Botany

Botany

second edition

Randy Moore
University of Louisville

W. Dennis Clark
Arizona State University

Darrell S. Vodopich
Baylor University

Contributing Authors

Kingsley R. Stern
California State University-Chico

Ricki Lewis
State University of New York at Albany

WCB/McGraw-Hill

A Division of The **McGraw·Hill** Companies

WCB/McGraw-Hill

A Division of The **McGraw·Hill** *Companies*

Botany 2/e

Copyright © 1998, by The McGraw-Hill Companies, Inc. All rights reserved.
Previous edition 1995 by Wm. C. Brown Communications, Inc.
Printed in the United States of America. Except as permitted under the United States
Copyright Act of 1976, no part of this publication may be reproduced or distributed in any
form or by any means, or stored in a data base or retrieval system, without the prior written
permission of the publisher.

 Recycled/acid-free paper

1 2 3 4 5 6 7 8 9 0 QPD/QPD 9 0 9 8 7
Library of Congress Catalog Number: 96–60294

ISBN 0–697–28623–1

Editorial director: *Kevin T. Kane*
Publisher: *Michael D. Lange*
Sponsoring editor: *Margaret J. Kemp*
Developmental editor: *Kathleen R. Loewenberg*
Marketing manager: *Thomas C. Lyon*
Senior project manager: *Kay J. Brimeyer*
Production supervisor: *Cheryl R. Horch*
Designer: *Katherine Farmer*
Photo research coordinator: *Janice Hancock*
Cover photo: *© David Muench 1996*
Compositor: *Shepherd, Inc.*
Typeface: *9½/12 Sabon*
Printer: *Quebecor, Inc.*

INTERNATIONAL EDITION Copyright 1998. Exclusive rights by the McGraw-Hill Companies, Inc.
for manufacture and export. This book cannot be re-exported from the country to which it is consigned
by McGraw-Hill. The International Edition is not available in North America.

1 2 3 4 5 6 7 8 9 0 QPD/QPD 9 0 9 8 7

When ordering this title use ISBN 0–07–115404–3

brief table of contents

table of contents

unit four

The Form and Function of Plants. . . 259

chapter 13

Primary Growth: Cells and Tissues 282

chapter 14

Primary Growth: Stems and Leaves 306

chapter 15

Primary Growth: Roots 334

chapter 16

Secondary Growth 356

unit six

Nutrition and Transport. . . 467

unit seven

Evolution. . . 521

unit eight

Diversity. . . 571

chapter 24

chapter 25

chapter 26

unit nine

Ecology. . . 757

list of readings

list of tables

preface

he American public is continually bombarded by articles in the news media that disparage the sad state of science education in the United States. These articles are usually based on educational research that emphasizes the need to improve the critical thinking skills of students at all levels. As classroom teachers, we work hard to address this need in our own courses.

We've written the second edition of *Botany* to help teachers develop students' critical thinking skills in a scientific context. We do this by showing students how scientists generate knowledge about the biology of plants: how we know what plants are, how we know about the functions of plants, how we know about the interactions of plants with each other and with their environment, and how we know where plants came from.

As in the first edition, *Botany* contains many interesting facts, but we have continued to de-emphasize the details of some scientific knowledge in favor of explanations involving the process of science. The result is that we have reduced the overwhelming number of new terms that usually appear in a textbook. This strategy follows the recommendations of educational research that show an improvement in students' thinking skills upon reducing the overburden of new terminology.

Our textbook mirrors the nature of botany: botanists advance our knowledge by posing and testing hypotheses. We lead students into this process by posing many questions, some of which we don't answer and others of which we answer by showing how botanists have addressed them. In so doing, we take students to the edge of scientific knowledge, where the answers to many questions remain unknown. By repeatedly seeing how this process works for different topics in each chapter, students will, indeed, be practicing the skills of critical thinking.

In addition to emphasizing *how* we know about the biology of plants, throughout this book we also show *what* we know. We present this information in the context of guiding principles (e.g., the relation of structure and function) and integrate it into an informational framework. We created that framework in the second edition of *Botany* by using the following themes:

The importance of plants. Plants affect virtually everything that we do. If students can understand plants, they can better appreciate the influence of plants on their lives.

The process of science. We stress the process of science throughout this textbook. We not only describe the products of research in all areas of botany, but we also explain the dynamics of scientific thought in each area. Moreover, we show how botanists are advancing science by applying modern research techniques to test competing hypotheses within current topics of interest.

Interactive learning. We designed this book to be a private tutor that will help students learn effectively. Consequently, we have included a variety of learning aids that link concepts and themes within chapters that show the beauty and excitement of studying plant biology, and that guide students in their own online explorations of botany using worldwide information resources. We believe that these features will encourage students to learn by making them active participants in their own learning and by enabling them to develop a strong sense of ownership of their botanical knowledge.

We've used these features to present information with authority, breadth, and quality. Although we've treated some topics in depth, we've not written a botanical encyclopedia. That is, this text is not a compendium, but rather a stimulus. Because the first step to knowing is the curiosity of a question, we hope that students will generate their own questions for study. Indeed, we'll help develop some of those questions in every chapter.

Unit Openers, Chapter Outlines, and **Chapter Openers** stimulate the students' curiosity and prepare them for what they will learn and why. These features will also pique their curiosity and alert them to the content that follows.

Readings expand the scope of botany to include people, controversies, curiosities, and new developments.

Chapter Summaries help students retain the main points of each chapter and integrate that knowledge with what they've learned in previous chapters.

Questions for Further Thought and Study help students apply what they have learned and stimulate additional thought.

These are questions that emphasize critical thinking, not true-false or multiple-choice questions that focus on the mere recall of isolated "facts." We ask students to put their ideas together, not just fill in the blank of someone else's thinking.

Internet Sites are windows to worldwide information about plants. The sites listed in each chapter expand the material of that chapter, lead students to discover what other people are doing in botany and in related disciplines, provide links to other universities and institutions, and describe where to get science-related software, and how to download and use it.

Suggested Readings provide sources of additional information that support or expand on the topics of each chapter.

The Lore of Plants tells about the fascinating—and sometimes even bizarre—history of plants, and about their uses by humans.

What are Botanists Doing? challenges students to learn the newest ideas about various botanical subjects.

Writing to Learn Botany encourages students to use writing as a tool to learn botany. This feature aids the development of communication skills, which are crucial to success in virtually any profession.

We have kept the strengths of the first edition while making the second edition stronger. We've done this with the help of excellent suggestions and recommendations from users and reviewers of the first edition. The second edition continues to be up-to-date with the latest findings in fast-developing areas such as molecular genetics. We have also bolstered Unit Eight (Diversity) by including more full-color life cycles of representative organisms and by expanding our presentation of the fungi and separating them into their own chapter (Chapter 26). This organizational change allowed us to put more focus on bacteria and viruses in Chapter 25, without making that chapter too large and unwieldy.

Our textbook is entertaining and accessible because we have combined the fun and fascination of botany with the process of scientific thinking. We have also rejected the awkward, stoic writing style typical of "scientific writing" and in its place have put together a readable scientific guide to botany that stresses communication and understanding. We believe the combination of our process-oriented approach and our engaging style is a critically important feature for making students better learners, and instructors better teachers.

Additional Learning Aids

Instructor's Manual and Test Item File
Prepared by Rebecca McBride DiLiddo, University of Illinois, Chicago, and Allan Nelson, Texas A & M University, Kingsville, this manual features Chapter Outlines, Teaching Goals, Vocabulary Words, Mastery Concepts, Teaching Tips, Puzzles, and Projects. The Test Item File contains 40–50 objective questions/answers per chapter for the instructor's use in preparing exams. Both the Instructors Manual and Test Item File have been updated to correspond to the changes in the second edition of the main text.

Classroom Testing Software
This helpful, easy-to-use, computerized test-generating software is available to instructors for testing and grading. It is available in Windows and Macintosh formats.

Transparencies
A set of 100 key figures from the text are available free to instructors.

Slides
A set of 200 additional images not found in the book are available free to instructors.

Student Study Guide
Written by Rebecca McBride DiLiddo, this guide helps students test their comprehension of important concepts and principles of botany. The study guide contains learning objectives, concept maps, multiple-choice, and fill-in-the blank questions.

Laboratory Manual
Prepared by Randy Moore and Darrell Vodopich, this unique manual is written in a straightforward style and covers basic biological principles in its 32 exercises. Updated in the second edition, it features exercises on molecular biology, a carefully developed learning system, numerous illustrations and diagrams to aid in lab explorations, and an appendix on "How to Write a Scientific Paper."

Laboratory Resource Guide
This useful guide contains a list of equipment needs for each exercise in the Laboratory Manual. Helpful hints for completing the labs safely and correctly are also included.

Net Quest: Exploring Botany
A unique supplement for any student of botany. This Internet activities booklet includes 17 chapters of basic botanical concepts and topics with overviews, URLs (Internet addresses) of interesting and informative sites, and related activities.

Videotapes
Tapes One, Two, and Five in the *Life Science Animation Videotape Series* will provide you with more than 30 animations of the most important concepts found in a botany course.

Life Sciences Living Lexicon
This interactive CD-ROM includes complete glossaries for all life science disciplines, a section describing the classification system, an overview of word construction, and more than 500 vivid illustrations and animations of key processes.

How to Study Science
Written by Fred Drewes, Suffolk County Community College, this workbook is an excellent guide for students enrolled in a science course. It offers tips on how to take notes, how to utilize laboratory time, and how to overcome science anxiety.

acknowledgments

The second edition of *Botany* required much work, and I am grateful to many people for their help. I especially thank Kathy Loewenberg, Kay Brimeyer, Marge Kemp, and Michael Lange for their ongoing encouragement, undeserved patience, and steadfast commitment to making this book the best available; all authors should work with such helpful editors. I also thank my many colleagues who reviewed chapters of the first edition; their excellent advice helped improve the book significantly. My students were also instrumental in this revision; I thank them for teaching me so much about botany. All of these people helped shape *Botany* in important ways.

I remain deeply grateful to my parents and wife; without their support I could not have produced this revision. And finally, I thank Stella and Lucy for never saying anything bad about me.

R.M.

List of Reviewers

Fred L. Arthaud	*North Central Missouri College*
D. K. Bagga	*University of Montevallo*
William F. Beasley, Jr.	*Paducah Community College*
Dorothea Bedigian	*Antioch College*
Jack M. Bostrack	*University of Wisconsin*
Richard G. Bowmer	*Idaho State University*
James A. Brenneman	*University of Evansville*
C. Evert Brown	*Casper College*
Paul J. Bybee	*Utah Valley State College*
Richard Churchill	*Southern Maine Technical College*
Jerry A. Clonts	*Anderson College*
John Cruzan	*Geneva College*
David B. Czarnecki	*Loras College*
Robert R. Dockery	*Horticulture Technology*
Ben L. Dolbeare	*Lincoln Laud Community College*
Roland Dute	*Auburn University*
Patrick E. Elvander	*University of California–Santa Cruz*
Sharon Eversman	*Montana State University*
Mary B. Fields	*Ursinus College*
Daniel E. Flisser	*Cazenovia College*
Donna Ford	*West Virginia University*
Bernard L. Frye	*University of Texas–Arlington*
Jean M. Gerrath	*University of Northern Iowa*
Sibdas Ghosh	*University of Wisconsin–Whitewater*
John J. Gillen	*Hostos Community College, CUNY*
Canzater C. Gillespie	*Milwaukee Area Technical College*
Holly L. Gorton	*St. Mary's College of Maryland*
Dalton Gossett	*Louisiana State University –Shreveport*
Suzanne M. Harley	*Weber State University*
Elizabeth M. Harris	*Illinois University*
Richard L. Hauke	*Georgia State University*
Donna Hazelwood	*Dakota State University*
Brian T. Hazlett	*Briar Cliff College*
Glen D. Hegstad	*Western Iowa Tech Community College*
Hon H. Ho	*SUNY–New Paltz*
Doug Jeffries	*University of the Ozarks*
Clarence Johnson, Jr.	*Fort Valley State College*
Carol Kasper	*McMurray College*
Robert W. Kingsolver	*Kentucky Wesleyan College*
Alan K. Knapp	*Kansas State University*
James Marshall Knapp	*Onondaga Community College*
Mark Knauss	*Shorter College*
Randy Landren	*Middlebum College*
Alfred R. Loeblich III	*University of Houston*
Anne C. Lund	*Hampden-Sydney College*
Chuck Lura	*North Dakota State University –Bottineau*
Paul C. Makareweez	*Three Rivers Community –Technical College*
Stephen D. Manning	*Arkansas State University–Beebe*
Craig E. Martin	*University of Kansas*
Jerri Dawn Martin	*Sue Bennett College*
William J. Mathena	*Kaskaskia College*

W. D. McBryde — Central Texas College
Susan B. Mitchell — Benedict College
Patricia J. Moore — Transylvania University
J. J. Muchovej — Florida A&M University
Lytton J. Musselman — Old Dominion University
Richard A. Niesenbaum — Muhlenberg College
Norine E. Nolan — Milwaukee Area Technical College
Jeanette C. Oliver — Flathead Valley Community College
Larry Peterson — University of Guelph
Donald H. Pfister — Harvard University
William J. Pietraface — SUNY–Oneonta
Paul Przybylowicz — The Evergreen State College
David M. Reid — The University of Calgary
H. Bruce Reid — Kean College of New Jersey
Stanley Rice — Southwestern State University
Dennis F. Ringling — Pennsylvania College of Technology

Herbert C. Robbins — Dallas Baptist University
Donald R. Roeder — Simon's Rock College of Bard

Sabine Rundle — Western Carolina University
A. Edward Salgado — Christian Brothers University
Renee M. Schloupt — Delaware Valley College
Fayla Schwartz — Everett Community College
Wanda C. Schwarz — Georgia Military College
Bruce S. Serlin — DePaul University
Brian R. Shmaefsky — Kingwood College
Richard Sims — Jones County Junior College
David Starrett — USDA/ARS
Paul Stutzman — Lehigh University
Max R. Terman — Tabor College
Jun Tsuji — Siena Heights College
Douglas C. Ure — Northeastern Junior College
Pamela J. Weathers — Worcester Polytechnic Institute
David B. Wing — New Mexico Tech
Robert Winget — Brigham Young University–Hawaii
Ian A. Worley — University of Vermont
James C. Zech — Sul Ross State University

Botany

Butterfly weed (*Asclepias tuberosa*), also called pleurisy root, is a member of the milkweed family. Powdered roots of these plants have been used to treat wounds, pulmonary disease, and rheumatism.

An Introduction to Botany

1

Chapter Outline

Chapter Overview

This chapter describes what botany is and why you'll enjoy studying it. You'll learn about the importance of plants, the lore of plants, and the concepts that unify the study of plants. More important, you'll learn about the scientific method, which underlies the process of scientific research. This information prepares you for the remaining chapters in this book, which integrate scientific knowledge about plants and discuss how we've applied the scientific method to understand botany.

Plants and Life

For many centuries, humans have relied on plants for survival and pleasure. Asian civilizations were based on rice, Middle Eastern civilizations were based on wheat and barley, and American civilizations were based on corn. Today, our dependence on plants persists: almost everything we do is influenced, either directly or indirectly, by plants. We've even fought wars over plants. For example, the First Opium War (1839–1842) was fought over opium, an extract of the opium poppy, *Papaver somniferum* (Britain won that war and, as part of the settlement, claimed Hong Kong). Today we use plants to make clothes, string, rope, resin, lumber, musical instruments, sports equipment, furniture, fabrics, cardboard, medicines, and explosives. Even these many uses of plants, however, are only a small part of their importance to us (see fig. 1.1). Consider the following examples:

- Green plants and algae generate the oxygen and sugars that sustain life on earth.

- Plants supply our food and many of our drinks. For example, about 95% of our food comes from only 20 species of plants; 80% of our food comes from six of those species. We've made tequila from the century plant (*Agave tequilana*) for hundreds of years, and Egyptians made beer from barley as early as 500 B.C. Tea and coffee, the world's two most popular beverages, are also made from plants (see reading 1.1, "Coffee: Even If You Don't Know Beans").

- We use extracts of plants and plantlike organisms to make paint, plastics, soap, oils, adhesives, natural rubber, waxes, dyes, spices, and drugs such as morphine, cocaine, aspirin, caffeine, codeine, digitoxin, quinine, vinblastine, and most antibiotics. We use flowers for decorations, perfumes, and to express our feelings.

- About 33 million people in the United States have home gardens. Thirty percent of these people use their gardens as a source of fresh fruits and vegetables, 25% as a source of better-tasting vegetables, 15% as a source of income, and a lighthearted 22%—the croquet and lawn tennis group—use them as a source of fun. These gardens provide $16 billion worth of food; that's about 23 kg (50 lb) of food per gardener, down from 57 kg (126 lb) per gardener in 1910. This decrease in production over the last 80 years is due to our increasing urbanization. Today, the people who don't have gardens blame a lack of space (35%) or a lack of time (28%), followed closely by the hammock-lovers who say that gardening is too much work (27%).

- On the negative side, plants clog rivers, damage crops, cause allergies, and poison us. Socrates was poisoned with poison hemlock, and Claudius, the father of Nero, was poisoned with mushrooms and monkshood juice. Many children are poisoned each year from eating poisonous plants.

Today, plants dominate our lives and economy, just as they have in all civilizations.

c o n c e p t

Plants affect virtually everything that we do.

Our need-driven uses of plants, along with our curiosity about our world, have spawned **botany**, the scientific study of plants. Today, botany is a thriving, exciting science that, directly or indirectly, deals with the largest part of our national and world economy.

But what exactly is a plant? Although you probably already have some notion as to what a plant is—a quiet, green organism that we eat, mow, and use for decoration and shade—it's difficult to come up with a complete definition for the word *plant*. For example, not all plants are green, and some plants consume animals. Indeed, many plants do not look or act like plants at all. Nevertheless, if you're like most people, you've probably not given plants much thought. You might have assumed that they're all basically alike and, because they don't move around, that they're uninteresting. To dispel this notion, consider the giant sequoia (*Sequoiadendron giganteum*) and its taller but slimmer relative, the coastal redwood (*Sequoia sempervirens*) (fig. 1.2).

- These cloud-piercing trees are the world's tallest organisms: they would tower above our nation's Capitol, dwarf the Saturn V rockets that flew astronauts to the moon, and reach from home plate to the outfield bleachers of any baseball park in the country. A 20-year-old sapling, a mere sprig of a thing, is often more than 15 m (50 ft) high.

- The General Sherman Tree in California weighs 1,400 tons—as much as 13 space shuttles, 20 million boxes of toothpicks, 200 elephants, or 10 blue whales, the animal kingdom's largest representative. These giant trees are supported by roots that seldom go deeper than 1.8 m (6 ft) into the soil.

- The oldest giant sequoias are about 3,200 years old, meaning that the trees that grow today were already 200 years old when King David ruled the Israelites.[1] In its seventies—the sunset years for humans—a sequoia is still a teenager and bears its first seeds. A mature tree produces about 2,000 cones per year, each of which contains about 200 seeds. Each of these seeds weighs less than 0.01 g (i.e., about eight-thousandths of an ounce), a mere one-hundred-billionth of the weight of a full-size tree.

---

1. Even the huge redwoods seem like youngsters when compared to bristlecone pines—gnarled, low-lying conifers that seldom grow higher than about 9 m (30 ft). Many of these pines are more than 5,000 years old, meaning that they were growing for 500 years before the pyramids were built.

Foxglove (*Digitalis purpurea*) contains cardiac glycosides used to treat congestive heart failure. These plants, along with *Digitalis lanata*, have saved the lives of many heart-attack victims and are helping millions of people with heart problems to lead normal lives.

Chocolate is made from seeds of cocoa (*Theobroma cacao*). Because these seeds were believed to be of divine origin, botanist Carolus Linnaeus named the plant *Theobroma*, meaning "food of the gods." Most famous as the main ingredient of the drink given to Cortés by Montezuma in 1519, cocoa beans were also used as currency by the Aztecs, who paid their taxes with them until 1887. Chocolate has long been considered an aphrodisiac, and Casanova is said to have preferred chocolate to champagne as an inducement to romance. Today, chocolate remains a popular gift for romantics.

Piper nigrum is the source of black pepper, the most commonly used spice. The lure of pepper, the only spice that could make decaying or heavily salted meat edible, drew Columbus and medieval merchants to "discover" the rain-forested areas of Earth. For Europeans, North America was a by-product of the maritime search for pepper.

Flax (*Linum usitatissimum*), with its wiry stems and distinctive flowers, has been used for centuries for its fibers (from stems, used to make linen) and linseed oil (crushed from its seeds). Flax fibers are about three times stronger than cotton fibers. The luster of linen results from the fibers' capacity to reflect light.

Tea is made from leaves of *Camellia sinensis,* a small evergreen shrub related to the garden camellia. Each year more than 2 million tons of tea are produced from about 25 countries. North Americans drink about 40 billion cups of tea per year.

Money is one of the most important things we make from plants. Paper money is made from fibers of flax.

figure 1.1

Plants affect virtually all aspects of our lives. Many plants are economically important; our paper money is even made from plants.

COFFEE: EVEN IF YOU DON'T KNOW BEANS

Throughout this book you'll see essays such as this that describe the excitement, lore, and recent discoveries of botany. Here, in the first of them, is "the rest of the story" about a plant extract that millions of people enjoy every day: coffee.

Coffee originated in the Middle East more than 1,000 years ago. Since then, it's been used as food, medicine, wine, and even as an aphrodisiac. Ever since the Boston Tea Party in 1773, coffee has been America's most popular beverage. Today, many people treasure quality coffees and pay more than $100 per pound for gourmet brands. Such value isn't a recent phenomenon—coffee beans are used as currency in some isolated parts of Africa, and the failure of a Turkish husband to provide his wife with coffee was once considered as "grounds" for divorce.

Coffee (*Coffea arabica*) is grown from seeds. After two years, the seedlings are about 0.5 m tall and are transplanted to large plantations. Two to three years later, the plants produce white, jasmine-scented flowers that form fruit. When ripe, the fruits are red and sweet. Each fruit contains two green coffee beans that are harvested, dried, and sold.

Different kinds of coffee plants produce different kinds of beans. Most plants produce robusta beans, which are used to make grocery-store varieties of coffee. Each of these trees produces about 5,000 fruits—1.4 kg (3 lbs) of coffee—per year. Robusta coffee is about 2.5% caffeine.

Gourmet coffees are made from arabica beans, which are more susceptible to disease and more costly to harvest. These

reading figure 1.1
Coffee berries mature in 7 to 9 months, turning from dark green to yellow, then red.

plants grow at elevations exceeding 1,800 m (6,000 ft) and produce only about 2,500 fruits—0.7 kg (1.5 lb) of coffee—per plant per year. Arabica coffee is about 1% caffeine. Expresso, the darkest roast, has even less because some caffeine burns off in the roasting process.

Given the potentially harmful effects of caffeine, several companies have marketed substitutes for coffee. The most popular of

these was Postum, a nutritious beverage of wheat, molasses, and wheat bran sold by Charles W. Post in 1893. Post got the idea for this beverage while being treated for ulcers at the Western Health Reform Institute, operated by John Kellogg in Battle Creek, Michigan. Although Post left the institute uncured, he successfully marketed Postum with a clever advertising campaign that declared that Postum "makes your blood red" (to learn more about Post and Kellogg, see reading 1.2, "Breakfast at the Sanitarium"). Nevertheless, the public continued to want coffee. Sales increased greatly when Boston's Chase and Sanborn began selling roasted coffee in cans in 1878, and hoarding of coffee in the United States led to rationing in November 1942. Coffee's popularity has increased ever since.

Today, half of the people in the United States won't start the day without a cup of coffee; this amounts to an average consumption of about 13 lbs per person (in world-leading Finland, 38 lbs per year is the norm). Worldwide, the coffee industry generates $16 billion per year, and the more than 15 billion lbs of coffee produced annually in 50 countries provide more than 21 million jobs. Coffee shops such as Starbucks* are immensely popular throughout most parts of the world. Brazil produces more than one-third of the world's coffee, helping to make coffee the world's second-largest commodity—second only to oil—in international trade.

*Starbucks is named after the coffee-loving first mate in Herman Melville's classic adventure tale, *Moby Dick.*

- The average sequoia needs more than 1,100 l of water per day. The energy required to lift this water to the tree's leaves each day is enough to launch a can of Pepsi into a low orbit of earth.

Clearly, these giants belie the notion that plants are too static or too quiet to be as thrilling as stampeding elephants or tail-pounding dinosaurs. However, the stories of fascinating and useful plants don't stop with sequoias and redwoods. Indeed, plants have a fascinating lore:

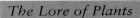

The Lore of Plants

Sweet potatoes were once considered strong aphrodisiacs. When Shakespeare's Falstaff shouted, "Let the sky rain potatoes!" in *The Merry Wives of Windsor*, he was hoping for sweet potatoes, not the spuds we use to make French fries.

figure 1.2

The sequoias are among the most impressive members of the plant kingdom. The volume of a mature sequoia's trunk is almost twice that of a typical two-story house.

figure 1.3

Dead American elm trees ravaged by Dutch elm disease.

- Although George Washington never chopped down his father's cherry tree, others have used cherry trees for many purposes. Cherry bark has been used to make a tea to ease the pains of childbirth and to cure coughs, colds, and dysentery. Queen Victoria and others since her time have enjoyed cherry brandy. Daniel Boone carved several coffins out of cherry and always kept one under his bed. He gave away all but the last one.

- The Republican Party was founded under a white oak in Jackson, Michigan in 1856. Oaks have also played prominent roles in some of Hollywood's greatest movies. For example, Twelve Oaks—Ashley Wilkes's estate in *Gone with the Wind*—was named for its dozen kinds of live oaks. Incidentally, the Spanish moss that covered those oaks is not a moss but a bizarre relative of pineapple. Similarly, poison oak is not an oak at all, but rather is a cousin of poison ivy, mango, and cashew.

- Most wood connoisseurs don't like elm. The wood is hard to split, too wet to burn, and usually smells bad—hence the name *piss elm*. Nevertheless, elms have witnessed many important events of history. For example, physicists at the University of Chicago met

beneath an elm in 1943 to discuss the experiments that led to the atomic bomb. Today, most elms in the United States are being killed by Dutch elm disease, a disease caused by the fungus *Ceratocystis ulmi*[2] (fig. 1.3).

- Despite warnings to the contrary, Americans eat only about 8.1 kg (18 lb) of apples per person per year—far less than the one per day to keep the doctor away. Italians do better—25 kg (56 lb) per year. Only the Dutch, who each eat an average of 45 kg (100 lb) per year, keep the doctor away.

---

2. To try to help save these majestic trees, botanists have recruited an unlikely soldier: the bacterium *Pseudomonas syringae*. Some strains of this bacterium produce a compound that kills *C. ulmi*. In 1987 Gary Strobel and his colleagues at Montana State University discovered a genetically altered strain of *P. syringae* that made large amounts of the bacterial toxin. When preliminary experiments with greenhouse-grown elms worked well—injecting the bacterium into the trees protected the trees from later infections—Strobel and his colleagues injected fourteen 15-year-old elms on their campus with the genetically altered fungus. However, Strobel did this without the approval of the U.S. Environmental Protection Agency, whose mission includes preventing scientists from releasing unauthorized organisms into the environment. Strobel's release of *P. syringae* was the first deliberate release of genetically altered organisms without the government's approval. The furor that resulted from Strobel's experiment caused the project to be put on hold. On 3 September 1987, Strobel ended the experiment by cutting down all of the experimental elms.

- We use spruce to make the 23,000 tons of newsprint needed each day to produce the 65 million newspapers in the United States. A bumper issue of the *New York Times* clears about 400 hectares (990 acres) of trees, and the average tree produces enough newsprint to produce about 400 copies of a 40-page tabloid. Each person in the United States uses an average of about 290 kg (640 lbs) of paper each year—that's roughly equivalent to 0.8 cubic meters (25 cubic feet) of wood. Much of that paper produces the dreaded "paperwork" of many people's jobs. Indeed, 2 trillion pieces of paper flow through offices each year; the 120 billion sheets that are filed pack about 5 million filing cabinets.

- Not all products of spruce are as inexpensive as newsprint. One of the most valuable is the spruce wood that's part of a 300-year-old violin made by Antonio Stradivari. If you're lucky enough to find one of those violins in your attic, be sure to play it often; vibrations keep the wood elastic, while neglect stiffens it. Don't let your violin end up like Paganini's: his was kept locked away in a vault for 50 years, after which it was useless.

- Although cartoons and nagging parents try to convince you otherwise, spinach contains little that is nutritious except vitamin A.[3] Spinach might have given Popeye the Sailor an edge as a night pilot, but it probably did him little good in his dockside brawls.

- Some of Isaac Newton's great ideas were triggered when he saw an apple fall from a tree. Other important ideas have been stimulated by plants; for example, Velcro was invented by George deMestral after he examined the structure of a cocklebur clinging to his clothes.

- Plants have been the star witnesses in several criminal trials, most notably in the 1932 trial of Bruno Hauptmann. In a trial based largely on botanical evidence, Hauptmann was convicted of kidnapping the baby boy of Charles and Anne Lindbergh and was sentenced to death. Similar evidence has been used to convict rapists, murderers, and thieves.

- The preserving effects of peat moss have shown us the gruesome details of ancient religious ceremonies and have provided us with ancient human DNA (see reading 28.2, "Secrets of the Bog").

- Cox's orange pippin apples can be traced to a neglected apple tree that Richard Cox discovered while walking across an abandoned farm. Similarly, the seedless navel orange industry can be traced to an orange tree growing near the corner of Magnolia and Arlington Avenues in Riverside, California.

- The Union army used onion juice to cleanse gunshot wounds. General Ulysses Grant, deprived of onions in one campaign, sent a terse memo to the War Department saying, "I will not move my troops without onions." The War Department sent him three cartloads.

- People have used fibers for more than 10,000 years. We use hemp to make rope, sisal to make brooms and brushes, ramie to make textiles, and flax to make linen, wrappings for Egyptian mummies, cigarette paper, and money (fig. 1.1).

- People have shown their fascination with plants by sculpting them into various shapes, incorporating them into logos and names, and naming their children after them. We've also done some rather bizarre things with plants. For example, people in Charlevoix, Michigan—presumably taking a break from their diets—baked a 7-ton cherry pie, and a fellow in Tulsa claimed to have four rooms of his house filled with turnips. However, nothing beats the story of Jay Gwaltney of Chicago who, in 1980, ate a 3.4 m (11 ft) birch tree. It took him just over 89 hours, but he ate the *whole plant*.

concept

Botany is the scientific study of plants. Plants are the basis for our national and world economy and have a fascinating history and lore.

A Botanist's View of Life

Botanists, other scientists, and nonscientists often share an almost insatiable curiosity about life and its diverse but related forms. We watch nature shows on television, visit zoos and botanical gardens, buy pets, and tend our gardens. Botanists, however, also often watch life in more exotic places—for example, while perched in a treetop of a tropical rain forest or while in a greenhouse surrounded by a variety of plants bred specially for experiments. Such intense observation has generated many questions about plants, most of which can be grouped as follows:

- How are plants constructed?

- How do plants work?

- How did plants get here?

- Why are plants important?

Botanists have answered these and other questions by using an experiment-based process called the **scientific method,** a systematic way to describe and explain the universe based on observing, comparing, reasoning, predicting, testing, concluding, and interpreting. All in all, it's not very different from the way that most nonscientists think about this fascinating thing we call life.

---

3. Although spinach contains much iron, humans do not absorb it during digestion.

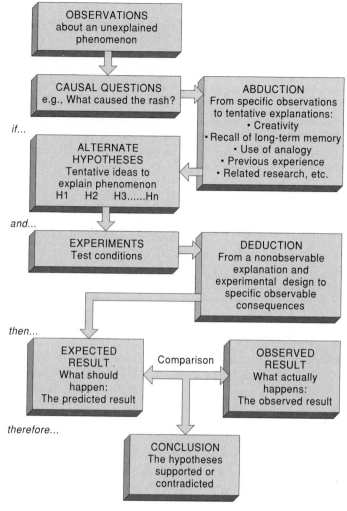

figure 1.4

The scientific method.

The Scientific Method

Superficially, botany is a collection of "facts" that describe and explain the workings of plants and the living world. Although these facts are interesting—even fascinating—they are not the essence of botany. Rather, the excitement of botany lies in the intriguing observations and the carefully crafted experiments that botanists and others have devised to help us learn about plants. The important thing here is the *process* of botany—not just knowing the facts, but appreciating *how* botanists discovered (and continue to discover) those facts. This process is the distinguishing feature of science that you'll read about throughout this book.

The scientific method begins with things that we are all familiar with: observation and curiosity (fig. 1.4). Such observations can happen just about anywhere—in a research laboratory, garden, or while you're watching a television program. For example, the observation that leaves of many plants can follow the movement of the sun catches our at-

tention and piques our curiosity. This may prompt us to ask, "What enables the leaves of some plants to follow the movement of the sun?" Similarly, the observation that some plants in our garden are not attacked by Polish whiteflies while others are infested may lead us to ask, "What enables resistance to Polish whiteflies in some plants but not others?" These are the kinds of **causal questions** that are at the heart of the scientific method. Science is fundamentally about finding answers to such questions. To find these answers, botanists use past experiences, ideas, and observations to propose **hypotheses** that may produce predictions. To determine whether predictions are accurate, botanists do experiments. If experimental results match the predictions of a hypothesis, the hypothesis is accepted; if results do not match the predictions, the hypothesis is rejected. The effect of all this is to make scientific progress by revealing answers piece by piece.

Posing hypotheses is perhaps the most creative step in the scientific method. It requires a type of reasoning called **abduction,** which is the process of devising explanations for observations. Abduction involves sensing ways in which a new situation is somehow similar (analogous) to other known situations and using this similarity to make hypotheses about the new situation.

Let's see how the scientific method works in real life. Suppose that you wake up one night with a bothersome rash (observation). You want to know what caused this rash (causal question). Your first try to figure out how this situation is similar to other known situations. Perhaps a classmate had a similar rash that was caused by an allergy to the grass at the neighborhood baseball park. You realize that your rash has appeared after all of the baseball games that you've played at that park. The similarity (analogy) between your friend's allergy and your rash leads to the explanation (hypothesis) that your rash is caused by an allergy to the grass, too.

If your hypothesis is correct, you would not expect to get a rash from other kinds of grass (prediction). When you play baseball at several other parks, all with different kinds of grass than your neighborhood park (experiment), you do not get a rash (results). Your results match the prediction of the grass-allergy hypothesis, from which you decide that your rash is caused by the grass at your neighborhood park (conclusion). However, if this was your only hypothesis, you have not actually identified the cause. Perhaps your rash was caused by the paint used in the dugout. Or perhaps it was caused by fertilizers, pesticides, bacteria, molds, or other agents associated with the grass or soil at one baseball park and not another. By testing a single hypothesis, you have not ruled out any of these other possibilities. To do so, you would have to devise alternative hypotheses, make predictions from them, and obtain experimental results to compare with the predictions. By this process, you may be able to reject all hypotheses, even the grass-allergy hypothesis. Either way, you make progress by testing several hypotheses, not just one.[4] If you can reject all but the grass-allergy hypothesis, you can more confidently conclude that the grass caused your rash. Furthermore, if the grass is in the outfield but not in the infield, you may decide to become an infielder so that you can avoid the irritating plant (use of new knowledge).

If all of this seems like common sense, it is. As Thomas Henry Huxley said, "Science is nothing but trained and organized common sense." Although a conclusion marks an end to the scientific method for a particular experiment, it seldom ends the process of scientific inquiry. Any conclusion must be placed in perspective with existing knowledge. Moreover, an underlying characteristic of scientific thinking is that the sequence of observing, comparing, reasoning, predicting, testing, concluding, and interpreting is a cycle, with new ideas spawned at every step. To the curious scientific mind, a conclusion is never the final answer. There is always something more to study, something new to learn.

Using the Scientific Method

In science, as well as life, it is important to keep an open mind—to maintain the objectivity that the scientific method requires in drawing conclusions and designing experiments—and not allow biases or expectations to cloud the interpretation of results. This is sometimes difficult, because it is human nature to be cautious in accepting an observation that does not "fit" existing knowledge. For example, for centuries people believed that life arose from nonliving matter. Several studies—especially those involving exposing decaying meat to the air and observing what grew on it—"proved" that life arises spontaneously from the nonliving, a conclusion that brilliant scientists such as Newton, Harvey, and Linnaeus never questioned. However, spontaneous generation was finally rejected by another set of different, more critical experiments done by Louis Pasteur. The conclusion from those experiments—namely, that life does not arise from decaying meat—surprised many who believed that mice were created by mud, that flies came from rotting beef, and that beetles sprouted from cow dung. Similarly, biologists initially rejected the idea that DNA is the genetic material, thinking that protein was a more likely candidate.

Perhaps no one knew the conflicting feelings of joy and frustration that accompany a scientist's discovery of a quirk of nature better than the late Barbara McClintock (fig. 1.5). In the 1940s, McClintock studied the inheritance of kernel color in corn—by, as she put it, "Asking the maize (corn) plant to solve specific problems and then watching it respond." While watching her carefully bred plants respond, McClintock noticed that some kernels had a peculiar pattern of spotting.

McClintock concluded that the units of inheritance (genes) moved around in corn cells. This seemed preposterous at the time, for genes were thought to be immovable parts of larger structures called chromosomes. Despite McClintock's evidence, the scientific community refused to accept her conclusion. However, after a decade of additional discoveries of mobile genes in other organisms, Nobel Prize winner Barbara McClintock was finally recognized as the brilliant botanist that she had always been. Her observation of "jumping genes" in corn plants almost five decades ago may ultimately help explain how roving genes cause certain cancers in humans.

Is the Scientific Method Foolproof?

Although the scientific method is a powerful tool for answering some kinds of questions, it is not foolproof. For example, several studies have shown that animals that are fed large doses of vitamin E live longer than those receiving only average amounts of the vitamin. From those studies, many concluded that vitamin E retards aging, thereby promoting all sorts of claims by skin-cream manufacturers. However, the animals fed large amounts of vitamin E also lost weight, another factor that correlates positively with increased life expectancy. The original experiment did not distinguish these possible interpretations. As is usually the case, more experiments are necessary to distinguish the other possible interpretations.

Many experiments that biologists do are difficult to validate. For example, consider the experiments of Stanley Miller, who sought to re-create the chemical reactions that might, on the primitive earth, have formed the compounds found in living organisms today. Miller's experiments on the origin of life are difficult to validate. Although he provided interesting results and inspired much more research, we can't really know whether he successfully re-created the conditions and events of the planet's early history—unless, of course, we invent a time machine to go back and see.

Creativity in Science

Science is not always as planned and organized as the scientific method suggests. Sometimes discoveries involve luck and creativity, as with Barbara McClintock. The creative side of science involves making the mental connections to take advantage of accidental observations. For example, consider the case of Alexander Fleming. In 1928 Fleming left a culture dish of disease-causing *Staphylococcus* bacteria uncovered in his lab. Other organisms quickly contaminated the culture. Just before throwing out the contaminated plate, Fleming noticed several

4. The danger of trying to do science armed with just one hypothesis is that the scientific method rapidly degenerates into a hunt only for supporting evidence. In the example, if the rash happened to be caused by a mold that infects only the kind of grass at your neighborhood park, then predicting that you would not get a rash at a park with different grass is correct even though the grass-allergy hypothesis is false. You would come to the wrong conclusion because you did not examine other possible explanations (alternative hypotheses).

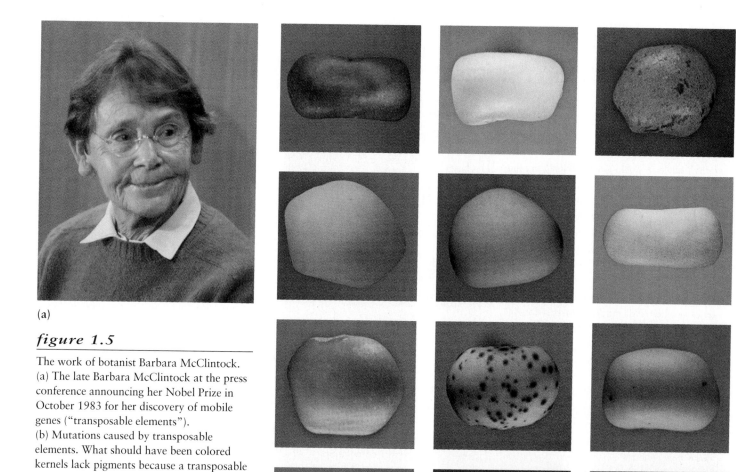

(a)

figure 1.5

The work of botanist Barbara McClintock.
(a) The late Barbara McClintock at the press conference announcing her Nobel Prize in October 1983 for her discovery of mobile genes ("transposable elements").
(b) Mutations caused by transposable elements. What should have been colored kernels lack pigments because a transposable element lies in or near the gene for pigment production. During development, the element sometimes leaves the gene, restoring pigmentation to that cell and all of its descendants. This produces the color variations of the kernels.

(b)

clear areas on the plate where the bacteria were not growing. This keen observation suggested to Fleming that some contaminant from the air had stopped the growth of the bacteria. Fleming tried to isolate and characterize the substance responsible for this inhibition of growth but failed because he was a poor chemist. However, other scientists later identified and purified penicillin, a product of the fungus whose effects Fleming had accidentally discovered. Fleming's observation, along with the follow-up work, saved millions of lives and eased much pain.

Was Fleming lucky? Perhaps—but he would not have discovered penicillin if he had not been observant and able to recognize the potential importance of what he saw. His discovery, like so many others, exemplifies a famous saying: "Luck is when preparation meets opportunity." Most others would not even have recognized the clear areas on the plate, much less thought that those areas could be important. Such observations are critical to science.

The Role of Technology in Scientific Discovery: A Case Study of the Cell Theory[5]

Discoveries are determined by our ability to observe, which often depends on technology. To appreciate this, consider the development of the cell theory, the set of postulates that holds the cell to be the fundamental unit of life. Before the middle of the seventeenth century, no one even knew that cells existed. Because most cells are too small to be seen with the naked eye, their discovery awaited the invention of microscopes. In 1665, the English scientist Robert Hooke used a crude, hand-built microscope to study a thin slice of cork. To his surprise, Hooke saw a honeycomb of compartments that he described and named *cells*, meaning "little rooms." Although Hooke's discovery was a landmark event in the history of biology,

---

5. A theory is a collection of assumptions (postulates) that, taken together, try to explain a set of related phenomena.

BREAKFAST AT THE SANITARIUM: THE STORY OF BREAKFAST CEREALS

Cornflakes and several other breakfast cereals can be traced to Seventh-Day Adventists, an American religious sect. In 1866, the Adventists organized the Western Health Reform Institute in Battle Creek, Michigan. The institute was popular: patients such as Henry Ford, John D. Rockefeller, and Harvey Firestone flocked to the institute for recuperation and rejuvenation. The institute was later renamed the Battle Creek Sanitarium and, due to the sect's religious beliefs, served only vegetarian meals. Each morning, the flamboyant John Kellogg (a physician) and his quiet brother Will made breakfast for the sanitarium's patients by running a potful of cooked wheat through some rollers to form sheets of wheat. These sheets, when toasted and ground up, became a sort of cereal.

One night in 1895, the Kellogg brothers were called away on an emergency and left some wheat soaking for more hours than usual. When they tried to make their wheat-sheets the next day, each wheat grain formed its own flake. Patients at the sanitarium had a new breakfast food: wheat flakes. Four years later, the Kellogg

reading figure 1.2

brothers tried their new approach with corn, added some malt flavoring, and produced the first cornflakes. Patients at the sanitarium liked the flakes, especially ones made of corn, and thus was born Kellogg's Corn Flakes of the Kelloggs' cereal empire, which still exists in Battle Creek.* Although the Kellogg brothers later had a falling out (Will won use of the family name, which explains

why his famous signature appears on the boxes of cereal), their product remained immensely popular. Indeed, within 15 years, more than 40 cereal companies were centered in or near Battle Creek, some of them operating out of tents.

Interestingly, one of the patients (a suspender salesman) at the sanitarium saw how the Kelloggs made their cornflakes. He later began a competing business that produced breakfast cereals, naming his first cornflakes "Elijah's Manna." However, the American clergy considered that name blasphemous, prompting the former patient in 1908 to change the name of the flakes. That patient was C. W. Post, who named his flakes Post Toasties.

---

*Interestingly, wheat flakes (e.g., Wheaties) weren't marketed until 26 years after cornflakes were sold. The process used by the Kelloggs to make flakes was soon modified to produce grains, shreds, and puffs. Puffs are made by heating the grains, causing pressure to build. When this pressure is released, the water vapor in the grain explodes the grain into a puff.

Hooke thought that these cells were unique to cork and therefore never realized the full importance of his discovery. Several years later, Dutch naturalist Anton van Leeuwenhoek built several tiny microscopes, each of which would fit easily in the palm of your hand. With these microscopes, van Leeuwenhoek studied pond water and became the first to describe microscopic organisms (he also described other organisms and discovered red blood cells). However, like Hooke, van Leeuwenhoek did not fully appreciate the importance of his own work. Neither did other biologists, for the discoveries of Hooke and van Leeuwenhoek were ignored for almost two centuries.

By the early 1800s, technology had produced stronger and easier-to-use microscopes. In 1838, German botanist Matthias Jakob Schleiden used one of these microscopes to study a variety of plants. He concluded that all plants "are aggregates of fully individualized, independent, separate beings, namely the cells themselves." This was the first postulate of what we now call the **cell theory** (see Chapter 3) and was a brilliant example of inductive reasoning—making a generalization based on concurring observations. The next year, German zoologist Theodor Schwann reported that animals are made of cells and proposed a cellular basis of life. Later, in 1855, the great German pathologist Rudolf Virchow completed the

modern cell theory when he induced that "the animal arises only from an animal and the plant only from a plant." This strong statement refuted the idea of spontaneous generation—that life could arise from nonliving matter—that had been suggested by Aristotle and had persisted for many centuries. Today, the cell theory is tremendously important because it emphasizes the similarity of all living systems and thereby unites the many studies of different kinds of organisms. The cell theory guides the work of many biologists and physicians as they study disease and food production. However, the development of the cell theory was limited by technology (the availability of microscopes) and depended on keen observation and inductive reasoning.

Today, much of science is driven by new technology. Just as microscopes revolutionized our understanding of cells, so too will new technology produce remarkable discoveries and new insights. That technology will overturn many of the "facts" that you read about in this or any other science book. This is why *Botany* concentrates so strongly on the *process* of science; the facts and interpretations may change, but the process used to discover them—that is, the process that enables you to think critically, solve problems, and even become a professional scientist—remains the same.

We All Think Like Scientists at Times

The cycle of questioning and answering that defines the scientific method is second nature to a practicing scientist; it's more a philosophy than a set of rules. However, the use of the scientific method is not restricted to scientists: we all observe, compare, guess, plan, do experiments, and interpret and use the results of experiments. For example, our ancestors used experimental trial and error to determine which plants were edible. In a specific experiment in 1820, a New Jersey colonel named Robert Johnson bravely ascended the steps of a county courthouse and ate a basketful of tomatoes, much to the amazement of the 2,000 townspeople who thought that tomatoes were poisonous. His bold experiment and its results (his survival) showed the townspeople that tomatoes are not poisonous. Parents do a similar experiment when they introduce food one item at a time to their children to check for allergic reactions. They also experiment—by tasting the food in question—to check its safety or taste.

Our use of the scientific method has greatly advanced knowledge and improved our quality of life. For example, how would Gregor Mendel, a monk who described the principles of heredity more than a century ago, feel if he knew that his observations of pea plants would one day help to explain how a perfectly healthy couple could have a child afflicted with a crippling disease?[6] Similarly, it took keen observation, experimentation, and reasoning to discover that the mild and supposedly harmless tranquilizer called thalidomide could cause 10,000 European children to be born without arms and legs, and that a tiny virus could destroy the immune system and cause AIDS.

Many students mistakenly believe that botanical research is done only by professors, graduate students, and full-time technicians. Not true! Indeed, many undergraduates—that is, students just like you—do important botanical research. To convince you of this and to stimulate your interest in research, we'll periodically include essays by undergraduates about their research. Our first essay is by Aaron Ball, an undergraduate who works in Judy Verbeke's laboratory at the University of Arizona.

---

6. Mendel did his famous experiments in the monastery garden at Brünn, Austria, in the mid-1800s. Although Mendel also taught natural science at Brünn high school, he never managed to pass the exams necessary to get a teaching certificate.

focus on student research

Aaron Ball
University of Arizona

My research began with a curious "mutant" plant that was discovered before I began working in the lab. The mutant was found in a seed lot of normal *Catharanthus roseus*. It is a variety that has no petals.

The goal of my project is to determine if the mutant variety is genetically stable. To do this, I must show that the mutant plants can produce offspring. The challenge is that since the mutant has no male parts, cross-pollination with pollen from normal plants is necessary. My initial experiments involved cross-pollination within the same variety, because I wanted to first understand the mechanisms and techniques of cross-pollination within a variety before applying these techniques to pollinate the mutants. *Catharanthus roseus* is a highly self-pollinated plant, and its floral structures support this reproductive strategy. These plants, therefore, are difficult to emasculate, before introducing new pollen.

Another stage of this experiment included testing pollen viability. After several tests, I found that an extract made from stigmas from the same variety that the pollen would come from was necessary to facilitate pollen germination. After much trial and error, I was able to apply the techniques learned in the initial cross-pollination to crossing the mutant. We're now waiting for these results. If my tests show this trait is stable, the next step will be to isolate those genes responsible for the abnormality and determine how they work.

This research has been very important to me during my undergraduate career. It has exposed me to the workings of a true research lab. This has given me valuable experience I will be able to use after I graduate and begin to look for a job. But I believe there has been an even more important aspect to my work in this lab. This research has allowed me to use the knowledge I have learned in my classes because I have had to use things I learned in chemistry, biology, and plant sciences. This research has reassured me that I have learned valuable information as an undergraduate. It has also taught me problem-solving skills. As an undergraduate, I have been exposed to a recipe approach in a lot of my lab classes. This research project has allowed me to break from these recipes and use my own knowledge and creativity to solve real research problems.

(a)   (b)   (c)   (d)   (e)

figure 1.6

The unifying themes of botany. (a) **Plants consist of organized parts.** The cells in the center of this cross section of a buttercup (*Ranunculus*) root conduct fluids, while those surrounding the center of the root store food. (b) **Plants exchange energy with their environment.** Virtually all energy that powers life is captured from sunlight. (c) **Plants respond and adapt to their environment.** The growth of these stems toward light increases the chances of the leaves being able to absorb light for photosynthesis. (d) **Plants reproduce.** Flowering plants such as this Easter lily (*Lilium longiflorum*) reproduce sexually by producing flowers. (e) **Plants share parts of a common ancestry and have coevolved with other organisms.** Here, an Anna's hummingbird visits a flower. Both of these organisms evolved with compatible anatomy.

The Unifying Themes of Botany

Our studies of plants have shown us that plants, like all organisms, share several important features. Just as you appreciate an opera or symphony if you know its themes, so too will you better understand plants if you understand their repeating themes (fig. 1.6). These themes are integrated throughout this book.

Plants consist of organized parts.

Just as builders use the same kinds of bricks to make large or small houses, plants use the same building blocks to make all of their parts. Building all of these parts from the same building blocks is analogous to making thousands of words from the 26 letters of the alphabet— the secret is in the arrangement of the building blocks. The differences we see in organs such as leaves and roots result primarily from differing arrangements of similar tissues rather than from the presence of unique tissues.

Plants exchange energy with their environment.

Although organization is a requirement for life, it alone does not make something alive. Plants, like all organisms, absorb energy from their surroundings and, in turn, impact their environment as they use that energy for their activities. This conversion of energy from one form to another occurs via a set of chemical reactions collectively called *metabolism*.

Plants respond and adapt to their environment.

Such adaptations are inherited characteristics that help to ensure survival. These responses include the ability of plants to detect and respond to stimuli such as gravity; to convert light energy to chemical energy; to gather nutrients; and to lure animals into helping pollinate, defend, and disperse the plants. All of these adaptations and responses are important because they increase a plant's chances of survival and reproduction. Today's more than 260,000 species of plants show that there are at least 260,000 ways that plants adapt to their environment.

Plants reproduce.

Their reproduction can be as "simple" as splitting a cell or as complex as orchestrating flower production, mobilizing energy reserves, or luring insects, birds, and other helpers. Although most plants can reproduce sexually or asexually, the greatest variation results from sexual reproduction, in which pieces of DNA called *genes* move from one generation to the next. This explains why plants resemble their parents, just as children resemble theirs. Genes, combined with the influence of the environment, produce the characteristic features of plants. The diversity of plants results from their diverse environments and differences in their DNA, and is a product of evolution. Understanding the language of DNA—that is, how cells read DNA and translate its instructions into metabolism and growth—is a major goal of many biologists and the basis of an exciting business: biotechnology.

Plants share parts of a common ancestry.

This ancestry was best explained in the most important science book ever written: Charles Darwin's *On the Origin of Species,* published in 1859. This book, "one long argument" for the diversity of life, used evidence, logic, and analogy to explain the central role of variation in the evolution of life. Although Darwin's ideas were based largely on the ideas of other scientists, no one had ever produced such a convincing argument. Darwin's argument forced scientists to accept his theory of natural selection despite all opposing arguments, primarily those of religious dogma. Darwin's book shook the foundations of biology and remains one of the most important books ever written.

The products of evolution surround us, and we've used geological and biochemical evidence to infer how and when life's diverse forms appeared on earth. Life on earth probably began about 3.6 billion years ago as a simple cell. Since then, natural selection has produced a variety of plants and plant-like organisms. For example, the algae are about a billion years old, and ferns appeared on earth at least 350 million years ago. Cone-bearing trees such as pines may have appeared 300 million years ago, and the flowering plants, which are relative newcomers, appeared a mere 200 million years ago. These flowering plants are the most diverse group and include monocots (plants such as grasses, lilies, and orchids, whose leaves have parallel veins) and dicots (plants such as apples, oaks, and sunflowers, whose leaves have netted veins). Today, we use and study all of these kinds of plants.

c o n c e p t

Plants consist of organized parts, exchange energy with their environment, respond and adapt to their environment, reproduce, and share parts of a common ancestry.

Our Goals

Botany discusses these themes in several contexts, including the relation of structure and function, the evolution and diversity of plants, and the lore, importance, and uses of plants. Throughout this book, you'll also learn about a new, exciting, $2-billion business: plant biotechnology. Biotechnology is a way of using organisms to make commercial products. Although we've used biotechnology for several decades, our ability to directly manipulate a plant's genome began in 1983 when botanists transferred a gene from a bacterium into a plant. Today, biotechnology and molecular biology are revolutionizing biology and biology-based industries; by the year 2005, plant biotechnology is predicted to be an $8-billion industry. Here are a few of the ways that we use biotechnology:

- **To make vitamins and drugs.** For example, just as pigs are now being used to make human hemoglobin, plants are being transformed into factories that make drugs and oils. A promising new vaccine for hepatitis B is made with baker's yeast, and plants are now used to make serum albumen, which is used for fluid replacement in burn victims. Botanists at Calgene have produced canola oil containing 40% laurate (an important ingredient of soaps and shampoos), as well as tomatoes that contain a gene that delays softening and reduces spoilage, thereby allowing supermarkets to sell full-flavored, vine-ripened tomatoes even in winter. You'll learn more about these Flavr Savr™ tomatoes in Chapter 11.

- **To produce plants that resist drought, disease, and insects.** Botanists at Monsanto have used genetic engineering to produce disease-resistant plants. Others have transferred a gene from *Bacillus thuringiensis*, a common soil bacterium, into cotton plants. This gene produces a protein that, when eaten by pests such as the tomato hornworm, induces paralysis and death (fig. 1.7). The protein has no effect on other organisms, making it environmentally safe. Although most *B. thuringiensis* is used to fight caterpillars, other strains of the bacterium are being developed to fight beetles and mosquitoes.[7] Botanists are searching for ways to help plants resist diseases that lay waste to, on average, 12% of crops worldwide.

- **To produce plants resistant to herbicides.** Genetically engineered plants can now tolerate glyphosate, the active ingredient in Roundup, a nonspecific herbicide that degrades quickly in the environment. Consequently,

---

7. Our use of *B. thuringiensis* as a soldier to fight pests has been significant. For example, 90% of the pesticide used today to control spruce budworm in Canada is *B. thuringiensis,* up from 10% in 1979. This amounts to an annual release of about 2.3 million kg of *B. thuringiensis*—that's about 4.5×10^{20} bacteria, a number roughly equivalent to the estimated number of stars in the universe.

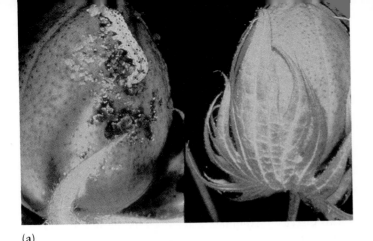

(a)

(b)

(c)

figure 1.7

Monsanto's use of genetic engineering to improve crops. (a) The improved cotton boll on the right with no insect damage is the result of combining traditional breeding techniques with the science of biotechnology. Insect control is achieved when a gene from a common soil bacteria, *Bacillus thuringiensis*, or *Bt*, is placed into plants, causing plants to produce a protein that acts as a natural insecticide against the tobacco budworm, pink bollworm, and cotton bollworm. An untreated boll on the left shows typical insect damage. (b) The large boll of cotton on the left is the product of a plant that has insect protection with the Bollgard™ gene. Plants not weakened by insects are healthier and produce more cotton. The boll on the right was grown in the same field but comes from an unprotected plant. (c) The potential increase in yield becomes evident when the leaves are stripped from a plant damaged by insects (right) and a plant that is insect-protected with the Bollgard™ gene (left).

farmers can spray fields with Roundup and kill all the plants except the crop plant. This could increase yields significantly, because weeds reduce agricultural productivity by more than 10%.

- **To make foods and beverages such as yogurt, cheese, bread, and beer.** Botanists at Quaker Oats are using biotechnology to increase the protein content of their oats, and botanists at Kraft are using similar techniques to decrease the amount of saturated fat in the soybean oil used in their products. Enzymes from engineered tobacco plants are used to make bread and low-calorie beer and to clarify wine and juices. Genetically engineered potatoes containing increased amounts of starch are easier and cheaper to process for fries and chips.

- **To make better-tasting food, clean the environment, recycle wastes, prevent tooth decay, and produce antibiotics, biodegradable plastics, and fragrances.** Microbiologists currently use about a dozen genera of fungi and bacteria to make about 4,500 antibiotics.

c o n c e p t

Plant biotechnology is a multibillion-dollar industry that is changing our lives and society.

Throughout this book you'll learn about the biotechnological revolution, as well as its more unusual applications. For example, genetically engineered tobacco plants now produce human antibodies useful for diagnosing and treating disease. Similarly, biologists have cloned and transferred a gene for an unusual protein—the one that keeps winter flounder from freezing—into tomato and tobacco plants. This "antifreeze gene" could help extend the growing range of these plants.

Another important goal of *Botany* is to show you the process of science and the doings of scientists. You'll see the scientific method in action—scientists formulating hypotheses, doing experiments, gathering data, and drawing conclusions. You'll also see that scientists, like other people, are often passionate about their work. For example, Rachel Carson wrote *Silent Spring* in response to a letter that she received in January 1958 from Olga Huckins describing how a small part of the world had been made lifeless by pesticides. Carson's response was to say, "There would be no peace for me if I kept silent."[8] Such passion fuels the work of many successful botanists.

8. Chemical companies and several manufacturers spent thousands of dollars to try to block the publication of *Silent Spring*, and Carson was maligned as "ignorant" and "hysterical." Nevertheless, *Silent Spring* became a best-seller that prompted President John F. Kennedy to create a special panel of the Science Advisory Committee to study the use of pesticides. That panel's report vindicated Carson and stimulated the government to act against pollution. *Silent Spring* launched the environmental movement.

A final goal of *Botany* is to help you appreciate living organisms. This is perhaps the most important goal of any biology book and any biologist's teaching, for such an appreciation of life is the first step toward respecting and conserving life. Appreciation and respect for life are critical, especially in light of our neglect of the environment for many decades—a neglect that is now costing us money and killing earth's organisms. For example:

- In 1996 businesses spent more than $81 billion to try to clean up their messes in the environment. All of these costs are passed to consumers as higher prices. Pollutants such as lead, mercury, and DDT kill many organisms.

- Producing 1,000 tons of low-grade paper from virgin pulp produces more than 40 tons of pollutants and requires 1,700 trees, 25 million gallons of water, and 20 billion joules of energy. That amount of energy is equivalent to the energy released during the explosion of 5 tons of TNT.

- In 1980 we cleared about 11.3 million hectares of rain forest—an area equivalent to that of Pennsylvania. By clearing these forests, we drive animals and plants to extinction and thereby lose many potentially valuable drugs—drugs that could save lives. Botanists estimate that as many as 50,000 species of tropical plants may be extinct within the next few decades because of the destruction of rain forests.

- More than 200 species of plants are thought to have become extinct during the past 200 years. About 900 species are endangered (i.e., in danger of becoming extinct in part of or throughout the world), and more than 1,400 are threatened (i.e., likely to become endangered). About 5% of the 80,000 species of plants native to temperate regions are near extinction. This has been caused by habitat destruction, overgrazing by domestic animals, the introduction of foreign plants, and the destruction of pollinators.

- Red spruce on Mount Mitchell, North Carolina, are dying because of acid rain. Similarly, almost 40% of the trees in Germany are affected by acid rain, and many ponds in the Adirondack Mountains of upstate New York are nearly lifeless because of low pH caused by acid rain. Acid rain has dropped the pH of some of these ponds to about 4.1—far too low for most organisms. These trees and ponds are like the canaries carried into mines to warn of natural gas: unless we heed their message, we'll continue to pay an ever-increasing price.

Although people today are more environmentally aware than people of previous generations, our environmental problems don't seem to be improving. For example, in 1980 most environmentalists talked about the "dangerous" rate of deforestation. By 1990, that rate had nearly doubled. This year, we'll clear rain forests having an area equal to that of Florida *plus*

Maine. Similarly, our land, air, and water are becoming increasingly polluted. Food production is higher than ever, yet millions of people are starving. What are we to do?

Such problems will not be solved by governmental panels, presidential commissions, blue-ribbon committees, or expert testimony before Congress. Such groups have been convened for decades, during which time our problems have only worsened. These problems will be solved only by a scientifically literate public. This textbook uses botany to teach such literacy; we hope it will enable you to help solve the many problems that we face.

Botany will help you appreciate what botanists know, what they don't know, and what they do. You'll learn not only the "facts," but the *process* of doing botany. Your observations and ideas concerning life will be more meaningful the more experience and learning you have on which to build. This book and your course will give you some of this valuable background, which should enable you to apply the scientific method to your own observations of the living world. There's no reason why you can't be the next person to make a major scientific discovery.

Chapter Summary

Botany is the scientific study of plants. Plants have a fascinating history and lore, and our national and world economies are based on plants. Virtually everything we do is influenced by plants.

Botanists learn about plants and the world by using the scientific method, a systematic way to describe and explain the universe, based on observing, comparing, reasoning, predicting, testing, and interpreting. Everyone uses the scientific method at times.

Plants consist of organized parts, exchange energy with their surroundings, respond and adapt to the environment, reproduce, and share a common ancestry. Plant biotechnology is a multibillion-dollar industry that is changing our lives and society.

 What Are Botanists Doing?

Botanists have used biotechnology to produce a variety of genetically engineered plants. Many of these plants could greatly benefit society—for example, by capturing heavy metals from groundwater and locking them in their tissues.

Go to a library and read two articles that describe uses of genetically engineered plants. How could these plants improve the quality of our lives? What restrictions, if any, should be placed on botanists who try to engineer such plants? Explain your reasoning.

 Writing to Learn Botany

What are the strengths and weaknesses of the scientific method?

Questions for Further Thought and Study

1. Why study plants?

2. Interferon is a protein that animals make to fight viral infections. Botanists recently inserted the gene for interferon into turnips, hoping the protein would make the plants more resistant to viral diseases. The experiment failed; engineered plants were no more resistant than were untreated plants. However, the interferon from the turnips appears to be active in animals, including humans. What does this tell you about the process of science?

3. How do you use the scientific method in your daily life?

4. Is the scientific method foolproof? Explain your answer.

5. How has our understanding of life been affected by technology?

6. Can the scientific method be used to make any judgments? Explain your answer.

7. How do plants influence your life?

8. Each year the National Cancer Institute tests extracts from about 5,000 plants for their effects on 60 kinds of human cancers and the AIDS virus. Fewer than 1% of these extracts are selectively toxic against these cancers. Indeed, these tests have produced only four marketable drugs (taxol is one of the program's successes). Given this low percentage and the program's high costs, should such tests be abandoned? Why or why not?

Web Sites

The World Wide Web—usually referred to as "The Web"—offers an astounding amount of information and is rapidly joining libraries as a primary place for researching a topic. Throughout *Botany*, we'll present a variety of Web sites at the end of each chapter so that you can use the Web as a way of learning about plants. Here are a few to get you started:

http://www.mhhe.com/moore

This site is devoted to helping you use this textbook to learn about plants. Check it often; you'll find interesting updates, information, and stories about plants.

http://www.scisoc.org/aps/careers/

This site, sponsored by the American Phytopathological Society, describes some of the exciting careers available to botanists.

http://www.indirect.com/www/bazza/cps/faq/faq.html

This site contains photographs of, and information about, the ecology, growth, and taxonomy of some of the world's most bizarre organisms: carnivorous plants.

http://www.lowes.com/howto/plants.html

Here's information about how to grow plants in your apartment or garden. Whether you're pruning a plant or trying to find a pretty plant that will grow in your apartment, you'll find helpful information here.

http://www.uaex.edu/vegfacs/tomatoes.html

This site is devoted to tomatoes, the most popular plant in most people's gardens.

Although you'll enjoy these sites, don't limit yourself—there's no telling what you'll learn about botany, and careers in botany, when you use the Web.

Suggested Readings

Articles

Darley, W. M. 1990. The essence of "plantness." *The American Biology Teacher* 52:354–357.

Gibbons, Ann. 1990. Biotechnology takes root in the Third World. *Science* 248:962–963.

Gibbs, A., and A. E. Lawson. 1992. The nature of scientific thinking as reflected by the work of biologists and by biology textbooks. *The American Biology Teacher* 54:137–152.

Moffat, A. S. 1992. Plant biotechnology: High-tech plants promise a bumper crop of new products. *Science* 256:770–771.

Roberts, Leslie. 1988. Extinction imminent for native plants. *Science* 242:1508.

Stern, W. L. 1988. Wood in the courtroom. *World of Wood* 41:6–9.

Books

Imes, Rick. 1990. *The Practical Botanist*. New York: Simon and Schuster.

Morton, A. G. 1981. *History of Botanical Science. An Account of the Development of Botany from Ancient Times to the Present Day*. New York: Academic Press.

Overfield, R. A. 1993. *Science with Practice: Charles E. Bessey and the Maturing of American Botany*. Ames: Iowa State University Press.

Simpson, B. B., and M. Conner-Ogorzaly. 1995. *Economic Botany: Plants in Our World*, 2d ed. New York: McGraw-Hill.

Talalaj, S., D. Talalaj, and J. Talalaj. 1991. *The Strangest Plants in the World*. Melbourne, Australia: Hill of Content Publishing.

Cells as the Fundamental Units of Life . . .

One of the unifying themes of botany, as mentioned in Chapter 1, is that plants consist of organized parts. This construction includes several levels of organization, from atoms and molecules to cells, tissues, organs, and whole plants. Specialties within botany often follow these organizational lines. Biochemistry involves molecules, physiology covers biochemical pathways, anatomy seeks information about structure, and morphology compares variation among whole plants and their different parts.

The most common approach to learning botany begins at the molecular level and moves up the organizational ladder to whole plants, as we do in this textbook. We begin in Unit 1 by discussing the molecular aspects of plant design (Chapter 2); then we describe the structure of cells (Chapter 3) and the main features of membranes and how they work (Chapter 4). Unit 1 should therefore give you a good foundation for understanding the key cellular processes of plants that are presented in Unit 2 ("Plants and Energy") and Unit 3 ("Genetics"). Unit 1 will also help you understand how plants grow and reproduce, which are the main subjects of Unit 4 ("The Form and Function of Plants"), Unit 5 ("Regulating Growth and Development"), and Unit 6 ("Nutrition and Transport").

We could just as easily have saved Unit 1 for the end of the text and instead presented botany from the top; that is, we could have begun with whole plants and gone down the organizational ladder. This descending order of presentation is justifiable because the organization of plants may be controlled at the organism level, not at the cellular or molecular level. Evidence for this point of view is discussed in Chapter 3. This issue is significant, because a "molecules-to-cells-to-tissues" approach implies that this is the only way plants can be organized. However, remember that there is an alternative hypothesis. We present botany in ascending order more as a matter of convenience and because of the historical tradition of our own professional training.

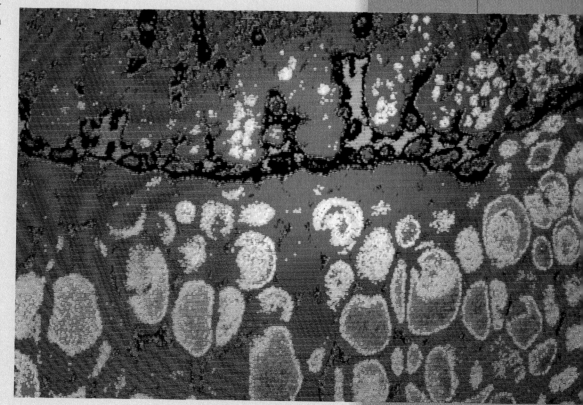

Molecular model of the amino acid L-alanine.

Atoms and Molecules: The Building Blocks of Life

2

Chapter Outline

Chapter Overview

Organisms are a succession of increasingly complex levels of organization, from molecules to individuals and communities. The traits that distinguish a leaf from a root, or a giant redwood from a dandelion, emerge from the type and arrangement of their molecules. The tree you see out the window consists of thousands of different kinds of molecules, some of which are shared by all life, and some of which are unique to that kind of tree. These molecules are joined to form cellular structures such as membranes, cell walls, and other cell parts. Different types of cells are grouped into tissues that form the leaves, roots, and branches of the tree. The properties of each level of complexity emerge from its components.

In this chapter we present the major kinds of molecules that form living systems. We emphasize the kinds of molecules common to all life, the kinds of molecules common to plants, and the kinds of molecules that distinguish one plant from another. These molecules include proteins, carbohydrates, nucleic acids, lipids, and secondary metabolites. The study of these molecules and how they are made provides the first insight into how plants live.

The physical and chemical behavior of matter can be anticipated, at least partly, by knowing its behavior at the atomic level. Similarly, cellular processes can be understood by knowing how biological molecules interact. For instance, our knowledge of chlorophyll helps us understand how plants harvest and use the energy of light to make sugars during photosynthesis. Similarly, we get basic information about the mechanisms of inheritance from studies of DNA. We understand how plants control the uptake of foreign substances because we know something about the chemistry of cell membranes. Such **reductionism**—studying simpler components to understand the functions of complex systems—is the most dominant strategy in biological research.

In this chapter we use the reductionist strategy to explain the properties of the main ingredients in living cells. However, this perspective cannot explain what life is, because a cell is more than a bag of chemicals. Nevertheless, the study of cellular components is a good place to start our study of living organisms. Note that a basic background in chemistry is necessary before reading this chapter. An overview of some of the most important background information is provided in Appendix A, "Fundamentals of Chemistry."

Elements of Life

Only about 16 of the 92 naturally occurring elements are essential to most plants. Four of these elements—hydrogen, carbon, oxygen, and nitrogen—make up more than 99% of the mass of most plant tissues (see table 2.1).

Most molecules that contain carbon are referred to as **organic** compounds. Almost all organic compounds also include hydrogen, and most include oxygen. However, pure carbon (graphite, diamond, buckminsterfullerene) and carbon compounds lacking hydrogen (carbonates, carbon dioxide, carbon monoxide, cyanide salts) are classified with noncarbon compounds as **inorganic** chemicals. **Biochemicals** are organic and inorganic molecules that occur in living organisms.

The major elements of living cells occur in thousands of combinations, each combination forming a different kind of molecule. Except for water, the mass of a plant or other organism consists primarily of four kinds of molecules: carbohydrates, proteins, nucleic acids, and lipids. However, plants also make many other kinds of compounds, usually in lesser amounts, including chemicals having such exotic names as phenolics, alkaloids, terpenoids, sterols, and flavonoids.

Large Molecules: Polymers and Their Monomers

Many biochemicals exist individually as small, single-unit molecules called **monomers**; however, most plant tissues consist of **polymers**, which are compounds made of many identical or similar monomers assembled into large molecules. Examples of polymers include cellulose, starches, enzymes, DNA, waxes, lignins, and tannins. Glucose is the only monomer in cellulose and starch. In contrast, enzymes and other proteins consist of twenty different amino acid monomers; there are probably more than 50,000 unique polymers made from amino acids in every plant. Only four nucleotides make a polymer of DNA, but every organism has a different set of DNA molecules. Variations in these and other classes of polymers account for the main differences we see among plants and animals that exist now or have existed in the past. In addition to the major classes of large molecules common to all organisms—carbohydrates, proteins, nucleic acids, and lipids—plants may also contain such polymers as lignins, tannins, polyterpenes, and polyacetylenes.

This section presents the main features of the major classes of biological polymers. Polymers unique to plants are discussed with other kinds of plant chemicals later in the chapter.

Most plant tissues are made of large molecules that are polymers of simpler molecules. The major biological polymers are divided into four classes: carbohydrates, proteins, nucleic acids, and lipids.

Carbohydrates

Glucose and other sugars, as well as their polymers, are called **carbohydrates**. Carbohydrates generally contain one oxygen and two hydrogens for every carbon. For example, glucose and fructose consist of six carbons, twelve hydrogens, and six oxygens, and have the formula $C_6H_{12}O_6$. Galactose, mannose, and many other monomers have this same formula, differing only in the arrangement of the elements (see reading 2.1, "Reading Chemical Structures"). This means they are **isomers**, which are different compounds with the same formula. Structural differences between isomers can impart significantly different chemical properties to the molecules. In the example of

table 2.1

The Most Common Elements in Plants

Element	Symbol	Atomic Mass	Average Percent of Plant Tissue	Example of Role in Plants
Hydrogen	H	1	47.4	One of the two components of water; also occurs in all organic molecules
Carbon	C	12	27.6	Backbone of all organic molecules
Oxygen	O	16	23.7	The other of the two components of water; also occurs in nearly all organic molecules; molecular oxygen is used in aerobic respiration
Nitrogen	N	14	0.8	Component of proteins, nucleic acids, chlorophylls, and alkaloids
Potassium	K	39.1	0.2	Prevalent ion in plants; regulates water uptake; activates certain enzymes
Calcium	Ca	40.1	0.1	Important in synthesizing pectin in cell walls; activates enzymes involved in chemical communication in cells
Magnesium	Mg	24.3	0.06	Part of chlorophyll and of some enzymes; helps to stabilize ribosomes
Phosphorous	P	31	0.05	Part of phosphate in energy-transfer molecules, nucleic acids, coenzymes, and phospholipids
Sulfur	S	32	0.02	Ingredient of most proteins and of some enzyme cofactors
Chlorine	Cl	35.4	0.002	Possible role in some reactions of photosynthesis
Iron	Fe	55.8	0.0016	Chlorophyll synthesis; part of active site of many important oxidation-reduction enzymes
Boron	B	10.8	0.0016	Unknown; may work in translocation of sugars
Manganese	Mn	54.9	0.0008	Prevalent enzyme-activating metal in plants
Zinc	Zn	65.4	0.0002	Activates many enzymes; involved in synthesis of the plant hormone indoleacetic acid
Copper	Cu	63.5	0.00008	Activates many enzymes; occurs in plastocyanin, an electron carrier of photosynthesis
Molybdenum	Mo	95.9	0.0000007	Important in nitrate reduction

Note: Several other elements, such as cobalt, either occur in smaller amounts or are important only in some plants. This latter group includes, for example, sodium, selenium, and silicon. Still other elements occur sporadically in plants because they happen to be in the soil. These include metal ions (gold, aluminum, lead, mercury), fluorine, and even uranium. Such elements are often toxic to the plants. We'll discuss minerals and plant nutrition in more detail in Chapter 20.

glucose and fructose,[1] one has little flavor (glucose), while the other (fructose) is the sweetest sugar known. Common carbohydrate monomers having different chemical formulas include ribose, xylose, and arabinose ($C_5H_{10}O_5$); ribose ($C_5H_{10}O_5$) and deoxyribose ($C_5H_{10}O_4$) (see p. 31); glucuronic acid and galacturonic acid ($C_6H_{10}O_7$); and rhamnose ($C_6H_{12}O_5$).

Carbohydrate monomers rarely occur as free sugars in plants; rather, they are usually bound to other kinds of molecules or linked together into larger carbohydrates. To distinguish different sizes of carbohydrates, monomers and dimers (molecules with two monomers) are called **monosaccharides** and **disaccharides,** respectively; polymers are called **polysaccharides.** The most common disaccharide in plants is sucrose, which is a glucose and a fructose linked into a dimer (fig. 2.1). Sucrose is common table sugar, also called *cane sugar* or *beet sugar.* This disaccharide is the major form of carbohydrate that moves in plants. People exploit such transport when they collect sap from sugar maples (*Acer saccharum*) to make maple syrup (see reading 21.3, "Making Maple Syrup"). People also use maltose, a disaccharide of glucose, to make beer (see reading 4.1, "Membrane Transport and Making Beer").

You are now prepared for an interesting and important exercise in carbohydrate chemistry: reading food labels for their sugar content. The different forms of sugar listed on food labels include: sugar, sucrose, glucose, dextrose, fructose, and high-fructose corn syrup. Sugar is an ingredient of many foods in one or more of these forms. Labels on the sweetest foods (e.g., candy, cookies, breakfast cereals) include several of these names. Even ketchup, salad dressings, bread, and other non-sweet foods often include sweeteners.

The Lore of Plants

Ancient Egyptians used dill (*Anethum graveolens*) as a soothing medicine, and the Greeks knew that "dill stayeth the hickets" (hiccups). During the Middle Ages, dill was prized as protection against witchcraft; magicians used dill in their spells, while lesser mortals added it to wine to enhance passion. Early settlers took dill to North America, where it became known as "meetin' seed," because children were given dill seed to chew during long sermons.

---

1. Glucose is also called *grape sugar, blood sugar,* or *dextrose.* Fructose is often referred to as *fruit sugar* because of its abundance in ripening fruits.

READING CHEMICAL STRUCTURES

At first encounter, chemical structures are often a pesky problem for students, because different chemicals are represented in variable and seemingly inconsistent ways. The common goal of all structural representations, however, is to portray chemicals simply but informatively. Such simplification is important because complete structures are cluttered and usually too tedious to read or write. For example, examine the complete structures of some common laboratory chemicals—glucose, cyclohexane, and toluene—shown on the left in the accompanying diagrams. These structures include a C for each carbon, an H for each hydrogen, and an O for each oxygen. Each chemical bond is shown as a single line. Contrast these three structures with the streamlined versions of each compound shown on the right. These are the "bare-bones" versions that eliminate the need to write some of the carbons, hydrogens, and chemical bonds. The unwritten atoms and bonds are still there, but by general agreement they are not shown; it is up to the reader to mentally fill in the appropriate atoms or bonds.

reading figure 2.1

Diagrams of three complete chemical structures (left) and their abbreviations (right).

There are no rules that encompass all structural simplifications. Generally, however, the streamlining involves only carbon and hydrogen. For example, unwritten carbons are the corners of rings or the joints of chains. Hydrogens attached to carbons are either not written, as in cyclohexane and toluene, or they are written without bonds to the right of the carbon they belong to, as in the CH_2 of glucose. Some inconsistencies do occur, however, such as between carbohydrates and other organic chemicals. The hydrogens on the carbons of the glucose ring are abbreviated as lines (bonds) without Hs, but those of cyclohexane are not shown in any way. Moreover, the bond sticking out of toluene represents a methyl group ($-CH_3$) and not a hydrogen as it does in glucose.

Perhaps the best advice on reading simplified chemical structures is to satisfy carbon—that is, to make sure that each carbon bonds to four other atoms. In cyclohexane, for example, where each carbon is shown to have two bonds, the two unwritten bonds must be to hydrogens. As a brief exercise, apply this principle to fill in the unwritten hydrogens and carbons for strychnine (fig. 2.13). (Hint: The formula for strychnine is $C_{21}H_{22}N_2O_2$.) The structure of this compound would be very messy if all the carbons and hydrogens were written.

Structural Polysaccharides

Polysaccharides that hold cells and organisms together are called **structural polysaccharides.** Cellulose is the most abundant structural polysaccharide in plants and the most abundant polymer on earth; between 40% and 60% of a cell wall is cellulose. Cellulose also occurs in almost pure form in some plants—for example, in the fibers surrounding cotton seeds (*Gossypium hirsutum*).

The strength of cellulose comes from its organization (see fig. 2.2). **Beta-glucose** units are linked by oxygen between the number 1 carbon of one glucose and the number 4 carbon of the next glucose. (*Beta* refers to a monomer whose hydroxyl at the number 1 carbon points up. When the hydroxyl points down, it is the *alpha* configuration, as in **alpha-glucose.**) This beta-1,4 linkage is repeated to make a linear molecule consisting of 100–15,000 glucose units (fig. 2.2). A higher level of organization occurs in cellulose when a thousand or more polymers twist together into **microfibrils.** Microfibrils are like strong, tiny cables. The tension that microfibrils can withstand before breaking—that is, their *tensile strength*—is as high as 110 kilograms per square millimeter. This means that if a microfibril could be made with a diameter about the size of this letter o, it could suspend your instructor. This tensile strength is about 2.5 times that of the strongest steel.

Several microfibrils intertwine to make cellulose fibrils. Layers of fibrils are cemented together into strong, three-dimensional

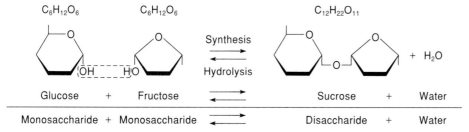

$$C_6H_{12}O_6 \qquad C_6H_{12}O_6 \qquad\qquad C_{12}H_{22}O_{11}$$

Synthesis
Hydrolysis

$+ H_2O$

| Glucose | + | Fructose | | Sucrose | + | Water |

| Monosaccharide | + | Monosaccharide | | Disaccharide | + | Water |

figure 2.1

Diagram showing the synthesis and hydrolysis of sucrose. During synthesis, a bond forms between glucose and fructose, and a molecule of water is removed. (Water is not actually released into the aqueous medium of the cell. See Appendix A for a discussion of where this water molecule goes.) Hydrolysis occurs as a molecule of water is added and the bond between glucose and fructose is broken. The enzyme that drives hydrolysis is sucrase; the synthesis of sucrose is actually a multistep reaction that involves several enzymes.

Alpha-glucose

Amylose

Beta-glucose

Cellulose

figure 2.2

Line diagrams of alpha-glucose and beta-glucose, plus primary structures of amylose and cellulose. The main difference is that amylose is made of alpha-glucose, and cellulose is made of beta-glucose.

grids by other kinds of structural polysaccharides (fig. 2.3). The gluey polysaccharides that hold cellulose fibrils together are called **pectins** and **hemicelluloses.** Cell-wall pectins are mostly polymers of galacturonic acid with alpha-1,4 linkages. Hemicelluloses are not related to cellulose; they are made of different kinds of monomers. Linkages in hemicelluloses are variable and complicated, but the composition of these polysaccharides can be studied by hydrolyzing them in acid and determining their component monomers. For example, hemicelluloses in grasses consist mostly of xylose, with lesser amounts of arabinose, galactose, and uronic acids. In contrast, hemicelluloses in legumes (the pea family) are high in uronic acids, galactose, and arabinose, with little xylose. Some of the most interesting hemicelluloses may not be structural polysaccharides at all. These are usually exuded from stems, roots, leaves, or fruits in a sticky mixture called **gum.** Gums are complex, branched polysaccharides consisting of several kinds of monomers. For example, a gum called **gum arabic** (from *Acacia senegal*) consists of arabinose, galactose, glucose, and rhamnose. Gum arabic is almost everywhere in our daily lives: it is used to stabilize postage-stamp glue, beer suds, hand lotions, and liquid soaps.

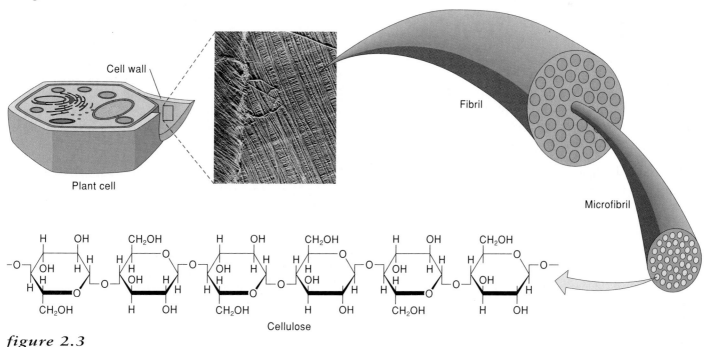

Cell wall

Plant cell

Fibril

Microfibril

Cellulose

figure 2.3

Model for the arrangement of fibrils, microfibrils, and cellulose in cell walls. The scanning electron micrograph shows the fibrils in a cell wall of the green alga *Chaetomorpha*, ×30,000.

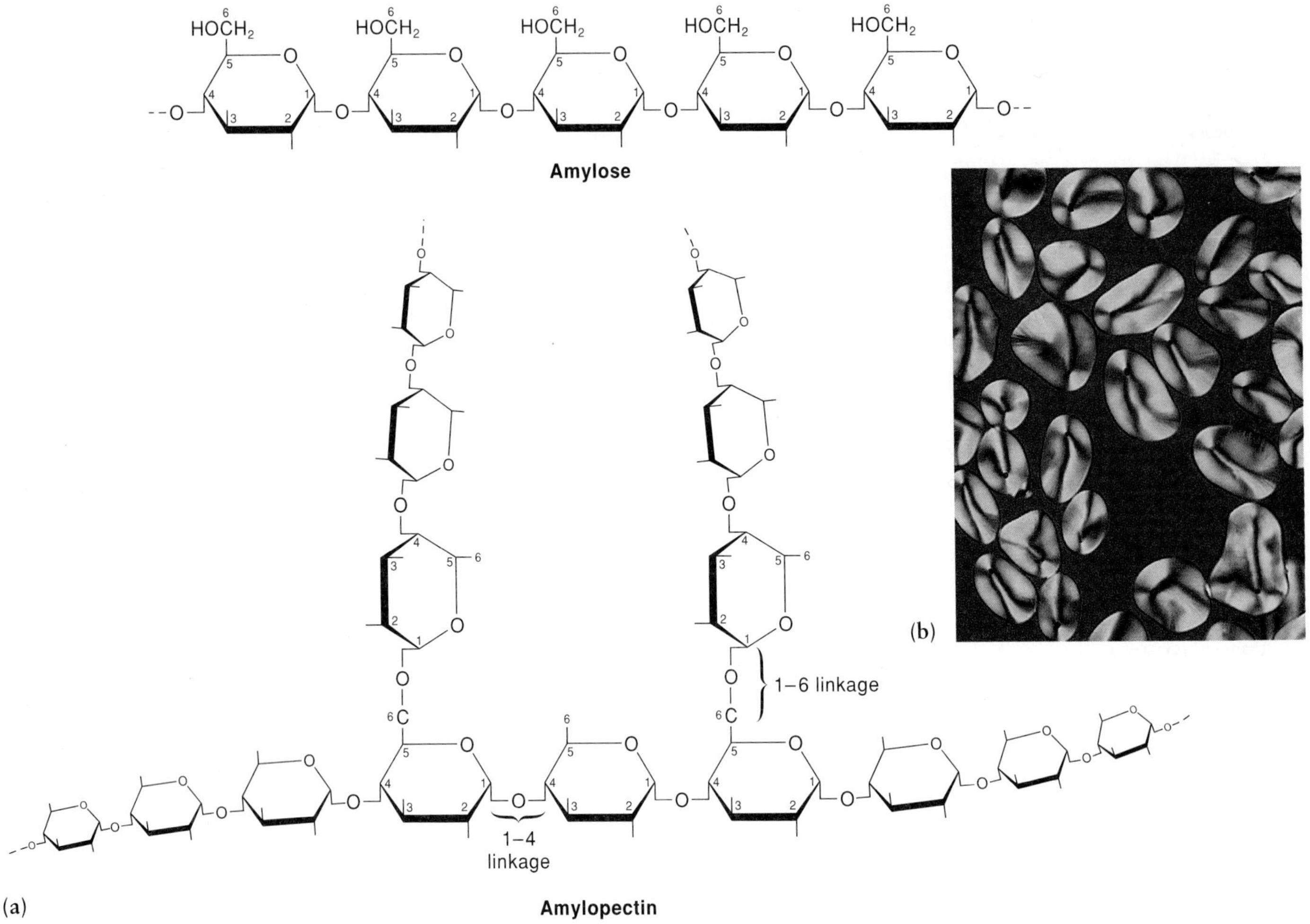

Amylose

Amylopectin

1–6 linkage

1–4 linkage

(a)

(b)

figure 2.4

Amylose and amylopectin are two forms of glucose-containing storage polymers in plants. (a) Amylopectin is made of short, helical chains that consist of branched amylose molecules. (b) Molecules of amylopectin aggregate into starch grains, as shown here from potato tubers, ×83.

Two commercially important polysaccharides of algae are **agar** and **carrageenan.** These polymers are the slimy substances that surround the cellulosic cell walls of certain red algae. Each polymer is a different mix of alpha-galactose sulfates having 1,4 and 1,6 linkages. Agar, which is harvested mostly from *Gelidium robustum* in the United States, is used to make drug capsules, cosmetics, gelatin desserts, and temporary preservatives; it is also used in scientific research as a medium for growing microorganisms and plant tissue cultures. Carrageenan, which comes mostly from species of the genus *Chondrus,* is used primarily as a stabilizer in paints and cosmetics and in foods such as salad dressings and dairy products.

Storage Polysaccharides

Two of the most common storage polysaccharides in plants are **amylose** and **amylopectin** (fig. 2.4a); they are the polymers of alpha-glucose that make up starch. Amylose is the smaller and simpler of the two: it consists of chains of a hundred to several thousand monomers with 1,4 linkages. In contrast, amylopectin is a highly branched polymer of up to 50,000 monomers. This polymer consists of short chains with 1,4 linkages, which are

cross-linked to other chains. Most starch grains are about 20% amylose and 80% amylopectin, but this depends on the plant. For example, starches in some types of corn (*Zea mays*) and rice (*Oryza sativa*) are almost entirely amylopectin. At the other extreme, a variety of wrinkled pea (*Pisum sativum* var. Steadfast) makes starch that is 80% amylose.

Starch grains have many shapes and structures (fig. 2.4b). These shapes depend not only on the linkages of amylose and amylopectin but also on how these molecules are folded into secondary structures. In general, amylose twists into coils, groups of which are surrounded by the larger amylopectin molecules.

Although starches are the most common storage polysaccharides in plants, they are not the only ones. Many plants make **inulin** instead of or in addition to starch. Inulin is a polymer of fructose having beta-2,1 linkages (the number 2 carbon of fructose reacts like the number 1 carbon of glucose). Inulin is a storage polysaccharide in dahlias (*Dahlia* species), Jerusalem artichoke (*Helianthus tuberosum*), globe artichoke (*Cynara scolymus*), chicory (*Cichorium intybus*), and sweet corn. The storage organs of these plants (e.g., tubers of Jerusalem artichoke, kernels of sweet corn) taste sweet because inulin releases fructose.

Amylose Versus Cellulose: Feast or Famine

As you learned from the preceding discussion, there is only a slight difference between alpha-glucose and beta-glucose (fig. 2.2); however, this difference gives the molecule significantly different properties. Cellulose is a strong structural polysaccharide because it forms microfibrils and fibrils. More important, cellulose is impervious to the enzymes that degrade starch. One such enzyme, called **alpha-amylase**, digests only alpha-glucose linkages in amylose. Thus, the calories stored in cellulose are unavailable for plant growth or for herbivorous animals. This is why plants cannot use cellulose as a storage reserve and why we get little energy from eating a stalk of celery or a spoonful of bran. Conversely, plants produce alpha-amylase in seeds, roots, and tubers when they harvest the energy from stored starch. Because human saliva contains alpha-amylase, the digestion of starch in our food begins even before it is swallowed.

Some organisms can digest cellulose because they make enzymes called **cellulases**. Examples include some fungi and bacteria, earthworms, and certain insects. Cows, sheep, termites, and other herbivores digest cellulose indirectly by housing cellulase-producing microorganisms in their digestive systems. The microorganisms digest the cellulose, and the animals absorb the glucose released from it.

---

c o n c e p t

Carbohydrates include monomers called *monosaccharides* and polymers called *polysaccharides*. Polysaccharides differ in the kind of monomer they contain and in the linkages between monomers. The two most widespread polysaccharides in plants are cellulose and starch. Cellulose is a structural polysaccharide, and starch is a storage polysaccharide.

Proteins

After cellulose, proteins make up most of the remaining biomass of living plant cells. A protein consists of one or more **polypeptides** and may also include sugars or other kinds of small molecules. A polypeptide is a chain of amino acids linked together by carbon-nitrogen bonds called **peptide bonds** (fig. 2.5). The smallest polypeptides have fewer than 100 amino acids, and the largest have several thousand. There are thousands of different kinds of proteins.

Like polysaccharides, proteins are important in cell structure and as storage reserves. Many proteins are also enzymes that catalyze biochemical reactions. Unlike polysaccharides, proteins do not occur as chains of the same monomer with different linkages. Instead, each protein contains different amino acids linked by the same peptide bonds. The diversity of proteins results from different sequences of the amino acids. Thus, each amino acid is like a letter in an alphabet, and each polypeptide is like a long word that usually contains the entire alphabet, with some letters repeated many times.

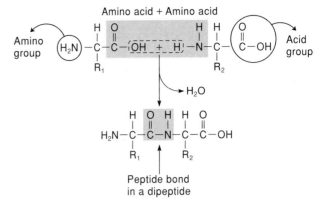

figure 2.5

Peptide synthesis occurs as a bond forms between the amine-nitrogen of one amino acid and the carboxyl-carbon of another, with the removal of a molecule of water. (Water is not actually released into the aqueous medium of the cell. See Appendix A for a discussion of where this water molecule goes.)

Each of the 20 amino acids in plants has two main parts. The first part, which all amino acids share, consists of a carbon with both a carboxylic acid group ($-COOH$) and an amino group ($-NH_2$ or $-NH-$) attached to it. The second part, which is attached to the same carbon as the amino and carboxyl groups, is the side chain. The side chain, or *R group*, differs for each amino acid. The R groups range in complexity from the simple hydrogen atom, as in glycine, to a ring structure, as in tryptophan (fig. 2.6).

Protein Structure

The sequence of amino acids in a protein is the protein's **primary structure** (fig. 2.7). However, proteins are not simply long molecules that wave randomly in the cell; rather, they fold and twist into three-dimensional shapes. The shape of a protein is determined by bonds and other interactions between amino acids at different positions in the primary structure. Hydrogen bonds occur at regular intervals and hold specific parts of a polypeptide together in a **secondary structure**. The secondary structure is shaped like either a coil or a pleated sheet (fig. 2.7).

Disulfide bonds, which are covalent bonds between the sulfurs of two cysteines, maintain the rigid **tertiary structure** of a protein. The tertiary structure is also maintained by weak bonds between ionized carboxyl and amino groups, and by adjacent nonpolar amino acids that aggregate by excluding water from their vicinity. The result of all of these interactions is a protein with a unique three-dimensional structure.

Many proteins consist of two or more polypeptide chains called **subunits**. For example, the plant enzyme ribulose-1, 5-bisphosphate carboxylase/oxygenase (known as *rubisco*) has sixteen subunits. In this enzyme, one polypeptide makes up eight large subunits, and another polypeptide makes up eight small subunits. The **quaternary structure** of this protein refers to the way that the different subunits attach to each other. Only proteins having more than one subunit have a quaternary structure.

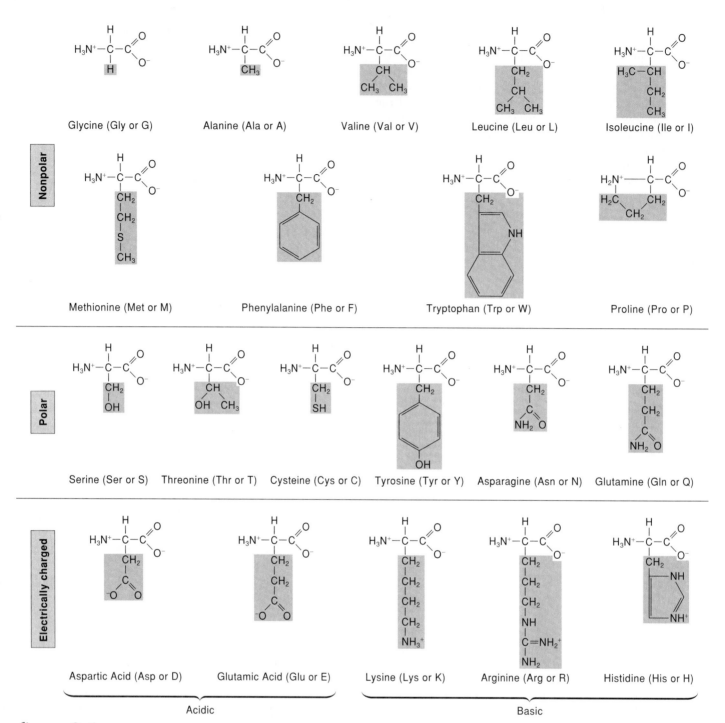

Nonpolar

Glycine (Gly or G)   Alanine (Ala or A)   Valine (Val or V)   Leucine (Leu or L)   Isoleucine (Ile or I)

Methionine (Met or M)   Phenylalanine (Phe or F)   Tryptophan (Trp or W)   Proline (Pro or P)

Polar

Serine (Ser or S)   Threonine (Thr or T)   Cysteine (Cys or C)   Tyrosine (Tyr or Y)   Asparagine (Asn or N)   Glutamine (Gln or Q)

Electrically charged

Aspartic Acid (Asp or D)   Glutamic Acid (Glu or E)   Lysine (Lys or K)   Arginine (Arg or R)   Histidine (His or H)

Acidic   Basic

figure 2.6

Structures of the twenty amino acids that occur most commonly in proteins. Each amino acid has one carbon that is bonded to both an amino group (H_2N—) and a carboxyl group (—COOH). At neutral pH (pH 7), the prevailing form of the amine is H_3N^+ and of the carboxyl is COO^-. The amino acids are arranged according to the main properties of their side chains. Each R group is highlighted by a colored background.

In solution, the three-dimensional structure of a protein may be somewhat flexible. Proteins can also be **denatured** in laboratory experiments by adding urea or other chemicals. Urea inhibits hydrogen bonds that contribute to secondary structure, thereby causing a protein to lose its shape and function. However, denaturation may be reversible, since proteins often regain their original shapes when the urea is removed from the solu-

tion. Proteins may also be irreversibly denatured by heat or harsher chemicals. For example, a cooked egg, which is mostly denatured protein, cannot be "uncooked" (i.e., renatured).

Proteins are difficult to classify because of their great diversity. The most commonly used schemes of protein classification are based primarily on a protein's solubility in water or detergent solutions and on its functions. For simplicity, we will

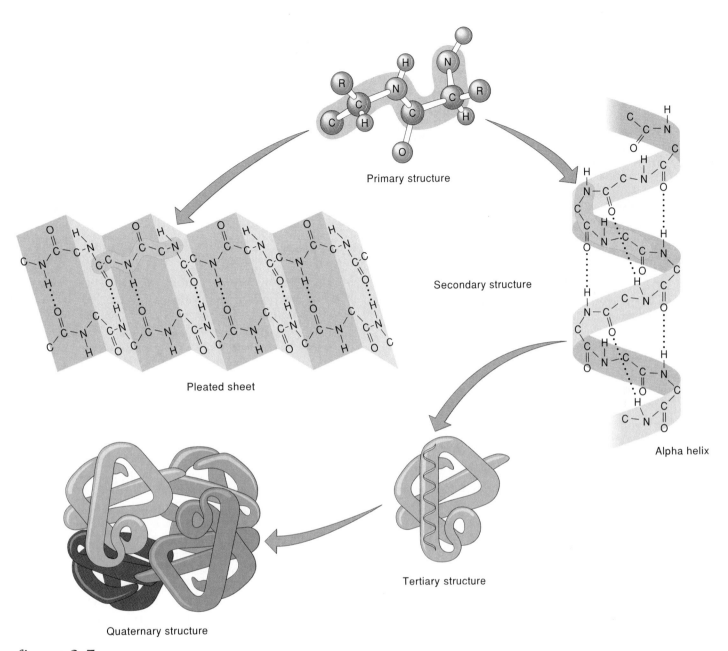

Primary structure

Secondary structure

Pleated sheet

Alpha helix

Tertiary structure

Quaternary structure

figure 2.7

Models of the primary, secondary, tertiary, and quaternary structures of a protein. The primary structure is the sequence of amino acids. The secondary structure may consist of coils (alpha helices) or folds (pleated sheets), depending on which amino acids form hydrogen bonds with each other. (For simplicity, R groups are omitted from the pleated sheet and the alpha helix.) Tertiary structure is maintained by different kinds of chemical interactions within the protein. Quaternary structure occurs between different subunits.

discuss proteins relative to their three major functions: structural proteins, storage proteins, and enzymes. Most of the proteins mentioned in this book fit into one of these categories.

Structural Proteins in Cell Walls and Membranes

In addition to containing carbohydrates, cell walls also include from 2% to 10% protein. Among these proteins are **expansins** that help cell walls increase their surface area. Expansins do this by causing slippage between the polysaccharides that form the wall. This slippage, in turn, results from the breakage and formation of hydrogen bonds between the polysaccharides. Several of the structural proteins in cell walls were originally thought to play a role in the expansion of cell walls, and were therefore named **extensin.** The synthesis of extensin is induced when cells are damaged by wounding, infection, or freezing, suggesting that extensins somehow help protect or repair damaged cells.

Extensins also consist of up to 30% carbohydrate, which makes them **glycoproteins**—that is, "sugar-proteins." The carbohydrates of extensins include mostly arabinose and galactose. Although the exact structure of the extensin-polysaccharide

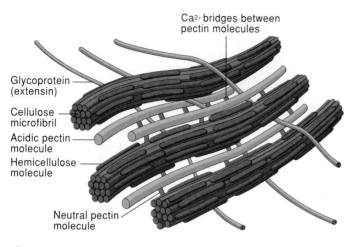

Ca²⁺ bridges between pectin molecules

Glycoprotein (extensin)

Cellulose microfibril

Acidic pectin molecule

Hemicellulose molecule

Neutral pectin molecule

figure 2.8

Model of the interconnections among major components of primary cell walls. Hemicellulose molecules bind to the surface of cellulose microfibrils by hydrogen bonds and to acidic pectin molecules by cross-links with neutral pectin molecules. Calcium (Ca^{2+}) holds acidic pectin molecules to one another. Glycoproteins (e.g., extensins and expansins) are tightly woven into the cell-wall matrix. Since the polysaccharides are covalently bonded to one another, we can interpret the cell wall as one giant macromolecule.

complex in cell walls is unknown, the arabinose or galactose of extensins probably attaches to one or more kinds of cell-wall polysaccharides (fig. 2.8).

Structural proteins also occur in all of the membranes of a cell, and each kind of membrane has a different protein composition. For example, the internal membranes of mitochondria and chloroplasts are about 75% protein, whereas the membrane that surrounds the cell is usually about 50% protein. Lipids comprise most of the nonprotein portion of membranes. The composition, structure, and functions of different membranes are discussed in more detail in Chapters 3, 4, 6, and 7.

Storage Proteins

Storage proteins are stored mostly in seeds and are used as a source of nutrition for the early development of seedlings. The composition of seed proteins depends on the plant species. For example, corn produces a storage protein called **zein,** which consists of nearly thirty polypeptides that occur in two major subunits and one minor subunit. In comparison, wheat produces a storage protein called **gliadin,** which has four major subunits that are made up of at least forty-six polypeptides. The most complex storage proteins may be the glutenins of wheat, which consist of up to fifteen different proteins, ranging in size from 1,100 to more than 13,300 **kilodaltons** (1 kilodalton is equal to a molecular mass of 1,000).

Seed storage proteins in soybean and cereal grains (oats, rice, wheat, corn, barley) are especially important because they are a major source of nutrition for humans and cattle.

However, corn and barley are low in the amino acids lysine, threonine, and tryptophan. Soybean and other legume seeds are deficient in the sulfur amino acids (cysteine and methionine), but their lysine content is adequate.

Seeds of some plants also contain proteins that have undesirable nutritional effects on the animals that eat them. For example, up to 10% of the protein in many cereal grains inhibits certain digestive enzymes of animals. These proteins are called **protease inhibitors** because they inhibit proteases, the enzymes that digest proteins. The seed proteins of a few plants are toxic. These include glycoproteins such as **ricin D** from the castor bean (*Ricinus communis*) and **abrin** from the rosary pea (*Abrus precatorius*).

Enzymes

Most proteins in a living cell are **enzymes,** which are the catalysts for biochemical reactions. This means that enzymes speed up reactions without being consumed in the process.

Enzymes usually have flexible, globular shapes. A specific place on each enzyme, called an **active site,** binds to one or more substrates (reactants). Enzymes are often named by adding the ending *-ase* to the name of a reactant or product of the reaction that the enzyme catalyzes. One example is alpha-amylase, the enzyme that digests amylose. Enzymes also catalyze reactions that make or digest cell walls, membranes, storage polymers, proteins, DNA, RNA, pollen walls, seed coats, chlorophyll, amino acids, and other plant metabolites. Examples of enzymes and their roles in the lives of plants are described in almost every chapter in this textbook.

Enzymes often remain active even after they are removed from the cell. Alpha-amylase, for example, digests amylose in a test tube if it is provided with the appropriate temperature, pH, and cofactors. Pure enzymes that maintain their activity are commercially important. These enzymes include the proteases **papain** and **chymopapain** from papaya (*Carica papaya*). Papain digests the muscle tissue of animals, which is why it is a major ingredient in meat tenderizers. Chymopapain is a drug used to treat the slippage of a disk in the spinal column. Chymopapain is injected directly into the problem area, where it dissolves the proteinaceous cartilage of which the disk is made. You will learn more about how enzymes work in Chapter 5.

c o n c e p t

Proteins consist of chains of up to twenty different amino acids bonded into polypeptides. The primary structure of a polypeptide is the amino acid sequence. Secondary and tertiary structures are the three-dimensional shapes created by various nonpeptide bonds and other interactions among amino acids within the polypeptide. Quaternary structure is the arrangement of more than one polypeptide in a protein. Like carbohydrates, proteins have roles in cell structure and in storage. Unlike carbohydrates, proteins may also be enzymes.

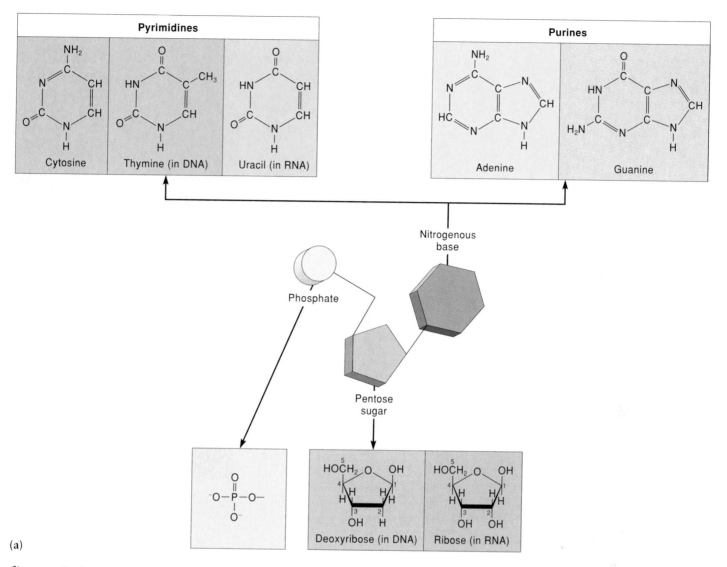

figure 2.9

The structure of DNA. (a) Each nucleotide monomer consists of three smaller building blocks: a nitrogenous base, a pentose sugar, and a phosphate group. (b) Nucleotide monomers are bonded to each other by covalent bonds between the phosphate of one nucleotide and the sugar of the next nucleotide. (c) DNA is usually a double strand held together by hydrogen bonds between nitrogenous bases: Adenine (A) pairs only with thymine (T), and cytosine (C) pairs only with guanine (G). The double strand is twisted spirally into a double helix. (d) A space-filling model shows the close stacking between nitrogenous bases. Color code for atoms: yellow = P, dark blue = C, red = O, turquoise = N, and white = H.

The Lore of Plants

There are no pineapple-flavored gelatin desserts, because pineapples contain a protease (bromelain) that hydrolyzes gelatin. Hydrolyzed gelatin cannot form a gel, so gelatin desserts containing pineapple juice would be too runny to appeal to the typical consumer.

Nucleic Acids

The most complex biological polymers are **nucleic acids.** Their complexity emerges from a primary structure that contains mostly four different monomers, each of which is called a **nucleotide.** Each nucleotide consists of a phosphate group, a simple sugar, and a nitrogen-containing base (fig. 2.9). The simple sugar is either ribose, which occurs in **ribonucleic acid (RNA),** or deoxyribose, which is the sugar in **deoxyribonucleic acid (DNA).** There are two classes of nitrogenous bases: pyrimidines, which have one ring, and purines, which have two rings. The pyrimidines in DNA are cytosine and thymine; the purines are guanine and adenine. The base composition of RNA is the same as that of DNA, except that RNA includes uracil instead of thymine. Furthermore, RNA occurs as a single strand, whereas DNA occurs as a two-stranded spiral called a **double helix.**

Nucleic acid molecules vary in size from about thirty nucleotides to several million nucleotides. The smaller nucleic acids are primarily RNA. Even the smallest nucleic acids, however,

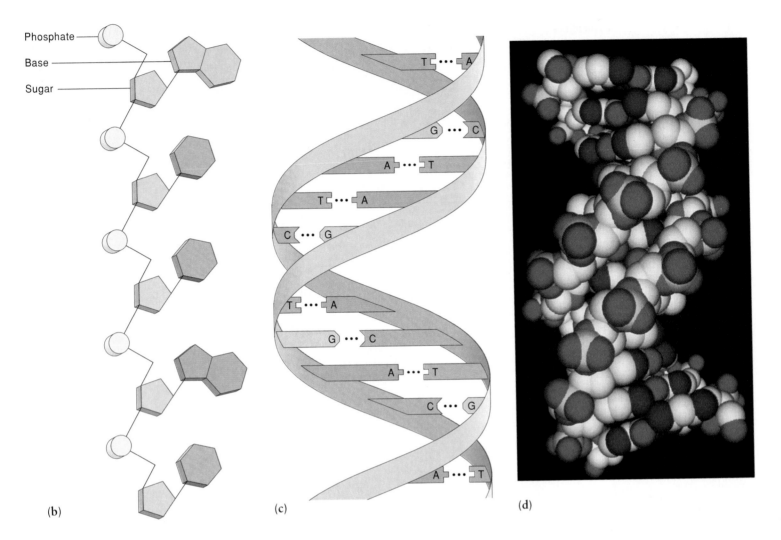

Phosphate
Base
Sugar

(b)   (c)   (d)

figure 2.9   (continued)

have a seemingly infinite number of nucleotide sequences. For example, four different nucleotides can be arranged in a ten-monomer sequence in 4^{10} (1,048,576) different ways. This is analogous to using an alphabet with four letters to write more than a million ten-letter words.

Although DNA consists of the same four nucleotides in all organisms, the amounts of different nucleotides are variable. These amounts are generally measured as the percentage of G (guanine) + C (cytosine) since guanine and cytosine always complement each other, and adenine and thymine always complement each other (the reason that G/C and A/T are complementary is explained in Chapter 9). For example, the GC content of plants is usually between 36% and 43%, but in the grass family (*Poaceae*) it exceeds 48%.

Nucleic acids are unique because they can replicate themselves. Furthermore, DNA can make RNA, which guides the assembly of proteins. Nucleic acids form the molecular foundation for every living organism. The roles of nucleic acids are so central to life that all of Unit 3 in this text ("Genetics") is devoted to their biology, chemistry, and importance.

Other Nucleotides

Nucleotides other than adenine, guanine, cytosine, and thymine occasionally occur in DNA in trace amounts. One that is abundant, however, is 5-methylcytosine, an otherwise normal cytosine that has a methyl group (—CH$_3$) added to it. The seemingly minor change from cytosine to 5-methylcytosine inactivates the DNA.

As much as 30% of the cytosine in plant DNA is replaced with 5-methylcytosine. Since cytosine is methylated and demethylated at different times during the life of a cell, these reactions probably regulate the activity of genes.

c o n c e p t

Nucleic acids are chains of nucleotides. Different nucleic acids vary in their nucleotide composition and sequence. The two main types of nucleic acids, RNA and DNA, also contain different sugars: RNA contains ribose, and DNA contains deoxyribose.

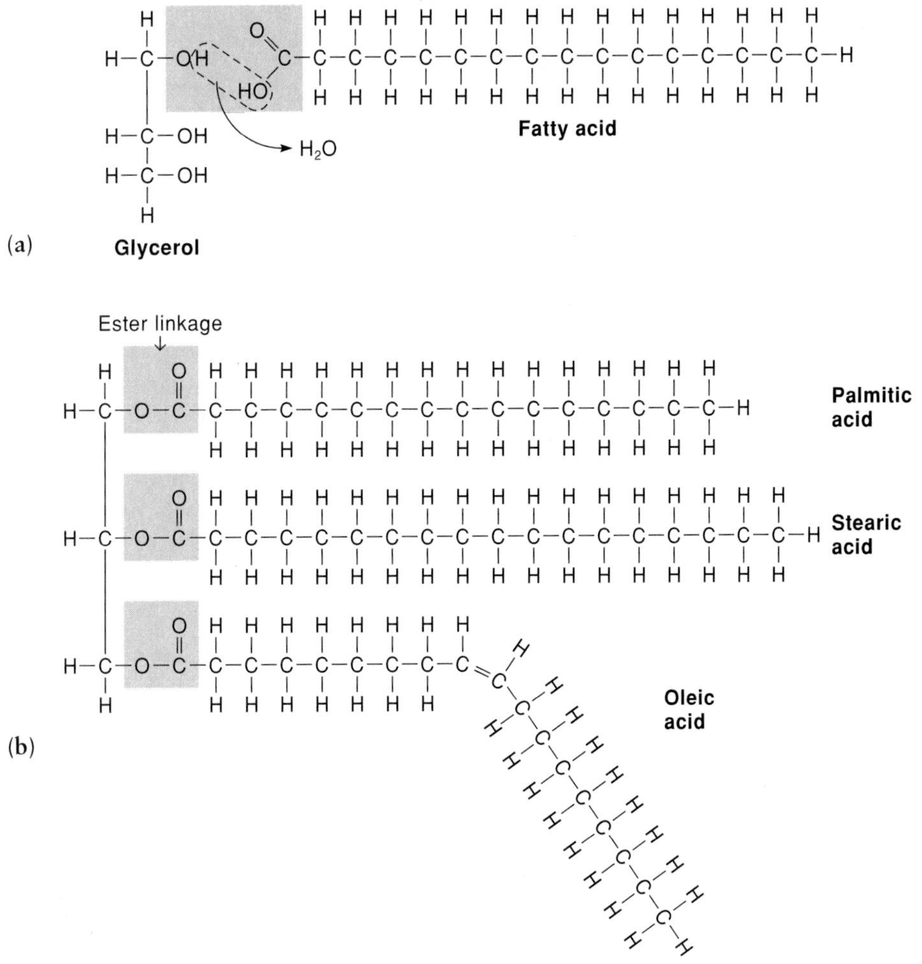

figure 2.10

The structure of a fat. (a) An acylglyceride bond forms when the carboxyl group of a fatty acid links to the hydroxyl group of a glycerol, with the removal of a water molecule. (b) Fats are triacylglycerides whose fatty acids vary in length and also in the presence and location of carbon-carbon double bonds.

Lipids

Unlike other biological polymers, lipids are not defined by specific, repeating, monomeric subunits. Rather, they are defined by their water-repellent property (*lipid* means "fat"). The only common structural theme shared by all lipids is a large proportion of nonpolar hydrocarbon groups (—CH₃, —CH₂, or —CH). These hydrocarbon groups are often made from polymers of a two-carbon compound called *acetate*. The major plant lipids with acetate-derived hydrocarbon chains include oils, phospholipids, and waxes.

Oils

Oils are fats that are liquid at room temperature. A fat is a molecule of glycerol with three long-chain organic acids, called **fatty acids,** linked to it (fig. 2.10). The linkage between a fatty acid and glycerol is called an *acylglyceride linkage.* Thus, a fat is a **triacylglyceride.**

Fatty acids of triacylglycerides may differ in length and in the number and placement of double bonds in the chain. A fatty acid having no carbon-carbon double bonds is **saturated.**

Oils are liquid because their fatty acids are **unsaturated;** that is, they have double bonds between one or more pairs of carbon atoms. Double bonds provide molecular rigidity at sharp angles, which prevents the molecules from packing tightly into a solid. This is why unsaturated lipids such as corn oil and peanut oil are liquids at room temperature.

Although triacylglycerides occur in all parts of a plant, they are most abundant in seeds. Like carbohydrates, seed oils are a form of chemical energy that is harvested when the seed germinates. Some seeds contain enough oil to be commercially valuable. The best known of these are cotton, sesame, safflower, sunflower, olive, coconut, peanut, corn, castor bean, and soybean. You can buy most of these oils at grocery stores. Indeed, more than 90% of the oils that we harvest from these plants are used for products such as margarines, shortenings, salad oils, and frying oils; the rest is used for nonfood products such as lubricants, fuels, coatings, and soaps.

Fatty acids in plant oils are usually either monounsaturated or polyunsaturated with two or more double bonds. The most common fatty acids are oleic acid (one double bond), linoleic acid (two double bonds), and linolenic acid (three double bonds).

WHAT'S WRONG WITH TROPICAL OILS?

In most people, saturated fats raise blood-cholesterol levels more than do monounsaturated or polyunsaturated fats. In fact, blood-cholesterol levels are raised more by saturated fats than by cholesterol in the diet. High blood cholesterol has been linked with an increased risk of heart disease. Thus, it is important to minimize saturated fats in our diets.

How does this often-repeated advice relate to tropical oils? Tropical oils are the most common "hidden" source of saturated fats. They have long been used to make a variety of processed foods, including breakfast cereals, crackers, cookies, coffee creamers, baked goods, and microwave popcorn. Palm kernel oil and coconut oil are also used in cocoa mixes, diet desserts, instant soup, and chocolate.

It seems like a simple task to avoid tropical oils, because all you have to do is read the ingredients list on a food label. Some manufacturers are already making this easier by printing "NO TROPICAL OILS" in bold, colorful letters on the fronts of packages. However, many such products still contain saturated fats in the form of "hydrogenated vegetable oil." When hydrogen is added to the double bonds of a polyunsaturated fat, the fat becomes saturated. This means that linolenic acid, with three double bonds, becomes palmitic acid when it is hydrogenated. Since palmitic acid from linolenic acid does not come from palms or coconuts, the "no tropical oils" claim is legitimate; however, the saturated fat content may still be high.

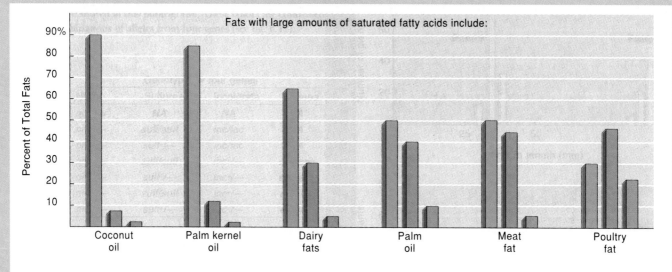

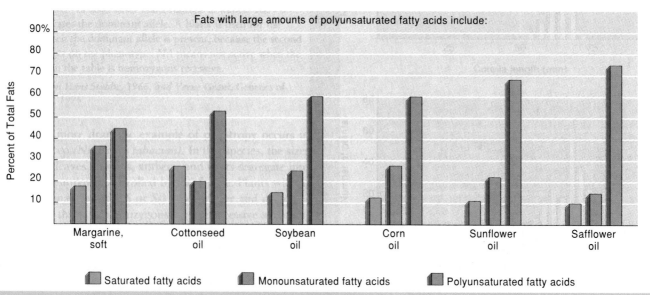

reading figure 2.2

Dietary fat and fatty acid proportions in different foods.

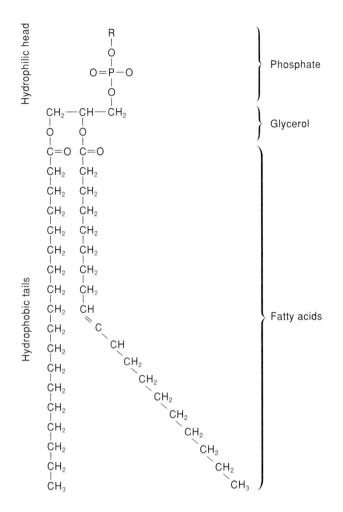

figure 2.11

The structure of a phospholipid. A phospholipid consists of glycerol that is bonded to two fatty acids, which are hydrophobic, and to one phosphate group, which is hydrophilic. Phospholipids vary by their fatty acids and by side chains (R groups) that are attached to the phosphate. The R groups include glycerol, sugars, and amine-containing carbon chains.

Linoleic acid and linolenic acid are essential fatty acids. This means that they are necessary for human growth and development, but our bodies cannot synthesize them.

Seed oils from palms, coconuts, and other tropical plants contain mostly palmitic acid, which is saturated. Reading 2.2, "What's Wrong with Tropical Oils?" explains how the composition of different plant oils affects human health.

Phospholipids

Membranes contain lipids in which a phosphate group is substituted for one of the fatty acids (fig. 2.11). Such lipids are called **phospholipids.** The phosphate group gives the compound a polar end that dissolves in water or forms a covalent bond with a membrane protein. This property causes phospholipids to have

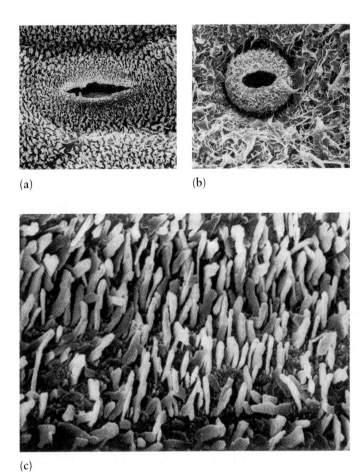

figure 2.12

Scanning electron micrographs of leaf surfaces, showing the different structures of epicuticular waxes. These are all from different orchids. (a) *Cleistes rosea*, ×700. (b) *Stelis gemma*, ×550. (c) *Eulophia paivaeana*, ×1,400.

a dual solubility, since the phosphate end is water-soluble (hydrophilic) and the fatty end is water-repellent (hydrophobic). Consequently, phospholipids interact with the polar groups of proteins at one end and form an oily matrix at the other. This versatility enables membranes to control the passage of polar and nonpolar substances through them. Chapter 4, "Membranes and Membrane Transport," presents more details about the role of phospholipids in membrane structure and function.

Waxes and Waxlike Substances

Waxes are complex mixtures of fatty acids linked to long-chain alcohols. This mixture also contains free fatty acids, fatty acids with hydroxyl groups (e.g., –OH), and long-chain hydrocarbons. Waxes that comprise the outermost layer (*cuticle*) of leaves, fruits, and herbaceous stems are called **epicuticular wax.** Waxes embedded in the cuticle are called **cuticular wax.** The structures of different waxes vary depending on the plants that produce them (fig. 2.12).

Waxes are usually harder and more water-repellent than other fats, which is why you don't "fat" your car or kitchen floor. Wax from plants such as the carnauba palm (*Copernicia cerifera*, the source of carnauba wax) is especially hard. Other commercial waxes come from the bayberry (*Myrica pensylvanica*), which is used to make novelty candles, and from candellila (*Euphorbia antisyphilitica*). Candellila wax, which is often substituted for carnauba wax, was once the main component of the hard "melts in your mouth, not in your hands" coating of M&M chocolate candies. (The *antisyphilitica* part of the scientific name implies another use for this plant, but there is no evidence that it can be used against sexually transmitted diseases.) Unfortunately, candellila is near extinction because the plants have been over-collected for their wax. Collecting this plant is forbidden in the United States, but it is still heavily collected in Mexico.

Although most waxes are hard, one significant exception is the wax from jojoba seed (*Simmondsia chinensis*). About half of a jojoba seed is liquid wax. This wax is sometimes called an oil because it is liquid at room temperature, but it has the chemical makeup of a wax. Jojoba "oil" is used in cosmetics because it penetrates the outer layer of human skin. It is also used as a lubricant. Jojoba oil is similar to the oil of sperm whales, which are an endangered species. Botanists and chemists are trying to modify jojoba oil to substitute for sperm whale oil as a lubricant for high-speed industrial machinery.

Cuticular wax is also associated with **cutin**, another waxy substance, which makes up most of the cuticle. Cutin consists of a variety of hydroxylated fatty acids linked together. A similar substance, called **suberin**, occurs in cork cells in bark and in the same cells of underground plant parts. Cutin and suberin differ mainly in the kinds of fatty acids they contain, but both function as barriers to water loss.

c o n c e p t

Lipids are water-repellent compounds made mostly of carbon and hydrogen. The major lipids of plants are triacylglycerides, phospholipids, and waxes, which all contain fatty acids. Triacylglycerides function mainly as storage in oilseeds. Phospholipids are important components of membranes. Waxes and waxlike substances prevent water loss from plants.

Secondary Metabolites

Plants make a variety of less widely distributed compounds, such as morphine, caffeine, nicotine, menthol, and rubber. These compounds are the products of **secondary metabolism,** which is the metabolism of chemicals that occur irregularly or rarely among plants and that have no known general metabolic role in cells.

Most plants have not yet been examined for their secondary products, and new compounds are discovered almost daily. Most secondary products can be grouped into classes based on structural similarities, biosynthetic pathways, or the kinds of plants that make them. The largest such classes are the alkaloids, the terpenoids, and the phenolics. Examples of several kinds of secondary metabolites are presented in table 2.2.

Secondary metabolites often occur in plants in combination with one or more sugars. Such combination molecules are called **glycosides**. The most common sugars in glycosidic secondary products are glucose, galactose, and rhamnose. Several other sugars also occur in glycosides, some of which are rare. For example, digitoxose is known only from the genus of the purple foxglove (*Digitalis purpurea*; fig. 2.14c), and apiose is unique to parsley (*Petroselinum*) and its relatives.

Functions of Secondary Metabolites

The most common roles suggested for secondary products are ecological roles that govern the interactions between plants and other organisms. For example, many secondary products are brightly colored pigments that attract insects and other animals for pollination or dispersal of fruits and seeds. Conversely, nicotine and other toxic chemicals may protect plants against attacks by hungry insects or invasion by pathogenic microbes. This defensive role seems to be true in some cases but not in others. For example, phenolics in soybeans provide resistance to fungal infection, and nicotine from tobacco is toxic to many insects. Conversely, tobacco hornworms have little difficulty eating tobacco leaves that contain up to 30% nicotine, and monarch butterflies are unaffected by the cardiac glycosides of milkweeds. Instead of protecting the plants from being eaten by butterfly caterpillars, these toxins remain in the adult butterflies and protect them from insect-eating birds. Although there are many examples of the ecological functions of these kinds of chemicals, most secondary metabolites have not been examined for their potential ecological roles in plants.

Alkaloids

Alkaloids generally include alkaline substances that contain nitrogen as part of a ring structure. The alkaloids, of which more than 6,500 are known, comprise the largest class of secondary metabolites. They occur in several plant families, especially the pea, sunflower, poppy, citrus, and potato families. Alkaloids are unknown in mosses, ferns, conifers, and most families of flowering plants.

Alkaloids are a diverse group of secondary products, ranging from simple compounds like coniine to complex compounds like strychnine and tomatine (fig. 2.13). They often produce dramatic physiological effects in humans and other animals. For example, coniine, strychnine, and tubocurarine are infamous toxins, while morphine, codeine, atropine, and vincristine are important therapeutic drugs. Alkaloids are often bitter; one of the most bitter substances known is the alkaloid quinine.

table 2.2

Examples of Well-Known Secondary Metabolites

Compound	Example Source	Comment
Alkaloids		
coniine	poison hemlock	nerve toxin; killer of Socrates
strychnine	strychnine tree	potent nerve stimulant and convulsant
tubocurarine	curare tree	component of arrow poisons; used as muscle relaxant during surgery
tomatine	tomato leaves	inhibits feeding by Colorado potato beetles, which are also pests of tomatoes
morphine	opium poppy	main painkiller used worldwide
codeine	opium poppy	cough suppressant
atropine	belladonna	used in eye exams to dilate pupils and as an antidote to nerve gas
vincristine	Madagascar periwinkle	main treatment for certain kinds of leukemia
quinine	quinine tree	bitter flavor of tonic in gin and tonic; used to prevent malaria

Opium poppy (*Papaver somniferum*)

Compound	Example Source	Comment
Other Nitrogen-Containing Compounds		
canavanine	jack beans	toxic nonprotein amino acid
betanin	roots of garden beet	red pigment
sinigrin	horseradish	"burning" spice of mustards
amygdalin	apricot seeds	releases cyanide; may be active ingredient of unapproved cancer drug called Laetrile

Garden beet (*Beta vulgaris*)

Compound	Example Source	Comment
Terpenoids		
menthol	mints and eucalyptus tree	strong aroma; used in cough medicines
camphor	camphor tree	component of disinfectants and plasticizers
nepetalactone	catnip	very attractive to cats
smilagenin	sarsaparilla	steroidal glycoside
digitalin	purple foxglove	cardiotonic used to stimulate heart action
oleandrin	oleander	heart poison (similar to digitalin)
lycopene	tomatoes	red/orange pigment
rubber	rubber tree	component of rubber tires
taxol	Pacific yew	drug that inhibits cancerous tumors, especially of ovarian cancer

Pacific yew (*Taxus brevifolia*)

Compound	Example Source	Comment
Phenolics		
salicin	willow tree	folk medicine against headaches and fever; precursor of aspirin*
gallic acid	walnut husks	main component of some tannins
myristicin	nutmeg	main flavor of the spice
rutin	buckwheat	common "bioflavonoid" sold in nutrition stores
cyanidin glucoside	chrysanthemums	deep red pigment
limonin	grapefruit	bitter flavor

Nutmeg (*Myristica fragrans*)

*The medicinal reputation of willow (*Salix alba*) stems from salicin, a chemical first isolated from willow bark in 1827. Although salicin was medically useless because of its severe side effects, related compounds were later isolated from other plants (e.g., salicylic acid from meadowsweet, *Spiraea ulmaria*). Finally, in 1899, Felix Hoffman produced acetylsalicylic acid to help his father deal with arthritis. Acetylsalicylic acid, which produced no bad side effects, was later mass-produced by Friedrich Bayer & Co. Bayer named the compound "aspirin"—*a* for acetyl and *spirin* for *Spiraea*—which, although catchy, is misleading, because *Salix* was there first. Today, aspirin is the world's most frequently taken medicine; Americans swallow almost 40 tons of it every day.

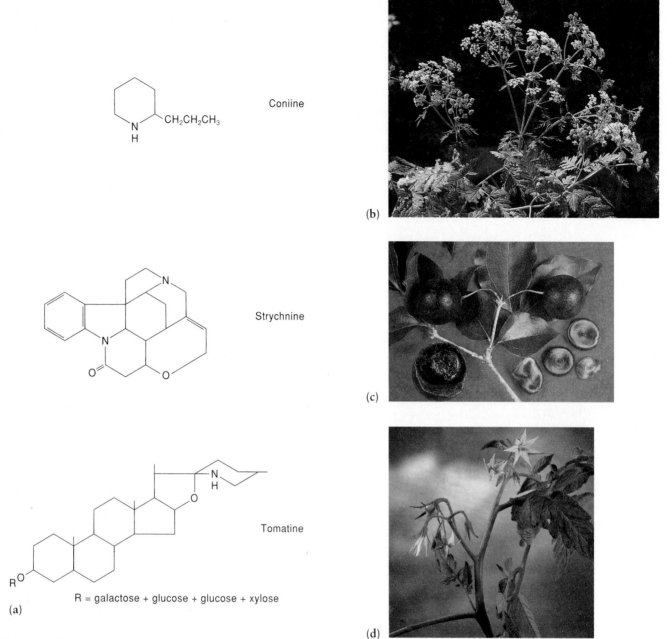

Coniine

Strychnine

Tomatine

R = galactose + glucose + glucose + xylose

(a)

(b)

(c)

(d)

figure 2.13

Secondary metabolites in plants. (a) Structures of three alkaloids. (b) Poison hemlock (*Conium maculatum*) produces coniine in its leaves. (c) Strychnine plant (*Strychnos nux-vomica*) produces strychnine in seed coats. (d) Tomato leaves (*Lycopersicon esculentum*) are a source of tomatine.

Terpenoids

Terpenoids are dimers and polymers of five-carbon precursors called **isoprene** units (C_5H_8; fig. 2.14). The production of these isoprene units, which often evaporate from plants and contribute to the haze that we see on hot sunny days, is an expensive proposition for plants: isoprene production typically siphons off 2% of the carbon fixed in photosynthesis—carbon that could otherwise be converted to sugars. The smallest terpenoids, the **monoterpenes,** have two isoprene units. Monoter-

penes such as geraniol and menthol, from the leaves of mints and eucalyptus, are volatile and usually have strong odors. Pines and other resinous plants also synthesize terpenoids made from four isoprene units (**diterpenes**) or six isoprene units (**triterpenes**).

One subclass of triterpenes is the **sterols,** which are chemically similar to the steroidal hormones of animals. Sterols may be combined with nitrogen to form alkaloids, as in tomatine (fig. 2.13), or with sugars in steroidal glycosides

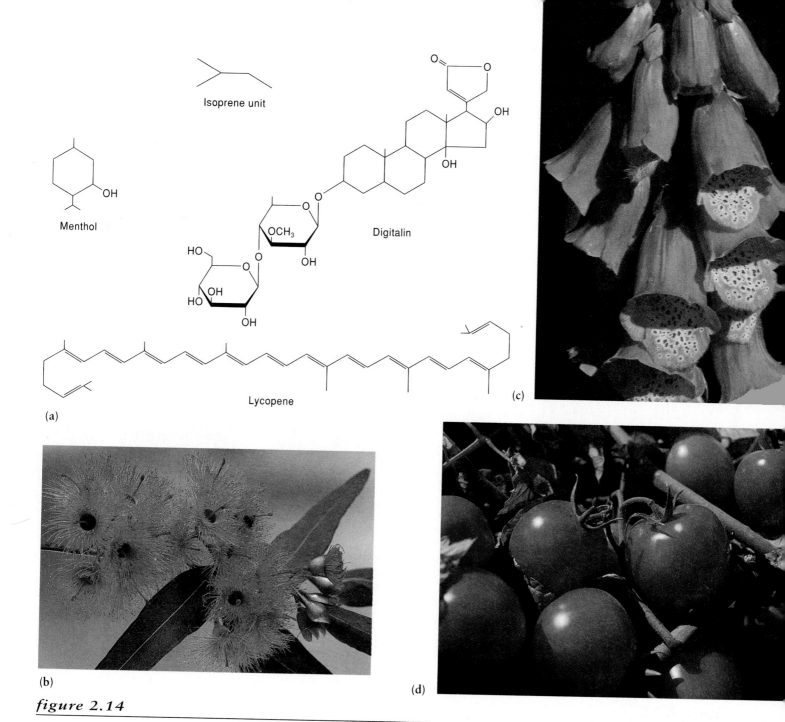

figure 2.14

Terpenoids. (a) Structures of three terpenoids. (b) *Eucalyptus* species produce menthol. (c) The purple foxglove (*Digitalis purpurea*) is a source of digitalin. (d) Lycopene is the main red pigment of tomatoes.

like digitalin (fig. 2.14). Terpenoids having eight isoprene units form a class of yellow to red pigments called **carotenoids** (see Chapter 7). The largest terpenoids, which can have more than 6,000 isoprene units in a single molecule, are known as **rubber.** Rubber is made by about 2,000 plant species, but most of them make this substance in amounts too small for commercial use.

Although terpenoids are considered secondary metabolites, they include some compounds that have clear roles in plants. The most prominent terpenoids are **abscisic acid,** which is made from three isoprene units; the **gibberellins,** which are diterpenes; sterols; and the carotenoid **beta-carotene.** Abscisic acid and the gibberellins are important plant hormones that regulate plant growth and development; Chapter 18 discusses these and other plant hormones. Plants need sterols for maintaining membrane function. Beta-carotene is a universally occurring accessory pigment in photosynthetic organisms. Abscisic acid, the

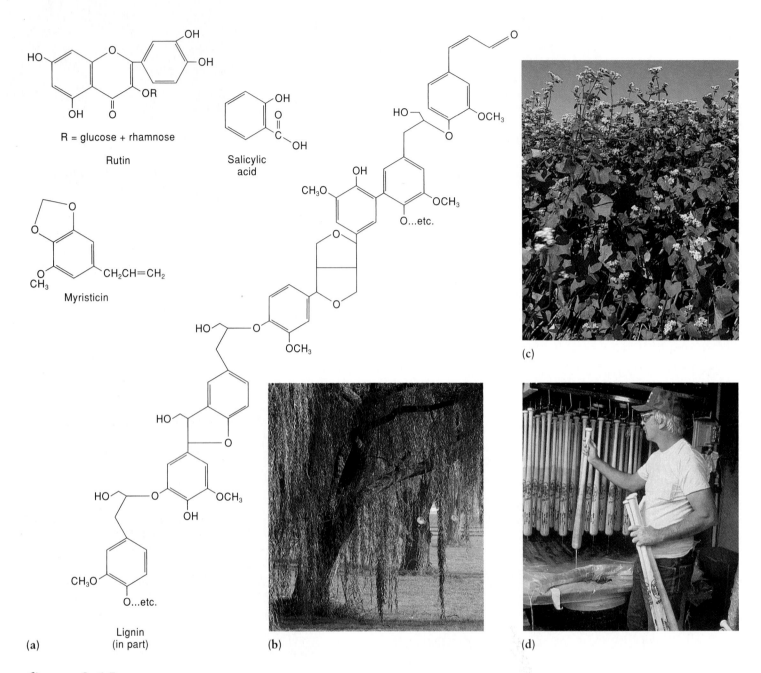

figure 2.15

(a) Structures of four phenolics. (b) Willows (*Salix* species) accumulate salicylic acid in their bark. (c) Rutin is produced abundantly in buckwheat (*Fagopyrum esculentum*) and many other plants. (d) Lignin is the strengthening polymer that makes wood valuable commercially. These baseball bats are made of the wood of white ash (*Fraxinus americana*). To learn how these bats are made, see reading 16.2, "The Bats of Summer: Botany and Our National Pastime," on page 370.

gibberellins, sterols, and beta-carotene are examples of basic metabolites that are derived from secondary metabolic pathways.

Phenolics

Compounds that contain a fully unsaturated, six-carbon ring linked to an oxygen are called **phenolics.** Simple phenolics, such as salicylic acid, are single-ring chemicals adorned by simple side groups (fig. 2.15). Complex phenolics that have a three-carbon side chain are called **phenylpropanoids** (*phenyl* refers to the ring; *prop* refers to the three-carbon side chain); they are derived from tyrosine and phenylalanine, which are phenylpropanoid amino acids (fig. 2.6). Myristicin, the main flavor of nutmeg (table 2.2), is an example of a familiar phenylpropanoid (fig. 2.15).

Phenylpropanoids commonly occur as parts of more complex molecules. For example, an entire subclass of secondary metabolites, called **flavonoids,** includes phenylpropanoids that are condensed into complex three-ringed structures. Certain flavonoids, called **anthocyanins,** are the red and blue pigments of many flowers; most others are colorless except in ultraviolet light. Flavonoids are sold in nutrition centers and health-food stores, usually as supplements with vitamin C. The most commonly available flavonoid is rutin, which is easily obtained from eucalyptus or buckwheat leaves (fig. 2.15c).

Like the terpenoids, the phenolics also form polymers. Flavonoid polymers, called **tannins,** can tan leather and are astringent to the taste—they are the compounds that impart the "dryness" (astringency) to dry wines. Perhaps the most significant phenolic polymer is **lignin,** a major structural component of wood. Lignin consists of polymeric phenylpropanoids whose combined structure is unknown.

Minor Classes of Secondary Metabolites

Besides the "big three" (alkaloids, terpenoids, phenolics), plants also make several minor classes of secondary metabolites. For example, **mustard oil glycosides** are nitrogen-sulfur compounds that occur in cabbage, broccoli, horseradish, watercress, and other members of the mustard family. These compounds produce the main flavor and odor of plants in this family.

Members of the pea family produce amino acids that are not incorporated into proteins. About 400 of these so-called **nonprotein amino acids** have been discovered.

Other minor classes of secondary products, such as **polyacetylenes** and **cyanogenic glycosides,** occur in several families of plants. Polyacetylenes are long-chain derivatives of fatty acids that contain carbon-carbon triple bonds. Most known polyacetylenes are from the sunflower family or the magnolia family. Cyanogenic glycosides are sugar-containing compounds that release cyanide gas when they are hydrolyzed. They occur in many plant families, but they are especially common in the pea and rose families. Cyanide from such plants is not usually fatal, but two cups of well-chewed apple seeds would probably kill an adult human. The seeds would have to be well chewed because cyanogenic glycosides are hydrolyzed only when the cells are damaged.

There are many other minor classes of secondary products. Many are familiar to us as the unique flavors and odors of common edible plants or the colors of garden flowers. New secondary products from these and other plants are reported in scientific journals almost daily. The tremendous diversity of secondary metabolites shows that plants are amazing biochemical factories.

c o n c e p t

Plants have common metabolic processes that involve carbohydrates, proteins, nucleic acids, and lipids. They also have a highly diverse secondary metabolism that makes many rare chemicals. Most secondary products belong to one of three main groups: terpenoids, phenolics, or alkaloids. The functions of secondary metabolites are generally unknown for most plants.

Chapter Summary

Organisms are organized into a complex hierarchy of atoms, molecules, and cells. Beginning with the structure of atoms, the successively higher levels of complexity depend on the properties of their simpler components. Elements comprise the simplest level that cannot be broken down by ordinary chemical means. The main elements in organisms are carbon, hydrogen, oxygen, and nitrogen. These and other elements bond in a variety of combinations to make carbohydrates, proteins, nucleic acids, lipids, and many other classes of plant chemicals.

Carbohydrates are monomers and polymers of sugars. The most abundant carbohydrate polymers are cellulose and starch, which exhibit the two main functions of carbohydrates: structure (cellulose) and storage (starch). Cellulose is a major component of cell walls, and starch is a common energy reserve. The structures of carbohydrate polymers depend on linkages between monomers and on hydrogen bonding between different chains of the polymer.

Proteins, like carbohydrates, are also important in structure and storage. Many proteins are enzymes that catalyze the chemical reactions of cells. Structural proteins occur mostly in membranes, but certain glycoproteins, such as the extensins, occur in cell walls. Storage proteins are common in seeds but not in other parts of plants.

The primary structure of a protein is determined by its sequence of amino acids. Secondary and tertiary structures depend on bonding and interactions between amino acids in the protein. Proteins consisting of more than one polypeptide have quaternary structure, which is the configuration of the subunits relative to each other. Nucleic acids are the largest and most complex polymers in cells. The two kinds of nucleic acids, DNA and RNA, differ in the type of sugar they contain and in one of their four nitrogen-containing bases. The basic monomer of each type of nucleic acid includes a sugar, a nitrogenous base, and a phosphate group.

Lipids are water-repellent compounds such as fats. Fats consist of hydrocarbon chains of fatty acids that are attached three at a time to glycerol to make triacylglycerides. Phospholipids consist of two fatty acids plus a phosphate group attached to glycerol. Phospholipids are important parts of membranes, whereas triacylglycerides are primarily storage reserves in seeds. Waxes are combinations of fatty acids and long-chain alcohols mixed with other kinds of long-chain hydrocarbons.

Plants also make many other classes of chemicals. The occurrence of particular compounds in these other classes is often limited to just a few plants, and such chemicals in them have no obvious or universal function in plant metabolism. Thus, these chemicals are called secondary metabolites to distinguish them from the widespread, basic classes of plant compounds (i.e., carbohydrates, proteins, nucleic acids, lipids). Nevertheless, some products of secondary metabolism are widespread and important in basic plant metabolism, and others seem to be defensive chemicals against parasites or herbivores.

 ### What Are Botanists Doing?

Use the computerized database or the journal abstracts (e.g., *Biological Abstracts*) in your library to find out what kinds of studies have been done on flavonoids during the past twelve months.

 ### Writing to Learn Botany

Enzymes that are beta-glucosidases hydrolyze bonds between beta-glucose molecules. What nutritional benefits would animals get from having this kind of enzyme? What nutritional problems would humans have if we had this kind of enzyme?

Questions for Further Thought and Study

1. Biochemical reactions form or break bonds between fatty acids and glycerol (ester bonds), between amino acids (peptide bonds), and between carbohydrates (glycosidic bonds). Suggest reasons why the pathways of synthesis and breakdown are different.

2. How do the chemical and physical properties of sunflower oil, jojoba oil, and petroleum oil differ? How are they alike?

3. One of the earliest explanations for the function of secondary metabolites was that they were waste products of plant metabolism. Suggest reasons why this might be a good explanation; also suggest reasons why it might not be a good one. How could you test each idea?

4. How do the properties of amino acids contribute to the three-dimensional structures of proteins?

Web Sites

http://www.ebi.ac.uk/htbin/bwurld.pl
Bio-wURLd, in the United Kingdom, is a user-maintained collection of URLs related to biochemistry and molecular biology.

http://clas.www.pdx.edu/Algalphysiology/biochem.html
Portland State University Algal Physiology Home Page presents a colorful and informative overview of the biochemical composition of algae and plants.

http://www.ifrn.bbsrc.ac.uk/gm/lab/docs/iftmb.html
The Internet for the Molecular Biologist provides information on all aspects of the Internet (WWW, FTP, gopher, databases, e-mail, newsgroups, etc.) that are related to molecular biology and other life sciences.

http://molbio.info.nih.gov/modeling/
This is the Molecular Modeling Home Page of the National Institutes of Health. It includes modeling software called *Molecules R Us* that allows you to design, rotate, and color molecules. The page also includes many links to universities.

http://www.acs.org
This web site, sponsored by The American Chemical Society (membership 160,000), promotes scientific research and education for all levels. It includes internship programs and the new Experimental Programs in chemistry section, which describes ways you can gain real-life experience in the field of chemistry.

http://www.msi.com
At this web site you'll find information about software for molecular simulations in life sciences, as well as a chemistry archive and the WWW Virtual Library of Chemistry.

http://www.camsci.com
The Cambridge Soft Web site includes a database of more than 3,500 compounds searchable by name and chemical formula. Each compound has links to related information on the Web. You can also download chemical viewing software from this site.

Suggested Readings

Articles

Brown, K. S. 1995. The green clean. *BioScience* 45:579–582.

Duchesne, L. C., and D. W. Larson. 1989. Cellulose and the evolution of plant life. *BioScience* 39:238–241.

Patton, A. 1991. Tropical oils. *Diabetes Forecast* 44(4):56–59.

Richards, F. M. 1991. The protein folding problem. *Scientific American* 264(January):54–65.

Welch, R. M. 1995. Micronutrient nutrition of plants. *Critical Reviews in Plant Science* 14:49–82.

Books

Campbell, M. K. 1995. *Biochemistry*. 2d ed. New York: Saunders.

Harborne, J. B. 1988. *Introduction to Ecological Biochemistry*. 3d ed. New York: Academic Press.

Sackheim, G. 1996. *Introduction to Chemistry for Biology Students*. 5th ed. Redwood City, CA: Benjamin/Cummings.

Synder, C. 1995. *The Extraordinary Chemistry of Ordinary Things*. 2d ed. New York: Wiley.

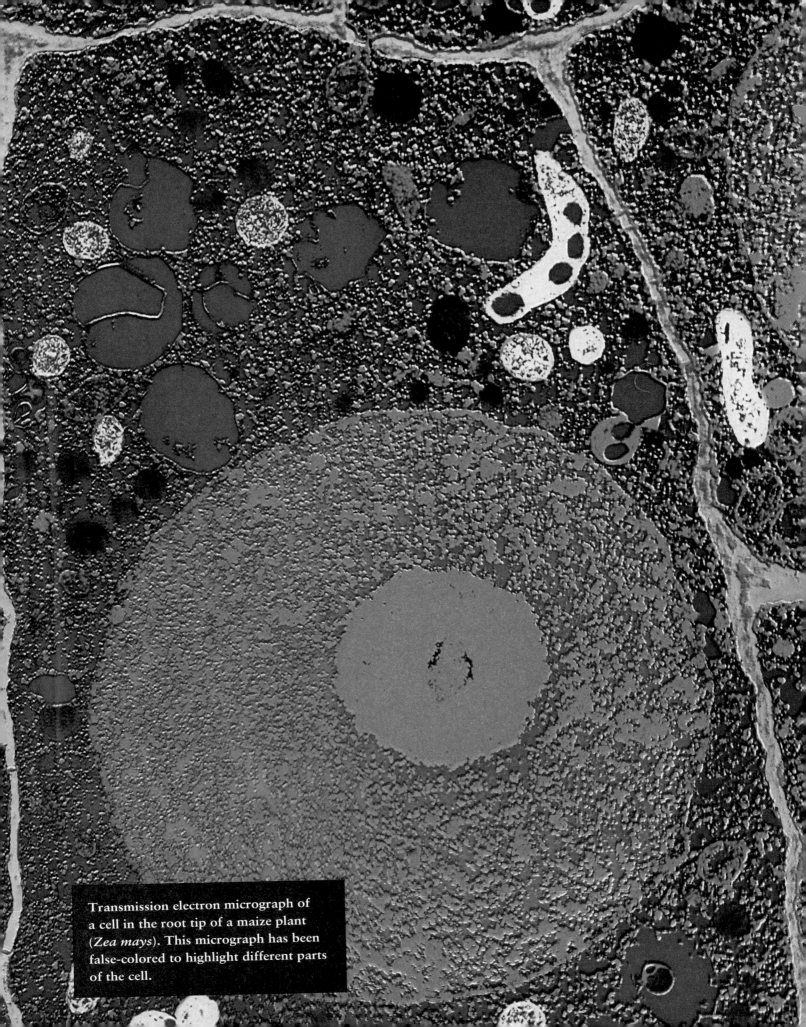

Transmission electron micrograph of a cell in the root tip of a maize plant (*Zea mays*). This micrograph has been false-colored to highlight different parts of the cell.

chapter

Structure and Function of Plant Cells

3

Chapter Outline

Chapter Overview

After atoms and molecules, the next higher level of complexity in living organisms includes cells and their components. All plants are made of cells. Some cell components occur in all living cells, while others occur only in the cells of leaves, roots, or other parts of plants. Depending on their components, cells can divide, grow, transport sugar and water, photosynthesize, secrete nectar or resin, or harvest energy from organic molecules. Most types of cells also contain genetic material that controls the activities of the cell. This genetic material is inherited by new cells after cell division.

In this chapter we introduce the main components of cells and the primary functions of these components in plant structure and metabolism. We emphasize the common components of cells and present examples to show that these components vary among different cell types and different plants. For example, components such as the plasma membrane occur in all living cells, while chloroplasts occur only in cells of green tissue. These and other examples discussed in this chapter show how tissues can differ at the cellular level from one part of a plant to another.

All plants consist of cells (fig. 3.1a), which are the simplest units of a plant that can live independently. The smallest organisms are single cells (fig. 3.1b), but plants are made of billions of cells. Plant cells have many shapes and sizes, the smallest of which are the dividing cells at the tips of roots and stems. These cells are usually about 12 micrometers (μm) in diameter; it would take about 100 of these cells to equal the thickness of a dime. Conversely, the largest cells are long, thin fiber cells. Fibers of jute (*Corchorus capsularis*), which are used to make burlap and rope, can be more than 2.3 m long (fig. 3.1c). In comparison, your height is probably between 1.5 and 2.0 m.

Each cell in a plant consists of a cell wall that surrounds a **plasma membrane,** which encloses many smaller parts called **organelles.** Because each organelle has its own set of functions, the job of each cell in a plant is determined by how many and which organelles the cell contains and what the organelles do. For example, leaf cells contain chloroplasts, nectar-secreting cells contain many dictyosomes, and the storage cells of oil-containing seeds have many glyoxysomes. These organelles have roles in photosynthesis, secretion, and oil metabolism, respectively. They and the other organelles of plant cells are discussed further in later sections of this chapter.

Like plant chemistry, plant cell biology can be reduced to the study of the structure and function of smaller components. By analyzing the anatomy of a cell, we can find clues to how the cell works. In this chapter we present an overview of plant cell structure and how it controls cellular processes. Many of these processes are discussed in greater detail in later chapters.

Scales of Microscopic Observation

Before reading about cells, you should have a good perspective on the sizes of cells and organelles relative to what you can see with your unaided eyes. Sizes are described either in metric units (see Appendix B) or in **magnifications.** Figure 3.2 shows metric units based on the meter, which is slightly longer than a yard.

The power of the naked eye is 1, or 1×. An ordinary magnifying lens can give a 5× magnification, meaning that it can make an object five times larger than the object appears to the naked eye (fig. 3.3). Magnifying power can be increased further by combining two or more lenses, as is done in a **light microscope,** which consists of a series of lenses. A light microscope that has two 5× lenses has a magnifying power of 25× (5× · 5× = 25×; fig. 3.4). Modern light microscopes generally offer an optional combination of lenses that magnify from 100× to 1,000×.

Another factor important in microscopy is **resolving power,** which is the minimum distance necessary to distinguish two points from one another. For example, the resolving power of the human eye is about 100 micrometers (μm), meaning that two dots less than 100 μm apart appear to us as one dot. The best light microscopes can distinguish objects that are about 0.2 μm apart; this is about 500 times more resolving power than the human eye and is sufficient to see plant cells and some of their parts. The resolving power of a microscope also determines the sharpness of the magnified image.

In light microscopy, resolving power is limited by the wavelengths of visible light used to create the image. The shortest wavelength of light visible to the human eye is around 400 nanometers (nm), which is 0.4 μm. However, greater resolution is obtained by using beams of electrons, which have shorter wavelengths. A microscope that focuses an electron beam instead of light is called an **electron microscope** (fig. 3.5). The wavelengths of electron beams depend on the amount of voltage used to generate them. For example, with 60,000 volts, an electron microscope can generate a beam of electrons having a wavelength of about 0.005 nm. This wavelength can achieve a resolution of up to 0.4 nm, which is about 1,000 times more resolution than that of a light microscope. At this level of resolution, sharp images may be obtained at magnifications exceeding 100,000×.

This chapter contains many photographs of plant cells and their parts. It is important that you keep a clear perspective on the scales of measurement in these photographs so that you can better understand how these structures fit together to make a cell. As an aid to your perspective of scales, figure 3.2 shows the size range of familiar objects relative to molecules and to cells and some of their components.

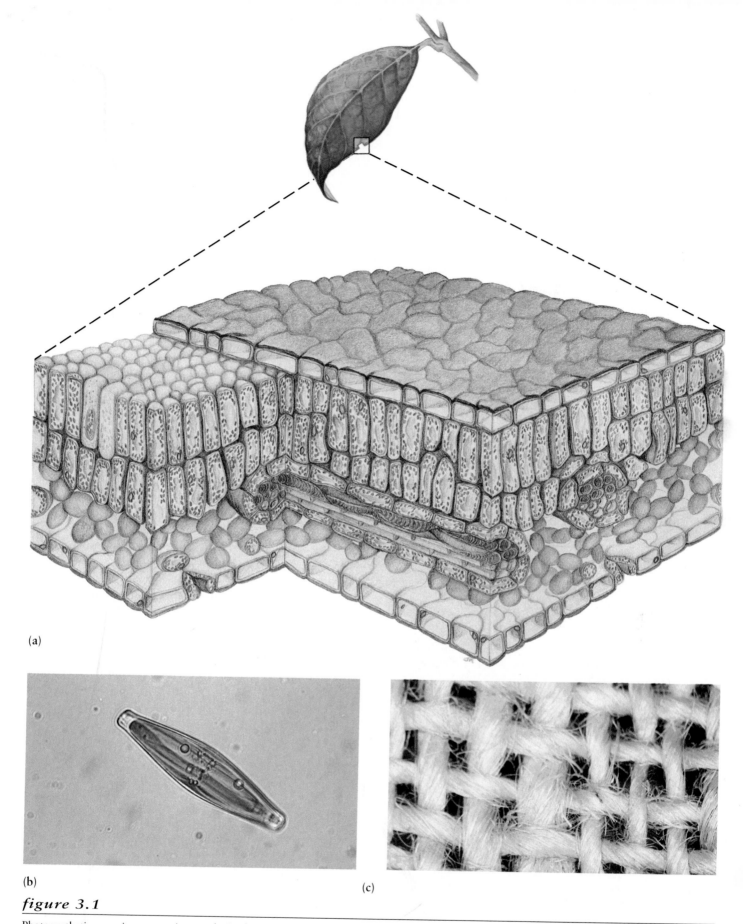

(a)

(b)

(c)

figure 3.1

Photosynthetic organisms range in complexity from single-celled algae to plants that are made of billions of cells. (a) Diagram of a leaf with small portion enlarged and sectioned to show cellularity. (b) Light micrograph of a diatom. The entire organism is a single cell, ×450. (c) Photo of jute fibers woven into burlap, ×10.

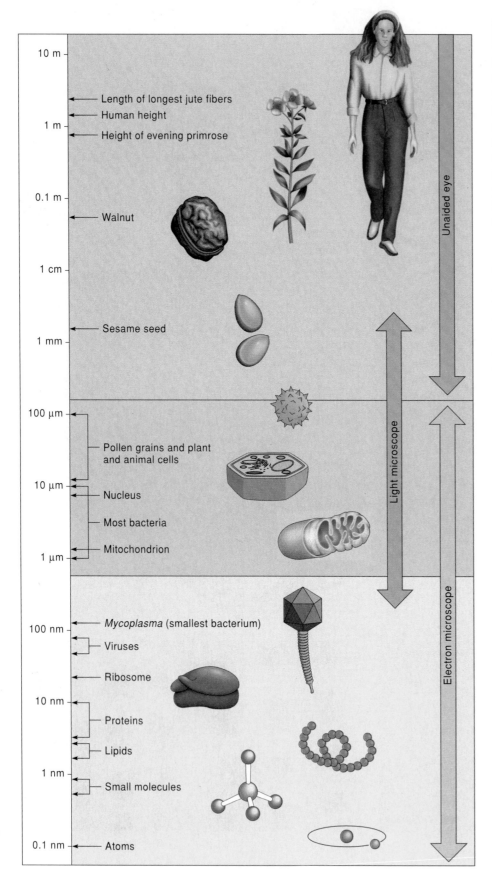

10 m — ← Length of longest jute fibers
1 m — ← Human height
← Height of evening primrose

0.1 m — ← Walnut

1 cm —

1 mm — ← Sesame seed

100 μm — Pollen grains and plant and animal cells

10 μm — ← Nucleus

← Most bacteria

1 μm — ← Mitochondrion

100 nm — ← *Mycoplasma* (smallest bacterium)
← Viruses

← Ribosome

10 nm — ← Proteins

← Lipids

1 nm — ← Small molecules

0.1 nm — ← Atoms

Unaided eye

Light microscope

Electron microscope

Most cells cannot be seen with the unaided eye. However, cells and their components can be seen with light and electron microscopes. Light microscopes magnify objects up to about 1,000×; electron microscopy magnifies them to more than 100,000×. The quality of the magnified image is determined by resolving power, which is limited by the wavelength of the light or electron beam that is used for observation.

Methods of Cytology

Cytology is the study of cell structure and function. Cytologists use light microscopy, electron microscopy, and cell chemistry to learn information about cells. Several kinds of light microscopy can enhance images of cells; for example, differential-interference microscopy exaggerates differences in density within a specimen. Figure 3.6 shows images obtained by this and other kinds of light microscopy. Light microscopy is also aided by chemicals that stain specific parts of the cell; for example, acetocarmine stains the nucleus deep red or purple (fig. 3.7). The next few paragraphs describe how electron microscopy and cell chemistry also provide information about cell structure and function. Before you read about these methods, see reading 3.1, "How to Prepare Cells for Microscopy."

figure 3.2

The relative sizes of cells and cellular parts, in metric units. Cells and certain cellular components are so small that they are at the limit of visibility by light microscopy. For more information about measurements in science, see Appendix B, "The Metric System."

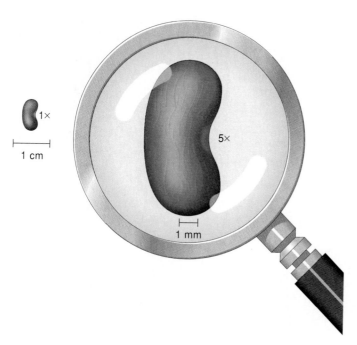

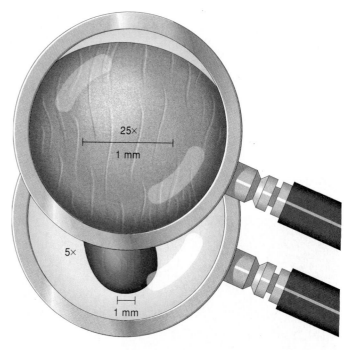

figure 3.3

A 5× magnifying lens makes this kidney bean look five times larger than its actual size.

figure 3.4

The kidney bean looks 25 times larger than normal under two 5× magnifying lenses (5× · 5× = 25×).

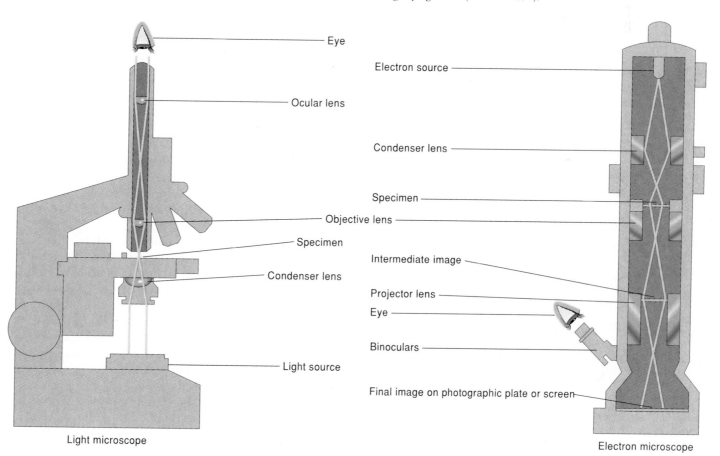

Light microscope

Electron microscope

figure 3.5

Comparison of light microscopes and transmission electron microscopes. In standard light microscopy, light is focused on the specimen by a condenser lens. The image is subsequently magnified by an objective lens and an ocular lens. All lenses are made of glass. In contrast, electron beams are focused by electromagnetic lenses in transmission electron microscopy. Electrons cannot pass through glass lenses.

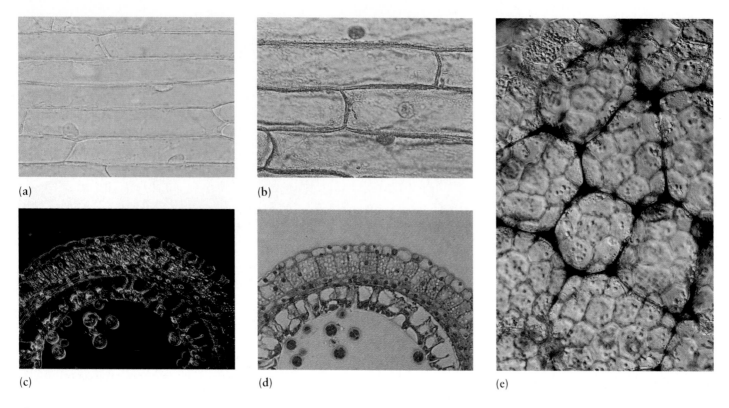

(a)　　　　　　　　　　(b)

(c)　　　　　　　　　　(d)　　　　　　　　　　(e)

figure 3.6

Comparison of different types of light microscopy. (a) Standard (brightfield), with fresh or living specimen. Light passes directly through the specimen, but details are washed out where there is no natural pigment. (b) Standard, with specimen that has been fixed and stained. Staining with different dyes enhances contrast. (c) Darkfield. Light passes through the specimen at an angle, so only scattered light is seen. (d) Phase-contrast. This method amplifies different densities within the specimen and is especially useful for unpigmented structures in living cells. (e) Differential-interference. This technique intensifies differences in density within the specimen.

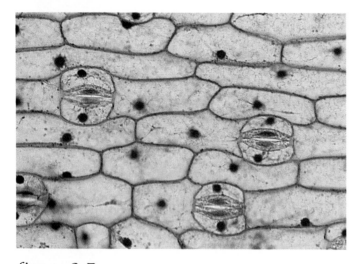

figure 3.7

Light micrograph of leaf cells. Nuclei (dark red) were stained by acetocarmine.

Electron Microscopy

The pattern of contrast in an electron micrograph reveals the ultrastructure of the cell—that is, the structure that can be seen only with an electron microscope. For example, a membrane 10 nm thick can be seen with an electron microscope but not with a light microscope (fig. 3.8).

Electron microscopy that uses transmitted electrons is called **transmission electron microscopy (TEM)**. Observations obtained with this type of microscopy are limited to nonliving specimens that are sliced into thin sections and treated with stains containing elements that deflect the beam of electrons (fig. 3.8a). Images from TEM come from the differences between areas of the specimen that deflect the electrons (dark areas) and areas that do not (light areas). By contrast, in **scanning electron microscopy (SEM)**, bright areas (high spots) and dark areas (low spots) show the actual surface structure of the specimen. Thus, SEM specimens appear as three-dimensional images (figs. 2.12 and 3.9). Unlike TEM, SEM does not require that tissues be sliced into thin sections and stained.

HOW TO PREPARE CELLS FOR MICROSCOPY

Although living cells can be studied by light microscopy, much greater detail about cellular structure can be obtained from cells that are preserved and stained. Tissue preserved in alcohol, for example, is generally soaked in a dye that stains a specific cellular component. One of the most widely used dyes is leucofuchsin, which stains the nucleus because the dye is specific for DNA. Other common dyes are periodic acid for carbohydrates and Sudan red for lipids. Specimens for the electron microscope must be stained with materials that deflect electrons. The most frequently used stain for electron microscopy is a heavy metal such as osmium tetroxide, which creates dark areas wherever the electron-dense metal binds in the cell.

For best results, cells are generally cut into thin sections before staining. Before being sliced, they must first be embedded in wax (light microscopy) or plastic (electron microscopy) so they will be rigid and not collapse when they are cut. Sections are cut with a microtome, which resembles a small-scale meat saw. Steel knives are used to make sections for light microscopy. For electron microscopy, sharp glass or diamond knives are used to cut thin sections through cells embedded in hard plastic.

A remarkable array of subcellular structures can be seen after cells are preserved, embedded, sectioned, and stained. However, two kinds of difficulties arise from preparing cells for microscopy. One is that cellular structure may change during preparation of the fixed cells; for example, proteins that seem to connect cytoskeletal filaments may be artifacts caused by cell preservation. The other kind of difficulty is that thin sections give a static, two-dimensional view of cellular components that were formerly living, three-dimensional structures. A mitochondrion, for example, may appear round or cylindrical, depending on the angle of the section through it. This second problem can be partially overcome by the use of computers with graphic imaging programs that add images together from many thin sections. In this way a simulated three-dimensional structure can be reconstructed from two-dimensional photographs.

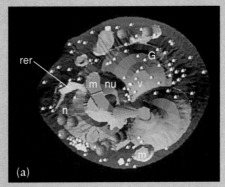
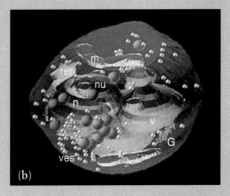

reading figure 3.1

Computerized three-dimensional reconstruction of a yeast cell. (a) Top view. (b) Side view. (G = dictyosome; m = mitochondrion; n = nucleus; nu = nucleolus; rer = rough endoplasmic reticulum; v = vacuole; ves = membrane vesicle).

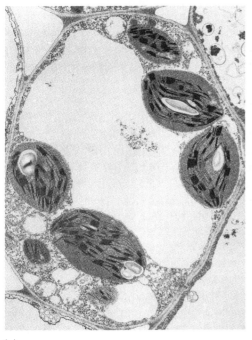

(a)

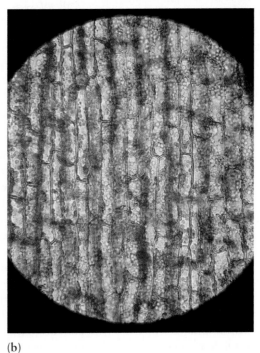

(b)

figure 3.8

Micrographs of plant cells.
(a) Transmission electron micrograph of a plant cell, showing organelles and membranes, ×16,000. (b) Light micrograph of plant cells. Membranes cannot be seen at this magnification and resolution, ×225.

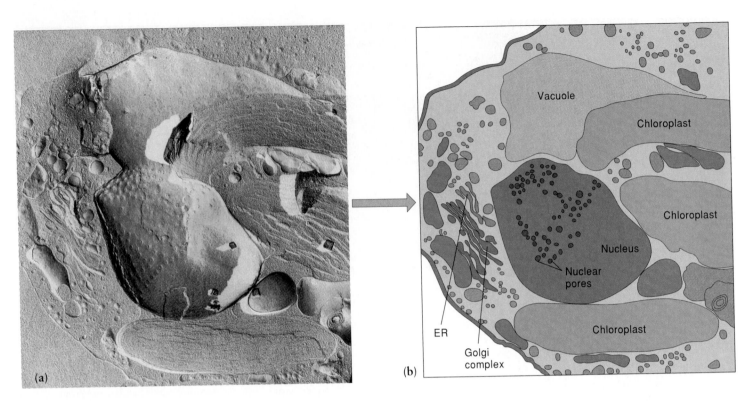

figure 3.9

Freeze-fracturing of plant cells. (a) Scanning electron micrograph of freeze-fractured cell from an onion root tip, showing organelles and the surface of a membrane, ×20,200. As in all micrographs of freeze-fractured specimens, this is not an image of the original specimen but a carbon-platinum replica of it. (b) Diagram showing interpretation of structures in the micrograph in (a). The identities of the brown-colored structures cannot be confirmed from this micrograph, but they are probably a collection of vesicles and small pieces of mitochondria.

Freeze-Fracturing and Electron Microscopy

While SEM is useful for studying surfaces, cells must be broken open before internal structures can be studied with scanning electron microscopy. One method of doing this is **freeze-fracture electron microscopy.** In this method, a block of cells is first frozen to -196° C using liquid nitrogen in the presence of an antifreeze. The frozen block is then cracked open using a knife blade, thereby causing the cells to fracture along lines where there was little water to form strong ice. Such regions include the lipid-rich parts of membranes, which break apart to reveal inner membrane surfaces. The resulting fracture-faces are shadowed with platinum, the cellular material is removed, and the platinum replicas are viewed in an electron microscope (fig. 3.9).

Cell Fractionation

The functions of organelles, which are the membrane-bound components of a cell, are not usually obvious from their structure. Therefore, botanists often purify and concentrate each type of organelle so that its function can be determined and correlated with its structure. This is the purpose of **cell fractionation:** to obtain large quantities of organelles so their functions can be determined more easily.

In preparation for cell fractionation, cells are first homogenized into small fragments, and the homogenate is laid atop a sucrose solution in a test tube. The test tube is then spun at high speed in an instrument called an **ultracentrifuge.** An ultracentrifuge can spin at more than 100,000 revolutions per minute, which is more than 50 times faster than an idling car engine. During spinning, different organelles separate on the basis of their buoyancy in the sucrose solution. Nuclei, which are the densest organelles, sink the farthest. Chloroplasts are usually intermediate in density and therefore form a bright green band near the middle of the test tube.

Biochemical Cytology

Specific substances in cells can be stained with chemicals that bind to the substances but not to other parts of the cell. Such selective staining is useful for finding specific kinds of molecules in cells prepared for microscopy (see fig. 3.7). This method is one of the tools used in a subdiscipline of cytology called **biochemical cytology** or **cytochemistry.** Biochemical cytology uses the biochemical properties of cell components in conjunction with techniques of microscopy to unravel the details of cellular structure and function.

One of the most sophisticated and informative techniques of biochemical cytology uses the immune systems of animals. Large molecules from plant cells, especially proteins, become **antigens** (antibody inducers) when they are

(a)

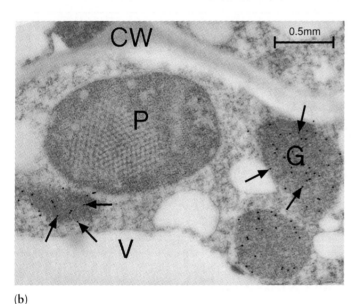

(b)

figure 3.10

Using biochemical cytology to study cells. (a) Light micrograph of a cell that has been immunofluorescent-stained twice—once for the proteins of microtubules (green) and once for the proteins of actin filaments (red) (see fig. 3.12). (b) Electron micrograph of section through a cotton (*Gossypium*) cotyledon, showing staining by antibody-gold complex. Colloidal gold particles (arrows) are bound to malate synthase in glyoxysomes (G). Lack of gold particles in the cell wall (CW), plastid (P), and vacuole (V) indicates an absence of the enzyme in these parts of the cell.

injected into an animal, usually a rabbit or a mouse. The animal makes **antibodies** (protein complexes that bind to antigens), and the antibodies induced by the plant protein are then used to locate the antigen in the plant cell. The antigen-antibody complex is visible because it has been linked to another complex that includes a dye, such as fluorescein. The location of the antigen shows as a fluorescent area when it is exposed to ultraviolet light. An example of this method is shown in figure 3.10a; in this example, protein from microtubules was used as the antigen.

Immunological cytochemistry is also useful in electron microscopy, although antibody complexes used in electron microscopy must be linked to a heavy metal that deflects electrons. For example, an antibody-gold complex linked to malate synthase, the enzyme that converts glyoxylic acid to malic acid, shows where this protein occurs in cells of cotton seedlings (fig. 3.10b)

<table>
<tr><td align="center">c   o   n   c   e   p   t</td></tr>
</table>

Cytology uses microscopy and cell chemistry to study cellular structure and function. Cell chemistry is useful in staining specific parts of the cell to enhance their visibility by microscopy. Organelles or other parts of cells can be separated by cell fractionation and studied as isolated components.

Functional Organization of Cells

Why Are Cells So Small?

For convenience, most of the contents of a cell are referred to as its **cytoplasm;** the only component of the cell that is not part of the cytoplasm is the nucleus. The plasma membrane, which surrounds the cytoplasm, is a barrier that protects the cell from harmful substances. It has the consistency of salad oil and must allow the passage of gases and nutrients into and out of the cell. However, the surface area of the membrane can service only so much cellular volume—that is, the surface-to-volume ratio must have a lower limit. Multicellular organisms avoid this limit by making more but smaller cells for a given volume. Consider the example of a cube-shaped cell that is 250 μm on each side; it has a volume of 1.56×10^7 μm^3 (250 μm × 250 μm × 250 μm) surrounded by 3.75×10^5 μm^2 (250 μm × 250 μm × 6 sides) of plasma membrane (fig. 3.11). The surface-to-volume ratio is 0.024, which is unacceptably small for most cells. This means that such a cell would be too big for its plasma membrane. However, if this single large cell were to be divided into cells 25 μm on a side, the total volume would be the same, but the surface area of all plasma membranes would be 3.75×10^6 μm^2 (25 × 25 μm per side × 6 sides per cell × 1,000 cells). The surface-to-volume ratio would be 0.24, which is 10 times more membrane for the same volume as the single large cell. A ratio of 0.24 is about what many plant cells have for the relative area of plasma membrane surface to cell volume.

The need for a sufficient amount of plasma membrane for the volume of the cell only partially explains why most cells are so small. Other important factors include the limits on rates of synthesis and transport of molecules within the cell and the requirement that a single nucleus manage the metabolism of the entire cell.

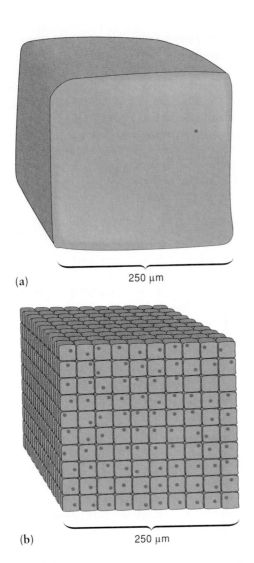

(a) 250 µm

(b) 250 µm

figure 3.11

Diagrams showing (a) a single large cell compared with (b), a cube ten cells across having the same total volume as the single cell. Although the volume is the same for both, the surface-to-volume ratio for the large cell is much less than for the sum of the small cells.

<table>
<tr><td>c o n c e p t</td></tr>
</table>

The small size of cells is partially explained by the amount of plasma membrane that can service the cell's volume. A group of smaller cells has a greater ratio of plasma membrane area to cell volume than does a single large cell.

Membranes and Cell Compartments

Many metabolic functions in a cell occur in or on membranes. The plasma membrane alone is inadequate for all of these processes, regardless of cell size. Additional internal membranes, either attached to the plasma membrane or included in organelles, compensate for the insufficiency of the plasma membrane. These other membranes also form compartments

that maintain different environments within the cell. These compartments allow many different kinds of reactions to occur in the cell simultaneously.

Biological membranes are usually 7–10 nm thick and consist mostly of phospholipids and proteins. In addition to the plasma membrane, there are membranes that surround nuclei, mitochondria, chloroplasts, vacuoles, and microbodies. Membranes also pervade the cell between organelles in a complex system that complements the functions of the organellar membranes. Further characteristics of organelles and the internal membrane systems of cells are presented in later sections of this chapter.

<table>
<tr><td>c o n c e p t</td></tr>
</table>

The plasma membrane is supplemented by internal membranes that provide additional areas for enzymatic reactions and divide the cell into compartments.

The Lore of Plants

Cooks are the plant cell biologists of the kitchen. Every cookbook describes how to prevent browning of fruits and vegetables before they are canned or frozen. Browning is caused by the enzyme-catalyzed polymerization of phenolics in damaged cells. The enzymes (phenolic oxidases) and the phenolics are kept in separate compartments in intact cells. The polymerization reaction can be stopped by covering the freshly cut tissue with ascorbic acid (vitamin C), as crystals or in lemon juice, or with vinegar, which contains acetic acid. These organic acids inhibit the reactions that are driven by phenolic oxidases.

The Cytoskeleton

The **cytoskeleton** is a network of filaments that forms a mechanical support system in the cell. These filaments also help maintain organelle position, direct cell expansion, and control the movement of chromosomes during nuclear division. Groups of filaments may also form channels for transporting large molecules within the cell.

There are at least three kinds of filaments that comprise the cytoskeletons of cells in plants and animals (fig. 3.12). The largest filaments are **microtubules,** which are hollow tubes about 18–25 nm in diameter. Microtubules are made of two types of globular proteins, **alpha tubulin** and **beta tubulin.** These proteins are bound to each other as dimers, which aggregate into microtubules.

The smallest filaments in the cytoskeleton are **actin filaments,** which consist of two intertwined strands of the globular protein **actin.** Actin filaments are 4–7 nm in diameter. The proteins of microtubules and actin filaments are the same in plants and animals. This similarity suggests that these filaments work the same way in both kingdoms, even though the tissues that contain them, such as muscle tissue and leaf tissue, are different.

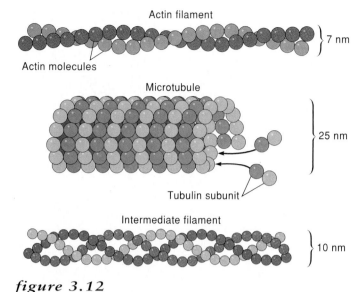

Actin filament

Actin molecules

7 nm

Microtubule

25 nm

Tubulin subunit

Intermediate filament

10 nm

figure 3.12

Structural models of the three kinds of cytoskeletal filaments.

The third kind of cytoskeletal filaments are intermediate in size between microtubules and actin filaments; for this reason, they are called **intermediate filaments.** These filaments are made of fibrous proteins wound into coils having diameters of 8–12 nm. Several kinds of proteins in intermediate filaments occur in animals, but little is known about them in plants. Unlike the proteins of microtubules and actin filaments, the proteins of intermediate filaments in animals are fibrous and more numerous. Moreover, different proteins occur in different types of cells.

The proteins of intermediate filaments are probably similar in both plants and animals. The evidence for this suggestion includes cytochemical studies showing that antibodies made against intermediate filaments from animals bind to intermediate filaments from plants. At least three kinds of proteins of intermediate filaments have been identified in plants by this method.

How Does the Cytoskeleton Work?

It is easy to imagine the cytoskeleton as a stationary lattice of filaments that holds cells in specific shapes and keeps cell components in specific places; the name *cytoskeleton* implies such a static function. However, dividing cells often change size and shape, and organelles and other internal components of all living cells are in almost constant motion, which means that the cytoskeleton cannot be a rigid structure. This assertion is confirmed by observations that expansion of and internal movements in cells occur in conjunction with the growth and breakdown of microtubules and actin filaments. Such observations are indirect evidence that the cytoskeleton has many dynamic functions in the cell. Experimental evidence comes from treating cells with colchicine, an alkaloid from the meadow saffron (*Colchicum autumnale*). Colchicine disrupts the assembly of microtubules. The formation of new cell walls during cell division is inhibited by colchicine, which suggests that cell division requires microtubule synthesis.

The least understood parts of the cytoskeleton are the intermediate filaments. Because their proteins are fibrous, intermediate filaments do not have globular subunits that are easily assembled and disassembled like those of microtubules and actin filaments. Thus, intermediate filaments may be relatively static, tension-bearing structures that are not as dynamic as their larger and smaller counterparts in the cytoskeleton. Although the various filaments of the cytoskeleton must somehow cooperate to control the overall organization of the cell, we do not yet know how their functions are integrated.

c o n c e p t

The cytoskeleton consists of microtubules, actin filaments, and intermediate filaments. Each type of filament is made of different kinds of proteins. The cytoskeleton functions not only as a structural support for the cell but also as a dynamic network that controls certain aspects of cell growth and internal movement.

The Cell Wall

The most easily observed part of a plant cell is the cell wall. In some cells, such as the cork cells in bark, the cell wall is the only remnant of a formerly living cell. Cell walls are dynamic parts of cells that can grow and change their shape and composition. Their composition varies in different cell types and from one species to another. Up to 60% of a cell wall may be cellulose; other components include hemicelluloses, pectins, lignins, and proteins. Almost all plant cells have cellulose-containing cell walls.

Young cells and cells in actively growing areas have **primary cell walls** that are relatively thin and flexible. Examples of such cells include the dividing cells at tips of roots and shoots. The primary wall is usually less than 25% cellulose, the remainder being hemicelluloses, pectins, and glycoproteins. Some primary cell walls also contain small amounts of lignin. However, many cells having only primary cell walls can change shape, divide, or differentiate into other kinds of cells.

Certain kinds of cells stop growing when they reach maturity. When this occurs, these cells form a **secondary cell wall** inside the primary cell wall (fig. 3.13). The secondary cell wall is more rigid than the primary cell wall and therefore functions as a strong support structure. Although cellulose is one of their main components, the secondary walls of cells in wood are up to 25% lignin, which adds hardness and resists decay. Because of its lignin content, wood is one of the strongest materials known.

Some cell walls, such as those of cork cells, also contain suberin. Suberized tissues inhibit water loss through bark, which is why cork from the cork oak (*Quercus suber*) is useful in making stoppers for wine bottles (see reading 16.4, "Cork"). Unlike primary cell walls, secondary cell walls are rigid and lack glycoproteins. Most types of cells that have secondary cell walls die when they reach maturity.

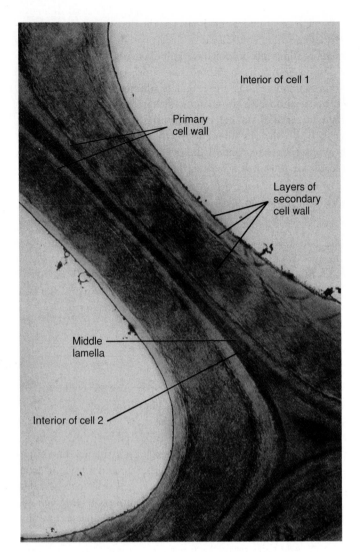

figure 3.13

Transmission electron micrograph of cell walls. The primary wall is constructed when the cell is young. Thicker secondary walls are added later when the cell stops growing, ×3,000.

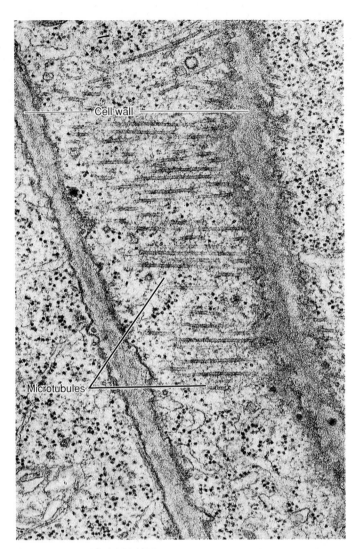

figure 3.14

Transmission electron micrograph of a growing section of a root tip cell, showing orientation of microtubules, ×25,000.

Cells that adjoin one another are probably held together by pectins. The pectic layer between cells is called the **middle lamella.** Some tissues, such as the flesh of apples, are so rich in pectins that these polysaccharides are extracted for use as thickening agents in making jams and jellies.

How Cell Walls Grow

All materials necessary for making cell walls come from inside the cell. Dictyosomes play a role in this process by transporting pectins, hemicelluloses, and glycoproteins to the plasma membrane. These substances pass through the membrane in the region of cell-wall synthesis, where they are assembled inside the existing wall. This is how cell walls thicken.

Cellular expansion requires the cell wall to stretch or deform as the cell absorbs more water. As the wall enlarges, the existing cellulose microfibrils must separate because they cannot stretch. During expansion, new cellulose microfibrils are deposited inside the loosened cell wall in different patterns, depending on the cell type. In stem cells, microfibrils are oriented mostly perpendicular to the direction of cell expansion. Thus, cells may grow 20 times their original length, with little growth in width. In comparison, microfibrils are deposited in random arrays in cells of storage tissues and tissue cultures. This pattern enables growth in these cells to be more or less uniform in all directions.

As cell walls grow, the sites and orientation of new microfibrils are controlled by arrays of microtubules that occur immediately inside the plasma membrane. These microtubules are arranged in the same orientation as the microfibrils (fig. 3.14). Evidence for the role of microtubules in this process is based on the growth of cells that are treated with colchicine. For example, in water-conducting cells of stems, colchicine changes the normal development of regular wall

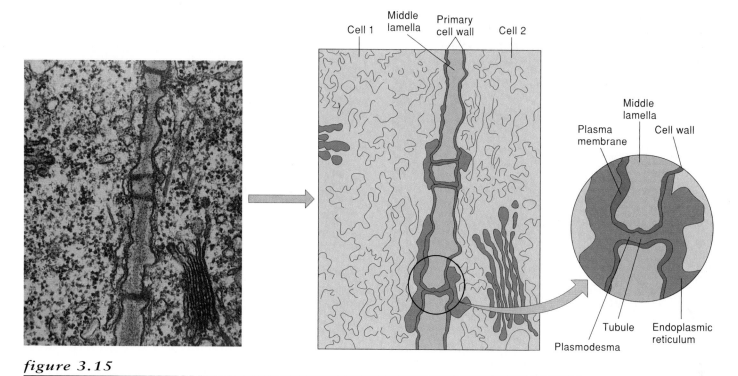

figure 3.15

This transmission electron micrograph, ×17,000, and accompanying drawing of a primary cell wall show the plasmodesmata (which contain plasma membrane) and a tubule that connects the endoplasmic reticulum between adjacent cells.

thickenings to a disorganized smear of wall material. Despite this effect, colchicine apparently does not affect the synthesis of new microfibrils, and in some cases colchicine has little or no effect on the orientation of new microfibrils that are deposited in a preexisting pattern. This means that the role of microtubules in cell-wall growth is not as straightforward as might have been expected. Although results from experiments with colchicine suggest that cell-wall components outside the cell connect directly or indirectly to microtubules across the plasma membrane, there is no direct evidence for such connections, and their molecular basis is unknown.

c o n c e p t

Primary and secondary cell walls are made mostly of a variety of carbohydrates, but the most rigid walls also contain lignin. Secondary cell walls are deposited inside primary cell walls. Microtubules have a role in the pattern of cell-wall deposition.

Connections between Cells

Primary cell walls have thin areas where many tiny connections, called **plasmodesmata** (singular, **plasmodesma**), occur between adjacent cells. Plasmodesmata are lined by the plasma membrane, thereby forming an uninterrupted channel for the movement of materials from one cell to another. This means that all cells in a plant are interconnected and have the potential to exchange substances through plasmodesmata.

Plasmodesmata often occur in clusters where primary cell walls are particularly thin. These regions are called primary pit-fields (fig. 3.15). Primary pit-fields and plasmodesmata are abundant in conducting cells and secretory cells, such as those in nectar glands or oil glands. Plasmodesmatal channels are usually about 30 nm in diameter, but they can be much larger in some cells.

The structure of plasmodesmata and the frequency of their occurrence in conducting and glandular cells suggest that these connections function in transport between cells. Direct evidence for this function comes from experiments with dyes and electric currents. When a dye that does not easily cross the plasma membrane is injected into one cell, it quickly passes into neighboring cells. Similarly, the high electrical resistance of the plasma membrane can be bypassed by electric currents that pass through plasmodesmata. In addition, virus particles pass between cells through plasmodesmata. Despite such evidence, cells probably do not exchange all materials freely; neighboring cells can differentiate into different cell types and maintain different internal concentrations of various chemicals.

Water-conducting tissue is an important exception to the general occurrence of plasmodesmata. Cells of this tissue die as they mature, so they have no living material to share between them. Instead, they function as inanimate "straws" formed by many cells. In flowering plants, water moves through perforations in the primary cell wall, around which there is no secondary cell wall (fig. 3.16). The movement of water through plants is discussed in more detail in Chapter 21.

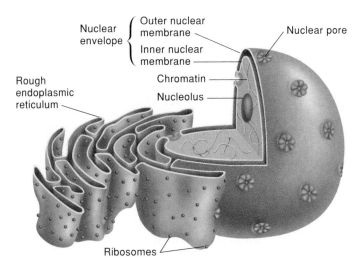

figure 3.17

The nuclear envelope has pores that link the cytoplasm with the inside of the nucleus. The outer nuclear membrane is continuous with the endoplasmic reticulum.

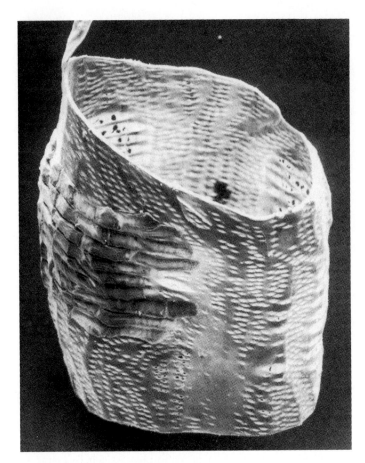

figure 3.16

Scanning electron micrograph of a water-conducting vessel element, showing perforation and pitting, ×325.

c o n c e p t

All living cells in a plant are interconnected by plasmodesmata, which are lined with a continuous plasma membrane. Some substances can move from cell to cell through plasmodesmata. Despite these intercellular connections, neighboring cells can maintain different chemical composition.

The Nucleus

The **nucleus** is usually the most conspicuous organelle in a cell; when stained, it can be seen easily with a light microscope (fig. 3.7). The nucleus contains most of a cell's DNA, which occurs with proteins in threadlike chromosomes (see Chapter 10). The nucleus is surrounded by two membranes, together called the **nuclear envelope** (fig. 3.17). The outer membrane is continuous with the membrane of the endoplasmic reticulum, which is discussed later in this chapter. The inner and outer nuclear membranes are separated by a space of 20–40 nm, except where they fuse to form pores in the envelope. These nu-

clear pores are small circular openings, 30–100 nm in diameter, bordered by proteins that probably influence the passage of molecules between the nucleus and the rest of the cell. For example, certain proteins move into the nucleus, where they join with ribosomal RNA to make the subunits of ribosomes (see next section). In turn, ribosomal subunits and other RNA-containing molecules that are made in the nucleus move out into the cytosol through the nuclear pores. The **cytosol** is the semifluid matrix between organelles.

The complexity and importance of the nucleus and its activities are the focus of modern research in genetics. An entire unit of this text (Unit 3, "Genetics") is devoted to this branch of biology.

Ribosomes

Ribosomes are the sites of protein synthesis. They are about 20 nm in diameter and consist of approximately equal amounts of protein and ribosomal RNA (**rRNA**). Each ribosome is assembled from two subunits that are produced in the nucleus and exported to the cytosol. The two subunits are joined when they attach to a molecule of messenger RNA (**mRNA**). Ribosomes usually occur in clusters, called **polysomes,** on a single molecule of mRNA (fig. 3.18). Unlike the nucleus and other organelles, ribosomes are not surrounded by membranes.

Ribosomes are either attached to membranes or move freely in the cytosol. Proteins made by cytosolic ribosomes are usually also cytosolic; that is, they are not associated with membranes. These proteins include enzymes that degrade sucrose and glucose in the cytosol. Conversely, membrane-bound ribosomes usually make proteins that will be attached

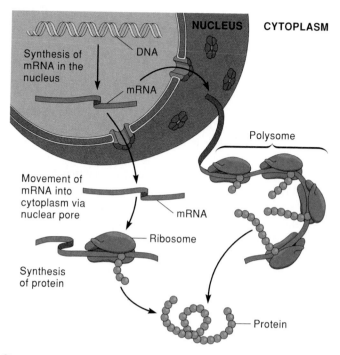

figure 3.18

Ribosomes play a central role in protein synthesis. Unlike other organelles, ribosomes do not have membranes.

to or embedded in membranes. Examples of such proteins are those that control the transport of ions and other substances through the plasma membrane.

The number of ribosomes varies among cell types and in different stages of cell development: A growing cell can make about 10,000 ribosomes per minute. Ribosomes are especially abundant in dividing cells because these cells make large amounts of protein. The mechanisms of protein synthesis by ribosomes and the nucleus are discussed in detail in Chapter 11, "Molecular Genetics and Gene Technology."

The Membrane System

All of the membranes in a cell are connected either by direct contact with each other or by the exchange of membrane segments. These interconnected membranes function together as a **membrane system** that includes the plasma membrane and the various organellar membranes.

Many biochemical processes occur in or on membranes. For example, the enzymes involved in photosynthesis and in ATP synthesis are embedded in membranes. Membranes also provide a framework for making more membranes.

All membranes have a similar structure, which consists of a double layer of phospholipids that is impregnated with proteins (see Chapter 4, fig. 4.4). However, not all membranes have exactly the same structure and function; for example, the permeability of the plasma membrane differs from that of the nuclear envelope. Each membrane controls

how much and what kinds of materials pass through it; that is, each membrane has a different **membrane selectivity**, which depends on its composition. The composition and physical properties of a membrane can change during cell development.

In principle, internal membranes increase the ratio of membrane surface to cell volume. However, internal membranes do more than this. Indeed, they are physically and chemically distinct from the plasma membrane and from each other. Furthermore, they divide a cell into functionally distinct compartments (i.e., organelles) that are bounded by their own differentially permeable membranes. Thus, a cell is separated into a set of individual reaction vessels. Each reaction vessel has its own specialized functions, which are performed by a unique set of enzymes bound to or held within the membrane.

The Plasma Membrane

The plasma membrane is a remarkably changeable, multipurpose membrane. As already mentioned, it acts as a differentially permeable barrier to substances that enter and leave the cell. Some substances can pass through it; others cannot.

Partly because of its outer position, the plasma membrane receives and translates chemical and environmental signals from outside the cell. Signals translated by the plasma membrane change cellular metabolism. For example, hormones received by the plasma membrane can initiate a series of enzymatic reactions that cause the cell to enlarge. The plasma membrane also accepts packets of raw materials from other membranes inside the cell and directs the assembly of these materials into cell-wall microfibrils.

The Lore of Plants

Readers of A. A. Milne's *Winnie the Pooh* know how fond donkeys are of thistle; indeed, cotton thistle (*Onopordum acanthium*) gets its name from the Greek *onos*, meaning "an ass," and *perdon*, meaning "I disperse wind." This may be why the character Eeyore had such a sorrowful and solitary existence.

The Endoplasmic Reticulum

Most of the membrane surface area inside cells occurs in the **endoplasmic reticulum (ER)**, which is an extensive network of sheetlike membranes distributed throughout the cytosol. When viewed with an electron microscope, the ER appears to meander throughout the cell (fig. 3.19a). Seen in three dimensions, however, the ER is a system of flattened tubes and sacs (fig. 3.19b) that is continuous between the plasma membrane and the outer membrane of the nuclear envelope. Internal compartments of the ER are isolated from the rest of the

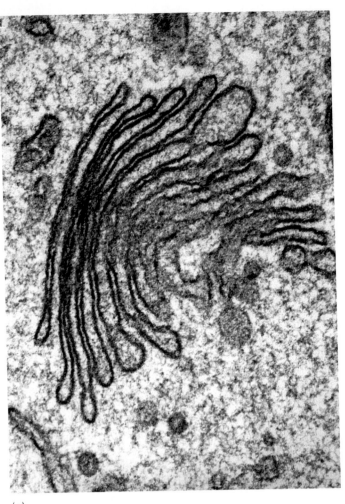

(a)

cytoplasm but in contact with the space between the two membranes of the nuclear envelope. Plasmodesmata also contain portions of ER, which form a continuous internal membrane between cells.

Two regions of ER can be distinguished in electron micrographs. One region is called the *rough* ER because the many ribosomes attached to it give it a rough appearance (fig. 3.19b, fig. 3.20). In contrast, the other region is called the *smooth* ER because it has no ribosomes attached to it (fig. 3.19b).

Biochemical cytology and studies of isolated ER membranes have shown that the rough ER is the major region of protein synthesis in the cell. The smooth ER is involved with the synthesis of phospholipids and the assembly of new membranes. Both types of ER form vesicles that break away and fuse with other membranes. When one of these vesicles fuses with the plasma membrane, its contents are secreted from the cell; cell-wall components, waxes, lipids, and mucilages are secreted by such vesicles. Other vesicles can fuse with organellar membranes and become part of them.

c o n c e p t

The endoplasmic reticulum is a complex internal membrane that has many functions. It contains enzymes that catalyze phospholipid synthesis, the construction of other membranes, and the synthesis of cell-wall precursors. The rough endoplasmic reticulum also bears ribosomes for protein synthesis.

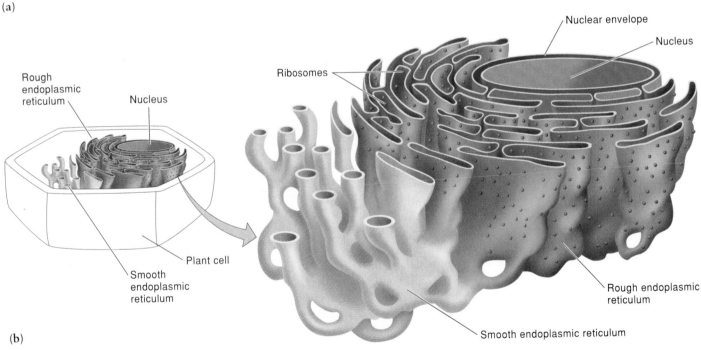

(b)

figure 3.19

(a) Transmission electron micrograph of a dictyosome, ×75,000. (b) The three-dimensional character of the ER, as shown here, cannot be seen in the electron micrograph.

3–17   *Unit One*   Cells as the Fundamental Units of Life

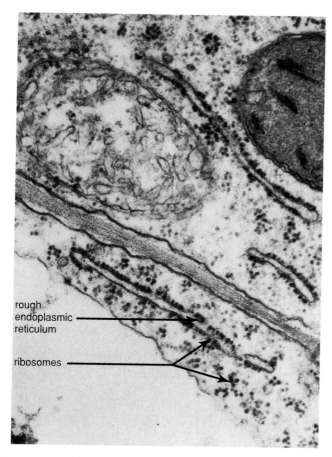

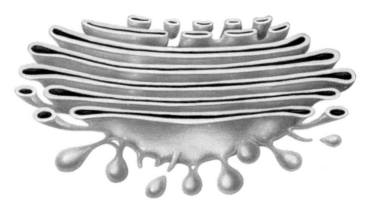

rough
endoplasmic
reticulum

ribosomes

figure 3.20

Transmission electron micrograph from a young leaf cell of maize, ×30,000. Areas of the ER with numerous ribosomes attached are the rough endoplasmic reticulum.

Dictyosomes

Stacks of flattened, membranous vesicles are called **dictyosomes,** or sometimes **Golgi bodies** (fig. 3.21). Dictyosomes are usually two-sided, with one side facing the smooth ER and one side facing the plasma membrane. Membrane-bound vesicles occur near the edges of the dictyosomes and are considered to be part of them.

Dictyosomes receive material from the smooth ER, either through direct connections or in vesicles released by the ER. Transport vesicles from the ER fuse with the inner face of the dictyosome and release their contents into its interior. These vesicles contain proteins, lipids, and other substances, which are often chemically modified in the dictyosome and then sorted into separate packets. For example, sugars in dictyosomes become linked to proteins, converting them into glycoproteins. Packets of glycoproteins or other materials eventually move to the edge of the dictyosome near the outer face, where the dictyosome membrane is pinched off into another vesicle. This new vesicle moves to the plasma membrane or to other sites in the cell.

Vesicles that move to the plasma membrane are secretory vesicles, because they fuse with plasma membrane and secrete their contents to the exterior of the cell. This type of

figure 3.21

Three-dimensional representation of a dictyosome (Golgi body). Vesicles near the stacked membrane are considered to be part of the dictyosome.

secretion is called **exocytosis.** Polysaccharides secreted by exocytosis become part of the cell wall. Other substances are secreted from specialized cells with the aid of dictyosomes. These substances include nectar that flowers secrete as a reward for hummingbirds or other pollinators, and the oils or similar resinous chemicals secreted from the glands on leaves and stems of mints and many other fragrant plants.

In dividing cells, dictyosomes help build new primary walls after the nuclei have divided. Wall construction occurs when dictyosome vesicles are guided by microtubules to a region between nuclei. Vesicles containing cell-wall precursors fuse in this region to form a disk-shaped structure called a **cell plate.** The cell plate grows by the addition of more vesicles until it fuses with the parent cell wall, which completes the partition between the two new cells.

A logistical problem arises in some cells as a consequence of building secondary walls. As secretory vesicles continually fuse with the plasma membrane, large amounts of new membrane are added to it (estimates are that active cells can double the amount of plasma membrane every 20 minutes), but plasma membrane has little room to grow in cells that are forming rigid secondary walls. Cells must therefore have a system for recycling excess plasma membrane. In animal cells, certain regions of the plasma membrane become coated with bristlelike structures. These regions, called **coated pits,** form vesicles that pinch off into the cell, thereby removing a portion of the plasma membrane. Coated pits have also been discovered in plant cells, which suggests that the mechanism for recycling plasma membrane may be the same in plants as it is in animals.

c o n c e p t

Dictyosomes are membranous organelles that modify chemicals from the endoplasmic reticulum, build primary cell walls between newly divided nuclei, secrete substances through the plasma membrane, and probably help to recycle excess plasma membrane.

Vacuoles

Vesicles from the ER and dictyosomes often fuse to form larger sacs called **vacuoles.** Immature cells of plants and animals may contain several small vacuoles, but in most plant cells these small vacuoles fuse into larger ones as the cell matures. A mature plant cell typically has one large vacuole that can occupy up to 95% of the cell's volume (fig. 3.22a). The membrane of the central vacuole has its own name, the **tonoplast.**

As plant cells grow, most of their enlargement results from the absorption of water by vacuoles (fig. 3.22b). This absorption of water by the vacuole expands and pushes the rest of the cell's contents into a thin layer against the cell wall. Vacuoles that are filled with water create pressure, called **turgor pressure,** on the cell walls, which contributes to structural rigidity of the cell. When a plant receives too little water, turgor pressure decreases and the plant wilts. We can see the effects of turgor pressure by letting carrots or celery dry out and become flaccid; we can make them crisp again by putting them in water. This reacquired crispness (turgor) is caused by vacuoles that have refilled with water.

Vacuoles are versatile organelles, as indicated by the diversity of substances that occur in them. In addition to water, vacuoles contain enzymes and other proteins, water-soluble pigments, growth hormones, and ions. Vacuolar enzymes digest storage materials and components from other organelles for recycling into the cytosol. Pigments in vacuoles, especially red and blue anthocyanins, impart bright colors to flowers, fruits, and other plant parts. Some plants harbor toxic alkaloids or other secondary products in their vacuoles. These alkaloids, which are water-soluble at the acidic pH of vacuoles, may deter insects and other animals from eating the plants that contain them.

Ions such as potassium and chloride are stored in vacuoles for easy retrieval to the cytosol when needed for cellular metabolism. In plants specialized for high-salt habitats, such as those along coastlines and near marine estuaries, vacuoles can accumulate chloride salts to concentrations several thousand times greater than in the cytosol. The cytoplasm of these plant cells is thereby protected from salt toxicity, enabling the plants to thrive in their harsh environment. In other plants, such as rhubarb (*Rheum rhabarbarum*), oxalic acid accumulates in the vacuoles and forms crystals of calcium oxalate. The tartness of these plants comes from these oxalates.

c o n c e p t

Most plant cells have a large central vacuole that arises from the fusion of vesicles and many smaller vacuoles. The central vacuole contains mostly water, but it may also contain enzymes, salts, pigments, alkaloids, and other kinds of chemicals.

Microbodies

The smallest membrane-bound organelles in a cell are called **microbodies.** Microbodies, which are bound by a single membrane, are usually spherical and 0.5–1.5 µm in diameter

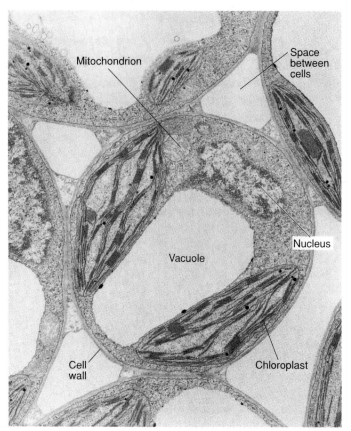

(a)

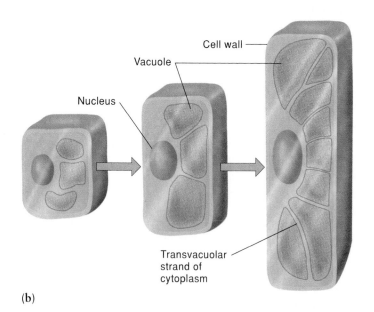

(b)

figure 3.22

Vacuoles. (a) Transmission electron micrograph from a coleus leaf (*Coleus blumei*). The expanding central vacuole compresses the rest of the cell contents into a small space against the cell wall, ×20,000. (b) Schematic diagram of cell growth, showing how an increase in cellular volume can occur without a large increase in the amount of cytoplasm. The peripheral layer of cytoplasm may be interconnected through the vacuole by cytoplasmic strands that radiate from the region of the nucleus.

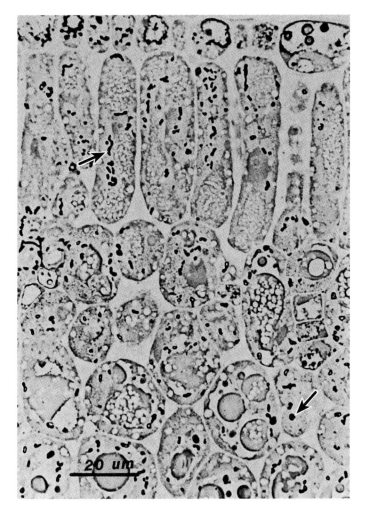

figure 3.23

Phase-contrast light micrograph of a section through a cotyledon of cotton (*Gossypium*). Cells were stained to localize the enzyme catalase (arrows), an enzyme in microbodies that degrades hydrogen peroxide. The localization of catalase shows where microbodies occur. The microbodies in these cells are peroxisomes.

(fig. 3.23). These tiny organelles are often associated with membranes of the ER, but they may also be closely associated with chloroplasts and mitochondria. Different types of microbodies have specific enzymes for certain metabolic pathways. Two of the most important kinds of microbodies are **peroxisomes,** which occur primarily in leaves, and **glyoxysomes,** which are common in germinating oil-bearing seeds and the young seedlings that grow from them.

Peroxisomes are so named because they metabolize hydrogen peroxide (H_2O_2). The enzyme that catalyzes this breakdown is **catalase.** However, peroxisomes also contain **oxidases** that catalyze the production of H_2O_2. The importance of having these oxidases and catalase in the same organelle is a mystery. It might seem logical to suggest that peroxisomes protect cells by destroying peroxide, but this hypothesis does not explain why peroxisomes make peroxide, too. Furthermore, the cytosol and mitochondria also contain H_2O_2-producing oxidases, which enable these cell compartments to make peroxide

that is not metabolized in peroxisomes. Moreover, many cells that make H_2O_2 lack peroxisomes. Thus, it seems that peroxisomes are not just peroxide-detoxifying organelles; their importance may be in their use of H_2O_2 to oxidize other cellular toxins, such as ethanol and nitrites.

Glyoxysomes also contain enzymes that catalyze the breakdown of fatty acids into acetyl-CoA. This acetyl-CoA is used to make organic acids that can be exported from the glyoxysome and used in other metabolic pathways, such as respiration and sucrose synthesis. Unlike peroxisomes, glyoxysomes rarely occur in animals. Consequently, plants can convert lipids to carbohydrates, but animals generally cannot.

c o n c e p t

Microbodies include peroxisomes and glyoxysomes. Peroxisomes break down some of the toxic products of cell metabolism by oxidizing them with hydrogen peroxide. Glyoxysomes link the breakdown of fatty acids to the synthesis of carbohydrates.

Organelles for Energy Conversion

Cells thrive on the energy of ATP. Two kinds of organelles, **chloroplasts** and **mitochondria,** produce most of the ATP needed for cellular metabolism. These organelles are similar in several respects. For example, both are bounded by two membranes, and much of their internal membranes is folded and stacked to form complex compartments. Their internal membranes contain the enzyme ATPase, which uses the electrochemical energy of protons to phosphorylate ADP into ATP. Chloroplasts and mitochondria also contain DNA that controls the synthesis of some of the enzymes necessary for their respective metabolic pathways (nuclear DNA provides instructions to make additional enzymes). Finally, chloroplasts and mitochondria are semiautonomous: they grow and divide in the cell on their own (see reading 3.2, "Cellular Invasion: Origin of Chloroplasts and Mitochondria").

The differences between chloroplasts and mitochondria include their respective sources of energy for making ATP, their appearance, and their composition. Chloroplasts use the energy of light to make ATP, whereas mitochondria use the energy of chemical bonds. Chloroplasts contain chlorophyll, which makes them green, while mitochondria are colorless. Photosynthesis occurs in chloroplasts, and most respiration occurs in mitochondria. Each process requires a different set of enzymes. Chloroplasts have many shapes and sizes, but they are generally larger than mitochondria, which are often cigar-shaped. A brief overview of these two organelles follows; details of their functions are presented in Chapters 6 and 7.

Chloroplasts

The fluid inside chloroplasts is called the **stroma** (fig. 3.24a). Membranes occurring throughout the stroma are called **thylakoids.** Thylakoids are either aggregated into stacks, called **grana,**

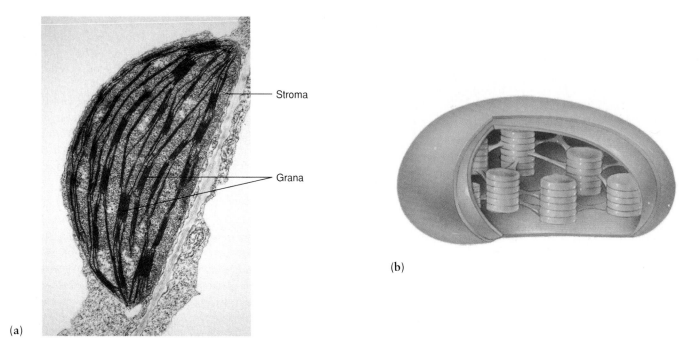

- Stroma

- Grana

(a)

(b)

figure 3.24

Chloroplasts. (a) Transmission electron micrograph of a chloroplast, showing stroma, thylakoids, and thylakoid stacks (grana), ×35,000. Also see the chloroplasts in figure 3.8a. (b) Three-dimensional model of chloroplast membranes, showing vesicle-like thylakoids in grana and continuous thylakoid connections between grana.

or they form connections between stacks (fig. 3.24b). Because these connections are so common, all thylakoids in a chloroplast are probably formed by the same, continuous membrane system.

Thylakoid membranes contain a rich diversity of enzymes and pigments that characterize chloroplasts. Because these enzymes catalyze the reactions of photosynthesis, they are distinct from the enzymes of mitochondria. Pigments in chloroplasts include the chlorophylls, which create the green color of leaves and other green organs, and the carotenoids, which are the yellow, orange, or red colors of autumn leaves, tomatoes, and carrots. In addition to enzymes and pigments, chloroplasts often contain starch or oil.

The greenest cells of a leaf may each contain more than 50 chloroplasts. However, chloroplasts are just one type of **plastid.** Other plastids are classified according to the kinds of pigments or storage products they contain; for example, amyloplasts store starch, and elaioplasts store oils. Colored, nongreen plastids are chromoplasts and are usually red, orange, or yellow.

All plastids develop from proplastids in young, unspecialized cells, and each type of plastid inherits the same genetic material from its proplastid precursors. Depending on environmental conditions, such as light versus dark, plastids may change. When green tissues are kept in the dark, chloroplasts become colorless, but upon reexposure to light they become green again.

Mitochondria

A mitochondrion consists of a smooth outer membrane and an inner membrane that is folded into tubular or vesicle-shaped cristae (fig. 3.25). The internal membrane system of mitochondria arises from the inner membrane of the mitochondrial envelope, whereas the outer membrane of the envelope is smooth. This arrangement of membranes creates two compartments within the mitochondrion: one is the space between the two membranes, and the other is enclosed by the inner membrane.

Many of the reactions of aerobic respiration are catalyzed by enzymes bound to mitochondrial membranes. Other reactions occur either in the space between the inner and outer membranes or in the matrix that is enclosed by the inner membrane (fig. 3.25).

A cell may contain several hundred mitochondria; the actual number of mitochondria in a cell is usually proportional to its requirements for ATP. Dividing cells and cells that are metabolically active need large amounts of ATP and usually have the largest numbers of mitochondria. You will learn more about the structure and function of mitochondria in Chapter 6 ("Respiration").

Cell Movements

Cells prepared for study with microscopes are usually arrested in static positions that conceal their dynamic nature. However, the contents of cells are mobile; internal movements can be seen directly in living cells with a light microscope, or they can be tracked by time-lapse photography. Neither of these techniques is possible with electron microscopy because specimens

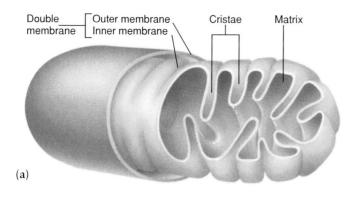

(a)

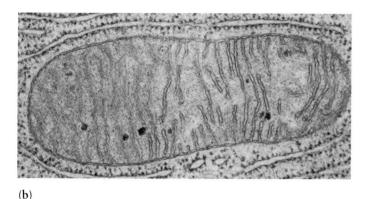

(b)

figure 3.25

Mitochondria. (a) Three-dimensional model of a mitochondrion. The inner membrane is folded into numerous cristae. (b) Transmission electron micrograph of a mitochondrion, ×20,000.

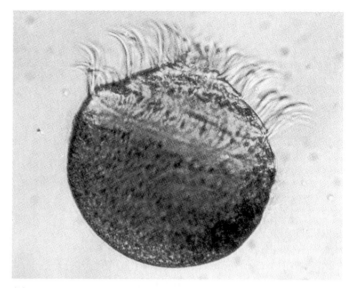

(a)

(b)

figure 3.26

Cells that swim. (a) Light micrograph of sperm cell from a cycad (*Zamia*), showing numerous flagella, ×250. (b) Photo of a cycad that produces flagellated sperm like that shown in (a).

must be killed and fixed before they can be studied with the electron microscope. By means of three-dimensional reconstructions from numerous serial sections, however, electron microscopy can also reveal internal cell movements. Unlike observations with light microscopy, serial reconstructions from electron micrographs reveal the changeability of small organelles and membranes.

In addition to internal movements, some entire cells are motile; that is, they can swim. In plants, only sperm cells swim. Only seedless plants and a few seed plants have swimming sperm cells. The sperm cells of flowering plants cannot swim.

Internal Movements

When we observe living plant cells by light microscopy, we see that their cytoplasm moves constantly. Organelles and other particles usually move in a circle around the central vacuole. This streaming movement is called **cyclosis** because of its circular path.

Chloroplasts and mitochondria tumble along definite paths that are associated with actin filaments and microtubules of the cytoskeleton. The outermost region of the cytoplasm is more anchored and relatively immobile, whereas the innermost region is more fluid. Rows of actin filaments occur between the two regions and induce streaming. Cyclosis enhances the exchange of materials among organelles, between membranes and organelles, and even between cells.

Cells That Swim

Cells that swim have hairlike locomotor organelles that protrude into the medium surrounding the cell. In plant cells these hairs can be up to 150 μm long and are called **flagella** (singular, **flagellum**). The only such swimming cells in plants are sperm cells, which may have two flagella, as in mosses, or several thousand flagella, as in cycads (fig. 3.26).

CELLULAR INVASION: ORIGIN OF CHLOROPLASTS AND MITOCHONDRIA

Chloroplasts and mitochondria are similar to bacteria: they are about the same size, they contain DNA arranged in circular form, they have ribosomes that are smaller than those made in nuclei, and they have a common gene structure that differs from that of nuclear genes. Furthermore, the inner membranes of chloroplasts and mitochondria are similar to the plasma membranes of bacteria. Protein synthesis in chloroplasts and mitochondria is also inhibited by antibiotics that have similar effects on ribosomes in bacteria but not those made in nuclei. Such similarities between these organelles and bacteria are evidence for the endosymbiotic hypothesis of the origin of chloroplasts and mitochondria. According to this hypothesis, small bacteria invaded or were engulfed by larger bacteria. The smaller bacteria, now represented by chloroplasts and mitochondria, developed symbiotic relationships with their larger hosts. The mitochondrial precursors may have provided aerobic (oxygen-using) metabolism for a host that was probably unable to use oxygen. The photosynthetic precursors of chloroplasts provided the ability to make carbohydrates.

Bacterialike organisms probably originated more than 3.5 billion years ago. The first endosymbiotic fossils are about 1.5 billion years old. Nevertheless, modern organisms provide good clues about the bacterial relatives of chloroplasts and mitochondria. Physiological and genetic similarities point to the largest group of bacteria as a source of ancestors for these organelles. For example, one extant member of this group, *Prochlorothrix*, contains chlorophyll *b*, which is otherwise known only in plants and some algae. Other photosynthetic bacteria lack this pigment. However, certain other bacteria have pigments similar to those in chloroplasts of the red algae. This may indicate that chloroplasts originated at least twice, from two different kinds of bacteria. Nevertheless, botanists do not generally agree on how many times chloroplasts arose or which kinds of bacteria gave rise to them.

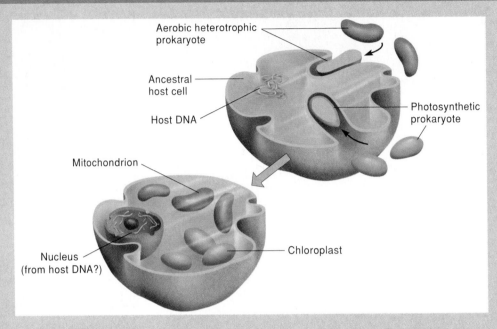

reading figure 3.2

Model of cellular invasion to illustrate the endosymbiotic hypothesis of the origin of chloroplasts and mitochondria. The origin of the nucleus is not explained by this hypothesis.

Other plants that have flagellated sperm cells are ferns and other seedless vascular plants and the maidenhair tree (*Ginkgo biloba*), which is a seed plant. Some algae, water molds, and animals also have flagellated sperm cells. Some water molds and algae have other kinds of cells that swim by flagella as well.

Regardless of their occurrence in diverse organisms, all flagella in plants, animals, fungi, and protists have the same internal structure and the same mechanism of action. Each flagellum consists of a membrane that surrounds ten pairs of microtubules. One pair occupies the center of the flagellum, and nine pairs occur in a ring around the central pair (fig. 3.27a). Each outer pair of microtubules is connected to its neighboring pairs by sidearms that are evenly spaced along the length of the flagellum. Similarly, spokelike extensions from the outer microtubules connect to the inner microtubules (fig. 3.27b). The sidearms are made of a protein called **dynein.** Dynein arms use ATP as a source of energy for their movement, which entails bending and sliding between pairs of microtubules. This pattern of dynein movement is coordinated among nine pairs of microtubules, which enables the flagellum to whip in a powerful wavelike motion.

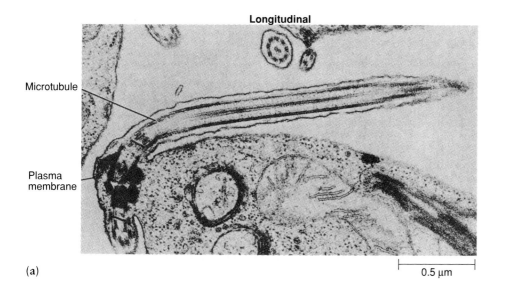

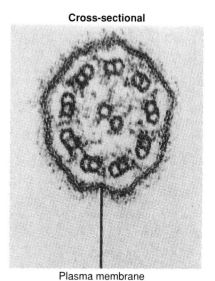

Microtubule

Plasma membrane

(a)

0.5 µm

Plasma membrane

figure 3.27

Flagella. (a) Transmission electron micrographs of a flagellum in the reproductive cell of the green alga *Ulvaria*. Longitudinal section shows that the membrane surrounding the flagellum is continuous with the plasma membrane. Cross section shows 10 pairs of microtubules, one central pair surrounded by nine peripheral pairs in a ring. (b) Model of a flagellum, showing the three-dimensional arrangement of internal components. Microtubules are attached to each other by dynein arms.

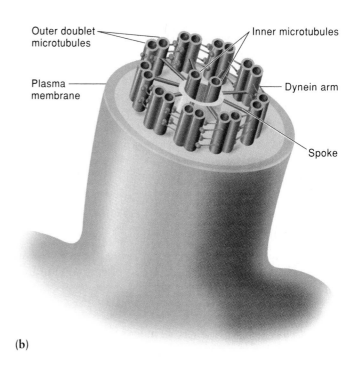

Outer doublet microtubules

Plasma membrane

Inner microtubules

Dynein arm

Spoke

(b)

c o n c e p t

The nucleus and cytoplasm of most plant cells are in constant motion. Organelles and other particles move by cyclosis, which is associated with microtubules and actin filaments in the cytoskeleton. Another kind of cellular movement occurs in the sperm cells of some plants, which have flagella that enable the sperm cells to swim.

Cell Theory: The Dogma and an Alternative

The history of cell science, like that of scientific development in general, parallels the invention of instruments that expanded our observations beyond their previous limits. For example, the discovery of cells was made possible by the invention of the light microscope. In 1665, Robert Hooke used a handmade microscope to see cells in the dead outer bark of a cork oak. He made thin slices of cork tissue and observed the cell walls. Cork tissue was ideal for Hooke because it is rigid enough to slice without collapsing and because it has a simple structure due to the absence of living matter. These advantages do not exist for most other tissues, which may help explain why it was 150 years after Hooke's discovery before any significant advances were made in the study of cells.

The most significant advance in cytology was based on the idea that cells are the fundamental units of life, not just an

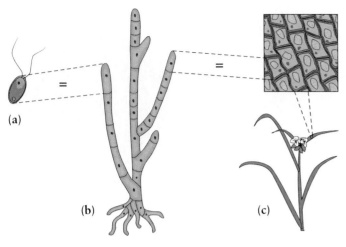

figure 3.28

Equivalency of plant organization according to the cell theory.
(a) Vegetative cell of the single-celled green alga *Chlamydomonas*.
(b) Branched filament of the green alga *Spongomorpha*. (c) Leaf cells
from the leaf of spiderwort (*Tradescantia*).

oddity of cork. This idea is a basic postulate of the **cell theory,**
which was proposed in 1838 by Matthias Schleiden and
Theodor Schwann and became generally accepted by the
1860s (see Chapter 1). The main evidence for the cell theory
was the observation that organisms are made of cells. However, there are many exceptions to the cell theory and its consequences for biological organization.

The Cell Theory: Postulates and Problems

As the cell theory was developed, several principal postulates were proposed for explaining the organization of living
organisms:

1. All living substance is concentrated in cells.

2. Cells in an organism are all individuals of the same
   organizational rank.

3. The cell is the basic unit of structure and function.

4. An organism is an aggregate of cells, which are its
   building blocks.

5. The action of an organism is the sum of many actions of
   different kinds of collaborating cells.

6. The appearance of an organism depends on
   multicellularity.

In sum, these postulates explain that an organism is
analogous to a "cell republic": each cell in an organism is at
the same level of organization as each cell in another organism. The equivalency of cells in different organisms, according
to the cell theory, is shown in figure 3.28.

One organism whose organization is consistent with all of
the postulates of the cell theory is the slime mold, *Dictyostelium
discoideum,* which exists as an aggregation of independent,
amoebalike cells for most of its life history. When it begins to
reproduce, the individual cells come together to form a spore-
containing organ. Similar examples occur in certain green algae.
Such organisms provide support for the cell theory. However,
many biologists question the cell theory because this kind of organization is rare and because it does not occur in plants or animals. Flowers and leaves, for example, are not aggregations of
formerly dissociated cells.

Further doubt arises when we compare a wider variety
of organisms than is shown in figure 3.28. Such a comparison is presented in figure 3.29, which shows that the equivalent level of organization among green algae and flowering
plants may be the whole organism, not the cell. Moreover,
the noncellular body of *Derbesia* in figure 3.29 shows that
appearance does not depend on multicellularity, at least in
green algae. Similar noncellularity also occurs in certain developmental stages of seed plants. For example, the nutritional tissue of many seeds is multinucleate cytoplasm before
it becomes cellular; coconut milk is a common example of
such noncellular tissue. The embryos of pines and other
cone-bearing trees are also noncellular in their early development. Thus, their development is independent of multicellularity. Similar examples of noncellularity also occur in fungi,
protozoans, and animals.

The cell theory was derived primarily from animal development. However, the multicellular construction of plants is
distinctly different from that of animals. Unlike animals,
plants are characterized by incomplete cleavage between dividing cells, continuous cytoplasmic connections (i.e., plasmodesmata) between cells, and the absence of cellular motility during development. The fixed position and interconnectedness of
all cells in a plant provide evidence for the concept of the **symplast.** This term describes the entire living mass of the plant as
a continuous unit, contrary to the idea that cells are separate
individuals.

An Alternative: The Organismal Theory

If cells are not the basic units of biological organization, then
what is the alternative? Perhaps the best alternative comes
from the **organismal theory,** which holds that the whole organism is the basic unit of organization. This means that an
entire plant is like a large cell that is compartmentalized into
many parts. According to this theory, a whole plant constitutes the same level of organization as a single-celled green
alga (fig. 3.29). One of the main postulates of this theory is
that the appearance of an organism is independent of multicellularity; that is, development is not coupled to cell division.
Derbesia presents observational evidence for this suggestion
(fig. 3.29).

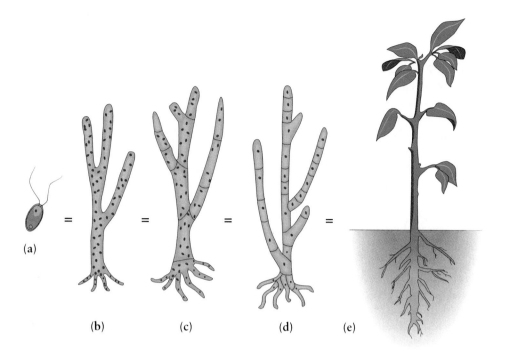

figure 3.29

Equivalency of plant organization according to the organismal theory. (a) *Chlamydomonas*. (b) Branched filament of the noncellular green alga *Derbesia*. (c) Branched filament of the multinucleate but cellular green alga *Cladophora*. (d) *Spongomorpha*. (e) Diagram of *Vicia faba* (broad bean).

Botanists draw additional support for the organismal theory from the lemon tree, in which they have observed that the cuticle is first formed around the zygote, which is the first cell of the tree. The cuticle expands and maintains its coverage of the embryo, the seedling, and the tree as the plant grows. This observation suggests that the cuticle is a characteristic of the whole organism rather than of a particular cell type. This means that the plant differentiates by compartmentation of the zygote and all other cells within the same cuticle and not by the aggregation of separate cells, each with its own cuticle.

Support for the organismal theory also comes from experimental evidence that shows a lack of a cause-and-effect relation between cell division and development. The experiment involved wheat grains irradiated with gamma rays, which suppressed cell division in the seedlings. In spite of this treatment, the wheat grew to normal size but more slowly and with fewer cells than untreated wheat (see Chapter 12). This experiment provides dramatic evidence that cell division does not cause development, a result that supports the organismal theory. Although the focus here is on plants, support for the organismal theory has also been found in animals.

c o n c e p t

The cell theory states that the cell is the basic unit of organization of living organisms. In contrast, the organismal theory states that the whole organism is the basic unit of living organisms. The cell theory is best represented in slime molds and certain algae, but observations and experiments supporting the organismal theory come from plants and animals.

Chapter Summary

Cells are the simplest units of a plant that can live independently. Each plant cell consists of a cell wall that surrounds a plasma membrane, which encloses the contents of the cell. The contents of a plant cell usually include a nucleus and the cytoplasm. The cytoplasm includes all organelles except the nucleus, plus all internal membranes and the cytosol.

The study of cells is aided by instruments such as the light microscope and the electron microscope. The quality of microscopic observation depends on both magnification and resolving power. Resolving power is limited by the wavelength of the light or the electron beam that is used for observation. Our study of plant cells has been aided by examining the functions of organelles that have been isolated by cell fractionation and by selectively staining different parts of cells prior to microscopy.

Membranes divide cells into many interconnected compartments. These compartments are connected by membranes that either move between or are attached to the nucleus and other organelles. All membranes consist of a double layer of phospholipids that has enzymes attached to or embedded in it. The composition of membranes varies from one organelle to another.

The distribution and movement of membranes and other cell compartments are affected by the cytoskeleton, a network of filaments that includes microtubules, actin filaments, and intermediate filaments. All three types of cytoskeletal filaments are made of different kinds of proteins.

Young cells and actively growing cells have flexible primary cell walls. As cells mature, however, they often form rigid secondary cell walls just inside the primary walls. Cell-wall synthesis is controlled by arrays of microtubules just inside the plasma membrane.

Plant cells are connected to one another by plasmodesmata. These connections contain cytoplasm and a plasma membrane that is continuous between cells.

The nucleus, chloroplasts, and mitochondria are organelles that are surrounded by double membranes. Microbodies have single membranes, and ribosomes have no membranes. Dictyosomes and the endoplasmic reticulum are composed mostly of membranes that enclose relatively little internal fluid. Many of the chemical reactions in a cell are catalyzed by enzymes that are in or on membranes, including the plasma membrane.

Sperm cells in plants and animals, as well as other kinds of cells in algae and water molds, have flagella that enable them to swim. The structure and function of the flagella are identical in these organisms.

The cell theory has been the dominant theory of how living things are organized. It explains how cells are the basic units of organization. An alternative to this idea is the organismal theory, which holds that the basic unit of organization is the whole organism. Some evidence supports both theories, but observational and experimental evidence supports the organismal theory better in plants.

 ### *What Are Botanists Doing?*

Use the reference section of your library to find articles that have been published in the past twelve months on work that involves the use of colchicine in plant cell biology. Answer the question, "What is the prevalent use of colchicine in plant cell biology today?"

 ### *Writing to Learn Botany*

What might be the advantages and disadvantages of studying isolated organelles?

Questions for Further Thought and Study

1. Discuss the importance of having both proteins and lipids as components of membranes.

2. How does the function of rough ER differ from that of smooth ER?

3. Where are microtubules in cells? What are their roles?

4. How does microscopy provide evidence for the functions of cellular organelles?

5. What evidence supports the organismal theory?

6. The cell theory is an excellent example of inductive reasoning. How do you use inductive reasoning in your life?

Web Sites

http://ampere.scale.uiuc.edu/~m-lexa/scripts/cell.cgi
At this site you can tour a three-dimensional model of a plant cell. Cut, magnify, and rotate the cell as you learn about its organelles and their functions.

http://www.mblab.gla.ac.uk:80/~julian/Dict.html
Need to research a definition of a term in cell biology? Try this online dictionary.

http://sunsite.unc.edu/herbmed/
This site presents an interesting array of medicinal and culinary herbs, pictures of Finnish plants, the Herblist logs, herb programs, and pictures by Marco Bleekers.

Suggested Readings

Articles

Albersheim, P. 1975. The walls of growing plant cells. *Scientific American* 232 (April):81–95.

de Duve, C. 1983. Microbodies in the living cell. *Scientific American* 248 (May):74–84.

Kaplan, D. R., and W. Hagemann. 1991. The relationship of cell and organism in vascular plants. *BioScience* 41:693–703.

Lane, M. A. et al. 1990. Forensic botany. *BioScience* 40:34–39.

Niklas, K. J. 1989. The cellular mechanisms of plants. *American Scientist* 77:344–349.

Rothman, J. E. 1985. The compartmental organization of the Golgi apparatus. *Scientific American* 253 (September):74–89.

Storey, R. D. 1990. Textbook errors and misconceptions in biology: Cell structure. *The American Biology Teacher* 52:213–218.

Symmons, M. et al. 1989. The shifting scaffolds of the cell. *New Scientist* (18 February):44–47.

Vogel, S. 1987. Mythology in introductory biology. *BioScience* 37:611–614.

Zimmer, C. 1995. First cell. *Discover* 16(11):70–78.

Books

Becker, W. M., and D. W. Deamer. 1991. *The World of the Cell.* 2d ed. Redwood City, CA: Benjamin/Cummings.

de Duve, C. 1984. *A Guided Tour of the Living Cell.* Scientific American Library. New York: W. H. Freeman.

Gunning, B. E. S., and M. W. Steer. 1986. *Plant Cell Biology: An Ultrastructural Approach.* Copyright M. W. Steer.

Tobin, A. K. *Plant Organelles.* New York: Cambridge University Press.

The movement of water and nutrients between photosynthetic cells (green) and vascular tissues (red) entails transport across membranes.

Membranes and Membrane Transport

4

Chapter Outline

Chapter Overview

In Chapter 3 you learned that all living cells have membranes. Membranes surround cells, connect cells to one another, form complex internal networks, and divide cells into distinct compartments. The abundance of membranes in cells underscores their importance, but it does not begin to show the diversity of their functions. Membranes are active and changeable participants in cellular metabolism: everything that happens in a cell is directly or indirectly associated with a membrane.

In this chapter we explain more about membrane-dependent metabolism and how membranes control the life of a cell. You will also see how, in spite of their wide range of capabilities, all membranes have the same basic structure. Moreover, you will learn how plasma membranes enable organisms to grow in salty soil or water, how organellar membranes harvest energy from ion movement, and how membranes help cells communicate. All of these functions of membranes are essential to life. In fact, one of the criteria for describing life is the presence of a dynamic membrane system.

In the preceding chapter you learned that membranes have various functions in cellular metabolism. Most of the important activities of cells are associated with membranes. For example, proteins destined for secretion or for insertion into cell membranes are made by ribosomes that are attached to a membrane system called the endoplasmic reticulum. The separation of cytosol and organellar contents from one another is maintained by membranes.

Suggestions about the structure of membranes appeared in the 1920s, long before membranes were seen with the electron microscope. The first ideas about membrane structure were based on the soaplike properties of phospholipids in artificial membranes. Phospholipids, like soaps, have a dual solubility. Their long hydrocarbon "tails" are nonpolar and hydrophobic. *Hydrophobic* literally means "water-fearing" and refers to chemicals that do not dissolve in water. In contrast, the ionic phosphate "head" of a phospholipid is polar and hydrophilic. *Hydrophilic* means "water-loving" and refers to chemicals that can dissolve in water. In water, phospholipids aggregate spontaneously into a **bilayer,** which is a double membrane with an interior of hydrophobic hydrocarbons and an exterior of hydrophilic phosphates (fig. 4.1). As you will learn in this chapter, this artificial membrane resembles the phospholipid backbone of biological membranes. However, the great diversity of membrane functions cannot be explained by such a simple structural model. The properties of membranes also depend on proteins, which are their other main ingredient.

Many functions of membranes relate to the general properties of phospholipids and proteins. However, different membranes are made of different proteins and lipids. In this chapter we first discuss the basic functions of membranes in terms of the general characteristics of proteins and lipids. Later, we examine different kinds of membranes and their functions.

Overview of Membrane Structure and Function

Early ideas about the structure of phospholipid bilayers explained how some molecules, including nonpolar molecules of gases (N_2 and O_2) and small polar molecules such as water could flow across membranes (fig. 4.2). However, lipid bilayers could not account for the ready passage of larger polar molecules such as monosaccharides and amino acids. The passage of these molecules through membranes was first explained in the 1930s by H. Davson and J. F. Danielli, who suggested that the lipid bilayer was coated on both sides with hydrophilic proteins that were attached to the polar heads of phospholipids. According to this model, the hydrophilic proteins absorbed polar molecules and somehow eased their passage through the nonpolar layer of the membrane. In the 1950s, the first electron micrographs of membranes seemed to confirm this model (fig. 4.3). These pictures showed membranes to have an electron-transparent inner region sandwiched between two electron-dense outer layers. The outer layers were assumed to be made of proteins and phospholipid heads, and the inner region was believed to be made of the hydrocarbons in phospholipid tails.

Structure of Membranes

Despite the apparent support from electron microscopy, flaws in the Davson-Danielli model began to accumulate. For example, cell biologists found that this model could not explain

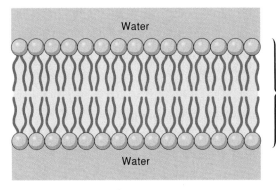

Water

Water

Phospholipid bilayer

figure 4.1

Model of an artificial membrane consisting of phospholipids. Phospholipids aggregate spontaneously into a bilayer that consists of inner fatty acids and outer phosphates.

structural and biochemical variations among different kinds of membranes. Mitochondrial membranes, for instance, are thinner than plasma membranes and contain a much higher proportion of protein. This and other findings contradicted the Davson-Danielli model, which held that membrane proteins must be hydrophilic and that all membranes have the general structure of a protein-lipid-protein "sandwich."

The current view of membrane structure entails several modifications of the Davson-Danielli model. This newer version, proposed in 1972, is called the **fluid mosaic model** because it holds that proteins occur as a mosaic in a fluid bilayer of phospholipids (fig. 4.4). Visual evidence for the fluid mosaic model comes from scanning electron microscopy

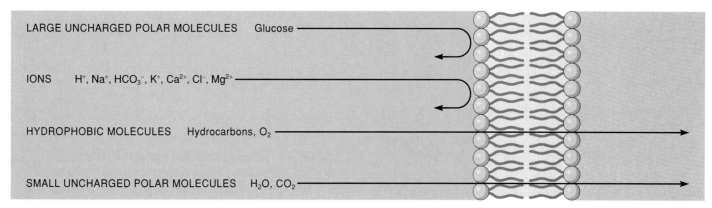

figure 4.2

Differential permeability of a phospholipid bilayer. The bilayer is much more permeable to hydrophobic molecules and small uncharged polar molecules than it is to ions and large polar molecules.

LARGE UNCHARGED POLAR MOLECULES — Glucose

IONS — H^+, Na^+, HCO_3^-, K^+, Ca^{2+}, Cl^-, Mg^{2+}

HYDROPHOBIC MOLECULES — Hydrocarbons, O_2

SMALL UNCHARGED POLAR MOLECULES — H_2O, CO_2

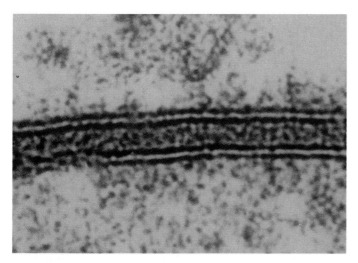

figure 4.3

Transmission electron micrograph of two membranes. Each membrane appears as two dark lines that are separated by a lighter region. At first, micrographs of this sort seemed to confirm the Davson-Danielli model of membrane structure, but this interpretation was later found to be incorrect.

of freeze-fractured membranes. Recall from Chapter 3 that freeze-fracturing often splits membranes between the two layers of phospholipids. The interior of the membrane is then seen as a dotted plain: the plain is a sea of lipids, and the dots are proteins inserted into them (fig. 4.5). The proteins are partly hydrophobic and partly hydrophilic, and are anchored at the protein-lipid interface by hydrophobic amino acids. Hydrophilic amino acids occur inside the proteins and also where the proteins protrude from the membrane into the aqueous environment on either side of the membrane (fig. 4.6).

The name *fluid mosaic model* implies that the membrane contains liquid. In fact, the lipid bilayer is oily; its fluidity results from the loose packing of the fatty acid tails of the phospholipids. The mosaic of proteins actually floats through the lipids. The mobility of proteins is a significant feature of the fluid mosaic model; it accounts for the movement and intermingling of proteins that must touch each other to function.

Another important feature of membrane structure, one that is not immediately obvious from the fluid mosaic

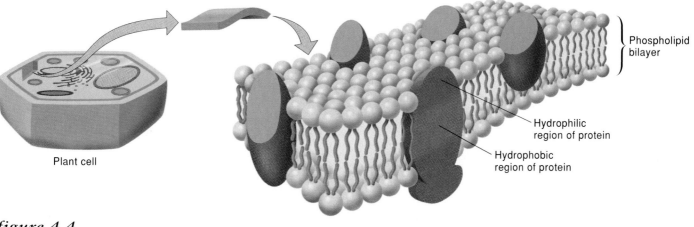

Plant cell

Phospholipid bilayer

Hydrophilic region of protein

Hydrophobic region of protein

figure 4.4

The fluid mosaic model of membrane structure. According to this model, proteins are dispersed in the phospholipid bilayer.

figure 4.5

Scanning electron micrograph of a freeze-fractured chloroplast membrane. Proteins appear as dots embedded in the lipid matrix.

model, is its asymmetry; that is, one side of a membrane is different from the other. This difference comes mostly from the carbohydrates that are attached to proteins on the outside surface of the membrane. Proteins with carbohydrates attached to them are called **glycoproteins,** and they do not usually occur on the inner surfaces of membranes. Examples of the roles of glycoproteins are discussed later in this chapter in the section on "Cellular Communication."

c o n c e p t

Membranes are a mosaic of proteins that move around in a fluid double layer of phospholipids. This is called the *fluid mosaic model* of membrane structure. Proteins that protrude from the outer surface of membranes are often glycoproteins.

Functions of Membranes

Proteins control most of the functions of membranes. There may be fifty or more different kinds of proteins in a plasma membrane and perhaps as many in the tonoplast and other organellar membranes. This diversity of proteins is reflected in the enormous range of activities associated with membranes. Some of the more important of these activities are discussed in the next few sections of this chapter and summarized below:

Movement of water and solutes. The plasma membrane generally allows the unrestricted movement of water and certain dissolved substances into or out of the cell (fig. 4.2). Water balance is crucial for maintaining turgor pressure, which makes the cell rigid and drives cellular expansion during growth.

Differential permeability. Membranes control or block the passage of some kinds of molecules (fig. 4.2); such membranes are referred to as *differentially permeable* membranes. Differential permeability is different for different membranes.

Ion pumps. Certain ions, such as K^+ and H^+, are pumped through membranes. Ion pumps in the plasma membrane use energy from ATP to move ions from the cell, while ion pumps in mitochondrial and chloroplast membranes are important for making ATP.

Enzyme activity. Enzymes that cooperate in multistep processes, such as ATP synthesis or the absorption of light energy, often occur together in a particular spot on a membrane.

Cellular communication. The plasma membrane contains proteins that bind molecules released from other cells. Once bound to an external molecule, these proteins activate other proteins in the membrane that cause metabolic changes in the cell.

c o n c e p t

Membranes have many roles in a cell. Besides blocking or facilitating the entry of substances into the cell, membranes are also the sites of enzyme activity, ATP synthesis, and cellular communication.

The Movement of Water and Other Molecules through Membranes

The most common molecule in cells is water. Ions and other polar molecules are dissolved in this water; that is, they are **solutes,** and water is the **solvent** that dissolves them. The

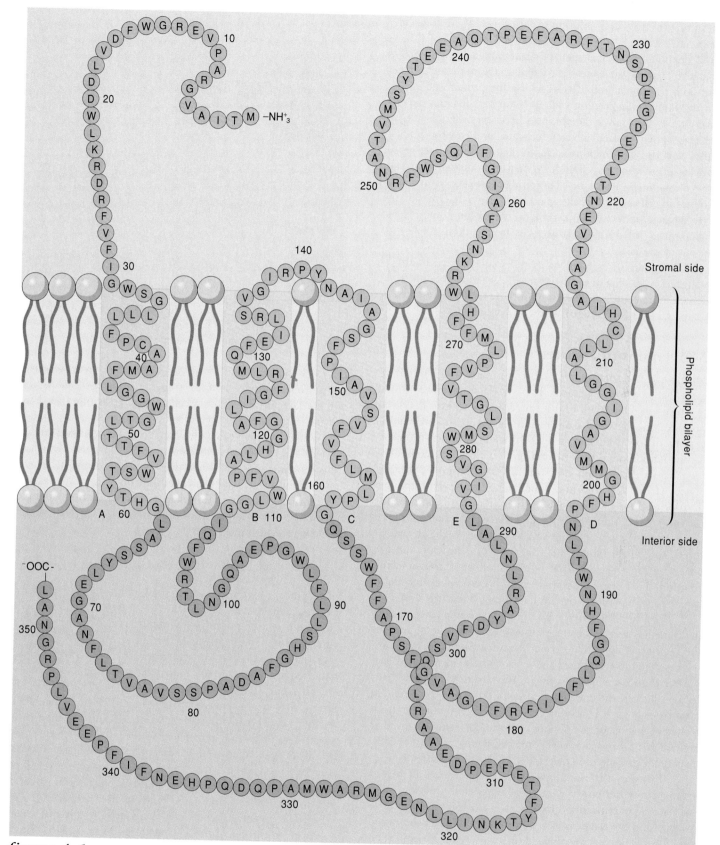

figure 4.6

Primary structure of a thylakoid protein. Five regions of the protein occur in the membrane (boxes A–E); the remaining regions protrude either into the stroma or into the interior of the thylakoid. This membrane protein consists of 352 amino acids, each designated by a one-letter abbreviation (see Chapter 2, fig. 2.6).

Source: Wim F. J. Vermaas.

resulting mixture is a **solution.** Solutes include protons (H^+), mineral ions such as potassium (K^+) and magnesium (Mg^{2+}), and organic compounds such as sugars and amino acids. The passage of these substances through membranes is determined by both the phospholipid bilayer and the proteins embedded in it. Nonpolar molecules, such as hydrocarbons and oxygen, pass easily through a membrane's lipid core. Small and uncharged polar molecules, such as water and carbon dioxide, also pass through membrane lipids without hindrance. Ions, however, are almost entirely prevented from diffusing through the phospholipid bilayer. The concentrations of different solutes in cells and organelles are closely regulated, despite the free passage of water and certain solutes through the plasma membranes. This inhibits substances from leaking freely into or out of the cell.

Some terms involving water and solute movement are introduced in this chapter because they relate to membrane function. A more complete discussion of these terms is presented in Chapter 21, "Movement of Water and Solutes."

Movement of Solutes

All molecules display random thermal motion, or kinetic energy; that is, a solute molecule has a tendency to move around in a solution. One result of this random movement is that molecules diffuse outward from regions of high concentration to regions of lower concentration. **Diffusion** by random movement continues until the distribution of molecules becomes homogeneous throughout the solution. For example, when a crystal of dye is placed in a container of pure water, the dye diffuses as the crystal dissolves. The rate of diffusion depends on the size of the dye molecules (larger molecules move slower) and the temperature of the solution (higher temperatures cause faster movement). Regardless of this rate, the dye concentration will eventually become uniform throughout the solution. This phenomenon is easily illustrated by placing a drop of a water-soluble dye into a glass of water (fig. 4.7).

Because diffusion is based on random movements, each molecule has an equal chance of moving toward or away from the region of high solute concentration. However, there are more solute molecules near the crystal, which means there is a greater chance for **net movement** away from the source. Molecular movement continues after homogeneity is reached, even though there is no longer a net change in concentration from one region of the solution to another. This means that after such a balance is reached, for every molecule that vacates a spot, another molecule takes its place just by chance.

Diffusion is the net movement of solutes from an area of greater concentration to an area of lesser concentration—that is, down a **concentration gradient.** We can think of solutes as marbles and the gradient as a hill; marbles rolling down a hill would be analogous to solutes diffusing down a concentration gradient. Furthermore, a steeper hill would be analogous to the steeper gradient caused by a higher concentration of solutes. Thus, a steeper gradient causes a higher net rate of solute movement. Just as marbles can move up the hill if there

figure 4.7

Beakers of water before and after diffusion of the dye bromthymol blue. Random movements of water and dye molecules cause diffusion, eventually resulting in a uniform concentration of the dye.

is a force to push them, so too can solutes move from lower to higher concentrations if there is energy to push them. As you will see later in this chapter, cellular energy is often used for moving solutes up their concentration gradients.

Anything that can fall, change, or flow from a higher level to a lower one has the potential to do work, which is called **potential energy.** When the solute is moving, it has energy of movement, which is called **kinetic energy.** We discuss these forms of energy in this chapter in relation to membrane function, but a broader discussion of energy and its use by plants is presented in Chapter 5.

Water Potential

Like solutes, water also has potential energy to flow to where it is less concentrated. The potential energy of water has a special name: **water potential.** Water tends to move down a water-potential gradient—that is, from a region of high water potential to a region of low water potential. Also, like solutes, water requires energy to move up a water-potential gradient.

By general agreement, the water potential of pure water is zero. This means that the water potential of a solution has a negative value because the water is less concentrated than pure water. Also by general agreement, water potential is expressed in units of pressure instead of units of energy, because pressure is simpler to measure. Thus, water potential may be expressed in **bars** or **megapascals (MPa).** One bar equals atmospheric pressure at sea level and room temperature, and 0.1 MPa is 1 bar. The potential of seawater is about -25 bars, the water potential in the cells of freshly watered herbs is more than -1 bar (meaning closer to zero), and the water potential in dry seeds is about -200 bars. For comparison, the recommended air pressure for most car tires is about 2 bars. We'll discuss water potential in more detail in Chapter 21.

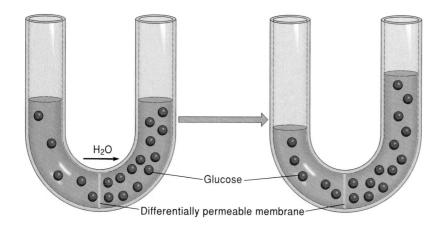

figure 4.8

Osmosis is demonstrated by the movement of water through a differentially permeable membrane in a U-tube. Glucose, which cannot pass through the membrane, is more concentrated in the right-hand part of the tube than in the left-hand part. Net movement of water into the more concentrated glucose solution causes the volume of the right-hand solution to increase.

c o n c e p t

The random thermal motion of all molecules causes them to move about in a solution. Because of this motion, molecules move from a region of higher concentration to a region of lower concentration. This net movement of molecules down a concentration gradient is called *diffusion*. The potential for solutes to move down a gradient is called *potential energy*. The potential energy of water to move down a gradient has a special name, *water potential*.

Osmosis

Many substances, including water, move through biological membranes as freely as they move through an aqueous solution. Such unrestricted movement of a substance through a biological membrane is called **passive transport.** The energy for passive transport is the kinetic energy that is inherent in all molecules. That is, passive transport does not require energy from cellular metabolism.

The passive transport of water influences many activities of the cell, including cell growth, structural rigidity, and photosynthesis. Because of its roles in so many processes, the diffusion of water through a selectively permeable membrane has a special name: **osmosis.**

Osmosis is influenced by different water potentials on either side of a membrane, as in the following example: Consider a membrane that is permeable to water but impermeable to glucose. When such a membrane separates two halves of a container, each having a different concentration of glucose, water diffuses by osmosis into the side having the higher glucose concentration (i.e., lower water potential) (fig. 4.8). The net movement of water will stop either when both sides of the container have the same concentration of glucose or when the force of gravity equals the force of water movement, whichever occurs first. Note that the side that began with a higher concentration of glucose increases in volume.

The foregoing example of osmosis also hints at another feature of this process. Immediately before osmosis begins, water on the **hypotonic** side (low solute concentration) of the

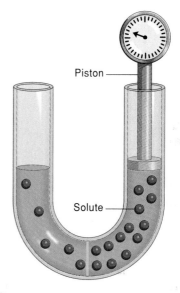

figure 4.9

Higher water potential in the left-hand solution causes pressure for water to move into the right-hand solution. The amount of pressure that is necessary to maintain constant volume on the right equals the force of water movement across the membrane.

membrane has a higher water potential, which means a greater potential to move. Its movement can be prevented, however, if a piston is placed on the **hypertonic** side (high solute concentration), with just enough downward pressure to keep the volume constant. A pressure gauge on this piston measures the force required to maintain a constant volume (fig. 4.9).

The potential of pure water to move into a solution on the other side of a membrane is called **osmotic pressure** (fig. 4.10). Conversely, the potential of the solution to cause osmotic pressure is called **osmotic potential.** Solutions with high concentrations of solutes (i.e., more negative osmotic potential) cause high osmotic pressure. Osmotic potential and water potential are expressed in the same units of measurement and may be easily confused. However, although osmotic potential occurs only across a membrane, water potential has no such limitation.

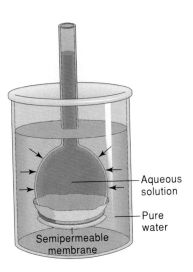

figure 4.10

Osmotic pressure (arrows) is the pressure of pure water to move into a solution on the other side of a membrane. The osmotic potential of the solution is the potential to cause osmotic pressure.

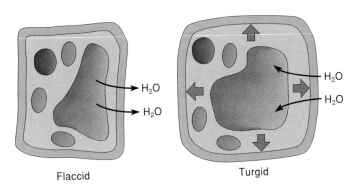

Flaccid   Turgid

figure 4.11

Flaccid and turgid cells. The plumpness of the turgid cell is maintained by turgor pressure.

c o n c e p t

Osmosis is the diffusion of water across a differentially permeable membrane. The potential for water movement into a cell is continuous because the concentrations of solution are generally higher in the cytoplasm than in the matrix outside the cell. This potential causes pressure, called *osmotic pressure*, from outside the cell. Inside the cell, the counterpart of osmotic pressure is called *osmotic potential*.

Turgor

Most plant cells are surrounded by a hypotonic environment. As a result, cells absorb as much water as they can hold. The outward pressure of the plasma membrane against the cell walls is called **turgor pressure** because it keeps the cells turgid (fig. 4.11).

figure 4.12

Scanning electron micrograph of an open stoma, ×640. Stomata are open when guard cells are turgid; they are closed when guard cells lose turgidity (see Chapter 21).

Turgor pressure is vital to plants in many ways. During growth, cell expansion is caused by turgor pressure on cell walls that have become relaxed. Turgor pressure also keeps herbaceous (nonwoody) plants upright, supports the fleshy stalks and leaves of trees and shrubs, and keeps supermarket vegetables crisp when they are sprayed with water. Changes in turgor also cause movements in plants, such as the opening and closing of stomata and the curling of grass leaves (fig. 4.12). Some movements caused by changes in turgor are dramatic, such as leaf movement in the sensitive plant (see Chapter 19).

Cells lose turgor when they are placed in a hypertonic solution. The continued loss of turgor causes the cytoplasm to shrink away from the cell wall (fig. 4.13). Osmotically induced shrinkage of the cytoplasm is called **plasmolysis**. This phenomenon occurs in crop and garden plants when salt accumulates in the soil from extensive use of hard (i.e., mineral-rich) water. It also occurs when people apply too much fertilizer. The loss of turgor in these plants causes their leaves and stems to wilt.

Inducing Osmosis: The Control of Turgor in Plants

Although water moves freely across biological membranes, plants can control osmosis by regulating the concentrations of solutes in their cells. Loss of turgor causes the uptake of potassium ions (K^+), which are **osmotically active** because they change the cell's osmotic potential. As the concentration of K^+ increases, the osmotic potential increases, and more water enters the cell. Conversely, cells reduce their osmotic potential by secreting K^+. This causes water to leave the cell.

         4–9   *Unit One*   Cells as the Fundamental Units of Life

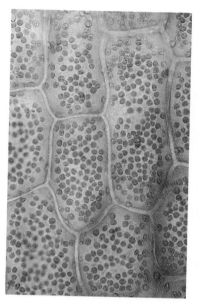
(a)

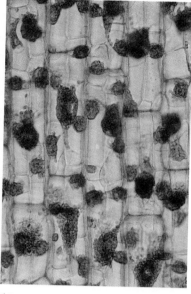
(b)

figure 4.13

Light micrographs of (a) turgid cells, ×400, and (b) plasmolyzed cells, ×100, showing the effects of plasmolysis as the cell contents shrink away from the cell wall.

figure 4.14

The salt bush (*Atriplex occidentalis*) grows in salt flats along the Pacific Coast. These plants produce large amounts of osmotically active compounds that enable them to absorb water from the dry, salty soil.

The uptake of K⁺ occurs against its concentration gradient; that is, K⁺ moves from a region of low concentration outside the cell to a region of high concentration inside the cell. This process, which is called **active transport,** requires metabolic energy (see following page).

Plants that live in high-salt environments such as salt flats, near ocean bays, or inland where oceans once were accumulate large amounts of osmotically active solutes such as the amino acid proline and the carbohydrate mannitol (fig. 4.14). These organic solutes help the plant absorb water (via osmosis) from dry, salty soil.

Osmotic potential causes water pressure to push the plasma membrane against the cell wall. This pressure, called *turgor pressure*, keeps the cell turgid. Decreased osmotic potential causes a loss of water and, consequently, a loss of turgor. When turgor is lost, the plasma membrane shrinks from the cell wall in a process called *plasmolysis*. Although membranes allow the free passage of water, turgor pressure is controlled by the active transport of ions or the synthesis of other osmotically active solutes. These processes require energy from cellular metabolism.

Differential Permeability of Membranes

As already mentioned in this chapter, biological membranes are differentially permeable. This is one of their most important properties, because it keeps metabolically important substances inside the cell or organelle and prevents inappropriate or toxic substances from entering. Membranes also enable ions and larger polar molecules, such as sugars, to pass into the cell through specific membrane proteins called **transport proteins.**

There are two main types of transport proteins. One type includes proteins that work by active transport, which requires energy from ATP to move solutes up a concentration gradient; the other includes passive transport proteins, which do not require metabolic energy. In passive transport, proteins merely act as channels for the diffusion of certain molecules down their concentration gradients. Each passive transport channel is specific for one or two solutes. In contrast to simple diffusion through the phospholipid bilayer, passive transport through protein channels is called **facilitated diffusion.** As in simple diffusion, potential energy is also released during facilitated diffusion.

Facilitated Diffusion

Like simple diffusion, facilitated diffusion is driven by a concentration gradient. Solutes move through transport proteins from the hypertonic side of the membrane to the hypotonic side of the membrane (fig. 4.15). Each transport protein forms a continuous, hydrophilic pathway for polar molecules. Some proteins allow only one solute to diffuse at a time, whereas others only work when two solutes move at the same time, by cotransport.

Little is known about how transport proteins work. The best guess is that they alternate between two forms. One form of the protein accepts a solute molecule on one side of the membrane, which changes the protein to the other form. That second form of the transport protein releases the solute on the other side of the membrane. According to this model, we must also assume that "empty" proteins flip-flop randomly between the two forms. Because of their greater number, molecules on the hypertonic side of the membrane would have more frequent contact with transport proteins than those on the hypotonic side of the membrane; thus, solutes would move down their concentration gradient.

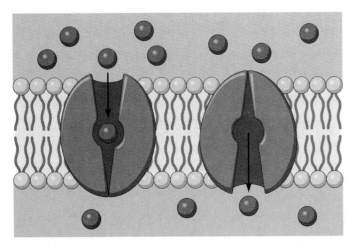

figure 4.15

A possible mechanism for facilitated diffusion. The transport protein (purple) accepts solute molecules (red spheres) on one side of the membrane and releases them on the other side. The transport protein alternates between two forms, depending on whether it is "open" to one side of the membrane or the other.

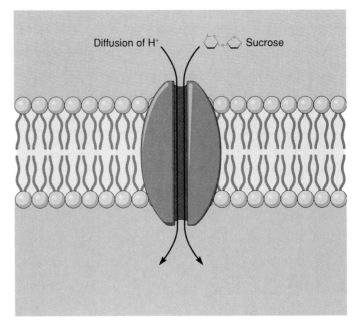

Diffusion of H⁺   Sucrose

figure 4.16

Cotransport across membranes uses energy from the force of diffusion of one solute (H^+) to move another solute (sucrose) against its concentration gradient.

Sugars typically move by facilitated diffusion that involves cotransport with another solute. For example, sucrose moves into conducting cells of leaf veins by hitching a ride with hydrogen ions (fig. 4.16): the energy for sucrose transport comes from the force of H^+ diffusion. The force of H^+ diffusion can be powerful enough to move sucrose against its own concentration gradient.

Active Transport

Many substances move into or out of cells and organelles against a concentration gradient without the aid of facilitated diffusion by cotransport. The uptake of potassium, mentioned earlier in this chapter, is one example. Likewise, marine algae secrete sodium, even though the seawater surrounding them is much saltier than their cytoplasm. In both cases, the transport of solutes requires energy from the cell to overcome the energy of thermal motion that drives passive transport. This energy-requiring process is called *active transport.*

The energy required for active transport usually comes from the hydrolysis of ATP. This reaction is catalyzed by membrane-bound enzymes called *ATP phosphohydrolases* (*ATPases*), which are transport proteins that use the energy of ATP hydrolysis (fig. 4.17). Many ATPases actively transport ions against the ion's concentration gradient, thereby creating potential energy for the passive cotransport of other solutes back across the membranes. The cotransport of sucrose with H^+ depends on a higher concentration of H^+ outside the cell. This gradient is maintained by the active transport of H^+ across the plasma membrane. This is an example of a **coupled cotransport system,** so-called because it uses energy from active transport to create a gradient that drives the passive cotransport of two solutes.

In addition to a concentration gradient, H^+ and other ions also have an electrical gradient because they are charged particles. Thus, ion transport is also influenced by an electrical gradient. The combination of the concentration gradient and the electrical gradient of ions is called an **electrochemical gradient.** The effects of electrochemical gradients on ion transport are described more fully later in this chapter.

c o n c e p t

Ions and large polar molecules are prevented from diffusing through the lipid bilayer of membranes. Instead, membranes selectively allow these kinds of solutes to pass through transport proteins. The movement of solutes through transport proteins, down a concentration gradient, is called *facilitated diffusion.* When solutes move against a concentration gradient, metabolic energy from ATP is required for *active transport.*

The Lore of Plants

Leonardo da Vinci and Michelangelo both claimed that rue (*Ruta graveolens*) enhanced their eyesight and creative power. Robbers who stripped plague victims protected themselves with "vinegar of the four thieves," of which rue was an ingredient.

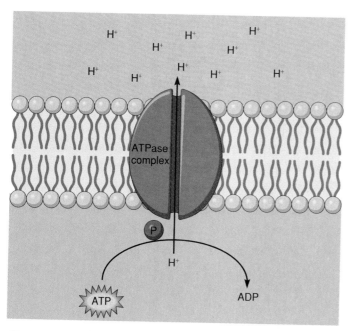

figure 4.17

Active transport uses energy released by the hydrolysis of ATP by ATPases in the membrane. This energy is used to transport ions (H⁺) against their concentration gradient. Phosphate (P) from ATP binds to ATPases during hydrolysis.

Bypassing Membrane Transport

Simple diffusion, facilitated diffusion, and active transport all entail direct movement through the phospholipid bilayer or through proteins embedded in it. Nevertheless, membrane transport is often bypassed by **exocytosis** (fig. 4.18).

Plant cells secrete polysaccharides and proteins across the plasma membrane for assembly into cell walls. Moreover, cells of root tips secrete a slimy polysaccharide that lubricates their passage through soil as they grow, and cells covering leaves exude waxy substances onto their surfaces to inhibit water loss. Leaves of the Venus's-flytrap and other insectivorous plants secrete enzymes that digest insects. However, unlike the exocytosis of cell-wall materials, which occurs via dictyosome vesicles, the secretion of digestive enzymes relies on vesicles derived from the endoplasmic reticulum.

The movement of starch-digesting enzymes in cereal grains has probably been studied more thoroughly than any other example of exocytosis in plants. The physiology of cereal grains is important to brewers of beer and beverages. Reading 4.1, "Membrane Transport and Making Beer," presents a brief discussion of membrane transport and starch digestion in cereal grains.

Cells can also bypass membrane transport into the cytoplasm by pinching off small coated pits in the plasma membrane. This process, called **endocytosis,** is common in animal cells, but it is not readily observed in plants. Plant cells do have coated pits (fig. 4.19), however, which is indirect evidence for endocytosis.

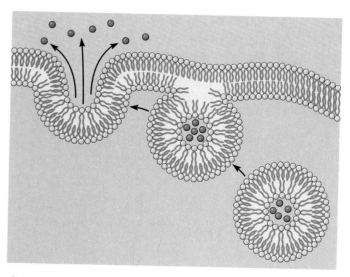

figure 4.18

Exocytosis transports large molecules out of cells. This kind of transport involves membrane-bound vesicles that fuse with the plasma membrane.

Endocytosis is apparently more difficult in plants than in animals because the plasma membrane of plant cells is usually pressed against the cell wall by turgor pressure. This turgor pressure hinders the plasma membrane from invaginating into the cytoplasm.

c o n c e p t

The fusion of membrane-bound vesicles to the plasma membrane and the expulsion of their contents is called *exocytosis*. Exocytosis bypasses the diffusion and transport mechanisms of membranes. The reverse process, where the plasma membrane invaginates to release vesicles into the cytoplasm, is called *endocytosis*.

Movement of Ions across Membranes

Plasma and organellar membranes have unequal concentrations of negatively charged ions (*anions*) and positively charged ions (*cations*) on one side versus the other. For example, the cytoplasm has a higher concentration of anions and a lower concentration of cations than does the matrix of the cell wall. This unequal distribution of ions creates an electrical gradient that is analogous to a concentration gradient. However, because an electrical gradient is based on electrical charge, the diffusion force of a charge is an electrical potential instead of a chemical potential. Because membranes selectively control the passage of ions, this electrical potential is called the **membrane potential.** Like any other electrical potential,

MEMBRANE TRANSPORT AND MAKING BEER

Malting is the process of soaking grains of barley (*Hordeum vulgare*) in water to induce seed germination and starch hydrolysis. Malting is of interest to plant biologists because of its importance in brewing beer and because it is a model system for studying membrane transport. During malting, different solutes move by a variety of transport mechanisms. The result is that glucose is provided to the rapidly growing embryo. Reading figure 4.1 shows where the principal steps of malting occur in a barley grain.

The primary enzyme that digests starch is alpha-amylase. Synthesis of this enzyme begins when the embryo sends a chemical signal to the aleurone layer of the grain. This chemical signal is a gibberellin, a plant hormone (see Chapter 18). Thus, the role of gibberellin in seed germination is to "tell" aleurone cells to make alpha-amylase. The enzyme is then transported to the endosperm, where starch is stored. Finally, starch there is hydrolyzed by alpha-amylase into maltose units (dimers of glucose) and then by maltase into glucose units, which are carried to the embryo.

Membrane transport moves gibberellin from the embryo to the aleurone layer, alpha-amylase from aleurone cells to the endosperm, and glucose from the endosperm to the embryo. The movement of alpha-amylase and glucose requires energy, but little is known about the movement of gibberellin. Studies of barley-seed germination generally begin when gibberellin is applied to grains whose embryos have already been removed.

When gibberellin induces the synthesis of alpha-amylase, aleurone cells show an increase in the amount of endoplasmic

(a)

reading figure 4.1

Barley. (a) A germinating barley grain. (b) Alpha-amylase is synthesized in the aleurone layer of a barley grain and then transported to the endosperm, where it hydrolyzes starch.

reticulum. Vesicles from the ER move the newly synthesized enzyme to the plasma membrane, where it is secreted by exocytosis. The enzyme must then be transported into cells of the endosperm and into amyloplasts (i.e., where starch is stored). When starch is completely hydrolyzed, the osmotic potential of the amyloplasts increases because, unlike starch, glucose is an osmotically active solute. Glucose probably moves out of the amyloplasts and into the cytoplasm by facilitated diffusion, but its transport from the endosperm to the embryo is hastened by a cotransport system that uses energy from H+-ATPase.

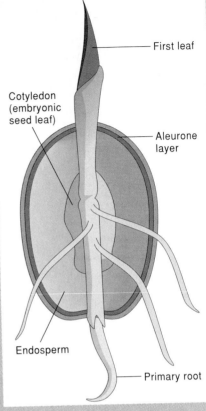

First leaf

Cotyledon (embryonic seed leaf)

Aleurone layer

Endosperm

Primary root

(b)

During malting, the temporary buildup of glucose in endosperm cells has a regulatory effect on starch hydrolysis, because glucose inhibits the secretion of alpha-amylase from aleurone cells. This inhibition leads to incomplete malting, which affects the flavor of beer adversely. Commercial breweries add gibberellin during malting to stimulate the production of alpha-amylase, which drives the malting to completion and, ultimately, makes a better-tasting beer.

---

membrane potential is measured in volts. Figure 4.20 shows some typical membrane potentials for a leaf cell. Note that membrane potentials are negative to indicate that the cytoplasm or organelle has more anions than does the matrix outside the cell or organelle.

Ion Pumps

Membrane potentials are maintained by proteins that actively transport ions. A membrane protein that pumps ions is called an **electrogenic pump** because it generates voltage across a membrane. Different ions are pumped by different proteins, but the main electrogenic pumps of plants are proton pumps, the H+-ATPases (fig. 4.17). One function of proton pumps, discussed earlier in this chapter, is to provide energy for the coupled cotransport of uncharged solutes such as sucrose. Another function is to regulate pH. Chemical reactions in a cell often incorporate or release ions that affect the pH of cells; the uptake of ions from soil also affects pH. Pumping protons out of the cell keeps the cytoplasm at a constant pH of about 7.4.

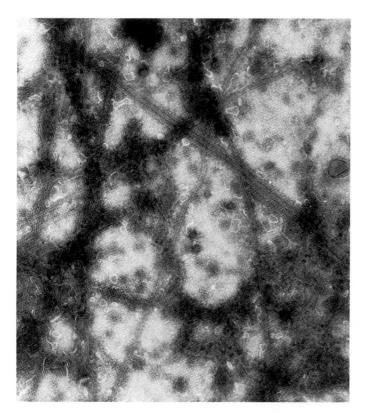

figure 4.19

Transmission electron micrograph of the plasma membrane from an unspecialized tobacco cell in culture, showing coated pits, ×67,000. The occurrence of coated pits is indirect evidence for endocytosis in plant cells.

Similarly, pumping protons into vacuoles keeps the pH there at about 5.0. This low pH is ideal for enzymes that break down organic compounds that are dumped into vacuoles for disposal. Proton pumps also regulate the pH of other organelles (fig. 4.21).

Proton pumps also influence cellular elongation. When H^+-ATPases in the plasma membrane are stimulated, the outward transport of hydrogen ions decreases the pH in the surrounding cell wall; this causes certain enzymes in the cell wall, which are activated at a lower pH, to begin to degrade cellulose microfibrils. This degradation loosens the cell wall, thereby allowing the cell to expand because of turgor pressure. Loosening of the cell wall can also be induced by applying **auxin,** a plant hormone. Physiologists suspect that auxin stimulates cellular elongation by stimulating the proton pump. Auxin and other hormones are discussed in more detail in Chapter 18, "Plant Hormones."

ATP Synthesis

There are two types of proton pumps. One type uses ATP and occurs mainly in the plasma membrane and in the tonoplast; the other produces ATP and occurs in the membranes of mitochondria and chloroplasts. In these ATP-synthesizing organelles, ATP is made from ADP and a phosphate group

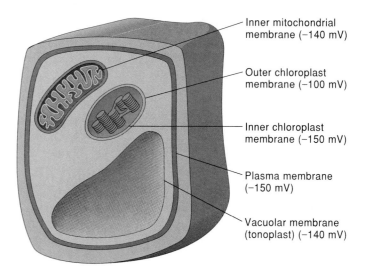

figure 4.20

Membrane potentials of different membranes in a leaf cell. Potentials are relative to the extracellular matrix at zero millivolts (mV; 1mV = 10^{-3} volt). For comparison, the potential between the two terminals of a typical flashlight battery is 1.5 volts (i.e., 1,500 mV).

when the diffusion of H^+ down its electrochemical gradient releases energy (fig. 4.22). This is the opposite of what happens in a proton pump that is driven by ATP. However, ATP is made only when a gradient of H^+ already exists; therefore, energy must be used to maintain this gradient. In chloroplasts, the energy for such a gradient comes from light energy during photosynthesis. In mitochondria, the energy comes from the rearrangement of chemical bonds during respiration. Both of these processes are explained more fully in Unit 2, "Plants and Energy."

c o n c e p t

The concentrations of ions are different on either side of a membrane. This difference produces an electrical potential, called *membrane potential,* which is measured in volts. Membrane potential is maintained by proteins in the membrane that actively transport ions against their electrical gradient. Protons are pumped by ATPases that use the energy of ATP in the plasma membrane and tonoplast. In contrast, proton pumps in mitochondrial and chloroplast membranes produce ATP.

Cellular Communication

Cells in a complex organism interact with their environment (e.g., gravity—see Chapter 19), with one another, and with the cells of other organisms. Cell-to-cell interactions occur when chemical or electrical signals released from one cell are received by another, where they change some aspect of metabolism. Auxin is an example of an internal chemical signal— that is, a signal that moves from cell to cell in the same

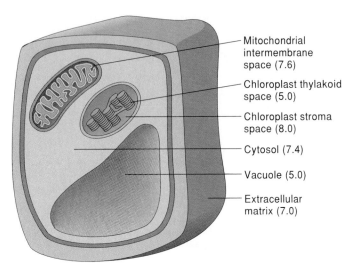

figure 4.21

The pH of different compartments in a leaf cell. Proton pumps use energy from the hydrolysis of ATP to move protons out of the cell (raises pH of the cytosol) or into the vacuole (lowers pH).

Mitochondrial intermembrane space (7.6)
Chloroplast thylakoid space (5.0)
Chloroplast stroma space (8.0)
Cytosol (7.4)
Vacuole (5.0)
Extracellular matrix (7.0)

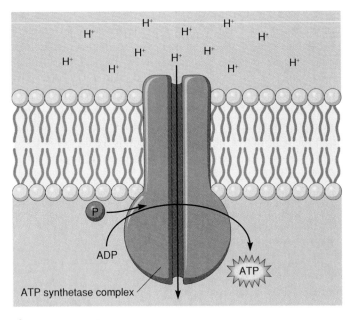

figure 4.22

Membranes of chloroplasts and mitochondria use energy stored in a proton gradient to make ATP from ADP and phosphate (P). Energy is harvested from protons as they diffuse down their concentration gradient through ATP synthase complexes in the membranes.

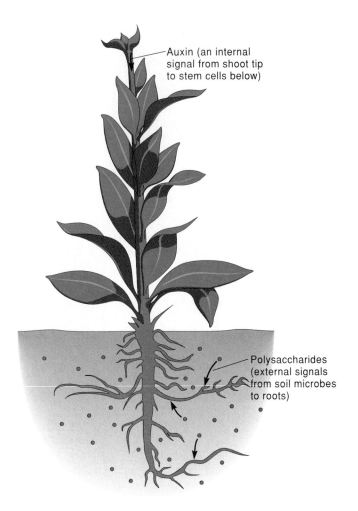

Auxin (an internal signal from shoot tip to stem cells below)

Polysaccharides (external signals from soil microbes to roots)

figure 4.23

Auxin is an example of an internal signal between cells. Microbial polysaccharides may function as a signal from bacteria to roots.

plant. External signals are those that pass between different organisms—for example, between plants and bacteria or fungi (fig. 4.23).

The reception of chemical signals and the transmission of their messages are important functions of proteins in membranes. Studies of signal transduction in plants have focused on the role of calcium ions (Ca^{2+}) and **calmodulin,** a protein that is activated when it binds to calcium. In its active form, the Ca^{2+}-calmodulin complex activates enzymes in membranes, essentially telling them to get to work. As much as 2% of the plasma membrane may be calmodulin.

An example of how calmodulin works is its role in transmitting a chemical "message" from auxin. The message causes the endoplasmic reticulum to release Ca^{2+}, which activates calmodulin. Calmodulin, in turn, activates protein kinase, which is an enzyme that activates or inactivates still other enzymes by transferring phosphates to and from them. Protein kinase removes phosphate groups from, for example, ATPases. The ATPases can then accept more phosphates and hydrolyze more molecules of ATP to ADP. In this case, the result is that energy is harvested to fuel the proton pump. This series of steps in signal transduction may seem unreasonably complicated, but it is effective in amplifying proton pumping. In this example, a single molecule of auxin can, by causing the activation of a series of enzymes, influence many protons to be transported across the membrane (fig. 4.24).

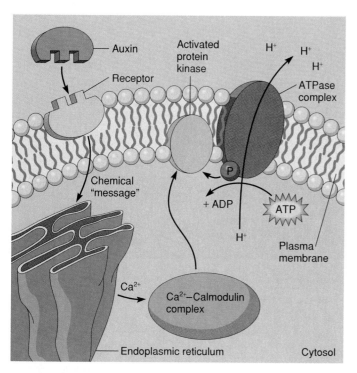

figure 4.24

Auxin can induce and amplify proton pumping. One molecule of auxin starts a series of signals that results in the activation of protein kinase. One activated molecule of protein kinase removes several phosphates from ATPases, thereby enabling hydrolysis of more ATP molecules.

Hormone Receptors

Signal transduction in plant cells begins when a hormone binds to a receptor protein on the plasma membrane. Plants make several different hormones, each of which must be recognized by a different receptor. Studies of plant hormone receptors have concentrated on auxin receptors because auxin has so many effects on plant growth and development (see Chapter 18). Each auxin receptor causes different metabolic changes, depending on where it occurs in a plant. Furthermore, the amount of binding varies from one tissue to the next; for example, auxin receptors in leaf stalks bind more than 100 times more auxin than do receptors in fruits. The multiple characteristics of hormone binding mean that there are probably many different receptors for auxin, as well as for each of the other plant hormones. Little is known about how they work or what they have in common, but the subject of hormone reception and signal transduction is an active area of botanical research.

Membrane Interactions with Other Organisms

Each plant is surrounded by other organisms, including animals, bacteria, fungi, and other plants, and interactions are common among plants and many of the organisms in their environment. Most of our knowledge about these kinds of interactions concerns the effects of microbes that cause plant diseases.

Two kinds of molecules are commonly secreted by microbes that cause plant diseases: small polypeptides and small polysaccharides. For example, a toxic polypeptide called syringomycin is secreted by *Pseudomonas syringae*, a species of bacteria that infects corn, beans, stone fruits (cherries, peaches, etc.), and many other plants. Syringomycin alters the membrane potential of the plasma membrane, probably by inducing protein kinase to phosphorylate the ATPase. This saturates ATPase with phosphates and inactivates the proton pump. Inactivated proton pumps allow protons to diffuse back into the cytoplasm. These and other effects of syringomycin are poorly understood, but they ultimately cause the membrane to leak and disintegrate. Enzymes from the bacteria digest the cell as it disintegrates.

To fight infections, the plasma membranes of many plant cells can send signals into their own cytoplasm that induce a defensive response. For example, polysaccharides secreted by the fungus *Colletotrichum lindemuthianum* cause cells in kidney bean leaves to make several different kinds of chemicals that inhibit the growth and spread of the fungus. Among these chemicals are phenolics that inhibit fungal growth. Cells at the edge of the infection also make lignin, which is indigestible to the fungus and limits its spread. Although the mechanisms for this response are not well understood, the process probably starts when the plasma membrane recognizes the fungal polysaccharides and sends a signal to the nucleus. This signal activates genes that control the synthesis of lignin, phenolics, and many other products that help to defend the plant. The result is that the leaf inhibits the growth of the fungus in the infected cells and blocks fungal growth at the edge of the infection.

Lectins and Glycoproteins

Proteins that bind to carbohydrates on cell surfaces are called **lectins.** Many lectins are glycoproteins. Lectins occur in all parts of the cell and in different kinds of tissues, and are usually associated with the endoplasmic reticulum and other membranes, including the plasma membrane. A lectin recognizes a specific carbohydrate on the basis of which monomers the carbohydrate contains and the linkages between them. Because of this specificity, plant lectins can recognize the cell-wall carbohydrates of specific bacteria and fungi.

Cellular recognition is perhaps best understood among plants in the legume family and bacterial species of the genus *Rhizobium*. For example, lectins in the root hairs of white clover (*Trifolium repens*) bind only to the cell walls of *Rhizobium trifolii*. Once bound to the root hairs, these bacteria infect the roots and cause them to swell into nodules (fig. 4.25). The interaction between root-nodule bacteria and plants is mutually beneficial because the organisms trade metabolites: the plants supply carbohydrates for bacterial respiration, and the bacteria convert nitrogen from the air into ammonia, a metabolically useful form of nitrogen for plants. This process, called nitrogen fixation, is discussed in more detail in Chapter 20.

figure 4.25

Photograph of roots of pea plants with bacteria-containing nodules. In these nodules, atmospheric nitrogen is converted to ammonia via a process called nitrogen fixation. The infection of roots by bacteria is enabled by root-hair lectins that bind to bacterial cell walls.

Proteins, most often glycoproteins, also have important roles in sexual reproduction. For example, they enable the reproductive cells of some algae to recognize their appropriate mates for fertilization. In flowering plants, reproductive compatibility is controlled by interactions between the glycoproteins of pollen, which carry the sperm, and the glycoproteins of stigmas, which are borne on the seed-producing organ. If their respective glycoproteins fit together appropriately, pollen tubes grow normally, and fertilization can occur; if they do not fit together, the pollen tubes grow irregularly and incompletely, and fertilization does not occur. In some cases, glycoproteins of the pollen and stigma within the same flower do not fit together, which makes the plant self-incompatible (fig. 4. 26).

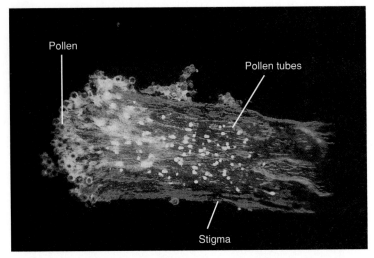

figure 4.26

Light micrograph of fluorescent-stained pollen tubes growing through the stigma, ×20.

Cells communicate with other cells by chemical signals. Receptor proteins on plasma membranes are activated by binding to hormones and other chemicals. The activated proteins induce enzymes and other proteins to change the metabolism of the cell. Membrane surfaces also recognize specific secretions or cell-wall components from other organisms. Glycoproteins and other proteins are probably the most common receptors for recognizing other cells by their carbohydrates. Glycoproteins also have a role in controlling sexual reproduction.

The Lore of Plants

Plants can be used to determine your blood type. Lectins from jack beans, lima beans, and lotus bind to glycoproteins on the plasma membranes of red blood cells. Because the cells of different blood types have different glycoproteins, cells of each blood type bind to a specific lectin. This is one example of the many clinical and research applications of plant lectins in human medicine.

Chapter Summary

Membranes consist of two main components: phospholipids and proteins. The structure of membranes is best described by the fluid mosaic model. The phospholipids form a fluid bilayer, with the hydrophobic tails of fatty acids at its core and the hydrophilic heads of phospholipids on both sides. Proteins move in the fluid. The lipid layer allows the unrestricted diffusion of small molecules across the plasma membrane, but membrane lipids block the diffusion of ions and large polar molecules. This property of membranes is called *differential permeability*. The diffusion of hydrophilic solutes is facilitated by proteins embedded in the membranes.

The diffusion of water through a differentially permeable membrane is called *osmosis*. Plants regulate osmosis by controlling the uptake of ions or by making osmotically active solutes. In so doing, cells maintain turgor pressure against the cell walls. Because turgor pressure prevents further uptake of water, the water outside the cell exerts pressure, called *osmotic pressure*. Its counterpart inside the cell is the osmotic potential of the cell, which is expressed as a negative number.

Osmotic potential results from the potential of water to diffuse down its concentration gradient. This concentration gradient of water is controlled by the uptake or synthesis of osmotically active solutes, which requires energy from the hydrolysis of ATP. The energy-requiring transport of solutes is called *active transport*. The most common actively transported solutes are ions, especially hydrogen ions. Transport proteins for hydrogen ions are also called *ATPases* because they use energy from the hydrolysis of ATP. The lipid bilayer and membrane proteins are bypassed by exocytosis, the process by which proteins and other large molecules are secreted in vesicles that fuse with the plasma membrane.

Ion transport is affected by the concentration gradient of ions and by the electrical gradient of their charges. Together, these two gradients create an electrochemical gradient across a membrane. The electrical component of this gradient is measured in volts and is called the *membrane potential*. Ion transport proteins, called *electrogenic pumps*, maintain the membrane potential. The most common electrogenic pumps are proton pumps. In the plasma membrane and tonoplast, proton pumps use the energy of ATP to move protons against an electrochemical gradient. In the membranes of mitochondria and chloroplasts, proton pumps are reversed—that is, they produce ATP. The ATP required to run these pumps comes from respiration (in mitochondria) or photosynthesis (in chloroplasts).

Membrane proteins are also receptors for hormones and other chemicals from other cells or organisms. As receptors, these proteins signal other parts of the cell to change their metabolism. These changes contribute to cell growth, defense against disease, recognition between mating cells, and many other aspects of plant life.

 What Are Botanists Doing?

Look through two or three general biology textbooks for their discussions of membrane potential. Then use reference materials at your library to get a general idea of why botanists are interested in membrane potential. Compare the textbook coverage with the information you obtained from the library.

 Writing to Learn Botany

Suppose you wanted to measure the osmotic pressure of a living cell. How would you do it? Explain the rationale for the technique you've decided to use. What would be the strengths and shortcomings of the technique you chose?

Questions for Further Thought and Study

1. How does the outside surface of the plasma membrane differ from the inside surface? In what ways does this difference reflect the functions of the plasma membrane?

2. How is the Davson-Danielli model of membrane structure deficient in explaining what we now know about membranes?

3. What features of phospholipids account for the following characteristics of a biological membrane: (a) its double layer; (b) its fluidity; (c) its hydrophobic and hydrophilic properties; (d) its differential permeability?

4. Can you explain why glucose is osmotically active and starch is not?

5. Certain brands of water purifiers for the home clean water by "reverse osmosis." How do you think this process might relate to osmosis?

6. Why do you think ATP synthetase complexes are often referred to as ATPases?

7. Some botanists think that the response of kidney beans to infection by *Colletotrichum* is evidence that plants have an immune system. How can you support or refute that idea?

Web Sites

Review the "Doing Botany Yourself" essay and assignments for Chapter 4 on the *Botany Home Page*. What experiments would you do to test the hypotheses? What data can you gather on the Web to help you refine your experiments? Here are some other sites that you may find interesting:

http://kauai.cudenver.edu:3010/0/nutrition.dir/membrane.html

This site provides basic information about the structure of membranes. It's also linked to The Cell, a series of pages that describe the structure and function of cells.

http://www.interedusoft.com/scientia/caduceus/cards/biology/
plasmembrane.html

This site reviews the structure and function of membranes, as well as how substances move across membranes (e.g., diffusion, active transport).

http://www.dcn.davis.ca.us/~carl/diffuse.html

This site discusses diffusion, osmosis, facilitated diffusion, and active transport.

Suggested Readings

Articles

Bretscher, M. S. 1985. The molecules of the cell membrane. *Scientific American* 253 (October) :100-108.

Lavenda, B. H. 1985. Brownian motion. *Scientific American* 252 (February) :70-84.

Longenecker, N. E., and E. T. Hibbs. 1986. Active transport. *The American Biology Teacher* 48:304-306.

Maloney, P. C., and T. H. Wilson. 1985. The evolution of ion pumps. *BioScience* 35:43-48.

Nelson, N., and L. Taiz. 1989. The evolution of H^+-ATPases. *Trends in Biological Sciences* 14:113-116.

Satir, B. 1975. The final stage in secretion. *Scientific American* 233 (October) :28-37.

Slayman, C. L. 1985. Proton chemistry and the ubiquity of proton pumps. *BioScience* 35:16-17.

———. 1985. Plasma membrane proton pumps in plants and fungi. *BioScience* 35:34-37.

Unwin, N., and R. Henderson. 1984. The structure of proteins in biological membranes. *Scientific American* 250 (February) :78-94.

Books

Gennis, R. 1989. *Biomembranes: Molecular Structure and Function.* New York: Springer-Verlag.

Robinson, D. G. 1985. *Plant Membranes: Endo- and Plasma Membranes of Plant Cells.* New York: John Wiley & Sons.

Tosteson, D. 1989. *Membrane Transport: People and Ideas.* London: Oxford University Press.

Plants and Energy . . .

C orn is the most widely planted crop in the United States. To people who have driven across the midwestern United States (also called the Corn Belt) during the summer, this is no surprise, for in many areas there is little except mile after monotonous mile of corn. These crops cover a combined area about the size of Arizona and produce about 8 billion bushels of corn per year. According to one mathematically minded urbanite, that's enough corn to bury all of Manhattan Island about 5 m deep.

Corn is the most efficient of all major grain crops; that is, corn is the champion at converting sunlight to sugars that the plant stores as starch in grain. We use these grains for a variety of purposes. For example, we feed about half of our corn to livestock; another 25% of the crop is exported. About 10% of the crop—optimally within 5 minutes of picking—appears on our dinner tables. The rest of the crop passes into an enormous number of peripheral products, including corn syrup, cornmeal, corn oil, cardboard, crayons, firecrackers, aspirin, wallpaper, gasohol, pancake mix, shoe polish, ketchup, chewing gum, marshmallows, soap, and corn whisky.

Nearly a fourth of the Calories in people's diets can be traced to corn. Where do those Calories come from? How does corn convert this energy to sugars? How do corn and other organisms extract this energy from the sugars?

In this unit you'll learn about energy and its use by plants. After an introduction to the laws governing energy conversions, you'll learn about the most important set of chemical reactions on earth: photosynthesis. These are the reactions that convert sunlight to chemical energy, thus providing fuel for virtually all other organisms. You'll also learn how organisms—plants included—extract the energy trapped in sugars and other compounds for work. In the process, you'll come to appreciate one of the themes of life: that all organisms, in addition to requiring energy to stay alive, exchange energy with their environment.

Like all organisms, plants convert energy from one form to another. This corn (*Zea mays*) plant uses photosynthesis to convert sunlight to chemical energy. This chemical energy is then used to fuel the plant's growth and development.

Energy and Its Use by Plants

5

Chapter Outline

Chapter Overview

Although the word *energy* is common today, it was not coined until about 200 years ago when the Industrial Revolution shifted our concept of energy from horses (i.e., horsepower) and falling water to combustion engines. Thereafter, a quantitative understanding of the concepts of work and energy became not only practical but essential. Biologists soon realized that these concepts could help them understand how organisms function. Applying the concepts of energy to living organisms began the study of **bioenergetics**, a fascinating discipline that helps us understand life.

Today, few things are more important than energy: countries fight over oil supplies, and accidents at nuclear power plants endanger lives. Everything involves energy, including music, games, tides, seasons, and the orbits of planets. Plants and animals are not exceptions: all aspects of their lives, ranging from maintenance and repair to exotic tasks such as producing beautiful flowers, require energy. We can't understand or appreciate plants unless we know something about energy, including what it is and how it is used.

In this chapter, you'll learn how organisms use energy for their various activities. These principles will help you understand the topics of the following two chapters, respiration and photosynthesis.

Metabolism is a bit like a capitalist economy: labor is hired and paid for, products are made from raw materials that have to be bought, and goods are packaged and shipped to different locations by a service industry that requires compensation. The gold standard of metabolism is energy: plants harvest energy from the sun, convert it to a metabolically useful form, and move it in different forms ("currencies") within cells according to an energy budget. Energy is needed to transport ions and molecules across membranes, to make large molecules from smaller ones, and to move molecules through the cytosol and into and out of organelles and cells.

In this chapter we look at what energy is and the different forms it takes in plants. Some of the discussion is general, and some applies only to living organisms. The latter is referred to as *bioenergetics* because it involves the compounds of energy metabolism.

What Is Energy?

Energy is the ability to do work—that is, to bring about change or move matter against an opposing force such as gravity or friction. Because energy is an *ability* to do work, it is not always as obvious to us as matter, which has mass and occupies space. We describe energy according to how it affects matter. Humans use energy for conspicuous activities such as dancing, mowing the lawn, playing baseball, and studying plants. However, plants expend most of their energy in subtle, nearly unrecognizable ways. For example, consider the philodendron (*Philodendron*) plants shown in figure 5.1. Their large leaves slowly gather the energy available in sunlight and use it to fuel their metabolism and growth. However, the calm existence of these plants changes when they reproduce. Their large flowers open for only a couple of days. At night, when air temperatures often hover near freezing, their flowers can reach temperatures exceeding 46° C/115° F (by comparison, butter melts at 30° C/88° F). These furnace-like flowers maintain their high temperature for many hours in the cold night air. Plants such as skunk cabbage and voodoo lily also generate large amounts of heat. This heat may be used to help the plants reproduce. For example, the heat produced by voodoo lily helps to disperse compounds that smell like dung or rotting flesh. The insects attracted to the plants by these odors assist in pollination. Clearly, understanding the bioenergetics of plant growth helps us understand how plants live and reproduce.

Measuring Energy

Since energy exists in many forms, it's not surprising that energy is described with many units. Most scientists measure energy in calories (cal) or joules (J). A **calorie** (**cal**; note the small *c*) is the amount of energy required to raise the temperature of 1 g of water by 1° C. The most common unit for measuring the energy content of food and the heat

figure 5.1

The clublike spadix in the center of this *Philodendron* flower heats up to temperatures as high as 46° C. This heat helps the plant disperse compounds that attract pollinators.

EARTH'S ENERGY

The energy available on earth can be traced to the sun (e.g., fossil fuels, biomass, wind, and incoming radiation), cosmic evolution preceding the origin of our solar system (nuclear power), lunar motion (tidal power), or the earth's core (geothermal power). Of these, sunlight provides the most energy for organisms. Sunlight originates inside the sun, where temperatures of 10,000,000° C fuse hydrogen to form helium and release gamma rays, which then produce electrons and photons:

$$H + H \rightarrow He + \text{gamma rays} + \text{energy}$$

$$\text{gamma rays} \rightarrow \text{electrons} + \text{photons} + \text{energy}$$

Life depends on transforming the energy contained in photons of sunlight into chemical energy before it is transformed to heat. The sun's total energy output is about 3.8 sextillion (3,800,000,000,000,000,000,000) megawatts of electricity. Earth intercepts only about 1/2,000,000,000 of this energy; this amounts to the energy-equivalent of about 2×10^{14} tons of coal per year. Of this,

- Thirty percent is reflected back to space. This is what astronauts see while orbiting the earth.

- Fifty percent is absorbed, converted to heat, and reradiated.

Nineteen percent powers the hydrologic cycle, creates winds, and drives photosynthesis. Of this 19%, only 0.05% to 1.5% is incorporated into plant material. However, this relatively small amount of energy sustains life. Without photosynthesis, almost all life would quickly disappear from the earth.

The amount of sunlight energy at the earth's surface is about 2 million trillion calories per second, an amount of energy equivalent to that of 400 million atomic bombs of the size dropped during World War II.* Stated another way, the annual amount of energy that strikes the earth as sunlight is 15,000 times greater than the world's present supply of energy. This is more than 20 billion times greater than our present rate of energy consumption.

Despite this vast resource of energy, we've learned relatively little about how to use renewable sources of energy such as sunlight. These so-called "alternate sources" of energy differ from oil, gas, and nuclear power because they are plentiful and free of pollution. Today, renewable forms of energy (e.g., hydropower and biomass) supply only 18% of the world's energy needs, while nuclear power supplies 4%. The rest of our energy demands are met by increasingly scarce, nonrenewable sources such as fossil fuel. Our demands are huge. For example, to exert the power equivalent to the amount of electricity used by U.S. manufacturers in one week, one worker, doing average manual labor for fifty 40-hour weeks per year, would have to work for 145,193,740 years.

*Every day, the earth receives about 1.5 billion times more energy than is contained in all of the electricity used in the United States each year.

output of organisms is the **Calorie** (**Cal**; note the large C), which is the energy required to raise the temperature of 1 liter of water by 1° C.[1] A **joule** (**J**) is the amount of energy needed to move 1 kilogram through 1 meter with an acceleration of 1 meter per second per second (1 m/sec^2; for comparison purposes, 1 cal = 4.12 J).[2] To help you put these units into better perspective, consider that a slice of apple pie provides enough energy (1.5×10^6 J, or 365 Cal) for a woman to run for an hour or for a typist to enter about 15 million characters on a manual typewriter (almost 11,000 typewritten pages). Here's the energy involved in some common (and a few not-so-common) events:

Severe earthquake (Richter 8)	10^{18}J
Atomic bomb dropped on Hiroshima	8.4×10^{13} J
Man running for 1 h	2.5×10^6 J
Woman running for 1 h	1.8×10^6 J
Lethal dose of X rays	7×10^2 J
Depressing a key of a manual typewriter	10^{-1} J
Turning this page	10^{-3} J
Chirrup of a cricket	9×10^{-4} J
Wingbeat of a honeybee	8×10^{-4} J
Moonlight on a person's face for 1 sec	8×10^{-5} J
Energy released by splitting one uranium atom	4×10^{-11} J
Energy of a photon within the visible range	$2.5\text{-}5.1 \times 10^{-19}$ J

Organisms must continually obtain energy to stay alive—heterotrophs such as humans must eat food, and plants must absorb light for photosynthesis. These transformations of food and light into usable energy occur via energy conversions.

---

1. The Calorie (written with a capital C) used to measure the energy content of food is equivalent to 1,000 calories (written with a lowercase c), or 1 kcal.

2. Units to measure energy and power throughout the world (but not necessarily in the United States) include ergs, British thermal units (Btu), watts (W), kilowatt hours (kW h), and horsepower. An erg is the energy needed to move 1 g of mass 1 cm with an acceleration of 1 cm sec^{-2}. A Btu is the energy needed to raise the temperature of 1 lb of water 1° F. A watt is the energy provided when 1 J is used for 1 sec. A kilowatt hour is the energy used when 1 kW is available for 1 h. A horsepower is the energy needed to raise 550 lb 1 ft in 1 sec. Although these different units often bewilder consumers, they are interconvertible. Here are the conversion factors for these units: 1 erg = 10^{-7} J; 1 Btu = 1055 J = 252 cal = the energy given off by a wooden kitchen match; 1 cal = 4.1 J = 0.001 kcal = 0.001 Cal; 1 hp = 746 W; 1 W = 0.001341 hp.

Energy Conversions

All activities—ranging from cellular division and heat production by flowers to home runs and nuclear explosions—involve converting energy from one form to another. For example, we convert the energy contained in coal and oil to electricity, and then convert the electricity into light energy to illuminate our homes, streets, and parks. Similarly, plants convert sunlight into chemical energy that they use to reproduce, repair their DNA, and build new parts. This conversion of light energy to chemical energy is **photosynthesis,** which sustains almost all life on earth. Animals stay alive by eating other animals and/or plants and their stored energy. All aspects of the lives of organisms center on energy and energy conversions.

There are two types of energy: *potential energy* and *kinetic energy* (fig. 5.2). **Potential energy** is stored energy—that is, energy available to do work. Examples of potential energy include a teaspoon of sugar, an unlit firecracker, and a rock atop a hill. Potential energy is determined by the position (e.g., water held at an altitude behind a dam) or arrangement (e.g., the type of chemical bonds) of matter. In organisms, potential (i.e., latent) energy is stored subtly in chemical bonds such as those in sugars, starch, and fats.

Kinetic energy is energy being used to do work. Examples of kinetic energy include burning sugar, an exploding firecracker, a rock rolling down a hill, light emitted by a firefly, and a root forcing its way through soil. Kinetic energy affects matter by transferring motion to other matter, much as a moving billiard ball transfers its kinetic energy to other billiard balls. Similarly, flowing water can be used to turn a turbine, and a growing root can break a concrete sidewalk. Kinetic energy moves objects, whether they be mountains, molehills, or molecules.

Heat is kinetic energy because it involves the movement of molecules. Again study the example shown in figure 5.2. The rock atop the hill contains much potential energy—the *capacity* to do work—because of its position, but since it is at rest, it has no kinetic energy. If the rock were given a nudge, it would spontaneously roll down the hill, transforming its potential energy into kinetic energy, which could be used to do work.

Just as the mainspring of a tightly wound watch can propel the hands around the dial, potential energy can be converted to kinetic energy to do work. Likewise, winding a watch transforms kinetic energy from the person winding the watch into potential energy stored in the watch's mainspring. Under most conditions, potential and kinetic energy are freely interconvertible, though not at 100% efficiency.

c o n c e p t

Energy is the ability to do work. Potential energy is energy available to do work, and kinetic energy is energy being used to do work. Under most conditions, kinetic energy and potential energy are freely interconvertible, though not at 100% efficiency.

The Laws of Thermodynamics

Life depends on energy transformations. For example, our bodies transform the chemical energy in food to mechanical energy that enables us to study, play, and dance; and our appliances convert electrical energy to light for reading and to heat for warming our homes. Combustion engines convert the chemical energy in gasoline to mechanical energy that runs our automobiles and mows our lawns. Plants convert sunlight to chemical energy that sustains life on the planet.

Energy transformations are regulated by laws of thermodynamics. These laws involve a system and its surroundings. The collection of matter being studied is called the *system,* and the rest of the universe is referred to as the *surroundings.* A closed system, such as that approximated by a thermos bottle, is isolated from (i.e., does not exchange energy with) its surroundings; conversely, an open system exchanges energy with its surroundings. Don't let all of this intimidate you; the laws of thermodynamics are rather simple and are based on common sense. More important, they are unbreakable laws that apply to *all* energy transformations, whether they be the combustion of gasoline in an engine, the breakdown of glucose in a cell, the closing of a Venus's-flytrap, or the generation of heat by the philodendron flowers shown in figure 5.1. The laws of thermodynamics are important because they govern the existence of all organisms.

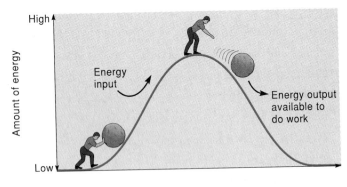

figure 5.2

Pushing a boulder to the top of a hill requires an input of energy because it increases the potential energy of the boulder. The boulder atop the hill has potential energy because it can do work as it rolls down the hill. As the boulder rolls down the hill, potential energy is converted to kinetic energy.

From Postlethwait and Hopson, Nature of Life, 2d ed. Copyright © 1992 McGraw-Hill, Inc., New York, NY. Reproduced with permission of McGraw-Hill, Inc.

The First Law of Thermodynamics

The **first law of thermodynamics** is the law of conservation of energy. This law states that energy cannot be created or destroyed but only converted to other forms. It can also be stated in other ways:

- In any process, the total amount of energy in a system and its surroundings remains constant.

- The total amount of energy in any isolated (i.e., closed) system is constant.

- The amount of energy in the universe is constant.

- You can't get something for nothing.

For example, the energy used to wind a watch comes from the person winding the watch. Similarly, a power plant does not create energy—it merely transforms energy from one form (e.g., fossil fuel) to another (e.g., electricity). Likewise, green plants are not energy producers—they merely trap and convert the energy in sunlight into chemical bonds. The first law asserts that the energy in sunlight that warms our planet and drives photosynthesis must come from somewhere else in the system. In this case, it comes from the sun.

Energy conversions often generate much heat. For example, as you read this book, your body generates the heat of a 100 W lightbulb (a class of ten students would produce the heat equivalent of a kilowatt heater). However, your body temperature is neither increasing nor decreasing because the heat generated by your body is radiated into your surroundings. That is, there is no change in the total amount of energy in the system—the energy radiated from your body that heats the room you're in can be traced to the energy contained in the food that you ate. According to the first law of thermodynamics, the amount of energy released by your body cannot exceed the amount of energy contained in the food that you eat. If you stop eating, you'll eventually run out of energy, die, and stop releasing heat.

The first law of thermodynamics has tremendous implications for everyday life. For example, it explains why an automobile can go only a limited distance, regardless of its fuel efficiency or mileage rating: the energy used to move the car cannot exceed that contained in the chemical bonds of its fuel. When the car runs out of gas, it can go no farther until more energy (i.e., gasoline) is added to the system. This addition of energy to the car corresponds to a loss of energy from somewhere else—the storage tank at the gas station.

The first law of thermodynamics also dictates that the energy trapped by leaves during photosynthesis cannot exceed the energy of the absorbed light. For example, if 100 units of light energy strike a leaf, no more than 100 units of energy can be trapped in carbohydrates produced by photosynthesis. No matter how hard you try, you can't get more energy out of a system than you put in. There are no exceptions to this law.

The Second Law of Thermodynamics

The **second law of thermodynamics** is the law of **entropy**, or disorder. This law states that all energy transformations are inefficient; that is, the amount of concentrated, useful energy decreases in all energy transformations. The following statements express the second law in other ways:

- Systems tend toward increasing entropy (disorder).

- Any system tends spontaneously to become disorganized.

- In all energy transformations, some usable energy is lost as heat.

- Any spontaneous change decreases the amount of usable energy.

- A perpetual motion machine cannot exist; that is, no real process can be 100% efficient.

- You can't break even.

These restatements of the second law are based on our world being overwhelmingly irreversible. This irreversibility is all around us, including our aging, the futility of trying to unscramble an egg, and the impracticality of coasting downhill in a car to refuel its tank. Irreversibility results from the message of the second law of thermodynamics—namely, the loss of *usable* energy as heat during an energy transformation.

To better understand this, consider a person throwing a ball or heating a cup of tea. Both of these processes require energy. According to the first law, no energy has been created or destroyed in moving the ball or heating the tea: the energy used to heat the tea probably came from breaking the chemical bonds of natural gas, while that used to throw the ball came from energy in food used to contract muscles. However, the energetics of the moving ball and of heating the tea are drastically different. The moving ball heats the air and any object that it strikes, and is in *coherent* motion: all of its parts move together in an orderly way. In contrast, energy in the heated tea is contained in the random, *incoherent* motion of its molecules. There's no order to it; the heat energy results only from random molecular motion. This randomness distinguishes heat from all other forms of energy: any other form of energy can be converted completely to heat, but heat cannot be completely converted to other forms of energy. This again goes back to our everyday experience: the thrown ball heats the air and the glove that it hits, but applying the same amount of heat to the air doesn't move the ball. All energy is ultimately converted to heat, and heat is not usable energy.

Consider again the example of a combustion engine in a car. Gasoline is a concentrated, orderly source of energy—its energy resides in the covalent bonds of octane. However, when these bonds break and release energy in the car's engine, less than one-fourth of the energy is used to move the automobile (i.e., most combustion engines are less than 25% efficient). According to the first law of thermodynamics, no energy was lost when the car moved: the amount of energy used to move the car, heat the engine block (and the air around it), power the radio, and heat the tires equals that originally contained in the gasoline. However, applying heat to a car does not move the car. That is, the energy contained in the heated tires, pavement, air, and engine block cannot be recycled to run the car, because heat energy resides in randomly moving molecules and is therefore not in a concentrated, useful form. This heat represents the inefficiency inherent in any energy transformation and is the basis for the second law of thermodynamics.[3] Because all energy transformations produce heat (i.e., an unusable form of energy), all things naturally become more disorganized.

The consequences of the second law of thermodynamics are important and familiar to everyone. For example, rocks tumble downhill rather than uphill, and pieces of a jigsaw puzzle never spontaneously fall into place when they're poured from the box. In short, the second law of thermodynamics explains why disorder in the universe is increasing continually. All naturally occurring processes tend to reach maximum entropy (i.e., maximum disorder). Life exists at the expense of the universe, which is constantly running down.[4]

Cells derive energy from sugars and fats for growth, repair, and reproduction. The chemical reactions that free this energy are inefficient and release much heat; indeed, the cells of most organisms extract less than half of their fuel's energy for useful work (e.g., the energy used to power your brain while you sleep is equivalent to that of a 60 W bulb). Thus, although organisms can channel the transformation of energy from one form to another—they can hoard the energy in reserves (e.g., fat or starch) or use it for repair, movement, or reproduction—these diversions are only temporary. Eventually, all energy is transformed to heat.

The loss of useful energy as heat during energy transformations increases the entropy in a system, and only processes that decrease the amount of useful energy occur sponta-

neously. Therefore, there is a natural tendency for things to become less organized. Although the entropy of one system, such as an organism or cell, may decrease (i.e., the system may become more organized), the entropy of the universe is always increasing. On a more local level, our apartments and desks quickly become messy unless we periodically straighten them, and parents often point out that children are "entropy's little helpers."

Because organisms are highly ordered, it might seem that they are exceptions to the laws of thermodynamics. However, the second law applies only to closed (i.e., isolated) systems. Organisms remain organized because they are *not* closed systems: they use inputs of matter and energy such as food and sunlight to reduce randomness (i.e., decrease entropy) and stay alive. The energy that keeps organisms alive comes ultimately from the sun—that is, plants transform sunlight into the chemical bonds of carbohydrates, which humans and other organisms use as an energy source. Life is possible only because organisms temporarily store and later use some of the energy flowing through the system. Plants and plantlike organisms (such as algae) are the first and most important part of the scheme (see fig. 5.3).

The Lore of Plants

Valerian (*Valeriana sylvatica*) is an ancient medicinal herb whose name is derived from the Latin *valere*, meaning "to be in health." Despite its awful odor, dried valerian root was always in the medicine bag of Canadian Indian warriors, who used it as a wound antiseptic. People used valerian during World Wars I and II for shell shock and nervous stress. Valerian's old name, *V. phu*, could be the origin of our expression for bad odor.

We're most familiar with energy transformations such as explosions, home runs, moving vehicles, and electricity, which release large amounts of energy at once. However, energy transformations in cells each involve small amounts of energy. It is the cumulative effect of these many small transformations that we see as growth and development.

The two primary energy transformations in plants are photosynthesis and cellular respiration (fig. 5.4). Photosynthesis uses light energy to convert CO_2 and H_2O—both of which are energy-poor—into sugars. In the process, oxygen gas (O_2) is released. Cells extract energy from sugars via cellular respiration. Some of this energy is stored in molecules such as ATP (discussed later in this chapter). If all of the energy in the chemical bonds of sugars were released at once (as in an explosion), the energy would be converted mostly to heat and produce lethally high temperatures. To avoid these problems, cells extract energy from glucose and other molecules by

---

3. Heat can be used to do work only when there is a steep temperature gradient, as in a steam engine. These gradients do not exist in living organisms, and nature has never evolved a steam engine. Indeed, such gradients would quickly denature enzymes and kill cells. In the uniform conditions characteristic of living organisms, heat is useless except for warming.

4. The laws of thermodynamics determine the energy relationships of cells and organisms. They also predict the fate of our universe. For example, the first law of thermodynamics indicates that the energy present when the universe formed 14 billion years ago is all that can ever exist. Similarly, the second law means that our sun (and all other suns) will eventually burn out (i.e., all of its energy will be transformed to heat), thus leaving the earth without an energy source. When this occurs, all of the energy originally present when the universe formed will remain present, but there will be no usable energy for work. Everything will be at the same temperature, and there will be no more energy transformations; only small, randomly moving molecules will remain in the universe. All usable energy will have been frittered away into heat, and all life in the universe will end, a scenario described by biochemist and science fiction writer Isaac Asimov in *The Last Question*. In short, the lights will be out and the party will be over. But don't be overly concerned—even the most pessimistic folks don't expect this to happen for several billion years.

figure 5.3

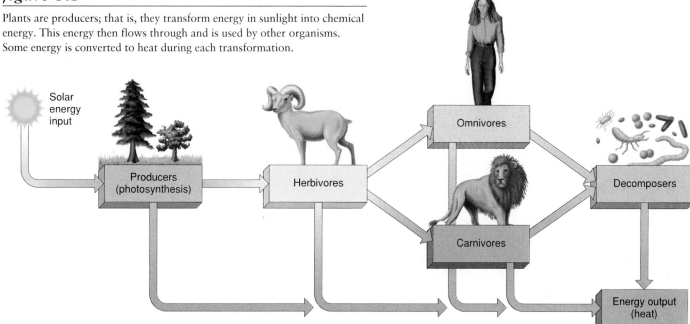

Plants are producers; that is, they transform energy in sunlight into chemical energy. This energy then flows through and is used by other organisms. Some energy is converted to heat during each transformation.

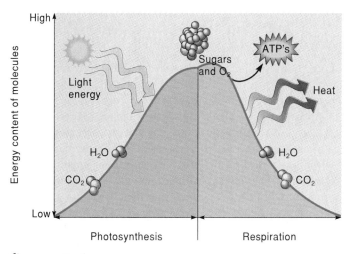

figure 5.4

Plants use photosynthesis, shown here as an uphill reaction, to make energy-rich sugars (and O_2) from energy-poor CO_2 and water. Plants and other organisms use respiration, shown here as a downhill reaction, to convert sugars to CO_2 and water. Some of the energy released during respiration is conserved in chemical bonds of other molecules, especially ATP.

slowly oxidizing the molecule in a *series* of chemical reactions. During each reaction, there is a drop in the potential energy of the molecule. Some of this energy is lost as heat; however, much of it is trapped in the chemical bonds of other molecules in the cell. The chemical reactions that transform energy in cells are collectively called *metabolism.*

c o n c e p t

All of life's activities involve changing energy from one form to another. These energy transformations are governed by the laws of thermodynamics. The first law states that energy cannot be created or destroyed but only converted to other forms. The second law states that all energy transformations are inefficient because they involve the loss of usable energy as heat. The quantitative measure of disorder created by any spontaneous reaction is entropy. All spontaneous reactions increase entropy.

Metabolism: Energy for Life's Work

Metabolism (from the Greek word meaning "change") is a fundamental property of life arising from energy transformations in cells; it is the sum of the vast array of chemical reactions that occur in an organism. These reactions do not occur randomly; rather, they occur in step-by-step sequences called **metabolic pathways,** in which the product of one reaction becomes the substrate (i.e., the starting point) for another. The various metabolic pathways in a cell are much like the roads on a map (fig. 5.5).

Each reaction in a metabolic pathway rearranges atoms into new compounds, and each one either absorbs or releases energy. The net release or uptake of energy equals

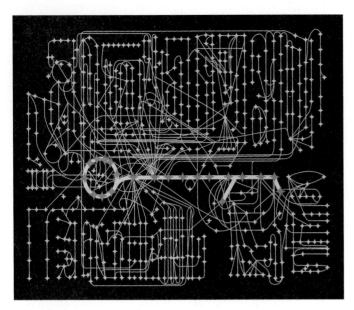

figure 5.5

Cellular metabolism. This diagram traces some of the metabolic reactions that occur in a cell. Dots represent molecules, and lines represent reactions—each catalyzed by a specific enzyme—that change the molecules. The stepwise sequences of reactions are called *metabolic pathways*. The pathway shown in yellow is central to most other pathways.

the difference of energy released and energy consumed. For example, burning a mole (16 g) of methane releases 160 kcal of energy:

$$CH_4 + 2O_2 \rightarrow CO_2 + 2H_2O + energy$$

This heat of reaction is the heat that you feel from the stove (i.e., the net amount of energy released by the reaction) and is represented by ΔH (delta H). It is derived from the total potential energy of the molecules, a measure called **enthalpy**. Therefore, we say that the heat released into the surroundings comes from the enthalpy of the reacting molecules. In the case of burning methane, the products (CO_2 and H_2O) have 160 kcal less enthalpy than the reactant (CH_4). Such heat-releasing reactions are called **exothermic** reactions and change the molecules so that their energy content decreases.

---

c o n c e p t

Metabolism is the sum of the vast array of chemical reactions that occur in an organism. Metabolism occurs in stepwise sequences called *metabolic pathways*. Each reaction of a metabolic pathway rearranges atoms into new compounds.

Although most exothermic reactions are spontaneous, some are not. To accurately predict whether a reaction will occur spontaneously, we must introduce a new concept: free energy.

Free Energy

The potential energy of a compound is contained in its chemical bonds. When these bonds break, the energy that is released can be used to do work, such as form other bonds. The amount of energy available to form other bonds is the **free energy** of the molecule and is represented by G (for its discoverer, Yale University physicist Josiah Gibbs).

Chemical reactions change the amount of free energy available for work. This change in free energy is called ΔG (delta G) and is the most fundamental property of a chemical reaction: it is equivalent to the change in the heat content minus the change in entropy. These relationships are represented by the following formula:

$$\Delta G = \Delta H - T\Delta S$$

where:

ΔG is the change in free energy of the reaction and is the part of the potential energy that can do work. The remaining energy is not available for work because of entropy.

ΔH is the change in enthalpy (heat content), which is the energy in chemical bonds.

T is the temperature measured on a scale of ° C above absolute zero.[5]

ΔS is the change in entropy, or disorder.

Entropy is amplified at higher temperatures because temperature measures random molecular motion (i.e., the intensity or potential of heat), which increases disorder. Therefore, higher temperatures speed reactions and increase disorder. This is also why water evaporates faster at higher temperatures.

Spontaneous reactions release heat.[6] These reactions usually increase entropy and are called **exergonic** (energy outward) reactions. They have a ΔG less than zero and therefore form products with less free energy than their reactants; this energy is potentially available for cellular work. All spontaneous reactions decrease the amount of free energy because some energy is dissipated, thereby increasing entropy. Because this energy can do no work, life is a constant struggle against entropy.

Not all reactions are spontaneous. For example, consider the formation of sucrose (table sugar) and water from glucose and fructose:

$$glucose + fructose \rightarrow sucrose + H_2O$$

---

5. Temperature in the equation for calculating the change in free energy is expressed in degrees Kelvin (K), which are units above absolute heatlessness (i.e., where molecular motion stops). Since temperature measures the intensity of molecular motion, all Kelvin temperatures exceed zero. A kelvin (the unit) is the same as a degree centigrade; to get from centigrade to Kelvin, add 273 (i.e., K = ° C + 273). For example, water freezes at 273 K, room temperature is ~ 293 K, and body temperature is ~ 310 K. Note that there is no degree sign used with Kelvin measurements.

6. *Spontaneous* indicates only that the reaction will occur, not its rate.

This reaction has a ΔG of +5.5 kcal, meaning that its products have 5.5 kcal mol^{-1} *more* energy than its reactants. This reaction absorbs energy from the surroundings and is not spontaneous. Such reactions are called **endergonic** (energy inward) reactions and will not occur without a net input of energy.

The free energy of a particular reaction determines many of a reaction's properties. Most important, the ΔG of a reaction dictates how much work a reaction can perform. For example, consider the oxidation of a mole of glucose to carbon dioxide and water:

$$C_6H_{12}O_6 + 6O_2 \rightarrow 6CO_2 + 6H_2O + energy$$

$$\Delta G = -686 \text{ kcal mol}^{-1}$$

Since ΔG for this reaction is less than zero, this reaction is exergonic and spontaneous. The carbon dioxide and water it produces store 686 fewer kcal than does glucose.

Free Energy and Chemical Equilibrium

When a chemical reaction reaches equilibrium, ΔG equals zero. Similarly, ΔG increases as one moves away from equilibrium. Because all of life's processes require work, *cells must remain far from equilibrium to stay alive.* They accomplish this by continually preventing the accumulation of any of the reactants of metabolic pathways. For example, the huge difference in free energy between glucose and its oxidation products (carbon dioxide and water) pulls cellular metabolism strongly in one direction; as soon as reactants form, they are quickly converted to new compounds by other reactions.

c o n c e p t

Chemical reactions change the amount of free energy available for work. This change in free energy is the amount of energy available to form other bonds.

Oxidation, Reduction, and Energy Content

Most energy transformations in organisms involve chemical reactions called *oxidations* and *reductions* (fig. 5.6). **Oxidation** is the loss of electrons, either alone or with hydrogen, from a molecule. This is equivalent to adding oxygen because oxygen is strongly electronegative and therefore attracts electrons from the original atom. Oxidative reactions such as the breakdown of glucose to carbon dioxide and water degrade molecules into simpler products and are therefore examples of **catabolism.** They are "downhill" reactions that release energy.

Reduction is the addition of electrons, either alone or with hydrogen, to a molecule. Reduction changes the chemical properties of a molecule, not necessarily its size. Electrons removed from a molecule during oxidation reduce another mol-

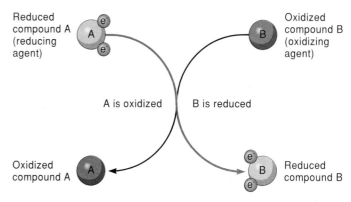

figure 5.6

Oxidation and reduction occur simultaneously; as compound A is oxidized (i.e., donates electrons), compound B is reduced (i.e., accepts electrons). Oxidized compounds such as CO_2 contain less potential energy than do reduced compounds such as CH_4.

ecule. Reduction reactions such as the formation of lipids usually involve synthesis of more complex molecules and are therefore examples of **anabolism.** They are "uphill" reactions that require a net input of energy. Oxidation and reduction reactions always occur simultaneously; if something is reduced, something else is oxidized.

Many energy transformations in living systems involve oxidation and reduction of carbon. Reduced carbon contains more energy than does oxidized carbon. This explains why reduced molecules such as methane (CH_4) are explosive, while oxidized molecules such as carbon dioxide (CO_2) are not. The same principle applies to other compounds: the more reduced they are, the more energy they contain. This principle also applies to food. For example,

Substance	Energy Content (J kg^{-1})
Hydrogen	122,000,000
Natural gas	55,000,000
Gasoline	42,000,000
Wood (dry)	17,000,000
Starch	17,000,000
Sucrose	17,000,000
Wheat flour	15,000,000 (mostly starch)
Lima beans, raw	5,000,000 (67% water)
Potatoes, raw	3,000,000 (80% water)

This example should help you better appreciate how the different energy content of foods affects our lives. Most college students require about 3,000 Calories per day—that's equivalent to the energy in a pound of butter. If you had to obtain all of that energy from lettuce or celery, you'd have to eat about 60 pounds of lettuce or about 450 celery sticks every day. Although your friends might start mistaking you for a rabbit, an "Eat All the Lettuce You Want and Still Lose Weight" diet would be effective.

figure 5.7

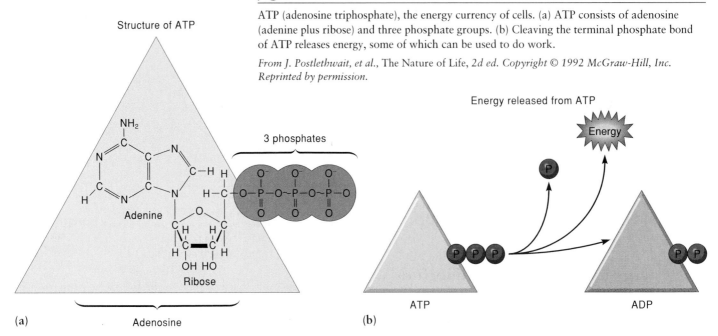

ATP (adenosine triphosphate), the energy currency of cells. (a) ATP consists of adenosine (adenine plus ribose) and three phosphate groups. (b) Cleaving the terminal phosphate bond of ATP releases energy, some of which can be used to do work.

From J. Postlethwait, et al., The Nature of Life, 2d ed. Copyright © 1992 McGraw-Hill, Inc. Reprinted by permission.

c o n c e p t

Many energy transformations in organisms involve chemical reactions called *oxidations* and *reductions*. Oxidation is the loss of electrons from a molecule, and reduction is the addition of electrons to a molecule. Oxidation and reduction reactions always occur simultaneously.

Organisms extract energy from energy-rich compounds such as sugars and fats via catabolic reactions collectively called **cellular respiration.** They use this energy to drive anabolic reactions that do work—for example, moving flagella, building cell walls, and replicating genetic information. An important hub through which energy passes during cellular metabolism is adenosine triphosphate, a compound more commonly known as *ATP.*

ATP: The Energy Currency of Cells

When cells need energy, they hydrolyze **adenosine triphosphate,** or **ATP.** ATP is a nucleoside triphosphate made of adenine (a nitrogen-containing base), ribose (a five-carbon sugar), and three phosphate (HPO_4^{2-}) groups (fig. 5.7). ATP molecules contain much energy that is released when the terminal phosphate group (represented by a squiggly line, ~) is cleaved from the molecule.[7] Because the breakdown of ATP links energy ex-

changes in cells, ATP is the energy currency of the cell: when cells need energy to do something, they "spend" ATP by converting it to adenosine diphosphate (ADP), inorganic phosphate (Pi), and energy (fig. 5.7):

$$\text{ATP} + H_2O \rightarrow \text{ADP} + \text{Pi} + \text{energy}$$
$$\Delta G = -7.3 \text{ kcal mol}^{-1}$$

Several properties of ATP make it ideally suited as the energy currency of cells:

- The amount of energy released by converting ATP to ADP + Pi (7.3 kcal mol^{-1}) is about twice as much as is needed to drive most cellular reactions. The rest of the energy is dissipated as heat.

- Much of its energy is immediately available to cells. Although fats and starch also store large amounts of energy, their energy must first be converted to ATP before it can be used. ATP represents the readily available energy "cash" of a cell; fats and starch are analogous to energy stocks and bonds.

- Unlike the covalent bonds linking carbon and hydrogen in molecules such as methane and glucose, the terminal phosphate bond of ATP is unstable—that's why it breaks so easily.

ATP is a common energy currency—all cells of all organisms use ATP for energy transformations. Like all of the different appliances that can plug into an electrical outlet and do different things, so too can different chemical reactions use a cell's ATP to do different kinds of work. Organisms use ATP for virtually all of their work, including making new cells and macromolecules, pumping materials, and moving materials through cells and throughout the organism. Accomplishing all

---

7. The terminal phosphate bond is often called a *high-energy* bond. This is incorrect. The energy contained in ATP is not simply in this phosphate bond; it resides in the complex interaction of atoms in the entire molecule.

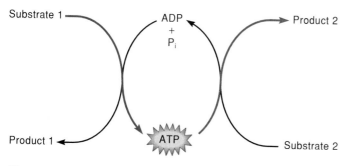

figure 5.8

Coupled reactions. Energy released by an exergonic reaction can be used to make ATP from ADP and Pi. This ATP can then be used to drive an endergonic reaction.

of this work requires huge amounts of ATP. For example, a typical adult uses the equivalent of about 200 kg (440 lb) of ATP per day, but has only a few grams of ATP on hand at any one time. Even the hungriest of us doesn't eat 200 kg of food in a week, much less in a day. Nevertheless, organisms that run out of ATP—that is, those that go "bankrupt"—immediately die. Where does all this ATP come from? Obviously, we do not make 200 kg of ATP from scratch each day. Rather, we recycle our ATP at a furious pace, turning over the entire supply every minute or so:

$$ADP + Pi + energy \rightarrow ATP$$

$$\Delta G = +7.3 \text{ kcal mol}^{-1}$$

Since the ΔG for this reaction is positive, each reaction requires an input of energy, which comes from molecules that are broken down in other reactions that are coupled to the synthesis of ATP.

Coupled Reactions

Just as cells couple the breakdown of food to the production of ATP, so too do they couple the breakdown of ATP to other reactions that occur at the same time and place in the cell. These **coupled reactions** drive other reactions that do work or make other molecules (fig. 5.8). For example, consider again the formation of sucrose (table sugar) and water from glucose and fructose:

$$\text{glucose} + \text{fructose} \rightarrow \text{sucrose} + H_2O$$

$$\Delta G = +5.5 \text{ kcal mol}^{-1}$$

Because the ΔG for this reaction exceeds zero, the reaction is not spontaneous and will not occur without a net input of energy. This input of energy is provided by two moles of ATP, each of which provides 7.3 kcal mol^{-1} of energy:

$$ATP \rightarrow ADP + Pi + energy$$

$$\Delta G = -7.3 \text{ kcal mol}^{-1}$$

This changes the equation for sucrose synthesis to

$$\text{glucose} + \text{fructose} + 2ATP$$
$$\downarrow$$
$$\text{sucrose} + H_2O + 2ADP + 2Pi + energy$$

and makes the $\Delta G = 5.5 - 14.6 = -9.1$ kcal mol^{-1}. Thanks to the expenditure of energy by the cell (i.e., as ATP), the reaction proceeds because its overall ΔG is negative. In this reaction, the breakdown of ATP is coupled to the formation of sucrose.

ATP accomplishes much of its work by transferring its phosphate group to another molecule in a process called **phosphorylation**. Phosphorylations energize the molecules receiving the phosphate group, so that they can be used in later reactions. The original "cost" of this phosphorylation is returned in subsequent reactions.

concept

Adenosine triphosphate, or ATP, is the energy currency of cells. That is, cells use ATP to do work. ATP releases about twice as much energy as is needed to drive most cellular reactions. ATP is used by all cells for energy transformations. Many of these transformations involve coupled reactions in which the energy released by ATP drives other reactions.

Other Compounds Involved in Energy Metabolism

Several other compounds besides ATP affect energy transformations in plant cells. Nonprotein helpers such as Mn^{2+} and Na^+ aid cellular metabolism and are called **cofactors**. Cofactors are often ions; for example, Mg^{2+} is a cofactor required to transfer phosphate groups between molecules. Organic cofactors are called **coenzymes** and usually carry protons or electrons. These are often nucleotides; unlike ATP, their energy content depends on their oxidation state, not on the presence or absence of a particular phosphate bond. Coenzymes are vitamins that occur in all cells. Humans and other animals must obtain vitamins from food; plants produce their own vitamins.

NAD+: Nicotinamide Adenine Dinucleotide

NAD^+ is similar to ATP in that it is made of adenine, ribose, and phosphate groups (fig. 5.9a). However, the active part of NAD^+ is a nitrogen-containing ring, called *nicotinamide*, which is a derivative of nicotinic acid (niacin, or vitamin B_3, one of the compounds added to products such as cornflakes to make them "vitamin fortified"). NAD^+ is reduced when it accepts two electrons and a proton from the active site of an enzyme or from a substrate. NAD^+ removes two protons and two electrons from a substrate, and both electrons and one proton are added to the NAD^+:

$$NAD^+ + 2H^+ + 2e^- \rightarrow NADH + H^+$$

$$\Delta G = -52.6 \text{ kcal mol}^{-1}$$

$NADH + H^+$ is fully reduced and is therefore energy-rich. It is used to make ATP and to reduce other compounds in cells. You'll learn in Chapter 6 how other pathways "cash in" the energy contained in $NADH + H^+$ for ATP.

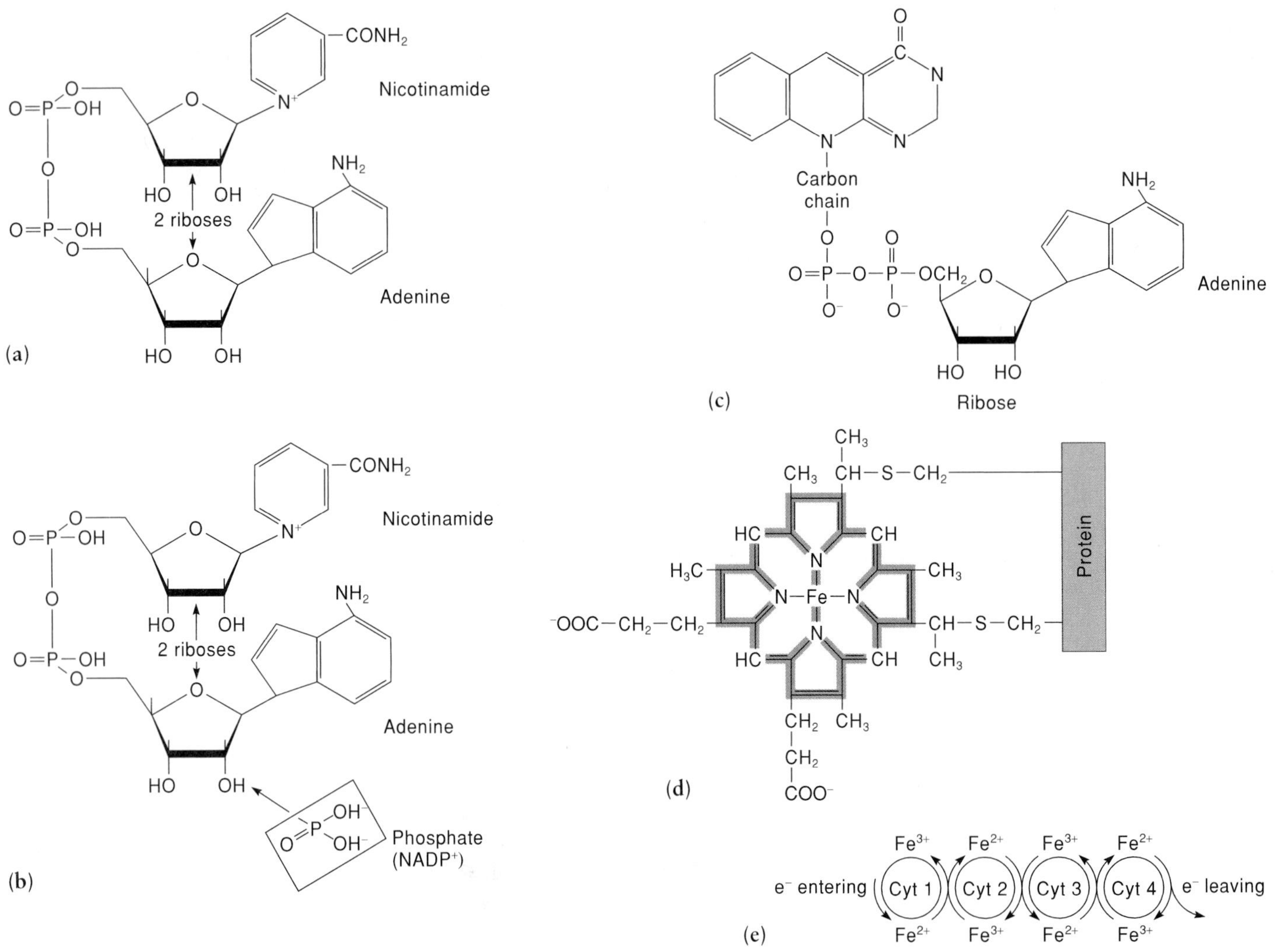

figure 5.9

Structures of electron carriers in plant cells. (a) NAD+ (nicotinamide adenine dinucleotide). The nicotinic acid ring is the part of NAD+ that is oxidized or reduced during energy transformations. (b) NADP+ (nicotinamide adenine dinucleotide phosphate). NADP+ differs from NAD+ only in that NADP+ has a phosphate group. (c) FAD (flavin adenine dinucleotide). (d) Cytochrome c, one of several kinds of cytochromes in cells. Cytochromes contain a heme group (outlined in blue), which is a ring of nitrogens with an iron atom at its center. (e) The iron atoms of cytochromes alternate between a reduced (Fe^{2+}) and an oxidized (Fe^{3+}) state during electron transport. At each step there is a drop in free energy; some of the energy that is released is used to make ATP.

NADP+: Nicotinamide Adenine Dinucleotide Phosphate

NADP+ has a structure similar to NAD+ with an added phosphate group (fig. 5.9b). NADPH supplies the hydrogen that reduces CO_2 to carbohydrate during photosynthesis. NADPH also supplies the hydrogen used to reduce nitrate to ammonia.

FAD: Flavin Adenine Dinucleotide

FAD is one of the coenzyme forms of riboflavin (vitamin B_2; fig. 5.9c). FAD, like NAD, carries two electrons; however, FAD accepts both protons to become $FADH_2$. FAD functions in cellular respiration.

Other Nucleoside Triphosphates

Several other nucleoside triphosphates function in cellular metabolism. For example, uridine triphosphate (UTP) is involved in making cell walls, guanosine triphosphate (GTP) is involved in protein synthesis, and cytidine triphosphate (CTP) is involved in membrane production.

Cytochromes

Like chlorophyll and hemoglobin, cytochromes are a group of metal-containing molecules that participate in metabolism by transferring electrons (fig. 5.9d and e). When oxidized, the iron in cytochromes exists as Fe^{3+}. When this iron accepts an

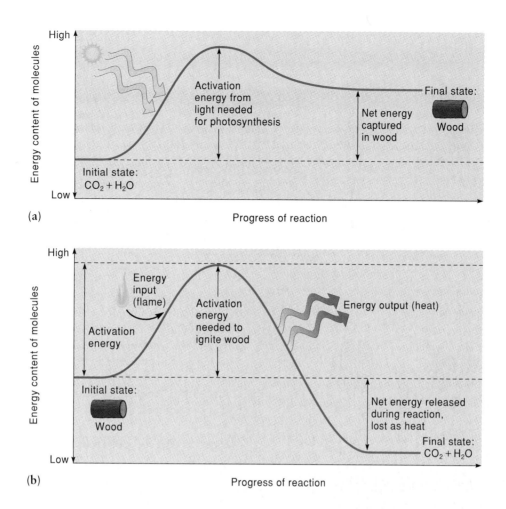

figure 5.10

Energy of activation. (a) Plants use photosynthesis to make sugars that are, in turn, used to make products such as wood. The energy to do this—that is, the activation energy for photosynthesis—comes from sunlight. (b) The energy captured in wood is released when wood burns. To start this reaction, activation energy must be provided to ignite the wood.

electron, it is reduced to Fe^{2+}. There are several types of cytochromes, all of which carry electrons in cells. These molecules participate in virtually all energy transformations in living organisms, suggesting that the machinery for energy transformation in all organisms has a common heritage.

Enzymes and Energy

You learned earlier in this chapter that exergonic reactions (e.g., the combustion of methane) are spontaneous. If this is true, then why haven't all of these reactions already occurred? The answer is that these reactions require an energy input to start. To understand this, consider again the example shown in figure 5.2. The rock atop the hill contains potential energy but is at rest. If given a nudge, the rock will roll down the hill and release its kinetic energy. The nudge is the energy input required to start the reaction. Once started, the rock rolls to the bottom of the hill spontaneously (i.e., on its own).

Many chemical reactions are similar to the rock atop the hill: they proceed only when activated by an energy input, called the **energy of activation** (E_{act}). In photosynthesis, the energy of activation is provided by sunlight, and the net gain of energy is trapped in products such as wood (fig. 5.10a). Although this wood contains much energy, the wood is too stable to burn spontaneously. That is, although the wood contains much potential energy, it will not burn unless it is first heated (fig. 5.10b). The heat necessary to ignite the wood is the energy of activation. Once ignited, the wood will continue to burn and release energy.

Although heat is generally an effective catalyst, it is usually not effective for cells. Indeed, the heat needed to activate most metabolic reactions would quickly kill most cells. Thus, cells rely on biocatalysts called *enzymes* to start their reactions.

Enzymes are globular proteins that speed reactions by decreasing the energy of activation of a reaction (fig. 5.11). They do this by binding the substrate so that the reaction can occur. Decreasing the energy of activation greatly speeds the

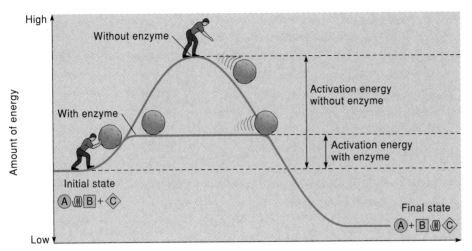

figure 5.11

Enzymes speed the rate of spontaneous reactions by lowering the energy of activation of the reaction.

From J. Postlethwait, et al., The Nature of Life, 2d ed. Copyright © 1992 McGraw-Hill, Inc. Reprinted by permission.

However, in the presence of carbonic anhydrase, H_2CO_3 forms at a rate of about 600,000 molecules per second, an increase of more than 10^7. This emphasizes the importance of enzymes: they are critical to life because they speed spontaneous reactions to a biologically useful rate.

Higher temperatures typically increase enzymatic activity. Indeed, up to a point, enzymatic activity typically doubles for every increase of 10° C. However, beyond about 60° C, entropy wins out as the protein is denatured and the reaction stops. Although a few organisms can tolerate high temperatures (e.g., bacteria in hot springs of Yellowstone National Park thrive at 70° C), most enzymes work best at much lower temperatures. For example, most enzymes in our bodies function best near body temperature (37° C/98.6° F).

Regulating Metabolism

Enzymes regulate energy transformations by controlling metabolic reactions. If metabolism is likened to a series of interconnected roads (fig. 5.5), then enzymes would function as traffic lights that control the flow of energy in cells. How do enzymes do this?

HYPOTHESIS 1:

Enzymes decrease the ΔG of a chemical reaction.
No; enzymes do not change the ΔG of a chemical reaction.

HYPOTHESIS 2:

Enzymes heat cells to increase reaction rates.
No; enzymes do not heat cells. The heat required to speed most metabolic reactions to useful rates would kill the cells. How, then, do enzymes regulate metabolism? The products of a pathway often affect the activity of the pathway. For example, plants use a five-step pathway to make isoleucine from threonine (fig. 5.12). When isoleucine accumulates, it inhibits the first enzyme of the pathway, thereby decreasing production of isoleucine until the current supply is used. This means of slowing a pathway when its products aren't needed is called **feedback inhibition** and is common in plants and animals. Feedback inhibition balances supply and demand in cells, thereby averting unnecessary excesses and deficiencies.

One means of feedback inhibition involves **allosteric regulation,** in which the product binds weakly to a receptor site on the enzyme that differs from the active site. Allosteric regulation is common in enzymes having more than one subunit. The binding of a molecule to an allosteric site (usually located where subunits join) changes the enzyme's activity, thereby affecting

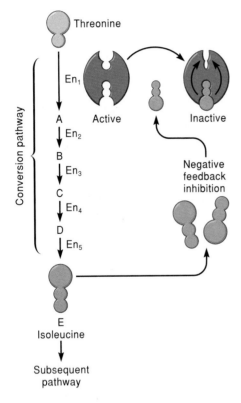

figure 5.12

Feedback inhibition. The metabolic conversion of threonine to isoleucine consists of five steps, each controlled by an enzyme (En_1, En_2, En_3, . . .). An accumulation of isoleucine inhibits the first enzyme of the pathway (En_1), thereby preventing an unnecessary buildup of isoleucine.

reaction. For example, consider the formation of H_2CO_3 from CO_2 and H_2O, a reaction involved in gas exchange and catalyzed by carbonic anhydrase:

$$CO_2 + H_2O \xrightarrow[\textit{carbonic anhydrase}]{} H_2CO_3$$

Without carbonic anhydrase, only about 1 molecule of H_2CO_3 forms per second. This is far too slow to be useful to organisms.

the cell's metabolism. For example, isoleucine allosterically inhibits the first enzyme of its synthetic pathway and thus prevents the unnecessary buildup of isoleucine.

Many enzymes are also inhibited by other molecules that compete for the enzyme's active site. These so-called **competitive inhibitors** mimic the substrate and can be overcome by increasing the concentration of substrate (i.e., diluting the concentration of the inhibitor). Drugs such as sulfanilamide competitively inhibit enzymes.

Other compounds inactivate enzymes by binding to parts of the enzyme that are different from the active site, thereby preventing the enzyme from binding the substrate at its active site. Compounds that do this, such as lead and nerve gas, are called **noncompetitive inhibitors.** Although many noncompetitive inhibitors bind reversibly to an enzyme, many do not. For example, penicillin is a noncompetitive inhibitor that binds irreversibly to an enzyme that makes cell walls in bacteria. This blockage of cell-wall synthesis ultimately kills the bacteria, thus accounting for the antibiotic effect of penicillin.

c o n c e p t

Enzymes catalyze biological reactions by lowering the energy of activation. Enzymes are critical to life because they speed spontaneous reactions to a biologically useful rate. Enzymes, and therefore metabolism, are controlled by feedback inhibition, competitive inhibition, and noncompetitive inhibition.

The Major Energy Transformations in Plants: Photosynthesis and Respiration

The energy transformations that sustain life occur similarly in all organisms. The most important energy transformations in plants are photosynthesis and respiration, the topics of the next two chapters. To appreciate these reactions, again examine figure 5.4. The energy-requiring uphill stage of the process is photosynthesis. During this process, light energy absorbed by chloroplasts is used to release oxygen and reduce carbon dioxide (a low-energy compound) to carbohydrate (a high-energy compound):

carbon dioxide + water + light

↓

carbohydrate + oxygen

Carbohydrate fuels the activities of plants and all other organisms.

The energy-releasing (i.e., exergonic), downhill stage of the process is cellular respiration. During this process, energy-rich molecules such as sugars are oxidized to carbon dioxide and water:

carbohydrate + oxygen

↓

carbon dioxide + water + energy

Respiration drives the cellular economy of most organisms.

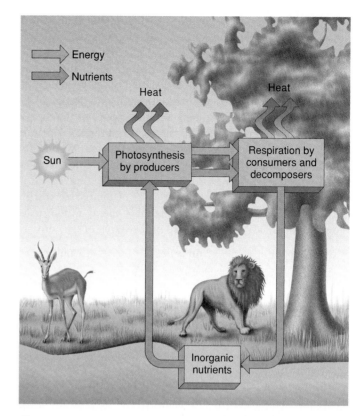

figure 5.13

Photosynthesis and respiration, the major transformations of energy by organisms. Photosynthesis provides food for virtually all organisms on earth. Organisms use cellular respiration to extract energy from food. Although nutrients cycle in an ecosystem, energy flows through an ecosystem. All energy is eventually converted to heat.

The Flow of Energy

Examine again the equations for photosynthesis and respiration. Each process involves oxidation or reduction of carbon by the addition or removal of hydrogen. Photosynthesis removes hydrogen from water and adds it to carbon; thus, *carbon is reduced during photosynthesis.* Cellular respiration removes hydrogen from carbon and adds it to oxygen to form water; thus *carbon is oxidized during respiration.* Although one equation is the reverse of the other, they proceed by different mechanisms.

Now examine figure 5.13, which shows how photosynthesis and respiration are integrated in nature. This diagram illustrates an important concept about energy: *Energy flows through a system and is ultimately converted to heat.* For example, sunlight is the energy input that sustains life on earth. This is possible only because plants use photosynthesis to convert sunlight into chemical energy (e.g., sugars). Animals stay alive by eating plants (or by eating other animals that ate plants) and using the stored energy to power their activities. These animals are then eaten by other animals to fuel their activities. Ultimately, the organisms die and are decomposed by bacteria and fungi. In the process, the energy stored temporarily by organisms in this so-called "food chain" is released as heat. Organisms may temporarily store various amounts of the energy, but the net effect is that it flows through the system and is ultimately transformed to heat.

According to the second law of thermodynamics, all of the energy transformations at every step in the food chain are less than 100% efficient. Indeed, there is a 90% loss of usable energy at each stage. This has tremendous implications. For example, consider the crop of corn shown at the beginning of this unit. Most of the energy striking the plants' leaves is reflected or converted to heat; only a small amount of the energy is trapped by photosynthesis in sugars. When these sugars are converted to starch, more of the energy is lost. When this starch is fed to a cow, only about 10% of the usable energy is stored by the cow; the rest is lost as heat. By the time we eat a steak made from the cow, another 90% of the usable energy has been lost. Thus, the amount of usable energy in the steak is only about 1% of that contained in the starch of the corn.

100 units of energy in corn	→	10 units of energy in an herbivore	→	1 unit of energy in a carnivore

These energy transformations, as predicted by the second law of thermodynamics, are inefficient because much of the energy is converted to heat. By inserting an extra energy transformation between the corn plant and ourselves (i.e., the cow), we lose much of the usable energy.

Look again at figure 5.13. Notice that although energy moves through the system in a one-way flow (i.e., toward heat), nutrients are cycled. That is, photosynthesis produces sugars and oxygen from carbon dioxide, light, and water, and the sugars and oxygen are recycled to carbon dioxide and water during cellular respiration. Thus, *nutrients cycle in an ecosystem.* You'll learn more about these nutrients in Chapter 20.

c o n c e p t

Photosynthesis and respiration represent the major energy transformations in organisms. Photosynthesis uses light energy to reduce carbon dioxide and water to carbohydrate, while respiration oxidizes carbohydrates, producing ATP in the process. Together, photosynthesis and respiration compose a system by which energy flows through an ecosystem and is ultimately converted to heat. Unlike energy, nutrients cycle in an ecosystem.

The Lore of Plants

On average, vegetables and fruits make up about 10% of our intake of calories, 7% of our intake of protein, and an unimpressive 1% of our intake of fat.

Chapter Summary

Energy is the ability to do work. It is measured with a variety of units, the most common of which are calories and joules. Potential energy is energy available to do work, while kinetic energy is energy being used to do work.

All of life's activities require that energy be transformed from one form to another. These energy transformations are governed by the laws of thermodynamics. The first law of thermodynamics is the law of conservation of energy; it states that in all chemical and physical changes, matter and energy cannot be created or destroyed, only converted to other forms. The second law of thermodynamics states that all energy transformations are less than 100% efficient. This inefficiency results from the loss of usable energy in the form of heat. The quantitative measure of the disorder created by any spontaneous reaction is entropy. All spontaneous reactions increase entropy.

Metabolism is the sum of the vast array of energy and matter transformations in cells, which occur in step-by-step sequences called *metabolic pathways*. Each reaction of a metabolic pathway rearranges atoms into new compounds and changes the amount of free energy available for work. This change in free energy is the amount of energy available to form other bonds.

Many energy transformations in organisms involve chemical reactions called oxidations and reductions. Oxidation is the loss of electrons from a molecule; reduction is the addition of electrons to a molecule. Oxidation and reduction reactions always occur simultaneously.

Adenosine triphosphate, or ATP, is the energy currency of cells, which use it to do work. ATP is suited for this function because (1) most of its energy is available immediately to cells, and (2) it releases about twice as much energy as is needed to drive most cellular reactions. ATP is used by all cells for energy transformations. Many of these transformations involve coupled reactions in which the energy released by ATP is used to drive other reactions. Other compounds involved in cellular metabolism include NAD, NADP, FAD, and cytochromes.

Enzymes catalyze biological reactions by lowering the energy of activation of the reaction. Enzymes are critical to life because they speed spontaneous reactions to a biologically useful rate. Enzymes, and therefore metabolism, are controlled by feedback inhibition, competitive inhibition, and noncompetitive inhibition.

Photosynthesis and respiration are the major energy transformations in organisms. Photosynthesis uses light energy to reduce carbon dioxide to carbohydrate, while respiration converts carbohydrate to ATP. Together, photosynthesis and respiration comprise a system by which energy flows through an ecosystem and is ultimately converted to heat. Unlike energy, nutrients cycle in an ecosystem.

 ### What Are Botanists Doing?

Go to the library and browse through a botany journal such as *Plant Physiology, American Journal of Botany,* or *Planta.* Read an article that describes a study of metabolism or bioenergetics. What did the authors do? What did they conclude? What experiments would you do if you wanted to extend their work?

Creationists frequently claim that the evolution of increasingly complex organisms during the history of life contradicts the second law of thermodynamics, which is an unbreakable rule. Therefore, they claim, biological evolution is invalid. How would you respond to this argument?

Questions for Further Thought and Study

1. Why are the laws of thermodynamics important?

2. Does gravity affect the potential energy of an object? If so, how?

3. What's the difference between potential and kinetic energy?

4. What is entropy? How does the second law of thermodynamics affect plants?

5. Why isn't heat considered usable energy?

6. Why haven't all spontaneous reactions already occurred?

7. Which contains more energy, ATP or reduced NAD? How do you know?

8. How are ATP and NADP similar? How are they different?

9. What are coenzymes, and why are they important?

10. What is energy, and why is it important?

11. What are coupled reactions? Why are they important?

12. Several catabolic enzymes involved in cellular respiration have allosteric sites that fit ATP and ADP. How could this regulate cellular respiration?

13. What does ΔG tell us about a chemical reaction?

14. How could a cell minimize the amount of enzyme needed to produce or concentrate a metabolic product?

15. As you read this book, your body uses about 25 cal sec^{-1}. Your brain alone consumes the equivalent of 250 M&Ms worth of sugar per day. Where does this energy ultimately come from?

16. We use oxidation and reduction reactions for a variety of purposes, including in breathalyzers to test for drunk drivers, in bleaches to clean our clothes, and as food preservatives (antioxidants). How do you use oxidation and reduction reactions?

Web Sites

Review the "Doing Botany Yourself" essay and assignments for Chapter 5 on the *Botany Home Page*. What experiments would you do to test the hypotheses? What data can you gather on the Web to help you refine your experiments?

Here are some other sites that you may find interesting:

http://nasacci.cs.olemiss.edu/photo/

This site, sponsored by the National Aeronautics and Space Administration (NASA), includes an overview of photosynthesis, a short quiz, and an experiment that you can do on-line.

http://clas.www.pdx.edu/Algalphysiology/photomemb.html

Review this site for research being done on photosynthetic membranes in bacteria and algae.

http://mss.scbe.on.ca/DSRESPIR.htm

Use this site to review the similarities and differences between cellular respiration and photosynthesis.

http://mss.scbe.on.ca.DSPHOTOS.HTM
http://esg-www.mit.edu:8001:esgbio/ps/intro.html

These sites include excellent reviews of the history of photosynthesis, as well as current information summarizing what we know about plants and energy.

http://www.life.uiuc.edu/bio100/lessons/photosynthesis_links.html

At this site you'll find The Photosynthesis Center, which includes summaries of the most recent research about photosynthesis. This site also includes electron micrographs and links to *Virtual Cell*.

Suggested Readings

Articles

Davis, Ged R. 1990. Energy for planet earth. *Scientific American* 263 (September):54–62.

Seymour, R. S. 1997. Plants that warm themselves. *Scientific American* (March 1997) 276:104–109.

Storey, Richard D. 1992. Textbook errors and misconceptions in biology: Cell energetics. *The American Biology Teacher* 54:161–166.

Books

Becker, W. M. 1986. *The World of the Cell*. Menlo Park, CA: Benjamin/Cummings.

Galston, A. W. 1994. *Life Processes of Plants*. New York: Freeman.

Harold, F. M. 1986. *The Vital Force: A Study of Bioenergetics*. New York: W. H. Freeman.

Stryer, L. 1989. *Biochemistry*. 3d ed. San Francisco: W. H. Freeman.

Walker, D. 1992. *Energy, Plants, and Man*. 2d ed. Brighton, U.K: Oxygraphics.

The starch-rich endosperm of this corn kernel has been stained blue-black with iodine. Starch is fuel for the emergent seedling.

Respiration

6

Chapter Outline

Chapter Overview

All organisms, including plants, harvest energy from stored chemicals in much the same way. The metabolic pathways by which organisms liberate stored energy are referred to as *cellular respiration*. This process breaks down complex molecules into simpler chemicals and harvests the energy stored in them.

The products of the complete oxidation of glucose are the same whether the glucose is burned in the laboratory or respired in a cell: CO_2, H_2O, and energy. When it is burned, a mole of glucose (180 g) makes a small fire that lasts for several minutes. In a cell, respiration controls this process, so the heat does not destroy the tissue. Instead, some of the energy from glucose is trapped in ATP. The heat lost during respiration never causes smoke or fire because it is released slowly by many chemical reactions.

There are several kinds of respiration, but the most common ones require oxygen and produce ATP. Other kinds of respiration make little or no ATP; rather, they lose most of the energy as heat or transfer energy into the synthesis of organic molecules other than ATP.

In the previous chapter you learned that all organisms convert energy from one form to another. Plants convert light energy to chemical energy by photosynthesis, a process that you will learn more about in the next chapter. Plants and animals convert stored chemical energy to ATP by respiration. The ATP harvested from energy stored in carbohydrates and other organic molecules fuels the energy-requiring activities of the cell, including the movement of organelles and small particles within a cell, active transport across membranes, and the transport of organic compounds from one part of a plant to another. The carbon from storage compounds goes into carbohydrates, proteins, fats, nucleic acids, and other kinds of organic molecules.

Beginning with glucose, respiration harvests energy in three stages. The first stage is **glycolysis**, in which glucose is split into smaller molecules in the cytosol. The second stage, called the **Krebs cycle**, completely metabolizes the products of glycolysis. In the third stage several electron carriers, together called the **electron transport chain**, harvest energy from the movement of electrons (fig. 6.1). The Krebs cycle and the electron transport chain occur in mitochondria.

All three stages of respiration produce ATP. In this chapter you will learn how respiration accomplishes this task. Although the process is complicated, remember the theme of the chapter: Cells make ATP from energy stored in organic molecules.

c o n c e p t

Energy stored in glucose is retrieved in the three stages of respiration: glycolysis, the Krebs cycle, and electron transport. Part of this energy is converted into ATP, which can be used for cellular work.

Retrieving Glucose from Other Molecules

Most discussions of respiration focus on what happens to glucose, probably because the metabolism of glucose is so similar in all organisms. However, glucose is usually not abundant in cells; thus, respiration begins by converting storage compounds to glucose, which can then be shunted into glycolysis. Such storage compounds vary from one organism to another or, in plants, from one organ to another. For example, humans and other vertebrates store the polysaccharide glycogen; potatoes and bananas store starch; and Jerusalem artichokes and onions store polymers of fructose (fig. 6.2). Furthermore, sucrose is the common starter molecule for respiration in leaves, but in the same plant, respiration may begin with starch in nongreen storage organs such as roots and stems. In other plants, respiration may begin with different carbohydrate polymers, different sugars, fats, organic acids, or proteins. Nevertheless, the most well-known and probably most common sources for glucose in plants are sucrose and starch.

Retrieval from Sucrose

Sucrose, abundant in sugarcane stems and sugar beet roots, is common in plants and is commercially important because humans consume so much of it as table sugar. Besides being a common energy source in photosynthetic cells, sucrose is the carbohydrate that moves most readily through conducting cells to growing tissues.

In slow-growing and mature cells, the degradation of sucrose is mostly an irreversible hydrolysis to its component monomers, glucose and fructose:

$$sucrose + H_2O \rightarrow glucose + fructose$$

Enzymes that catalyze this reaction are called **sucrases**; they occur in the cytosol, in the central vacuole, and sometimes in cell walls. Hydrolysis by sucrases frees glucose and fructose for glycolysis.

In starch-storing organs and rapidly growing cells, sucrose from starch degradation is usually hydrolyzed by the enzyme **sucrose synthase** in a reversible reaction that releases fructose but binds glucose to a carrier molecule, **uridine diphosphate (UDP)**:

$$sucrose + UDP + H_2O \rightleftharpoons fructose + UDP\text{-}glucose$$

Fructose from this reaction is available for glycolysis, but glucose is either used in cell-wall synthesis or released from the UDP-glucose complex for further respiration.

Retrieval from Starch

Starch is a complex branched polymer made of glucose monomers (Chapter 2). Because of the complexity of starch, its breakdown requires several enzymes for the complete retrieval of glucose. These enzymes attach directly to starch grains in chloroplasts and amyloplasts. Three general kinds of enzymes degrade starch. One kind is the **amylases**, which hydrolyze alpha-1,4 glucose linkages. Another kind of enzyme

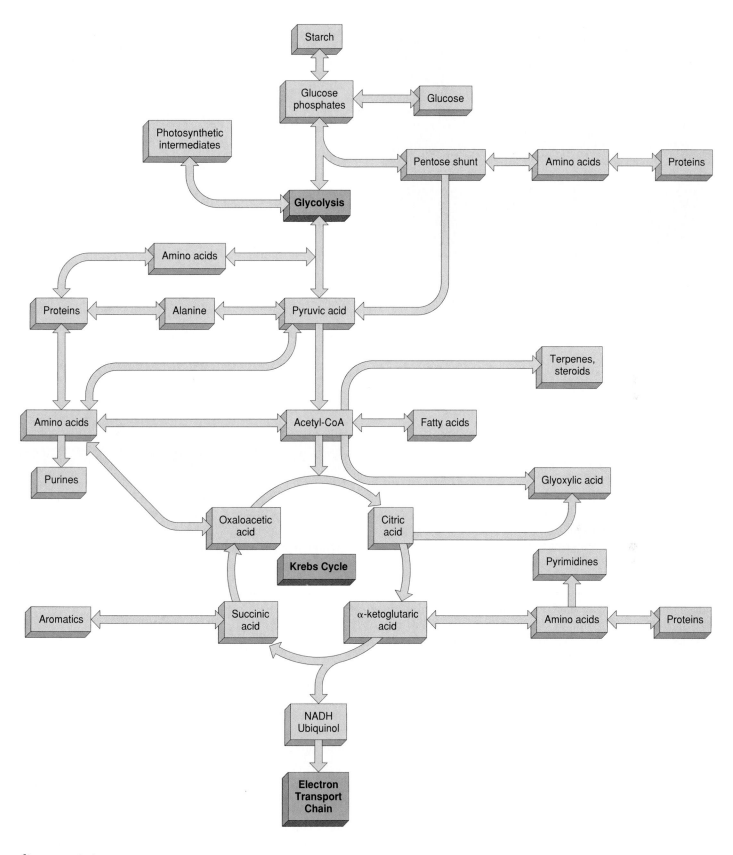

figure 6.1

Respiration and its major metabolic relationships. The main pathway of respiration begins with the breakdown of carbohydrates, continues through the Krebs cycle, and ends with the electron transport chain. Many kinds of organic compounds are made from metabolic spinoffs of respiration. The metabolic relationships shown here represent many of the hundreds of reactions that are depicted in fig. 5.5.

figure 6.2

Different organisms store polysaccharides in different forms, including starch (potatoes, bananas) and polymers of fructose (onions, Jerusalem artichokes).

is **starch phosphorylase,** which cleaves glucose at one end of a glucose polymer by adding a phosphate to it:

$$\text{starch} + H_2PO_4 \rightleftharpoons \text{glucose-1-phosphate}$$

Although this reaction is reversible, starch synthesis from it is negligible. Also, glucose-1-phosphate from this reaction bypasses the need for an ATP in the first step of glycolysis, as discussed later in this chapter.

Neither the amylases nor starch phosphorylase can break the alpha-1,6 glucose linkages at the branch points in starch. Only a third group of enzymes, called **debranching enzymes,** can hydrolyze starch at branch points.

Although glucose is retrieved from starch in plastids, it must be exported to the cytosol for glycolysis. However, intact glucose does not move out of plastids readily. Therefore, glucose in plastids must be modified before it is exported to the cytosol. This usually entails breaking the glucose into two smaller molecules, exporting them, and then either reassembling them into glucose in the cytosol or shunting them directly into glycolysis.

Harvesting Energy from Glucose: An Overview

The overall equation for the respiration of glucose is as follows:

glucose   + oxygen → carbon dioxide + water + energy
$C_6H_{12}O_6$ + O_2 → CO_2 + H_2O + energy

This reaction links the solar energy obtained during photosynthesis, which is in glucose in this equation, to the energy that is needed by cells in all organisms. Some of the energy on the right side of this equation ends up in ATP; the rest either resides in CO_2 and H_2O or is lost as heat. The efficiency of the overall process is the topic of reading 6.2.

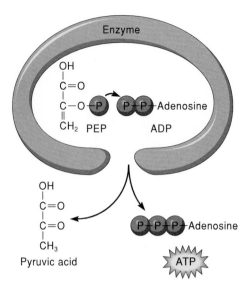

figure 6.3

Substrate-level phosphorylation. Phosphate is enzymatically transferred from a substrate, in this case phosphoenolpyruvic acid (PEP), to ADP. The reaction proceeds because the phosphate bonds of ATP are more stable than the phosphate bond of PEP.

The Lore of Plants

The amount of CO_2 released by running a car engine is about 100 kg per 40 l tank of gasoline. If 1 mole of glucose (180 g) yields 6 moles of CO_2 (264 g) during aerobic respiration, then burning a tank of gasoline is equivalent to the respiration of about 68 kg of glucose, or the mass of the average adult human. This is roughly the amount of glucose respired by 10,000 young sunflower plants during a warm summer night.

Synthesis of ATP occurs in two ways. The first involves the transfer of phosphate from organic compounds to ADP. Because a substrate provides the phosphate for ATP directly, this reaction is called **substrate-level phosphorylation** (fig. 6.3). The transfer of phosphate is catalyzed by an enzyme that binds the substrate and ADP.

The energy of substrate-level phosphorylation comes from the phosphate-containing substrate. Regardless of which substrate is used, the removal of a phosphate from it must release more energy than necessary for making ATP from ADP. For example, removal of a phosphate from phosphoenolpyruvic acid (PEP) releases 14.8 kcal mol^{-1} (fig. 6.3). This is more than enough energy to phosphorylate ADP, which requires 7.3 kcal mol^{-1}. The remaining energy (7.5 kcal mol^{-1}) is lost as heat.

The second way that ATP is made involves coupling energy from an electron donor to an electrochemical gradient that spans the inner mitochondrial membrane. The electron donor is usually NADH, a molecule that you learned about in Chapter 5. Electrons from NADH are passed through a chain of electron carriers by a series of oxidation-reduction reactions. The energy from this movement of electrons is used by proton pumps to generate a proton electrochemical gradient that fuels the phosphorylation of ADP to ATP. The overall process is called **oxidative phosphorylation.**

Membranes play an important role in coupling energy between the electron transport chain and the phosphorylation of ADP. This energy-coupling is usually referred to as **chemiosmosis** because it involves chemical reactions and transport across membranes. Chemiosmosis is also used to make ATP in chloroplasts during photosynthesis (see Chapter 7). The mechanism of chemiosmosis in mitochondria is described later in this chapter.

Of the two ways that ATP is made during respiration, substrate-level phosphorylation is the simpler, but it accounts for only a small percentage of ATP synthesis. Most ATP is made by oxidative phosphorylation.

c o n c e p t

In glycolysis and the Krebs cycle, chemical bond energy from different substrates is used to bond phosphate to ADP to make ATP. This is called *substrate-level phosphorylation*. In electron transport, energy for phosphorylating ADP comes from a series of oxidation-reduction reactions. This is called *oxidative phosphorylation*, or phosphorylation by *chemiosmosis*.

Glycolysis

Potential Energy of Glucose

As you learned in Chapter 5, glucose contains 686 kcal mol^{-1}. This means that with the addition of some heat (i.e., energy of activation, E_{act}) to get it going, 1 mole of glucose will yield 686,000 calories when it is completely oxidized to CO_2 and H_2O. However, cells recover only a small amount of this energy during glycolysis; the rest is lost as heat. Because the energy is released in small increments, the excess heat does not damage the cell.

Breakdown of Glucose to Pyruvic Acid

During glycolysis, glucose is split into two three-carbon compounds. The entire process requires ten steps, all of which occur in the cytosol (fig. 6.4). The first step phosphorylates glucose by using one ATP. This phosphorylation activates glucose so that the appropriate enzyme can do the next step. Phosphorylated glucose is rearranged and phosphorylated again by a second ATP before it is split in half. After it is split, each of the three-carbon products has one phosphate. One of the products is glyceraldehyde-3-phosphate, which is further metabolized in glycolysis. The other product is dihydroxyacetone phosphate, which is not metabolized directly but is instead converted to glyceraldehyde-3-phosphate. These products represent the halfway point of glycolysis (see steps 1–5 in fig. 6.4). By this point, energy has been used but not yet harvested.

The first energy-harvesting step of glycolysis occurs in the second half of the pathway. This step is the reduction of NAD$^+$ to NADH by the oxidation of glyceraldehyde-3-phosphate. After this step, some of the energy from glucose is stored in the energy-rich electrons of NADH. This oxidation also releases enough energy to add a second phosphate group to glyceraldehyde-3-phosphate, converting it to 1,3-bisphosphoglyceric acid. At this point, glycolysis is finally ready to make ATP. Substrate-level phosphorylation occurs when one of the phosphates of 1,3-bisphosphoglyceric acid is transferred to ADP. The three-carbon molecule that remains is then rearranged to form phosphoenolpyruvic acid, which becomes pyruvic acid when it gives up its phosphate to a second ADP. In this way, each glyceraldehyde-3-phosphate from the first half of glycolysis is used to make two molecules of ATP and one molecule of pyruvic acid.

Since two molecules of glyceraldehyde-3-phosphate can be made from one molecule of glucose, each glucose yields four ATPs and two pyruvic acids from glycolysis. However, two ATPs are used in the first half of glycolysis, so the net gain is two ATPs for each glucose.

By the end of glycolysis, a small amount of the chemical energy that started out in glucose ends up in ATP and NADH. Most of the energy of glucose remains in pyruvic acid. The energy stored in pyruvic acid is used to make more ATP in mitochondria. The synthesis of ATP from the energy of pyruvic acid occurs in the second phase of respiration, the Krebs cycle.

c o n c e p t

During the first half of glycolysis, glucose is phosphorylated twice and split into two three-carbon compounds. This first half of glycolysis uses ATP. The three-carbon compounds are oxidized to make pyruvic acid during the second half of glycolysis, which also makes ATP and NADH.

The Krebs Cycle

Pyruvic acid that is transported into the mitochondrion is not used in the Krebs cycle directly. Instead, pyruvic acid first loses a molecule of carbon dioxide. The remaining acetyl group is attached to a coenzyme to form **acetyl coenzyme A (acetyl-CoA).** Figure 6.5 outlines the conversion of pyruvic acid to acetyl-CoA in the mitochondrion. Note that when carbon dioxide is released from pyruvic acid, NAD$^+$ is reduced to NADH.

Glycolysis and the Krebs cycle are linked by the conversion of pyruvic acid to acetyl-CoA. Pyruvic acid is the final product of glycolysis, and acetyl-CoA is the compound that enters the Krebs cycle.

Steps in the Krebs Cycle

The Krebs cycle, named in honor of Sir Hans Krebs, is a cycle because the last step regenerates the starter chemicals for the first step (fig. 6.6; also see reading 6.1, "Botany and Politics"). In all, there are eight steps, seven of which occur in the mitochondrial matrix. Step 6 occurs in the mitochondrial membrane.

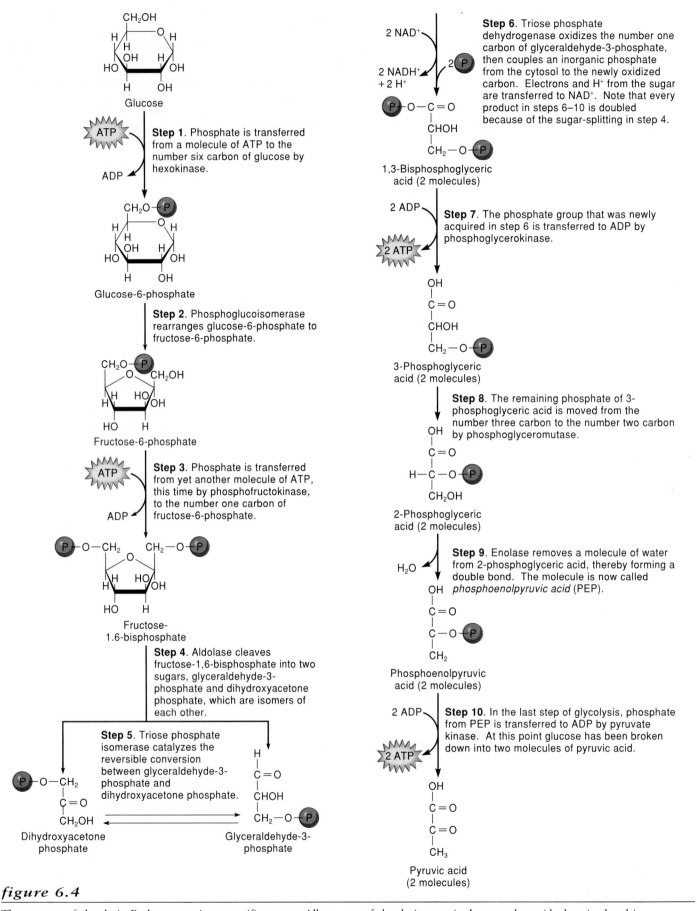

figure 6.4

The ten steps of glycolysis. Each step requires a specific enzyme. All enzymes of glycolysis occur in the cytosol, outside the mitochondria.

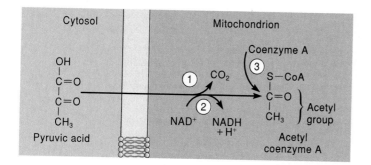

figure 6.5

Conversion of pyruvic acid to acetyl-CoA. (1) CO_2 is removed from pyruvic acid that has been imported into the mitochondrion. (2) The remaining two-carbon fragment is oxidized to an acetyl group, while NAD^+ is reduced to NADH. (3) The acetyl group is attached to the sulfur atom of coenzyme A (CoA).

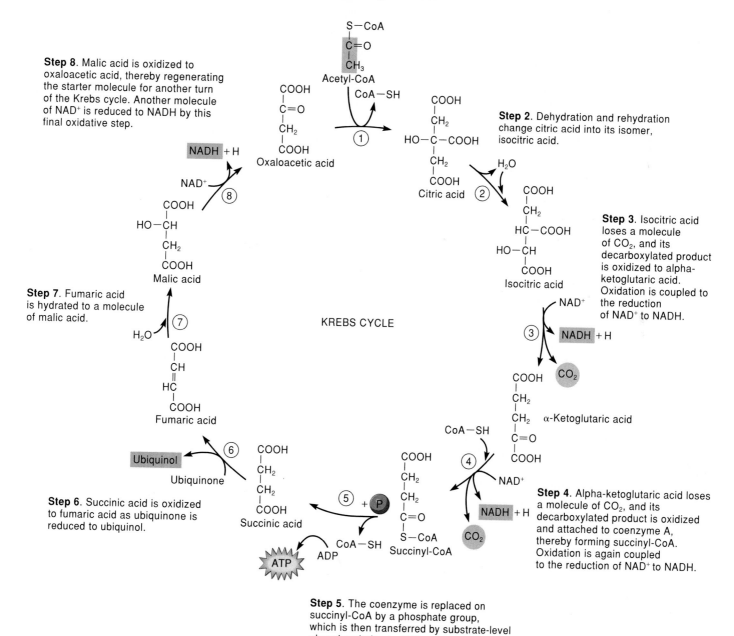

Step 1. The acetyl group of acetyl-CoA is attached to oxaloacetic acid, thereby making citric acid.

Step 8. Malic acid is oxidized to oxaloacetic acid, thereby regenerating the starter molecule for another turn of the Krebs cycle. Another molecule of NAD^+ is reduced to NADH by this final oxidative step.

Step 2. Dehydration and rehydration change citric acid into its isomer, isocitric acid.

Step 3. Isocitric acid loses a molecule of CO_2, and its decarboxylated product is oxidized to alpha-ketoglutaric acid. Oxidation is coupled to the reduction of NAD^+ to NADH.

Step 7. Fumaric acid is hydrated to a molecule of malic acid.

KREBS CYCLE

Step 4. Alpha-ketoglutaric acid loses a molecule of CO_2, and its decarboxylated product is oxidized and attached to coenzyme A, thereby forming succinyl-CoA. Oxidation is again coupled to the reduction of NAD^+ to NADH.

Step 6. Succinic acid is oxidized to fumaric acid as ubiquinone is reduced to ubiquinol.

Step 5. The coenzyme is replaced on succinyl-CoA by a phosphate group, which is then transferred by substrate-level phosphorylation to ADP.

figure 6.6

The Krebs cycle. Purple carbons are either added from acetyl-CoA or removed as CO_2. Orange boxes contain reduction products whose formation depends on oxidative steps in the cycle.

BOTANY AND POLITICS

Science is a search for truth. Although "truth" is not affected by politics, the business of science often is. Consider, for example, Otto Heinrich Warburg, a German biochemist. Warburg began his scientific career in the early 1900s and in 1924 announced his discovery of "iron oxygenase," the oxygen-carrier that we now call cytochrome oxidase. In 1931, Warburg was awarded a Nobel Prize for this work. However, less than two years later, Hitler was appointed chancellor of Nazi Germany. Although Warburg was half-Jewish, he was unharmed by the Nazis because he studied cancer. Indeed, when Nazis removed Warburg from his position in 1941, he was personally ordered back to work on cancer by Hitler, who feared dying of cancer. Warburg stayed in Germany because he thought that his work would be destroyed if he left Germany. Although Warburg's pact with the Nazis incensed his colleagues outside of Germany, his research was brilliant. For example, Warburg's 1935 discovery that nicotinamide is the active group of hydrogen-transferring enzyme would have earned him a second Nobel Prize, but Hitler forbade German citizens from accepting the award. Nevertheless, Warburg continued to study cancer, photosynthesis, and respiration until his death in 1970.

Nazi policies also affected other German scientists. One of these scientists was Hans Adolf Krebs, who worked as an assistant with Warburg in Berlin from 1926–1930. In 1932, using much of what he learned from Warburg, Krebs announced the first metabolic cycle ever to be described. However, the next year (1933) Krebs was fired by the Nazis and fled from Nazi Germany to Britain. Only four years after arriving in Britain, Krebs announced his discovery of the "tricarboxylic acid cycle," a pathway used by almost all organisms to convert two-carbon fragments to carbon dioxide, water, and energy. This pathway, which later became known as the Krebs cycle, earned Krebs a Nobel Prize in 1953. Krebs was knighted in 1958, making him Sir Hans Krebs. Krebs worked in Britain until his death in 1981.

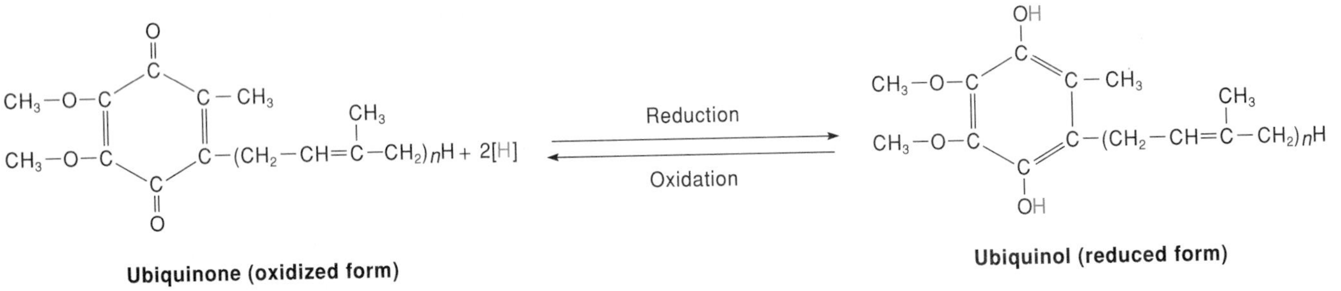

figure 6.7

Reversible conversions between ubiquinone and ubiquinol. The n in each formula stands for the number of times the five-carbon side chain is repeated. Naturally occurring ubiquinones have 6–10 of these five-carbon units (30–50 carbons in the side chain).

In the first step, coenzyme A is cleaved from acetyl-CoA, and the acetyl group is attached to **oxaloacetic acid,** which has four carbons. The resulting six-carbon compound is **citric acid.** (The Krebs cycle is also known as the **citric acid cycle** because it begins with citric acid or as the **tricarboxylic acid cycle** because citric acid has three carboxyl groups.) Citric acid is then rearranged to **isocitric acid** (step 2), which is the beginning substrate for two oxidative steps that entail the removal of two molecules of carbon dioxide. These steps also reduce two molecules of NAD+ to NADH.

In the first oxidative step from isocitric acid (step 3), the removal of carbon dioxide yields **alpha-ketoglutaric acid.** In the second oxidative step (step 4), carbon dioxide is removed from alpha-ketoglutaric acid, and the product is bound to coenzyme A to make **succinyl-CoA.** Each step also reduces a molecule of NAD+ to NADH. Coenzyme A is then cleaved from succinyl-CoA, thereby producing **succinic acid** and providing energy for the substrate-level phosphorylation of ADP to ATP (step 5).

Following the formation of succinic acid, three more oxidative steps occur (steps 6–8): succinic acid is oxidized to **fumaric acid,** fumaric acid to **malic acid,** and malic acid to oxaloacetic acid. The oxidation of succinic acid also reduces the electron carrier **ubiquinone** to **ubiquinol** (fig. 6.7);[1] the oxidation of malic acid to oxaloacetic acid reduces NAD+ to NADH. Thus, for each acetyl-CoA that enters the Krebs cycle, only one ATP is made by substrate-level phosphorylation. Most of the energy derived from the oxidative steps of the Krebs cycle is stored in the high-energy electrons of NADH and ubiquinol. The energy in these molecules is harvested in the third phase of respiration, oxidative phosphorylation.

---

1. Most textbooks show that the oxidation of succinic acid reduces FAD to FADH$_2$. (Recall that FADH$_2$ is similar to NADH; see Chapter 5.) Although this reaction does occur, FAD and FADH$_2$ are highly transient, enzyme-bound intermediates during the reduction of ubiquinone to ubiquinol; therefore, ubiquinol is the mobile product from this step in the Krebs cycle, not FADH$_2$.

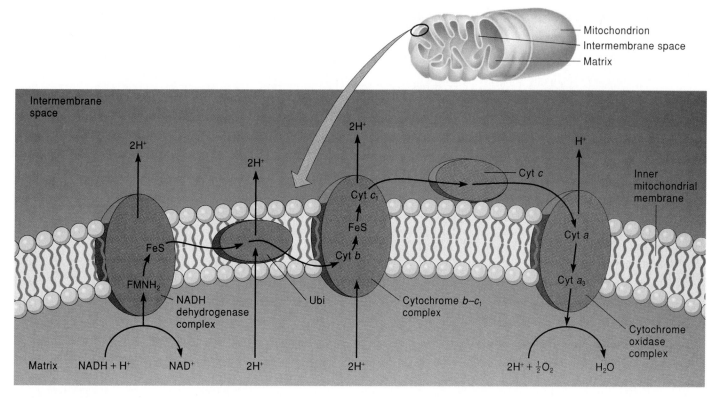

figure 6.8

Model for the sequence of steps in the electron transport chain. Two mobile electron carriers, ubiquinone (Ubi) and cytochrome c (Cyt c), transfer electrons between three electron-carrier complexes that are embedded in the membrane: the NADH dehydrogenase complex, the cytochrome $b-c_1$ complex, and the cytochrome oxidase complex. Electron flow along the chain causes proton pumping from the mitochondrial matrix into the intermembrane space. This model shows that one NADH fuels the transport of six protons, although the number of protons is variable and can be as high as eleven.

c o n c e p t

Pyruvic acid from glycolysis is moved into mitochondria and converted into acetyl-CoA, which enters the Krebs cycle. Each turn of the Krebs cycle uses one molecule of acetyl-CoA and produces one molecule of ATP, one molecule of ubiquinol, and three molecules of NADH. The acetyl carbons are released as carbon dioxide.

Electron Transport and Oxidative Phosphorylation

Relative to glycolysis and the Krebs cycle, oxidative phosphorylation is a bonanza for ATP synthesis. Energy for oxidative phosphorylation comes from NADH and ubiquinol, which are produced by the first two phases of respiration. The high-energy electrons of NADH and ubiquinol are not used directly for ATP synthesis; instead, they start a series of oxidation-reduction reactions that move electrons through several carriers. The sequence of electron carriers is known, appropriately enough, as the **electron transport chain.** Energy from electron flow through these carriers maintains a proton gradient across the inner mitochondrial membrane, which drives the synthesis of ATP. *Oxidative phosphorylation* refers to the combination of the oxidation-reduction reactions of electron transport that enable a cell to use the energy in NADH and ubiquinol to phosphorylate ADP to ATP.

Electron Transport

The electron transport chain is like a series of tiny, successively stronger magnets; that is, each has a higher potential than its predecessor. This means that each component of the chain pulls electrons away from a weaker neighbor and gives them up to a stronger one. Thus, the strongest acceptor in the chain is the terminal acceptor, oxygen, which has a potential of 0.82 volts. This is slightly more than half of the potential between the two terminals of a typical flashlight battery.

There are at least nine electron carriers in the series, most of which are proteins. Some of these carriers also accept protons and release them into the intermembrane space. Although scientists are uncertain how these components work together for electron and proton transport, figure 6.8 shows a widely accepted model for the sequence of steps in the chain.

How Efficient Is Respiration?

An age-old question in biology and botany classes is, "How many ATPs can a cell make from a molecule of glucose?" This seems like a simple but important question that can be answered by adding up the number of times ADP is phosphorylated between glucose and the end of the electron transport chain. If we knew this number, we could calculate respiratory efficiency by comparing the number of calories in glucose to the number of calories in ATP made by respiration. However, such calculations require some simple and probably incorrect assumptions.

The Dogma of Calculating Respiratory Efficiency

The usual calculation shows that a maximum of 36 molecules of ATP can be made from one molecule of glucose (see reading table 6.2). This number is obtained by adding the presumed maximum net number of ATPs from glycolysis, the Krebs cycle, and oxidative phosphorylation. For example, substrate-level phosphorylation yields two ATPs from glycolysis and two ATPs from the Krebs cycle (one ATP each from two turns of the cycle). But these are the only steps where ATP can be counted directly. Most ATP from respiration is made by the electron transport chain via oxidative phosphorylation. To estimate the number of ATPs made by the electron transport chain, we assume that one ATP is produced for each pair of actively transported protons. This means that, based on the model in figure 6.8, three ATPs would be produced from each NADH and two ATPs from each ubiquinol. Thus, we calculate ATP production from the number of NADHs and ubiquinols that enter the chain. Hence, one molecule of glucose yields two NADHs from glycolysis, two NADHs from decarboxylating two molecules of pyruvic acid to acetyl-CoA, and six NADHs and two ubiquinols from two turns of the Krebs cycle. The 10 NADHs would yield 30 ATPs, and the two ubiquinols would yield four more ATPs, which when added to the four ATPs from substrate-level phosphoryla-

tion, would yield a total of 38 ATPs. However, NADH made from glycolysis must be actively transported into mitochondria, perhaps using as many as two ATPs. This would reduce the net production of ATPs from a molecule of glucose to 36. Glucose contains 686 kcal mol^{-1} of potential energy, and ATP contains 7.3 kcal mol^{-1} of potential energy. Based on the ratio of 36 ATPs to one glucose, the maximum efficiency of respiration is $(36 \times 7.3) / 686 = 0.383$, or about 38%.

An Alternative View

The ledger for ATP synthesis is simple for substrate-level phosphorylation. However, such ATP-counting is much more complex for oxidative phosphorylation than our ledger shows. The reason is that NADH production does not translate directly into a specific number of ATPs by chemiosmosis.

Some variability in the energy harvest comes from the use of NADH from glycolysis. This variability occurs because NADH made in the cytosol cannot cross the inner mitochondrial membrane. Instead, NADH from the cytosol may trigger a series of oxidation-reduction reactions that span the membrane. The final reactant produces a new molecule of NADH in the mitochondrial matrix (see reading figure 6.2a). Because some energy is lost in this process, each original NADH cannot power the complete reduction of another NAD$^+$. The amount of lost energy probably varies among different cell types and at different stages of cell growth, so the amount of energy from each NADH from glycolysis that actually gets to the electron transport chain cannot be calculated accurately.

Another difficulty in ATP-counting is that the amount of proton transport caused by each NADH is probably variable. For example, laboratory measurements show that each NADH can transport up to eleven protons from the mitochondrial matrix to the intermembrane space, not six as predicted by the model in figure 6.8. If a pair of protons could fuel the production of one ATP, then one NADH could produce 5.5 ATPs instead of three, as previously assumed. Conversely, recent measurements of chemiosmosis show

reading table 6.2

Potential Production of ATP by Respiration of One Molecule of Glucose

				ATPs
I.	Glycolysis			
	Substrate-level phosphorylation:	2 ATP	→	2 ATP
	Reduction of NAD$^+$:	2 NADH*		
		(to mitochondrion)		
II.	Pyruvic acid → Acetyl-CoA (×2)			
	Reduction of NAD$^+$:	2 NADH*		
III.	Krebs cycle (×2)			
	Substrate-level phosphorylation:	2 ATP	→	2 ATP
	Reduction of NAD$^+$:	6 NADH*		
	Reduction of ubiquinone:	2 ubiquinols**		
IV.	Electron transport			
	*Oxidation of 10 NADH:			30 ATP
	**Oxidation of 2 ubiquinols:			4 ATP
				38 ATP
	Minus energy for active transport of NADH from glycolysis into mitochondrion			–2 ATP
			TOTAL:	36 ATP

that the energy of at least four protons is needed to make each ATP. Thus, the maximum number of ATPs from NADH would be 2.75—that is, eleven protons divided by four protons per ATP. Ignore for the moment that three-quarters of a phosphate bond makes little sense. The main problem with this kind of counting is that it implicitly assumes that all protons diffusing back through the inner mitochondrial membrane go through ATP syntheses. This is incorrect; the proton gradient is used for other things besides making ATP (see reading figure 6.2b). For example, inorganic phosphate and other metabolites are actively transported into the mitochondrial matrix by the proton gradient. The proton gradient is also used to export ATP from the mitochondrial matrix (and simultaneously import ADP), so that it can be used for metabolism elsewhere in the cell. Finally, some protons leak through the membrane; their energy is not harvested at all.

We cannot calculate the number of ATPs made by the proton gradient because energy from the gradient is used for so many other processes at the same time. The energy demand by these processes varies among cell types and at different stages of

cell development; therefore, the amount of ATP synthesis also varies under the same circumstances.

Regardless of the complexity of respiration, trying to balance the books on respiratory efficiency is still one of the favorite activities of biochemists. Current speculation is that a minimum of 21 ATPs is produced

by one glucose and that the maximum of 36 ATPs is probably unreachable. Thus, average respiratory efficiency is probably between 22%–38% (21–36 ATPs per glucose). For comparison, the lower limit of respiratory efficiency is probably equivalent to the efficiency of your car in harvesting energy from gasoline—about 20%–25%.

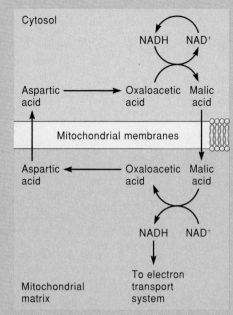

reading figure 6.2a

One of the pathways for energy transfer from cytosolic NADH to mitochondrial NADH. In this pathway, cytosolic NADH reduces oxaloacetic acid to malic acid, which is imported into the mitochondrion. Oxidation of malic acid to oxaloacetic acid regenerates NADH in the mitochondrion by reducing NAD^+. The transfer cycle begins again when oxaloacetic acid is converted to the amino acid aspartic acid, which is exported to the cytosol and converted back to oxaloacetic acid. Energy is lost during this transfer cycle, so more than one cytosolic NADH is required to regenerate each mitochondrial NADH.

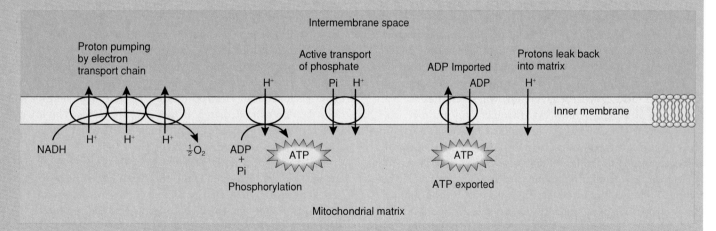

reading figure 6.2b

Uses of the proton gradient generated by electron transport energy. The amount of energy consumed by phosphorylation and active transport or lost by ATP export and proton leakage is unknown.

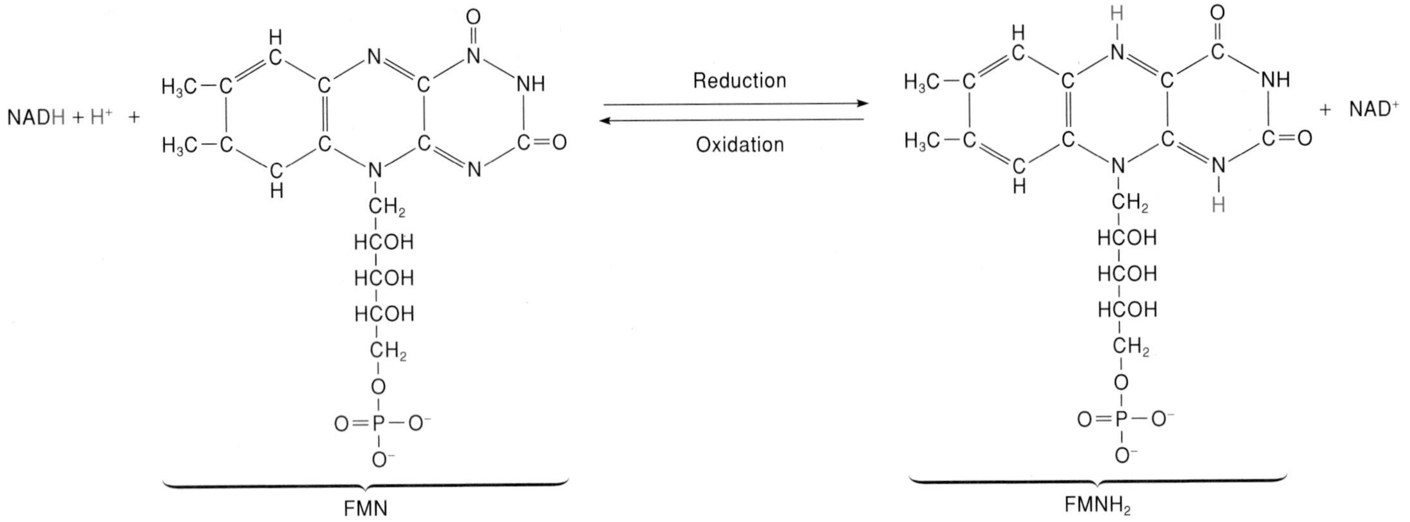

figure 6.9

In the first step of the electron transport chain, flavin mononucleotide (FMN) is reduced in a reversible reaction to FMNH$_2$ by NADH and H$^+$ from the mitochondrial matrix. (FMN is also called *riboflavin* or *vitamin B$_2$*.)

According to the model, in the first step of electron transport, **flavin mononucleotide (FMN)** takes electrons from NADH in the mitochondrial matrix. FMN also takes two protons (H$^+$), one from NADH and one from the aqueous matrix (fig. 6.8). In this step, NADH is oxidized to NAD$^+$, and FMN is reduced to FMNH$_2$ (fig. 6.9). In the second step an iron- and sulfur-containing protein is reduced as FMNH$_2$ is oxidized back to FMN (i.e., the iron-sulfur protein pulls electrons from FMNH$_2$). Meanwhile, the protons from FMNH$_2$ are released into the intermembrane space. The net result of these first two steps is that energy from oxidation-reduction reactions drives the transport of protons from one side of the inner mitochondrial membrane to the other. The carriers responsible for these steps are called the **NADH dehydrogenase complex,** named after the enzyme that removes hydrogens from NADH.

In the third step of electron transport, electrons are pulled from the NADH dehydrogenase complex by ubiquinone (also called **coenzyme Q**), which shuttles the electrons to a complex of two **cytochromes** and another iron-sulfur protein. This complex is called the **cytochrome *b-c$_1$* complex.** Protons are again transported from the matrix side of the inner membrane to the intermembrane space, using the energy of electron flow from ubiquinone through the cytochrome *b-c$_1$* complex. Thus, ubiquinone acts as a second proton pump in the electron transport system, and the cytochrome *b-c$_1$* complex acts as a third proton pump. Note that ubiquinone can also accept electrons directly from ubiquinol in the matrix.

Finally, electrons from the cytochrome *b-c$_1$* complex are pulled away by a mobile cytochrome, cytochrome *c,* and shuttled to the terminal carrier complex. The terminal complex consists of two more cytochromes, cytochrome *a* and cytochrome *a$_3$*, which form the **cytochrome oxidase complex.** Cytochrome *a$_3$* donates its electrons to oxygen, which is the terminal electron acceptor in the electron transport chain. The reduced oxygen then combines with protons in the mitochondrial matrix, producing water.

One of the key features of the electron transport chain is that it involves iron. This element occurs not only in the iron-sulfur proteins but also in the cytochromes. Thus, the oxidation and reduction of these electron carriers entails interchanges between the reduced (Fe^{2+}) and the oxidized (Fe^{3+}) forms of iron. In cytochromes, this change occurs in a complex ring group, called a **heme,** that is attached to the protein (fig. 6.10). The heme of cytochromes is similar to the heme of hemoglobin in the blood of animals.

Note that the transport of protons from one side of the inner mitochondrial membrane to the other requires energy, which is provided by the movement of electrons through the electron transport chain. Electron movement yields energy because each reduced electron carrier has less free energy than does the donor immediately before it in the series. The lowest level of free energy in the chain occurs in the water that forms when oxygen is reduced in the last step.

Between the NADH at the start of electron transport and the oxygen at the end, the total change in free energy is −53 kcal mol^{-1}. The energy change between ubiquinol and oxygen is about one-third less because ubiquinol enters the chain at a lower energy level (fig. 6.11).

Why not have NADH and ubiquinol simply donate electrons directly to oxygen? If this happened, the total free-energy change would be the same. However, there are at least two reasons why this does not occur. The first is that such a single-step reaction would probably release too much heat for

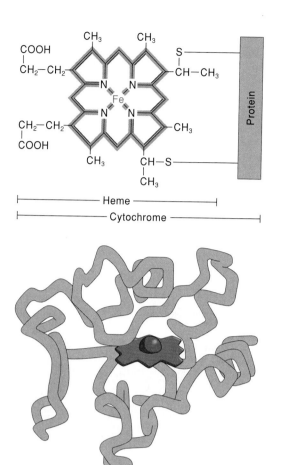

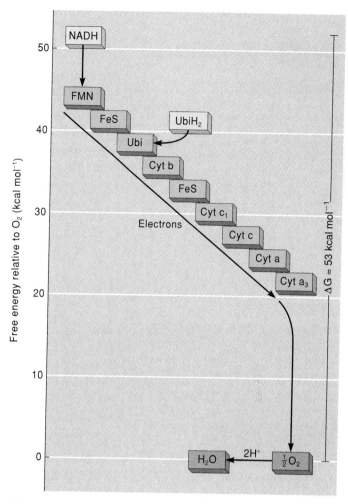

figure 6.10

Structural models of cytochrome c, showing the iron-containing heme group. The heme group is blue and the iron is red. During electron transport, each iron accepts one electron and is reduced from Fe^{3+} to Fe^{2+}.

figure 6.11

Free-energy scale of the electron transport chain. The energy drop for electrons from NADH to oxygen is 53 kcal mol^{-1}. The energy drop from ubiquinol is about one-third less.

the cell to absorb without damage. The second is that a single step from NADH or ubiquinol to oxygen would skip the proton pumping that is necessary to make ATP.

c o n c e p t

Electrons from NADH and ubiquinol reduce electron carriers that pass electrons to other carriers. The drop in free energy from one carrier to the next provides energy for transporting protons from the mitochondrial matrix into the intermembrane space. Thus, mitochondrial proton pumps are fueled by the energy of electron transport.

Chemiosmosis and ATP Synthesis

How do the protons that are pumped by the electron transport chain provide energy for ATP synthesis? While many scientists searched for high-energy compounds formed directly by the electron transport chain, British scientist Peter Mitchell suspected that he could understand *how* ATP was made if he knew *where* it was made. Mitchell and others knew that all of the carriers of the electron transport chain were part of the inner mitochondrial membrane. Consequently, Mitchell suspected that the membrane must have a role in ATP synthesis. To test this idea, he first isolated the inner membranes of mitochondria and then made vesicles of the membranes. When he added oxygen and NADH to the medium containing the vesicles, the electron transport chain functioned as it does in intact cells. However, something else also happened: the pH of the medium surrounding the vesicles decreased (recall that a lower pH indicates a greater concentration of protons). Mitchell concluded that the protons were pumped out of the vesicles during electron transport. Scientists then used electron microscopy to show that **ATP synthases,** the enzymes that make ATP, protruded from one side of the vesicles (fig. 6.12).

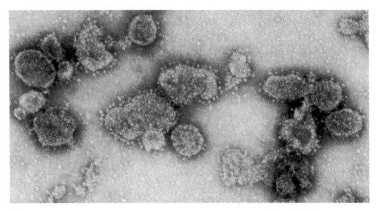

(a)

figure 6.12

The chemiosmotic theory for ATP synthesis. (a) Transmission electron micrograph of vesicles made of inner membrane fragments from mitochondria. Bulblike protrusions on these vesicles are portions of ATP synthase complexes on the matrix side of the membrane, ×145,000. (b) Model of ATP synthase complex in the inner mitochondrial membrane. Energy from electron transport pumps protons out of the matrix. Energy from the proton gradient is used for phosphorylating ADP when protons diffuse back into the matrix through ATP synthase complexes.

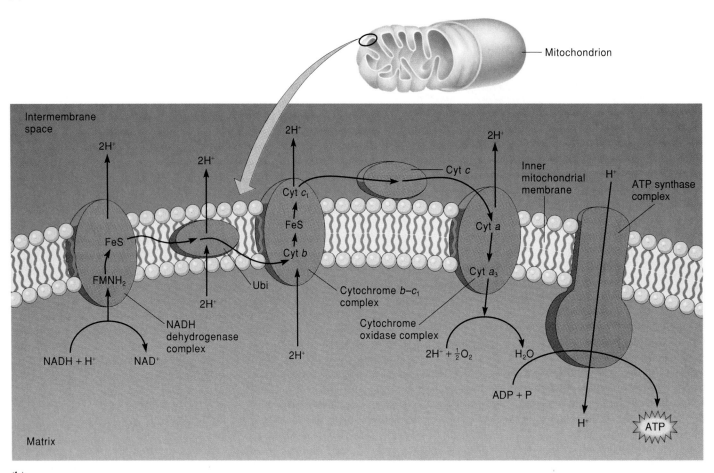

(b)

Mitchell used all this information to formulate his chemiosmotic theory of ATP synthesis:[2]

- Carriers of the electron transport chain are embedded in the inner mitochondrial membrane. The flow of electrons through this series of carriers is used to pump protons across the membrane. Because the membrane is not very permeable to protons, relatively few protons leak back across the membrane.

- Continued electron transport pumps more and more protons into the space between the inner and outer membranes of the mitochondria. This creates a gradient of protons across the membrane.

2. Mitchell won a Nobel Prize in 1978 for his study of chemiosmosis.

- The proton gradient serves as a battery to do work, including making ATP.

According to Mitchell's model, which is supported by much experimental evidence, the interaction between electron transport and ATP synthesis is indirect: electron transport produces a proton gradient, and that gradient drives the synthesis of ATP.

The inner membrane of a mitochondrion is folded many times, so that much surface area is available for membrane-dependent metabolism. Thousands of copies of the ATP synthase complex are embedded throughout the membrane, making the inner mitochondrial membrane a powerhouse for making ATP.

c o n c e p t

The proton gradient that is maintained by electron transport provides the energy necessary for making ATP. Protons diffuse down their electrochemical gradient through enzyme complexes that harness the energy of proton flow for the phosphorylation of ADP.

Other Types of Respiration

The foregoing discussion of respiration focuses on the main type of respiration in most organisms, which is called **aerobic respiration** because it requires oxygen as the terminal electron acceptor. When oxygen is not available, other electron acceptors can be used in a process called **anaerobic respiration**. This is one of several alternatives to aerobic respiration that are discussed in the next few paragraphs.

Anaerobic Respiration

Although oxygen is required only at the end of electron transport during aerobic respiration, electron transport and the Krebs cycle are both inhibited when oxygen is not available. Glycolysis works normally in an oxygen-free environment, but energy stored in the electrons of the NADH from glycolysis can become a problem if the NADH cannot be used fast enough elsewhere in the cell or if the supply of NAD^+ is depleted. In both cases, there would be no NAD^+ available to oxidize glyceraldehyde-3-phosphate; consequently, glycolysis would stop. In animals, certain fungi, and bacteria, pyruvic acid is reduced to lactic acid by excess NADH. In other fungi and in plants, the abundance of NADH is relieved when pyruvic acid is converted to acetaldehyde, which is then reduced by NADH to ethanol (fig. 6.13). Together, these anaerobic reactions are called **fermentation**. Unless they have adaptations (such as aerenchyma) to facilitate diffusion of oxygen to their roots, many plants ferment when they grow in mud or in oxygen-poor water; this is why most houseplants die when they are overwatered. Anaerobic respiration is inefficient because it produces little or no ATP.

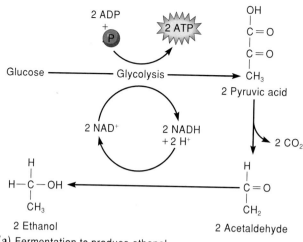

(a) Fermentation to produce ethanol

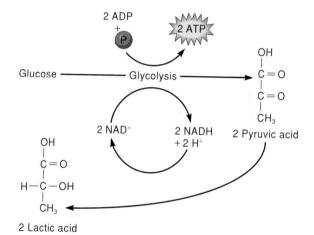

(b) Fermentation to produce lactic acid

figure 6.13

Anaerobic respiration usually produces ethanol and carbon dioxide in plants and certain fungi (a); in other fungi and in bacteria and animals, it produces lactic acid (b).

Anaerobic respiration is economically important. For example, fermentation by bacteria and fungi is important for flavoring cheese and yogurt. Similarly, fermentation by yeast is important for making alcoholic beverages and bread; wine, beer, and bread are made by different strains of brewer's yeast (*Saccharomyces cerevisiae*). The carbon dioxide produced during fermentation makes bread rise, but the alcohol from fermentation evaporates when the bread is baked. Wine is usually made by yeast that grows on grapes, although fermented honey (mead), apples (cider), and other carbohydrate-rich substrates are also used to make wine or winelike beverages. Likewise, although beer is usually made from barley, rice beer is common in the Orient, and corn beer is made by the Tarahumara tribe of northern Mexico.

Respiration of Pentose Sugars

Glucose-6-phosphate from glycolysis is often redirected to a series of reactions that involve several five-carbon sugars (pentoses) in what is usually called the **pentose phosphate pathway** (fig. 6.14). Like glycolysis, the pentose phosphate pathway releases carbon dioxide and reduces an electron carrier—in this case $NADP^+$ instead of NAD^+. The NADPH is not used in electron transport, and no other useful energy comes from this pathway. However, NADPH is used in certain reactions that do not use NADH, such as the reduction of nitrate to ammonia. Nitrogen in ammonia can then be incorporated into the amino groups of amino acids. Of equal importance are the intermediates of pentose metabolism that are precursors of other kinds of molecules. These include ribose, which is used to make DNA and RNA, and erythrose, which is a precursor of many phenolic compounds, including lignin.

Respiration of Lipids

Lipids that include fatty acids are important storage compounds in oil-containing seeds. When such seeds germinate, their fatty acids are metabolized in glyoxysomes to recover the energy stored in them. Fatty acids are first removed from glycerol in storage triacylglycerides, then snipped into two-carbon pieces that are released as acetyl-CoA (fig. 6.15). This reaction, called **beta-oxidation,** is repeated for every pair of carbons until all of the fatty acid is converted into acetyl-CoA molecules. Thus, a molecule of stearic acid, which has eighteen carbons, yields nine molecules of acetyl-CoA. Beta-oxidation occurs either in the cytosol or in glyoxysomes. The release of acetyl-CoA drives the reduction of NAD^+ and FAD to NADH and $FADH_2$, respectively.

Acetyl-CoA from beta-oxidation can be used in the Krebs cycle for further recovery of the energy stored in fatty acids, or it can be routed to other metabolic pathways (see fig. 6.1). NADH and $FADH_2$ may be transported to mitochondria, where their electrons can be used in the electron transport chain. ($FADH_2$ donates electrons to an iron-sulfur protein that donates to ubiquinone.)

Cyanide-Resistant Respiration

Aerobic respiration may be inhibited when the terminal electron carrier, cytochrome oxidase, combines with cyanide

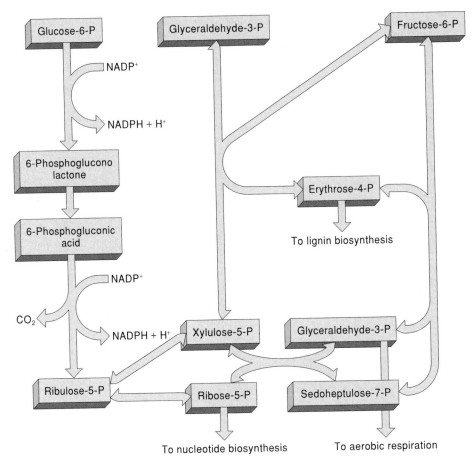

figure 6.14

The major intermediates of the pentose phosphate pathway. Single-headed arrows indicate reversible conversions whose direction depends on metabolic demand for nucleotide synthesis, lignin synthesis, and aerobic respiration.

(CN^-), azide (N_3^-), or certain other negatively charged ions. These ions complex with iron in the oxidase, thereby poisoning the enzyme and halting electron transport. However, in many plant tissues, respiration continues despite such respiratory poisons. Such respiration is called **cyanide-resistant respiration.** Certain plants, fungi, and bacteria have cyanide-resistant respiration, but it is rare in animals.

The reason respiration does not stop when cytochrome oxidase is poisoned is that mitochondria have an alternative short chain of electron carriers that branches off at ubiquinone. Because this chain also ends in an oxidase, it is aerobic—that is, oxygen is the terminal electron acceptor (fig. 6.16). Little or no oxidative phosphorylation is coupled to this chain, however, so the energy from the oxidation of NADH in this pathway produces heat instead of ATP.

Cyanide-resistant respiration is most active in sugar-rich cells, where glycolysis and the Krebs cycle occur unusually fast. This observation is indirect evidence that the main electron transport chain becomes saturated and the cyanide-resistant pathway takes up the overflow of electrons. Thus, the main benefit of this pathway for plants is probably that it dissipates excess energy from respiration.

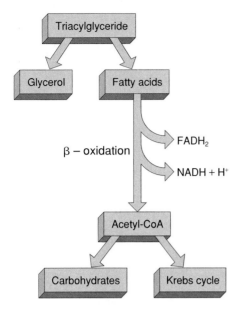

figure 6.15

Respiration of lipids. Triacylglycerides are hydrolyzed into glycerol and fatty acids. Fatty acids are degraded by beta-oxidation into acetyl groups that are attached to coenzyme A, which can be used in respiration or for carbohydrate synthesis. FAD and NAD^+ are also reduced during beta-oxidation.

In some plants, such as the aroid lilies, the amount of heat generated from cyanide-resistant respiration is remarkable. For example, in the eastern skunk cabbage (*Symplocarpus foetidus*), this heat can raise temperatures in some parts of the plant more than 20° C above the air temperature. Such extreme temperatures are important in the ecology of these plants (see reading 6.3; also see Chapter 5).

Photorespiration

Some plants increase their use of oxygen when available CO_2 diminishes. This light-dependent process, called **photorespiration,** is discussed in more detail in Chapter 7. Photorespiration, which interferes with photosynthesis, causes large losses in the yield of many crops.

Chapter Summary

Respiration harvests energy from organic molecules. There are several kinds of respiration, but the most common kind is aerobic respiration, which requires oxygen and produces ATP. Respiration occurs in three stages. The first stage, glycolysis, entails the breakdown of glucose into smaller organic compounds; it occurs exclusively in the cytosol and converts some of the energy stored in glucose to ATP and NADH. During glycolysis, ATP is made by substrate-level phosphorylation. The remainder of the energy from glucose is either lost as heat

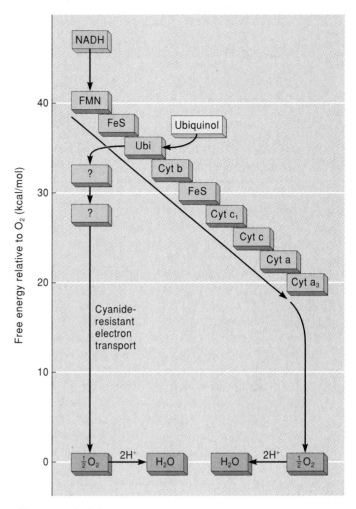

figure 6.16

Model of electron transport chain for cyanide-resistant respiration. Question marks represent unknown electron carriers. As in cyanide-susceptible electron transport (fig. 6.8), the final carrier of cyanide-resistant electron transport is an oxidase, because the terminal electron acceptor is oxygen.

or remains in pyruvic acid, which is shunted into the mitochondria. In the mitochondria, pyruvic acid is converted into acetyl-CoA in a reaction that also reduces NAD^+ to NADH.

Acetyl-CoA enters the second stage of respiration, the Krebs cycle, which is a series of oxidation-reduction reactions that produce ATP, NADH, ubiquinol, and CO_2. As in glycolysis, ATP is made in the Krebs cycle by substrate-level phosphorylation.

Energy-rich electrons from NADH and ubiquinol fuel the electron transport chain, which is the third stage of respiration. Electrons move through a series of carriers that release energy at each step. The terminal electron acceptor is oxygen, which is reduced to form water. The energy of electron transport is used to pump protons from the mitochondrial matrix into the intermembrane space. As protons diffuse by chemiosmosis back into the matrix through ATP synthase complexes,

WHY SKUNK CABBAGE GETS SO HOT

reading figure 6.3

Skunk cabbage spends the spring and summer packing starch into its fleshy rhizome. The root then acts as a furnace during late winter by providing the energy needed to heat the flowering shoot and melt the snow around it as it grows up through the snow (see reading figure 6.3). Skunk cabbage burns this fuel and uses oxygen at a rate comparable to that of a hummingbird. Heat is conserved by a thick, spongy structure that acts as an insulator around the flowering shoot. The overall effect is that the flowers develop in a near-tropical climate even though the plant is buried in snow.

Respiration that generates so much heat is costly to the plant; moreover, it makes little or no ATP. There is also no evidence that getting such an early jump on spring gives skunk cabbage an advantage over other plants. Why, then, does skunk cabbage get so hot? One possible explanation is an evolutionary one—that is, heat-generating respiration was passed down unchanged from the ancestors of skunk cabbage.

To study this hypothesis, botanists have compared skunk cabbage with its tropical relatives, which include many well-known houseplants, such as caladiums, philodendrons, voodoo lilies, and dumbcanes. These plants, as well as skunk cabbage, belong to the Arum family (*Araceae*), most of whose members still live in the tropics. When some aroids heat up, they vaporize foul odors that diffuse into the air around the flowering shoot. These odors include such chemicals as putrescine, cadaverine, and skatole. They smell like dung or decaying flesh, which attracts flies and other insects seeking suitable places for laying their eggs. Instead of finding places to lay their eggs, however, the flies and insects are often trapped temporarily by the plant and become covered with pollen. After they are released, they may get trapped again and leave pollen in another flowering shoot. In this way they transfer pollen from one flower to another.

Unlike its tropical relatives, skunk cabbage does not seem to benefit from its vaporized fragrance by tempting pollinators. For one thing, few insects are active when skunk cabbage is hot. Also, although skunk cabbage smells a lot like its tropical relatives, it has no mechanism for trapping insects.

The heat-generating ability of skunk cabbage and some of its relatives has intrigued botanists for decades. Is the heat-generating process an adaptation for surviving cold weather, or is it merely an evolutionary remnant of a feature of some tropical plants? What kinds of information and experiments would help you answer this question?

---

their energy is coupled to the phosphorylation of ADP to ATP. This synthesis of ATP from the energy of electron transport is called *oxidative phosphorylation.*

There are several kinds of respiration. Anaerobic respiration occurs in the absence of oxygen. The pentose phosphate pathway uses products from the breakdown of glucose to synthesize other organic molecules. Lipids are respired by beta-oxidation to yield acetyl units that enter the Krebs cycle. Cyanide-resistant respiration uses the energy of NADH to generate heat. Photorespiration occurs when photosynthesis cannot proceed due to the depletion of CO_2 in leaves.

 ## What Are Botanists Doing?

Plant scientists are finding that an atmosphere of artificially enriched CO_2 affects plant respiration. Using the reference resources at your library, find out how plant respiration is affected by a high-CO_2 environment. What hypotheses, if any, have plant scientists proposed to explain these effects?

 ## Writing to Learn Botany

What are the advantages and disadvantages of aerobic versus anaerobic respiration?

Questions for Further Thought and Study

1. If the proton gradient in mitochondria worked only to make ATP, how many ATPs could be made from each NADH? Explain why such maximum ATP production cannot be attained.

2. Suggest reasons why the absence of oxygen inhibits both electron transport and the Krebs cycle but not glycolysis.

3. Without distillation, wine and other fermentation products do not exceed about 12% alcohol. Suggest reasons why anaerobic respiration does not continue after this concentration of alcohol is reached.

4. How does aerobic respiration differ from cyanide-resistant respiration?

5. What might the similarities and differences be between the inhibition of respiration by carbon monoxide versus carbon dioxide?

6. Where could an organism live if it could only respire anaerobically?

7. Too much oxygen is toxic to aerobic organisms. Explain how too much oxygen might affect respiration.

8. Re-read reading 6.1, "Botany and Politics." How do politics and governmental policies affect botany today?

Web Sites

Review the "Doing Botany Yourself" essay and assignments for Chapter 6 on the *Botany Home Page*. What experiments would you do to test the hypotheses? What data can you gather on the Web to help you refine your experiments?

Here are some other sites that you may find interesting:

http://mss.scbe.on.ca/DSRESPIR.htm

This site includes descriptions of membranes, cellular respiration, and fermentation, as well as micrographs and a laboratory experiment. You can also review the similarities and differences between cellular respiration and photosynthesis.

http://www.life.uiuc.edu/bio100/lessons/respiration_links.html

Use this site to review the principles of cellular respiration. This site also includes electron micrographs and links to *Virtual Cell*.

Suggested Readings

Articles

Amthor, J. S. 1991. Respiration in a future, higher-CO_2 world: Opinion. *Plant, Cell and Environment* 14:13–20.

Babcock, G. T., and M. Wikstrom. 1992. Oxygen activation and the conservation of energy in cell respiration. *Nature* 356:301–308.

Cammack, R. 1987. $FADH_2$ as a "product" of the citric acid cycle. *Trends in Biochemical Sciences* 12:377.

Davies, D. D. 1987. Introduction: A history of the biochemistry of plant respiration. In *The Biochemistry of Plants*, vol. 11, D. D. Davies, ed. New York: Academic Press, 1–38.

McCarty, R. E. 1985. H^+-ATPases in oxidative and photosynthetic phosphorylation. *BioScience* 35:27–33.

McIntosh, L. 1994. Molecular biology of the alternative oxidase. *Science* 105:781–786.

Meeuse, B. J. D., and I. Raskin. 1988. Sexual reproduction in the arum lily family, with emphasis on thermogenicity. *Sexual Plant Reproduction* 1:3–15.

Prince, R. C. 1988. The proton pump of cytochrome oxidase. *Trends in Biochemical Sciences* 13:159–160.

Ryan, M. 1991. Effects of climate change on plant respiration. *Ecological Applications* 1:157–167.

Storey, R. D. 1991. Textbook errors and misconceptions in biology: Cell metabolism. *The American Biology Teacher* 53:339–343.

Wivagg, D. 1987. Research reviews: How many ATPs per glucose molecule? *The American Biology Teacher* 49:113–114.

Books

Amthor, J. S. 1989. *Respiration and Crop Productivity.* New York: Springer-Verlag.

Douce, R., and D. A. Day, eds. 1985. Higher Plant Cell Respiration. *Encyclopedia of Plant Physiology.* Vol. 18. New York: Springer-Verlag.

Egginston, S., and H. F. Ross, eds. 1992. *Oxygen Transport in Biological Systems: Modelling of Pathways from Environment to Cell.* New York: Cambridge University Press.

Kramer, S. P. 1986. *Getting Oxygen: What Do You Do If You're Cell Twenty-Two?* New York: Crowell.

Salisbury, F. B., and C. W. Ross. 1992. *Plant Physiology.* 4th ed. Belmont, CA: Wadsworth.

Plants use photosynthesis to convert sunlight to chemical energy.

Photosynthesis

7

Chapter Outline

Chapter Overview

Plants surround us and enrich our lives. Plants also sustain ecosystems with photosynthesis.

Photosynthesis is a light-driven series of reactions that convert energy-poor compounds such as carbon dioxide and water to energy-rich sugars. In plants, photosynthesis also oxidizes water and releases oxygen. Over time, the oxygen released by photosynthesis has dramatically changed the earth's atmosphere and enabled the evolution of aerobic respiration. Today, virtually all life depends on photosynthesis.

In Chapter 5 you learned that cells are open systems—systems that can absorb, but not create, energy. Before the evolution of photosynthesis, virtually all organisms used organic compounds in the planet's "primeval broth" as sources of energy; those organisms were *heterotrophs*—that is, organisms that make organic compounds from other energy-rich organic compounds that they absorb. As these organisms oxidized the carbon compounds from the primeval broth, they released carbon dioxide into the environment. This life-style was adequate in the short term, but there was a long-term problem: the primeval broth was nonrenewable. This created a constant competition for the limited and ever-dwindling supply of carbon compounds.

The evolution of photosynthesis about 3 billion years ago gave many organisms a new source of energy. Photosynthetic organisms, rather than relying on the ever-dwindling amount of primeval broth for energy, began to use sunlight as an energy source. In doing so, they became solar-powered and could exploit a reliable and abundant supply of energy. These organisms were the earth's first *photosynthetic autotrophs*—organisms that use light energy to make organic compounds from inorganic compounds such as water and carbon dioxide. Their ability to convert light-energy to chemical energy soon supported almost all other forms of life on the planet. These autotrophs also changed the planet and its remaining organisms by decreasing the atmospheric concentration of carbon dioxide, diminishing the greenhouse effect, and filling the atmosphere with a waste product that some of the other organisms ultimately found essential for life: oxygen. All of the oxygen in air that we breathe has been cycled through plants via photosynthesis, "life's grand device." This oxygen allowed the evolution of aerobic respiration and higher life-forms.

Today, only about 500,000 of the earth's 3–5 million species of organisms are photosynthetic. However, the importance of these organisms cannot be overstated: without photosynthesis, humans and virtually all other animals would become extinct.

Photosynthesis packages light into chemical bonds. Indeed, each year photosynthesis produces about 1.4×10^{14} kg (3.1×10^{14} lb) of carbohydrates—enough sugar to fill a string of boxcars reaching to the moon and back 50 times. Here's another way of appreciating the annual contribution of photosynthesis: if all of the sugars made by photosynthesis were converted to an equivalent amount of coal loaded into standard railroad cars (each holding about 50 tons of coal), then the earth's photosynthesis would fill more than 100 cars per second with coal. Photosynthesis is the reaction of life.

How We Learned about Photosynthesis

Considering how much we know about biology (witness the size of this and other biology books), it's easy to forget that for most of recorded history, scientists had no idea that the sun supplies the earth with virtually all of its energy or that green plants trap energy and produce the invisible gases that we breathe. Ancient Greeks, noting that fertilizing the soil increases plant growth and that the lives of animals depend on the food that they eat, reasoned that plant growth must result from "food" that the plants "eat" from the soil. This concept of plants as soil-eaters went unchallenged until 1648, when the Dutch physician named Jan-Baptista van Helmont reported a simple but elegant experiment (fig. 7.1):

> That all vegetable matter immediately and materially arises from the element of water alone I learned from this experiment. I took an earthenware pot, placed in it 200 lb of earth dried in an oven, soaked this with water, and placed in it a willow shoot weighing 5 lb. After five years had passed the tree growth therefrom weighed 169 lb and about 3 oz. But the earthenware pot was constantly wet only with rain or (when necessary) distilled water; and it was ample in size and imbedded in the ground; and, to prevent dust flying around from mixing with the earth, the top of the pot was kept covered with an iron plate coated with tin and pierced with many holes. I did not compute the weight of the deciduous leaves of the four autumns. Finally, I again dried the earth of the pot, and it was found to be the same 200 lb minus about 2 oz. Therefore, 164 lb of wood, bark, and root had arisen from the water alone.

Although van Helmont's conclusion would later be proven wrong, he did show that plants are not soil-eaters. His careful measurements also set the stage for similar studies of how plants grow. Near the same time, other scientists were studying combustion, a topic that interested not only medieval alchemists but also their successors, who laid the foundation of modern chemistry. One of the people interested in the changes made in air by combustion was an English chemist and radical, nonconformist Unitarian minister named Joseph Priestley (1733–1824). Priestley lived above a brewery and was impressed by the bubbling of the fermenting vats. He became interested in gases and how they affect animals. He designed a series of elegant experiments that showed that combustion somehow "injured" the air: if a candle were burned in a closed container, it would go out (fig. 7.2). If a mouse were then put in the container, the mouse would die (fig. 7.2). Similarly, when the contents of bubbles from beer fermentation were substituted for pure air in the container, the mouse would die. In 1771 Priestley described his findings:

> I flatter myself that I have accidentally hit upon a method of restoring air which has been injured by the burning of candles, and that I have discovered at least one of the restoratives which nature employs for this purpose. It is vegetation. In what manner this process in nature operates, to produce so remarkable an effect, I do not pretend to have discovered; but a number of facts declare in favour of this hypothesis. I shall introduce my account of them, by reciting some of the observations which I made on the growing of plants in confined air, which led to this discovery. One might have imagined that, since common air is necessary to vegetable, as well as to animal life, both plants and animals had affected it in the same manner, and I own I had that expectation, when I first put a sprig of mint into a glass-jar, standing inverted in a vessel of water; but when it had continued growing there for some months, I found that the air would neither extinguish a candle, nor was it all inconvenient to a mouse, which I put into it.

> Finding that candles burn very well in air in which plants had grown a long time, and having had some reason to think, that there was something attending vegetation, which restored air that had been injured by respiration, I thought it was possible that the same process might also restore the air that had been injured by the burning of candles.

> Accordingly, on the 17th of August, 1771, I put a sprig of mint into a quantity of air, in which a wax candle had burned out, and found that, on the 27th of the same month, another candle burned perfectly well in it. This experiment I repeated, without the least variation in the event, not less than eight or ten times in the remainder of the summer. Several times I divided the quantity of air in which the candle had burned out, into two parts, and putting the plant into one of them, left the other in the same exposure, contained, also, in a glass vessel immersed in water, but without any plant; and never failed to find, that a candle would burn in the former, but not in the latter. I generally found that five or six days were sufficient to restore this air, when the plant was in its vigour; whereas I have kept this kind of air in glass vessels immersed in water many months without being able to perceive the least alteration had been made in it.

Young willow tree
2.25 kg (5 lbs)

Soil
90 kg (200 lbs)
dry weight

+

Rain
water

Lid

5 years

Leaves shed in
4 autumns
(not weighed)

76.1 kg (169.2 lbs)
of trunk, roots,
and branches

Soil
89.9 kg (199.8 lbs)
dry weight

+

figure 7.1

J. B. van Helmont's willow experiments. Van Helmont concluded from these experiments that 164 lb of wood, bark, and root had arisen from water alone.

(a)   (b)   (c)

figure 7.2

Joseph Priestley's experiments showed that (a) a candle burning in an airtight jar went out, (b) a mouse died if kept in an airtight jar, and (c) the mouse lived if a plant were placed in the airtight jar with the mouse. Today, we explain Priestley's results by saying that plants use CO_2 produced by combustion or exhaled by animals and that animals inhale and use the O_2 released by plants.

Today we explain Priestley's results by saying that plants use CO_2 produced by combustion or exhaled by animals and that animals inhale and use the O_2 released by plants.

Priestley's experiments offered the first logical explanation of how air remained "pure" and able to support a mouse despite the burning of candles and the breathing of animals. Although he did not realize it, Priestley's experiments were the first demonstration that plants produce oxygen. Nor did Priestley realize that light was essential for photosynthesis. Nevertheless, he received a medal and a citation that read, in part, as follows: "For these discoveries we are assured that no vegetable grows in vain . . . but cleanses and purifies our atmosphere." Priestley implicated oxygen (*dephlogisticated air*, as he called it) in photosynthesis by showing that green plants could renew air made bad by the breathing of animals. However, a serious problem appeared: others could not repeat Priestley's work. Indeed, even Priestley could not get the same results when he repeated his experiments (he had probably moved the experimental apparatus to a dark part of his lab). Seven years later, however, the Dutch physician Jan Ingenhousz confirmed Priestley's work and made an important addition by showing that light is necessary for plants to release oxygen (although, like Priestley, he knew nothing about oxygen at the time and explained it in other terms). Ingenhousz reported that plants in the dark "contaminate the air" and make it "harmful to animals." He also made a bold and accurate statement: "The sun by itself has no power to mend air without the concurrence of plants." Ingenhousz published a wonderfully titled book, *Experiments upon Vegetables, Discovering Their Great Power of Purifying the Common Air in the Sun-Shine, and of Injuring It in the Shade and at Night*, in which he also reported that only the green parts of a plant could photosynthesize and that plants, too, "injure" air when kept in darkness. Ingenhousz must have been rather alarmed by his findings, because he recommended that plants be removed from houses at night to avoid poisoning the occupants.

In 1782 a Swiss preacher and part-time scientist named Jean Senebier showed that photosynthesis depended on a particular kind of gas, which he called *fixed air* (and we call CO_2). Senebier also claimed that this air (CO_2) produced by animals and plants in darkness stimulated production of purified air (O_2) by plants in light. Thus, by the late 1700s biologists knew that at least two gases participate in photosynthesis. Work done by Lavoisier and others (especially P. S. Laplace, a French mathematician) showed that these gases were CO_2 and O_2.[1] Ingenhousz adapted the ideas of Lavoisier and suggested that plants do not just exchange "good air" for "bad air." Rather, he suggested that plants absorb carbon from carbon dioxide, "throwing out at that time the oxygen alone, and keeping the carbon to itself as nourishment." Ingenhousz's work was extended in 1804 by the Swiss botanist and physician Nicholas de Saussure, who noted that approximately equal volumes of CO_2 and O_2 are exchanged during photosynthesis. He added the final component of the overall photosynthetic reaction when he showed that photosynthesis requires water:

carbon dioxide + water + light

$$\downarrow$$

organic material + oxygen

By this time, many botanists were studying photosynthesis. Among the most clever was T. W. Engelmann, who in 1883 designed an elegant experiment that simultaneously studied the light requirements and biochemistry of photosynthesis (fig. 7.3). Engelmann studied photosynthesis in *Spirogyra*, a green alga

1. Lavoisier was a French chemist who named oxygen and hydrogen and wrote the first modern textbook of modern chemistry. He was also a tax farmer—a term used to designate a tax collector empowered to make a profit. Although Lavoisier used his profits to fund his research, such activities were not popular with the general public. Soon after the French Revolution, Lavoisier's career ended at the guillotine.

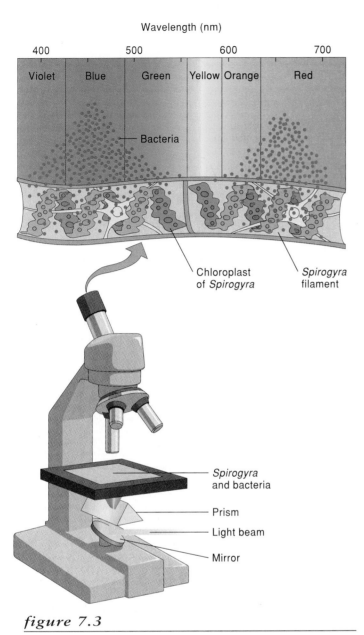

Wavelength (nm)

400   500   600   700

| Violet | Blue | Green | Yellow | Orange | Red |

Bacteria

Chloroplast of *Spirogyra*

Spirogyra filament

Spirogyra and bacteria

Prism

Light beam

Mirror

figure 7.3

T. W. Engelmann's experiment. Engelmann illuminated a filament of *Spirogyra* with light dispersed into a spectrum by a prism inserted into a beam of light. Oxygen-requiring bacteria gathered in the blue and red light, suggesting that these colors of light are most effective for photosynthesis.

that has a long, spiral chloroplast in each of its cells. He placed the alga in a drop of water containing an oxygen-requiring bacterium and used a prism to illuminate different parts of the chloroplast with different colors of light. When Engelmann viewed the *Spirogyra* through a microscope, he saw that the bacteria clustered around parts of the chloroplast illuminated by red and blue light. He concluded that red and blue light are most effective for producing oxygen during photosynthesis.

Other biologists began to follow up Ingenhousz's finding that light was required for the release of O_2. Julius Sachs reported that chlorophyll, the photosynthetic pigment, occurs in chloro-

plasts and that photosynthesis forms carbohydrates only in light. Thus, Sachs revised the overall reaction of photosynthesis:

$$n\text{CO}_2 + n\text{H}_2\text{O} + \textbf{light}$$

$$\downarrow \textit{chlorophyll}$$

$$(\text{CH}_2\text{O}) + n\text{O}_2$$

In this reaction, CH_2O is an abbreviation for starch or other carbohydrates. Further research showed that Sachs's conclusion was correct: there is no known exception to the linkage of chlorophyll with oxygen production.

By the turn of the twentieth century, most biologists accepted Ingenhousz's suggestion that the oxygen released during photosynthesis came from carbon dioxide. Others, however, questioned this assumption. In the 1920s, a Stanford University graduate student named C. B. van Niel began a study that would resolve this question and become a milestone in biological research. Van Niel studied a photosynthetic bacterium that uses H_2S as an electron source and deposits sulfur as a byproduct. Photosynthesis in these bacteria occurred as follows:

$$\text{CO}_2 + 2\text{H}_2\text{S} \overset{\textit{light}}{\longrightarrow} \text{CH}_2\text{O} + \text{H}_2\text{O} + 2\text{S}$$

Van Niel's work didn't attract much attention until he made a bold extrapolation: he asserted that the oxygen released during photosynthesis came from water, not carbon dioxide. His reasoning was based on analogies between the roles of H_2S and H_2O, and of O_2 and sulfur.

Sulfur bacteria:
$$\text{CO}_2 + 2\text{H}_2\text{S} \overset{\textit{light}}{\longrightarrow} \text{CH}_2\text{O} + \text{H}_2\text{O} + 2\text{S}$$

General equation:
$$\text{CO}_2 + 2\text{H}_2\text{X} \overset{\textit{light}}{\longrightarrow} \text{CH}_2\text{O} + \text{H}_2\text{O} + 2\text{X}$$

Green plants:
$$\text{CO}_2 + 2\text{H}_2\text{X} \overset{\textit{light}}{\longrightarrow} \text{CH}_2\text{O} + \text{H}_2\text{O} + \text{O}_2$$

Van Niel's conclusion that O_2 released by plants comes from water rather than CO_2 was tested in 1941 by Samuel Ruben and Martin Kamen, who exposed cultures of *Chlorella* (a green alga) to H_2O labeled with ^{18}O, a nonradioactive isotope of oxygen that they could detect with a mass spectrometer. Ruben and Kamen reasoned that if oxygen released during photosynthesis came from water, then the oxygen would be tagged with ^{18}O. Conversely, if the oxygen were derived from carbon dioxide, it would not be labeled with ^{18}O. Their results were striking: the oxygen, not the carbohydrates, were labeled with $^{18}O_2$:

$$\text{CO}_2 + 2\text{H}_2^{18}\text{O} \overset{\textit{light}}{\longrightarrow} \text{CH}_2\text{O} + \text{H}_2\text{O} + ^{18}\text{O}_2$$

These results confirmed van Niel's claim that oxygen released during photosynthesis comes from water, not carbon dioxide.

Further evidence for this conclusion was provided by Robin Hill and his coworkers, who discovered that isolated chloroplasts could release O_2 in the absence of CO_2 if given a suitable electron acceptor for the electrons removed from water. This light-driven splitting of water in the absence of CO_2 became known as the **Hill reaction** and showed that (1) whole cells are unnecessary for some of the reactions of photosynthesis, and (2) the light-driven release of O_2 during photosynthesis is not linked directly to the fixation of CO_2.

In 1951 botanists discovered that the electron acceptor in chloroplasts is $NADP^+$, a coenzyme that can accept electrons. Later studies showed that $NADP^+$ reduced during photosynthesis is used to reduce CO_2 during photosynthesis. Similarly, light drives the reduction of $NADP^+$, thereby converting it to an oxidizable fuel that can release energy to power the growth and development of plants.

c o n c e p t

Botanists have studied photosynthesis for hundreds of years. During that time, they have discovered that (1) photosynthesis uses carbon dioxide; (2) photosynthesis requires light and water and releases oxygen; (3) light for photosynthesis in plants is absorbed by chlorophyll; and (4) oxygen released during photosynthesis comes from water.

The Nature of Light

The sun is a giant thermonuclear reactor—each minute more than 120 million metric tons of solar matter are converted to radiant energy. Much of this energy is released as radiation that moves in rhythmic waves similar to those created by pebbles dropped in a pond (however, light moves as disturbances of electric and magnetic fields, not as disturbances of the surface of water; see fig. 7.4a). Eight minutes later, about two-billionths (5×10^{24} kcal; 1.73×10^{17} W) of this energy has traveled 160 million km and hit the earth's upper atmosphere. About one-third of the light hitting the atmosphere is reflected back to space (fig. 7.4b), and only about 1% of the light is used for photosynthesis. Since light powers photosynthesis, we must know something about light to understand photosynthesis.

Light is the part of the electromagnetic spectrum having wavelengths visible to the human eye (about 390–760 nm). Virtually all life depends on this light. Our understanding of light began about 300 years ago, when Isaac Newton showed that white light that passed through a prism, droplet of water, or soap bubble would separate into a band of colors that, if passed through another prism, could be recombined to form white light. Based on that experiment, Newton proposed that white light is actually a spectrum of colors ranging from violet to red (fig. 7.4b). Two hundred years later, in the 1860s, a Scottish mathematician named James Maxwell showed that the visible light that Newton separated into a spectrum of colors is only a small part of a much larger spectrum of radiation (fig. 7.4b).

In 1905, Einstein linked the ideas of Newton and Maxwell by proposing that light consists of packets of energy called **photons,** which are the smallest divisible units of light. The intensity (i.e., brightness) of light depends on the number of photons (i.e., the amount of energy) absorbed per unit of

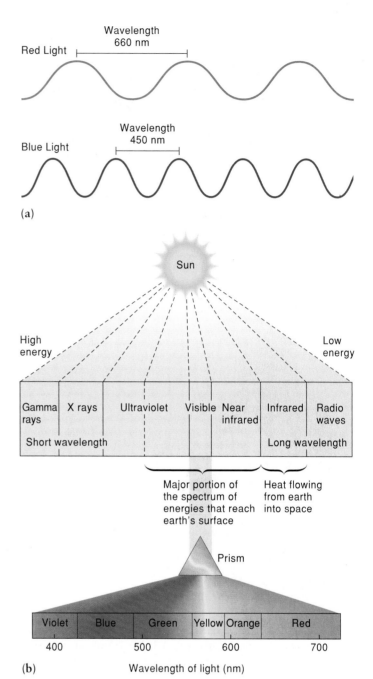

(a)

(b)

figure 7.4

Properties of light. (a) The wavelike nature of light. (b) Visible light is only a small part of the electromagnetic spectrum. Visible light consists of a rainbow of colors ranging from violet to red.

time. Each photon carries a fixed amount of energy that is determined by how the photon vibrates: the slower the vibration, the less energy carried by the photon (table 7.1). The distance moved by the photon during a complete vibration is referred to as the photon's **wavelength** (fig. 7.4a). Stated another way, the wavelength is the distance between vibrational crests of a photon. The wavelengths of visible light are measured in nanometers (nm), or billionths (10^{-9}) of a meter. We perceive the different wavelengths as different colors. For example, violet light

table 7.1

Radiant Energies of Different Wavelengths of Light

Wavelength (nm)	Color	Energy (joules μmol^{-1})
300		0.399
400	violet	0.299
450	blue	0.277
500	green	0.239
600	orange	0.199
700	red	0.171

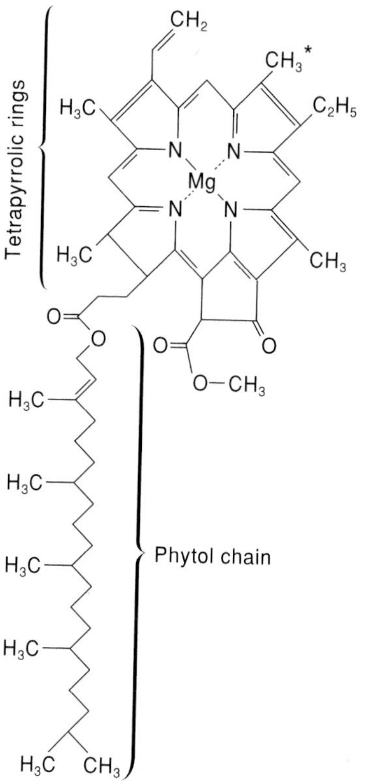

figure 7.5

Chlorophyll *a*, the primary pigment of photosynthesis. If an aldehyde (—CHO) replaces the methyl (—CH$_3$) group that is marked with an asterisk, the molecule is chlorophyll *b*.

has a wavelength of about 400 nm, which is about one-fortieth the thickness of this page. The energy of a photon is a **quantum** and is inversely proportional to the wavelength of the light: the longer the wavelength (i.e., the longer the distance traveled during a vibration), the less energy per photon (table 7.1).

Sunlight consists of about 4% ultraviolet radiation, 52% infrared radiation, and 44% visible light. Each of these kinds of light has different energy and affects organisms differently.

Ultraviolet radiation (UV) contains too much energy for most biological systems. Indeed, its high-energy photons often drive electrons from molecules, thus explaining why UV is also called **ionizing radiation.** UV breaks weak bonds and causes sunburn. It is absorbed by O$_2$, ozone (O$_3$), and glass. You won't get a tan by sitting in front of a glass window.

Infrared radiation (IR) doesn't contain enough energy per photon to be useful to living systems. Cells absorb IR radiation, but this energy is insufficient to excite electrons. Consequently, most of the energy of IR is converted immediately to heat. IR is absorbed by water and carbon dioxide but goes through glass. Thus, although you won't tan in front of a window, you will warm up.

Visible light contains just the right amount of energy for biological reactions such as photosynthesis. To have an effect, however, light must first be absorbed. Light is absorbed by pigments.

c o n c e p t

Virtually all life depends on light, which powers photosynthesis. Light moves in waves, and its energy is contained in packets called *photons.* The energy of a photon is inversely proportional to the wavelength of the light: the longer the wavelength, the less energy per photon. Sunlight consists of a spectrum of colors of light. Red light and blue light are the most effective for photosynthesis.

Pigments

Light striking an object is either reflected, transmitted, or absorbed. Only light that is absorbed can have an effect. Light is absorbed by molecules called **pigments,** which are colored because they transmit particular colors of light. Black pigments absorb all wavelengths of light, while white pigments absorb no wavelengths of light.

Pigments in Plants

Evolution repeats its inventions and, in the process, adapts existing mechanisms to new and different purposes. Many kinds of pigments are similar but have different functions. For example, certain pigments are derived from tetrapyrroles, a group of compounds present in all organisms. Tetrapyrroles are large rings made of four smaller rings, each of which is called a *pyrrole ring* and consists of four carbons and one nitrogen (fig. 7.5). The four pyrrole rings of the tetrapyrrole are linked by one-carbon bridges that sequester a metal atom. If the metal is iron, the pigment is a heme, such as in hemoglobin or cytochrome (iron in hemoglobin is always reduced as Fe^{2+}, while in cytochromes the Fe shuttles between Fe^{2+} and Fe^{3+}; see fig. 5.9e). If the metal is copper, the pigment is turacin, the pigment in many feathers. If the metal is magnesium, the pigment is chlorophyll (or the more reduced bacteriochlorophyll of bacteria), the primary pigment of photosynthesis (fig. 7.5).

Chlorophylls are hydrophobic pigments that occur in plants, algae, and all but one primitive group of photosynthetic bacteria. Unlike the color of hemoglobin, which is insignificant for the molecule's function as an oxygen carrier, the color of chlorophyll *is* significant. Chlorophyll absorbs maximally at wavelengths of 400–500 nm (violet-blue) and 600–700 nm (orange-red) (fig. 7.6). The synthesis of chlorophyll and several other pigments in plants is stimulated by light, which explains why plants grown in the dark (i.e., etiolated plants) contain no chlorophyll. Similarly, the light-stimulated synthesis of anthocyanin, another plant pigment, explains why apples are always redder on the sunny side of an apple tree.

The Lore of Plants

Chlorophyll, the photosynthetic pigment that makes plants green, was wildly popular as a deodorizer in the 1950s. Pepsodent marketed a toothpaste named Chlorodent in 1950 and soon thereafter started selling a chlorophyll-based dog food. Other chlorophyll-based products included gum, mouthwash, deodorant, diapers, popcorn, and cologne, and there were plans for chlorophyll-based salami and beer. Although the chlorophyll industry was damaged by tests done by the U.S. Food and Drug Administration showing that chlorophyll was a poor deodorizer, the folly of the craze was summarized by a magazine article that pointed out that although goats eat almost nothing but green grass, they nevertheless stink. Soon thereafter, chlorophyll returned to its everyday job.

There are several types of chlorophyll, the most important of which is **chlorophyll *a*,** the primary photosynthetic pigment. Chlorophyll *a* ($C_{55}H_{72}O_5N_4Mg$) is a grass-green pigment whose structure includes an atom of magnesium (Mg) (fig. 7.5). It occurs in all photosynthetic organisms except photosynthetic bacteria and absorbs maximally at 430 and 662 nm (fig. 7.6).

The relationship of light absorption by chlorophyll *a* versus wavelength is shown in figure 7.7. This **absorption spectrum** shows that chlorophyll *a* absorbs all visible light except green light; it reflects and transmits green light, and therefore appears green. This absorption spectrum for chlorophyll *a* closely matches the graph showing how photosynthesis varies with different wavelengths of light (fig. 7.7). This so-called **action spectrum** shows that the pattern of photosynthesis closely follows that of chlorophyll *a*, suggesting that chlorophyll *a* is the primary pigment in photosynthesis. However, the absorption spectrum of chlorophyll *a* does not *perfectly* match the action spectrum for photosynthesis. Therefore, we conclude that other pigments must be involved in photosynthesis. Those pigments are called *accessory pigments.*

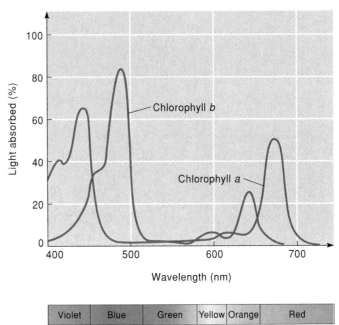

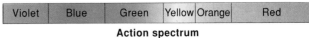

Absorption spectra of chlorophylls *a* and *b*.

figure 7.6

The absorption of light by chlorophylls *a* and *b*. Chlorophylls absorb maximally at wavelengths of 400–500 nm (violet-blue) and 600–700 nm (orange-red).

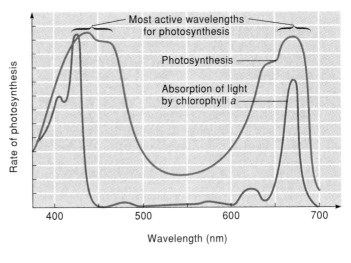

Action spectrum

figure 7.7

The absorption spectrum of chlorophyll *a* shows that chlorophyll *a* absorbs maximally at 430 nm and 662 nm. The rate of photosynthesis in different wavelengths of light is similar to the ability of chlorophyll *a* to absorb those wavelengths. The similarity of the action spectrum of photosynthesis and absorption spectrum of chlorophyll *a* suggests that chlorophyll *a* is the primary photosynthetic pigment and that light absorbed by chlorophyll *a* drives photosynthesis.

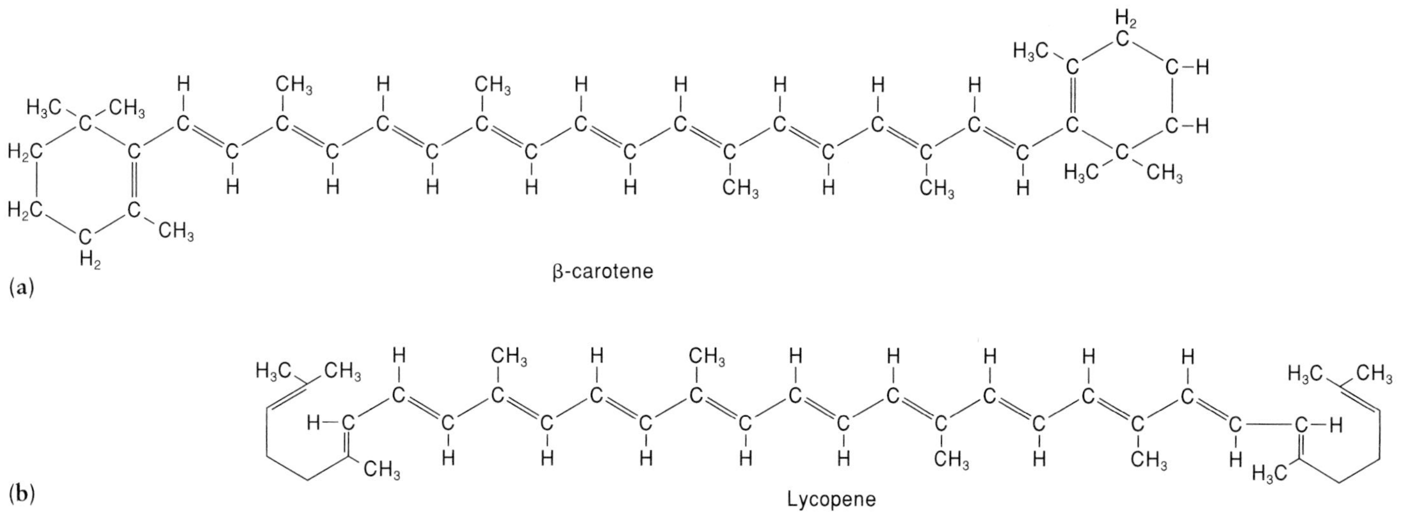

figure 7.8

Carotenoids. (a) Beta-carotene, a reddish-yellow carotenoid. (b) Lycopene, a carotenoid that colors tomatoes. Both beta-carotene and lycopene have the empirical formula of $C_{40}H_{56}$.

Accessory Pigments

Plants also contain a rainbow of other pigments. Those pigments extend the range of light useful for photosynthesis by absorbing photons not absorbed by chlorophyll *a* and transmitting that energy to chlorophyll *a*. The most common of these **accessory pigments** in plants are chlorophyll *b* and carotenoids.

Chlorophyll *b* ($C_{55}H_{70}O_6N_4Mg$) is a bluish-green pigment that absorbs maximally at 453 and 642 nm (figs. 7.5, 7.6). It occurs in all plants, green algae, and some prokaryotes. Plants usually contain about half as much chlorophyll *b* as chlorophyll *a*.

Carotenoids are accessory pigments that occur in all photosynthetic organisms. They contain forty carbons, are fat-soluble, and often contain no oxygen (fig. 7.8). Carotenoids absorb maximally at wavelengths between 460 and 550 nm; therefore, they are red, orange, and yellow. Carotenoids are chemically unrelated to chlorophylls and consist of carbon rings linked by long carbon chains having alternating single and double bonds.

Like other accessory pigments such as chlorophyll *b*, carotenoids extend the range of photosynthesis by absorbing light that is not absorbed by chlorophyll *a*. Carotenoids also protect plants against photo-oxidation, which occurs when excited chlorophyll transforms oxygen into high-energy radicals. These radicals can attract hydrogen from nearby molecules, thereby destroying the molecules and killing the cells. Mutants that lack carotenoids are susceptible to such radicals, which explains why they are bleach-white when grown in light and soon die (fig. 7.9). Many herbicides kill plants by blocking the synthesis of carotenoids, thereby causing the plant to photo-oxidize itself.

figure 7.9

Carotenoids protect against photo-oxidation. The plants on the left were treated with an herbicide that blocks synthesis of carotenoids. Consequently, the plants became photo-oxidized (bleached) and died when grown in light. Untreated plants (right) make carotenoids and therefore are not photo-oxidized by light.

The most common carotenoid is **beta-carotene**, a reddish-yellow pigment consisting of two six-carbon rings connected by an eighteen-carbon chain (fig. 7.8). Beta-carotene absorbs maximally at wavelengths between 400 and 500 nm. When split in half, beta-carotene becomes two molecules of vitamin A, which is a precursor of retinal, a pigment essential for human vision. Carotenoids occur throughout the plant kingdom and produce the colors of tomatoes, carrots, squash, bananas, avocados, and autumn's colorful leaves.

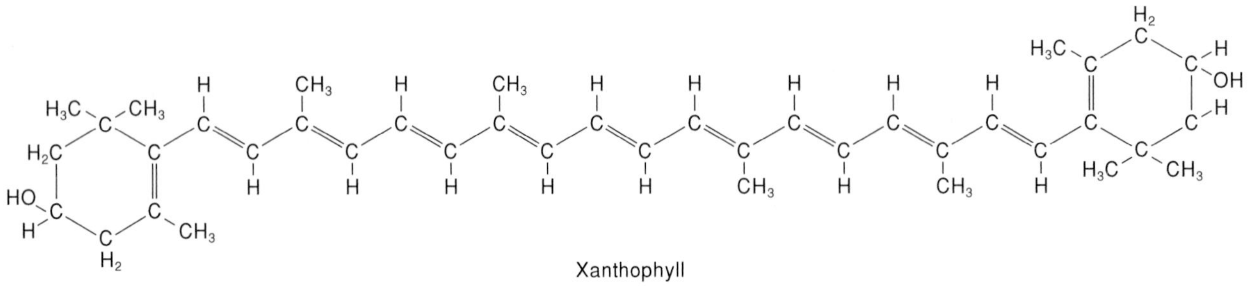

figure 7.10

Xanthophyll. Xanthophylls are abundant in carrots, leaves, and many algae.

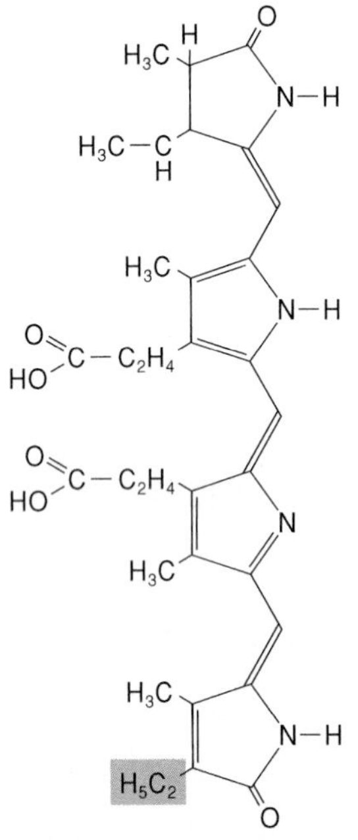

Phycocyanin

Phycoerythrin component

figure 7.11

Phycocyanin and phycoerythrin are accessory pigments in red algae and cyanobacteria. Phycoerythrin is obtained by exchanging the highlighted groups.

Unlike chlorophyll, carotenoids also occur in animals. Animals cannot make carotenoids, but they can metabolize and use carotenoids from plants. For example, carotenoids color egg yolks, flamingo feathers, insect wings, and the black ink released by squid, as well as the bodies of corals, fish, and amphibians. The dazzling array of colors of carotenoids includes blue, green, red, violet, gray, chocolate, and black; they result largely from proteins that are attached to the carotenoid. When heated, this protein breaks off and frees the carotenoid (this accounts for the red pigment released when we cook lobsters).

The Lore of Plants

Tomatoes are red because of the carotenoid lycopene, the same pigment that colors the flesh of pink grapefruits and watermelon. White tomatoes lack carotenoids, and yellow tomatoes make yellow carotenoids but no red carotenoids. There is even a variety of tomato that ripens but does not break down its chlorophyll. This variety, which stays confusingly green, is called Evergreen.

Oxidizing carotenes produces **xanthophylls,** which are red and yellow pigments in tomatoes, carrots, leaves, algae (e.g., fucoxanthin in brown algae), and photoautotrophic bacteria (fig. 7.10). One xanthophyll, zeaxanthin, is the pigment that absorbs the blue light that induces curvature of some plants toward blue light (you'll learn more about this phenomenon in Chapter 19). Xanthophylls are less efficient at transferring energy during photosynthesis than are carotenoids.

Other accessory pigments include chlorophylls *c* and *d,* phycoerythrin (a red pigment), and phycocyanin (a blue pigment; fig. 7.11). Different forms of phycoerythrin and phycocyanin occur in red algae and cyanobacteria.

c o n c e p t

Light is absorbed by pigments. Chlorophyll *a* is the primary pigment for photosynthesis and occurs in all photosynthetic organisms except photosynthetic bacteria. Accessory pigments such as carotenoids and chlorophyll *b* absorb light that chlorophyll *a* cannot absorb, thereby extending the range of light useful for photosynthesis.

Making and Destroying Pigments

Most plants regularly destroy and resynthesize their chlorophyll. Indeed, each year the 300 million tons of chlorophyll on earth are turned over an average of three times. When the rate of chlorophyll synthesis lags behind that of its breakdown, leaves begin a photochemical suicide. The decreasing amounts

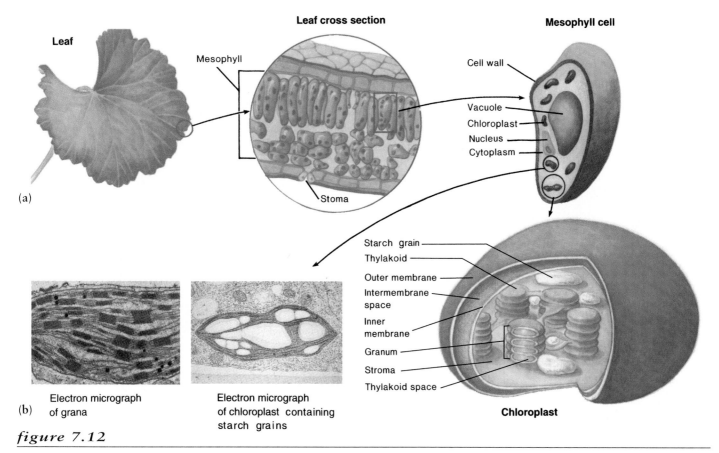

Leaf

Leaf cross section

Mesophyll

Stoma

Mesophyll cell

Cell wall

Vacuole

Chloroplast

Nucleus

Cytoplasm

(a)

Starch grain

Thylakoid

Outer membrane

Intermembrane space

Inner membrane

Granum

Stroma

Thylakoid space

Chloroplast

(b) Electron micrograph of grana

Electron micrograph of chloroplast containing starch grains

figure 7.12

In plants, photosynthesis occurs in chloroplasts. Light is absorbed and converted to chemical energy in thylakoids and grana; this chemical energy is then used in the stroma to make sugars.

of chlorophyll no longer mask the other pigments, which results in the colors of autumn that are praised by poets and songwriters. In many trees, this "turning of the leaves" is spectacular. Tourists who visit New England to see the carotenoids of trees generate a billion-dollar-per-year industry, most of which happens in just 3 weeks. States such as New Hampshire even have a Fall Foliage Hotline to keep people up-to-date on the breakdown of chlorophyll.

Chloroplasts

The green color of leaves is due to solar chemical factories called **chloroplasts,** the site of photosynthesis in eukaryotes (fig. 7.12). Chloroplasts in plants are usually shaped like footballs, with a diameter of 5–10 μm and a depth of 3–4 μm. Most photosynthetic cells have 40–200 chloroplasts, which amounts to about 500,000 chloroplasts per square millimeter of leaf area. To help you put the size of a chloroplast in perspective, consider that it would take about 2,000 chloroplasts to stretch across your thumbnail.

Each chloroplast is surrounded by two membranes that enclose a gelatinous matrix called the **stroma** (fig. 7.12). The stroma contains ribosomes, DNA, and enzymes that make the sugars involved in photosynthesis. Suspended in the stroma are neatly folded sacs of membranes called **thylakoids** (fig. 7.12). These membranes are unique to chloroplasts. In some parts of chloroplasts, 10–20 thylakoids are stacked into **grana** (fig. 7.12). Thylakoids and grana contain chlorophyll and are where light is absorbed during photosynthesis. Most of the proteins in chloroplasts are coded by nuclear genes, produced in the cytoplasm, and then shipped to the chloroplast.

Complexes of Pigments in Chloroplasts

In the early 1950s Robert Emerson at the University of Illinois made an unusual observation: he noted that red light having wavelengths exceeding 690 nm was ineffective for photosynthesis despite the fact that it was absorbed by chlorophyll. However, when this light was supplemented by light having a shorter wavelength, photosynthesis occurred faster than it did in either light alone. This effect, which became known as the *Emerson enhancement effect,* showed that plants contain two light-harvesting systems.

In all but the most primitive bacteria, light is captured by a network of chloroplast pigments arranged in aggregates on thylakoids. These aggregates, called **antennae complexes** (fig. 7.13), are anchored in a protein matrix and consist of proteins, about 300 molecules of chlorophyll *a,* and about 50 molecules

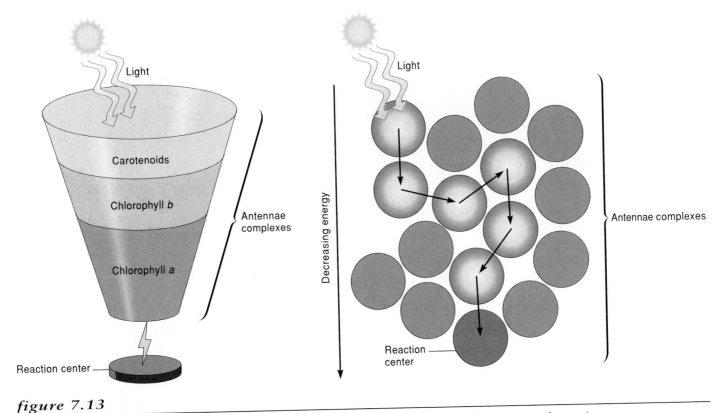

figure 7.13

The light-harvesting system in chloroplasts functions like a funnel; it collects photons and passes their energy to the reaction center.

of carotenoids and other accessory pigments that gather light. Energy absorbed by antennae complexes flows energetically "downhill" to a special pair of energy-collecting molecules of chlorophyll *a* and associated proteins called a **reaction center** (fig. 7.13). Although reaction centers comprise less than 1% of the chlorophyll in plants, chlorophyll *a* in the reaction center is the electron acceptor that participates directly in photosynthesis; all other photosynthetic pigments function as antennae.

There are two kinds of reaction-center chlorophylls, each having light-harvesters in their antennae. One of these chlorophylls absorbs maximally at 700 nm and is called **P700** (for pigment 700) (fig. 7.15). The complex containing P700 is called *Photosystem I*. The core of Photosystem I consists of about forty molecules of chlorophyll *a*, several molecules of beta-carotene, lipids, four manganese, one iron, several calcium, several chlorine, two molecules of plastoquinone, and two molecules of pheophytin, a colorless form of chlorophyll *a*.

The chlorophyll *a* molecule in *Photosystem I* resonates from energy transmitted by about 100 molecules of chlorophyll *a* and *b* (in a 4:1 ratio) bound to nuclear-coded proteins in the antennae. Three peripheral polypeptides bind calcium and chlorine, which explains why these nutrients are essential for photosynthesis. The chlorophyll *a* molecule in Photosystem II is identical to that of Photosystem I, but it is associated with different proteins; in Photosystem I, the reaction center is bound to a large protein (molecular weight 110,000), while that in Photosystem II is bound to a smaller protein (molecular weight 47,000). The associated protein in Photosystem II shifts the maximal absorption to about 680 nm. Consequently, the reaction center in Photosystem II is called **P680** (fig. 7.15). P680 resonates from energy transmitted by about 250 molecules of chlorophyll *a* and *b* (in equal numbers). Its core includes numerous xanthophylls (but no beta-carotenes) and integral proteins coded in the nucleus and made in the cytoplasm.

What Happens When Pigments Absorb Light?

Light must be absorbed by a pigment before it can have an effect. When light is absorbed, the energy of its photons is captured by the pigment and is used to boost the energy of electrons—that is, it changes the configuration of electrons by boosting them to higher-energy orbitals. These excited electrons can have several fates (fig. 7.14):

- The energy can be released as heat (i.e., molecular motion; fig. 7.14a). This is what is happening to the electrons in the pigment (i.e., the ink) that you're looking at now.

- The energy can be released as an afterglow of light via a process called **fluorescence**. Light released during fluorescence has a longer wavelength (and therefore, less energy) than the light that excited the pigment. Chlorophyll fluoresces deep red. Isolated chlorophyll fluoresces because no molecules are nearby to accept the energized electrons (fig. 7.14b).

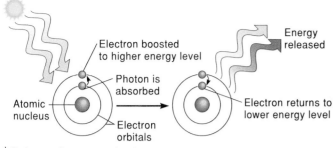

(a) Release of energy as heat

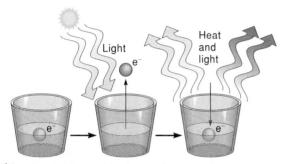

(b) Isolated chlorophyll molecule: Release of energy as heat and light

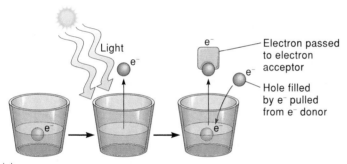

(c) Chlorophyll in intact chloroplast: Passage of energy to other molecules

figure 7.14

Fates of energy in energized electrons of chlorophyll. (a) Release of energy as heat. (b) Isolated chlorophyll: electrons that had been raised fall to original energy level; light energy absorbed is released as heat and light (fluorescence). (c) Intact chloroplast: energy passes to electron acceptors.

- The energy can be passed to a neighboring molecule. Chlorophyll in thylakoids is surrounded by other molecules that trap the energy of the excited electrons (fig. 7.14c).

The light-driven reactions of photosynthesis, which pass the energy of photons to other molecules, occur on photosynthetic membranes. In plants and algae, these photosynthetic membranes are enclosed in chloroplasts, the evolutionary descendants of photosynthetic prokaryotes (which lack chloroplasts but do have membranes similar to thylakoids;

see reading 3.2 on p. 66). In these organisms, chlorophyll is in membranes, in vesicles, or—as in cyanobacteria—in parallel stacks of flattened sacs.

Photophosphorylation: Chemiosmosis in Chloroplasts

Photons absorbed by pigments energize electrons. Plants pass that energy through a series of molecules called an *electron transport chain*. This transfer of electrons involves reduction and oxidation, or *redox*, reactions: the electron donor is oxidized as the electron acceptor is reduced. For electrons to flow "down" this chain, each receiver must attract the electrons more strongly than does the donor.

The potential energy of the electrons drops at each step of the electron transport chain. Plants couple this exergonic flow of electrons to an endergonic reaction that makes ATP. This light-driven production of ATP from electron transport is called **photophosphorylation** and depends upon a proton gradient (fig. 7.15). Photons captured by pigments excite electrons, which are shuttled along carriers embedded in the thylakoid membrane. When these electrons reach transmembrane H^+-pumps, their arrival induces transport of H^+ across the membrane, thereby creating the proton gradient that drives the synthesis of ATP.

As you've probably noticed, chemiosmosis in chloroplasts (photophosphorylation) resembles chemiosmosis (oxidative phosphorylation) in mitochondria (see Chapter 6). In both cases, (1) protons are pumped through a series of carriers that are progressively more electronegative, and (2) the free energy released by electron transport generates and maintains a proton gradient across a membrane. However, there are some differences. For example, in mitochondria, energized electrons moving through the electron transport chain are extracted by the oxidation of food; in chloroplasts, light energizes the electrons, and no food is necessary. In mitochondria, the inner membrane pumps protons from the matrix out to the intermembrane space; this is the reservoir of protons that powers the synthesis of ATP. In chloroplasts, electron carriers in the thylakoid pump protons from the stroma into the lumen (i.e., thylakoid space), which is the proton reservoir. This light-driven pumping of protons into the lumen decreases the pH there to about 5, while the pH of the stroma increases to 8 (i.e., a thousandfold difference in the concentration of protons). Light is required to generate this proton gradient; the difference in pH across the thylakoid membrane disappears quickly in the dark. The pH gradient is discharged during the synthesis of ATP as protons flow into the stroma across channels whose catalytic heads protrude like knobs into the stroma (fig. 7.15). Coupling-factor proteins in these channels include ATP synthase, an enzyme that harnesses the flow of protons to make ATP. The ATP produced by the ATP-synthase complexes is released into the stroma, where it is used to make carbohydrates.

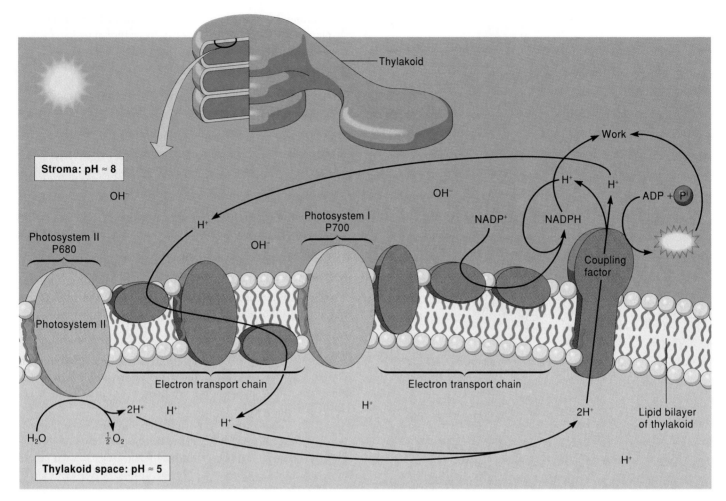

figure 7.15

The light-driven transport of electrons during photosynthesis creates a proton gradient across the thylakoid membrane. This proton gradient drives the formation of ATP.

c o n c e p t

Photosynthesis in eukaryotes occurs in chloroplasts. Chloroplasts consist of a gelatinous matrix called the *stroma* and stacks of membranes (thylakoids) called *grana*. Aggregates of pigments in grana absorb light and funnel its energy to special pairs of chlorophyll and proteins called *reaction centers*. The energy of the light is used to energize electrons, which, in turn, are used to pump protons from the stroma into the thylakoid space. The resulting pH gradient is discharged during the synthesis of ATP as protons flow back into the stroma.

The Photochemical and Biochemical Reactions of Photosynthesis

In 1905 F. F. Blackman reported an interesting set of experiments showing that photosynthesis is more complex than most botanists had realized. Blackman reported that in dim light,

increasing the intensity of light increased the rate of photosynthesis but increasing the temperature did not. Thus, these reactions were light-dependent and temperature-independent. However, in bright light, increasing the temperature increased the rate of photosynthesis, while further increases in light intensity did not. Thus, these reactions were temperature-dependent and light-independent.

Blackman's experiments provided the first evidence that photosynthesis is a two-stage process (fig. 7.16). The light-dependent reactions became known as the **photochemical** (or **light) reactions** of photosynthesis; these reactions are insensitive to changes in temperature. The temperature-dependent reactions became known as the **biochemical** (or **light-independent** or **dark) reactions** of photosynthesis because of their insensitivity to light; *dark* means only that light is not directly involved in the reactions. Indeed, these reactions will occur in the light or dark, provided that ATP, NADPH, and CO_2 are present. The photochemical reactions oxidize water, release oxygen, and produce ATP and reduced NADP+ that are used in the biochemical reactions, which reduce carbon dioxide to carbohydrate (fig. 7.16).

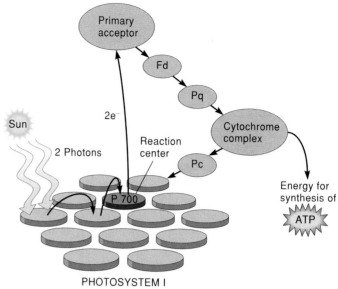

figure 7.17

Cyclic flow of electrons. Electrons ejected from Photosystem I are passed through a series of electron carriers: ferredoxin (Fd, an iron-containing protein), plastoquinone (Pq, a mobile electron carrier similar to ubiquinone in the electron transport chain of mitochondria), a complex of two cytochromes, and plastocyanin (Pc, a copper-containing protein). The cycle is completed when plastocyanin returns the electron to P700. The loss of potential energy at each step is used to create a proton gradient that, in turn, is used to generate ATP (see fig. 7.15).

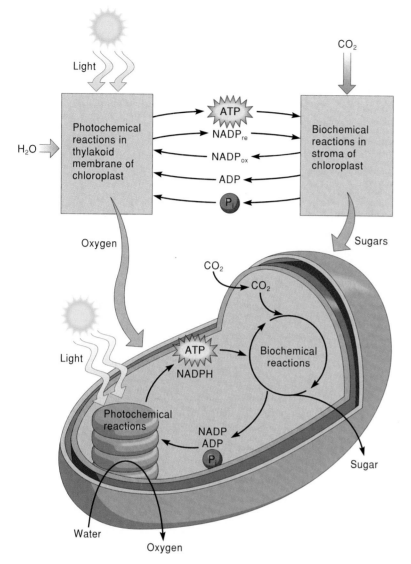

figure 7.16

Photosynthesis consists of photochemical and biochemical reactions. The photochemical reactions convert light energy to chemical energy: ATP and NADPH. The biochemical reactions use the ATP and NADPH produced by the photochemical reactions to reduce CO_2 to sugars. The photochemical reactions occur on thylakoid membranes, whereas the biochemical reactions occur in the stroma.

The Photochemical Reactions of Photosynthesis

Cyclic Electron Flow

Photosynthetic prokaryotes use only **cyclic photophosphorylation,** a process geared to energy production rather than to biosynthesis. Photophosphorylation is relatively simple and involves only Photosystem I (fig. 7.17). Electrons energized by light are recycled through the electron transport chain and return to the same reaction center from which they originated. Electron carriers move the electrons only in the company of

hydrogen ions, which are transported across the membrane as the electrons move to the next member of the chain. In the process, the energy-releasing flow of electrons is coupled to the energy-requiring manufacture of ATP from ADP and Pi. The coupling mechanism is chemiosmosis, as in mitochondria. At each step of the electron transport chain, electrons lose potential energy, finally returning to their ground-state energy in P700. Absorption of more light excites the reaction center again, restarting the cycle. Electrons that return to P700 have only about half of the energy they had when they left; the remainder of the energy is the photosynthetic payoff.

Cyclic phosphorylation short-circuits production of reduced NADP+ by shuttling the energized electrons from P700 to ferredoxin and then to plastoquinone. As the electrons again move down the electron transport chain, ATP forms. This process generates no oxygen or reduced NADP+.

Noncyclic Electron Flow

Cyclic photophosphorylation alone occurred for more than a billion years and is geared to energy production, not biosynthesis. It suffices for prokaryotes that use molecules such as H_2S as an electron source. However, plants strip electrons from water. This requires a huge change in biochemistry, because the voltage from one light-energized reaction center of P700 doesn't have enough energy to strip electrons from water and give them to carbon

Karin Brueschweiler
Arizona State University

My research is to find genes encoding polypeptides that are involved in Photosystem I assembly. Photosystem I consists of at least 14 subunits, 5 of which are encoded by chloroplast genes and 9 by nuclear genes. The functions of the chloroplast-encoded proteins of Photosystem I are well characterized, but little is known about the functions of the nuclear gene products. My project involves tagging and identifying these unknown nuclear genes by randomly inserting a known plasmid into their nuclear DNA.

The organism I use is a mutant of *Chlamydomonas reinhardtii*. This unicellular green alga cannot grow on a culture medium that lacks the amino acid arginine. I randomly inserted a plasmid, called *arg-7*, that contains the gene for arginine synthesis into the nuclear DNA of the mutants. Algae cells that are successfully transformed can subsequently grow in the absence of supplemental arginine and also contain a nuclear gene that is interrupted by the *arg-7* plasmid (insertional mutants). To identify a nuclear Photosystem I gene, I look for mutants that are nonphotosynthetic and fail to assemble the Photosystem I complex.

Currently, I have isolated five mutants that are impaired in accumulation of Photosystem I. My future research will focus on mapping, identifying, and cloning the tagged genes.

From: Karin Brueschweiler, Arizona State University. Reprinted by permission.

dioxide. Plants overcame this problem by grafting onto the bacterial system a second, more powerful photosystem that enabled a linear, **noncyclic electron flow** (fig. 7.18). These two pumps are connected in series, much as we use two batteries to increase the voltage in a flashlight. The second pump is Photosystem II, whose unique arrangement of pigments allows it to harvest shorter wavelengths of light. This photosystem acts first in what is now called the *Z scheme*, a model for noncyclic flow of electrons (and energy) during the photochemical reactions of photosynthesis (see fig. 7.18). Both cyclic and noncyclic photophosphorylation occurs in green plants, algae, and photosynthetic bacteria, and involve two photosystems that (1) make ATP in the electron transport chain linking Photosystems I and II, and (2) reduce $NADP^+$. Plants and algae can switch between cyclic and noncyclic electron flow as needed to balance metabolic demands for ATP and NADPH.

A Summary of the Photochemical Reactions

The photochemical reactions of photosynthesis start at P680 and P700, and end at $NADP^+$. The pathway for electron flow in this process is shown in figure 7.18a and is referred to as the *Z scheme*, because the pathway resembles a Z on its side. These reactions are linear rather than cyclic and produce ATP and NADPH. In figure 7.18 follow how they occur:

- A photon hits the end of the reaction center nearest the inner surface of the membrane, exciting an electron. This electron carries its energy to the other end of the reaction center (the end nearest the outer surface of the membrane).

- Three more photons cause other electrons to do the same thing. Thus, four photons strike P680 of Photosystem II, which functions like a tiny capacitor by

creating a separation of charge. Although the reaction center captures 98%–99% of the photons that it absorbs, only about half of their energy is stored in the charge separation. Thus, Photosystem II is only about half as efficient as a battery.

- Like all energy-rich substances, the excited electrons are unstable and release much of their energy as heat. The electrons ejected from P680 leave "holes" that are filled by electrons from a compound called Z, which gets its electrons from water. Thus, Photosystem II removes electrons from water, releases oxygen, and uses the electrons to replace those ejected by photons absorbed by the reaction center. All of this occurs in less than a billionth (10^{-9}) of a second.

- Splitting two molecules of water releases a molecule of O_2 and four electrons, all at once. However, only one electron at a time can be accepted by P680. The electrons stripped from water are managed by manganese, whose stable oxidation states, ranging from +2 to +7, suit it ideally for its function as a charge accumulator. At saturating intensities of light, one molecule of oxygen is released per about 2,500 molecules of chlorophyll.

- Within a few trillionths of a second, the electrons ejected from P680 of Photosystem II reduce pheophyton, a pigment that accepts electrons from chlorophyll. The resulting separation of charge widens as the electrons move across the thylakoid membrane and are accepted by plastoquinone (Pq), which is embedded on the outer, stromal side of the membrane.

- The electrons then descend an electron transport chain of molecules (such as cytochromes) in thylakoids linking

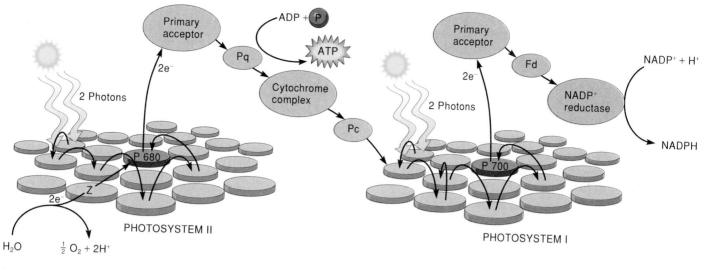

(a)

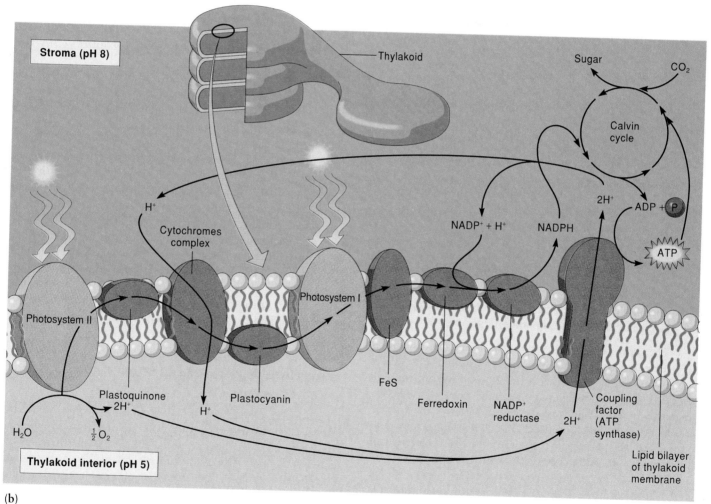

(b)

figure 7.18

(a) Summary of noncyclic flow of electrons. When Photosystems I and II are illuminated, electrons flow from water to NADP⁺. Electrons ejected from P680 are replaced by electrons from water; a by-product of this process is O_2. The energized electrons from P680 flow through an electron transport chain to P700, generating an ATP in the process. Illumination of P700 energizes the electrons again; these electrons are used to reduce NADP⁺.
(b) Photosystems I and II are linked in series in the thylakoid membrane. Noncyclic flow of electrons produces ATP, reduced NADP⁺, and O_2.

Photosystem II and Photosystem I,[2] which are connected in series in thylakoid membranes. As the electrons move down this chain, they form a proton gradient that generates ATP. Herbicides such as monuron (CMU) and diuron (DCMU) kill plants by blocking this electron transport.

- The final electron acceptor in the electron transport chain is P700, the reaction center of Photosystem I. There, the electrons have a higher energy than when they left P680 a split second earlier. At P700, four more photons absorbed by antennae transfer their energy to the reaction center, where the energy is used to eject electrons from P700's electron donor, located on the thylakoid space side of the membrane. These energized electrons cross the membrane and reduce ferredoxin, a small, iron-containing protein that accepts electrons from Photosystem I. Ferredoxin then reduces $NADP^+$ to NADPH. The herbicides diquat and paraquat kill plants by accepting electrons from Photosystem I before ferredoxin. They transfer these electrons to oxygen and make superoxide (O_2^-), a free radical that destroys unsaturated fatty acids and kills the cells.

- Because ferredoxin is on the outer, stromal side of the membrane, NADPH forms in the stroma, where it is used in biochemical reactions to reduce carbon dioxide to carbohydrate, the basic nutrient of life.

- The eight photons[3] used in the photochemical reactions make three molecules of ATP and reduce two molecules of $NADP^+$.

$$2H_2O + 2NADP^+ + 3ADP + 3Pi$$
$$\downarrow 8 \; photons$$
$$O_2 + 2NADPH + 3ATP + 4e^- + 2H^+$$

- These ATPs and NADPHs are used to reduce CO_2 in the biochemical reactions of photosynthesis.

The entire system of electron flow during photosynthesis has been compared to water flowing between hills and valleys in a hydroelectric system: as water flows downhill through successive dams and turbines, its potential energy is converted to electrical energy that can be stored in batteries and later be used to do work. In chloroplasts, the electrons are analogous to the water, the photons to the pumps, the electron carriers to the turbines, and ATP and NADPH to the batteries.

The evolution of Photosystem II forever changed life on earth. For example, it made photosynthetic organisms

---

2. These photosystems were named in their order of discovery, not their organization in chloroplasts. This explains why electrons move from Photosystem II to Photosystem I, rather than from I to II.

3. Theoretically, plants need only eight photons to reduce one molecule of carbon dioxide. Most plants need 15–20 photons to do this in field conditions. In ideal lab conditions, plants need 9–12 photons to reduce one CO_2.

independent of H_2S from decaying organic compounds as sources of electrons. Moreover, it liberated oxygen into the atmosphere, thereby allowing the evolution of aerobic respiration.

c o n c e p t

The photochemical reactions of photosynthesis convert light-energy into chemical energy. In plants, the photochemical reactions oxidize water, produce ATP, reduce $NADP^+$, and involve two photosystems connected in series. The ATP and reduced $NADP^+$ made in the photochemical reactions are used in the biochemical reactions of photosynthesis to reduce carbon dioxide to carbohydrate.

Are Both Photosystems Necessary?

For nearly 30 years, botanists have been confident that they understand the major aspects of photosynthesis. Specifically, botanists believe that plants require two photosystems: Photosystem I is essential for producing the reduced $NADP^+$ used to reduce CO_2, whereas Photosystem II oxidizes water and releases O_2.

But as this book went to press in early 1997, a group of botanists at Oak Ridge National Laboratory in Tennessee questioned this long-held idea. These botanists studied mutants of *Chlamydomonas* (a single-celled alga) in which Photosystem I does not operate. Nevertheless, the algae fix CO_2, release O_2, and grow. These results suggest that the mutants can perform photosynthesis without Photosystem I (i.e., with only Photosystem II). While not all researchers are ready to accept this conclusion, the study may force botanists to consider alternatives to the model presented in figure 7.18. The study may also help us understand how photosynthesis evolved. The mutant grows better in the absence of oxygen, suggesting that Photosystem II evolved first under anaerobic conditions that existed before photosynthesis filled the atmosphere with O_2. That is, the earliest forms of photosynthesis utilized only Photosystem II; Photosystem I may have evolved later to stabilize Photosystem II and help the plants cope with O_2.

The Biochemical Reactions of Photosynthesis

The biochemical reactions of photosynthesis reduce carbon dioxide to carbohydrate, a storage depot of chemical energy. Straightforward as this may sound, the actual mechanisms of photosynthesis have been difficult to explain. The most important contribution to our understanding of how plants reduce carbon dioxide to carbohydrate was provided by Melvin Calvin and his colleagues at the University of California at Berkeley in the early 1950s. These botanists studied photosynthesis by using a product of World War II: ^{14}C, or radioactive carbon (^{14}C has identical chemical properties to the common ^{12}C but is radioactive because it has more neutrons). Calvin first exposed a dense culture of *Chlorella* (a green alga) growing in a lollipoplike chamber to radioactively labeled CO_2 (fig. 7.19). Then, using

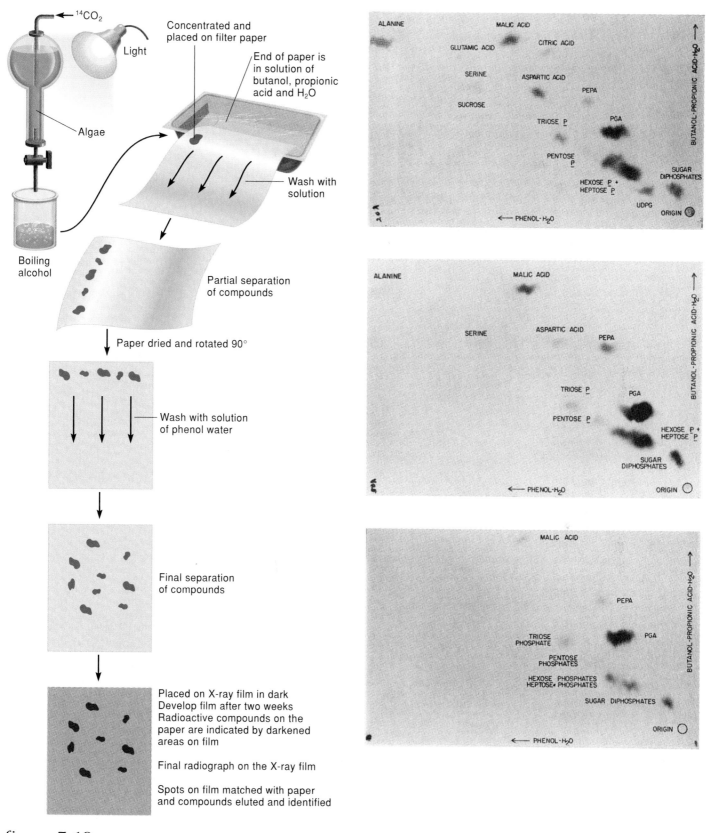

figure 7.19

A summary of Calvin's experiments and results. Calvin exposed algae to radioactive CO_2 for various times (shown here are results for 2, 7, and 60 sec.) and then killed the algae at different times by placing them in boiling ethanol. Calvin then applied an extract of the algae onto pieces of paper and ran solvents through the paper from one direction. This process, called *chromatography*, separated the compounds in the extract: different compounds moved different distances in the solvent moving through the paper. Calvin then placed film against the pieces of chromatography paper. The radioactive compounds in the original extract—now at different points on the paper—exposed the film. Calvin then identified the compounds by repeating the process with known standards. This technique is known as *autoradiography*. Calvin used chromatography and autoradiography to decipher the biochemical reactions of photosynthesis.

chromatography (to separate the compounds) and autoradiography (to identify the separated compounds), Calvin traced the path of the ^{14}C as it moved from CO_2 to carbohydrate (fig. 7.19). When Calvin exposed the alga to labeled CO_2 for 60 seconds, he found that many compounds in the alga were labeled with ^{14}C. Calvin reasoned that he needed to modify the experiment and expose the alga to the $^{14}CO_2$ for less time. When he repeated the experiment with a 7-second exposure, most of the radioactive carbon appeared in 3-phosphoglyceric acid (3-PGA), a three-carbon molecule (fig. 7.19). This suggested to Calvin that the molecule that joined with CO_2 (i.e., the CO_2 acceptor) was a two-carbon compound, since combining such a two-carbon acceptor with CO_2 (a one-carbon molecule) would produce a three-carbon compound. Calvin searched for this two-carbon acceptor but did not find it.

Faced with these results, Calvin began searching for another molecule that, if combined with CO_2, would produce a three-carbon product. He found the elusive acceptor in ribulose-1,5-bisphosphate (RuBP), a five-carbon sugar. The product of RuBP and CO_2 is an unstable six-carbon molecule that immediately breaks down into two molecules of 3-phosphoglyceric acid (3-PGA):

$$CO_2 + RuBP \rightarrow 2 \ 3\text{-PGA}$$

Calvin used a similar approach to figure out the rest of the biochemical reactions of photosynthesis. Follow in figure 7.20 how these reactions occur:

- CO_2 diffuses into the chloroplast stroma and is fixed; that is, it is incorporated into an organic compound.[4] This occurs by the covalent bonding of CO_2 to RuBP. The enzyme that catalyzes this reaction is *RuBP carboxylase/oxygenase* (also called *rubisco*), a large (molecular weight 490,000) enzyme consisting of eight large subunits and eight small subunits. The large subunits are coded in the chloroplast; the mRNA for these subunits is part of the chloroplast RNA and is made on ribosomes in the chloroplast.[5] The eight small subunits are coded in the nucleus; their mRNA is associated with cytoplasmic ribosomes and includes a polypeptide leader that directs the mRNA into the chloroplast. The assembly of the large and small subunits into rubisco is controlled by light.

- Rubisco occurs in all autotrophs except a few species of bacteria and is the most important and abundant protein on earth. It makes up about 20%–25% of the soluble protein in leaves and is made at a rate of 4×10^{10} g yr^{-1},

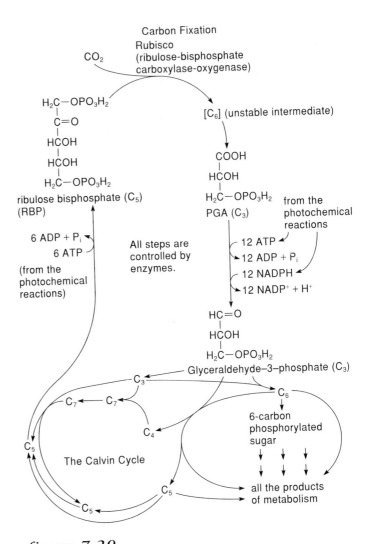

figure 7.20

The Calvin (C_3) cycle.

which is the equivalent of about 10^6 g sec^{-1}. Every person on earth is supported by about 44 kg of rubisco.

- The reaction of CO_2 with RuBP produces an unstable six-carbon compound that immediately breaks into two molecules of 3-phosphoglyceric acid (3-PGA) (fig. 7.21). This three-carbon molecule is the first stable product of photosynthesis, which explains why botanists refer to this series of reactions as the C_3 **cycle** (this cycle is also referred to as the **Calvin cycle** and the **reductive pentose cycle**).[6] Similarly, plants that use only the Calvin cycle to fix carbon dioxide are called C_3 **plants**. About 85% of plant species are C_3 plants, including cereals (e.g., barley, oats, rice, wheat), peanuts, cotton, sugar beets, tobacco, spinach, soybeans, most trees, and lawn grasses such as rye, fescue, and Kentucky bluegrass. The features of C_3 plants are summarized in tables 7.2 and 7.3 later in this chapter.

---

4. CO_2 actually moves into cells as bicarbonate ion (HCO_3^-). CO_2 dissolves in water on the wet surface of mesophyll cells. This forms carbonic acid (H_2CO_3), which then breaks down into a proton and a bicarbonate ion:

$$CO_2 + H_2O \rightleftharpoons H_2CO_3 \rightleftharpoons H^+ + HCO_3^-$$

5. Each photosynthetic cell has several hundred to several thousand copies of the gene to make the large subunit of rubisco. This results not only from the 40–200 chloroplasts per cell but also from the polyploid nature of chloroplast DNA; there are 50–1,000 copies of the genome per chloroplast.

6. Calvin's research earned him a Nobel Prize in 1961.

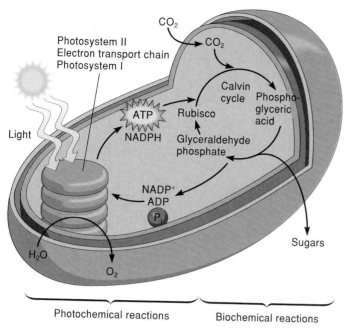

figure 7.21

A summary of photosynthesis. The photochemical reactions, which occur in thylakoids and grana, produce O_2, ATP, and reduced NADP⁺. The Calvin cycle (i.e., biochemical reactions), which occurs in the stroma, uses the ATP and reduced NADP⁺ to reduce CO_2 to carbohydrate (three key intermediates are shown here). The by-products ADP, Pi, and NADP⁺ are returned from the Calvin cycle to the photochemical reactions.

- A phosphate group is cleaved from each molecule of 3-PGA, and ATP and electrons from NADPH (produced by the photochemical reactions) are then used to reduce diphosphoglycerate (DPGA) to glyceraldehyde-3-phosphate (G-3-P; also referred to as PGAL—phosphoglyceraldehyde—to emphasize its similarity to PGA), a three-carbon sugar. This sugar, not glucose, is the carbohydrate produced by the Calvin cycle and is the starting point for many other metabolic pathways in the plant.

- Some of the G-3-P is used to re-form RuBP. This can occur in the dark and requires ATP made in the photochemical reactions. The rest of the G-3-P moves through a series of chemical reactions to form fructose diphosphate, which is used to make glucose, sucrose, starch, and other compounds needed by the plant.

A Summary of the Biochemical Reactions

The first steps of the Calvin cycle make precursors for glucose and other carbohydrates, while the later steps regenerate RuBP. Since the Calvin cycle takes in only one carbon (as CO_2) at a time, it takes six turns of the cycle to produce a net gain of six carbons (i.e., two molecules of glyceraldehyde-3-phosphate, a three-carbon sugar). These six turns of the cycle require eighteen ATPs (three per carbon; one-third of these ATPs are used to regenerate RUBP) and twelve NADPHs (two

per carbon), all of which come from the photochemical reactions of photosynthesis (fig. 7.21):

$$6CO_2 + 18ATP + 12NADPH + 12H_2O$$

$$\downarrow$$

$$C_6H_{12}O_6 + 18ADP + 18Pi + 12NADP^+ + 6H_2O + 6O_2$$

Although this summary equation is convenient for understanding the basics of photosynthesis, it is imperfect; some of the intermediates of the Calvin cycle are siphoned off to make other compounds (e.g., amino acids).

c o n c e p t

The biochemical reactions of photosynthesis use ATP and reduced NADP⁺ made in the photochemical reactions to reduce carbon dioxide to carbohydrate. These reactions are collectively referred to as the *Calvin*, or C_3, *cycle* and begin with the fixation of carbon dioxide by an enzyme called *RuBP carboxylase/oxygenase*. The product of the biochemical reactions is glyceraldehyde-3-phosphate, a sugar that is the starting point for several metabolic pathways. Plants that use only the Calvin cycle to fix CO_2 are called C_3 *plants*.

The Efficiency of Photosynthesis

How efficient is photosynthesis? To reduce two molecules of NADP⁺ requires that Photosystems I and II each absorb four photons (i.e., a total of eight photons). Since 1 mole of photons having a wavelength of 600 nm carries 47.6 kcal of energy, 8 moles of such photons carry 381 kcal. It takes 114 kcal of energy to reduce 1 mole (44 g) of carbon dioxide to hexose. Therefore, the theoretical efficiency of photosynthesis is 114/381 = 30%. On cloudy days, field measurements of the photosynthetic efficiency of individual plants average about 0.1%, while those for intensively cultivated plants average about 3%. Most crops range from 1%–4% (as do algae), and values as high as 25% have been recorded in laboratory conditions. A photosynthetic superstar of field-growing plants is *Oenothera claviformis*, an annual winter-evening-primrose that grows in Death Valley: its value of 8% is the highest of all plants studied in natural conditions. Sugarcane is a close second at about 7%.

Photosynthesis Is Not Perfect: Photorespiration

Photosynthesis took a long time to evolve, but it still has many shortcomings. For example, it wastes much light energy (see reading 7.1) and often does a poor job of fixing carbon. The first biologist to show this imperfection of photosynthesis was Otto Warburg, who in 1920 demonstrated that increased amounts of oxygen inhibit photosynthesis in C_3 plants (to learn more about Warburg, see reading 6.1, "Botany and Politics" on p. 118). This effect became known as the Warburg effect.

THE EVOLUTION OF PHOTOSYNTHESIS: WHY AREN'T PLANTS BLACK?

Photosynthetic autotrophs use light-energy to reduce carbon dioxide to carbohydrate. Because the amount of energy absorbed by an organism largely determines its rate of photosynthesis, why then aren't plants black? Stated another way, why do plants reflect, and therefore waste, green light? Why don't they use all of the spectrum of visible light for photosynthesis? The answers to these questions lie in the evolution of photosynthesis.

The earliest photosynthetic organisms were aquatic bacteria, several of whose descendents are around today. Chief among these is *Halobacterium halobium*, a purple bacterium that grows only in extremely salty water (3–4 times saltier than seawater). No other organisms can tolerate this environment.

Since water absorbs most light outside the visible part of the spectrum, most ultraviolet and infrared light was unavailable to the first photosynthetic organisms. Natural selection therefore favored the evolution of pigments such as bacteriorhodopsin, the photosynthetic pigment in *Halobacterium*. Bacteriorhodopsin, a purple pigment that resembles rhodopsin (the light-sensing pigment in our eyes), absorbs broadly in the middle of the visible spectrum of light (reading fig. 7.1). When it absorbs light, it pumps protons out of the cell, thereby creating a proton gradient that the cell uses to make ATP. That ATP powers *Halobacterium*.

Organisms living on the surface of the sediment faced a different problem. The bacteriorhodopsin in the *Halobacterium* swimming above the sediments absorbed most of the green light. Consequently, the light that reached the bottom-dwelling organisms was poorly suited for another bacteriorhodopsin-based type of photosynthesis. Similarly, the surrounding water absorbed all of the infrared and ultraviolet light of sunlight. Natural selection thus favored the evolution of a pigment that could absorb the remaining wavelengths—red and blue. This pigment was chlorophyll, which—in addition to pumping protons out of the cell—also pumped electrons into the cell. Their ability

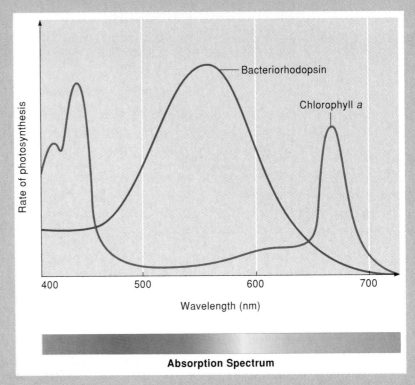

Absorption Spectrum

reading figure 7.1

The absorption spectrum of bacteriorhodopsin peaks in the green wavelengths, while that of chlorophyll *a* peaks on either side, in the red and blue wavelengths.

to pump electrons enabled these organisms to reduce carbon dioxide to sugars.

Plants and plantlike organisms have evolved other pigments to help compensate for their poor use of light. Accessory pigments absorb wavelengths near the center of the visible spectrum and pass this energy to chlorophyll. Cyanobacteria and red algae have phycocyanin and allophycocyanin as accessory pigments, which absorb orange light but don't cover all of the wavelengths missed by chlorophyll. They also contain a red pigment called phycoerythrin, which absorbs green light and thus extends the range of photosynthesis beyond that of most plants. Indeed, despite their name, many red algae are dark-colored, approaching black, making them ideally suited for photosynthesis. Similarly, the dark colors of brown algae result from chlorophyll and fucoxanthin, an accessory pigment. Not surprisingly, many red and brown algae grow in deep water, where

light is extremely dim. Some darkly colored red algae in the western Atlantic live 268 m (884 ft) down, where the light is less than 1% of full sunlight.

And what about land plants? Since light is abundant on land, there has been relatively little selection pressure for the evolution of a wider range of pigments. In other words, unlike red and brown algae, land plants grow in adequate light and do not have to extract as much light as possible. Consequently, land plants depend almost entirely on chlorophyll for photosynthesis. Carotenoids partly fill the gap, especially toward blue, but they do not pass this energy efficiently to chlorophyll. Since no pigment in land plants absorbs significantly in the green part of the spectrum, this light is wasted. It also creates a paradox: the most advanced plants have the least advanced system for absorbing light. Nevertheless, this is not all bad—few of us would trade a green countryside for a black one in the name of photosynthetic efficiency.

You may wonder how this is relevant to plants, since the amounts of CO_2 and O_2 in air are relatively constant. The answer is that it depends on what air you're talking about. Indeed, the composition of air outside a leaf often differs significantly from that inside a leaf where photosynthesis occurs.

During photosynthesis, plants fix CO_2 from the atmosphere. Although the concentration of CO_2 in this air is only about 0.034%, this represents an almost endless supply of CO_2 for photosynthesis—indeed, there are about 7×10^{11} tons of carbon (as carbon dioxide) in the atmosphere (for comparison, there are about 4.5×10^{11} tons of carbon in vegetation as carbohydrate).[7] Consequently, you wouldn't think that plants would have much trouble obtaining enough carbon dioxide for photosynthesis. However, all of this carbon dioxide is available to a plant only when the plant's stomata are open. As CO_2 diffuses into a leaf for photosynthesis, water diffuses out through the stomata. As long as the plant has plenty of water, this is not a serious problem, but on hot, dry days, plants close their stomata to conserve water. When this occurs, the continued fixation of carbon dioxide (via the Calvin cycle) decreases the concentration of carbon dioxide inside the leaf, which greatly increases the relative amount of O_2. This has a tremendous effect on the plant: when the concentration of carbon dioxide inside the leaf dwindles to about 50 parts per million (ppm),[8] rubisco stops fixing CO_2 and starts fixing oxygen. The products of this oxygenase reaction (remember that the enzyme is ribulose-1,5-bisphosphate carboxylase/*oxygenase*) are phosphoglycolic acid (a two-carbon compound), and PGA (a three-carbon compound):

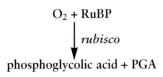

$$O_2 + RuBP$$
$$\downarrow rubisco$$
$$\textbf{phosphoglycolic acid + PGA}$$

The PGA stays in the Calvin cycle, but the phosphoglycolic acid leaves the cycle. In a complex series of reactions occurring in mitochondria and peroxisomes (fig. 7.22), the phosphoglycolic acid is hydrolyzed to glycine, a two-carbon amino acid. Two molecules of glycine then combine to form CO_2 and serine, a three-carbon amino acid. The serine is converted to 3-PGA (at the expense of ATP), which can enter the Calvin cycle. However, the CO_2 is often lost from the plant.

The reactions initiated by rubisco's fixation of oxygen rather than carbon dioxide are collectively called **photorespiration** because they occur only in light, consume oxygen, and release carbon dioxide. Unlike respiration, however, photorespiration produces no ATP. That is, photorespiration squanders large amounts of carbon originally fixed at great expense by the Calvin cycle. Although some of the CO_2 released during photorespiration is recycled by the Calvin cycle, as much as half of the carbon fixed in photosynthesis by C_3 plants is reoxidized and lost as CO_2. Thus, photorespiration undoes photosynthesis.

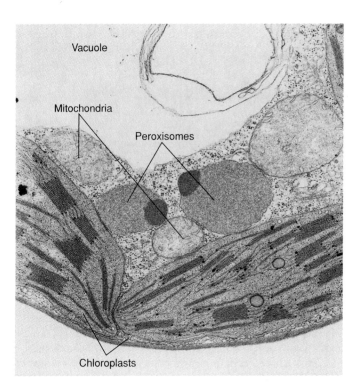

figure 7.22

Electron micrograph showing the close association of chloroplasts, peroxisomes, and mitochondria, ×58,000. All of these organelles participate in photorespiration.

How did plants evolve a wasteful process such as photorespiration? Some botanists suggest that the oxygenase reaction of rubisco is inevitable because of rubisco's structure: its active site makes it impossible for the enzyme to discriminate between CO_2 and O_2. Another suggestion is that photorespiration represents evolutionary baggage—a metabolic relic from when the atmosphere contained no free oxygen and competition for CO_2 didn't matter. The buildup of oxygen derived from photosynthesis led to photorespiration after millions of years.

c o n c e p t

Photorespiration occurs in C_3 plants when the internal concentration of carbon dioxide becomes low, such as when stomata close during drought. During photorespiration, rubisco fixes oxygen, and carbon dioxide is later released. Consequently, photorespiration undoes photosynthesis: as much as half of the carbon fixed in the Calvin cycle is released by photorespiration during hot, dry days.

Photorespiration peaks in hot, dry conditions (especially at temperatures above 28° C) when plants begin to close their stomata to conserve water. When this occurs, photorespiration causes plants to lose carbon. Clearly, plants that could avoid photorespiration would have a significant advantage. That advantage was provided by the evolution of C_4 photosynthesis.

---

7. About 80% of the earth's carbon is in rocks, and the remainder is in organic compounds such as cellulose and oil. The atmosphere contains about 1/1000 of 1% of the total, or 720 billion metric tons of carbon. About 560 billion metric tons of carbon are in living organisms.

8. One part per million is analogous to a second in 277 hours.

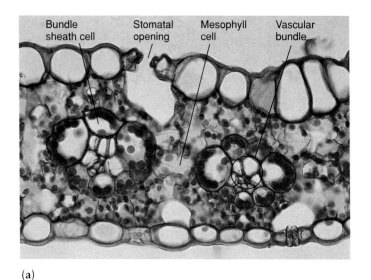 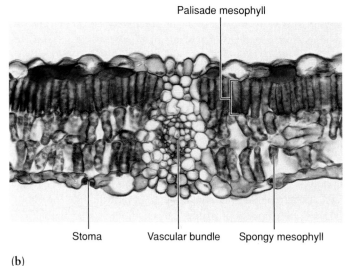

(a)   (b)

figure 7.23

Structure of leaves of C_4 and C_3 plants. (a) Leaves of C_4 plants have a characteristic Kranz anatomy: veins are tightly surrounded by photosynthetic bundle-sheath cells, which, in turn, are surrounded by photosynthetic mesophyll cells. This is a light micrograph of a cross section of a leaf of corn (*Zea mays*). (b) Leaves of C_3 plants lack a prominent and photosynthetic bundle sheath. Leaves of C_3 plants have a palisade mesophyll (of densely packed photosynthetic cells) and a spongy mesophyll (of loosely packed photosynthetic cells). You'll learn more about leaf anatomy in Chapter 14.

C_4 Photosynthesis

Soon after Calvin reported his brilliant study, other biologists began to repeat Calvin's study with other plants. Most of these studies confirmed Calvin's findings, but one study of sugarcane in 1965 stood out. In this plant, a 1-second exposure to radioactive CO_2 resulted in 80% of the radioactivity being found not in a three-carbon compound such as PGA but in a four-carbon acid called malic acid. Moreover, most of this malic acid appeared only in thin-walled mesophyll cells which, strangely enough, lack most of the enzymes of the Calvin cycle. Only after about 10 seconds did the label appear in PGA in bundle-sheath cells, which are thick-walled cells surrounding the vascular bundles (fig. 7.23). This kind of leaf anatomy—veins encased by thick-walled photosynthetic bundle-sheath cells that are, in turn, surrounded by thin-walled mesophyll cells—is characteristic of tropical grasses such as sugarcane and is referred to as **Kranz** (**halo** or **wreath**) **anatomy** (fig. 7.23). More interesting, at least initially, was the fact that CO_2 being fixed into malic acid was inconsistent with the idea that the Calvin cycle was the only means of fixing CO_2 in plants. What was going on here?

Careful studies of the structure and function of these leaves finally unlocked the puzzle—sugarcane *does* fix CO_2 via the Calvin cycle but only *after* fixing it via another set of reactions that preface the Calvin cycle. Here's the explanation (fig. 7.24):

- Carbon dioxide diffuses into the leaf through stomata and is fixed in mesophyll cells. These cells lack rubisco; during greening, light blocks the transcription of the mRNA to make the large subunit of rubisco.

- Carbon dioxide in mesophyll cells combines with a three-carbon compound called *phosphoenolpyruvic acid* (*PEP*), a respiratory intermediate. This reaction is catalyzed by PEP carboxylase and produces oxaloacetic acid (OAA), or oxaloacetate, a four-carbon acid (and an intermediate in the Krebs cycle):

$$CO_2 + PEP$$
$$\downarrow PEP\ carboxylase$$
$$oxaloacetic\ acid\ (OAA)$$

- The four-carbon acid is quickly converted to malic acid (malate) or aspartic acid (aspartate), which then moves through plasmodesmata to the adjacent bundle-sheath cell. There, these acids are split into CO_2 and a three-carbon compound. Thus, malic acid and aspartic acid function as short-lived reservoirs of CO_2; they turn over in about 10 seconds.

- The pumping by C_4 plants of CO_2 into bundle-sheath cells keeps the internal concentration of CO_2 in bundle-sheath cells 20–120 times greater than normal. This high concentration of CO_2 allows rubisco to fix CO_2 at maximal rates (recall that in C_3 plants, rubisco operates at only about 25% of its maximum rate). This eliminates photorespiration, thereby enabling C_4 plants to fix CO_2 more efficiently than C_3 plants do.

- The CO_2 released in the bundle-sheath cells is fixed by the Calvin cycle. Meanwhile, the three-carbon compound moves back to the mesophyll cell, where it is converted to PEP (at the expense of an additional ATP), the initial CO_2 acceptor in C_4 photosynthesis.

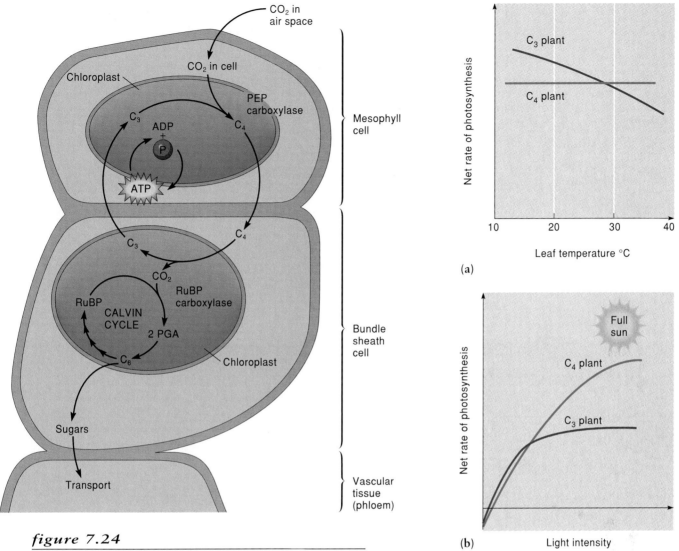

figure 7.24

Summary of C_4 photosynthesis. Mesophyll cells fix CO_2 into four-carbon acids that move into bundle-sheath cells. There, the acids are decarboxylated; the CO_2 released by this decarboxylation is fixed via the Calvin cycle in the bundle-sheath cell. The three-carbon compound moves back to the mesophyll cell, where it is converted to PEP, the CO_2 acceptor.

figure 7.25

C_4 plants are photosynthetically more efficient than C_3 plants only in hot, dry, bright conditions. Their increased efficiency in these conditions is due to the absence of photorespiration in C_4 plants.

Plants such as sugarcane are called C_4 plants because a four-carbon acid is the first stable product of their photosynthesis. Although none of the photosynthetic reactions of C_4 plants is unique, they produce dramatic differences in plant growth. The advantage of the added set of reactions in C_4 photosynthesis is that PEP carboxylase scavenges CO_2 and does not react with oxygen. This, combined with the pumping of CO_2 by surrounding mesophyll cells into bundle-sheath cells, enables C_4 plants to fix CO_2 with rubisco (in bundle-sheath cells) that is insulated from high concentrations of oxygen (fig. 7.24). Consequently, C_4 plants exhibit no photorespiration; they can fix CO_2 until the internal concentration of CO_2 reaches zero. This allows them to continue to fix CO_2 even in hot, dry weather when stomata begin to close (fig. 7.25a). As a result, they need only about half as much water as C_3 plants for photosynthesis (table 7.2).

C_4 plants also use nitrogen more efficiently than do C_3 plants, largely because they saturate their rubisco (a nitrogen-rich protein) with CO_2, thereby maximizing its efficiency. This allows C_4 plants to make less rubisco than C_3 plants and offsets the expense of the additional reactions of the C_4 cycle. Moreover, it is almost impossible to light-saturate C_4 plants (fig. 7.25b). Consequently, they usually outcompete C_3 plants in hot, dry weather. However, if given enough water, C_3 plants such as peanut and sunflower fix carbon at rates comparable to those of C_4 plants. Other features of C_4 plants are summarized in tables 7.2 and 7.3.

C_4 photosynthesis evolved in the hot conditions of the tropics. The primary selecting mechanism for C_4 photosynthesis was probably the decrease in CO_2 concentrations during the last 50 to 60 million years. Indeed, hot and arid conditions

table 7.2

Photosynthetic Characteristics of C₃, C₄, and CAM Plants

Characteristic	C₃	C₄	CAM
Leaf anatomy	Bundle-sheath cells without dense arrangement of chloroplasts	Bundle-sheath cells with dense arrangement of chloroplasts; no distinctive palisade cells	Large vacuoles
Primary carboxylating enzyme	Ribulose bisphosphate carboxylase/oxygenase (rubisco)	PEP carboxylase, then rubisco	PEP carboxylase at night, rubisco during the day
Grams of water required to produce 1 g of dry matter	450 to 950	250 to 350	50 to 55
Requires sodium as micronutrient	No	Yes	Probably
CO_2 compensation point (ppm)	30–70	0–10	0–5
Photorespiration detectable	Yes	Only in isolated bundle-sheath cells	No
Optimum temperature for photosynthesis	15°–25° C	30°–47° C	About 35° C
Approximate tons of dry matter produced per hectare per year	20–25	35–40	Usually low and variable

table 7.3

Maximum Photosynthetic Rates of Major Types of Plants in Natural Conditions

Type of Plant	Example	Maximum Photosynthesis mg CO_2 dm^{-2} h^{-1}
CAM	*Agave americana* (century plant)	1–4
C₃		
Tropical, subtropical, and Mediterranean evergreen trees and shrubs; temperate zone evergreen conifers	*Pinus sylvestris* (Scotch pine)	5–15
Temperate zone deciduous trees and shrubs	*Fagus sylvatica* (European beech)	5–20
Temperate zone herbs and C₃ pathway crop plants	*Glycine max* (soybean)	15–30
C₄		
Tropical grasses, dicots, and sedges with C₄ pathway	*Zea mays* (corn)	35–70

have been common throughout history, but only during the last 50 to 60 million years have CO_2 concentrations dropped to levels that give C₄ plants a selective advantage over C₃ plants.

Today, all known C₄ plants are angiosperms and are most common in hot, open ecosystems. They occur in at least seventeen families, none of which includes only C₄ plants. Because these families are diverse, distantly related, and have no common C₄ ancestors, C₄ photosynthesis probably evolved independently several times. Although most C₄ plants are monocots (especially grasses and sedges), more than 300 are dicots (including some trees and shrubs). Examples of C₄ plants include members of the Poaceae (corn, sorghum, sugarcane, millet, crabgrass, and Bermuda grass), pigweed (*Amaranthus*), and *Atriplex*. There are no known C₄ gymnosperms, bryophytes, or algae.

c o n c e p t

C₄ photosynthesis occurs in many plant species and fixes carbon dioxide into a four-carbon acid in mesophyll cells. This acid then moves to bundle-sheath cells, where it is broken down to carbon dioxide and a three-carbon molecule. The three-carbon molecule moves back to the mesophyll cell to accept another molecule of carbon dioxide. Meanwhile, the carbon dioxide is fixed by the Calvin cycle in bundle-sheath cells. This pumping of carbon dioxide into bundle-sheath cells eliminates photorespiration, which makes C₄ plants more efficient than C₃ plants during hot, dry conditions.

Why Don't C₄ Plants Dominate the Landscape?

If C₄ plants are more efficient than C₃ plants, why then don't they dominate our landscape? The answer is that C₄ plants are more efficient than C₃ plants *only in bright, hot conditions* (fig. 7.25). To best appreciate this, examine the summary equation for C₄ photosynthesis:

$$6CO_2 + 30ATP + 12NADPH + 12H_2O$$

$$\downarrow$$

$$C_6H_{12}O_6 + 30ADP + 30Pi + 12NADP^+ + 6H_2O + 6O_2$$

Compare this reaction with that of C₃ photosynthesis:

$$6CO_2 + 18ATP + 12NADPH + 12H_2O$$

$$\downarrow$$

$$C_6H_{12}O_6 + 18ADP + 18Pi + 12NADP^+ + 6H_2O + 6O_2$$

When it's not hot and dry, C₄ plants are *less* efficient than C₃ plants because of the extra ATPs they expend to run the C₄ part of their photosynthesis. However, in a hot, dry environment in which photorespiration would otherwise remove much of the carbon fixed in photosynthesis, the additional expense of C₄ photosynthesis is the best compromise available. Indeed, many of the plants that grow in hot climates are C₄ plants. However, this added expense is the reason they cannot compete effectively with C₃ plants in wet environments when temperatures are less than 25° C or in habitats such as dark forest floors.

Only about 0.4% of the 260,000 known species of plants are C₄ plants. So why all the fuss about C₄ photosynthesis? The answer is that several economically important plants such as corn and sorghum are C₄ plants.[9] Moreover, in hot, dry conditions they produce more biomass than do C₃ plants. Thus, the C₄ machinery represents a potential way for botanists to increase crop yields.

C₃–C₄ Intermediates

Genera such as *Flaveria* (Asteraceae), *Panicum* (Poaceae), and *Alternanthera* (Amaranthaceae) contain C₃ and C₄ species as well as species intermediate between C₃ and C₄ photosynthesis. These so-called C₃–C₄ intermediates have the following characteristics:

- *Intermediate leaf anatomy.* Their leaves contain bundle-sheath cells, but they are indistinct and poorly developed.

---

9. Each year the United States produces 7 million bushels of corn from about 700 billion plants. These plants cover about 70 million acres, a cornfield the size of Arizona. On a smaller scale, the 10,000 corn plants that grow on 1 acre fix 2,500 kg (5,512 lbs) of carbon in one growing season. This requires about 10 metric tons (11 tons) of CO_2.

reading 7.2

ATMOSPHERIC CO₂ AND PHOTOSYNTHESIS

The atmospheric concentration of CO_2 has increased from about 270 ppm in 1870 to 365 ppm today. Although this increase in atmospheric CO_2 could significantly enhance the greenhouse effect and promote global warming, it also provides the carbon that plants use in photosynthesis to make sugars, starch, and other compounds. How will this increase in atmospheric CO_2 affect plants?

The increased amount of CO_2 in the atmosphere could be a boon to agriculture. Although C₄ photosynthesis doesn't change significantly in response to increases in atmospheric CO_2, C₃ plants such as wheat and rice respond with a burst of photosynthesis. This increase in photosynthesis usually increases yields by as much as 60%.

Although increased amounts of atmospheric CO_2 increase photosynthesis of crops, we do not understand how increased amounts of CO_2 affect natural ecosystems. Indeed, in some systems, the initial burst of photosynthesis levels off after a few weeks. Despite this uncertainty, most botanists are convinced that increasing the amount of CO_2 in an ecosystem will affect plants and, in the process, be a potent force of global change.

- *Reduced photorespiration.* The CO_2 compensation point is the concentration of CO_2 at which net photosynthesis stops. In C₃ plants it ranges from 30 to 70 ppm (1 ppm = 1 mmol mol^{-1}), while in C₄ plants it is usually less than 5 ppm. The CO_2 compensation point for C₃–C₄ intermediates ranges from 7 to 28 ppm.

- Interestingly, closely related taxa often have different types of photosynthesis. For example, diploids of *Alloteropsis semialata* are C₃, while polyploids are often C₄.

Crassulacean Acid Metabolism (CAM)

The renewed interest in photosynthesis stimulated by Calvin in the 1950s prompted many botanists to begin studying the biochemistry of a variety of plants. Biologists familiar with the literature knew that their predecessors in the nineteenth century had shown that some members of the Crassulaceae family become acidic at night and progressively more basic during the day. The significance of these diurnal changes was not appreciated until 1958, when botanists reported that these plants open their stomata at night and fix CO_2 into

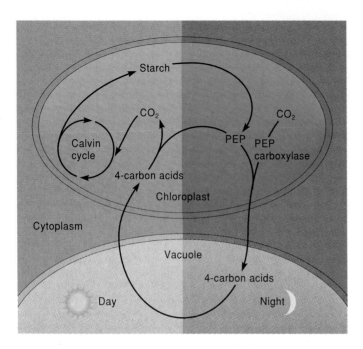

figure 7.26

Photosynthesis in CAM plants. At night, CAM plants fix CO_2 into four-carbon acids, which are stored in the vacuole. During the following day, these acids are decarboxylated, and the CO_2 is fixed via the Calvin cycle. CAM photosynthesis conserves water, thus enabling CAM plants to grow in dry environments (e.g., deserts) where most other kinds of plants cannot grow.

malic acid that is stored overnight in vacuoles of the large, succulent photosynthetic cells (fig. 7.26); the concentration of malic acid in the vacuole can reach 0.3 M, dropping the pH to as low as 4. The next day, the stomata close, thereby conserving water. Plants exhibiting this diurnal acidity and nocturnal fixation of CO_2 are called **Crassulacean acid metabolism (CAM) plants** because their unusual photosynthesis was discovered in a member of the Crassulaceae.

During the day, the acids made the previous night are decarboxylated, and the liberated CO_2 is fixed via the Calvin cycle in chloroplasts (fig. 7.26). Unlike C_4 plants, in which the initial fixation of CO_2 occurs in a different cell than does the Calvin cycle, CAM plants perform all of their photosynthesis in the same cell. Therefore, while the reactions of C_4 plants are separated spatially (i.e., they occur in different cells), the reactions of the Crassulaceae are separated both spatially (i.e., in different parts of the cell) *and* temporally (i.e., at different times of the day) (see figs. 7.24 and 7.26). The PEP that reacts with CO_2 during the night comes from the breakdown of starch, which is resynthesized during the day (fig. 7.26).

CAM plants open their stomata at night when temperatures drop and relative humidity increases. This enables CAM plants to use water more efficiently than do C_3 plants. Indeed, although CAM plants grow slowly, they need much less water

than do either C_3 or C_4 plants. Other features of CAM plants are summarized in tables 7.2 and 7.3.

CAM is more widespread than C_4 photosynthesis: it occurs in more than twenty families that include monocots, dicots, and primitive plants. Most CAM plants are succulents (fleshy plants having a low surface-to-volume ratio), but not all succulents are CAM plants. For example, pineapple is a CAM plant that is not succulent, and many halophytes (plants that grow in salty soil) are succulents but are not CAM. Examples of CAM plants include cacti, orchids, maternity plant (*Kalanchoë daigremontiana*), wax plant (*Hoya carnosa*), pineapple (*Ananas comosus*), Spanish moss (*Tillandsia usneoides*), and at least two genera of ferns. About 10% of plant species are CAM plants, but the only ones that are agriculturally significant are pineapple and an *Agave* species used to make tequila and as a source of fibers.

There are several variations of CAM. Some CAM plants revert to C_3 photosynthesis at the end of the day when their stored acids are gone and when given adequate water. Many CAM plants revert to C_3 photosynthesis when they're given large amounts of water, and many C_3 plants become CAM plants during drought. Plants such as *Peperomia camptotricha* lead double lives—the seedlings of these plants use C_3 photosynthesis, whereas mature plants use CAM.

c o n c e p t

Crassulacean acid metabolism (CAM) occurs in some plants that live in extremely dry habitats. CAM plants fix carbon dioxide at night into a four-carbon acid that is stored in the vacuole until daylight. The acid is then decarboxylated, and the carbon dioxide is used in the Calvin cycle. All of these reactions occur in the same cells (in C_4 plants they occur in different cells). Fixation of carbon dioxide at night (i.e., when the relative humidity increases) helps CAM plants conserve water.

Control of Photosynthesis

The previous sections discussed one factor that controls photosynthesis: temperature. Other factors that control photosynthesis include light, the concentration of carbon dioxide, the availability of water, and metabolic sinks.

Light Chlorophyll absorbs red and blue light strongly; therefore, these wavelengths stimulate photosynthesis more than does green light, which is reflected or transmitted (fig. 7.6). Maximum rates of photosynthesis occur near noon when light is brightest; at these times, plants usually fix about eight times more carbon dioxide than is released during respiration. Averaged over a day, photosynthesis fixes about six times more carbon dioxide than is released by respiration.

The Calvin cycle includes twelve enzymes, at least five of which are rate-limiting to photosynthesis. These same five enzymes are all activated by light.

In darkness, there is no photosynthesis, and respiration produces carbon dioxide that is released from the plant. As light intensity increases, so too does photosynthesis, eventually reaching a point at which the rate of photosynthesis equals the rate of respiration. This **light-compensation point** varies in different plants. "Sun plants," such as corn and sugarcane that grow in bright sun, have a high light-compensation point and require much light to saturate photosynthesis. Conversely, plants that grow in low light (e.g., on forest floors) have a low light-compensation point and saturate at low light intensities. The rates of photosynthesis in these plants also differ significantly: shade plants fix about 5 mg C dm^{-2} h^{-1}, while those of sun plants fix about 20 mg C dm^{-2} h^{-1}. You'll learn more about sun and shade plants in Chapter 14.

Concentration of Carbon Dioxide The concentration of carbon dioxide at which plants show no net fixation of carbon dioxide is the CO_2 **compensation point.** In C_3 plants such as wheat and rice, the CO_2 compensation point is about 50 ppm (at 25° C and 21% oxygen). For C_4 plants, the CO_2 compensation point is 0–5 ppm.

Increasing the concentration of carbon dioxide in the air to 0.10% doubles the rate of photosynthesis in C_3 plants by reducing photorespiration. As you'd expect, such "fertilizing" of the air with CO_2 does not significantly affect the rate of photosynthesis in C_4 plants—they saturate at CO_2 concentrations near 0.04%, just above normal. The benefit of increased concentrations of CO_2 is limited, however, because stomata close and photosynthesis stops at CO_2 concentrations exceeding about 0.15% (see reading 7.2, "Atmospheric CO_2 and Photosynthesis").

Water Availability The inability of plants to maintain enough water in their leaves causes stomata to close, thereby inhibiting photosynthesis. The lack of water is the most limiting factor controlling photosynthesis.

Metabolic Sinks Metabolic sinks are parts of a plant where photosynthate is stored or used, such as roots and growing areas of the plant. Photosynthesis decreases when sinks are removed and increases when sinks are created (e.g., by wounds or infections). In general, high rates of photosynthesis require high rates of translocation of the photosynthate to sinks.

c o n c e p t

The primary factors that control photosynthesis are temperature, water availability, light, the concentration of carbon dioxide, and metabolic sinks.

What Happens to the Products of Photosynthesis?

The products of photosynthesis are called photosynthate and have many fates:

- About half of all photosynthate is used as fuel for cellular respiration and photorespiration.

- Some of the 3-PGA moves into the cytoplasm, where it is used to make amino acids.

- The PGAL is used to make fructose-1,6-diphosphate, which, in turn, is used to make other sugars and starch. Fructose-6-phosphate is used to regenerate RuBP.

- Chloroplasts use ATP to reduce sulfate to sulfhydryl groups. Ferredoxin and NADPH in chloroplasts are used to reduce nitrite to ammonia for incorporation into amino acids (nitrate is reduced to nitrite in the cytoplasm).

- Some of the photosynthate is used to make glucose. Molecules of glucose are smaller and store more energy than does ATP, which is reactive, polar, and difficult to move across membranes.

- Some of the photosynthate is used to make sucrose, a disaccharide that is shipped throughout the plant. Glucose is often stockpiled as dense, inert granules of starch, which is stored in roots and stems. Small amounts of starch are stored in chloroplasts (fig. 7.12). Unlike glucose, starch has little effect on the osmotic properties of the cell and serves as a starting point for making other compounds that the cell needs.

- Much of the photosynthate is used to make cellulose, the most abundant organic compound on earth. Cellulose, you'll recall, is a primary ingredient in cell walls (see Chapter 3, pp. 55–58).

- Plants use some of their photosynthate to make secondary metabolites such as latex. Some of these metabolites resemble crude oil; for example, the grape-size fruits of *Pittosporum undulatum* have been used for centuries as torches. Melvin Calvin and other botanists are studying how to use the hydrocarbons in the latex of some plants (e.g., members of the spurge family and relatives of poinsettia) as fuel. Some of the results are promising. For example, extracts from the Amazonian copa iba tree (*Copaifera langsdorfii*) are a diesel fuel that can be burned in automobile engines with no further refining. Similarly, 1 acre of gopher plants (*Euphorbia lathyris*) can produce enough latex to make twelve barrels of oil. Plantations of *Euphorbia* would be economically feasible when oil prices reach about $31 per barrel.

- We use photosynthate of corn to make almost 90% of the ethanol used in the United States, some of which is mixed with gasoline (in a 1:9 ratio) to make gasohol. Most cars made in Brazil run on 100% ethanol, and in the United States, gasohol accounts for about 10% of gas sales. One bushel of corn will make about 3 gallons of ethanol.

- Every year we use photosynthesis to produce about 1.5 billion tons of grain, the staple of the world's diet.

Each year, photosynthesis fixes about 10% of the carbon in the atmosphere. Humans, either directly or indirectly, use about 40% of the net products of photosynthesis from terrestrial ecosystems. These uses affect all aspects of our lives: textiles, food (e.g., cereals such as oats and wheat provide half of the world's food), fuels such as coal and oil (formed millions of years ago), wood (the United States uses more than 3 billion cubic feet of wood per year to make paper), drugs (e.g., caffeine, quinine, cocaine, nicotine), waxes, oils (e.g., rose, jasmine), rubber, spices (e.g., pepper, cinnamon, peppermint), and even the perfumes that Cleopatra used to entice Marc Antony and Julius Caesar. This page, the oxygen you breathe, and all of the atoms in your body were once parts of plants.

No process can match the importance or magnitude of photosynthesis. It sustains virtually all life on earth. Without photosynthesis, all other biological reactions would be irrelevant.

Chapter Summary

Photosynthesis is the light-driven conversion of water and carbon dioxide to carbohydrates. Virtually all life depends on light. Light moves in waves, and its energy is contained in packets called *photons*. The energy of a photon is inversely proportional to the wavelength of the light: the longer the wavelength, the less energy per photon. Light occurs in a spectrum of colors. Red and blue light are most effective for photosynthesis.

Only light that is absorbed can have an effect. Light is absorbed by pigments. Chlorophyll *a* is the primary pigment for photosynthesis and occurs in all photosynthetic organisms except photosynthetic bacteria. Accessory pigments such as carotenoids and chlorophyll *b* absorb light that chlorophyll *a* cannot absorb, thereby extending the range of light used for photosynthesis.

Photosynthesis in eukaryotes occurs only in chloroplasts, which consist of a gelatinous matrix called the *stroma* and stacks of membranes (thylakoids) called *grana*. Aggregates of pigments in grana absorb light and funnel its energy to special pairs of chlorophyll and proteins called *reaction centers*. The energy of the light energizes electrons, which are used to pump protons from the stroma into the thylakoid space. The resulting pH gradient is discharged during the synthesis of ATP as protons flow back into the stroma.

The photochemical reactions of photosynthesis convert light-energy into chemical energy. In plants, the photochemical reactions oxidize water, produce ATP, reduce $NADP^+$, and involve two photosystems connected in series. The ATP and reduced $NADP^+$ made in the photochemical reactions are used in the biochemical reactions of photosynthesis to reduce carbon dioxide to carbohydrate. The biochemical reactions are collectively referred to as the *Calvin*, or C_3, *cycle* and begin with the fixation of carbon dioxide by an enzyme called *ribulose bisphosphate carboxylase/oxygenase* (*rubisco*). The product of the biochemical reactions is glyceraldehyde-3-phosphate, a sugar that is the starting point for several metabolic pathways. Fixing each molecule of carbon dioxide and regenerating RUBP requires three ATPs and two reduced $NADP^+$s. Plants that use only the Calvin cycle are called C_3 *plants*.

Photorespiration occurs in C_3 plants when the internal concentration of carbon dioxide becomes low, such as when stomata close during drought. During photorespiration, rubisco fixes oxygen, and carbon dioxide is later released. Consequently, photorespiration undoes photosynthesis: as much as half of the carbon fixed in the Calvin cycle is released by photorespiration during hot, dry days.

C_4 photosynthesis occurs in many plant species and involves fixation of carbon dioxide into a four-carbon acid in mesophyll cells. The acid then moves to bundle-sheath cells, where it is broken down to carbon dioxide and a three-carbon molecule. The three-carbon molecule moves back to the mesophyll cell to accept another molecule of carbon dioxide. Meanwhile, the carbon dioxide is fixed by the Calvin cycle in bundle-sheath cells. This pumping of carbon dioxide concentrates carbon dioxide in bundle-sheath cells, thereby eliminating photorespiration. This, in turn, makes C_4 plants more efficient during hot conditions.

Crassulacean acid metabolism (CAM) occurs in many plants that live in dry habitats. CAM plants fix carbon dioxide at night into a four-carbon acid that is stored in the vacuole until daylight. The acid is then decarboxylated, and the carbon dioxide is used in the Calvin cycle. Fixation of carbon dioxide at night helps CAM plants conserve water. The primary factors that control photosynthesis are temperature, water availability, light, the concentration of carbon dioxide, and metabolic sinks.

 What Are Botanists Doing?

Use a reference source at the library to document the global distributions of C_3 and C_4 plants. What do these distributions tell you about the evolution of these plants?

 Writing to Learn Botany

How does photosynthesis differ from respiration?

Questions for Further Thought and Study

1. You learned in Chapters 5 and 6 that the complete combustion of 1 mole of glucose to carbon dioxide and water releases 686 kcal of energy (i.e., a mole of glucose stores about 686 kcal of energy in its chemical bonds). However, making glucose requires almost 2,000 kcal. What happens to the rest of the energy?

2. What is the significance of using NAD^+ in respiration and $NADP^+$ in photosynthesis?

3. Geldikkop is a rare disease in sheep in which chlorophyll that gets into the blood is energized by light and causes lesions. Why would chlorophyll in blood cause such a problem?

4. Zooplankton are transparent and eat phytoplankton. Most zooplankton avoid bright light by migrating down during the day and up during the evening and night. They also contain large amounts of carotenoids. What is the significance of this?

5. The chloroplast genome is the same in the mesophyll and bundle-sheath cells of C_4 plants. How, then, can these chloroplasts be so different?

6. Explain how the adaptations of CAM and C_4 plants enhance photosynthesis in hot, dry environments.

7. Life has been defined as the heat generated between photosynthesis and respiration. Explain the basis for this statement.

8. How does visible light differ from ultraviolet light? From infrared light?

9. How is an absorption spectrum different from an action spectrum?

10. How does cyclic electron flow differ from noncyclic electron flow? What is the significance of these differences?

11. What is photorespiration? Why is it significant?

12. Why don't C_4 plants dominate the landscape?

13. What are some of the things that we do not know about photosynthesis? What experiments would you do to learn more about these mysteries?

14. Use a reference source at the library to document the effect of temperature on photosynthesis. What does this information tell you about the process?

Web Sites

Review the "Doing Botany Yourself" essay and assignments for Chapter 7 on the *Botany Home Page*. What experiments would you do to test the hypotheses? What data can you gather on the Web to help you refine your experiments?

Here are some other sites that you may find interesting:

http://www.public.asa.edu/~larryorr/photosyn/study/html
This site includes a review of how photosynthesis relates to agriculture energy.

http://www.bio.net/hypermail/photosynthesis
At this site you can join a discussion group about photosynthesis. Ask questions or learn from the questions and comments of others.

http://www.bio.net/archives.html
At this site, you'll find links to other biology-related pages, including those about botanical research, forestry, and software for biologists.

http://copernicus.bbn.com/www/CoE/photo/PandR.html
This site includes a review of photosynthesis and an interactive simulation to test your understanding of photosynthesis.

http://mnonline.org/ktca/newtons/a/phytosy.html
This site links you to a summary of photosynthesis in *Newton's Apple*, a popular television show.

http://www.nyu.edu/pages/mathmol/modules/photosynthesis/photo.html
This site shows a color photograph of the arrangement of pigments in a reaction center.

http://mss.schbe.on.ca/DSRESPIR.htm
At this site you can review the similarities and differences between cellular respiration and photosynthesis.

http://www.asu.edu/clas/photosyn/
The Arizona State University Center for the Study of Early Events in Photosynthesis was established in 1988 as part of a joint grant program of the Department of Energy, the National Science Foundation, and the Department of Agriculture. It reports on all aspects of photosynthesis.

Suggested Readings

Articles

Andrews, T. J., and G. H. Lorimer. 1987. Rubisco: Structure, mechanisms, and prospects for improvement. In *The Biochemistry of Plants*, vol. 10, M. D. Hatch and N. K. Boardman, eds. San Diego: Academic Press, 131–218.

Bazzaz, F. A., and E. D. Jajer. 1992. Plant life in a CO_2-rich world. *Scientific American* 266 (January):68–74.

Culotta, Elizabeth. 1995. Will plants profit from high CO_2? *Science* 268:654–656.

Ehleringer, J. R., R. F. Sage, L. B. Flanagan, and R. W. Pearcy. 1991. Climate change and the evolution of C_4 photosynthesis. *Trends in Ecology and Evolution* 6:95–99.

Govindjee, and William J. Coleman. 1990. How plants make oxygen. *Scientific American* (February):50–58.

Hendry, George. 1990. Making, breaking, and remaking chlorophyll. *Natural History* 90(5):36–41.

Storey, Richard D. 1989. Textbook errors and misconceptions in biology: Photosynthesis. *The American Biology Teacher* 51:271–274.

Williams, Nigel. 1996. Mutant alga blurs classic picture of photosynthesis. *Science* 273:310.

Zelitch, Israel. 1992. Control of plant productivity by regulation of photorespiration. *BioScience* 42:510–516.

Books

Baker, N. R., and H. Thomas. 1992. *Crop Photosynthesis: Spatial and Temporal Determinants*. New York: Elsevier Science Publishers.

Barber, J. 1992. *The Photosystems: Structure, Function, and Molecular Biology*. New York: Elsevier Science Publishers.

Galston, A. W. 1994. *Life Processes of Plants*. New York: Scientific American.

Lawlor, D. W. 1993. *Photosynthesis: Molecular, Physiological, and Environmental Processes*. New York: J. Wiley & Sons.

Genetics . . .

"Like begets like." This saying has been around for thousands of years because people have long recognized that, for example, only roses can make more roses and only camels can make more camels. This means that information passes from one generation to another, thereby defining what each organism can be. The inheritance of such information is the foundation for the discipline of genetics, the subject of this unit.

Plants manage inherited information at the molecular level. The management of this information is intertwined with how plants obtain and use energy for cellular metabolism, which was discussed in the previous unit. Reproduction requires energy from photosynthesis and respiration, and the synthesis of molecules that guide metabolism relies on directions from information inherited from previous generations.

The most sophisticated technology in biological research is now used to understand the molecular basis of inheritance. New discoveries are being made faster in genetics than in any other discipline. These discoveries underscore the importance of genetics to science and society in the late twentieth century. They also mean that some of the information presented in this text is probably changing as you read it. Some of the controversies in genetics that are discussed here are being resolved even as others are being raised.

unit three

Garden pea (*Pisum sativum*) with flowers and pods. Gregor Mendel used the garden pea to establish the basic postulates of the theory of inheritance.

chapter

Patterns of Inheritance

8

Chapter Outline

Chapter Overview

Every organism is the expression of thousands of genes. Such genetic complexity would seem to prevent an understanding of how inheritance works. Nevertheless, the current theory of inheritance was proposed more than a century ago. The nineteenth-century assumptions of this theory are still correct, at least for characteristics that are under simple genetic control. However, many apparent exceptions and complex patterns of inheritance have been discovered since the beginning of the twentieth century. The search for explanations of complex inheritance is one of the primary subjects of modern genetics. However, unlike the earliest geneticists, we have a better understanding of what genes are and how they work.

In this chapter we present a historical perspective on the theory of inheritance and how early experiments laid the foundation for explaining heredity. We also discuss how different patterns of inheritance arise and how they relate to our current understanding of genetics. Much of this chapter is a springboard for more detailed discussions of cell reproduction, the mechanisms of heredity, and how genes work, which are presented in subsequent chapters.

People have always been interested in heredity and its importance in their families, crop production, and animal husbandry. We have known for centuries that all organisms inherit traits from their parents. The earliest farmers, thousands of years ago, selected bigger, more desirable seeds for continued cultivation. By so doing, they selected heritable characteristics, even though they had little or no understanding of inheritance.

Many theories on the nature of heredity, some based on religion or mythology and some based on factual observations, have held sway at one time or another during the past few centuries. The modern theory of inheritance was established in the mid-nineteenth century. The basic assumptions of this theory were first made by Gregor Mendel, an obscure amateur botanist working in a monastery garden in Austria (fig. 8.1). Several decades passed before biologists fully appreciated Mendel's work. Once they did, these botanists made many discoveries involving the chemical and biological nature of heredity. Among the more recent of these discoveries are that DNA is the hereditary material and that genes are the basic units of inheritance.

The Theory of Inheritance: Gregor Mendel's Discovery

Gregor Mendel (1822–1884) studied a simple genetic system and recognized the importance of his results for understanding patterns of inheritance that are basic to all organisms. Remarkably, Mendel chose an appropriate plant species, designed powerful experiments, and interpreted his results without knowing about chromosomes or genes. By doing so, he derived a set of correct assumptions about heredity in the garden pea *(Pisum sativum)* that form the theory of inheritance. Subsequent sections of this chapter describe how Mendel achieved such a significant scientific advancement and how his work fits into the framework of modern genetics.

Setting the Scene: Plant Hybridization before Mendel

The first plant breeders were probably ancient Babylonians and Assyrians who cultivated date palms *(Phoenix dactylifera)* in the region of modern-day Iraq more than 4,000 years ago. Like many plants, date palms are **dioecious,** meaning that each plant produces either pollen or fruits, but not both. Date growers discovered that pollen from just a few trees could be transferred by hand to several hundred fruit-producing trees, causing them to make dates. This practice was important economically because only a few fruitless (pollen) trees had to be maintained in each plantation. Unknown to the growers, pollen transfer was also important biologically because it promoted variation in the size, flavor, shape, and color of the dates. As a result, desirable variations were selected for replanting through many generations, producing more than 400 varieties of dates (fig. 8.2) Even though the patterns of inheritance were exploited, how plants inherited traits from their parents remained unknown for many centuries.

figure 8.1

Gregor Mendel (1822–1884). Mendel's experiments with the garden pea established the basic postulates for the theory of inheritance.

figure 8.2

Some of the more than 400 different varieties of dates. Selection of dates for different traits probably began more than 4,000 years ago.

By the end of the seventeenth century, the need for pollen in sexual reproduction had been shown experimentally in several species. During the eighteenth and nineteenth centuries, European botanists began to study heredity by artificial hybridization (see reading 8.1, "Artificial Hybridization in Plants"). The most notable early experiments were done in the 1760s by Josef Kölreuter of Germany, who discovered that hybrid offspring had features that were either intermediate between their parents or from only one parent. He also showed that continued self-pollination of hybrids over several generations produced much variability, including the reappearance of features that had disappeared in earlier generations. Hence, Kölreuter discovered a crucial piece in the puzzle of inheritance: that parental traits could be absent in one generation of offspring but reappear in a subsequent generation.

The most extensive studies of plant hybridization were done by Karl Freidrich von Gaertner, who was from the same district in Germany as Kölreuter. Beginning in the 1820s, Gaertner did nearly 10,000 experiments involving artificial hybridization that established the role of pollen in transmitting traits of the pollen parent to the hybrid offspring. Also, Gaertner discovered that certain traits are expressed preferentially over others. For example, purple flowers are **dominant** over white flowers. A dominant trait is one that masks the alternative trait for a particular feature. The masked trait is said to be **recessive**. Gaertner determined the dominant and recessive forms of several traits, including flower color (purple over white), seed shape (round over wrinkled), and pod color (green over yellow) in the garden pea. By confirming the tendency of these dominant traits to be more commonly expressed among offspring, Gaertner further illuminated the theory of inheritance.

Extensive genetic studies of plants were done a century before Mendel began his experiments with garden peas. These studies showed that pollen transmits traits to offspring. Dominance was known, and the reappearance of recessive traits in offspring had been observed. This knowledge provided important background information for Mendel's studies.

How Did Mendel Begin?

The voluminous work of Gaertner and his predecessors was a rich resource for Mendel. The challenge for Mendel was to find order in seemingly overwhelming complexity. The problem, as he perceived it, was that

> Whoever surveys the work in this field will come to the conviction that among the numerous experiments not one has been carried out to an extent or in a manner that would make it possible to determine the number of different forms in which hybrid progeny [offspring] appear, permit classification of these forms in each generation with certainty, and ascertain their numerical interrelationships.

Armed with a strong background in biology and mathematics, Mendel tried to measure the results of hybridization experiments—to "ascertain their numerical interrelationships." His work is an excellent example of how expertise in one discipline (mathematics) can provide insight into another (biology). Although many people had done the same experiments as Mendel, it was his use of mathematics that led to his great discoveries.

Mendel studied several kinds of plants, mostly from the pea family, that were easy to grow and manipulate in hybridization experiments. The focus of the following discussion is on one of these, the garden pea, because it is the only one that showed patterns of inheritance that Mendel could fully explain.

Before making hybrids, Mendel grew his pea plants for 2 years, through several generations, until he could select **true-breeding** (purebred) strains for each of several traits. He chose nine sets of traits based on the ease of distinguishing one trait from another. Three of these (flower color, seedling axil color, seed coat color) were so well correlated, however, that he regarded them as expressions of the same factor. Thus, he followed the inheritance of seven separate pairs of traits, shown in figure 8.3. By focusing on these simple and easily distinguished traits, Mendel eliminated the potential for complex variability.

In his initial experiments, Mendel crossed (hybridized) pairs of plants that differed in only one trait. For example, he crossed purple-flowered plants with white-flowered plants and wrinkled-seeded plants with smooth-seeded plants. In so doing, he made reciprocal crosses between parental strains— that is, he transferred pollen from purple flowers to white flowers and pollen from white flowers to purple flowers.[1] He

---

1. Reciprocal crosses occur when one plant is the pollen parent and the other plant is the seed parent for the first cross, and the pollen parent and seed parent are reversed for the second cross.

ARTIFICIAL HYBRIDIZATION IN PLANTS

The offspring of parents with different traits is called a **hybrid.** Reproductive structures of many kinds of flowers allow easy manipulation of the pollen-containing anthers, which means that plant hybrids can be made artificially—that is, by human hands. In hybridization experiments, anthers are typically removed from each flower to prevent self-pollination. This is necessary only when a plant is self-compatible—that is, when it can fertilize itself. Not all plants are self-compatible. Mendel took the precaution of removing anthers because the garden pea is a self-compatible species (reading fig. 8.1a).

After the anthers are removed, pollen from another individual is brushed onto the stigma, which is the receptive surface of the seed-producing organ (carpel). The pollen can be from any other plant, but successful fertilization depends upon genetic compatibility between the pollen parent and the seed parent. It is impossible to fertilize a garden pea with pollen from an orchid, because peas and orchids are too dissimilar genetically. However, pollen from one variety of garden pea can fertilize another variety of garden pea (reading fig. 8.1b). The limits of interfertility are inconsistent from one type of plant to another. For example, many species of pines cannot be hybridized artificially, but hybrids of orchids, mustards, carnations, and many other plants have been made.

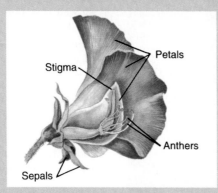

reading figure 8.1a

A pea flower (longitudinal section). The petals enclose the anthers and stigma, thereby promoting self-pollination.

Plant breeders develop many new kinds of plants by selecting flowers having the most desirable traits (size, color, etc.) for making artificial hybrids. Most commercially available garden plants are the products of many generations of hybridization. These are referred to as *cultivars* to denote cultivated varieties that are not found in nature. As a commercial enterprise, the business of making new cultivars has produced thousands of particularly remarkable or unique forms. The most valuable of these are patented.

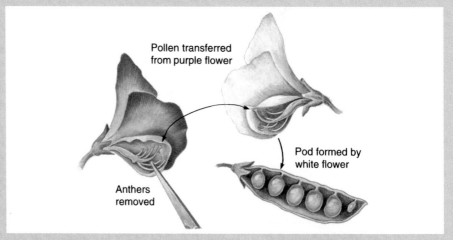

reading figure 8.1b

Artificial hybridization in pea flowers. Cross-pollination occurs when anthers from one flower are brushed against the stigma of another flower.

then allowed the offspring of the various crosses to self-pollinate for several generations and counted the plants with each trait in every generation.

Mendel's Results: A Theory of Inheritance

The first generation of offspring is called the F_1 generation. (The *F* comes from **filial,** which is derived from the Latin word for daughter.) In every experiment that Mendel did, only the dominant trait was expressed in the F_1 generation. However, when F_1 plants were allowed to self-pollinate, the recessive trait reappeared in the F_2 generation. By counting the offspring in the F_2 generation, Mendel found that the ratios of dominant to recessive traits were the same for all seven pairs of traits, about 3:1

(table 8.1) Mendel concluded that "factors" from each parent somehow controlled the traits he observed. Moreover, he suggested that each offspring gets one of two factors from each parent. In the F_1 generation all of the offspring have both factors, but the factor for the dominant trait masks the factor for the recessive trait. Mendel also predicted that any plant having two different factors would donate the recessive factor to one-half of the offspring and the dominant factor to the other half. Thus, the probability for F_2 plants to receive two recessive factors would be one-half times one-half, or one-quarter. This prediction matches the observed ratio of three dominant traits (3/4) to one recessive trait (1/4) in the F_2 generation (fig. 8.4).

Mendel accounted for heritable factors and the probabilities of inheritance for each trait *only because he counted the offspring* of each cross. Several assumptions, or *postulates,* can

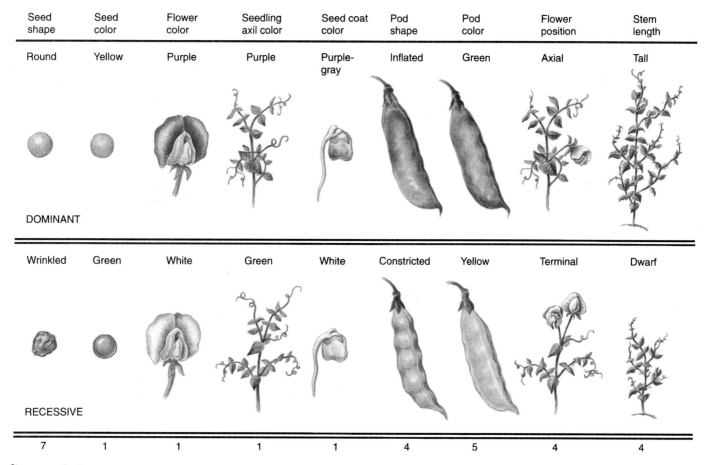

Seed shape	Seed color	Flower color	Seedling axil color	Seed coat color	Pod shape	Pod color	Flower position	Stem length
Round	Yellow	Purple	Purple	Purple-gray	Inflated	Green	Axial	Tall

DOMINANT

Wrinkled	Green	White	Green	White	Constricted	Yellow	Terminal	Dwarf

RECESSIVE

| 7 | 1 | 1 | 1 | 1 | 4 | 5 | 4 | 4 |

figure 8.3

Characteristics of the garden pea studied by Mendel. Mendel regarded flower color, seedling axil color, and seed coat color as expressions of the same factor. The number below each characteristic is the number of the chromosome that bears the gene for the characteristic. This information was not discovered until the twentieth century.

table 8.1

Mendel's Experimental Results for Seven Characteristics of Garden Pea (These Data Represent the F_2 Generation)

	Sample Size	Dominant Form	Recessive Form	Ratio
	7,324 seeds	5,474 round	1,850 wrinkled	2.96:1
	8,023 seeds	6,022 yellow	2,001 green	3.01:1
	929 plants	705 purple-flowered	224 white-flowered	3.15:1
	1,181 plants	882 with inflated pods	299 with constricted pods	2.95:1
	580 plants	428 with green pods	152 with yellow pods	2.82:1
	858 plants	651 with axial flowers	207 with terminal flowers	3.14:1
	1,064 plants	787 with long stems	277 with short stems	2.84:1
TOTAL:	19,959	14,949	5,010	2.98:1

Note: Two additional characters, seedling axil color and seed coat color, were also examined by Mendel. However, purple seedling axils and purple-gray seed coats were always associated with purple flowers, so Mendel regarded these three features as expressions of the same factor. He therefore recorded data for flower color and disregarded the other two.

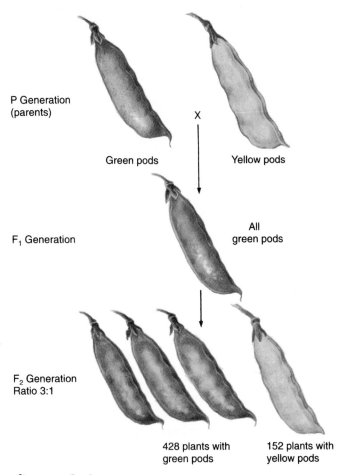

P Generation (parents)

Green pods   X   Yellow pods

F₁ Generation

All green pods

F₂ Generation Ratio 3:1

428 plants with green pods   152 plants with yellow pods

figure 8.4

The inheritance pattern of one of Mendel's crosses (green versus yellow pods) through two generations. All plants in the F_1 generation had green pods. About three-fourths of the F_2 generation had green pods, and about one-fourth had yellow pods.

be derived from his initial experiments. Taken together, these postulates are referred to as Mendel's **theory of inheritance.** Note, however, that the postulates of this theory, which are listed next, are modernized versions of Mendel's ideas; Mendel did not actually state them as postulates, nor did he propose a theory of inheritance. In fact, Mendel barely mentioned the ideas behind postulates 7 and 8.

1. Traits are controlled by heritable factors.

2. Factors are passed from parent to offspring in reproductive cells.

3. Each individual contains pairs of factors in every cell except reproductive cells.

4. Paired factors segregate during the formation of reproductive cells so that each reproductive cell gets one of the factors of a pair.

5. There is an equal chance that a reproductive cell will get one or the other factor of a pair.

6. Each factor from one parent has an equal chance of combining either with the identical factor or with the other factor from the other parent during fertilization.

7. Sometimes one factor dominates the other factor; in such cases, the dominant factor controls that feature of the plant.

8. When two or more traits are under consideration, the factors for each trait assort independently to the reproductive cells.

The significance of Mendel's work is clear now, but it remained obscure to science until the beginning of the twentieth century. Since their rediscovery, postulates 4, 7, and 8 have often been called Mendel's **laws of inheritance:**[2] the **law of segregation,** the **law of dominance,** and the **law of independent assortment,** respectively. Unfortunately, the exalted status of these postulates as laws came from twentieth-century geneticists who were a little overzealous in clarifying Mendel's ideas. Mendel proposed no such laws. On the contrary, Mendel did not regard the segregation of factors as a law because he could not observe it directly. He did not believe dominance to be a law because he knew of exceptions to this pattern, and he only did two sets of experiments that showed independent assortment, which he barely mentioned. (See "Experiments Using Multiple Traits" later in this chapter.) Nevertheless, many modern biology textbooks still perpetuate these myths about Mendel's experiments.

In modern genetics, each of Mendel's paired factors is called a **gene.** Each factor of a pair is an **allele,** which is an alternative form of the gene. This means that in true-breeding parents, the genes controlling flower color or other features used by Mendel had two identical alleles for each gene. A plant that has the same alleles for a gene is said to be **homozygous** for that gene. In contrast, a plant that has different alleles for a gene is **heterozygous** for that gene. Thus, the parental generation in Mendel's experiments was homozygous for each gene, either dominant or recessive. The F_1 generation was therefore heterozygous for each gene, and the F_2 generation included both homozygous and heterozygous plants.

Genotypes and Phenotypes

For clarity, we distinguish between an organism's genes, individually or collectively, which we call its **genotype,** and the observable characters they control, which we call its **phenotype.** For example, in the garden pea, phenotypes for flower color are purple and white. The genotype can be designated in several ways, but the most common way is by the first letter of the

---

2. The term *law* is derived from a time when naturalists believed they were observing the universal truths of God's plan. Scientists no longer make this assumption. On the contrary, our modern view of science prohibits us from ever finding ultimate or universal truth (see the discussion in Chapter 1 on the scientific method). In our opinion, therefore, the term *law* is no longer useful. (Physicists would disagree, however, since they still refer to many natural phenomena as laws, such as the laws of thermodynamics that were discussed in Chapter 5.)

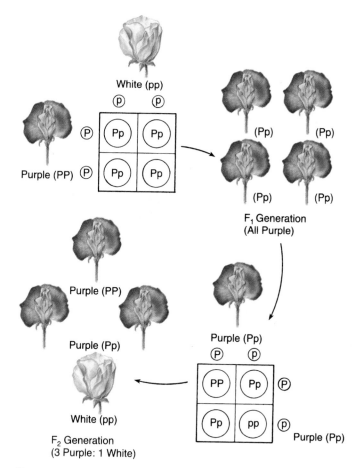

White (pp)

Purple (PP)

F₁ Generation
(All Purple)

Purple (PP)

Purple (Pp)

White (pp)

Purple (Pp)

Purple (Pp)

F₂ Generation
(3 Purple: 1 White)

figure 8.5

Use of a Punnett square to show the inheritance of phenotypes and underlying genotypes in Mendel's cross with flower color. *P* and *p* at top and side of the Punnett squares represent gametes that have dominant (*P*) or recessive (*p*) alleles of the gene for flower color. The genotypes of the offspring are predicted by combining alleles from different gametes into the boxes of the Punnett square.

dominant trait. Accordingly, the gene for flower color is *P* for the dominant allele and *p* for the recessive allele (fig. 8.5). Consequently, the genotype for flower color is *PP* (homozygous) or *Pp* (heterozygous) for purple and *pp* (homozygous) for white.

Note that Mendel counted phenotypes and calculated phenotypic ratios. The phenotypic ratios for all seven features of garden pea in the F₂ generation were the same: 3 dominant to 1 recessive (fig. 8.4). However, what do we know about the genotypic ratios? We know that the genotype for white flowers *(pp)* occurs 25% of the time because of the probability of matching two recessive alleles from heterozygous parents *(Pp)*. Similarly, the probability of matching two dominant alleles *(PP)* would be the same. The remaining 50% of the genotypes are therefore heterozygous. Thus, the 3:1 phenotypic ratio in the F₂ generation is based on a genotypic ratio of 1 *PP* to 2 *Pp* to 1 *pp* (1:2:1).

Genotypic and phenotypic ratios can be conveniently calculated by using a gametic grid called a **Punnett square,**

table 8.2

Mendel's Results from a Dihybrid Cross

	Sample Size	Phenotypes	Ratio
Seed shape/Seed color	556 seeds	315 round/yellow	9.84
		108 round/green	3.38
		101 wrinkled/yellow	3.16
		32 wrinkled/green	1.00

Note: Ratios are calculated by using the wrinkled/green phenotype as the common denominator.

named for R. C. Punnett. This is a checkerboardlike diagram that has the gametic genotype of one parent across the top and the gametic genotype of the other parent down one side (fig. 8.5). In this way, we show that the **gametes** (i.e., sperm and egg) from a cross between two heterozygotes for flower color *(Pp × Pp)* would be *P* and *p* across the top and *P* and *p* down the side. When the gametes of such a cross fuse (fertilize), we add them into an expected ratio of 1:2:1 in the genotypes of the offspring. This ratio is perhaps as simple to obtain by direct inspection as it is by using a Punnett square. Nevertheless, the Punnett square is indispensable for calculating ratios from multiple traits, as discussed next.

Experiments Using Multiple Traits

Mendel also did experiments using two or three pairs of traits at a time. For example, he studied combined inheritance of seed shape (dominant round versus recessive wrinkled) and seed color (dominant yellow versus recessive green) in the same plants. One set of parents was homozygous for both dominant traits *(RRYY)*, and another set of parents was homozygous for both recessive traits *(rryy)*; this means he crossed plants that were true-breeding for round and yellow seeds with plants that were true-breeding for wrinkled and green seeds. All of the seeds produced from this cross were round and yellow. Moreover, since they received dominant alleles from one parent and recessive alleles from the other parent, these seeds (F₁) were heterozygous for both genes *(RrYy)*. This pattern of inheritance conforms to predictions from experiments using one pair of traits at a time. (Note that a Punnett square of this cross would be simple: One parent has only gametes with the genotype *RY*, and the other parent has only gametes with the genotype *ry*. Thus, the sole genotype of their offspring is *RrYy*.)

When the hybrid seeds were grown into mature F₁ offspring and allowed to self-pollinate, they produced the F₂ generation of seeds, which included four phenotypes in the quantities listed in table 8.2. The approximate ratios among these phenotypes are 9 round-yellow to 3 round-green to 3 wrinkled yellow to 1 wrinkled-green (9:3:3:1). This is the expected result of a **dihybrid cross,** which follows two genes that are both heterozygous (fig. 8.6). Using the pea seed

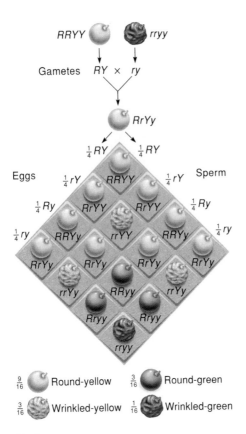

RRYY ○ ● rryy

Gametes RY × ry

RrYy

¼ RY   ¼ RY

Eggs   ¼ rY   RRYY   ¼ rY   Sperm

¼ Ry   RrYY   RrYY   ¼ Ry

¼ ry   RRYy   rrYY   RRYy   ¼ ry

RrYy   RrYy   RrYy   RrYy

rrYy   RRyy   rryy

Rryy   Rryy

rryy

$\frac{9}{16}$ ○ Round-yellow $\frac{3}{16}$ ● Round-green

$\frac{3}{16}$ Wrinkled-yellow $\frac{1}{16}$ Wrinkled-green

figure 8.6

Inheritance in the garden pea. This Punnett square shows the expected pattern of inheritance in a dihybrid cross according to independent assortment of alleles for seed color and seed shape. It predicts four phenotypes in the F_2 generation in a ratio of 9:3:3:1.

example, a dihybrid cross is written *RrYy* × *RrYy*. In a Punnett square, each parent has four gametic genotypes: *RY, Ry, rY,* and *ry* (fig. 8.6). This means that the Punnett square of a dihybrid cross will have 16 boxes (4 × 4). These 16 boxes will contain 9 different genotypes that underlie the 4 phenotypes of the F_2 generation (fig. 8.6).

The phenotypic ratios from Mendel's dihybrid cross can be explained only if the segregation of one pair of traits (green versus yellow seeds) is not influenced by the other pair of traits (round versus wrinkled seeds)—that is, if different genes are inherited independently of each other. Although Mendel did not recognize the 9:3:3:1 ratio of F_2 phenotypes, he did note that different pairs of traits were inherited independently of one another. From this observation we derive postulate 8 (independent assortment) of Mendel's theory of inheritance.

c o n c e p t

Twentieth-century geneticists derived several postulates from Mendel's experiments on inheritance in garden peas. These postulates make up Mendel's theory of inheritance.

Mendel was not only an amateur botanist but also an active member of the local beekeeper's society. Some historians have suggested that he was inspired to begin his genetic studies by a paper published on inheritance in bees. We can imagine, therefore, that he at least considered using bees in his experiments. If he had done so, the theory of inheritance might have been based on Mendel's bees instead of Mendel's peas.

A Note about Meiosis and Chromosomes

Meiosis is a type of nuclear division that produces daughter nuclei that have half of the alleles of the parent nucleus. This feature of meiosis explains how **haploid** cells are produced in certain parts of flowers. Haploid cells have half the amount of nuclear genetic material that their parent cells have. Cells that have the full amount of nuclear genetic material are **diploid** cells. Such a simple description of meiosis accounts for the segregation of alleles into haploid cells that will become gamete-producing structures. Hence, Mendel's peas had two factors for each feature of the plant because they were diploid, and because of meiosis, each diploid parent passed on only one of the factors in each haploid gamete.

Chromosomes were seen as threadlike structures in plant nuclei as early as the 1830s. By the start of the twentieth century, biologists knew that chromosomes existed in pairs and that the pairs separated during meiosis. The parallel separation between pairs of chromosomes and between pairs of alleles during sexual reproduction was the first indication that genes were associated with chromosomes. This was the main postulate of the **chromosomal theory of heredity.** This postulate accounts for the separation of two alleles for each gene on **homologous** chromosomes, which are chromosome pairs that have alleles for the same genes. Our current view on the meiotic separation of homologous chromosomes and their alleles is shown in figure 8.7 for one pair of traits from the garden pea. The figure shows that chromosomes bear genes, that alleles separate on homologous chromosomes during meiosis, and that homologous chromosomes reunite during fertilization.

c o n c e p t

Genes occur on chromosomes. Chromosomes and alleles segregate during meiosis. Haploid nuclei have one set of chromosomes and one allele for each gene.

Meiosis is necessary for sexual reproduction, which means that it has many important consequences for inheritance. A more complete description of the roles of meiosis and chromosomes in heredity is presented in Chapter 10, "Meiosis, Chromosomes, and the Mechanism of Heredity."

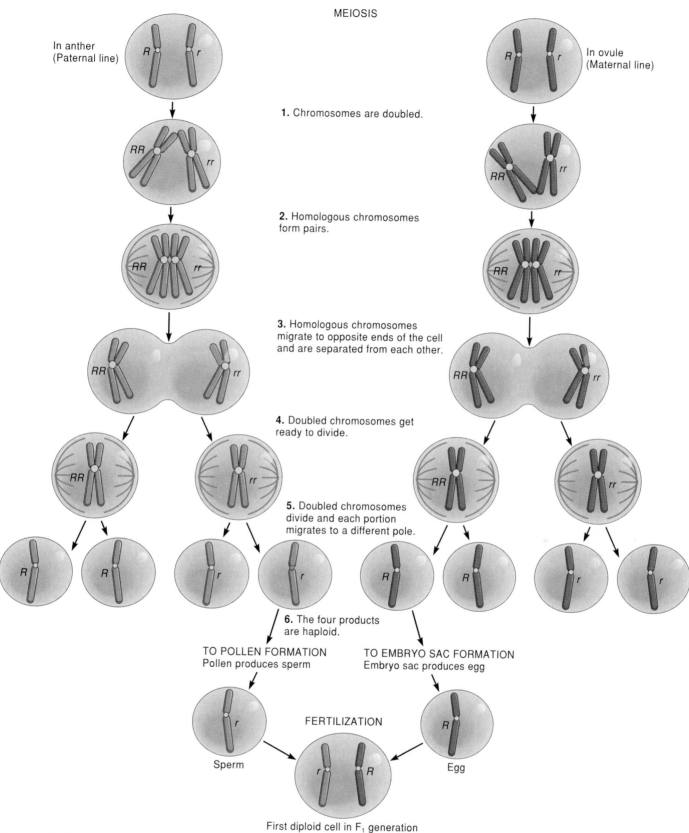

MEIOSIS

In anther
(Paternal line)

In ovule
(Maternal line)

1. Chromosomes are doubled.

2. Homologous chromosomes form pairs.

3. Homologous chromosomes migrate to opposite ends of the cell and are separated from each other.

4. Doubled chromosomes get ready to divide.

5. Doubled chromosomes divide and each portion migrates to a different pole.

6. The four products are haploid.

TO POLLEN FORMATION
Pollen produces sperm

TO EMBRYO SAC FORMATION
Embryo sac produces egg

Sperm

FERTILIZATION

Egg

First diploid cell in F₁ generation

figure 8.7

In anthers, alleles separate during meiosis and form the paternal genotype in the sperm. In ovules, alleles separate during meiosis and form the maternal genotype in the egg. Maternal and paternal alleles are reunited by fertilization. The parental genotype *(Rr)* is reproduced in the F₁ cell in this example, but different combinations of sperm and egg alleles will yield *RR* and *rr* genotypes in other offspring of the same parents.

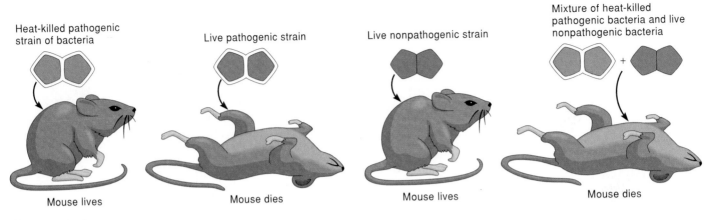

figure 8.8

Evidence for bacterial transformation. Bacteria must be alive and have a polysaccharide coat to be pathogenic. Dead cells or strains without the polysaccharide coat *(blue)* are not pathogenic. However, a mixture of living, coat-free cells (harmless) and dead, coated cells (harmless) kills mice. Bacterial cells cultured from those mice have coats, which means that the living cells are "transformed" by the dead ones.

The Search for the Hereditary Material

In the decades following Mendel's experiments, biologists made important discoveries about the cellular and chemical basis of heredity. New methods and improvements in light microscopy permitted detailed studies of the nucleus and of chromosomes, while advances in chemistry led to the discovery of DNA and nuclear proteins. Although the role of chromosomes in heredity was suspected, this role was not demonstrated until after 1900. The importance of DNA in inheritance was not confirmed until halfway into the twentieth century.

The Importance of the Nucleus in Heredity

The nucleus was the only subcellular body that could easily be seen with a light microscope in the 1800s. Because of this limitation, early studies of the substances of inheritance focused on the potential role of the nucleus. But it was not until the 1870s that scientists agreed that the nucleus was separate from the cytoplasm and that fertilization involved the fusion of two nuclei.

Choosing between Protein and DNA

By the 1880s, nuclei were known to consist mostly of proteins and DNA, together referred to as **chromatin** because this material can be stained with various dyes (see fig. 9.3). The dual chemical nature of chromatin created a difficult puzzle: Which of the two components are genes made of? Without evidence one way or another, the logical choice seemed to be proteins. They consist of complex arrays of amino acids that form many different kinds of proteins. Complex genetic processes, it was thought, must be controlled by complex protein molecules. Conversely, since there were only four different nucleotides in DNA, chromosomal DNA was thought to be too simple a molecule to meet complex cellular demands.

Resolving the issue of DNA versus proteins was difficult because nuclear DNA is associated with proteins in chromosomes. In bacteria and viruses, however, DNA does not occur in chromosomes and is not associated with proteins. Two experiments, one using bacteria and one using viruses, provided substantial support for DNA as the hereditary material.

By 1928, a British microbiologist, Frederick Griffith, had discovered that pathogenic (i.e., disease-causing) strains of the bacterium now called *Streptococcus pneumoniae* could transform nonpathogenic strains into infectious strains (fig. 8.8). Later, in 1944, a group of researchers led by Oswald Avery discovered that the "transforming principle" of this bacterium is DNA. They showed this by introducing DNA from killed, pathogenic strains into cultures of nonpathogenic strains. As a result, the nonpathogenic strains became infectious because of the DNA they absorbed. This result showed that DNA had some influence on regulating the cell like a hormone, although not necessarily as the substance that controlled the cell. In hindsight, Avery's experiments were the first indication that DNA is the hereditary substance. General acceptance of this notion, however, did not come until nearly a decade later, when direct evidence for the hereditary role of DNA was obtained in experiments using viruses.

Viruses have even simpler genetic systems than do bacteria. By the 1950s, viruses were already characterized as being made of nucleic acids surrounded by a protein coat (see Chapter 25). Certain types of viruses, called **bacteriophages**, parasitize bacteria: they attach to bacterial cells, inject their genes into the host, and cause the host cell to make more viruses. In other words, viruses transform bacterial cells into miniature factories for making more viruses. In 1952, Alfred Hershey and Martha Chase grew a strain of virus, called bacteriophage T2, with radioactive isotopes of either sulfur (^{35}S) or phosphorus (^{32}P). The ^{35}S was incorporated into the protein coat of one culture, and the ^{32}P was incorporated into the DNA of the other culture. Each radiolabeled phage was then allowed to infect cells of *E. coli*, its host. However, before the bacterial cells were destroyed, they were agitated to remove the viral coats

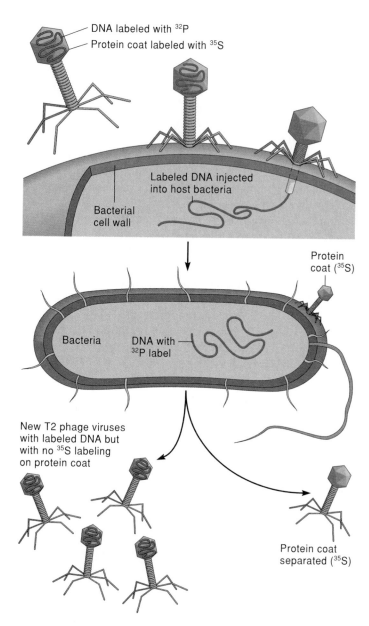

Genes

Mendel's results can be readily appreciated in light of the accepted role of DNA as the hereditary substance and of the nature of genes as sequences of nucleotides. We now know that each of the seven pairs of traits that Mendel studied in garden peas is controlled by a unique sequence of nucleotides—that is, by a single gene. The position of a gene on a chromosome is referred to as that gene's **locus** (plural, **loci**). Thus, Mendel's experiments dealt with seven loci.

At the molecular level, genes are codes for making different molecules, including proteins or protein subunits and different kinds of RNA. The size of a locus ranges from a few dozen to several thousand bases for different genes. Details about how DNA works as a code for proteins and RNA are discussed in Chapter 11.

How does a gene produce a phenotype? Although many genes have been described both by their base sequences and their coded products, we do not understand how the sum of this molecular information becomes a complex organism. Our understanding is generally limited to biochemical reactions catalyzed by enzymes, such as when a single enzyme in flowers of garden pea makes a colored pigment from a colorless precursor. This reaction, controlled by the product of one gene, explains how that gene influences flower color. However, we have no idea how other genes made the flower in the first place.

c o n c e p t

Genes are codes for making other molecules, which make up the phenotype. Exactly how they interact to make a complex organism is mostly unknown.

Complex Inheritance

The traits Mendel used and the patterns of inheritance he discovered are often described as **Mendelian inheritance.** Many other traits in Mendel's peas, however, are inherited differently than the ones he studied. Moreover, the same kinds of traits that Mendel studied in peas may be inherited differently in other plants, and some hereditary patterns do not conform to Mendelian predictions at all. Such complex and non-Mendelian patterns of inheritance are discussed in the next few sections of this chapter. As you will see, Mendel was lucky to have avoided some pitfalls that would have greatly complicated his task.

Types of Dominance

The seven genes studied by Mendel all exhibit **complete dominance,** which is a relatively rare type of inheritance. Complete dominance occurs when one trait completely masks its recessive allele. More frequently, the phenotype for one allele is only partly masked by the other, a condition called **incomplete dominance.** Incomplete dominance occurs when hybrids have a

figure 8.9

Evidence for the role of DNA in heredity. Bacteriophage T2 is radiolabeled with ^{35}S in its protein coat and ^{32}P in its DNA. The bacteriophage reproduces after infecting bacterial cells. New bacteriophages contain radioactive DNA but not radioactive protein, which means that the DNA was inherited, but the protein was not.

attached to their cell walls. After they were separated from the cells, the empty viral coats contained ^{35}S, which meant that the coats contained only protein (recall that DNA does not contain sulfur). Conversely, the bacterial cytoplasm contained ^{32}P that had been incorporated into the viral DNA (proteins lack phosphorus). Thus, only genes made of DNA could have entered the bacteria and caused their transformation to produce more viruses (fig. 8.9). These bacteriophage experiments confirmed the role of DNA in heredity and added to our rapidly growing knowledge of this molecule.

figure 8.10

Camellias *(Camellia japonica)* show incomplete dominance in flower color. Pink flowers are heterozygous for flower color *(Rr)*; red and white are homozygous for flower color *(RR* and *rr*, respectively).

phenotype intermediate between those of the two parents. For example, the allele for red flowers in camellia *(Camellia japonica)* is incompletely dominant over the allele for white flowers. As a result, the F_1 progeny from a cross between red-flowered and white-flowered camellias all have pink flowers. The phenotypic ratio in the F_2 offspring is 1:2:1 (25% red, 50% pink, 25% white) (fig 8.10). Accordingly, in cases of incomplete dominance, the phenotypic and genotypic ratios are the same.

Codominance occurs when both alleles of a heterozygote are expressed equally, so there is really no dominance at all. Codominance is common for heterozygous genes that code for two equally functional enzymes. This means that there is more than one form of the same enzyme. The differing forms of enzymes made by different alleles of the same locus are called **allozymes**. Although allozymes catalyze the same reaction, they differ from each other by one or a few amino acids, which makes them slightly different from each other in size and overall electric charge. For example, in wild sunflower *(Helianthus debilis)*, there are allozymes of phosphoglucoisomerase, which catalyzes one of the first reactions in glycolysis (Chapter 6). Heterozygotes produce both forms of the enzyme, but homozygotes produce only one or the other (see reading 8.2).

Multiple Alleles of the Same Gene

Diploid plants have only two alleles at a single locus. The two allozymes at a locus are usually referred to as *fast* or *slow* forms, depending on how far they move on an electrophoresis gel (see reading 8.2). Also, some populations of plants have more than two allozymes for some loci. For example, most

---

wild sunflowers have fast or slow forms of the allozymes of the phosphoglucomutase gene, *Pgm-3*, but other individuals of this species have additional forms that replace either the fast or slow allozyme. This means that there are more than two allozymes for *Pgm-3*. This example shows that even though each diploid plant can have a maximum of two allozymes, a population of plants can have more than two allozymes.

Botanists are not sure why populations of plants have multiple forms of the same enzyme, but such variation may be adaptive. There are two kinds of indirect evidence for this explanation. One is that allozymes work at different optimum pHs, temperatures, or other conditions. We imply from this evidence that allozyme variation enables plants to thrive under a range of environmental conditions. The second kind of evidence is that certain allozymes occur more frequently in populations, for example, at higher elevations, in wetter soils, or within shadier forests. In this case, allozymes are thought to be adaptive because their occurrence is correlated with where certain populations live. Nevertheless, we have no strong direct evidence for the adaptiveness of allozymes.

Multiple Genes

Allelic variation of a single gene is often complicated by the presence of more than one gene for the same enzyme. The same enzymes from different genes are called **isozymes** to distinguish them from allozymes. Multiple isozymes are common in plants. For example, *Helianthus debilis* has two nuclear genes and two chloroplast genes for phosphoglucomutase (phosphoglucomutase and other glycolytic enzymes in chloroplasts are thought to be holdovers from the prokaryotic ancestors of these organelles). The four genes for phosphoglucomutase in this sunflower make a total of nine forms of the enzyme.

The most frequent application of isozyme/allozyme studies involves population genetics. Botanists routinely study the variability patterns of more than two dozen types of enzymes, many having isozymic and allozymic forms. By studying the distribution of these enzymes in plant populations, we can learn about the reproductive biology of plants. For example, populations with relatively high numbers of heterozygous loci indicate a high level of cross-pollination (i.e., pollination between plants). Fewer heterozygous loci are associated with a high level of self-pollination (i.e., pollination within a single flower or between flowers of the same plant). Further, when all samples from a population have the same allozymes and isozymes, the "population" is probably a clone of the same genotype—that is, sexual reproduction is absent. In contrast, a relatively high level of allozyme/isozyme variability at the edge of a population may indicate hybridization with a nearby species.

Serial Gene Systems

Unlike the independently expressed genes of Mendel's peas, genes often act together to control one or more characteristics.

This phenomenon can occur, for example, in the multistep biosynthesis of a complex molecule. Such a serial system of genes consists of two or more Mendelian genes acting together to complete a developmental sequence. This interaction of two or more genes is called **epistasis.**

In most plant species, flower color is controlled by the epistatic effects of several genes in a series. In the snapdragon (*Antirrhinum majus*), seven different flower colors are controlled by a series of four genes (fig. 8.11). One of these, called the *nivea* gene (abbreviated *niv*), blocks the synthesis of a precursor molecule when both of the gene's alleles are recessive. When this occurs, there is no pigment, and the flowers are white. When at least one dominant allele occurs at the *niv* locus, flower color depends on the genotypes of one or more of the other three genes. Figure 8.12 shows the genotypes of flower colors in snapdragons.

Polygenic Inheritance

In his less well-known experiments with garden beans *(Phaseolus multiflorus)*, Mendel noticed that flowers of F_2 plants exhibit a continuous series of colors ranging from purple to white. He suggested that two or three independent factors control a single color trait, but he provided no further evidence to support this hypothesis. We now know that heritable variation in many characteristics appears to be continuous. Body weight, plant height, quantity of seed oil, and length of corn ears are examples of continuous characters.

Multiple genes can act additively, without dominance, to control a continuous trait. Such genes are referred to as **polygenes.** For example, polygenes control the inheritance of corolla length in *Nicotiana longiflora* (fig. 8.13a), a species of wild tobacco. When long-corolla plants were crossed with short-corolla plants, measurements of this character in the F_1 generation yielded the distribution of phenotypes shown in figure 8.13b. The F_2 generation showed a similar distribution of phenotypes. This pattern of inheritance is attributed to the polygenic control of corolla length.

Expression by polygenes, especially those involving size, can be modified by temperature, the availability of water and nutrients, or by other environmental factors. Because of the potential complexity of interactions between polygenes and the environment, the phenomenon of polygenic control is poorly understood. We can only assume that traits that vary continuously are at least partially controlled by polygenes.

Pleiotropy

Genes that affect more than one phenotypic characteristic are called **pleiotropic genes.** Mendel may have encountered a pleiotropic gene in the garden pea—the one that influences the color of flowers and seed coats (fig. 8.3). Purple or purplish coloration occurs in all of the organs or in none of them, thereby behaving like a group trait controlled by a single gene.

figure 8.11

Snapdragons *(Antirrhinum majus)* have seven different flower colors, most of which are shown in this photograph. These colors are controlled by different combinations of alleles from four genes (see fig. 8.12).

Color	Genotype for four genes			
	Nivea	Sulfurea	Incolorata	Eosinea
white	*niv/niv*	*NA*	*NA*	*NA*
yellow	*niv⁺/—*	*sulf/sulf*	*inc/inc*	*NA*
ivory	*niv⁺/—*	*sulf⁺/—*	*inc/inc*	*NA*
bronze	*niv⁺/—*	*sulf/sulf*	*inc⁺/—*	*eos/eos*
pink	*niv⁺/—*	*sulf⁺/—*	*inc⁺/—*	*eos/eos*
crimson	*niv⁺/—*	*sulf/sulf*	*inc⁺/—*	*eos⁺/—*
magenta	*niv⁺/—*	*sulf⁺/—*	*inc⁺/—*	*eos⁺/—*

figure 8.12

Control of flower color in snapdragons by four genes. Gene loci are designated by three- or four-letter abbreviations in italics. The "+" superscript indicates the dominant allele. A blank is used for the second allele when the dominant allele is present, because the second allele has no effect on the phenotype. *NA* means not active when the preceding gene in the table is homozygous recessive.

Source: Data from Hans Stubbe, 1966, and Verne Grant, Genetics of Flowering Plants, *1975.*

A clearer and more dramatic example of pleiotropy occurs in common tobacco *(Nicotiana tabacum)*. In this species, the sizes and shapes of leaves, flowers, anthers, and fruits segregate into two sets of phenotypes controlled by the *S* gene. Plants with at least one dominant allele (*SS* or *Ss*) grow longer and narrower organs. Plants that are homozygous for the recessive allele *(ss)* have shorter and broader structures (fig. 8.14).

A pleiotropic gene also influences development in the European columbine *(Aquilegia vulgaris)*, which has a dwarf form in which secondary cell walls thicken earlier than normal. Because of this early wall development, dwarfs are shorter than normal plants and have brittle stems, erect flower buds, smaller flower parts, and more branches. This set of traits is controlled by a single homozygous recessive gene.

(a)

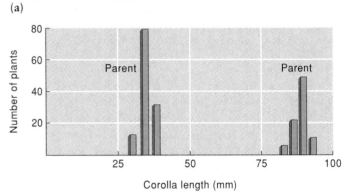

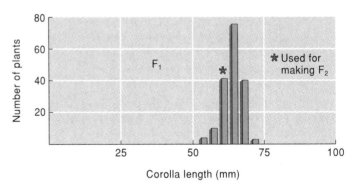

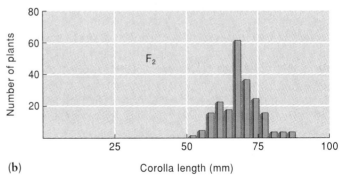

(b)

figure 8.13

(a) Flowers of *Nicotiana longiflora*. (b) Frequency distribution of corolla lengths in *Nicotiana longiflora*. Both the F_1 and F_2 generations show continuous variation of corolla length, which indicates polygenic inheritance.

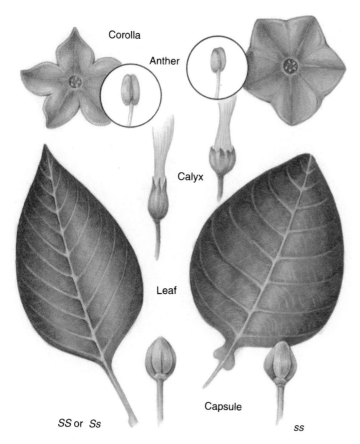

figure 8.14

The pleiotropic effects of two alleles of the *S* gene on various parts of the tobacco plant. The *S* allele is dominant; genotypes *SS* and *Ss* have longer and narrower organs, whereas organs of the *ss* genotype are shorter and broader.

---

c o n c e p t

Complex inheritance is revealed by various combinations of Mendelian genes. These include polygenic control of a single trait, multiple enzymic phenotypes from different loci, and single genes that control more than one phenotypic characteristic.

Non-Mendelian Inheritance

Aside from such complicating factors as polygenic control or pleiotropy, even the inheritance of many single-gene traits does not yield Mendelian ratios. The most common causes of non-Mendelian inheritance are linkage, cytoplasmic inheritance, mutations, and transposable elements.

Linkage

Many genes occur on each chromosome. When two or more genes occur on the same chromosome, it seems as if they should be inherited together, not independently. The simulta-neous inheritance of genes on the same chromosome is called genetic **linkage**. Linkage violates the postulate of independent assortment of genes.

Linkage was first reported in the sweet pea *(Lathyrus odoratus)*, a relative of the garden pea. In 1906, geneticists at Cambridge University discovered that genes for flower color and pollen shape did not assort into the expected 9:3:3:1 phenotypic ratio in the F_2 generation. Instead, the ratio was 7:1:1:7 (fig. 8.15). This means that almost 44% (7/16) of the F_2 plants had the dominant flower color (purple) and the dominant pollen shape (oblong), while an equal proportion had the recessive flower color (red) and the recessive pollen shape (spherical). This pattern indicates that the purple-oblong and red-spherical phenotypes are inherited together, which means that the genes for these traits are linked.

The main puzzle of linkage, however, was that some of the F_2 plants were red-oblong and some were purple-spherical, unlike the purebred parents or the hybrids of the F_1 generation. If linkage is perfect, it seems that these phenotypes should not occur. Imperfect linkage occurs when chromosomes exchange complementary fragments during meiosis. As a result, fragments that have different alleles for linked genes may be rearranged to produce nonparental combinations of alleles. This process is called **crossing-over** and is discussed more completely in Chapter 10, "Meiosis, Chromosomes, and the Mechanism of Heredity."

How did Mendel avoid results influenced by linkage? He noted that traits in a dihybrid combination and a trihybrid (three-character) combination showed independent inheritance. The simplest explanation for this pattern is that each of the genes used in Mendel's experiments occurs on a separate chromosome. This reasoning is correct for his dihybrid cross, because the gene for seed shape is on chromosome 7 and the gene for seed color is on chromosome 1 (fig. 8.3). However, this reasoning is incorrect for color of flowers, seedling axils, and seed coat because genes for these features are all located on chromosome 1. Nevertheless, we can explain independent assortment of linked genes on chromosome 1 by frequent crossing-over between them. Such frequent crossing-over causes genes to assort independently even though they are on the same chromosome.

Our assumption of frequent crossing-over can account for all but one of the pairs of linked genes in Mendel's peas: the genes that control pod shape and plant height. Modern geneticists have shown that these genes do not assort independently because they are too close to each other for frequent crossing-over to occur between them. Fortunately for Mendel, his experiments were not confounded by this result because he apparently did not make hybrids with this gene combination.

---

c o n c e p t

Each chromosome has many genes. Genes on the same chromosome may not assort independently because they are linked. Because of crossing-over, however, linked genes can be rearranged into nonparental combinations of alleles and mimic independent assortment.

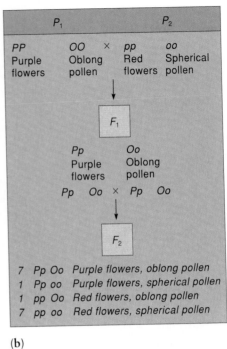

P_1			P_2	
PP	*OO*	×	*pp*	*oo*
Purple flowers	Oblong pollen		Red flowers	Spherical pollen

F_1

Pp	*Oo*
Purple flowers	Oblong pollen

Pp   Oo   ×   *Pp   Oo*

F_2

7	*Pp Oo*	Purple flowers, oblong pollen
1	*Pp oo*	Purple flowers, spherical pollen
1	*pp Oo*	Red flowers, oblong pollen
7	*pp oo*	Red flowers, spherical pollen

(a)

(b)

figure 8.15

(a) Sweet pea *(Lathyrus odoratus)* in bloom. (b) A dihybrid experiment with sweet pea does not yield the expected 9:3:3:1 ratio of phenotypes in the F_2 generation because of linkage between the loci for flower color and pollen shape. Instead, most of the offspring are like the parents (P_1 generation). However, linkage is not perfect, since a small number of nonparental combinations (red-oblong and purple-spherical) appear in the F_2 generation.

Cytoplasmic Inheritance

You learned in Chapter 3 that chloroplasts and mitochondria contain DNA. Genes in these organelles control certain aspects of photosynthesis and respiration, respectively. Inheritance of these genes is independent of sexual reproduction because they are transmitted to offspring with the cytoplasm, often that of the maternal parent (see reading 8.3, "Cytoplasmic Male Sterility in Corn").

One example of cytoplasmic gene control occurs in certain forms of the cultivated four-o'clock *(Mirabilis jalapa)* that have yellowish white leaves instead of green leaves. This difference in leaf color is caused by defective chloroplast genes. Phenotypic expression depends solely on the seed parent. Thus, when pollen from a white-leaved plant is transferred to a green-leaved plant, all the offspring have green leaves. In contrast, all of the offspring of the reciprocal cross have white leaves. This is an example of the **cytoplasmic inheritance** of non-nuclear genes.

The cooperation of organellar and nuclear genes is often necessary for normal metabolism. For example, the photosynthetic enzyme ribulose-1,5-bisphosphate carboxylase/oxygenase (rubisco; Chapter 7) has two subunits, one derived from a nuclear gene and one from a chloroplast gene. Similarly, some ATPases (Chapter 4) have a dual origin between the nucleus and mitochondria. In each case, the final product—that is, a complete and functional enzyme—depends on genes from two sources in the same cell.

Mutations

Dutch botanist Hugo de Vries was one of the rediscoverers of Mendel's work. De Vries studied patterns of inheritance in several kinds of evening primrose *(Oenothera* species). Most of his results conformed to Mendelian inheritance, but occasional characteristics appeared that were not present in either parent. De Vries called these spontaneous hereditary changes **mutations,** a term that remains in widespread use today. In modern genetics, the term *mutation* includes a variety of genetic changes, including chromosomal rearrangements and sequence changes in DNA. Details of both types of mutations are presented in Chapters 9 and 10.

Transposable Elements

One of the most important kinds of mutations involves sequences of DNA that seem to multiply and move spontaneously among an organism's chromosomes. These movable pieces of DNA are called **transposable elements** and have been estimated to account for as many as half of the plant's spontaneous mutations. Such fragments of DNA may contain one or more genes that can function as inhibitors, modifiers, or mutators, thereby affecting the action and phenotypic expression of ordinary genes in many ways. Inserting transposable elements into or near functional genes produces unpredictable patterns of inheritance.

Transposable elements were first identified in the 1940s by Barbara McClintock (see fig. 1.5), who observed

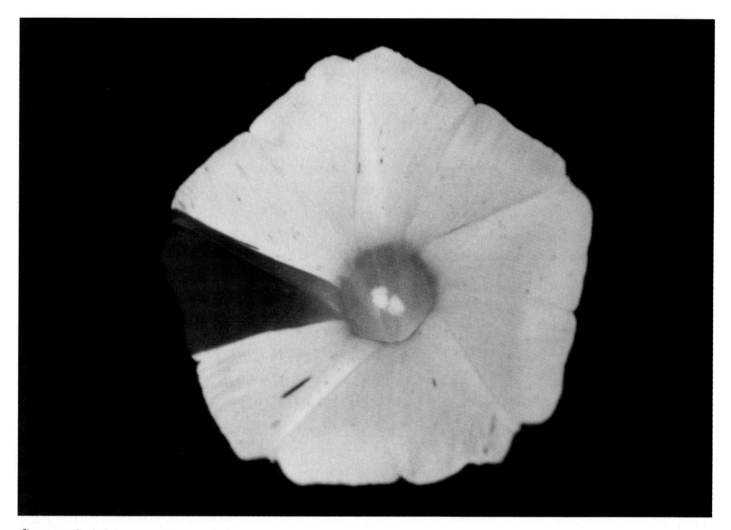

figure 8.16

Photograph of a morning glory flower shows the effects of transposable elements on flower color. White sections of the flower occur where transposable elements disrupt the biosynthesis of pigments.

© *Evelyne Cudel-Epperson, University of California, Riverside, 1992.*

their effects on pigment patterns in corn kernels. Specifically, McClintock noticed that a fully pigmented grain was often flanked by a totally unpigmented grain on one side and on the other side by a grain having pigment over only a part of its area (fig. 1.5b). If all of the cells in, for example, the endosperm of a given grain are derived from a single cell, why weren't they all alike? McClintock proposed that portions of kernels remained white, even though genes for pigment synthesis were present, because the pigment genes were disrupted by "controlling [transposable] elements." McClintock noted that reversion to the wild, nondisrupted genotype occurred when these elements moved away from the affected genes. Normal pigmentation appeared in cells in which reversion occurred. The pigment-inhibiting mutation was so unstable that many groups of cells reverted to the wild-type pigmentation as each kernel developed. This produced a patchwork of colored spots and streaks, distributed in seemingly random patterns among colorless portions of the kernel (fig. 1.5). For McClintock, examining an ear's variously colored grains was like reading the history of the movement of transposable elements in cells.

Recent research has shown that coloration patterns in snapdragons and morning glories *(Ipomoea purpurea)* are also caused by transposable elements. Variegated flower colors result from complex interactions between mobile genes and genes for pigment synthesis (fig. 8.16). Similar color patterns are also probably caused by transposable elements in many other species of plants.

Transposable elements are now accepted as a general feature of many organisms and are widely used to induce mutations. Scientists believe, however, that since not all genes are active at the same time, the potential effects of transposable elements on gene regulation vary. For example, transposable elements may help regulate cancer-causing genes, which are

CYTOPLASMIC MALE STERILITY IN CORN

The development of anthers and pollen in hybrids is often sensitive to interactions between nuclear and cytoplasmic genes. In many species, certain nuclear genotypes interact with specific mitochondrial genes to produce sterile pollen. The phenotype is then said to show **cytoplasmic male sterility (cms).** Crop-plant breeders can take advantage of male-sterile plants in controlled hybridizations. One crop that has been used in this manner is corn *(Zea mays).* Corn plants have two kinds of flowers, one for pollen and one for kernels, both occurring on the same plant. The pollen flowers occur in **tassels** at the tops of each plant, so that most of the pollen falls downward to fertilize the kernels on the same plant (reading fig. 8.3a). To make hybrids, the tassels must be removed. This step becomes very expensive when whole fields must be cross-pollinated to produce large quantities of hybrid corn. Fortunately, certain varieties of corn have cytoplasmic male sterility (i.e., produce sterile pollen). Consequently, it is unnecessary to remove tassels before making hybrid kernels.

One particular variety of corn, which contains what is called Texas cytoplasm, or *cms-T,* was widely used for the commercial production of hybrid seed corn before about 1970. At its peak use, it was estimated that more than 85% of the corn grown in the United States contained *cms-T.* Unfortunately, heavy reliance on *cms-T* became

reading figure 8.3a

Corn plant *(Zea mays)* in full bloom. Upper arrow: tassels of pollen flowers. Lower arrow: husks containing cobs of seed flowers. Cornsilk threads protruding from the husks are stigmas of the seed flowers.

reading figure 8.3b

Corn husks infected with southern corn blight *(Helminthosporium maydis).* This corn has a strain of cytoplasmic male sterility called *cms-T,* which is susceptible to infection by the fungus.

disastrous for growers, beginning in 1970. At that time, all *cms-T* varieties became susceptible to southern corn blight, a disease caused by the fungus *Helminthosporium maydis* (reading fig. 8.3b). This fungus destroyed *cms-T* varieties of crops in epidemic proportions. The susceptibility factor was subsequently found to be controlled by mitochondrial genes associated with the T-cytoplasm.

Male-sterile corn continues to be widely used for making hybrid corn. This means that there is a continuing need for studying the interactions between cytoplasmic and nuclear genes. With the advent of molecular biology, cytoplasmic male sterility has become one of the most intensely studied subjects in plant genetics.

---

widespread among humans but are normally not activated. The importance of transposable elements led to the awarding of a Nobel Prize to Barbara McClintock in 1983.

c o n c e p t

Many kinds of non-Mendelian inheritance patterns cannot be explained by linkage. Changes in the base sequences of DNA, either by random mutation or by the insertion of transposable elements, alter inheritance patterns in unpredictable ways. Also, the inheritance of organellar genes does not conform to Mendelian predictions. In many plants, plastid and mitochondrial chromosomes are inherited with the cytoplasm of the female gamete.

Chapter Summary

The basic postulates for the theory of inheritance were derived from experiments by Gregor Mendel. However, we now know that the heredity of an organism is much more complex than it initially appeared. Exceptions to the basic postulates occur for several reasons, including linkage, cytoplasmic inheritance, mutations, and transposable elements. Moreover, complex patterns of inheritance arise from multiple genes, serial genes, or interactions between genes.

Chromosomes are made of DNA and proteins. Genes were originally thought to be made of proteins, but experiments with bacteria and viruses, which do not have DNA associated with protein, showed that the hereditary substance is DNA.

Different alleles control the production of alternative forms of an enzyme. These enzymes are called *allozymes* when they are made by the same gene locus and *isozymes* when they are produced by different loci. Several allozymes often occur for a single locus, indicating that multiple alleles of the locus exist among the individuals of a population.

Interactions of many genes can contribute to the expression of a quantitative characteristic. The expression of such polygenic characters (such as plant height) can be plotted on a continuous curve. Polygenic inheritance is usually influenced by environmental factors.

Genes that occur on mitochondrial or chloroplast chromosomes control cytoplasmic inheritance. The expression of nuclear genes occasionally depends on these organellar genes. Examples include enzymes (such as rubisco) having at least two subunits, one derived from a nuclear gene and one from a mitochondrial or chloroplast gene.

Mutations are changes in the positions or composition of genes. One type of mutation that occurs frequently involves spontaneous movement of DNA segments called *transposable elements*. The expression of ordinary genes is often influenced by their proximity to these mobile elements.

What Are Botanists Doing?

Compile a list of plants that are known to have transposable elements. What kinds of observations do botanists use to indicate the presence of transposable elements? What kinds of evidence do they use to confirm such a suggestion?

Writing to Learn Botany

Hybridization experiments should ideally include reciprocal crosses. Why?

Questions for Further Thought and Study

1. Suppose you have a purple-flowered garden pea. What would its genotype be if it were crossed with a white-flowered individual and the phenotypic ratio of F_1 plants was 1:1?

2. Assume that a red and a white allele exist for flower color and that a blue and a yellow allele exist for pollen color. What phenotypic ratios would be predicted in a dihybrid cross in which both traits showed incomplete dominance?

3. What would the genotypic ratios be in the cross described in question 2?

4. Why was DNA discounted as the hereditary substance for so long?

5. Avery's results in 1944, from which he suggested that hereditary material is made of DNA and not protein, were not widely accepted until 1952. What objections to his experiments do you think might have caused this delay?

6. What happened in 1952 that confirmed Avery's findings?

7. Explain why a 9:7 phenotypic ratio in the F_2 generation indicates control by two genes.

8. Why is the inheritance of chloroplast and mitochondrial genes non-Mendelian?

9. How can one parent plant reproduce sexually?

10. Describe how the movement of transposable elements could produce the patchwork pigmentation of the corn kernels shown in figure 1.5. Account for fully pigmented kernels, unpigmented kernels, and kernels having varying degrees of pigmentation.

11. Barbara McClintock's work was either neglected or disbelieved for almost 40 years, after which microbiologists obtained independent evidence for the existence of transposable elements in bacteria. Well into her eighties, McClintock was awarded a Nobel Prize and saw her work embraced and extended by a new generation of biologists. McClintock, who died in 1992 at the age of ninety, lived to witness the results of the revolution she had wrought. What does her story tell you about science and scientists?

Web Sites

Review the "Doing Botany Yourself" essay and assignments for Chapter 8 on the *Botany Home Page*. What experiments would you do to test the hypotheses? What data can you gather on the Web to help you refine your experiments? Here are some other sites that you may find interesting:

http://www.netspace.org/MendelWeb/
- *MendelWeb*
  http://hermes.astro.washinghton.edu:80/mirrors/MendelWeb/
- *MendelWeb*
  http://www.stg.brown.edu/MendelWeb/

MendelWeb is an educational resource for teachers and students interested in the origins of classical genetics, introductory data analysis, elementary plant science, and the history and literature of science.

http://www.botany.duke.edu/DCMB/Chlamy.htm

The *Chlamydomonas* Genetics Center is maintained at Duke University.

http://www.vxzbr.cz/mendel.html

This site, which is maintained by the Mendel University of Agriculture and Forestry, includes a short biography of Mendel.

http://nasc.nott.ac.uk:8200/nasc.html

This site includes links to a variety of genetics-related sites throughout the world.

Suggested Readings

Articles

Blixt, S. 1975. Why didn't Mendel find linkage? *Nature* 256:206.

Corcos, A., and F. Monaghan. 1985. Some myths about Mendel's experiments. *The American Biology Teacher* 47:233–236.

Corcos, A., and F. Monaghan. 1990. Mendel's work and its rediscovery: A new perspective. *Critical Reviews in Plant Sciences* 9:197–212.

Dahl, Hans-Henrik M. 1993. Things Mendel never dreamed of. *Medical Journal of Australia* 158:247–254.

Federoff, N. 1984. Transposable genetic elements in maize. *Scientific American* 249 (June):85–98.

Hartl, D. L. 1992. What did Mendel think he discovered? *Genetics* 131:245–254.

Huckabee, C. J. 1989. Influences on Mendel. *The American Biology Teacher* 51:84–88.

Janick, J. 1990. Gregor (Johann) Mendel (1822–1884). *HortScience* 25:1211–1213.

Laughnan, J. R., and S. Gabay-Laughnan. 1983. Cytoplasmic male sterility in maize. *Annual Review of Genetics* 17:27–48.

Rogers, J. 1991. Mechanisms Mendel never knew. *Mosaic* 22(3):2–7.

von Tschermak-Seysenegg, E. 1951. The rediscovery of Mendel's work. *Journal of Heredity* 42:163–71.

Books

Corcos, A. F., and F. V. Monaghan. 1993. *Gregor Mendel's Experiments on Plant Hybrids: A Guided Study.* New Brunswick, NJ: Rutgers University Press.

Fowler, C., and P. Mooney. 1990. *Shattering: Food, Politics, and the Loss of Genetic Diversity.* Tucson: University of Arizona Press.

Galston, A. W. 1994. *Life Processes of Plants.* New York: Scientific American Library.

Grant, V. 1975. *Genetics of Flowering Plants.* New York: Columbia University Press.

John, B., and G. Miklos. 1988. *The Eukaryote Genome in Development and Evolution.* Boston: Allen and Unwin.

Klug, W. S., and M. R. Cummings. 1995. *Concepts of Genetics.* New York: Macmillan.

Murphy, T. M., and W. F. Thompson. 1988. *Molecular Plant Development.* Englewood Cliffs, NJ: Prentice-Hall.

Olby, R. C. 1985. *The Origins of Mendelism.* 2d ed. Chicago: University of Chicago Press.

Roberts, H. F. 1929. *Plant Hybridization before Mendel.* Princeton, NJ: Princeton University Press.

Watson, J. D. 1968. *The Double Helix.* New York: Atheneum.

The Cape Sundew (*Drosera capensis*) on the left and the Boston Fern (*Nephrolepis exaltata*) on the right were both grown from cultured meristematic tissue.

chapter

The Cell Cycle

9

Chapter Outline

Chapter Overview

Every plant begins as a single cell and grows as cells repeatedly expand and divide. As the plant matures, however, only a few cells continue to divide. Most cells instead become specialized for a limited set of functions, such as photosynthesis, storage, and transport. Cells that continue to divide occur only in specific regions of the plant, such as stem tips, root tips, and buds; in these regions, cellular expansion and division can theoretically continue indefinitely. The uninterrupted repetition of cell expansion and division, together called the cell cycle, gives plants the potential for unlimited growth. Some plants, such as giant redwoods, come closer to this potential than others.

Most of the attention devoted to the cell cycle involves the mechanisms of DNA synthesis and the behavior of chromosomes during mitosis. These subjects are the focus of this chapter.

Virtually all living cells in a plant have identical sets of chromosomes in their nuclei. This observation leads to the prediction that each cell is **totipotent**—that is, that each cell has the same genes and therefore the same genetic potential to make all other cell types. The first confirmation of this prediction came from experiments in the 1950s by F. C. Steward and his colleagues at Cornell University (fig. 9.1). They began by growing small pieces of tissue from carrots in a nutrient broth. Cells that broke free from the fragments dedifferentiated, meaning that they reverted to unspecialized cells. As these unspecialized cells grew, however, they divided and redifferentiated back into specialized cell types. Eventually, cell division and redifferentiation produced new plants. Each unspecialized cell from the nutrient broth expressed its genetic potential to make all the other cell types in a plant (fig. 9.1).

Whole plants of many species are now routinely regenerated from cell cultures. This process is used, for example, to grow clones of orchids and other ornamental plants inexpensively and to make disease-free clones of potatoes and other crop plants. Cloning by cell culture is also important in genetic engineering of plants, which is discussed in Chapter 11.

Experiments that show totipotency led to one of the biggest puzzles in genetics: how do identical genomes dictate the differential expression of the same genotype in different cells? In other words, how can some genes be active while others are not? Solutions to such a complex puzzle lie in the answers to a multitude of questions about the functions of chromosomes and their DNA. Some questions have been answered, but many seemingly acceptable hypotheses still lack experimental support. Most processes remain largely unknown. Much of our understanding of genetic mechanisms comes from studying the changing structure and behavior of chromosomes during cellular growth and division, especially during DNA synthesis.

The repeating processes of cellular growth and division are known collectively as the **cell cycle.** These processes occur in specific tissues during normal plant growth. In this chapter we introduce the concept of the cell cycle briefly from a historical perspective and then present the major events of cellular growth and reproduction in plants.

The Cell Cycle in Plants

The cell cycle is complete only in cells that divide. The first dividing cell in the life history of a plant is the **zygote,** which forms by fusion of a sperm and an egg. The zygote grows into a pre-embryo by cellular division and expansion. The pre-embryo becomes an embryo that has specialized regions, called **meristems,** that undergo cell division. (Chapter 12 presents a complete discussion of meristems.) As a seed germinates, cells at the growing tips of embryonic roots and shoots never stop dividing; the cell cycle in these cells is complete. In contrast, cells in leaves and certain tissues of stems and roots stop dividing and become specialized; the cell cycle in these cells is arrested.

Cells also divide and grow in several other sites in plants. For example, flowers, fruits, and branches arise by cell division in buds, which contain meristems that are active only during specific periods of plant development. Such buds usually occur in the axils of leaves. Branches in roots also arise from meristems, but lateral meristems in roots are internal and are not derived from axillary buds (see Chapter 12). Furthermore, many kinds of nonmeristematic cells in stems and roots and, in the leaves of some plants, can be stimulated to divide when they are damaged. Cell division in damaged tissues continues until the wound is sealed with a layer of protective cork. Thus, the cell cycle is complete, when needed, for the formation of reproductive organs, stem and root branches, and wound-repair tissue.

Overview of the Cell Cycle

In 1858, nearly 200 years after cells were first seen through a microscope, the German physiologist Rudolf Virchow proposed that all cells must come from preexisting cells. Since then, we have discovered that the formation of new cells includes both nuclear and cytoplasmic processes. During the past century, however, much more attention has been given to events that occur in the nucleus. Emphasis on the nucleus began because chromosomes were relatively easy to see with light microscopy. Nuclear processes still comprise the most intensely studied aspects of the cell cycle.

The cell cycle has two main parts: cell growth and cell division. The phases of cell growth are collectively called **interphase,** and the phases of cell division are collectively called **mitosis** and **cytokinesis.** Mitosis is the division of nuclei,

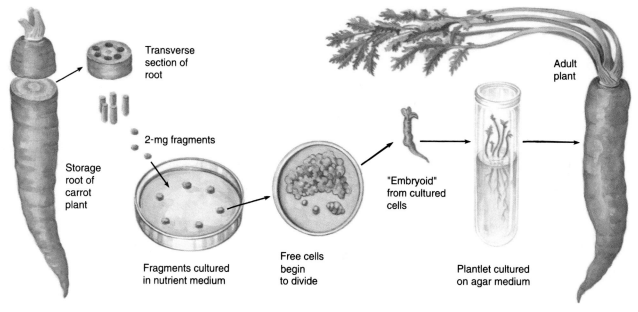

figure 9.1

The main steps involved in culturing vegetative cells and regenerating a new plant from cell culture. Unspecialized cells in suspension divide and redifferentiate into all of the cell types needed to grow into another plant. The results of this experiment are evidence that each cell in a carrot has the genetic potential to make all other cell types.

and cytokinesis is the division of cytoplasm. Several shorter phases can be distinguished in each of the two stages of the cell cycle. Interphase includes three phases. The first is the G_1 phase (*first gap*), which occurs between the end of mitosis and the onset of DNA synthesis. This is followed by the **S phase**, which is when DNA is synthesized. The third part of interphase is the G_2 phase (*second gap*), which begins at the end of the S phase (i.e., when DNA synthesis is complete) and lasts until the beginning of mitosis.

Mitosis is a continuous process that is divided for convenience into four phases: **prophase, metaphase, anaphase,** and **telophase.** Cytokinesis usually follows telophase but may begin before telophase is completed. In certain tissues of plants and other organisms, cytokinesis is delayed or does not occur at all. Cells in these tissues are therefore multinucleate, either temporarily or permanently (see fig. 3.29).

concept

The two main stages of the cell cycle are growth (interphase) and division (mitosis and cytokinesis). The cell cycle is a continuous process, but several phases can be recognized in each stage.

Interphase

Most meristematic cells spend about 90% of their time in interphase (fig. 9.2). However, throughout the first half of this century, little was known about what happens during interphase,

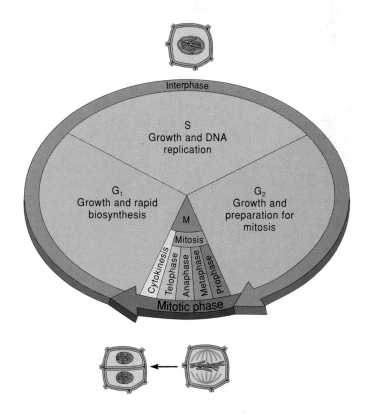

figure 9.2

Periods of the cell cycle. Most meristematic cells spend about 90% of their time in interphase and about 10% of their time in mitosis and cytokinesis.

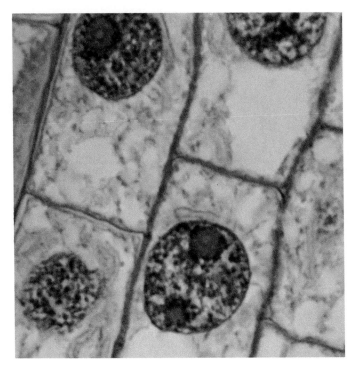

figure 9.3

Interphase nucleus from a cell of an onion root tip *(Allium cepa),* stained to show diffuse chromatin, ×500. DNA synthesis and cell growth occur while chromatin remains diffuse.

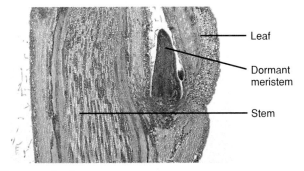

figure 9.4

Dormant axillary bud in a longitudinal section of a pear stem *(Pyrus communis),* ×40. Cells in these meristems are usually arrested in the G_1 phase of the cell cycle.

since nothing but diffuse chromatin could be seen by light microscopy for this part of the cell cycle (fig. 9.3). It was instead believed, incorrectly, that everything important in the cell cycle happened immediately before and during mitosis. This included chromosome doubling and other mysterious processes, both nuclear and cytoplasmic, that prepare the cell for division.

As techniques such as light microscopy, cell preparation, and chemical staining improved, biologists made discoveries that contradicted the initial assumptions about interphase nuclei. One of these discoveries came by using the now famous Feulgen stain, a chemical mixture that forms a purple complex with DNA. The absorbance intensity of this complex, which is proportional to the amount of DNA, was used by Hewson Swift in 1950 to estimate the relative amounts of DNA in meristematic cells of corn *(Zea mays)* and spiderwort *(Tradescantia* species). By making such determinations of many individual nuclei, Swift discovered that the amount of DNA increases during interphase, long before mitosis begins. This was the first indication that interphase nuclei are active and that they play a key role in preparing for nuclear division.

In 1953, Alma Howard and S. R. Pelc determined precisely when DNA is synthesized during interphase and how long it takes. They did this by measuring the incorporation of radioactive phosphorus (^{32}P) into the growing root tips of broad beans *(Vicia faba).* Since phosphate is a major ingredient of DNA, the uptake of ^{32}P indicates that DNA is being made. After extracting other phosphorus-containing compounds (e.g., ATP, inorganic phosphate, phosphorylated proteins), Howard and Pelc carefully measured changes in radioactivity caused by DNA synthesis. They found that DNA synthesis in broad beans (1) begins about 12 hours after the previous mitosis, (2) requires about 6 hours for completion, and (3) stops about 8 hours before the next mitosis. Accordingly, about 87% of the 30-hour cell cycle in broad beans is spent in interphase, and about 20% of the cell cycle is spent in DNA synthesis. Howard and Pelc were the first to designate the phases of interphase as G_1, S, and G_2. Their work is the foundation for our current model of the cell cycle.

c o n c e p t

The longest period of the cell cycle is interphase. Most studies of interphase involve the S phase, when DNA is synthesized.

The G_1 Phase

Cellular activities that are slowed or arrested during mitosis are reactivated during the G_1 phase. For example, cytoskeletal microtubules reassemble to support the growing cell. Also, the cell enlarges, organelles multiply, mitochondrial and plastid DNA increases manyfold, and many enzymes and structural proteins are made during the G_1 phase. One of the main tasks of G_1 is the synthesis of nucleotides that will be used to make DNA in the S phase to follow.

G_1 is the most variable phase in the cell cycle; it can be almost absent in rapidly dividing cells, or it can last for hours or days in slowly dividing cells. When cell cycles are suspended during winter dormancy, they are usually arrested in a G_1 phase that lasts for several months (fig. 9.4). The cell cycle in dormant cells speeds up when the cells resume activity in the spring. Conversely, the G_1 phase is permanent in cells that are differentiated. Such cells can continue the cell cycle only if they are induced to do so by things such as wounds, infections, or gall-forming organisms (fig. 9.5).

figure 9.5

"Sea urchin" galls on an oak leaf. Leaf cells are normally arrested in the G_1 phase of the cell cycle. Gall-forming insects induce the cells to continue the cycle so that the leaf will make new tissue to house the parasite.

Beginning a New Era: The Theory of DNA Structure and Duplication

Progress in understanding the cell cycle, especially the S phase, was accelerated in the 1950s by a new theory of the structure of DNA. At that time, interest in the structure of DNA was at its peak. Recent evidence indicated that DNA was the substance of genes (see Chapter 8), and several different kinds of studies began to reveal the complexity of the DNA molecule. These studies were like the pieces of a puzzle—the structure of DNA being the puzzle. The relevance of various kinds of information to understanding the structure of DNA was first correctly recognized by James Watson and Francis Crick of Cambridge University. They are therefore credited with discovering the structure of DNA, the most significant discovery in biology of this century. Watson and Crick accomplished their feat by comparing the structure of proteins to a possible structure of DNA and by rationalizing all of the chemical and physical data on DNA into a unified structural proposal. The pieces for the puzzle were available, and Watson and Crick put them together.

Watson and Crick were aided in their quest for the structure of DNA by results from two other laboratories. One of these was at the California Institute of Technology, where Linus Pauling had found that parts of a protein form a **helix** (spiral) in three dimensions. Pauling's interpretation of protein structure was derived from X-ray diffraction patterns of a protein crystal (such patterns are obtained when X rays are shot through a crystal and bent according to the distances between different parts of the molecule). Pauling further explained that the twists of a protein helix are held together by hydrogen bonds between different parts of the molecule. These results led to the second significant piece in the DNA puzzle, which was provided in 1953 by Rosalind Franklin and Maurice Wilkins at King's College in London. They showed that X-ray diffraction

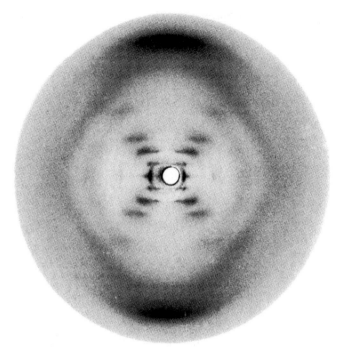

figure 9.6

Photograph of the X-ray diffraction pattern of DNA. This photograph, which was made by Rosalind Franklin, shows the characteristic X-shaped pattern of helical structures.

patterns of DNA resemble those of proteins, thereby also indicating a helical structure for DNA (fig. 9.6).[1] Based on this information, several models of DNA structure were proposed, but only the one by Watson and Crick reconciled all of the available data. Their model was a **double helix**—that is, a two-stranded spiral (fig. 9.7). The use of the term *model* in this case is not abstract. Watson and Crick actually built a model of a DNA molecule by putting together pieces of metal to represent the chemical components of DNA (fig. 9.8).

The double-helix model incorporated the property of macromolecules, such as Pauling's protein, to form hydrogen bonds at normal cytoplasmic pH. Specifically, Watson and Crick showed that hydrogen bonding could pair only certain nucleotides with each other (fig. 9.9). Furthermore, in summing all of the appropriate electronic interactions and bond angles for each string of nucleotides (base + deoxyribose sugar + phosphate; see Chapter 2), Watson and Crick accounted for the twisting forces that form the helix itself. When assembled, the model immediately revealed that the two strands are oriented in opposite directions—that is, they are **antiparallel.** By convention, the directionality of a DNA strand derives from the orientation of its deoxyribose sugars. Each sugar has a phosphate at its 3´ (read "3-prime") -carbon and one at its 5´-carbon. Thus, one strand of the double helix is said to be in the 3´ to 5´ direction, and the other is in the 5´ to 3´ direction.

---

1. Franklin and Wilkins were the first to publish the suggestion that DNA is a helix. Before publication, Wilkins gave the necessary X-ray data to Watson and Crick, without Franklin's knowledge. Franklin thought Watson and Crick had simply beat her to the punch on figuring out the structure of DNA.

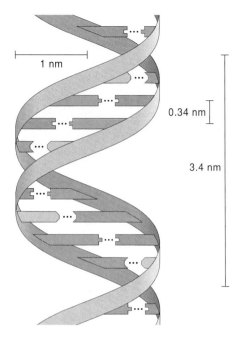

figure 9.7

Diagrammatic representation of the double helix of DNA. Ribbons represent the deoxyribose-phosphate backbones of each strand. The bars between ribbons represent paired nitrogenous bases.

figure 9.8

Watson and Crick with their model of DNA in 1953. The pieces of metal in this model mimic the sizes, shapes, and relative positions of the chemical components of DNA.

c o n c e p t

Discovery of the helical structure of DNA was aided by its similarity with a known protein structure; however, DNA is a double helix, unlike the single helix of the protein.

The impact of the Watson-Crick model was immediate. It explained how DNA could exist in many different sequences of nucleotide bases, thereby being sufficiently complex to produce genes. The model hinted that genes could exist as base sequences of defined length. It also accounted for the constant ratios between cytosine and guanine and between adenine and thymine that had been found in all organisms. Finally, the complementary nature of opposite strands suggested a potential mechanism for self-replication of the molecule. The postulates made by Watson and Crick make up the theory of DNA structure and duplication, which is summarized in table 9.1. Explanations of DNA structure and synthesis presented in this chapter are based on this theory.

In 1962, Watson, Crick, and Wilkins received a Nobel Prize for their landmark studies. Franklin did not receive the prize because she died before the award was made (the Nobel Prize is not awarded posthumously).

c o n c e p t

The Watson-Crick model for DNA structure explains how DNA can be complex enough to accommodate a large number of genes in a multitude of nucleotide sequences. The pairing of adenine with thymine and cytosine with guanine also explains why base composition ratios are constant in each organism. The model also suggests how DNA can replicate itself.

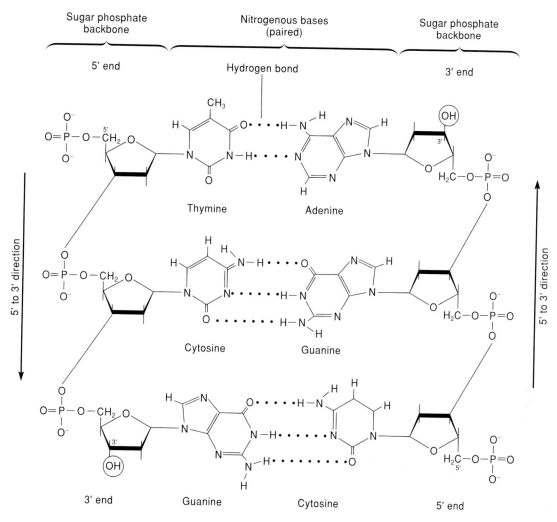

figure 9.9

Components of a portion of a DNA molecule. The two strands are held together by hydrogen bonds between pairs of nucleotides, adenine pairing with thymine and cytosine pairing with guanine. The 3´-carbon of each sugar is bound to a phosphate group that is attached to the 5´-carbon of the next sugar.

The Structure of Chromosomes during Interphase

When the chromatin of an interphase nucleus is examined by electron microscopy, it appears as regularly spaced beads on a thin string (fig. 9.10e, k). The beads are called **nucleosomes** and are the basic structural unit of the chromosome. Nucleosomes are stable complexes of negatively charged DNA and positively charged proteins called **histones**. Histones are positively charged because they contain large amounts of arginine and lysine, which have positively charged amino groups in their side chains. DNA packs tightly around the histones in a form that does not react with most enzymes (fig. 9.10d).

A nucleosome usually contains about 150 pairs of nucleotides, but the exact number can be different for different organisms. There are eight histones in each nucleosome, each composed of two copies of four different proteins. The arrangement of histones and DNA in the nucleosome is not known precisely, but DNA probably wraps around the histones (fig. 9.10e). A double-helical strand that is 50 nm long and 2 nm across is packaged into a nearly spherical nucleosome having a diameter of about 10 nm. Nucleosomes are separated by spacer DNA, usually 40–60 nucleotides long, which is associated with another histone (H1). This histone binds nucleosomes together in a regular, repeating array that is stabilized in a **chromatin fiber** about 30 nm in diameter (fig. 9.10d).

Chromatin fibers are further packed into folds called **looped domains** (fig. 9.10c), which may consist of 20,000 to 100,000 nucleotide pairs. The loops extend from the main axis of the chromosome, but the mechanisms of loop formation and maintenance are unknown. One hypothesis is that

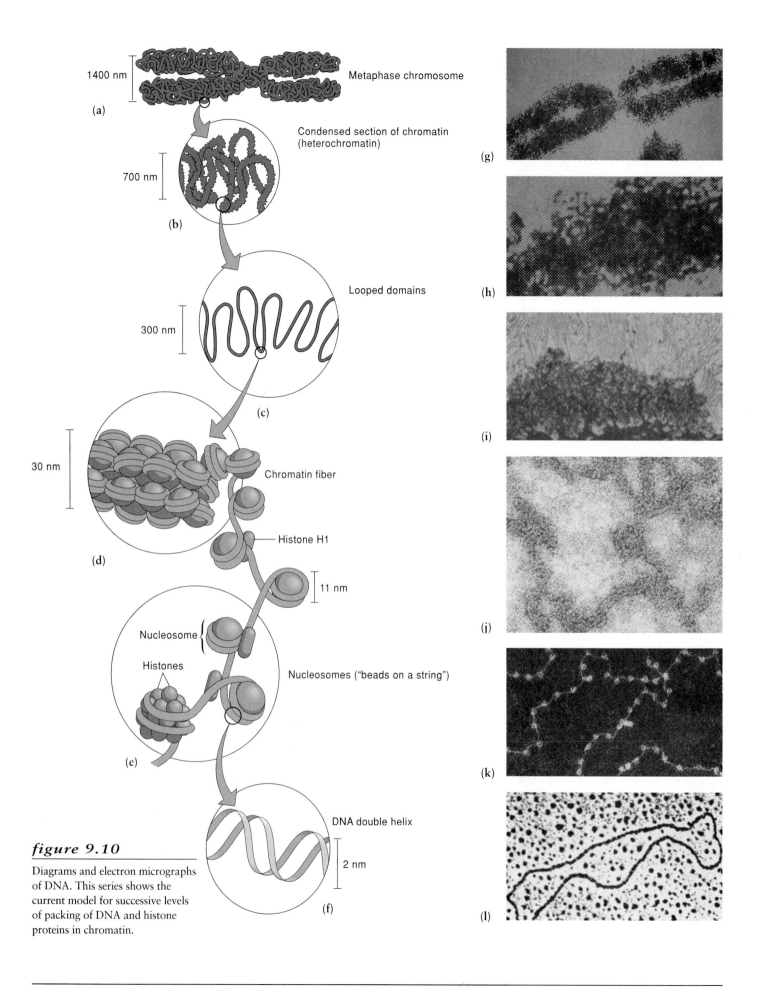

1400 nm

(a)

Metaphase chromosome

700 nm

(b)

Condensed section of chromatin
(heterochromatin)

300 nm

(c)

Looped domains

30 nm

(d)

Chromatin fiber

Histone H1

11 nm

Nucleosome

Histones

(e)

Nucleosomes ("beads on a string")

DNA double helix

2 nm

(f)

figure 9.10

Diagrams and electron micrographs
of DNA. This series shows the
current model for successive levels
of packing of DNA and histone
proteins in chromatin.

(g)

(h)

(i)

(j)

(k)

(l)

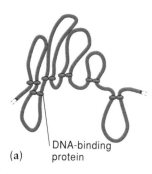

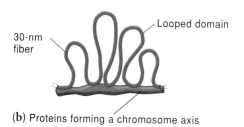

(a) DNA-binding protein

30-nm fiber

Looped domain

(b) Proteins forming a chromosome axis

figure 9.11

Two models for the maintenance of looped domains. (*a*) DNA-binding proteins hold each end of a loop together. (*b*) The ends of loops are all attached to a chromosome axis of proteins.

Source: Alberts, Molecular Biology of the Cell, *2d ed. Copyright © Garland Publishing Company, New York, NY.*

DNA-binding proteins hold each end of a loop (fig. 9.11). Another hypothesis is that the ends of loops attach to proteins in a chromosome axis. There is no direct evidence for either of these hypotheses, but regions near the ends of loops are enriched for certain enzymes.

Some looped domains coil and fold even further, until stained chromatin becomes visible by light microscopy during interphase. This visible chromatin is called **heterochromatin.** The more open form of chromatin, which cannot be seen easily by light microscopy, is called **euchromatin.** Heterochromatin is more condensed than euchromatin, which may mean that the genes in heterochromatin are packed too tightly to be functional. This is one explanation for how some genes can be inactive at different times during the cell cycle. Chromosomes condense even further during mitosis, which is discussed later in this chapter.

c o n c e p t

Chromatin contains DNA and proteins that are arranged in a succession of levels of organization. The basic level of organization is the chromatin fiber, which may be further packed into looped domains and heterochromatin.

Making Chromatin: The Major Task of the S Phase

Chromatin synthesis dictates that approximately equal amounts of DNA and histones be made during the S phase.

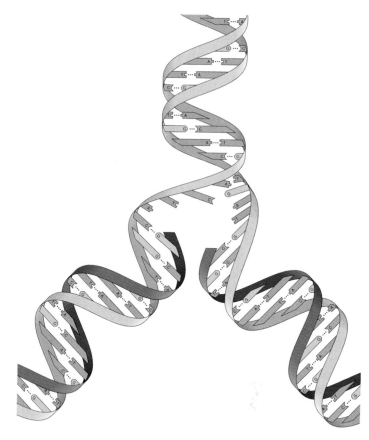

figure 9.12

Semiconservative replication means that half of each parent molecule is conserved in each daughter molecule; the other half is newly synthesized. After replication, the two daughter molecules are identical to the parental double helix.

This means that two different products must be made: DNA and protein. Thus, DNA must not only duplicate itself but also direct the synthesis of histones. Most studies of chromatin synthesis have been devoted to the DNA portion of chromosomes. The synthesis of DNA is called **self-replication,** because DNA makes exact copies of itself. The involvement of DNA in protein synthesis is a function of genes, which will be discussed in Chapter 11.

When Watson and Crick proposed the double-helix model for the structure of DNA, they immediately saw that this structure revealed a potential mechanism for self-replication. They envisioned that the two strands would "unzip" between base-pairs by breaking the hydrogen bonds between them. Each single strand could then act as a template for guiding the formation of a new chain onto the old one (fig. 9.12). According to this postulate, two double-stranded molecules would be produced, each made of one parent strand and one new strand. This is referred to as **semiconservative replication,** because half of each daughter molecule is conserved from one strand of the parent double helix. The semiconservative model seemed logical, but it lacked experimental support for about 5 years after it was proposed.

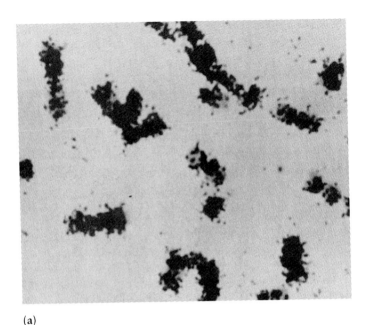

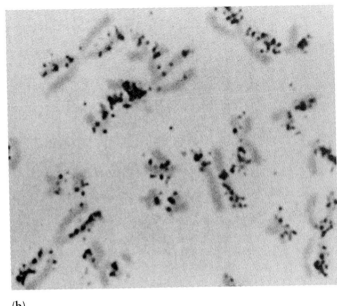

(a)   (b)

figure 9.13

Semiconservative replication as shown by radioactive labeling. (a) Metaphase chromosomes after DNA synthesis in the presence of tritium-labeled thymidine (^{3}H-thymidine), which is radioactive. Radioactivity is shown by dark areas. (b) One cell cycle later, after DNA synthesis in the absence of ^{3}H-thymidine. The radioactive parent DNA remains, and the newly synthesized DNA is not radioactive.

Evidence for Semiconservative Replication

In 1957, J. Herbert Taylor and others at Columbia University grew seedlings of the broad bean in a medium containing thymidine bound to tritium (^{3}H), a radioactive isotope of hydrogen. Because emissions from radioisotopes react with photographic film, Taylor could photograph radioactive chromosomes in root-tip cells. After one cell cycle, the DNA in each chromosome was radioactive throughout (fig. 9.13). After the next cell cycle, which occurred in the absence of the ^{3}H-thymidine, the parent DNA remained radioactive, but the newly made DNA was not radioactive. This was the first evidence that replication was semiconservative.

Further support for semiconservative replication of DNA was obtained in 1958 by Matthew Meselson and Franklin Stahl at the California Institute of Technology. They devised a simple but ingenious method involving the incorporation of heavy nitrogen (^{15}N) into cells of *Escherichia coli*, a common gut bacterium. After they grew the bacteria for many generations in a medium containing ^{15}N-ammonium chloride, the DNA of virtually all cells was heavier than ordinary. Meselson and Stahl showed that heavy DNA moves farther than normal DNA when centrifuged at high speed in a gradient of cesium chloride (fig. 9.14; see Chapter 3 for an explanation of cell fractionation). They then found that *E. coli* containing only heavy DNA would make DNA of intermediate density after growing for one cell cycle in a medium with normal ammonium chloride. This intermediate band consisted of hybrid DNA that contained equal amounts of ^{14}N and ^{15}N. Furthermore, after one more cycle in a normal medium, two bands appeared, one at intermediate density and one at normal den-

sity. This means that each bout of DNA synthesis conserved one old strand for each new strand.

c o n c e p t

DNA synthesis is semiconservative, meaning that one strand of a new double helix is newly synthesized and the other strand is conserved from the double helix of the parent cell.

Enzymes of Replication: Keys to Understanding the S Phase

Although the concept of semiconservative replication is relatively simple, the process is a complex set of biochemical reactions. Each step is catalyzed by a separate enzyme that works with other enzymes to accomplish the overall task.

As figure 9.15 shows, several reactions occur more or less simultaneously. The DNA double helix must first be untwisted and split into two single strands. We do not know how this starts in plants, but in bacteria the initial untwisting of DNA is associated with special enzymes, called **helicases,** which open the helix by breaking hydrogen bonds between complementary base-pairs. As the strands unwind and separate, their natural tendency to fuse back together and rewind is controlled by **single-strand binding proteins.** These proteins prevent re-fusion and keep the single strands from becoming tangled.

Synthesis occurs when **DNA polymerases** catalyze the co-valent bonding of newly made nucleotides into a new strand. The order of nucleotides in the new strand is directed by the

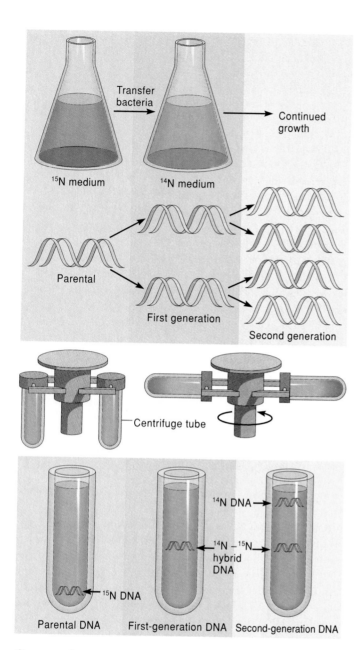

figure 9.14

Incorporation of heavy nitrogen (^{15}N) into DNA makes the DNA heavier. Upon centrifugation, heavy parental DNA sinks farther into a gradient of cesium chloride (CsCl). After one generation in a normal culture medium containing ^{14}N, DNA becomes more buoyant. One strand in each double helix still has ^{15}N because of semiconservative replication; the other strand of the "hybrid" DNA has ^{14}N. After the second generation in a normal culture medium, half of the DNA in the cell is "hybrid" DNA, and half contains only ^{14}N (i.e., is most buoyant).

sequence of the single-stranded DNA template in the parent strand.[2] Thus, only bases complementary to the template can be

2. DNA can also be made in a test tube by mixing isolated DNA with purified DNA polymerase. A short fragment (usually 10–30 nucleotides) of artificially made DNA is added as a primer to get the reaction started. This procedure, which is called the **polymerase chain reaction (PCR)**, and its application to molecular biology are discussed in Chapter 11.

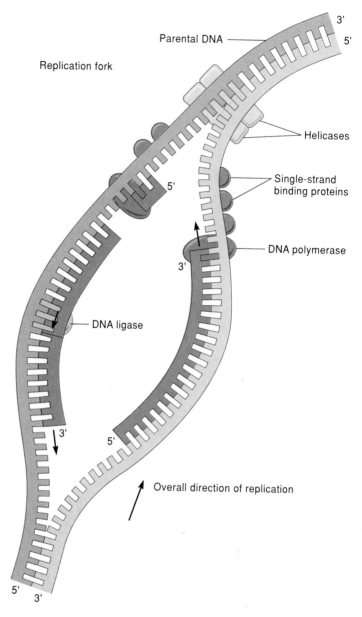

figure 9.15

DNA replication requires helicases to unwind the double helix, single-strand binding proteins to stabilize the unzipped double strand, DNA polymerase to assemble nucleotides into a new strand, and DNA ligase to bind fragments together from discontinuous synthesis.

added to the growing chain, which keeps adenines opposite thymines and cytosines opposite guanines in the developing double strand (but see reading 9.1, "Mutations and DNA Repair: Do Mistakes Have to Be Corrected?"). Together, the processes of unwinding the helix and separating the strands create a moving **replication fork,** so named because of its forklike appearance in electron micrographs (fig. 9.16). Two replication forks move in opposite directions from a **replication origin,** which is the point of initiation for DNA synthesis. These moving forks form an expanding **replication bubble** between them.

MUTATIONS AND DNA REPAIR: DO MISTAKES HAVE TO BE CORRECTED?

Mutations are permanent alterations in chromosomal DNA. Many types of changes fit this definition. For example, in Chapter 8, we discussed mutations caused by inserting mobile sequences (transposable elements) into genes. Sequences may also be duplicated or deleted. Finally, mutations may involve the substitution of one nucleotide for another, or the gain or loss of one or more nucleotides.

An essential function of DNA polymerases is to add nucleotides in a specific direction to a growing chain of DNA during replication. Polymerases follow the lead of primers, which are short pieces of RNA complementary to the DNA template. Because the synthesis of RNA primers does not also depend upon primers, their sequences are relatively error-prone. Errors in RNA primers cause about one in every 10,000 bases to be mismatched. Nevertheless, repairing these "mistakes" is possible because they can be recognized by DNA polymerases during replication. When a mismatch is encountered in the primer, polymerase backs up and other enzymes cut out the offending base. This means that DNA polymerases "proofread" replication. Imagine what havoc there would be if mistakes in RNA primers were not corrected. Even in the smallest of plant genomes, that of the common wall cress *Arabidopsis thaliana*, there could be as many as 280,000 mutations in every cell cycle! Nevertheless, we can find evidence that mutations have occurred and not been repaired in different genomes, although not at such an enormous rate. This evidence comes from comparing DNA sequences for the same genes among different organisms.

The advent of modern molecular biology has enabled biologists to determine nucleotide sequences of specific genes. DNA sequences vary in different species yet remain functional in their respective plants. Such comparisons enable biologists to determine both the magnitude and the rate of mutations.

One of the more popular genes for study in this regard is *rbcL*, which occurs in the chloroplast genome and is responsible for the large protein subunit of rubisco (ribulose-1, 5-bisphosphate carboxylase/oxygenase; see Chapter 7). This gene is studied intensively because it is important for photosynthesis, is relatively easy to extract from many plants, and because its inheritance is not confounded by sexual reproduction. Comparisons between the *rbcL* of maize (*Zea mays*) and tobacco (*Nicotiana tabacum*) show that more than 20% of the nucleotide positions differ in the *rbcL* of these plants. This comparison shows that even though DNA repair is reasonably dependable, it is not perfect. Some mutations escape the process and, if not lethal, are maintained without apparent harm to the organism.

When DNA polymerases were first studied, they were found to have a seemingly severe limitation: they could make chains of DNA in only one direction, from 5′ to 3′. This means that uninterrupted synthesis (**continuous synthesis**) can occur on only one of the two parent strands. It was later found, however, that synthesis on the other strand occurs when polymerase starts closer to the replication fork and then builds the new chain backward in the 5′ to 3′ direction. Pieces of this new chain are built in the opposite direction of the growing replication fork. Because it occurs in fragments, this interrupted synthesis is called **discontinuous synthesis**. Nevertheless, the net growth of a new chain by discontinuous synthesis occurs in the same direction as continuous synthesis.

More than a dozen other kinds of enzymes are involved in DNA replication. Discussion of the roles of all of these would be exceedingly complicated, but one additional type of enzyme has attracted considerable interest. This is the type called **topoisomerases,** so named because they influence the architecture (topology) of DNA. In the nucleus, DNA is often so severely twisted that it is like an overwound rubber band. Topoisomerases relieve the formation of kinks that would otherwise block the movement of replication forks. Topoisomerases can break one or both strands, allowing them to uncoil by swiveling around one another (fig. 9.17).

After swiveling, DNA is then reconnected in its original form by the same enzymes. In this manner, topoisomerases loosen knots of DNA and allow replication enzymes access to the newly exposed double strands. Such access is crucial for replication, since enzymes cannot react with segments of DNA that are tied in knots. Furthermore, by breaking and swiveling, DNA does not have to uncoil as replication occurs through a chromosome. If DNA did not break and swivel, then uncoiling would have to stay ahead of a replication rate of about 50 nucleotides per second, which converts to 5 revolutions per second (10 nucleotides per turn of the helix). This means that parts of chromosomes would be spinning at about 300 revolutions per minute, or about half the speed of a car engine at low idle. The havoc from such chromosome spinning would be hard to imagine.

Interestingly, topoisomerases can also *form* knots. This process hides DNA segments from biochemical reactions, perhaps in the form of heterochromatin. Furthermore, one type of topoisomerase is abundant at the ends of loops in looped domains, which may indicate a role for this enzyme in making loops. Because of their many roles in controlling chromatin structure, topoisomerases are studied intensely for their potential involvement in regulating gene expression. These studies underscore the importance of chromosome structure for controlling the functions of DNA.

9–13

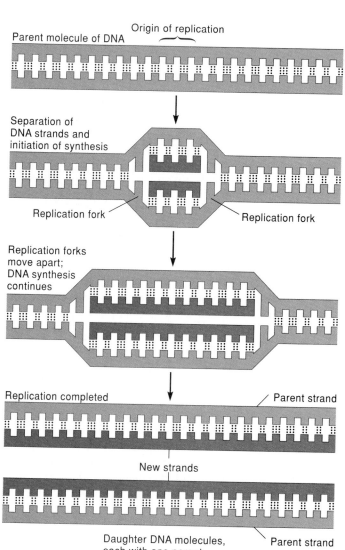

Parent molecule of DNA

Origin of replication

Separation of
DNA strands and
initiation of synthesis

Replication fork   Replication fork

Replication forks
move apart;
DNA synthesis
continues

Replication completed   Parent strand

New strands

Daughter DNA molecules,   Parent strand
each with one parent
strand and one new strand

(a)

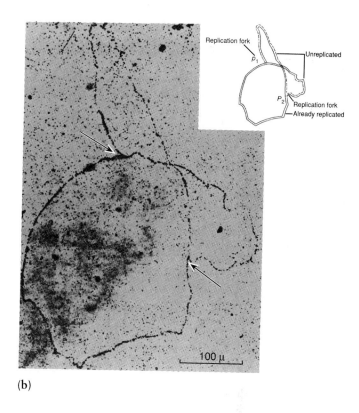

Replication fork   Unreplicated

P_1

P_2 Replication fork
Already replicated

100 μ

(b)

figure 9.16

Origin of replication. (a) Model of the initiation and expansion of a
replication bubble for DNA synthesis. (b) Electron micrograph of a
replication bubble. Arrows point to replication forks at either end of
the bubble.

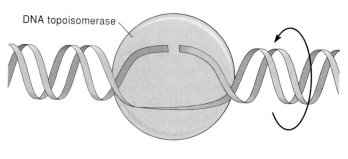

Strained double helix

DNA topoisomerase

Relaxed double helix

figure 9.17

A strained double helix can form coils and kinks, but DNA
topoisomerases can break one strand and allow the strained region to
swivel. Once the strain is relieved, the topoisomerases also repair the
broken strand.

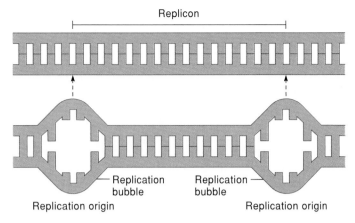

Clues to the mechanisms of DNA replication come from identifying the enzymes involved and determining their functions. Enzymes must stabilize and unwind the double helix, temporarily break hydrogen bonds between strands, and build a new chain along the parent DNA template.

The Lore of Plants

As often happens in science, "normal" processes, such as those of the cell cycle, have many exceptions. One common exception to the cell cycle is that, in many organisms, the S phase can occur many times before mitosis. The result is that the nuclei of some cells can have much more than the expected amount of DNA. For example, hair cells and glandular cells may have from 16 to more than 4,000 times the haploid amount of DNA. The record for plants is probably in certain cells of the seeds of the lords-and-ladies arum *(Arum maculatum)*, which have more than 24,000-fold the haploid amount of DNA. But this number pales in comparison with the giant neurons of a certain species of mollusk, which have at least 75,000 times the haploid amount of DNA.

Replicons

Replication starts at replication origins, and it stops where the forks of adjacent replication bubbles meet. The block of DNA between two replication origins is a **replicon** (fig. 9.18). Replicons vary in size from 30,000 to 300,000 pairs of nucleotides. This means that, based on an average replication rate of 50 nucleotides per second, it takes from 10 to 100 minutes to complete one replicon. Replication of a whole genome, therefore, requires that some replicons be active simultaneously and others be active at different times during the several hours of the S phase.

Different patterns of replicon activation may explain why genomes of vastly different sizes do not have S phases of vastly different lengths of time. For example, the haploid genome of common wall cress *(Arabidopsis thaliana)* is about 70 million nucleotide pairs, which is the smallest plant genome known. In contrast, onion *(Allium cepa)* has about 4.4 billion nucleotide pairs in its haploid genome, nearly 63 times greater than that of wall cress (see readings 9.2 and 9.3, "Determining the Size and Composition of Genomes" and "An Important Little Weed"), yet its S phase lasts only 3.9 times longer than that of wall cress; the S phase in wall cress lasts 2.8 hours whereas in onion it lasts 10.9 hours.

Based on observations of a few organisms, including wall cress, replicons are apparently activated along a chromosome

figure 9.18

A replicon is the block of DNA between two replication origins. Replicons vary in size from 30,000 to 300,000 pairs of nucleotides. Many replicons can be active simultaneously throughout a genome.

in clusters of 50–80 at a time. Many clusters operate on different chromosomes at the same time, with some finishing before others start. Furthermore, families of replicon clusters are active throughout the genome at specific times during the S phase. In wall cress, two replicon families accommodate all 10 chromosomes; clusters in the first family activate 36 minutes before synthesis begins in the second. In comparison, DNA synthesis in onion probably depends on many more replicons per cluster and larger replicon families than it does in wall cress.

Genomes are duplicated in specific and highly organized patterns. The basic unit of replication is the replicon, which is the sequence between adjacent replication origins. Replicons are activated simultaneously in clusters. Replicon clusters operate on different chromosomes at specified times during the S phase of the cell cycle.

The G_2 Phase

The G_2 phase begins after chromatin synthesis is complete, when newly replicated chromatin gradually begins to coil and condense into a compact form. During G_2, the cell finishes preparing for mitosis. This preparation includes making tubulins for mitotic microtubules, making proteins for processing chromosomes, and breaking down the nuclear envelope.

The average length of a G_2 phase is 3–5 hours, and it continues until mitosis begins. The end of G_2 is the end of interphase.

DETERMINING THE SIZE AND COMPOSITION OF GENOMES

Scientists have long been interested in variations among the genome sizes of different organisms. The first measurements of genome size used cytophotometry to estimate the mass of DNA in a nucleus. This method entails staining the chromatin of cells in the G_1 phase and then recording its absorbance by a photometer through a light microscope. The density of the darkened nucleus is proportional to the size of the genome, which is usually expressed in picograms (pg; 1 pg = 10^{-12} g) per nucleus. Cytophotometry is relatively simple and can be used to measure genome sizes of many plants in a short time. The genomes of several hundred species have been measured in this manner. The results reveal a tremendous variation in haploid genome sizes among plants, ranging from about 0.26 pg in common wall cress (*Arabidopsis thaliana*) to more than 340 pg in a species of lily (*Lilium longiflorum*) (reading fig. 9.2a). A picogram is equivalent to approximately 269 million pairs of nucleotides, so the genome of *Arabidopsis* is about 70 million base-pairs and that of the lily is more than 90 billion base-pairs (for comparison, the human genome is almost 12 pg, or about 3 billion base-pairs). The massive genomes of some species are due to polyploidy, wherein many copies of the same genome occur in a single nucleus. However, most variation cannot be explained so easily. For example, the amount of DNA in diploid species of the bean genus *Vicia* varies sevenfold.

The phenomenon of different genome sizes has fostered much interest in the molecular biology, ecology, and evolution

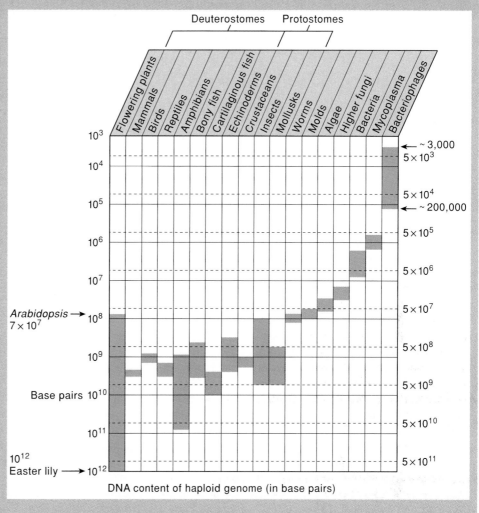

reading figure 9.2a

The range of amounts of DNA among plants and other organisms.

Source: Fristrom and Clegg, Principles of Genetics, *2d ed. Copyright © Chiron Press, Cambridge, MA.*

Continued on next page

Mitosis and Cytokinesis

In one of the earliest descriptions of dividing nuclei, a German cytologist named Walther Flemming referred to the various stages of chromatin appearance as *mitosen* (threads), giving rise to the name **mitosis** for the process of nuclear division. Flemming's work in the 1870s inspired many studies of mitosis. By the start of the twentieth century, biologists knew that this process was the foundation for nuclear division in all plants and animals.

Mitosis refers to the separation of chromosomes and the formation of two genetically identical daughter nuclei. It is usually followed by cytokinesis, but there are many exceptions to this pattern. For example, in seed development, the nuclei of some tissues divide many times before they become separated by cell walls. Furthermore, in many species of algae, cytokinesis occurs only during sexual reproduction. Vegetative stages of these algae are permanently multinucleate.

Mitosis can be studied in several types of meristematic cells, either in whole plants, in cell cultures, or in wound or

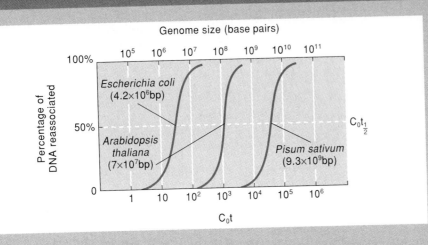

reading figure 9.2b

Examples of $C_o t$ curves. The point at which half of the DNA is reassociated ($C_o t_{1/2}$) is correlated with the amount of DNA in the genome. *Escherichia coli* is a bacterium; *Arabidopsis thaliana* and *Pisum sativum* are flowering plants (bp = base-pairs).

of genome size variation. For example, as discussed in this chapter, the mechanics of the cell cycle must account for the amount of DNA made in the S phase. Accordingly, the number, replicon size, and activity patterns have been studied intensively at the molecular level. At the organismic level, however, the reasons for excessive amounts of DNA in species with large genomes have been elusive.

One of the most informative methods for measuring the size of a genome exploits the property of DNA strands to reassociate after they are physically separated. Separation is induced by heating native (i.e., double-stranded) DNA, usually to about 90° C, until hydrogen bonds between the two strands break, thus forming **denatured** (i.e., single-stranded) DNA. When cooled, these single strands reassociate into native DNA. In this renaturation process, DNA strands collide randomly until a small region on one strand bumps into a complementary region on another and forms hydrogen bonds with it. Then the two strands zip in both directions from that point until the native DNA completely re-forms. The extent of strand reassociation depends upon the initial DNA concentration (C_o) and the incubation time (t) required to complete it, $C_o \times t$. When plotted on a logarithmic scale, the reassociation graph, called a **cot curve,** has a sigmoidal shape with an inflection point that is halfway to complete reassociation. This halfway point, referred to as the $C_o t_{1/2}$ value, is compared with a standard of known genome size, such as that of *E. coli*. The genome

sizes of plants can be extrapolated from such comparisons (reading fig. 9.2b).

Early measurements of genome size also revealed some unexpected surprises about genome composition. Much of the DNA in plants and other eukaryotes reassociates faster than expected, based on comparisons with prokaryotes and viruses. Since rates depend on concentration, some portions of a genome occur in higher concentrations than others. That is, some DNA sequences are repeated many times. The portion of DNA that reassociates faster than expected is the percentage of repetitive DNA in the genome. Species with larger genomes usually have higher proportions of repetitive DNA. For example, the small genome of *Arabidopsis* has almost no repetitive DNA, but the genomes of garden peas and mung beans (*Vigna radiata*) are about 75% repetitive. Repeated sequences can account for more than 90% of the genome of the lily, *Trillium erectum*. For comparison, the human genome consists of at least 90% repetitive DNA.

Reassociation occurs a million times faster in some sequences than in others. This means that these DNA sequences are present in a million (10^6) copies per nucleus. Such highly repetitive DNA accounts for the massive genomes found in some plants.

The presence of huge amounts of DNA in some but not all organisms raises the question about the necessity of repetitive DNA. If many organisms have little repetitive DNA, why do other genomes maintain large amounts of it? There is as yet no convincing answer for this question.

tumor tissues. For example, cells undergoing mitosis can be readily observed in onion root tips, as illustrated in figure 9.19. At any one time, most nuclei are in interphase, and the remainder are in various stages of nuclear division.

Before Prophase

Before prophase, microtubules and actin filaments that are just inside the plasma membrane begin to form a narrow bundle, called the **preprophase band,** around the nucleus. The plane of the preprophase band is perpendicular to the direction of chromosome movement during mitosis (fig. 9.20). As mitosis begins, the microtubules disappear, but the actin filaments remain and provide a circular support network between the dividing nucleus and the plasma membrane. Toward the end of mitosis, this network guides the deposition of precursors for the new cell wall. This means that the radial strands of actin act as a "memory" of the preprophase band, thereby predetermining where cytokinesis will occur.

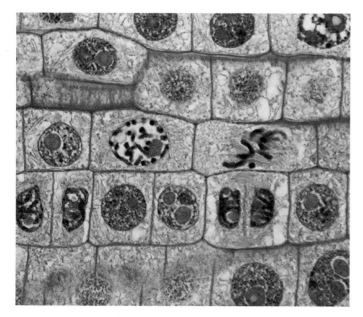

figure 9.19

Mitosis can be seen easily among the nuclei of actively dividing cells, such as those of onion root tip (*Allium cepa*). At any one time, most nuclei are in interphase, ×280.

Chromosome Condensation: Prophase

Chromatin of the interphase nucleus condenses rapidly during early prophase. As condensation proceeds, chromatin appears as a mass of elongated threads. By late prophase, individual chromosomes are visible, appearing as two parallel threads attached at a constriction point called the **centromere** (fig. 9.21a). The centromere consists of a specific sequence of DNA that is required for segregation of chromosomes later in mitosis. In this double-thread form of the chromosome, each thread is called a **chromatid**. The double chromatid is still considered to be one chromosome because it has just one centromere.

The end of prophase is marked by the disappearance of nucleoli and the nuclear envelope. Without a nuclear boundary, chromosomes protrude into the cytoplasm, and their behavior appears to be chaotic and uncontrolled. As we shall see, this could not be further from the truth.

Chromosomes are at their shortest near the end of prophase and the beginning of metaphase. In the garden pea, the seven pairs of chromosomes have a combined length of about 335 µm at their shortest stage. This represents a reduction from a total DNA double-strand length of about 7 m for the 20.4 billion nucleotide pairs that are in the diploid nucleus after the S phase. This is more than a 20,000-fold shrinkage from the double-helical strands of DNA to the most highly coiled form of chromatin. Such shrinkage would be equivalent to reducing the length of a football field to about the height of this line of type.

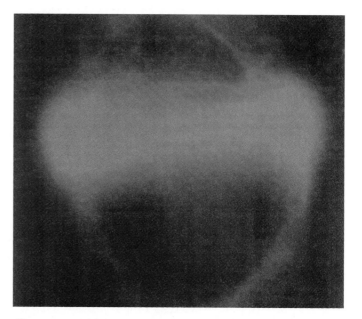

figure 9.20

Photomicrograph of a cell about to undergo mitosis, viewed with immunofluorescence microscopy to show the microtubules in the preprophase band and those near the nuclear surface. Brackets indicate where cell boundaries are. The plane of the preprophase band is perpendicular to the direction the chromosomes will move in mitosis and parallel to the plane of the new cell wall that will form to divide this cell into two cells at cytokinesis.

Chromosome Alignment: Metaphase

After the nuclear envelope dissolves, several nearby parallel fibers, called **spindle fibers,** accumulate near the chromosomes in early metaphase. Spindle fibers, which are made of microtubules, are collectively called the **spindle apparatus** (*spindle* for short). The mature spindle apparatus is elongated, with its ends pointing to opposite poles of the cell (fig. 9.21b). Fibers from opposite poles of the spindle attach to each chromosome on either side of its centromere, to a disc-shaped structure called the **kinetochore** (fig. 9.22), a complex of proteins that binds to the centromere. Most fibers are not attached to kinetochores, because there are usually many more fibers than chromosomes. The unattached fibers are called *polar fibers,* to distinguish them from kinetochore-bound fibers.

As viewed by light microscopy, the metaphase chromosomes line up independently of one another in a circle around the circumference of the spindle apparatus. The plane of this circle, called the **metaphase plate,** is perpendicular to the axis of the spindle. The position of chromosomes along the metaphase plate is maintained by the balance of forces between two sets of kinetochore-bound fibers, one set from each pole of the spindle.

By the end of metaphase, the chromosomes have all been forced into a tight plane at a specific position in the cell. Disorder has become order. The chromosomes are now ready for the splitting of their centromeres and the separation of their chromatids.

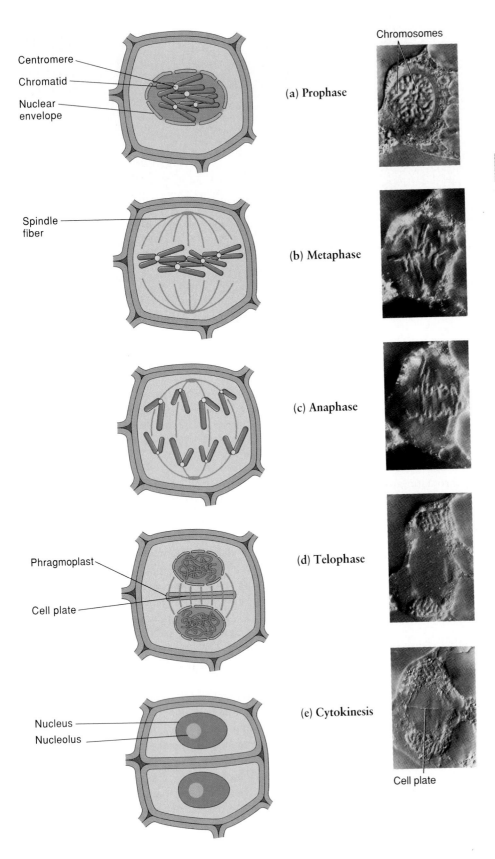

Centromere
Chromatid
Nuclear envelope

(a) Prophase

Chromosomes

Spindle fiber

(b) Metaphase

(c) Anaphase

Phragmoplast
Cell plate

(d) Telophase

Nucleus
Nucleolus

(e) Cytokinesis

Cell plate

figure 9.21

Mitosis and cytokinesis. (a) Chromosome condensation: prophase. (b) Chromosome alignment: metaphase. (c) Chromosome migration: anaphase. (d) Chromosomes decondense and new nuclei are formed: telophase. The cell plate forms at this point if telophase is to be followed by cytokinesis. (e) Cells divide: cytokinesis.

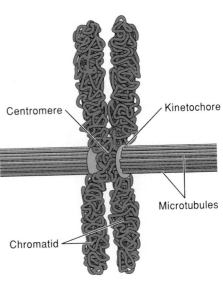

Centromere
Kinetochore
Microtubules
Chromatid

figure 9.22

Diagram of a metaphase chromosome with spindle fibers from opposite ends of the cell attached to opposing kinetochores. A kinetochore is a complex of proteins that binds to the centromere.

Chromosome Separation: Anaphase

During anaphase, which is the shortest stage of mitosis, chromatids attached to the same centromere (i.e., *sister chromatids*) separate and move to opposite poles of the spindle (fig. 9.21c). Centromeres split before the chromatids separate and are pulled along the axis of the spindle. When the centromere is near the center of a chromosome, the two chromosome arms form a V as they are dragged through the cytoplasm. Upon separation, the term *chromatid* no longer applies; each structure is considered a separate (i.e., daughter) chromosome because it has its own centromere.

In most cells, the spindle elongates during anaphase, thereby increasing the distance between the poles. This movement further separates the two sets of chromosomes as they move along the spindle. By the end of anaphase, chromosomes have been separated into two genetically identical nuclei.

An Important Little Weed

More than a half century ago, German botanist Friedrich Laibach claimed that he had found the perfect plant for studying genetics: *Arabidopsis thaliana*, a commercially useless member of the mustard family (reading fig. 9.3). Although Laibach's advice was initially ignored, today *Arabidopsis* is one of the most prized organisms for studying how genes control development. Here's why:

Arabidopsis is small and easy to grow. Thousands of plants can be grown on a lab bench under fluorescent lights. No tractors or overalls are required.

Arabidopsis has a short generation time; its entire life cycle takes only about a month. Moreover, each plant produces thousands of seeds.

Arabidopsis has the smallest known genome of any flowering plant. Indeed, its 30,000 genes are on only five pairs of chromosomes. For comparison, the genome of corn is 30–40 times larger than that of *Arabidopsis*. The genome of *Arabidopsis* is only about 20 times larger than that of the common bacterium *Escherichia coli*.

Unlike most other plants, *Arabidopsis* has almost no repetitive DNA; this is why plants such as wheat and tomato—which have about the same number of genes as *Arabidopsis*—have so much more DNA than does *Arabidopsis*. The relatively simple genome of *Arabidopsis* makes it easier for botanists to associate specific genetic functions with certain regions of DNA.

Today, botanists are using *Arabidopsis* to answer a variety of questions about plant growth and development, evolution, and genetics. Studies of *Arabidopsis* have uncovered genes that regulate processes ranging from photosynthesis to disease resistance and flowering; one group of botanists has inserted bacterial genes into *Arabidopsis* to create plants that produce biodegradable plastics. There's even an international program to sequence the genome of *Arabidopsis*.

You'll read about studies of *Arabidopsis* throughout this book. For more up-to-the-minute information about *Arabidopsis*, check out the following sites on the World Wide Web:

Arabidopsis Monograph Review
(http://genome-www.stanford.edu/Arabidopsis)
Arabidopsis thaliana Genome
(wais://weeds.mgh.harvard.edu:210/Arabidopsis_thaliana_Genome)
Teaching with *Arabidopsis*
(http://www.arabidopsis.com/ltc.html)

reading figure 9.3

Arabidopsis thaliana, a tiny weed that botanists are using to answer important questions about plant development.

The Lore of Plants

To appreciate the ability of DNA to store information, consider an organism such as humans whose genome is about 3 billion base-pairs. If this DNA were stretched out, there would be about 2 m of DNA per cell. However, all of this DNA is stored in a nucleus, which is only about 0.005 mm (0.002 in) in diameter, or about 1/500 the thickness of a dime. These 3 billion base-pairs—that is, 6 billion bits of information—store 21 times more information than is found in *The Encyclopedia Britannica*, 79 times more information than *The Oxford English Dictionary*, and 5,000 times more information than is found in the Bible. A cubic centimeter of DNA contains as much information as 57 billion *Encyclopedia Britannicas*; one cubic inch contains as much information as about 930 billion *Britannicas*.

Formation of Daughter Nuclei: Telophase

During telophase, daughter nuclei seem to mimic prophase in reverse. The spindle apparatus disappears, a nuclear envelope forms around each of the two sets of chromosomes, and nucleoli appear in each new nucleus (fig. 9.21d). The chromosomes steadily elongate and decondense once again into diffuse chromatin.

concept

Mitosis is a continuous process that produces two genetically identical nuclei. For convenience, biologists recognize several stages of the process: chromosome condensation (prophase), alignment (metaphase), separation (anaphase), and formation into daughter nuclei (telophase).

Division into Cells: Cytokinesis

As early as late anaphase or telophase, the cell begins to prepare for cytokinesis. As part of this preparation, a cylindrical system of short microtubules starts to form between the daughter nuclei, parallel to the spindle. This system of microtubules, called a **phragmoplast,** is divided by the plane that was established by the preprophase band, which is also where the metaphase plate had been. This plane, which is perpendicular to the spindle apparatus, will be the division plane of the cytoplasm. As cytokinesis continues, dictyosome vesicles laden with cell-wall precursors become trapped by the phragmoplast. These vesicles fuse into a single platelike vesicle in which the new cell walls of the daughter cells begin to form. Together, the phragmoplast, central vesicle, and new walls become the first structure of cytokinesis that is easily visible by light microscopy. This structure is called the **cell plate** (fig. 9.21d).

The cell plate grows as dictyosome vesicles are added to the outer edge of the central vesicle. The new walls in the central vesicle grow outward along their edges. The cell plate continues to grow in this manner until the central vesicle fuses with the plasma membrane of the parent cell, thereby becoming part of the plasma membranes of the daughter cells. At the same time, new cell walls in the central vesicle join the wall of the parent cell. Fusion of the cell walls is the last step in cytokinesis (fig. 9.21e). As cytokinesis nears completion, the wall that surrounds the parent cell loosens and stretches, thereby enabling subsequent growth of the daughter cells by cell expansion. Shortly after cytokinesis, the daughter cells deposit additional primary cell-wall materials on both sides of the newly formed cell wall and on all of the preexisting walls as well.

Two features of the cytokinesis just described characterize plants and a few algae. One is the formation of a cell plate that involves a phragmoplast. In most other organisms having rigid cell walls, cytokinesis occurs by constriction of the parent wall and plasma membrane. In animal cells, which lack cell walls, cells are pinched off by constriction of the parent plasma membrane. Neither of these nonplant types of cytokinesis involves a cell plate. In some algae, however, a cell plate forms from microtubules oriented perpendicular to the spindle apparatus, not parallel as in a phragmoplast. In such algae, the cell-plate forming structure is called a **phycoplast** (*phyco* refers to algae).

The second feature of cytokinesis that characterizes cell division in plants and some algae is the formation of plasmodesmata (see Chapter 3). As dictyosome vesicles accumulate in the growing cell plate, they trap parts of the endoplasmic reticulum. After cytokinesis, parts of ER that were caught between fused vesicles connect the cytoplasms of the daughter cells. Although the formation of plasmodesmata by this mechanism seems to be random, it is probably controlled. Indirect evidence for this suggestion is that the number of plasmodesmata varies with cell type (see Chapter 3).

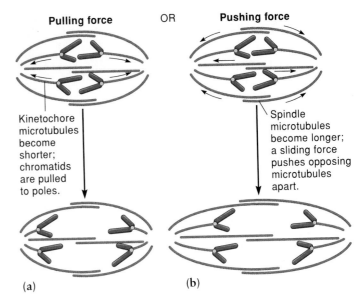

(a) **(b)**

figure 9.23

Forces that move chromosomes. (a) Pulling force on chromosomes is created by shortening of kinetochore-bound microtubules. (b) Pushing force against poles of the spindle apparatus comes from sliding action between opposite polar microtubules. Arrows indicate the location and direction of forces during anaphase.

c o n c e p t

Cytokinesis splits the cytoplasm of a dividing cell into daughter cells. In plants and a few algae, cytokinesis involves a phragmoplast and cell plate. The cell plate from a phragmoplast forms in a division plane that is perpendicular to the mitotic spindle. The new cell wall and plasma membranes of daughter cells are derived from the cell plate.

Mysteries of the Mitotic Spindle: How Are Chromosomes Moved?

Mitosis is dominated by the spindle apparatus. The mechanisms of spindle function have received much attention in studies of mitosis, but the precise way that forces are formed by the spindle is still not fully understood.

Recall from earlier in this chapter that two kinds of forces contribute to chromosome movement during anaphase (fig. 9.23). One force is the *pull* by kinetochore-bound microtubules on either side of each centromere, which moves sister chromatids in opposite directions. The other force is the *push* by microtubules that converge at the poles of the spindle but are not attached to kinetochores. This pushing force moves the poles of the spindle apparatus in opposite directions from each other. How can some microtubules pull while others push?

9–21

The Push of Polar Microtubules

Polar spindle fibers from one pole overlap those of the other pole near the middle of the spindle apparatus. This suggests that opposing fibers somehow slide past one another, thereby pushing their respective poles in opposite directions. This *sliding-microtubules hypothesis* is supported by studies of isolated spindles of *Stephanopyxis turris*, a freshwater diatom (a type of alga). Unlike the spindles of plants, those isolated from *Stephanopyxis* remain active after being removed from the organism. In isolation, opposing spindle fibers separate until the overlap zone disappears. Once the overlap zone is gone, spindle movement stops, suggesting that movement depends on contact between opposite microtubules. In contrast, the overlap zone is not exhausted in living cells, presumably because polar microtubules continue to be made.

The mechanism by which microtubules slide during mitosis is unknown, but one suggestion is that it resembles the movement of flagella. Spindle movement and flagellar movement are similar because they can both be blocked by inhibiting the hydrolysis of ATP. This means that ATP provides energy for microtubule sliding and for flagella movement (see Chapter 3). Unlike spindles, however, flagellar contain **dynein**, a large protein complex that converts energy from ATP into a sliding force between flagellar microtubules. Dynein has not yet been found in the mitotic spindle. Furthermore, spindle microtubules are oriented in directions opposite to their slide mates, while flagellar microtubules are oriented in the same direction.

The Pull of Kinetochore Microtubules

The pulling force of microtubules that are attached to a chromosome is probably caused by disassembly of microtubules. Partial evidence for this hypothesis is that the alkaloid taxol, which suppresses microtubule disassembly, inhibits chromosome movement. Further evidence is that the alkaloid colchicine, which inhibits assembly of microtubules, causes spindle fibers to shorten faster than normal; this accelerates the movement of chromosomes.

It is not known whether the disassembly of microtubules occurs only at the kinetochores or both at the kinetochores and at the poles (fig. 9.24). The model for microtubule disassembly is also incomplete, because the force generated by kinetochore-attached fibers is proportional to fiber length. This suggests that the pulling force is distributed along the entire length of the fiber, rather than only at the polar ends where disassembly occurs. Consequently, we still cannot fully explain how chromosomes are separated during mitosis.

c o n c e p t

Microtubules create both pulling and pushing forces that separate chromosomes during mitosis.

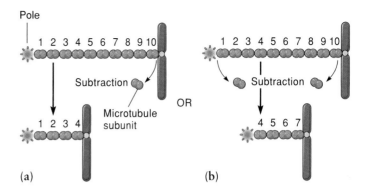

figure 9.24

Possible patterns of disassembly of kinetochore-bound microtubules. (a) Disassembly at kinetochore end only. (b) Simultaneous disassembly at pole and at kinetochore.

Control of the Cell Cycle

During the past 10 years, research on the cell cycle has focused on the properties of **cyclins**, which are proteins that activate enzymes of the cell cycle, and on **cell division cycle (cdc) genes**, primarily in yeasts. More recently, *cdc* genes similar to those in yeasts have also been found in plants and animals. Most of the *cdc* genes that have been discovered so far regulate transitions from the G_1 phase to the S phase or from the G_2 phase to mitosis.

Studies of one of the yeast genes, called a **mitotic inducer gene** (*cdc25*), have revealed how plant *cdc* genes might function. This discovery was made when *cdc25* was transferred to tobacco (Chapter 11 describes how genes can be transferred between organisms). Tobacco plants containing *cdc25* underwent faster cell division in some cells of their flower buds, thereby causing their daughter cells to be smaller than normal. Such plants also produced flowers faster but without petals. This experiment shows how a gene that controls the timing of cell division affects cell size and floral development. Plant scientists believe that the action of *cdc25* in tobacco provides clues about how *cdc* genes work in plants. This is the first step toward understanding the molecular regulation that links the cell cycle with cell size and plant development.

Chapter Summary

The cell cycle is a set of repeated processes that foster cell growth and division. When cell growth and division stop in mature, nonmeristematic cells, the cell cycle also stops. Such mature cells remain in the phase that precedes DNA synthesis, unless they are induced to divide by wounding or some other external influence.

The two main stages of the cell cycle are interphase, which mostly entails cell expansion and DNA synthesis, and cell division. Interphase is further recognized to have three phases: (1) the gap phase between the end of the previous mitosis and DNA synthesis (G_1 phase), (2) the phase involving DNA synthesis (S phase), and (3) the gap phase between the end of DNA synthesis and the beginning of mitosis (G_2 phase). In comparison, cell division has several phases involving chromosome behavior, together called *mitosis*, and a phase called *cytokinesis* that divides cytoplasm into daughter cells.

Most of the interest in interphase is directed at the S phase. Our knowledge of the S phase relies heavily on understanding the structure of DNA and chromatin, which is the DNA-protein complex that makes up chromosomes. DNA synthesis occurs only after chromosomes are unknotted and their DNA helices are unwound and split into single strands. Such manipulations of chromosomes require several kinds of enzymes that work together to replicate DNA. These enzymes are referred to as *replication enzymes*.

DNA synthesis is referred to as *semiconservative replication* because each new double helix contains one strand from the previous double helix. Just one strand is newly synthesized. Semiconservative replication uses a single strand of parent DNA as a template for guiding the synthesis of the new strand. The sequence of nucleotides in the new strand complements that of the parent strand.

The main phases of mitosis are easily observable by light microscopy. Chromosomes condense and become visible in prophase, align along a division plane in metaphase, split into daughter chromosomes and move to opposite poles of the cell in anaphase, and organize into separate daughter nuclei in telophase. Mitosis is usually followed by cytokinesis. When it occurs, cytokinesis begins as early as late anaphase by forming a microtubule-containing structure called a *phragmoplast*. The phragmoplast guides the accumulation of dictyosome vesicles into a large central vesicle that becomes the new cell walls and plasma membranes between daughter cells. Plasmodesmata are formed by pieces of the endoplasmic reticulum that are trapped between dictyosome vesicles. Cytokinesis that involves a phragmoplast, cell plate, and the formation of plasmodesmata is unique to plants and a few algae.

Chromosome movement during mitosis is caused by the behavior of microtubules in the spindle apparatus. Microtubules generate two kinds of forces: one pulls chromosomes to opposite poles of the spindle, and one pushes the poles of the spindle apart.

 What Are Botanists Doing?

Look through several journals related to cell biology or conduct a computer search of the scientific literature in your library to estimate how many research articles have been published in the past 12 months on the cell cycle or related topics. What proportion of these articles is based on plants? How can you explain the difference in the number of botanical versus nonbotanical studies on the cell cycle?

 Writing to Learn Botany

At different times during the cell cycle, microtubules seem to appear, disappear, and reappear at specific places in the cell. This pattern can be explained by the hypothesis that microtubules in all of these places are derived from the same tubulin subunits, which are disassembled in one place and reassembled where they are needed. How could you test this hypothesis?

Questions for Further Thought and Study

1. Before 1950, why did scientists believe that chromosomes doubled immediately before prophase? How was chromatin synthesis discovered to occur several hours before the start of mitosis?

2. What function could centromeres have in nuclear division?

3. Mitosis is said to be "observable" with a light microscope. However, meristematic cells are killed before they are mounted onto a glass microscope slide for study. How, then, do we observe this living process by examining dead cells?

4. DNA polymerases catalyze the assembly of DNA chains in one direction only. Can you suggest reasons for this seemingly severe limitation?

5. Organellar and nuclear genomes must be passed on to daughter cells during cell reproduction. During which period(s) of the cell cycle are plastids and mitochondria most likely to be reproduced? Why?

6. Measurements of genome sizes have been obtained by measuring rates of reassociation of denatured DNA. How can reassociation rates be used to determine genome size?

7. By examining *cot* curves, we can learn about the composition of chromatin. The earliest comparisons of *cot* curves between prokaryotes and eukaryotes yielded some surprising results. What were these results, and why were they unexpected?

8. Onion and wall cress have a 63-fold difference in their respective genome sizes, but less than a 4-fold difference in S-phase duration in their root tip cells. How can we account for the comparatively much faster duplication of the onion genome?

9. What properties of histones do you think are especially important for the function of these proteins in maintaining chromatin structure? Why?

10. How could heterochromatin function in gene regulation?

11. Suggest how a topoisomerase could sometimes unkink chromatin fibers and at other times tie them in knots.

12. The genomes of organisms contain gigantic amounts of information. For example, each human parent contributes to an embryo a set of DNA consisting of almost 3.3 billion nucleotides arranged in 23 chromosomes. If such a DNA sequence were typed in a 10-point font on a continuous ribbon, the ribbon would stretch from San Francisco to Chicago, then to Baltimore, Houston, Los Angeles, and finally back to San Francisco. Even individual genes consist of huge amounts of information (recall that one looped domain consists of 20,000 to 100,000 nucleotide pairs). How would you organize that much information so that you could compare it with the findings of other biologists?

Web Sites

Review the "Doing Botany Yourself" essay and assignments for Chapter 9 on the *Botany Home Page*. What experiments would you do to test the hypotheses? What data can you gather on the Web to help you refine your experiments? Here are some other sites that you may find interesting:

http://www.gene.com/ae/index.hmtl

This page is maintained by *Access Excellence,* Genentech's educational site. Here you'll find links to several content-rich pages, including one that describes the experiments leading to the discovery of the structure of DNA. Another site includes an interview with Dr. Francis Crick (one of the co-discoverers of the structure of DNA). While you're at this site, be sure to check out the links to other pages.

http://xenon.chem.uidaho.edu/hypermail/talkback/0002.htme

At this site, you can participate in discussions about mitosis, DNA, and the cell cycle.

Suggested Readings

Articles

Allen, R. D. 1987. The microtubule as an intracellular engine. *Scientific American* 256 (February): 42–49.

Andrews, B. J., and S. W. Mason. 1993. Gene expression and the cell cycle: A family affair. *Science* 261: 1543–1544.

Chang, F., and P. Nurse. 1993. Finishing the cell cycle: Control of mitosis and cytokinesis in fission yeast. *Trends in Genetics* 9: 333–335.

Francis, D., and N. G. Halford. 1995. The plant cell cycle. *Physiologia Plantarum* 93: 365–374.

Grime, J. P., and M. A. Mowforth. 1981. Variation in genome size—An ecological interpretation. *Nature* 299: 151–153.

Hartwell, L. H., and T. A. Weinert. 1989. Checkpoints: Controls that ensure the order of cell cycle events. *Science* 246: 629–634.

Hollenbeck, P. J. 1985. Mitotic spindles in isolation. *Nature* 316: 393–394.

Koshland, D. E., Jr. 1989. The cell cycle [editorial]. *Science* 261: 545.

Koshland, D. E., T. J. Mitchison, and M. W. Kirschner. 1988. Polewards chromosome movement driven by microtubule depolymerization *in vitro*. *Nature* 331: 499–504.

Murray, A. W. 1989. Dominoes and clocks: The union of two views of the cell cycle. *Science* 246: 614–621.

Murray, A. W. 1993. Cell-cycle control: Turning on mitosis. *Current Biology* 3: 291–293.

Murray, A., and M. Kirschner. 1991. What controls the cell cycle? *Scientific American* 264 (March): 56–63.

North, G. 1985. Eukaryotic topoisomerases come into the limelight. *Nature* 316: 394–395.

Pickett-Heaps, J. D., D. H. Tippit, and K. R. Porter. 1982. Rethinking mitosis. *Cell* 29: 729–744.

Sloboda, R. D. 1980. The role of microtubules in cell structure and cell division. *American Scientist* 68: 290–298.

Ward, G. 1988. How-to-do-it: A handy model for mitosis. *The American Biology Teacher* 50: 170–172.

Books

Alberts, B., D. Bray, J. Lewis, M. Raff, K. Roberts, and J. D. Watson. 1994. *Molecular Biology of the Cell*. 3d ed. New York: Garland.

Anderson, J. W., and J. Beardall. 1991. *Molecular Activities of Plant Cells*. Oxford: Blackwell Scientific Publications.

Becker, W. M., and D. W. Deamer. 1991. *The World of the Cell*. 2d ed. Menlo Park, CA: Benjamin/Cummings.

Bernstein, J. 1978. *Experiencing Science*. New York: Basic Books. (See Chapter 4, "A Sorrow and a Pity: Rosalind Franklin and the Double Helix.")

Bryant, J. A., and D. Francis, eds. 1985. *The Cell Division Cycle in Plants*. New York: Cambridge University Press.

Lehninger, A. L., D. L. Nelson, and M. M. Cox. 1992. *Principles of Biochemistry*. 2d ed. New York: Worth.

Portugal, F. H., and J. S. Cohen. 1977. *A Century of DNA*. Cambridge, MA: MIT Press.

Pollen grains contain cells that are haploid. These cells descended from spores that developed by meiosis in diploid parent cells.

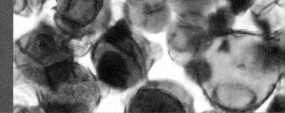

Meiosis, Chromosomes, and the Mechanism of Heredity

10

Chapter Outline

Chapter Overview

All types of organisms, except bacteria, undergo sexual reproduction that involves a regular alternation between meiosis and fertilization. Meiosis produces haploid nuclei from diploid nuclei, and fertilization combines two haploid nuclei into a new diploid nucleus. Offspring that arise from new diploid nuclei usually differ genetically from each other and from their parents. We assume that sexual reproduction and genetic variation are advantageous, since most plants and animals have developed it. However, sex is not necessary for reproduction; many organisms do without it, either periodically or permanently. For example, plants may propagate vegetatively by sprouting from roots or stems. Aspens, dandelions, strawberries, and cholla cacti are all examples of species that reproduce asexually. Unlike organisms that reproduce sexually, those that reproduce asexually rely solely on mitosis for producing offspring. Asexually produced offspring are genetically identical clones of the parent organism.

Sexual reproduction is fundamentally different in plants and animals because plants have multicellular haploid and diploid phases of their life cycle. In contrast, gametes are the only haploid cells in an animal's life cycle. Nevertheless, all sexual organisms rely on meiosis as a major source of genetic variation.

This chapter focuses on the basic features of the meiotic cell cycle and how genetic variation arises from it. It expands on the classical genetics presented in Chapter 8 and shows how the nuclear division of sex cells differs from the asexual nuclear division that was discussed in Chapter 9. This chapter will also help you understand how genes work, which is the main subject of Chapter 11.

The discovery in the 1880s that gametes of a particular worm are haploid led to the idea that some cells are produced by a special type of nuclear division, which was named **meiosis** (from the Greek word *meion,* meaning "less"). It was not until about 1900 that meiosis was described in plants. At first, meiosis was thought to be a simple process whereby half the chromosomes in a diploid nucleus migrated to one side of the cell and half to the other, followed by cytokinesis to produce two haploid cells. The behavior of chromosomes during meiosis turned out to be much more complex than expected, even more so than during mitosis.

The major features of meiosis were described by the 1930s. By that time the mechanisms of meiosis had become a central theme in cytogenetic studies of plants. The role of meiosis in generating genetic variation was established as a result of such work.

In this chapter we first describe the role of meiosis for sexual reproduction in plants and then provide an overview of chromosome behavior during meiosis. These first sections give you a foundation for understanding later discussions of how meiosis produces genetic variation. By the end of this chapter you will be prepared to read Chapter 11, which discusses how genes work and how we can manipulate them by genetic engineering.

Sexual Reproduction in Flowering Plants

Sexual reproduction in all types of organisms entails meiosis, which produces haploid cells from diploid cells, and fertilization, which restores diploidy by the fusion of two haploid cells (gametes) into a diploid cell. This process is called a *life cycle* because it is repeated every generation. In plants, the haploid cells produced by meiosis are called *spores,* which are cells that can divide by mitosis to produce a multicellular haploid organism. This haploid plant makes gametes, either eggs or sperm, so it is called the **gametophyte.** The diploid plant that bears meiotic, spore-producing cells is called the **sporophyte.** Such a life cycle is known as an **alternation of generations** because it alternates between multicellular haploid and diploid "generations," which are phases of the same life cycle (fig. 10.1).[1]

The following paragraphs present a brief overview of sexual reproduction in flowering plants, emphasizing the features of reproduction that they all share. We chose the flowering plants for this discussion because they are the most diverse and widespread group of plants on earth. We present reproduction in other plant groups and in other types of organisms in Chapters 25–30. The diversity of reproductive structures in this plant group is described in detail in Chapter 17.

Spores and Gametes

Meiosis produces two kinds of spores in flowering plants. Spores produced in the anthers of flowers are called **microspores** (fig. 10.2). Microspores are produced from specialized cells called **microspore mother cells.** Each microspore mother cell is diploid and divides meiotically to form four haploid microspores. Following meiosis, each microspore divides by mitosis and cytokinesis inside the spore wall and produces an immature male gametophyte. At this stage, the combined structure that includes a microspore wall and a male gametophyte is called a **pollen grain.** The male gametophyte matures when one of the cells later divides into two sperm cells, which are the male gametes. This means that the sexually mature, full-fledged male plant consists of three cells.

Spores that are produced in immature seeds (**ovules**) are called **megaspores.** In each ovule only one of the diploid cells, the **megaspore mother cell,** undergoes meiosis to form four haploid megaspores. The fate of the megaspores varies among different flowering plants (see Chapter 17). The following description is based on lily (*Lilium* sp.), whose development of the female gametophyte is more commonly known than it is for other plants. Three of the four megaspores in lily disintegrate, leaving the ovule with only one megaspore, the **functional megaspore.** The nucleus of the functional megaspore usually divides by mitosis three times, which produces eight free nuclei. These eight nuclei are then partitioned into seven cells: six have one nucleus, whereas the other cell contains two nuclei. The seven-celled, eight-nucleate structure is the female gametophyte, also called the **embryo sac** (fig. 10.3). Hence, the sexually mature, full-fledged female plant consists of seven cells. The two nuclei that remain free are called the **polar nuclei** because they come from opposite ends of the embryo sac.

---

1. Note that botanists are stuck with a terminological difficulty regarding the word *generation.* This word refers to diploid offspring of diploid parents (e.g., the F_1 generation), as we discussed regarding Mendel's peas in Chapter 8. It also refers to the sporophyte and gametophyte generations (i.e., phases) of the plant life cycle. Because of this difficulty, some botanists prefer to call the plant life cycle an "alternation of phases" instead of an alternation of generations. We use the latter in this text merely because it still is more widely accepted at this time.

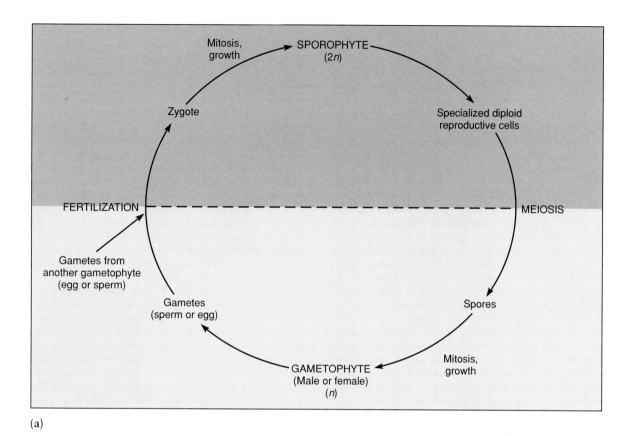

(a)

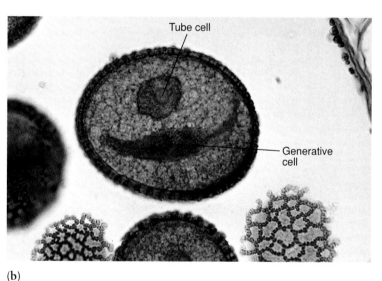

(b)

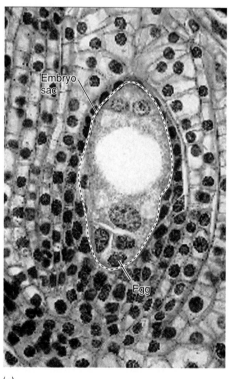

(c)

figure 10.1

Alternation of generations in plants. (a) The plant life cycle is an alternation of
generations between diploid sporophytes and haploid gametophytes. Each sporophyte
produces haploid spores by meiosis. Spores divide and grow into gametophytes, which
produce gametes. The union of the two gametes (i.e., fertilization between sperm and
egg) forms a diploid cell, the zygote, which restores the sporophyte generation.
(b) Immature pollen of lily. These pollen consist of a two-celled male gametophyte; the
generative cell will divide to become two sperm cells; the tube cell will produce the
pollen tube that grows to the female gametophyte, where the sperm are released for
fertilization, ×450. (c) Section through an immature lily seed (ovule) with an embryo sac
containing an egg cell (not all cells are visible in this section). The development of male
and female gametophytes is shown in figures 10.2 and 10.3.

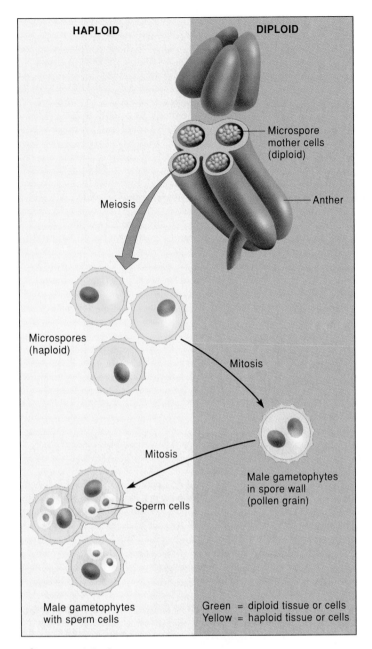

Microspore mother cells (diploid)

Anther

Meiosis

Microspores (haploid)

Mitosis

Mitosis

Male gametophytes in spore wall (pollen grain)

Sperm cells

Male gametophytes with sperm cells

Green = diploid tissue or cells
Yellow = haploid tissue or cells

figure 10.2

Development of the male gametophyte. Microspore mother cells in the anther undergo meiosis. Microspores grow into two-celled male gametophytes, each producing two sperm cells.

Pollination and Fertilization

Sexual reproduction in flowering plants depends on **pollination,** which is the transfer of pollen grains from an anther to the receptive region of a **pistil.** Pistils are the ovule-bearing structures of flowers. Each pistil has three main parts: (1) the **stigma,** which is the receptive surface for pollen grains; (2) the **style,** which conducts pollen tubes to the ovary and raises the stigma to a receptive position; and (3) the **ovary,** where ovules form. In nature, pollen is usually transferred from anther to stigma by animals or wind.

After a pollen grain reaches a stigma, it germinates, forming a **pollen tube.** The cell that forms the pollen tube is the **tube**

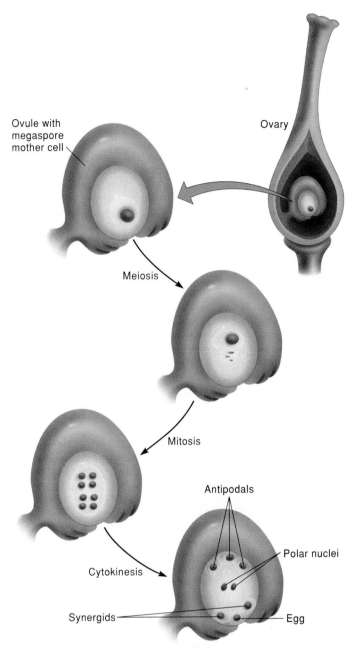

Ovule with megaspore mother cell

Ovary

Meiosis

Mitosis

Antipodals

Polar nuclei

Cytokinesis

Synergids

Egg

Embryo sac (7 cells, 8 nuclei)

figure 10.3

Development of the *Lilium*-type female gametophyte. A single megaspore mother cell in each ovule undergoes meiosis. Three of the megaspores disintegrate, leaving one functional megaspore. Three mitotic divisions in the functional megaspore produce the embryo sac, which consists of eight nuclei that are walled off into seven cells. One of the cells is the egg, which is flanked by two synergids. Three cells are called antipodals, and the remaining large cell contains two polar nuclei.

cell. The second cell in the pollen grain is called the **generative cell** because it divides and generates two sperm cells. This cell divides as the pollen tube grows, and the two sperm cells move through the tube to a small opening in an ovule (fig. 10.4). The two sperm cells enter the embryo sac through one of the synergids, which are the cells next to the egg. The first sperm then

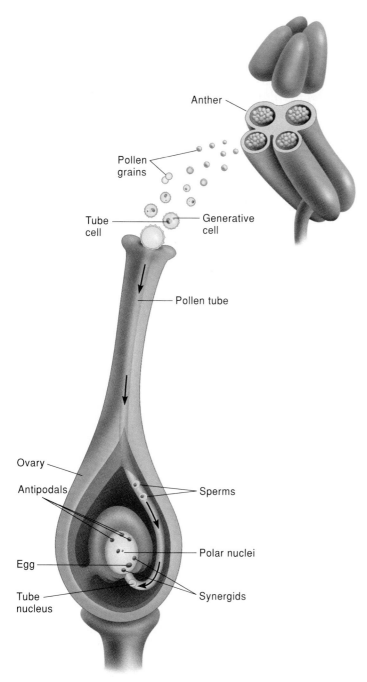

Anther

Pollen grains

Tube cell

Generative cell

Pollen tube

Ovary

Antipodals

Sperms

Egg

Polar nuclei

Tube nucleus

Synergids

figure 10.4

Pollination and fertilization. During pollination, pollen is transferred from the anther to the stigma of the pistil. Two sperms move from the pollen grain through the pollen tube to an opening in the ovule. One sperm fertilizes the egg, and the other fertilizes the polar nuclei.

fertilizes the egg, and the second sperm moves to the polar nuclei and fuses with them. This process is called **double fertilization** because both sperm cells undergo fusion. The diploid cell produced by the fusion of sperm and egg is the **zygote,** which will become the embryo (fig. 10.5). The zygote is therefore the first cell of the new sporophyte generation. The triploid cell formed by the fusion of sperm and polar nuclei will usually form the **endosperm,** a nutritive tissue for the developing embryo. After fertilization, the ovule matures into a seed.

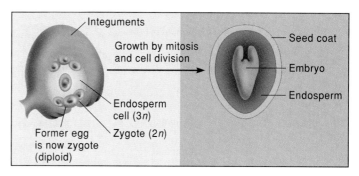

Integuments

Growth by mitosis and cell division

Seed coat

Embryo

Endosperm

Endosperm cell (3n)

Zygote (2n)

Former egg is now zygote (diploid)

figure 10.5

The zygote forms the embryo, which is the beginning of the new sporophyte generation. The product of the second fertilization in this embryo sac develops into a triploid (i.e., 3n) endosperm. Integuments harden and darken into a seed coat.

Some of the genetic variation among offspring of the new sporophyte generation comes from mixing the genomes of different parents. For example, pistils from several flowers of the same plant may receive nonidentical pollen from many other plants. Variation occurs because pollen from different plants contains different genomes. Thus, 10 sources of pollen will produce embryos with at least 10 different genotypes. However, 10 pollen-producing plants will produce many more than 10 different pollen genotypes. Such additional genetic variation develops as a result of meiosis. The remainder of this chapter describes how meiosis works and how it produces genetic variation.

c o n c e p t

Sexual reproduction in plants involves regular alternation between diploid and haploid generations. The diploid generation produces the haploid generation by meiosis. The haploid generation restores the diploid generation by fertilization.

Meiosis

Chromosomes are duplicated during the interphase portion of the meiotic cell cycle. After duplication, chromosomes enter the first of two successive nuclear divisions: **meiosis I** and **meiosis II.** The phases of meiosis are distinguished by a *I* for the first division and a *II* for the second. Accordingly, meiosis I consists of **prophase I, metaphase I, anaphase I,** and **telophase I;** in meiosis II nuclei go through **prophase II, metaphase II, anaphase II,** and **telophase II.** The names of the phases in meiosis are the same as those of mitosis because chromosomes behave similarly in both types of nuclear division. Nevertheless, starting with prophase I, meiosis exhibits several major differences from mitosis. The most obvious distinction is that **homologous** chromosomes form pairs in meiosis but not in mitosis. Chromosomes are homologous to one another when they have the same genes (but not necessarily the same alleles). Starting with chromosome pairing in prophase I, meiosis is a much more complicated and time-consuming process than mitosis.

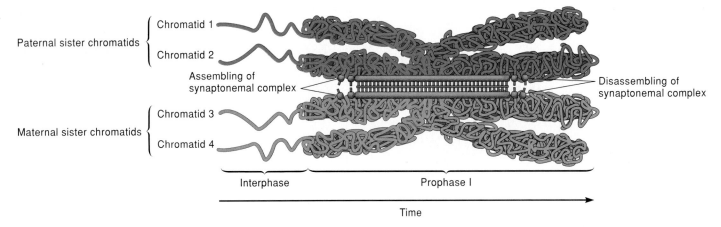

figure 10.6

Prophase I. Homologous chromosomes form pairs that attach to a synaptonemal complex. The synaptonemal complex disintegrates during chromosomal unpairing (desynapsis) in late prophase I.

Prophase I

Prophase I is the longest and most complex stage of meiosis; it typically lasts about 90% or more of the total time required for meiosis. Prophase I involves several stages, which include chromosome condensation, pairing (**synapsis**), and unpairing (**desynapsis**). The major events of prophase I are summarized in figure 10.6. Pairing between homologous chromosomes is achieved by a series of proteins, collectively called the **synaptonemal complex.** Details about how the synaptonemal complex works and its importance in fostering genetic variation appear later in this chapter.

As prophase I begins, the chromosomes condense into long threads. In contrast to mitotic prophase, chromosomes in prophase I of meiosis appear to consist of single chromatids (fig. 10.7). Evidence from cytophotometry, however (see Chapter 9), shows that they have a doubled amount of DNA. This means that each chromosome is actually a pair of sister chromatids, even though it looks like one chromatid. Paired chromosomes in prophase I are called **bivalents** (fig. 10.8a). Because each member of a bivalent has doubled during interphase, a bivalent consists of four chromatids and two centromeres (fig. 10.8b).

The Lore of Plants

One of the oddest modifications of meiosis in plants occurs in a group of North American evening primroses (*Oenothera* sp.). They are all diploid, with a diploid chromosome number of 14. Instead of forming seven bivalents in synapsis, however, they form a ring of up to fourteen chromosomes, seemingly arranged end to end. Botanists discovered this oddity in the early nineteenth century, and they are still trying to figure out the hows and whys of sexual reproduction in these plants.

Soon after chromosomes pair, nonsister chromatids may twist around each other and exchange genetic material by **crossing-over.** Crossing-over occurs when two intertwined, nonsister

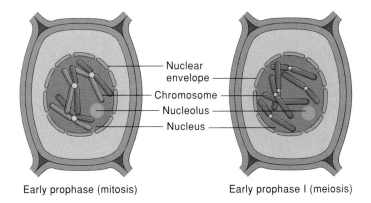

figure 10.7

Comparison of early prophase in mitosis and meiosis. Sister chromatids are distinguishable from each other in early mitosis but not in early meiosis.

chromatids break, and each broken end re-fuses with its nonsister chromatid instead of with its original chromatid (fig. 10.9). Thus, fragments of chromosomes switch places between homologs. Synapsis and crossing-over can therefore change the genetic makeup of two chromatids when the alleles on one fragment differ from those on the other. Such chromosomal rearrangements during crossing-over are a type of **genetic recombination,** which is the general term for producing offspring that are genetically different from the parents. Several kinds of genetic recombination are derived from meiosis in addition to chromosomal rearrangements (see "Genetic Recombination" later in this chapter).

As prophase I continues, the synaptonemal complex dissolves and the nuclear envelope disintegrates. At this time the arms of homologous chromosomes seem to repel each other, but they are still attached at their centromeres and at attachment points called **chiasmata** (singular, **chiasma**) (fig. 10.9). Chiasmata occur where two of the four chromatids have crossed over. Near the end of prophase I, chiasmata move to the ends of the chromosomes; the chiasmata eventually all disappear, but homologs are still held together at their centromeres.

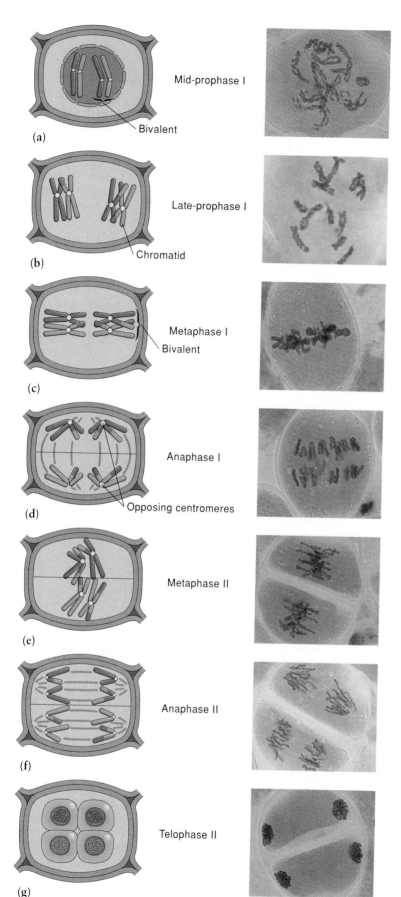

Mid-prophase I

Bivalent

(a)

Late-prophase I

Chromatid

(b)

Metaphase I

Bivalent

(c)

Anaphase I

Opposing centromeres

(d)

Metaphase II

(e)

Anaphase II

(f)

Telophase II

(g)

Prophase I begins when already replicated, homologous chromosomes synapse along their entire lengths. While fused, chromosomes exchange genetic material by crossing-over. By the end of prophase I, homologs are paired only at their centromeres.

The Remainder of the First Meiotic Division

Metaphase I is characterized by the formation of a meiotic spindle apparatus, which moves the bivalents to a metaphase plate at the center of the cell (fig. 10.8c). Microtubules attach to the kinetochore on only one side of each centromere in a bivalent. Because of this pattern of spindle attachment, opposing centromeres are pulled apart during anaphase I (fig. 10.8d). This means that homologous chromosomes separate into different nuclei. The separation of centromeres in meiosis I differs from the splitting of centromeres in mitosis. By splitting mitotic centromeres, mitosis maintains diploidy in the daughter nuclei. However, because intact centromeres separate in meiosis I, the daughter nuclei are haploid. This is why meiosis I is called the **reduction division** of meiosis. Each daughter cell has a haploid set of chromosomes, but each chromosome still consists of two chromatids.

Chromosomes reach their destinations and either begin to decondense in telophase I or continue directly to the second meiotic division. In either case, nuclei do not enter another interphase. If nuclei enter telophase I, a new nuclear envelope forms around the daughter nuclei. This envelope disintegrates in the second meiotic division. When telophase I is skipped, the already condensed chromatin of meiosis I skips prophase II and

figure 10.8

Stages in meiosis. (a) Mid-prophase I. Sister chromatids are distinguishable. (b) Late-prophase I. Nonsister chromatids cross over between homologous chromosomes, making it possible to exchange genetic material between homologs. (c) Metaphase I. Chromosomes align along the metaphase plate. (d) Anaphase I. Homologs separate and move to opposite ends of the spindle. Chromosome fragments that were exchanged during prophase move with recombined chromosomes. (e) Metaphase II. Chromosomes align along metaphase plates in a plane that is perpendicular to the direction of the plate in metaphase I. (f) Anaphase II. Centromeres divide and former sister chromatids move to opposite poles of the spindle. (g) Late telophase II. Chromosomes decondense and nuclear envelopes form around each haploid daughter nucleus. At this stage the products of meiosis are called spores, which form spore walls and are later released from the parent cell. Light micrographs are of meiosis in the anthers of Easter lily (*Lilium longiflorum*).

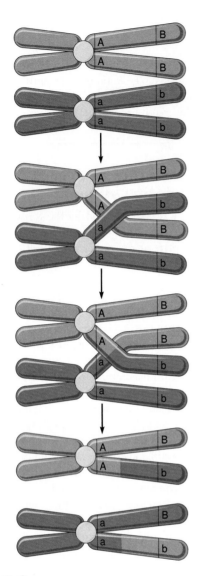

figure 10.9

Homologous chromosomes align during prophase I, enabling nonsister chromatids (indicated by different colors) to cross over and form chiasmata. After breakage and refusion, fragments of chromosomes switch places. In this way, nonsister chromatids exchange genetic material, resulting in genetic recombination.

enters metaphase II directly. The shortcut from anaphase I to metaphase II does not involve the formation and disintegration of a nuclear envelope.

The Second Meiotic Division

The second meiotic division is the same as in mitosis. Chromosomes either condense in prophase II or, if telophase I was bypassed, they go straight to metaphase II (fig. 10.8e), where the spindle apparatus forms and aligns chromosomes in a plane. Once they are aligned, chromosomes enter anaphase II, when centromeres divide and sister chromatids become individual chromosomes. The new centromeres are pulled apart during anaphase II, thereby separating duplicate chromosomes from

each other (fig. 10.8f). In telophase II, chromosomes decondense and a nuclear envelope forms around each of the four daughter nuclei (fig. 10.8g).

By the end of meiosis II, two divisions have been completed. In sum, the diploid parent nucleus that entered meiosis has divided twice to form four haploid nuclei. After the second meiotic division, cytokinesis divides the cytoplasm into four spores. Cytokinesis may occur after meiosis I and after meiosis II or only after meiosis II.

Meiosis is usually an orderly process that produces four haploid nuclei from one diploid nucleus. However, mistakes in meiosis have apparently occurred many times during the evolution of plants. The evidence for such meiotic errors is that many plants are **polyploid,** which means that they have more than two sets of chromosomes. Polyploidy can be induced by treating cells in prophase I with colchicine, a drug that disrupts the assembly of microtubules in the spindle apparatus; the products of meiosis in colchicine-treated plants are therefore diploid instead of haploid. This artificial treatment shows how meiosis may have failed in the past. As explained in reading 10.1, "Polyploidy in Plants," meiotic mistakes have played an important role in plant evolution and, more recently, in the origin of cultivated plants.

c o n c e p t

After prophase I, homologous chromosomes align on a metaphase plate and then segregate into two haploid nuclei for the remainder of meiosis. If necessary after meiosis I, chromosomes condense once again as they enter meiosis II. They then align on a metaphase plate, their centromeres split, and the former sister chromatids move to opposite poles of the cell. The net result of the two meiotic divisions is four haploid nuclei from one diploid nucleus.

Synaptonemal Complex

The synaptonemal complex, which forms during synapsis, begins as a protein axis along each chromosome. The synaptonemal complex links homologous parts of chromosomes, allele for allele (fig. 10.10). Such alignment can be best seen with a light microscope when homologs differ in allele sequence. For example, when one chromosome has an inverted segment relative to the other, the bivalent makes a loop to ensure proper pairing of chromosomes (fig. 10.11). Early evidence that allelic matching occurs between homologs came from studies of such inversion loops.

In contrast to the fidelity of allele matching, exact base-pairing is not required along the entire length of each interacting pair of chromosomes. Homologous chromatin strands, which are at least 100 nm apart, are too widely separated by the synaptonemal proteins for base-pair recognition to occur. Also, the synaptonemal complex can join regions of chromosomes that are different; this is crucial for allowing nonidentical alleles at the same locus to fuse and exchange different genetic material (see fig. 10.16).

POLYPLOIDY IN PLANTS

About 35% of the species of flowering plants are polyploid. The most common level of polyploidy is tetraploidy, which means that plants have four sets of chromosomes instead of two. This also means that the gametes are diploid instead of haploid. Most of the easily observed effects of polyploidy are associated with increased size, and they are usually observed in plants that have been induced to become polyploid by treatment with colchicine. In nature, however, polyploid plants are often indistinguishable from diploid plants, at least upon initial inspection. Instead, for example, polyploids may differ less obviously by being better adapted to temperature or water stress than their diploid relatives. Thus, polyploids can survive in habitats that are not as suitable for their diploid counterparts.

Many cultivated plants are polyploids. One of these is bread wheat (*Triticum aestivum*), the most commonly cultivated crop in the world. Bread wheat probably

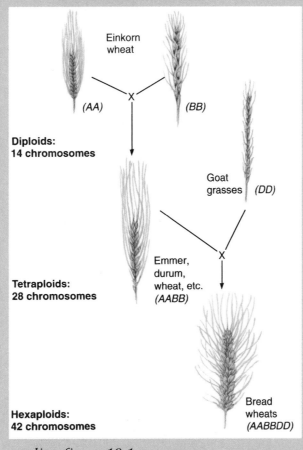

reading figure 10.1

Domesticated wheats evolved through hybridization and polyploidy. A, B, and D refer to different haploid genomes.

Adapted from "Wheat" by Paul C. Mangelsdorf. Copyright © 1953 by Scientific American, Inc. All rights reserved.

arose in the Middle East at least 8,000 years ago. It evolved by hybridization between a durum wheat, which is a tetraploid that has 28 chromosomes, and one of the goat grasses, which are wild diploid relatives of wheat that have 14 chromosomes. Thus, bread wheat has 42 chromosomes, which means that it is hexaploid. Durum wheat arose as the polyploid descendant of a diploid hybrid between two kinds of einkorn wheat and is still cultivated as the principal grain used in macaroni products (to learn more about wheat, see the Epilogue starting on page 806).

In a polyploid series such as that of wheat, distinct genomes can be distinguished by light microscopy. Thus, haploid genomes of the two einkorn wheats are designated *A* and *B*, respectively, and the haploid genome of goat grasses is designated *D*. The tetraploid genome of durum wheat is therefore *AABB*, and that of bread wheat is *AABBDD* (reading fig. 10.1).

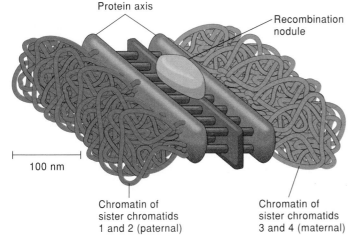

figure 10.10

Model of a synaptonemal complex. A synaptonemal complex consists of a protein axis that binds homologous chromosomes together at alleles of the same gene locus.

Recombination Nodules

Chromosome exchange occurs in specific regions of the synaptonemal complex that are called **recombination nodules;** these regions are so named because of the large, protein-containing spheres associated with them. Although the synaptonemal complex provides a structural framework for chromosomal rearrangements, the actual process is thought to be mediated by recombination nodules. Each nodule has a diameter of about 90 nm, which almost completely spans the distance between homologous chromosomes in a synaptonemal complex. Each recombination nodule is probably made of several enzymes that act in concert to bring together homologous segments of DNA from opposite chromatids.

Chiasmata and Chromosome Segregation

As described previously, chiasmata in late prophase I indicate that crossing-over has probably occurred. In addition to

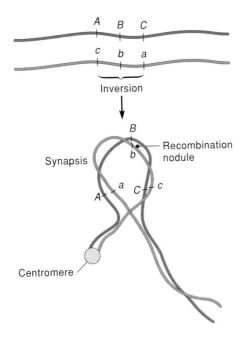

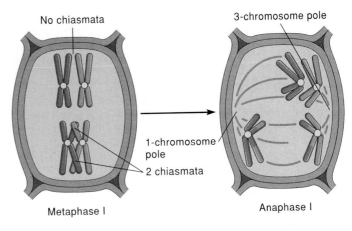

figure 10.11

Drawing of synapsed chromosomes with an inversion loop. An inversion loop forms when the order of alleles in a segment of one homolog is inverted relative to the other homolog. The loop enables the appropriate matching of alleles during synapsis.

recombining genes, crossing-over helps to segregate chromosomes during the first division of meiosis. This is because chiasmata continue to join homologs on the spindle after desynapsis. In mutant or hybrid plants that have fewer chiasmata than normal, however, homologs often fail to segregate normally. The products of such an abnormal meiotic division receive the wrong number of chromosomes and usually either die or are sterile (fig. 10.12).

After bivalents align in metaphase I, chiasmata begin to dissolve. Their dissolution coincides with the detachment of the fused kinetochores of sister chromatids. The attraction between the arms of sister chromatids is also disrupted, giving them a splayed appearance as they enter anaphase I (fig. 10.8d).

c o n c e p t

Chiasmata usually indicate that crossing-over has occurred. Chiasmata ensure correct chromosome segregation because they hold chromosomes together until bivalents are properly aligned on the metaphase I plate.

Genetic Recombination

Meiosis always produces genetic recombination, much of which comes from the exchange of chromosomal fragments during prophase I. However, genetic recombination would

figure 10.12

Chiasmata in prophase I are associated with proper separation and migration of homologs in anaphase I. Homologs may not separate when there are no chiasmata. Too few or too many chromosomes usually cause the new cells to die or to be sterile.

occur even without crossing-over. New combinations of traits would still appear in offspring because of chromosomal separation in meiosis.

Recombination by Chromosomal Assortment

Homologs are not oriented along the metaphase plate until spindle fibers attach to each centromere. This means that, before spindle attachment, each homolog has an equal chance of moving to either pole. Furthermore, the direction of movement of each homolog is independent of all other bivalents. This is analogous to independent assortment in Mendelian genes, but for entire chromosomes.

How does chromosomal separation cause genetic recombination? Imagine that all the chromosomes of a sperm cell are of one color and all the chromosomes of an egg cell are of another color. The zygote from their fertilization would have chromosome sets of both colors (fig. 10.13). When the mature plant from this zygote reproduces sexually, all the bivalents in meiosis would have the color of one sperm-derived homolog and the color of one egg-derived homolog. If, during metaphase I, some of the sperm-derived chromosomes moved to the same pole as some of the egg-derived chromosomes, then each haploid nucleus would have chromosomes of both colors. This means that the gametophytes and the gametes produced by them would have new combinations of chromosomes and that the offspring would therefore differ genetically from the parents.

To illustrate the potential for new genetic combinations based on chromosome separation, consider the gametes of the desert sunflower, *Machaeranthera gracilis*. Cells of the sporophyte have only four chromosomes, which is the smallest chromosome number known in plants. This means that there are only two bivalents in prophase I. If we color-code the homologs of these bivalents, we see that they can be oriented in

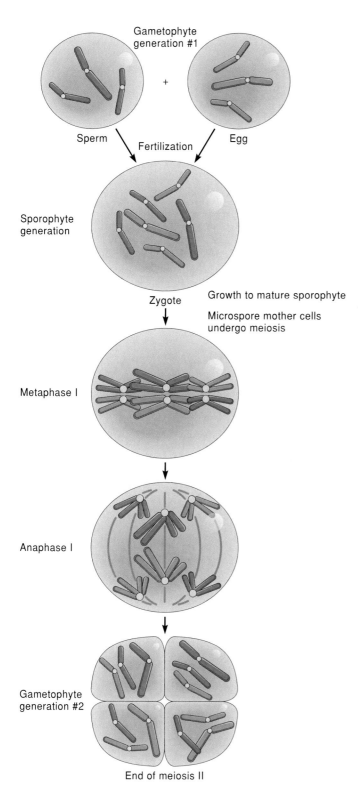

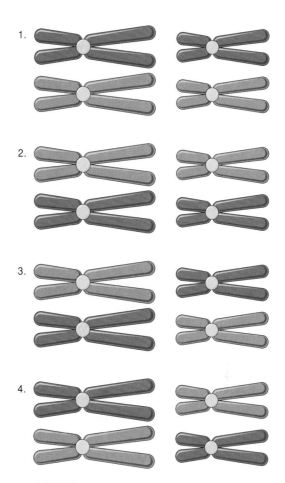

figure 10.13

The pattern of separation of meiotic chromosomes is a source of genetic combination. Differently colored chromosomes from the sperm and egg in the first gametophyte generation are mixed together in new combinations in the second gametophyte generation. Mixing comes from the orientations of homologous pairs in metaphase I.

figure 10.14

There are four possible orientations of two bivalents in metaphase I. Two of them (1 and 2) will segregate into parental combinations in the gametophyte generation. The other two (3 and 4) will result in new combinations of chromosomes in the gametophyte generation.

four (i.e., 2^2) combinations in metaphase I (fig. 10.14). In anaphase I, two of the combinations will keep same-colored chromosomes together, whereas the other two combinations will produce nuclei with mixed colors. Thus, by chance, half of the spores in the desert sunflower will have different combinations of chromosomes; that is, they will differ genetically from their parents.

The number of orientations of chromosomes in metaphase I increases dramatically for larger numbers of chromosomes. For example, there are 8 ways to orient three bivalents (i.e., $2^3 = 8$). As before, 2 of these are parental combinations (all one color), so 6 are new combinations. Thus, based on random chromosomal separation alone, 75% of the spores will differ genetically from the previous gametophytic generation. However, most organisms have more than three kinds of chromosomes. For example, the garden pea has seven pairs of chromosomes, which have 128 (i.e., 2^7) possible orientations of bivalents in metaphase I. Thus, 1.6% (2/128) are parental, and 98.4% (126/128) are new combinations. Similarly, the 23 pairs of chromosomes in humans have more than 8 million possible patterns of chromosome separation ($2^{23} = 8,388,608$), but only two of these are parental combinations.

Remember that (1) the random separation of chromosomes during meiosis produces microspores in anthers and megaspores in ovules, and (2) the chance of reconstituting the parental chromosome combination in the microspore is independent of that in the megaspore. In the garden pea, for instance, this means that the probability of reconstituting the parental combination in a sperm (i.e., 2/128 = 0.0078) and in an egg (i.e., 0.0078) is 0.0078 × 0.0078 = 0.00006. In other words, only 6 of every 100,000 fertilizations have a chance of producing offspring without new combinations of chromosomes. Even without considering how two such gametes might get together, these simple calculations show that the odds of maintaining any one combination of chromosomes from parent to offspring are minuscule. The probability of genetic recombination by chromosome assortment is overwhelming.

c o n c e p t

Homologs separate independently of each other during meiosis. Spores are therefore likely to have new combinations of chromosomes, some from the maternal parent of the sporophyte and some from the paternal parent. The number of new combinations of chromosomes is greater for higher chromosome numbers.

Evidence for Chromosomal Exchange during Crossing-Over

The first experimental evidence in plants that genetic recombination occurs during meiosis came from breeding experiments with corn (*Zea mays*) performed by H. S. Creighton and Barbara McClintock in the 1930s (to learn more about Barbara McClintock, see Chapter 1). Their evidence came from the inheritance of linked genes on a chromosome with unusual morphology. The two linked genes of interest control kernel color and texture, respectively. Kernels are either colored (C) or colorless (c) and have either standard texture (Wx) or waxy texture (wx). The chromosome that bears these genes sometimes has a darkly stained knob at one end and a piece of another chromosome at the other end. In the experimental plants, a chromosome of normal appearance carried alleles for c and Wx, and its homolog, with both the knob and the extra segment, carried C and wx.

The results of the physical exchange between homologous chromosomes were seen with a light microscope and were associated with recombination of the two gene loci. Recombined chromosomes have the knob on one homolog and the extra piece on the other (fig. 10.15). Such switching of chromosome segments explains why offspring from this cross can have colored kernels with standard texture or colorless kernels with waxy texture, even though the parents had neither of these combinations of traits.

Crossing-over is more easily observed in the bread mold *Neurospora crassa*. Spores are haploid, and therefore show the results of meiotic chromosome behavior directly, without having to wait for the next diploid generation, as in plants and

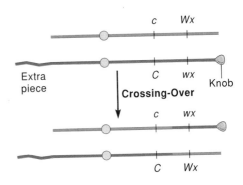

figure 10.15

Evidence for crossing-over in corn (*Zea mays*). When nonsister chromatids cross over in prophase I, the chromatid bearing the allele for waxy kernel texture (*wx*) is exchanged with its homologous chromosome bearing the allele for normal kernel texture (*Wx*). Although the alleles themselves cannot be seen, the new position of the knob is direct evidence for chromosomal exchange.

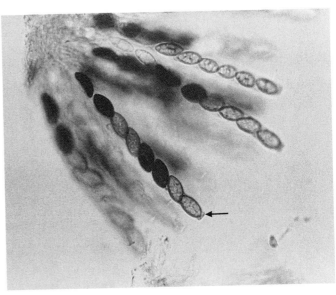

figure 10.16

Light micrograph of spore sacs from the bread mold *Neurospora crassa*. Each spore sac contains eight spores in single file. When crossing-over occurs between chromosome segments that bear the gene for spore color, the sac contains an alternating pattern of spore colors (arrow).

animals. Also, in contrast to what happens in plants, spores form in single file in small sacs in *Neurospora*, and each spore undergoes mitosis immediately after meiosis, thereby making 8 spores in each sac (fig. 10.16). Spores can be either brown or gray, depending on one or the other allele of a single gene. After meiosis and mitosis, spore sacs usually have 4 brown spores in a row, followed by 4 gray spores in a row (fig. 10.17a). However, some sacs have the following pattern: 2 brown, 2 gray, 2 brown, 2 gray. This is evidence that crossing-over occurred between nonsister chromatids bearing the gene for spore color, as diagrammed in figure 10.17b.

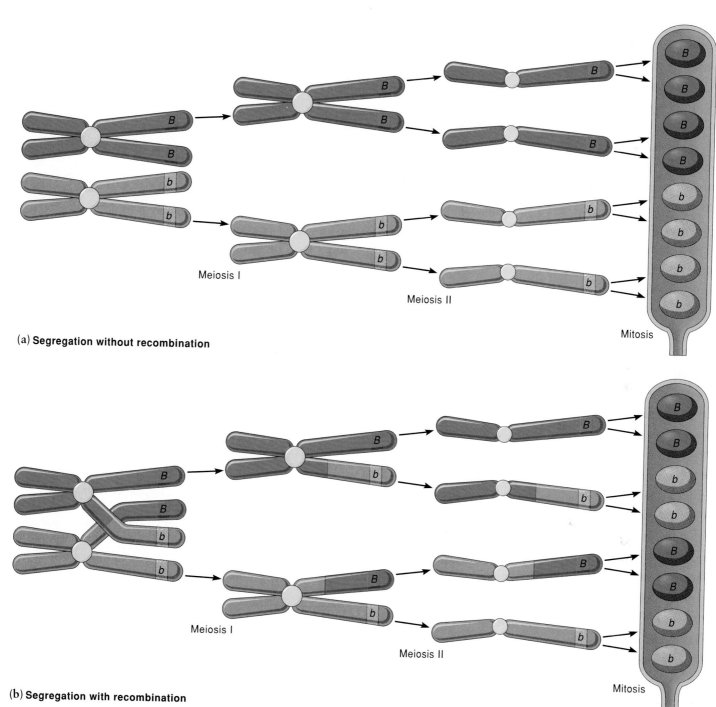

(a) **Segregation without recombination**

Meiosis I

Meiosis II

Mitosis

(b) **Segregation with recombination**

Meiosis I

Meiosis II

Mitosis

figure 10.17

Evidence for crossing-over from the segregation of spore colors in bread mold. (a) Without crossing-over, brown spores are together at one end of the sac and gray spores are together at the other end. (b) After crossing-over, brown and gray spores alternate in pairs.

Gene Conversion

Besides promoting recombination from crossing-over and chromosome separation, meiosis may also expose mistakes during DNA replication in spore mother cells. One such change was discovered in *Neurospora crassa* by Mary Mitchell in 1955. She found that one allele can apparently be changed to an alternate allele during spore production by this fungus. This change is called **gene conversion.**

Like crossing-over, gene conversion in *Neurospora* can be shown by following the inheritance of spore color (fig. 10.18). Instead of containing 4 brown and 4 gray spores, spore sacs occasionally contain 2 brown and 6 gray spores or 6 brown and 2 gray spores. This ratio (3:1) cannot be the result of allelic interactions such as dominance, because the spores are haploid. Mitchell interpreted this pattern to mean that one allele is changed to the other, either brown to gray or gray to brown, during meiosis.

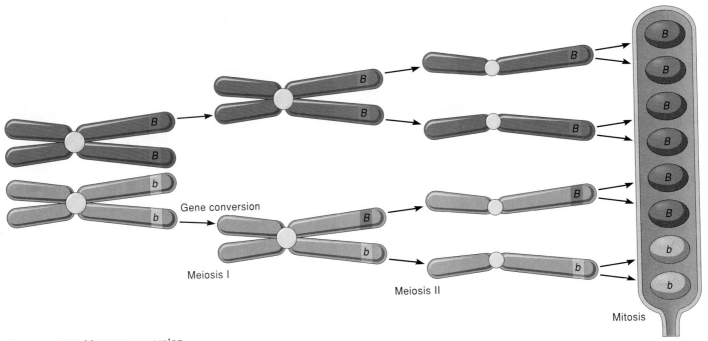

Segregation with gene conversion

figure 10.18

Evidence for gene conversion in bread mold. When one allele is changed to the other, the sac contains six spores of one color and two of the other (3:1 ratio), instead of four of each (1:1 ratio).

In the figure: Meiosis I, Meiosis II, Mitosis, Gene conversion

Although gene conversion probably develops by different mechanisms in different organisms, all mechanisms are believed to involve recombination and DNA repair between mismatched homologs. In one of the models for this process, gene conversion begins when a single strand in one double helix breaks and unzips from its complementary strand. The single-stranded "whisker" then invades a double strand on the other homolog, thereby displacing part of the other double helix (fig. 10.19).

When the invading DNA whisker is not perfectly complementary with its homolog, any mismatches are recognized and repaired by DNA repair enzymes. If the invader is used as the template, then the complement is changed to it, thereby undergoing gene conversion. If the complement is used as the template, then the invader is changed, which means that no gene conversion occurs. Meanwhile, the single strand that was left behind by the departing whisker is used as a template to replace the missing DNA fragment. The sequence that was displaced by the whisker on the invaded strand is removed and broken down.

The enzymes that initiate single-strand breakage of DNA and promote recombination are well known only in bacteria, which are incapable of meiosis. Nevertheless, gene conversion probably occurs in all organisms that can repair DNA. The small amount of DNA synthesis that occurs during prophase I in eukaryotes is considered to be indirect evidence for gene conversion in these organisms.

c o n c e p t

Gene conversion is a process by which the DNA sequence of one allele is converted to that of the other allele. The process entails breakage and repair of DNA. Although gene conversion cannot be readily seen in most organisms, it is believed to occur when DNA is repaired after a single strand from one chromosome invades a complementary but nonidentical region of its homolog.

Unequal Crossing-Over and Gene Duplication

The discussion to this point has focused on the exchange of equal amounts of DNA by crossing-over. However, unequal amounts of DNA may be exchanged when homologs are not aligned perfectly (fig. 10.20). In such cases, crossing-over is called **unequal crossing-over,** and it probably occurs most often when a gene has already been duplicated at least once. The mechanisms for making an extra copy of a single gene are unknown, but they seem to be associated with the activity of transposable elements (see Chapter 8) and the irregular fusion of chromosome fragments during synapsis. One of the more common outcomes of such duplications is a **tandem repeat,** which is the occurrence of two or more copies of a gene in a row. Figure 10.21 shows how a tandem repeat might form during crossing-over.

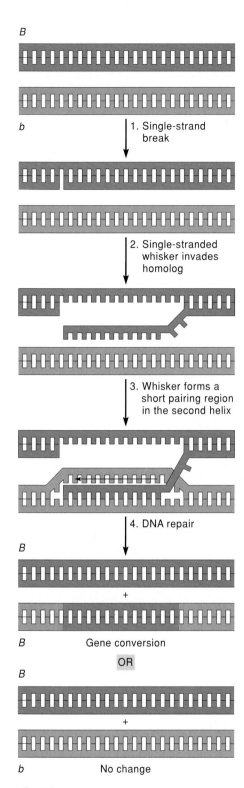

B

b

1. Single-strand break

2. Single-stranded whisker invades homolog

3. Whisker forms a short pairing region in the second helix

4. DNA repair

B

+

B   Gene conversion

OR

B

+

b   No change

figure 10.19

Model for gene conversion: (1) a single strand of one homolog breaks; (2) the whisker of single-stranded DNA invades the double strand on a nonsister chromatid; (3) the invading whisker forms a short pairing region in the second helix; (4) the mismatched region between the invading whisker and the invaded chromatid undergoes DNA repair to match both strands in the region. DNA displaced by the invasion is degraded; the gap left by the whisker is filled in by new DNA. Gene conversion occurs when the invading whisker is used as the template for DNA repair (*b* allele becomes *B* allele in second helix). There is no change when the second helix is used as the template for DNA repair.

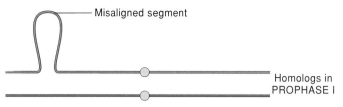

Misaligned segment

Homologs in PROPHASE I

figure 10.20

A chromosome may have an extra segment due to unequal crossing-over in a previous generation. During synapsis, the extra segment protrudes as a loop because it has no matching sequence on its homolog.

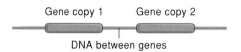

Gene copy 1   Gene copy 2

DNA between genes

figure 10.21

Model of a tandem repeat. Two or more copies in a row of the same gene are said to be in tandem.

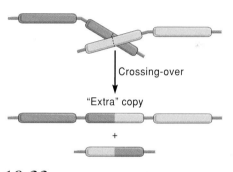

Crossing-over

"Extra" copy

+

figure 10.22

Model of gene duplication. Tandemly repeated genes may misalign during synapsis and exchange unequal-sized fragments. As a result one of the homologs ends up with an "extra" copy, and the other homolog has one copy fewer than it had before.

The most likely mechanism for making a third copy of two tandem genes is shown in figure 10.22. According to this model, chromosomes first misalign during synapsis so that one member of the tandem is paired and one is unpaired on each homolog. When crossing-over occurs in the paired region, genetic exchange may produce a "hybrid" gene between the tandem genes on one of the homologs, but the other homolog loses one member of the tandem at the same time. Thus, after unequal crossing-over, one homolog has three copies of the gene in a row, and the other homolog has one gene left from the original tandem.

Tandem repeats are common, which is considered to be indirect evidence for gene duplication by unequal crossing-over. For example, some strains of corn have more than 9,000 tandemly repeated genes that code for ribosomal RNA (rRNA). Tobacco has about 1,000 copies of rRNA genes, and humans have only 50 copies. These genes are the most highly duplicated in all organisms because of the high demand for ribosomes. Nevertheless, many other kinds of genes also occur in duplicate, but usually in only 5–10 copies. In addition,

MULTIGENE FAMILIES

A set of duplicated genes is referred to as a multigene family. Although most genes are single-copy genes, multigene families probably occur in all plants. The number and diversity of genes in a multigene family represent the historical record of gene duplication in the genome. Scientists are fascinated by this phenomenon because it has the potential for revealing insights into the evolution of new genes.

Genes associated with photosynthesis are of particular interest to scientists in studying molecular evolution in plants. For example, polypeptides from two kinds of genes combine to form rubisco, the primary CO_2-fixing enzyme of photosynthesis (see Chapter 7). The genes for this enzyme have received more attention for evolutionary studies than any other genes.

One of the genes for rubisco occurs in chloroplasts and makes the large subunit of the enzyme. The other gene occurs in the nucleus and makes the small subunit. In all plants that have been examined, the gene for the small subunit, called *rbc*S, occurs in a multigene family. The tomato, for example, has five copies in its *rbc*S family, and the petunia has eight copies in its *rbc*S family. In tomatoes, three of the five copies occur in a head-to-tail sequence, close to each other on one chromosome. These are tandem repeats. A fourth copy is farther away on the same chromosome, and the fifth copy is on another

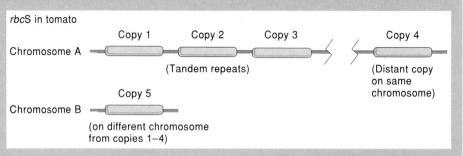

reading figure 10.2

Distribution of *rbc*S genes of tomato (*Lycopersicon esculentum*). Three copies of this gene are tandem repeats, one is distant from the tandem repeats on the same chromosome, and one is on a different chromosome than the other four copies.

chromosome entirely (reading fig. 10.2). Comparing the DNA sequences of each copy shows the amount of similarity among the different gene copies. In so doing, evolutionary biologists have discovered that the three head-to-tail genes are almost identical to each other (99% similarity), but the other two differ by at least 10%. This means that the nontandem copies of the gene have evolved so that they differ slightly from their counterparts in tandem repeats.

Duplicated genes from unequal crossing-over should, when first formed, be identical and in the same place on one chromosome. So how can we explain the diversity of sequences and the different genomic locations of genes in the same multigene family? The best explanation for their different locations relies on the behavior of transposable elements. When mobile sequences of DNA jump from one location to another, they often take nearby DNA with them. In so doing, a transposable element can move the duplicated copy of a gene, which was originally formed by unequal crossing-over, to another location. The new copies can then accumulate random mutations, becoming different from the DNA sequence of the parent gene. In contrast, the tandem copies may remain nearly identical because gene conversion "homogenizes" them—that is, one copy acts as a template to correct any mutations that arise in another copy. Tandem genes may also be nearly identical because gene duplication was recent, and the newest copies have not yet had time to accumulate many mutations.

some copies of duplicated genes have apparently mutated independently from other copies (see reading 10.2, "Multigene Families").

c o n c e p t

Many genes occur in multiple copies that probably arose by genetic exchange during unequal crossing-over. Some genes, such as those for ribosomal RNA, may be present in thousands of copies. However, most duplicate genes have ten or fewer copies.

Why Sex?

The cycle of meiosis and fertilization depends on so many variables that it seems to flirt with failure every time sexual reproduction is attempted. For instance, most male gametes are wasted, since most pollen grains do not arrive at the stigma of a flower. This is especially true of wind-pollinated species, such as oaks, pines, walnuts, and birches. In addition, sexual reproductive structures require substantial amounts of energy. In some species, such as the small monkey flower (*Mimulus kelloggii*), more than 20% of the energy from photosynthesis is used to make flowers. Moreover, genetic recombination disrupts adaptive gene combinations. This means that although an individual plant is well adapted to its environment, it is unlikely that its gene combination will remain intact in the offspring. In a constant, unchanging environment, it is also unlikely that genetic recombination will produce offspring that are more fit than the parents, since the parents are already well adapted.

Sexual reproduction generates genetic variability, which is important for the evolutionary adaptation of a specific to new or changing environments. However, if genetic recombination is often detrimental to the fitness of an individual's offspring, and if plants can be successful asexually, why is there sex at all? In trying to answer this question, biologists have focused most of their attention on the evolutionary consequences of sex. The assumption is that sexual reproduction is advantageous because it is so common in virtually all kinds of organisms. Although most scientists agree that there is some underlying evolutionary benefit from sexual reproduction, they do not agree on what the benefit is. The two main hypotheses to account for the significance of sexual reproduction at the molecular level are the DNA repair hypothesis and the transposon hypothesis. (See Chapter 22 for a discussion of the significance of sexual reproduction at the population level.)

DNA Repair Hypothesis

Crossing-over does not occur in certain mutant strains of yeast (*Saccharomyces cerevisiae*). These strains also cannot repair double-stranded DNA. This observation is evidence that the mechanisms of crossing-over are shared with those of the DNA repair system, at least partially. Furthermore, such evidence may mean that the primary function of crossing-over is to repair DNA (see fig. 10.19).

A damaged chromosome can be repaired if it has an appropriate template to guide the repair. Fusion to such a template occurs only during synapsis in prophase I, when gene conversion can change a defective chromosome back to its undamaged state. According to this reasoning, DNA repair is the main function of meiosis I.

According to the DNA repair hypothesis, damage in one haploid generation should be repaired before another haploid generation is produced. From this perspective, the diploid generation need only be transient, existing just long enough to bring homologs together, repair any damage, and segregate chromosomes back into haploid cells. The first eukaryotic organisms to reproduce sexually were haploid and had short-lived diploid generations. Among extant organisms, many algae and fungi still have this type of life cycle. The only cell in the diploid generation is the zygote, and meiosis occurs at that stage. Moreover, fertilization in these organisms is often induced by low or high temperatures, desiccation, nutrient depletion, or some other kind of stress. This may be additional indirect evidence for the DNA repair hypothesis, because stressed haploid cells are more likely to suffer lethal damage to

DNA than diploid cells. However, such damage can be repaired when the cells fuse and then undergo synapsis and gene conversion.

Although the DNA repair hypothesis helps to explain how sexual reproduction may have started, most of the evidence for it is circumstantial. Nevertheless, studies of mutant strains of yeast and of life cycles in single-celled eukaryotes provide evidence that is consistent with the hypothesis.

Transposon Hypothesis

Transposable elements, also called *transposons*, were discussed in Chapter 8 with regard to their ability to disrupt gene expression. Such disruption occurs when a mobile sequence of DNA moves into a gene, thereby inhibiting its function. Besides their mobility, many transposable elements are apparently duplicated during transposition; the parental copy stays in place while a duplicate copy is inserted elsewhere in the genome. Furthermore, newly made copies of transposons are believed to spread, like viruses, from one genome to another by meiosis and fertilization. That is, transposons probably spread throughout populations and species because of sexual reproduction.

The main function of transposons seems to be to make more transposons. They are therefore considered to be genomic parasites. They are efficient parasites, however, because they exploit sexual processes for their own reproduction. For this reason, they have been referred to as **selfish DNA.** Because the spread of such selfish DNA is enhanced by sexual reproduction, transposon sequences are most abundant in sexual species, especially those that rely predominantly on crossbreeding between different individuals. Indeed, in some species of plants, transposons make up about 10% of the genome. The abundance of transposon-derived DNA sequences explains, at least partially, why so many sexually reproducing organisms have large amounts of repetitive DNA (Chapter 9).

The transposon hypothesis asserts that the driving force for the origin of sex came from parasitic DNA sequences that could enhance their own fitness by promoting cycles of chromosome fusion and segregation. This hypothesis is supported by the behavior and widespread occurrence of transposons in sexual organisms. Furthermore, it eliminates the need to explain how cells or organisms can derive benefits from sexual reproduction. Such benefits, if any, are irrelevant to the individual plant or animal; benefits accrue instead to the genomic parasites.

Evidence that sex is an adaptation of parasitic DNA is also known in prokaryotes. In the bacterium *Escherichia coli*, cell fusion and genetic recombination result from interactions between the main bacterial genome and DNA that occurs in plasmids, which are self-replicating DNA sequences that are not part of the primary bacterial genome. In addition, the DNA of bacteriophages, which are all parasitic, also causes genetic change in the host bacterium. Thus, both kinds of genomic parasites—plasmids and phages—provide forms of sexuality among prokaryotes. Thus, eukaryotic sex may have begun with a type of intracellular genetic parasitism that arose from prokaryotic ancestors.

Chapter Summary

The life cycle of plants is an alternation between diploid and haploid generations. The diploid generation produces the haploid generation by meiosis; the haploid generation produces the diploid generation by fertilization. The stages of meiosis have the same names as the stages of mitosis: prophase, metaphase, anaphase, and telophase. However, meiosis consists of two divisions. The first division is a reduction division, and the second is like mitosis. Meiosis is also fundamentally different from mitosis because, in meiosis, homologous pairs of chromosomes fuse at the beginning of the first prophase. Thus, paired chromosomes separate into haploid nuclei in the first meiotic division, and centromeres divide and separate in the second meiotic division.

During the first prophase, chromosomes exchange genetic material across the synaptonemal complex that holds them together. Chiasmata, the apparent physical indicators of chromosomal crossing-over, ensure correct chromosome segregation because they hold chromosomes together until bivalents align properly on the metaphase plate.

New genetic variation occurs during meiosis because of chromosomal separation and genetic exchange between homologs. Chromosomal separation produces new combinations of parental chromosomes in the offspring. Genetic exchange also occurs because of gene conversion by DNA repair processes. In gene conversion, a strand from one chromosome functions as a template for repairing a strand on its homolog. When crossing-over occurs between imperfectly aligned homologs, the exchange of genetic material can be unbalanced. Through such repeated unequal crossing-over, entire genes can be duplicated on one homolog and deleted from another. This process of gene duplication also explains how the number of genes for ribosomal RNA can be in the thousands in some plants. Many other kinds of genes occur in multiple copies, but the usual number of copies is ten or fewer.

Sexual reproduction benefits species and populations because it produces genetic variation, but it is not clear how individuals or their progeny benefit from sexual processes. Two hypotheses have been advanced to explain the origin and maintenance of sexual reproduction at the molecular level. One states that chromosome fusion and segregation evolved as a mechanism to repair DNA. An alternative hypothesis for the origin of sex is based on the evolution and spread of parasitic DNA sequences, or transposable elements.

What Are Botanists Doing?

Use the reference resources in your library to answer the question, "How many multigene families have been discovered in plants during the past twelve months?" As you compile your list, note the plant names and the names or descriptions of the genes or gene products on the list.

Writing to Learn Botany

Instead of undergoing meiosis, diploid nuclei could simply segregate their chromosomes into equal parts. This occurs in some fungi. How would the haploid nuclei of such a process differ from haploid nuclei produced by meiosis? After fertilization occurred between gametes from these "haploidized" cells, how would the new diploid generation differ from offspring produced by normal sexual reproduction?

Questions for Further Thought and Study

1. What are the major differences and similarities between meiosis and mitosis? What is the significance of each process?

2. If diploidy evolved because of its potential to repair DNA, it seems that a single-celled diploid stage would suffice to accomplish it. Yet most organisms are dominated by a complex, multicellular diploid generation. How can this discrepancy be explained?

3. Genes in a multigene family can vary from one copy to another. What are the possible reasons for this variation?

4. What evolutionary implications can you think of for the occurrence of multigene families?

5. Two races of *Machaeranthera gracilis* exist, one with $n = 2$ chromosomes and one with $n = 3$ chromosomes. Assume that gametes from the two races are interfertile and that the hybrid from such a cross can grow into a normal-looking plant. Would you expect this hybrid to be fertile or sterile? Why?

6. Some cholla cacti are triploid with a chromosome number of 33. How might prophase I look when synapsis occurs in cells of these plants?

Web Sites

Review the "Doing Botany Yourself" essay and assignments for Chapter 10 on the *Botany Home Page.* What experiments would you do to test the hypotheses? What data can you gather on the Web to help you refine your experiments?
Here are some other sites that you may find interesting:

http://www.mcl.ucsb.edu/classes/bio4a/meiosis.htm
This site includes a short tutorial on the relationship between the movements of chromosomes during meiosis and the segregations of alleles during meiosis. You can also download some movies from this site.

http://esg-www.mit.edu:8001/esbgio/mg/meiosis.html

This site includes "Meiosis and Genetic Recombination," a section from the Mendelian Genetics chapter of the 7.001 Hypertextbook.

http://nasc.nott.ac.uk:8200/nasc.html

This site includes links to a variety of genetics-related sites throughout the world.

Suggested Readings

Articles

Anderson, A. 1992. The evolution of sexes. *Science* 257:324–326.

Atkins, T., and J. M. Roderick. 1991. "Dropping your genes." A genetics simulation in meiosis, fertilization, and reproduction. *The American Biology Teacher* 52:164–169.

Brown, C. R. 1990. Some misconceptions in meiosis shown by students responding to an advanced level practical examination question in biology. *Journal of Biological Education* 24:182–185.

Maguire, M. P. 1992. The evolution of meiosis. *Journal of Theoretical Biology* 154:43–55.

Petrusky, B. 1990. DNA on target: Homologous recombination. *Mosaic* 21(May):44–52.

Rose, M. R. 1983. The contagion mechanism for the origin of sex. *Journal of Theoretical Biology* 101:137–146.

Simchen, G., and Y. Hugerat. 1993. What determines whether chromosomes segregate reductionally or equationally in meiosis? *BioEssays* 15:1–8.

Stahl, F. 1987. Genetic recombination. *Scientific American* (February):90–101.

Stewart, J., B. Hafner, and M. Dale. 1990. Students' alternate views of meiosis. *The American Biology Teacher* 52:228–232.

Tanksley, S. D., and E. Pichersky. 1988. Organization and evolution of sequences in the plant nuclear genome. *Plant Evolutionary Biology*, L. D. Gottlieb and S. K. Jain, eds. New York: Chapman and Hall.

Troyer, J. R. 1989. John Henry Schaffner (1866–1939) and reduction division in plants: Legend and fact. *American Journal of Botany* 76:1229–1246.

Books

Alberts, B., D. Bray, J. Lewis, M. Raff, K. Roberts, and J. D. Watson. 1994. *Molecular Biology of the Cell*. 3d ed. New York: Garland.

Farley, J. 1982. *Gametes and Spores: Ideas about Sexual Reproduction, 1750–1914*. Baltimore: Johns Hopkins University Press.

John, B. 1990. *Meiosis*. New York: Cambridge University Press.

Klug, W. S., and M. R. Cummings. 1994. *Concepts of Genetics*. New York: Macmillan.

Marguilis, L., and D. Sagan. 1986. *Origins of Sex*. New Haven, CT: Yale University Press.

Michod, R. E., and B. R. Levin, eds. 1988. *The Evolution of Sex: An Examination of Current Ideas*. Sunderland, MA: Sinauer.

Moens, P. B., ed. 1987. *Meiosis*. New York: Academic Press.

The fertile heads of these genetically transformed wheat plants are silhouetted against a Southern blot that shows the unique features of the transformed plants. In this chapter you'll learn about genetic engineering and its associated techniques (e.g., Southern blotting).

chapter

Molecular Genetics and Gene Technology

11

Chapter Outline

Chapter Overview

Imagine that the nucleus is an enormous dictionary containing the words that define life but that the words are written in a microscopic code. As recently as the 1950s, this code seemed so complicated that scientists thought it might be beyond human comprehension. Furthermore, the code was so large for any given organism that if its letters could be printed in the same size as this type, the dictionary would fill several hundred volumes the size of this book. Now we know that the code is made of DNA and that DNA is organized into "words" called *genes*. But how is the code translated into organisms? There is still no good answer for this question, although pieces of the answer have been discovered by finding out how genes work at the cellular level. Studies of the mechanisms of gene action constitute a relatively new subdiscipline of genetics known as *molecular genetics*, the topic of this final chapter in the unit on genetics.

The methods and discoveries of molecular genetics have greatly improved crop productivity, gene therapy, and other aspects of applied biology, which rely on our knowing how to make new genes, change existing genes, or transfer genes from one organism to another. This work is the basis for a highly sophisticated, multibillion-dollar industry called *gene technology*, which is also discussed in this chapter.

The first trait of the garden pea that was described in 1865 by Mendel was the round versus wrinkled shape of mature seeds (fig. 11.1). The genetic basis for this trait was not described until 1990, however, by Madan Bhattacharyya and several colleagues at the John Innes Institute in Norwich, England. This group identified the enzyme that controls roundness, found the gene that makes it, and determined the differences between the dominant-round allele (*R*) and the recessive-wrinkled allele (*r*). Roundness is apparently controlled by one form of **starch-branching enzymes** (**SBEI**; I = isoform), which converts the straight chains of amylose to the branched polymers of amylopectin (see Chapter 2). This form of the enzyme is active in the early development of round seeds but absent in wrinkled seeds. Seeds with either *RR* or *Rr* genotypes have large amounts of starch with high ratios of amylopectin. Conversely, seeds that are *rr* have less starch

and proportionately less amylopectin; instead, they have more sucrose. Wrinkles occur because sucrose is osmotically active, which causes *rr* seeds to absorb water. When *rr* seeds dry as they mature, the loss of water makes them shrivel. In contrast, since starch is not osmotically active, *RR* and *Rr* seeds contain less water. Hence, when these seeds mature, they lose little water and remain full and round.

Bhattacharyya's study is an example of how genes are found and how they influence phenotypes. However, a complete explanation of the molecular genetics of wrinkled seeds requires an understanding of the mechanisms of gene action and how they can be exploited to find and manipulate genes of interest in the laboratory. These are the subjects of this chapter. We therefore postpone further discussion of the genetic details of wrinkled peas until you have read about how genes work.

How Genes Work

Proteins are made in the cytoplasm, not in the nucleus where the genes (DNA) occur. This observation was one of the first clues that genes do not make proteins directly. Another clue was that polypeptide synthesis is associated with ribosomes, which occur outside the nucleus. These clues led to the discovery that polypeptides are made in ribosomes. Protein synthesis is a multistep process that begins when DNA is used as a template for making RNA by a process called **DNA transcription.** While still in the nucleus, RNA is edited (cut, spliced, or otherwise modified) and then exported into the cytoplasm. There are three main kinds of RNA in the cytoplasm. **Messenger RNA (mRNA)** is the actual message for a polypeptide. **Transfer RNA (tRNA)** binds to amino acids, so that they can be brought together during polypeptide synthesis. **Ribosomal RNA (rRNA)** is one of the main ingredients of ribosomes. The process of making a polypeptide from mRNA is called **translation.** Once translation is complete, polypeptides are often assembled into multipolypeptide units or otherwise modified before becoming functional proteins.

Overall, the way genes work (**gene expression**) to transfer information from DNA to protein can be summarized as follows:

DNA (Genes)

↓ *Transcription*

Nuclear RNA

↓ *Editing and Export*

Cytoplasmic RNA

↓ *Translation*

Polypeptides

↓ *Assembly, Modification*

Proteins

figure 11.1

The first trait that was described by Mendel in 1865 for his experiments with the garden pea was seed shape (round versus wrinkled). The molecular genetic basis for this trait was not discovered until 1990.

The major exception to this overall process occurs in viruses, most of which have genomes of RNA. In such viruses, RNA genes are first reverse-transcribed to DNA, which then undergoes transcription.

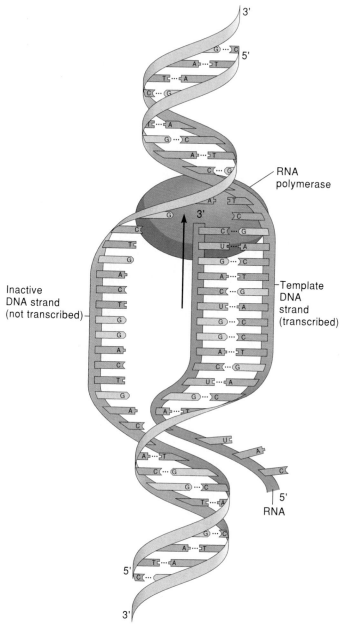

RNA polymerase

Inactive DNA strand (not transcribed)

Template DNA strand (transcribed)

RNA

figure 11.2

Diagrammatic representation of transcription. RNA polymerase assembles nucleotides in a sequence that is complementary to the DNA template. RNA is therefore identical to the nontranscribed strand of DNA, except for the substitution of uracil (U) for thymine (T) in RNA.

RNA Synthesis: Transcription

During transcription, DNA serves as a template for making complementary RNA copies of itself. The entire gene is transcribed into molecules of RNA that are about the same length as the DNA template (fig. 11.2). Unlike DNA, RNA contains ribose instead of deoxyribose and uracil (U) instead of thymine (T). Accordingly, a DNA sequence of ATGCCTGGA will be transcribed into an RNA sequence of UACGGACCU.

The main features of DNA transcription are similar to those of DNA replication. As in DNA replication, DNA transcription proceeds from the 5′ end to the 3′ end of the new strand, so that the complementary RNA strand is assembled antiparallel to the template DNA strand. The oldest (first-made) end of the RNA is the 5′ end.

Steps in Transcription

Transcription occurs in three main steps. The initiation step of transcription begins when special enzymes, including **RNA polymerase,** attach to the DNA sequence (fig. 11.2). These enzymes attach to the DNA at a promoter site, which is a special sequence of DNA that is the initiation signal for transcription. Only one strand of DNA is the coding strand for each gene—that is, only one strand contains the coding sequence for a polypeptide. The other strand is the noncoding strand and is not transcribed. How does RNA polymerase recognize the coding strand versus the noncoding strand of double-stranded DNA? The key to recognizing the coding strand is the promoter site. Sequences vary among different promoters, but almost all of them include two specific sets of six nucleotides, TATAAT and TTGACA, to which transcription enzymes bind. The noncoding strand contains their complements (ATATTA and AACTGT, respectively), which are not recognized by transcription enzymes. The coding strand for some genes is the noncoding strand for other genes, which means that genes may occur on either strand throughout an entire chromosome.

In the second step in transcription, the elongation step, the polymerase enzyme moves along the strand of DNA, adding nucleotides to the growing chain of RNA (fig. 11.3). The transcribed region may also include a **leader sequence** of variable length—that is, a noncoding sequence between the promoter and the beginning of the gene. After about thirty bases have been transcribed, a chemical **cap** is attached to the 5′ end of the RNA. This cap is a nucleotide triphosphate, which is analogous to ATP, called 7-methylguanosine triphosphate (7mG). The 7mG cap protects the RNA molecule from degradation and later serves as the initiation signal for translation.

During the termination step, transcription stops and the new RNA molecule is released from its DNA template. After RNA polymerase reaches the end of the gene, it continues transcription until it arrives at a termination signal. The extra amount of RNA beyond the end of the gene is called a **trailer sequence.** Like the leader sequence, the trailer sequence does not function as a coding region; it is a signal for another polymerase to cut the RNA molecule and release it from the DNA template. Immediately after cleavage, 100–200 adenylic acid molecules are added to the 3′ end of the RNA. With the addition of this **poly-A tail,** the complete transcription product is called the **primary RNA transcript.**

The demand for certain polypeptides may be immediate and crucial for cellular metabolism. To satisfy this demand, a single gene may be transcribed by many RNA polymerases at the same time (fig. 11.4).

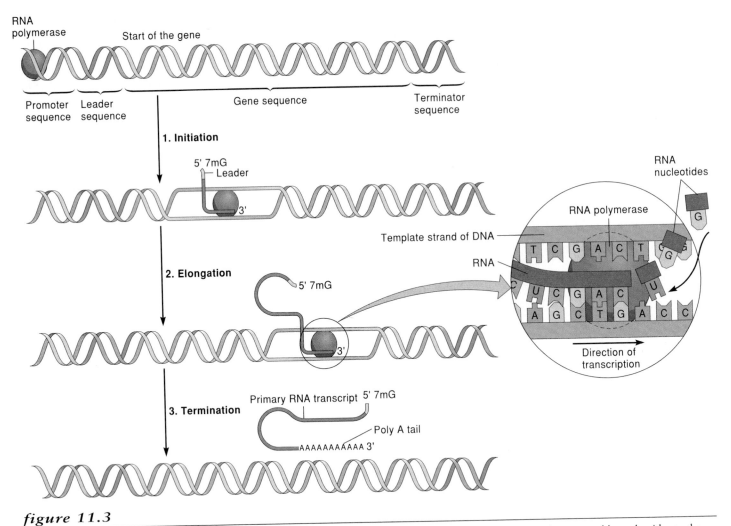

figure 11.3

Steps in RNA transcription. (1) RNA polymerase attaches to the DNA sequence at a promoter site. (2) RNA polymerase adds nucleotides to the growing chain of RNA as the DNA unwinds ahead of it. (3) Transcription stops, and the new RNA molecule is cleaved from its template.

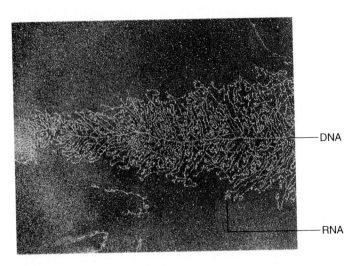

figure 11.4

Electron micrograph of simultaneous transcription of a single gene by several RNA polymerases. Each branch from the central thread of DNA is a strand of RNA with its own RNA polymerase.

c o n c e p t

Transcription entails three main steps: initiation, elongation, and termination of RNA synthesis. The coding strand of DNA contains sequences for initiation and termination that are recognized by the enzyme RNA polymerase. In addition to transcribing the gene, RNA polymerase also transcribes a noncoding leader sequence and a noncoding trailer sequence. The primary RNA transcript is completed when it is bound to a long poly-A tail and released from the DNA template.

Different Kinds of RNA

All RNA comes from the transcription of DNA during gene expression. But RNA occurs in several forms, some of which stay in the nucleus and some of which move into the cytoplasm. In this chapter we focus on nuclear RNAs that help make polypeptides and on cytoplasmic RNAs (mRNA, tRNA,

11–5

RIBOZYMES AND THE ORIGIN OF LIFE

In 1981, Thomas Cech discovered that primary RNA transcripts from *Tetrahymena*, a single-celled protist, can be processed in the absence of enzymes. Processing occurs because one of the introns is self-splicing, which means that it acts like an enzyme to catalyze a chemical reaction. Unlike an enzyme, however, the intron works on itself (reading fig. 11.1). To acknowledge this difference, Cech named this intron a ribozyme. His work on ribozymes earned Cech a Nobel Prize in 1986.

Cech's discovery sparked a search for ribozymes in other organisms. So far, they have been found in the nuclei and organelles of plants, animals, and fungi, and in the genomes of bacteria. Additional functions have also been discovered for ribozymes. For example, they work like nucleases, which are enzymes that digest nucleic acids. Ribozymes also mimic polymerases by joining nucleotides in short chains and by joining short RNA sequences into longer ones.

What do ribozymes have to do with the origin of life? One possibility is that they may be the answer to the "chicken versus egg" paradox between nucleic acids and proteins. The paradox is that if proteins came first, how could they be encoded without DNA or RNA? Conversely, if nucleic acids came first, how could they be replicated without enzymes? This paradox rests on the dogma that only proteins can be enzymes and only nucleic acids can contain genetic information. However, we now know that RNA can play both roles. Thus, the first precellular organisms probably contained RNA long before DNA or proteins evolved.

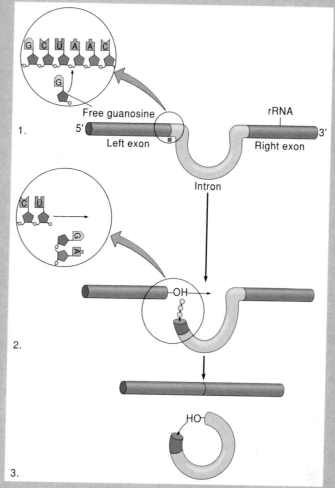

reading figure 11.1

Model of RNA splicing by intron.

rRNA). Certain kinds of RNA molecules called **ribozymes** can also function as enzymes. These RNAs are discussed in reading 11.1, "Ribozymes and the Origin of Life."

RNA in the Nucleus

Initially, each molecule of RNA is about the same size as the DNA sequence that serves as its template. A gene of 8,000 nucleotides is copied into an RNA molecule of 8,000 nucleotides, including leader and trailer sequences. Primary RNA transcripts from different genes occur in different sizes, from a few thousand nucleotides to more than 20,000 nucleotides. The pool of

large RNA transcripts in the nucleus is called **heterogeneous nuclear RNA (hnRNA)** because of the heterogeneity of sizes among RNA molecules for the same gene. Large transcripts must be trimmed by **RNA processing**, also called **RNA splicing**, before they leave the nucleus. The splicing of RNA transcripts involves small molecules of RNA that remain in the nucleus (**snRNA**), which are about 100–200 nucleotides each. These small RNA molecules condense with proteins, which are called **small nuclear ribonucleoproteins** (**snRNPs**). A complex of numerous snRNPs is called a **spliceosome.** Each spliceosome binds to a large primary RNA transcript, cuts out certain parts of the transcript, and splices the rest of the RNA back into a continuous strand

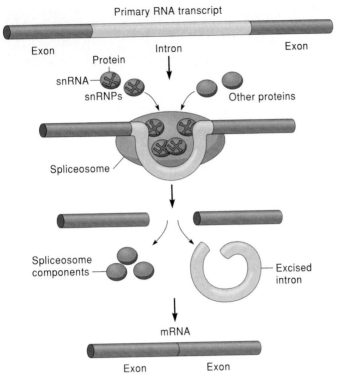

Primary RNA transcript

Exon | Intron | Exon

Protein
snRNA
snRNPs
Other proteins

Spliceosome

Spliceosome components

Excised intron

mRNA

Exon | Exon

figure 11.5

RNA processing. Portions of a primary RNA transcript are removed before the RNA is exported from the nucleus. Complexes of proteins and snRNA (small nuclear RNA) bind to segments to be excised (introns), cut them out, and splice the functional segments (exons) together.

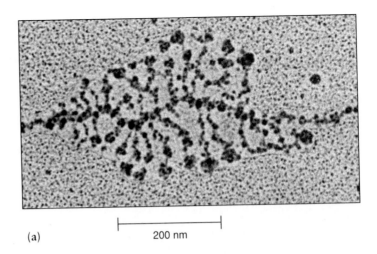

(a)

200 nm

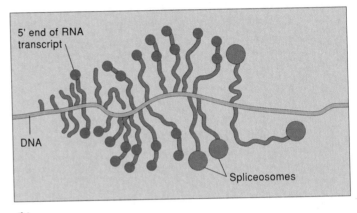

5' end of RNA transcript

DNA

Spliceosomes

(b)

figure 11.6

(a) Electron micrograph of RNA processing during transcription. (b) The larger circles represent whole spliceosomes. The smaller circles are probably snRNA-protein particles at the ends of introns. They will become spliceosomes as they get larger.

(b) Source: Alberts, Molecular Biology of the Cell, *2d ed. Copyright © Garland Publishing Company, New York, NY.*

(fig. 11.5). The parts of the primary transcript that are cut out are called **introns;** the parts that are spliced together are called **exons.** Several spliceosomes can attach to an RNA transcript simultaneously. Spliceosomes begin processing RNA immediately after transcription, even as some parts of the RNA molecule are still being transcribed (fig. 11.6). After processing, the final form of the RNA molecule is often thousands of nucleotides smaller than the primary transcript. For example, it is common for a primary transcript of 8,000 nucleotides to be processed into an RNA molecule of about 1,200 nucleotides.

There are often parts of a primary RNA transcript that do not function in polypeptide synthesis because they are cut out before the RNA leaves the nucleus. This means that primary RNA transcripts and the genes that code for them are interrupted by noncoding sequences. The functions of interruptions in genes are still unknown, but they may have roles in gene regulation or genetic recombination. The experiments that led to the discovery of interrupted genes are described in reading 11.2.

Messenger RNA

Messenger RNA (mRNA) gets it name because it contains the "message" (i.e., coding sequence) for a polypeptide. The mRNA molecule has four main parts: the 7mG cap, the leader

sequence, the coding sequence, and the trailer sequence with a poly-A tail (fig. 11.7). The coding sequence usually ranges between a few hundred nucleotides to about 2,000 nucleotides among mRNA molecules from different genes.

Each polypeptide in a cell is usually coded by a separate mRNA. This means that there are thousands of different kinds of mRNA. Cells maintain at least 5–10 copies of most kinds of mRNA, but a few kinds of mRNA occur in 10,000 or more copies at a time. The number of copies of mRNA depends on the metabolic, regulatory, and structural needs for particular polypeptides. However, cells usually contain only 3%–5% of their total RNA as mRNA.

Transfer RNA

The *transfer* in transfer RNA (tRNA) refers to its role in bringing amino acids together during translation. Whereas there are thousands of distinct molecules of mRNA in each cell, there

11–7

THE DISCOVERY OF INTERRUPTED GENES

Until the 1970s, most studies of the structure and function of genes involved bacteria. Bacterial genes were found to be made of continuous coding sequences, and there seemed to be no obvious reason why genes of other organisms should be organized differently. However, by the early 1970s, hnRNAs (heterogeneous nuclear RNAs) had been discovered as the precursors of mRNAs in eukaryotes. Because hnRNAs and mRNAs have poly-A tails at their 3' ends, geneticists assumed that mRNAs were derived by extensive degradation of a long leader sequence at the 5' ends of hnRNAs. This hypothesis was rejected when biologists discovered that hnRNAs have 5' caps and leader sequences that were preserved during the conversion of hnRNA to mRNA. Hindsight makes it seem obvious that nucleotides *within* the hnRNA molecule must be removed to make mRNA, thereby leaving the 5' and 3' ends intact. This hypothesis seemed absurd at the time, but by 1977 molecular biologists had evidence from two kinds of experiments to support it.

The first evidence for interrupted genes came from studies of a human virus called *adenovirus-2*. In these experiments, a segment of double-stranded viral DNA was denatured in a mixture that included mRNA from a gene in the segment. The mixture

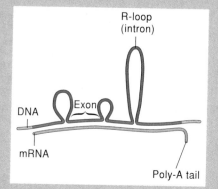

reading figure 11.2

Evidence for introns came from DNA-mRNA binding experiments. Binding only occurs where both sequences are complementary, thereby excluding noncomplementary sequences of DNA (introns) as R-loops.

was then cooled to a temperature that promotes annealing between RNA and DNA but not between two strands of DNA. However, "hybrids" between mRNA and its DNA template occurred only in regions where the two were complementary. Sequences of DNA within the gene that did not match the mRNA were displaced into loops, called **R-loops,** which could be seen by electron microscopy (reading fig. 11.2). Because they do not match the coding strand of mRNA, R-loops are noncoding sequences that interrupt the gene. Later, R-loops were

also discovered in genes from plants and fungi, which means that interrupted genes are a general feature of eukaryotes.

At about the same time of the discovery of R-loops, molecular biologists discovered different but comparably surprising evidence for interrupted genes in chickens, rabbits, and mice. In these experiments, DNA was made by reverse transcription (RNA → DNA) from an mRNA template using a special polymerase from viruses. Evidence for gene interruptions came from comparisons of this DNA with its original gene. Enzymes that cut DNA at specific sequences did not cut the reverse-transcribed DNA, but they did cut the original gene into several pieces. This observation is evidence that native genes contain sequences that are absent in the mRNA that is transcribed from them. Furthermore, since the cut sequences do not occur in mRNA, they must occur in noncoding regions—that is, in sequences that interrupt the gene. We now know that the noncoding portions of genes correspond to introns, which are removed during the processing of primary RNA transcripts into mRNA.

The discovery of introns was a major breakthrough toward understanding how genes work. In 1993 several scientists were awarded a Nobel Prize for their contribution to this discovery.

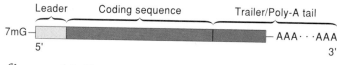

figure 11.7

Model of mRNA that is exported from the nucleus. It consists of a 7mG cap, a leader sequence, a coding sequence for a polypeptide, and a poly-A tail.

are only about sixty different kinds of tRNA. Although the average size of a molecule of tRNA is only 80 nucleotides, tRNA makes up about 15% of total cellular RNA.

The probable three-dimensional structure of a tRNA molecule is shown by computer simulation in figure 11.8. Note that the molecule is twisted and folded back on itself, so that four double-stranded regions and three loops are formed. This is the characteristic form of most tRNA molecules.

Each tRNA molecule recognizes a certain sequence of nucleotides on an mRNA molecule and attaches to it. The recognition site on an mRNA molecule is called a **codon,** and the attachment site on the tRNA molecule is called the **anticodon.** The anticodon sequence, which occurs in one of the loops in the tRNA molecule, is complementary to the codon sequence. Each tRNA molecule also contains a specific sequence at its 3′ end, called the **amino acid acceptor site,** which recognizes and binds to one of the twenty amino acids that make up proteins.

Ribosomal RNA

Ribosomes are the sites of protein synthesis. Each ribosome consists of a complex of ribosomal RNA (rRNA) and proteins, which are organized into a large subunit and a small subunit (fig. 11.9). In an average eukaryotic ribosome, the large subunit contains about 50 proteins and three kinds of

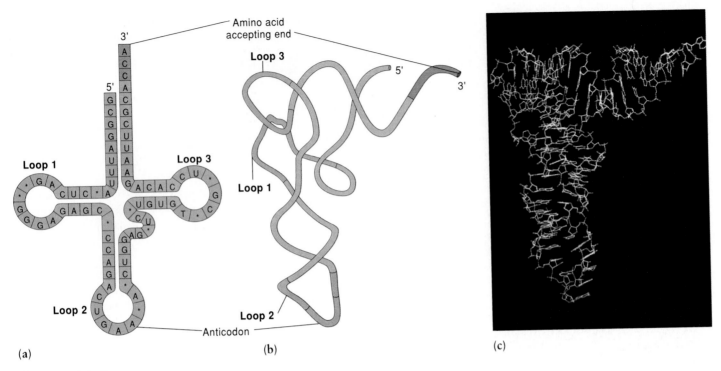

figure 11.8

The structure of transfer RNA. (a) Two-dimensional representation, showing three loops and four regions of internally complementary nucleotides that form double strands. At one end of the molecule is the amino acid attachment site, and at the middle loop (Loop 2) is the mRNA recognition site (anticodon). Asterisks denote nucleotides that are changed chemically (e.g., two hydrogens added to uracil, two methyls added to guanine, etc.). (b) Diagram of three-dimensional representation. (c) Computer-generated three-dimensional structure of tRNA.

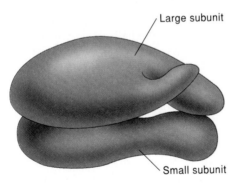

figure 11.9

Structure of a eukaryotic ribosome. Each ribosome consists of a large subunit and a small subunit. The two subunits join together for protein synthesis.

single-stranded RNA: one that is about 4,700 nucleotides, one of about 160 nucleotides, and one that is about 120 nucleotides. The small subunit contains more than 30 proteins and one kind of single-stranded RNA of about 1,900 nucleotides. More than half of the mass of a ribosome is RNA.

Transcription of rRNA molecules is similar to transcription of other kinds of RNA, except that rRNA transcripts are made by a different kind of RNA polymerase. Also, genes for rRNA occur in a string of tandem repeats (see Chapter 10). The number of each repeating unit varies in different plants,

even within the same species. For example, rRNA genes range from as few as 3,000 copies to more than 9,000 copies in different strains of corn (*Zea mays*).

Some rRNA may be transcribed from tandem three-gene clusters on regions of chromosomes in the nucleolus (fig. 11.10). Other rRNA is transcribed from single genes in tandem but not in the nucleolus. This means that the construction of ribosomes is complicated by having to unite products from two separate gene clusters. The process is further slowed because transcribed regions are long, the primary transcripts are made of multiple subunits, and the completed rRNA must be packaged with many proteins in the cytoplasm. It may take up to an hour to construct one ribosome. Nevertheless, several million ribosomes are made during each cell cycle. Such large numbers of ribosomes can be made only because the cell contains many copies of ribosomal RNA genes and because each gene is transcribed simultaneously by many molecules of RNA polymerase. At least 80% of the total cellular RNA is rRNA.

c o n c e p t

Three main kinds of RNA work together to make polypeptides. Messenger RNA carries the polypeptide code from the gene. Transfer RNA is directed by mRNA to bring amino acids together in the proper order. Ribosomes, which contain protein and ribosomal RNA, are where mRNA and tRNA come together to make polypeptides in the cytoplasm.

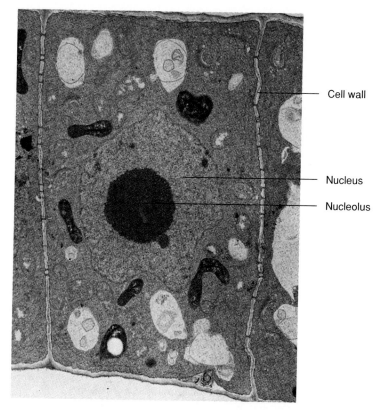

figure 11.10

Electron micrograph of a plant cell showing the nucleus and a nucleolus, ×14,000. The nucleolus is where some of the cell's rRNA is synthesized.

Polypeptide Synthesis: Translation

Translation is like transcription because it uses enzymes to make a polymer. Both processes occur in stages that include initiation, elongation, and termination. Translation differs, however, because it makes amino acid chains, not nucleotide chains.

Before Translation Starts

Molecules of tRNA must bind their appropriate amino acids before translation can start. Enzymes called **aminoacyl-tRNA synthetases** ensure that each tRNA molecule bonds covalently to a specific amino acid. There is a different synthetase for each of the twenty amino acids used in translation. Each aminoacyl-tRNA is made in a process that utilizes energy from ATP and includes an intermediate that consists of an amino acid and adenosine monophosphate (AMP) (fig. 11.11). In this way, for example, methionine is attached only to molecules of tRNA^{Met}, and valine is attached only to molecules of tRNA^{Val}.

Every aminoacyl-tRNA molecule has two main functions. One function is as a decoder of nucleotide sequences. The decoder function comes from its dual codes: the anticodon that reads a codon and the amino acid acceptor site that reads an amino acid. The second function is to activate the carboxyl group of an amino acid so that it reacts with the amino group of another amino acid to make a peptide bond (see Chapter 2).

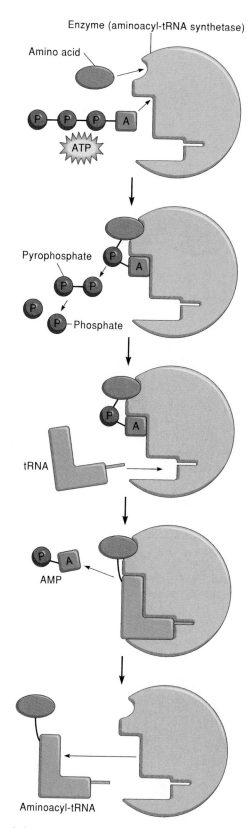

figure 11.11

Synthesis of aminoacyl-tRNA. Aminoacyl-tRNA synthetase binds an amino acid and a molecule of ATP, using the energy of two phosphate bonds. The remaining AMP is replaced at the enzyme by the appropriate tRNA, which is then bound to the amino acid; AMP is released from the enzyme. The aminoacyl-tRNA complex is then released from the enzyme and can be used in protein synthesis.

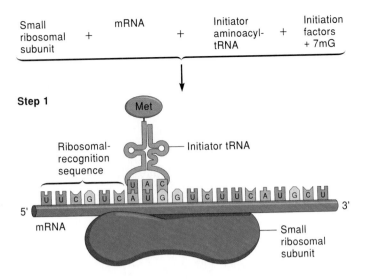

Step 1

Ribosomal-recognition sequence

Met

Initiator tRNA

mRNA

5'

3'

Small ribosomal subunit

Step 2

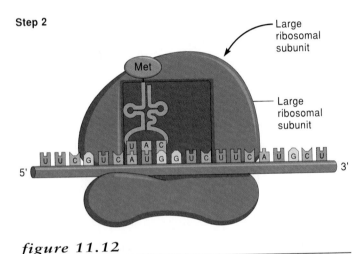

Met

Large ribosomal subunit

Large ribosomal subunit

5'

3'

figure 11.12

Initiation of protein synthesis. An initiator aminoacyl-tRNA binds to the start codon on mRNA, in a complex together with the small ribosomal subunit. Protein synthesis begins after the large ribosomal subunit is attached to this complex.

Steps in Translation

Translation starts when mRNA and an initiator tRNA bind to a small ribosomal subunit. Binding is aided by proteins in the ribosomal subunit that recognize the 7mG cap at the 5′ end of mRNA. The initiator tRNA is the molecule that carries the first amino acid for a polypeptide, which is usually methionine. The anticodon of the initiator tRNA is UAC, which is the complement of the **start codon** (AUG) on the mRNA. After the initiator tRNA, the mRNA, and the small subunit are in place, the large ribosomal subunit binds to the small subunit. This phase of initiation completes the assembly of a functional ribosome (fig. 11.12).

Elongation continues when a second tRNA molecule is attached next to the initiator tRNA on the large ribosomal subunit. The second tRNA is determined by the second codon on the mRNA. For example, the codon GUC will bind only to its complementary anticodon, CAG. The tRNA with the anticodon CAG carries the amino acid valine, which would therefore be the second amino acid in the growing polypeptide. The two amino acids are joined covalently when the enzyme **peptidyl transferase** moves the methionine from the initiator tRNA to the valine on the next tRNA (fig. 11.13). After methionine is bound to valine, tRNA^{Met} is released from the ribosome. The ribosome then moves forward one codon along the mRNA molecule. The movement of the ribosome brings the tRNA^{Val} molecule into the place vacated by the tRNA^{Met} molecule. A third tRNA then binds to the mRNA next to the tRNA^{Val}, bringing another amino acid with it. This process continues to link amino acids, one by one, until the polypeptide is complete (fig. 11.14).

Translation stops when the ribosome arrives at a **stop codon** on the mRNA molecule. The stop codon does not attach to tRNA; instead, it binds to special cytoplasmic proteins, called **release-factors.** Release-factors interrupt translation and hydrolyze the bond between the final amino acid in the new polypeptide and its tRNA.

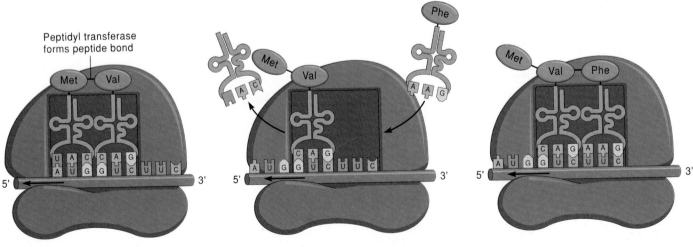

Peptidyl transferase forms peptide bond

Met   Val

Met   Val

Phe

Met   Val   Phe

5'   3'   5'   3'   5'   3'

figure 11.13

Steps in elongation. A second aminoacyl-tRNA binds to the ribosome next to the initiator tRNA, upon which the two amino acids are joined by a peptide bond. The initiator tRNA is then released, the mRNA advances one codon through the ribosome, and a third aminoacyl-tRNA binds to the spot on the ribosome that was vacated by the second aminoacyl-tRNA. A peptide bond forms between the second and third amino acids. These steps are repeated for each additional amino acid in the polypeptide.

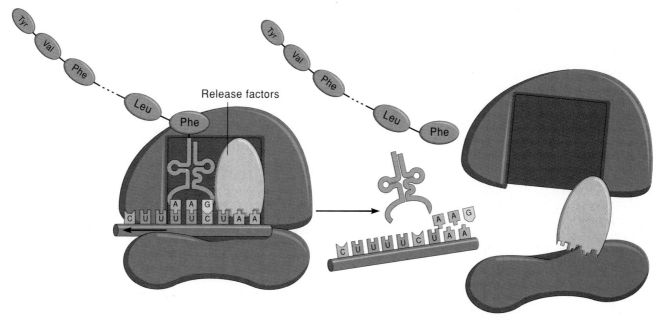

figure 11.14

Translation stops when the ribosome reaches a stop codon. The stop codon binds to release-factors that hydrolyze the final aminoacyl-tRNA, thereby interrupting translation.

Speed of Translation

The amount of polypeptide synthesis is the result of two levels of activity. The first level is the transcription of a gene to produce a large number of mRNA molecules. The second level is the repeated translation of each mRNA molecule. It takes less than a minute to translate mRNA into a polypeptide of 400 amino acids. Simultaneous translation by many ribosomes (*polyribosomes*) allows each mRNA molecule to make as many as ten polypeptides per minute. By using several thousand mRNA molecules at the same time, a cell can translate tens of thousands of polypeptides per minute. Continued translation at this rate produces millions of polypeptides during a single cell cycle.

Making the Finished Product: Polypeptide to Protein

Polypeptides must often be modified before they become functional proteins. For example, a polypeptide can be joined to one or more other polypeptides to make a protein having multiple subunits, as in the enzyme rubisco. Other changes occur in polypeptides that contain the amino acid cysteine. In these polypeptides, two cysteine molecules form a bond, called a *disulfide bond*, between their respective sulfur atoms. Also, membrane polypeptides bond covalently to sugars or lipids, thereby converting these polypeptides into glycoproteins or lipoproteins, respectively. Polypeptides that will help make cell walls or plasma membranes are often converted to such proteins in dictyosomes. Other kinds of polypeptides must bind to metal ions before becoming functional proteins. Examples include metalloenzymes such as iron- and copper-containing cytochrome oxidases and zinc-containing DNA polymerases and NADH-dehydrogenases.

The Genetic Code

The **genetic code** is a set of nucleotide messages that specify amino acids during translation. The smallest unit of the code is a codon. Each codon is three nucleotides long, collectively called a **triplet.** By using three nucleotides, the genetic code has 64 possible codons (4^3). (A two-nucleotide codon has only sixteen—that is, 4^2—possible combinations of four nucleotides, which would not be enough for 20 amino acids.) The presence of 64 possible codons means that there are more triplets than needed for 20 amino acids and a stop signal. What do the other 43 codons do? As explained next, it took a long time to answer this question, but eventually, all 64 combinations were found to be useful in the genetic code.

Although the size of a codon was easy to calculate, codon messages were not deciphered until the 1960s. The first triplet to be decoded was UUU. In 1961, Marshall Nirenberg made artificial mRNA with U as its only nucleotide. When he added this mRNA to the appropriate ingredients for protein synthesis, he obtained a polypeptide that contained only

Second Base

	U	C	A	G	
U	UUU⎫ Phe UUC⎭ F UUA⎫ Leu UUG⎭	UCU⎫ UCC⎮ Ser UCA⎮ S UCG⎭	UAU⎫ Tyr UAC⎭ Y UAA Stop UAG Stop	UGU⎫ Cys UGC⎭ C UGA Stop UGG Trp W	U C A G
C	CUU⎫ CUC⎮ Leu CUA⎮ L CUG⎭	CCU⎫ CCC⎮ Pro CCA⎮ P CCG⎭	CAU⎫ His CAC⎭ H CAA⎫ Gln CAG⎭ Q	CGU⎫ CGC⎮ Arg CGA⎮ R CGG⎭	U C A G
A	AUU⎫ AUC⎮ Ile AUA⎮ I AUG Met or M Start	ACU⎫ ACC⎮ Thr ACA⎮ T ACG⎭	AAU⎫ Asn AAC⎭ N AAA⎫ Lys AAG⎭ K	AGU⎫ Ser AGC⎭ S AGA⎫ Arg AGG⎭ R	U C A G
G	GUU⎫ GUC⎮ Val GUA⎮ V GUG⎭	GCU⎫ GCC⎮ Ala GCA⎮ A GCG⎭	GAU⎫ Asp GAC⎭ D GAA⎫ Glu GAG⎭ E	GGU⎫ GGC⎮ Gly GGA⎮ G GGG⎭	U C A G

(First Base — left axis; Third Base — right axis)

figure 11.15

The genetic code for mRNA. Read each codon in the 5′ to 3′ direction, starting with the first base on the chart, to find the message for each amino acid. Both the three-letter and one-letter abbreviations are given for each amino acid. Three of the sixty-four codons are stop signals; they do not translate into amino acids.

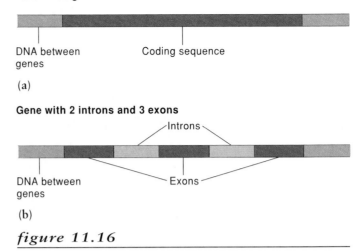

Intron-free gene

DNA between genes   Coding sequence

(a)

Gene with 2 introns and 3 exons

Introns

DNA between genes   Exons

(b)

figure 11.16

Diagram of (a) intron-free gene structure and (b) intron-containing gene structure. Exons encode mRNA sequences that will be translated into polypeptides, but introns are noncoding sequences.

---

phenylalanine. In this way he discovered that the codon UUU designates the amino acid phenylalanine. In similar experiments, the other three uniform codons were also quickly deciphered: AAA for lysine, CCC for proline, and GGG for glycine. Codons for the other sixteen amino acids were more difficult to determine, but all codons had been deciphered by 1966, 13 years after Watson and Crick proposed their model for the structure of DNA. The messages of all codons are listed in figure 11.15.

Note that all 64 triplets are used in the genetic code; 43 codons are not "left over" after 20 amino acids and a stop signal are coded. Instead, most amino acids are encoded by more than one codon. For example, lysine is coded by both AAA and AAG, and leucine is coded by six different triplets.

Codons that specify the same amino acid are called **synonymous codons** because they translate to the same product. The word *synonymous* may be misleading, however, because nobody knows if different codons for the same amino acid are identical metabolically. Although their products are the same, synonymous codons may be translated at different rates. Nevertheless, in all, there are 61 codons for 20 amino acids, and three codons (UAG, UAA, UGA) are stop signals. One of the coding triplets, AUG (methionine), is also a start codon.

Knowing that the genetic code consists of triplet codons, we can see how a molecule of mRNA that is 1,200 nucleotides long makes a polypeptide having 399 amino acids. The first 1,197 nucleotides comprise 399 triplets for amino acids, and

the last 3 nucleotides are a nontranslated stop codon. Also, since the start codon is usually AUG, the first amino acid in most polypeptides is methionine.

The Structure of Eukaryotic Genes

Genes that are made entirely of codons, such as genes of histone proteins, are the simplest genes because they do not contain introns (fig. 11.16a). However, many eukaryotic genes contain exons that are separated by introns (fig. 11.16b; look again at reading 11.2, "The Discovery of Interrupted Genes").

The number of introns varies from one to several among different genes, but the number is usually stable in each gene for each organism. Conversely, the same gene may have different numbers of introns among different organisms. For example, one of the enzymes of glycolysis, triosephosphate isomerase, is coded by a gene that has eight introns in maize (*Zea mays*), six introns in the domestic chicken (*Gallus domesticus*), and five introns in the fungus *Aspergillus*. This gene has no introns in brewer's yeast (*Saccharomyces cerevisiae*) or in the bacterium *Escherichia coli* (fig. 11.17).

How many genes are interrupted? This question cannot be answered at present, but introns are continually being discovered in different genes. Genomes with the highest proportion of interrupted genes occur in the nuclei of complex eukaryotes—that is, in plants and animals. Fewer interrupted genes occur in genomes of chloroplasts and mitochondria and in the nuclei of simpler eukaryotes such as yeasts. Introns are unknown in most prokaryotes.

Roles of Gene Interruptions

Although the functions of introns are unclear, scientists speculate that they have two possible roles. One is in regulating development. In this role, long or frequent introns slow gene expression because they take longer to process.

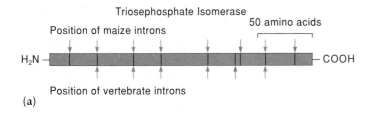

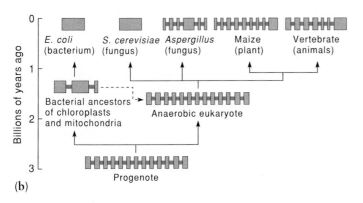

figure 11.17

Evolution of interrupted genes. (a) A comparison of the enzyme triosephosphate isomerase in maize (*Zea mays*) and vertebrate animals reveals the protein to be translated from several exons that are shared between plants and animals. Five identical intron positions are indicated by red arrows; unshared intron positions are marked by green arrows. Shared intron positions are indirect evidence that these introns arose before the evolutionary divergence of plants and animals, more than 1 billion years ago. (b) Outline of the most likely path of evolution of the triosephosphate isomerase gene in different organisms. Exons are in blue and introns are in red. The common ancestor for all organisms is called a *progenote*. The dashed arrow marks the endosymbiotic invasions of prokaryotes that gave rise to chloroplasts and mitochondria (see reading 3.2, p. 66).

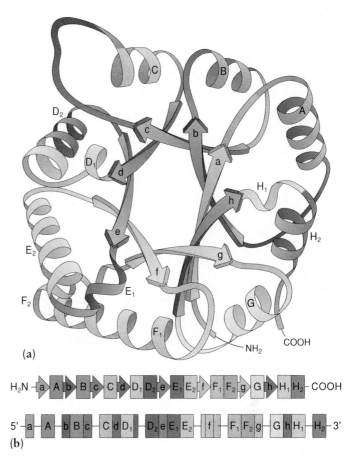

figure 11.18

Computer-graphic model of triosephosphate isomerase enzyme, color-coded to show domains (*a*). Domains generally correspond to specific exons (*b*).

The Exon-Shuffling Hypothesis

The genetic recombination hypothesis for the role of introns implies that each exon in a gene is somehow independent because it can move to another gene and still function. This means that structural and functional portions of a polypeptide, called **domains,** may be encoded separately by specific exons. Evidence for this hypothesis comes from comparing protein structures with gene structures. For example, the domains of triosephosphate isomerase generally correspond to specific exons (fig. 11.18).

The potential mobility of exons has led to the **exon-shuffling hypothesis,** which explains the origin of complex new genes by the joining of independent exons into new combinations. Exon shuffling may therefore produce new proteins from already existing exons. Circumstantial evidence for this hypothesis comes from the similarity of exons in different genes. For example, exon 4 of the chloroplast gene, *psb*A, and exon 17 of the mouse band 3 protein gene both encode similar amino acid sequences.[1] Although they

Hence, different genes are expressed at different rates, depending on the size and number of their introns. The idea of a regulatory role for introns is supported by circumstantial evidence: introns occur mainly in complex, multicellular organisms whose development is carefully regulated, including plants and animals. In contrast, single-celled organisms that grow as fast as possible, without complex developmental control, have few or no introns at all. Bacteria that lack introns are in this group. Moreover, introns are rare in *Saccharomyces*, a single-celled fungus, but they are common in *Aspergillus,* a multicellular fungus.

Introns may also enhance genetic recombination during crossing-over in meiosis (see Chapter 10). According to this hypothesis, when unequal crossing-over occurs within a gene, homologs that break and reconnect in introns would not affect the coding sequence of the gene. In contrast, breakage and nonreciprocal exchange within an exon would probably disrupt the coding sequence and ruin the polypeptide coded by it. This means that genes with long or frequent introns have a better chance of successful genetic exchange than do genes with short or few introns.

---

1. The name of a gene usually has a logical derivation. The gene *psb*A, for example, includes "ps" for photosystem, "b" for photosystem II, and "A" for the first enzyme of its type to be discovered (there are now more than a dozen known). These enzymes have a role in photosynthesis (see Chapter 7). Some genes, however, are nameless. For example, the gene for the third protein band on an electrophoresis gel, in an extract of mouse membrane proteins, is simply referred to as the gene for the mouse band 3 protein.

(a)

(b)

(c)

figure 11.19

Examples of genetically engineered plants. (a) Herbicide-resistant (right) and normal cotton. (b) Insect-resistant (left) and normal tobacco. (c) Spoilage-resistant (left) and normal tomatoes.

are in different proteins, both amino acid sequences function as membrane-spanning domains. These exons seem to be descendants of a single, ancestral exon that was reshuffled into different genes before the evolutionary divergence of plants and animals.

The idea that ancestral exons can be recombined to make new proteins has led a team of scientists, headed by Walter Gilbert of Harvard University, to suggest that existing genes consist of a much smaller number of reusable exons. By comparing amino acid sequences from all known exons, they determined which exons were most likely to come from the same ancestor. From these comparisons, they estimated that only 1,000–7,000 original exons were needed to make the hundreds of thousands of proteins that exist today. This proposal, however, is controversial.

c o n c e p t

Genes begin with a start codon and end with a stop codon. Codons for amino acids and the start and stop signals are three nucleotides long. Many amino acids have more than one codon. In eukaryotic genes, the protein-coding sequence between the start and stop signals is often interrupted by noncoding sequences.

Genetic Engineering

Can you imagine a blue rose, a potato that makes plastic, a tobacco plant that glows in the dark, or a truly tasty tomato from the supermarket? All of these products have already been made by **genetic engineering**, which is the artificial manipulation of genes, or the transfer of genes from one organism to another (fig. 11.19). Such artificial (i.e., human-directed) exploitation of genes is also called **recombinant DNA technology**. In these examples, the rose has genes from the petunia that control the synthesis of blue floral pigments. The potato has bacterial genes that make polymers that can be used to make biodegradable plastics. The luciferase gene, which causes fireflies to glow, has been transferred to tobacco plants as an easily identifiable marker gene for the detection of successful gene transfers; if the plants glow, then the transfer of the luciferase gene and other genes linked to it was successful. These plants are all **transgenic,** which means that they contain genes from other organisms. Conversely, tomatoes are genetically engineered when the tomato gene for a specific carbohydrate-degrading enzyme is altered in the laboratory and reinserted into the tomato plant. The vine-ripened fruits from these plants, which have all their natural flavor, stay firm during shipping and storage (see fig. 11.19c).

Recombinant DNA technology relies on **vectors** (DNA carriers), such as viruses or bacterial plasmids, which are independent molecules of DNA apart from the primary bacterial genome. They are vectors because they carry inserted DNA as their own DNA, which is then replicated. In this way, DNA from plants or animals is **cloned** by inserting it into bacteria or viruses. As discussed next, DNA cloning is an important tool in gene technology. It is used to find and study important genes and to make genetically engineered plants.

11–15

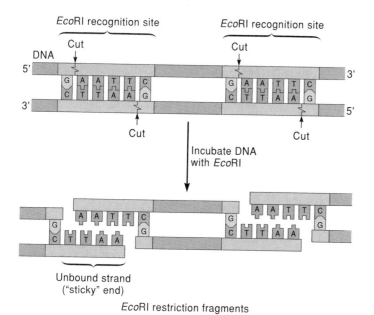

figure 11.20

Like most restriction enzymes, *Eco*RI makes a zigzag cut wherever it finds a recognition site. Single-stranded portions of restriction fragments are called *sticky ends* because they can form hydrogen bonds with complementary sticky ends on other molecules of DNA.

DNA Cloning

Before recombination, both the foreign DNA and the vector DNA must be made receptive to each other. This job is done by bacterial **restriction enzymes,** which cut (i.e., restrict) DNA. Their use in DNA cloning is discussed in the next few paragraphs.

Restriction Enzymes

Different bacteria make hundreds of restriction enzymes, each of which recognizes a specific DNA sequence of 4–8 nucleotides. In nature, these enzymes protect bacteria by destroying foreign DNA. To prevent self-destruction, bacterial DNA is chemically modified to be inert to its own restriction enzymes.

An example of a restriction enzyme is **Eco RI,** which is harvested for commercial sale from *Escherichia coli*. *Eco*RI recognizes the nucleotide sequence GAATTC and then cuts the DNA between the guanine and the adenine. The six-nucleotide *Eco*RI site has identical sequences on both DNA strands, because its antiparallel complement is CTTAAG. This means that by cutting between guanine and adenine on each strand, the enzyme makes a zigzag cut that leaves short, single-stranded ends on each DNA double strand (fig. 11.20). Such single-stranded ends are called *sticky ends* because they can be "glued" by hydrogen bonding to complementary sticky ends of other DNA molecules.

Cloning in Plasmids

After cutting up a whole genome with a restriction enzyme, the next step in cloning DNA is to insert the DNA into a vector. One

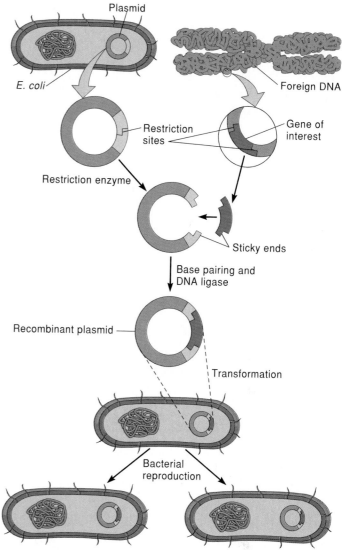

figure 11.21

Gene cloning by bacteria. The gene of interest from another organism is cut out of its genome by a restriction enzyme and inserted into a bacterial plasmid that has been cut with the same restriction enzyme. Sticky ends of the foreign fragment bind to the complementary sticky ends of the plasmid, followed by the formation of new sugar-phosphate bonds through the action of DNA ligase. Finally, the altered plasmid is taken up by the bacteria, which make many copies of the foreign gene by duplicating the plasmid during reproduction.

type of vector is a bacterial plasmid, which is a small, circular molecule of DNA (fig. 11.21). After the foreign DNA is inserted into isolated plasmids, the plasmids are absorbed back into bacteria and reproduced (i.e., cloned) during normal DNA replication.

Foreign DNA and plasmid DNA are receptive to each other when they are cut by the same restriction enzyme. For example, an enzyme such as *Eco*RI makes the same sticky ends on both the foreign DNA and the plasmid DNA. When the two kinds of DNA are mixed, they anneal because their sticky ends complement each other.

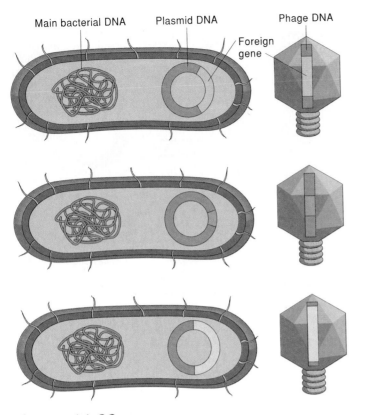

figure 11.22

Genomic libraries are clones of DNA fragments, usually consisting of a single gene, that are stored in vectors. Cloning may be by plasmids or by phages. The bacterial cells on the left and the viruses on the right represent the cloning of three foreign genes by each type of vector.

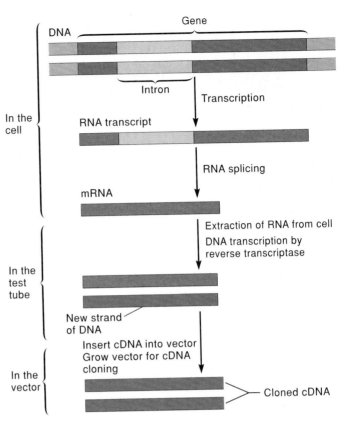

figure 11.23

Steps for making complementary DNA. After transcription and RNA editing in the cell, mRNA is extracted and used as a template for making DNA by reverse transcription in a test tube. The cDNA from reverse transcription is then inserted into a vector for cloning. Unlike native genes, complementary DNA lacks introns because it is made from processed RNA.

At first, plasmid DNA is weakly held to foreign DNA by hydrogen bonds between complementary nucleotides. The linkage between them is strengthened when sugar-phosphate bonds form between two molecules of DNA. Sugar-phosphate bonds are made by the enzyme **DNA ligase.** When this enzyme is added to the mixture, it joins (i.e., ligates) phosphate groups to deoxyribose between adjacent strands of DNA.

Cloning in Viruses

Methods for cloning DNA in plasmids also work with viruses. Like plasmid DNA, viral DNA is cut by restriction enzymes, and foreign DNA is inserted into it. Since viruses are parasites, however, they must be reintroduced into their hosts before they can replicate the recombined DNA. For cloning DNA in viruses, the most convenient hosts are bacteria such as *E. coli* that can be easily cultured in the laboratory. The most commonly used viruses for cloning DNA are bacteriophages.

Libraries of Genes

A whole genome can be cut by restriction enzymes and many of the resulting DNA fragments inserted into vectors. This has been done for many plants, animals, and fungi. When inserted into a vector, a fragment or gene of interest from any of these

organisms can be retrieved from storage by culturing the appropriate vector. This storage and retrieval of genetic information is like having a library of genes, so scientists call the set of cloned fragments of a genome a **genomic library** (fig. 11.22); such libraries may have several thousand to several million entries, each a recombined plasmid or virus that contains a different restriction fragment.

Libraries are also made using complementary DNA (**cDNA**). To make a cDNA library, mRNA is reverse-transcribed to make cDNA. This reaction from RNA back to DNA is catalyzed by the enzyme **reverse transcriptase,** a viral enzyme that is available commercially (fig. 11.23). The cDNA is later modified to have the appropriate sticky ends, after which it is annealed and ligated to a vector. Libraries of cDNA are always incomplete because they contain only genes whose mRNA has been reverse-transcribed. However, this can also be an advantage of cDNA libraries, because it results in a library of genes that are expressed in a single type of tissue or organ. Leaf mRNA or root mRNA, for example, would yield a leaf cDNA library or a root cDNA library, respectively. In addition, cDNA genes are easier to manipulate because they are generally smaller than in their native state, since their introns are missing.

In addition to being clonable, some genes can make polypeptides in their host vectors. This means that intron-free genes in a genomic library and genes from a cDNA library may be translated in the vector. The products of these genes can therefore be made in large amounts for research or industrial uses by culturing the vector. The first such product to be available commercially was human insulin, but many proteins are now being made by cloning. Unfortunately, however, intron-containing genes in genomic libraries cannot be translated in a bacterial vector, because prokaryotes lack the enzymes to process RNA.

Making and screening gene libraries is a tedious and time-consuming procedure. A plasmid or a virus can replicate only a few thousand nucleotides of foreign DNA, so many clones are needed to replicate whole genomes. For example, by chance alone, $EcoRI$ will find the sequence GAATTC once out of every 4,096 nucleotides. This is 4^6, which is the number of ways four different nucleotides can be arranged in a six-base sequence. This means that the genome of $Arabidopsis\ thaliana$, the smallest known in plants (70 million nucleotide pairs), will be cut into more than 17,000 fragments with an average size of 4,096 nucleotides each by $EcoRI$ (70,000,000/4,096 = 17,089). Thus, $Arabidopsis$ would require more than 17,000 clones in its complete genomic library. By the same calculation, genomes of the garden pea (10.2 billion nucleotide pairs) would require about 2.5 million clones, and that of Easter lily (90 billion nucleotide pairs) would require about 22 million clones! However, gene libraries are incomplete because only part of an organism's genome can be efficiently taken up by vectors.

Making Artificial Chromosomes

One of the most important recent advances in genetic engineering is the construction of artificial chromosomes that can be cloned in yeast (*Saccharomyces cerevisiae*). These **yeast artificial chromosomes (YACs)** are made by putting fragments of foreign DNA into native yeast chromosomes. The YAC contains nonyeast DNA, a centromere, at least one replication origin, and two **telomeres,** which are DNA sequences that counteract chromosome condensation before mitosis (fig. 11.24). When such artificial chromosomes are introduced into yeast cells, they are replicated like native chromosomes.

Unlike bacterial or viral vectors, YACs can unite with and replicate millions of nucleotides of foreign DNA. Large fragments of foreign DNA are obtained by lightly digesting genomic DNA from the source organism, so that only some of the restriction sites are cut. The average size of restriction fragments made by $EcoRI$ may therefore be in the millions of nucleotides instead of 4,096. A complete genomic library of 3 billion nucleotide pairs may require about a thousand YACs. Besides requiring fewer clones by this method, the construction of YACs has the advantage that intron-containing genes from the source organism can be expressed, because yeast (a eukaryote) has the appropriate enzymes for processing RNA. However, the use of YACs is so complicated and expensive that the only extensive work currently being done with them involves the human genome.

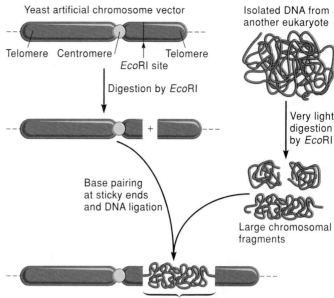

figure 11.24

Construction of a yeast artificial chromosome (YAC) with inserted foreign DNA. Once a YAC is made, it can be taken up by a yeast cell and replicated like a normal chromosome when the cell reproduces. YACs can replicate fragments of foreign DNA that are millions of nucleotides long.

concept

Bacteria and viruses can incorporate foreign DNA into their own. The entire genome of a plant can be cut into fragments and inserted into bacterial or viral vectors. By replicating foreign DNA with their own DNA, these vectors make clones of the genes inserted into them. The collection of bacterial or viral clones of plant DNA is like a genomic library. Genomic libraries can also be cloned in artificial chromosomes that are replicated in yeast.

Finding the Gene of Interest

Perhaps the most difficult and time-consuming step in genetic engineering is finding the gene of interest. One method is to isolate specific mRNA and then make a cDNA copy of it for cloning. For example, fungal infection stimulates soybeans to make large amounts of mRNA from the chalcone isomerase gene, a gene that is part of a biosynthetic pathway that makes compounds to fight the infection. After it is induced to make mRNA, the chalcone isomerase gene of the soybean can be cloned as cDNA.

Genes are found in gene libraries by using **probes** to search for them. A probe is a sequence of radioactive DNA that matches part of the gene of interest. In such cases, the amino acids are used to predict the DNA sequence that might

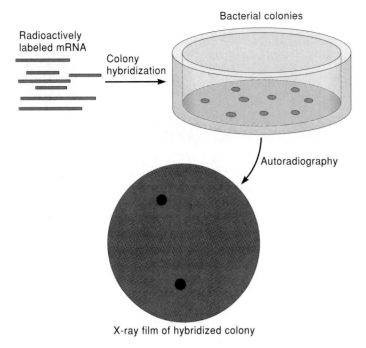

Radioactively labeled mRNA

Colony hybridization

Bacterial colonies

Autoradiography

X-ray film of hybridized colony

figure 11.25

Colony hybridization. Only those colonies containing DNA of the target gene, which is complementary to the radioactive mRNA, will show up on X-ray film. This diagram shows that two colonies have the target gene.

code for the polypeptide. A short fragment containing this DNA sequence is made chemically and radiolabeled, and then used as a probe.

By using a probe, we can detect the gene of interest in bacterial cultures by **colony hybridization.** In this procedure, an agar plate containing bacterial cultures is blotted with filter paper to make a replica of the pattern of colonies. The replica is then covered in a solution containing the probe, which hybridizes (i.e., anneals) to its DNA complement in those colonies that have the target gene (fig. 11.25). Only those colonies that have the target gene will bind to the probe and become radioactive. Their location is marked by exposing the filter paper to X-ray film, which reacts with radiation from the probe-bound colonies. The radioactive colonies are then picked out of the culture and grown individually, thereby cloning the gene of interest for further study.

Gene libraries can be screened more directly by searching for specific polypeptides. This method is similar to colony hybridization, except that instead of probing for a DNA sequence, antibodies are used to probe for a protein. For example, in the molecular genetic study of round versus wrinkled seeds in the garden pea, the first step was to make an antibody against the purified starch-branching enzyme. When a solution of the antibody was infused into the vector culture of the gene library from *RR* plants, it reacted only in colonies containing the starch-branching enzyme, SBEI, thereby revealing the presence of the gene.

Polymerase Chain Reaction

The newest technique for both finding and cloning a gene of interest is the **polymerase chain reaction (PCR),** in which DNA polymerase assembles free nucleotides, thereby making a complementary sequence from a strand of template DNA.[2] In cells, the DNA polymerase replicates the entire genome during a cell cycle (see Chapter 9), but the goal of PCR is to replicate one gene, not the entire genome. Therefore, a specific gene is targeted by including **primers** for it in the reaction mixture. A primer is a small sequence of DNA, usually 10–30 nucleotides long, that is complementary to one end of the target gene. Once it is annealed to the gene, the DNA polymerase adds nucleotides to it along the template. Both strands of the target gene are copied by using two primers, one for the 5′ end of the coding strand and one for the 5′ end of the complementary strand.

One cycle of the PCR takes about 10 minutes and requires three steps:

1. Denature the DNA at high temperature.

2. Cool the mixture to allow the primers to anneal to the target gene.

3. Incubate the mixture to allow the polymerase to make new DNA (fig. 11.26).

The amount of a target gene doubles during each cycle. Normally, the PCR runs for 30 cycles, which means that several million copies of a gene are made in a few hours. By comparison, a plasmid or a virus takes several weeks to make the same amount of a specific gene.[3]

Industrial Uses of Genetically Engineered Bacteria

As described earlier, bacterial vectors are used to store and retrieve genes in modern genetic research. Genetically engineered bacteria are also important for making foreign gene products on a commercial scale. The polypeptide of interest is either harvested from bacteria that are grown in large vats, or the live bacteria are used directly (fig. 11.27). The synthesis of a eukaryotic polypeptide in a vector is often enhanced by linking the gene to a highly active promoter, which helps to make large amounts of the polypeptide at relatively low cost.

---

2. Kary B. Mullis is given credit for inventing or discovering PCR, but what he actually did was show how the mechanisms of DNA synthesis in the nucleus could be imitated in a test tube. He devised such a method in 1983 and was awarded a Nobel Prize for this work in 1993.

3. Techniques such as PCR and the analysis of DNA fragments from restriction enzymes are the basis for DNA typing that is now used in criminal trials, paternity suits, and the identification of abandoned babies. Gel electrophoresis of PCR products or restriction fragments looks like a bar code on a grocery store product, and each person (except identical twins) has a unique bar code. Such *DNA fingerprinting* was used recently to identify a plant in a murder case in Arizona. Seeds from a palo verde tree (*Cercidium floridum*) were found in the back of a suspect's truck. DNA from the seeds matched that of a tree near the scene of the crime. This information was used as evidence to place the suspect's truck at the scene of the crime, which helped the jury reach a verdict of guilty.

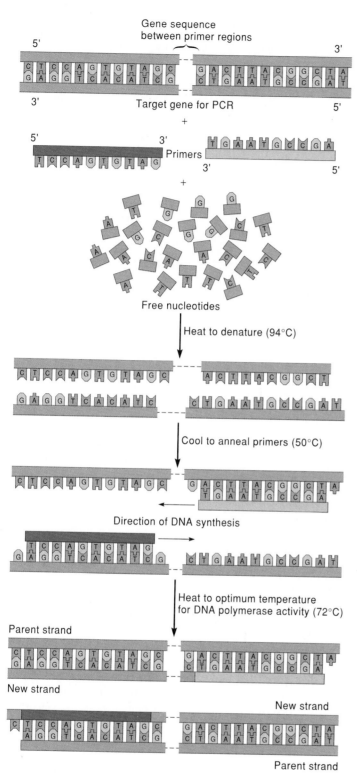

figure 11.26

The polymerase chain reaction. Target DNA is amplified by DNA polymerase after primers anneal to their complementary sequences in the target. The cycle is repeated until several million copies of the target gene have been made. Temperatures in each cycle and the number of cycles vary for different target genes and primers.

figure 11.27

An industrial fermenter used to grow genetically engineered bacteria. Such large-scale bacterial cultures produce large amounts of the foreign polypeptide at relatively low cost.

Most commercial uses of genetically altered bacteria involve agriculture and medicine. In agriculture, bacteria that contain foreign genes are used directly on crops. For example, bacteria in the genus *Pseudomonas* live on plants and normally produce a protein that acts as a "seed crystal" around which ice crystals form. The ice crystals damage the plant and nourish the bacteria, often destroying the crop. To combat this problem, farmers can now spray their crops with **ice-minus bacteria,** which are genetically engineered *Pseudomonas* that have had the gene for the ice-inducing protein removed. The ice-minus bacteria compete with the normal, ice-inducing bacteria, thereby displacing them and reducing crop loss. In another example, root-colonizing bacteria have been altered to make a protein toxin that protects crops from root pests. These bacteria contain the toxin gene from another species of bacteria, *Bacillus thuringiensis,* which does not colonize roots.

In medicine, genes that code for the synthesis of several polypeptide drugs are now produced by bacteria. Unlike the bacteria used in agriculture, which contain genes from other prokaryotes, drug-producing bacteria contain genes from eukaryotes. These genes make pharmaceutically important proteins, including insulin used to treat diabetes, human growth hormone used to treat dwarfism, and interferons and interleukins used to stimulate the immune system. Before these genetically engineered products were available, all such proteins had to be obtained from animals. Human growth hormone was especially scarce because it had to come from human cadavers.

figure 11.28

The tumorous growths on these *Gladiolus* bulbs arose from cells that were induced to become meristematic by the bacterium *Agrobacterium tumefaciens*. Such growth is known as *crown gall disease*.

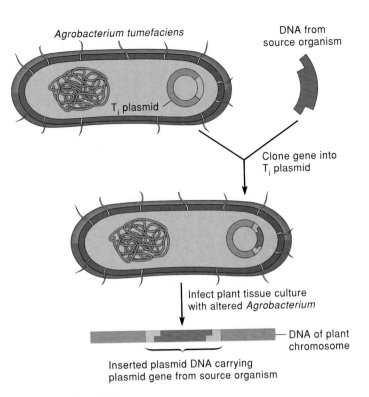

figure 11.29

Inserting foreign genes by using the *Agrobacterium* T_i plasmid. First, the bacterium is transformed by the insertion of a foreign gene into its T_i plasmid, which carries genes for inserting itself into plant chromosomes. Then a plant cell that is infected with the plasmid receives the foreign gene into its genome.

In genetic engineering, the product of interest is not always a protein. For example, genetically engineered plants are now used to make drugs (e.g., morphine, quinine, codeine), food additives (e.g., vanillin, mint oil), and fragrances (e.g., jasmine oil, rose oil). Similarly, potato plants are now used to make a type of plastic polymer. Genes for this polymer come from certain bacteria that are difficult to grow and that make only small amounts of the polymer. The polymer genes have now been inserted into *E. coli*, which is easy to grow, but polymer production is still low. Scientists are searching for promoters that will enhance expression of the polymer genes. If this work is successful, biodegradable plastics from bacteria may replace petroleum-based plastics, which remain in the environment too long.

Making Transgenic Plants: Problems and Solutions

The main problem in making transgenic plants is that, unlike bacterial plasmids, plant DNA cannot be recombined with foreign DNA by the direct use of restriction enzymes, nor can plant cells be transformed easily by absorbing DNA. Nevertheless, nature has its own genetic engineers that can overcome these difficulties: parasites that have special genes for inserting their DNA into plant genomes. One such parasite is the bacterium *Agrobacterium tumefaciens*, which causes plants to grow galls (tumors) (fig. 11.28). *Agrobacterium* has a plasmid, called the T_i plasmid, that is transferred into the host plant's DNA (fig. 11.29). In doing so, *Agrobacterium* can genetically transform the plant cells.

The most common way to make a transgenic plant is to use bacteria whose T_i plasmid contains an inserted gene of interest. When such bacteria infect plant cells, recombined T_i plasmids combine with the plant's chromosomes. The foreign gene is then replicated in each cell cycle as if it were a normal part of the plant genome. Ideally, the newly inserted gene is also expressed in all of the cells that contain it.

When these methods are used on whole plants or tissues, only some cells are infected with recombined T_i plasmids and become transgenic. To make an entire plant transgenic, however, all of its cells must be genetically altered. For many plants, this is a minor problem that is solved by genetically engineering their cells in culture and growing the cultured cells back into whole plants (fig. 11.30). The genes of this plasmid are then replicated and expressed as if they belonged to the plant's genome. Such genes reactivate the cell cycle in cells that had stopped in the G_1 phase (see Chapter 9). These cells then divide rapidly and form plant tumors.

Antisense Technology

Although particular genes can be inactivated by inserting transposons or by mutagenic chemicals, these processes are random and inefficient, and can alter genes that we do not want to change. We can avoid this problem by using **antisense technology**, which inactivates specific genes. To appreciate this technique, recall that active genes are codes for polypeptides. When one copy of a strand of DNA is transcribed, the

11–21

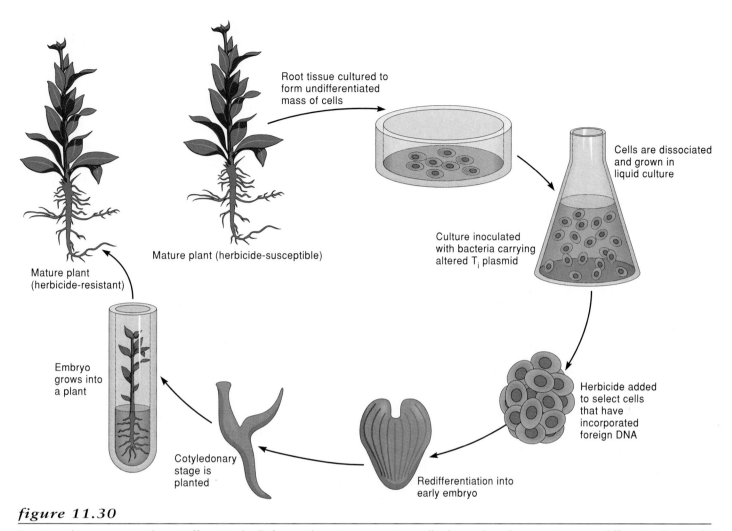

figure 11.30

Steps in making transgenic plants. Differentiated cells from a plant tissue are put into cell culture, where they grow into an undifferentiated mass of cells. The culture is transferred to a liquid medium and infected with T$_i$ plasmids carrying a gene for herbicide resistance. Cells that survive herbicide treatment are transgenic. Plants that are regenerated from these cells are also transgenic, making them resistant to the herbicide in the garden.

sequence of bases in the DNA is preserved, except that the sequence is exactly complementary to the original code. The mRNA, now containing instructions from DNA in its own language, is then translated into a polypeptide. If done correctly, the message is said to be in correct **sense.** However, if an exactly complementary (i.e., antisense) RNA were present to combine with the sense RNA, the message of the sense RNA would be nullified.

To make an RNA complement, an antisense message can be made from the "other" strand of DNA and introduced into a cell, or it can be transcribed by the cell after an antisense gene is introduced. When the sense RNA couples with its antisense RNA, it cannot combine with a ribosome and is inactivated. Thus, the original gene is blocked by its own antisense RNA, without affecting other genes.

Antisense technology has been used to control a variety of plant processes, including fruit ripening (see "Improvement

in Fruit Storage," p. 253). Because of its specificity and simplicity, antisense technology has tremendous potential for use in agriculture.

c o n c e p t

Transgenic bacteria are made directly by inserting foreign genes into their plasmids, but transgenic plants must be grown from cell cultures that are infected by parasitic bacteria carrying foreign genes.

Uses of Transgenic Plants

Table 11.1 lists several goals of plant genetic engineering. Of these, the development of herbicide resistance has been the most successful. Transgenic plants that resist certain insects or

PUBLIC ISSUES IN PLANT GENETIC ENGINEERING

Regardless of the seemingly unlimited potential for human benefit from transgenic plants, the scientific and commercial aspects of plant genetic engineering bring forth some emotional, ethical, and legal issues for the public. One of the main developments giving rise to these issues is that the natural barriers to transferring genes from one organism to another are rapidly disappearing. Unlike breeding experiments, where hybridization transfers genes between closely related species, genetic engineering enables the transfer of genes between organisms from different kingdoms. Public opinion of this kind of work was revealed in a recent survey of consumers regarding genetically engineered food products. About 33% of the people disapproved of foods made by transferring genes from one plant to another, but 75% could not stomach the idea of foods that are genetically engineered by mixing genes between plants and animals. In anticipating that genetically engineered foods will continue to gain government approval for commercial sale, some 1,500 well-known chefs across the United States have joined a "Pure Foods Campaign" to boycott them.

Organizations have already been formed by people who object to genetic engineering. Some groups believe that it is too dangerous to make transgenic organisms because we might accidentally make and release super-resistant disease bacteria or uncontrollable weeds into the environment. Other groups believe that it is morally or ethically wrong to "play God" by manipulating genes artificially. These are just some of the social issues that have been explored in conferences, books, articles, and government reports over the past several years. These kinds of issues are certain to remain at the forefront of public consciousness for years to come.

The development of transgenic plants as valuable financial commodities is also at the heart of recent legal battles over the patent rights to the procedures and products of genetic engineering. Since the early 1980s, the U.S. Patent and Trademark Office has granted more than 100 patents for transgenic plants or technological developments that are used to make them. Three particular patents lead the way in the legal battles that have

followed. The most stunning of these is a 1992 patent held by Agracetus Inc. of Middleton, Wisconsin, for rights to all forms of genetically engineered cotton. This patent is remarkable because it grants a 17-year ownership to one company for all transgenic strains of an entire species, no matter what genes or techniques are used to make them. A similarly noteworthy patent was granted to Agracetus by the European Patent Office for all forms of genetically engineered soybeans. The third sweeping patent of note was granted in the United States to Enzo Biochemical of New York City for antisense RNA technology. Enzo is already suing Calgene Inc., a Davis, California-based company that allegedly infringed on this patent by using an antisense gene to produce the Flavr Savr™ tomato (see p. 254). All three of these patents are under legal challenges that are likely to last for years. It will take that long before we know whether or to what extent any single company can own the rights to genetically engineering an entire species.

viruses have also been made. Most of the goals listed, however, have not yet been accomplished. Some examples of those that have been achieved, at least experimentally, are discussed in the following paragraphs.

Herbicide Resistance

Most plants are killed by **glyphosate,** a herbicide that inhibits an enzyme called *EPSP synthetase* that is required for making aromatic amino acids. (EPSP stands for the product made by EPSP synthetase, which is 5-EnolPyruvyl-Shikimic acid-3-Phosphate.) When plants cannot make aromatic amino acids, their metabolism stops and they die.

Glyphosate-resistant petunias have been made by inserting extra copies of the EPSP synthetase genes into them. These petunias make enough enzyme to overcome inhibition by glyphosate. Plants that contain a bacterial form of EPSP synthetase also resist glyphosate (fig. 11.31).

Ideally, crops that resist glyphosate do not have to be weeded. Treating a field with the herbicide kills the weeds but does not affect the genetically engineered crop.

Disease Resistance

Botanists have discovered a variety of genes involved in disease resistance by plants. For example, the RPS_2 gene in *Arabidopsis thaliana* confers resistance to bacteria such as *Pseudomas syringae*. Similarly, tobacco mosaic virus (TMV) causes a mosaic pattern of infection on tobacco leaves (fig. 11.32). When the gene for the protein coat of TMV is transferred to tobacco, the plants become resistant to the virus. The TMV responds to the genetically engineered tobacco cells as if they were already infected, and since the virus cannot infect cells that are already infected, the plant is immune to infection.

table 11.1

Goals of Plant Genetic Engineering

1. Crop resistance to herbicides, insects, diseases, and drought
2. Reduction of photorespiration in C_3 crops
3. Atmospheric nitrogen fixation by crop plants
4. Tolerance to high-salt soils and to flooding in crops
5. Increased nutritional value of plant storage proteins
6. Cold-tolerance in tropical and subtropical crops
7. Longer storage life of fruits and vegetables
8. Enhanced productivity in ornamental and food plants
9. Lipids for nutrition and industrial use
10. Vaccines against human disease
11. Cotton-polyester fibers
12. Biodegradable plastics
13. Industrial lubricants
14. Components of soaps and detergents
15. Drugs and other pharmaceuticals

Source: Data from D. Grierson and S. N. Covey, *Plant Molecular Biology*, 2d ed. Blackie Academic and Professional, imprint of Chapman & Hall, 1988.

figure 11.32

This tobacco leaf is infected with tobacco mosaic virus (TMV). Resistance to this virus can be engineered into tobacco by using a gene from the virus itself.

figure 11.31

Normal tomato plants (*right*) and transgenic plants containing a gene for herbicide resistance (*middle*) were treated with the herbicide glyphosate. All of the normal plants died. Plants in the left row were not treated with herbicide.

Insect Resistance

As discussed previously in this chapter, crops can often be protected against insects by being sprayed with bacteria that contain the toxin gene from *Bacillus thuringiensis*. This gene has also been inserted into tomato, potato, and tobacco plants. Caterpillars that normally eat the leaves of these species do not eat plants that contain the toxin made by this gene (fig. 11.33a).

Perhaps the most important crop for genetically engineering resistance to insects is cotton; indeed, most of the chemical pesticides used in the United States are used on cotton. Experimental cotton plants that contain the toxin gene from *B. thuringiensis* typically have better resistance to insects and require fewer chemical pesticides.

Drought Resistance

Genes from desert petunias have been transferred into cultivated petunias. The transgenic petunias require much less water than the normal plants. If crop plants likewise can be altered genetically to require less water, there will be less need for irrigation. Such crops will become more important as the demand for water increases in semiarid agricultural regions of the southwestern United States.

Changing Storage Proteins

Grains from cereal grasses are deficient in lysine, which is an essential amino acid in our diet. However, induced mutations in cultured rice cells have increased their production of lysine and protein. Plants from these mutant cell cultures produce grains containing more lysine and protein than do normal rice plants. Botanists hope to transfer the gene for the high-lysine mutation to other cereal grasses.

Improvement in Fruit Storage

One of the main problems with fruits is that when they are left on the plant too long, they get mushy during shipping and storage. To overcome this problem, fruits are picked and shipped while they are still immature; they are ripened later by exposure to the hormone ethylene. Unfortunately, such artificially ripened fruits often have little flavor.

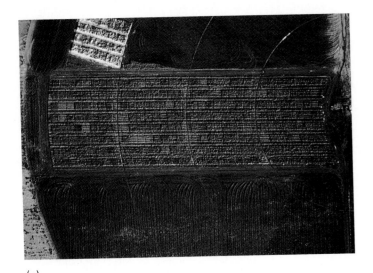

(a)

(b)

figure 11.33

Genetic engineering and plants' resistance to pests. (a) Infrared aerial image (enhanced by false color) of potato fields. Normal potatoes were defoliated by the Colorado potato beetle, thereby allowing wet ground to be exposed (*green*). Beetles avoided transgenic potatoes that contained an insecticide-producing gene (*red*). (White areas are wheat fields.) (b) The peas on the left have been genetically engineered for resistance to weevil larvae. These were the first pest-resistant fruits produced by genetic engineering.

Genetic engineering has been used to solve this problem for tomatoes, which become mushy because a carbohydrate-digesting enzyme breaks down cell walls in the fruit. Specifically, botanists have introduced an antisense message for the enzyme system that affects the mushiness of the fruit. Consequently, these plants produce fruits that do not become mushy. Such genetically engineered tomatoes that are left to ripen on the parent plant maintain their firmness (and flavor) during storage. The first such genetically engineered tomato, called the Flavr Savr™ tomato, became available in grocery stores in the United States in 1994.

In countries such as Brazil, farmers often lose 20%–40% of their crops to storage pests such as weevils. This problem is being addressed by engineering plants whose fruits are pest-resistant. For example, botanists have produced bean plants having fruit that produces an inhibitor of α-amylase. This inhibitor blocks the α-amylase secreted by attacking weevils, thereby reducing the harvest's susceptibility to the pest (fig. 11.33b).

Production of Polymers

Sales of nonfood products from genetically transformed plants will grow from about $21 million per year in 1997 to $320 million or more by 2005. The first nonfood commercial product of plant genetic engineering is lauric acid, a 12-carbon fatty acid that is used in making soaps and detergents. The transgenic plant that makes this chemical is rapeseed (*Brassica napus*). Normally, fatty acids in this plant are 18 carbons long, but scientists at Calgene Inc. have inserted into rapeseed a gene from California bay tree (*Umbellularia californica*) that shuts off fatty acid synthesis after 12 carbons. Oil from a few thousand acres of transgenic rapeseed, which is easy to grow and a good oil-producer, was marketed in 1995. Other transgenic rapeseed lines, in different stages of development, will be producing lubricants, nylon, and fatty acids for making margarine and shortening.

Besides fatty acids, other polymers from transgenic plants include a biodegradable plastic that is made by bacterial genes inserted into *Arabidopsis thaliana* and polyester-like fibers from similar genes inserted into cotton (*Gossypium hirsutum*). Neither of these polymers has yet reached commercial production.

The Lore of Plants

Biological vectors are not the only carriers for inserting foreign genes into plant cells. Molecular geneticists have several nonbiological tools for this purpose, the most interesting of which is the gene gun. With this device, DNA is placed on tiny metal beads that are loaded into the barrel of a .22-caliber gun and literally shot into cells. The cells that are hit just right by the metal beads integrate the foreign DNA into their chromosomes about as efficiently as cells that are infected with biological vectors.

The Molecular Genetic Basis of Round versus Wrinkled Peas

Let's now return to our round and wrinkled peas that we discussed at the beginning of this unit. So far, you have learned that the wrinkled-pea phenotype is probably caused by the absence of a certain starch-branching enzyme (p. 232) and that the gene for this enzyme was detected by colony screening with an antibody against the enzyme in a gene library of *RR* plants (p. 248). The question remains, What is the difference between the *R* allele and the *r* allele?

The first part of the answer is that the antibody reacts to the gene library from *RR* plants but not to the gene library from *rr* plants. This does not mean that the gene locus is missing in *rr* plants, but it does mean that the *r* allele is not translated. The next part of the answer was found when the *R* allele from the *RR* gene library was used to probe the *rr* genome.

The gene library that was used to find the *R* allele came from a digestion of the *RR* genome by *Eco*RI. This digestion gave two fragments that together contained the whole *R* allele, which means that the allele had an *Eco*RI site (GAATTC) in it. Clones containing these *Eco*RI fragments were used as the probes.

The first probe analysis was done by a procedure called **Northern blotting.** For this procedure, all of the RNA is extracted from the plant and separated by gel electrophoresis (Chapter 8 describes gel electrophoresis of proteins). Nucleic acids, like proteins, are ionic and therefore separable in an electric field, and like proteins, they are separated on the basis of different sizes: larger fragments are slowed more by the gel, so they do not move as far. For a Northern blot analysis, after electrophoresis the RNA is transferred to filter paper (*blotting*), and the filter paper is immersed in a solution that contains the radioactive probe. The probe then binds to sequences that are complementary to it wherever they occur on the filter paper. Sequences that bind to the radioactive probe are then detected by exposing the probed filter paper to X-ray film (fig. 11.34).

The RNA extracts from *RR* and *rr* plants showed a significant difference when probed by the *Eco*RI fragment of the *R* allele. The *R* allele made large amounts of an mRNA molecule that was about 3,300 nucleotides long, but the *r* allele made much smaller amounts of an mRNA molecule that was about 4,100 nucleotides long (fig. 11.35). The first probe analysis showed that mRNA from the *r* allele is about 800 nucleotides larger than mRNA from the *R* allele. Also, the *r* allele is not transcribed as much as the *R* allele. This is evidence that the 800 extra nucleotides have been inserted into an exon and that this insert blocks translation. If the insert had been in an intron, it would have been removed by RNA processing and therefore would not have appeared in the mRNA.

In the second analysis, *RR* and *rr* genomes were digested with different restriction enzymes, and the DNA was probed directly. This type of analysis is called **Southern blotting,** which is named for its discoverer, E. M. Southern (RNA detection is called *Northern blotting* simply to distinguish it from Southern blotting). Analysis by Southern blotting showed that the *r* allele is larger than the *R* allele by about 800 nucleotide pairs, thereby confirming the results from Northern blotting.

At this point, the explanation of the molecular genetic basis for wrinkled seeds is nearly complete. The *r* allele contains an 800-nucleotide insert in a coding region, and this insert inhibits translation of mRNA into a functional starch-branching enzyme. But how did this insert get into the *r* allele in the first place? To address this question, Bhattacharyya's group used a clone containing the insert to probe the garden pea genome. They found that this probe binds to many regions of both the *RR* and the *rr* genomes, indicating that the insert is repeated many times throughout the genome of the garden pea.

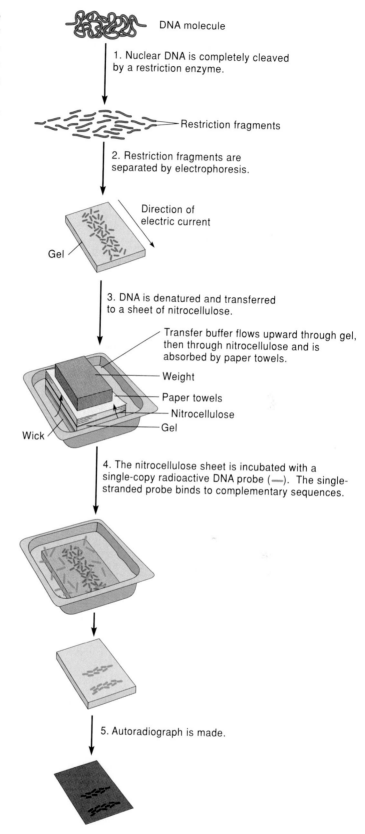

1. Nuclear DNA is completely cleaved by a restriction enzyme.

Restriction fragments

2. Restriction fragments are separated by electrophoresis.

Direction of electric current

Gel

3. DNA is denatured and transferred to a sheet of nitrocellulose.

Transfer buffer flows upward through gel, then through nitrocellulose and is absorbed by paper towels.

Weight
Paper towels
Nitrocellulose
Gel
Wick

4. The nitrocellulose sheet is incubated with a single-copy radioactive DNA probe (—). The single-stranded probe binds to complementary sequences.

5. Autoradiograph is made.

figure 11.34

Detection of target DNA by Southern blotting and target mRNA by Northern blotting. Different names are used for the detection of each type of nucleic acid, although the blotting procedure is the same.

Source: Fristrom and Clegg, Principles of Genetics, *2d ed. Copyright © Chiron Press, Cambridge, MA.*

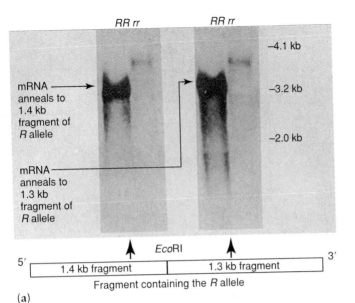

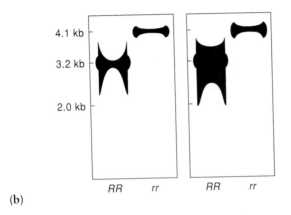

figure 11.35

Understanding the molecular basis of round versus wrinkled peas.
(a) Northern blots of mRNA extracts from garden pea embryos with
either the genotype *RR* or *rr*. Blots were made using radiolabeled
fragments of the *R* allele. (b) The *RR* genotype makes large amounts of
mRNA that is about 3,300 nucleotides; the *rr* genotype makes small
amounts of a larger mRNA molecule (about 4,100 nucleotides). This is
evidence that the *r* allele contains an insertion that interrupts translation.

Furthermore, the sequence of nucleotides in the pea insert was
found to be similar to sequences from snapdragon, maize, and
parsley that correspond to parts of transposable elements.
Therefore, the insert in the *r* allele probably came from the activity of a transposable element that is widespread in the garden pea and other plants.

Chapter Summary

Genes work as templates for making RNA during transcription. In eukaryotes, genes are interrupted by nonsense DNA,
which must be removed from RNA transcripts before they are
transported out of the nucleus. The removal of nonsense sequences from RNA is called RNA processing and is done by
spliceosomes.

After messenger RNA is processed and transported out
of the nucleus, it is translated into polypeptides in the cytoplasm. Translation also requires transfer RNA and ribosomal
RNA. Transfer RNA brings amino acids together. Ribosomal
RNA is a component of ribosomes, which are the sites of
polypeptide synthesis.

The unit of DNA that codes for an amino acid is a
codon, which is a triplet of three nucleotides. Most amino
acids have more than one codon. Codons that code for the
same amino acid are called synonymous codons.

Coding sequences in interrupted genes are called exons,
and noncoding sequences are called introns. Genes that contain introns occur almost exclusively in eukaryotes. Introns are
rare in simple eukaryotes, such as yeasts, but they are common
in complex eukaryotes, such as plants and animals. Many
exons code for separate domains in polypeptides.

Whole genomes can be cut into fragments by bacterial
restriction enzymes and inserted into vectors, such as bacterial
plasmids or viruses. Once inserted into a vector, the genome
can be cloned by culturing the vector, thereby maintaining a library of the genes from that organism.

Individual genes are also produced in test tubes by DNA
polymerase in the polymerase chain reaction. Unlike cloning in
vectors, which may take weeks to produce enough of the target gene for further study, the polymerase chain reaction takes
only a few hours.

Genetic engineering, a subfield of *recombinant DNA
technology*, is the transfer of genes from one organism to another. Transgenic organisms are those whose genomes contain
a foreign gene. Transgenic bacteria are made by inserting foreign DNA into plasmids that are incorporated back into the
bacteria. Making transgenic plants requires extra steps, because plants lack plasmids. However, foreign DNA can be inserted into plant genomes by parasitic bacteria. Cultured plant
cells that are infected by such a transgenic parasite can be regenerated into transgenic plants.

The molecular genetic study of round versus wrinkled
seeds in the garden pea is an example of how genes can be
studied by using modern laboratory techniques. The transcription product, the translation product, and the gene itself were

all manipulated in this study, which showed that the *r* allele is defective because of a DNA insert that blocks translation of the mRNA. The absence of the translation product, a starch-branching enzyme, apparently causes metabolic changes in seeds with the *rr* genotype, thereby producing the wrinkled phenotype when the seeds mature.

 ## What Are Botanists Doing?

Read at least three journal articles about using radioactive probes to find a plant gene of interest. Examine the articles to find out how the probes were obtained. Based on this list, what is the most common method for obtaining probes for molecular genetic studies of plants?

 ## Writing to Learn Botany

Should we be concerned about releasing artificially produced genes into nature? Why or why not?

Questions for Further Thought and Study

1. How may introns regulate genes?

2. What are the advantages and disadvantages of a genomic library versus a cDNA library?

3. Most research in plant genetic engineering involves tobacco, tomatoes, or petunias. Why?

4. Starting with the gene, what are the fastest and slowest steps in protein synthesis? What makes the fast steps fast and the slow steps slow?

5. Compare the polymerase chain reaction with DNA replication.

6. What are the natural functions of restriction enzymes? How are these enzymes used in gene technology?

7. What is the minimum number of different mRNAs, tRNAs, rRNAs, codons, and amino acids for making all of the proteins in a cell? How many of each are known? Explain the discrepancy between the minimum and the actual numbers if they are different from each other.

8. Why do you suppose the 800-nucleotide insert in the *r* allele of the garden pea blocks translation? That is, why does this allele not translate into a protein that is 267 amino acids longer than SBEI from the *R* allele?

9. What is the importance of reverse transcriptase in gene technology?

10. Some people claim that the release of genetically transformed plants in the wild could upset ecosystems—for example, that a gene for herbicide resistance could be transferred to wild relatives, thereby producing herbicide-resistant weeds. Is this a reasonable concern? If so, how should it be addressed?

Web Sites

Review the "Doing Botany Yourself" essay and assignments for Chapter 11 on the *Botany Home Page*. What experiments would you do to test the hypotheses? What data can you gather on the Web to help you refine your experiments? Here are some other sites that you may find interesting:

http://ophelia.com/Ophelia/pgr/
This site, which is maintained by the American Society of Plant Physiologists, is the *Plant Gene Register*. It includes a searchable index to articles about plant genetics.

http://members.gnn.com/kalimantan/mw4/page13.htm
This site is devoted to transgenic plants. Here you can read about the history of plant molecular biology, discover links to related sites, and read on-line articles.

http://www.gene.com/ae/AB/BC/index.html
This site includes *The Biotech Chronicles*, a brief history of the applications of biotechnology in medicine, farming, and other industries.

http://www.gene.com:80/ae/AB/index.html
This site includes information about the history, tools, ethical concerns, and environmental concerns of biotechnology.

http://www.ifrn.bbsrc.ac.uk/gm/lab/docs/iftmb.html
The Internet for the Molecular Biologist provides information on all aspects of the Internet (e.g., databases, e-mail, newsgroups) that are related to molecular biology.

http://muse.bio.cornell.edu/
http://muse.bio.cornell.edu/cgi-bin/hl?botany
This collection of botany resources spans informational databases to high-resolution images.

http://www.ncbi.nlm.nih.gov/
Welcome to the National Center for Biotechnology Information. The NCBI builds, maintains, and distributes GenBank, the National Institute of Health genetic sequence database that collects all known DNA sequences from scientists worldwide.

Suggested Readings

Articles

Barton, J. H. 1991. Patenting life. *Scientific American* 264(March):40–46.

Boulter, D. 1995. Plant biotechnology: Facts and public perceptions. *Phytochemistry* 40:1–9.

Cech, T. R. 1986. RNA as an enzyme. *Scientific American* 255(November):64–75.

Chambon, P. 1981. Split genes. *Scientific American* 244(May):60–71.

Clegg, M. T. 1990. Dating the monocot-dicot divergence. *Trends in Ecology and Evolution* 5:1–2.

Daviss, B. 1990. Super spud. *Discover* 11(5):30.

Gasser, C. S., and R. T. Fraley. 1992. Transgenic crops. *Scientific American* 266(June):62–69.

Holliday, R. 1989. A different kind of inheritance. *Scientific American* 260(June):60–73.

McKnight, S. L. 1991. Molecular zippers in gene regulation. *Scientific American* 264(April):54–64.

Moffat, A. S. 1995. Exploring transgenic plants as a new vaccine source. *Science* 268:658–660.

Mullis, K. B. 1990. The unusual origin of the polymerase chain reaction. *Scientific American* 262(April):56–65.

Ogden, R., and D. A. Adams. 1989. Recombinant DNA technology: Basic techniques. *Carolina Tips* 52(4):14–16. Burlington, NC: Carolina Biological Supply Co.

Pace, N. R., and T. L. Marsh. 1985. RNA catalysis and the origin of life. *Origins of Life* 16:97–116.

Powledge, F. 1995. Who owns rice and beans? *BioScience* 45:440–444.

Stone, R. 1995. Sweeping patents put biotech companies on the warpath. *Science* 268:656–658.

Strange, C. 1990. Cereal progress via biotechnology. *BioScience* 40:5–9, 14.

White, R., and J. M. LaLouel. 1988. Chromosome mapping with DNA markers. *Scientific American* 258(February):40–49.

Books

Grierson, D., and S. N. Covey. 1988. *Plant Molecular Biology.* 2d ed. New York: Chapman and Hall.

Juma, C. 1989. *The Gene Hunters: Biotechnology and the Scramble for Seeds.* Princeton, NJ: Princeton University Press.

Klug, W. S., and M. R. Cummings. 1994. *Concepts of Genetics.* New York: Macmillan.

Lewin, B. 1990. *Genes IV.* New York: Oxford University Press.

Murphy, T. M., and W. F. Thompson. 1988. *Molecular Plant Development.* Englewood Cliffs, NJ: Prentice-Hall.

Watson, J. D., J. Tooze, and D. T. Kurtz. 1989. *Recombinant DNA: A Short Course.* 2d ed. New York: Scientific American Books.

Wilson, T. M. A., and J. W. Davies, eds. 1992. *Genetic Engineering with Plant Viruses.* Boca Raton, FL: CRC Press.

The Form and Function of Plants

Plants have a long and interesting evolutionary history. They evolved from algae 450 million years ago and then diversified in a way that forever changed our planet: they invaded land, thereby paving the way for the subsequent colonization of land by animals. Because the terrestrial habitats encountered by plants were more diverse than their ancestral aquatic environments, land plants faced many new problems and selective pressures. As a result, they evolved rapidly and became more complex and diverse than did their algal ancestors.

In this unit we present generalities, explanations, and examples to help you understand the form and function of vascular plants. Vascular plants (such as ferns, conifers, and flowering plants) have vascular tissues specialized to transport water and dissolved substances throughout the plant. Nonvascular plants such as mosses and liverworts lack vascular tissues. In this unit we emphasize vascular plants for two reasons:

1. Vascular plants represent several closely related lines of evolution and share a common form and function.

2. Vascular plants dominate our terrestrial ecology and our economy; they affect almost every aspect of our lives.

Since there are probably more than 300,000 species of vascular plants, you should expect some variety. We will also introduce a few of the exceptions to our generalities, because unusual plants often help us understand the norm.

The most diverse group of vascular plants is the flowering plants, which include more than a quarter of a million species. Because we know more about this group than any other, the focus in this unit will be on the flowering plants. Nevertheless, vegetative tissues develop and are organized similarly in almost all vascular plants.

Leaves of this flame bromeliad (*Neoregelia carolinae*) form a cup that gathers rain and nutrients.

Plant Growth and Development

12

Chapter Outline

Chapter Overview

Plant growth and development result from interactions of cellular division, cellular enlargement, and the patterned formation of tissues and organs. Roots and shoots grow from meristems, which are embryonic regions of cellular division and expansion. Although this growth enlarges a plant, it doesn't necessarily adapt a plant to its environment. Rather, adaptation results from differentiation, whereby new cells are organized into tissues and organs specialized for different functions. Because organs such as leaves, roots, and stems are produced continually, each adapts to changing environmental conditions. Growth and development are limited by a plant's genome and are integrated with environmental signals such as light and gravity.

Roots absorb water and minerals from soil and, in doing so, anchor a plant in place. This sedentary life-style has profound consequences for plant growth and development. Unlike advanced animals, whose development is controlled internally and usually isn't affected much by the environment, plants use environmental signals such as light and gravity to direct their growth and development. Growth and development are ultimately limited by a plant's genome, but the genome isn't a strict developmental blueprint. Rather, plants constantly adjust their growth and development to changing environmental conditions.

Living in one place makes plants unable to flee danger or pursue food. As a result, they are constantly threatened by damage and starvation. Natural selection has helped plants to cope with these threats by producing parts that are totipotent, controlled locally, and able to revert to an earlier stage of development, if necessary. This developmental flexibility adapts plants to their environment; indeed, in the worst of times, plants can become dormant or regenerate themselves. However, this adaptability has a price: plant structure must remain relatively simple (fig. 12.1). For example, seedlings consist largely of only three kinds of cells: parenchyma, collenchyma, and sclerenchyma. **Collenchyma** and **sclerenchyma** are thick-walled cells that support the plant. **Parenchyma** cells are more versatile: their functions range from protection and secretion to photosynthesis and storage.

The simple design of plants also extends to tissues and organs:

- Simplicity of plant tissues. Plants consist primarily of three kinds of tissues: epidermis, vascular tissue, and ground tissue. Examine the locations of these tissues in figure 12.1. **Epidermis** covers the plant, protecting the underlying tissues and regulating the movement of gases between the plant and the atmosphere. Just interior to the epidermis is **ground tissue,** which is usually modified for photosynthesis or storage. The various parts of a plant are linked by **xylem** and **phloem,** which are **vascular tissues** that conduct water and dissolved solutes.

- Simplicity of plant organs. Epidermis, ground tissue, and vascular tissue are arranged into three kinds of vegetative organs: stems, leaves, and roots (fig. 12.1). Leaves are usually the primary photosynthetic organs and are arranged in spirals on the stem. Above the point at which a leaf attaches to a stem is an axillary bud. Although axillary buds are usually dormant, under certain circumstances they can form branches, flowers, or specialized shoots such as tendrils. Shoots, like roots, grow at their tips.

This chapter introduces the principles of development common to roots and shoots. It is meant to be an overview, so concentrate on the principles, not the specifics. In the other chapters of this unit you'll learn how these principles apply to roots and shoots.

A Perspective on Plant Form and Function

Many aspects of plant structure were discovered long ago, when people believed that all living organisms were created as a series of divine "types." Because there was no reason to expect intermediates between these divinely created types, most botany of the nineteenth century involved describing and categorizing the types of organs, tissues, and cells in various groups of plants. Had this approach continued, botany would have been little more than a description of the different plant types and lists of families and genera.

Today we explain plant structure and function not with types and divine creation, but with biochemistry, physiology, and genetics, all of which are guided by evolution. Thus, in this textbook we do not discuss types; rather, we present plant structure and development in relation to its function. Instead of reading lists of types and families, you will learn how plants are adapted to diverse environments, how plant parts perform various functions, and why each function is important for a plant's survival.

The Lives of Plants

Multicellular plants and animals grow differently. Most animals are embryonic for only a short time, during which all parts of their bodies grow simultaneously. This *closed growth*, or *determinate growth*, produces an animal whose shape doesn't change much throughout the rest of its life. Conversely, most plants never attain a fixed size because they grow indefinitely. This *open growth*, or *indeterminate growth*, is peculiar to plants; with the exception of a few organisms, such as fish, nothing like it occurs in higher animals.[1] As a result, the life spans of plants are often determined by environmental factors such as drought and disease, rather than by genetic limitations. For example, annuals usually complete their life cycle in one year, but many of them become perennials when protected from cold and drought. Similarly, perennials often reach astounding ages: many trees are

---

1. Usually only roots and stems grow indeterminately. Leaves, fruits, and flowers grow determinately and therefore reach a mature size and shape.

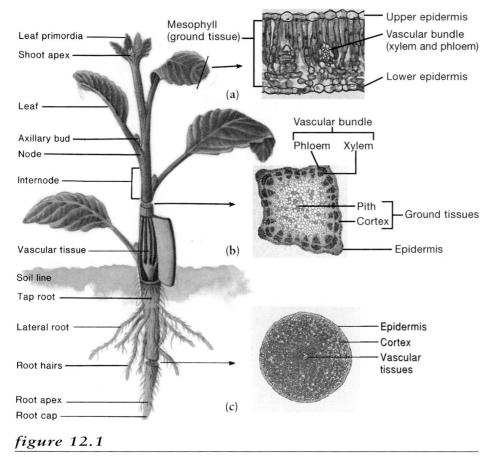

Leaf primordia
Shoot apex
Leaf
Axillary bud
Node
Internode
Vascular tissue
Soil line
Tap root
Lateral root
Root hairs
Root apex
Root cap

Mesophyll (ground tissue)
Upper epidermis
Vascular bundle (xylem and phloem)
Lower epidermis
(a)

Vascular bundle
Phloem   Xylem
Pith
Cortex — Ground tissues
Epidermis
(b)

Epidermis
Cortex
Vascular tissues
(c)

figure 12.1

The tissues and organs of a herbaceous plant. Cross sections of a (a) leaf, (b) stem, and (c) root.

thousands of years old. Few plants die of old age. Rather, they succumb to disease, drought, or even their own success—some plants become so large that their transport systems can't service all of their parts.

Not all parts of a plant grow at the same time. Rather, plant growth is restricted to perpetually embryonic regions called **meristems**. Plants have four types of meristems (figs. 12.1, 12.2):

1. **Apical meristems** occur near the tips of roots and shoots, and produce *primary tissues*. These meristems account for *primary growth*, which is elongation of roots and shoots. Primary growth is important because it enables a plant to explore new environments for light, water, and nutrients.

2. **Axillary buds** are reiterations of the apical meristem left behind by primary growth. Axillary buds occur in the axil of a leaf and typically undergo a dormant period (fig. 12.1). Axillary buds are important because they are a shoot's insurance policy: they are inactive cells that can form a branch or a flower.

3. **Lateral meristems** are cylindrical meristems that form in subapical regions of the roots and shoots of woody plants. Lateral meristems produce *secondary growth*, which increases the girth of the plant. Secondary growth is important because it makes a plant sturdier. This, in turn, enables the plant to grow taller and intercept light.

4. **Intercalary meristems** occur between mature tissues. These meristems are most common in grasses, where they occur at the bases of nodes. Intercalary meristems are important because they help regenerate parts removed by grazing herbivores (or lawn mowers).

Localized growth in meristems has important implications for plant growth and development: it means that plants are a mixture of young dividing cells, maturing cells, and mature cells, all of which are derived from meristems. As a result, plants function as mature organisms while still growing. Plants are especially sensitive to damaged meristems, however, because such damage often stops growth. In shoots, plants minimize this risk by protecting their meristems with young leaves and by forming dormant reserve meristems (i.e., buds) that can take over if the apical meristem is damaged. Similarly, roots protect their apical meristems with a root cap. If the primary root is damaged, its function is often assumed by a lateral root.

We begin this chapter by discussing apical meristems, because they occur in all plants and ultimately produce the cells that form mature plants. The other kinds of meristems are discussed later in this unit.

Meristems

Functions of Apical Meristems

Roots and shoots enlarge by primary growth from apical meristems located near their tips. Apical meristems produce cells that, when they expand, lengthen the root or shoot. The cells of apical meristems usually have thin walls, prominent nuclei, and small vacuoles. Compared to other types of plant cells, meristematic cells appear rather simple and are therefore sometimes referred to as *undifferentiated*. However, this is incorrect: distinct types of meristematic cells have unique structures, even though they may be in the same plant. Moreover, cells in different zones of an apical meristem change structure as the meristem changes function. Thus, meristematic cells are unique and differentiated; none should be considered unspecialized.

Apical meristems consist of meristematic initials and their immediate derivatives. As these cells divide, they leave a

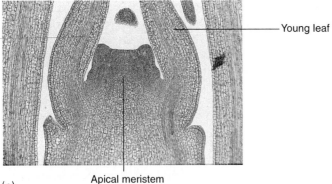

Young leaf

Apical meristem

(a)

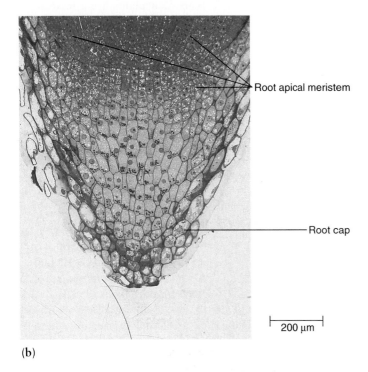

Root apical meristem

Root cap

⊢——— 200 µm ——⊣

(b)

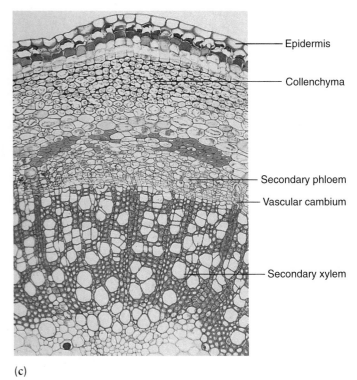

Epidermis

Collenchyma

Secondary phloem

Vascular cambium

Secondary xylem

(c)

figure 12.2

Meristems in plants. Apical meristems in shoots (a) and roots (b) produce *primary growth*, which elongates the organs. The shoot apical meristem (*a*, shown in longitudinal section) produces young leaves, which usually overarch the meristem. The root apical meristem (*b*, also shown in longitudinal section) is covered by a thimble-shaped root cap that protects the meristem as the root grows through the soil. The vascular cambium (*c*, shown in cross section) is a lateral meristem that produces *secondary growth*, or thickening, of the root or shoot of a woody plant. The vascular cambium produces wood (secondary xylem, which conducts water and dissolved minerals) to the inside and secondary phloem (which conducts dissolved organic compounds) to the outside. The young stem shown in *c* is covered by a protective epidermis, under which is collenchyma tissue that helps support the stem.

cylindrical root or shoot behind them. Apical meristems have two primary functions: (1) establishing patterns, and (2) producing new, genetically healthy cells.

Establishing Patterns

Apical meristems establish many patterns in plants (e.g., the arrangement of leaves). When an apical meristem is damaged, the remaining tissues continue to grow. However, only those tissues whose patterns were already set by the apical meristem develop normally. Continued patterned development is delayed until a meristem is regenerated.

Providing New, Genetically Healthy Cells

As you learned in Chapter 9, meristematic cells divide cyclically: divisions are followed by cellular enlargement, after which the cells again divide. These divisions push cells out of the apical

meristem, where they become specialized *while remaining meristematic.* That is, cells continue to divide after they're pushed out of the meristem. This is important because it allows meristems to function as a reserve of mitotically young and genetically healthy cells. To understand the significance of this, recall that DNA replication can make mistakes that result in mutations. If uncorrected, these errors would accumulate to lethal levels within a few mitotic cycles. Although cells minimize these mistakes with elaborate proofreading and correction procedures, mutations nevertheless appear at rates of approximately one base-pair per billion base-pairs replicated. If plants grew only from small meristems, the millions or billions of replications necessary to produce all of its cells would produce a useless genome long before the plant could reproduce.

Plants and animals have solved the problem of accumulating mutations in different ways. In plants, apical meristems are

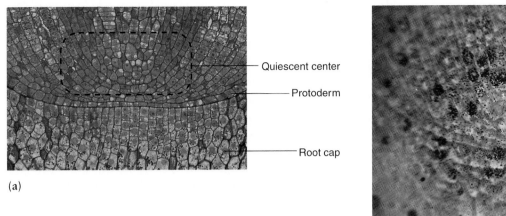

- Quiescent center
- Protoderm
- Root cap

(a)

(b)

100 µm

figure 12.3

The root tip of corn (*Zea mays*). (a) The apical meristem and quiescent center (outlined) are located just behind the root cap. (b) The quiescent center consists of 500–1,000 apparently inactive cells. The apparent inactivity of these cells was tested by botanists using a technique called *autoradiography*. Dividing cells exposed to radioactive thymidine incorporate the thymidine into their DNA. When sections of the labeled root are covered by a photographic emulsion, producing dark grains over dividing cells, the cells of the quiescent center exposed few grains. This indicates that cells of the quiescent center divide much less frequently than do the surrounding cells of the meristem.

---

reading 12.1

MAKING APPLES AND ORANGES

Humans have long exploited the apical meristem's immortality and ability to produce genetically healthy cells. For example, in 1825 a retired brewer named Richard Cox plucked an apple from a tree while walking across an abandoned farm. He was so enamored with the apple's taste that he grafted buds of the apple tree onto other "normal" apple trees to propagate his prized discovery. Subsequent growth of these grafted buds formed stems that produced apples identical to the one he'd discovered during his walk. Similar grafts have been made thousands of times. Today, Cox's orange pippin apples can be traced to the neglected tree that Richard Cox discovered more than 160 years ago. Similarly, the seedless navel orange of California can be traced to a tree that today grows in Riverside, California. This tree was sent to California as a bud-grafted tree discovered by a missionary in Bahia, Brazil, in 1870.

When horticulturists realized that Cox's apples were best-sellers, they tried to hasten the cloning via tissue culture. These experiments failed because the cultured tissues produced numerous new "sports" and "mutants," each having features different from those of Cox's original apple. Why don't similar changes occur in intact meristems? Recent research suggests that mutations *do* arise in meristems; however, the intact meristem recognizes and inhibits the growth of these "rogue" cells. As a result, the apical meristem remains genetically intact and continues to produce clones of the original apple. Apparently this censoring that occurs in intact meristems doesn't occur in cultured tissues, suggesting that recognizing mutant cells depends on the presence of an intact meristem.

---

multistep meristems, in which a small group of initials divides occasionally to provide cells for the rest of the meristematic region (fig. 12.3). These derivatives, rather than cells of the meristem itself, continue to divide for several weeks or months and form the various tissues of the plant body. By not dividing rapidly, the initials ensure that the adjacent meristematic region is constantly supplied with new, genetically healthy cells (see reading 12.1, "Making Apples and Oranges"). In contrast, determinate growth in animals minimizes the effects of these mutations by producing adults without any cells that go through multiple mitotic cycles. Animals set aside their sex cells as reserves early in development; because these cells do not go through many mitotic cycles, their genome usually remains error-free.

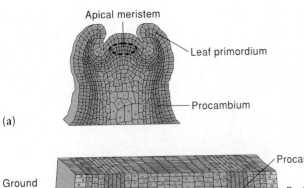

(a)

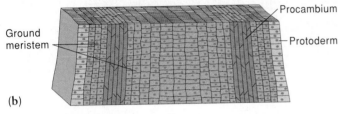

(b)

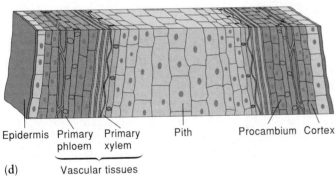

(c)

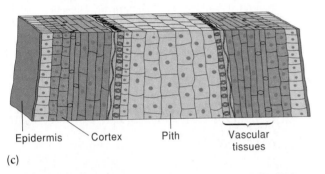

Epidermis   Primary   Primary   Pith   Procambium   Cortex
            phloem   xylem

(d)   Vascular tissues

figure 12.4

The three kinds of transitional meristems in plants. (a)–(d) The protoderm produces the epidermis, while the ground meristem forms the cortex and pith. The procambium forms the primary vascular tissues: primary xylem and primary phloem.

Derivatives of Meristems

Most cell divisions in tips of roots and shoots occur in derivatives of meristems rather than in the meristem's initials. These derivatives of meristems form regions of cellular division and specialization called **transitional meristems.** There are three transitional meristems in plants (fig. 12.4):

1. **Protoderm** is the transitional meristem that forms epidermis. Protodermal cells usually divide only

anticlinally (i.e., perpendicular to the surface) and form a sheet of new epidermal cells that covers the growing root or shoot. Interestingly, the protoderm is irreplaceable: if damaged, no other cells can assume its functions.

2. **Procambium** is the transitional meristem that produces vascular tissues. Procambium differentiates as strands connected to the plant's mature vascular tissues.

3. **Ground meristem** produces ground tissue and consists of two other meristems: the flank meristem and the rib meristem. The *flank meristem* forms the cortex, and the *rib meristem* forms the pith. Cells of the ground meristem divide perpendicular to the stem's or root's axis and form longitudinal files (i.e., ribs) of cells. In shoots, division and subsequent elongation of these cells expand internodes (i.e., lengths of stems between leaves or branches) and separate the shoot's leaves. Dwarf and rosette plants, such as dandelion (*Taraxacum*) and *Agave*, don't have a rib meristem and thus have short internodes and compacted leaves. The hormone gibberellin strongly influences the activity of the rib meristem; applying gibberellin to many rosettes and dwarfs induces internodal elongation (see Chapter 18).

Transitional meristems aren't the only means by which specialized cells can form. For example, masses of callus tissue grown *in vitro* via tissue culture (see fig. 12.8) lack procambium but contain isolated clumps of vascular tissue. Thus, procambium is not necessary for the differentiation of vascular cells. Rather, procambium is important because it forms vascular tissue in *patterned strands* that link the growing apex with mature vascular tissues.

The procambium and ground meristem form in response to developmental signals received from nearby organs and tissues. For example, young leaves control the differentiation of procambium: removing leaves from a shoot apex stops procambial differentiation, and grafting leaves to a mass of callus tissue induces formation of procambium and vascular tissue.

c o n c e p t

Apical meristems are multistep meristems that produce orderly arrangements of new, genetically healthy cells. Most divisions occur in derivatives of meristematic initials called *transitional meristems.* Protoderm, procambium, and ground meristem are transitional meristems that produce epidermis, vascular tissues, and ground tissue, respectively.

How Apical Meristems Are Organized

Botanists have devised several models to describe the organization of apical meristems. One of the earliest models suggested that the three primary tissues of a plant (i.e., epidermal, vascular, and ground tissues) originate from distinct zones in the apex called *histogens.* This model also stated that the developmental fate of a plant cell, like that of many animal cells, is controlled by its lineage. A simple and convincing experiment discredited the histogen model: when regions of the meristem designated as histogens were surgically removed and rearranged, the root or shoot developed normally. Thus, the fate of a plant cell usually isn't determined by its lineage. That is, a cortical cell differentiates from a meristem not because the cell is unique, but because it receives developmental signals that change its metabolism to that of a cortical cell. When the histogen model failed to account for the development of meristems, botanists developed new models to explain how shoots and roots are organized.

Shoot Apical Meristems

Examine the shoot apical meristem shown in figure 12.5. The shoot apical meristem produces a stem and several kinds of lateral appendages, including photosynthetic leaves and modified stems such as thorns and tendrils.

The first step toward our current understanding of shoot apical meristems came in 1924, with the formulation of the **tunica-corpus model.** According to this model, a shoot apex consists of two zones—a tunica and a corpus—each of which is distinguished by its planes of cellular division (fig. 12.6). The **tunica** is the outermost layer(s) of the shoot apex; these layers divide only anticlinally. That is, these divisions occur in one plane and create sheets of cells that cover the apex of the shoot (fig. 12.6). Most angiosperms have one or two tunica layers, the outermost of which becomes protoderm. The **corpus** includes cells of the shoot tip below the tunica. These cells divide in all planes and produce bulk growth. In contrast to the histogen model, the tunica-corpus model is purely descriptive; it cannot be used to predict the developmental fate of individual cells.

The next step in understanding the organization of apical meristems was the discovery that the corpus is not homogeneous. Rather, it contains several zones with distinctive cellular features. To appreciate this, examine figure 12.6, which shows the positions of these zones and their relation to the protoderm, procambium, and ground meristem. The *central mother cells* are large, cuboidal cells in the uppermost zone of the corpus. These cells divide in all planes. Flanking the central mother cells is a cone-shaped *peripheral meristem.* These cells are small and densely cytoplasmic. The peripheral meristem, along with the protoderm and procambium, forms leaves. The leaves form in regular patterns so rapidly that nodes (i.e., points on stems where leaves or branches attach) and internodes are indistinguishable at the apex. Internodal expansion is the function of the peripheral meristem and the *pith-rib meristem,* which forms

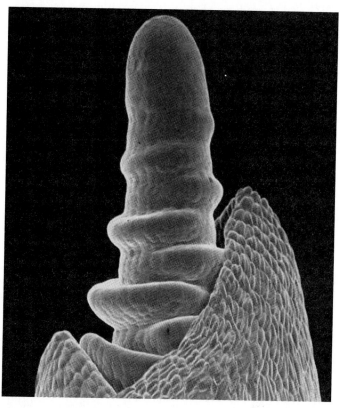

figure 12.5

The shoot apex of wheat (*Triticum*). The apical meristem produces the stem and leaves, which first appear as bulges on the side of the apex. (Scanning electron micrograph, ×200.)

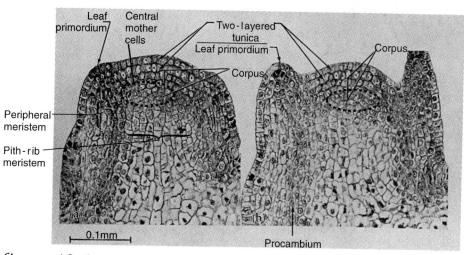

figure 12.6

The shoot apex of potato. (a) The tunica-corpus organization. The two-layered tunica overlies the corpus. (b) The central mother cells form the upper zone of the corpus. Leaves are formed by the procambium and peripheral meristem. The pith-rib meristem (along with the peripheral meristem) is responsible for internodal expansion.

just below the central mother cells. Most cellular divisions in shoot tips occur in expanding internodes.

In plants such as corn (*Zea mays*), the entire shoot apex is mitotically active, and cells flow from the central mother cell region to the peripheral meristem. In other plants, however, the meristems have a central region of inactivity bordered by a ring of mitotically active cells that produces cells for vegetative growth. Cells in the center of the meristem remain inactive and therefore genetically healthy as they wait for an environmental signal (e.g., changing day-length) to trigger their activation.

Root Apical Meristems

The root tip is covered by a slimy *root cap* that protects the root apical meristem and helps the root move through the soil. Behind the root cap is the root apical meristem, at the center of which is the *quiescent center*, which consists of 500–1,000 cells (fig. 12.3). As the name suggests, these cells are relatively inactive. Indeed, cells in root caps divide fifteen times more often than do cells in the quiescent center. The functional basis for these quiescent cells is that the root cap is effective but imperfect at protecting the apical meristem as the root forces its way through soil. As a result, the meristem is often damaged during growth. When this occurs, cells of the quiescent center divide and re-form the meristem and root cap. Thus, the quiescent center, like the central mother cells of plants such as sunflower, is a reservoir of genetically healthy cells. Surrounding the quiescent center are the root's transitional meristems, which produce longitudinal files of cells that differentiate into vascular tissues, cortex, and epidermis.

---

c o n c e p t

Apical meristems are collections of several types of cells that function together to produce roots and shoots. Shoot apical meristems are protected by young leaves, while root apical meristems are protected by the root cap.

Lateral Meristems

Woody plants such as pines, oaks, and magnolias have lateral meristems that produce secondary growth, which increases the girth of the plant. There are two kinds of lateral meristems: the vascular cambium and phellogen (cork cambium). These meristems form a woody secondary body consisting of wood and bark (i.e., all tissues outside the vascular cambium). Wood (secondary xylem) and secondary phloem are derived from the vascular cambium; phellogen forms the periderm. The periderm forms a layer of dead, suberized cells that protects the inner tissues of the secondary plant body.

Secondary growth is important because it provides support that enables plants to grow taller, thereby increasing their chances of intercepting light for photosynthesis. You'll learn more about lateral meristems and secondary growth in Chapter 16.

Plant Growth and Development

Plant growth and development typically involve quantitative increases in the *number* and *size* of cells and qualitative differences in *types* of cells.

Growth

What Is Growth?

Growth can mean several different things, such as an increase in size, volume, weight, or number of cells. For example, a piece of dry wood "grows" (i.e., swells) when placed in water, and a child who is "growing up" is becoming taller. We'll define growth *as any irreversible increase in size of an organism or its parts.* Growth is accompanied by metabolic processes that occur at the expense of metabolic energy. Thus, the production of a leaf is growth, whereas the water-induced swelling of dead wood is not.

Cellular Division and Enlargement

Cellular division usually accompanies cellular enlargement. Meristematic cells repeatedly enlarge and divide, thereby maintaining an average cell-volume in the meristem. However, cellular expansion doesn't trigger division, and cellular division doesn't always accompany enlargement. Therefore, cellular division and enlargement must be regulated separately. Furthermore, cellular division is not necessary for cellular enlargement. This conclusion is supported by experiments involving irradiated wheat seedlings called *gamma plantlets*. You'll recall that such plants were discussed in Chapter 3 with regard to the organismic theory. Irradiation stops DNA synthesis and cellular division but doesn't affect seed germination or growth of the seedlings. Thus, gamma plantlets develop without cellular division; they grow only by cellular enlargement.

Two events affect cellular enlargement:

1. **"Loosening" the cell wall so that it can stretch.** One means by which walls are loosened is by secretion of acids into the cell walls. These acids loosen the cell wall by activating pH-dependent enzymes that break bonds between cellulose molecules in the wall. Cell-wall acidification associated with cellular elongation is strongly influenced by auxin, a plant hormone. You'll learn more about auxin in Chapter 18.

2. **Positive turgor pressure.** Turgor pressure is pressure in a cell resulting from the osmotically driven uptake of water; it occurs primarily in response to increasing concentrations of solutes in the vacuole. Thus, filling the vacuole with water is what "fills up" an expanding cell; for example, cellular expansion in onion roots corresponds to a 30- to 150-fold increase in the size of the vacuole. This means of enlargement is more efficient in terms of energy use than synthesizing an equal volume of protein, enzymes, and organelles. It can also be fast; for example, petioles of a tropical water lily (*Victoria regia*) can lengthen more than 2 cm h^{-1}.

Plants can't grow without cellular enlargement. Since enlargement requires turgor, which, in turn, requires water, plant growth and development are intimately linked with a plant's water status. Fluctuations in water availability also influence metabolism, which controls events such as seed

CELL DIVISION AND PLANT GROWTH

In plants, certain proteins encoded by a group of genes called *cyclin* trigger cell division. To determine how cell division affects the rate of root growth, Peter Doerner and his colleagues at the Salk Institute for Biological Studies devised a clever experiment: they spliced *cyclin1* into a regulatory sequence of DNA, thereby forcing the gene to make its protein throughout the cell cycle instead of just before cell division. They then transferred the gene into *Arabidopsis* and observed the results (reading fig. 12.2):

> The transgenic plants reached a standard size; however, their roots grew 40% faster than did roots of unaltered plants.

> Both transgenic and unaltered plants had normal root patterns.

> Having the *cyclin1* gene active throughout the cell cycle reduced the effectiveness of the mechanism that normally stops growth.

Doerner and his colleagues changed the growth rates of the plants by stimulating the *cyclin1* gene, prompting them to conclude that plant growth depends entirely on cell division. Is this conclusion consistent with data from experiments involving gamma plantlets? Why do you think that only the roots of the transgenic plants grew faster?

 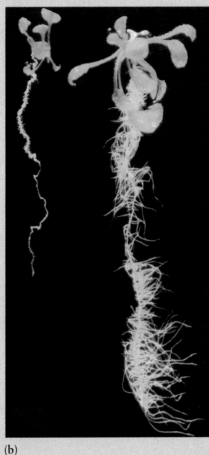

(a)   (b)

reading figure 12.2

Cell division and plant growth. Left to right: (a) Normal and transgenic 7-day-old *Arabidopsis*: (b) 10-day-old normal and transgenic plants treated with a growth promoter.

germination, bud growth, and the cellular elongation that causes roots and shoots to respond to environmental signals such as light and gravity.

Cells often enlarge in only one direction. The direction in which a cell elongates is determined by which cell wall is most elastic, and that, in turn, is determined by the orientation of the wall's cellulose microfibrils (fig. 12.7). We don't know what determines how these microfibrils are deposited in the cell wall, but we do know that microtubules are intimately involved. For example, microtubules near the plasmalemma are oriented in the same direction as microfibrils deposited in the cell wall (fig. 12.7a). Other evidence supporting the dependence of microfibril deposition on microtubules involves experiments using colchicine, a drug that disrupts microtubules but not the deposition of cellulose microfibrils. Cells treated with colchicine lack microtubules, and their cellulose microfibrils are deposited randomly in the cell wall. As a result, these cells become spherical rather than elongate (fig. 12.7b).

Cellular division and enlargement are important for growth but do not by themselves constitute development. To understand this, consider the callus masses growing *in vitro* shown in figure 12.8. These callus masses divide and expand rapidly but do not resemble a plant, because they lack roots, stems, and leaves. Producing these characteristic parts of a plant is referred to as *plant development* (fig. 12.8) and requires cellular specialization; that is, it requires that cells differentiate.

concept

Growth is an irreversible increase in size of an organism or its parts and results from increases in the number and size of cells. Cellular division and elongation are controlled separately. Cellular elongation accounts for most primary growth and requires "loose" cell walls and positive turgor pressure.

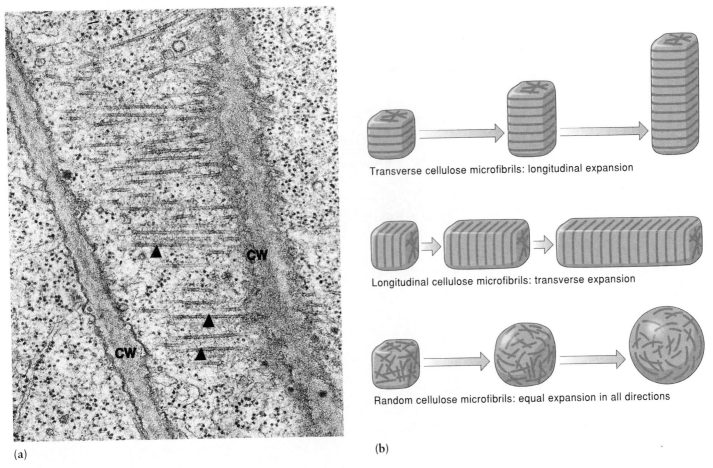

(a)

(b)

figure 12.7

Microfibrils and cellular shape. (a) The microtubules (arrowheads) shown in this electron micrograph are involved in the deposition of cellulose microfibrils in the cell wall (cw). In this cell, how would the microfibrils be arranged? What are the consequences of such an arrangement? (b) How the orientation of microfibrils affects cellular shape.

Development

Roots, stems, and leaves are organized similarly—so much so that their organizing influences must be similar in different groups of plants. For example, vascular tissues consist of the same types of cells in a wide variety of plants. Furthermore, each tissue forms in a characteristic position related to its function. How do cells having similar origins (i.e., from the same meristem) become different?

Cellular Differentiation

Differentiation is an orderly process by which the structure and function of genetically identical cells become different. It occurs throughout a plant's life and is always preceded by complex biochemical changes. Differentiation produces cells, tissues, and organs specialized for different functions.

Much of what we know about differentiation in plants comes from studies using **tissue culture,** a technique for growing fragments of plants in an artificial medium. The basis for tissue culture was proposed in 1902 by the German botanist Gottleib Haberlandt, who suggested that plant cells are totipotent; that is,

each living cell has the full genetic potential of the organism. Gardeners already knew that the organs of many plants were totipotent; they had cultivated potatoes and several other plants for centuries with cuttings that, when planted, grew into entire plants. However, Haberlandt's suggestion extended totipotency to individual cells. Botanists soon began testing Haberlandt's idea: evidence for totipotency would be the regeneration of an entire plant from one or a few nonzygotic cells. The earliest attempts failed; cultured cells remained alive for a while but didn't divide and soon died. There were two possible explanations for these failures:

HYPOTHESIS 1

Plant cells are not totipotent. That is, the cells were genetically incompetent to regenerate a new plant.

HYPOTHESIS 2

The workers' failures were due not to the cells' genetic incompetence, but rather to other factors such as inadequate nutrients and growth conditions. Were the cultured cells being given everything they needed?

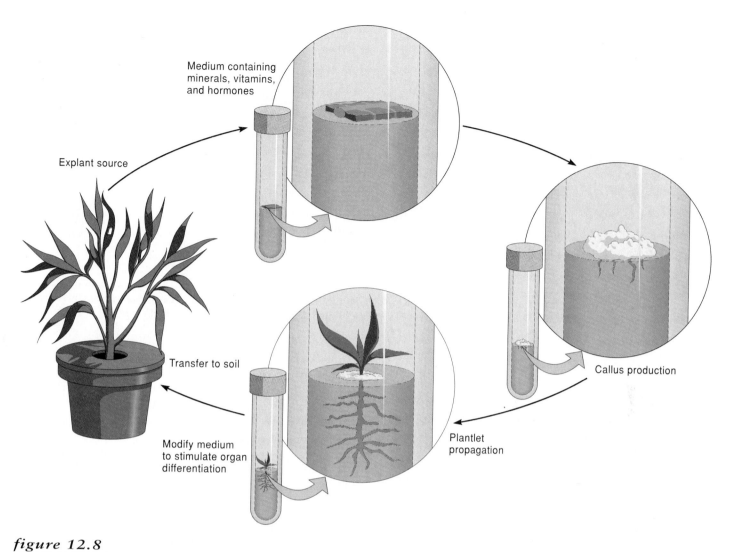

Medium containing
minerals, vitamins,
and hormones

Explant source

Transfer to soil

Modify medium
to stimulate organ
differentiation

Plantlet
propagation

Callus production

figure 12.8

Plant development and propagation by tissue culture.

Continued research finally produced a breakthrough when, in 1958, Cornell University botanist F. C. Steward regenerated a carrot plant from a tiny piece of phloem (see Chapter 9). Why was Steward successful? Like previous workers, he supplied the cultured cells with sugars (i.e., sources of reduced carbon), minerals, and vitamins, but he also added a new ingredient: coconut milk. Coconut milk contains, among other things, a substance that induces cellular division (subsequent research identified this substance as cytokinin, a plant hormone). Once the cultured cells began dividing, they were transplanted to new media, where they formed roots and shoots and developed into plants (fig. 12.8).

Steward's work was important because it showed that plant cells are totipotent and that, given the proper environment, a zygote can be replaced by other types of cells. Today, tissue culture is an important tool for plant biotechnology and

crop improvement. Tissue culture is also an important technique for studying cellular differentiation because it is relatively simple and gives a researcher a great deal of experimental control. Many plants produce a mass of callus tissue when grown in culture; the effects of compounds such as minerals, sugars, and hormones can be tested by observing what happens to the callus tissue when these compounds are added or removed from the growth medium.

c o n c e p t

Differentiation is an orderly process by which the structure and function of genetically identical cells become different. Differentiation results from the differential activation of a cell's genome. Differentiation is important because it specializes cells for different functions.

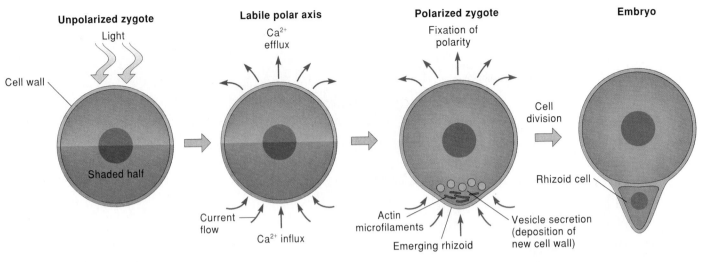

figure 12.9

The development of environmentally controlled polarity in the brown alga *Fucus*. Undirectional light induces a polarity characterized by an influx and efflux of Ca^{2+} in different halves of the zygote. The rootlike rhizoid forms where there was an influx of Ca^{2+} into the zygote.

Dedifferentiation

Wounds disrupt structural patterns in plants. For example, a grazing herbivore might sever the vascular tissue in a stem or leaf, or strong winds could break branches from a tree. When this occurs, cells near the wound respond to the disruption by **dedifferentiating,** meaning that they become meristematic, and prepare for differentiation. Severing a vascular bundle induces divisions of parenchyma cells near the wound, thereby forming a layer of parenchyma that links the severed ends of the vascular bundle. This new parenchyma then differentiates into vascular tissues that reconnect the severed strands. Dedifferentiation is important because it enables plants to repair wounds and to re-form disrupted patterns.

Polarity

Many of the environmental signals that direct plant growth and development are polar (i.e., directional). For example, gravity is a unidirectional force, and light for photosynthesis often comes primarily from one direction. Since these directional signals control plant growth and development, it is not surprising that plants and their parts also exhibit a marked **polarity,** which is a directionality expressed as differences between different sides or ends of a cell, tissue, organ, or organism. For example, the root and shoot ends of a plant's axis are radically different, as are the upper and lower sides of a leaf.

Polarity is controlled by the environment and the plant's genome. The plant's genome ultimately limits growth and development; that is, genes are the ultimate arbiters of the cellular division, enlargement, and differentiation that create polarity. However, these genes are answerable to and activated by environmental signals such as light, gravity, touch, and temperature. For example, aerial roots of *Cissus*, a relative of grapes, don't branch until their tips touch the soil; similarly, the spores of many ferns require light to germinate and grow.

Another example of environmental control of polarity occurs in the zygotes of *Fucus*, a brown alga. *Fucus* zygotes are an ideal experimental system because they are relatively easy to obtain and because they develop freely and individually in seawater. When released, *Fucus* zygotes have a random distribution of organelles (i.e., they are nonpolar). However, when illuminated by light from one direction, a cascade of events soon occurs that produces a remarkable polarity. Here's what happens (fig. 12.9):

1. Half of the zygote becomes densely cytoplasmic and electronegative relative to the other half. This electrical polarity involves an influx of Ca^{2+} at the densely cytoplasmic end and an efflux of Ca^{2+} from the other end, which drives an electrical current across the zygote.

2. When the cell divides, the daughter cell derived from the densely cytoplasmic half of the zygote (i.e., where Ca^{2+} enters) becomes a rootlike rhizoid, while the daughter cell derived from the half of the cell characterized by Ca^{2+} efflux forms a shootlike thallus. These electrical currents and calcium asymmetries determine polarity in *Fucus* zygotes. Rhizoids can be induced to grow from the half of a zygote to which calcium is applied, and placing zygotes in an electrical current determines the site of rhizoid formation (fig. 12.9)

A similar environmentally determined polarity also occurs in other plants; for example, the first cellular divisions in spores of horsetails (*Equisetum*) are always perpendicular to the direction of incoming light. Environmental signals determine the polarity of many plants and their parts.

Polarity can also be inherent and independent of environmental signals. For example, filamentous algae such as *Cladophora* and *Griffithsia* form rhizoids only at their physiological base, regardless of the filament's orientation. If the filament is broken into smaller pieces, each piece

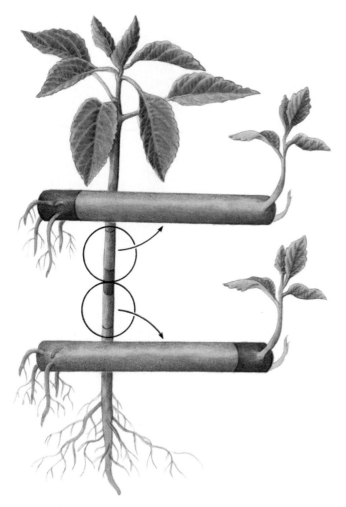

figure 12.10

Polarity of plant tissues. Roots grow from the base of a cutting, while buds grow from the apex, even when the cuttings are kept horizontal and in uniform conditions. Development is determined by the location on the cutting, not on the intact plant.

forms rhizoids only at its physiological base. This polarity extends even to individual cells of the filament, which form a "shoot" cell at their apex and rhizoids at their base. Thus, these filaments are analogous to a bar magnet: they retain their original polarity no matter how many times they are subdivided. This cellular polarity can be changed by centrifuging the cells, suggesting that it results from a differential distribution of a cell's contents.

Polarity in angiosperms is often fixed and difficult to change. To appreciate this, consider the experiment using internodal segments shown in figure 12.10. Although the physiological basal ("root") and apical ("shoot") ends of these segments are structurally indistinguishable, they respond differently during organogenesis: roots always form at the basal end, and shoots always form at the apical end of the cutting, even if the cutting is inverted. Furthermore, this polarity doesn't depend on having an entire internode, since cuttings maintain their original polarity regardless of how many times

the internode is divided. These results indicate that (1) root-shoot polarity is an inherent property of a plant axis and (2) the same part of a plant can produce different apices, depending on its location relative to the rest of the cutting.

What could account for the inherent polarity of angiosperm tissues? Several botanists have suggested that the polarity results from metabolic gradients. For example, the hormone auxin moves polarly in stems from the apex toward the base, regardless of the stem's orientation. The resulting accumulation of auxin (or other metabolite) at the base could induce formation of roots and inhibit formation of buds, and therefore could account for the root-shoot polarity of angiosperms. However, the polarity persists from one season to the next, as well as through dormant periods when concentration gradients would be unlikely to persist. Therefore, phenomena such as polar auxin transport and gradients may *result from* rather than *cause* polarity. Today, many botanists believe that polarity results from a combination of (1) polar transport of morphological signals (e.g., auxin) and (2) formation of a developmental gradient along the plant's axis (e.g., root versus shoot).

c o n c e p t

Plants and their parts are polar, meaning that they have directional differences. In many instances, polarity is determined by environmental signals such as light and gravity, while in others it is inherent and independent of the environment.

The polarity of a tissue or organ reflects the polarity of its individual cells, which often begins with a choreographed rearrangement of the cell's parts. This rearrangement of organelles often precedes another polar event that influences differentiation: asymmetric cellular division.

Asymmetric Cellular Division and Differentiation

Cellular differentiation is often initiated by an asymmetric division, in which the smaller of the two daughter cells differentiates into a specialized structure. For example, asymmetric divisions of epidermal cells of leaves produce a large, vacuolate cell and a smaller, densely cytoplasmic cell (fig. 12.11). The large cell becomes an ordinary epidermal cell. The smaller cell becomes a stomatal mother cell, which then divides equally at a right angle to the first division to produce the two identical guard cells that form the stoma. Stomata regulate gas exchange in leaves.

Similar asymmetric divisions occur throughout plant growth and development. For example, the larger cell from asymmetric divisions that form pollen grains becomes the vegetative cell that forms the pollen tube. The smaller cell becomes the generative cell, which eventually divides equally to produce two identically shaped sperm cells. In asymmetric divisions of the epidermal cells of roots, the larger cell becomes an ordinary epidermal cell, while the smaller cell becomes a root hair.

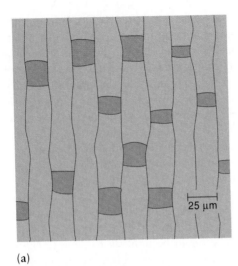

(a)

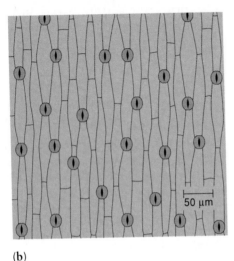

(b)

figure 12.11

Stomatal differentiation in leaves. (a) Early in development, epidermal cells divide asymmetrically. The small cells (shaded) divide to form stomatal mother cells, while the larger cells become ordinary epidermal cells. (b) This type of cell lineage produces regularly spaced stomata. Stomata regulate gas exchange.

It is interesting to note that structures such as stomata and root hairs, whose differentiation is preceded by asymmetric divisions, don't form in gamma plantlets, indicating that asymmetric divisions may be critical to differentiation and polarity. We know little about what controls asymmetric cellular division, but it is strongly influenced by Ca^{2+}. Asymmetric divisions are often preceded by an accumulation of Ca^{2+} in the cytoplasm and cell wall where the division will occur.

Although asymmetric cellular divisions often initiate differentiation, they don't necessarily determine the developmental fate of cells. For example, asymmetric cellular divisions in ferns produce a large and a small cell. The smaller cell becomes a reproductive structure if gibberellins are present, but it becomes a hairlike rhizoid if gibberellins are absent. Therefore, although asymmetric divisions sometimes

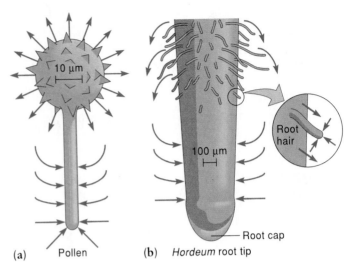

(a) Pollen (b) *Hordeum* root tip

figure 12.12

Growing plants generate electrical currents. (a) Currents enter tips of growing pollen grains and leave laterally behind their tips. (b) Similarly, currents enter tips of growing roots and root hairs. Abolishing these currents abolishes the polarized growth typical of the organ or cell.

may be essential for differentiation, they may not be sufficient for complete differentiation. Additional developmental signals are necessary to direct the process.

Signals That Regulate Plant Growth and Development

In addition to totipotency, Haberlandt suggested that cellular differentiation resulted from differential expression of the cell's genetic potential. This suggestion, like that of totipotency, has been supported by many experiments and today is a foundation of molecular biology and genetics. But if all plant cells have the same potential and if specialization results from differential expression of this potential, what controls which part of a cell's potential is expressed? Stated another way, what internal signals determine the developmental fate of a cell?

The signals that control plant growth and development have intrigued botanists for centuries. Today, we know that these signals are complex and involve electrical, hormonal, positional, biophysical, and genetic controls.

Electrical Currents

Electrical currents influence several aspects of plant development, ranging from the closure of Venus's-flytraps to cellular elongation and polarity. For example, currents enter pollen tubes, roots, and root hairs at their tips and exit laterally behind their tips (fig. 12.12). When these currents are abolished, polarized growth is also abolished. Similarly, growth *table 12.1*

Functions of Plant Hormones

Hormone	Major Functions	Where Produced or Found in Plant
Auxin (such as IAA)	Stimulates stem elongation, root growth, differentiation and branching, apical dominance, development of fruit; instrumental in phototropism and gravitropism	Endosperm and embryo of seed; meristems of apical buds and young leaves
Cytokinins (such as kinetin)	Affect root growth and differentiation; stimulate cell division and growth, germination, and flowering; delay senescence	Synthesized in roots and transported to other organs
Gibberellins (such as GA$_1$)	Promote seed and bud germination, stem elongation, leaf growth; stimulate flowering and development of fruit; affect root growth and differentiation	Meristems of apical buds, roots, and young leaves; embryo
Abscisic acid	Inhibits growth; closes stomata during water stress; counteracts breaking of dormancy	Leaves, stems, green fruit
Ethylene	Promotes fruit ripening; opposes or reduces some effects of auxin; promotes or inhibits growth and development of roots, leaves, flowers, depending on species	Tissues of ripening fruits, nodes of stems, senescent leaves

and development usually change when plants are grown in electrical fields, indicating that electrical currents influence development.

Chemical Signals

Botanists have long known that one part of a plant affects other parts of the plant. For example, removing the shoot apex usually stimulates bud growth, and seeds often germinate faster when they're removed from fruit. These effects have often been attributed to chemical signals called **hormones,** which are organic molecules made in one part of a plant and transported to another part, where they elicit a physiological response.

There are five major groups of plant hormones: auxins, gibberellins, cytokinins, abscisic acid, and ethylene. Several aspects of growth and development are influenced by these hormones (table 12.1). For example, auxin strongly affects the differentiation of procambium into young leaves. Here's how we think it happens:

1. Young leaves produce large amounts of auxin.

2. This auxin moves polarly down the stem.

3. Auxin moving down the stem induces differentiation of vascular tissues.

As you would predict, replacing leaves of a shoot apex with auxin stimulates differentiation of procambium. Furthermore, applying auxin to a callus mass induces differentiation of procambium at the site of application. These observations suggest that auxin formed in leaves and transported down the stem stimulates the differentiation of procambium into the leaf. Thus, young leaves use auxin production to control the differentiation of vascular tissues for service and support.

Each of the five types of hormones has many different and redundant effects in different tissues (e.g., auxin can stimulate or inhibit cellular enlargement and, like cytokinin, often stimulates cellular division). Thus, plant hormones are not specific like animal hormones,[2] and probably no phase of plant growth and development is controlled exclusively by one hormone. Rather, hormones are probably integrating agents that are necessary for but do not control a particular response. For example, cytokinins are necessary to break the dormancy of buds, but they do not control the subsequent growth of the bud.

The influences of hormones are nonspecific and are influenced by other hormones and other substances, such as calcium. For example, asymmetries of Ca^{2+} affect events ranging from cellular elongation to responses to light and gravity. Similarly, pollen tubes have a tip-to-base gradient of Ca^{2+}, with the highest concentration of Ca^{2+} in the growing tip. Disrupting this gradient of Ca^{2+} abolishes polarity and polar growth of the pollen tube. Similarly, artificially establishing a gradient of Ca^{2+} across tips of roots or shoots induces differential growth and curvature.[3]

The effects of hormones are largely a function of their targets: plant hormones affect growth and differentiation of cells that are programmed to differentiate in a certain way. This programming is usually a function of the cell's position and biophysical constraints.

Positional Controls

Cellular differentiation and polarity often create beautiful patterns in plants (fig. 12.13). Stomata in many plants are arranged in regular arrays, and leaves form in spirals so uniform that they've intrigued botanists, mathematicians, philosophers, and artists for centuries. How do these patterns form? Since plant cells don't move after they're produced, they must receive *positional signals* that inform them of their position so that they "know" where they are in the developing pattern. Cells then respond to these signals by becoming specialized (i.e., by differentiating) for their particular position in the pattern.

A cell's position also influences its rate of division and enlargement. Cells of growing carrot embryos, for instance, divide approximately twice per day. If they maintained this rate of cell division, after four weeks there would be about 2^{28} cells (about the number of cells in a mature carrot); if it continued another two weeks, the carrot would have about 2^{42} cells and weigh 2.5 tons. Obviously, carrots don't mature in only a month, and they seldom weigh more than about 150 g. The reason for these differences is that cells usually divide only when they're in an embryo or a meristem. Thus, although restricting divisions to embryos and meristems limits growth, it also ensures that other cells are present to protect and service the fragile (and vulnerable) dividing cells.

You've probably already thought of several questions about the positional signals that control differentiation. For example, what are the signals? How do they arise? And how is positional information transmitted? Unfortunately, we have few answers to these questions.

Biophysical Controls

Physical pressure generated by growing organs affects plant growth and development. For example, roots and shoots are cylinders bound by an epidermis. Cellulose microfibrils in the outer walls of epidermal cells function as reinforcing hoops and direct the growth of these cylindrical organs (fig. 12.14). Altering the arrangement of cellulose microfibrils in epidermal cells alters the directional pressure, which, in turn, alters growth and development. For example, one of the earliest signs of leaf initiation is altered deposition of microfibrils in protodermal cells of the shoot apex. This alteration creates a "weak spot" from which the leaf primordium bulges and grows.

Genetic Controls

Although botanists have long assumed that genetic controls underlie the specialization of cells in meristems, until recently no such developmental genes had been discovered. In 1989, however, a team of scientists led by Sarah Hake discovered a gene that causes odd growths (knots) on the leaves of maize (*Zea mays*). To Hake's surprise, the DNA sequence of this gene, called *knotted-1* or *kn1*, included a segment that was already known in the developmental genes of fruit flies. By early 1994, Hake and her colleagues had found several other maize genes containing such segments.

In fruit flies, developmental genes determine which cells in the early embryo will form legs, antennae, or other body parts. By analogy, Hake suggests that *kn1* and similar genes in maize dictate which cells in a meristem will form leaves or stems. Hake's evidence for this hypothesis is that when the maize *kn1* gene is inserted into a tobacco plant, the tobacco leaf cells behave like meristems and sprout new stems and leaves. However, just how the *kn1* gene controls the meristem and how such genes are activated only in meristems remain to be explained.

---

2. This lack of specificity has prompted many botanists to question the validity of plant hormones as regulators of plant growth and development. This and other aspects of plant hormones are discussed in Chapters 18 and 19.

3. Calcium ions (Ca^{2+}) regulate many aspects of growth and development, including responses to gravity and light. Ca^{2+} exerts these effects by activating calmodulin, a small protein that affects several enzymes (see the discussion on p. 86).

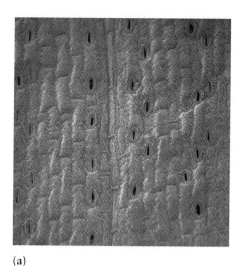  

(a)   (b)   (c)

figure 12.13

Patterns in plants resulting from cellular differentiation and polarity. (a) The regular arrangement of stomata on leaves of corn (*Zea mays*). (b) Spiral arrangement of leaves in sedge. (c) Sunflower.

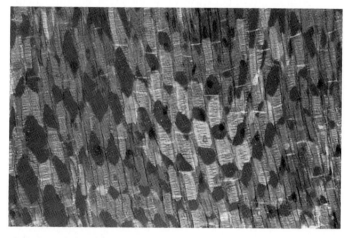

figure 12.14

Hoop reinforcement in roots of *Sprekelia* as seen with polarized light. In this longitudinal section, the files of cells are sectioned obliquely, so that bands of cell wall alternate with bands of cytoplasm. The cell walls are lighter than the background, indicating that most of their cellulose microfibrils are oriented transversely.

c o n c e p t

Electrical currents, hormones, position, and biophysical constraints are developmental signals that control growth and development of plant cells. Genetic controls of plant development resemble those of animals.

Competence of Cells to Respond to Developmental Signals

Although presumably totipotent, not all plant cells are competent to respond to developmental signals. The particular effect on any of these signals is also influenced by neighboring cells and the timing of developmental signals.

Neighboring Cells

Isolated cells and small pieces of callus (i.e., pieces less than 100–200 μm in diameter) cannot differentiate, regardless of what's in the growth medium. Competence to form roots and shoots usually develops only after a callus mass enlarges to a minimum size. Therefore, *competence to differentiate is an acquired condition and typically requires the support of neighboring cells.*

Timing of Developmental Signals

The timing of developmental signals strongly influences plant growth and development. For example, there are two meristems at the base of each leaf of a spike moss (*Selaginella*): the ventral meristem typically forms a root, and the dorsal meristem forms a shoot. Applying auxin to the dorsal meristem early in development causes it to produce a root, but similar applications late in development have little effect. Furthermore, an auxin-transformed meristem reverts to producing a shoot unless it is continually supplied with auxin. Therefore, *there is a window of time during which developmental signals can be received; these signals fix the developmental fate in a stable but not necessarily irreversible way.*

c o n c e p t

The competence of a cell to respond to developmental signals is influenced by neighboring cells and the timing of developmental signals.

The Lore of Plants

Ornamental plants are often prized for their oddities. One such oddity comes from unusual meristematic activity in cacti that causes the formation of broad crests at the tops of stems that make the plants look like some sort of spiny-headed aliens. These crests arise because of multiple apical meristems, which cause divisions to occur primarily in one direction instead of two. The origin of multiple apical meristems is one of the unsolved puzzles of plant development.

Modular Plant Growth

Plants are modular organisms made of standardized parts: roots, shoots, and leaves. These modules are produced over and over by meristems, linked by vascular tissue, and protected by a waterproof epidermis. Once formed, each module follows a do-it-yourself strategy: each develops somewhat autonomously, which enables it to adapt to its particular environment.

Modules of plants cooperate and help plants exploit a patchy environment. For example, the twinflower (*Linnaea borealis*) is a pretty plant that grows in northern forests. As it grows across the forest floor, cooperation between modules allows roots to develop in a nutrient-rich hollow or leaves to exploit a spot of sun. Such cooperation occurs within as well as between plants. For example, Douglas fir trees and fig trees—even those of different species—graft together to form a giant supertree in which each module (i.e., tree) shares its resources with adjacent grafted trees. This cooperation accounts for why, after a patch of Douglas firs is cut down, many of the stumps remain alive for more than 20 years, sustained by root grafts to the remaining trees.

Although cooperation between plant modules helps plants survive, it has a price: exploitation by individuals that break developmental "rules." For example, mistletoes such as *Phoradendron* and *Viscum*[4] are parasitic plants that tap a host's vascular tissue, much like a developing leaf. The host plant can't recognize the incursion as that of a parasite, and so the mistletoe is accepted as another module of the plant. However, the mistletoe has a genome different from its host and therefore doesn't behave according to the host's developmental rules. Mistletoe leaves open their stomata very wide—much wider than stomata of their host—and therefore lose water faster than leaves of the host. As a result, the parasite receives a disproportionate share of the host's xylem sap and, with it, a disproportionate share of the nutrients carried in the sap. The vulnerability of vascular tissues to parasites such as mistletoe suggests that a plant's vascular system has few if any mechanisms for excluding parasites. However, the stability that plants gain from a modular strategy apparently outweighs the risks associated with this vulnerability.

---

4. *Viscum* was worshipped by Druids, an order of priests assisting in religious rites in ancient Gaul, Ireland, and Britain. At certain times—probably at Midsummer Eve—Druids harvested *Viscum*. The mistletoe was cut from the parasitized tree with a golden sickle and caught in a white cloth; only then were the animals (and even humans) killed and burned. In contemporary Europe, there are many remnants of the Druidic ceremony, although they are rarely recognized for what they are. The Druidic rituals have been replaced by the newer and more sophisticated mythology of Christmas, in which the pagan mistletoe often plays a part.

Tabasco (meaning "land where soil is humid") brand pepper sauce is made at Avery Island, Louisiana. Peppers that are hand-picked before 3:30 P.M. each day are mashed and mixed with Avery Island salt and fermented in white oak barrels for 3 years. The fermented peppers are then mixed with white vinegar, the seeds are strained out, and the sauce is put into bottles. (The sauce was originally sold in cast-off cologne bottles.)

The Avery Island factory produces almost 80 million 2 oz bottles of Tabasco sauce per year, generating more than $50 million in sales throughout the world. Labels for Tabasco sauce have been printed in fifteen different languages, and seeds extracted from the original lineage of pepper plants are kept in a bank vault.

If you're ever near Avery Island, stop by the factory for a free tour.

Chapter Summary

Plants have a simple structure. They consist of three kinds of tissues: epidermal, vascular, and ground tissues. These tissues form leaves, roots, and stems, the vegetative organs of plants. Organs cooperate with each other and are adapted to their particular environment.

Plants grow throughout their lives from perpetually embryonic regions called meristems. Apical meristems occur near tips of roots and shoots. Elongation of cells produced by apical meristems produces primary growth. Axillary buds are reiterations of the apical meristem left behind by primary growth; they occur in leaf axils and, after a period of dormancy, can form a branch or flower. Lateral meristems occur only in woody plants and increase the plant's girth.

Apical meristems establish patterns and produce new, genetically healthy cells. Apical meristems are multistep meristems, meaning that a small group of meristematic initials divides and provides cells for the rest of the meristem. Most divisions occur in derivatives of meristematic initials called transitional meristems. Protoderm, procambium, and ground meristem are transitional meristems that produce epidermis, vascular tissues, and ground tissue, respectively.

Apical meristems are collections of several distinct meristems that function together to produce roots and shoots. Botanists use the tunica-corpus model to describe the organization of shoot apical meristems. The tunica is the outermost layer(s) of the meristem; these cells divide only anticlinally (i.e., perpendicular to the surface). The corpus includes cells of the shoot apex below the tunica and is not homogeneous. Central mother cells occur in the uppermost part of the corpus and are surrounded by a cone-shaped peripheral meristem that, with protoderm, produces leaves. Internodal expansion is a function of the pith-rib meristem, which is located just below the central mother cells. The shoot apical meristem is protected by leaf primordia, while the root apical meristem is covered by a protective root cap. Cells in the center of the root apical meristem are usually inactive and comprise the quiescent center. Transitional meristems adjacent to the quiescent center produce the epidermis, vascular tissues, and ground tissue (i.e., cortex) of the root.

Growth is an irreversible increase in the size of an organism or its parts. In multicellular plants, growth results primarily from cellular elongation, which requires a "loosened" cell wall and positive turgor pressure. Cells often elongate primarily in one direction, which is determined by the orientation of cellulose microfibrils in the cell wall.

Differentiation is an orderly process by which the structure and function of genetically identical cells become different and results from the differential activation of a cell's genome. Asymmetric divisions often precede cellular differentiation, which depends more on a cell's position than its lineage. Wounds typically induce dedifferentiation, meaning that the cells resume meristematic activity, reprogram their genome, and prepare to differentiate. Dedifferentiation is important because it enables plants to repair wounds and to re-form disrupted patterns.

Plants and their parts are polar, meaning that they have directional differences. Polarity in plants is often determined by environmental signals such as light and gravity, while in other instances it is inherent and independent of the environment.

Electrical currents, hormones, a cell's position, biophysical conditions, and genes control plant development. The competence of a cell to respond to these signals is influenced by neighboring cells and the timing of the signal.

What Are Botanists Doing?

Conduct a computer-aided search of scientific journals to find what research has been done in the past 5 years on the effects of irradiation on plants. Does any of this research relate to earlier experiments involving gamma plantlets of wheat? If so, how?

 Writing to Learn Botany

In the 1800s, German plant anatomist Anton de Bary said, "The plant forms cells, not cells the plant." What does this mean, and how is it related to plant growth and development?

Questions for Further Thought and Study

1. A primary function of apical meristems is to provide new, mitotically healthy cells. How do apical meristems accomplish this?

2. Is pattern formation unique to living organisms? Give examples to support your answer.

3. What are meristems, and why are they important?

4. What is indeterminate growth, and why is it important for plant growth and development?

5. The differentiation of procambium is influenced strongly by young leaves. Why is this significant?

6. What are transitional meristems, and why are they important?

7. What is the role of asymmetric cellular division in cellular differentiation?

8. What is polarity? Discuss several examples of polarity in plants.

9. Discuss how electrical, hormonal, positional, and biophysical controls affect plant development.

10. Discuss the concept of modular growth in plants.

11. How is a callus mass like a meristem? How is it different?

12. What is dedifferentiation, and why is it important?

13. Leaves of many grasses grow from meristems located at the base of the leaf. Why don't trees grow from basal meristems at the base of their trunks?

14. The idea that the body plan of plants is established in the embryo, as occurs in animals, contradicts the traditional view that plants have an "open" and "plastic" type of development. At first glance, plants grow by adding new structures produced by apical meristems located at opposite ends of the plant's axis. However, recent evidence suggests that the primary plant body plan is generated in the embryo and that meristems are merely terminal elements of the embryonic axis. How would you test this idea?

Web Sites

Review the "Doing Botany Yourself" essay and assignments for Chapter 12 on the *Botany Home Page*. What experiments would you do to test the hypotheses? What data can you gather on the Web to help you refine your experiments? Here are some other sites that you may find interesting:

http://www.ub.gu.se/Gb/res/rese.html

This site includes a variety of links to sites containing information about plant hormones and research data (e.g., the American Society of Plant Physiology).

http://www.yahoo.com/Science/Biology/Botany

This site includes links to a variety of botany-related sites, including botanical gardens, databases, software, and research journals.

http://mollie.berkeley.edu/NSF_Center/

The National Science Foundation's Center of Plant Developmental Biology is an interdisciplinary facility at the University of California at Berkeley.

News:sci.bio.botany

Exchange ideas with botanists all over the world in this newsgroup. Post your own questions and see who responds!

Suggested Readings

Articles

Aloni, Roni. 1987. Differentiation of vascular tissues. *Annual Review of Plant Physiology* 38:179–204.

Cosgrove, Daniel. 1986. Biophysical control of plant cell growth. *Annual Review of Plant Physiology* 37:377–405.

Hardwick, Richard. 1986. Construction kits for modular plants. *New Scientists* (April 10):39–42.

Jürgens, G. 1992. Genes to greens: Embryonic pattern formation in plants. *Science* 256:487–488.

Schnepf, E. 1986. Cellular polarity. *Annual Review of Plant Physiology* 37:23–47.

Books

Barlow, P. W., and D. J. Carr. 1984. *Positional Controls in Plant Development.* Cambridge: Cambridge University Press.

Galston, Arthur W. 1994. *Life Processes of Plants.* New York: Scientific American.

Lyndon, R. F. 1990. *Plant Development: The Cellular Basis.* London: Unwin Hyman.

Mauseth, James D. 1988. *Plant Anatomy.* Menlo Park, CA: Benjamin/Cummings.

Sachs, T. 1991. *Pattern Formation in Plant Tissues.* Cambridge: Cambridge University Press.

Smith, R. 1992. *Plant Tissue Culture: Techniques and Experiments.* San Diego: Academic Press.

Steeves, T. A., and I. M. Sussex. 1989. *Patterns in Plant Development,* 2d ed. Cambridge: Cambridge University Press.

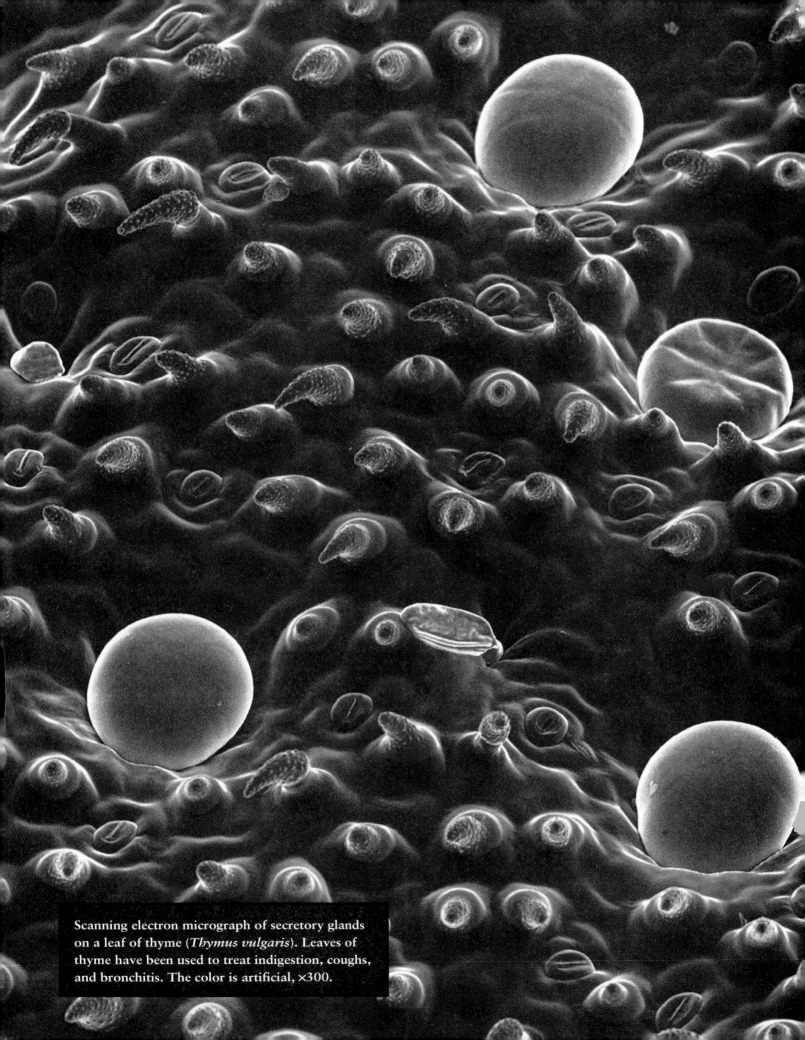

Scanning electron micrograph of secretory glands on a leaf of thyme (*Thymus vulgaris*). Leaves of thyme have been used to treat indigestion, coughs, and bronchitis. The color is artificial, ×300.

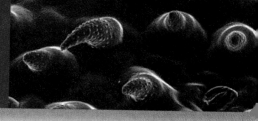

c h a p t e r

Primary Growth: Cells and Tissues

13

Chapter Outline

Chapter Overview

Photosynthetic cells of land plants have requirements similar to those of the green algae from which they evolved. Foremost among these requirements is an aquatic cellular environment, since unprotected cells, whether of algae or land plants, desiccate and die within minutes after being exposed to dry air. Thus, the survival of plants on land depends on their ability to establish an aquatic cellular environment in the dry land environment. Establishing this environment is a complex problem and requires specializations for gas exchange, ventilation, support, water retention, and protection. Plant tissues are the evolutionary solutions to these problems. For example, the epidermis waterproofs the plant and regulates gas exchange, vascular tissues move water and solutes throughout the plant, and secretory structures seal wounds and deter herbivores. Together, these tissues enable plants to grow in the hostile land environment.

Chapter 12 stressed two basic features of plants: (1) their localized growth via meristems, and (2) structural and functional specializations that occur via cellular differentiation. Taken together, these processes produce the **primary body** of a plant, which is an axis consisting of a root and shoot. This axis is made of **primary tissues,** which are groups of cells having a common structure or function. In this and the following two chapters, we'll examine the structure and function of these various cells and tissues, and how they are integrated in the primary body of a plant. We'll consider secondary growth in a separate chapter because it begins only after the primary plant body has formed. Secondary growth produces unique cells and tissues, redirects the plant's resources for growth, and does not occur in all plants.

The primary body of plants consists of four tissues: meristems, ground tissue, dermal tissue, and vascular tissues. We've already discussed meristems (see Chapter 12), which are localized regions of cellular division that form a plant's cells. Like meristems, the ground, dermal, and vascular tissues consist of distinctive types of cells and usually have more than one function. For example, epidermal tissue absorbs nutrients and protects plants from desiccation, and vascular tissues provide support and move water and solutes throughout the plant. The functions of each of these tissues represent evolutionary adaptations to life on land. Indeed, plants are designed much like buildings: they have supporting structures, ventilating and plumbing systems, storage areas, and protective coverings. These life-support activities in plants are performed by the ground, vascular, and dermal tissues.

Ground Tissue

Ground tissue differentiates from the ground meristem and constitutes most of the primary body of a plant. Ground tissue occurs throughout a plant, including other tissues such as vascular tissues. The cortex and pith of stems and roots consist almost entirely of ground tissue. Ground tissue has several functions, including storage, basic metabolism, and support. These functions are performed by its three kinds of cells: parenchyma, collenchyma, and sclerenchyma.

Parenchyma

Parenchyma (from the Greek words *para,* meaning "beside," and *en + chein,* meaning "to pour in") cells are the most abundant and versatile cells in plants. They are identified by their relatively inconspicuous structure; parenchyma cells have few distinctive structural characteristics. In practice, botanists classify as parenchyma any cell not assignable to any other structural or functional class. The shortcomings of this classification scheme become apparent when we consider some of the diverse functions of parenchyma cells:

- **Storage.** Plants store nutrients in parenchyma cells. Most nutrients in plants such as corn and potatoes are contained in starch-laden parenchyma cells (fig. 13.1). Similarly, the thick primary walls of storage parenchyma cells in the seeds of plants such as persimmon (*Diospyros virginiana*), date palm (*Phoenix dactylifera*), and coffee (*Coffea arabica*) contain hemicellulose, which is used as an energy source by germinating embryos.

- **Basic metabolism.** Parenchyma cells are the primary sites of the metabolic functions that you learned about in earlier chapters of this textbook, including photosynthesis, respiration, and protein synthesis.

Clearly, all parenchyma cells are not homogeneous; it is only our lack of understanding that prompts us to group such functionally diverse cells as a single cell type.

The more specialized cells of plants evolved from parenchyma cells, which are therefore considered to be phylogenetically primitive. As their name suggests, parenchyma cells usually form the matrix in which the other, more specialized cells are embedded. However, parenchyma cells are not merely filler, as evidenced by their many functions. Moreover, parenchyma cells can **dedifferentiate** and then **redifferentiate,** meaning that they can change activities and become more specialized (see Chapter 12). For example, wounding often stimulates parenchyma cells to divide and form masses of undifferentiated cells from which roots develop on a cutting (fig. 13.2). The ability of parenchyma cells to dedifferentiate and redifferentiate is important because it is a primary way that plants develop and adapt to various influences such as wounding and changing environments. Indeed, parenchyma cells are the "ready reserves" from which a plant makes specialized cells to meet its changing needs.

Parenchyma cells are made by all of a plant's meristems and occur throughout the plant body. They comprise the photosynthetic tissue of a leaf, the flesh of fruit, and the storage tissue of roots and seeds. Parenchyma cells are alive at maturity and usually have only a primary cell wall. Mature parenchyma cells are more or less isodiametric (i.e., have equal diameters in all directions; see fig. 13.1b). Finally, parenchyma cells often have large vacuoles. When full of water, these vacuoles make the cells turgid—this is what makes the lettuce in your salad crisp (see Chapter 4).

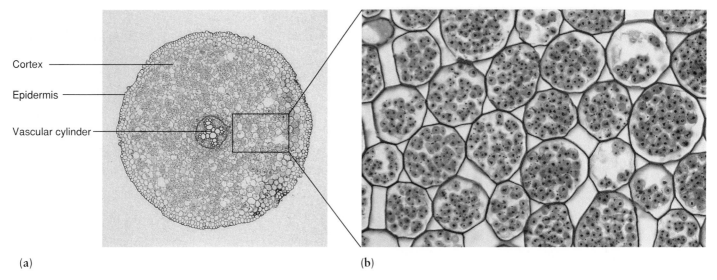

Cortex

Epidermis

Vascular cylinder

(a)

(b)

figure 13.1

Transverse sections of the root of a buttercup (*Ranunculus*). (a) Overall view of mature root. The vascular cylinder includes tissues specialized for long-distance transport of water and solutes, while the epidermis forms a protective outer layer of the root, ×16. (b) Detail of cortex, ×250. Each parenchyma cell in the cortex contains many amyloplasts, which store starch.

Chlorenchyma cells are chloroplast-containing parenchyma cells specialized for photosynthesis (fig. 13.3). Another kind of specialized parenchyma tissue is *aerenchyma* (fig. 13.4), which is characterized by prominent intercellular spaces. These spaces improve the gas-exchange capacity of the tissue and, not surprisingly, usually occur in or near metabolically active tissues. For example, the spongy mesophyll of leaves is aerenchyma that promotes the gas exchange needed for photosynthesis (fig. 13.4b). From a structural perspective, aerenchyma like that in *Juncus* stems (fig. 13.4a) provides maximum support with a minimum metabolic requirement.

Transfer cells are parenchyma cells specialized for short-distance transport of solutes. To complement this function, transfer cells have highly convoluted, nonlignified, cell walls (fig. 13.5). Because the plasmalemma conforms to the shape of the cell wall, these invaginations of the cell walls increase the surface area of the plasmalemma as much as twentyfold, thereby increasing the cell's capacity for transport. Transfer cells occur in areas of high solute transport, such as in secretory glands of carnivorous plants, in nectar-secreting tissues of flowers, and along the conducting cells of xylem and phloem.

c o n c e p t

Parenchyma cells are the primary components of ground tissue and are the most abundant and least structurally specialized cells in plants. Parenchyma cells are the site of the basic functions of plants. Parenchyma cells can form other, more specialized tissues.

Collenchyma

Collenchyma (from the Greek word *kolla*, meaning "glue") cells are elongate (up to 2 mm long) cells having unevenly thickened primary cell walls (fig. 13.6). They support growing regions of

figure 13.2

Roots at the base of a plant cutting. Wounding induces parenchyma cells near the cut surface to dedifferentiate into meristematic cells. These meristematic cells then divide and produce root apical meristems that form roots on the cutting.

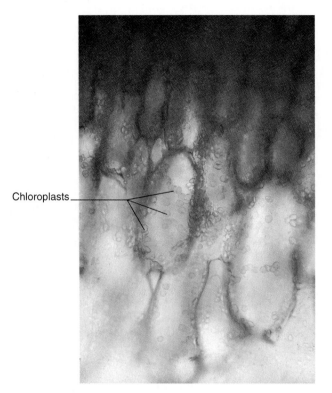

Chloroplasts

figure 13.3

Chloroplast-containing parenchyma cells are termed *chlorenchyma*. This light micrograph of a cross section of a leaf of *Frithia* shows chloroplasts in chlorenchyma cells.

(a)

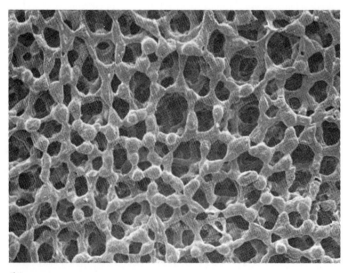

(b)

figure 13.4

Aerenchyma tissue, a type of parenchyma specialized for gas exchange. (a) Light micrograph of aerenchyma from *Juncus*, a plant that often grows along the edges of ponds and lakes, ×400. Note the large amount of intercellular space. Aerenchyma in this plant provides support while maximizing gas exchange. (b) Aerenchyma tissue in leaves (shown here in the broad bean) is spongy mesophyll, a type of chlorenchyma. The numerous intercellular spaces increase gas exchange for photosynthesis in the chlorenchyma cells.

shoots and are therefore common in expanding leaves, petioles, and elongating stems. Collenchyma cells are exquisitely adapted for support: their nonlignified cell walls can stretch, thereby enabling the cells to elongate as the tissue they are a part of elongates. Furthermore, collenchyma cells often differentiate in strands or as a cylinder beneath the epidermis. This arrangement maximizes support, because a cylinder provides more support than does a rod located in the center of a stem or petiole. You're already familiar with strands of collenchyma—they are the resilient strings in petioles of celery (*Apium graveolens*).

Collenchyma cells differentiate from parenchyma cells and are alive at maturity. Their differentiation is strongly influenced by mechanical stress. For example, celery plants shaken for 9 hours per day for a month have twice as much collenchyma as do unshaken controls. Furthermore, the walls of collenchyma cells in shaken plants are 40%–100% thicker than those of controls. This stimulation of collenchyma differentiation by mechanical stress helps a plant organ remain upright despite environmental disturbances such as wind and rain.

Sclerenchyma

Sclerenchyma (from the Greek word *skleros*, meaning "hard") cells are rigid and have thick, nonstretchable secondary cell walls (figs. 13.7–13.10). They support and strengthen nonextending regions of plants such as mature stems and are usually dead at maturity. Thus, the support provided by sclerenchyma

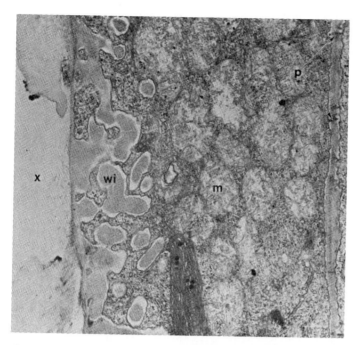

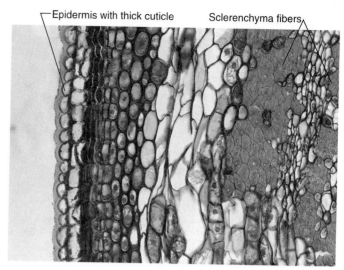

figure 13.7

Sclerenchyma fibers from a stem of basswood (*Tilia americana*) seen in cross-sectional view, ×250. Sclerenchyma fibers have thick walls and usually occur in groups, as shown here.

figure 13.5

Part of a transfer cell in the shoot of *Galium aparine,* showing wall ingrowths (wi). These ingrowths increase the surface area available for transport. (x = xylary element, a cell that conducts water and dissolved minerals; m = mitochondrion; p = plasmodesmata.)

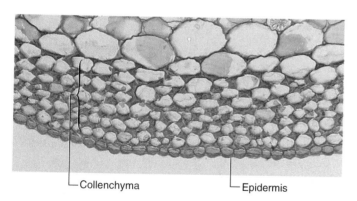

figure 13.6

Light micrograph of a transverse section of collenchyma tissue from a petiole of rhubarb (*Rheum rhaponticum*). In fresh tissue like this, the unevenly thickened collenchyma cell walls have a glistening appearance. Collenchyma supports growing regions of plants.

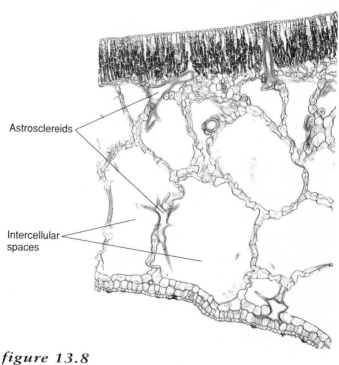

figure 13.8

Branched sclereid (also called an astrosclereid) from a leaf of water lily (*Nymphaea odorata*), ×250. Branched sclereids such as this are common in petioles and leaves of many plants.

cells is attributable to its cell-wall "skeleton," which is produced before the cell dies. Sclerenchyma cells occur in all mature parts of plants, including leaves, stems, roots, and bark.

There are two types of sclerenchyma cells: sclereids and fibers, both of which differentiate from parenchyma. Sclereids and fibers are distinguishable on the basis of cellular shape and grouping. Sclereids are relatively short, have variable shapes, and usually occur singly or in small groups (figs. 13.8–13.9), while fibers are long, slender cells typically occurring in strands (fig. 13.7).

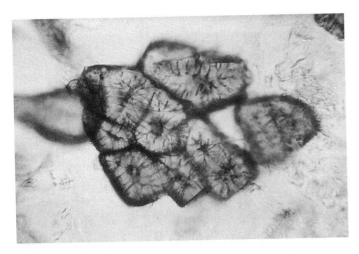

figure 13.9

Sclereids in fruits of pear (*Pyrus communis*) are called brachysclereids. These sclereids give pears their gritty texture.

Sclereids occur throughout plants, including in roots, leaves, stems, seed coats, and even in the hulls of peanuts (*Arachis*). They often form hard layers; for example, the tough core of an apple (*Malus*) consists mostly of sclereids, as does the shell of a walnut (*Juglans*); sclereids also produce the gritty texture of pears (*Pyrus*; fig. 13.9). The differentiation of sclereids is influenced strongly by wounding and by the position of the cell. Sclereids often occur at vein endings and near the edges of leaves.

Fibers are long, slender cells occurring in single strands or bundles often associated with vascular tissue (figs. 13.7, 13.10). They differentiate from parenchyma or the vascular cambium and vary in length; for example, fibers in sisal (*Agave sisalana*) are 1–8 mm long, while those of ramie (*Boehmeria nivea*) may be over half a meter long. Fibers in stems elongate as internodes grow, after which they deposit a thick secondary wall that occupies as much as 90% of the cell's volume.

Several factors influence the differentiation of fibers. For example, fibers do not form in stems or petioles when leaves are removed, suggesting that their differentiation is promoted by a transmittable substance emanating from leaves. This substance is probably not auxin (see Chapter 18), since applying auxin does not replace the differentiating effects of leaves. Rather, gibberellins (a type of plant hormone that you'll learn more about in Chapter 18) may influence fiber differentiation. Jute (*Corchorus capsularis*) and hemp (*Cannabis sativa*) plants treated with gibberellin have more fibers than do untreated controls; these fibers are thicker and up to four times longer than those in untreated plants. As is true for collenchyma, mechanical stress promotes the differentiation of fibers; plants grown on shakers contain more fibers than do unshaken controls.

Classification of Fibers

Fibers are classified in several ways. One common classification scheme is based on their location. Fibers in xylem are called

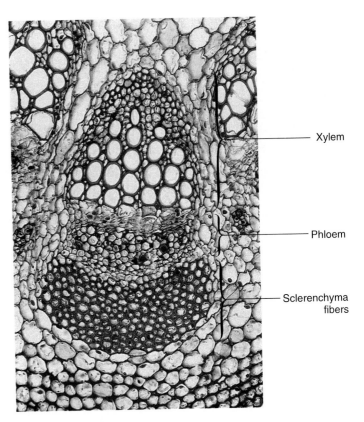

figure 13.10

Cross-sectional view of fibers associated with the vascular tissue of a sunflower stem, ×485. Note that the walls of fibers are much thicker than those of adjacent cells.

xylary fibers, while those that occur in tissues other than the xylem are called *extraxylary fibers*. Extraxylary fibers are usually longer than xylary fibers and are used in fabrics such as linen.

Commercial fibers are also classified according to their hardness. Fibers harvested from most monocots also include xylem, thereby making them lignified and stiff. Consequently, these fibers are called *hard fibers*. Conversely, fibers harvested from many dicots are called *soft fibers* because they lack lignin or contain only small amounts of it, which makes them more flexible than hard fibers. The consequences of this difference in cell-wall structure are easily appreciated by comparing two products made of fibers: coarse rope and linen cloth. Coarse rope is made from hard, lignified fibers of sisal (*Agave sisalana*), whereas fine linen is woven from phloem fibers of flax (*Linum usitatissimum*), which are almost exclusively cellulose. In general, dicot fibers are stronger and more durable than monocot fibers.

Uses of Fibers

Humans have used fibers for more than 10,000 years. People living in the northwestern United States extracted and bound fibers into cords as early as 8,000 B.C., and a complete bag made of fibers dates to 5,000 B.C. Flax and hemp have been cultivated for

13–7

figure 13.11

Harvesting *Musa textilis*, commonly known as abaca, or Manila hemp. The trunk, which consists of leaf stalks, is used for its fibers; these fibers make high-grade cordage.

fibers for more than 5,000 years; in fact, hemp was one of the largest crops grown by George Washington on his Virginia plantation (see reading 13.1, "Hemp"). Today, humans cultivate more than forty families of plants for fibers (fig. 13.11), many of which you encounter every day. For example,

Musa textilis (Manila hemp) is used to make ropes and cords (fig. 13.11).

Agave sisalana (sisal; century plant) is used to make coarse ropes and twines.

Furcraea gigantea (Mauritius hemp) is used to make ropes, cords, and coarse fabrics.

Cannabis sativa (hemp; marijuana) is used to make twine and rope (see reading 13.1, "Hemp").

Linum usitatissimum (flax) is used to make linen. Linen is flexible, soft, and lustrous but not as elastic and flexible as cotton.

Boehmeria nivea (ramie) is used to make fine Oriental textiles and, more recently, Western clothing.

Corchorus capsularis (jute) is used to make coarse fabrics, bags, burlap, and sacks. After cotton, jute is the most important textile fiber in world commerce.

By now you're probably wondering, "What about cotton fibers?" Cotton (*Gossypium*) is the most important commercial fiber, but it is not sclerenchyma. Rather, cotton fibers are *trichomes*, which are outgrowths of epidermal cells. You'll learn about cotton fibers and trichomes later in this chapter.

Dermal Tissue: The Epidermis

Dermal tissues cover the plant body. The dermal tissue that covers the primary body of plants is the **epidermis**. The epidermis has several functions, including the absorption of water and minerals, secretion of cuticle, protection against herbivores, and control of gas exchange. Each of these functions is attributable to one or more of the unique features of the epidermis, such as the presence of a cuticle, few intercellular spaces, and multifunctional outgrowths called trichomes.

Cuticle

The outer walls of epidermal cells are covered with a waterproof **cuticle** made of a fatty material called *cutin* (fig. 13.12). The cuticle protects the plant from desiccation by helping to maintain a watery, aquatic environment inside the plant. The thickness of the cuticle, and hence the ability of a plant to retain water, is strongly influenced by the environment: plants growing in dry environments typically have thick cuticles, while those growing in wet environments have thin cuticles. The cuticle also protects a plant from microbes because it resists microbial infection and degradation. This resilience also accounts for why cuticles have been recovered from fossils millions of years old.

The cuticle is often covered by **epicuticular wax,** which is deposited as smooth sheets, rods, or filaments (figs. 13.13, 2.12). This wax imparts a whitish coloring to the surface of many leaves and fruits. On the undersides of leaves of wax palm, *Copernicia cerifera* (from the Latin word *cerifera*, meaning "wax bearer"), the whitish layer of epicuticular wax is often more than 5 mm thick. Wax from leaves of this plant is called *carnauba wax* and is used to make polishes, candles, and lipstick. It takes about 300 large leaves to produce a kilogram of wax.

The cuticle and its thick underlying cell wall protect the plant from stresses such as wind, desiccation, predators, and abrasion. Some plants also use debris from the environment as a protective coating. For example, *Cerochlamys pachyphylla* is a small desert succulent whose epidermal cells secrete a sticky, gluelike substance. Windblown sand adheres to this glue and protects the plant from predators and the hot sun.

Hemp, a fabric made from fibers of *Cannabis sativa,* was once as common as cotton is today. Thomas Jefferson farmed hemp, and George Washington proclaimed, "Sow it everywhere!" Washington's soldiers wore hemp uniforms, and the first two drafts of the Declaration of Independence were written on hemp paper. Even Betsy Ross's American flag was red, white, and blue hemp.

Hemp is a low-maintenance crop: it requires little water and no fertilizers or pesticides. Its deep roots prevent erosion. Hemp is also a superior fabric. It looks and feels like its pedigreed cousin, linen; only an expert armed with a microscope can distinguish between the two fibers. Despite these advantages, hemp was eventually replaced by cheap cotton and wood pulp. In 1937, a federal tax—which recognized the use of the plant as a drug, marijuana—raised prices of hemp to prohibitive levels, thereby ending production and making the fabric obsolete (reading fig. 13.1).

reading figure 13.1

Marijuana (*Cannabis sativa*), the source of hemp.

Today, the prospects of growing hemp again in the United States are unclear. In 1971, Congress repealed the 1937 tax, so it is now up to the states to decide whether to permit production. But because the cultivation of hemp involves growing marijuana, the federal Drug Enforcement Agency can still intervene. To avoid drug-related problems, hemp merchants must import hemp from countries where its cultivation is legal, primarily China and Hungary. France, which views hemp as a huge growth market, has developed cultivars of *Cannabis sativa* that lack the psychoactive agent tetrahydrocannabinol (THC), precluding it use as a drug.

Today, hemp is an expensive fabric—on average, it costs about six times as much as cotton: a pair of hemp jeans costs about $70. Despite its cost, hemp is becoming increasingly popular: products made of hemp are sold by companies such as J. Crew, Ralph Lauren, and Calvin Klein. At Walt Disney World, you can buy a hemp hat at the Indiana Jones Gift Shop.

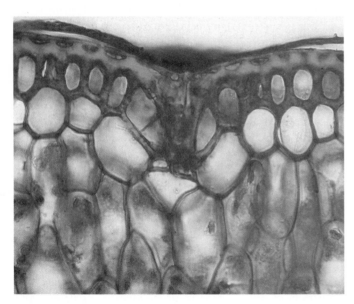

figure 13.12

Epidermal cells cover the primary body of a plant. These cells are covered by a water-resistant cuticle that protects the underlying cells and tissues.

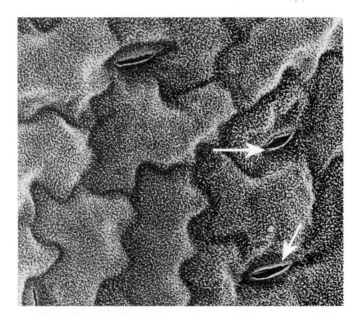

figure 13.13

Scanning electron micrograph of epidermal cells on the upper surface of a leaf of *Pisum sativum* (garden pea). Epidermal cells are packed closely together; the only spaces occur between specialized cells that form stomata (arrowheads). Stomata allow for gas exchange for photosynthesis, ×100.

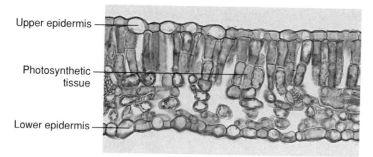

figure 13.14

Cross section of a leaf, showing the upper and lower epidermis. In most plants the epidermis is one cell thick.

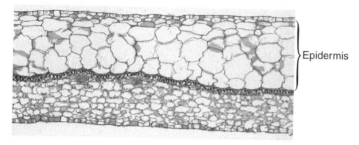

figure 13.15

The epidermis of a leaf of *Peperomia* is several cells thick and is used to store water.

Epidermal Cells and Gas Exchange

Most epidermal cells are flat, tilelike cells packed together like bricks in a brick wall (fig. 13.13). These cells typically lack chloroplasts and are transparent; it is the underlying chlorenchyma cells that give leaves their green color. However, the vacuoles of epidermal cells occasionally contain pigments. For example, red cabbage is colored by anthocyanins in epidermal cells, as are flowers and colored parts of variegated leaves of plants such as *Coleus*.

The epidermis differentiates from protoderm and is usually only one cell thick (fig. 13.14). However, the epidermis of leaves of *Peperomia* and the rubber plant (*Ficus elastica*) is typically several cells thick; the additional layers of cells store water (fig. 13.15). Similarly, the roots of many epiphytic orchids have a multicellular epidermis called a *velamen* that helps the plant retain water (fig. 13.16).

The only intercellular spaces in the epidermis are **stomata** (singular: **stoma**). Each stoma (from the Greek word *stoma,* meaning "mouth") is surrounded by two *guard cells* (fig. 13.17). Stomata occur on stems, leaves, flowers, and fruits but tend to be most abundant on the undersides of leaves. Stomata usually cover less than 1% of the epidermal surface, yet they are numerous: the leaves of most plants have 10,000–80,000 stomata cm^{-2}. Stomata are often surrounded by distinctively shaped cells called *subsidiary cells,* which function as reservoirs for water and ions that enter and leave guard

(a)

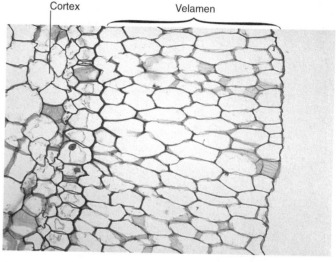

(b)

figure 13.16

The specialized epidermis of epiphytic orchids. (a) Roots of an epiphytic orchid. Epiphytes are plants that grow not in soil, but on other objects such as trees. (b) Cross section of the outer portion of an orchid root, showing the velamen, ×100. The velamen helps the orchid conserve water.

cells (fig. 13.18). Although substomatal cavities form after stomata differentiate, not all guard cells open to a cavity.

Guard cells in dicots are kidney-shaped, while those of monocots are shaped like dumbbells (fig. 13.19). Despite their different structures, guard cells in all plants have the same function: they regulate the exchange of gases (e.g., CO_2, O_2, H_2O) by opening and closing the stomatal pore. Guard cells regulate gas exchange; for example, the opening and closing of stomata regulate the diffusion of CO_2 into the leaf for photosynthesis, which occurs in underlying chlorenchyma cells. When CO_2 enters a leaf through stomata, water leaves, indeed, for every gram of carbon fixed via photosynthesis, the plant loses from 250

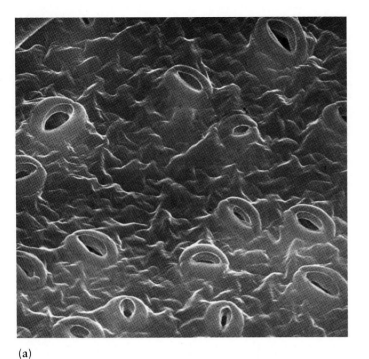

(a)

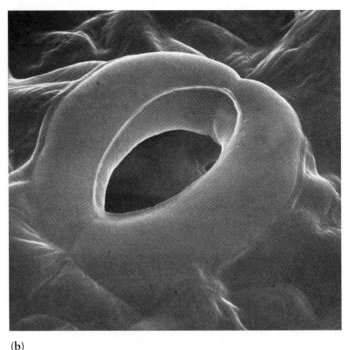

(b)

figure 13.17

Stomata. (a) Stomata on lower surface of a cucumber leaf, ×1,400. The pore formed by the guard cells allows for gas exchange between the atmosphere and the underlying cells. (b) Stoma in epidermis of a cucumber leaf. Two guard cells surround the pore.

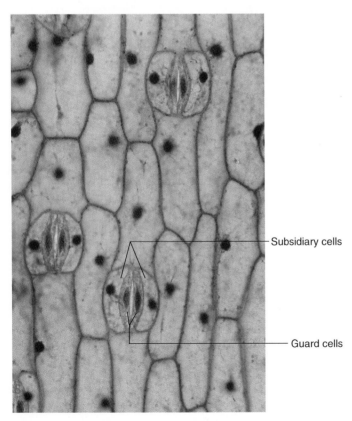

Subsidiary cells

Guard cells

figure 13.18

Epidermis of *Zebrina*. The large polygonal cells are ordinary epidermal cells. Each pair of guard cells is bordered by subsidiary cells, ×150.

to 600 g of water, depending on the type of photosynthesis (e.g., C₃, C₄, or CAM; see Chapter 7). Such large losses of water through stomata are an inevitable cost of photosynthesis.

Stomata open and close in response to environmentally induced changes in the turgor pressure of guard cells (fig. 13.19). Light and other environmental signals cause K^+ to enter guard cells, thereby drawing water into them via osmosis. This influx of water nearly doubles the volume of the guard cells and causes them to elongate. As they elongate, they bow apart and open the stomatal pore. Similarly, stomata close in the dark when K^+ and water leave the guard cells. We'll discuss the details of the structure and function of stomata in Chapter 21. For now, remember that stomata are a critical adaptation for conserving water and thus for maintaining life on land.

The epidermis coats the entire primary body of plants and is therefore a plant's first line of defense against herbivores. Although generally effective, the epidermis is not impenetrable; many pathogens invade plants through stomata. For example, *Pseudoperonospora humuli* and *Plasmopara viticola* are fungal pathogens that cause downy mildew on hop (*Humulus lupulus*) and grape (*Vitis*). Zoospores of these pathogens settle on stomata in light. Having located an entry into the plant, the zoospores then germinate and grow into the vulnerable inner tissues of the leaf. Many fungal pathogens locate stomata by tactile stimulation (touch).

Guard cells are not the only specialized cells in the epidermis. The upper epidermis of the leaves of many grasses contains longitudinal rows of large, vacuolate cells called *bulliform cells*. When it is hot and dry, bulliform cells rapidly lose

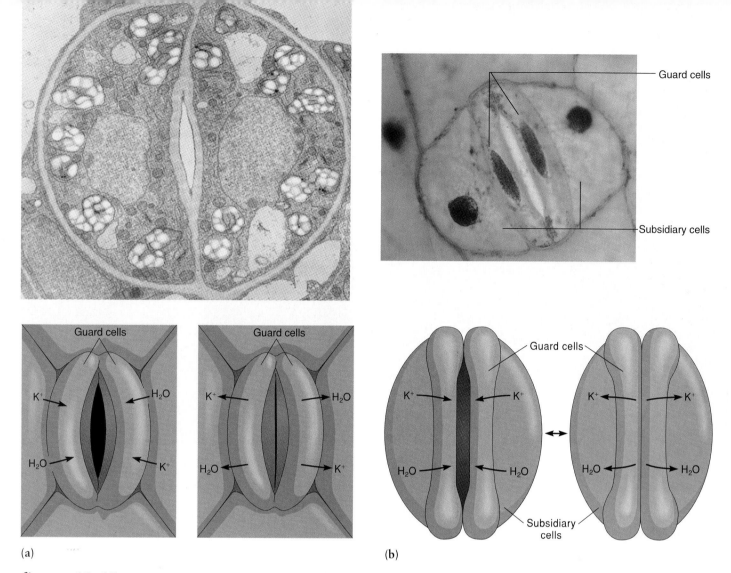

figure 13.19

Stomata of (a) cow pea (*Vigna sinensis*), a dicot, ×13,000, and (b) corn (*Zea mays*), a monocot, ×400. Both kinds of stomata function similarly (see drawings). Light triggers an influx of K⁺ into guard cells, which causes an influx of water (via osmosis) into the guard cells. As the cells elongate, they bow apart and form the stomatal pore. The pore closes via the reverse mechanism: K⁺ exits the cells, causing water to move out of them via osmosis. This, in turn, shrinks the cells and closes the pore.

water from their vacuoles and shrink, causing the leaf to roll into a protected cylinder. Leaf rolling reduces the surface area that is exposed to the environment, thereby helping the leaves conserve water.

Trichomes

Trichomes are single-celled or multicellular outgrowths of epidermal cells. The most economically important trichomes are cotton fibers, which are up to 6 cm long and are made by the epidermis of cotton seeds. The flexibility and soft texture of cotton are due to the absence of lignin—indeed, cotton fibers are 95% cellulose. However, trichomes on other seeds may be hard and lignified, such as those on the seed coat of *Strychnos nux-vomica*, the source of strychnine.

The contents of trichomes of some plants are used by humans. For example, menthol is a volatile oil collected from trichomes of peppermint (*Mentha piperita*), and trichomes of *Cannabis sativa* produce the tetrahydrocannabinols (THC) responsible for marijuana's soothing effects. The purified resin from *Cannabis* trichomes is called *hashish* and is a powerful drug.

Functions of Trichomes

Nutrition and Absorption Trichomes of carnivorous plants such as sundew (*Drosera*) and butterwort (*Pinguicula*) secrete enzymes that help the plant liquefy its trapped prey (fig. 13.20; you'll learn more about carnivorous plants in Chapter 20). Trichomes also absorb nutrients released as the prey are dissolved in traps.

Root hairs are outgrowths of epidermal cells specialized for absorbing water and minerals from soil (fig. 13.21). They

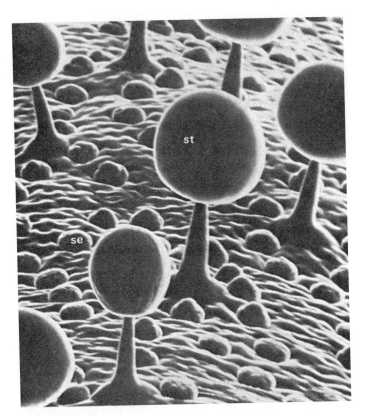

figure 13.20

Scanning electron micrograph of part of a leaf of butterwort (*Pinguicula grandiflora*), a carnivorous plant. The stalked (st) and smaller sessile (se) trichomes secrete enzymes and other compounds that help trap and digest animals.

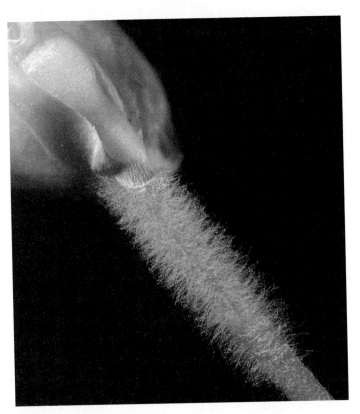

figure 13.21

Corn seedling. Note the many root hairs, which increase the surface area for absorption of water and dissolved minerals. Root hairs form just behind the elongating tip of the root.

occur near the tips of roots, where they are abundant (e.g., many plants have as many as 40,000 root hairs cm^{-2}). Root hairs increase the absorptive surface area of roots several thousandfold, thereby enabling the plant to extract water and dissolved minerals more effectively from nooks and crannies in the soil.

Protection   In many plants, trichomes deter marauding animals. For example, the resistance of cotton and soybean to leafhoppers is proportional to the density of trichomes on their leaves. Similarly, the resistance of *Passiflora adenopoda* to larvae of *Heliconius* butterflies is due to the presence of hook-shaped trichomes; larvae move only a few millimeters before being impaled by these trichomes. Similarly, aphids often inadvertently break off the heads of trichomes, thereby releasing their sticky contents. The aphids become trapped in this sticky secretion and soon starve to death.

Some trichomes protect plants from larger prey, including humans. For example, the leaves of stinging nettle (*Urtica*) have brittle, vase-shaped trichomes (fig. 13.22). When touched, the tip of the trichome breaks off, leaving behind a sharp, syringe-shaped cell that readily penetrates your skin. Pressure from the contact also injects the trichome's irritating chemicals into your skin. Ouch!

The Lore of Plants

The protective function of trichomes on sunflower leaves has apparently been overcome by *Chlosyne* butterflies. Female butterflies are attracted to chemicals in the trichomes as cues for where to lay their eggs. Larvae from these eggs eat the leaves, apparently unaffected by the chemicals in the trichomes.

Epidermal Cells and Cellular Recognition

Epidermal cells often recognize other organisms, such as pathogenic and nonpathogenic bacteria. Recognition typically involves interactions between molecules called *lectins* on the surfaces of epidermal cells and the contacting organism; it is referred to as **cellular recognition**. These chemical identification tags on a plant's epidermis are important because they direct many aspects of plant growth and development, including germination of pollen grains, formation of nitrogen-fixing nodules in roots, and whether a pathogen can successfully invade a plant (see Chapter 4).

figure 13.22

Stinging hairs of *Urtica dioica* (stinging nettle) are large cells filled with toxin. The tip is pointed and glassy, which makes it brittle and easily broken. The small tip allows it to catch in an animal's skin and break off, thus releasing the toxin, much like a microscopic hypodermic needle.

c o n c e p t

The epidermis covers the primary body of a plant. It is covered with a waxy cuticle that decreases water loss and is perforated by pores, called stomata, that control gas exchange. Outgrowths of epidermal cells called trichomes have many functions, including absorption and protection. The epidermis is also the site of cellular recognition, by which a plant chemically recognizes and responds to other organisms.

Fate of the Epidermis

The epidermis is short-lived in many plants. Radial expansion of stems caused by activation of the vascular cambium usually ruptures the epidermis during the first year of growth. When this occurs, the epidermis is replaced by a secondary dermal tissue called the *periderm*. You'll learn about the periderm when we discuss secondary growth in Chapter 16.

Vascular Tissues: Xylem and Phloem

Vascular tissues are specialized for long-distance transport of water and dissolved solutes. They ramify throughout the plant and are easily seen as veins in leaves. Vascular tissues typically contain transfer cells, secretory cells, and fibers in addition to parenchyma and conducting cells.

Xylem and phloem are the two kinds of vascular tissues in plants. In this section we'll concentrate on their differing structures and functions, and on the ways in which these properties adapt plants to life on land. You'll learn more about transport in the xylem and phloem in Chapter 21.

Xylem

Xylem (from the Greek word *xylos*, meaning "wood") transports water and dissolved nutrients in an unbroken stream from the roots to all parts of a plant. The water transported in xylem replaces that lost via evaporation through stomata. There are two types of xylem: primary xylem and secondary xylem.

Primary and Secondary Xylem

Primary xylem differentiates from procambium in the apical meristem and occurs throughout the primary body of a plant. Xylem in the elongating regions of plants is called *protoxylem*. During growth, protoxylem is stretched and replaced by newly differentiated protoxylem. When elongation stops, the procambium forms *metaxylem*, a relatively permanent type of primary xylem. The secondary cell walls of protoxylem and metaxylem are elaborately sculptured into hoops, bands, or springlike helices that allow elongation and prevent the cells from being crushed by adjacent cells or by internal tensions generated by water transport (fig. 13.23).

Secondary xylem differentiates from vascular cambium and is commonly called *wood*. Secondary xylem contains the same types of cells as primary xylem: the main difference is that these cells are more abundant and occur in different frequencies in secondary xylem than in primary xylem. Such differences vary in different species and account for the diverse properties of different kinds of wood. For example, balsa (*Ochroma*) wood contains many large, thin-walled parenchyma cells and relatively few thick-walled fibers. This combination of cells gives balsa wood a low specific gravity (0.10–0.16), making it lightweight and ideally suited as an insulator and component of model airplanes and lifeboats. Conversely, oak (*Quercus*) wood contains many thick-walled fibers and relatively few thin-walled parenchyma cells; thus, it is much denser (its specific gravity is approximately 0.6) than balsa wood. You'll learn more about wood and its characteristics in Chapter 16 when we examine secondary growth.

Conducting Cells in Xylem

Conducting cells in xylem are called **xylary elements.** There are two kinds of xylary elements: *tracheids* and *vessel elements* (figs. 13.24, 13.25). Both are elongate, are dead at maturity, and have thick, lignified secondary cell walls. Because water is typically pulled through xylary elements at a negative pressure (i.e., tension; lower than atmospheric pressure), these thick walls help prevent the cells from collapsing. The thick walls of xylary elements also help support the plant.

Tracheids are the most primitive (i.e., least specialized) xylary elements and are the only water-conducting cells in

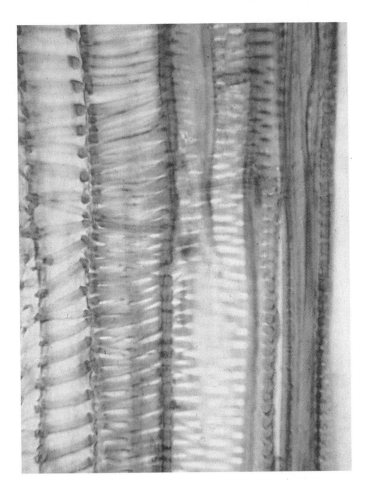

figure 13.23

The secondary walls of some of the first-formed xylary elements of the protoxylem are often deposited in rings or spirals. These thickenings strengthen the tracheary element while enabling the tracheary element to elongate during growth.

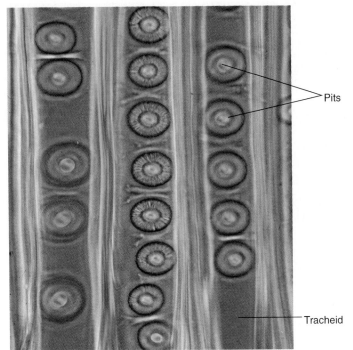

figure 13.24

Light micrograph of tracheids in wood of pine (*Pinus*), ×163. Tracheids are connected by bordered pits, which are thin areas in adjacent cell walls. The wood of most gymnosperms (e.g., conifers) resembles that of pine; the primary conducting cells are tracheids.

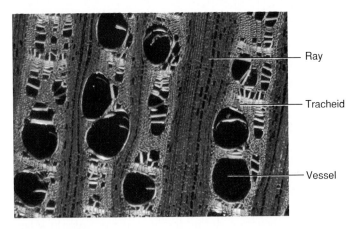

figure 13.25

Light micrograph of a cross section of wood of a flowering plant, showing tracheids, vessels, and multiserate rays. The wood of most flowering plants (e.g., oak) has tracheids and vessels to conduct water and dissolved materials from the roots to leaves. You'll learn more about wood in Chapter 16.

most woody, nonflowering plants such as pine. Tracheids are long, slender cells with tapered, overlapping ends. Water moves upward from tracheid to tracheid through thin areas in cell walls called *bordered pits* (fig. 13.24).

Vessel elements are more evolutionarily advanced than tracheids and occur in several groups of plants, including angiosperms. With only a few exceptions, all angiosperms contain vessels and tracheids (fig. 13.25). The vessel elements of many plants are barrel-shaped; that is, they are relatively short and wide. Vessel elements, which are shorter and wider than tracheids, are arranged end-to-end. Their transverse walls are partially or wholly dissolved, forming long, hollow vessels through which water can move as much as 3 m without having to move through a pit (fig. 13.26). This feature of vessels, along with their relatively large diameter, enables them to transport water more rapidly than can smaller, thinner tracheids. The evolution of vessels in angiosperms and their greater water-transport capacity is a major reason why angiosperms dominate today's landscapes.

c o n c e p t

Xylem conducts water and dissolved minerals from roots to leaves. Water moves through xylary elements called *tracheids* and *vessel elements*, which are dead, hollow cells having thick secondary cell walls.

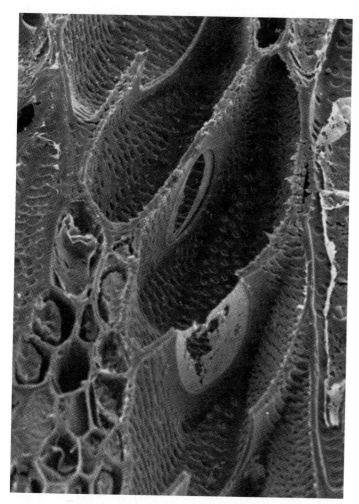

figure 13.26

Vessel elements in wood of *Knightia excelsa*, an angiosperm. The vessel elements form in long rows, are wider than they are long, and have thick, reinforced walls. The transverse walls of adjacent vessel elements are partly or wholly dissolved, forming long, hollow vessels through which water and dissolved minerals flow.

Tracheids have also evolved into thick-walled, fiberlike cells called *fiber-tracheids*. Unlike vessel elements, fiber-tracheids are specialized for support rather than water transport.

Phloem

Phloem (from the Greek word *phloios*, meaning "bark") transports dissolved organic materials (especially sucrose) throughout the plant. Unlike xylem, which transports water only upward, solutes in the phloem move in all directions. Furthermore, dissolved solutes move under a positive pressure in the living conducting cells of phloem, whereas they move at a negative pressure in the dead conducting cells of xylem.

Primary and Secondary Phloem

Primary phloem differentiates from procambium and extends throughout the primary body of a plant. Primary phloem in

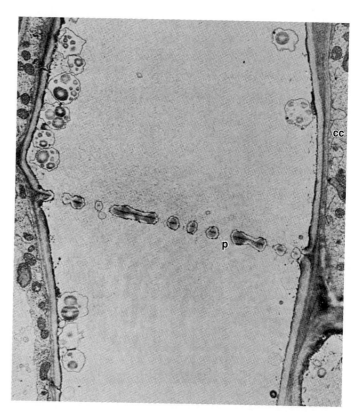

figure 13.27

Sieve elements of tobacco (*Nicotiana tabacum*). Note that the pores (p) of the sieve plate are open. Sieve elements conduct water and dissolved organic matter (especially sugars) in all directions in the plant (cc = companion cell).

elongating regions of the plant is called *protophloem*, while that in nonelongating regions is called *metaphloem*. Cells of primary phloem have primary cell walls that are not elaborately sculptured.

Secondary phloem differentiates from the vascular cambium and constitutes the inner layer of bark. Like xylem, phloem consists of several types of cells, including transfer cells and other parenchyma cells, fibers, conducting cells, and secretory structures.

Conducting Cells of Phloem

Conducting cells of phloem are called **sieve elements.** Sieve elements lack nuclei and are alive at maturity. They also have thin areas along their cell walls called *sieve areas* that are perforated by *sieve pores* (fig. 13.27); solutes move from sieve element to sieve element through these pores. Sieve elements have thin, primary cell walls and are the most delicate cells in plants. They usually live for less than a year, although in palms they can live for more than a century.

There are two kinds of sieve elements: sieve cells and sieve tube members. **Sieve cells,** which are more primitive than sieve tube members, occur in nonflowering plants. They are long and have tapered, overlapping ends. Sieve areas occur in equal frequencies all over the surface of sieve cells, including

figure 13.28

Sieve tube members in the phloem of squash (*Cucurbita*). The dark regions near the center of the photo are side views of sieve plates.

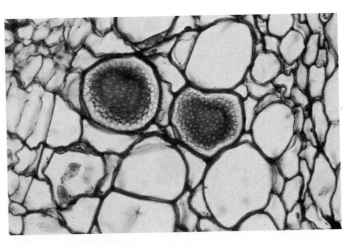

figure 13.29

Face view of sieve plates of *Cucurbita*. During peak periods of transport, enough water and solutes move through a sieve plate to empty and refill a 0.5 mm-long sieve tube member every 2 sec.

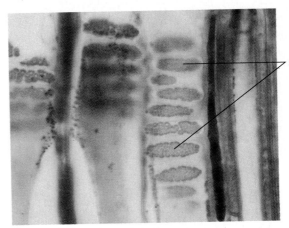

Sieve areas

figure 13.30

Sieve areas of sieve tube members of grape (*Vitis* sp.). Each sieve area consists of pores bordered by callose, which is stained blue in this preparation, ×400. Callose helps to seal wounded sieve tube members.

along their longitudinal walls. Sieve cells may be associated with specialized parenchyma cells called *albuminous cells,* which help regulate the sieve cells' activities. Despite this interdependency, however, sieve cells and albuminous cells do not arise from the same mother cell.

Sieve tube members occur in all but a few of the most primitive angiosperms. They are usually shorter and wider than sieve cells and are arranged end-to-end into sieve tubes (fig. 13.28). Their sieve areas have larger sieve pores than do sieve cells, and they are concentrated along the contacting endwalls of adjacent sieve tube members (fig. 13.29). These specializations allow solutes to move through sieve tubes faster than through sieve cells.

Mature sieve tube members are rather strange: they are the only cells in plants that lack nuclei but contain living cytoplasm. Furthermore, each sieve tube member is associated with at least one specialized cell called a **companion cell,** which regulates many of its activities (fig. 13.27). Companion cells help regulate the loading and unloading of carbohydrates from sieve tube members. Sieve tube members and companion cells arise from the same cell and are linked via numerous plasmodesmata. Their interdependence is apparently absolute: the death of a sieve tube member also causes the demise of its companion cell.

The absence of nuclei is not the only unusual thing about sieve tube members. Indeed, they also contain a proteinaceous substance called *P-protein* (for *p*hloem-protein) along their longitudinal walls. Sometimes they also contain a spirally wound polymer of glucose called *callose* (fig. 13.30). P-protein and callose seal wounded sieve tubes.

c o n c e p t

Phloem conducts water and organic solutes (especially sucrose) in all directions in plants. Solutes move through sieve elements called *sieve cells* and *sieve tube members,* which are living at maturity.

13–17

Secretory Structures

Plants secrete a variety of substances from structures called **secretory structures**. Secretory structures are seldom classified as a separate type of plant tissue because they often intergrade with other tissues. For example, some secretory structures are epidermal trichomes, while others are parts of the ground and vascular tissues. Despite their various origins in plants, secretory structures typically meet the requirement for classification as a plant tissue: they form a group of cells having a common function.

There are two types of secretory structures: external secretory structures and internal secretory structures.

External Secretory Structures

Nectaries

Nectaries are structures that secrete *nectar,* a sugary exudate that attracts insects, birds, or other animals. Most nectaries are associated with flowers and are called *floral nectaries.* Their nectar is 10%–50% sugar (especially sucrose, glucose, and fructose) and also contains amino acids. Nectar typically is pushed or diffuses through the walls of secretory cells, but it may also ooze from stomata.

Nectar in floral nectaries attracts insects and other animals that pollinate the plant. Plants usually secrete small amounts of nectar, which forces foraging animals to visit several flowers before getting a full meal. Thus, a single animal can pollinate tens or hundreds of plants.

Nectaries that occur on vegetative parts of plants are called *extrafloral nectaries* (fig. 13.31). These nectaries often attract animals that defend the plant. For example, the extrafloral nectaries of plants such as trumpet creeper (*Campsis radicans*) and *Costus* attract ants that eat the nectar and, in return, defend the plant from leaf-eating insects. These tiny defenders are surprisingly effective: *Costus* plants deprived of ants are quickly devastated by fly larvae and produce only one-third as many seeds as plants protected by ants.

Hydathodes

Hydathodes are tissues that secrete water via a process called *guttation.* Hydathodes are loose arrangements of parenchyma cells usually located at vein endings along the edges of leaves; they occur in plants ranging from ferns to flowering plants. Guttation from hydathodes occurs through modified stomata that are always open. The significance of hydathodes is unknown, but they may help relieve pressure caused by the uptake of excessive water by roots at night. You'll learn more about the uptake and transport of water in Chapter 21.

Digestive Glands of Carnivorous Plants

Carnivorous plants have trichomes that secrete enzymes and digest trapped prey (fig. 13.20). For example, trichomes on the

Extrafloral nectaries

figure 13.31

Extrafloral nectaries of passion flower (*Passiflora incarnata*). What do you suppose is the primary function of these glands? How would you test your hypothesis?

sticky "flypaper" leaves of *Pinguicula* secrete digestive enzymes within an hour after trapping a victim; the secretion is so abundant that pools of the enzymes form on the leaf within 2–3 hours. These enzymes help gather nitrogen from animal protein, thereby allowing the plants to survive in nitrogen-deficient soil.

Salt Glands

Plants such as *Atriplex, Spartina,* and *Tamarix* that grow in salty soil have secretory structures called *salt glands.* These glands often form a glistening crust of salt that covers the leaves of the plant. In effect, salt glands are dump sites for the excess salt absorbed in water from the soil. Salt glands help plants adapt to life in saline environments.

Internal Secretory Structures

Secretory Cells

Secretory cells are large cells containing substances such as oils, tannins, resins, mucilage, and crystals. They often occur in groups and have several functions, including the storage and production of chemical deterrents to foraging animals. Secretory cells are also a source of many oils used by humans. For example, peanut oil comes from secretory cells in cotyledons of peanut (*Arachis hypogaea*), safflower oil from seeds of safflower (*Carthamus tinctorius*), clove oil from flowers of clove (*Eugenia caryophyllata*), palm oil from fruits of palms (such as *Elaeis guineensis*), and cinnamon oil from the secondary phloem of *Cinnamomum zeylanicum.*

Canals, Ducts, and Cavities

Many plants secrete oils and resins into internal canals, ducts, and cavities. For example, myrrh, frankincense, *Citrus* oils,

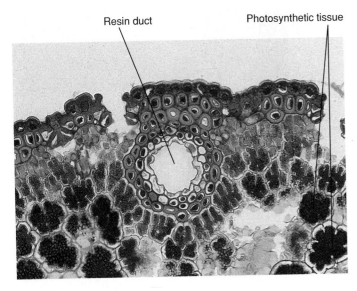

figure 13.32

Resin duct from a cross section of a leaf of *Pinus resinosa*. When ruptured, these ducts release resin, which deters herbivores and seals the wound.

Eucalyptus oil, and pine (*Pinus*) resin are extracted from internal cavities, ducts, and canals (fig. 13.32); these oils deter grazing animals, and resin rapidly seals wounds.

As already mentioned, resins and oils help plants defend themselves. However, some animals have evolved means of not only tolerating these resins but also using them for their own defense. Sawfly larvae, for example, eat pine and store its resin in their bodies. When disturbed by another animal, the larva rears up and secretes a drop of resin onto the intruder. This typically troubles the rude visitor and ensures privacy for the larva.

Laticifers

Laticifers are secretory structures that contain *latex*, which is cytoplasm containing a hodgepodge of carbohydrates, organic acids, alkaloids, terpenes, oils, resins, enzymes, and rubber (fig. 13.33). Laticifers occur in several families of plants either as long, single cells called *nonarticulated laticifers* or as a series of fused cells called *articulated laticifers*. Laticifers have only a primary cell wall and can be several centimeters long. Because the contents of laticifers are under positive pressure, latex oozes from wounded surfaces of plants such as *Euphorbia* and dandelion (*Taraxacum*). It may be white (as in *Euphorbia*), colorless (as in some species of *Morus*), or orange-yellow (as in *Cannabis*). Latex is important because it deters grazing animals and helps seal wounds.

Humans have many uses for latex. For example, opium, morphine, and codeine are derived from latex of *Papaver somniferum*, the opium poppy (see reading E.1, "Poppies, Opium, and Heroin"). Latex from the rubber tree (*Hevea brasiliensis*) is 30%–50% rubber; Amazon Indians used this latex to make rubber balls and containers more than 450 years ago. Today,

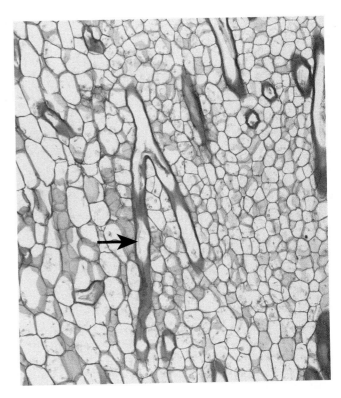

figure 13.33

Longitudinal section of shoot tip of *Nerium oleander*, showing a branched, nonarticulated laticifer (arrow). Laticifers contain latex under positive pressure. When these cells are ruptured, latex oozes from the wound, deterring herbivores and helping to seal the wound.

Hevea remains the primary source of natural rubber: yields at Central American plantations often exceed 2,000 kg of rubber per hectare per year (about 0.7 liters per week from a 5-year-old plant). The latex of several other plants contains rubberlike particles: gutta percha used to make dentures and golf balls is derived from the latex of *Palaquium*, and chicle (as in Chiclets) gum comes from *Achras sapota*.

c o n c e p t

Plants secrete many different kinds of substances, including resins, nectar, and digestive enzymes. These compounds are secreted internally and externally, and are important because they protect plants, attract pollinators, and seal wounds. Secretions such as rubber and oils are economically important.

Plant Tissues and Life on Land

The tissues that you learned about in this chapter adapt plants for life on land. In upcoming chapters you will learn how plants use these tissues to build organs such as leaves, stems, and roots. However, we need to look beyond organs to understand the significance of plant tissues; that is, we need to understand tissues in the context of the entire plant and its habitat. To appreciate this perspective, consider a unicellular alga and an oak tree. Despite having drastically different sizes and origins separated by hundred of millions of years of evolution, the basic life processes of these organisms are similar: both have similar cellular organelles, trap chemical energy via photosynthesis, use respiration to free this energy for work, and have nearly identical metabolic pathways. But if these cells are so similar, why are oaks so different from algae?

The answer to this question lies not in *how* the cells function but *where* they function. Algae live in aquatic environments where they are protected from desiccation. When removed from such environments, unprotected algal cells desiccate and die in less than a minute. Similarly, unprotected cells of an oak tree also die when placed in air. Thus, all cells, whether of a towering oak or a microscopic alga, are strictly aquatic.

The similar life processes of oak and algal cells require a delicate balance of several factors, such as light, nutrients, and gases; excesses or deficiencies of any of these quickly kill each organism. Algae live in aquatic systems where these delicate balances occur naturally, such as in moist soil and surface waters. There, they conveniently exchange gases, absorb nutrients, and secrete waste products directly into the surrounding aquatic environment. Terrestrial plants such as oaks, however, face a dilemma: they extend far beyond the areas where these delicate balances naturally occur. To live outside these areas, plants must re-create an aquatic oasis in the hostile atmosphere. Plants can survive on land because they maintain an internal aquatic environment that is protected from the harsh atmosphere.

With the exception of chlorenchyma cells and a few reproductive cells, all of a plant's tissues function to create the aquatic environment necessary for life. Foremost among these tissues is the epidermis, which forms a waterproof, protective shell. In leaves, the epidermis encases millions of algae-like photosynthetic cells. Thus, the epidermis maintains the oak tree's aquatic environment, despite the fact that the deadly, desiccating atmosphere is only fractions of a centimeter away.

The epidermis establishes and protects the leaf's aquatic environment. However, this creates other problems for the plant. As a consequence of having a protective epidermis, plants must import and export gases, nutrients, and waste products. These "services" are the functions of other plant tissues. For example, xylem supports the plant and transports water and dissolved minerals to leaves to replace that lost by evaporation. Similarly, phloem transports energy-rich sugars to nonphotosynthetic parts of the plant for storage or to enable continued growth. Finally, plants form secretory structures, produce chemicals that deter herbivores, and use other compounds to seal wounds and attract animals that pollinate and protect the plant. Although these life-support systems seem rather diverse, they share a common function: to establish, maintain, and protect the internal aquatic environment necessary for life.

table 13.1

Plant Tissues and Their Cell Types

Tissue	Cell Types
Parenchyma	Parenchyma cells
Collenchyma	Collenchyma cells
Sclerenchyma	Fibers or sclereids
Epidermis	Usually parenchyma cells; guard cells and trichomes; sclerenchyma cells
Phloem	Sieve cells or sieve tube members; albuminous cells or companion cells; parenchyma cells; sclerenchyma cells
Xylem	Tracheids; vessel elements; sclerenchyma cells; parenchyma cells
Periderm	Usually parenchyma cells; sclerenchyma cells

From Peter H. Raven, Ray F. Evert, and Susan E. Eichhorn, *Biology of Plants,* 5th edition, Worth Publishers, New York, 1992. Reprinted with permission.

Despite this commitment, however, the aquatic environment inside terrestrial plants is often threatened because plants must exchange gases directly with the atmosphere. In doing so, they lose large amounts of water through their stomata. Because photosynthesis and water conservation are diametrically opposed events, there is no perfect solution to the gas-exchange dilemma faced by plants. However, stomata are an adequate, albeit imperfect, solution to the problem. Stomata use internal and external cues to balance the photosynthesis-desiccation compromise. As a result, photosynthesis occurs only when and to the extent that the internal aquatic environment is not threatened.

Chapter Summary

Table 13.1 summarizes the plant tissues and their cell types; Table 13.2 describes the features, locations, and functions of these types of plant cells. Refer to these tables as you read this summary of Chapter 13.

The primary body of plants consists of four kinds of primary tissues: meristems, ground tissue, dermal tissue, and vascular tissues. These tissues contain distinctive types of cells and are typically multifunctional. Meristems are regions of localized cellular division and growth. They produce all of the cells of the plant body and are responsible for many instances of pattern formation in plants.

Ground tissue differentiates from the ground meristem and constitutes most of the primary body of a plant. Ground tissue consists of three kinds of cells: parenchyma, collenchyma, and sclerenchyma cells.

Parenchyma cells are the most abundant and least structurally specialized cells in plants. Parenchyma cells have many functions, including storage, photosynthesis, and basic *table 13.2*

Cell Types in Plants

Cell Type	Location	Features	Function
Parenchyma	Throughout the plant (e.g., parenchyma tissue in cortex, as pith and pith rays)	Usually, many-sided Cell wall: primary, or primary and secondary; may be lignified, suberized, or cutinized Alive at maturity	Metabolic processes as respiration and photosynthesis; storage and conduction; wound healing and regeneration
Collenchyma	Beneath the epidermis in young elongating stems; often as a cylinder of tissue or only in patches; in ribs along veins in some leaves	Elongate shape Cell wall: unevenly thickened; primary only—nonlignified Alive at maturity	Support in primary plant body
Fibers	Usually associated with xylem and phloem; in leaves of monocotyledons; sometimes in cortex of stems	Usually very long Cell wall: primary and thick secondary—often lignified Often (not always) dead at maturity	Support
Sclereids	Throughout plant	Variable shape; shorter than fibers Cell wall: primary and thick secondary; usually lignified May be alive or dead at maturity	Protective; mechanical
Tracheid	Xylem	Elongate and tapered shape Cell wall: primary and secondary; lignified; contains pits but not perforations Dead at maturity	Primary water-conducting element in gymnosperms and seedless vascular plants; also found in angiosperms
Vessel element	Xylem	Elongate; usually not as long as tracheids Cell wall: primary and secondary; lignified; contains pits and perforations; several vessel elements end-to-end form a vessel Dead at maturity	Primary water-conducting element in angiosperms
Sieve cell	Phloem	Elongate and tapered shape Cell wall: primary in most species; with sieve areas; callose often associated with wall and pores Alive at maturity; either lacks or contains remnants of a nucleus at maturity	Primary food-conducting element in gymnosperms and seedless vascular plants
Albuminous cell	Phloem	Elongate shape Cell wall: primary Alive at maturity; associated with sieve cell; has many connections with sieve cell	Movement of food into and out of sieve cells
Sieve tube member	Phloem	Elongate shape Cell wall: primary, with sieve areas; sieve areas on end-wall with much larger pores than those on side walls Alive at maturity; either lacks a nucleus at maturity or contains only remnants of nucleus; contains a proteinaceous substance known as slime, or P-protein, in dicots and some monocots; several sieve tube members are arranged vertically to form a sieve tube	Primary food-conducting element in angiosperms
Companion cell	Phloem	Usually elongate Cell wall: primary Alive at maturity; associated with sieve tube members; has many connections with sieve tube member	Movement of food into and out of the sieve tube member

From Peter H. Raven, Ray F. Evert, and Susan E. Eichhorn, *Biology of Plants*, 5th edition, Worth Publishers, New York, 1992. Reprinted with permission.

metabolism. Parenchyma cells can dedifferentiate, meaning that they can change activities. The ability of parenchyma cells to dedifferentiate helps plants adapt to changing environmental demands.

Chlorenchyma, aerenchyma, and transfer cells are special types of parenchyma cells. Chlorenchyma cells are chloroplast-containing parenchyma cells specialized for photosynthesis.

Aerenchyma tissue contains large intercellular spaces and is specialized for gas exchange. Transfer cells have elaborate ingrowths of their cell walls and are specialized for short-distance transport of solutes.

Collenchyma cells are long, living cells with thickened primary cell walls. These walls can expand and support growing regions of the plant.

Sclerenchyma cells have thick, nonstretchable secondary cell walls that support nonextending regions of plants. There are two kinds of sclerenchyma cells: sclereids and fibers. Sclereids are relatively short cells that have a variety of shapes and occur singly or in small groups. Fibers are long, slender cells occurring as single strands or bundles in mature parts of the plant. Humans use the fibers of many plants to make textiles.

Epidermis is the dermal tissue that covers the primary body of a plant. The epidermis has several functions, including absorption, secretion, digestion, protection, cellular recognition, and the control of gas exchange. These functions are attributable to the unique features of the epidermis, including the presence of a cuticle, few intercellular spaces, and the presence of trichomes. The outer walls of epidermal cells are covered by a waterproof cuticle that protects the internal tissues from desiccation.

Stomata are direct avenues of gas exchange between the environment and the interior of the plant. Each stoma consists of two guard cells that surround a stomatal pore. This pore opens and closes as guard cells enlarge and shrink, thereby controlling gas exchange.

Trichomes are single-celled or multicellular outgrowths of epidermal cells. They have several functions, including absorbing water and dissolved nutrients, and protecting the plant from predators.

Vascular tissues are specialized for long-distance transport of water and dissolved solutes. Xylem and phloem are the two vascular tissues in plants. Both tissues occur throughout a plant and contain parenchyma cells, secretory cells, fibers, and conducting cells.

Xylem transports water and dissolved minerals from roots to leaves. Water and dissolved minerals move through xylem under negative pressure. Primary xylem differentiates from procambium. The first primary xylem that forms is called *protoxylem,* whereas that formed in more mature regions is called *metaxylem.* Secondary xylem, or *wood,* is produced by the vascular cambium, a lateral meristem.

There are two kinds of conducting cells in xylem: tracheids and vessel elements. Both are dead cells having thick, lignified, secondary cell walls. Tracheids are long, narrow cells with tapering, overlapping ends. Water moves upward from tracheid to tracheid through thin areas called pits. Tracheids occur in all vascular plants.

Vessel elements are more evolutionarily advanced than tracheids and occur mostly in flowering plants. Vessel elements are arranged end-to-end and are shorter and wider than tracheids. The transverse walls of vessel elements are partially or wholly dissolved, thereby forming long, hollow pipes called vessels. Vessels can transport water more rapidly than can tracheids because of their width and dissolved transverse walls.

Phloem transports dissolved sugars in all directions in plants. Substances in phloem, unlike those in xylem, move under positive pressure through living cells. Primary phloem differentiates from the procambium. The primary phloem formed first is called *protophloem;* that formed later in the mature region of the plant is called *metaphloem.* Secondary phloem is the inner layer of bark and is made by the vascular cambium.

There are two kinds of conducting cells in phloem: sieve cells and sieve tube members. Both are living cells that have only a primary cell wall. Water and dissolved solutes move between adjacent sieve tube members through sieve pores of sieve plates. Sieve cells occur in all vascular plants except angiosperms. Sieve areas occur all over sieve cells and may be associated with albuminous cells. Sieve tube members occur only in angiosperms and are arranged end-to-end in sieve tubes. Their sieve areas are concentrated in sieve plates occurring along their transverse walls. Sieve tube members lack nuclei and are controlled by adjacent parenchyma cells called companion cells.

Secretory structures occur in ground, vascular, and dermal tissues. Secretory structures secrete substances internally into secretory cells, laticifers, or ducts and externally from nectaries, hydathodes, salt glands, and digestive glands. Secretory structures protect plants, attract beneficial animals, and aid in the plant's nutrition.

Plant tissues have enabled vascular plants to successfully invade the land. With the exception of chlorenchyma and reproductive cells, all plant tissues serve as life-support systems, maintaining the aquatic environment necessary for life. The epidermis minimizes desiccation, xylem transports water to replace that lost by evaporation, and secretory structures seal wounds and deter grazing animals.

 What Are Botanists Doing?

The ultrastructures of cells have become important for plant classification. Find a recent article on the use of ultrastructure in classification and summarize the results that it presents.

 Writing to Learn Botany

Collenchyma is common in shoots but relatively rare in roots. What could account for this?

Questions for Further Thought and Study

1. Justify why parenchyma can be considered a cell type or a tissue.

2. Discuss the structural and functional differences between (a) parenchyma, collenchyma, and sclerenchyma; and (b) xylem and phloem.

3. Discuss how plant tissues establish, maintain, or protect an aquatic environment in terrestrial plants.

4. What are the functions of trichomes? Secretory structures? Epidermis?

5. How would you distinguish a vessel element from a sieve tube member? A vessel element from a tracheid?

6. List the unique properties of the epidermis. Why is each important for plant growth?

7. Discuss how the structure of xylem could be used to identify plants.

8. What is redifferentiation? Why is it important in plants?

9. Nectaries typically contain large amounts of phloem, whereas hydathodes contain large amounts of xylem. Discuss how these structural differences relate to the functions of these secretory structures.

10. Vessel elements evolved from tracheids in several groups of angiosperms. What do you conclude from this observation?

11. Discuss the structure and function of (a) fibers, (b) guard cells, (c) tracheids, (d) sieve cells, and (e) companion cells. What is the adaptive significance of each?

12. Of what adaptive significance is it that the differentiation of sclerenchyma and collenchyma is induced to form by mechanical stress?

13. How are nectaries important to plants?

14. What might be the adaptive advantage of sclereids differentiating near wounds?

15. Late in 1996, Philip Becraft and his colleagues reported that the *crinkly* 4 (*cr* 4) mutation changes the structure and function of epidermal cells. These changes also correspond with the occurrence of graftlike fusions between organs of mutated plants. What does this suggest about the role of the epidermis in maintaining the structural integrity of a plant organ?

Web Sites

Review the "Doing Botany Yourself" essay and assignments for Chapter 13 on the *Botany Home Page.* What experiments would you do to test the hypotheses? What data can you gather on the Web to help you refine your experiments?

Here are some other sites that you may find interesting:

http://www.axis-net.com/pfaf/

This site houses *The Plant Tracker*, a searchable database of almost 7,000 species of plants and their edible, medicinal, and other uses.

http://www.whidbey.com/mvg/plant.htm

At this site, you'll learn about the Plant-of-the-Month. This site is updated monthly, so check it often to learn about unusual plants.

Suggested Readings

Articles

Walker, D. B. 1978. Plants in the hostile environment. *Natural History* 87:74–81.

Books

Burgess, J. 1985. *An Introduction to Plant Cell Development.* Cambridge: Cambridge University Press.

Buvat, R. 1989. *Ontogeny, Cell Differentiation, and Structure of Vascular Plants.* New York: Springer-Verlag.

Cutter, E. G. 1978. *Plant Anatomy. Part 1: Cells and Tissues.* 2d ed. Boston, Mass.: Addison-Wesley.

Fahn, A. 1979. *Secretory Tissues in Plants.* London: Academic Press.

Hall, J. L., and C. R. Hawes. 1991. *Electron Microscopy of Plant Cells.* San Diego: Academic Press.

Lydon, R. F. 1990. *Plant Development: The Cellular Basis.* London: Unwin Human.

Mauseth, J. D. 1988. *Plant Anatomy.* Menlo Park, CA: Benjamin/Cummings.

Sachs, T. 1991. *Pattern Formation in Plant Tissues.* New York: Cambridge University Press.

Leaves of cottonwood (*Populus fremontii*).

Primary Growth: Stems and Leaves

14

Chapter Outline

Chapter Overview

The autotrophic life-style of plants requires that they expose large amounts of surface area to the environment. The most conspicuous surface area is the plant's shoot, which is an integrated system of stems and leaves. Stems support leaves—that is, stems are the "hat racks" on which plants hang their solar collecting leaves. The most important functions of leaves and stems are photosynthesis and support. Some plants modify stems and leaves for climbing, trapping insects, and reproduction. The structure and function of stems and leaves are sensitive to changes in light and moisture, which allows plants to adapt to different and changing environments.

Stems and leaves are the most conspicuous and diverse organs of plants. The different shapes and structures of stems and leaves do not result from the presence of different or unique tissues but from different *arrangements* of the tissues discussed in the previous chapter. That is, what we call different organs are merely different combinations of the same building blocks. Leaves and stems are important to us because they make our food, oxygen, building materials, and many of our drugs.

We usually think of stems and leaves as different organs; after all, they usually look different, and we can easily distinguish them from each other. However, this distinction is not always as clear as you may think. Indeed, stems form leaves, and there is often no clear-cut point at which a stem ends and a leaf begins. Furthermore, the leaves of many plants look like stems, while the stems of other plants look like leaves.

Although we'll discuss leaves and stems separately in this chapter, remember that this distinction is somewhat artificial. The structure and function of stems and leaves can best be appreciated by viewing them as parts of a single dynamic shoot system.

Stems

Stems and Their Functions

A stem is a collection of integrated tissues arranged as nodes and internodes. **Nodes** are regions where leaves attach to stems, and **internodes** are the parts of stems between nodes (fig. 14.1). Together, nodes and internodes perform several important functions:

- **Stems support leaves,** the solar collectors of plants. Turgor pressure in stems provides a hydrostatic skeleton that supports the young plant. Leaves are also supported by a stem's internal skeleton of collenchyma and sclerenchyma.

- **Stems produce carbohydrates.** Stems of plants such as *Salicornia* are green and photosynthetic. Although photosynthesis in stems is usually insignificant compared to that in leaves, in plants such as cacti it accounts for most of the plant's carbon fixation.

- **Stems store materials.** Parenchyma cells in stems store large amounts of starch and water. For example, water accounts for as much as 98% of the weight of many cactus stems.

- **Stems transport water and solutes between roots and leaves.** The vascular system of stems maintains an aquatic environment in leaves and transports sugars and other solutes between leaves and roots. Stems link leaves with the water and dissolved nutrients of the soil.

Control of Stem Growth

The growth of stems is controlled by several factors, of which leaves are perhaps the most important. Leaf primordia control the differentiation of procambium in stems. As a result, there is no procambium above the youngest primordium, and vascular tissues in developing leaves align and connect with mature vascular tissues of the stem. Light also controls stem growth: the stems of seedlings growing in darkness elongate much faster than those growing in light.

Stems elongate in subapical regions in response to auxin and gibberellins, which are made in young leaves. These hormones elongate stems by stimulating cellular elongation and division. In many plants, elongation occurs throughout the internode; in others, it proceeds as a wave originating at the base of the internode. You'll learn more about plant hormones in Chapter 18.

Grasses such as bamboo elongate at meristems intercalated between mature tissues at the bases of their nodes and leaf sheaths (fig. 14.2). These meristems are called **intercalary meristems** and remain meristematic long after tissues in the rest of the internode are fully differentiated. Intercalary meristems are important because they re-form the stem or leaf of a grass when its tip is torn off by a grazing animal or lawn mower. Because grasses are a primary food for grazing animals, and because humans and other predators depend to a greater or lesser extent on grazers for food, intercalary meristems are largely responsible for keeping our meat markets open.

Intercalary meristems can produce phenomenal rates of growth in many plants. For example, bamboo stems can elongate almost a meter per day (that's more than an average child grows between birth and its tenth birthday) and can grow to be over 35 m tall. Intercalary meristems also have a direct impact on us: no matter how many afternoons we spend vibrating behind our lawn mowers, we can be sure that the grass blades will soon grow back, thanks to intercalary meristems.

Plants whose stems do not elongate are called **rosette plants**; these plants have short internodes with tightly packed leaves. The stems of rosette plants are short and made almost entirely of overlapping leaf bases. These structures can reach impressive proportions. For example, although the large, overlapping leaf bases of a banana tree form a "trunk" that can raise leaves more than 10 m, its pyramid-shaped stem barely reaches

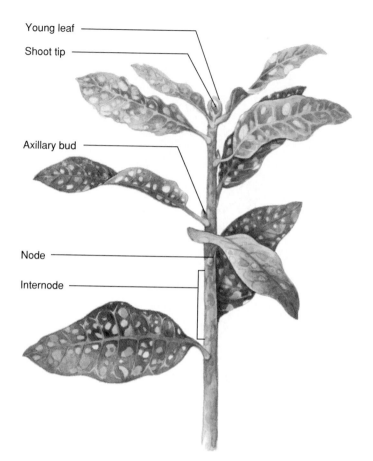

Young leaf

Shoot tip

Axillary bud

Node

Internode

figure 14.1

Stems consist of nodes and internodes. Nodes are regions where leaves attach to stems, whereas internodes are the parts of stems between nodes.

above ground. Most rosettes fail to elongate because they do not make enough physiologically active gibberellins. Thus, treating rosettes with gibberellins typically causes internodal elongation.

c o n c e p t

Stems support leaves, produce and store carbohydrates, and transport materials between roots and leaves. Stem elongation occurs in subapical regions, is influenced by hormones, and helps plants intercept light.

The Structure of Stems

Stems are made of three tissues: epidermal tissue, ground tissue, and vascular tissue (fig. 14.3).

Epidermal Tissue

Stems are surrounded by a transparent epidermis that is usually one cell thick and often bears trichomes. The trichomes of tomato plants secrete an irritating juice that deters hungry insects, while hook-shaped trichomes often entangle insects and prevent them from feeding while they struggle to free themselves.

figure 14.2

Bamboo growing in Indonesia. These plants elongate rapidly from intercalary meristems at the bases of their nodes.

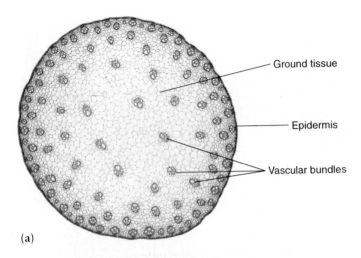

(a)

Ground tissue

Epidermis

Vascular bundles

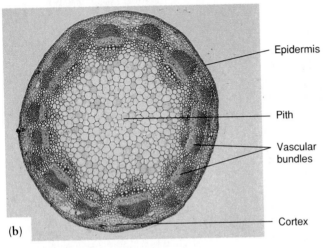

(b)

Epidermis

Pith

Vascular
bundles

Cortex

figure 14.3

Cross section of a stem of (a) corn (*Zea mays*), a monocot, ×5, and
(b) sunflower (*Helianthus annuus*), a dicot, ×10. In monocots,
vascular bundles (see fig. 14.4) occur throughout the ground tissue. In
dicots, vascular bundles typically occur in a ring surrounding a
parenchymatous pith.

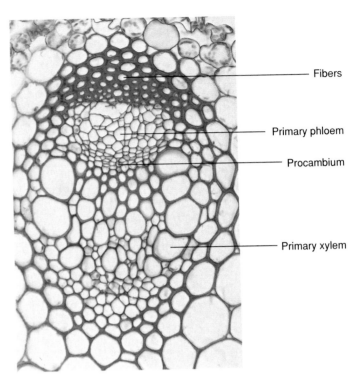

Fibers

Primary phloem

Procambium

Primary xylem

figure 14.4

Vascular bundle from a stem of *Ranunculus*, ×400. A bundle consists
of primary phloem, primary xylem, and procambium. Vascular
bundles are often associated with bundles of fibers.

Vascular Tissue

Substances coming from young leaves control the differentiation of procambium and vascular tissue in stems. Indeed, there
is no procambium above the youngest leaf primordium, and removing a plant's leaf primordia stops vascular differentiation.
The xylem-inducing effect of a leaf primordium can be produced by auxin, suggesting that auxin from leaf primordia controls the differentiation of vascular tissue in stems. Because vascular tissue differentiates in response to leaf primordia, it is not
surprising that *vascular tissue in stems is arranged relative to
leaves.* This is most obvious at nodes, where one or more vascular bundles branch into leaves.

Xylem and phloem in stems occur in vascular bundles
(fig. 14.4). Phloem forms before xylem and differentiates on the
outside of the bundle. Xylem forms on the inside of the bundle.

Vascular bundles are often enclosed by or associated
with sclerenchyma fibers that differentiate after the internode
has finished elongating. However, a layer of cells between the
xylem and phloem remains meristematic (fig. 14.4). In woody
plants and some herbaceous dicots, this layer of cells later becomes part of the vascular cambium, which is the lateral
meristem that produces secondary growth (see Chapter 16).
The absence of a vascular cambium in monocots, however, is
an important feature of this group of plants.

Vascular bundles in stems are arranged differently in different groups of plants (fig. 14.3):

- **Monocots** such as corn (*Zea mays*) have vascular
  bundles embedded throughout the ground tissue.

- **Most dicots** such as alfalfa (*Medicago sativa*) and
  sunflower (*Helianthus annuus*) have a single ring of
  vascular bundles embedded in ground tissue.

- **Many nonflowering plants and a few dicots** such as
  linden (*Tilia americana*) have concentric cylinders of
  xylem and phloem. The cylinder of phloem surrounds an
  inner cylinder of xylem.

These generalities are not absolute: monocots such as Job's
tears (*Coix lacryma-jobi*) have a ring of vascular bundles,
while dicots such as *Bougainvillea* have vascular bundles scattered throughout the stem.

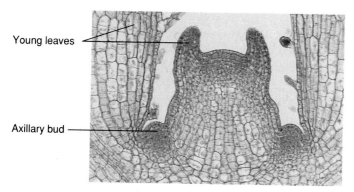

figure 14.5

Shoot tip of *Coleus*. Axillary buds form in the axil where a leaf attaches to the stem. The products of the apical meristem of a *Coleus* plant are shown in figure 14.10.

Ground Tissue

Between the epidermis and ring of vascular tissue in dicots is the **cortex** (fig. 14.3b). Most cortical cells are parenchyma. Cortical cells are photosynthetic in plants such as *Pelargonium* and often store starch.

In dicots, the parenchymatous ground tissue in the center of the stem is specialized for storage and is called **pith** (fig. 14.3b). Pith cells are often lignified and arranged loosely, and the pith may contain secretory structures such as laticifers. In some plants, stresses induced by growth destroy the central part of the pith, thereby forming a hollow stem. Because monocots have vascular bundles throughout their ground tissue, their stems do not have a readily discernable pith or cortex (fig. 14.3a); the parenchyma cells in monocot stems are referred to simply as ground tissue.

---

c o n c e p t

Stems consist of epidermal, ground, and vascular tissue. The vascular tissue is embedded in ground tissue, which often produces or stores food. Stems are covered by a protective epidermis.

Axillary Buds and Branching

Early in leaf development, a small island of meristematic cells forms in the axil where a leaf attaches to the stem. These cells rapidly form an **axillary bud** (fig. 14.5), which typically undergoes a dormant period controlled by hormones made by the shoot apex. In most plants, axillary buds near the shoot apex remain dormant, while those progressively farther away from the tip start to grow. This dominating effect of the shoot apex on growth of axillary buds is called *apical dominance* and strongly influences the symmetry of the shoot. Apical dominance is strong in plants such as pine and spruce, and accounts for their tiered, Christmas-tree shape. Conversely, plants with weak apical dominance have a shrublike shape. Axillary buds are important because they are a shoot's insurance policy: they are inactive (i.e., genetically resting) cells that can form a branch or flower. As such, axillary buds can replace or modify the roles of the shoot apical meristem.

(a)

(b)

figure 14.6

Modified stems that grow aboveground. (a) Stolons of beach strawberry (*Fragaria chilensis*). Roots that form at intervals provide new sources of water and minerals for the plant. Rooted parts of the plant can live independently if the stolon is cut. (b) Tendrils help a plant cling to other objects.

Modified Stems

Plants often modify their stems for special functions. Such modified stems can have unusual and even bizarre shapes, and grow aboveground and belowground. Refer to figures 14.6 and 14.7 as you learn about these unusual stems.

Modified Stems That Grow Aboveground

Stolons or **runners** are horizontally oriented stems that grow along the soil surface. Their function is vegetative reproduction.

figure 14.7

Modified stems that grow belowground. (a) Onions are bulbs consisting of layered, fleshy leaf bases attached to a short stem. Roots form on the underside of the stem. (b) The Irish potato is a tuber. Sprouts growing from the "eyes" of this potato will produce new plants.

For example, buds at nodes of strawberry stolons produce shoots and roots that eventually form new plants (fig. 14.6a). Other plants having stolons include Boston fern (*Nephrolepis exaltata*), Bermuda grass (*Cynodon dactylon*), crabgrass (*Digitaria sanguinalis*), and spider plant (*Chlorophytum*).

Tendrils and **twining shoots** of plants such as morning glory and sweet potato coil around objects and help support the plant. Tips of tendrils of plants such as Virginia creeper (*Parthenocissus*) have adhesive pads that stick to nearby objects. Tendrils and twining shoots are important because they help support plants (fig. 14.6b).

Searcher shoots are stems with long internodes that move in circles through the air, thereby increasing the probability that they will contact a supportive structure. Searcher shoots may be more than 3 m long and often die if they do not find a substrate. Examples of plants with searcher shoots include honeysuckle (*Lonicera*) and *Wisteria*.

Cladodes or **cladophylls** are flat, leaflike stems modified for photosynthesis. Examples of plants with cladodes include butcher's broom (*Ruscus*), greenbrier (*Smilax*), asparagus (*Asparagus officinalis*), and orchids (e.g., *Epidendrum*).

Thorns are modified stems that protect plants from grazing animals. Examples of plants with thorns are honey locust (*Gleditsia triacanthos*), fire thorn (*Pyracantha*), hawthorn (*Crataegus*), and bougainvillea (*Bougainvillea spectabilis*).

Short and long shoots are distinguished by internodal elongation: long shoots have long internodes, while short shoots are rosettes that have short internodes. This difference in shoot length usually underlies a functional division of labor. For example, long shoots of pine (*Pinus*) produce protective scale leaves, while short shoots produce foliage leaves. Examples of plants with short and long shoots include cedar (*Cedrus*), maidenhair tree (*Ginkgo*), and barberry (*Berberis*).

Succulent stems of plants such as cacti have a low surface-to-volume ratio. Succulent stems store large amounts of water and are common in plants growing in deserts.

Modified Stems That Grow Belowground

Bulbs are rosette stems surrounded by fleshy leaves that store nutrients. When these nutrients are removed (such as during flowering), the fleshy leaves collapse into a papery scale leaf surrounding the stem. Onion, lily, hyacinth, tulip, narcissus, and daffodil are examples of plants that produce bulbs (fig. 14.7a). Bulbs are important because they enable many plants to survive stressful periods, such as winter.

Rhizomes are underground stems that grow near the soil surface. They typically have short internodes and scale leaves, and produce roots along their lower surface. Rhizomes store food for renewing growth of the shoot after periods of stress, such as cold winters. This explains why plants such as Johnson grass (*Sorghum halepense*) are so difficult to eradicate—although their aerial shoots die back during winter, their dormant rhizomes renew growth the following spring. Other examples of plants having rhizomes include quack grass (*Agropyron repens*), *Iris, Canna, Begonia,* and ginger (*Zingiber officinalis*).

Corms are stubby, vertically oriented stems that grow underground. Corms, which have only a few thin leaves, store nutrients. Like rhizomes, corms enable many plants to survive winter. Examples of plants having corms are *Gladiolus*, *Cyclamen*, and *Crocus*.

Tubers are swollen regions of stems that store food for subsequent growth. For example, Irish potatoes (*Solanum tuberosum*) are tubers produced on stolons that burrow into the soil (fig. 14.7b). The eyes of potatoes are buds in the axils of small, scalelike leaves; the areas between a potato's eyes are internodes.

c o n c e p t

Plants use modified stems for reproduction, climbing, photosynthesis, protection, and storage. These modified stems adapt plants to different environments.

The Economic Importance of Stems

We have many uses for stems. Sugar and molasses from sugarcane (*Saccharum officinalis*) go in our drinks and over our pancakes, and stem fibers are used to make several fabrics. We also eat a few stems, such as tubers of potatoes. In some parts of the world, potato tubers are the most important part of the diet. Andean tribes mash and dry tubers to form *chuño*, which is added to just about everything they eat. Indeed, members of the tribes say that "stew without *chuño* is like life without love." Although we don't expect you to take your potatoes quite that seriously, we do hope you'll remember what you're eating next time you order some fries.

Lumber, pulp for paper, charcoal, corks, insulation, life preservers, quinine, cinnamon, and even a diesel-like fuel are derived from stems. However, these products come mainly from wood or bark and will be discussed in Chapter 16 when we discuss secondary growth.

Leaves

Leaves are the most active and conspicuous organs of plants. The most important of their functions is absorbing sunlight for photosynthesis. To do this, they expose large amounts of surface area to the environment. For example, a maple tree (*Acer*) with a trunk 1 m wide has approximately 100,000 leaves with a combined surface area exceeding 2,000 m²—that's roughly the area of six basketball courts. Oak trees have approximately 700,000 leaves, and you'd better think twice before agreeing to rake a lawn shaded by American elm trees (*Ulmus americana*)—when mature, these trees can each produce more than 5 million leaves per season. On a global basis, leaves produce more than 200 billion tons of sugars per year. Those sugars sustain virtually all life on this planet.

How Leaves Form

Leaves are the most diverse of all plant organs—they can be tubular, needlelike, feathery, cupped, sticky, fragrant, smooth, or waxy. Leaves range in size from the pinhead-sized leaves of watermeal (*Wolffia*) to the 20 m fronds of tropical palms, and they range in number from the millions of leaves on American elms to those of *Welwitschia mirabilis*, a desert plant of southwest Africa that grows only two leaves during its lifetime (several hundred years). However, regardless of their number or size, leaves are formed by the coordinated efforts of several meristems, each of which is named for its position.

Although mature leaves often consist of millions of cells (e.g., leaves of broad bean contain about 40 million cells), the earliest stage of leaf development is a small bulge at the shoot apex called a **leaf buttress** or **leaf primordium** that consists of only 100 to 300 cells. This buttress is formed by cellular divisions one to three cell layers below the overlying protoderm. Continued cellular divisions and cellular expansion produce a symmetrical cone called an **apical peg**. This peg has an apical meristem and a procambial strand that, in most species, forms the leaf's midrib. The adaxial surface (i.e., the surface closest to the internode above it) of the apical peg elongates slower than the abaxial surface, thereby arching the leaf primordium over the shoot apical meristem. This differential growth is important because it positions the primordium to protect the apical meristem.

The leaf then forms an **adaxial meristem** that thickens the leaf. Soon thereafter (i.e., when the leaf primordium is about 0.2–0.5 mm high) it forms an **upper leaf zone** and a **lower leaf zone**. The upper leaf zone contains **marginal meristems** that form the flattened blade and stalklike petiole that attaches the leaf to the stem. In plants such as cocklebur (*Xanthium*), marginal growth continues for approximately 3 weeks and produces a blade about six cells thick. The lower leaf zone forms the leaf base.

Continued growth of a leaf involves cellular expansion and division. For example, leaf formation in cocklebur involves almost thirty generations of cells; these divisions continue until the leaf is one-half to three-fourths grown. Cellular expansion forms most of the intercellular spaces in a leaf. Stomata differentiate soon after intercellular spaces form. Except for vascular tissue, which differentiates acropetally (i.e., from the base into the tip of the leaf), other tissues in leaves differentiate basipetally (i.e., from the tip toward the base). Until it is 30%–40% of its final size, a growing leaf depends on the rest of the plant for its nutrition.

Developmental abnormalities that alter cellular expansion and division also affect a leaf's morphology. For example, tobacco mosaic virus inhibits the marginal meristems of developing tobacco leaves. Consequently, infected leaves have deformed blades or no blades at all.

Fern leaves are called *fronds*. Unlike the leaves of other plants, fronds usually form in a curled structure called a *fiddlehead* (fig. 14.8). During development, the adaxial side of the fiddlehead elongates faster than the abaxial side, causing the fiddleheads to unroll and form a frond. Fronds of some ferns can live more than 50 years and be more than 15 m long. We even eat some of them: for example, fiddleheads of ostrich fern (*Matteuccia struthiopteris*) are an asparagus-like delicacy enjoyed by many New Englanders.

figure 14.8

Leaves of ferns are called *fronds*. Fronds unroll from base to tip. When they are young and coiled as shown in this photo, they are referred to as fiddleheads.

Phyllotaxis

Phyllotaxis (from the Greek words *phyllon*, meaning "leaf," and *taxis*, meaning "arrangement") is the arrangement of leaves on a stem. It is determined at the shoot apex and is species-specific.

- Most plants have **spiral**, or **alternate**, **phyllotaxis** (fig. 14.9), meaning that they have one leaf per node. Birch (*Fagus*), oak (*Quercus*), and *Agave* have a spiral phyllotaxis. If the leaves form two parallel ranks along the stem, the phyllotaxis is *distichous*, as in ginger and pea (*Pisum sativum*).

- Plants with **opposite phyllotaxis** have two leaves per node, as in maple (*Acer*) and ash (*Fraxinus*). When pairs of leaves at successive nodes form at right angles to each other, the phyllotaxis is *decussate*, as in *Coleus* (fig. 14.10).

- Plants with **whorled phyllotaxis** have three or more (and as many as twenty-five) leaves per node. Oleander (*Nerium oleander*), *Peperomia*, and horsetail (*Equisetum*) have a whorled phyllotaxis.

figure 14.9

Alternate (spiral) phyllotaxis in poplar.

figure 14.10

Coleus has decussate phyllotaxis. It has two leaves per node; these pairs of leaves at successive nodes form at right angles to each other. This view is from above the plant. The meristem that produced these leaves is shown in figure 14.5.

Phyllotaxis is independent of leaf shape and can be described mathematically. The leaves of plants with alternate phyllotaxis are arranged in spirals around the stem, and each leaf is vertically superimposed on other leaves of the spiral. That is, the leaves are arranged in vertical files on stems, which are called **orthostichies**. Botanists use orthostichies to describe phyllotaxis with the following equation:

$$\text{phyllotaxis} = \frac{\textbf{(number of turns of spiral)}}{\substack{\textbf{(number of leaves between} \\ \textbf{successive leaves of an orthostichy)}^*}}$$

*Counting the first leaf, but not the last leaf directly above or below the first.

figure 14.11

Whorls of spines (modified leaves) on cactus. Each spine is part of two opposing spirals called parastichies. This cactus has a phyllotaxis of (13 + 21).

Plants with decussate leaves, such as *Coleus* (fig. 14.10), have a phyllotaxis of (1 + 2).* This means that you must circle the stem once to find two leaves that are superimposed vertically. There are two leaves on this spiral, meaning that each leaf is halfway around the stem. Many botanists also quantify phyllotaxis based on the position of leaves at the apex. For example, examine the shoot apex of the cactus shown in figure 14.11. Each leaf is part of two opposing spirals called **parastichies**. Botanists describe a plant's phyllotaxis by counting these spirals. The cactus shown in figure 14.11 has a phyllotaxis of (13 + 21). Other common plants and their phyllotaxis include cabbage (*Brassica oleracea*: 1 + 2), flax (*Linum usitatissimum*: 3 + 5), and sunflower (*Helianthus annuus*: 34 + 55). The (2 + 3) phyllotaxis is most common in angiosperms, and almost half of all nonflowering seed plants have a (3 + 5) phyllotaxis. As you might have already noticed, each of the numbers that describe phyllotaxis is part of a mathematical progression called the *Fibonacci series*, a numerical series in which each number (except the first two) is the sum of the two preceding numbers: 1, 2, 3, 5, 8, 13, 21, 34, 55, etc. (see reading 14.1, "Leonardo the Blockhead"). Leaf formation can also be described temporally. A **plastochron** is the time required to form two successive leaves at the shoot apex. Plastochrons in most plants vary considerably, depending on the plant's growth rate and nutritional status. Plastochrons in flax (*Linum*) and spruce (*Picea*) range from 3 to 22 hours, while in clover (*Trifolium*) they range from 1.2 to 2 days.

---

*Phyllotaxis is sometimes presented as a fraction. In this example, it would be ½.

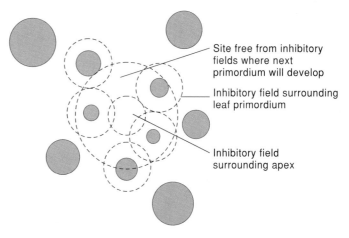

figure 14.12

The position of a leaf on the mature plant is determined largely by where it forms in the apical meristem. One hypothesis to explain where leaves will form in the apex is the *inhibitor field* hypothesis, which suggests that a new leaf forms in the next available space that is free of the inhibitory influences of other leaves.

What Controls Phyllotaxis?

Although phyllotaxis has been studied by botanists and philosophers alike, we still know relatively little about the process. Since phyllotaxis is usually unaltered by changes in daylength, light intensity, or moisture, what causes leaves to form at precisely the "right" place? One hypothesis is that older primordia determine the sites where new primordia form by releasing compounds that inhibit the formation of other primordia. As a result, each older primordium is surrounded by an inhibitor field in which new primordia cannot form (fig. 14.12). Older leaves produce progressively smaller amounts of the inhibitor, thereby forming a moving field. New primordia form where these fields are weakest. Although we do not know what these inhibitors are or if they actually exist, plant hormones can sometimes alter phyllotaxis. For example, applying gibberellic acid to shoot tips of cocklebur changes the phyllotaxis from (2 + 3) to (3 + 5).

Another hypothesis is that phyllotaxis is controlled physically by differential reinforcement in the outermost layers of the shoot apex. According to this hypothesis, leaves form at points of least reinforcement, much as bulges form at the weakest areas of tires (see Chapter 12). Although surgical experiments to test these and other ideas have produced interesting results, we still do not understand exactly what controls phyllotaxis.

c o n c e p t

Leaves are the primary photosynthetic organs of most plants. Leaves are formed by several meristems in specific positions. The arrangement of leaves on a stem is called *phyllotaxis* and is usually species-specific.

LEONARDO THE BLOCKHEAD

His real name was Leonardo, but his friends in Pisa called him "the Blockhead." History speaks of him simply as Fibonacci, a thirteenth-century mathematician who discovered one of the great mysteries of the universe. In 1202, at the age of twenty-seven, Fibonacci published *Liber Abaci (The Book of the Abaci)*, a historic book that introduced Europeans to Arabic numbers. A small section of the book contained a puzzling problem that has fascinated mathematicians and other scientists for centuries. Fibonacci wrote:

> Someone placed a pair of rabbits in a certain place, enclosed on all sides by a wall, to find out how many pairs will be born in the course of one year, it being assumed that every month a pair of rabbits produces another pair, and that rabbits begin to bear young two months after their own birth.

Fibonacci calculated the total pairs of rabbits at the end of each month and came up with the following sequence of numbers: 1, 2, 3, 5, 8, 13, 21, 34, 55, 89, 144, 233, 377, 610, and so on. Except for the first two, each of these numbers is the sum of the two preceding numbers. Interestingly, these numbers appear throughout the plant kingdom (e.g., see fig. 14.11 and reading fig. 14.1). For example,

- You'd better check out the Fibonacci series before trying the "loves me, loves me not" routine—daisies usually have 21, 34, 55, or 89 petals.

- Most sunflowers have spirals of 34 and 55 seeds. Smaller sunflowers have spirals of 21 and 34 seeds, and a giant sunflower grown in the former Soviet Union reportedly had spirals containing 89 and 144 seeds.

reading figure 14.1

Part of a branch of *Prunus*. The line that passes from leaf to leaf shows a (2 + 5) phyllotaxis: you must circle the stem twice and pass five leaves before you find two leaves that are superimposed vertically.

- Pine needles usually grow in clusters of 2, 3, or 5, depending on the species. Furthermore, the modified leaves of pine cones are arranged in spirals based on Fibonacci numbers.

- There are usually 13 buds arranged between 2 vertical lines on stems of pussy willow. To count these buds, you must circle the stem 5 times.

The Fibonacci series extends far beyond botany. Look again at the Fibonacci numbers. Each number has a special relationship to the numbers surrounding it—if you divide a Fibonacci number by the next higher number, you'll see that it is precisely 0.618034 times as large as the number that follows. (This figure is valid only when the Fibonacci numbers are large enough to be precise—it works for all numbers after the fourteenth in the sequence.) This number oc-

curs throughout nature with amazing precision. For example, it is the basis for Greek architecture (including the Parthenon), the Egyptian pyramids, playing cards, and the work of countless artists, including another Leonardo—Leonardo da Vinci. However, even these examples hardly touch the importance of Fibonacci's numbers. To understand this, consider the so-called golden rectangle having a length:width ratio of 0.618034. If you draw a square into one end of the rectangle, you are left with another, smaller golden rectangle. Repeating this process produces a succession of golden rectangles. If you connect the centers of the successive squares that you draw, you'll have a spiral, any part of which is 0.618034 times larger than the remainder of the spiral. This "golden spiral" underlies the design of mollusk shells, ram horns, parrot beaks, elephant tusks, lion claws, breaking waves, comet tails, spider webs, meteor craters, the shoreline of Cape Cod, and even the spiral galaxies of the universe.

Finally, the rabbits of Fibonacci pop up in music, a universal language that we hear and feel. For example, Western music is based on an 8 note octave, which, on a piano keyboard, is represented by 5 black and 8 white keys—a total of 13 keys. Furthermore, the major third is the interval that our ears like best; in the major third from note C to note E, for example, the E vibrates at a ratio of 0.625000 to the C, only 0.006966 away from the golden mean. These notes produce good vibrations in your inner ear's cochlea, a spiral-shaped organ. So next time you catch yourself dancing or tapping your toes to the beat of your favorite song, think of Leonardo the Blockhead and how his magic numbers describe what we like to build, look at, feel, and hear.

The Structure of Leaves

External Structure

Although leaf morphology can be affected by the environment, it can often be used as a taxonomic tool to identify plants; maple leaves, for example, are easily distinguished from leaves of, say, walnut. Despite this strong genetic control of leaf morphology, no two leaves on a plant are identical.

There are four basic kinds of leaves: simple, compound, peltate, and perfoliate leaves. Refer to figures 14.13–14.15 as you learn about these kinds of leaves:

figure 14.13

Simple leaves of *Thottea*. Simple leaves have a flat, undivided blade attached to the stem by a stalk called a petiole.

Simple leaves have a flat, undivided *blade* that is supported by a stalk called a *petiole* (fig. 14.13). The petiole is typically supported by collenchyma and sclerenchyma fibers. In plants such as silk tree (*Albizia*), the petiole includes a jointlike swelling called a *pulvinus* that enables the leaf to respond to environmental stimuli such as light and gravity (see Chapter 19). Leaves of plants such as *Zinnia* that lack petioles are called *sessile* leaves. Redbud, elm, and maple have simple leaves.

Compound leaves have blades divided into *leaflets* that form in one plane (fig. 14.14). Leaflets lack axillary buds, but each compound leaf has a single bud at the base of its petiole. There are two kinds of compound leaves: pinnately compound leaves and palmately compound leaves.

Leaflets of **pinnately compound leaves** form in pairs along a central, stalklike *rachis* (fig. 14.14a). Ash, walnut, and rose have pinnately compound leaves.

Leaflets of **palmately compound leaves** attach at the same point, much as fingers are attached to your palm (fig. 14.14b). Examples of palmately compound leaves

(a)

(b)

figure 14.14

Compound leaves. (a) Pinnately compound leaves of this black walnut (*Juglans nigra*) form in pairs along a central rachis. (b) Palmately compound leaf of Buckeye. This leaf of *Aesculus* is palmately compound, or fan-shaped.

include horse chestnut (*Aesculus*), marijuana (*Cannabis sativa*), and lupine (*Lupinus*). The most famous palmately compound leaves are those of the trifoliate shamrock (*Trifolium*) that St. Patrick used to explain the doctrine of the Trinity. Today, the shamrock is the national plant of Ireland and is traditionally worn on St. Patrick's Day. Similarly, the "leaves" of a four-leaf-clover are actually leaflets of a palmately compound leaf.

Peltate leaves have petioles that attach to the middle of the blade (fig. 14.15b). Mayapple (*Podophyllum*) is an example of a plant with peltate leaves. The most extreme examples of peltate leaves are the tubular leaves of carnivorous and other plants.

(a)

(b)

figure 14.15

Perfoliate leaf (a) and peltate leaf (b).

Perfoliate leaves are sessile leaves that surround and are pierced by stems (fig. 14.15a). Perfoliate leaves of yellow-wort (*Blackstonia perfoliata*) and thoroughwort (*Eupatorium perfoliatum*) form by the fusion of two opposing leaves.

Many plants have adult and juvenile leaves that form on adult and juvenile parts of the plant. For example, juvenile leaves of some species of *Acacia* are compound, whereas adult leaves are simple. Similarly, elm produces progressively more teeth along the edges of its leaves as the tree ages.

Although leaf shape is characteristic for many species, leaf size can vary greatly. Clearly, the number and size of epidermal cells determine leaf area; some evidence suggests that the ratio of epidermal cells to the total number of cells in a leaf is relatively constant after cellular divisions stop. Because cellular divisions stop first in the epidermis, the epidermis may govern leaf size.

Internal Structure

Leaves consist of epidermal, ground, and vascular tissues.

Epidermis The epidermis of most leaves is compact, transparent, and usually not photosynthetic. It also contains numerous stomata (e.g., a cabbage leaf typically has more than 11 million stomata). In horizontally oriented leaves, there are usually more stomata on the protected lower side than on the exposed upper side. Conversely, as shown in the following table, vertically oriented leaves usually have similar numbers of stomata on the different sides of their leaves:

Plant and Leaf Orientation	Stomatal Frequency (stomata cm^{-2})	
	Upper Epidermis	Lower Epidermis
Horizontally oriented leaves		
Apple (*Malus sylvestris*)	0	38,760
Bean (*Phaseolus vulgaris*)	4,031	24,806
Oak (*Quercus velutina*)	0	58,140
Pumpkin (*Cucurbita pepo*)	2,791	27,132
Vertically oriented leaves		
Corn (*Zea mays*)	9,800	10,800
Pine (*Pinus sylvestris*)	12,000	12,000
Onion (*Allium cepa*)	17,500	17,500

The frequency and distribution of stomata vary in different species and in different parts of individual leaves. For example, stomata differentiate in parallel rows in monocot leaves, while in dicots they are more scattered.

Although stomata usually occupy less than 1% of the leaf surface, they lose huge amounts of water to the atmosphere. The evaporation of water from leaves into the atmosphere is called *transpiration* and can influence patterns of rainfall. For example, about half of the moisture in rainfall in Amazon rain forests originates from transpiration. On the more local scene, your neighbor's 1-acre lawn can transpire more than 100,000 l of water per week during the summer.

Vascular Tissues Xylem and phloem in leaves form in strands called **veins.** Xylem forms on the upper side of a vein (on a vertical leaf, on the side next to the stem), and phloem forms on the lower side (on a vertical leaf, on the side away from the stem) (fig. 14.16). Veins are often supported by fibers and are usually surrounded by a layer of parenchyma cells called the *bundle sheath.* The bundle sheath often extends to the epidermis of the leaf. These *bundle-sheath extensions* help support the veins and may conduct water to epidermal cells.

Veins are a leaf's "fingerprint" and can be used to identify plants. Most dicots and some nonflowering plants have *netted venation,* meaning that they have one or a few prominent *midveins* from which smaller *minor veins* branch into a meshed network (fig. 14.17a). The leaves of most monocots have *parallel venation,* meaning that several prominent and parallel veins interconnect with smaller, inconspicuous veins (fig. 14.17b). Minor veins are extensive; for example, there are about 70 cm of minor veins per cm^2 in the leaves of sugar beet

14–13 *Unit Four* The Form and Function of Plants

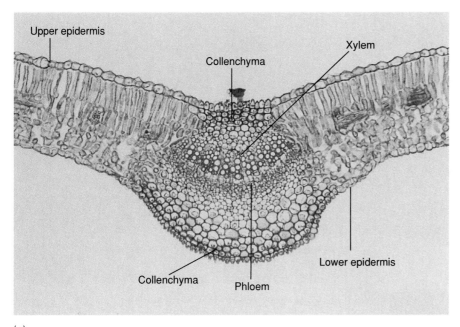

(a)

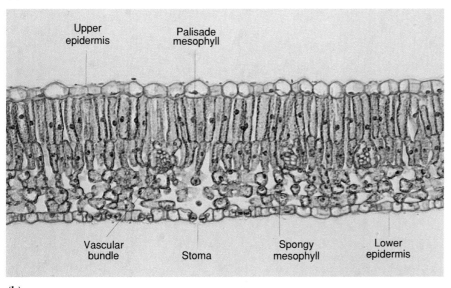

(b)

figure 14.16

Cross sections of a leaf of lilac (*Syringa vulgaris*) through (a) the midvein and (b) the blade. Minor veins are visible in the blade.

Although veins are prominent in most leaves, in no plant are they more obvious than in Madagascar lattice-leaf (*Aponogeton fenestrale*), an aquatic plant with large, frilly leaves. Leaves of these plants have only a few chlorenchyma cells surrounding their veins; that is, there are no chlorenchyma cells linking adjacent veins. As a result, the leaves look like a skeleton of veins and are a natural demonstration of venation.

Ground Tissue The ground tissue of leaves is called **mesophyll.** It contains several types of cells, including sclerenchyma (usually sclereids), storage parenchyma, and chlorenchyma. Chlorenchyma cells are photosynthetic cells that, as you learned in Chapter 7, can have differing arrangements and metabolisms. In general, the arrangement of chlorenchyma is determined genetically and is influenced by whether the leaf is oriented horizontally or vertically when it forms.

Horizontally oriented leaves. Examine the cross section of the privet (*Ligustrum*) leaf shown in figure 14.19. These leaves have two different kinds of chlorenchyma tissue. Along the upper side of the leaf are one or more layers of long, columnar chlorenchyma cells called **palisade mesophyll** cells, which are densely packed (i.e., have small intercellular spaces) and perform as much as 90% of the leaf's photosynthesis. Palisade cells contain large amounts of chlorophyll and are thus specialized for the light-absorption and carbon-fixation portions of photosynthesis.

Along the lower side of horizontally oriented leaves are **spongy mesophyll** cells, which are a type of photosynthetic aerenchyma tissue. Spongy mesophyll cells are green, irregularly shaped cells separated by large intercellular spaces connected to stomata. Depending on the species, these air spaces account for 10%–70% of the leaf volume, thereby increasing the amount of internal surface area for gas exchange. Spongy mesophyll is specialized for the gas-exchange portion of photosynthesis.

Vertically oriented leaves. Plants such as corn and other monocots form leaves vertically. These leaves intercept light from all directions, and their chlorenchyma is usually arranged differently than that of horizontally oriented leaves, which intercept light primarily on their upper surfaces. To appreciate this difference in leaf anatomy, examine the cross section of the corn leaf shown in figure 14.20a.

(*Beta*). In many species these minor veins end blindly in the mesophyll as *vein endings*, which are the "business end" of the leaf's vascular system (fig. 14.18). Each vein ending services a small neighborhood of cells and is where most water and solutes are exchanged with cells of the leaf. Each vein ending in leaves of sugar beet services about 30 cells, each of which is within about two cell diameters (70 μm) of the vein.

Several factors influence the formation of veins, including the presence of other veins. For example, new veins form in blades of *Narcissus* as soon as cellular divisions separate existing veins by more than about eleven cells.

(a)

figure 14.18

Vein endings in *Diospyros kaki*. Most water and solutes are exchanged between the veins and other cells of the leaf at vein endings.

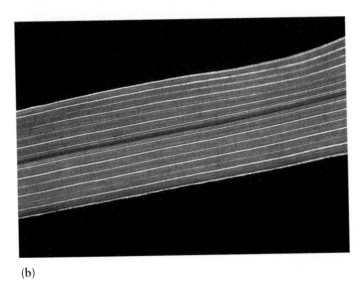

(b)

figure 14.17

Venation in leaves. (a) Leaves of dicots such as pumpkin (shown here) have a netted venation. (b) Leaves of monocots such as lily have parallel venation.

Vertically oriented leaves lack palisade and spongy layers. Rather, most of their chlorenchyma cells appear similar; this is why these cells are called uniform mesophyll cells.

The ground tissue of many tropical grasses is arranged concentrically (fig. 14.20). Uniform chlorenchyma surrounds a photosynthetic bundle sheath containing large, active chloroplasts. This combination of uniform chlorenchyma and photosynthetic bundle-sheath cells is referred to as *Kranz* (German for "wreath") *anatomy* and underlies a metabolic division of labor called C_4 photosynthesis (see Chapter 7). Pigweed (*Chenopodium*), corn (*Zea*), Bermuda grass (*Cynodon*), crabgrass (*Digitaria*), and nutsedge (*Cyperus*) are examples of C_4 plants that have vertically oriented leaves with Kranz anatomy. However, not all grasses are C_4 plants; fescue (*Festuca*) and bluegrass (*Poa*) are C_3 plants. As you might guess, the bundle sheaths of these C_3 plants are not photosynthetic.

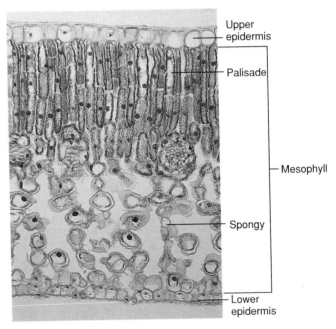

figure 14.19

Cross section of a leaf of the common hedge privet (*Ligustrum*), ×50. The photosynthetic tissue consists of a densely packed palisade mesophyll and a loosely packed spongy mesophyll.

c o n c e p t

Leaves have many shapes and sizes, and are made of ground tissue, veins, and epidermis. Chlorenchyma cells of the ground tissue produce sugars and receive nutrients and water from vascular tissues of veins. Leaves are covered by epidermis containing stomata that regulate gas exchange.

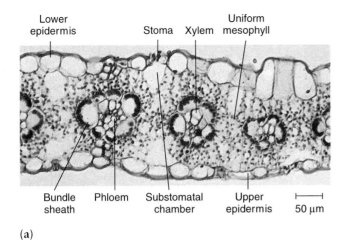

Lower epidermis   Stoma   Xylem   Uniform mesophyll

Bundle sheath   Phloem   Substomatal chamber   Upper epidermis   50 µm

(a)

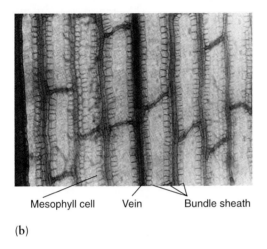

Mesophyll cell   Vein   Bundle sheath

(b)

figure 14.20

(a) Cross section of a leaf of corn (*Zea mays*), a C₄ plant. Vascular bundles are surrounded by photosynthetic bundle-sheath cells, which, in turn, are surrounded by photosynthetic uniform mesophyll. This arrangement of chlorenchyma is typical of C₄ plants and underlies a photosynthetic division of labor (see Chapter 7). (b) Veins (i.e., vascular bundles) of C₄ plants are tightly surrounded by bundle-sheath cells. Although the bundle-sheath cells appear to be on only two sides of the vein, they actually encircle the vein.

Environmental Control of Leaf Variation

The morphological differences that distinguish plants growing in different habitats are most striking in leaves. Indeed, leaf development is affected by several environmental factors, the most influential of which are light and moisture.

Light

Daylength, light intensity, and the presence or absence of light strongly affect leaf development.

The Presence or Absence of Light   Leaves of most dicots require light to expand and produce chlorophyll. Light that controls leaf expansion is absorbed by phytochrome, a family of

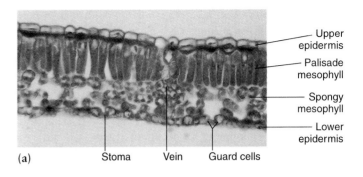

Upper epidermis — Palisade mesophyll — Spongy mesophyll — Lower epidermis

(a)   Stoma   Vein   Guard cells

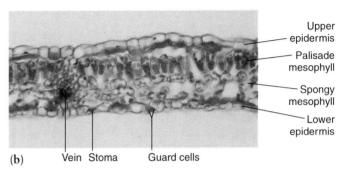

Upper epidermis — Palisade mesophyll — Spongy mesophyll — Lower epidermis

(b)   Vein   Stoma   Guard cells

figure 14.21

Cross section of (a) sun versus (b) shade leaves of maple. Sun leaves have larger mesophyll cells and more chloroplasts than do shade leaves.

pigments that also influences several other aspects of plant growth and development (see Chapter 19). This light-controlled expansion of leaves is important because it ensures that a plant will not form leaves unless light is present for photosynthesis.

Daylength   Leaf development is also affected by daylength, which is the amount of light per day. During short days, plants such as *Kalanchoë* produce small, succulent leaves that are sessile and not lobed. During long days, *Kalanchoë* produces large, thin, lobed leaves with petioles. Daylength also affects a variety of other processes in plants, including flowering and seed germination (see Chapter 19).

Light Intensity   The leaves of many plants respond to differing intensities of light. To best appreciate this, consider the plants growing in a dense rain forest. Leaves atop the plant canopy are bathed in intense light and are called **sun leaves.** These leaves have significantly different structures than do **shade leaves,** which grow in the dim light on the forest floor. Look at the sun and shade leaves shown in figure 14.21.

Sun leaves are smaller and thicker than shade leaves. For example, leaves of *Plectranthus* grown in intense light are three times thicker than leaves grown in dim light.

In intense light, sun leaves fix carbon faster than do shade leaves. For example, sun leaves fix 16–20 mg C dm⁻² h⁻¹, while shade leaves fix only 3–5 mg C dm⁻² h⁻¹. Similarly, sun leaves respire 3–5 times faster than shade leaves.

*Sun leaves have smaller and more numerous chloroplasts than
do shade leaves.* Chloroplasts in sun leaves have fewer
grana than do chloroplasts in shade leaves. Epidermal
cells of shade leaves often contain many chloroplasts.

Sun and shade leaves can form on the same plant; leaves
near the shoot apex are often sun leaves, while leaves in the
dimmer light of the lower canopy are shade leaves. This flexibility in leaf development allows plants to exploit different
and changing environments.

Moisture

As mentioned in the previous chapter, obtaining enough water
is the biggest challenge faced by land plants. Therefore, it is
not surprising that the availability of water strongly influences
leaf development. For example, the features described earlier
in this chapter typify **mesophytes,** which are plants that grow
best in moist but not wet environments (i.e., environments
having intermediate amounts of water). Most plants you are
familiar with are mesophytes. The structural features of mesophyte leaves intergrade with those of plants that live in more
extreme environments: xerophytes and hydrophytes.

Xerophytes   The leaf in figure 14.22a is from a **xerophyte,**
which is a plant that grows in habitats characterized by seasonal or persistent drought. Since these environments are usually bright deserts that support relatively little vegetation, the
availability of light seldom limits the growth of xerophytes.
Rather, their growth is limited by how efficiently they use
water. The leaves of xerophytes usually have one or more of
the following modifications that help conserve water:

*Xerophytes have small, thick leaves with well-developed
spongy and palisade layers.* These leaves are often
modified for storing water and typically contain
relatively few intercellular spaces. Xerophytes such as
sagebrush (*Artemisia*) that grow in seasonally dry
habitats produce relatively large leaves during the wet
season; when drought begins, these leaves are replaced
by smaller leaves. Conversely, some plants abort all of
their leaves during drought. When this occurs, the
survival problem is no longer transpiration but rather
how to stay alive until foliage leaves form in response to
the next rainy season. Plants solve this problem by
producing photosynthetic stems and storing water in
their succulent stems.

*Xerophytes are covered by an epidermis having thick cell
walls, numerous stomata, and a thick cuticle.* For
example, leaves of mesquite (*Prosopis*) growing in dry
areas have cuticles ten times thicker than those growing
in wet soil. The stomata of xerophytes are often sunken
and overlaid with trichomes. The large number of
stomata in leaves of xerophytes increases their
photosynthetic rates during rare wet periods—a
"get-it-while-you-can" strategy.

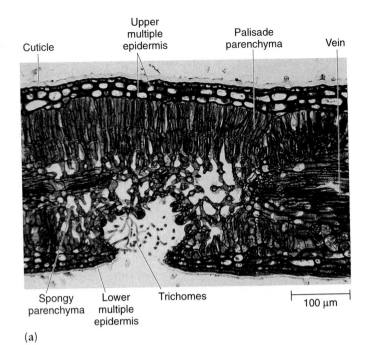

Cuticle   Upper multiple epidermis   Palisade parenchyma   Vein

Spongy parenchyma   Lower multiple epidermis   Trichomes   100 μm

(a)

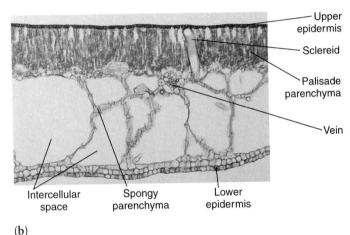

Upper epidermis
Sclereid
Palisade parenchyma
Vein

Intercellular space   Spongy parenchyma   Lower epidermis

(b)

figure 14.22

(a) Cross section of a leaf of oleander (*Nerium oleander*), a xerophyte,
×55. Note the thick cuticle and multiple epidermis. Stomata and
trichomes are sunken into chambers called stomatal crypts. (b) Cross
section of a leaf of water lily (*Nymphaea odorata*), a hydrophyte,
×100. Stomata form only on the upper surface of these floating leaves,
and the vascular tissue (especially the xylem) is reduced. Large
intercellular spaces add buoyancy to the leaves.

*Xerophytes have large amounts of supporting tissues in
their leaves.* Turgor pressure supports the leaves of
plants growing in most environments. Since turgor
cannot always be maintained in xerophytes, their
leaves usually contain large amounts of sclerenchyma
for support.

Not all botanists think that these leaf modifications are
adaptations for conserving water. Indeed, transpiration does
not always decrease when stomata are sunken and covered
with trichomes. Such observations have prompted several

CLOGGING THE WATERWAYS

Most hydrophytes contain large amounts of aerenchyma, which enhances gas exchange between the atmosphere and submerged tissues. Some hydrophytes use aerenchyma for support. For example, petioles of water hyacinth (*Eichhornia crassipes*) have bladders filled with aerenchyma that buoys the plant's leaves.

Water hyacinth was introduced to North America in 1864 at a horticultural exhibit in Louisiana. One exhibitor, especially attracted to the plant's luscious leaves and elegant flowers, took some of the plants and began cultivating them near his home in Florida. There, without its natural herbivores, water hyacinth spread wildly throughout waterways, where today it clogs ponds and drainage canals (reading figs. 14.2a and 14.2b). Similar problems caused by introduced water hyacinth now also plague India and the Nile and Congo Rivers of Africa.

Transpiration from dense mats of water hyacinth can drain a pond at an incredible rate; ponds covered by *Eichhornia* lose water almost 8 times faster than uncovered surfaces. Although *Eichhornia's* leaves can be harvested to feed cattle and generate methane, its clogging of waterways costs millions of dollars to control.

Water hyacinth is not a problem in the Amazon River, its natural habitat. There, herbivores and periodic flooding prevent it from clogging the river.

reading figure 14.2a

Water hyacinth was introduced into Florida in the 1860s. These plants, which can double their population in only two weeks, clog many of the waterways in Florida and other southeastern states.

reading figure 14.2b

Close-up of water hyacinth. Note the air-filled bladders that help buoy the plant.

---

botanists to suggest that xerophytic traits may be caused by adaptations to intense light. Most xerophytic modifications can also be induced in well-watered plants by cold temperatures or nutrient deficiencies. For example, plants given only small amounts of nitrogen often develop more striking xerophytic features than do those deprived of water.

Hydrophytes   Look again at figure 14.22. The leaf in figure 14.22b is from a **hydrophyte,** which is a plant that grows in habitually wet environments (see reading 14.2, "Clogging the Waterways"). Aquatic plants that grow partly or completely submerged in fresh water are hydrophytes and often have unusual leaves. *Wolffia,* a hydrophyte, is the smallest flowering plant; it is common in ponds and slow-moving streams and has leaves only 1 mm wide. Conversely, the leaves of royal water lily (*Victoria amazonica*) are large floating discs up to 2 m wide. Although their undersides have protective spines, their upper surfaces are smooth and are used as floats by insects, frogs, and children. Leaves of royal water lily inspired the architecture of the Crystal Palace of London.

Obtaining enough water is never a problem for hydrophytes; however, the water surrounding submerged plants such as *Elodea* reduces the intensity of light reaching submerged leaves and severely limits gas exchange (remember that diffusion through liquid is much slower than through air). Thus, the watery environment solves one problem (i.e., desiccation) but creates two others: absorbing enough light and exchanging gases. Most leaf modifications in hydrophytes enhance light absorption and gas exchange.

Hydrophytes have large, thin leaves with poorly developed spongy and palisade layers. Epidermal cells typically contain chloroplasts and are photosynthetic, while ground tissue is modified for storage. Leaves of hydrophytes also contain large amounts of aerenchyma for gas exchange and support.

Hydrophytes are covered by a thin cuticle and have thin cell walls. Submerged leaves lack stomata, while floating leaves have stomata only on their upper surface.

Hydrophytes contain relatively little xylem and supporting tissue. Submerged leaves of hydrophytes are supported by aerenchyma and the buoyancy of the surrounding water.

As you can see, the characteristics that distinguish the leaves of hydrophytes are the opposites of those of xerophytes. Hydrophytes also have reduced root systems; nutrients are absorbed by epidermal cells, which are often modified into transfer cells.

The submerged leaves of aquatic plants often have different shapes than the floating leaves. For example, floating leaves of buttercup (*Ranunculus*) are large and flat, while submerged leaves are highly dissected and lacelike. Similarly, submerged leaves of American pondweed (*Potamogeton nodosus*) are long, narrow, and thin (3–4 cells thick); have a uniform chlorenchyma; and lack stomata, cuticle, and large air spaces. Floating leaves are elliptical and somewhat similar to the leaves of many terrestrial plants; that is, they have a cuticle, numerous stomata on their upper surface, large air spaces, and palisade and spongy chlorenchyma. The occurrence of morphologically distinct leaves on the same plant is called **leaf dimorphism.** Plant hormones influence the formation of dimorphic leaves. For example, pondweed produces floating leaves instead of submerged leaves when exposed to abscisic acid, a plant hormone. This effect can be overcome by simultaneously exposing the leaf to other plant hormones such as gibberellins and cytokinins. You'll learn more about plant hormones in Chapter 18.

c o n c e p t

Light and the availability of water strongly modify leaf structure. This plasticity in leaf development adapts plants to different and changing environments.

Leaf Movements

The leaves of many plants move. For example, leaves often orient themselves perpendicularly to sunlight, thereby increasing the amount of light they absorb for photosynthesis. As a result of this light-directed movement, leaves form **mosaics** that minimize the shading of leaves by each other (fig. 14.23). Similarly, the leaves of many desert plants often orient themselves parallel to sunlight, thereby decreasing their heat load. However, movements and arrangements of leaves often go beyond even these generalizations. For example, leaves of the Ceará rubber tree (*Manihot glaziovii*) of northeastern Brazil fold like an umbrella, while those of some species of rhododendron roll into a cylinder when it's cold. In contrast, leaves of the compass plant (*Silphium laciniatum*) do not move but use light for orientation—half of the plant's leaves point to the east and half point to the west. A most unusual kind of leaf movement occurs in the telegraph plant (*Desmodium gyrans*), a member of the

figure 14.23

Leaves of plants such as this Boston ivy (*Parthenocissus tricuspidata*) form mosaics that help ensure that almost all leaves are exposed to light. Leaf mosaics often form in houseplants exposed to light from one direction.

legume family. Its leaves sometimes move in circles; those on one side of the plant circle in one direction, and the rest circle in the opposite direction. At other times, the leaves seem to go berserk: they jerk and twitch wildly, move up and down, or move in circles—with rest periods between the spasms. We do not know the basis for these bizarre movements.

Modified Leaves

Like other organs, leaves are often modified for functions other than photosynthesis. Refer to figure 14.24 as you read about these modifications.

Tendrils Tendrils of plants such as sweet pea (*Lathyrus odoratus*) and trumpet flower (*Bignonia capreolata*) are leaves modified for support (fig. 14.24a). In garden pea (*Pisum*

(a)

(b)

(c)

(d)

figure 14.24

Modified leaves. (a) Tendrils on this garden pea enable the plant to cling to other objects for support. (b) In window leaves of plants such as *Frithia pulchra*, the tip of the leaf is transparent (i.e., is a "window"); this allows light to reach the photosynthetic tissue, which is below the soil level. (c) Leaves of *Sarracenia* are modified into pitchers that trap and kill insects. Digested insects are an important source of nitrogen for the plant. (d) Leaves of *Kalanchoë* form tiny plants at their margins. When separated from the parent, these tiny plants can become new individuals.

sativum), the *afila* gene converts leaflets to tendrils. In plants such as yellow vetchling (*Lathyrus pratensis*), the entire leaf is a tendril; photosynthesis in these plants is delegated to leaflike structures called *stipules* at the base of each leaf. Conversely, only the petiole is a tendril in potato vine (*Solanum jasminoides*) and garden nasturtium (*Tropaeolum majus*). Tendrils of many plants may be up to 30 cm long.

Stipules Stipules are small, leaflike structures at the base of petioles. Stipules have a variety of functions. For example, stipules of woodruff (*Asperula*) and sweet pea are photosynthetic, while those of black locust (*Robinia pseudoacacia*) and spurge (*Euphorbia*) form protective spines. Stipules protect buds in oak and beech, and in plants such as *Smilax* they become tendrils that coil around objects that they touch. These tendrils are extremely sensitive: they'll coil around a wire weighing only 1.23 mg—that's about 1/50 the weight of a paper clip.

Spines Spines of plants such as ocotillo (*Fouquieria splendens*) and cacti are leaves modified for protection.

Bud Scales Bud scales are tough, overlapping, waterproof leaves that protect buds from frost, desiccation, and pathogens. Bud scales form before the onset of unfavorable growing seasons such as winter.

Window Leaves Window leaves are common in many desert plants such as fairy-elephant's-feet (*Frithia pulchra*) (fig. 14.24b). These leaves are shaped like tiny ice cream cones and grow mostly underground, with only a small, transparent "window" tip protruding above soil level. The covering soil shields window plants from the desert's drying winds and increases their chances of being overlooked by grazing herbivores. Similarly, their windows allow light to penetrate and illuminate the chlorenchyma tissue, which is below soil level. Underground photosynthesis!

Bracts Bracts are floral leaves that form at the base of a flower or flower stalk. They are usually small and scalelike, and protect developing flowers. Some plants have colorful bracts; for example, the colorful portions of the flowers and inflorescences of poinsettia, Indian paintbrush (*Castilleja*), bird-of-paradise (*Strelitzia*), *Bougainvillea*, and chaconia (*Warszewiczia coccinea*, the national flower of Trinidad) are bracts. In these plants, bracts replace petals and attract pollinators (the petals of these plants are small and inconspicuous). The tiny flowers of dogwood (*Cornus*) are a drab, inconspicuous yellow-green and are arranged in a circle (i.e., the "eye" of the flower); the large bracts are pink or white and attract pollinating insects.

Storage Leaves Storage leaves of plants such as onion (*Allium cepa*) and lily are fleshy, concentric leaves modified to store food (fig. 14.7a). Onions have tubular leaves, the white bases of which form the bulb. The leaves of most bulbs store sugar or starch.

Flowerpot Leaves Leaves of flowerpot plants are packed tightly into a flowerpot-like structure that catches falling water and debris. Many epiphytes (i.e., plants that grow on other objects rather than soil) grow roots among the bases of their flowerpot leaves to absorb nutrients collected by the flowerpot. More ingenious plants such as *Dischidia* have hollow leaves that function as flowerpots. These leaves do not catch falling debris; rather, they are homes for ants that bring soil and debris into the leaf. Almost one-third of the plant's nitrogen comes from feces and decaying debris deposited by ants; nitrogen from these wastes is mined by roots that form at the node and grow into the hollow leaf. Some species of *Dischidia* also get almost 40% of their carbon from carbon dioxide released by ants living in the flowerpot leaves.

Insect-Trapping Leaves In carnivorous plants, insect-trapping leaves are modified for attracting, trapping, and digesting animals. These adaptations range from sticky "flypaper" surfaces of leaves such as those of butterwort (*Pinguicula*) to the vatlike leaves of pitcher plants such as *Sarracenia* and *Nepenthes* (fig. 14.24c). Pitcher plants typically use nectar to lure insects into the leaf chamber. About halfway down the pitcher of many pitcher plants, the epidermal surface abruptly becomes flaky wax. When insects step onto this wax, their legs become covered with the wax, and their delicate feet are transformed into unwieldy clodhoppers. Unable to cling to the flaky wax on the side of the pitcher, the insects then slip into the pitcher's vat and die. They are then digested, and their nutrients are absorbed by the leaf.

Leaves Modified for Reproduction Succulent plants such as *Peperomia*, *Begonia*, rock-lettuce (*Sedum*), and maternity plant (*Kalanchoë*) commonly have leaves that are modified for reproduction (fig. 14.24d). Leaves of these plants produce tiny plants that become new individuals when they are shed from the parent leaves.

Cotyledons Cotyledons are embryonic leaves. Monocots, such as corn, usually have one cotyledon, while dicots have two. However, there are some exceptions; for example, *Degeneria vitiensis*, one of the most primitive dicots, has three or four cotyledons. Cotyledons have several functions. In beans they absorb the endosperm and therefore store energy used for germination. Storage products in cotyledons are usually carbohydrates, but they may also be oils, as in peanuts (*Arachis hypogaea*). The cotyledons of the cacahuanache tree (*Licania arborea*) contain large amounts of flammable oils—enough that the seeds can be strung on sticks and used as torches. Oils extracted from *Licania* cotyledons are also used to produce candles, soaps, and grease. Although the cotyledons of some plants are green and grow aboveground, they usually lack stomata and are nonphotosynthetic.

Prophylls Prophylls are the first leaves to form on axillary buds. Monocots usually have one prophyll, whereas dicots have two, which suggests that these tiny leaves may be analogous to cotyledons. Prophylls protect axillary buds.

LEAVES AND WIND

The deep shadows in a forest attest to trees' ability to absorb sunlight efficiently. This ability results primarily from the positioning of leaves. Although the design of leaves is primarily for the absorption of light, it is also influenced by wind. The large surface area of leaves—a tremendous asset for absorbing light—produces much drag in wind, thereby becoming a major liability. Why, then, don't more trees blow over in storms?

To survive winds, trees must have a strong trunk, flexible branches, and a sturdy root system. They also need leaves that have minimal drag in wind. Leaves do this in a variety of ways. For example, leaves of an American holly clump together as the speed of the wind increases; in winds of 30 mph, the leaves are flat against each other, thereby minimizing drag. Similarly, leaflets of a black locust roll together into a tight cylinder. In tulip trees, winds cause leaves to gradually roll themselves into tight cones that point into the wind. This minimizes drag, thereby reducing the probability that the tree will be blown over by the wind (reading fig. 14.3).

A variety of leaves roll up when in strong winds; examples include red maple, sycamore, sweetgum, and redbud. All of these leaves have long petioles and wide blades that flare backward toward the stem. These unremarkable features, which may seem trivial in everyday circumstances, may be crucial to survival in a storm.

0 mph

5 mph

10 mph

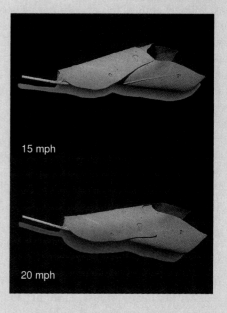

15 mph

20 mph

reading figure 14.3

As wind speed increases, leaves of tulip trees roll into a cone that points into the wind. This decreases the drag, thereby keeping the tree from blowing over during a storm.

c o n c e p t

Plants use modified leaves for obtaining and storing nutrients, climbing, protection, and reproduction.

How Leaves Defend Themselves

Since they cannot run away or physically defend themselves from herbivores, the delicate, nutrient-rich leaves of most plants would apparently be easy prey for the 300,000 known species of herbivorous insects. However, leaves are not defenseless against the prospect of being eaten. The defenses of some plants are obvious: their lignin makes them hard to chew, and trichomes deter many herbivores. In other plants, an attacker's first nibbles unleash a barrage of chemicals that deter animals.

Plants That Poison Their Attackers

Plants such as white clover (*Trifolium repens*) and bird's-foot trefoil (*Lotus corniculatus*) are chemical minefields containing cyanogenic compounds, and merely touching the leaves of a plant such as stinging nettle (*Urtica*) will quickly get the attention of most animals. Even more powerful are the poisons of plants such as poison hemlock (*Conium maculatum*, an herb not to be confused with the hemlock tree, *Tsuga*, whose leaves are not poisonous). The parsleylike leaves of poison hemlock contain coniine, a deadly alkaloid. The ancient Greeks knew about this powerful poison and used it to kill prisoners, including Socrates. Poisons of other plants are still used. For example, the Maku Indians of Colombia and Brazil use extracts from *Euphorbia cotinifolia* and *Phyllanthus brasiliensis* to kill fish. Shoots of these plants are placed on bridges over streams and beaten, so that their juices trickle into the stream and suffocate the fish. The fish are then gathered a few hundred yards downstream.

Interestingly, several insects use plant poisons for defense. For example, the grasshopper *Poekilocerus bufonius* eats only milkweeds that contain poisonous cardenolides. The grasshopper stores these poisons in special glands. When attacked, the grasshopper sprays the poisons on its predator, killing the predator. Eating milkweed is a critical part of the grasshopper's defense; grasshoppers fed diets lacking milkweed secrete 90% less cardenolides and are relatively defenseless against their attackers.

Many plants produce chemicals called *photosensitizers* that are toxic only when insects are in light. Photosensitizers produce highly reactive molecules that literally burn up insects. Some insects avoid the effects of photosensitizers by rolling themselves in leaves when they eat.

Plants That Change the Life Cycles of Their Attackers

Ecdysteroids and juvenile hormone are hormones that regulate insect development. Many plants produce these hormones, and when insects eat them, their life cycle is altered, usually to the benefit of the plant. For example, swarms of locusts can denude large prairies in only a few hours, but they do not usually attack bugleweed (*Ajuga remota*). This plant laces its leaves with large amounts of ecdysone-like compounds, many of which are more potent than those made by insects. Insects that eat bugleweed develop several head capsules when they change from larvae to adults. These extra head capsules block their mouthparts, and the insects starve to death.

The discovery of substances similar to juvenile hormone in plants is an excellent example of the scientific method in action. Two researchers, one in the United States and one in Europe, were collaborating on a research project that involved growing an insect called *Pyrrhocoris apterus*. These insects developed normally when grown in Europe, but those reared in the United States failed to become adults. Puzzled, the researchers tried to discover why they could not get the same results. They used carefully designed control experiments to sequentially eliminate several factors, including temperature, light, and humidity. Finally, they discovered the culprit: the filter paper on which the insects were being grown. The active ingredient of the paper was traced to the pulp of balsam fir, a primary source of North American (but not European) paper products, including the filter paper used in laboratories. This compound, which was named *juvabione*, is also abundant in leaves; it arrests insect development and enables fir trees to eliminate many herbivores.

Plants also produce chemicals that alter the life cycles of vertebrates. For example, chemicals in pine leaves (i.e., needles) eaten by grazing cattle double the concentration of progesterone, a hormone needed to maintain normal pregnancy. As a result, pregnant cattle that eat as little as 0.7 kg of needles per day usually abort their calves.

Plants That Make Themselves Less Digestible

Many plants fight their attackers by making themselves less digestible. For example, leaves of tomato and potato attacked by chewing insects release substances that move through the plant from the site of wounding and induce the formation of enzymes that interfere with insects' digestion. As a result, tomato plants become less nutritious, and insects tend to dine elsewhere.

Plants also warn each other of impending attacks by insects. Sitka willows (*Salix sitchensis*) being attacked by herbivores produce compounds that decrease the digestibility of their leaves. Uninfested plants up to 60 m away also produce these chemicals, suggesting that willows use airborne signals to warn others that the herbivore is coming.

Finally, the leaves of many plants use animals for defense. For example, stipules of bull's-horn acacia (*Acacia cornigera*) are modified as hollow spines that resemble bull horns and serve as homes for ants. The plant's nectaries produce sugars and fatty globules, which are eaten by aggressive ants living in the spines. In return for this food, the ants defend the plant fiercely: an intruder landing on the plant is soon covered by a swarm of ants. These armies of ants are important, since *Acacia* plants without ants suffer more damage by herbivores than do plants protected by ants.

Plants That Shift Their Resources

Plants can also defend themselves by changing their production of certain compounds. One group of leaves on a maple tree, for example, may contain large amounts of toxins, while another group may contain relatively little nitrogen. Plants change the concentrations of these chemicals regularly and in response to insect attacks. This shell-game strategy forces insects to move around constantly, thereby exposing them to *their* predators.

c o n c e p t

Leaves protect themselves by poisoning attackers, changing the life cycle of their attackers, making themselves less digestible to attackers, and shifting their resources.

Many of the chemical defenses of plants are elicited by attacks. For example, phytoalexins are antimicrobial compounds made in response to carbohydrates released from the cell walls of attacking bacteria and fungi. Phytoalexins have several functions, including preventing germination of spores and inhibiting the enzymes that degrade cell walls. Plants make large amounts of phytoalexins when attacked by pests. Many botanists hope to develop resistant hybrids by using naturally occurring elicitors to turn on a plant's defenses *before* an attack—something similar to vaccinating plants against their enemies.

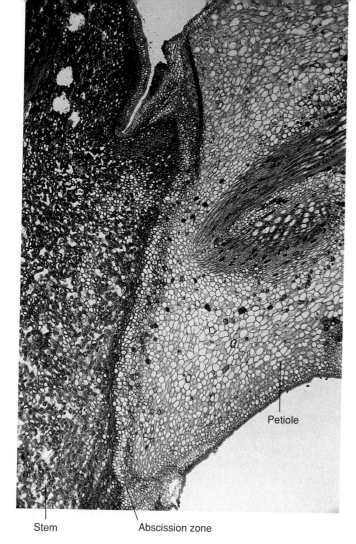

Petiole

Stem   Abscission zone

figure 14.25

Leaf abscission occurs when an abscission zone forms near the base of the petiole. This zone minimizes infection and loss of nutrients.

The associations described between ants and plants are examples of **coevolution,** a reciprocal process in which characteristics of one organism evolve in response to specific characteristics of another. The interactions between the two organisms can be remarkably specific: many plants are defended by only one kind of ant, and many plants are pollinated by only one kind of insect. As a reward for their fidelity, insects get food and, occasionally, a place to live.

Leaf Abscission

Leaves typically have a limited life span, after which they are shed from a plant via a complex process called **leaf abscission.** Abscission can result from injury or seasonal changes in climate such as drought or the shortening days of autumn. Plants such as oak and pecan (*Carya*) that abscise their leaves in the fall are called *deciduous.* Leaves of *evergreens* such as pine and oleander live 3–5 years and are shed year-round.

Leaves abscise at a predetermined region at the base of the petiole called the *abscission zone* (fig. 14.25). In most leaves, this zone contains reduced amounts of sclerenchyma,

few intercellular spaces, and thin-walled cells. Injury or short days alter the production of auxin and ethylene, which are plant hormones that control abscission. As a result, a *separation layer* forms in the abscission zone. Cells along the separation layer suberize (i.e., form suberin) and, in doing so, isolate the senescing leaf from the stem. Once the separation layer forms, wind or other disturbances break the dead leaf from the stem. What's left on the stem is a suberized *leaf scar* that resists infection and desiccation. Leaf abscission is important because it prunes plants and rids them of injured or dying leaves. You'll learn more about leaf abscission in Chapter 19.

The Economic Importance of Leaves

We use leaves and their products every day. Most important, chlorenchyma cells produce the sugars and oxygen that we and most other organisms need to live. Leaves also have other important, though less critical, uses.

Food, Spices, and Drinks   The leaves of plants such as cabbage (*Brassica oleracea*), lettuce (*Lactuca sativa*), spinach (*Spinacia oleracea*), chard (*Beta vulgaris*), sage (*Salvia*), thyme (*Thymus vulgaris*), celery (*Apium graveolens*), parsley (*Petroselinum crispum*), and bay laurel (*Laurus nobilis*) have long been parts of our diets and favorite items in our salads. Many of these leaves are especially nutritious; for example, spinach contains large amounts of vitamin A. Conversely, other plants are inedible in the wild; for example, wild lettuce contains large amounts of a bitter and narcotic latex.

Edible leaves of some other plants have mythical stories associated with them. Peppermint (*Mentha piperita*), for instance, derives its name from Menthe, a nymph of Greek mythology.[1] Thyme, oregano (*Origanum vulgare*), peppermint, spearmint (*Mentha spicata*), wintergreen (*Gaultheria procumbens*), basil (*Ocimum basilicum*), and sage are all derived from leaves. Similarly, tea is extracted from leaves of a relative of the garden *Camellia* (fig. 1.1), and *Agave* leaves are used to make tequila and mescal. It takes about 50 lb of *Agave* leaves to produce a gallon of mescal.

Dyes   Although most dyes are now made from coal tar, plants were the original sources of most colorings. The leaves of bearberry (*Arctostaphylos uva-ursi*) contain a yellow dye, while those of henna (*Lawsonia inermis*) contain a red dye that was used to stain fingernails and the cloth used to swath Egyptian mummies.

Fibers   You learned in Chapter 13 that fibers from plants such as flax, sisal, and Manila hemp are woven into ropes and fabrics prized by clothiers and sailors. Palm leaves are used to make Panama hats, clothing, brooms, and thatched huts in the tropics.

---

1. Menthe was a nymph beloved by Pluto, alias Hades (god of the dead). She was trampled by Persephone (daughter of Zeus and the symbol of rebirth of crops in spring) and turned into mint. As was true for Menthe, the more you crush mint, the sweeter it gets.

Fuel   The leaves of plants such as yareta (*Azorella yareta*) contain flammable resins that can be used as fuel.

Drugs   Many leaves contain poisons that, when administered in small amounts, are useful drugs. For example, digitoxin and digitalis are popular drugs extracted from foxglove (*Digitalis purpurea*) (fig. 1.1). These heart stimulants increase the relaxation of the heart without affecting its contraction, thereby increasing blood circulation. Hyoscyamine, atropine, and scopolamine are derived from deadly nightshade (*Atropa belladonna*), a poisonous plant named after Atropos, the Greek god of fate who held the shears to cut the thread of human life. *Atropa* was a savior for Macbeth, whose soldiers used it to poison the Danish army during peace talks. The *belladonna* epithet is Italian for "beautiful woman," a reference to Italian women's use of *Atropa belladonna* to dilate their eyes and make them brighter and more attractive. *Aloe* leaves are used to treat burns. The lobeline sulfate used in drugs for quitting smoking is derived from leaves of a relative of the garden lobelia. Coffee and tea leaves (fig. 1.1; reading 1.1) contain as much as 5% caffeine, which, like drugs from nightshade, is a heart stimulant. Trichomes on *Cannabis* leaves contain narcotic tetrahydrocannabinols. Cocaine, a drug extracted from coca (*Erythroxylon*) leaves, is a dangerous yet popular drug (fig. E.5d).

Tobacco leaves contain large amounts of drugs such as nicotine that cause cardiovascular problems. Thanks to billions of dollars in federal subsidies, we produce almost a billion kilograms of tobacco leaves annually. The return for this investment of tax dollars is that smoking kills 17.2% of the people in the United States (alcohol abuse, motor vehicle accidents, and illicit drugs account for only 4.9%, 2.3%, and 0.2% of U.S. deaths, respectively). But the social costs of tobacco use do not end with funerals. At last count, smoking costs Americans $53.7 billion in absenteeism, lost production, and health care—a staggering $1.79 per pack of cigarettes.

These examples underscore the indispensable roles of plants in society and modern medicine. Drugs derived from plants account for more than 20% of all prescriptions dispensed in the United States. These prescriptions do not come cheaply; in 1994, Americans spent more than $13 billion for prescription drugs derived from plants.

Other Uses of Leaves   Carnauba wax is derived from leaves of the carnauba palm, and extracts from the lancelike leaves of *Aloe* are ingredients of medicated soaps and creams. *Aloe* is sacred to many Muslims, who often hang its leaves above their doors to show that they have made a pilgrimage to Mecca. We also sculpt leaves into unusual shapes and use them as names for sports teams (e.g., the Toronto Maple Leafs hockey team), on national flags (e.g., the Canadian flag), and even as parts of corporate logos. For example, the shamrock on the Boston Celtics logo is a palmately compound leaf.

Finally, do not overlook the elegant beauty and aesthetic importance of leaves. They provide shade for us and other organisms, creating a cool oasis that we often seek during summer heat waves.

The Lore of Plants

The trees at about ten suburban homes can produce a ton of dead leaves each fall. Burning these leaves produces smoke that is not as nice as it smells. Indeed, burning these leaves releases 90 kg of soot, 260 kg of carbon monoxide (a poison), and various amounts of hydrocarbons that can irritate the eyes, nose, throat, and lungs. Some of these hydrocarbons are carcinogens.

Chapter Summary

The shoot system of a plant consists of leaves and stems. Both are made of the same tissues; the structural and functional differences between them result from different arrangements of these tissues. Stems are collections of nodes and internodes. Nodes are areas where leaves attach to stems, and internodes are the portions of stems between nodes. Stems support leaves, produce and store food, and transport water and solutes between roots and leaves.

Stem elongation occurs in subapical regions. Many grasses elongate via the activities of intercalary meristems, which are meristems intercalated between mature tissues at the bases of their nodes. Internodal elongation helps plants intercept light. Plants whose stems do not elongate are called rosette plants. The internodes of these plants do not elongate, leaving the leaves packed tightly on a short stem.

Stems are made of epidermal, ground, and vascular tissues. The epidermis of stems is made of tightly packed and cutinized cells that prevent desiccation. Vascular tissues are arranged in vascular bundles: in monocots these bundles are scattered throughout the stem's ground tissue, whereas in most dicots they are arranged in a single ring. Ground tissue inside this ring is called the pith and is often modified for storage. Between the vascular bundles and the epidermis is the cortex, which stores food or is photosynthetic.

Stolons, tendrils, searcher shoots, cladodes, thorns, short and long shoots, succulent stems, bulbs, rhizomes, corms, and tubers are stems modified for reproduction, climbing, photosynthesis, protection, and storage. We use stems for food, paper, drugs, spices, and lumber.

Leaves are the most active and diverse organs on plants. Their formation depends on several meristems, and their final shape results from cellular division and expansion. Simple leaves have a flattened blade and a stalklike petiole. Blades of compound leaves are divided into leaflets. Leaflets of pinnately compound leaves form in pairs along a central rachis, while those of palmately compound leaves form at the same point. Axillary buds, which form in the axil between the petiole and stem, form flowers or branches that modify that plant's shape and ability to intercept light.

Phyllotaxis is the arrangement of leaves on stems. Plants with one leaf per node have spiral phyllotaxis, while those with two leaves per node have opposite phyllotaxis. Plants with three or more leaves per node have whorled phyllotaxis. Phyllotaxis is probably controlled by fields of inhibitors at the shoot apex or differential reinforcement of cells of the tunica layers.

Leaves are made of epidermal, ground, and vascular tissues. The epidermis is usually transparent and contains numerous stomata. Vascular tissues in leaves are arranged in veins: xylem forms on the upper side of the vein, and phloem forms on the lower side. Leaves of dicots have netted venation, in which a prominent midvein connects with a meshlike network of minor veins. Leaves of monocots have parallel venation, meaning that the major veins are oriented parallel to each other and are connected by smaller minor veins. Vein endings in the ground tissue are where water and solutes move into and out of veins.

The ground tissue in leaves is called mesophyll. Horizontally oriented leaves have densely packed, columnar palisade mesophyll cells along the upper side and spongy mesophyll cells along the lower side. Most light absorption and carbon fixation occurs in the palisade mesophyll, and most gas exchange occurs in the spongy mesophyll. Vertically oriented leaves have a uniform mesophyll. Mesophyll cells of grasses such as corn are arranged concentrically around a photosynthetic bundle sheath. This arrangement of mesophyll underlies C_4 photosynthesis, which in hot, dry environments is more efficient than C_3 photosynthesis.

The structure and function of leaves are strongly influenced by light and moisture. Leaves require light to expand, and daylength often determines what kind of leaves a plant produces. Leaves in intense light are sun leaves and are smaller, thicker, and capable of higher rates of photosynthesis than shade leaves, which grow in dim light.

Xerophytes grow best in dry environments and have small, thick leaves with numerous stomata, a thick cuticle, and well-developed palisade and spongy mesophyll layers. Hydrophytes grow best in wet environments and have large, thin leaves with a thin cuticle and few stomata. The environmental control of leaf formation adapts leaves to different and changing environments.

Tendrils, spines, bud scales, window leaves, bracts, storage leaves, flowerpot leaves, insect-trapping leaves, leaves modified for reproduction, stipules, prophylls, and cotyledons are leaves modified for obtaining and storing nutrients, climbing, protection, and reproduction. Leaves defend themselves by poisoning their attackers, altering the life cycles of their attackers, making themselves less digestible, and shifting their resources.

Old and injured leaves abscise from plants at a predetermined abscission zone. Leaves of evergreens live several years and are shed year-round. Deciduous plants shed their leaves seasonally, usually in response to drought or the shortening days of autumn. We use leaves for food and to make spices, drinks, dyes, fibers, and drugs.

 What Are Botanists Doing?

Potatoes are one of our most important food crops, so studies on how to improve them are always underway. Go to the library and read about some of this research. How are botanists trying to improve potatoes?

Writing to Learn Botany

Discuss the various ways in which leaves defend themselves. How do we exploit these potential defenses?

Questions for Further Thought and Study

1. Explain why submerged leaves often resemble shade leaves, whereas the leaves of xerophytes often resemble sun leaves.

2. How can you distinguish a leaf from a leaflet?

3. The tiny plants that form along the edges of leaves of plants such as *Kalanchoë* (fig. 14.24d) are genetically identical to the parent. Why?

4. How does the ability of a plant to form (a) tendrils, (b) thorns, (c) bulbs, and (d) rhizomes help a plant survive?

5. What is the significance of having more stomata on the lower side of a leaf than on the upper side?

6. Describe three ways that light affects leaf development. What is the significance of each?

7. How do sun leaves differ from shade leaves? What is the significance of these differences?

8. How do the leaves of xerophytes differ from those of hydrophytes? What is the significance of these differences?

9. List six kinds of modified leaves and the functional significance of each.

10. What advantages do climbing plants have over erect plants? What disadvantages?

11. What are the advantages of producing many small leaves instead of fewer large leaves? What are the disadvantages?

12. Plasmodesmata link all cells of leaves. Why, then, do leaves have veins for transporting water and solutes?

13. What are the advantages of having a palisade layer on the upper surface of a leaf?

14. When attacked by caterpillars, some plants release chemicals that attract wasps—the natural enemies of the invading caterpillars. What would be the selection pressure for the evolution of such a defense mechanism?

Web Sites

Review the "Doing Botany Yourself" essay and assignments for Chapter 14 on the *Botany Home Page*. What experiments would you do to test the hypotheses? What data can you gather on the Web to help you refine your experiments?

Here are some other sites that you may find interesting:

http://www.hpl.hp.com/bot/cp_home

This database of over 3,000 entries provides information about a variety of carnivorous plants. Many of the entries are accompanied by photographs.

http://www.cco.caltech.edu/~aquaria/Krib/Plants/

This site presents descriptions and photographs of a variety of hydrophytes. You'll also learn how to grow many of these plants in aquaria.

http://aquat1.ifas.ufl.edu/

Welcome to the Center for Aquatic Plants and to APIRS, the Aquatic Plant Information Retrieval System. If you want information about hydrophytes, this is an excellent place to start your search.

http://www.graylab.ac.uk/usr/hodgkiss/succule.html

This site is a resource for people who collect, grow, propagate, and conserve xerophytes.

http://chipmunk.apgea.army.mil/ento/PLANT.HTML

Many plants defend themselves by producing toxins that deter herbivores. Information from this Guide to Poisonous and Toxic Plants might save your life one day.

Missouri Botanical Garden
http://www.mobot.org/

The Missouri Botanical Garden is one of the best in the world and features a Center for Plant Conservation. Visit this conveniently searchable web site.

WWW Journal of Biology
http://epress.com/w3jbio/

Try surfing through the subheadings of the World Wide Web *Journal of Biology*. Successful communication and publication is a key to success in science.

Suggested Readings

Articles

Bolz, D. M., and K. B. Sandved. 1987. A world of leaves: Familiar forms and surprising twists. *Smithsonian* 16:150–155.

Dale, J. E. 1992. How do leaves grow? *BioScience* 42:323–332.

Rosenthal, G. A. 1986. The chemical defenses of higher plants. *Scientific American* 254 (January):94–99.

Vogel, S. 1993. When leaves save the tree. *Natural History* (September): 59–63.

Books

Jean, R. V. 1994. *Phyllotaxis*. Cambridge: Cambridge University Press.

Sandved, K. B., and G. T. Prance. 1985. *Leaves*. New York: Crown Publishers.

Woods, R. K. S., ed. 1982. *Active Defense Mechanisms in Plants*. New York: Plenum Press.

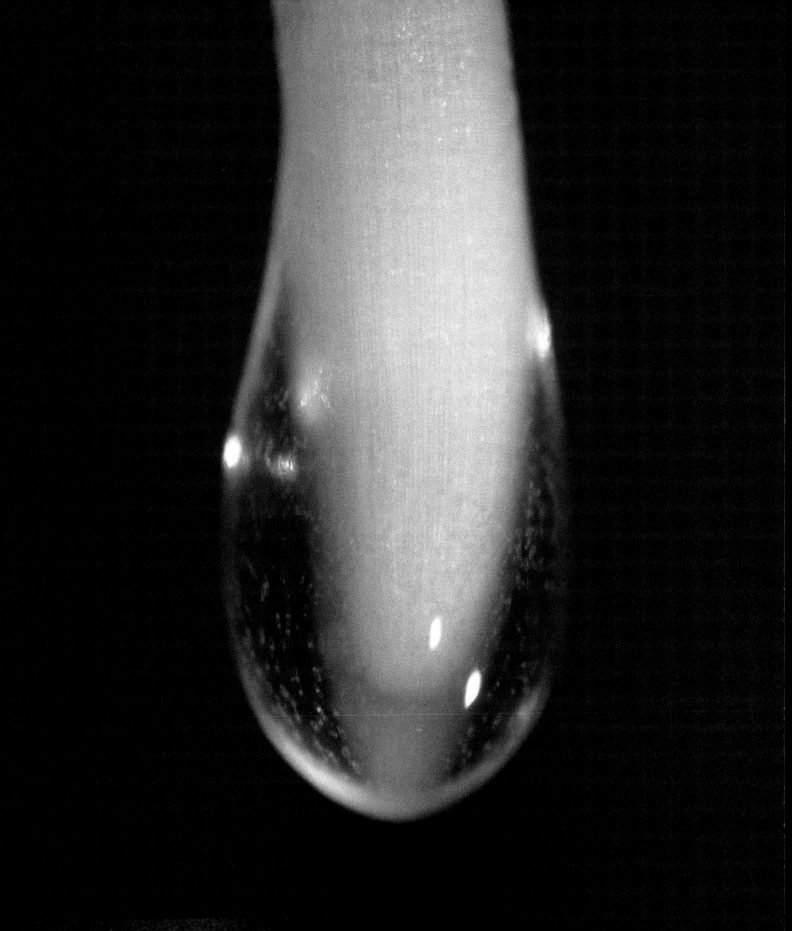

Tip of a primary root of corn (*Zea mays*). Root
tips produce large amounts of mucigel that help
roots force their way through the soil.

Primary Growth: Roots

15

Chapter Outline

Chapter Overview

We're all familiar with stems and leaves because they grow aboveground and are conspicuous. Although we do not see much of roots, they are equally important to plant growth because they provide the photosynthetic cells of stems and leaves with water and dissolved minerals. Roots constantly grow into new territory, where they absorb and transport water and minerals from the soil to the shoot, thereby linking the plant's photosynthetic cells with the soil's moisture and nutrients. Simultaneously, roots receive sugars and other organic compounds from the shoot. These compounds are stored, used for growth, and released to the soil's microbes. As a result, much of a plant's photosynthate supports an extensive underground mining operation. The ability of roots to extract water and minerals from the soil is affected by many of the same environmental factors that affect the growth of shoots, including light and water. Some roots are modified for functions such as support, movement, propagation, aeration, and parasitism.

Plants and animals have different strategies for growth. Animals usually expose only enough surface area to sense the environment and move about to gather food. This strategy is successful because it makes animals more efficient, decreases injury, and helps them obtain concentrated sources of nutrients. Conversely, plants absorb dilute nutrients from the environment. For example, the CO_2 essential for photosynthesis accounts for only 0.035% of air, and sunlight is often at nonsaturating intensities. Thus, exposing large amounts of surface area increases a plant's chances of absorbing light and nutrients from the environment. For the surface area of the shoot, there's usually some light available each day, and the breeze constantly bathes leaves in a fresh supply of CO_2. Thus, the millions of leaves on a mature elm tree usually have no problem producing enough sugar for growth, as long as they get enough water and nutrients from roots. Roots grow in unstirred soil, however, which makes it necessary for them to grow *to* water and nutrients.

The growth necessary to locate dispersed water and nutrients produces large roots and extensive root systems. To appreciate this, consider the simple but tedious experiment performed in the 1930s by Howard Dittmer. Dittmer grew a winter rye plant (*Secale cereale*) in a shoebox-sized container full of fertile soil. After 4 months he carefully unearthed the plant and measured its root and shoot systems. The shoot was 8 cm high and consisted of 80 leaves covering 5 m², an area about the size of a Ping-Pong table. However, the situation was different underground. The root system consisted of more than 13 million roots and 1.4×10^{10} root hairs having a surface area larger than two-and-a-half tennis courts. The combined length of these roots and their root hairs spanned more than 11,000 km, a distance almost one-third that of the earth's circumference.

Dittmer's experiment emphasizes an important principle of plant growth—namely, that plants invest heavily in roots. Although roots seldom occupy more than 5% of the soil's volume, they often consume more than half of a plant's net primary production each year. Furthermore, roots account for more than 80% of plant biomass in ecosystems such as short-grass prairies and tundra. In trees such as pecan (*Carya illinoensis*), roots are usually longer and spread wider than the shoot. As a result, roots can effectively mine dispersed water and minerals from the soil. Roots are also important to animals because they unlock the soil's store of nutrients and move elements from the soil into the food chain. Thanks to roots, animals can obtain essential elements such as calcium and sulfur by eating plants instead of soil.

Kinds of Root Systems

Taproot System

The first root to emerge from a seed is the **radicle**, or **primary root** (fig. 15.1). In most dicots,[1] the radicle enlarges to form a prominent **taproot** that persists throughout the life of the plant. Many progressively smaller **branch roots** grow from the taproot. This type of root system consisting of a large taproot and smaller branch roots is called a **taproot system** and is common in conifers and many dicots (fig. 15.2). In plants such as sugar beet and carrot, fleshy taproots are the plant's food pantry: they store large reserves of food, usually as carbohydrates. Not all taproots are modified for storage, however; for example, the long taproots of poison ivy (*Rhus toxicodendron*) and mesquite (*Prosopis*) are modified for reaching water deep in the ground.

Taproots usually control the growth and development of branch roots, much as the shoot apex controls the growth of axillary buds. For example, if the radicle is damaged, a branch root enlarges and assumes the dominating role of the taproot. Because taproots of most dicots grow faster than branch roots throughout the life of the plant, many plants have long taproots. Indeed, engineers digging a mine in the southwestern United States uncovered a mesquite root 53 m down. If you'd like to see firsthand how big roots can get, visit the Roto-Rooter Monster Root Hall of Fame in Des Moines, Iowa. There you can see Moby Root, a 31 m root pulled out of a drainage pipe of a parking garage in 1994.

Fibrous Root System

Most monocots (including grasses) have a **fibrous root system** consisting of an extensive mass of similarly sized roots (fig. 15.3). In these plants, the radicle is short-lived and is replaced by a mass of **adventitious roots** (from the Latin word *adventicius*, meaning "not belonging to"), which are roots that form on organs other than roots. Fibrous roots of a few plants are edible; for example, sweet potatoes are fleshy parts of fibrous root systems of *Ipomoea batatas*. Because the adventitious roots of monocots are so extensive and cling tenaciously to soil particles, such plants are excellent for preventing erosion.

---

1. Recall that dicots are flowering plants that have two embryonic leaves (cotyledons) in their seeds; examples include rose, pea, sunflower, and magnolia. Monocots are flowering plants that have one embryonic leaf in their seeds; examples include corn, lily, pineapple, and banana.

figure 15.1

Primary root of a corn seedling.

Shoot

Primary root with root hairs

figure 15.2

Taproot system of dandelion (*Taraxacum*). Taproot systems consist of a prominent taproot and smaller lateral roots.

figure 15.3

Fibrous root system of a grass. Fibrous root systems consist of many similarly sized roots roots that form extensive networks in the soil.

Adventitious Roots

The adventitious roots of most monocots begin growing soon after the seed germinates. Each node in the embryo usually produces 2–6 *seminal roots* that grow from the seed. These roots are soon supplemented by an extensive system of *crown roots* that grow from nodes of the growing shoot. Within a few weeks, a corn plant's root system consists of seminal and crown roots that form an extensive network in the soil. In only 4 weeks, the hemispherical root system of a corn seedling is more than 45 cm deep and 60 cm in diameter. By the time the plant forms fruit, the root system spans more than 1.5 m.

Recent studies of monocot root systems tell us that they are more complex than we originally thought. For example, corn plants have two types of adventitious roots, one of which is usually unbranched and has actively growing tips. These *feeder roots* are associated with a 1 mm thick soil sheath permeated by root hairs and cemented together by secretions of the root and the soil's microbes. Other roots of the root system are long, branched, and lack an encasing soil sheath. These roots lack actively growing tips; that is, they grow determinately. Some plants have roots with characteristics of both tap and fibrous root systems. For example, clover (*Trifolium*) has a taproot and an extensive fibrous system of roots produced at nodes of its stolons.

There are several types of adventitious roots besides those of monocots. For example, adventitious roots are common along rhizomes of ferns, club mosses (*Lycopodium*), and horsetails (*Equisetum*). In many plants, adventitious roots are a primary means of vegetative reproduction; forests of quaking aspen (*Populus tremuloides*) are often a single clone spread by adventitious roots. We use adventitious roots on cuttings to propagate many plants (see fig. 13.2), including raspberries (*Rubus*), apples (*Malus*), cabbage (*Brassica*), and brussels sprouts (*Brassica*).

Adventitious roots form in all sorts of places on plants, including leaves, petioles, and stems. The formation of adventitious roots is controlled by hormones such as auxin, which are the active ingredients in the rooting compounds sold in stores. Although they arise at different locations, primary roots and adventitious roots have similar structures and functions.

c o n c e p t

Most dicots have a taproot system consisting of a large taproot and smaller branch roots. Taproot systems maximize support and storage. Monocots have fibrous root systems consisting of similarly sized roots that maximize absorption. Adventitious roots are roots that form on organs other than roots.

Functions and Structure of Roots

We have long known that roots are critical for plant growth: Aristotle taught about the importance of roots in the fourth century B.C., and you do not have to be a botany whiz to know that most plants die when separated from their roots. Roots have four primary functions:

Anchorage. To locate water and minerals, roots permeate the soil. In doing so, they anchor the plant in one place for its entire life.

Storage. Roots store large amounts of energy reserves. In biennials (i.e., plants that complete their life cycle in 2 years) such as carrot and sugar beet, these reserves are concentrated in only one or a few roots. We harvest these roots after the first year of growth, before the plant uses the stored energy for vegetative growth and reproduction.

Absorption. Roots absorb large amounts of water and dissolved minerals from the soil. For example, the roots of a corn plant absorb more than 2 l of water per day.

Conduction. Roots transport water and dissolved nutrients to and from the shoot. The roots of plants such as quillwort (*Isoetes*) and shoreweed (*Littorella*) even transport CO_2 to leaves for photosynthesis.

c o n c e p t

Roots anchor plants, store reserves, absorb water and dissolved minerals, and transport materials to and from the shoot. Each of these functions is linked to the unique structure of roots.

Root Tip

Root Cap

The tips of roots are covered by a thimble-shaped **root cap,** which has its own meristem that pushes cells forward into the cap (fig. 15.4a). As they move through the cap, these cells differentiate into elongate *columella cells,* so named because they are arranged in longitudinal columns (fig. 15.4b). Each columella cell contains 15–30 amyloplasts that sediment in response to gravity to the lower side of the cell (fig. 15.4b). Many botanists suspect that the sedimentation of these amyloplasts is how roots perceive gravity (although other botanists now question this interpretation; see Chapter 19). Besides protecting the growing root tip and its meristem, the root cap senses light and pressure exerted by soil particles. There is no structure in shoots that corresponds to a root cap.

Within 2–3 days, continued cellular divisions in the root-cap meristem push columella cells to the periphery of the root cap, where they differentiate into *peripheral cells* (figs. 15.4b, 15.5). Thus, the root cap is in constant flux: new cells

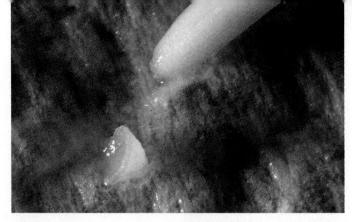

(a)

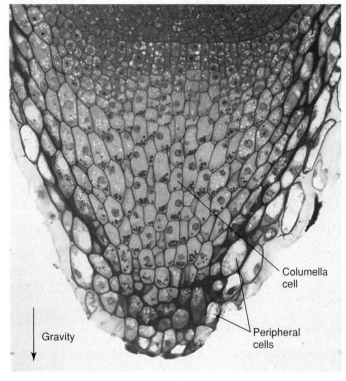

Columella cell

Peripheral cells

Gravity

(b)

figure 15.4

The root tip and root cap. (a) Root tip (upper right) with its thimble-shaped root cap (lower left) removed. (b) Light micrograph of cap of primary root of corn (*Zea mays*). The cells in the center of the cap are arranged in columns and are called *columella cells.* Note the dense amyloplasts at the lower side of the columella cells; the sedimentation of these amyloplasts may be involved in the perception of gravity by roots (see Chapter 19). Surrounding the columella cells are peripheral cells. These cells secrete mucigel, which lubricates the root tip and eases growth through the soil (also see fig. 15.5 and the photo at the beginning of this chapter).

constantly move through the cap and replace the thousands of peripheral cells that are shed from the cap as the root pushes its way through the soil. The root caps of corn shed as many as 10,000 peripheral cells per day.

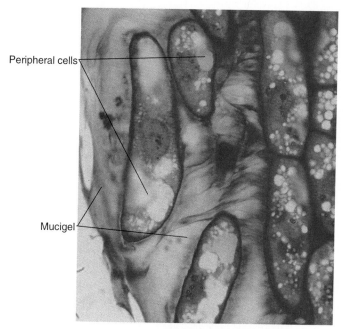

figure 15.5

The outermost peripheral cells of root caps secrete mucigel, a hydrated polysaccharide that lubricates the tip of the root as it moves through the soil. In this micrograph, mucigel surrounds the outermost peripheral cells of a root cap of corn (*Zea mays*).

The peripheral cells of root caps secrete large amounts of **mucigel,** a slimy substance made by dictyosomes (fig. 15.5; also see the photo that opens this chapter). Mucigel is a hydrated polysaccharide containing sugars, organic acids, vitamins, enzymes, and amino acids. A root weighing 1 g can secrete as much as 100 mg of mucigel per day. Although this rate of secretion may seem rather insignificant, the total amount of mucigel secreted by an actively growing group of plants can reach impressive proportions: for example, the roots of one hectare (10,000 m²) of corn secrete more than 1,000 m³ of mucigel during a growing season—that's enough mucigel to fill a typical two-story, four-bedroom house.

Mucigel has several important functions:

Protection. Mucigel protects roots from desiccation and contains compounds that diffuse into the soil and inhibit growth of other roots. For example, the mucigel of giant foxtail decreases the growth of nearby corn roots by 35%.

Lubrication. Mucigel lubricates roots as they force their way between soil particles.

Water absorption. Soil particles cling to mucigel, thereby increasing the root's contact with the soil. The water-absorbing properties of mucigel help maintain the continuity between roots and soil water.

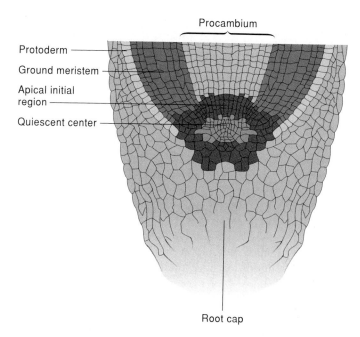

figure 15.6

Diagram of median longitudinal section through a hypothetical root. Cells in the quiescent center divide much less frequently than do other cells in the root tip. The protoderm, ground meristem, procambium, and root cap are derived ultimately from the apical initials.

Nutrient absorption. Carboxyl groups in mucigel influence ion uptake, and organic acids in mucigel make certain ions more available to plants. Also, fatty acids, lectins, and sterols in mucigel may help establish beneficial symbioses with soil microbes. For example, nitrogen-fixing bacteria such as *Azospirillum* are attracted to and inhabit the mucigel of plants such as corn.

Quiescent Center

Just behind the root cap is the **quiescent center,** which consists of 500–1,000 seemingly inactive cells (fig. 15.6). These cells are typically arrested in the G_1 phase of the cell cycle and divide only about once every 15–20 days (in comparison, cells of the adjacent meristem divide more than once per day). Quiescent and meristematic cells are differentially sensitive to environmental perturbations such as radiation. For example, meristematic cells stop dividing when exposed to intense X rays, while quiescent cells are unaffected by radiation and soon begin dividing to re-form the meristem. Thus, cells of the quiescent center are not inherently quiescent; rather, they function as a reservoir to replace damaged cells of the meristem. The quiescent center is also important because it organizes the patterns of primary growth in roots.

Tips of roots are covered by a root cap, which senses environmental stimuli such as gravity. The root cap also produces mucigel that protects, lubricates, and helps absorb materials from the soil. The quiescent center is located just behind the root cap and is made of seemingly inactive cells. The quiescent center organizes patterns of primary growth and replaces damaged cells of the adjacent meristem.

Subapical Region

The subapical region of roots has traditionally been divided into three regions: the zones of cellular division, cellular elongation, and cellular maturation (fig. 15.7). Although such divisions are useful for teaching, these regions intergrade and are not sharply defined. Moreover, they do not always accurately describe what is happening in a particular region of the root. For example, cells of some tissues elongate in the zone of cellular division, while those of others mature in the zone of cellular elongation. Concentrate on how structure correlates with function as we examine each of the subapical regions of a root tip.

Zone of Cellular Division

Surrounding the quiescent center is a dome-shaped apical meristem located 0.5–1.5 mm behind the root tip (fig. 15.6). This is the first of four unique features of roots as compared to shoots—namely, that *the apical meristem of a root is subterminal*. This meristematic region is the **zone of cellular division** and is made of small (diameter = 10–20 μm), densely cytoplasmic cells. Meristematic cells in roots divide every 12–36 hours; in some plants, the meristem produces almost 20,000 new cells each day. Divisions rarely occur past 1 cm behind the root tip except when new lateral root primordia are formed.

Zone of Cellular Elongation

The **zone of cellular elongation** occurs 4–10 mm behind the root tip (fig. 15.7). Cells in this zone elongate by as much as 150-fold, primarily by filling their vacuoles with water. Thus, this zone is easily distinguished from the root cap and zone of cellular division by its long, vacuolate cells. Cellular elongation in the elongating zone shoves the root cap and apical meristem through the soil at rates as high as 4 cm per day (secondary roots grow progressively slower: 0.05–0.5 cm per day). Cells behind the elongating zone do not elongate.

Zone of Cellular Maturation

Cellular elongation typically begins the process of cellular differentiation. Differentiation is completed in the **zone of cellular maturation**, which occurs 1–5 cm behind the root tip (fig. 15.7). The maturation zone is easily distinguished by the presence of many ephemeral root hairs—as many as 40,000 cm^{-2}. Root hairs, which increase the absorptive surface area of the root several thousandfold, are usually less than a millimeter long (figs. 15.8, 13.21). In most plants they form from asymmetric divisions of the protoderm (see p. 274) and

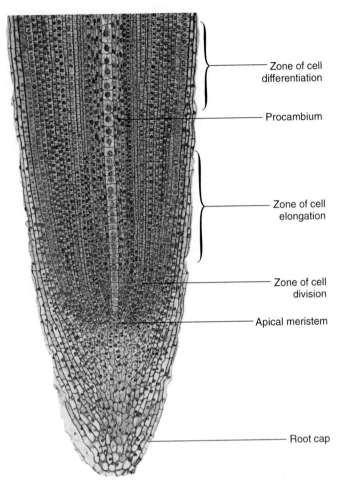

figure 15.7

The subapical region of roots includes the zone of cellular division, zone of cellular elongation, and zone of cellular differentiation. Although these designations are useful for teaching, the regions intergrade and seldom describe what occurs in a particular part of a root.

usually live only a few days, with old hairs farthest from the tip constantly being replaced by new ones closer to the tip.

Root hairs form only in the maturing, nonelongating region of the root. This gives root tips a simpler structure than the tips of shoots and accounts for the second unique feature of roots: *roots have no lateral appendages at their tips*. Because root hairs are fragile extensions of epidermal cells, they usually break off when plants are transplanted. Therefore, do not wash off the adhering soil when you transplant your plants.

The subapical region of roots can be divided into the zones of cellular division, elongation, and maturation. The zone of cellular division includes the root apical meristem, which is subterminal. Roots elongate in the zone of cellular elongation, and cellular differentiation is completed in the zone of maturation. Root hairs form in the zone of cellular maturation and greatly increase the root's absorptive surface area. Roots have no lateral appendages at their tips.

Mature Region

Primary tissues complete differentiation in or just "behind" (i.e., distally to) the zone of cellular maturation. A cross section through this region gives us our first look at the primary structure of a root (fig. 15.9).

Epidermis

The root is surrounded by an *epidermis*, which is usually one cell thick. Epidermal cells differentiate from protoderm and usually either lack a cuticle or have a thin cuticle that does not significantly affect water absorption. The epidermis covers all of the root except the root cap and usually lacks stomata.

Cortex

Just interior to the epidermis is the *cortex*, which is formed by the ground meristem. The cortex usually occupies the largest cross-sectional area of a root (fig. 15.9; also see fig. 13.1) and consists of three concentric layers: the hypodermis, storage parenchyma cells, and endodermis.

Hypodermis In many plants, the outermost layer(s) of the cortex is a suberized, protective layer called the *hypodermis* (fig. 15.10), which is usually most prominent in roots growing in arid soil or near the soil's surface. Hypodermal cells are lined with suberin (a waxy substance that is impervious to water) and complete their differentiation well behind the root-hair zone. Suberin in the hypodermis slows the outward movement of water and dissolved nutrients by as much as 200-fold, thereby helping roots retain water and nutrients that they've absorbed.

Storage Parenchyma Cells Most of the cortex consists of thin-walled storage *parenchyma cells* (figs. 15.9, 15.10). These cells often contain starch and are separated by large intercellular spaces that can occupy as much as 30% of the root's volume.

Endodermis The innermost layer of the cortex is the *endodermis* and differentiates 5–8 mm from the root tip. Unlike other cortical cells, endodermal cells are packed tightly together and lack intercellular spaces. Furthermore, their radial and transverse walls are impregnated with a *Casparian strip* made of lignin and suberin and arranged similarly to a rubber band around a rectangular box (fig. 15.11). This may help you visualize the structure of the endodermis: if

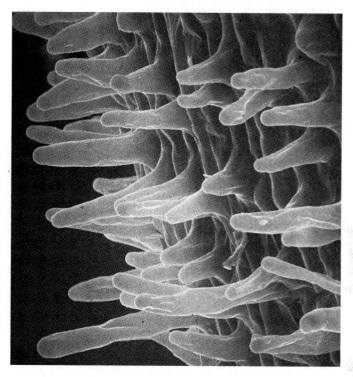

figure 15.8

Scanning electron micrograph of root hairs on a primary root of a radish (*Raphanus sativus*), ×840. Root hairs, which are extensions of epidermal cells, greatly increase the absorptive surface area of the root. For another view of root hairs, see figure 13.21.

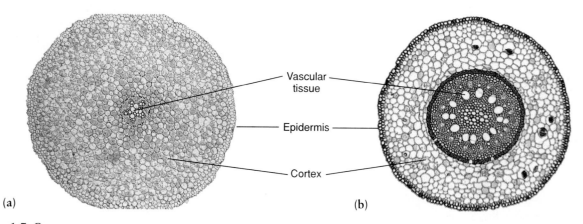

(a)   (b)

Vascular tissue

Epidermis

Cortex

figure 15.9

Cross section of roots of (a) buttercup (*Ranunculus* sp.), a dicot, and (b) greenbrier *(Smilax)*, a monocot. At this magnification, the most obvious tissues are the epidermis, cortex, and the vascular tissues of the vascular cylinder.

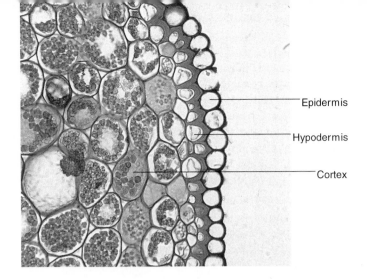

Epidermis

Hypodermis

Cortex

figure 15.10

Cortical cells contain many amyloplasts (stained darkly in this micrograph) and are usually separated by intercellular spaces (also see fig. 13.1). In many plants, the outermost layer of the cortex is specialized as a hypodermis, which retards the loss of water and dissolved nutrients. Surrounding the root is the epidermis.

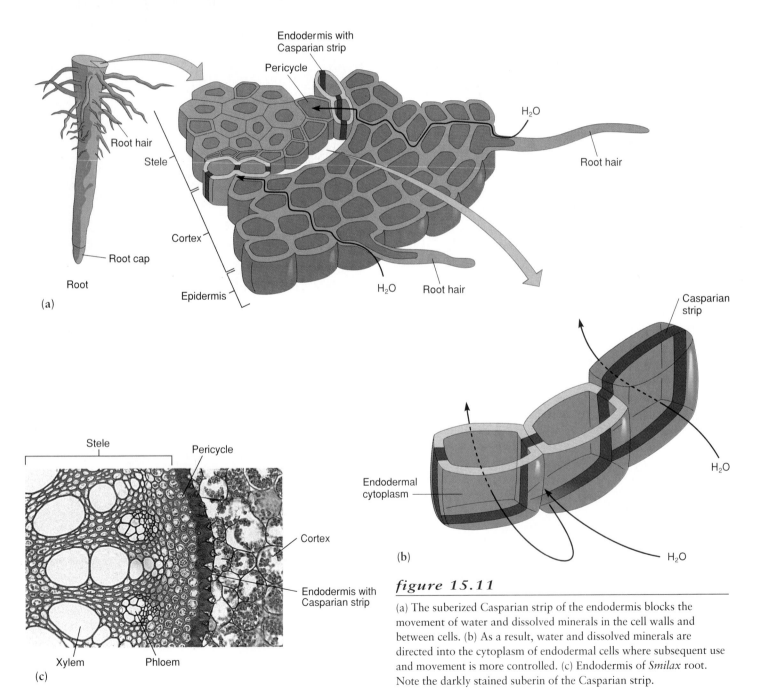

Endodermis with Casparian strip

Pericycle

H_2O

Root hair

Root hair

Stele

Cortex

Root cap

Root

Epidermis

(a)

Root hair

H_2O

Root hair

Casparian strip

H_2O

Endodermal cytoplasm

H_2O

(b)

Stele

Pericycle

Cortex

Endodermis with Casparian strip

Xylem

Phloem

(c)

figure 15.11

(a) The suberized Casparian strip of the endodermis blocks the movement of water and dissolved minerals in the cell walls and between cells. (b) As a result, water and dissolved minerals are directed into the cytoplasm of endodermal cells where subsequent use and movement is more controlled. (c) Endodermis of *Smilax* root. Note the darkly stained suberin of the Casparian strip.

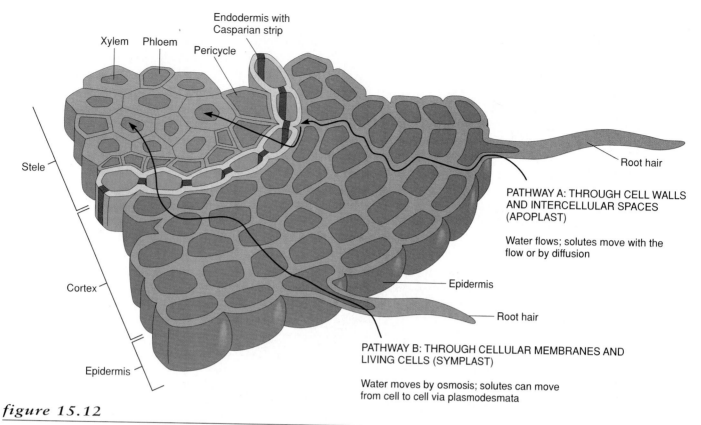

Xylem Phloem
Endodermis with
Casparian strip
Pericycle
Stele
Cortex
Epidermis

Root hair

PATHWAY A: THROUGH CELL WALLS
AND INTERCELLULAR SPACES
(APOPLAST)

Water flows; solutes move with the
flow or by diffusion

Epidermis

Root hair

PATHWAY B: THROUGH CELLULAR MEMBRANES AND
LIVING CELLS (SYMPLAST)

Water moves by osmosis; solutes can move
from cell to cell via plasmodesmata

figure 15.12

Diagram of cross section of a root, showing the alternative routes for uptake of water and dissolved nutrients. Note that the suberin of the Casparian strip blocks the movement of water through intercellular spaces and between cells. How would the presence of a suberized hypodermis affect the absorption of water via these two pathways?

endodermal cells were likened to bricks in a brick wall, then the Casparian strip would be analogous to the mortar surrounding each brick. The Casparian strip differentiates around cells approximately 1 mm from the apical meristem, which is just prior to where root hairs form. It is also tightly fused with the plasmalemmas of endodermal cells, thereby preventing the inward movement of water and nutrients through the cell wall and intercellular space. Thus, water and dissolved minerals must pass through the plasmalemmas of endodermal cells to reach the vascular tissues of the root (fig. 15.12). As a result, the endodermis functions somewhat as a valve that regulates the movement of nutrients into the vascular tissue via the **symplast** (i.e., membranes and living cells). This arrangement is critical for helping to eliminate leaks and for conserving ions in the vascular tissue. No such control can be exerted over transport through the **apoplast** (i.e., through cell walls and intercellular spaces).

Suberin is deposited continually in endodermal cells as a root ages. Beyond 5 cm or so from the root tip, suberin completely blocks the movement of water and solutes through the endodermis. Thus, these mature zones of the root absorb little water. Rather, they store reserves and anchor the plant in the

soil. Although the precise pathway for exchange between the cortex and vascular tissues in these parts of roots is unknown, most botanists think that it occurs through *passage cells,* which are endodermal cells that do not produce excess suberin. However, this hypothesis is controversial.

concept

The epidermis surrounds the mature region of the root. Interior to the epidermis is the cortex, which has three layers: hypodermis, storage parenchyma, and endodermis. The hypodermis protects roots, and storage parenchyma tissue stores reserves for subsequent use. The endodermis is lined with a strip of suberin called the Casparian strip, which diverts water and dissolved minerals into the cytoplasm of endodermal cells.

Stele

The *stele* includes all of the tissues inside the cortex. It consists of the pericycle, vascular tissues, and sometimes a parenchymatous pith.

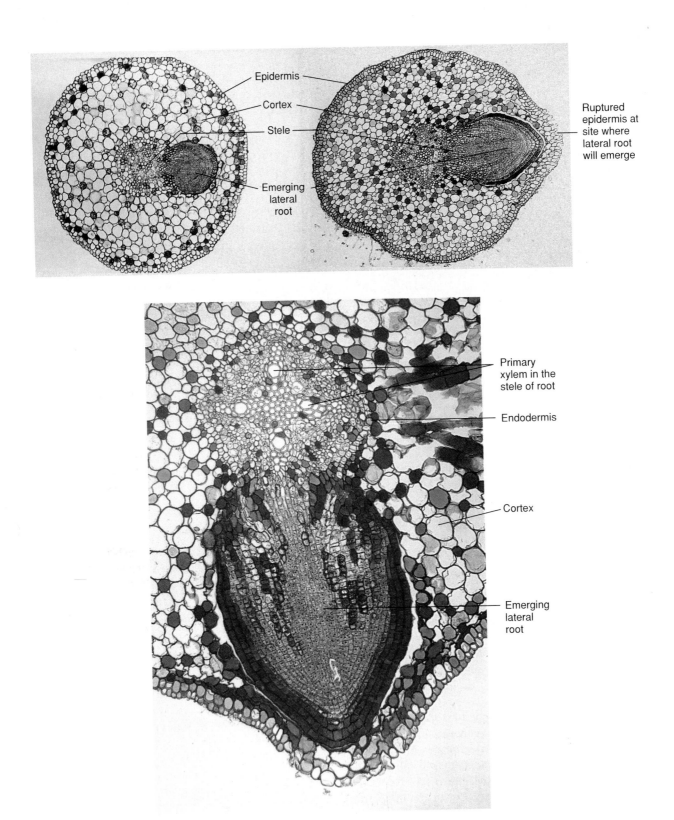

figure 15.13

Formation of lateral roots. Lateral roots are formed by the pericycle, the outermost layer of the stele. The growth of lateral roots eventually ruptures the epidermis of the parent root.

15–11 *Unit Four* The Form and Function of Plants

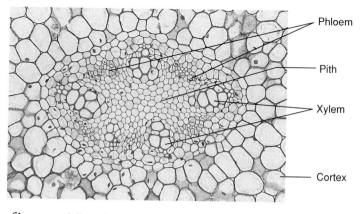

figure 15.14

In siphonostelic roots such as this one of mayapple (*Podophyllum*), a ring of vascular tissue surrounds a parenchymatous pith, ×100.

Pericycle The outermost layer of the stele is the *pericycle*, a meristematic layer of thin-walled parenchyma cells one to several cells thick. The pericycle is important because it produces *branch roots* (also called *secondary* or *lateral roots*). This is the third unique feature of roots: *lateral appendages form endogenously* (fig. 15.13). Branch roots typically form 8–20 mm from the root tip.

The earliest sign of branch-root formation is cellular divisions in the pericycle (fig. 15.13). Soon thereafter, a root cap and primary tissues form. The branch root then forces its way through the cortex and epidermis of the parent root, much as it will later force its way through the soil. The vascular tissues of branch roots link with those of the parent root. Older branch roots form near the root-shoot junction; young ones form near the root tip.

Each branch root services a small area of the shoot; that is, certain parts of a shoot are targeted for deliveries from specific branch roots. This is best seen when examining plants infected with the fungus *Cephalosporium*. This pathogen typically enters the xylem of branch roots; soon thereafter, symptomatic yellow stripes form in the shoot at the delivery sites of that root.

Vascular Tissue and Pith Inside the pericycle is the root's vascular tissue. Roots of most dicots and some monocots (e.g., wheat and barley) are **protostelic,** meaning that the procambium forms a lobed, solid core of primary xylem in the center of the root. Roots with two lobes of xylem are called *diarch* (*arch* here means "first," referring to the first cells of the lobe), those with three lobes are called *triarch,* and those with many lobes are called *polyarch.* However, the number of lobes of xylem is a poor means of classifying plants because it is highly variable, even in the same plant.

The roots of many monocots and a few dicots are **siphonostelic,** meaning that a ring of vascular tissue surrounds a parenchymatous pith (fig. 15.14). In these roots, primary xylem typically forms in isolated rows that radiate toward the periphery of the root. The protoxylem of roots, unlike that in shoots, differentiates centripetally—that is, from the outside

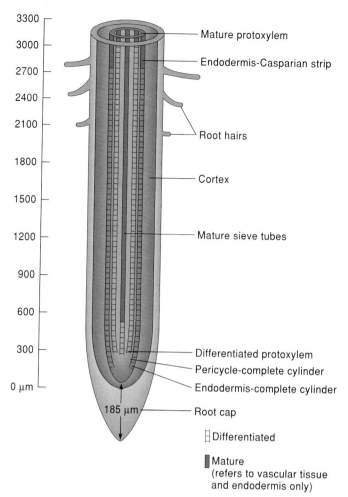

figure 15.15

Diagram of root tip of mustard (*Sinapis alba*), showing where different tissues differentiate relative to the root apex. These data emphasize the artificial nature of labeling discrete zones of cellular division, elongation, and differentiation (see fig. 15.7).

Source: Data from Peterson, Canadian Journal of Botany, *45:319–331, National Research Council of Canada.*

toward the inside of the root. As a result, protoxylem forms next to the pericycle on the outer side of the metaxylem (fig. 15.15). This type of differentiation is called *exarch.* Thus, roots have an exarch arrangement of protoxylem. You'll recall from the previous chapter that the arrangement of protoxylem in shoots is *endarch.*

In roots, bundles of primary phloem differentiate between lobes of xylem. This is the final unique feature of roots: *xylem and phloem in roots alternate with each other.* Sieve elements typically differentiate nearer the root tip than do xylary elements (fig. 15.15). The precise position where vascular tissue differentiates depends on several factors, including the root's growth rate, type, developmental stage, and external influences such as temperature, aeration, and soil texture. In many dicots a vascular cambium later forms between the xylem and phloem, and produces secondary growth. The pericycle also contributes to secondary growth in roots of some plants. You'll learn more about secondary growth in roots in Chapter 16.

The stele includes all of the tissues inside the cortex, including the pericycle and vascular tissues. The pericycle produces branch roots, which form endogenously. Xylem and phloem form in alternating strands interior to the pericycle. The roots of most dicots have a solid core of xylem, while those of many monocots have a parenchymatous pith.

Transition Region between the Root and the Shoot

We mentioned in the previous section that vascular tissues in roots and shoots are arranged differently. For example, the vascular tissue in dicot roots is surrounded by a cortex (i.e., there is no pith), whereas in stems it usually encircles a pith. Moreover, strands of xylem and phloem alternate in roots, but they are arranged opposite each other in the vascular bundles of shoots.

The vascular tissues of roots and shoots join in the **transition region** located between the root and the lower internodes of the shoot. The transition region is different in different plants; for example, it may be abrupt or extend for some distance. Although terms such as *inverted*, *twisted*, and *rotated* are often used to describe the connection between the different vascular systems, cells in the transition region do not move. Rather, the transition between the vascular tissues of roots and shoots is established early in plant development by differentiation of the procambium.

The Root-Soil Interface

The narrow zone of soil surrounding a root and subject to its influence is called the **rhizosphere.** It extends up to 5 mm from the root's surface and is a complex and ever-changing environment. Growth and metabolism of roots modify the rhizosphere in several ways:

Roots enrich the soil with organic matter. Plants transport as much as 60% of their net photosynthate to roots, which deposit more than 30% of this material in the soil as mucigel and other compounds. In plants such as wheat, the amount of carbohydrate deposited in the soil often exceeds that stored in the plant's fruits. Roots leave massive amounts of organic matter in the soil when they die and decay.

Roots compress the soil. Roots compress the soil as they force their way through crevices and between soil particles. Later, when the roots die and decay, they leave open channels that help aerate the soil.

Absorption of nutrients by roots alters the chemical composition of the soil. Roots secrete bicarbonate when they absorb anions such as nitrate (NO_3^-) and secrete H^+ when they absorb cations such as ammonium (NH_4^+). As a result, the pH of the rhizosphere and the surrounding soil can differ by more than

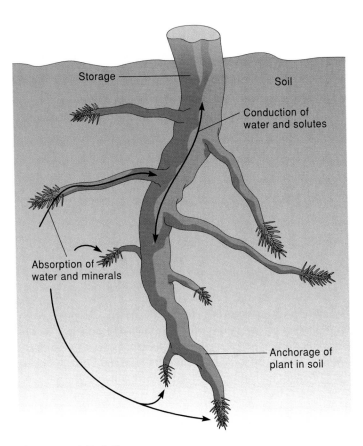

figure 15.16

Diagram summarizing the major functions of roots.

2 pH units. Hydrogen ions alter the availability of other minerals. For example, decreasing the pH increases the availability of aluminum, often to toxic amounts. The respiration of roots decreases the concentration of oxygen and increases the concentration of CO_2 in the rhizosphere. You'll learn more about how roots absorb nutrients from the soil in Chapter 20.

As a result of these influences, the rhizosphere differs significantly from bulk soil. For example, the rhizosphere usually contains large amounts of energy-rich molecules. These molecules feed microbial floras often exceeding 10^{10} organisms cm^{-3} of soil—populations that are 10–100 times more dense than those in bulk soil.

The narrow zone of soil surrounding a root and subject to its influence is called the rhizosphere. Roots change the rhizosphere by enriching it, compressing it, and absorbing nutrients from it. Many microbes live in the rhizosphere.

Earlier in this chapter you learned the four primary functions of roots: absorption, anchorage, conduction, and storage. Let's now consider how the structure of a root relates to these four functions (fig. 15.16).

figure 15.17

The interior of the rhizotron at the University of Michigan Biological Laboratory. Professor John Lussenhop is studying roots, fungi, and soil invertebrates through a microscope. The underground laboratory is lined with 34 windows that contain removable glass panes that provide researchers with direct access to the soil. Roots that press against the glass panes can be studied relatively easily.

Absorption

Most water and nutrients are absorbed by root hairs in the zone of maturation (fig. 15.8). Water enters the root via two pathways: an apoplastic route consisting of intercellular spaces and the cell wall, and a symplastic route involving movement across the plasmalemma and into the cytoplasm of cells (fig. 15.12). Water and nutrients moving through the epidermis in the apoplast are absorbed by cells in the cortex; these cells form a vast collecting system that gathers water and nutrients from the cell walls and intercellular spaces. Nutrients seeping through the apoplast toward the stele finally encounter the endodermis, which is the primary barrier to absorption. The Casparian strip in the endodermis ensures that water and nutrients enter the stele via the symplast. Most nutrients are absorbed and accumulate in the apical 3–5 cm of the root, where most growth occurs.

Anchorage

Relatively little absorption occurs past a few centimeters beyond the root tip, because these parts of the root lack root hairs and have a heavily suberized endodermis. These nonabsorptive regions of roots anchor plants and may later produce branch roots.

Conduction and Storage

Water and dissolved minerals absorbed by roots move to the shoot in xylary elements. Similarly, roots are het-erotrophic and therefore must receive sugars and other compounds from the shoot via the phloem. These nutrients are either used immediately for growth or are stored in cortical cells.

Factors Controlling the Growth and Distribution of Roots

We do not know as much about roots as we do about shoots, primarily because roots grow underground and are more difficult to study than shoots. Much of what we know comes from studies of potted plants or seeds germinated in artificial environments such as sterile dishes or moist paper towels. However, there is little evidence that these roots behave like those growing in field conditions. Understanding how roots grow in their natural environment requires much unusual work. Some of it has been ingenious as well as tedious, and has involved methods ranging from underground cameras and radioactive tracers to painstaking excavations. The newest and most elaborate methods for studying roots involve laboratories called *rhizotrons*, which are underground walkways with glass walls (fig. 15.17). As they grow, many roots press against these glass walls and can be easily studied. Observations in rhizotrons have been supplemented with other studies to reveal that the growth and distribution of roots are controlled by several factors.

Temperature

Roots usually become dormant when it gets cold. Dormancy involves sealing the root until warmer temperatures and more favorable growing conditions return.

Other Organisms

Most microbes in soil live near roots. These microbes secrete compounds that affect the growth and distribution of roots. Equally important, they increase the uptake and translocation of minerals from the soil. Plants growing in sterile soil absorb fewer minerals than do those growing in rhizospheres that include microbes.

    Competition for water and nutrients in soil is often fierce; indeed, a single gram of fertile soil contains approximately 10^9 bacteria, 10^6 actinomycetes, 10^5 fungi, 10^3 algae, and several millimeters of roots. Although we do not understand what characteristics enable the roots of some species to outcompete those of others, different plants have evolved different strategies for competing with their neighbors:

Plants produce many roots. Plants must produce extensive root systems that locate water and minerals dispersed in the soil. In only 6 weeks a corn seedling produces more than 2,000 roots, a 100-year-old Scotch pine (*Pinus sylvestris*) more than 5 million roots, and a mature red oak (*Quercus rubra*) more than 500 million roots.

Roots permeate the soil. Roots of plants such as salt cedar (*Tamarix*) and *Retama* spread tens of meters and burrow more than 30 m deep. The taproot of a 12 m oak tree goes down more than 5 m, while its branch roots span more than 40 m. Most roots grow throughout the upper 3 m of soil, where nutrients are most abundant.

Roots of different plants often grow in different zones of the soil. Plants minimize competition by growing into different areas of the soil. Consider the root systems of mesquite (*Prosopis*) and a saguaro cactus (*Carnegiea gigantea*), two plants that grow in dry environments. Mesquite produces long taproots that obtain water from deep underground. As a result, mesquite grows as a mesophyte in arid environments. Saguaro cacti survive in the same environment by producing an extensive mass of shallow roots that spread as far as 30 m. These roots maximize water absorption after infrequent rains.

Roots protect themselves from other organisms. Since they usually grow underground, roots are protected from many of the herbivores, winds, and lawn mowers that plague stems and leaves. Thus, it's not surprising that roots have fewer obvious adaptations for protection than do shoots. Most of a root's defenses against soil pathogens are chemical rather than

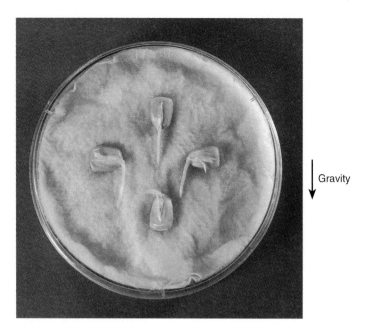

figure 15.18

Primary roots of corn (*Zea mays*) are positively gravitropic; that is, they grow downward. They do so whether the grain germinates in a vertical, horizontal, or inverted position.

structural. Roots often secrete phytoalexins and other obnoxious chemicals that inhibit the growth of pathogens and other organisms.

Roots form partnerships with other organisms. Roots of legumes often form partnerships with bacteria that can provide nitrogen-containing compounds to the plant. Similarly, roots of most plants establish partnerships with fungi that help the plant obtain nutrients. These partnerships are extremely beneficial: in a forest, most of the available nitrogen and phosphorus is contained in these root-fungi associations. You'll learn more about these partnerships later in this chapter and in Chapter 20.

Light

Roots of many plants grow away from light, and light inhibits root growth in corn, wheat, peas, and rice. Light is sensed by the root cap and inhibits growth by slowing the rates of cellular division and elongation.

Gravity

Different roots respond differently to gravity. For example, primary roots grow down; that is, they are *positively gravitropic* (fig. 15.18). Branch roots growing out of primary roots often do not respond to gravity and grow in whatever direction they happen to diverge from the primary root. In plants such as castor bean (*Ricinus communis*), branch roots grow laterally for several centimeters, after which they become graviresponsive and grow down. As a result of this differential responsiveness to gravity, roots permeate the soil at various angles and efficiently absorb its water and dissolved nutrients.

Genetic Differences

The growth and distribution of roots are also controlled genetically. For example, plants such as locoweed (*Astragalus*) always grow deep taproots, whereas most grasses produce shallow, fibrous root systems. Similarly, the root systems of corn are denser near the soil surface than those of soybeans (*Glycine max*), regardless of the type of soil they are growing in.

Stage of Plant Development

Prior to and during fruit formation, most of a plant's resources are used for shoot rather than root growth. Thus, root growth typically slows during flowering and fruit formation. This shift in the allocation of resources is most obvious in some monocots, in which much of the root system dies before harvest.

Soil Properties

Soil Texture   Roots usually grow best in loosely packed soil. For example, roots of wheat grow four times faster in loose sand than in tightly packed clay. Roots that grow in tightly packed soil are usually shorter and thicker than those that grow in loosely packed soil.

Moisture and Air   Most roots grow best in moist but not wet soil. Similarly, roots usually grow deeper in moist, aerated soil than in soaked, poorly aerated soil. The slow growth of roots in poorly aerated soil may also be due to the accumulation of the plant hormone ethylene, which diffuses 10^4 times slower from roots in flooded soil than in well-aerated soil. As a result, ethylene accumulates in flooded soil and slows root growth. This ethylene also stimulates the formation of aerenchyma tissue that improves aeration in the root. We'll discuss other effects of ethylene in Chapter 18.

The roots of hydrophytes (i.e., plants that live in wet areas) are usually small and modified for gas exchange rather than water absorption. They contain relatively small amounts of xylem, usually lack root hairs, and are often filled with aerenchyma, which improves ventilation. Conversely, the roots of xerophytes (i.e., plants that live in arid areas) are often extensive and modified for rapid transport of water. They often have a thin cortex, which shortens the distance between the soil and vascular tissue. They also contain large amounts of well-developed xylem that allows them to move water rapidly to the shoot after a rain.

Nutrients   Roots tend to proliferate in pockets of nutrient-rich soil.

Other Factors   Root growth is also influenced by the presence of other roots, the pH of the soil, and the amount of ethylene in the soil. Ethylene levels in some soils can reach concentrations of 10 parts per million, which would probably slow root extension.[2]

c   o   n   c   e   p   t

The growth and distribution of roots are controlled by temperature, other organisms, light, gravity, genetic differences, the stage of plant development, and properties of the soil. Plants compete for the soil's nutrients by producing many roots, permeating the soil, and growing roots in places different from those of other plants. Most of a root's defenses are chemical rather than structural.

The Lore of Plants

Some epiphytes can be cultivated only in acidic soil (pH 4.0; that's about the pH of tomato juice). This requirement is due to the adaptation of these plants to form roots in arboreal ant nests, which are very acidic environments.

The Growth of Roots versus Shoots

All of the preceding factors strongly influence the major management problem faced by a plant—namely, how best to allocate its resources between roots and shoots. Plants produce shoots that intercept light for photosynthesis, but efficient photosynthesis requires an equally efficient root system for gathering water and nutrients from soil. Clearly, there is a trade-off, and efficient growth requires a mechanism that effectively allocates energy for the growth of roots versus shoots. Botanists study this problem in a variety of ways, including by determining the ratio of the weights of roots and shoots in plants subjected to differing environmental conditions. This *root-shoot ratio* tells us how the environment affects the growth of roots and shoots. The root-shoot ratio is relatively large for seedlings and decreases gradually as a plant ages. Decreasing the amount of nutrients available to roots decreases the root-shoot ratio, as does diminishing light. This tells us that roots are affected more strongly by nutrients and light than are shoots. If stressed, most plants route proportionally move energy and materials to their shoots.

---

2. One part per million equals 0.0001%; it is equivalent to a second in 277 hours (11.5 days).

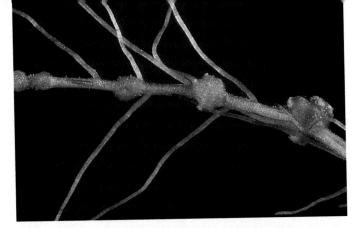

(a)

(b)

(c)

figure 15.19

Modified roots. (a) Nodules on the roots of these soybeans (*Glycine max*) consist of cortical cells infected with nitrogen-fixing bacteria. The red color of the nodules is due to leghemoglobin, a pigment that is similar to the hemoglobin in red blood cells of vertebrates. Much of the nitrogen fixed by the bacteria is used by the host plant for growth. (b) Prop roots on corn (*Zea mays*) are adventitious roots that arise from the stem. Prop roots help support the plant. (c) Some tropical trees produce planklike buttress roots at the base of their trunk. Buttress roots help stabilize and support the tree.

Modified Roots

Like stems and leaves, roots are often modified for special functions (fig. 15.19).

Storage

In plants such as beets (*Beta vulgaris*), turnips (*Brassica rapa*), radish (*Raphanus sativus*), dandelion (*Taraxacum officinale*), and cassava (*Manihot esculenta*), roots store large amounts of starch. Roots of other plants store carbohydrates as sugars; for example, the roots of sweet potato (*Ipomoea batatas*) contain 15%–20% sucrose. Roots can also store large amounts of water; the taproots of some desert plants store more than 70 kg of water.

Propagation

The roots of cherry (*Prunus*), pear (*Pyrus*), apple (*Malus*), and teak (*Tectona grandis*) produce adventitious buds, which form aerial shoots called (rather disrespectfully) *suckers*. When separated from the parent plant, suckers become new individuals. Adventitious buds are a common means of propagating many other plants. For example, most groups of creosote bushes (*Larrea tridentata*) are clones derived from a single plant. Some of these clones are more than 12,000 years old; this means that the first seed germinated approximately 4,000 years before humans began writing.

The most massive organism in the world is believed to be a quaking aspen (*Populus tremuloides*) that has grown thousands of suckers from the same root system. This plant, which grows in the Wasatch Mountains of Utah, consists of more than 47,000 tree trunks, each with the usual complement of leaves and branches. It covers almost 43 hectares, and its mass has been estimated to be about 6 million metric tons. The discoverers of this aspen named it *Pando*, a Latin word meaning "I spread."

Aeration

Many plants grow in stagnant water and mud that contains less than 3% of the oxygen in air. Plants such as black mangrove (*Avicennia germinans*) that grow in these environments avoid suffocation by producing specialized roots called *pneumatophores* that import oxygen from the atmosphere (see fig. 21.14). Pneumatophores contain as much as 80% aerenchyma and grow up into the air, where they function like snorkels through which oxygen diffuses to submerged roots.

Movement

Contractile roots are common on corm and bulb-forming plants such as lily and *Gladiolus*, as well as on nonbulbous plants such as ginseng (*Panax*) and dandelion. These roots contract by shrinking their cortical cells, which twists the

 Unit Four The Form and Function of Plants

Witchweed (*Striga asiatica*) is a parasitic plant that uses its roots to infect grain crops and legumes in sub-Saharan Africa. *Striga* infections can be devastating; for example, *Striga* reduces yields by as much as 70% and is now the leading cause of famine in Africa. According to the United Nations' Food and Agricultural Organization, witchweeds cause more than 100 million farmers to lose at least half of their already reduced crops (reading fig. 15.1).

The key to survival for the parasite's seedlings is to find a host quickly. To accomplish this, its seeds typically germinate only in the presence of a signal released by a host. For example, a hydroquinone released by sorghum roots triggers germination of *Striga* seeds that are within a few millimeters of sorghum roots; the seeds then grow toward increasing concentrations of the hydroquinone and, in the process, toward their

reading figure 15.1

This red-flowered *Striga* is parasitizing a corn (*Zea mays*) plant. Worldwide, *Striga* causes millions of dollars worth of damage to crops.

nearby meal ticket. Seeds farther from host roots do not germinate because the host's hydroquinone is transformed by oxygen into an inactive compound that does not stimulate germination. Strigol, another compound that is released by sorghum plants and stimulates germination of *Striga* seeds, is active at concentrations as low as 10^{-15} M.

Plant biotechnologists are now providing hope for farmers with witchweed-infested land. For example, a special hybrid corn grain developed by Pioneer HiBred International has been biologically engineered for resistance to the imidazolinone herbicides. When farmers soaked these grains with the herbicides and planted them in infested fields, the plants from herbicide-treated grains did well; untreated plants were quickly destroyed by the witchweeds. Less than 0.5 mg herbicide per corn grain tripled yields in the most heavily infested areas.

---

xylary elements into a corkscrew, wrinkles the surface of the root, and contracts it by as much as 3 mm d^{-1}. As a result, contractile roots can shrink more than 50% in only a few weeks. Contractile roots are important because they can pull a plant into the more stable environment of deeper soil.

Nutrition

Parasitism

Plants such as witchweed (*Striga*) and broomrape (*Orobanche*) use their roots to parasitize other plants. Losses caused by these parasitic plants often devastate a crop (see reading 15.1, "Finding a Host").

Mycorrhizae

The epidermis and cortex of many roots are associated with beneficial fungi. Such associations between a root and a fungus occur throughout the plant kingdom and are called **mycorrhizae**. In *ectomycorrhizal roots* of plants such as pine (*Pinus*) and oak (*Quercus*), the fungus produces a mass of filaments on the surface of the root. These filaments invade the root and form an extensive netlike structure between the cortical cells. In *endomycorrhizal roots* the fungus forms inconspicuous filaments on the root's surface and invades cortical cells. Orchids

have specialized mycorrhizae called *orchidaceous mycorrhizae*. Mycorrhizal roots often lack root hairs, suggesting that the fungi replace the absorptive functions of root hairs. A mature tree has thousands of mycorrhizae.

Mycorrhizae are a type of *mutualism*, meaning that both the plant and the fungus benefit from the association. The fungus serves as a subcontractor that absorbs nutrients from the soil; in return, the host plant provides the fungus with carbohydrates, amino acids, vitamins, and other organic substances.

Plants infected with mycorrhizal fungi tolerate stress better than do uninfected plants. This is not surprising, since mycorrhizae absorb nutrients for the plant. A benefit of mycorrhizal associations was demonstrated in a set of experiments done in the rhizotron operated at the University of Michigan Biological Laboratory (fig. 15.17): tree roots having mycorrhizae grew toward water to which nitrogen had been added but not toward pure water alone.

Mycorrhizae are critical to plant growth. Mycorrhizal influence whether plants can invade disturbed areas. Moreover, of all the photosynthesis in a temperate forest, 75% of the carbon is stored underground, and 50% of it ends up in fungi. Mycorrhizae also support a variety of other organisms: for example, a white, grublike insect called *Collembola* dines exclusively on mycorrhizae. You'll learn more about mycorrhizae and plant nutrition in Chapter 20.

Nodules

In the roots of legumes, cortical cells are often infected with *Rhizobium*, a nitrogen-fixing bacterium. This bacterium infects the roots through root hairs and, once inside, forms infection threads that permeate the root. Localized swellings in response to these infections are called **nodules** and represent another type of mutualism between roots and microbes (fig. 15.19a). Bacteria receive carbohydrates and other substances from the host, while the host plant receives nitrogen-containing ions from the bacteria.

Aerial Roots

Aerial roots are adventitious roots that are formed by aboveground structures such as stems. In plants such as mangrove, they account for more than 25% of the aboveground biomass. Aerial roots have different functions in different kinds of plants:

Water retention. The multiple epidermis of roots of some epiphytic orchids, called a **velamen** (fig. 13.16), was originally thought to be a modification for absorbing water, but this conclusion was premature. Subsequent studies have shown that a velamen is almost impermeable to water. The primary functions of the velamen are mechanical protection and the *retention* of water in the root.

Photosynthesis. In plants such as *Philodendron*, vanilla orchid (*Vanilla planifolia*), and several aquatic plants, the roots are photosynthetic. In orchids such as *Microcoelia smithii*, aerial roots are flat and green, and look remarkably like shoots.

Support. *Prop roots* are supportive aerial roots that grow into the soil; they are common in plants such as banyan trees (*Ficus benghalensis*) that grow in mudflats, as well as in plants such as corn (fig. 15.19b). Banyan trees may produce thousands of prop roots that grow down from horizontally oriented stems and form pillarlike supports. Similarly, roots of ball moss (*Tillandsia recurvata*), Spanish moss (*Tillandsia usneoides*), and English ivy (*Hedera helix*) anchor the plant's shoot to its substrate.

Shallow-rooted tropical trees such as *Khaya* and fig (*Ficus*) produce remarkable *buttress roots* at the base of their trunks (fig. 15.19c). These planklike roots are often more than 4 m high and are specialized for support: they contain large amounts of fibers and relatively small amounts of xylary elements.

c o n c e p t

Modified roots are used for storage, propagation, aeration, movement, parasitism, nutrition, and support.

The Economic Importance of Roots

Roots have long been among our favorite foods. For example, we've cultivated carrots (*Daucus carota*) for more than 2,000 years, and sugar beets are a common source of table sugar. Similarly, radish (*Raphanus sativus*), horseradish (*Rorippa armoracia*), sweet potatoes (*Ipomoea*), and turnips appear regularly on our tables. Of these roots, sweet potato is the most nutritious: it is about 5% protein and contains large amounts of calcium, iron, and other minerals. Conversely, cassava, which is used to make tapioca, contains almost no protein, yet it provides more starch per hectare than any other cultivated crop. Cassava also can grow in nutrient-poor soil. Spices such as licorice (*Glycyrrhiza glabra*), sassafras (*Sassafras albidum*), and sarsaparilla (*Smilax;* the flavoring used to make root beer) are derived from roots.

Roots also provide several other economically important products. Drugs such as aconite, gentian, ipecac, ginseng, the tranquilizer reserpine, and the heart-relaxant protoveratrine are extracted from roots. Members of the coffee family provide several dyes, as do carrots: their carotene is sometimes used to color butter.

Finally, a woodland shrub called the wahoo plant (*Euonymus*) is sold in some novelty shops as a cure to "uncross" victims of witches' spells. Folklore has it that the victim will be saved from the curse by holding a piece of the plant's root overhead and screaming "wahoo" seven times.

Chapter Summary

Plants invest heavily in roots. The first root to emerge from a seed is the radicle. Most dicots produce a taproot system consisting of a prominent taproot and numerous smaller branch roots. Taproot systems maximize support and absorption of materials from deep in the soil. Monocots produce a fibrous root system made of similarly sized roots. Fibrous root systems are relatively shallow and rapidly absorb materials from near the soil surface. Root systems, especially of monocots, also include adventitious roots, which are roots that form on organs other than roots. Roots anchor plants, store food, absorb water and dissolved minerals, and transport materials to and from the shoot.

Root tips are covered by a root cap that senses light, pressure, and perhaps gravity. Root caps produce and secrete mucigel, which protects and lubricates roots and helps them absorb water and nutrients. Just behind the root cap is the quiescent center, which is a cluster of seemingly inactive cells. The quiescent center serves as a reservoir of cells that replace damaged cells of the adjacent meristem.

The subapical region of a root can be loosely divided into the zones of cellular division, cellular elongation, and cellular maturation. The zone of cellular division includes the root apical meristem and is characterized by high rates of mitosis. Cells produced by the meristem elongate in the zone of cellular elongation, which is located 4–10 mm behind the root tip. This elongation accounts for most of root extension. Cells complete differentiation in the zone of cellular maturation, which is located 1–5 cm behind the root tip. Root hairs and mature tissues form in the zone of cellular maturation. Roots have no apical appendages.

The primary body of roots is enclosed by an epidermis. The epidermis surrounds the cortex, which consists of three concentric layers: hypodermis, storage parenchyma, and endodermis. The hypodermis is the protective outer layer(s) of the cortex. Storage parenchyma, which comprises most of the cortex, stores starch that is used for subsequent growth. The endodermis is the inner layer of cortex and the primary barrier to absorption. Its tangential walls are surrounded by a Casparian strip that forces water and dissolved minerals into the cytoplasm of endodermal cells. The pericycle and vascular tissues are located inside the cortex and form the stele. Branch roots form endogenously from the pericycle. Xylem and phloem form in alternating strands interior to the pericycle. Roots of most monocots have a pith; those of most dicots have a solid core of xylem.

The narrow zone of soil surrounding a root and subject to its influence is called the *rhizosphere*. Roots change the rhizosphere by enriching the soil, compressing soil particles, and absorbing nutrients. The rhizosphere houses numerous microbes.

Plants compete with other organisms for the soil's water and nutrients. Most of a root's defenses are chemical rather than structural.

The growth and distribution of roots are controlled by temperature, other organisms, light, gravity, genetic differences, the stage of plant development, and properties of the soil. Modified roots are used for storage, propagation, aeration, movement, parasitism, nutrition, and support. Roots of many plants produce economically important products, such as food, spices, and drugs.

 ### What Are Botanists Doing?

Botanists have discovered that when seedlings are infected with *Agrobacterium rhizogenes*, the plants form root hairs over the entire length of each rootlet. What does recent scientific literature tell us about the potential economic importance of these so-called hairy roots?

 ### Writing to Learn Botany

Explain how roots can have such large areas and yet occupy less than 5% of the soil's volume.

Questions for Further Thought and Study

1. Discuss how roots are structurally and functionally modified for storage, anchorage, and absorption.

2. Of what adaptive value are the compounds that inhibit the growth of other plants' roots?

3. Suppose you are given a slide of a root cross section and a slide of a stem cross section. How would you distinguish the two?

4. What's the difference between a small root and a root hair?

5. List six types of modified roots and the functional significance of each.

6. List and discuss the functional significance of four unique features of roots.

7. What is the functional importance of mucigel?

8. List the unique features of the endodermis. How do these features relate to the absorption of water and ions by roots?

9. Pneumatophores typically have diaphragms located throughout the length of the root. What might be the functional significance of these diaphragms?

10. What are the advantages and disadvantages of a fibrous root system? A taproot system?

11. Describe seven factors that control the growth and distribution of roots.

12. What is the adaptive significance of the fact that most annuals have fibrous root systems, while many perennials have taproot systems?

13. Why are roots important to plants? Why are they important to animals?

14. Discuss the economic importance of roots.

15. What characteristics distinguish the zone of cellular division? Cellular elongation? Cellular maturation?

16. Describe how roots modify the soil.

Web Sites

Review the "Doing Botany Yourself" essay and assignments for Chapter 15 on the *Botany Home Page*. What experiments would you do to test the hypotheses? What data can you gather on the Web to help you refine your experiments?

Here are some other sites that you may find interesting:

http://herb.biol. uregina. ca/liu/bio/idb.shtml

At this site you'll find the Internet Directory for Botany, an alphabetical list of botany-related links to databases and other sites. Visit this site to learn more about museums, botany-related organizations, and other resources.

http://130.17.2.208/

- *CSUBOWEB (California State University Biological Sciences World Wide Web Server)*
  http://arnica.csustan.edu

The California State University Biological Sciences World Wide Web Server is a great place to browse in all areas of biology.

http://www.uoguelph.ca/CBS/Botany/roots.html

Root Biology and Mycorrhiza Research Group is maintained at the Department of Botany, University of Guelph, Ontario, Canada. Research focuses on developmental biology of plant roots and the association of plant roots with soil microorganisms, particularly mycorrhizae.

Suggested Readings

Articles

Davies, W. J., and Z. Zhang. 1991. Root signals and the regulation of growth and development of plants in drying soil. *Annual Review of Plant Physiology and Plant Molecular Biology* 42:55–76.

Feldman, L. J. 1984. Regulation of root development. *Annual Review of Plant Physiology* 35:223–242.

Gressel, J. 1996. Plant biotechnology can quickly offer solutions to hunger in Africa. *The Scientist* (30 September):10.

Stewart, G. 1990. Witchweed: A parasitic weed of grain crops. *Outlook on Agriculture* 19:115–117.

Books

Box, J. R., and L. C. Hammond, eds. 1990. *Rhizosphere Dynamics*. Boulder, CO: Westview Press.

Elliott, Douglas B. 1976. *Roots: An Underground Botany and Forager's Guide*. Old Greenwich, CT: Chatham Press.

Givnish, Thomas J. 1986. *On the Economy of Plant Form and Function*. Cambridge: Cambridge University Press.

Gregory, P. J., J. V. Lake, and D. A. Rose, eds. 1987. *Root Development and Function*. Society for Experimental Biology, Seminar Series, vol. 30. Cambridge: Cambridge University Press.

Iqbal, M. 1995. *Growth Patterns in Vascular Plants*. Portland, OR: Timber Press.

Majestic trees, such as these adult and immature
sequoias, are produced by secondary growth.

c h a p t e r

Secondary Growth

16

Chapter Outline

Chapter Overview

Woody plants are characterized by secondary growth. Secondary growth results from lateral meristems that produce large amounts of secondary phloem and secondary xylem. These secondary tissues are protected by a suberized tissue called cork and increase the plant's girth, thereby making the plant sturdier and able to grow taller. This increased height decreases the chances that the plant's leaves will be shaded by other plants. Humans use secondary xylem (i.e., wood) to make a variety of products, most notably paper and lumber.

Competition among organisms for the environment's resources is fierce. In the previous chapter you learned how roots form alliances with other organisms to improve their chances of obtaining water and minerals from the soil. Competition among organisms is also keen aboveground. Although winds maintain a relatively constant concentration of CO_2 throughout the atmosphere, there is only a limited amount of light available for photosynthesis. Plants compete with each other to intercept this light and have evolved various adaptations for maximizing its absorption. For example, plants have elaborate guidance systems that direct them toward light. This light-directed growth, called phototropism, helps ensure that plants intercept light (see Chapter 19). Another means of intercepting light is simply to grow taller than nearby plants. However, primary tissues such as parenchyma and phloem cannot support a tall plant against the pull of gravity. Plants such as climbing vines have overcome this problem by clinging to other objects for support. This adaptation is relatively rare, however, and few herbaceous plants are more than a meter or two high. Clearly, a tall plant that could support itself against gravity would have a decided advantage: it could grow taller and therefore avoid being shaded by other plants or objects.

An improved system for supporting a plant against gravity's pull evolved 370 million years ago (i.e., in the Middle Devonian) and involved the formation of two lateral meristems: the vascular cambium and phellogen (cork cambium). Unlike apical meristems that increase the length of roots and shoots, the vascular cambium and phellogen increase the girth (i.e., thickness) of stems and roots by a process called **secondary growth.** Such growth makes plants sturdier and enables them to grow taller, thereby improving their chances of intercepting light. Secondary growth can produce huge plants. For example, the canopy of a banyan tree (*Ficus benghalensis*) near Gutibayalu, India, covers more than 21,000 m^2, an area about the size of five football fields.

Although secondary growth improves a plant's chances of intercepting light, it also means that the plant must maintain and defend a larger mass of permanent tissue. Increased thickness and height also put new demands on a plant's transport systems. Thus, tissues derived from the vascular cambium and phellogen have several functions besides support, the most important of which are conduction and protection. Support and conduction are functions of derivatives of the vascular cambium, and protection is a function of derivatives of the phellogen.

Tissues produced by lateral meristems form the **secondary body** of a plant. Most of the secondary body is made by the vascular cambium.

The Vascular Cambium

The vascular cambium is a thin band of cells arranged in an open-ended cylinder (fig. 16.1). Except for a few unusual monocots such as *Dracaena*, only dicots and nonflowering seed plants have secondary growth. There are two kinds of cells in the vascular cambium: ray initials and fusiform initials.

Ray initials are small cells oriented perpendicular to the axis of the stem (fig. 16.2). They form sclerenchyma cells or ray parenchyma (i.e., rays) that radially divide a stem, much like slices of a pie. Rays elongate perpendicular to the axis of the stem; they transport water and dissolved solutes radially.

Fusiform initials are tapered, prism-shaped cells oriented parallel to the axis of the stem (fig. 16.2). They produce secondary xylem and secondary phloem (i.e., the axial system of the stem) between the rays. The conducting cells of these tissues transport water and dissolved solutes longitudinally.

There are several differences between meristematic cells of the vascular cambium and those of apical meristems. Fusiform initials of the vascular cambium divide longitudinally toward their tips, whereas other meristematic cells usually divide along the shortest distance across the cell. Also, cells of the vascular cambium are highly vacuolate, while those of apical meristems are densely cytoplasmic. The vacuolation of cambial cells presumably enables them to withstand pressure generated by the expanding secondary xylem.

Most divisions of fusiform initials are periclinal (i.e., parallel to the nearest surface) and produce radial rows of secondary xylem (wood) to the inside and secondary phloem to the outside (fig. 16.3). Derivatives of fusiform initials usually divide several times before differentiating, thus producing a cambial *zone* rather than a single layer of meristematic cells. On average, the vascular cambium produces 4–10 times more xylem than phloem—a ratio that is unaffected by environmental changes such as rain and temperature. Occasional anticlinal divisions (i.e., perpendicular to the nearest surface) expand the vascular cambium to accommodate the increasing girth of the stem.

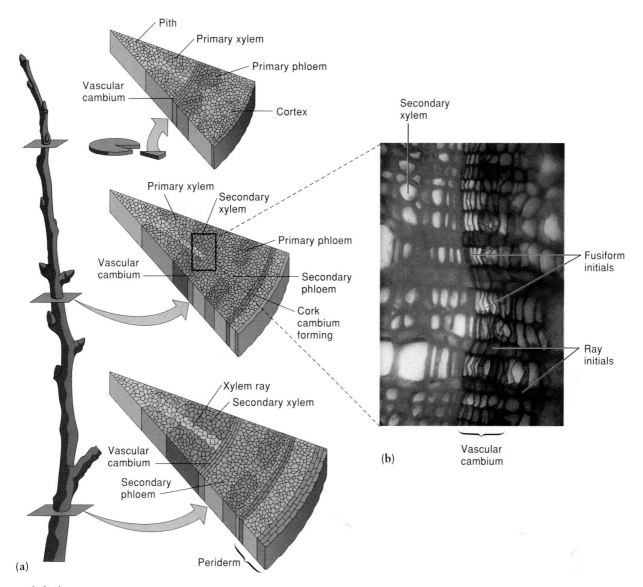

figure 16.1

Secondary growth of dicot stems. (a) The vascular cambium differentiates between the primary xylem and primary phloem, and forms an open-ended cylinder. The vascular cambium produces secondary phloem to the outside and secondary xylem (wood) to the inside. Derivatives of the secondary phloem form a protective layer called the *periderm*. (b) Cross section of the vascular cambium shows ray initials, which form rays that transport water and dissolved solutes radially. Fusiform initials form secondary xylem and secondary phloem between the rays.

c o n c e p t

The vascular cambium consists of fusiform initials and ray initials. Ray initials produce rays that transport fluids radially. Fusiform initials produce secondary xylem and secondary phloem, which are the conducting cells that transport liquids longitudinally.

Where and How the Vascular Cambium Forms

Examine figure 16.4, which shows where the vascular cambium forms in roots and shoots. In roots, the vascular cambium differentiates from latent procambium between xylem

and phloem. These regions of *fascicular cambium* differentiate simultaneously and are subsequently linked by *interfascicular cambium* that differentiates through the pericycle. Thus, the vascular cambium of roots is initially lobed and surrounds the spokes of primary xylem. The vascular cambium in roots differentiates acropetally (i.e., toward the apex) and becomes roughly circular after it begins to divide.

In roots, secondary growth is usually less extensive than in shoots. Nevertheless, secondary growth in roots has similar consequences as in shoots: it increases the girth of roots by producing secondary xylem, secondary phloem, and periderm. These tissues—especially the periderm—curtail the absorption of water. The portions of roots that have undergone secondary growth help anchor the plant.

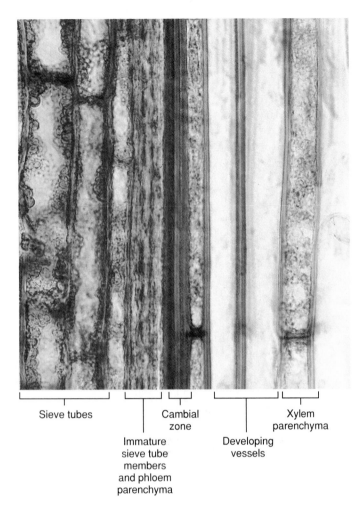

Sieve tubes | Cambial zone | Xylem parenchyma

Immature sieve tube members and phloem parenchyma | Developing vessels

figure 16.2

Longitudinal (tangential) section of a portion of the vascular cambium of an apple (*Malus sylvestris*) tree, showing the orientation of ray initials and fusiform initials, ×100. Compare this with the cross section of the vascular cambium shown in figure 16.1

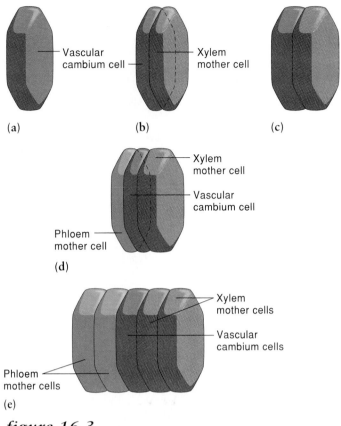

(a) (b) (c)

Vascular cambium cell — Xylem mother cell

Xylem mother cell

Vascular cambium cell

Phloem mother cell

(d)

Xylem mother cells

Vascular cambium cells

Phloem mother cells

(e)

figure 16.3

Differentiation of secondary xylem and secondary phloem from the vascular cambium. Derivatives of the vascular cambium—xylem mother cells and phloem mother cells—divide several times before differentiating, thereby forming a zone of meristematic cells.

In shoots, the fascicular cambium differentiates from latent procambium between xylem and phloem of the stem's ring of vascular bundles (fig. 16.4). The interfascicular cambium then differentiates from cortical cells separating the vascular bundles, thereby forming a cylindrical cambium. As in roots, the vascular cambium in shoots differentiates acropetally (i.e., from the base toward the tip) and is continuous with the vascular cambium of branches.

What controls the formation and polarity of the vascular cambium? Hormones such as auxin coming from young leaves are important, as is the predisposition of cambial cells. To understand this, consider the results of an elegant grafting experiment. When plugs of tissue were removed from the interfascicular region and reoriented 180° prior to differentiation of the vascular cambium (i.e., so that the end of the plug that originally was nearest the epidermis was now in the center of the stem), the cambium differentiated normally across the grafted tissue. However, it produced secondary tissues that were 180° off—that is, it produced secondary xylem to the *outside* and secondary phloem to the *inside*. These and other experiments show that the cortical cells that ultimately differentiate into vascular cambium are developmentally predetermined. The developmental basis for this predetermination of cells remains unknown.

What Controls the Activity of the Vascular Cambium?

During autumn and winter, the vascular cambium is inactive, and its dormant cells have thick cell walls. The moist, warmer, and longer days of spring reactivate the vascular cambium by thinning its walls and inducing cellular divisions. What controls this reactivation? The earliest observations suggested that the activity of the cambium is controlled by auxin made by and released from buds and young leaves. Indeed, reactivation of the cambium begins in buds and twigs, and moves progressively down the stem in a wave. These results prompted botanists to suggest that spring weather induces auxin synthesis in leaves and buds and that this auxin moves down the stem, activating the vascular cambium as it goes (fig. 16.5).

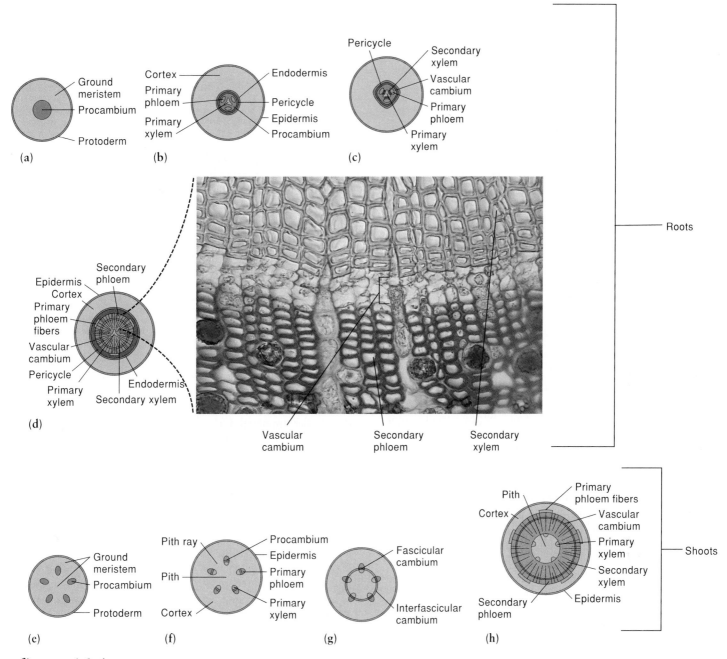

figure 16.4

Differentiation of the vascular cambium in dicot roots (a–d) and shoots (e–h). **Roots.** (a) Early stage of primary growth in roots, showing primary meristems. At this stage, the vascular cambium has not formed. (b) At the completion of primary growth, the meristematic procambium remains between the primary xylem and primary phloem. (c) In this root, the vascular cambium forms from procambium between the strands of primary xylem and primary phloem. The pericycle opposite the three poles of protoxylem also contributes to the vascular cambium. (d) The vascular cambium produces secondary phloem to the outside and secondary xylem to the inside. The radiating lines represent rays. **Shoots.** (e) Early stage of primary growth in a dicot stem, showing primary meristems. (f) After primary growth. (g) Initiation of the vascular cambium. The fascicular cambium (between the xylem and phloem) links with the interfascicular cambium (between the different bundles) to form a cylinder. (h) The fully differentiated vascular cambium produces secondary xylem to the inside and secondary phloem to the outside, ×400.

More recent evidence has prompted other botanists to question this model. Indeed, the inactivation of the cambium that occurs when the apex is removed cannot be replaced by auxin applied to the cut surface. Moreover, the concentration of auxin in the cambium remains high even when the cambium is dormant, suggesting that the onset and cessation of cambial activity are not due to changes in auxin concentration. Therefore, the dormancy of the vascular cambium is probably not due to a deficiency of auxin but rather to a decreased *sensitivity* to auxin: auxin affects the cambial cells in the spring only after the

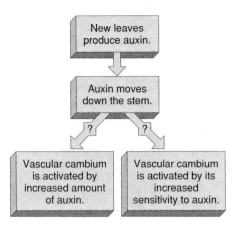

figure 16.5

Hypotheses to account for how auxin from young leaves activates and influences the activity of the vascular cambium.

cambial cells regain their ability to respond to the auxin. This increased sensitivity to auxin begins in buds and twigs, and moves progressively down the stem in a wave, possibly in response to something released by buds and young leaves. In both of these models, buds and young leaves control the formation of vascular tissues for service and support.

Wood (secondary xylem) formed by the vascular cambium during the moist days of spring (i.e., early in the growing season) is called **spring** or **early wood** and consists of large, thin-walled cells (fig. 16.6a). Later in the season, the drier days of summer gradually slow the activity of the vascular cambium and cause it to produce **summer** or **late wood** made of smaller cells with thicker walls (fig. 16.6a). This environmentally induced difference between spring and summer wood is influenced strongly by auxin: auxin stimulates the formation of spring wood, and applying auxin-transport inhibitors stimulates the formation of summer wood. Differences between spring and summer wood are abrupt in trees such as pine and are visible in most trees as **growth rings** (fig. 16.6b). Each growth increment increases the diameter of the tree trunk; once formed, each increment remains unchanged in size or position during the life of the tree.

In temperate regions, the predictable seasons usually produce one growth ring per year, thus accounting for the term *annual ring*. However, trees such as ebony (*Diospyros ebenum*) and jacaranda (e.g., *Jacaranda arborea*) that grow in the seasonless tropics lack growth rings, whereas plants growing in arid and semiarid regions often produce more than one growth ring per year in response to sporadic rains.

Growth rings appear as concentric circles in cross section and are usually 1–10 mm wide (fig. 16.6). Since climate (especially water availability) strongly influences the formation of growth rings, a cross section of wood is a diary of the climatic history of a particular region. The science of interpreting history by studying growth rings is called *dendrochronology* and is an important tool for meteorologists and anthropologists. For ex-

ample, the White Mountains of eastern California are home of bristlecone pines (*Pinus longaeva*), whose slow growth packs almost 1,000 growth rings into only 13 cm or so of wood. By overlapping the growth rings of living and dead trees, dendrochronologists have reconstructed the area's climate for the past 10,000 years. They've used this information to date cliff dwellings and to document droughts in 840, 1067, 1379, and 1632—each 200 to 300 years apart.

c o n c e p t

Water availability and the sensitivity of cambial cells to auxin control the activity of the vascular cambium. Wood made during the moist days of spring contains large xylary elements and is called *spring* or *early wood*. Wood made during the drier days of summer contains fewer large xylary elements and is called *summer* or *late wood*. Differences between early and late wood are visible as growth rings.

Extent of Secondary Growth

The vascular cambium is reactivated each year and is virtually immortal. Many plants are more than 1,000 years old, and a bristlecone pine named Methuselah growing in California is almost 5,000 years old; this means that the seed that produced this plant germinated more than 500 years before the Egyptian pyramids were completed. The perpetual reactivation of the vascular cambium for thousands of years produces impressive trees—the largest and most conspicuous plants ever to live. The fattest tree is a chestnut (*Castanea*) named the Tree of One Hundred Horses, which lives on Sicily's Mount Etna: this tree is 58 m around. A 2,000-year-old tule tree (*Taxodium mucronatum*) near Oaxaca, Mexico, is almost 45 m in circumference,[1] and at last measurement, a California redwood (*Sequoia sempervirens*) was more than 110 m tall. That redwood weighs approximately 1,600 tons, which is roughly equivalent to the weight of ten blue whales (see reading 16.1, "The Lore of Trees").

Trees have attained these gigantic sizes and old ages despite their lack of an immune system and their inability to flee enemies. Trees attacked by pathogens defend themselves by **compartmentalization,** which involves walling off infections, fortifying cell walls near wounds, and making antimicrobial compounds such as phenols. This strategy is effective but has important consequences for a tree. Phenols that kill invading microbes also damage the tree because they poison its tissues. Trees survive these sacrifices of their parts by continually producing new tissue to maintain life. However, the walled-off

---

1. The Mayas, who recognized divinities in trees, attached offerings to this tree. To deflect people's religious interest from this tree, Catholic missionaries built a church nearby. Today, the church's courtyard surrounds this sacred tree, which is visited by thousands of people every year.

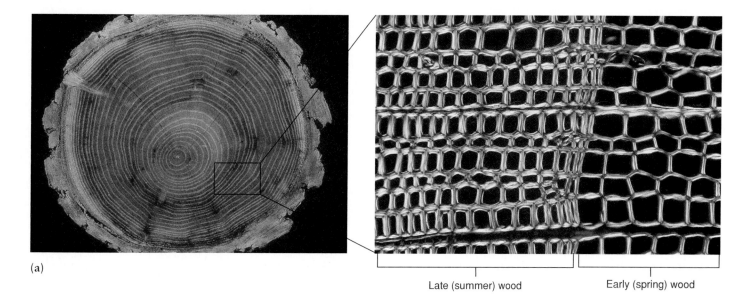

Late (summer) wood   Early (spring) wood

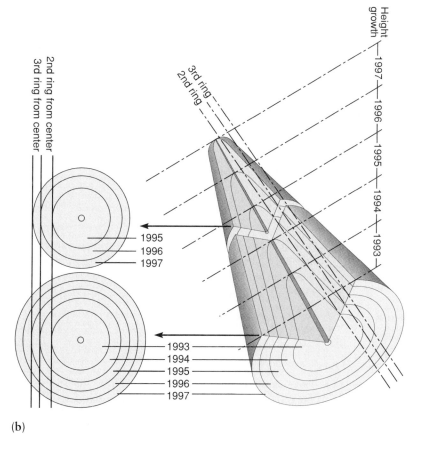

(b)

figure 16.6

Growth rings. (a) The seasonal activity of the vascular cambium in temperate trees produces growth rings in wood. Each ring consists of early (spring) wood made of larger, thin-walled cells, and late (summer) wood made of small, thick-walled cells. (b) This diagram of a tree trunk shows a hypothetical situation that would result from growth during average conditions (no floods or drought). In general, growth begins at the top of the tree and progresses toward the base. Although the width of each growth ring decreases as you move toward the edge of the trunk, the total cross-sectional area remains the same for each year.

compartments of a tree are inaccessible, which forces it to grow a "new" tree over the sealed compartments of the old tree. This new tree is visible as growth rings and provides new tissues to perform the tree's activities. Thus, a tree's survival depends on how fast and effectively it can react to and wall off its invaders— American chestnut (*Castanea dentata*) is nearly extinct, and American elm (*Ulmus Americana*) faces possible extinction because they cannot react fast enough to pathogens. Too many walled-off compartments (i.e., too many infections) stop the movement of a tree's vital fluids and thereby kill the tree.

Secondary Xylem: Wood

Secondary xylem, or wood, is the inner derivative of the vascular cambium and comprises about 90% of a typical tree. About one-third of the United States is covered by trees, of which about 65% are farmed commercially (see reading 16.3). We have all sorts of uses for the 1.7 million cubic meters of wood produced by these forests each day. Each person in the United States uses an average of 2.25 m^3 of wood per year. About 50% of this wood is used for lumber, 25% for pulp and paper products, 10% for plywood and veneer, and the rest for fuel and miscellaneous products such as fence posts, toothpicks, bowling pins, guitars, baskets, decks, and barrels. Each December we chop down about 50 million Christmas trees, and Americans use 250,000 tons of napkins and 2 million tons of newsprint and writing paper per year. These and other uses of wood make forestry a $39 billion-per-year industry.

Kinds of Wood

There are several kinds of wood. We classify them on the basis of the kind of plant that produces the wood (i.e., softwoods versus hardwoods) and the location and function of the wood in the plant (i.e., sapwood versus heartwood).

Softwoods and Hardwoods

Softwoods are nonflowering seed plants native to temperate zones and include pine, spruce, larch, fir, and redwood. Their wood is relatively homogeneous because it is about 90% tracheids and lacks vessels (fig. 16.7). Many softwoods, such as pine and spruce, also contain vertically oriented *resin canals* filled with resin secreted by living parenchyma cells that line the canals (fig. 16.7). Resin hardens when exposed to air and is an effective means of sealing wounds. Fossilized resin is *amber*, a substance described in some mineralogy books as a gem.

Most softwoods are relatively light (i.e., soft) and are easily penetrated by nails, which explains their widespread use in construction. The large amount of lignin in softwoods, such as white pine, makes them ideal for lumber, because lignin makes wood stable and unlikely to warp. Most commercial plantations grow pines or firs because these trees grow faster and are more profitable than dicots. Pulp and plywood come primarily from softwoods.

The uniseriate rays of softwoods are vertically oriented sheets of cells that are 1 cell wide, 1–20 cells high, and hundreds of cells long (fig. 16.7). Softwoods often contain resin canals that form naturally (fig. 16.7) as well as in response to injury (fig. 16.8).

Hardwoods are dicots native to temperate and tropical regions and include oak, maple, ash, walnut, and hickory. Unlike softwoods, which contain only tracheids and ray parenchyma cells, hardwoods contain fibers and vessels (fig. 16.9). Fibers make most hardwoods stronger and denser (i.e., harder) than softwoods, which explains why woods like oak, walnut, maple, and hickory are hard to nail and seldom used for construction.

The conducting cells in hardwoods are tracheids and vessels. Vessels are often visible to the naked eye and are frequently called *pores* by woodworkers. If the largest vessels occur only in early wood, the wood is called *ring-porous wood* (see fig. 16.9a). Because of their targetlike arrangement of vessels, ring-porous woods such as elm, chestnut, oak, and ash have easily discernible growth rings. Ring-porous wood is an evolutionary specialization and occurs only in a few species native to temperate regions. In most trees there is little contrast in the size or pattern of vessels across growth rings; that is, vessels in these trees are distributed uniformly in early and late wood (fig. 16.9b). Such wood (e.g., maple and birch) is called *diffuse-porous wood*.

Rays of hardwoods are more diverse than rays of softwoods and seldom contain resin canals. These rays may be uniseriate but are usually more than twenty cells wide (i.e., are multiseriate) and hundreds of cells long.

---

reading 16.1

THE LORE OF TREES

Trees, like all plants, have a fascinating history and lore:

- There are about 1,000 species of trees in the United States and more species of trees in the Appalachian Mountains than in all of Europe. However, there are more species of trees in one hectare (10,000 m^2) of Malaysian rain forest than in all of the United States.

- The average tree planted today along a street in New York City lives only about 7 years.

- A fully grown deciduous tree can pull more than a ton of water from the soil each day.

- Planting three or four trees around every American house would save 10%–50% on air-conditioning bills.

- The most isolated tree in the world is a Norwegian spruce growing in the wasteland of Campbell Island, Antarctica. Its nearest arboreal neighbor is 120 miles away in the Auckland Islands.

- The fastest-growing tree is *Albizia falcata*, a member of the pea family. One tree in Malaysia grew more than 10 m (in height) in only 13 months and more than 30 m (in height) in just over 5 years.

- The California Department of Forestry and Fire Protection estimates that a single tree that lives for 50 years provides about $200,000 worth of services to its community. These services include providing oxygen ($32,000), recycling water and regulating humidity ($37,000), producing protein ($2,500), controlling pollution ($62,500), providing shelter for animals ($31,000), and reducing erosion ($31,000).

- Each year, the average tree takes in about 12 kg of carbon dioxide, an amount equivalent to that emitted by a car on a 7,000 km trip. The tree also releases enough oxygen to keep a family of four breathing for a year.

- Last year, people in the United States planted more than 4 billion trees. However, if you live in an urban area, you might not have noticed: in some of these areas, as many as four trees die for each one that is planted.

- Several cities in the United States impose severe penalties for killing a tree. In New York City, anyone who cuts down one of the city's 2 million park trees or 700,000 street trees can be fined $1,000 and put in jail for 90 days. In New Jersey, anyone found guilty of cutting down a shade tree can be fined up to $1,500 and pay a "replacement fee" of up to $27 per square inch of tree. This means that someone convicted of cutting down a white oak 75 cm in diameter and 1.5 m tall would have to pay—in addition to the fine—a replacement fee of $19,085.

(a)

(c)

(b)

(d)

reading figure 16.1

(a) El Arbol del Tule, a Montezuma bald cypress (*Taxodium mucronatum*), is the oldest living organism in Mexico. (b) Red firs (*Abies magnifica*) on Mt. Shasta, California. Firs have been used to make telephone poles. (c) The world's largest Valley Oak (*Quercus lobata*) in Butte County, California. (d) Giant sequoia (*Sequoiadendron giganteum*) growing in Sequoia National Park in California. Adult trees are 10–20 times more massive than blue whales, the largest animals.

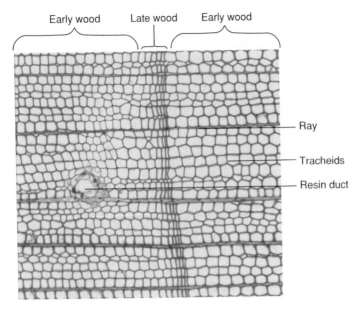

Early wood | Late wood | Early wood
Ray
Tracheids
Resin duct

figure 16.7

Cross section of softwood of eastern white pine (*Pinus strobus*), a gymnosperm. Softwood is relatively homogeneous because it lacks vessels. Note the uniseriate rays and resin ducts in the wood. The annual ring shown here results from differing diameters of tracheids produced at different times of the year.

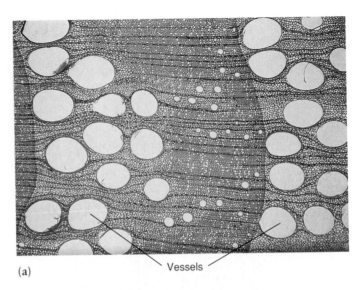

(a)

Vessels

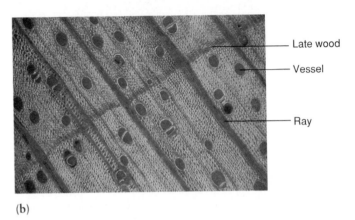

Late wood
Vessel
Ray

(b)

figure 16.9

Cross sections of wood from hardwoods, which contain tracheids and vessels. (a) Because the largest vessels in oak (*Quercus* sp.) occur only in early wood, oak wood is referred to as ring-porous wood. (b) Because the largest vessels in sugar maple (*Acer saccharum*) are distributed throughout the wood, maple wood is referred to as diffuse-porous wood.

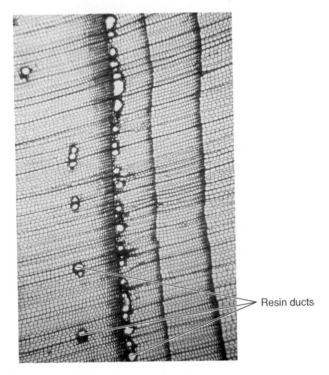

Resin ducts

figure 16.8

Resin ducts in pine (*Pinus*) wood. This band of ducts resulted from a forest fire. At what time of year did the fire occur?

Because of their many fibers, hardwoods are usually denser than softwoods. However, there are many exceptions to this generalization. For example, hardwoods such as poplar (*Populus*) and basswood (*Tilia*) are softer than softwoods such as hemlock (*Tsuga*) and yellow pine. Similarly, balsa (*Ochroma lagopus*) is a broadleaf hardwood whose wood has a specific gravity (0.12) only half that of cork. But even balsa (Spanish for "raft") is not the lightest wood—that distinction goes to the pith-plant (*Aeschynomene aspera*), whose wood has a specific gravity of only 0.04.

figure 16.10

This cross section of a locust tree shows the dark heartwood (inner 12–13 rings) and the lighter sapwood. The transport of water and dissolved minerals occurs only through the sapwood. The dark color of heartwood results from the presence of resins, gums, and other metabolites.

c o n c e p t

Softwoods are nonflowering seed plants native to temperate regions. Their wood is homogeneous and contains tracheids and uniseriate rays. Hardwoods are dicots whose wood contains multiseriate rays, tracheids, vessels specialized for transport, and fibers specialized for support.

Sapwood and Heartwood

We also classify wood as either sapwood or heartwood. This distinction involves differences in position, function, and appearance of a tree's wood and is based on different parts of a woody stem being specialized for different functions.

Water and dissolved nutrients move in the outer few centimeters of secondary xylem. This wood that transports sap is called **sapwood** and is usually light, pale, and relatively weak (fig. 16.10).

The dry wood in the heart (i.e., center) of a tree is called **heartwood** and is the dump site for some of the tree's waste products (fig. 16.10). As a tree ages, metabolites such as resins, gums, oils, and tannins are gradually deposited in heartwood, where they clog the xylary elements and eventu-

ally stop up the wood. Heartwood, which can be more than a meter wide in large trees, is darker, denser, more durable, and more aromatic than sapwood. Heartwood helps support a tree.

c o n c e p t

Sapwood transports water and dissolved nutrients, and comprises the outer few centimeters of wood. Heartwood occurs in the center of a tree and is where many waste products are stored. Heartwood does not transport water and minerals.

Characteristics of Wood

Interpreting the features of wood requires an understanding of how the wood you are examining was cut. Figure 16.11 shows the three ways that wood is cut:

Cross section. Growth rings of trunks cut in cross section (or *transverse* section) are arranged in concentric circles, much like the circles on a target. Rays radiate from the center of the section, like spokes from the center of a wheel. Cross sections of trees are seldom used because they often split after they're cut.

Radial section. A radial section is a longitudinal section that goes through the center of the stem. Boards made from radial sections are called *quartersaw cuts*. In these boards, growth rings appear as parallel lines oriented perpendicular to rays. Quartersawed boards are favorites from trees like oak, whose large rays add texture to the wood. However, only a few boards can be quartersawed from a tree trunk.

Tangential section. A tangential section is a longitudinal section that does not go through the center of the stem. Boards made from tangential sections are called *planesaw cuts*. Growth rings in these boards are arranged in large, irregular patterns of concentric Vs.

Keep these sections of wood in mind as you examine the following properties of wood, which are illustrated in figure 16.12.

Knots are bases of branches that have been covered by lateral growth of the main stem. Some knots fit tightly in wood, while others fit loosely and usually fall out (i.e., produce a "knothole" in the wood); the "fit" of a knot in wood depends on whether the branch was dead or alive when the trunk grew around it. Knots produced by dead limbs have no xylary continuity with the main stem and therefore fall out of lumber when the wood dries. Conversely, knots of living branches do not fall out because their xylem is continuous with that of the main axis. Knots usually weaken lumber and decrease its value.

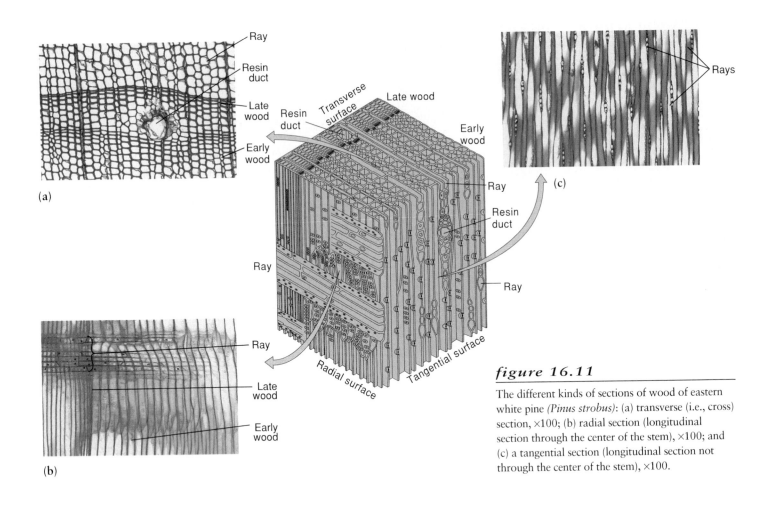

Ray

Resin duct

Late wood

Early wood

(a)

Resin duct

Transverse surface

Late wood

Early wood

Ray

Resin duct

Ray

Ray

Rays

(c)

Ray

Late wood

Early wood

(b)

Ray

Radial surface

Tangential surface

figure 16.11

The different kinds of sections of wood of eastern white pine *(Pinus strobus)*: (a) transverse (i.e., cross) section, ×100; (b) radial section (longitudinal section through the center of the stem), ×100; and (c) a tangential section (longitudinal section not through the center of the stem), ×100.

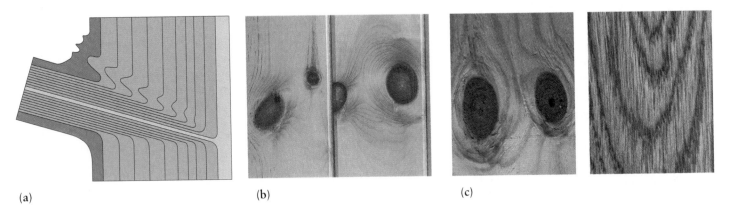

(a)

(b)

(c)

figure 16.12

Features of wood. (a) Radial section of trunk, showing a living branch whose cambium and growth rings are continuous with the branch. Lateral growth of the main stem eventually covers the branch. (b) Tangential section of a covered branch shows a knot, which is a cross section of the branch that was covered by lateral growth of the main stem. (c) Two prominent features of finished wood are grain and figure. These features vary, depending on how the wood was cut (see fig. 16.11).

Grain is the direction of axial cells relative to the longitudinal axis of a tree or a particular piece of wood. Grain may have regular patterns or, as in trees such as English elm (*Ulmus procera*), be irregular and lack patterns. In trees such as sycamore (*Platanus*), the grain changes every few years and is referred to as interlocked grain.

Texture refers to the arrangement and size of the pores in wood. Texture depends primarily on the size and distribution of vessels, tracheids, and, to a lesser extent, rays. Trees such as oak have large vessels and a coarse texture, whereas softwoods lack vessels and have a fine texture.

Figure refers to patterns produced by variations in color, arrangements of tissues, growth rings, grain, and knots.

Density of wood is expressed as *specific gravity*, which is the ratio of the density of the wood to that of water. In general, fast-growing conifers such as pine and diffuse, porous hardwoods are less dense (i.e., have a lower specific gravity) than slow-growing hardwoods. The specific gravity of wood is determined by the size of its cells, thickness of the cell walls, amount of lignin, and the proportions of early and late wood. Most woods have a specific gravity between 0.3 and 0.7; for example, hickory has a specific gravity of 0.65, white cedar (*Thuja*) has a specific gravity of only 0.30, and white ash (*Fraxinus americana*), the tree used to make baseball bats (see reading 16.2, "The Bats of Summer: Botany and Our National Pastime") has a specific gravity of 0.55. Thus, all of these woods float in water. Among the densest woods are lignum vitae (*Guaiacum officinale*) and black ironwood (*Krugiodendron ferreum*), which have specific gravities of 1.25 and 1.30, respectively. Blocks of these and a few other woods sink when placed in water.

Durability of wood refers to its ability to resist weathering and decay. Woods such as redwood (*Sequoia*), teak (*Tectona*), incense cedar (*Calocedrus*), and black locust (*Robinia*) contain large amounts of natural preservatives such as tannins and are extremely durable, while those such as cottonwood (*Populus*), willow (*Salix*), and fir (*Abies*) contain relatively few secondary products and decay rapidly. One of the most durable woods is lignum vitae, a hard and resinous wood that's been found in near-perfect condition in 400-year-old sunken Spanish galleons. Another is sequoia, which is rich in tannins. A well-preserved sequoia log was carbon-dated in 1964 and found to be 2,100 years old; this means that fungi and bacteria have been sidestepping it since the beginning of the Roman empire.

Water-content of wood accounts for as much as 75% of the weight of a living tree. Once harvested, tree trunks are dried (i.e., *cured* or *seasoned*) in large ovens that reduce

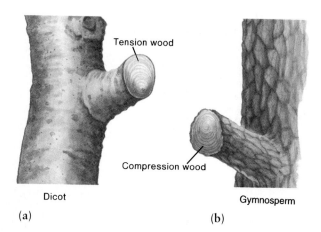

figure 16.13

There are two types of reaction wood: tension wood, which forms in angiosperms, and compression wood, which forms in gymnosperms. (a) In dicots (i.e., angiosperms), tension wood forms on the upper side of the stem. (b) In gymnosperms such as pine, compression wood forms on the lower side of the stem.

the water-content of the wood to between 5% and 20%. Dry wood is about 65% cellulose, 20% lignin, and 15% secondary products such as resins, gums, oils, and tannins (ash, the inorganic compound in wood, seldom accounts for more than 0.5% of wood). Drying wood is important because wood with a water-content less than 20% is essentially immune to attack by most fungi.

c o n c e p t

Figure, knots, grain, texture, durability, density, and water-content are distinguishing features of wood.

Reaction Wood

Horizontal stems and stems bent by wind or other disturbances right themselves by making **reaction wood,** which is wood produced in response to stress caused by gravity. Auxin influences the formation of reaction wood. Tilting a stem causes auxin to accumulate on the stem's lower side, where auxin triggers the vascular cambium to produce reaction wood.

In dicots, reaction wood forms on the upper side of the stem and is called *tension wood* (fig. 16.13a). Tension wood contains relatively few vessels and many gelatinous fibers made primarily of cellulose. These fibers resist cutting and make tension wood brittle. The gelatinous fibers of tension wood shrink and thus pull the stem upright.

Reaction wood in gymnosperms such as pine forms on the lower side of the stem and is called *compression wood*

THE BATS OF SUMMER: BOTANY AND OUR NATIONAL PASTIME

Baseball is a sport influenced heavily by plants. Pitchers use rosin to improve their grip, hitters coat their bat handles with pine tar, and an occasional cheater "corks" his bat by filling its barrel with cork to improve bat speed and drive. No part of baseball conjures up as much superstition and irrational behavior than the game's icon—the baseball bat. Players have prayed for them, coated them with manure, and even slept with them to improve their hitting. One aspiring minor leaguer in the Philadelphia Phillies' farm system even painted eyes and glasses on his bat after going 0 for 16, to help his bat see the ball better. He must have nearly blinded his bat that night when he hit a pair of home runs.

Baseball bats are made from white ash (*Fraxinus americana*) growing in the northeastern United States, especially New York and Pennsylvania (reading fig. 16.2). Ash is ideal for bats because it is tough, light, and will drive a ball. Ash trees harvested to make bats are each about 75 years old (40 cm in diameter) and produce about 60 bats per tree. Technicians use hand-turned lathes to produce one major league bat every 10–20 minutes. The wooden bats on sale at your local sporting goods store receive somewhat less attention—machine lathes produce a retail bat in only 8 seconds.

reading figure 16.2

Wooden baseball bats are made from wood of white ash (*Fraxinus*).

Following lathing, a bat's trademark is branded on the wood's tangential surface and against the grain. Thus, a hitter who follows the adage of "keeping the trademark up" will strike a ball with the grain and where the bat is strongest. When Yogi Berra dismissed this advice by saying he went to the plate "to hit, not to read," his bat makers moved the trademark of his bats a quarter of a turn, thus saving the lives of countless bats.

Although ash bats are durable and strong, they frequently break. Each major leaguer uses six to seven dozen bats per season—approximately one bat per base hit. Hillerich and Bradsby, the makers of Louisville Sluggers, cut down about 200,000 trees per year to meet major leaguers' demands for bats. Today, bats made of aluminum—"aerospace alloy," as their manufacturers say—have replaced wooden bats in amateur baseball. As a result, Hillerich and Bradsby produce fewer than 1 million bats per year, down from 6 million several years ago. They sell these bats to major league baseball teams for $24 to $30 each.

(fig. 16.13b). Compression wood contains intercellular space and large amounts of lignin, which causes it to shrink less than normal wood on the upper side of the stem. The differential shrinkage of compression wood as compared to normal wood lifts the stem.

Reaction wood is important because it helps establish a tree's architecture, reorients stems, and maximizes the interception of light. It is useless as lumber, however, because it shrinks unpredictably when dried.

concept

Reaction wood is abnormal wood that develops in leaning branches. In conifers it is called compression wood and forms on the lower side of a stem. In angiosperms it is called tension wood and forms on the upper side of a stem. Reaction wood is important because it helps establish a tree's architecture and reorients leaning stems.

CHUCK LEAVELL: WORLD-CLASS TREE FARMER, WORLD-CLASS MUSICIAN

When I met Chuck Leavell in the late 1970s, he was already a well-known musician. His previous group, the Allman Brothers Band, had disbanded, and Chuck had formed an eclectic band named Sea Level. I spent many an evening—including most of finals week one quarter—in a club called The Last Resort listening to Chuck play. He was the best I'd ever heard.

Through the years, I occasionally heard about Chuck. After Sea Level disbanded in 1980, Chuck played and toured with a variety of musicians, including Dr. John; Dion; Dave Edmunds; Bonnie Bramlett; the Black Crowes; Chuck Berry; Hank Williams, Jr.; Eric Clapton; and George Harrison. If you're one of the tens of millions of people who bought a Rolling Stones CD or who saw the Rolling Stones on any of their tours during the past 15 years or so, you've heard Chuck play (reading fig. 16.3). If you're one of the millions of people who bought "Eric Clapton Unplugged" (the Grammy's 1992 Album of the Year), you've heard Chuck play. Chuck is sought and admired by the best musicians throughout the world. However, Chuck is more than a great musician.

Chuck and his wife, Rose Lane (a successful businesswoman in her own right), are also world-class tree farmers. Charlane Plantation is their 1,500-acre award-winning tree farm in aptly named Twiggs County, Georgia (northeast of Macon). Unlike many tree farmers, Chuck and Rose Lane base their management plan on conservation, habitat preservation, and balance. As Chuck often says, "Biodiversity is the key." Instead of consisting of a sprawling monoculture of one kind of tree, Charlane Plantation includes a variety of hardwoods (e.g., red, white, and post oak, tupelo, black gum) and softwoods (e.g., loblolly, slash, long leaf, and short leaf pines), as well as an astonishing array of other plants, including established legumes (e.g., partridge pea), native grasses, and Egyptian wheat. Charlane Plantation also supports a diverse population of wildlife, including quail, deer, turkey, and even a group of grumpy alligators.

Chuck and Rose Lane's emphasis on conservation and habitat preservation

(a)

(b)

reading figure 16.3
(a) Chuck Leavell (right) with his musical colleague, Mick Jagger. (b) Chuck Leavell at his tree farm.

distinguishes their style of management from that of most other tree farmers. The Leavell's environmentally friendly approach has been remarkably successful: indeed, under the guidance of Chuck and Rose Lane, Charlane Plantation has become one of the most well-managed tree farms in the nation. The quality of their work is exemplified by their many awards: the American Forest Foundation's Georgia Tree Farmer of the Year, the American Pulpwood Association's Forestry Activist of the Year, the National Arbor Day Foundation's Good Steward Award, and the Central Georgia Soil and Water District's Conservationist of the Year, among others. These awards follow a family tradition: Mrs. A. V. White, Rose Lane's grandmother, also won conservation awards for her forestry work.

Although his work as a musician has made Chuck famous, Chuck and Rose Lane are equally proud of their forestry and conservation awards; in Chuck's words, "Awards like these mean every bit as much to me as a gold or platinum record." True to those words, Chuck proudly displays his conservation awards right beside his gold and platinum records.

Chuck and Rose Lane—each quiet and unassuming—are also strong advocates for conservation and science education. For example, Chuck participates in the Georgia Forestry Association's "Project Learning Tree," an award-winning, in-school environmental education program that helps children develop greater respect for and appreciation of forests and other natural resources. Chuck and Rose Lane regularly host field trips and workshops at Charlane Plantation for teachers and schoolchildren. They've installed—all at their own expense—a nature trail so that teachers and students can best appreciate the diversity of the forest.

Chuck and Rose Lane take an immediate, personal, and hands-on approach to their work as tree farmers and conservationists: on days when he's not touring (he often registers at hotels under his pseudonym, Mr. Forester), Chuck gets on his four-wheeler or old John Deere 4020 tractor and builds dams, plants trees, and gets dirty. He also sponsors fund-raising events that benefit conservation and habitat preservation, and has established a scholarship fund for students studying the environment, wildlife, and forestry (the Chuck Leavell Scholarship Fund, Daniel B. Warnell School of Forestry, University of Georgia, Athens, GA 30602).   —R. M.

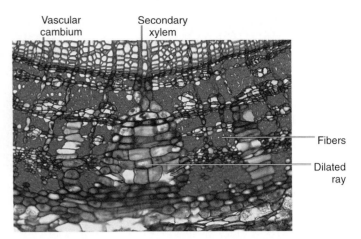

figure 16.15

Cross section of secondary phloem of basswood (*Tilia*), a dicot. Secondary phloem is the outer derivative of the vascular cambium. Where in this photo are the conducting cells of the secondary phloem?

Bark: Secondary Phloem and Periderm

All of the tissues outside of the vascular cambium constitute **bark** (fig. 16.14). In mature tress, bark consists of two tissues:

Secondary phloem transports water and organic solutes between roots and leaves.

Periderm is an outer suberized layer that protects and insulates underlying tissues.

Secondary Phloem

Secondary phloem is the outer derivative of the vascular cambium (fig. 16.15). It includes sieve elements and parenchyma cells that often alternate with bands of thick-walled fibers and prevent splitting of the inner bark. Only the inner centimeter or so of secondary phloem contains functional sieve elements; sieve elements in the outer parts of secondary phloem are dead and nonfunctional, and help protect the inner tissues.

Periderm

Radial expansion resulting from secondary growth eventually ruptures the epidermis of stems and roots. The ruptured epidermis is replaced by periderm, which protects the underlying tissues (fig. 16.16). Periderm consists of three tissues: phellogen (cork cambium), phellem (cork), and phelloderm (secondary cortex).

Phellogen

The **phellogen** (from the Greek words *phellos*, meaning "cork," and *genesis*, meaning "birth"), or cork cambium, is the meristem that produces the periderm (fig. 16.16).

(a)

(b)

figure 16.14

Varieties of bark. (a) Bark of Pacific yew (*Taxus brevifolia*). The bark of these trees contains a drug that is often effective at slowing or stopping the growth of some types of ovarian cancer. (b) Bark of aspen (*Populus tremuloides*). The markings on the bark are "graffiti" made by elk.

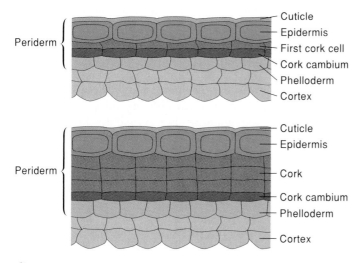

figure 16.16

Periderm. Diagrams showing (a) the origin of the first cork cambium and (b) the first layer of cork of the periderm.

Phellogen differentiates soon after the vascular cambium forms and originates from cellular divisions in the cortex, secondary phloem, or epidermis. Unlike the vascular cambium, the phellogen has only one type of initial.

Secondary growth in most trees splits the outer periderm. Trees cope with this problem by continually forming new layers of phellogen deeper and deeper in the cortex. When the cortex is gone, these replacement layers of phellogen form from parenchyma in secondary phloem. In trees such as paper birch (*Betula*), the replacement phellogen is cylindrical; thus, these trees shed their bark in strips or large, hollow cylinders and are called *ring bark* trees (fig. 16.17a). Trees that continually form overlapping plates of phellem (cork) are called *scale bark* trees and include sycamore, elm, maple, and pine (fig. 16.17b). As their name implies, these trees produce scales rather than large rings of bark.

Phellem

The outer derivatives of the phellogen are radial rows of densely packed **phellem** (cork) cells, which are dead, suberized, and lack intercellular spaces (fig. 16.16). In one season, the phellogen produces as many as forty layers of cork cells that waterproof, insulate, and protect the plant. Cork also forms from phellogen on many fruits and potatoes, over wounds, and in abscission zones.

Trees typically produce several overlapping periderms (fig. 16.18). Phellem made by the innermost periderm seals the outer tissues from the water and nutrients carried in secondary xylem and phloem. As a result, all of the tissues outside of the innermost periderm layer die and are called the *rhytidome*.

Gas exchange across the phellem occurs through **lenticels,** which are raised, localized areas of loosely packed cells (fig. 16.19). Lenticels are formed by the phellogen and

(a)

(b)

figure 16.17

Ring and scale bark. (a) The thin, paperlike ring bark of paper birch (*Betula papyrifera*). (b) The scale bark of sycamore (*Platanus occidentalis*). Scale bark is produced by overlapping plates of phellem (cork).

figure 16.18

Overlapping periderms in bark. The lightly colored layers of this piece of outer bark are phellem. The innermost layer of phellem isolates the outer layers from the rest of the plant.

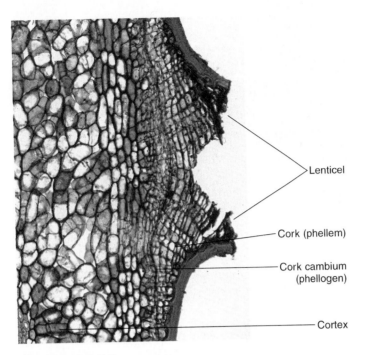

Lenticel

Cork (phellem)

Cork cambium
(phellogen)

Cortex

figure 16.19

Cross section of part of a young stem of elderberry (*Sambucus*). Gas exchange across the phellem occurs through lenticels.

range in shape from round to elliptical. Lenticels are easily visible on pears and apples and are the dark spots and streaks on corks (see reading 16.4, "Cork").

Phelloderm

The inner derivative of the phellogen is a parenchymatous **phelloderm**, or secondary cortex (fig. 16.16). Phelloderm cells are alive, are not suberized, and may be photosynthetic. Phelloderm contains many intercellular spaces for gas exchange.

c o n c e p t

All of the tissues outside of the vascular cambium constitute bark. Bark includes secondary phloem, which transports water and organic solutes, and periderm, which protects and insulates underlying tissues. Periderm is made by phellogen, or cork cambium. The outer derivative of phellogen is a suberized, protective tissue called phellem (cork), while the inner derivative is parenchymatous phelloderm. Gas exchange across the periderm occurs through lenticels.

Replacement of the epidermis by periderm during secondary growth has important consequences for plants. Epidermal cells are unique because they seldom dedifferentiate. Consequently, organs covered by an intact epidermis retain their individual integrity and do not graft, even when they touch each other for long periods. This is why our lawns consist of thousands of individual blades of grass rather than a gigantic, fused blob of photosynthetic astroturf. Unlike epidermal cells, certain

cells in the periderm can dedifferentiate and graft with members of the same or a related species. Grafts between adjacent organisms can be extensive; for example, more than half of the trees of a white-pine forest are often grafted together by secondary growth of their roots, thus creating a gigantic individual that "captures" other plants via more root grafts. These grafts establish close associations between different trees and radically change a forest. Root-grafted trees grow and develop as a single organism, and individual trees survive or die as part of a giant, interconnected system. Grafts also transmit pathogens and poisons among adjacent trees of the system; thus, applying poison to one tree often kills several adjacent trees.

Unusual Secondary Growth

Dicots

Not all dicots have the typical kind of secondary growth described earlier in this chapter. The unusual secondary growth of some dicots results from one of three anomalies:

1. *Unusual behavior of a typical vascular cambium.* In plants such as carrot (*Daucus*) and some cacti, the vascular cambium produces large amounts of storage parenchyma and only a few procambial strands. In cacti, these parenchyma cells store water, whereas in carrots they store carbohydrates.

2. *Presence of more than one cambium.* In plants such as beets (*Beta vulgaris*), cambia are arranged in concentric circles. These *supernumerary cambia* form phloem toward the outside and xylem toward the inside, along with large amounts of storage parenchyma (fig. 16.20a). Conversely, plants such as sweet potato (*Ipomoea batatas*) have a normal cambium and several *accessory cambia* that form around xylary elements. These accessory cambia produce xylem to the inside and phloem to the outside (fig. 16.20b).

3. *Differential activity of a vascular cambium.* In plants such as Dutchman's pipe (*Aristolochia*), parts of the cambium produce parenchyma, while other parts produce vascular tissue. As a result, the stems of these plants are often ridged. In some species of *Bauhinia*, only two opposite areas of the vascular cambium are active, which produces a flattened stem.

Monocots

Secondary growth by monocots is considered unusual because most monocots form only a primary body. However, some arborescent monocots such as *Dracaena* and *Cordyline* form a vascular cambium just outside their vascular bundles (fig. 16.21a). Unlike that of dicots, the monocot cambium consists of only one irregularly shaped type of cell. The outer derivative of this cambium is a parenchymatous *secondary cortex*, while the inner derivatives are procambial strands and lignified *conjunctive tissue*. Procambial strands later produce vascular bundles that connect with the plant's primary vascular system.

16–19

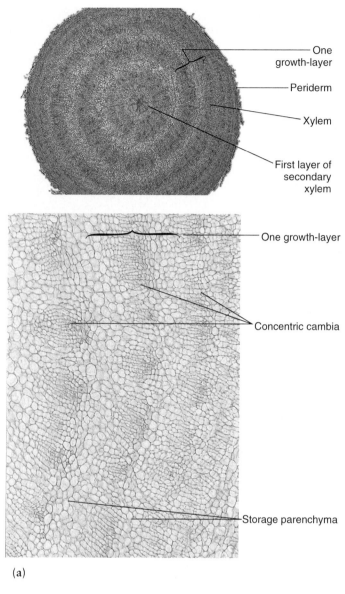

One growth-layer

Periderm

Xylem

First layer of
secondary
xylem

One growth-layer

Concentric cambia

Storage parenchyma

(a)

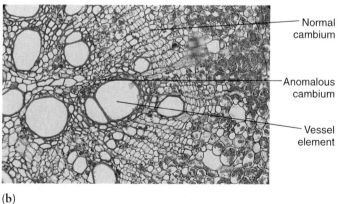

Normal
cambium

Anomalous
cambium

Vessel
element

(b)

figure 16.20

Unusual secondary growth. (a) In roots of sugar beet (*Beta vulgaris*), cambia form in concentric circles. Each cambium produces xylem, phloem, and storage parenchyma. (b) Roots of sweet potato (*Ipomoea batatas*) have a normal cambium plus several other anomalous cambia that form around vessels in secondary xylem. Each of these cambia produces phloem to the outside and xylem to the inside, ×100.

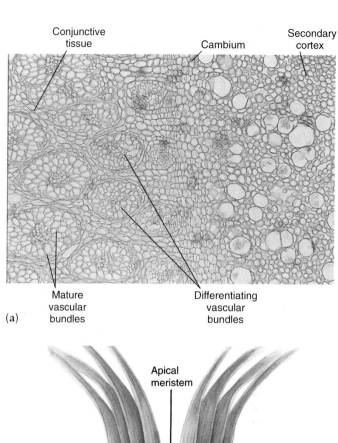

Conjunctive tissue

Cambium

Secondary cortex

Mature vascular bundles

Differentiating vascular bundles

(a)

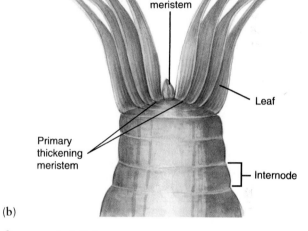

Apical meristem

Leaf

Primary thickening meristem

Internode

(b)

figure 16.21

Vascular cambia and primary thickening in monocots. (a) *Dracaena* is a monocot with a vascular cambium that produces a secondary cortex of parenchyma to the outside and conjunctive tissue and procambial strands to the inside. The procambial strands later form vascular bundles, ×100. (b) In palms, primary thickening occurs just below the apical meristem.

Secondary growth of some monocots can be impressive: for example, some species of *Dracaena* can be more than 18 m high and 12 m around. However, not all arborescent monocots exhibit secondary growth. For example, palms are overgrown herbs that become treelike by *primary thickening*, in which intense lateral expansion occurs just behind the apical meristem (fig. 16.21b). Cells in these plants expand radially before elongating.

Periderms of monocots also differ from those of dicots. All cells of a monocot's periderm are suberized, and there are no different layers such as phellem, phellogen, or phelloderm.

Each year we use almost 200,000 metric tons of commercial cork for things as diverse as stoppers for wine bottles, insulation for the space shuttle, and grips on symphony conductors' batons. Commercial cork is the outer bark of cork oak (*Quercus suber*), an evergreen oak grown in cork plantations in the western Mediterranean.

The first periderm, which forms from the epidermis, is commercially useless. Consequently, it is stripped and discarded when the tree is approximately 10 years old. A usable cork layer 3–10 cm thick can be harvested when a tree is 20–25 years old (i.e., when the tree is about 40 cm in diameter) and thereafter once per decade until the tree is approximately 150 years old. Since cork oaks grow crookedly, cork is harvested by hand rather than machine; workers use hatchets to peel it off in lumberlike slabs. The cork is then dried and boiled before being marketed. In contrast to that of most other trees, the cork of a cork oak breaks away at the cork cambium and can be peeled without harming the tree (reading fig. 16.4).

reading figure 16.4

The bark of this cork oak (*Quercus suber*) has just been harvested for its cork. The inner bark remains, so that the tree does not die when the cork is harvested.

Commercial cork has several properties that make it ideal for a variety of uses. It consists of densely packed, suberized cells (about 1 million cells cm^{-3}) that make it impermeable to liquids and gases. Half of a cork's volume is trapped air; thus, it is 4 times lighter than water. It is also compressible—cork can tolerate 400 kg cm^{-3} and expand to its original shape within one day. Cork is virtually indestructible, fire-resistant, and durable; resists friction; and absorbs vibration and sound.

Uses of Secondary Xylem and Phloem

We make more than 5,000 products from the secondary xylem and phloem of the 1,000 or so species of trees that grow in the United States.

Lumber

Most of the wood consumed in the United States is used for lumber and comes from about thirty-five species of trees. We use this lumber for all sorts of things, including fence posts, telephone poles, houses (white pine), furniture and paneling (black walnut), barrels and flooring (white oak), oars (white ash), skis (hickory), acoustic instruments such as violins (spruce, rosewood, maple), bowling pins, croquet balls, and cue sticks (hard maple). Since wood conducts heat poorly, wood floors are a good way to insulate a house. We also use lumber to make veneers, plywood, and particleboard:

Veneers are thin (about 1 mm thick) sheets of wood peeled from large logs. They are often glued over cheaper and less attractive wood to make cabinets and furniture.

Plywood is made by gluing several veneers together at right angles. These layered veneers make plywood strong, hard to split, and well suited for construction.

Particleboard is made by gluing and pressing together small particles (i.e., chips) of wood into flat sheets. Particleboard is used as insulation and to make boxes.

Pulp and Paper

Paper is made from wood pulp, which is pulverized wood.[2] To make pulp, paper mills first strip the bark from harvested trees. The trunks are then either pulverized on a grindstone or cooked in a batch of chemicals. Both of these processes degrade the wood into pulp, which is about 99% water and 1% fibers. The untreated pulp is fed into a Fourdrinier machine, which spreads and dries it into newsprint, an inexpensive type of paper containing only about 5% water. In only 1 minute, a papermaking machine can produce a sheet of paper covering more than 22,000 m², an area bigger than five football fields. An average tree harvested by a lumber company produces enough pulp to make about 400 copies of a forty-page newsprint tabloid.

Because newsprint is made from untreated pulp, it contains much lignin; this lignin contributes to the paper becoming yellow and brittle after a few days. To make higher quality paper, paper mills first remove the lignin from the pulp. They then add titanium dioxide to whiten the paper and, if necessary, china clay, latex, and alginates (compounds extracted from kelps) to produce glossy paper. All of these processes increase costs and decrease yield by 20%–50%, which is why fine stationery costs more than newsprint. Although more than 90% of paper is derived from wood, special types of paper are derived from a few other plants. For example, money and cigarette paper are made from flax (*Linum*) (fig. 1.1).

Most wood pulp is used to make paper, for which there is an ever-increasing demand. Each person in the United States uses an average of 681 g of paper per day for things ranging from advertisements and botany books to greeting cards, clothes, and general paperwork. This year, almost 2 trillion pieces of paper will move through offices.

Wood pulp is also used to make several other products, including cardboard, fiberboard, nitrocellulose explosives, synthetic cattle food, fillers in ice cream and bread, linoleum, and plastic films such as cellophane. In fact, you may be wearing a derivative of wood pulp: rayon (artificial silk) and several other synthetic fibers are made from the dissolved cellulose of wood pulp.

Fuel

More than half of all wood used worldwide is used for fuel: more than 1.5 billion people get at least 90% of their energy for heating and cooking by burning wood, while another 1 billion use wood for at least half of their heating and cooking needs. In developing countries, almost 90% of wood is used for fuel, while only 10% of the wood consumed in the United States is for fuel.

2. The writing material of the ancient world was woven reeds of papyrus, an aquatic plant that still grows in the delta of the Nile River. However, the Chinese began making paper in about A.D. 105. Before that, they wrote and drew on silk.

Charcoal and Related Products

Charcoal is made by burning blocks of hardwood in the presence of relatively little air. Subsequent distillation of the briquettes produces other economically important products, including methanol (wood alcohol), methane, acetic acid, and acetone. Distilling blocks of softwoods produces turpentine and a sticky rosin used to make paint, ink, lacquer, soap, and polishes. Rosin is also rubbed onto the bows of violins, the shoes of boxers and ballerinas, and the hands of baseball pitchers to increase friction.

Fabrics and Rope

Fibers from the inner bark of baobab (*Adansonia digitata*) trees are woven into fabric and rope. Baobab ropes are strong: as natives of tropical Africa say, "As secure as an elephant bound with baobab rope." Baobab trees are often more than 10 m wide, and their hollowed trunks have been used for everything from morgues to city jails.

Sugar and Spice

Cinnamon is the outer bark of *Cinnamomum zeylanicum;* it is sold either as a pulverized powder or as "sticks," which are the curled outer-bark segments split from twigs. Maple syrup originates from the wood of sugar maples (see reading 21.3, "Making Maple Syrup").

Dyes

Dyes extracted from the inner bark of alder (*Alnus*) are used to color wool and other fabrics.

Drugs

The bark of *Cinchona* trees contains quinine, a drug that kills the protozoan responsible for malaria, the world's leading killer. Although quinine has been a prized drug since the seventeenth century, it has now been replaced by the synthetic drug chloroquine. A more recent discovery is that the bark of the Pacific yew (fig. 16.14a) contains a promising drug (taxol) that arrests the growth of tumors; to extract only one kilogram of the drug requires that we cut down about 12,000 trees and strip from them more than 27,000 kg of bark. Botanists are now producing taxol in cells grown via tissue culture (see fig. 12.8 and accompanying discussion).

Other Uses of Secondary Xylem and Phloem

Tannins from chestnut and tan oak are used to prepare animal skins for sale, and natural chewing gum is extracted from sapodillo (*Manilkara zapota*) trees (most chewing gums sold today are synthetic polymers). *Hevea brasiliensis* produces natural rubber, and turpentine is distilled from longleaf and

All over the world, from the banyan groves of India to the dark forests of Scandinavia, trees have played important roles in ancient folklore, myths, and rituals. For example, Native Americans of the Pacific Northwest used Pacific yew to make totem poles that recorded the experiences and mythical adventures of their ancestors. Other trees have been worshiped as gods—for example, the Druids worshipped oaks, and Hindus revere the banyan. However, no tree has been more revered than cedars. For example, the Plains Indians (including the Shoshoni) regard red cedar (*Juniperus virginiana*) as the Tree of Life; its bark is burned to purify the atmosphere during religious ceremonies (they also made their teepee poles of red cedar because of its durability). Cedar was also revered in ancient civilizations. Indeed, in biblical times, cedar wood—like pine—was used for timber, ornaments, and masts. Its oil and resin were used to embalm the dead.

The cedar of Lebanon (*Cedrus libani*) is the symbol of Lebanon and is represented on the Lebanese flag. It is referred to in the Old Testament as being "excellent above all trees in the field" and was regarded as a symbol of

reading figure 16.5

Cedrus libani, the cedar of Lebanon.

prosperity, power, and long life (reading fig. 16.5). In ancient times, people used cedar wood to make perfumes and royal coffins. It was also considered to be imperishable; indeed, excavators of the Great Pyramid in Giza, Egypt, could, in 1954, still smell the aroma of cedars imported to Egypt over 4,000 years before, and a sacred boat made of cedar wood (buried near the Pharaoh's tomb by his son for use in the afterlife) remains intact some 4,600 years after it was built.

The wood of cedar of Lebanon is best known for having been used to make King Solomon's Temple on Jerusalem's Mount Moriah in about 1,000 B.C. Over a 20-year period, Solomon sent 30,000 Israelites, 150,000 slaves, and 3,300 officers to cut down huge forests of fir and cedar. Although known for his wisdom, Solomon apparently didn't know much about ecology. The shoots from the devastated forests were later eaten by farm animals, and the forests (especially the cedar forests) never regenerated. Except for a few isolated stands of cedar in remote parts of the country, the lush cedar forests that carpeted Lebanon in biblical times have vanished.

---

slash pines. Gum arabic from *Acacia* trees is used to glaze doughnuts and to make the adhesive on stamps and envelopes. Balsa wood is used to make model airplanes and lifejackets, and ambatch, which is lighter than balsa, is used to make pith helmets. The average American annually uses the equivalent of a tree more than 30 m tall and 45 cm in diameter.

c o n c e p t

We have many uses for secondary xylem and bark, including lumber, pulp, fuel, charcoal and related products, fabrics and rope, spices, dyes, and drugs.

Commemorating Trees

Trees are commemorated each year on Arbor Day, a largely ignored holiday that falls on the last Friday in April. Although

Arbor Day was established in 1872 by the Nebraska state legislature, it didn't go into effect until several years later, because overeager settlers had already cut down most trees in the state. The first year, repentant Nebraskans planted a million trees; within 16 years, citizens had planted 600 million trees. The rest of the country was apparently impressed and gradually acknowledged Arbor Day. Although Arbor Day is a feature of the federal calendar, it remains a fake holiday: no one gets the day off.

The Lore of Plants

Each year, people in the United States cut down about 3 billion trees. This harvest exceeds that of all other countries, including Brazil.

Chapter Summary

Secondary growth is growth in girth produced by two lateral meristems: the vascular cambium and phellogen (cork cambium). These meristems produce a secondary body consisting of secondary xylem (wood) and bark (secondary phloem and periderm). Secondary growth is important because it enables plants to grow taller and increases their chances of intercepting light for photosynthesis.

The vascular cambium is a meristematic cylinder made of ray initials and fusiform initials. Ray initials produce rays that transport fluids radially. Fusiform initials produce the secondary xylem and secondary phloem. The conducting cells of these tissues transport fluids longitudinally.

The vascular cambium differentiates from latent procambium between primary xylem and phloem. These regions of fascicular cambium are subsequently linked when the interfascicular cambium differentiates through the pericycle or cortex.

The activity of the vascular cambium is controlled by the availability of water and the sensitivity of cambial cells to auxin. Wood made during moist days of spring contains large xylary elements and is called *early* or *spring wood*. Wood produced during the drier days of summer contains few large vessels and is called *late* or *summer wood*. Differences between early and late wood are visible as growth rings.

Trees fight pathogens by compartmentalization, which involves walling off infections, fortifying cell walls near wounds, and producing antimicrobial compounds such as phenols. Trees survive without these damaged compartments by continually producing new tissues to maintain their activities.

Secondary xylem, or wood, is the inner derivative of the vascular cambium. Softwoods are nonflowering plants native to temperate regions; their wood is homogeneous and contains tracheids and uniseriate rays. Hardwoods are dicots whose wood contains multiseriate rays, tracheids, vessels, and fibers.

Sapwood transports water and dissolved nutrients, and comprises the outer few centimeters of wood. Heartwood is located in the center of a tree and is where many waste products are stored. Heartwood does not transport water and nutrients.

Figure, knots, grain, texture, durability, density, and water-content are distinguishing features of wood.

Reaction wood develops in leaning branches. In conifers it is called *compression wood* and forms on the lower side of the stem. Reaction wood in angiosperms is called *tension wood* and forms on the upper side of the stem. Auxin influences the formation of reaction wood. Reaction wood is important because it helps establish a tree's architecture.

All of the tissues outside of the vascular cambium constitute bark. Bark includes secondary phloem, which transports water and organic solutes, and periderm, which protects and insulates underlying tissues. Periderm is produced by phellogen, or cork cambium. The outer derivative of phellogen is a suberized, protective tissue called *phellem* (cork), while the inner derivative is parenchymatous phelloderm. Gas exchange across the periderm occurs through lenticels.

Unusual secondary growth in dicots results from the unusual behavior of a typical vascular cambium, the presence of more than one cambium, and differential activity of a vascular cambium. A few monocots have a vascular cambium that produces secondary cortex to the outside and procambial strands to the inside. These procambial strands later produce vascular bundles.

We use wood and bark to make lumber, pulp, fuel, paper, charcoal, fabrics, rope, spice, dyes, and drugs.

 ## What Are Botanists Doing?

Find an example of variation in wood structure that has been found in the past 3–5 years. How is this kind of variation useful in botanical research?

 ## Writing to Learn Botany

The large vessels in early ring-porous wood conduct water at 0.5–1 m h^{-1}, which is 5–20 times faster than in diffuse-porous wood. However, trees with ring-porous wood usually don't lose more water than trees with diffuse-porous wood. Explain.

Questions for Further Thought and Study

1. The Pacific yew tree was once thought to be economically worthless. Now, however, we prize it for an anticancer drug we get from it. What does this tell us about the possible consequences of our destruction of tropical rain forests?

2. Are there growth rings in the secondary phloem? Would the number of growth rings in roots always equal that in shoots? Explain your answers.

3. Which is denser—early wood or late wood? Why?

4. Which would produce more uniform paper—pulp from a softwood or from a hardwood? Why?

5. What has happened to the epidermis, cortex, and primary phloem of a 200-year-old oak tree?

6. How does secondary growth in monocots differ from that in dicots?

7. Why don't liquids ooze through lenticels of bottle corks?

8. Is formation of lateral roots an example of primary or secondary growth? Why?

9. Why would extensive secondary growth in leaves of annual plants be a "poor investment"?

10. Cell walls of most plants have a specific gravity of approximately 1.53. How, then, can woods of different species have different specific gravities?

11. Beached blue whales are crushed by the weight of their bodies. Why aren't trees crushed by their weight?

12. What is the significance of secondary growth?

13. Discuss the contribution of each of the following to secondary growth: vascular cambium, rays, periderm, secondary phloem, heartwood, sapwood, compartmentalization of damage.

14. How would wood produced during a wet year differ from wood produced during a dry year?

15. How could thick bark benefit a tree?

16. Discuss the following data relative to the energy costs and benefits of secondary growth:

Tissue	Respiratory Rate $(mm^3\ O_2\ h^{-1}\ gfw^{-1})$*
Secondary phloem	88
Vascular cambium	180
Sapwood	17
Heartwood	0.3

*Cubic millimeters of oxygen consumed per hour per gram fresh weight of tissue.

17. Many trees are considered to be sacred in some cultures. For example, people in the Valley of Mexico prayed to willow trees when they were threatened by storms. What other trees have been viewed as sacred throughout history?

18. The structural diversity of plants results from different arrangements and proportions of tissues rather than from the presence of different types of cells. Use specific examples to support this statement, and indicate why it is important for understanding how plants are designed.

19. Two major functions of a tree trunk are conduction and support. Describe the anatomical basis for each of these functions.

20. Why does removing a ring of bark from around a tree (i.e., girdling) kill a tree?

21. How does reaction wood affect the shape of a tree?

Web Sites

Review the "Doing Botany Yourself" essay and assignments for Chapter 16 on the *Botany Home Page*. What experiments would you do to test the hypotheses? What data can you gather on the Web to help you refine your experiments?

Here are some other sites that you may find interesting:

http://www.gardenweb.com/forums/trees/
At this site you can join discussions about all aspects of trees.

http://www.youra.com/trees.html
This site describes the world-record trees of the Olympic National Park. You'll be amazed by the descriptions of many of these trees.

http://piar.ltrr.arizona.edu/
The Laboratory of Tree-Ring Research was organized in 1937 as an outgrowth of the pioneering tree-ring studies initiated by Andrew Ellicott Douglass at the University of Arizona in 1906. A division of the College of Arts and Sciences, the Laboratory conducts a unique program of teaching and research in all aspects of dendrochronology.

http://www.rbgkew.org.uk/
Much research is conducted at the Royal Botanic Gardens at Kew in Great Britain. Visit the gardens and then review some of their latest research projects.

http://sol.uvic.ca/treephys
This site lets you preview *Tree Physiology*, a journal that publishes articles about all aspects of trees.

http://www.orst.edu/instruct/for241
This site is maintained at Oregon State University and describes trees of the Pacific Northwest.

http://pages.icacomp.com//~/runesmith/woods.html
This site contains interesting information about sacred woods and the lore of trees.

Suggested Readings

Articles

Hitch, Charles J. 1982. Dendrochronology and serendipity. *American Scientist* 70 (May–June):300–305.

Lauchaud, Suzanne. 1989. Participation of auxin and abscisic acid in the regulation of seasonal variations in cambial activity and xylogenesis. *Trees* 3:125–137.

Patton, Phil. 1984. Wooden bats still reign supreme at the old ball game. *Smithsonian* 15 (October):152–176.

Shigo, Alex. 1985. Compartmentalization of decay in trees. *Scientific American* 252 (April):96–103.

———. 1986. Journey to the center of a tree. *American Forests* 92 (June) :18–23.

Stewart, D. 1990. Green giants. *Discover* (April):61–64.

Wilson, B. F., and R. R. Archer. 1977. Reaction wood: Induction and mechanism of action. *Annual Review of Plant Physiology* 28:23–43.

Books

Adkins, Jan. 1980. *The Wood Book*. Boston: Little, Brown.

Altman, N. 1994. *Sacred Trees*. San Francisco: Sierra Club Books.

Burnie, D. 1988. *Eyewitness Books: Tree*. New York: Alfred A. Knopf.

Hoadley, R. B. 1980. *Understanding Wood*. Newtown, CT: Taunton Press.

Johnston, David. 1983. *The Wood Handbook for Craftsmen*. New York: Arco Publishing.

Menninger, E. A. 1995. *Fantastic Trees*. Portland, OR: Timber Press.

Prance, G. T., and A. E. Prance. 1993. *Bark*. Portland, OR: Timber Press.

Zimmerman, M. H. 1983. *Xylem Structure and the Ascent of Sap*. Berlin: Springer-Verlag.

Flowers of the scarlet gilia (*Ipomopsis aggregata*). Flowers are the reproductive organs of flowering plants.

c h a p t e r

Reproductive Morphology of Flowering Plants

17

Chapter Outline

Chapter Overview

Plants have evolved an extraordinary diversity of features for reproducing sexually. The most common reproductive organ of plants is the flower. Indeed, there are more plant species with flowers than with any other kind of reproductive organ. For this reason, the focus of this chapter is on the reproductive morphology of flowering plants. In addition, Chapter 31 provides information about the origin and diversity of flowering plants. We discuss the reproductive structures of other groups of plants in Chapters 28–30.

Successful reproduction in different plants depends on adaptations to various environmental factors. Much of the diversity among flowers comes from specializations for certain pollinators. Further diversity is based on adaptations of seeds for germination and of fruits and seeds for dispersal. This means that features such as the colors and odors of flowers, the size of pollen, and the number of seeds in a fruit are functionally important.

The adaptive point of view is standard for general botany textbooks, probably because explanations of plant morphology are more powerful when structure can be related to function. As you read through this chapter, however, keep in mind that many of the morphologically distinctive features of different plants are not clearly distinctive in their functions. Mustard flowers, for example, have four petals and six stamens, but wintergreen flowers have five petals and ten stamens. It is not clear why, or even if, these kinds of differences are important in the reproduction of plants. Nevertheless, it does show the important and dynamic role that reproductive morphology has in our attempts to understand why there are so many kinds of plants.

The most familiar reproductive structures of plants are flowers, fruits, and seeds. Variations in these structures are important in classifying and identifying flowering plants. This group of plants gets more attention than any other group, largely because of their economic importance, beauty, and diversity. Flowers are distinguished from other kinds of reproductive structures by having seeds enclosed in fruits. As a group, therefore, the flowering plants are commonly referred to as the **angiosperms,** which indicates that their seeds (*sperm*) are in vessels (*angio*) (fig. 17.1).

The role of reproductive structures is the same in all plants: to produce offspring. Moreover, the basic life cycle is the same for all plants (see Chapter 10). The sporophyte produces spore mother cells that undergo meiosis, thereby producing spores. These spores grow into gametophytes that produce gametes; male and female gametes then fuse into a zygote that becomes the next sporophyte stage. Thus, the life cycle of plants is an alternation between sporophytic and gametophytic stages.

Structures that house meiosis and fertilization range from tiny spore capsules in mosses and ferns to large cones in pines and firs. The most familiar reproductive organs of plants to you, however, are probably flowers, which are the focus of this chapter.

Flowers

Like other structures, flowers consist of different tissues and parts that collectively have one main function: in this case, to produce a new generation. A flower is essentially a short stem of several nodes with short internodes between them. However, unlike most other shoots, a floral shoot does not have an apical meristem that grows continuously, because the flower is at the tip of the shoot.

Study the flower shown in figure 17.2. In this generalized flower, the region of the floral shoot where the flower parts are attached is the **receptacle.** The various parts of flowers are considered by some botanists to be modified leaves, all of which are attached at nodes. The outermost whorl of floral parts is comprised of **sepals,** collectively called the **calyx.** **Petals,** together called the **corolla,** form the next whorl. Sepals and petals are the sterile parts of flowers. The fertile parts are the **stamens,** which form just inside the corolla, and the **pistil** in the center of the flower (fig. 17.2).

Botanists are interested in floral variation because features of flowers are important for plant classification. Floral variation also represents the evolutionary record of flowering plants. The next few sections of this chapter present a brief overview of floral variation and the nature of different parts of flowers.

Stamens

Each stamen consists of four pollen-containing chambers that are fused into an **anther,** which is often on a stalk called a **filament** (fig. 17.3). The chambers are named **microsporangia** (singular, **microsporangium**), because that is where microspores are produced. The stamens are collectively called the **androecium** (plural, **androecia**), which means "male house." *Male* refers to the male gametophytes that develop from the microspores.

figure 17.1

Angiosperms are plants whose seeds are enclosed in fruits, as shown here from papaya (*Carica papaya*).

Historically, the androecium was believed to be the male part of the flower; indeed, stamens are still sometimes called the male floral parts. Reference to the "maleness" of the androecium is misleading, however, since no part of the sporophyte can have a gender. Only the gametophytes can be male or female, because only they can produce male or female gametes. Accordingly, the "male" flowers of pumpkin, date palm, and corn are more correctly called *staminate flowers.* Likewise, "male" mulberry, pistachio, and jojoba plants, which have only staminate flowers, should instead be called *staminate plants.*

Variation among androecia comes mostly from the number and arrangement of stamens, the attachment of anthers to the filament, and the way the anthers open for releasing pollen. For example, the androecium consists of one stamen in

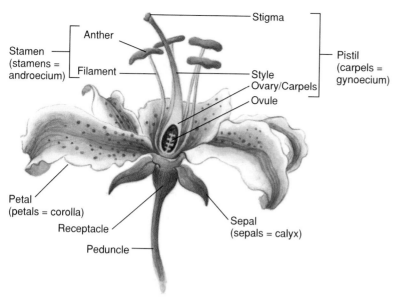

figure 17.2

The parts of a flower. This is a generalized flower that has four main kinds of parts: sepals, petals, stamens, and pistils.

some grasses, two in ash, four in fuchsia, six in mustard, and dozens in buttercups and poppies (fig. 17.4). In addition, the stamens of mustard, buttercup, and poppy are free from each other, but those of garden pea, cotton, and mallow are fused at their filaments (fig. 17.5). Although anthers usually release pollen by splitting longitudinally along a slit, rhododendrons and other members of the heath family release pollen through pores at the tip of the anther.

Pistils

Flowers bear one or more pistils, each consisting of an enlarged basal portion called the **ovary,** and the **stigma,** which is the receptive area for pollen (fig. 17.2). The stigma is often borne on a stalklike **style** that extends from the ovary, thereby elevating the stigma to a level that enhances pollination (see Chapter 10).

The ovary has one or more ovule-bearing units called **carpels.** In the simplest ovaries, there is one carpel with one or more ovules, as in garden pea (see Chapter 8, opening photograph, p. 164). The ovaries of other flowers, such as those of red larkspur, consist of two or more free carpels. Violets and many other flowers have carpels that coalesce at their edges.

Carpels are collectively called the **gynoecium** ("female house") in reference to the female gametophytes that are produced in this structure. Like the androecium, however, the gynoecium has no gender, because it is an organ of the sporophyte. Nevertheless, carpellate flowers, as well as plants that have only carpellate flowers, are often incorrectly referred to as female.

Areas where ovules are attached within the carpels are the **placentae** (singular, **placenta**). In garden pea and red larkspur,

(a)

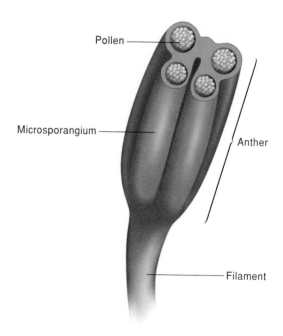

(b)

figure 17.3

(a) Flower of lily (*Lilium* sp.) showing six stamens, each with a white filament and golden brown anther, and a pistil. (b) A single stamen, showing pollen in four microsporangia.

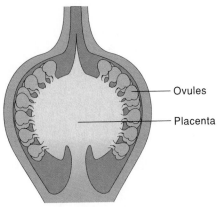

figure 17.4

The number of stamens in an androecium varies among different kinds of flowers. This photo shows dozens of stamens in St.-John's-Wort (*Hypericum* sp.).

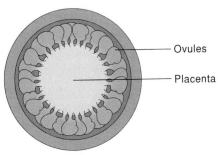

Free-central (longitudinal section)

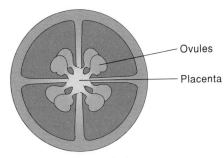

Free-central (cross section)

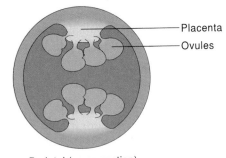

Axial (cross section)

figure 17.5

Stamens are fused in some flowers. Filaments in rose mallow (*Hibiscus* sp.), for example, are fused to the long style (arrow).

Parietal (cross section)

figure 17.6

Ovules are attached to different places in ovaries of different kinds of plants. The area where ovules attach to the ovary is called the placenta.

the placenta occurs along the margin—that is, on the opposite side of the pod from the main vein; this is referred to as **marginal placentation.** Ovaries derived from more than one carpel have more complex types of placentation (fig. 17.6). For example, when placentae are on the ovarian wall, as in violet, the attachment of ovules is called **parietal placentation.** In contrast, placentae on the central axis of the ovary are **axial** when there is more than one chamber, as in lily, or **free-central** when there is just one chamber, as in primrose (fig. 17.6).

17–5

figure 17.7

Flower of evening primrose (*Oenothera hookeri*), showing stigma with four lobes that indicate four fused carpels.

An ovary usually has several indicators for how many carpels it comprises. The most straightforward cases have a chamber for each carpel, as in citrus. This means that each section of an orange, for example, represents a carpel. In addition, stigmas or styles may also reflect the number of carpels. Accordingly, the lily has three seed-chambers and three stigmas, but it has one style that probably evolved by the fusion of three styles from the three carpels. Likewise, some kinds of evening primrose have four lobes on the stigma, signifying four carpels. Other kinds of evening primrose, however, have only one stigma but still have four seed-chambers (fig. 17.7). Conversely, in saguaro and other cacti, the ovary has one chamber and several stigmas, which suggests that the ovary evolved from several carpels. This suggestion is supported by the occurrence of several rows of ovules, each of which came from a separate placenta.

c o n c e p t

Flowers consist of two kinds of reproductive structures: the androecium, which produces the male gametophytes in pollen grains, and the gynoecium, which produces female gametophytes in ovules that are contained in carpels.

Petals

The corolla is usually the most noticeable part of a flower. Petals are often large and showy. Differences in color and odor distinguish many kinds of flowers. In addition, petals may be free (see figs. 17.2 and 17.7), fused into a short tube with large lobes (as in oleander; fig. 17.8a), or fused into a long tube that encompasses most of the corolla (as in honeysuckle; fig. 17.8b). Many of the modifications of corollas are important in pollination, which is discussed later in this chapter.

(a)

(b)

figure 17.8

Petals. (a) Flower of oleander (*Nerium oleander*) showing short tube of fused petals. (b) Flower of honeysuckle (*Lonicera* sp.) showing petals fused into a long tube.

In addition to color and odor, petal development also influences flower symmetry. When all petals develop equally, corollas are radially symmetrical (regular). However, when petals do not develop equally, corollas become bilaterally symmetrical (irregular). Mustard, lily, oleander, poppy, and stonecrop have regular corollas (fig. 17.9), whereas orchids, monkeyflower, garden pea, and honeysuckle have irregular corollas (fig. 17.8b).

figure 17.9

Flowers of stonecrop (*Sedum* sp.) have five sepals that alternate with the five petals. Each flower also has ten stamens and five carpels.

Each flower of a plant usually has a specific number of petals. For example, wild roses have five petals. The number of petals often corresponds to the number of stamens, carpels, and sepals. However, the flowers of cacti, buttercups, and magnolias have an indefinite number of petals. Furthermore, in cacti and magnolias, petals intergrade with sepals, thereby making it impossible to distinguish all petals from all sepals.

Sepals

Most sepals are leaflike, but they may resemble petals or intergrade with them. Sepals of lily are indistinguishable from the petals except for their location; both are referred to as *tepals*. The calyx in four-o'clocks looks like a corolla, but flowers of these plants have no petals. Like petals, sepals may be fused into a tube, and the calyx may be regular or irregular in its symmetry.

The number of sepals, like the number of petals, often corresponds to the number of other flower parts. In most flowers, the number of petals and the number of sepals are identical. Moreover, if you look directly at the center of a flower that has the same number of sepals and petals, the sepals appear to alternate with the petals (fig. 17.9).

Sepals protect the inner parts of a flower before it opens. The calyx is especially important for keeping the unopened flower from desiccating. However, sepals may fall off the flower when it matures. Poppies, for example, appear to have no sepals because they fall off just before the flower opens.

The Nature of Flower Parts

As mentioned, each part of a flower is thought by many botanists to be a specialized leaf. This idea is best supported for sepals because they are mostly leaflike. Botanists generally agree, therefore, that sepals are specialized leaves, but there is a long and fascinating history of debate over the origins of petals, stamens, and carpels. This disagreement stems from the

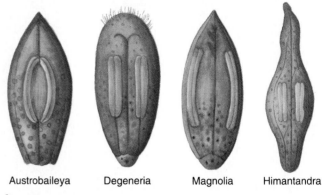

Austrobaileya Degeneria Magnolia Himantandra

figure 17.10

Stamens of *Austrobaileya* and certain other tropical trees look like leaves with microsporangia embedded in them. Such leaflike stamens are indirect evidence that stamens arose from fertile leaves.

absence of any direct evidence, either from fossils or from living plants, for their origins. Instead, different conclusions have been drawn from indirect evidence.

The possibility that stamens and carpels are derived from specialized leaves was so widely accepted at one time that the terms **microsporophyll** (microspore-leaf) and **megasporophyll** (megaspore-leaf), respectively, were used to describe them. These terms are still commonly used. Support for the origin of stamens from leaves comes from the stamen morphology of certain tropical trees, whose stamens look like leaves, each with several veins and four embedded microsporangia (fig. 17.10). If such stamens represent primitive types, then the common stamen-type evolved from them by loss of the lateral veins and by reduction of the leaf blade into a filament. However, in many flowers, the vascular bundles that lead to stamens originate in groups, not as single veins as if from a microsporophyll. This may suggest that each stamen came from a group of branching structures, which evolved into a solitary stamen by the loss of all but one member of the group. Such observations have prompted many botanists to question the leaf-origin hypothesis for stamens. Thus, the issue remains unsettled.

The argument for the origin of the carpel from a fertile leaf is the same as that for the origin of the stamen. A supportive example is the carpel of the genus *Drimys*, which is a tropical tree related to magnolia. The carpel of *Drimys* looks like it evolved from a leaf by folding in the middle and joining at the margins, with a stigma along part of the fused area (fig. 17.11). According to this model, carpels evolved further by sealing the margins and shrinking the stigmatic surface to a small area at or near the apex of the carpel. A pea pod represents the end point of such evolution.

Alternative views on the origin of the carpel come from fossil evidence. One view is that the carpel arose from the seed-bearing structure of an extinct group of plants called the *seed ferns*. These and other seed-bearing structures from fossil plants challenge the idea that carpels evolved from fertile leaves. Thus, like the origin of stamens, the origin of carpels remains controversial.

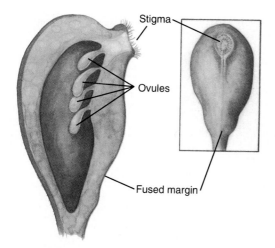

figure 17.11

Carpels of *Drimys* resemble leaves folded at the middle and fused at the edges, with a stigma along part of the fused area.

figure 17.12

"Double flowers" of cultivated roses arose by transformation of stamens into petals. Such transformations may indicate that petals evolved from stamens.

Petals are probably derived from stamens, according to evidence provided by abnormal flowers. For example, the "double flower" of cultivated roses and camellias appears to have risen by the transformation of stamens into petals (fig. 17.12). This transformation results from the broadening of the filament and anther of most or all of the stamens into petal-like structures, accompanied by a color change to that of the petals. In addition, such transformed stamens cannot produce pollen. Conversely, petals may have arisen from sepals. Many flowers, such as those of the saguaro cactus, lily, and magnolia, have petals that either intergrade with sepals or are identical with them. This observation supports the idea that both sepals and petals arose from specialized leaves.

Different parts of flowers may have arisen as modified leaves, or they may have arisen from the fertile structures of extinct seed plants. There is indirect evidence for both points of view.

Reproductive Morphology and Plant Diversity

The diversity of flowering plants is greater than that of any other group of plants; perhaps as many as 260,000 species of flowering plants have been discovered and named. The diversity of angiosperms is reflected in their different sizes, shapes, and forms. The basis for this diversity comes from the reproductive success of flowering plants in a wide variety of habitats. Reproductive success in angiosperms is based on the evolution of the flower, which allowed this plant group to diversify into new habitats. Such diversification was made possible by the protective ovary around the seeds and the potential for efficient pollination by insects and other animals.

The following sections describe some of the variations in flowers, fruits, and pollination mechanisms that represent angiosperm diversity. Understanding this diversity and knowing some of the terms that botanists use to describe it will help you to use field guides and other books to identify and appreciate plants in native habitats and gardens.

Flower Variation

A typical flower has several sepals, petals, stamens, and carpels. Much of the variation among flower types is based on variation of these basic parts. The showy, cream-colored flowers of the southern magnolia (*Magnolia grandiflora*), for example, have many carpels, stamens, and petals and usually three petal-like sepals (fig. 17.13). In contrast, the bright yellow flowers of stonecrop have five sepals, five petals, ten stamens, and five carpels (fig. 17.9). The highly simplified flowers of many grasses have three stamens, one functional carpel (and perhaps two nonfunctional ones), and no petals or sepals (fig. 17.14). Still other grass flowers have either stamens or a carpel, but not both. A flower that has all the major parts is a **complete flower**. An **incomplete flower** lacks one or more of the four whorls. However, even when sepals or petals are missing, a flower that has both an androecium and a gynoecium is called a **perfect flower**. In contrast, flowers that are only carpellate or staminate are **imperfect flowers**.

The position of the ovary also varies among different flower types. St.-John's-Wort, for example, has a **superior ovary**—that is, the base of the ovary is attached above (i.e., is superior to) the other three whorls (see fig. 17.4). In flowers such as the daffodil, the other three whorls grow from the top of the ovary, which is **inferior** to them

figure 17.13

Flowers of southern magnolia (*Magnolia grandiflora*) have many carpels, stamens, and petals, and usually three petal-like sepals.

(a)

figure 17.14

Grass flowers are highly simplified. The flower of this grass, for example, has three stamens, one functional carpel, and no petals or sepals.

(b)

figure 17.15

Longitudinal sections of flowers show differences in the positions of ovaries in different plants. (a) Daffodil (*Narcissus* sp.) has an inferior ovary. (b) In rose (*Rosa* sp.) the ovaries are surrounded by the receptacle.

(fig. 17.15a). Some members of the rose family have intermediate flowers. For example, flowers of the rose and cherry have a superior ovary, but it is surrounded by the receptacle; the corolla and androecium branch from the receptacle above the ovary (fig. 17.15b).

Pollen Development

As described in Chapter 10, pollen is formed by microspores that arise by meiosis in anthers. Variations on this theme occur early in development. In some plants, cell walls form after meiosis I and meiosis II, but cytokinesis in other plants is

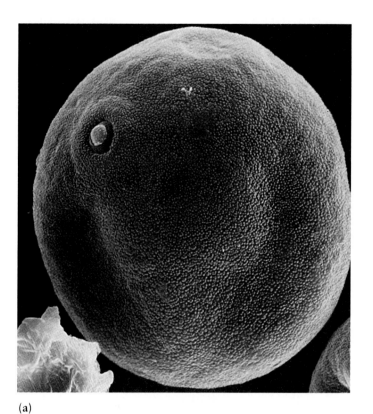

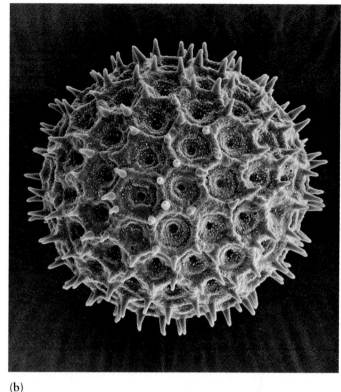

(a)   (b)

figure 17.16

Scanning electron micrographs of pollen from different plants, showing variation in pollen wall morphology. (a) Grass pollen has a smooth exine and a single pore. (b) Pollen from a morning glory has a spiny exine, ×1,160.

postponed until the end of meiosis II. Also early in development, sterile cells around the young microspores differentiate into a **tapetum,** which is the innermost layer of the pollen sac. The tapetum nourishes the developing pollen.

During or before the time that the microspores undergo mitosis and form two-celled pollen grains, each grain forms an outer wall, called the **exine,** and an inner wall, called the **intine.** The exine is made of a terpenoid polymer (see Chapter 2) called **sporopollenin** that protects the male gametophyte from desiccation. The intine comes from carbohydrate polymers that consist of cellulosic and pectic material, which is exported from the cytoplasm of the microspore.

Pollen grains shed from the anthers of most plants contain two cells: a **tube cell** that will form the pollen tube and a **generative cell** that will divide and form two sperm cells. In some plants, however, the generative cell divides to form two sperm cells before the pollen is released. Thus, depending on the plant species, some pollen is two-celled, and some is three-celled when it is transferred from the anther to the stigma. Furthermore, pollen may be shed in packets of two or more grains. Pollen is released in pairs from marsh arrow-grass (*Scheuchzeria palustris*), in tetrads from cattail (*Typha latifolia*), and in packets of sixteen grains from the silk tree (*Albizia julibrissin*). The largest packets occur in orchids, which shed all of the pollen from a flower in two or four masses that attach to the pollinator (see reading 17.1, "Orchids That Look Like Wasps").

The greatest variation among pollen occurs in the morphology of the pollen wall (fig. 17.16). The pollen of grasses, for example, has a relatively smooth exine, but the pollen of morning glories is spiny. Pollen size ranges from 250 μm long in the tropical cherimoya tree (*Annona cherimola*) to a diameter of less than 20 μm in Bermuda grass (*Cynodon dactylon*). In addition, pollen grains have pores or weak spots in the wall that enable the pollen tube to emerge from the grain when it germinates after pollination. Grass pollen has one such pore (fig. 17.16), and morning glory pollen has several, but most plants have three porelike or slitlike openings.

Development of the Embryo Sac

The embryo sac develops in the ovule, a relatively complex structure that consists of a stalk and one or two **integuments** that later develop into the seed coat (fig. 17.17). The integuments surround the megaspore-producing tissue, one cell of which undergoes meiosis in preparation for development of the embryo sac.

The most common type of development of embryo sacs was briefly described in Chapter 10. It is called the *Polygonum-*type of development because it was first found in knotweed (*Polygonum* sp.). About 70% of angiosperms have this type of development in their embryo sacs.

Development of the *Polygonum-*type embryo sac begins at meiosis, when the megaspore mother cell produces four megaspores (fig. 17.18). Three of them disintegrate, leaving

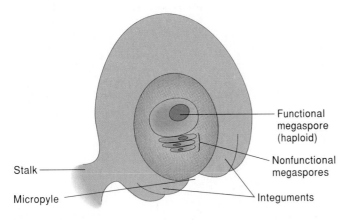

figure 17.17

An ovule consists of a stalk, integuments, and megaspore-producing tissue. Meiosis has already occurred in this ovule, but the embryo sac has not yet developed.

one **functional megaspore** that forms the female gametophyte. The nucleus of the functional megaspore divides by mitosis three times, thereby forming eight free nuclei. Three of the four nuclei nearest the **micropyle** (the opening in the ovule through which the pollen tube will enter) differentiate into the **egg apparatus.** The egg apparatus consists of three cells: the egg cell and two flanking cells, called **synergids.** Three of the four cells at the opposite end of the developing embryo sac, which is called the **chalazal pole,** are called the **antipodal cells.** The two remaining nuclei, one from each pole, migrate to the middle of the embryo sac and are called the **polar nuclei.**

Other plants form embryo sacs differently from the *Polygonum*-type of development. Some textbooks describe the development of the embryo sac in *Lilium* as being typical of angiosperms because the female gametophyte of lily is large and has large nuclei. However, the *Lilium*-type of embryo sac differs from that in *Polygonum* and is much rarer. The first difference in the *Lilium*-type is that meiosis is not followed by cytokinesis (fig. 17.19). Instead, all four spore nuclei remain in the cytoplasm of the megaspore mother cell and function in its development into an embryo sac. After meiosis, one nucleus migrates to the micropylar pole, and the remaining three nuclei migrate to the chalazal pole. The micropylar nucleus then undergoes two mitotic divisions, producing four haploid nuclei. Three of these nuclei differentiate into the egg apparatus, and one migrates to the middle of the sac and becomes a polar nucleus. Meanwhile, the three chalazal nuclei fuse into a triploid nucleus, which divides by two mitotic divisions into four triploid nuclei. Three of these nuclei become antipodal cells, and the fourth becomes another polar nucleus. Thus, the *Lilium*-type embryo sac looks like the *Polygonum* type, but the *Lilium* embryo sac has triploid antipodal cells; one of the polar nuclei is also triploid.

Several other types of embryo sacs are known in angiosperms, the development of which are summarized in

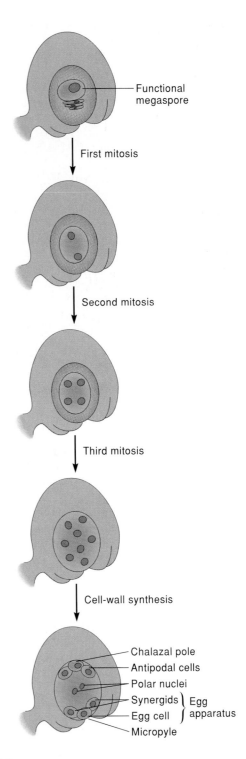

figure 17.18

Development of a *Polygonum*-type embryo sac. This type of development, which occurs in about 70% of angiosperms, involves a single functional megaspore that forms the embryo sac.

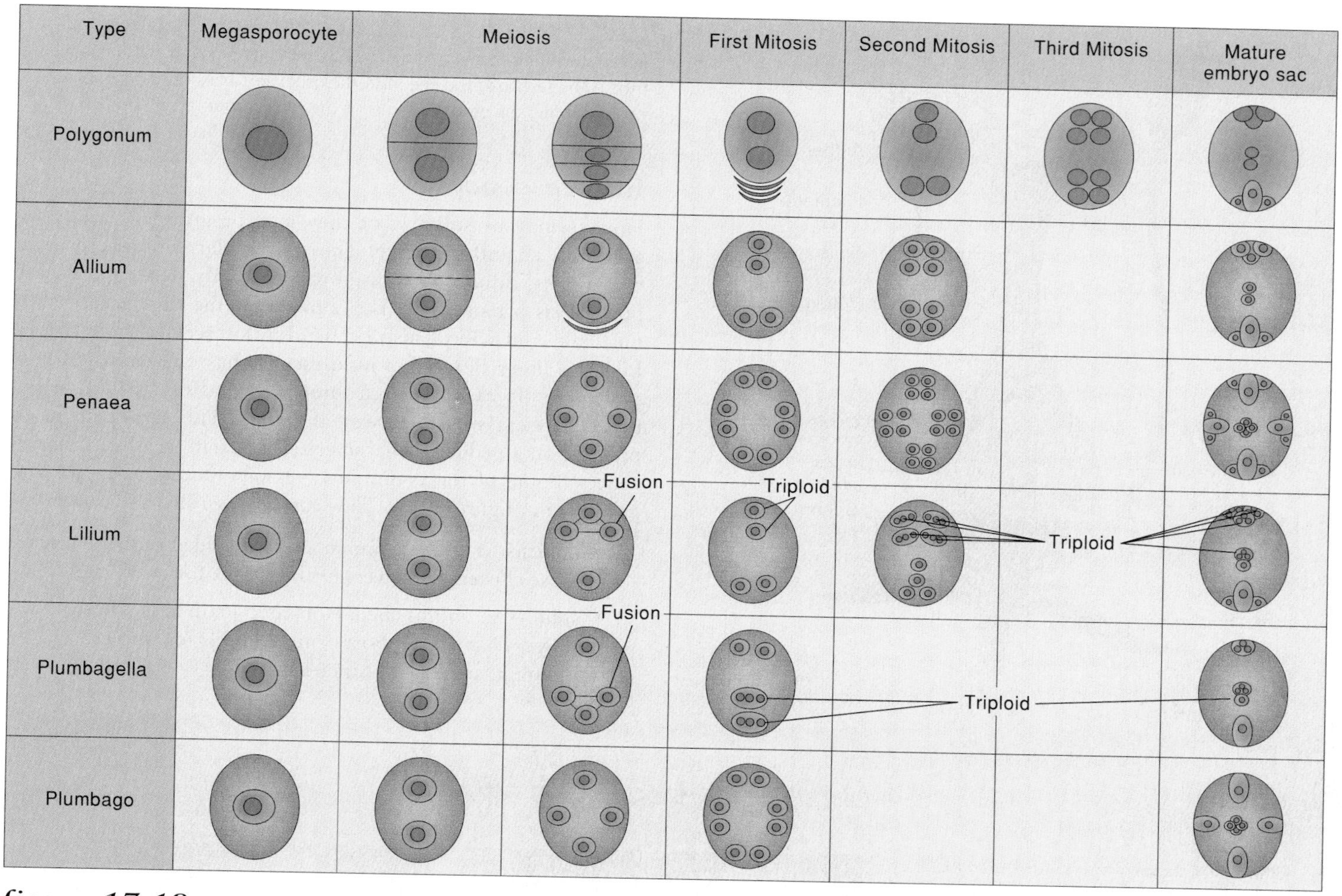

Type	Megasporocyte	Meiosis		First Mitosis	Second Mitosis	Third Mitosis	Mature embryo sac
Polygonum							
Allium							
Penaea							
Lilium			Fusion	Triploid	Triploid	Triploid	Triploid
Plumbagella			Fusion	Triploid		Triploid	
Plumbago							

figure 17.19

Variation in development of embryo sacs in angiosperms. Note that the embryo sac in *Polygonum* comes from a single megaspore nucleus, but the embryo sac in *Allium* comes from two megaspore nuclei, and the embryo sac in many other plants comes from four megaspore nuclei. Embryo sacs of *Penaea*-type and *Plumbago*-type plants apparently have four egg apparatuses, but only one egg per embryo sac normally participates in reproduction.

figure 17.19. They each have polar nuclei, antipodals, and an egg apparatus, with or without synergids. There is usually one egg, but the numbers of cells and polar nuclei vary. Nevertheless, *all types of embryo sacs function similarly during fertilization.* The pollen tube goes through the micropyle of the ovule to one of the synergids, if present. One sperm cell passes through the synergid and fertilizes the egg, and the other sperm cell fertilizes the polar nuclei (fig. 17.20). This is called **double fertilization.** The zygote that results from fusion of the egg and sperm is diploid ($2n$), but the endosperm that results from the union of sperm and polar nuclei may be diploid, triploid ($3n$), pentaploid ($5n$), or nonaploid ($9n$), depending on the type of development.

c o n c e p t

Flowers vary immensely in such features as numbers, sizes, shapes, and colors of flower parts. Different kinds of flowers also form different types of pollen and embryo sacs.

Monocots and Dicots

Floral variation and other aspects of reproductive morphology provide the basis for dividing the flowering plants into two major classes: the class Magnoliopsida (*dicots*) and the class Liliopsida (*monocots*). The informal name *dicot* refers to the presence of two embryonic leaves (**cotyledons**) in the seed; *monocot* refers to plants that have one embryonic leaf in the seed. In monocots, floral parts occur in multiples of three: three petals, three sepals, six stamens, and a carpel that has three chambers (see fig. 17.3). In dicots, flower parts usually occur in multiples of four or five (see figs. 17.7 and 17.9). Although many dicots and monocots have other numbers of floral parts, many other features are unique to each class. These are listed in table 17.1, along with several examples of familiar species in each class. Note that dicots include about 80% of all angiosperm species, including many herbaceous plants and all woody, flower-bearing trees and shrubs. Monocots are primarily herbaceous, but they also include nonwoody trees such as palms and Joshua trees (see Chapter 16 regarding why such trees are not woody).

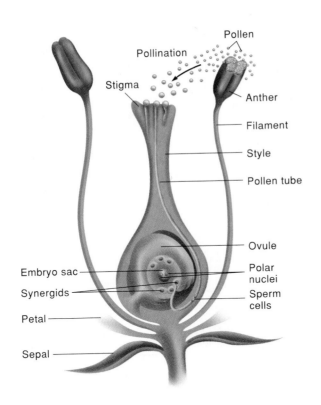

Pollen

Pollination

Stigma

Anther

Filament

Style

Pollen tube

Ovule

Embryo sac

Polar nuclei

Synergids

Sperm cells

Petal

Sepal

figure 17.20

Pollination and fertilization. Two sperm cells move down the pollen tube. One fertilizes the egg, and the other fertilizes the polar nuclei. In this example, the embryo sac is of the *Polygonum*-type (see p. 391).

Inflorescences

Flowers may be solitary, or they may be grouped closely together in an **inflorescence.** The spectacular inflorescences of urn plants, lupine, snapdragon, and many other plants are popular as ornamentals. Less obviously, the flowers of hazelnut, oak, and willow also occur in inflorescences (fig. 17.21). Like a solitary flower, an inflorescence has one main stalk, or **peduncle.** It also bears numerous smaller stalks, called **pedicels,** each with a flower at its tip. The arrangement of pedicels on a peduncle characterizes different kinds of inflorescences. Some of the common types of inflorescences are diagrammed in figure 17.22. They can be described as follows:

- **panicle**—a branched main axis with side branches bearing loose clusters of flowers (examples: oat, rice, fescue)

- **spike**—an unbranched, elongated main axis whose flowers have very short or no pedicels (examples: plantain, spearmint, tamarisk)

reading 17.1

ORCHIDS THAT LOOK LIKE WASPS

Although many flowers reward their pollinators, other flowers use deception. Perhaps the most sensational cases of pollinator deception occur in orchids. For example, flowers of *Ophrys* species imitate female wasps. These orchids have a wasplike shape, often including a lower lip that is fringed with red hairs like the abdomen of the female wasp. Chemicals in the flower's fragrance are similar to those secreted by the female wasp as a sexual attractant. These orchids bloom before most of the female wasps have emerged in the springtime.

Male wasps are attracted to the wasplike orchid and try to copulate with the flower (reading fig. 17.1). A large pollen packet is dislodged from the flower during the vigorous movements of this pseudocopulation. The pollen packet sticks to the wasp and is carried to another flower, where it detaches and nestles snugly among three stigmas. One of the stigmas is sterile and modified into a small outgrowth with a sticky area. The sticky spot ensures that the pollen packet stays on the stigmas after the wasp has left.

(a)

reading figure 17.1

(a) Wasplike flower of a species of *Ophrys*. (b) A wild bee (*Eucera longicornis*) during pseudocopulation with a flower of another species of *Ophrys*.

(b)

table 17.1

Main Differences between Monocots and Dicots

Characteristic	Dicots	Monocots
Flower parts	In fours or fives (usually)	In threes (usually)
Pollen	Usually having three furrows or pores	Usually having one furrow or pore
Cotyledons	Two	One
Leaf venation	Usually netlike	Usually parallel
Primary vascular bundles in stem	In a ring	Scattered arrangement
True secondary growth, with vascular cambium	Commonly present	Absent
Examples	Rose, pea, sunflower, magnolia, ash	Lily, corn, palms, pineapple, banana

(a)

- **raceme**—an unbranched, elongated main axis whose flowers have pedicels that are all about the same length (examples: lily of the valley, snapdragon, mustard)
- **corymb**—an unbranched, elongated main axis whose flowers have pedicels of unequal length, forming an inflorescence that appears flat-topped (examples: hawthorne, apple, dogwood)
- **simple umbel**—a peduncle bearing all of the pedicels at its apex (examples: onion, geranium, milkweed)
- **compound umbel**—a cluster of simple umbels at the apex of a main axis (examples: carrot, dill, parsley)
- **head**—a peduncle bearing flowers that have no pedicels (examples: sunflower, daisy, marigold)
- **catkin**—a spikelike inflorescence that bears only unisexual flowers; catkins occur only in woody plants (examples: hazelnut, willow, birch)

Some types of inflorescences characterize different plant groups. All members of the carrot family (Apiaceae; also called Umbelliferae) have compound umbels. This family includes celery, dill, parsley, and caraway. All members of the sunflower family (Asteraceae) have heads, including chrysanthemum, zinnia, marigold, and dandelion. Moreover, the heads often include two kinds of flowers. One kind is irregular and occurs at the periphery of the inflorescence; the other is regular and occurs in the center of the inflorescence. Thus, the "flower" of a sunflower is really many flowers packed into an inflorescence.

Breeding Systems

No discussion of reproductive morphology and plant diversity is complete without mention of how variation in flower structure can be related to its function. Reproductively, the most versatile

(b)

(c)

(d)

figure 17.21

The inflorescences of (a) urn plant (*Aechmea fasciata*) and (b) lupine (*Lupinus nootkatensis*) are spectacular because of their brightly colored flowers. The less obvious flowers of (c) hazelnut (*Corylus* sp.) and (d) willow (*Salix* sp.) are packed into catkins.

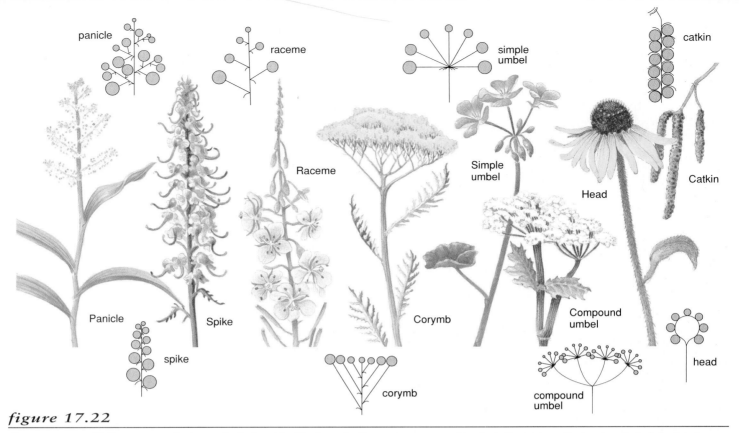

figure 17.22

Common types of inflorescences. Each type is distinguished by how the flowers are arranged in the inflorescence. Each colored circle in the diagrams represents an individual flower.

Insets: From Kingsley R. Stern, Introductory Plant Biology, *7th ed. Copyright © 1997 The McGraw-Hill Companies, Inc. All Rights Reserved. Reprinted by permission.*

plants have perfect flowers that are **self-compatible,** as in garden pea and snapdragon. This means that, in addition to cross-pollination, which is the transfer of pollen from the anthers of one plant to the stigma of another, successful reproduction can also occur following pollination within a flower or between flowers of the same plant. In contrast, corn and other plants that have imperfect flowers on the same sporophyte can only pollinate between flowers, either on the same sporophyte or between sporophytes. Such plants are called **monoecious** ("one house") because they have two kinds of flowers on the same sporophyte. Finally, mulberry, cottonwood, and willow are examples of plants that have staminate flowers and carpellate flowers on different sporophytes. These plants, therefore, are **dioecious;** that is, they have "two houses" (different plants) for reproduction. This means that reproduction in such plants can only occur by cross-pollination.

Mulberry, cottonwood, and willow are plants that depend on mating between gametophytes from different sporophytes; this is called **outcrossing.** In contrast, garden pea and snapdragon exemplify plants that are characterized by **inbreeding,** which in these plants is mating between gametophytes from the same sporophyte.

The breeding systems of plants range from complete outcrossing to complete inbreeding, but most plants use some combination of the two. For instance, flowers of the hoary plantain (*Plantago media*) only have carpels at first, the stamens forming after the carpels are mature (fig. 17.23). This breeding system enhances outcrossing because most pollen is shed from a particular flower before the carpel matures; however, self-pollination

figure 17.23

Inflorescence of the hoary plantain (*Plantago media*). Young flowers near the top of the inflorescence are carpellate only; the older flowers below are perfect.

is still possible as the flowers get older. In contrast, flowers of garden pea are equally receptive to self-pollination and cross-pollination throughout their development. This feature of garden pea enabled Mendel to produce pure-breeding plants for his experiments (see Chapter 8).

Some plants, such as the ground-cherry (*Physalis* sp.), produce perfect flowers that are **self-incompatible.** This means that even though self-pollination can occur, the pollen does not function properly with carpels of the same plant and no seeds are formed; thus, the pollination is unsuccessful. Self-incompatibility in the ground-cherry, as it probably is in most self-incompatible plants, is controlled by two genes. Depending on the plant species, these genes may work on the interaction between the pollen and stigma, thereby preventing the pollen tube from germinating. This is called *sporophytic self-incompatibility* because it is imposed by gene action in the stigma, a sporophytic structure. Conversely, in other plants the genes may work in the pollen grain as the pollen tube grows through the style. Self-incompatibility that is caused by genes in the male gametophyte is called *gametophytic self-incompatibility.*

Populations or species of plants usually have some combination of outcrossing and inbreeding. However, observations of thousands of different plants reveal that their reproductive features generally fall into one of the two categories shown in table 17.2. Some of the main features of outcrossing plants include many flowers per plant, long pedicels or peduncles, scented flowers, and many pollen grains per anther. Inbreeding plants generally have few flowers per plant, short pedicels or peduncles, unscented flowers, and fewer pollen grains per anther.

c o n c e p t

Breeding systems of flowering plants include self-compatibility, which is associated with inbreeding, and self-incompatibility, which confines plants to outcrossing.

Vegetative Reproduction

Plants are characterized by indeterminate vegetative growth; that is, they have meristematic cells that can divide indefinitely (see Chapter 12). Because of this property, tissues or cells can be removed from a plant and induced to grow into a new plant that is a **clone** of the old one (see Chapter 12). This kind of reproduction is **vegetative,** since meiosis and fertilization are not involved. Cloning is a modern term that simply refers to vegetative reproduction. Most plants reproduce vegetatively to some extent, and some species reproduce almost exclusively by vegetative means. Aspens, for example, grow in large stands that consist of just one or a few clones (see discussion in Chapter 15 on *Pando*, the giant quaking aspen). Humans exploit the clonability of plants by using cuttings to grow many kinds of houseplants, garden plants, and agricultural crops, including roses, African violets, potatoes, bananas, and oranges (see fig. 13.2).

In nature, the genotype of a clone may be long-lived. Clones of some grasses routinely live for a few hundred years. The oldest known living clone is a creosote bush (*Larrea tridentata*) in the Mojave Desert of California (fig. 17.24); this plant is more than 12,000 years old.

table 17.2

Features of Outcrossing versus Inbreeding Plants*

Outcrossing	Inbreeding
Self-incompatible	Self-compatible
Diploid sporophyte	Polyploid sporophyte
Many flowers per plant	Fewer flowers per plant
Long flower-stalks	Short flower-stalks
Large sepals	Smaller sepals
Large petals	Smaller petals
Often more than one high-contrast flower color	Usually one flower color; color of low contrast if more than one flower color
Nectaries present	Nectaries absent
Flowers scented	Flowers not scented
Nectar guides present	Nectar guides absent
Long carpel	Shorter carpel
Stamens longer or shorter than the carpel	Stamens the same length as the carpel
Anthers distant from stigma	Anthers close to stigma
Many pollen grains per anther	Fewer pollen grains per anther
Style protrudes from flower	Style does not protrude from flower
Anthers do not open when the stigma is receptive	Anthers open at the same time the stigma is receptive
Many ovules per flower or inflorescence	Few ovules per flower or inflorescence
Many ovules are not fertilized	All ovules fertilized
Some fruits do not mature	All fruits mature

*Note that dioecious plants, which can only outcross, often contradict some of these features. Such plants include walnut, cottonwood, birch, mulberry, willow, and hazelnut (see figs. 17.21c, d). Flowers of these plants have no petals.

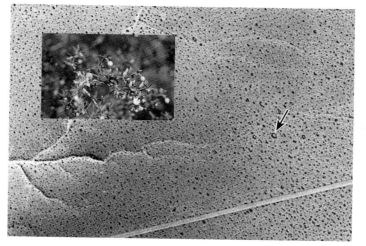

figure 17.24

The circle (arrow) of creosote bush (*Larrea tridentata*) in this aerial photo is a clone that is several meters in diameter. It is estimated to be at least 12,000 years old. Smaller circles nearby are of younger clones. The inset photo is a closer view of a creosote bush.

figure 17.25

Seedless fruits of the teddy-bear cholla (*Opuntia bigelovii*) fall to the ground and grow into clones of the parent plant.

Some plants seem to go through the motions of sexual reproduction without actually undergoing meiosis or fertilization. For example, many species of cholla cacti form flowers that develop fruits without seeds. These naturally seedless fruits fall to the ground and grow into clones of the parent plant (fig. 17.25).

Pollination Mechanisms

Pollination requires a *vector*, or carrier, to transfer the pollen from anther to stigma. Flowers are usually highly adapted for one type of vector, which may even be a single species of animal. Flowers and the animals that are attracted to them are often so closely coadapted that they are completely interdependent. As you read about pollination mechanisms, remember that pollination biology is a large and fascinating discipline that includes both botany and zoology. Pollination biologists are continually discovering new and complex interactions between flowers and their pollinators. Some of the major kinds of pollination mechanisms are described next.

Pollination by Wind and Water

The simplest method of pollination, and the most inefficient, occurs by wind. The main features of wind-pollinated angiosperms are as follows: they produce enormous amounts of lightweight, nonsticky pollen; they lack showy floral parts or strong fragrances; they have well-exposed stamens and large stigmas; they have a single ovule in each ovary; and they have many flowers packed into each inflorescence (fig. 17.26; see also fig. 17.21c,d).

Pollination by water is rare, simply because so few plants have flowers that are submerged underwater. Such plants include seagrasses (*Zostera* sp.), which release pollen that is carried passively by currents of water, much as wind-borne pollen is carried by wind. Other submerged aquatic plants, however, have more complicated pollination mechanisms. For example, pollen of ribbon weed (*Vallisneria spiralis*) is carried from one plant to another in "pollen boats" (fig. 17.27).

figure 17.26

Each flower of rye grass (*Lolium* sp.) produces about 50,000 dry, dustlike pollen grains, which readily take to the air in a light breeze.

figure 17.27

Staminate plants of the aquatic ribbon weed (*Vallisneria spiralis*) release their flowers as "pollen boats," which drift near the larger, surface-borne flowers of carpellate plants. Those pollen boats not eaten by fish may reach the edge of a dimple in the water that is created by surface tension around the carpellate flower. Once there, the pollen boats will slide down into the carpellate flower and be catapulted onto a receptive stigma.

(a)

(c)

(b)

(d)

Pollination by Insects

Insects are the most common group of animals that pollinate flowers. Figure 17.28 shows several types of flowers that are adapted for different insect pollinators. Although bees pollinate more kinds of flowers than any other type of insect, flowers can also be pollinated by wasps, flies, moths, butterflies, or beetles. There is no single set of characteristics for insect-pollinated flowers, because insects are such a large and diverse group of animals. Rather, each plant may have a set of reproductive features that attracts mostly one kind of insect, as summarized in table 17.3.

Many kinds of brightly colored flowers look like targets to insects because, in ultraviolet light, their petals are darker toward the center of the flower. Such targets, or **nectar guides,** are made by UV-absorbing pigments called *flavonoids,* which are visible to insects but invisible to other kinds of animals (fig. 17.29). Insect-pollinated flowers may also secrete strong fragrances.

(e)

figure 17.28

Insect-pollinated flowers. (a) Bumblebee gathering pollen from an aster. (b) Swallowtail butterfly on the inflorescence of an Indian paintbrush (*Castilleja* sp.). (c) The large nocturnal flowers of *Hydnora africana* emit foul odors that attract carrion beetles for pollination. (d) This elephant hawkmoth has a very long tongue that reaches deep into the narrow tube of honeysuckle flower (*Lonicera* sp.). (e) Ant pollinating a flower of *Orthocarpus pusillus.* This is one of the few examples of pollination by ants.

table 17.3

Floral Features Associated with Pollination by Different Kinds of Insects

Type of Insect	Flower Color	Flower Odor	Nectar Guides
Beetles	Dull	Strong and fruity	None
Carrion/ Dung flies	Reddish to purple-brown or greenish	Strong and foul	None
Bee-flies	Variable	Variable	None
Bees	Variable but not solid red	Usually sweet	Present
Hawkmoths	White or pale	Strong and sweet	None
Small moths	Variable but not solid red	Usually sweet	None
Butterflies	Variable, commonly pink	Moderately strong; sweet	Present

(a)

(a) (b)

figure 17.29

Flowers of the evening primrose (*Oenothera*) are uniform in color to the human eye (*a*), but insects see a different pattern in ultraviolet light (*b*).

(b)

figure 17.30

Nonpollinating flower visitors. (a) The hornet in this photo has bitten a hole in the corolla of an evening primrose flower (*Oenothera* sp.) and is drinking nectar through it. (b) This bright yellow crab spider seems conspicuous on the fuchsia-colored petals of this orchid flower, but it is camouflaged in UV light to pollinators who are its prey.

In contrast to flowers with bright colors or nectar guides, less showy flowers can also attract nocturnal insects. Some moths, for example, are attracted to strongly scented, night-blooming flowers, which are usually white or cream-colored (fig. 17.28d). Such flowers are often tubular or trumpet-shaped, which prevents all but the long-tongued moths from reaching the nectar.

Attractiveness and sweetness of fragrance are relative terms. Some "attractive" flowers are reddish brown and drab, and their fragrance is like that of rotting flesh (fig. 17.28c). These flowers are referred to as *carrion flowers* and are pollinated by carrion flies or beetles that are attracted to their foul odors.

Some animals get the rewards meant for pollinators by stealing the nectar without doing any work for the plant (fig. 17.30a). Some predatory animals use flowers as hunting grounds, where they prey on the pollinators attracted to the flowers (fig. 17.30b).

Pollination by Mammals

Like moth-pollinated flowers, flowers that attract bats and small rodents also open at night. Mammal-pollinated flowers are usually white and strongly scented, often with a fruity odor. Such flowers must be large and sturdy enough to bear the vigorous visits of these small mammals (fig. 17.31).

(a)

(a)

(b)

(b)

figure 17.31

Mammal-pollinated flowers. (a) A greater short-nosed bat feeds on the pollen and nectar of a banana plant (*Musa* sp.). (b) The tiny Australian honey-possum pollinates plants such as this coral gum (*Eucalyptus* sp.) as it forages for pollen and nectar.

figure 17.32

Bird-pollinated flowers. (a) A rufous hummingbird gets nectar from flowers of *Mimulus cardinalis*. (b) The yellow-plumed honeyeater has a brush-tipped tongue for lapping up nectar and pollen from the bell-fruited mallee (*Eucalyptus pressiana*).

Pollination by Birds

Hummingbirds are the most common group of flower-visiting birds in the Americas; honey creepers are common pollinators in Africa and Asia. The long beaks of both kinds of birds can reach to the base of long, tubular corollas to obtain nectar (fig. 17.32). Hummingbirds are mostly attracted to bright red and yellow flowers, colors that are not usually attractive to insects. Birds also have a relatively poor sense of smell, and hummingbird-pollinated flowers are generally odorless. Columbines, penstemons, and scarlet monkeyflowers are examples of flowers that attract hummingbirds.

Flowers are adapted for pollination mainly by wind or animals. Some kinds of flowers rely on pollination by water. Wind-pollinated flowers produce abundant pollen in nonshowy flowers. Animal-pollinated flowers are usually showy or strongly scented.

Mutualisms between plants and their pollinators are a precarious balance, with each participant trying to gain an advantage. Plants "win" when they attract pollinators, but do not deliver the promised reward (e.g., nectar). Interestingly, this "cheating" had evolved throughout the angiosperms. In some species, only a few flowers per plant lack rewards; in other species, all of the flowers are often rewardless.

table 17.4

Dichotomous Key to the Major Types of Fruit

Complex fruits (i.e., from more than one ovary)	
Fruit from many carpels on a single flower (magnolia, strawberry, blackberry):	**AGGREGATE FRUIT**
Fruit from carpels of many flowers fused together (pineapple, mulberry):	**MULTIPLE FRUIT**
Simple fruits (i.e., from a single ovary)	
Fleshy fruit	
Flesh derived from ovarian tissue	
Endocarp hard and stony; ovary superior and single-seeded (cherry, olive, coconut):	**DRUPE**
Endocarp fleshy or slimy; ovary usually many-seeded (tomato, grape, green pepper):	**BERRY**
Outer layer of berry a leathery skin containing oils (orange, grapefruit, lemon):	**HESPERIDIUM**
Outer layer of berry a thick rind not containing oils (watermelon, pumpkin, cucumber):	**PEPO**
Flesh derived from receptacle tissue (apple, pear, quince):	**POME**
Dry fruits	
Fruits that split open at maturity (usually more than one seed)	
Seeds released through longitudinal seams	
Split occurs along one seam in the ovary (magnolia, milkweed):	**FOLLICLE**
Split occurs along two seams in the ovary	
Seeds borne on one of the halves of the split ovary (pea and bean pods, peanuts):	**LEGUME**
Seeds borne on a partition between halves of the ovary (mustard, radish):	**SILIQUE**
Seeds released through pores or multiple seams (poppies, irises, lilies):	**CAPSULE**
Fruits that do not split open at maturity (usually one seed)	
Pericarp hard and thick, with a cup at its base (acorn, chestnut, hickory):	**NUT**
Pericarp soft and thin, without a cup	
Ovaries often together in pairs (parsley, carrot, dill):	**SCHIZOCARP**
Ovaries occur singly	
Pericarp winged (maple, ash, elm):	**SAMARA**
Pericarp not winged	
Single seed attached to pericarp only at its base (sunflower, buttercup):	**ACHENE**
Single seed fully fused to pericarp (cereal grains):	**CARYOPSIS**

Fruits

A fruit is a seed container derived from an ovary and any tissues that surround it. As such, fruits are products of flowers and therefore occur only in flowering plants. Beyond this simple definition, angiosperms have a remarkable diversity of fruit types. Fruits are classified on the basis of the characteristics of the mature ovary tissue—for example, whether the fruit is fleshy or dry, or whether the ovary is fused to other kinds of tissues. Examples of different types of fruit and their main features follow.

Types of Fruit

Ovaries that have matured into a fleshy fruit often consist of three regions: the skin (**exocarp**), the fleshy part (**mesocarp**), and the interior (**endocarp**) that surrounds the seeds. Stone fruits, such as apricots, fit this description, as do tomatoes. The main difference between an apricot and a tomato is that the endocarp in an apricot is hard and stony, whereas the endocarp of a tomato is soft and slimy. The mesocarp and endocarp are fused and indistinct from each other in other types of fruit. In dry fruits, all three layers are fused into one **pericarp,** which is often a thin layer around the seed.

Fruits are classified by several main features, including whether they are fleshy or dry, whether and how they split open at maturity, whether they are derived from one or more ovaries from a single flower or an inflorescence, and whether they consist solely of ovarian tissue or of ovarian tissue plus the tissue of other floral parts that may be fused to the ovary. Based on these and other differences, the major fruit types can be presented in a dichotomous key (table 17.4). By using this key, you can identify the major types of fruit by choosing between successive pairs of features (dichotomies) at each of several steps. Examples of each fruit type are listed in the key, and several are illustrated in figure 17.33. The key should enable you to determine the fruit types of, for example, bananas, peaches, and buckwheat.

Although the classification of fruits is informative, it is also inexact because so many variations in fruits do not fit into the key to major fruit types. For example, the fleshy tissue of strawberries is an expanded receptacle, which makes it an **accessory fruit**—that is, a fruit whose flesh is not derived from ovarian tissue. Pomes are also accessory fruits (see table 17.4). Other fruits, such as the coconut, are modified so that they partially fit the description of a certain fruit type but not perfectly. A coconut is a drupe, and drupes normally have a fleshy mesocarp; however, the mesocarp of a coconut is a fibrous husk (fig. 17.34). Only the bony endocarp containing the seed

figure 17.33

Examples of some of the types of fruit described in table 17.4: (a) grapes (berry), (b) apple (pome), (c) orange (hesperidium), and (d) acorns (nut).

figure 17.34

The fruit of a coconut (*Cocos nucifera*) is a drupe, but the mesocarp is fibrous instead of fleshy.

ends up on the grocery shelf. Blackberry and magnolia form aggregate fruits, although the individual ovaries in blackberry develop into tiny drupes and in magnolia into follicles. A cob of corn is a multiple fruit, but each kernel is a caryopsis.

Fruit Development

Fertilization is usually a prerequisite for the development of fruits. Chemical signals, called *hormones*, are secreted by seeds as they develop. These hormones induce the ovarian tissue to expand and mature into a fruit. Naturally produced hormones or their synthetic analogs are applied to some crops so that fruits will form and mature in synchrony, which makes harvesting more efficient and economical. Treatment of flowers with artificial hormones can also induce the formation of seedless fruits (e.g., seedless grapes) in the absence of fertilization. You'll learn more about the roles of plant hormones in fruit development in Chapter 18.

concept

Ovaries develop into fruits after fertilization. Fruits are either fleshy or dry. A fleshy fruit has three distinct regions of ovarian tissue, but some fleshy fruits are surrounded by accessory tissue that is not derived from the ovary. Dry fruits may remain tightly closed, or they may split open at maturity.

Seeds

Seed Structure

A distinguishing feature of a seed is the *seed coat,* which is the outer layer that develops from one or two integuments of the ovule and is often thin and papery. In addition, in some dicot seeds the embryo is surrounded by a nutrient endosperm; in others the endosperm is absorbed by the cotyledons (fig. 17.35). In the seeds of monocots, the embryo is either surrounded by or off to one side of the endosperm. The single cotyledon of monocot seeds is the absorptive organ that takes in nutrients from the digested endosperm. In grasses, which have the most complex seeds of monocots, the cotyledon is so highly modified for absorption that it has a special name, the **scutellum.** In the small seeds of some plants such as orchids, the endosperm does not develop.

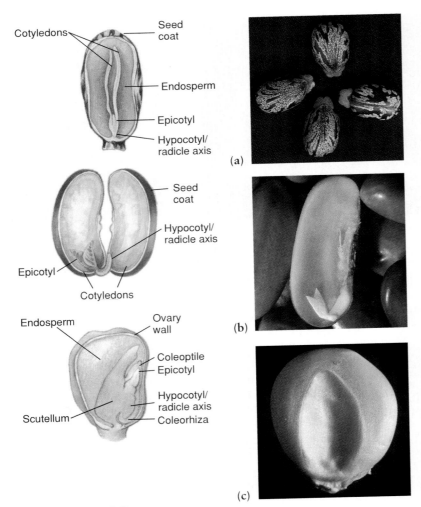

figure 17.35

Seeds. (a) Seeds of castor bean (*Ricinus communis*) have abundant endosperm that surrounds two thin cotyledons. (b) The two cotyledons in each seed of garden bean (*Phaseolus vulgaris*) absorb the endosperm before germination. (c) Corn (*Zea mays*) has seeds in kernels; the single cotyledon is an endosperm-absorbing structure called a scutellum.

The region of the embryo above the attachment point of cotyledons is the **epicotyl,** which gives rise to the shoot. The region below the attachment point is the **hypocotyl.** An embryonic root (**radicle**) is often distinguishable at the tip of the hypocotyl. The embryos of corn and other grass seeds are partially enclosed in protective sheaths. The sheath around the embryonic shoot is called the **coleoptile,** and the sheath around the radicle is called the **coleorhiza** (fig. 17.35).

Besides varying in structure, seeds also differ in their requirements for germination, as discussed next.

c o n c e p t

The main features of seeds are a seed coat and an embryo. Seeds may also contain endosperm or cotyledons that have already absorbed the endosperm. In angiosperms, the embryo consists of one or two cotyledons, an embryonic root, and an embryonic shoot.

Seed Germination

After fertilization and early development, the embryo usually stops growing. It begins to grow again when the seed germinates. This period of delayed germination is called **dormancy.** Different kinds of seeds require different internal or external stimuli to break dormancy and germinate. The most common internal factors include changes in the amount of the hormone *abscisic acid* (see Chapter 18) or other inhibitory chemicals. The most common external factors are water, temperature, and light.

Water Requirement

Perhaps the most common stimulus for breaking dormancy is water. A seed that absorbs water expands and bursts its seed coat. The seeds of many desert plants have a safeguard against premature germination after short rainfalls, which are common in deserts: they contain a germination inhibitor that must be leached out before they will germinate. Even though the seeds absorb water, the rainfall may be insufficient to wash out the inhibitor. This adaptation ensures that seeds will germinate only when there is enough water for the successful growth of the seedling. The seeds of rice and other aquatic or semiaquatic plants germinate only in soil that is submerged in water.

Temperature Requirement

The optimum temperature for the germination of most seeds is close to room temperature (25° C), although the seeds of plants adapted to colder or warmer climates can germinate at lower or higher temperatures, respectively. Germination can occur at temperatures as low as 0° C and as high as 45° C in different species. The seeds of woody plants in temperate climates often require a wet period that is followed by several weeks of a cold period before they will germinate; these conditions activate chemicals in the seed that stimulate germination. For example, the seeds of apple trees and most pines can be artificially induced to germinate by wetting them and placing them in a refrigerator. This process is called **stratification.**

Light Requirement

Not all seeds are affected by light. However, the seeds of birch, certain grasses, and some varieties of lettuce require light for germination. These seeds will not germinate in the dark because they contain an inhibitor that is broken down only in the presence of light. The inhibitor breaks down when *phytochrome*, which is a pigment that occurs in the seed, absorbs red light and sends chemical signals to other parts of the seed (you'll learn about this pigment in Chapter 19). In contrast, the seeds of other plants, such as geraniums and poppies, will germinate only in the dark. In these seeds, light stimulates the synthesis of compounds that inhibit germination.

17–23

THE DODO BIRD AND THE TAMBALACOQUE TREE

On the island of Mauritius in the western Indian Ocean, the seeds of the tambalacoque tree (*Calvaria major*) apparently required passage through the digestive tract of the now-extinct dodo bird (*Raphus cucullatus*) before they could germinate. This hypothesis is based on the observation that only about a dozen tambalacoque trees occur in natural stands, all from seeds that germinated at about the time the dodo went extinct (ca. 1680) (reading fig. 17.2a). There are no natural stands of seedlings or younger trees in spite of the abundant production of tambalacoque fruits for the past three centuries (reading fig. 17.2b).

The hypothesis of a dodo-tambalacoque interaction cannot be tested experimentally because the dodo is extinct, but much indirect evidence supports it. Tambalacoque fruits are fleshy drupes that are eaten by fruit-eating animals, such as the Mauritius parakeet and the Mauritius flying fox; these fruits were undoubtedly also eaten by dodo birds. Seeds germinate after the thick endocarp wall is abraded enough to allow the embryo to sprout through it. The endocarp resists crushing and destruction by forces that probably occurred in the gizzard of the dodo bird. When force-fed to turkeys, a small percentage of tambalacoque seeds

do·do \ˈdō(ˌ)dō\ *n, pl* **dodoes** *or* **dodos** [Pg *doudo*, fr. *doudo* silly, stupid] **1 a :** a large heavy flightless extinct bird (*Raphus cucullatus*, syn. *Didus ineptus*) related to the pigeons but larger than a turkey, that had dark ash-colored plumage with the breast and tail whitish, the rudimentary wings being yellowish white with black-tipped coverts, the bill blackish, and the legs yellow; that inhabited forests and laid a single large white egg in a nest of grass; and that was present in great numbers on the island of Mauritius prior to the arrival of European settlers but became extinct by 1681 **b :** a similar and apparently closely related bird of the neighboring island of Réunion that became extinct under similar circumstances at a slightly later date **2 a :** a person who is simplemindedly unaware of changing conditions and new ideas **:** a dull stupid person **b** *slang* **:** a flight cadet who has not yet soloed **3** *also*

reading figure 17.2a

The dodo bird has been extinct for more than 300 years.

can germinate, but turkeys are not attracted to tambalacoque fruits. The digestive systems of extant animals that do eat tambalacoque fruits do not abrade the endocarp enough for any seeds to germinate.

The dodo bird was driven to extinction by human activities. Perhaps due to the absence of this large, flightless bird, the extinction of the tambalacoque tree has also been accelerated. Although the loss of the dodo bird may have been a main factor, habitat de-

reading figure 17.2b

A three-centuries-old tambalacoque tree (*Calvaria major*). Seeds of the tambalacoque tree haven't germinated in the wild since the dodo bird became extinct.

struction and the introduction of weedy, non-native plants by humans have also contributed to the downfall of the tambalacoque tree.

Special Requirements

Seeds with thick or tough seed coats that do not split open may require special conditions for germination. A thick covering may reduce the uptake of water by the seed or prevent the embryo from expanding. For example, many legumes contain seeds with hard seed coats, and the seeds of some drupes are surrounded by a tightly sealed endocarp. Many such seeds will germinate only after the seed coat is scratched or cracked, or briefly soaked in concentrated acid. Such treatments are called **scarification.** In nature, seed coats may be scarified by bacterial action, freeze-thaw cycles, abrasive handling by rodents, or passing through the digestive tracts of animals (see reading 17.2, "The Dodo Bird and the Tambalacoque Tree").

Seed Banks

Many viable seeds do not germinate right after they are shed from the parent plant, nor do they germinate during the following growing season. Seeds can lie dormant for many years before conditions are suitable for their germination. Everywhere that seed plants grow, the soil contains viable, ungerminated seeds in natural storage—that is, a **seed bank.**[1] Seeds in a seed bank may be dormant because of their own inhibitors, as in many desert plants. In other habitats, seed germination may be inhibited by chemicals released from nearby plants or by a lack of nutrients in the soil.

Ecologists can sometimes determine what kinds of seeds are in the seed bank of a particular habitat by removing the shrubs from a small area. Shrubs often prevent seed germination because of the inhibitory chemicals that leach out of the leaves that they drop. Once the shrubs are removed, the seeds of many annual plants germinate during the next new growing season. Such experiments simulate what happens, in part, when fire sweeps through an area. In addition to eliminating the source of potential germination inhibitors, fire also releases the nutrients

---

1. Note that *seed bank* also refers to human-maintained seed collections (see discussion in the Epilogue).

figure 17.36

Plants that germinated after a fire. Most of these plants grew from seeds in the soil; these seeds were dormant until fire cleared the area.

figure 17.37

Arctic tundra lupine (*Lupinus arcticus*). Seeds of this species germinated after being frozen in a lemming burrow for about 10,000 years.

contained in plants. This is why annual plants grow abundantly in burned areas during the first growing season after a fire (fig. 17.36). As perennial plants become reestablished, the newly replenished seed bank of annual plants once again goes into natural storage until the next fire.

How long can seeds remain viable? The answer is that longevity varies among different kinds of plants and with environmental conditions. Some seeds, such as those of pumpkins, squashes, and other members of the melon family (*Cucurbitaceae*), can germinate after several years of storage. In some parts of the Mojave Desert, the seeds of weedy plants remain dormant for at least 15 years. In a seed-longevity experiment that began in the late 1800s at Michigan State University, the longest-lived seeds were of moth mullein, mullein, and mallow. Seeds from all of these plants germinated after 101 years of storage in buried jars. The record for longevity, however, belongs to the arctic tundra lupine (*Lupinus arcticus*) (fig. 17.37). Seeds of this species were frozen in a lemming burrow that has been estimated to be about 10,000 years old. Several of the seeds from this burrow germinated within 48 hours of planting, and one of the plants produced flowers within a year after planting.

c o n c e p t

Seed germination is often delayed in a period of dormancy. Germination may be induced by several factors, including changes in temperature, water availability, or light. Some seeds require scarification by animals, bacteria, or abrasion during repeated freeze-thaw cycles. In nature, seeds may remain dormant for several years before environmental conditions are favorable enough to induce germination.

Seedling Development

The first part of the embryo to emerge from the seed during germination is the radicle. This first root of the seedling increases the supply of water and nutrients to the shoot before the shoot breaks through the surface of the soil. The shoot may or may not bear cotyledons when it emerges, depending on which embryonic meristems are most active. In the garden bean, for example, the fastest growth occurs in the hypocotyl. As the new shoot pushes through the soil, the hypocotyl elongates and forces the embryonic apical meristem of the shoot, along with the cotyledons, above the surface (fig. 17.38). In contrast, cotyledons of the garden pea remain belowground because the fastest growth occurs at the embryonic shoot tip. In both cases, the cotyledons supply carbohydrates from the endosperm to the growing seedling. Aboveground cotyledons also become green and photosynthetic in garden bean. Cotyledons of garden pea, on the other hand, quickly shrivel as they are used up by the seedling (see discussion of cotyledons on p. 326).

Grasses have the most complex seedling development. The seed, which remains enclosed in the ovarian wall, germinates when the coleoptile that sheathes the shoot and the coleorhiza that sheathes the radicle both begin to grow. Each sheath grows several millimeters. The radicle then breaks through the tip of the coleorhiza and becomes the primary root. Similarly, when the coleoptile stops growing, the uppermost leaf pushes through it and becomes the first photosynthetic organ of the new seedling. During the germination of grass seeds, the cotyledon continues to absorb sugars from the endosperm until the new leaves make enough photosynthate for the plant to grow independently.

17–25

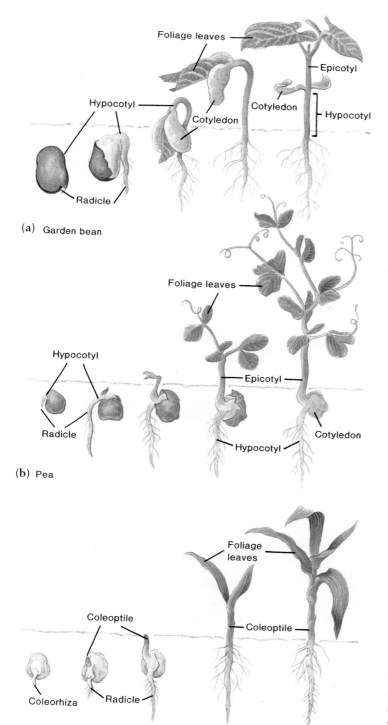

(a) Garden bean

(b) Pea

(c) Corn

figure 17.38

Seed germination. (a) Growth in the hypocotyl of garden bean forces the shoot apex and the cotyledons above the surface. (b) In garden pea, rapid growth at the shoot apex leaves the cotyledons belowground. (c) In corn and other grasses, the shoot apex grows out of the kernel through the tubelike coleoptile, and the root apex grows through the tubelike coleorhiza.

Seedling development varies depending on the function of the endosperm and the location of the greatest meristematic activity in the embryo. Faster growth in the hypocotyl pushes the shoot and cotyledons through the soil surface; faster growth at the shoot apex keeps the cotyledons belowground.

Dispersal of Fruits and Seeds

Seeds must get from the parent plant to a favorable place for germination. The best place for a new plant may be where the parent plant is already growing, had it not been there. When a seed germinates next to a mature plant, however, the young seedling may not get enough sunlight, water, or nutrients for continued growth, due to competition with the mature plant. The new generation has a greater chance of surviving if the seeds are moved away from the parent plant to a place that is favorable for seedling establishment, as well as away from herbivores that have found the mature plant.

Dispersal occurs either by physical or biological carriers. Wind and water are the physical carriers. The main biological carriers of fruits and seeds are animals, but some plants can forcefully eject their seeds away from the parent plant.

Dispersal by Wind and Water

The seeds of orchids and heaths are so small that they can be blown around by light winds. The familiar plumes of dandelions, willows, milkweeds, poplars, and some buttercups are examples of further modifications for seed or fruit dispersal by wind (fig. 17.39). A maple fruit has a curved wing that causes the fruit to spin as it floats through the air. Even in a light wind, maple fruits spin away from the parent plant. In arid areas where strong winds are common, tumbleweeds break off at their main stems and whole plants are blown around, releasing seeds as they tumble.

Coconuts are perhaps the most familiar example of adaptation to dispersal by water (fig. 17.40). The buoyancy of a coconut comes from its fibrous husk, which allows the fruit to stay afloat for days or weeks before too much saltwater seeps through the husk and kills the embryo. Many other plants that grow in or near water have fruits or seeds that float. Most sedges, for example, have air pockets around their seeds, so they can float for awhile before they absorb water and sink.

Dispersal by Animals

Fruits or seeds that are covered with a sticky substance (e.g., the seeds of mistletoes) or that have barbs or hooks (e.g., the fruits of bur clover and puncture vine) cling to beaks, fur, and skin (fig. 17.41). Such seeds or small fruits may be carried a few meters or many kilometers before they fall from the animal carrying them. The longest dispersal distances are probably achieved by seeds that get stuck to the muddy feet of migratory birds.

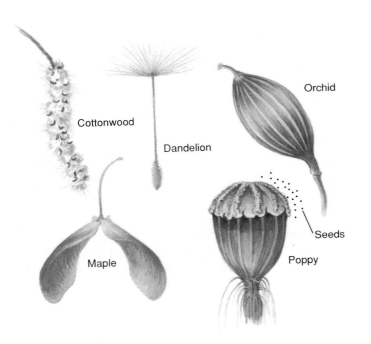

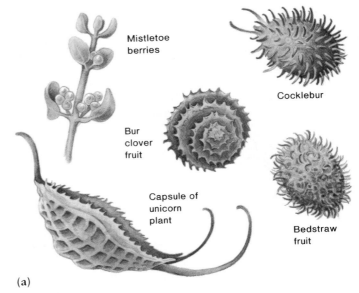

(a)

figure 17.39

Wind-dispersed fruits and seeds. Small seeds of orchids and poppies can be blown around by light breezes. Seeds or fruits of other plants have plumes or wings that catch the wind.

figure 17.40

Fruits of the coconut palm (*Cocos nucifera*) may float for weeks before reaching a beach where they can germinate.

(b)

figure 17.41

Seed dispersal by animals. (a) Sticky or prickly fruits adapted for dispersal by animals. (b) This elk cow has seeds stuck to her fur.

The colorful and sweet-tasting fruits of many plants attract fruit-eating animals. Seeds from such fruits usually pass unharmed through the digestive tracts of birds and mammals, or the animals may regurgitate them. A few plants, such as the mescal bean (*Sophora secundiflora*), have brightly colored seed coats that attract birds (fig. 17.42). However, birds drop these seeds after a short distance because the seed coats are too hard.

The most common seed dispersers in some habitats are ants. Seed-harvester ants obtain food from seeds, either from special oil bodies on the seeds or from the embryo. The ants remove the oil bodies and often leave the seeds unharmed. Many of these seeds are discarded by the ants after the seeds have been carried to the nest. When the seeds germinate, dense patches of plants grow around the entrance to the nest.

17–27

figure 17.42

Bright red seeds of the mescal bean (*Sophora secundiflora*) attract birds.

Self-Dispersal

The fruits of many plants disperse their own seeds by forceful ejection. For example, the seeds of touch-me-nots (*Impatiens*), witchhazel (*Hamamelis*), dwarf mistletoe (*Arceuthobium*), and some legumes (e.g., African *Acacia*) may be flung several meters from the fruit (fig. 17.43).

c o n c e p t

Successful reproduction involves transportation of seeds or fruits to favorable habitats. Seeds and fruits are dispersed either by wind, water, or animals. Some plants disperse their seeds by forceful ejection from the fruit.

(a)

(b)

figure 17.43

Self-propelled seed dispersal. (a) Fruits of the touch-me-not (*Impatiens glandulifera*) explode open, hurling seeds as far as 2 meters. (b) The green fruits of this Mediterranean cranesbill (*Erodium botrys*) still enclose seeds, but the dried-up fruits have already shed their seeds.

A single plant species, or two closely related ones, can be naturally separated by large distances (e.g., between North America and Asia, between South America and Africa). While some of these separations can be explained by continental drift, botanists believe that many were caused by long-distance dispersal. The mechanisms of ancient long-distance dispersal, however, have been harder to imagine; possible mechanisms have included such creative suggestions as seed transport by migratory, constipated birds or by plant migration over temporary land bridges that crisscrossed oceans.

Chapter Summary

Flowers are the reproductive structures of angiosperms. A flower is a stem tip that bears some or all of the following kinds of appendages on a receptacle: sepals, petals, stamens, and pistil. The pistil consists of an ovary, which includes one or more ovule-bearing carpels, and a stigma that is usually borne on a style. During pollination, pollen is transferred from an anther to a stigma. A pollen tube grows from the male gametophyte to the micropyle of an ovule and carries two sperms to the female gametophyte. After fertilization, ovules become seeds, and ovaries become fruits.

Both sperms from the male gametophyte fertilize cells in the female gametophyte. One sperm fertilizes the egg, which forms a zygote that develops into an embryo. The second sperm fuses with the polar nuclei, thereby forming the first cell of the endosperm. When the endosperm grows, it nourishes the embryo. The endosperm in some seeds is absorbed by the cotyledons, whereas in other seeds the endosperm is digested as the seed germinates.

Diversity in angiosperms is based largely on variation in reproductive morphology. Angiosperms are divided into two classes, the monocots and the dicots. Flowers in either class may be complete or incomplete, perfect or imperfect, regular or irregular. They also may be solitary or arranged in an inflorescence with several other flowers. Most flowers are adapted for pollination by wind or by different animals. Wind-pollinated flowers are usually incomplete and not showy, whereas insect-pollinated flowers and bird-pollinated flowers are often colorful. Night-blooming flowers attract nocturnal mammals or insects; these flowers are usually aromatic and white or cream-colored.

Fruits may be either fleshy or dry. Some kinds of dry fruits remain sealed at maturity, whereas others split open at maturity. Fleshy fruits that are colorful and sweet-tasting attract animals that eat them and disperse the seeds. Dry fruits, or the seeds from them, are also adapted for dispersal. These adaptations include plumes or wings that catch the wind, air sacs or spongy tissue for flotation, barbs or spines that attach to passing animals, and edible oil bodies that attract seed-harvester ants.

The distinguishing feature of a seed is the seed coat, which develops from one or two integuments. The embryo must expand and grow through the seed coat during germination. Seed germination is often delayed through a period of dormancy, which may be broken by changes in temperature, water availability, or light. Some seeds require special treatments, such as partial digestion by bacteria or animals or abrasive handling by rodents, before they will germinate.

Many seeds remain in natural storage in seed banks until dormancy is broken or environmental conditions are favorable for germination. The longevity of seeds in storage can be several years or several decades, depending on storage conditions. Seeds remain viable longer when they are kept cold and dry. The record for seed longevity is about 10,000 years for seeds of the arctic tundra lupine.

 ## What Are Botanists Doing?

Double fertilization is not unique to flowering plants; it also occurs in the genus *Ephedra*, a nonflowering seed plant. Recent studies of embryo sacs and fertilization in *Ephedra* help to answer the question, "Should *Ephedra* be classified as a flowering plant?" Find a journal article on this topic that was published within the past 5 years and see if the authors address this question, either directly or indirectly. How are the reproductive features of *Ephedra* used to support the inclusion of *Ephedra* in the flowering plants?

 ## Writing to Learn Botany

What are the possible advantages and disadvantages of vegetative reproduction?

Questions for Further Thought and Study

1. What are the components of a castor bean seed? What are the components of a garden bean? How do these two types of seeds differ in structure and function?

2. Seeds of orchids undergo double fertilization without developing endosperm. How might the absence of endosperm be adaptive for such seeds?

3. Why are strawberries, raspberries, and mulberries not berries? Since they are not berries, what are they?

4. Many seeds store their food reserves mostly as oil instead of carbohydrate. What are some examples of plants that produce oily seeds?

5. Why is it incorrect to refer to commercial sunflower "seeds" as seeds? Since they are not seeds, what are they?

6. Apples are fleshy, sweet-tasting fruits that attract and are eaten by fruit-eating animals that disperse their seeds. How could you explain the fact that the seeds of such an attractive fruit contain cyanide-producing chemicals?

7. What advantage is gained by seeds that contain germination inhibitors?

8. What might the advantages be for self-compatibility versus self-incompatibility?

9. Recent research suggests that the production of more flowers decreases a plant's chances of mating with others by increasing the likelihood of self-pollination. How would you test this hypothesis?

10. Do plants compete to attract pollinators? Write a short paragraph to explain your answer.

Web Sites

Review the "Doing Botany Yourself" essay and assignments for Chapter 17 on the *Botany Home Page*. What experiments would you do to test the hypotheses? What data can you gather on the Web to help you refine your experiments?

Here are some other sites that you may find interesting:

http://www.palomar.edu/Wayne/worthypl.html

At this site you'll find all sorts of interesting and unusual information about plants, including descriptions of the world's largest and smallest fruits, the world's largest vegetable, and "Supermarket Botany." In case you're wondering:

The title of "World's Largest Fruit" goes to the pumpkin: at the October 1996 Weigh-Off sponsored by the World Pumpkin Confederation, one monstrous cucurbit tipped the scales at 1,061 pounds. That pumpkin's growers received the grand prize check for $50,000.

The title of "World's Smallest Fruit" goes to fruits of *Wolffii globosa*, the world's smallest flowering plant. A mature plant is only about 0.6 mm long, 0.3 mm wide, and weighs about 150 mg: that weight equals the weight of two ordinary grains of table salt, and is seven-trillionths the weight of a sequoia. Each plant will slip through the eye of a needle, and a thimble will hold about 5,000 plants. And the fruits? Fruits of these tiny aquatic plants are about half as big as a grain of salt.

http://csdl.tamu.edu/FLORA/tfplab/reproch.htmReproductive

This site uses photos and text to discuss the reproductive structures of plants.

gopher://gopher.adp.wisc.edu/2070/11/.image/.bot/Fruits_130_d

This site, maintained by the University of Wisconsin, includes a variety of photos of fruits.

http://gears.tucson.ars.ag.gov/gears/nx/fossils/fossils.html

This site discusses which came first: bees or flowers? This evolutionary paradox provides an excellent example of the coevolution of two organisms.

http://www-wane.scri.fsu.edu/~mikems/
At this site you'll see the Florida Wildflower Showcase.

http://www.primenet.com/~tifehr/Gallery.html
These pages include descriptions and photos of daylilies. If you want to exchange information about daylilies, check out *http://a1.com/daylily/place.html*

http://www.ucr.ac.cr/lankester/albume.html
Lankester Botanical Garden presents its photos of Costa Rican Orchids. For more about orchids, visit *http://www.cfn.cs.dal.ca/Recreation/OrchidSNS/wwwsites.html*

http://rampages.onramp.net/~garylipe/
http://www.sccs.swarthmore.edu/~tkorn/wildflowers/
These sites organize information about wildflowers and identify other sites, references, and photos of wildflowers. If you still have questions about wildflowers, visit:
http://www.onr.com/wildflowers.html
a site maintained by the National Wildflower Research Center.

Suggested Readings

Articles

Beattie, A. J. 1990. Seed dispersal by ants. *Scientific American* 263(August):76.

Burd, M. 1994. Bateman's principle and plant reproduction: The role of pollen limitation in fruit and seed set. *The Botanical Review* 60:83–139.

Cook, R. E. 1983. Clonal plant populations. *American Scientist* 71:244–253.

Cox, P. A. 1993. Water-pollinated plants. *Scientific American* 269(October):68–74.

Fleming, T. H. 1993. Plant-visiting bats. *American Scientist* 81:460–467.

Keddy, P. A. 1981. Why gametophytes and sporophytes are different: Form and function in a terrestrial environment. *American Naturalist* 118:452–454.

Niklas, K. J. 1985. Wind pollination—A study in controlled chaos. *American Scientist* 37:462–470.

Owadally, A. W. 1979. The dodo and the tambalacoque tree. *Science* 203:1363–1364.

Postiglione, R. A. 1993. Velcro and seed dispersal. *The American Biology Teacher* 55:44.

Robacker, D. C., J. D. B. Meeuse, and E. H. Erickson. 1988. Floral aroma. *BioScience* 38:390–396.

Stebbins, G. L. 1981. Why are there so many species of flowering plants? *BioScience* 31:573–577.

Steele, L. C., and J. E. Keeley. 1991. Chaparral and fire ecology: Role of fire in seed germination. *The American Biology Teacher* 53:432.

Stein, B. A. 1992. Sicklebill hummingbirds, ants, and flowers. *BioScience* 42:27–33.

Wheelright, N. 1991. Frugivory and seed dispersal: "La coevolucion ha muerto—¡viva la coevolucion!" *Trends in Ecology and Evolution* 6:312.

Books

Barth, F. G. 1991. *Insects and Flowers: The Biology of a Partnership*. Princeton, NJ: Princeton University Press.

Bell, A. D. 1990. *Plant Forms: An Illustrated Guide to Flowering Plant Morphology*. New York: Oxford University Press.

Fenner, M., ed. 1992. *Seeds, the Ecology of Regeneration in Plant Communities*. Wallingford, England: CAB International.

Heywood, V. H., ed. 1985. *Flowering Plants of the World*. New York: Prentice-Hall.

Pijl, L. van der. 1982. *Principles of Dispersal in Higher Plants*. 2d ed. New York: Springer-Verlag.

Williams. E. G., A. E. Clarke, and R. B. Knox, eds. 1994. *Genetic Control of Self-Compatibility and Reproductive Development in Flowering Plants*. Boston: Kluwer Academic.

Willson, M. F. 1983. *Plant Reproductive Ecology*. New York: John Wiley and Sons.

Regulating Growth and Development

P lants seem to keep track of time because they do certain things at specific times of the day or year. For example, morning glories open their flowers in the morning, whereas flowers of the queen-of-the-night cactus open at night. Mulberries and aspens drop their leaves in the fall and grow new leaves in the spring; the timing is never the other way around. Plants also sense gravity, light, and touch. All of these activities show that plants sense their environment and respond to many different stimuli. Environmental stimuli vary, as do the responses of plants to these stimuli.

How do plants perceive a stimulus, communicate the stimulus to appropriate cells in the plant, and change their growth or development accordingly? This complex question can be addressed by learning about the roles of hormones (also called *plant growth-regulating substances*) and other chemical means of communication in plants and about the kinds of changes that plants undergo in response to different stimuli. These topics are the main subject of this unit.

As you read the chapters in this unit, keep in mind that the search for compounds that regulate plant growth and development is ongoing. New compounds that affect plant growth and development are discovered every year. Research on the established hormones continues because there are many unanswered questions about how they work.

Leaves of northern red oak. Leaf abscission is
strongly influenced by plant hormones.

Plant Hormones

18

Chapter Outline

Chapter Overview

Plant growth and development are strongly influenced by plant growth-regulating substances, which are organic compounds made in one part of a plant and transported to another part, where they elicit a response. These growth-regulating substances have traditionally been referred to as *hormones*. There are five major classes of plant hormones: auxin, gibberellins, cytokinins, abscisic acid, and ethylene. Each of these hormones elicits many responses and interacts in complex ways to stimulate or inhibit growth. Unlike most animal hormones, which have specific effects, each plant hormone is made in several parts of the plant and has several effects. The effects of hormones are governed by many factors, including the presence of other hormones, nonhormonal factors such as calcium ions, and the varying sensitivity of different tissues to hormones.

Plant growth and development result from complex, highly organized events. Those events involve more than forming masses of new cells or increasing the size of the plant. Rather, plants grow and develop by producing specialized cells, tissues, and organs having predictable shapes, locations, and functions. These specializations account for the form and function of a plant, which were the topics of the preceding unit. However, merely knowing something about the structure and function of specialized structures is not enough; we are also interested in how these specializations arise. For example, what determines that a cell differentiates into a vessel element and dies, while an adjacent cell remains meriste-matic or becomes a sieve element in the phloem? What controls if and when a leaf is shed by a tree? Why do the buds of most trees grow only during spring instead of all year long? In short, how do plants control all of the things that they do?

One of the earliest models of plant growth and development was proposed more than a century ago by Julius von Sachs, a German botanist who speculated that each plant organ resulted from the presence of a unique substance. Despite decades of intensive research, these substances have never been found. However, we *do* know that plant growth and development are controlled by internal signals, the most important being the information contained in DNA. This information governs the basic machinery for life, including growth and development, energy transformations, and reproduction. Since the signals for these processes are genetically predetermined, they provide a blueprint for many aspects of growth and development. The leaves of a corn plant, for example, always have the same basic shape, and seeds germinate in specific and predictable ways.

Plant growth is also influenced by external signals from the environment. Many genetic designs for growth and development come with several "options": different environmental conditions can activate different genes and produce different modes of growth. For example, specific genes stimulate growth during the favorable conditions of spring, whereas others induce dormancy during a harsh winter. These different modes of growth are important because they adapt plants to their changing environments. Although this adaptability is not without limits, it usually ensures growth and survival in the various conditions in which the species evolved.

The internal and external signals that regulate plant growth are mediated, at least in part, by plant growth-regulating substances, or **hormones** (from the Greek word *hormaein*, meaning "to excite"). Plant hormones are organic compounds that are made in small amounts in one part of the plant and transported to another part, where they initiate physiological responses. These responses are not always an "excitation" or stimulation; the regulation of bud dormancy, for example, results from inhibition of growth.

Botanists have identified five major classes of plant hormones: **auxin, gibberellins, cytokinins, abscisic acid,** and **ethylene.** According to our definition, these hormones have several characteristic features: they are made by the plant, are active in small quantities, are transported to other parts of the plant, and can cause a response. Consequently, in this chapter we'll discuss each of the following aspects of plant hormones:

- How the hormone was discovered
- How and where in plants the hormone is made
- How the hormone is transported
- Responses caused by the hormone

Despite the importance of hormones in plant growth and development, our knowledge of plant hormones is limited and, in many cases, controversial. We can give you only a few generalizations in this chapter. We'll begin with the three most important of them:

- Although a hormone has some characteristic effects, it may also have many other effects. That is, a single hormone can elicit many different responses.

- The particular effect elicited by a hormone depends on many factors, including the presence of other hormones, the amount of hormone present, and the sensitivity of the tissue to the hormone.

- Finally, it is difficult to predict a specific response, since the response changes under different conditions and in different plants. That is, the gene expression triggered by a plant hormone depends on the identity (i.e., message) of the hormone and its readability by a tissue.

You'll learn more about each of these generalizations throughout this chapter. For now, use them as a mental checklist as you learn how plant hormones were discovered and how all of the interest in them developed.

Plant hormones are organic compounds made in one part of a plant and transported to another part, where they elicit a response. Plant hormones are active in small concentrations. The five major classes of plant hormones are auxin, gibberellins, cytokinins, ethylene, and abscisic acid, each of which has many effects.

Auxin

Discovery

You're probably already familiar with Charles Darwin: his book *On the Origin of Species* formulated the theories of natural selection and descent with modification, and today stands as one of the most influential publications of all time. However, Darwin would have been famous even if he had never studied evolution.

During the late 1870s, Darwin and his son Francis began studying phototropism, which is the growth of stems and leaves toward light. The Darwins wanted to answer a seemingly simple question: how do plants grow toward light? They studied coleoptiles of canary grass (*Phalaris canariensis*) and oats (*Avena sativa*), both of which grow toward unilateral light (i.e., light coming from one direction; fig. 18.1a). In these plants, growth toward light occurs several millimeters below the tip of the coleoptile. One of the Darwins' first experiments involved blocking the incoming light by covering the tips of coleoptiles with metal foil. To the Darwins' surprise, these coleoptiles did not grow toward light, despite the fact that the growing region where curvature occurred was illuminated. Growth toward light resumed when the metal foil was removed or replaced with transparent glass. The Darwins then repeated the experiment, this time covering the growing region of the coleoptiles rather than their tips, and observed that the coleoptiles grew toward the light. They concluded that the growth of coleoptiles toward light was somehow controlled by the tip of the coleoptile.

In 1881, the Darwins published their findings in a book entitled *The Power of Movement of Plants*. They suggested that phototropism was due to an "influence" produced in the tip of a coleoptile that moved to the growing region, where it caused the coleoptile to grow toward light. Although the identification of plant hormones was decades away, you can already see the importance of the Darwins' work. The phototropic "influence" that they described was made in one part of the plant and transported to another, where it elicited a response. Their work was the first step toward discovering plant hormones.

The death of Charles Darwin in 1882 abruptly ended the Darwins' studies of phototropism. However, important questions remained, the most critical of which related to the nature of the "influence" that the Darwins had described. For example, was the "influence" a chemical signal, an electrical signal, or some other kind of stimulus?

In 1913, Danish plant physiologist Peter Boysen-Jensen began trying to answer this question (fig. 18.1c). When he cut off the tip of a coleoptile, the coleoptile stopped growing. This was consistent with the Darwins' observation that something from the tip of the coleoptile controlled growth. He then replaced the tip but separated it from the coleoptile with a tiny piece of agar, a gelatinous substance. Within a few minutes, the coleoptile resumed growth and curved toward the light (fig. 18.1c). Boysen-Jensen concluded that (1) the tips of coleoptiles do not have to be in their normal position to affect growth, and (2) the "influence" that controls phototropism can move through agar.

But what was the signal? Since it could move through agar, Boysen-Jensen reasoned that it probably was a water-soluble chemical. He tested this by replacing the agar blocks with butter. Because water is insoluble in butter, a water-soluble chemical coming from the tip would not move through the butter and reach the growing region of the coleoptile. Sure enough, the results were identical to those in which the tip was absent: there was no growth or curvature. Boysen-Jensen concluded that the "influence" controlling growth and phototropism of the coleoptiles was a water-soluble compound. But were electrical signals also involved? He tested this by replacing the agar blocks with pieces of platinum foil, which he expected would transmit electrical signals from the tip to the growing region of the coleoptile. Again, there was no growth or curvature, suggesting that the signal was not electrical.

In 1918, Hungarian plant physiologist Arpad Paál continued Boysen-Jensen's efforts to identify the Darwins' "influence" responsible for phototropism. Paál studied coleoptiles grown in the dark. He cut off the tips of dark-grown coleoptiles and asymmetrically replaced them on one side of the coleoptile's cut surface (fig. 18.1c). Surprisingly, these coleoptiles curved away from the side on which the tips were placed, despite the fact that the plants were in the dark. Furthermore, the curvature resembled that of plants growing toward light. These results suggested to Paál that the tip of the coleoptile produces a substance that moves down and stimulates growth and, more importantly, that light must cause it to accumulate on the shaded side of the coleoptile. By 1920, Paál's work had brought botanists tantalizingly close to finally accounting for the "influence" described by Charles and Francis Darwin some 40 years before.

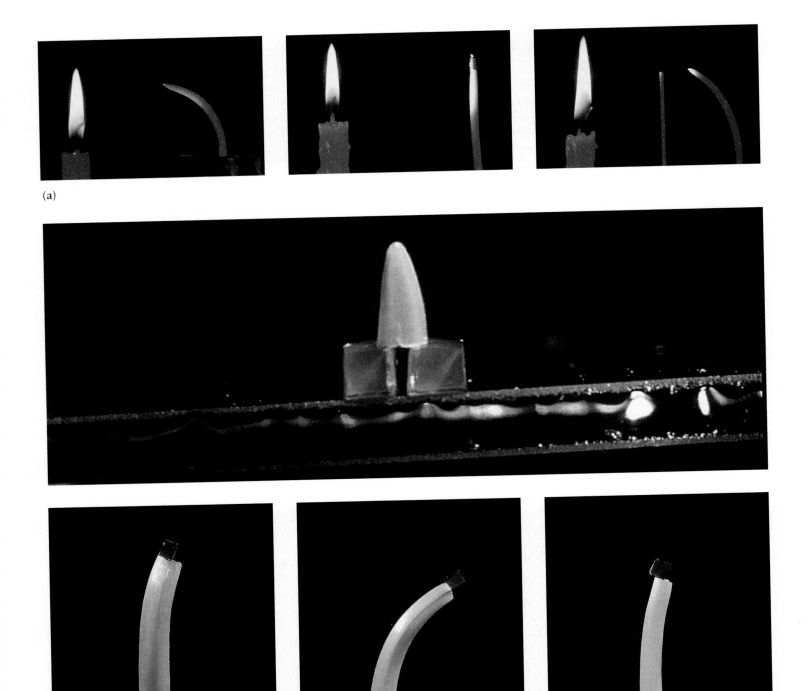

(a)

(b)

figure 18.1

(a) Darwin's study of phototropism. In the left photograph, the coleoptile is curving toward the light of a candle. On the coleoptile shown in the middle photograph, the top 2 mm of the plant has been covered by lightproof foil. The shoot does not bend, even though the growing zone of the coleoptile is illuminated. In the right photograph, the foil covers the growing zone but not the tip of the coleoptile. Curvature of the coleoptile toward the light shows that a signal must move from the light-sensitive tip to the growing zone, which is farther down. (b) Went's study of

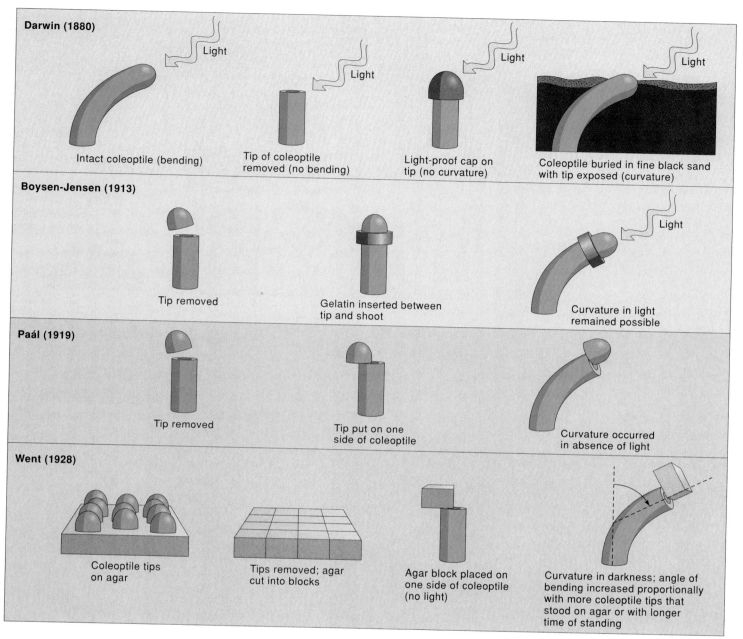

Darwin (1880)

Intact coleoptile (bending)

Tip of coleoptile removed (no bending)

Light-proof cap on tip (no curvature)

Coleoptile buried in fine black sand with tip exposed (curvature)

Boysen-Jensen (1913)

Tip removed

Gelatin inserted between tip and shoot

Curvature in light remained possible

Paál (1919)

Tip removed

Tip put on one side of coleoptile

Curvature occurred in absence of light

Went (1928)

Coleoptile tips on agar

Tips removed; agar cut into blocks

Agar block placed on one side of coleoptile (no light)

Curvature in darkness; angle of bending increased proportionally with more coleoptile tips that stood on agar or with longer time of standing

(c)

figure 18.1 (continued)

phototropism. These experiments were the first demonstration of a hormone in plants. Went gave the name *auxin* to the hormone involved in phototropism of coleoptiles. (c) A diagrammatic summary of some of the important experiments that demonstrated the existence of auxins. The experiments by Paál and Went were done in the dark.

MEASURING PLANT HORMONES

Botanists have traditionally thought that the amount of hormone controls the strength of hormonal responses. Thus, if a particular response is controlled by the amount of a hormone present, then the response must change as the amount of hormone changes. Although hormone-influenced responses such as leaf abscission, apical dominance, and phototropism are relatively easy to measure, measuring how much hormone is present is a more formidable problem. This is because hormones are active at minute concentrations (0.01 to 1 μM). Thus, any method for measuring a plant hormone must be sensitive. It must also be specific: the effect must be attributable to a specific hormone rather than to the influence of any other hormone or substance.

Hormones were originally measured with **bioassays,** which are procedures to quantify biologically active compounds by measuring their biological effects. Bioassays exhibit the extreme sensitivity and specificity of certain plant parts to particular hormones. For example, Frits Went showed that auxin could be collected from plants in agar and that placing this agar on one side of a decapitated *Avena* coleoptile would induce curvature. Since the curvature is proportional to the amount of indole-3-acetic acid (IAA) in the block, the amount of IAA can be measured by measuring the amount of curvature. This procedure, the *Avena* curvature test, became the standard bioassay for IAA. Similar bioassays were later developed for other hormones. For example, the classic bioassay for gibberellins is stimulation of growth in dwarf mutants (reading fig. 18.1).

Because they are inexpensive, relatively quick, and easy to do, bioassays are still used by many botanists to measure plant hormones. However, improved technology has resulted in more sensitive and specific techniques. Today, most hormones are measured with **gas chromatography,** a technique in which a sample is moved as a vapor through a column of liq-

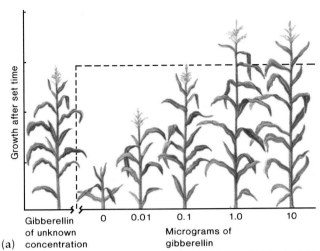
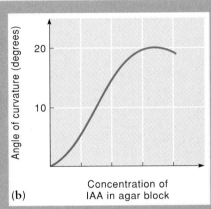

(a) Gibberellin of unknown concentration

Micrograms of gibberellin: 0   0.01   0.1   1.0   10

Growth after set time

(b) Angle of curvature (degrees) — Concentration of IAA in agar block

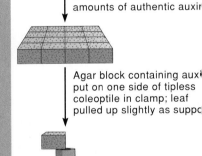

Coleoptile tip removed

Tips stood on agar for known periods

Tips removed from agar; agar cut into blocks; separate agar blocks made with known amounts of authentic auxin

Agar block containing auxin put on one side of tipless coleoptile in clamp; leaf pulled up slightly as support

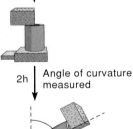

2h   Angle of curvature measured

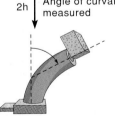

reading figure 18.1

Measuring plant hormones. (a) A plant's response to gibberellin can be used to measure the amount of hormone. The height of a gibberellin-treated plant is compared with that of dwarfs treated with a range of known concentrations of gibberellin. The technique of measuring the concentration of a chemical by measuring the response of living material to that chemical is called a bioassay. What was the unknown concentration of gibberellin in this bioassay? (b) The oat (*Avena*) coleoptile curvature bioassay is a popular bioassay for auxin. The amount of auxin that diffuses into an agar block can be estimated by comparing the results with a standard curve prepared from the responses of coleoptiles to known concentrations of auxin. Why must this bioassay be done in the dark?

uid or particulate solids. Components of the sample are differentially absorbed into the column, thereby separating the hormone from other parts of the sample.

A recently developed technique for measuring plant hormones is called **immunoassay.** This technique involves using animals to produce an antibody to the hormone and then using this antibody to select and bind the hormone from an extract. Thus, immunoassay uses the selectivity of antibodies to detect, identify, and measure quantities of plant hormones. Immunoassays of plant hormones are relatively inexpensive, speedy, extremely specific, and sensitive, and allow for convenient analysis of unpurified samples.

The Darwins' observations were finally explained in 1926 by Dutch plant physiologist Frits Went. While working in his father's laboratory, Went separated the "influence" for phototropism from the plants that produced it. His experiments were elegantly simple and were influenced strongly by the work of Boysen-Jensen and Paál. Went first cut off the tips of oat coleoptiles and placed their cut surfaces on agar. After about an hour, Went discarded the tips, and put pieces of agar that they had touched on the cut tips of decapitated coleoptiles grown in the dark. As you can see in figure 18.1b and 18.1c, Went's results were convincing:

- Decapitated coleoptiles without agar blocks did not grow. These results confirmed the Darwins' findings that the tips of coleoptiles produced something necessary for growth.

- Agar blocks that had not contacted cut tips of coleoptiles caused no response when they were placed on decapitated coleoptiles. Therefore, there was nothing in the agar that caused growth and curvature of the coleoptile.

- When agar blocks that had contacted cut tips were placed on the center of decapitated coleoptiles, they grew straight up. Thus, the coleoptile tips had produced a chemical that diffused into the agar, and this chemical stimulated the growth of coleoptiles.

- When agar blocks that had contacted cut tips were placed on one side of decapitated coleoptiles, they curved away from the agar blocks, just as Paál's seedlings did when the tip of a coleoptile was placed on one side of a decapitated coleoptile. This growth away from the agar blocks was similar to phototropic curvature, even though he kept the plants in the dark and the tips were absent. These results indicated that the agar blocks contained a chemical that stimulated the growth of coleoptiles.

Went concluded that phototropic curvature was not due to the mere presence of the coleoptile's tip but rather to a chemical coming from the coleoptile's tip that stimulated growth. Went named this chemical *auxin* (from the Greek word *auxein*, meaning "to grow"). Auxin does, in fact, influence phototropism: light striking one side of a coleoptile causes auxin to migrate to the shaded side of the coleoptile, where it stimulates growth toward light.

According to Went's explanation, auxin fit the definition of a hormone: it was made in one part of the plant (the tip of the coleoptile) and transported to another part (the growing region of the coleoptile), where it caused a response (increased growth). Thus, auxin was the first plant hormone to be discovered. Since its discovery, auxins have been found throughout the plant kingdom, in fungi, and in some bacteria.

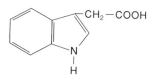

figure 18.2

The structure of indole-3-acetic acid (IAA).

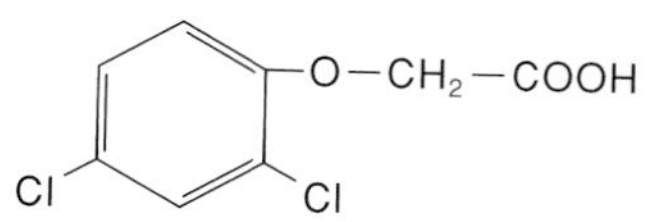

2,4-dichlorophenoxyacetic acid (2,4-D)

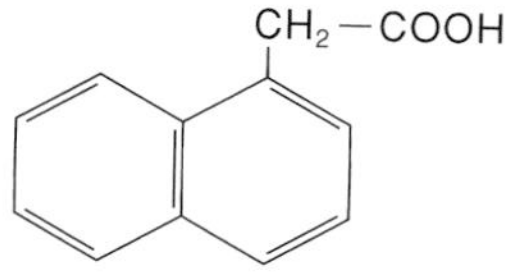

1-naphthaleneacetic acid (NAA)

figure 18.3

IAA is the most active naturally occurring auxin. Several synthetic compounds have auxinlike effects. Synthetic auxins such as 2,4-D and NAA have chemical structures similar to that of IAA.

Animals also produce auxins, although we do not know how or if they respond to them. Interestingly, the first auxin to be identified chemically was isolated from human urine.

Synthesis

The most active naturally occurring auxin in plants is **indole-3-acetic acid, or IAA** (fig. 18.2). Botanists thought for a long time that auxin was made from the amino acid tryptophan; however, a study in late 1991 using a mutant incapable of making tryptophan showed that plants can make IAA without tryptophan as an intermediate. Thus, the primary pathway for IAA synthesis may not involve tryptophan. The most active areas of IAA synthesis are shoot tips, embryos, young leaves, flowers, fruits, and pollen.

There are two other naturally occurring auxins: 4-chloro-IAA and phenylacetic acid. The precise roles of these auxins in plant growth and development are unknown; however, they are generally less active than IAA.

Synthetic Auxins

Although IAA is the most active naturally occurring auxin, several synthetic compounds have auxinlike effects. Synthetic auxins such as 2,4-D (2,4-dichlorophenoxyacetic acid) and NAA (naphthaleneacetic acid) have structures that resemble IAA (fig. 18.3).

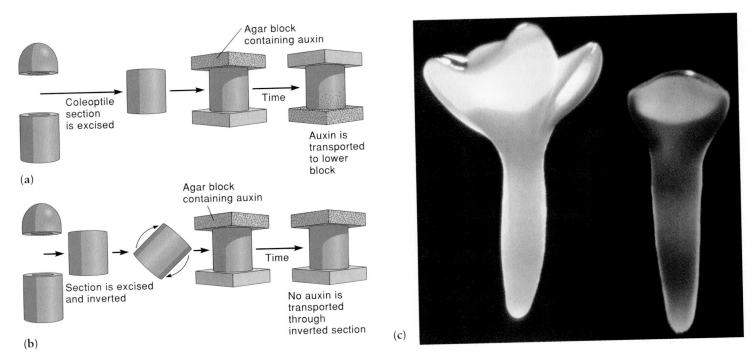

(a)

(b)

(c)

figure 18.4

The polar transport of auxin (shaded). (a) When the coleoptile section remains right-side-up, auxin moves through to the receptor block below. (b) No auxin moves through an inverted coleoptile section. (c) The influence of polar transport of auxin on embryo formation. When auxin transport is normal, the cotyledons develop normally (left). When the polar transport of auxin is blocked, the cotyledons fuse (right).

Synthetic auxins such as 2,4-D are used extensively as herbicides because they are inexpensive, relatively nontoxic to humans, and have a selective effect: they kill broadleaf dicots but not monocots. The exact mechanism for this selectivity is unknown, but it may involve the greater ability of broadleaf weeds to absorb and translocate the synthetic auxins. Synthetic auxins are also used to prevent preharvest dropping of fruit, to produce roots on cuttings, and to inhibit sprouting of lateral buds ("eyes") on Irish potatoes. Synthetic auxins have been used widely since the 1940s and have revolutionized weed-control in agriculture. Their use has decreased production costs by reducing the amount of labor and mechanical weeding needed to grow and efficiently harvest a crop.

Unfortunately, the effects of synthetic auxins on human physiology have not all been positive. Most of the negative effects can be traced to a defoliant used in the Vietnam War known by the code name Agent Orange. Agent Orange was a 1:1 mixture of 2,4,5-T (2,4,5-trichlorophenoxyacetic acid) and 2,4-D that was sprayed throughout the jungles of Vietnam and followed by napalm bombs. These treatments destroyed hundreds of square kilometers of forest. This defoliation was protested during the war by many botanists, but the problems have proven to be more serious than the destruction of the forests. Synthesis of the 2,4,5-T used in Agent Orange also produces dioxin (2,3,7,8-tetrachloro-dibenzo-*para*-dioxin), which is one of the most toxic synthetic chemicals known. Thousands of U.S. pilots and Vietnamese citizens exposed to Agent Orange (and often contaminated with dioxin) have had increased occurrences of

miscarriages, birth defects, leukemia, and other types of cancer. The Environmental Protection Agency banned the use of 2,4,5-T in the United States in 1979, but its effects continue to plague many Vietnam veterans and Vietnamese citizens.

Transport

IAA moves primarily through parenchyma cells of the cortex, pith, and vascular tissues. It moves slowly: typical rates of 1 cm hr^{-1} are slow compared to those of substances moving in xylem and phloem, but they are still about 10 times faster than those predicted by diffusion.

IAA moves polarly in roots and stems (fig. 18.4a,b). Polar transport requires energy; thus, inhibitors of ATP synthesis block the transport of IAA. In stems, IAA is transported *basipetally*, meaning that it moves toward the base. Basipetal transport in stems continues even if the stem is inverted, so that the apex is pointing down. In roots, IAA is transported *acropetally*, meaning that it moves toward the tip.

The polar transport of IAA has intrigued botanists for many years and may involve auxin carrier proteins in the plasmalemma (i.e., cell membrane). IAA made in mature leaves moves nonpolarly in the phloem.

Polar transport of auxin is also critical to normal growth and development of plants. To appreciate this, consider the embryos of Indian mustard (*Brassica juncea*) shown in figure 18.4c. During normal development (i.e., when auxin transport is polar), embryos with distinctive cotyledons (embryonic leaves) develop (fig. 18.4c, left). When polar transport of auxin is blocked (with chemical inhibitors), the embryos have fused cotyledons (fig. 18.4c, right). These and other observations suggest that polar auxin transport is critical for the initiation of bilateral symmetry in these embryos.

Most of the IAA in plants is joined to a variety of amino acids, proteins, and carbohydrates. These conjugates are believed to inactivate excess IAA, allow rapid changes in the concentration of free IAA, and transport IAA through the plant. The conjugation and deconjugation of IAA are probable mechanisms for controlling the concentration of free IAA and its effects on growth and development.

c o n c e p t

IAA is the most active naturally occurring auxin. It moves basipetally in stems and acropetally in roots. Synthetic auxins such as 2,4-D have widespread uses in modern agriculture.

Effects of Auxin

Cellular Elongation in Grass Seedlings and Herbs

Short-Term Effects There are two requirements for cellular elongation: (1) positive turgor pressure, and (2) increased plasticity (stretchability) of the cell wall. Turgor pressure in cells results primarily from the presence of dissolved solutes and is not significantly affected by IAA. However, IAA does increase the plasticity of the cell wall. One explanation of IAA-induced growth is the **acid-growth hypothesis,** which is summarized as follows:

1. IAA stimulates H^+ pumps in the plasmalemma.

2. Once activated, these pumps secrete H^+ into the cell wall, decreasing its pH to about 5.0 (for more about pH, see reading A.1, "Moles, Hydrogen Ions, and pH" in Appendix A).

3. This acidification of the cell wall activates pH-dependent enzymes such as expansins that break load-bearing bonds between cellulose microfibrils.

4. When these bonds break, the wall "loosens," and turgor pressure causes the cell to expand.

Several observations are consistent with the acid-growth hypothesis for IAA-induced growth. For example, applying acid to internodal segments induces growth similar to that resulting from applying IAA. Furthermore, IAA-induced cellular elongation is inhibited by bathing the tissue in neutral buffers, which prevent acidification of the cell walls.

IAA-induced growth is rapid: it occurs within as little as 3 minutes in grass seedlings. Because it occurs so fast, many botanists thought that it did not involve transcription or translation of genetic information. This conclusion was premature, however, for improved techniques have shown an accumulation of IAA-induced mRNA several minutes before H^+ secretion and wall acidification. Thus, IAA-induced growth may result from rapid gene activation, which, in turn, may activate H^+ pumps in the plasmalemma. Other effects of IAA may also alter mRNA synthesis, turnover, or activation.

Long-Term Effects According to the acid-growth hypothesis for IAA-induced growth, cellular elongation should continue as long as the cell wall is acidic. However, if this does not occur: growth induced by acid typically stops after 1–3 hours, while growth induced by IAA continues much longer. Therefore, acid-induced growth may account for only the early stages of IAA-induced growth, but the long-term growth induced by IAA may result from another mechanism. A clue to this other mechanism was provided in the 1960s, when researchers discovered that compounds that inhibit nucleic acid and protein synthesis also inhibit IAA-induced growth. This suggests that IAA may act at the gene level, possibly by activating a gene required for making a protein necessary for growth. Indeed, IAA stimulates the production of mRNA and ribosomes. Thus, long-term growth induced by IAA may result from selective activation of genes and altered patterns of protein synthesis. Among the genes turned on by IAA are those for the synthesis of cell-wall materials. These materials are made and selected by dictyosomes, and accommodate increases in cellular size that accompany growth.

Apical Dominance

Plants form axillary buds at the base of their petioles (see Chapter 14). In many plants, the growth of axillary buds is unequal along the stem: axillary buds near the shoot apex grow little, while those progressively farther from the tip grow much more. This inhibiting influence of the shoot apex on the growth of axillary buds in many plants is called **apical dominance** and produces a roughly cone-shaped plant.

If the shoot tip is removed, axillary buds near the tip of the stem begin growing within as little as 4 hours. Botanists long believed that apical dominance was due only to IAA coming from the shoot tip, since dominance could be restored by applying IAA to the cut surface of decapitated stems.

However, the mechanism for apical dominance is more complicated. For example, one cannot detect any radioactive IAA in axillary buds after applying radioactively labeled IAA (^{14}C-IAA) to the stem tip. Furthermore, applying IAA to buds does not inhibit growth; sometimes it even promotes bud growth. This may account for the less pronounced apical dominance typical of many plants.

The most recent hypothesis for apical dominance is based on the idea that high levels of IAA stimulate the production of ethylene, another plant hormone that you'll learn about later in this chapter. According to this explanation, IAA coming from the shoot tip stimulates cells around lateral buds to make ethylene, and it is the ethylene produced in response to IAA (rather than the IAA itself) that inhibits bud growth.

Cytokinins (other plant hormones) coming from roots also influence apical dominance. Indeed, cytokinins promote bud growth, even if the shoot apex is present. These observations emphasize another generalization we can make about plant hormones—namely, that *a single aspect of growth and development can be influenced by several hormones. A particular response probably results from changing ratios of hormones rather than from the presence or absence of an individual hormone.*

The Lore of Plants

The characteristic shape of most Christmas trees results from apical dominance. Except for a few nonconformists, such as pin oak, all Christmas-tree-shaped trees are conifers. Ever since President Franklin Pierce (a politician otherwise noted for putting stickum on the backs of postage stamps) first had a Christmas tree in the White House in 1856, people have been picky about their trees, preferring ones that are about 2 m high and 1.3 m wide at the base. Today, thanks largely to the efforts of a former car salesman named Fred Musser, Scotch pine is the most popular Christmas tree in the United States. Musser's marketing of Scotch pine (*Pinus sylvestris*) emphasized that the tree's needles, though long, stayed on the tree much longer than those of balsam fir or spruce, the previous best-sellers. The Scotch pine is also fast-growing; it takes only 8 years to grow a 2 m tree. To learn more about Christmas trees, see reading 30.4, "Christmas Trees."

Abscission

Another example of the interaction of IAA and other plant hormones is **abscission,** which is the shedding of leaves or fruit by a plant. Abscission of leaves is usually preceded by **senescence,** which is a collective term for the processes contributing to age-induced decline and death of a plant or plant part. During the final stages of senescence, leaves and fruit separate from the plant at the **abscission zone,** which is a layer of thin-walled parenchyma cells at the base of the organ (see fig. 14.25). Abscission occurs like this (also see discussion in Chapter 14):

1. Actively growing leaves and fruit produce large amounts of IAA, which is transported to the stem. This IAA, along with cytokinin and gibberellins from the roots, retards the onset of senescence and abscission.

2. Environmental stimuli such as the shorter days of fall, drought, wounding, or a nutrient deficiency cause decreased production of IAA. This triggers senescence.

3. Some signal, perhaps a "senescence factor," stimulates cells in the abscission zone to produce ethylene. This, in turn, stimulates cells of the abscission zone to expand, suberize, and produce cellulase and pectinase.

4. Cellulase and pectinase digest the middle lamella, which cements cells of the abscission zone together.

5. As a result of this wall digestion and concurrent cellular expansion, the cells separate and abscission occurs. Deciduous plants lose their leaves each fall; evergreens usually retain their leaves for 2–3 more years.

The inhibitory effect of IAA on abscission is used by horticulturists. For example, they apply synthetic auxins to orchards of apples, oranges, and grapefruits to prevent preharvest fruit drop. Similar applications of auxin are used to hold berries on holly, thereby making the holly marketable for longer periods of time. Conversely, applying large amounts of synthetic auxin stimulates the production of excess amounts of ethylene, which often hastens abscission. Heavy applications of synthetic auxins are used to coordinate the abscission of apples, olives, and other fruits, thereby allowing the crop to be harvested in a shorter period of time.

Differentiation of Vascular Tissue

The vascular cambium is activated in the spring by IAA produced by young, developed leaves (fig. 16.5). Gibberellin is also probably involved. Indeed, a high auxin/gibberellin ratio promotes xylem differentiation, while a low auxin/gibberellin ratio favors phloem differentiation. Nonhormonal factors such as sugars produced in leaves also influence the effect of IAA on cellular differentiation. For example,

- Auxin plus small amounts (2%) of sucrose favor differentiation of xylem.

- Auxin plus moderate amounts (3%) of sucrose favor differentiation of xylem and phloem.

- Auxin plus large amounts (4%) of sucrose favor differentiation of phloem.

These observations illustrate our next generalization about plant hormones: *Physiological responses elicited by hormones are strongly influenced by nonhormonal factors.*

Fruit Development

Seeds in fruit are the source of auxin that stimulates fruit development. For example, fruits of strawberry (*Fragaria ananassa*) are the small structures dispersed across the fleshy tissues of the strawberry: IAA from the seed in each fruit triggers development of the adjacent tissue (fig. 18.5). Thus, strawberries do not develop when the fruits (i.e., the sources of IAA) are removed. Similarly, when fruits are removed from half of the strawberry, only the remaining half develops normally. Commercially important seedless fruits of plants such as tomato and cucumber can be induced to form if you treat the unfertilized carpels with auxin. Such fruits are called **parthenocarpic** (from *parthenos,* meaning "virgin") fruit.

Formation of Adventitious Roots

Nurseries and amateur gardeners propagate plants by exploiting the ability of auxin to stimulate the formation of adventitious roots. The procedure is simple: cut surfaces of pieces of

figure 18.5

Auxin stimulates development of strawberries. IAA from the seeds in each fruit of the strawberry stimulates development of the adjacent tissue. Strawberries do not develop when fruits are removed. (a) Normal development. (b) All fruits removed. (c) All fruits but one removed. (d) All fruits removed and replaced by lanolin paste containing auxin.

a parent plant are dipped in a solution of synthetic auxin. The auxin stimulates the formation of adventitious roots at the cut surface (fig. 18.6), after which the cutting can be transplanted and grown into a new plant. However, it's not necessary to use synthetic auxins to stimulate formation of cuttings in all species. Some species transmit enough auxin from their stem apex to trigger formation of adventitious roots near the cut surface.

c o n c e p t

Auxin stimulates cellular elongation, differentiation of vascular tissue, fruit development, and formation of adventitious roots.

Auxin and Calcium

The responses elicited by IAA and other hormones depend on the presence of a nonhormonal factor, calcium ions (Ca^{2+}). For example, Ca^{2+} is required for the polar transport of IAA, and Ca^{2+}-deficient plants are usually not responsive to IAA. The most recent model accounting for the interaction of IAA and Ca^{2+} is as follows:

1. IAA stimulates the release of Ca^{2+} from the vacuole and endoplasmic reticulum of a cell.

2. This release of Ca^{2+} increases the concentration of Ca^{2+} in the cytoplasm. This has led botanists to suggest that Ca^{2+} acts as an intracellular messenger that conveys information about the nature of a stimulus to target proteins that guide the cellular response. Examples of target proteins activated by Ca^{2+} include calmodulin, Ca^{2+}-dependent protein kinases, phosphatases, and Ca^{2+}-stimulated phospholipases.

figure 18.6

Auxin promotes development of adventitious roots. To propagate plants, cut surfaces of parent plants are dipped in solutions of synthetic auxins, which promote the formation of adventitious roots at the cut surface. The stalk of the leaf on the right was placed in pure water.

3. Activated target proteins may then alter membrane transport, enzyme activity, and/or H^+ pumps and thereby produce the physiological response. According to this model, the wall acidification that accompanies growth occurs as a result of changes in the concentration of Ca^{2+} rather than changes in IAA itself.

Auxin increases the internal concentrations of Ca^{2+}. These auxin-induced increases in internal Ca^{2+} activate target proteins, which then may trigger the physiological response.

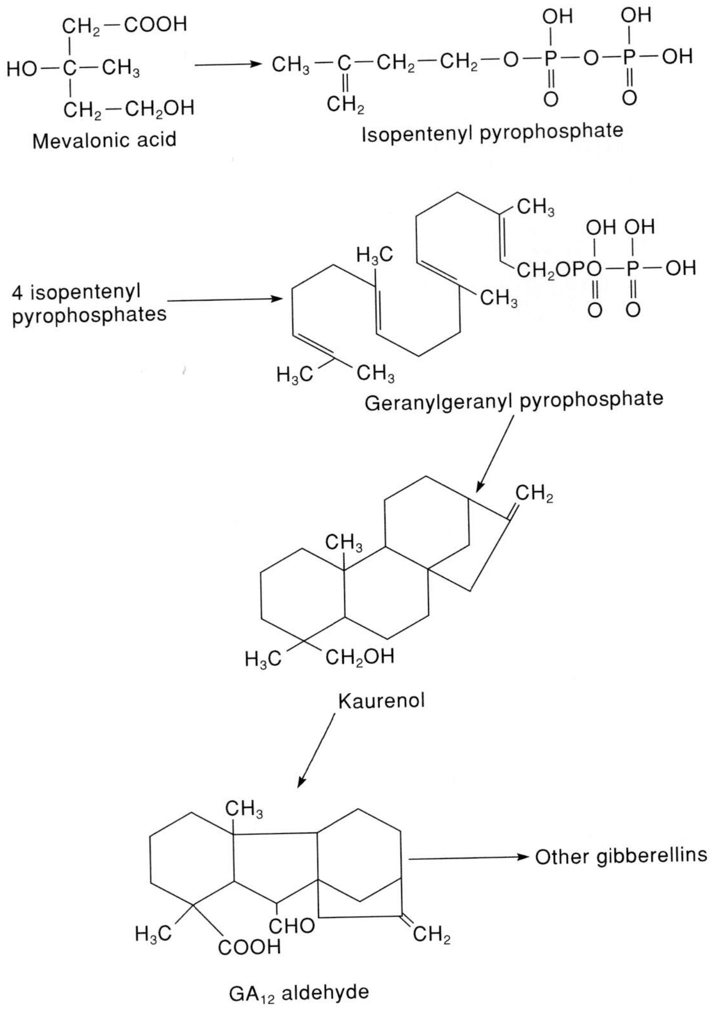

Gibberellins

Discovery

At about the time Frits Went published the results of his auxin experiments, botanists in Japan were beginning experiments that would ultimately lead to the discovery of another plant hormone: gibberellin. Ewiti Kurosawa and his colleagues were studying rice (*Oryza sativa*) plants suffering from *bakanae*, or "foolish seedling" disease. These plants were pale and spindly, and their excessively elongated growth caused them to fall over and die before they could be harvested. Kurosawa soon discovered that the diseased plants were infected by a fungus named *Gibberella fujikuroi*. Knowing this allowed him to design some clever experiments. First, he grew pure cultures of the fungus in his laboratory. When he ground up the fungus and applied the extract to healthy plants, the plants quickly developed symptoms characteristic of *bakanae*. Similar symptoms also developed when healthy plants were treated with media in which the fungus had grown. These results indicated that the fungus made and secreted a compound that caused *bakanae* in infected rice plants. Kurosawa named this compound *gibberellin*.

By 1939, Japanese botanists had isolated and chemically identified gibberellin. However, their work went largely unnoticed in the United States because of U.S. botanists' preoccupation with auxin, the lack of scientific communication with Japan, and World War II. Later studies in the United States showed that gibberellins are involved in many aspects of plant growth and development.

Synthesis and Transport

Gibberellin is made via the mevalonic acid pathway (fig. 18.7). Although more than 80 gibberellins have been isolated from various fungi and plants, no more than 15 have been found in a single species. Each of the gibberellins has an interlocking ring structure and one or more carboxyl groups that impart acidic properties to the molecule (fig. 18.8). Gibberellins are abbreviated **GA** (for gibberellic acid) and assigned subscript numbers that distinguish them from each other. GA_3 is isolated from *Gibberella fujikuroi* and is the most intensively studied gibberellin.

Several commercial compounds inhibit the synthesis of gibberellins. These inhibitors are called *growth retardants* and include Phosphon D, Cycocel (CCC), Amo-1618, and

figure 18.7

Gibberellin is made via the mevalonic acid pathway.

Ancymidol. Growth retardants inhibit stem elongation, thereby producing stunted plants. Growth retardants are frequently sprayed on growing chrysanthemums to produce flowers with thicker, sturdier stalks. People in the turf-grass industry often use growth retardants to control the development of seeds and stem elongation.

Why are there so many different gibberellins? Most gibberellins are inactive and are probably precursors of more active gibberellins. Indeed, GA_1 may be the only gibberellin that controls stem elongation in angiosperms.

Gibberellins occur in angiosperms, gymnosperms, mosses, ferns, algae, and fungi but are unknown in bacteria. In angiosperms, gibberellins occur in immature seeds, apices of roots and shoots, and young leaves. Unlike the transport of IAA, the transport of gibberellin is not polar: it moves in all directions in xylem and phloem.

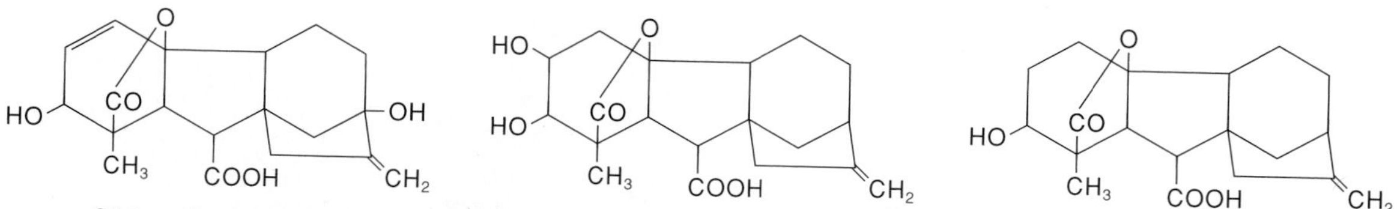

figure 18.8

Three of the more than eighty gibberellins that have been isolated from various fungi and plants. All gibberellins have an interlocking ring structure and one or more carboxyl groups (–COOH) that impart acidic properties to the molecules.

Effects of Gibberellins

Extensive Growth of Intact Plants

Many dwarf mutants grow normally if treated with GA. The sensitivity of these plants to GA is striking: for example, dwarf pea seedlings respond to as little as one-billionth of a gram of gibberellin. Dwarfism does not result from the absence of gibberellins: in fact, dwarf plants often contain *more* gibberellins than do normal plants. Rather, their stunted growth results from their inability to make enough *active* gibberellins, specifically GA_1.

Gibberellin is used commercially by Hawaii's sugarcane growers. Growing plants are sprayed with gibberellin, which stimulates shoot elongation and thereby increases the yield of sugar. However, the stimulation of shoot elongation by GA differs from that induced by IAA. For example,

- GA controls internodal elongation in mature regions of trees, shrubs, and a few grasses, whereas IAA regulates elongation in grass seedlings and herbs.

- GA induces cellular division *and* cellular elongation; IAA promotes cellular elongation alone.

- GA-stimulated elongation does not involve the cell wall acidification characteristic of IAA-induced elongation.

Thus, we identify another generalization about plant hormones: *different hormones can elicit similar effects via different mechanisms.*

Seed Germination

Gibberellins play a critical role in seed germination in cereal grasses such as barley. Here's what happens:

1. The sequence of events leading to seed germination begins with **imbibition,** which is the absorption of water by a seed. Imbibition causes the embryo to release gibberellins.

2. These gibberellins stimulate transcription of genes of hydrolytic enzymes in the seed's **aleurone layer,** a specialized layer of the endosperm 2–4 cells thick located just inside the seed coat. Gibberellin promotes the transcription of amylase mRNA.

3. These hydrolytic enzymes are secreted by dictyosomes into the seed's endosperm. One of the hydrolytic enzymes produced by the aleurone layer, amylase, catalyzes the conversion of starch to sugar, which is used as an energy source for the growing seedling.

Several relatively simple experiments provide evidence for the role of GA in the germination of barley seeds. For example, the endosperm is not broken down if the embryo (i.e., the source of GA) is removed. Similarly, the addition of GA to these embryoless seeds results in the production of amylase and the hydrolysis of endosperm starch to sugar. As is true for genetic dwarf plants, the endosperm of barley seeds is sensitive to GA: digestion begins when the endosperm is exposed to as little as nine-trillionths of a gram of GA.

As convincing as the preceding mechanism seems, it does not account for the germination of dicot seeds or even the germination of seeds of other cereal grains. For example, the aleurone layers of some cultivated oats and corn do not respond to added GA. In many of these seeds, the **scutellum** (i.e., cotyledon) rather than the aleurone layer is the primary source of enzymes that digest the endosperm. In the seeds of many dicots and gymnosperms, cytokinins may stimulate the breakdown of stored food reserves.

Many plants produce seeds that germinate only after exposure to cold and light. Gibberellins will substitute for this cold requirement in plants such as tobacco, lettuce, and oats. Gibberellins stimulate germination of these seeds by enhancing cellular elongation in the embryo, which causes the embryo to rupture the seed coat. The ability of gibberellins to stimulate germination of cereal seeds is exploited in making alcoholic beverages (see reading 4.1, "Membrane Transport and Making Beer"). Brewers treat barley grains with gibberellins to stimulate seed germination and speed the conversion of starch to sugar. We call these germinated seedlings *malt.* Brewing then converts the sugar to alcohol by anaerobic respiration (see Chapter 6).

Juvenility

Many plants have a juvenile stage and an adult stage of growth. For example, juvenile stages of eucalyptus (*Eucalyptus globulus*) have leaves shaped differently than those of adult stages (fig. 18.9). Gibberellins may help determine whether a particular part of a plant is juvenile or adult. For example, the buds of adult branches usually develop only into adult branches, but treating them with GA causes them to grow as juvenile branches.

Flowering

During their first year of growth, biennial plants such as cabbage have short internodes. These plants are called **rosettes** because their tightly packed leaves are arranged like the petals of a rose. Following winter's cold temperatures, internodes of these plants expand rapidly, and the plant forms flowers during the second year of growth. The rapid expansion of internodes and formation of flowers by rosette plants in response to cold is referred to as **bolting**. Applying GA to rosette plants induces bolting, suggesting that cold temperatures somehow stimulate the synthesis of GA during the following season (fig. 18.10).

GA-induced bolting in rosettes is exploited commercially to induce early production of seeds by some biennials. When these plants are treated with GA during their first year of growth, they form flowers and fruit a year early, thereby allowing growers to obtain seeds after only one season instead of the two seasons normally required by biennials. GA does not stimulate flowering in most other plants.

Finally, gibberellins are used by camellia fanciers to produce blooms of prize-winning size. The buds are treated with GA just before opening, resulting in larger petals. This treatment is called "gibbing the buds."

The Lore of Plants

In the 1920s lettuce was a seasonal crop. Thus, like corn and blackberries, it was available only during particular times of the year. However, one summer day, farmer W. Altee Burpee noticed an unusual lettuce plant growing in his garden. Unlike other varieties such as romaine and batavia lettuce, this unusual lettuce didn't bolt during hot weather, leading Burpee to describe it as "cool as an iceberg." Thus was born iceberg lettuce.

Iceberg lettuce is grown year-round and outsells all other varieties of lettuce combined. Indeed, Americans each eat an average of 28 lb of iceberg per year, 5 times more than we did in the 1920s (among vegetables, this rate of consumption ranks second only to potatoes). Contrary to popular belief, however, iceberg lettuce is nutritionally deficient: it contains only one-sixth the vitamin C as romaine, one-third the beta-carotene as bibb, and one-fourth the calcium as looseleaf lettuce. Nevertheless, it is immensely popular because of its crispness and ability to hold dressing. Almost 2 billion heads of iceberg are harvested per year (mostly in the Salinas Valley just south of San Francisco). This is why 75% of U.S. households have iceberg lettuce in the refrigerator right now.

(a)

(b)

figure 18.9

(a) Juvenile and (b) mature leaves in *Eucalyptus globulus* are shaped differently. Juvenile leaves are softer and opposite each other, with palisade parenchyma only on upper surfaces. Adult leaves are hard, spirally arranged, and have palisade on both sides. The white structures in (b) are flowers.

figure 18.10

Bolting in cabbage (*Brassica oleracea*). The plants on the right were untreated. The plants on the left were treated with gibberellin once a week for 8 weeks.

Fruit Formation

The most important commercial use of gibberellins involves their ability to increase the size of seedless grapes. Indeed, almost all vines of the "Thompson Seedless" grape grown in the Central and Imperial Valleys of California are sprayed with gibberellins each year. As a result of these treatments, the grapes increase in size almost threefold and are more loosely packed, making them less susceptible to fungal infections.

c o n c e p t

Gibberellins promote extensive growth of intact plants, stimulate seed germination in some cereal grasses, influence the formation of juvenile versus adult stages of growth, promote bolting of biennials, and increase fruit size in grapes.

Cytokinins

Discovery

The trail leading to the discovery of cytokinins can be traced to Gottlieb Haberlandt, who reported in 1913 that an unknown compound in vascular tissue stimulated cellular division. Like the Darwins' studies of the phototropic "influence," Haberlandt's work went largely unnoticed. However, Went's discovery of auxin in 1926 stimulated botanists to search for other plant hormones and later led to the rediscovery of Haberlandt's pioneering work. The search for the hormone controlling cellular division was a long and unusual one. It spanned several decades, involved many researchers, and included substances ranging from coconut milk to fish sperm.

In the 1940s, botanist Johannes van Overbeek made an interesting observation: plant embryos grew faster when they were supplied with coconut milk. Coconut milk is the liquid endosperm of coconut seeds and is rich in nucleic acids. This information was critical to the next set of experiments, which were done by Folke Skoog and his student Carlos Miller at the University of Wisconsin in the 1950s. Skoog and Miller were studying the influence of auxin on the growth of tobacco (*Nicotiana tabacum*) cells. They grew their cells by **tissue culture,** a technique for growing plants in artificial media in test tubes or beakers (you'll recall discussions of this technique in Chapters 9 and 12; see fig. 12.8). Growing plants in controlled cultures was important for Miller and Skoog because they could test the effects of individual compounds that they could add to or delete from the culture medium. When auxin alone was added to the medium, tobacco cells expanded but did not divide. Auxin therefore appeared to stimulate cellular enlargement, but that alone did not explain plant growth, which in intact plants involves cellular enlargement *and* division. Apparently, the signal for cellular division was missing.

One day in the lab, Miller noticed a jar containing something you probably don't have in your kitchen: herring-sperm DNA. Miller knew of van Overbeek's research showing that

PLANT HORMONES IN PLANT PATHOLOGY

Cytokinins and IAA play important roles in diseases caused by several plant pathogens. For example, *Corynebacterium fascians* is a bacterium that infects plants such as peas and chrysanthemums. Symptoms usually include flattened stems and extensive branching that forms broomlike structures called witch's brooms. Similar branching patterns can be induced in healthy plants by applying cytokinins. Thus, pathogenic symptoms can be induced by excess amounts of cytokinins. But where do these excess cytokinins come from?

Pathogenic strains of *C. fascians* contain a plasmid (i.e., a small circle of DNA that occurs independently of the organism's main DNA; see Chapter 11) having genes that produce cytokinins. Following infection, this plasmid is incorporated into the host's DNA. Subsequent transcription and translation of this plasmid results in excessive amounts of cytokinins, which cause extensive branching. Nonpathogenic strains of *C. fascians* lack this plasmid.

A similar mechanism is used by the bacterium *Agrobacterium tumefaciens* to produce crown gall, a cancerlike disease of dicots. During infection, *A. tumefaciens* transfers to its host a fragment of a large plasmid that carries genes responsible for making cytokinins and IAA. When these genes are expressed, the amounts of IAA

reading figure 18.2
Sawfly gall on willow.

and cytokinins increase, thereby producing the symptoms of crown gall (for a discussion of how botanists use *Agrobacterium* in genetic engineering, see Chapter 11).

Other plant pathogens do not cause their hosts to produce hormones that disrupt growth. Instead, the pathogens produce and release large amounts of IAA and cytokinins themselves. For example, the bacterium *Pseudomonas syringae* causes olive-knot disease in olive, oleander, and privet. Unlike *C. fascians* and *A. tumefaciens*, *P. syringae* does not transfer any of its DNA to the host. Instead, it exerts its pathogenic effects by se-

creting large amounts of IAA and cytokinins. Similarly, many rust fungi produce cytokinins that delay leaf senescence by establishing areas to which nutrients are transported. Consequently, necrotic lesions induced by the fungus are often surrounded by green islands of healthy tissue on which the fungus will subsequently feed. Witch's brooms are also common in trees infected by fungi such as *Exobasidium*. The degree of pathogenicity is roughly proportional to the amount of cytokinin produced and secreted by the fungus.

Plant abnormalities induced by excessive amounts of hormones are not limited to those produced by fungi and bacteria. For example, leaf galls form in response to parasitic insects such as mites and gall wasps (reading fig. 18.2). Female insects deposit their eggs in the leaf mesophyll. These eggs develop into larvae that produce cytokinins. When these cytokinins are released into the leaf, they induce cellular divisions that form a ball-like mass, or gall, that becomes the home for the developing insect. Because cytokinins also attract sugars and other nutrients, the gall provides shelter and a source of nutrition for the insect, even though the rest of the leaf may be undergoing senescence. During spring, the adult insect leaves the gall and lays eggs that infect other leaves, thereby completing its life cycle.

---

the DNA-rich endosperm of coconut stimulated plant growth, and he suspected that something in the DNA might cause the increased growth. When he added the herring-sperm DNA to the culture medium, the tobacco cells began dividing. Miller was excited about these results and quickly ordered another supply of herring-sperm DNA so that he could repeat and expand his experiments. To his chagrin, cells treated with the new batch of DNA did not divide: apparently, the growth factor present in the old DNA was not in the new DNA.

Miller reasoned that his original results might be due to the age of the original DNA; that is, the signal that triggered cellular division was probably a breakdown product of DNA rather than the DNA itself. Miller later discovered that he could also induce cellular division by supplementing the culture medium with adenine, a purine base in DNA. Similar results were later obtained by adding adeninelike compounds extracted from yeast. After painstaking work, Miller, Skoog, and their coworkers finally isolated the growth factor respon-

sible for cellular division from a DNA preparation. They named this substance *kinetin*, and they named the class to which kinetin belongs **cytokinins,** because these substances stimulated **cytokinesis,** or cellular division (fig. 18.11).

The first naturally occurring cytokinin was isolated in 1964 from corn (*Zea mays*) and was named *zeatin* (fig. 18.11). Soon thereafter, the influence of coconut milk on cellular division was explained when it was shown to contain zeatin and zeatin riboside, another cytokinin. Since then, botanists have isolated other naturally occurring cytokinins (e.g., dihydrozeatin and isopentenyl adenine) and have made several artificial cytokinins (e.g., kinetin and 6-benzylamino purine). All of these cytokinins have structures similar to adenine (fig. 18.11); that is, they have a side chain rich in carbon and hydrogen attached to nitrogen protruding from the top of the purine ring. Cytokinins are often minor components of RNA, but we do not know if these cytokinins are related to free cytokinins in cells.

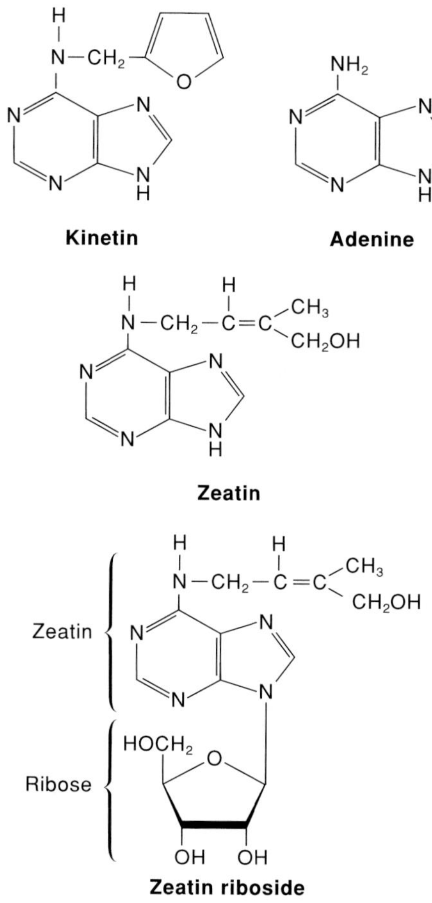

figure 18.11

Cytokinins. All cytokinins are structurally related to the purine adenine.

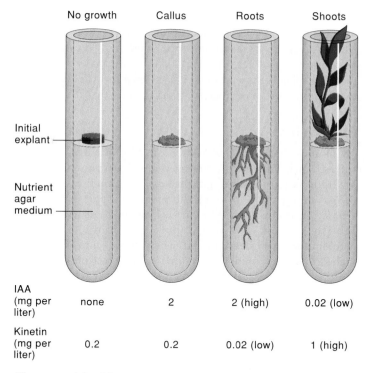

	No growth	Callus	Roots	Shoots
IAA (mg per liter)	none	2	2 (high)	0.02 (low)
Kinetin (mg per liter)	0.2	0.2	0.02 (low)	1 (high)

figure 18.12

The responses of plant tissue culture to kinetin and auxin. The initial explant is a small piece of sterile tissue cut from the pith of a tobacco plant. High ratios of auxin/cytokinin favor the formation of roots, whereas low ratios of auxin/cytokinin favor the formation of shoots.

Synthesis and Transport

Contrary to what was first believed, cytokinins are not the breakdown products of DNA. Rather, they are made via the mevalonate pathway, the same pathway used to make gibberellins. Cytokinins are widespread, if not universal, in plants: they have been isolated from angiosperms, gymnosperms, mosses, and ferns. In angiosperms, most cytokinins are made in roots and also occur in seeds, fruits, and young leaves. Cytokinins move nonpolarly in xylem, phloem, and parenchyma cells.

Effects of Cytokinins

Although cytokinins have relatively few uses in agriculture (e.g., turf-grass growers sometimes use them to promote root growth), they do strongly affect plant growth and development.

Cellular Division

Cytokinins stimulate cellular division by hastening the transition of cells from the G_2 phase (the growth phase following DNA replication) to the M phase (mitosis) of the cell cycle (see Chapter 9). This effect depends on the presence of auxin. For

example, cytokinins alone have no effect on cultured tobacco cells; cellular division begins only when auxin is added to the culture medium.

The discovery that cytokinins promote cellular division raised another interesting question: Could cytokinins be responsible for animal cancers, which result from uncontrolled cellular division? Research has demonstrated repeatedly that cytokinins have no effect on animal cells and that there is no direct link between animal cancers and cytokinins.

Effects on Cotyledons

Cytokinins promote cellular division and expansion in cotyledons. Cellular expansion results from cytokinin-induced increases in wall plasticity that do not involve wall acidification. Cytokinins also increase the amount of sugars (especially glucose and fructose) in cells, which may account for the osmotic influx of water and the resulting expansion of cytokinin-treated cells in cotyledons.

Organogenesis

Cytokinins and IAA affect **organogenesis**, which is the formation of organs. Figure 18.12 shows the influence of changing amounts of cytokinin and IAA on the formation of shoots and roots. Cultured cells grow only in the presence of cytokinin and

IAA. High cytokinin/auxin ratios favor the formation of roots, while low ratios favor the formation of shoots. Thus, a plant can be completely regenerated from single cells by varying the amounts of cytokinin and IAA. This hormonally controlled means of plant regeneration has been used to propagate plants that are resistant to pathogens, drought, and other stresses.

Senescence

Cytokinins delay the breakdown of chlorophyll in detached leaves, apparently by preventing genes that stimulate chlorophyll formation from being turned off. Cytokinin-treated areas of leaves remain healthy as the remaining parts of the leaf senesce. This effect of cytokinin may be due to its ability to establish a "sink" to which nutrients move. However, cytokinin-induced delay in leaf senescence occurs only in detached leaves; cytokinins have little or no effect on senescence in attached organs. Leaf senescence is also delayed by the formation of adventitious roots. Roots, you'll recall, are rich in cytokinins; the transport of these cytokinins to leaves could account for the delayed senescence.

Cytokinins are sometimes used commercially to maintain the greenness of excised plant parts, such as cut flowers. However, their use on edible crops such as broccoli is banned in the United States; any compound like cytokinin that resembles a nucleic acid component is automatically a suspected carcinogen.

c o n c e p t

Cytokinins are adeninelike compounds that stimulate cellular division, enlargement of cotyledons, and development of lateral buds. Cytokinins delay senescence of detached leaves and, with auxin, influence the formation of shoots and roots.

Cytokinins and Calcium

Small amounts of cytokinin do not stimulate cellular division of cultured cells, which merely enlarge due to the presence of auxin. However, adding calcium switches the growth pattern from cellular enlargement to cellular division. These results suggest that the presence of calcium increases a cell's sensitivity to cytokinin. Cytokinins also stimulate bud formation in mosses; substances that increase cells' permeability to Ca^{2+} mimic the effect of cytokinins, and cytokinin-induced formation of buds can be inhibited by applying inhibitors of calmodulin (see pp. 85–86). Similarly, the addition of calcium enhances the cytokinin-induced delay of leaf senescence. These effects may be mediated by cytokinin-induced changes in the concentration of Ca^{2+}.

Ethylene

Discovery

During the 1800s, the city streets of Germany were illuminated by lamps that burned "illuminating gas." Soon after these lamps were installed, city residents made a curious observation: plants growing near the lamps had shorter, thicker stems. Furthermore, the leaves fell off most of these plants. Were these effects caused by the lamp's light, heat, or some other factor?

This question was answered in 1901 when Soviet plant physiologist Dimitry Neljubow identified ethylene (C_2H_4) as the combustion product of "illuminating gas" that was responsible for defoliating and inhibiting the elongation of plants growing near the lamps. Neljubow also showed that these effects could be induced by as little as 0.06 parts per million (ppm) of ethylene (1 part per million = 1 milliliter in 1,000 liters = 1 minute in 2 years). Although botanists knew in 1901 that ethylene had many effects on plant growth, it would be more than three decades before they would learn that these effects were naturally occurring aspects of growth caused by ethylene made by plants.

In 1910 an annual report to the Japanese Department of Agriculture recommended that oranges not be stored with bananas, because oranges released something that caused premature ripening of the bananas. This "something" was not identified until 1934, when R. Gane showed that ethylene is made by plants and that it causes faster ripening of many fruits, including bananas. Subsequent research showed that ethylene met the requirements of a plant hormone; it is made in one part of a plant and transported to another, where it induces a physiological response. Thus was discovered the first gaseous plant hormone: ethylene.

Synthesis and Transport

Ethylene is made from methionine, an amino acid (fig. 18.13). Its synthesis is inhibited by CO_2 and requires oxygen. When plants are placed in pure CO_2 or O_2-free air, ethylene synthesis decreases dramatically.

All parts of angiosperms make ethylene, but especially large amounts are released into the air by roots, the shoot apical meristem, nodes, senescing flowers, and ripening fruits (e.g., the dark flecks on a ripening banana peel are patches of tissue that produce greater amounts of ethylene). Because most ethylene-induced effects result from ethylene in the air, the effects of ethylene can be "contagious." Ethylene made by one "bad" (overripe) apple *can* "spoil" (i.e., induce rapid ripening of) an entire bushel of apples.

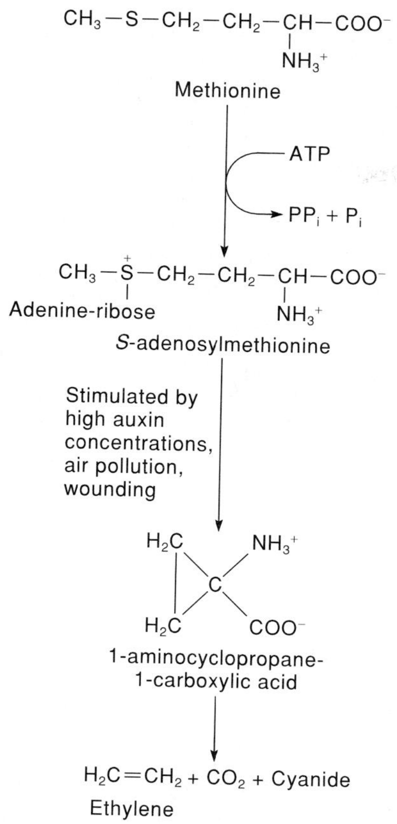

$$CH_3—S—CH_2—CH_2—CH—COO^-$$
$$\underset{NH_3^+}{|}$$

Methionine

ATP

$PP_i + P_i$

$$CH_3—\overset{+}{\underset{|}{S}}—CH_2—CH_2—CH—COO^-$$

Adenine-ribose $\qquad NH_3^+$

S-adenosylmethionine

Stimulated by
high auxin
concentrations,
air pollution,
wounding

$H_2C \qquad NH_3^+$

C

$H_2C \qquad COO^-$

1-aminocyclopropane-
1-carboxylic acid

$H_2C{=}CH_2 + CO_2 + Cyanide$

Ethylene

figure 18.13

Biosynthesis of ethylene from methionine, which serves as a precursor of ethylene in all higher plant tissues. The immediate precursor of ethylene is 1-amino-cyclopropane-1-carboxylic acid.

Effects of Ethylene

Fruit Ripening

The ancient Chinese knew that fruit would ripen faster if placed in rooms containing burning incense. The factor responsible for this hastened ripening was not heat but ethylene released as the incense burned. This stimulation of fruit-ripening by ethylene is multifaceted and includes the breakdown of chlorophyll and synthesis of other pigments (e.g., apples changing from green to red during ripening), fruit softening due to the breakdown of cell walls by cellulase and pectinase, production of volatile compounds associated with the scents of fruit, and conversion of starches and acids to sugars. Ethylene stimulates each of these aspects of fruit ripening.

In fruits such as tomatoes and apples, there is a conspicuous increase in respiration immediately before fruit ripening. This increase in respiration is called a **climacteric**, and fruits that display it are referred to as **climacteric fruits**. The climacteric begins just after a huge (up to 100-fold) increase in ethylene production. Thus, the climacteric and fruit ripening are triggered by ethylene. However, plants must be able to respond to ethylene for the ethylene to have an effect. The *Never-ripe* mutants of tomato—that is, mutants that are insensitive to ethylene—never ripen, even after being exposed to ethylene.

Fruit growers have often (and sometimes unknowingly) taken advantage of ethylene's stimulation of fruit ripening to have fruits available for sale out of season. For example, many apples are picked in September and October when they are green and immature. These apples are stored in rooms containing air that has small amounts (1%–3%) of O_2, large amounts (5%–10%) of CO_2, and no ethylene. Because large amounts of CO_2 and low levels of O_2 inhibit ethylene synthesis, the fruit can be stored without ripening. When these immature fruits are needed for sale, producers expose them to normal air containing about 1 ppm of ethylene. The ethylene induces the climacteric, and ripening soon occurs. Thus, the "fresh" apples you buy in March were probably harvested in September or October of the previous year. This "ripening on demand" is also used by producers of tomatoes, lemons, and oranges. Not all fruit can be ripened by exposure to ethylene, however. Such fruits (e.g., grapes) are called **nonclimacteric fruits** and are insensitive to ethylene.

Flowering

Ethylene inhibits flowering in most species but promotes it in a few plants, including mangos, pineapples, and some ornamental plants. This effect was known long ago by Puerto Rican pineapple growers and Filipino mango growers, who set bonfires near their crops. The fires produced ethylene, which initiated and synchronized flowering of their plants. In the 1950s, growers sprayed their crops with NAA, a synthetic auxin, which increased ethylene production and thereby stimulated flowering. Today, pineapple growers in Hawaii produce pineapple fruit all year by spraying plants with ethephon, a compound that breaks down in neutral and alkaline conditions to release ethylene.

Ethylene also promotes senescence of flowers. In plants such as orchids, the large amounts of auxin in pollen trigger ethylene synthesis in flowers; this ethylene triggers senescence of the flower parts.

Abscission

The increased production of ethylene at the abscission zone triggers the breakdown of the middle lamella, and thereby

LEAF ABSCISSION AND THE EXTERNAL FEATURES OF WOODY TWIGS

Twigs of deciduous woody plants such as horse chestnut (shown in reading fig. 18.3) have several important features, including buds. There are three main types of buds: (1) terminal buds at the tips of the twigs, (2) axillary buds in the axils of leaves, and (3) accessory buds, which occur in some species in pairs, one each on either side of axillary buds. In some species, accessory buds produce flowers and the axillary bud produces a leafy branch; in other species, the accessory buds do not develop if their associated axillary bud develops normally.

Leaf abscission leaves behind leaf scars (formed by the abscission zone) with their bundle scars. Since axillary buds form in the leaf axil above the leaf, leaf scars occur just below axillary buds. Bundle scars are the severed ends of vascular bundles that went from the stem to the leaf. The portion of stem between groups of bud-scale scars represents one year's growth.

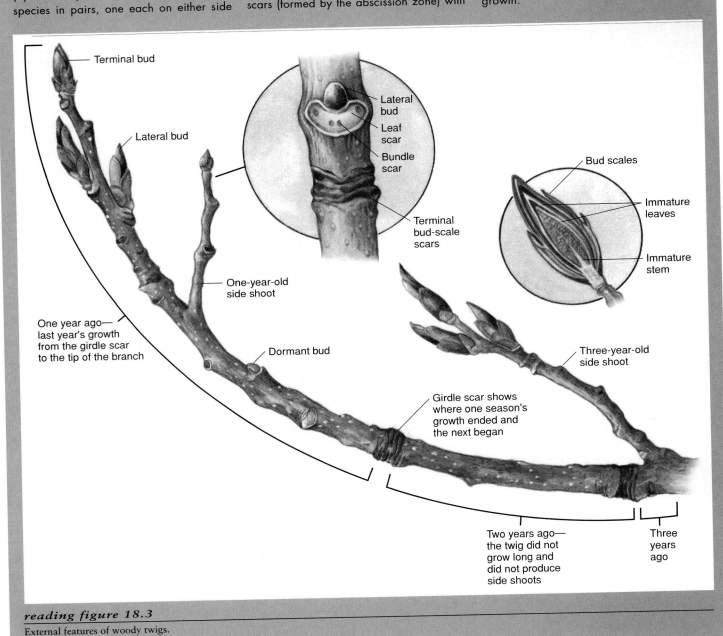

reading figure 18.3

External features of woody twigs.

18–22

Leaf abscission is strongly influenced by hormones, especially ethylene.

figure 18.15

Tomato (*Lycopersicon esculentum*) plants exposed to 28 consecutive days of brief, periodic shaking. Left: unstressed control; center: 30 seconds of stress once per day; right: 30 seconds of stress twice per day. Mechanical disturbances such as shaking decrease elongation.

initiates abscission (fig. 18.14; also see fig. 14.25). Like the influence of ethylene on flowering, this effect is used by horticulturists to increase the efficiency of harvesting fruit. Fruits such as cherries, grapes, and blueberries are sprayed with ethephon to coordinate abscission, thereby allowing growers to harvest crops in shorter periods of time.

Sex Expression

The sex of flowers on monoecious plants (i.e., plants that have male and female flowers on the same individual) is determined by ethylene and gibberellins. For example, cucumber (*Cucumis sativus*) buds treated with ethylene become female flowers, whereas those treated with gibberellins become male flowers. Buds that become female flowers produce more ethylene than do buds that become male flowers.

Stem Elongation

Mechanical disturbances such as shaking decrease elongation (fig. 18.15). This effect, called **thigmomorphogenesis,** is influenced by ethylene. Mechanical disturbances increase ethylene production severalfold, causing cells to arrange their cellulose microfibrils into longitudinal hoops. This lengthwise reinforcement inhibits cellular elongation, causing cells to expand radially and form shorter, thicker stems. This effect is opposite from that of auxin, which causes cells

to orient their microfibrils more transversely, thereby increasing cellular elongation.

Waterlogging

When a plant's roots are submerged for long periods, water eventually fills the intercellular spaces. Since these spaces are the primary routes of gas exchange with the atmosphere, the submerged roots become waterlogged and anaerobic. As you might predict, the symptoms of waterlogged plants (e.g., leaf chlorosis, shorter and thicker shoots, and wilting) can be induced by placing roots in oxygen-free air.

Because oxygen is needed to make ethylene, ethylene synthesis is greatly reduced in roots of waterlogged plants. The small amount of ethylene that is made in these roots is trapped, where it accumulates and eventually stimulates the activity of cellulase and pectinase. These enzymes break down the cell walls and, in doing so, form the many intercellular spaces characteristic of hydrophytes. Meanwhile, ethylene precursors in the shoot are converted to ethylene, which causes parenchyma cells on the upper side of the petiole to expand and point the leaf down, a response called **epinasty** (fig. 18.16).

Cell Differentiation

Ethylene may be a diffusible signal that regulates the differentiation of some types of cells. For example, root-hair formation is dramatically reduced in ethylene-insensitive mutants and by treatment with inhibitors of ethylene biosynthesis. Similarly, a root-hairless mutant of *Arabidopsis* forms hairs when exposed to ethylene.

figure 18.16

Leaf epinasty caused by ethylene. The *Coleus* plant on the right was exposed to ethylene for 2 days, whereas the plant on the left was untreated.

Ethylene Sensors

Plants sense ethylene via a protein kinase cascade. At least two genes are essential for ethylene signaling; these genes (*CTR1* and *ETR1*) code for two protein kinase sensors (recall from Chapter 4 that protein kinases are enzymes that activate or inactivate other enzymes by transferring phosphate groups to and from them). *ETR1* is the ethylene receptor in plants. Without functional *ETR1*, plants do not bind ethylene effectively. Another ethylene receptor, *ERS*, was discovered in 1995; the lack of ripening in the *Never-ripe* mutants of tomato results from a mutation in an *ERS*-like gene product. Interestingly, *ETR1* is expressed during flower and fruit senescence, while *ERS* is developmentally regulated during fruit ripening. Taken together, these results suggest that ethylene sensors are encoded by multigene families with members that are expressed differentially during plant growth and development.

c o n c e p t

Ethylene is a gaseous signaling hormone that stimulates fruit ripening, promotes abscission, affects sex expression, and regulates stem elongation.

Ethylene and Wounding

Wounding and stress such as chilling trigger the production of "wound ethylene." This production of ethylene is adaptive because the ethylene causes the senescence or abscission of the wounded part of the plant. This, in turn, removes the part from the plant's metabolic activities, thereby conserving resources (so they are not sent to a nonfunctional part of the plant) and preventing the entry of pathogens through damaged tissues.

Ethylene and Auxin

We mentioned earlier that high levels of IAA stimulate ethylene production, thereby linking the responses of these two hormones. But ethylene does not account for all of the effects elicited by applying IAA. Several responses of plants to IAA are unrelated to ethylene. For example, IAA's stimulation of cellular elongation and the production of lateral roots occur independently of ethylene. Conversely, leaf epinasty, decreased elongation of roots and shoots, and sex determination are responses to ethylene rather than to IAA.

Abscisic Acid

Discovery

The exciting discoveries of auxin, cytokinins, ethylene, and gibberellins offered new prospects for explaining plant growth and development. Most of the effects first discovered for plant hormones were stimulatory. For example, IAA stimulates cellular elongation, and cytokinins stimulate cellular division. But by the 1940s, botanists suspected that some aspects of plant growth and development resulted from inhibition rather than stimulation of growth. Near the end of that decade came a hint that this suspicion was correct: Torsten Hemberg of Sweden reported that dormant buds of ash and potato contained inhibitors that blocked the effects of IAA. When the buds germinated, the amount of these inhibitors decreased. Hemberg named these inhibitors *dormins*, because of their apparent influence on the dormancy of buds.

In the early 1960s, Philip Wareing confirmed Hemberg's findings and reported that applying dormin to a bud induced dormancy. At about the same time, F. T. Addicott discovered a compound that stimulated abscission of cotton fruit. He named this substance *abscisin*. Botanists later were surprised to discover that dormin and abscisin were the same compound. This compound was named **abscisic acid,** or **ABA**—an unfortunate name, because subsequent research has shown that ethylene rather than ABA controls abscission.

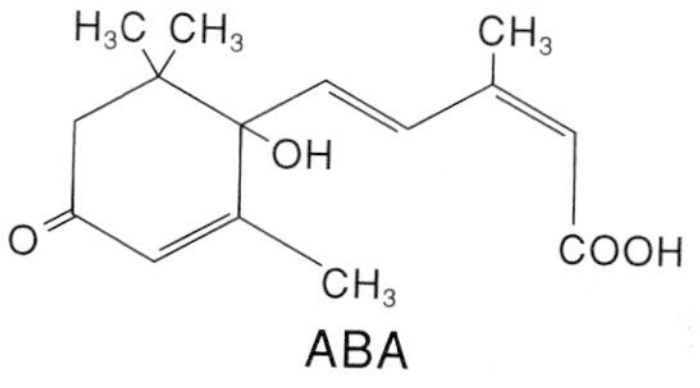

figure 18.17

The structure of abscisic acid (ABA).

Synthesis and Transport

ABA in plants is made from carotenoids (fig. 18.17). It occurs in angiosperms, gymnosperms, and mosses but apparently not in liverworts. Once synthesized, ABA moves throughout a plant in xylem, phloem, and parenchyma. Like gibberellins and cytokinins, ABA moves nonpolarly. There are no synthetic abscisic acids.

Effects of Abscisic Acid

Closure of Stomata

During drought, leaves make large amounts of ABA, which causes stomata to close. Thus, ABA functions as a messenger that enables plants to conserve water during drought. Because ABA closes stomata within 1–2 minutes, this effect probably occurs independently of protein synthesis. ABA probably exerts its effect by binding to proteins on the outer surface of the plasmalemma of guard cells. This makes the plasmalemma more positively charged, thereby stimulating transport of ions (especially K^+) from guard cells to epidermal cells. Because the loss of these ions causes water to leave guard cells via osmosis, the guard cells shrink, thereby closing the stomatal pore.

A plant's tolerance to drought may relate to its ability to make ABA. In some species, the levels of ABA increase by as much as tenfold within minutes after wilting occurs.

Bud Dormancy

ABA was initially thought to control bud dormancy, but recent evidence questions this conclusion. For example,

- After leaves are treated with radioactive ABA, no radioactivity can be detected in buds.

- In several plants, the induction and breaking of dormancy do not correlate with changes in endogenous amounts of ABA.

- Treatments that induce dormancy do not alter the amounts of ABA in buds.

These results suggest that bud dormancy is not controlled by ABA alone. Instead, dormancy is also probably influenced by cytokinins and IAA-induced synthesis of ethylene.

Seed Dormancy

Applying ABA delays seed germination in many species. Similarly, the amount of ABA in the seeds of many plants decreases when seeds germinate. Thus, ABA may control seed dormancy in certain species.

ABA Counteracts the Stimulatory Effects of Other Hormones

ABA usually inhibits the stimulatory effects of other hormones. For example, ABA (1) inhibits amylase produced by seeds treated with gibberellin, (2) promotes chlorosis that is inhibited by cytokinins, and (3) inhibits wall elasticity and cell growth promoted by IAA. How does ABA counteract these effects? One suggestion is that ABA is a Ca^{2+} antagonist; thus, its inhibition of the stimulatory effects of IAA and cytokinin may be due to its interference with Ca^{2+} metabolism.

Although ABA usually inhibits growth, it is not toxic to plants, as are inhibitors of energy metabolism and RNA/protein synthesis. Although ABA often decreases gene activity, there are instances of ABA stimulating genes. For example, ABA stimulates the synthesis of mRNAs for storage proteins in developing wheat grains.

c o n c e p t

Abscisic acid causes stomata to close, influences dormancy of some seeds, and generally inhibits the stimulatory effects of other hormones. These effects may involve changes in the plant's calcium metabolism.

Oligosaccharins and Other Plant Hormones

A growing number of reports indicate that plant hormones are not limited to auxin, gibberellins, cytokinins, ethylene, and abscisic acid. For example, fragments of cell walls called **oligosaccharins** influence plant growth, differentiation, reproduction, and defense against disease, and therefore function as plant hormones. However, oligosaccharins differ from other plant hormones because they elicit specific effects: different oligosaccharins can induce cultured cells to form undifferentiated callus, roots, shoots, or flowers in a variety of plants. Moreover, oligosaccharins that inhibit flowering and promote vegetative growth in one species have the same effects on other species. Furthermore, these effects are induced by impressively small amounts of the oligosaccharin: it takes 100 to 1,000 times less oligosaccharin than IAA or cytokinin to induce a response.

Oligosaccharins are released from cell walls by enzymes. Different oligosaccharins are released by different enzymes, and each oligosaccharin transmits a message that regulates a particular function. This specificity has prompted several botanists to suggest that the many effects of hormones like IAA and gibberellins may be due to the activation of enzymes that release specific oligosaccharins.

Other investigations have identified a variety of other compounds that function as hormones in various groups of plants. For example,

- Yams contain **batasins,** which induce dormancy of bulbils (vegetative reproductive structures) that form from lateral bulbs. Increased amounts of batasins induce dormancy, and cold treatments that break dormancy reduce the amounts of batasins.

- **Brassinosteroids** are steroid hormones that occur throughout the plant kingdom. These hormones may regulate the expression of some genes in light (such as the expression of light-regulated genes in photosynthesis), the promotion of cellular elongation, the senescence of leaves and chloroplasts, and the promotion of flowering.

Lore of Plants

Although hormones such as auxin are common signals in plants, so are pathogens. Indeed, a localized attack by a pathogen or pest causes a signal to spread systematically throughout the plant and induce the expression of defense genes. This long-lasting, broad-spectrum response is referred to as a *systemic acquired resistance* and is mediated by salicylic acid. Salicylic acid increases throughout a plant after only one part of the plant is attacked, and application of salicylic acid or certain analogs such as aspirin (acetylsalicylic acid) induces the rapid expression of pathogenesis-related genes acid (for more about plants and aspirin, see table 2.2). Moreover, plants unable to make salicylic acid cannot induce systemic acquired resistance and are susceptible to viral, fungal, and bacterial pathogens. The receptor for salicylic acid shares sequence identity with catalases. Catalases are enzymes that reduce and effectively inactivate hydrogen peroxides, a type of reactive oxygen compound.

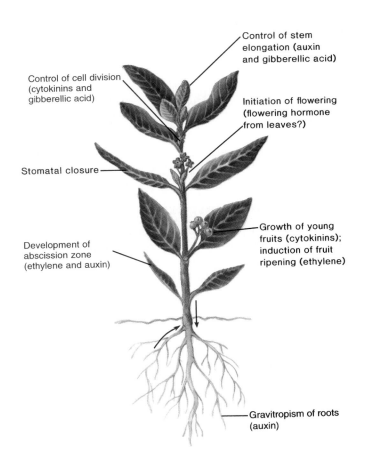

figure 18.18

Numerous hormonal interactions influence plant growth and development.

Controlling the Amounts of Plant Hormones: Hormonal Interactions

Hormones seldom, if ever, function alone; rather, plant growth and development usually result from interactions of plant hormones (fig. 18.18). We've already discussed several of these interactions, including those responsible for organogenesis and vascular differentiation. But what controls the amounts of hormones and thereby their ratios and resulting interactions? The amounts of hormones are controlled in two ways:

1. *Regulation of the rate of synthesis.* Several factors influence the rate of hormone production. For example, daylength can trigger the synthesis of IAA, and the synthesis of gibberellins in biennials is stimulated by cold temperatures.

18–26

2. *Regulation of the rate of breakdown or inactivation.* Inactivation of a hormone usually occurs either by oxidizing the hormone or by conjugating (i.e., combining) it with another compound. Coordinated synthesis and inactivation could control the amount of hormone present and thereby control the growth response.

Do Plant Hormones Really Exist?

Since plant hormones were discovered after animal hormones, most early research was based on the assumption that plant and animal hormones had fundamentally similar effects. However, there are important differences in plant and animal hormones:

- There is no evidence that the fundamental actions of plant and animal hormones are the same.

- Unlike animal hormones, plant hormones are not made in tissues specialized for hormone production.

- Animal hormones usually have specific effects, while those of plants seldom, if ever, have specific effects.

Despite these differences, botanists have traditionally suggested that plant hormones function like animal hormones; that is, the response elicited by a plant hormone is determined by the concentration of the hormone. Certain predictable responses often *can* be obtained by applying hormones to various parts of plants, but this does not necessarily mean that such responses are naturally controlled by the hormone. As a result, some botanists now question whether the concept of hormones applies to plants as it does to animals. These botanists—who emphasize these differences by referring to plant hormones as *plant growth regulators*—claim that the traditional view of plant hormones fails to explain plant growth and development. Rather, they suggest that plant hormones are integrating agents that are *necessary* for, but do not *control*, the response. For example, cytokinins regulate the transformation of buds from a dormant to nondormant state but do not control the subsequent growth of the bud. Thus, cytokinins are necessary for bud growth but do not control it. According to this perspective, the response elicited by a hormone is determined not by the *amount* of hormone but rather by the *sensitivity* of the tissue to the hormone.

Who's right? What *does* determine the response elicited by a hormone: its concentration or the tissue's sensitivity? There are several examples in which the amount of hormone determines the response. For example, inhibition of root growth is roughly proportional to the amount of IAA in the root. Conversely, other examples suggest that the sensitivity of a plant to a hormone changes with different environments and

developmental stages. For example, the sensitivity of corn coleoptile segments to IAA changes more than twofold during the first 3 hours after excision. Also, the responses characteristic of a particular hormone often occur without an accompanying increase in the amount of the hormone. For example, applying IAA to the elongating region of stems stimulates growth, but an increase in the amount of IAA has not been detected in rapidly growing stems.

c o n c e p t

Hormones are necessary for, but do not control, many aspects of plant growth and development. Plant growth and development are probably controlled by the amount of hormone present and the tissue's sensitivity to the hormone.

Chapter Summary

Plant hormones are organic compounds made in one part of a plant and transported to another part of the plant, where they cause a response. Plant hormones are active at small concentrations. Plant hormones and the responses that they elicit have the following characteristics:

- Although a hormone may have some characteristic effects, it also has many other effects. That is, a single hormone can elicit many different responses.

- The responses elicited by a hormone depend on many factors, including the presence of other hormones, the amount of hormone present, nonhormonal factors, and the sensitivity of the tissue to the hormone.

- Hormonal responses change under different conditions and in different plants.

- A single aspect of growth and development can be influenced by several hormones.

- Responses elicited by plant hormones probably result from changing ratios of hormones rather than from the presence or absence of any one hormone.

The five major classes of plant hormones are auxin, gibberellins, cytokinins, ethylene, and abscisic acid.

Auxin stimulates cellular elongation, differentiation of vascular tissue, fruit development, formation of adventitious roots, and production of ethylene. The most active naturally occurring auxin is indole-3-acetic acid (IAA), which is transported polarly in roots and stems. Synthetic auxins are used extensively in modern agriculture.

Gibberellins stimulate extensive growth of intact plants, the transition from juvenile to adult growth, bolting of biennials, fruit formation, and germination of some cereal grains.

Cytokinins stimulate cellular division, expansion of cotyledons, and growth of lateral buds. Cytokinins also delay senescence of detached leaves and, in combination with IAA, may influence formation of roots and shoots.

Ethylene is a gaseous hormone that influences fruit ripening, abscission, sex expression, and the radial expansion of cells. Ethylene also functions as a "wound hormone."

Abscisic acid (ABA) is an inhibitor that causes stomata to close, affects dormancy of some seeds, and, in general, counteracts the stimulatory effects of other hormones. These effects may occur because ABA is a calcium antagonist.

Oligosaccharins are fragments of cell wall that regulate plant growth and development. Unlike other plant hormones, oligosaccharins have specific effects.

Many botanists think that plant hormones are necessary for, but do not control, plant growth and development. According to this perspective, a plant's response to a hormone is not determined by the amount of hormone present but rather by the sensitivity of the tissue to the hormone.

 ### *What Are Botanists Doing?*

Go to the library and read a recent article about any of the plant growth-regulating substances mentioned in this chapter. Summarize the importance of the article. How does what you learned from the article relate to what you learned from this chapter?

 ### *Writing to Learn Botany*

What is the selective advantage to a plant of shedding its leaves? Its fruit?

Questions for Further Thought and Study

1. Calcium is often referred to as a "second messenger" in plant growth and development. What does this mean?

2. Calcium influences several effects of plant hormones. Why, then, isn't calcium considered a plant hormone?

3. Abscisic acid is a calcium antagonist. Discuss how this property could be the basis for inhibition by ABA of the stimulatory effects of other hormones.

4. How are oligosaccharins different from other plant hormones?

5. Some botanists refer to ABA as a "plant tranquilizer." Is this a good analogy? Why or why not?

6. Why are immature fruits stored in 1%–3% oxygen instead of 0% oxygen?

7. How might ethylene-producing fungi in soil affect plant growth and development?

8. Provide examples for each of the following generalizations about plant hormones:

   (a) A single plant hormone can produce many effects.
   (b) The effects elicited by a hormone depend on many factors, including the presence of other hormones.
   (c) Hormonal responses probably result from changing ratios of hormones rather than from the presence or absence of an individual hormone.

9. How is the transport of IAA different from that of other hormones?

10. What is the adaptive significance of the fact that auxin produced by young leaves influences vascular differentiation into the shoot apex?

11. What is the difference between a hormone and an enzyme?

12. Gibberellins, cytokinins, and IAA promote growth. How would you distinguish them from each other?

13. Morphactins are substances that influence plant growth and development by interfering with the transport of auxin. How would morphactins affect phototropism of coleoptiles?

14. The same concentration of auxin that stimulates formation of adventitious roots will subsequently stop the growth of these roots unless it is significantly reduced. Explain the significance of this observation.

Web Sites

Review the "Doing Botany Yourself" essay and assignments for Chapter 18 on the *Botany Home Page*. What experiments would you do to test the hypotheses? What data can you gather on the Web to help you refine your experiments? Here are some other sites that you may find interesting:

http://www.lars.bbsrc.ac.uk/

At this site you can learn how botanists in Japan used studies of the "foolish seedling" disease to isolate and identify gibberellic acid.

http://www.life.uiuc.edu/bio100/lessons/vegetable-growth.html

Here you'll learn about processes influenced by plant hormones, read discussions of how those hormones work, and review the major actions of plant hormones as well as experiments that show the effects of hormones.

http://www.plant-hormones.bbsrc.ac.uk/

This site is devoted to people interested in plant hormones. Here you'll find information about hormones, databases, links to other hormone-related sites, and more.

http://www.lars.bbsrc.ac.uk/plantsci/dormancy/dormancy.html

This site discusses the genetic basis of seed dormancy and its importance in agriculture.

Suggested Readings

Articles

Cline, M. 1991. Apical dominance. *The Botanical Review* 57:318–358.

Evans, M. L. 1984. The action of auxin on plant cell elongation. *CRC Critical Reviews in Plant Sciences* 2:317–365.

Huddart, H., R. J. Smith, P. D. Langton, A. M. Hetherington, and T. A. Mansfield. 1986. Is abscisic acid a universally active calcium antagonist? *New Phytologist* 104:161–173.

Kende, H. 1993. Ethylene biosynthesis. *Annual Review of Plant Physiology and Plant Molecular Biology* 44:283–308.

Russell, David W. 1996. Green light for steroid hormones. *Science* 272:370–371.

Ryan, C. A., and E. E. Farmer. 1991. Oligosaccharide signals in plants: A current assessment. *Annual Review of Plant Physiology and Plant Molecular Biology* 42:651–674.

Sisler, E. C., and S. F. Yand. 1984. Ethylene, the gaseous hormone. *BioScience* 33:233–338.

Theologis, A. 1995. Ethylene sensors: How perceptive! *Science* 270:1774.

Trewavas, A. 1991. How do plant growth substances work? *Plant, Cell and Environment* 14:1–12.

Books

Addicott, F. T., ed. 1983. *Abscisic Acid*. New York: Praeger.

Galston, A. W. 1994. *Life Processes of Plants*. New York: W. H. Freeman and Co.

Salisbury, F. B., and C. W. Ross. 1992. *Plant Physiology*. 4th ed. Belmont, CA: Wadsworth.

Wareing, P. F., and I. D. J. Phillips. 1981. *Growth and Differentiation in Plants*. 3d ed. Elmsford, NY: Pergamon Press.

Flowers of a lily (*Lilium superbum*). What controls flowering in plants?

How Plants Respond to Environmental Stimuli

19

Chapter Outline

Chapter Overview

Plant growth and development are strongly influenced by the environment. Some responses of plants to stimuli such as light, gravity, and touch are rapid; other responses, such as flowering, are relatively slow and are associated with changes of seasons. Regardless of their duration, most responses of plants to environmental signals are due to growth and are controlled, at least in part, by hormones. The coupling of plant growth and development with environmental signals determines the seasonality of growth, reproduction, and dormancy, ultimately fostering survival of the plant and the species in a given environment.

Plants have evolved an elaborate set of intricate and dramatic responses to the environment. These responses allow plants not only to survive adverse conditions that would kill most animals but also to coordinate their growth and development with appropriate environmental conditions. The degree to which plants rely on environmental cues to orchestrate development is unparalleled in the animal kingdom.

Responses of plants to environmental stimuli such as light, gravity, and touch occur in many ways and include such diverse events as flowering, the growth of stems toward light, and the trapping of food by carnivorous plants. Some of these behaviors are short-term responses. For example, a Venus's-flytrap takes less than a second to close, and curvature of a stem toward light is usually completed within a few hours. Other behaviors, such as flowering, take longer and are typically associated with changing seasons. Environmental signals trigger all of these seemingly unrelated responses: stems grow toward light, and rapid closure allows a Venus's-flytrap to feed on insects. The key element in these responses is *growth;* with few exceptions, plants respond to environmental stimuli by growing. This growth produces a variety of responses, among the most obvious of which are tropisms.

Tropisms

Plant growth toward or away from a stimulus such as light or gravity is called a **tropism**. There are several kinds of tropisms, each of which is named for the stimulus that causes the response. For example, phototropism is the growth of a stem or coleoptile toward light, and gravitropism is growth toward or away from gravity. Tropisms result from *differential growth,* meaning that one side of the responding organ elongates faster than the other side. This causes curvature of the organ toward or away from the stimulus. Growth of an organ toward a stimulus is called a *positive tropism*. Thus, stems that grow toward light are positively phototropic. Conversely, growth of an organ away from a stimulus is called a *negative tropism*. Roots, which grow away from light, are negatively phototropic.

Phototropism

Phototropism is a tropic response to unidirectional light (light coming from one direction; see fig. 19.1). As you learned in Chapter 18, Frits Went determined that during phototropism, the rapid elongation of cells along the shaded side of coleoptiles is controlled by IAA coming from the apex. Went formulated his hypothesis jointly with Russian plant physiologist Nicolai Cholodny, and this work became known as the Cholodny-Went hypothesis. Although this hypothesis was a major step in our understanding of phototropism, the mechanism by which IAA controlled phototropism remained a mystery. Either of two hypotheses could have accounted for the increased activity of IAA on the shaded side of the coleoptile:

Hypothesis 1

Light destroys IAA along the lighted side of the coleoptile.
Such a mechanism would result in more IAA along the shaded side of the coleoptile, which would account for phototropic curvature. **The evidence:** In the 1950s, Winslow Briggs and his colleagues determined that the amount of IAA produced by coleoptiles grown in light is the same as that made by coleoptiles grown in the dark (fig. 19.2). Thus, light does not destroy IAA.

Hypothesis 2

Light causes IAA to move to the shaded side of the coleoptile.
According to this hypothesis, the different concentrations of IAA in the lighted and shaded sides of the coleoptile would result from the movement of IAA rather than its destruction. **The evidence:** Briggs and his colleagues collected more IAA from the shaded side of coleoptiles than from the lighted side, suggesting that light causes IAA to move to the shaded side of the stem. This conclusion was confirmed by inserting impermeable barriers between the split tips of coleoptiles. These barriers blocked the movement of IAA to the shaded side of the coleoptiles (fig. 19.2). As a result, the coleoptiles did not curve toward the light. More recent experiments using IAA labeled with radioactive carbon (^{14}C) have confirmed that unidirectional light causes IAA to move to the shaded side of coleoptiles, where its increased concentration causes cells to elongate more rapidly than cells on the lighted side of the coleoptiles. As a result, the coleoptile curves toward the light.

Although the Cholodny-Went hypothesis adequately explains phototropism in coleoptiles, it often fails to account for phototropism in stems, which contain more chlorophyll than do coleoptiles. In stems, unidirectional light triggers production of an inhibitor that slows cellular elongation on the illuminated side of the stem. Because cellular elongation does not change on the shaded side of the stem, the stem curves toward the light (also see reading 19.1, "Tracking the Sun").

How does a plant sense unidirectional light? Botanists have determined that only blue light with wavelengths less than 500 nm effectively induces phototropism (fig. 19.3). Although flavoproteins (see Chapter 2) are probably the blue-light receptors for stomatal opening and the inhibition of stem elongation, a xanthophyll named zeaxanthin is probably the blue-light photoreceptor for phototropism. This pigment may alter the rate of transport of IAA, thereby speeding its movement to the shaded side of the stem or coleoptile.

c o n c e p t

Phototropism is a growth response to unidirectional light. Phototropism is probably influenced by IAA and results in stems growing toward light and roots growing away from light. This important response increases the probability that stems and leaves will intercept light for photosynthesis.

figure 19.1

Phototrophic curvature of corn shoots toward light. Unilateral light causes IAA to move to the shaded side of the stem. This IAA stimulates growth on the shaded side, causing the shoot to curve toward the light.

figure 19.2

Briggs's experiments (a) and (b) (done in the dark) showed that splitting the tip of a coleoptile did not significantly affect the total amount of auxin that was diffusing from the tip into the agar block. The degree of curvature produced when the agar block is applied to the side of a decapitated coleoptile is directly related to the amount of auxin produced and is shown by the numbers below the agar blocks. Comparing (a) and (b) with (c) and (d) shows that auxin production is not dependent on light. If a barrier is partially inserted into the tip, auxin is displaced from the lighted to the shaded side as shown in (e). Experiment (f) shows that actual displacement did occurr; the results shown in figure (e) do not reflect different rates of auxin production on the lighted and shaded sides.

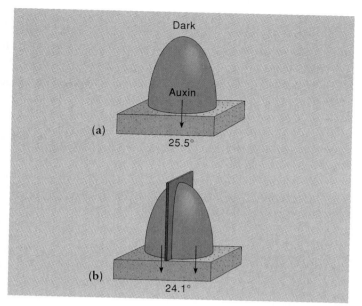

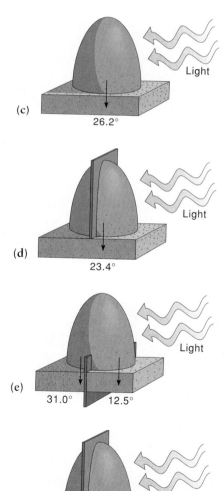

TRACKING THE SUN

One of the most unusual and adaptive movements in plants is solar tracking, also known as heliotropism (from the Greek word *helios*, meaning "sun"). Solar tracking is how the organs of plants track the sun across the sky, much as a radio telescope tracks a satellite (reading figs 19.1a & b). Like several other plant movements, solar tracking is caused by turgor changes in motor cells in pulvini located at the base of leaves and leaflets.

The tracking movements of leaves depend on environmental and physiological conditions. For example, the so-called compass plant (*Silphium laciniatum*) orients its leaves parallel to the sun's rays, thereby de-creasing leaf temperature and minimizing desiccation. Other plants often orient their leaves perpendicular to the sun's rays, thereby increasing the amount of light intercepted by the leaf for photosynthesis. Finally, some plants orient their leaves more obliquely to the sun when it is hot than when it is cooler. This variation in solar tracking helps keep the leaf temperature near the optimal temperature for photosynthesis.

Solar tracking occurs in many plants (e.g., cotton, alfalfa, beans, and soybeans) and is not restricted to leaves. What happens on cloudy, overcast days? On these days, leaves are oriented horizontally in their "rest-ing" position. If the sun appears from behind the clouds late in the day, leaves rapidly reorient themselves: they can move up to 60 degrees per hour, which is four times faster than the movement of the sun across the sky.

Solar tracking is controlled not only by the sun's position; leaves begin orienting themselves toward the direction of sunrise several hours before sunrise. How plants "remember" the position of the last sunrise is a mystery, but IAA may be involved; IAA from leaves can alter turgor of motor cells. Whatever the mechanism, plants are fast learners: only four sunrises are needed to entrain solar tracking.

reading figure 19.1a

Sunflowers (*Helianthus annuus*) oriented in the same direction, toward the sun. *Helianthus* means "sunflower."

reading figure 19.1b

Time-lapse photograph of a buttercup (*Ranunculus ficaria*) tracking a source of light.

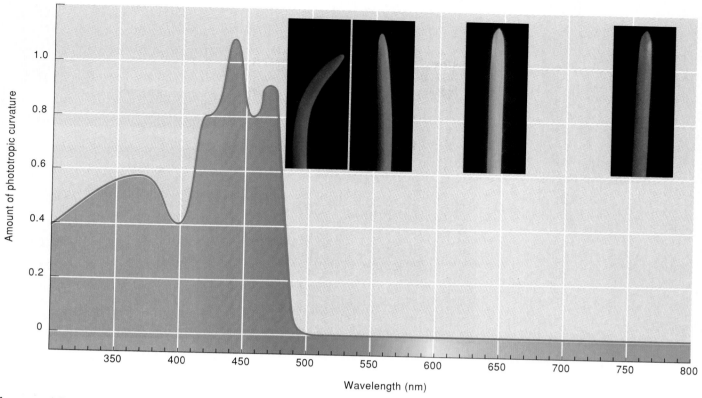

figure 19.3

Sensitivity of phototropism to different wavelengths of light. Phototropism is most sensitive to blue light having wavelengths less than 500 nm.
Source: Data from M. Wilkins, Plantwatching: How Plants Remember, Tell Time, Form Partnerships, and More, *Macmillan Publishing Company, 1988.*

Gravitropism

Charles Darwin was a phenomenally productive scientist. In addition to his landmark investigations of evolution, he and his son Francis also studied **gravitropism,** which is a growth response to gravity. One of their first experiments involved the responses of roots whose caps had been surgically removed. To the Darwins' surprise, these decapped roots continued to grow but did not respond to gravity. These clever experiments suggested that the root cap is necessary for root gravitropism.

Other researchers, following the Darwins' lead, began studying the root cap in hopes of understanding its strong influence on gravitropism. They soon made an exciting discovery: cells in the center of the cap contain numerous starch-laden amyloplasts that, under the influence of gravity, sink to the lower side of the cells (fig. 19.4). Could this gravity-dependent sedimentation of amyloplasts be how plants sense gravity? Several subsequent experiments suggested that this was true, which further intensified the study of amyloplasts as gravity sensors in plant roots. Despite over 50 years of intensive research, however, no convincing explanation has emerged for how the sedimentation of amyloplasts in the root cap could induce differential growth in the elongating zone of the root, which is 2–6 mm behind the root cap.

Studies by Timothy Caspar and his coworkers at Michigan State University suggest that research on amyloplasts as gravity

sensors in roots may have been misguided. Caspar took a unique approach to investigating the problem: he reasoned that a sure way to test if amyloplasts are gravity sensors in root caps would be to study gravitropism in a plant lacking amyloplasts. Although this reasoning was sound, it proved to be a formidable task, since all species that had been studied had amyloplasts in the cells of their root caps. To get around this problem, Caspar engineered his own experimental plant: a mutant of *Arabidopsis thaliana* that lacked starch and so also lacked amyloplasts (fig. 19.5; for more about *Arabidopsis,* see reading 9.3, "An Important Little Weed"). With this plant, the critical question and experiment became rather simple: do roots of mutant seedlings respond to gravity, even though they lack amyloplasts? When Caspar oriented roots of the mutant horizontally, they curved downward, indicating that amyloplasts are not necessary for root gravitropism. These results question the long-held assumption that amyloplasts constitute the gravity-sensing apparatus in roots.

More recent research by Randy Wayne and his colleagues at Cornell University suggests that roots respond to gravity by sensing gravitational pressure exerted by the protoplast, not because of the sedimentation of amyloplasts. Proteins at the interface between the cell membrane and cell wall may be required for sensing gravity. (The research of Cora Schmid, an undergraduate student who worked with Randy Wayne, is described on p. 449.)

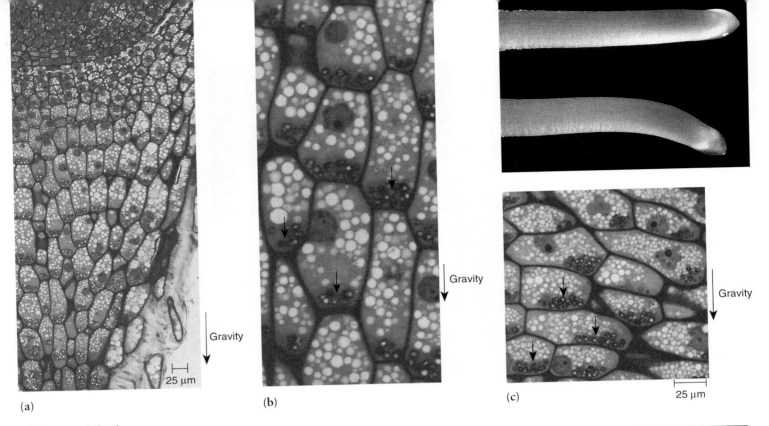

figure 19.4

The root cap and root gravitropism. (a) Longitudinal section of a root cap of corn. Cells in the center of the cap contain numerous starch-laden amyloplasts located in the lower part of the cells. (b) Magnified view of (a). Arrows indicate amyloplasts that have sedimented to the bottoms of the cells. (c) These amyloplasts (arrowed in lower photo) quickly sediment to the lower sides of cells when roots are oriented horizontally. This sedimentation of amyloplasts was long thought to influence how roots perceive gravity (upper photo).

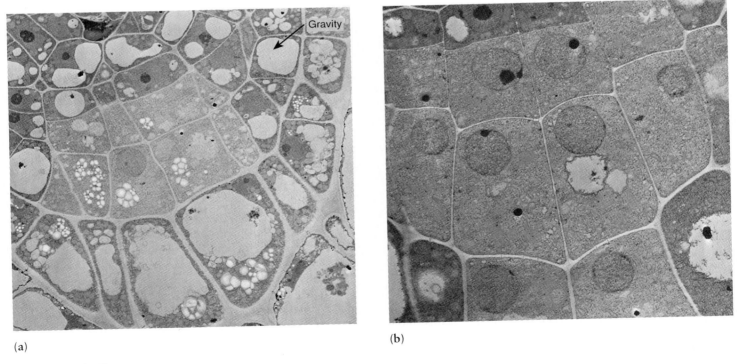

figure 19.5

Using mutants to understand how roots perceive and respond to gravity. These are electron micrographs of (a) normal (i.e., starchy), ×1,500, and (b) mutant (i.e., starchless) cells, ×3,000, of root caps of *Arabidopsis thaliana*, a relative of mustard. Roots of normal and mutant seedlings are graviresponsive, thus prompting many botanists to question the importance of root cap amyloplasts in root gravitropism.

Cora Schmid

If you have ever put a plant on a windowsill, you know that plants bend toward light. This phenomenon, called phototropism, is a plant's way of increasing the light it receives. This is important for a plant to do, because light is its energy source. The process that causes this bending is complex. The plant must first detect the direction of light and then transduce this information into a response that leads to bending. Although scientists have been studying phototropism ever since the great botanist Charles Darwin began his systematic study of movement in plants, the mechanism of this transduction process is not understood. Possible candidates for a photoreceptor have been determined, and the final mechanism of bending has been shown to be differential growth caused by differential auxin concentrations. However, the specific mechanism by which a plant uses the light signal to change concentrations of auxin remains a mystery.

We propose that the alga *Chara coralina* could serve as a model organism for the study of phototropism. This alga has toothpick-sized internodes, each of which is a single large cell. When subjected to directional light, *Chara* bends toward the light at the internodes. This results in visible cells that have been bent by unidirectional light. We believe that studying the mechanisms of phototropism may be easier in *Chara* than in larger, more complex higher plants. Of course, before we can use *Chara* as a model to study the signal transduction process of phototropism, we must first study the photoreceptor and differential growth process, which have not been studied previously in *Chara*. My project was to begin the characterization of the photoreceptor for phototropism in *Chara*.

To effect change in a biological system, light must first be absorbed by a pigment. Thus, only wavelengths of light that a specific pigment absorbs will effect a response. The level of response to different wavelengths will be identical to the wavelengths that the pigment absorbs. Because of this, studying the effects of different wavelengths on a system (action spectroscopy) can give insight into the chemical nature of the pigment, or photoreceptor, involved. We sought to develop an action spectrum for phototropism in *Chara*.

To determine the phototropic response to various wavelengths of light, we unilaterally illuminated plants with light of various wavelengths. We then measured the angle of bending periodically, calculating the rate of bending from averaged bending rates of multiple plants over 3 hours of illumination. While *Chara* showed no phototropic response to red light, it did respond to various wavelengths of blue and green light. We then needed a way to compare the relative phototropic response at specific wavelengths. We did this by measuring the amount (or intensity) of light required to elicit a half maximal bending response. Specifically, we varied the light intensity at each wavelength tested and then measured which light intensity caused plants to bend at a rate that was half as fast as the maximum rate measured for each wavelength. This enabled us to calculate the relative response (such that wavelengths requiring the least light for half maximal bending had the highest relative response).

The resulting spectrum (see illustration) showed a peak of activity (and hence absorption) around 450 nm with a possible peak next to it around 460 nm. There was a small bending response around 500 nm but no response at wavelengths longer than 550 nm. Further, though we only tested wavelengths in the visible range, the spectrum implies the beginning of a peak that may extend into the ultraviolet spectrum. This spectral pattern is consistent with action spectra for higher plants. While more work on the details of the spectra could be done, these results indicate that the photoreceptor for phototropism in *Chara coralina* is similar to that in higher plants.

—Cora Schmid

Cora Schmid did this research while an undergraduate at Cornell University. She worked under the direction of Randy Wayne.

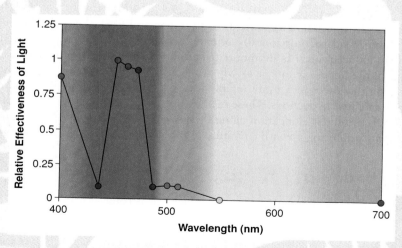

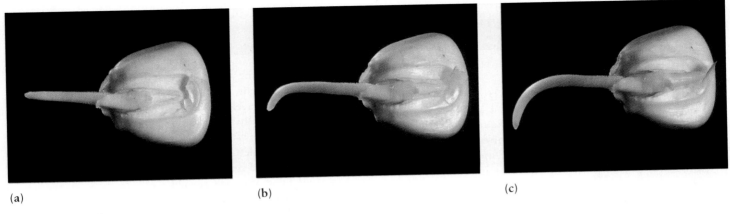

(a)   (b)   (c)

figure 19.6

Gravitropism by horizontally oriented roots of corn. (a, b)Downward curvature begins within 30 minutes and is completed (c) within a few hours. Curvature results from faster elongation of the upper side of the root than of the lower side.

Although we do not understand how roots *perceive* gravity, we are beginning to understand how they *respond* to gravity. When roots are oriented horizontally, growth slows along the lower side of the elongating zone, thereby causing the root to curve downward. One of the first events that ultimately causes this differential growth is the accumulation, not of a hormone, but of calcium ions (Ca^{2+}). Ca^{2+} moves to the lower side of the cap and elongating zone of horizontally oriented roots. This accumulation of Ca^{2+} along the lower side of the root triggers an accumulation of IAA along the lower side of the root tip. Since IAA inhibits cellular elongation in roots, the lower side of the root grows slower than the upper side of the root, and the root curves down (fig. 19.6). When the root reaches a vertical position, the lateral asymmetries of Ca^{2+} and IAA disappear, and straight growth resumes.

The *AUX1* gene plays a critical role in root gravitropism; indeed, mutations within this gene abolish root gravitropism. Thus, products of the *AUX1* gene mediate the hormonal control of root gravitropism, possibly by affecting the transport of an amino acid-like signaling molecule.

IAA and Ca^{2+} also direct the negative gravitropism of shoots. IAA accumulates along the lower side and Ca^{2+} along the upper side of horizontally oriented stems. Concurrently, auxin-induced mRNAs disappear from the cortex and epidermis of the upper (i.e., more slowly growing) side and accumulate on the lower (i.e., more rapidly growing) side of horizontally oriented hypocotyls. These mRNAs, or encoded proteins, stimulate cellular elongation along the lower side of the stem, thereby producing upward curvature of the stem (fig. 19.7).

c o n c e p t

Gravitropism is a growth response to gravity. It is controlled by calcium and IAA, and results in stems growing up and roots growing down. Gravitropism is important because it increases the probability that (1) roots will encounter water and minerals, and (2) that stems and leaves will intercept light for photosynthesis.

figure 19.7

Negative gravitropism by a stem of *Coleus*. This plant was placed on its side 24 hours before this picture was taken.

Thigmotropism

Thigmotropism is a growth response of plants to touch. The most common example of thigmotropism is the coiling of tendrils or of entire stems of plants such as morning glory (*Ipomoea*) and bindweed (*Convolvulus*) (fig. 19.8). Before touching an object, tendrils and twining stems often grow in a spiral pattern called *circumnutation* that increases their chances of contacting an object to which they can cling. Contact with an object is perceived by specialized epidermal cells, which induce differential growth in the tendril. Such growth can be extremely rapid: a tendril can encircle an object within 5–10 minutes. Furthermore, thigmotropism is often long lasting: stroking a tendril of garden pea for only a couple of minutes can induce a curling response that lasts for several days. Thigmotropism is probably influenced by IAA and ethylene; these hormones induce thigmotropic-like curvature of tendrils even in the absence of touch.

Tendrils can also store the "memory" of touch. Tendrils that are touched while growing in the dark do not respond until

figure 19.8

Thigmotropic coiling of a stem of bindweed.

after they are illuminated. Thus, although tendrils can store the sensory information received in the dark, light is required for the growth response to proceed. This light-induced expression of thigmotropism may be due to a requirement for ATP, since ATP will substitute for light in inducing thigmotropism of dark-stimulated tendrils. Various degrees of thigmotropism are exhibited not only by tendrils but also by leaves, stems, petioles, and roots.

c o n c e p t

Thigmotropism, a growth response to touch, is probably controlled by IAA and ethylene. Thigmotropism by tendrils allows plants to climb objects, thereby increasing the plants' chances of intercepting light for photosynthesis.

Nastic Movements

Unlike tropisms, nastic movements are independent of the direction of the stimulus: they occur in an anatomically predetermined direction, rather than toward or away from the stimulus. Nastic movements also occur in response to environmental stimuli. Nastic movements include some of the most unusual as well as spectacular responses in all of the plant kingdom.

Seismonasty

Seismonasty is a nastic movement resulting from contact or mechanical disturbances such as shaking. Seismonastic movements are based on a plant's ability to rapidly transmit a stimulus from touch-sensitive cells in one part of the plant to responding

cells located elsewhere. Among the most dramatic of these responses are those exhibited by the sensitive plant (*Mimosa pudica*): touching a leaf causes the leaflets to fold and the petiole to droop (fig. 19.9). Here's how the response occurs:

1. Touching a leaf generates an electrical signal that moves along the petiole.

2. This electrical signal is translated into a chemical signal that causes cell membranes to become more permeable to K$^+$ and other ions. The cells that are affected are called *motor cells;* these cells are large, thin-walled parenchyma cells located in a jointlike structure called a *pulvinus* (fig. 19.10). In the sensitive plant, a pulvinus is located at the base of each leaflet and petiole.

3. The movement of ions out of motor cells decreases the water potential in the surrounding extracellular space, which causes water to move out of motor cells via osmosis.

4. The loss of water causes the motor cells to shrink, thereby producing the seismonastic movement.

The unfolding of leaves takes 15–30 minutes and is accomplished by reversing this process. Motor cells take up K$^+$ and other ions, causing water to enter the cells via osmosis. This influx of water inflates the cells to their original size, thereby unfolding the leaves to their original positions.

Similar seismonastic movements occur in a few other plants. Such movements are interesting, but what is their adaptive significance? Seismonastic movements may scare insects off of plants, thereby decreasing a leaf's chances of being eaten. Presumably, folded leaves are more difficult to see, which may also divert a herbivore's attention to some other plant.

The most spectacular and macabre of plants' responses to environmental stimuli is the seismonastic response of Venus's-flytrap (*Dionaea muscipula*), a popular novelty plant. Although Venus's-flytraps have been studied since before the time of Charles Darwin, the mechanism responsible for trap-closure remains controversial. One hypothesis for trap-closure involves acid-induced growth. According to this hypothesis, movements of Venus's-flytrap involve increases in cellular size initiated by acidification of the trap's cell walls. Consistent with this hypothesis is the observation that traps will close if the walls of cells at the trap's "hinge" are acidified to pH 4.5 and below.

Leaves (i.e., traps) of Venus's-flytrap have two lobes, each of which has several sensitive "trigger" hairs overlying the motor cells. For the trap to close, at least two trap-hairs must be touched in sequence (the trap won't close if you merely touch the same hair twice). When a meandering animal touches these hairs, an electrical signal moves from the trigger hairs to the motor cells and initiates transport of H$^+$ to the walls of epidermal cells along the outer surface of the trap. The resulting acidification of these cell walls causes the outer epidermal cells along the central portion of the leaf to expand rapidly. Because epidermal cells along the inner surface of the leaf do not expand, the flytrap snaps shut (fig. 19.11). Closure of the trap takes 0.1 to 0.5 seconds and requires a large expenditure of ATP: motor cells use almost one-third of their ATP to pump the H$^+$ acidify the cell walls and close the trap.

(a)

(b)

(c)

figure 19.9

Seismonastic movement of leaves and leaflets of the sensitive plant (*Mimosa pudica*). In undisturbed plants (a), leaves are erect. Touching a leaf (b) causes leaflets to fold and the petiole to droop (c).

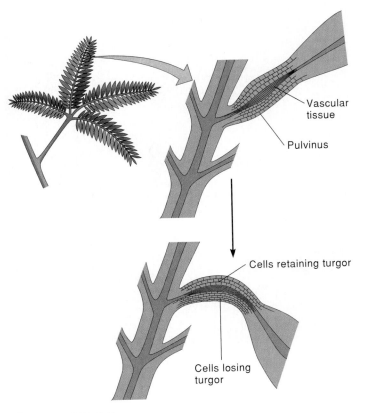

Vascular tissue

Pulvinus

Cells retaining turgor

Cells losing turgor

figure 19.10

Seismonasty in the sensitive plant is due to turgor changes in pulvini, which are swollen structures located at the bases of leaflets and petioles. Turgor changes on one side of the pulvinus cause cells there to shrink, thereby producing movement.

(a)

(b)

figure 19.11

Seismonasty enables a Venus's-flytrap to trap its prey. (a) The unsuspecting fly shown here was attracted to the trap by compounds secreted on the leaf's surface. (b) When this fly touched the trap's trigger hairs twice, it was snared.

figure 19.12

Seismonasty enables sundews to trap and digest their prey. Insects become trapped on the sticky mucilage secreted by glandular hairs. Seismonastic movement of these glandular hairs propels the insect to the center of the leaf, where it is digested.

(a)

(b)

figure 19.13

Nyctinastic movement in a prayer plant (*Maranta*). (a) During the day, leaves are oriented horizontally. (b) At night, the leaves become oriented in a vertical configuration resembling praying hands.

According to a competing hypothesis, trap closure results not from cell-wall acidification but rather from a release of tension. Mesophyll cells of a trap are extensible but are kept compressed in an open trap, thereby developing a tension in the tissue. This tension is somehow released by the action potential triggered by an animal that touches the trigger hairs, thereby causing the trap to close.

The trapping mechanism is quite sophisticated. Closure requires two stimuli, which are unlikely to result from inanimate objects such as falling leaves that touch the plant. The toothed edges of the trap mesh like a bear trap, pressing the captive animal against digestive glands along the inner surface of the trap. The prey's struggle to free itself stimulates the trap to close even tighter and glands in the trap to secrete digestive enzymes. Digestion lasts one to several days. The opening of empty traps usually requires 8–12 hours and results from the expansion of epidermal cells along the midrib of the inner surface of the leaf.

Seismonastic movements are also responsible for the feeding behavior of another carnivorous plant, the sundew (*Drosera* sp., fig. 19.12; also see photo that opens Chapter 20, p. 468). The club-shaped leaves of sundew are covered by glandular hairs that secrete nectar. Insects that land on the leaf become trapped in this nectar. The prey's contact with the leaf generates an electrical signal, which causes the surrounding tentacles to bend inward and carry the prey to the center of the leaf, where it is digested (fig. 19.12).

c o n c e p t

Seismonasty is a nastic response resulting from contact or mechanical disturbance. The mechanisms for this response include reversible changes in turgor.

Nyctinasty

Nyctinasty (from Greek *nyktos*, "night," and *nastos*, "pressed together"), or "sleep movement," is a nastic response caused by daily rhythms of light and dark. One of the most common nyctinastic responses occurs in the prayer plant (*Maranta* sp.), an ornamental houseplant. During the day, leaves of a prayer plant are horizontal, thereby maximizing their interception of sunlight. At night, the leaves fold vertically into a shape resembling a pair of praying hands (fig. 19.13). This movement of leaves in response to light and dark results from changes in the turgor of motor cells in a pulvinus located at the base of each leaf. In the dark, K^+ ions are transported from cells of the upper side of the pulvinus to cells along its lower side. This movement of ions causes water to move via osmosis into cells along the lower side of the pulvinus. This, in turn, causes cells along the upper side of the pulvinus to lose water and shrink as the cells along the lower side gain water and expand. Taken together, these changes in cellular volume move the leaf to a vertical position. At sunrise, the process is reversed, and the leaf again assumes its horizontal position.

Similar "sleep" movements occur in other plants, including sorrel (*Oxalis*) and legumes such as beans. In these plants, nyctinastic movements occur at regular times each day. A clever use of these regular movements was made by Carolus Linnaeus, a famous Swedish botanist. Linnaeus filled wedge-shaped portions of a circular garden with plants having sleep movements that occurred at different times. By seeing which plants of his so-called *horologium florae* (flower clock) were "asleep," Linnaeus could determine the time of day. You'll learn more about Linnaeus in Chapter 24.

c o n c e p t

Nyctinasty is a nastic response resulting from daily rhythms of light and dark. The mechanism underlying nyctinasty is turgor changes in pulvini.

The Lore of Plants

Thermonasty is a nastic response caused by changes in temperature. A popular example of thermonasty is the opening and closing of tulips. Another example is the curling of the edges of *Rhododendron* leaves in response to cold; some natives of the mountains in the Carolinas use this response as a means of estimating outdoor temperatures. Thermonasty has not been studied extensively, and its mechanism is unknown.

Thigmomorphogenesis

Plants growing in the protected environment of a greenhouse develop differently than do plants grown outside, because plants in natural environments are often disturbed by rain, wind, falling objects, and passing animals. These disturbances inhibit cellular elongation and produce shorter, stockier plants containing relatively large amounts of supportive tissue (i.e., collenchyma and sclerenchyma fibers). This response to mechanical disturbances is called **thigmomorphogenesis.** Thigmomorphogenesis is mediated by *touch-induced genes* that are activated by many stimuli, including wind and touch. Within 10–30 minutes after stimulation, mRNAs for these genes increase by as much as a hundredfold. One of these mRNAs encodes calmodulin, a calcium-binding protein that is involved in a variety of plant responses ranging from the secretion of enzymes to the actions of hormones. Ethylene is probably one of several factors involved in the control of thigmormophogenesis.

Seasonal Responses of Plants to the Environment

The behavior of plants is strongly influenced by seasonal changes in the environment. For example, the cooler nights and shorter days of autumn bring decreased growth, beautiful coloration of leaves, and dormancy of buds in preparation for the rigors of the upcoming winter. In the spring, buds resume growth and rapidly transform a barren forest into a dynamic, photosynthetic community. Clearly, plants must sense and anticipate seasonal changes. How do they do this? To answer this question, let's consider flowering, which is the basis for sexual reproduction in flowering plants.

Flowering

One of the most striking and predictable responses of most plants to changes of season is **flowering.** Gardeners have long known that many plants flower only during certain times of the year. For example, clover flowers during summer, and asters bloom during the short days of fall.

A study of flowering provided our first clues to how plants sense and anticipate changes of the season. In the early 1900s Wightman Garner and Henry Allard were studying tobacco at a research center in Beltsville, Maryland. Since tobacco plants flower during late summer in Maryland, Garner and Allard were not surprised when their tobacco plants began flowering in late August. However, one plant caught their eye: an oversized mutant that did not flower like the rest of the crop but continued to grow vegetatively into autumn. This mutant became much larger than other tobacco plants, leading Garner and Allard to name it Maryland Mammoth. Because the oversized mutant had the potential for increasing the yield of tobacco crops, Garner and Allard moved their Mammoth plant into the greenhouse to protect it from cold and continued to observe its growth. To their surprise, the mutant finally flowered in December. Their interest piqued, Garner and Allard began studying this "out-of-sync" response. First, they propagated the mutant so they would have more plants to work with. Plants grown from cuttings and seeds of the parent plant also flowered in December. After confirming their earlier observations, Garner and Allard set out to answer an important question: why didn't the mutant flower at the same time as other tobacco plants?

Photoperiodism

Garner and Allard began their inquiry by studying the influences of light, moisture, temperature, and nutrition on the flowering response. They discovered that the environmental signal for flowering is the **photoperiod,** which is the ratio of the length of day to the length of night. Most tobacco flowered only during the long days of summer. However, the Maryland Mammoth mutant was different: it flowered only during the short days of winter. These findings suggested to Garner and Allard that plants could measure daylength. They soon confirmed their hypothesis with experiments using soybeans. Garner and Allard set up several experimental plots, each planted a week or so apart. All of the plants flowered at the same time, even though the different planting times resulted in plants of different ages and sizes. These experiments convinced Garner and Allard that plants could measure seasonal changes in daylength. They called this phenomenon **photoperiodism.**

c o n c e p t

Photoperiodism is a response to changes in the relative lengths of night and day. Photoperiodism is important because it allows plants to measure and anticipate seasonal changes in climate.

Kinds of Flowering Responses in Plants

Subsequent studies of other flowering plants led botanists to classify plants into four groups, based on the plants' responses to photoperiod: day-neutral plants, short-day plants, long-day plants, and intermediate-day plants (fig. 19.14).

Day-neutral plants flower without regard to photoperiod; that is, daylength has no effect on their flowering. Examples of day-neutral plants include roses, snapdragons, cotton, carnations, dandelions, sunflowers, tomatoes, cucumbers, and many weeds.

Short-day plants flower only if light periods are shorter than some critical length. For example, ragweed plants flower only when exposed to 14 hours or less of light per day. Because of their light requirement and the need to accumulate photosynthate before flowering, short-day plants usually flower in late summer or fall. Asters, strawberries, dahlias, poinsettias, potatoes, soybeans, goldenrods, Christmas cactus, and some chrysanthemums are short-day plants.

Long-day plants usually flower in the spring or early summer; they flower only if light periods are longer than a critical length, which is usually 9–16 hours. For example, wheat plants flower only when light periods exceed 14 hours. Lettuce, spinach, radish, beet, clover, corn, gladiolus, and iris are long-day plants.

Intermediate-day plants flower only when exposed to days of intermediate length; they grow vegetatively if exposed to days that are either too long or too short. Examples of intermediate-day plants are sugarcane and purple nutsedge.

Do Plants Really Measure the Length of Day?

The findings of Garner and Allard stimulated many botanists to examine photoperiodism in other plants. Among these investigators were plant physiologists Karl Hamner and James Bonner. Their work is a model of how one set of observations can lead to new hypotheses and new knowledge.

Day-neutral plant

Long-day length   Short-day length

Short-day plant

Shorter than critical day length   Longer than critical day length

Long-day plant

Longer than critical day length   Shorter than critical day length

Intermediate-day plant

Longer or shorter than critical day length   Intermediate day length

figure 19.14

Kinds of flowering responses to daylength. Botanists originally thought that flowering was controlled by daylength. Flowering by day-neutral plants is not affected by daylength; short-day plants require a light period that is shorter than some critical length; long-day plants flower only if light periods are longer than some critical period; intermediate-day plants do not flower if the light period is too long or too short. Subsequent studies showed that the length of the *dark* period is what controls flowering in most plants.

Hamner and Bonner studied cocklebur (*Xanthium strumarium*), a short-day plant that requires 16 or fewer hours of light to flower. By growing plants in controlled-environment growth chambers, Hamner and Bonner could manipulate the photoperiods. To their surprise, they discovered that the length of the light period was unimportant. Rather, plants flowered only if the *dark* period exceeded 8 hours, regardless

IS THERE A FLOWERING HORMONE?

Many botanists have long suspected the existence of a hormone that controls flowering. One of the first botanists to provide evidence of such a hormone was Soviet scientist M. K. Chailakhyan, who studied short-day chrysanthemums in the 1930s. Chailakhyan separated the upper and lower halves of his chrysanthemums with a lightproof barrier so that he could expose the upper and lower halves of the plants to different photoperiods. When the upper half of the shoot was exposed to noninductive long days and the lower half to inductive short days, the entire plant flowered, even the upper portion that had been exposed to a noninductive photoperiod. Removing the leaves from the upper half of the plant did not change the results. However, flowering did not occur when leaves were stripped from the lower half of the plant (i.e., the part exposed to inductive short days) (reading fig. 19.2b). These results suggested that leaves perceive the photoperiod and that they produce a flower-inducing substance. Chailakhyan called this hypothetical substance **florigen** ("flower maker").

Subsequent research has provided additional information about flowering and florigen. For example, Karl Hamner and James Bonner showed that flowering would occur if only one leaf were exposed to an inductive photoperiod (reading fig. 19.2a). That is, the flowering stimulus can be transmitted from one leaf to the whole plant. However, the exposed leaf must be left on the plant for several hours after exposure for flowering to occur; these hours are presumably required for synthesis and transport of the stimulus from the leaf to the stem. Transport probably occurs in the phloem, since flowering does not occur if the phloem is stripped between the leaf and responding buds. Moreover, transport of the stimulus requires direct contact of living cells; flowering does not occur if the leaf and stem are separated by agar.

The most convincing evidence regarding florigen involves grafting of short-day and long-day plants. Short-day plants flower in long days if grafted to long-day plants that are flowering. Similarly, a long-day plant will flower in short days if grafted to a short-day plant that is flowering. In each instance, flowering begins nearest the graft union and then spreads progressively to the rest of the plant. These results suggest that there is a graft-transmittable substance that induces flowering and that this stimulus is similar in long-day and short-day plants.

Despite decades of intensive research, no one has yet isolated florigen. Thus, florigen remains the hypothetical flowering hormone. The inability to isolate florigen has prompted many botanists to suggest that flowering is induced not by a single hormone but by changes in ratios of other hormones. For example, plants such as pineapple flower when exposed to ethylene, and some long-day plants flower in short days when treated with gibberellic acid. However, short-day plants do not flower if treated with gibberellic acid. Thus, gibberellic acid is not florigen, although it may be involved in the flowering of some long-day plants.

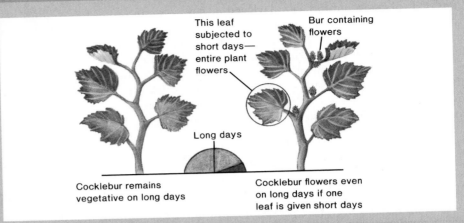

reading figure 19.2a

Cocklebur (*Xanthium strumarium*) is a short-day plant (left). Hamner and Bonner showed that exposure of just one leaf to short days induces flowering in the entire plant, even on long days (right).

reading figure 19.2b

Chailakhyan's experiments with short-day chrysanthemums (*Chrysanthemum* sp.). (a) Flowering occurred when leaves were subjected to short days. (b) Flowering did not occur when leaves were exposed to long days and upper, leafless parts were limited to short days.

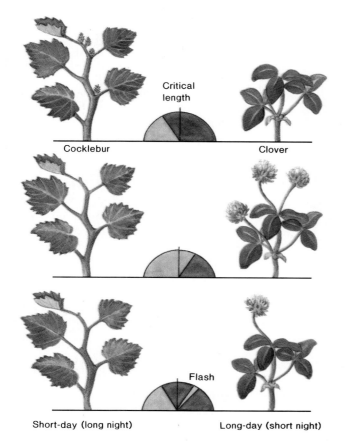

figure 19.15

The influence of daylength and night length on flowering by short-night plants (e.g., clover) and long-night plants (e.g., cocklebur). Long-night plants require an uninterrupted dark period longer than a critical period, while short-night plants require a dark period shorter than a critical length. Interrupting the dark period of a long-night plant inhibits flowering.

of the length of the light period. Hamner and Bonner then made another startling discovery: flowering did not occur if the dark period was interrupted by a 1-minute pulse of light, even if the regular light period remained less than 15 hours (fig. 19.15). Similar experiments in which the light period was interrupted with darkness had no effect on flowering. Other experiments with long- and short-day plants confirmed their finding: flowering requires a specific period of uninterrupted *dark* rather than uninterrupted light. Thus, short-day plants such as cocklebur are more accurately described as **long-night plants,** because they flower only if their uninterrupted dark period exceeds a critical length. Similarly, long-day plants such as clover are more accurately described as **short-night plants.**

Control and Sensitivity

The factor that determines if and when a plant flowers is not the absolute length of the photoperiod but rather whether the photoperiod is longer or shorter than the critical length

required for that species. For example, ragweed and spinach will both flower if exposed to 10 hours of dark: ragweed, a long-night plant, requires 10 hours *or more* of dark; spinach, a short-night plant, requires 10 hours *or less* of dark. Thus, spinach flowers in the short nights (i.e., long days) of summer, while ragweed blooms in the long nights (i.e., short days) of fall.

The measuring system in many plants is remarkably sensitive: henbane (*Hyoscyamus niger*), a short-night plant, will flower when exposed to dark periods of 13.7 hours but will not flower when exposed to dark periods of 14.0 hours. Plants stripped of their leaves are not responsive to changes in photoperiod, suggesting that the photoperiod is sensed by leaves.

How strict is the requirement for an inductive (flower-inducing) photoperiod? For some varieties of soybean (*Glycine max*), the requirement for an inductive photoperiod is absolute: these plants are long-night plants and will not flower unless exposed to long nights. This phenomenon is called *obligate photoperiodism.* Conversely, other plants such as marijuana (*Cannabis sativa*) and Christmas cactus (*Schlumbergera bridgesii*) will eventually flower without an inductive photoperiod. In these plants, the inductive photoperiod merely causes flowering to occur sooner. This response is called *facultative photoperiodism.* Finally, the photoperiodic requirement for flowering can be influenced by other features. For example, poinsettias (*Euphorbia pulcherrima*) are long-night plants at high temperatures and short-night plants at low temperatures. The mechanism underlying this temperature-controlled switch is unknown. Flowering of many plants is also influenced by moisture, soil conditions, nearby plants, and temperature (see reading 19.3, "Vernalization").

Significance of Photoperiodic Control of Flowering

The control of flowering by photoperiod influences the distribution of plants. For example, many short-night plants do not grow in the tropics because daylengths there are never long enough to induce flowering. Similarly, long-night plants such as ragweed do not grow far into northern areas because they do not have time to form seeds.

Photoperiodic initiation of flowering can occur only if a plant has passed from its juvenile stage into a phase designated as *reproductively mature,* or *ripe to flower.* For example, ragweed is a long-night plant that flowers only in the short days of autumn; why doesn't it also flower during the long nights of spring? The answer is that ragweed is not reproductively mature in the spring. The time required for reproductive maturation ranges from only a few days (as in the Japanese morning glory, *Ipomoea nil*) or weeks (as in annuals) to several years. Indeed, some species of *Agave* require decades before being ready to flower—hence their name, *century plant.* Similarly, the giant bamboos of Asia flower only every 33 or 66 years, even when transplanted to new environments. The delay in flowering caused by reproductive maturation ensures that the plant has stored enough food reserves to form and maintain flowers.

VERNALIZATION

Most people have never seen a carrot flower, even though they may buy carrots regularly at their local market. Why not? Don't carrots ever flower?

Yes, carrots do flower, but only during their second year of growth. Like other biennials such as turnips and beets, carrots spend their first year growing vegetatively and flower only after exposure to the cold temperatures of the ensuing winter (reading fig. 19.3a). Without this cold treatment, carrots would not flower; they would continue to grow vegetatively for an indefinite period of time. Of course, humans harvest carrots after the first year's growth (before flowering), when carbohydrates have accumulated in the root. As a result, carrots for sale in markets never have a flower stalk attached to them. Harvesting carrots in the spring of their second year of growth (when plants are flowering) would be pointless, because flowering consumes most of the food stored in the carrot.

The promotion of the springtime flowering of carrots and other plants in response to cold is called **vernalization** (from the Latin word *vernalis*, meaning "to make springlike"). Our understanding of this process is important for producing several kinds of crops, such as winter wheat, barley, and rye. For example, winter wheat is planted in autumn, and the seedlings are exposed to the cold temperatures of the following winter. These cold temperatures vernalize the wheat and cause it to flower in late winter or early spring, thereby producing an early crop. Such a crop is especially well suited to the Midwest, because it minimizes losses to pathogens and allows farmers

reading figure 19.3a

This sugar beet is 41 months old and was never exposed to low temperatures. Consequently, it has remained vegetative.

there to harvest before summer's drought. After harvesting winter wheat, farmers plant spring wheat, which does not require a cold treatment to flower. Spring wheat, like many other cereal crops, is a long-day plant and therefore flowers and produces fruit for a summer harvest (reading fig. 19.3b).

The vernalizing effects of winter can be simulated in the laboratory. For example, if moist seeds of winter rye are stored for a few weeks in a refrigerator, they will flower about 7 weeks after being planted in the spring. Seeds kept dry and warm throughout the winter will not flower until about 14 weeks after they are planted in the spring.

(a)

(b)

reading figure 19.3b

(a) Winter wheat in snow. (b) Spring wheat in summer.

Photoperiodism and Phytochrome

The experiments of Garner and Allard prompted botanists to search for the pigment that controls photoperiodism. Their search was aided by two clues:

1. Flowering is inhibited most effectively by interrupting the dark period with red light.
2. An interruption of the dark period with red light can be reversed if it is followed immediately by exposure to far-red light (the shortest wavelengths of infrared light).

These clues suggested that the pigment that was sensitive to daylength existed in two forms, one that absorbed red light and another that absorbed far-red light. The pigment was soon isolated and identified as **phytochrome.** Botanists later discovered that phytochrome is actually a family of pigments, with different forms predominating under different circumstances to control development throughout the life of the plant.

As suspected, phytochrome exists in two interconvertible forms, P_r and P_{fr}. Phytochrome is synthesized as P_r, which is the inactive form of the pigment. P_r absorbs red light and is

TURNING ON FLOWERING

Botanists have long known that genes such as *leafy* (*LFY*) and *apetala 1* (*AP1*) help determine the development of meristems. In late 1995, botanists discovered that these genes—each on its own—can trigger flower development. These genes are apparently part of a master switch: once they have flipped the flower-development switch, many other genes are expressed and flowering occurs (reading fig. 19.4).

The experiments demonstrating the roles of the *LFY* and *AP1* genes in flowering were remarkably clever:

- Botanists Alejandra Mandel and Martin Yanofsky genetically engineered mustard plants (*Arabidopsis*) so that the *AP1* gene

reading figure 19.4

Flowering can be promoted by triggering the *LFY* and *AP1* genes.

was continuously active. In these plants, shoots that normally produce stems and leaves form flowers instead in as little as one-third the time it takes normal plants to flower.

- Botanists Detlef Weigel and Ove Nilsson genetically engineered an aspen tree so that the *LFY* gene stayed active. The engineered tree produced flowers in 6 months, whereas natural aspens take 10–20 years.

These findings could make breeding and genetically altering plants and trees easier: if these discoveries can be applied to crops, they could produce plants with shorter growing seasons or shorten the time needed to improve crops and seeds.

---

converted to P_{fr}, the active form of phytochrome. P_{fr} promotes flowering of short-night plants and inhibits flowering of long-night plants. P_{fr} absorbs far-red light and is converted to P_r (fig. 19.16).

What is the significance of the interconvertibility of P_r and P_{fr} by red and far-red light? Sunlight contains proportionally more red light than far-red light. Thus, because P_r is converted to P_{fr} during the day, the presence of a large amount of P_{fr} could signal the plant that it is in light. When the plant is placed in darkness, P_{fr} is slowly converted to P_r. This conversion of P_{fr} to P_r was originally thought to start a set of reactions that enabled the plant to measure the length of darkness relative to the length of light. Indeed, this reasoning would explain why an uninterrupted period of dark (rather than light) controls photoperiodism. However, the dark-reversion of P_r to P_{fr} requires only 3–4 hours and therefore cannot fully account for the light-dark sensing mechanism that controls flowering. Rather, flowering is probably controlled by the combined actions of phytochrome and an internal clock.

To understand the coordination of phytochrome and this internal clock, consider a long-night plant in which flowering is inhibited by P_{fr}. This plant probably *measures* the longer-than-critical dark period with its internal clock, but *responds* (i.e., flowers) because the dark-reversion of P_{fr} to P_r removes an inhibition to flowering. Interrupting the dark period with red light would reconvert P_r to P_{fr}, thereby inhibiting flowering. Similarly, interrupting the dark period with a flash of red light followed by a flash of far-red light would leave most of the phytochrome in the P_r form, and flowering would occur. The internal clock involved in this process is discussed later in this chapter.

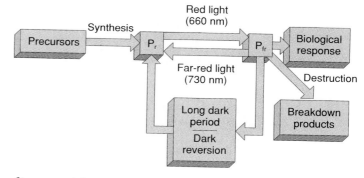

figure 19.16

Phytochrome is synthesized as P_r, which absorbs red light and is converted to P_{fr}. P_{fr} has several possible fates: it can absorb far-red light and be converted to P_r, initiate a biological response, be destroyed, or revert to P_r if maintained in darkness.

Other Responses Influenced by Photoperiod and Phytochrome

Phytochrome influences several responses of plants in addition to flowering. One of its most important roles is to control the early growth of seedlings. When a seed germinates underground or in darkness, the seedling has abnormally elongated stems, small roots and leaves, and a pale color, and appears spindly. This condition is termed **etiolated.** The rapid elongation of etiolated plants helps them reach light before exhausting their stored food reserves. Once a plant is in light, etiolated growth is replaced by normal growth. The light-controlled transformation from etiolated to normal growth is controlled by red light, which converts

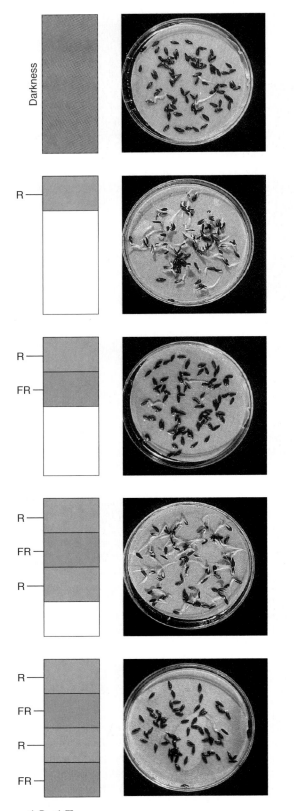

P_r to P_{fr} and thus brings about normal growth. Etiolated plants are sensitive to red light: exposing an etiolated plant to only 1 minute of red light will initiate normal growth. If this red light is followed immediately by exposure to far-red light, P_{fr} is converted to P_r, and etiolated growth continues.

Phytochrome may also provide plants with information regarding more subtle aspects of lighting, such as shading by plants overhead. Much of the red light of sunlight is absorbed by chlorophyll in a plant canopy. Underlying plants, such as those on a forest floor, therefore receive less red light, and less P_r is converted to P_{fr}. Because P_r promotes stem elongation (as in etiolated plants), information provided by the phytochrome system can help a plant reach sunlight more rapidly. Phytochrome may also help direct shoot phototropism. Unidirectional light could presumably create a P_r/P_{fr} gradient across the stem. Since P_r (abundant on the shaded side) promotes stem elongation, and P_{fr} (abundant on the illuminated side) inhibits stem elongation, this gradient of phytochrome could influence shoot phototropism.

Another aspect of plant development influenced by phytochrome is seed germination. In certain types of lettuce and weeds, red light stimulates germination, and far-red light inhibits germination. Alternating exposure of seeds to red and far-red light can be repeated indefinitely, but seed germination is affected only by the last exposure (fig. 19.17). Thus, germination

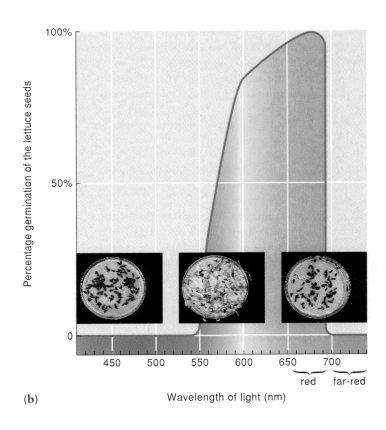

figure 19.17

Control of lettuce seed germination by red (R) and far-red (FR) light. (a) If the last exposure is to red light, most of the seeds germinate. The seeds remain dormant if the last exposure is to far-red light. This sensitivity to red and far-red light is controlled by phytochrome. (b) Lettuce seeds germinate in red light, but not in far-red light
Source: Data from M. Wilkins, Plantwatching: How Plants Remember, Tell Time, Form Partnerships, and More, *Macmillan Publishing Company, 1988.*

occurs after exposure to red/far-red/red/far-red/red light, just as it does after a single exposure to red light. Treatments with far-red and red/far-red/red/far-red/red/far-red light result in relatively little germination. Seeds of many species germinate when there is enough light to stimulate the conversion of P_r to P_{fr}. Most seeds with this sensitivity contain small amounts of stored food. It is unlikely that these seeds could survive germination from deep within the soil. Thus, the absence of P_{fr} (due to no sunlight) may account for the lack of germination by seeds buried deep in soil. This is highly adaptive, since it promotes the persistence of a "seed bank" in the soil.

Phytochrome and photoperiod also affect several other responses, including shoot gravitropism, stomatal formation, leaf abscission, chlorophyll synthesis, spore germination, sex expression, flower formation, and nyctinastic movements. In each of these responses, *phytochrome is the light receptor, not the effector.* How phytochrome induces these responses is unknown. It may alter membrane permeability, enzyme activity, gene expression, or the movement of hormones into and out of cells, thereby regulating cellular activities. Indeed, concentrations of most growth-promoting hormones increase immediately after exposure to red light.

There are many types of phytochrome, each of which may have a specific physiological role. We do not yet understand how P_{fr} induces the expression of responsive genes.

c o n c e p t

Phytochromes comprise a family of pigments that exist in two interconvertible forms. Phytochromes absorb red and far-red light, and influence many aspects of plant growth and development, including flowering. Phytochromes monitor the light environment and influence patterns of gene expression that enable plants to optimize growth and development in accordance with prevailing conditions.

Manipulating Photoperiodism

Horticulturists have long exploited our knowledge of photoperiodism to produce flowers when they are wanted for sale. For example, chrysanthemums are long-night plants that usually flower only in the fall. However, mums (i.e., flowers of chrysanthemums) can be made available year-round by using light-proof shades to create inductive, long-night photoperiods. Similarly, supplemental light can be used to induce out-of-season flowering in winter by short-night plants such as irises.

Senescence

Earlier in this chapter we stressed that plants usually respond to environmental stimuli by altering their growth patterns. However, some stimuli induce **senescence**, which is a collective term for the processes accompanying aging that lead to the death of a plant or plant part. In many instances, senescence is rapid; for example, flowers of plants such as wood sorrel (*Oxalis eu-*

ropaea) and heron's bill (*Erodium cicutarium*) shrivel and die only a few hours after being formed. Other examples of senescence, such as the colorful "turning" of leaves in autumn last longer and are triggered by changes in photoperiod (fig. 19.18). Whatever its duration, senescence is not merely a gradual cessation of growth; it is an energy-requiring process brought about by metabolic changes. For example, leaf senescence begins during the shortening days of late summer and involves mobilization of nutrients and breakdown of proteins. By the time a leaf is shed from a plant, it contains little more than cell walls and remnants of a nutrient-depleted protoplasm; most of its nutrients have long since been moved to the roots for storage. These nutrients are used in the spring when the plant resumes growth.

The most striking event in leaf senescence is the destruction or non-replacement of chlorophyll. When chlorophyll is degraded, the yellow and orange carotenoids previously masked by chlorophyll become visible. The senescing cells also produce brightly colored anthocyanins. These pigments cause leaves to change colors, often in spectacular fashion (fig. 19.18; see also "Making and Destroying Pigments" in Chapter 7).

What controls leaf senescence? Senescence is strongly influenced by plant hormones; for example, fruits of plants such as soybeans produce a "senescence factor" that is transported to leaves and induces senescence. In the laboratory, leaf senescence can be delayed by applying cytokinins, gibberellins, and/or IAA and can be promoted by applying abscisic acid or ethylene.

Dormancy

Leaf senescence is only one aspect of a plant's preparation for cold or drought. The shortening days of autumn induce **dormancy**, which is a period of decreased metabolism. Like senescence, dormancy involves structural and chemical changes within the plant. For example, cells make sugars and amino acids that function as antifreeze (i.e., they lower the freezing point in cells); these changes, as well as dehydration, thereby prevent or minimize damage from cold. Also, inhibitors accumulate in buds, transforming them into winter buds covered by thick, protective scales. These changes in preparation for new environmental conditions (such as winter) are called *acclimation.* As a result of acclimation, a plant can better survive a cold, dry winter and is said to be *cold-hardy.* Cold-hardiness plays a major role in determining the distribution of a species: the poor cold-hardiness of some plants restricts their growth to tropical or warm temperate regions.

Resumption of active growth in spring is usually influenced by photoperiod and/or temperature. For example, the lengthening days of spring can release dormancy in birch and red oak, whereas fruit trees such as apple and cherry resume growth only after exposure to winter's cold. This cold requirement for breaking dormancy presents problems for apple and cherry trees planted in warm climates. Since these climates usually do not have a cold or long winter, spring growth either does not occur or is greatly delayed.

In other plants, dormancy is released by factors unrelated to photoperiod or temperature. For example, rainfall

figure 19.18

Leaf senescence is a complex, energy-requiring process that is often controlled by photoperiod. The spectacular colors of leaves in the fall result from destruction of chlorophyll, which reveals the presence of other pigments such as carotenoids and anthocyanins.

alone releases dormancy in many desert plants, while plants such as potato require a dry period before renewing growth. The mechanisms by which photoperiod or cold break dormancy are unknown but probably involve changes in amounts of hormones or the sensitivity of tissues to hormones.

As mentioned in Chapter 18, seed dormancy in many plants is controlled by ABA. ABA begins forming in leaves when the plant senses that daylength is becoming progressively shorter. This ABA is transported to seeds, where it accumulates and prevents germination. During its dormancy, the seed slowly breaks down the ABA; when the amount of ABA falls to a permissive level, the seed can germinate. This usually occurs when the seasonal climate is best for the plant to begin growth.

Circadian Rhythms

Not all rhythmic responses of plants are seasonal. For example, the common four-o'clock (*Mirabilis jalapa*) opens its flowers only in late afternoon, and the yellow flowers of evening primrose (*Oenothera biennis*) open only in the evening. Similarly, nyctinastic movements of prayer plants occur at the same time every day, whereas some algae phosphoresce in warm ocean waters within a few minutes of midnight each night. These regular, daily rhythms are called **circadian rhythms** (from the Latin words *circa*, meaning "about," and *dies*, meaning "day"). Not all circadian rhythms are as obvious as the opening of flowers or the nyctinastic movements of

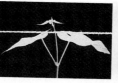

1:00 A.M.   6:00 A.M.   12:00 NOON   3:00 P.M.   10:00 P.M.   12:00 MIDNIGHT

figure 19.19

Leaf movements ("sleep movements") of bean leaves. A young bean plant was photographed at hourly intervals over a 24-hour period. The plant was held at constant temperature in the dark in front of a black velvet background and a horizontal string. An electronic flash was used to take the pictures. Six photographs were selected to produce the series shown.

leaves; subtle activities such as cellular division, stomatal opening, protein synthesis, secretion of nectar, hormone synthesis, and oscillations of the concentration of Ca^{2+} in cells and chloroplasts also occur in a daily rhythm. Circadian rhythms are widespread among eukaryotes.

What Controls Circadian Rhythms in Plants?

How do plants measure a day? Do circadian rhythms originate internally, or do plants depend on some external signal to measure a day? Some of the experiments that have helped answer this question have been elegantly simple:

Are circadian rhythms exact? Usually not; different species and different individuals of the same species have similar rhythms, but these rhythms may be a few hours longer or shorter than 24 hours.

What happens if the environment changes? Circadian rhythms, such as those of prayer plants, often occur even when the plants are kept under constant conditions and deprived of obvious external cues (i.e., information) about the time of day (fig. 19.19).

Some of the most interesting studies of circadian rhythms in plants have used a gene from the animal kingdom: the firefly luciferase gene. This gene codes for proteins that produce bioluminescence; this bioluminescence is then used to report the expression of nearby genes. In the experiments involving circadian rhythms, the luciferase gene was fused to the promoter of the *CAB2* gene, whose expression is regulated by light and the circadian clock. Expression of the gene was greatest in cotyledons and absent in roots (fig. 19.20). When the transgenic plants containing the gene were exposed to either red or blue light, the rhythmic period of bioluminescence was shortened. A phytochrome-deficient mutation lengthened the period in continuous red light but had little effect in continuous blue light. These results suggest that both phytochrome and blue-light receptors control the period of the circadian clock in plants. The phytochrome and blue-light receptors "reset the clock" by influencing the translation of mRNA. This resetting enables plants to keep up with seasonal changes in daylength.

Other experiments have been much more complicated and somewhat bizarre, such as launching plants into outer space, growing plants in the depths of mines, and observing plants grown at the South Pole. Taken together, these experiments have suggested that circadian rhythms are controlled internally rather than by external factors.

Biological Clocks and Entrainment

The internal timing mechanism that controls circadian rhythms is called the plant's **biological clock**; rhythmic events such as leaf movements in prayer plants can be viewed as the "hands" of this clock (also see the discussion of Linnaeus's *horologium florae* on p. 454). Interestingly, biological clocks do not speed up as temperature increases. This is somewhat surprising, because most biological processes are strongly influenced by temperature. Biological clocks must have some kind of feedback mechanism to compensate for changes in temperature. We know little about this feedback mechanism.

Although the control of circadian rhythms is internal, certain environmental factors can influence plants' rhythms. For example, slight modifications of the photoperiod can change the circadian rhythm. This resynchronization of the biological clock by the environment is called **entrainment.** However, entrainment to a new environment is limited: if the new photoperiod is too different from the plant's internal clock, the plant will revert to its own internal rhythms. Plants previously entrained to a new photoperiod will also revert to their natural rhythms when placed in constant light, regardless of how long they had been maintained in the modified photoperiod.

Of what significance is a biological clock? One important function of biological clocks is that they provide a timing mechanism for photoperiodism that allows plants to synchronize their activities. For example, biological clocks ensure that all flowers of a particular species are open at a particular time of day and are available for visits by pollinators. For some plants, this timing must be precise; flowers of *Cereus* cacti are pollinated by bats and must therefore open at night when bats are active. Furthermore, because flowers of some species of *Cereus* (e.g., queen-of-the-night) persist for only a night, different individuals must flower at the same time if they are to be cross-pollinated.

figure 19.20

Use of the firefly luciferase gene to study circadian rhythms in *Arabidopsis*. In these experiments, the luciferase gene was fused to the promoter region of a gene whose expression is regulated by light and the circadian clock. Expression is greatest (red) in the cotyledons and absent (blue) in the roots.

Biological clocks also allow plants to sense and anticipate changes in seasons. Phytochrome probably influences the timekeeping mechanism of biological clocks, as do changes in transport and metabolism in cells.

c o n c e p t

Circadian rhythms are rhythmic activities that occur daily. They are controlled by an internal biological clock and influence many aspects of plant growth and development.

Chapter Summary

Plants adjust their growth and development in response to environmental signals and rhythms.

Tropisms are short-term growth responses determined by the direction of an environmental stimulus. Tropisms result from differential growth and are important because they increase a plant's chances of intercepting more light for photosynthesis and of encountering water and minerals in soil. There are several types of tropisms:

- *Phototropism*, a growth response to unidirectional light, results from IAA moving to the shaded side of a coleoptile. The resulting accumulation of IAA on the shaded side stimulates cellular elongation there and causes the coleoptile to curve toward the light. Phototropism of stems involves production of inhibitors on the illuminated side of the stem. The inhibitors slow cellular elongation, thereby causing the stem to curve toward the light.

- *Gravitropism*, a growth response to gravity, results from an accumulation of IAA along the lower side of roots and stems. In stems, this accumulation of IAA stimulates cellular elongation, which causes the stems to curve up. In roots, the accumulation of IAA inhibits cellular elongation, which causes roots to curve down.

- *Thigmotropism* is a growth response to touch. Thigmotropism is common in tendrils, which coil around objects that they touch. Thigmotropism is controlled by IAA and ethylene.

Nastic movements occur in an anatomically predetermined direction. Like tropisms, nastic movements are widespread in plant growth and include activities such as leaf movements, trapping of prey by carnivorous plants, and defense. Nastic movements include:

- *Seismonasty* is a nastic movement resulting from contact or mechanical disturbances. These stimuli initiate electrical signals, which are subsequently translated into chemical signals that alter movement of ions. In some plants (e.g., sensitive plants), this movement of ions results in reversible turgor changes that cause movement.

- *Nyctinasty* is a nastic response resulting from daily rhythms of light and dark. It is caused by movements of ions, which cause turgor changes and movement.

Thigmomorphogenesis results in the stunted growth of plants due to mechanical disturbances such as wind and rain. It involves electrical signals and probably results from changes in ethylene production.

Photoperiodism is the ability of plants to sense changes in the relative lengths of night and day. One of the responses controlled by photoperiodism is flowering. The most critical factor for flowering is the length of the uninterrupted dark period.

Plants such as corn that flower only in response to long days and short nights are called *short-night* (i.e., *long-day*) *plants*, whereas those that flower only in response to short days and long nights are called *long-night* (i.e., *short-day*) *plants*. Intermediate-day plants such as sugarcane flower only when exposed to days of intermediate length. Day-neutral plants such as dandelions flower without regard to photoperiod.

Phytochrome is a group of pigments that influence many aspects of plant growth and development, including flowering. Phytochrome absorbs red and far-red light, and exists in two interconvertible forms, P_r and P_{fr}. P_{fr} forms upon exposure to red light and is the biologically active form of the pigment.

Senescence is the process that leads to the death of a plant or plant part. Leaf senescence in the fall is triggered by shortening days and involves mobilization of nutrients, destruction of chlorophyll, and abscission. Dormancy, a period of decreased activity, is usually controlled by photoperiod and temperature.

Circadian rhythms are regular rhythms of growth and activity that occur on an approximately 24-hour basis. They can be modified somewhat by the environment and are regulated by an internal biological clock.

What Are Botanists Doing?

Go to the library and read an article about flowering. What is the message of the article? What questions do you have about the work? Why are botanists so interested in flowering?

Writing to Learn Botany

What is the selective advantage of plants anticipating seasonal changes by measuring the photoperiod rather than rainfall, temperature, or some other weather factor?

Questions for Further Thought and Study

1. Would a short-night plant flower if its dark period were interrupted with light? Why or why not?

2. What would be the consequences of transplanting a long-night plant to the tropics? Explain your answer.

3. What would happen to IAA in a horizontally oriented stem that is illuminated from below?

4. Describe the possible significance (i.e., biological advantage) of each of the following: phototropism, thigmotropism, gravitropism, seismonasty, biological clocks, photoperiodism.

5. How does phytochrome differ from other plant pigments?

6. What is the difference between a tropism and a nastic movement? Of what value are nyctinastic movements?

7. Hoeing a garden or tilling a field often triggers the germination of many seeds buried in the soil. Why?

8. Why is it unnecessary to orient a seed in a particular way when you plant it?

9. Could florigen be an inhibitor rather than a promoter? Why or why not? What experiments would you do to try to isolate florigen?

10. Electrical signals are involved in several aspects of plant growth and development, including cellular differentiation, embryo development, growth of pollen tubes, attraction of symbionts, and resistance to pathogens. How are these signals similar to those in animals? How are they different?

11. In some mutants of corn and other plants, seeds germinate while still attached to the parent plant. These seedlings are referred to as being *viviparous*. Based on what you learned in this chapter, propose a hypothesis to account for viviparous seedlings.

12. How are the movements of pulvini and stomata similar? How are they different?

13. Only blue light with wavelengths less than 500 nm effectively induces phototropism (see fig. 19.3). Blue light also stimulates stomatal opening and inhibits stem elongation. What cellular mechanisms might these responses have in common?

Web Sites

Review the "Doing Botany Yourself" essay and assignments for Chapter 19 on the *Botany Home Page*. What experiments would you do to test the hypotheses? What data can you gather on the Web to help you refine your experiments?

Here are some other sites that you may find interesting:

http://liftoff.msfc.nasa.gov/spacelab/usml2/science/asc.html

This site, sponsored by National Aeronautics and Space Administration (NASA), describes how botanists are using the space shuttle to understand how plants perceive and respond to gravity.

http://www.reston.com/astro/biology/plant.plan.html

This site, also sponsored by NASA, describes the schedule for launching botany-related experiments into orbit, as well as the goals and accomplishments of NASA's Plant Biology Program.

Suggested Readings

Articles

Bennett, M. J. et al. 1996. *Arabidopsis Auk1* gene: A permease-like regulator of root gravitropism. *Science* 273:948–950

Cleland, C. F. 1978. The flowering enigma. *BioScience* 28:265.

Evans, M. L., R. Moore, and K. H. Hasenstein. 1986. How roots respond to gravity. *Scientific American* 255:112.

Lumsden, P. J. 1991. Circadian rhythms and phytochrome. *Annual Review of Plant Physiology and Plant Molecular Biology* 42:351–371.

Quail, Peter H. 1991. Phytochrome: A light-activated molecular switch that regulates plant gene expression. *Annual Review of Plant Physiology and Plant Molecular Biology* 42:389–409.

Salisbury, Frank B. 1993. Gravitropism: Changing ideas. *Horticultural Reviews* 15:233–278.

Smith, H. 1984. Plants that track the sun. *Nature* 308:774.

Sussman, Michael R. 1992. Shaking *Arabidopsis thaliana*. *Science* 256:619.

Wayne, Randy, Mark P. Staves, and A. Carl Leopold. 1992. The contribution of the extracellular matrix to gravisensing in characean cells. *Journal of Cell Science* 101:611–623.

Books

Galston, A. 1994. *Life Processes of Plants*. New York: W. H. Freeman and Co.

Kendrick, R. E., and B. Frankland. 1983. *Phytochrome and Plant Growth*. Studies in Biology no. 68. London: Edward Arnold.

Salisbury, F. B., and C. W. Ross. 1992. *Plant Physiology*. 4th ed. Belmont, CA: Wadsworth.

Wilkins, M. B. 1988. *Plant Watching: How Plants Live, Feel and Work*. London: Macmillan.

Nutrition and Transport

U p to this point, we have focused primarily on the organic nature of plants. Chromosomes, cell walls, cell membranes, hormones, wood, and ATP are all based on carbon; that is, they are organic. However, an introductory botany book would not be complete without discussing the role of inorganic compounds in plant growth. Inorganic ions, such as calcium and potassium, are indispensable as cofactors in many biochemical reactions and as parts of organic molecules. This means that the carbon-based metabolism of photosynthesis and respiration must be supplemented with ingredients that do not enter the plant by gas exchange; instead, they must enter the plant through the roots.

Understanding the inorganic side of plant metabolism comes from learning what nutrients are essential to plants, what minerals are available in the soil, and how plants absorb and move water and dissolved minerals into and throughout the plant. These are the topics that are discussed in this unit. Our discussion of plant nutrition and mineral transport will reveal unanswered questions as well as some surprises.

A sundew (*Drosera binata*), magnified to show gland-tipped hairs. These plants obtain much of their nitrogen by trapping and digesting insects.

Soils and Plant Nutrition

20

Chapter Outline

Chapter Overview

Life on earth is sustained by sunlight and soil. Although soils have differing properties, each influences the distribution of plants and animals on our planet. For example, animal nutrition depends on the ability of plants to find and absorb the sixteen elements essential for growth and reproduction. Gathering these elements is an elaborate and expensive venture that depends largely on root systems that permeate and mine the soil. The mineral nutrition of most plants also depends on other organisms with which plants establish relationships ranging from mutually beneficial symbioses to parasitisms. A few plants even get some of their essential nutrients by trapping and digesting animals. Nutrient absorption by plants is important because it links the living and nonliving parts of terrestrial and aquatic ecosystems.

Throughout this book you've seen examples of how various nutrients influence plant growth and development. For example, iron is an electron carrier in respiration, and calcium plays an important role in the responses of plants to environmental stimuli. Indeed, *all* aspects of plant growth and development depend on an adequate supply of nutrients.

But what do we mean by *adequate?* To answer this question, consider a harvest of 5 metric tons by an Idaho potato farmer. About 4 metric tons of the crop's weight is water—that is, hydrogen and oxygen. Almost 85% of the remaining "dry weight" is carbohydrates such as cellulose, starch, and sugars. The remainder includes nitrogen (2%) and minerals such as iron (0.01%), sulfur (0.15%), and chlorine (0.2%). Thus, plants require different amounts of different nutrients. Many of these "adequate" amounts of nutrients seem trivial. For example, all of the copper in the 5 tons of potatoes weighs only about 2.5 g, an amount roughly equal to that in a penny. However, no matter how small this amount of copper may seem, its role in plant growth is critical: plants *must* have copper to grow. Without that penny's worth of copper, there would be no potato plants and no harvest.

Where do nutrients come from? We have already studied an example of how plants obtain one of their nutrients from the air: carbon comes almost exclusively from atmospheric CO_2 fixed during photosynthesis. However, most nutrients are absorbed from soil by roots. Thus, roots are the interface between the living and nonliving parts of the terrestrial ecosystem: virtually all nutrients move from soil into food chains through roots. However, soil does more than store nutrients until they can be absorbed by a plant: it also determines the availability of nutrients and thus largely determines what plants and animals live in a particular area.

Soils are the central processing unit of earth's environment. You can't understand plants without knowing something about soil.

Soils

How Soils Form

The earth is more than 4.5 billion years old. In its youngest stages, it was a harsh place; its "land" was little more than a mixture of igneous, sedimentary, and metamorphic rocks. Once formed, these rocks were transformed into soil by glaciers, wind, and rain, and later also by organisms and their activities. This conversion of rock into soil is called *weathering* and is a slow process. For example, in eastern North America where rainfall is relatively abundant, it can take more than 200 years to form only 2 cm of topsoil. Although topsoil is usually less than a meter thick, it spans continents, bridging the rocks below with life above.

Soils have several layers, called **horizons,** each of which has several distinguishing characteristics (fig. 20.1):

The **O horizon** is the surface litter covering the soil. This horizon typically consists of fallen leaves and is only a few centimeters thick.

The **A horizon** is topsoil and usually extends 10–30 cm below the soil surface. In most fertile soils, the A horizon has a pH near 7 and contains 10%–15% organic matter, which gives this horizon a dark color.

The **B horizon** consists of larger soil particles than those in the A horizon and extends 30–60 cm below the soil surface. This horizon usually contains relatively little organic matter and is therefore lighter in color than the overlying A horizon. In many regions, the B horizon contains large amounts of minerals and clay particles washed by rainfall from the A horizon. Mature roots commonly extend into the B horizon, where minerals accumulate. The B horizon is often called subsoil.

The **C horizon** occurs 90–120 cm below the soil surface and consists primarily of partially altered to unaltered rock fragments and mineral grains. This horizon usually lacks organic matter and is often referred to as the parent material, since it is the raw material from which soil forms. The C horizon extends to the underlying and often impenetrable bedrock of igneous, sedimentary, or metamorphic rock.

Soil, the central processing unit of earth's environment, forms from rocks by a process called weathering. Soils are often arranged in distinctive layers called horizons, each of which has unique properties.

Soil with the horizons just described is called a *mollisol* and characterizes semiarid grasslands. Mollisols retain nutrients and are fertile. Indeed, mollisols of the Great Plains of the United States and the steppes of Eastern Europe (the Ukraine) are the earth's most fertile soils. The only problem with using mollisols for growing crops is that they often occur in semiarid climates and therefore must be irrigated frequently.

A
Topsoil

B
Subsoil

C
Weathering
bedrock

figure 20.1

Soil horizons. The A, B, and C horizons can sometimes be seen in road cuts such as this one in Australia. The upper layers developed from bedrock. The dark upper layer is home to most of the organisms that live in the soil.

Just as there is no such thing as a typical cell, there is no typical soil. Indeed, there are more than 70,000 kinds of soils in the United States alone, some of which identify places such as Redlands, California, and Black Earth, Wisconsin. Despite their diversity, however, these soils can be grouped into the following general categories:

Mollisols are nutrient-rich soils characteristic of semiarid grasslands. Most of the organic matter in a mollisol is in its A horizon.

Spodosols are light-colored soils characteristic of wet, temperate regions such as the coniferous forests of Canada. The A horizon of these soils contains small amounts of organic matter and may have a pH as low as 4. Most organic matter in a spodosol is in its B horizon.

Alfisols resemble spodosols, except that their A horizon contains more organic matter. Alfisols are relatively fertile, brown soils common in the eastern United States.

Aridosols occur in arid regions such as deserts. Since aridosols contain relatively little organic matter, they are usually light-colored.

Histosols are acidic soils typical of swamps and bogs. Histosols contain large amounts of organic matter and are usually dark brown or black.

Oxisols are acidic soils characteristic of rain forests. Despite the lush vegetation that they support, oxisols are nutrient-poor. As a result, only plants such as cassava and bananas that efficiently extract nutrients can grow well on oxisols. The lack of nutrients in oxisols makes them almost useless for modern agriculture. Indeed, the intense poverty of tropical South America and central Africa is due largely to the inability of their oxisols to sustain the agriculture necessary for increased amounts of food and industry.

Soils are sensitive to environmental changes. This is especially true of an iron-rich oxisol called *laterite* that occurs in many lowland rain forests. When shaded by the rain forest, laterite remains damp and soil-like. However, if the forest is cut and the exposed soil dries, the iron in laterite irreversibly cements the soil into hard clumps called *ironstone*. Once formed, ironstone cannot be reconverted to oxisol. Consequently, tropical rain forests often recover slowly, if at all, after being stripped of their vegetation.

c o n c e p t

There are many kinds of soils, each having different properties that support different plants. These soils range from the mollisols of fertile grasslands to the nutrient-depleted oxisols of tropical rain forests. Soils are sensitive to environmental disturbances.

Components of Soil

Regardless of the different properties of their horizons, all soils contain the same five components: mineral particles, decaying organic matter (humus), air, water, and living organisms. Differing amounts of these materials define the soil's properties and, therefore, the plants it can support.

Mineral Particles

Weathering breaks rocks into progressively smaller pieces, the smallest of which are called **soil particles.** These particles consist of **minerals,** which are naturally occurring inorganic compounds usually made of two or more elements. All soils contain three kinds of soil particles: sand, silt, and clay. Clays are the final product of weathering and the smallest of these particles:

Type of Soil Particle	Diameter of Soil Particle (mm)
Sand	0.02–2
Silt	0.02–0.002
Clay	<0.002

Different kinds of soils contain different proportions of sand, silt, and clay. Clays contain more than 30% clay particles, and sandy soils contain less than 20% silt and clay. Soils having

table 20.1

Physical and Chemical Properties of Soil as Affected by Soil Texture

Soil Texture	Water Infiltration	Water-Holding Capacity	Ion Exchange	Aeration	Workability	Root Penetration
Sand	Good	Poor	Poor	Good	Good	Good
Silt	Medium	Medium	Medium	Medium	Medium	Medium
Clay	Poor	Good	Good	Poor	Poor	Poor
Loam*	Medium	Medium	Medium	Medium	Medium	Medium

*Loam soil averages out as medium in all respects because it is of variable composition, depending on the actual proportions of sand, silt, and clay particles that are present.

approximately equal mixtures of sand, silt, and clay—that is, the soils in which most plants grow best—are called **loams**. Soil mineral particles have many effects on plant growth, including determining the availability of nutrients and water (see table 20.1).

How Soil Particles Influence the Availability of Nutrients
The ability of soil to retain nutrients depends primarily on the structure of its clay particles. Clay particles have a sheetlike structure and are called **micelles**. In their original form, micelles associate with large amounts of silica (Si^{4+}). Over time, however, the Si^{4+} of clay particles is often replaced by Al^{3+}, which, in turn, may be replaced by Mg^{2+} or Fe^{2+}. As a result of these substitutions, clay micelles become negatively charged, allowing them to reversibly bind cations such as Ca^{2+} and K^+ and prevent these cations from being washed from the soil by rainfall. This weak binding of cations by clay micelles is important because plants can later extract these cations by exchanging them for H^+ via a process called **cation exchange** (fig. 20.2). For example, roots can secrete H^+ that replaces and makes available the Ca^{2+} held by a clay micelle:

$$micelle{:}Ca^{2+} + 2H^+$$

$$\downarrow cation\ exchange$$

$$H^+{:}micelle{:}H^+ + Ca^{2+}$$

Cation exchange is also enhanced by the respiratory production of CO_2 by roots. CO_2 released by respiring roots dissolves in water to form carbonic acid (H_2CO_3), which then dissociates into HCO_3^- and H^+. This H^+ replaces cations held on clay micelles, thereby making the cations available to the plant.

Because of their negative charge, clays have a high **cation-exchange capacity.** Conversely, sand particles are not charged and therefore have a low cation-exchange capacity. Because cations do not bind effectively to sand particles, cations are rapidly leached from sandy soil after a rainfall. This explains why sandy soils are usually nutrient-deficient and poorly suited for growing crops. The high cation-exchange capacity of clays makes them well suited for plant growth. However, the cation-exchange capacity of a clay can also have negative consequences for a plant. For example, replacing clay micelles with H^+ acidifies the soil. If this process is unchecked, the pH of the soil can drop to as low as 3 or 4. This drop in the pH alters the availability of nutrients and thus can limit plant growth (see "Soil pH," p. 482).

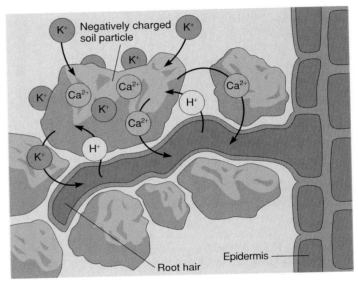

figure 20.2

Cation exchange in soil. Negatively charged clay micelles reversibly bind cations such as Ca^{2+} or K^+, thereby preventing these cations from being washed from soil by rainfall. Plants can extract these cations by exchanging them for H^+ via cation exchange. The secretion of H^+ by roots facilitates this exchange.

What about anions? Since soil particles are not positively charged, rainfall can rapidly leach anions such as SO_4^{2-} and NO_3^- from the soil. Because the lack of these anions frequently limits growth of plants, anions must be replenished by adding fertilizer to the soil (see reading 20.1, "Putting Things Back").

How Soil Particles Influence the Availability of Water The surfaces of soil particles adsorb much of the water in soil. The amount of water held in a soil is proportional to the surface area of its particles: the larger the surface area, the greater the retention of water. Because clay micelles have a sheetlike structure, they have a large surface area; indeed, the surface of clay particles in the upper few centimeters of soil in a 2-hectare cornfield equals the surface area of North America. Clay micelles are smaller than sand and therefore have a larger surface area per unit of soil volume than does sand. As a result, clay soils retain much more water than do sandy soils. This

PUTTING THINGS BACK

Plants extract large amounts of nutrients from soil. For example, during one growing season a wheat crop on 1 hectare of land removes 85 kg of nitrogen, 47 kg of potassium, and 17 kg of phosphorus from the soil. Some of these nutrients are replenished by decaying humus and by plowing under the remaining parts of the crop. However, such replenishment does not match what is lost when the crop is harvested if crops are grown repeatedly on the site. Consequently, these lost nutrients must be replenished with fertilizers. For example, the yield of an unfertilized soil that initially produced 100 bushels of corn per acre diminished to only 23 bushels per acre in 70 years. When this soil was fertilized, the yield increased to more than 130 bushels per acre (reading fig. 20.1). Although fertilization increases plant growth and crop yield, it rapidly reaches a point of diminishing returns: doubling the yield of already fertile soil often requires adding as much as 5 times more fertilizer.

CHEMICAL FERTILIZERS

Most chemical fertilizers have a rating that consists of three numbers, such as 12–6–6. These numbers refer to the amounts of nitrogen, phosphorus, and potassium, which are the three elements most likely to be deficient in soil. Thus, a 12–6–6 fertilizer contains 12% nitrogen (usually as ammonium salts), 6% phosphorus (as phosphoric acid), and 6% potassium (as potash).

Nitrogen, which is the most expensive of these elements to produce, is incorporated into fertilizer via the *Haber-Bosch* process:

$$3H_2 + N_2$$
$$\downarrow$$
$$500° C,$$
$$300 \text{ atm pressure}$$
$$\downarrow$$
$$2NH_3$$

reading figure 20.1

The application of fertilizer increases plant growth by increasing the availability of nutrients.

Nitrogen can either be added directly to the fertilizer as an ammonium salt or be converted to nitrate and then added as a nitrate salt (e.g., $NaNO_3$).

More than 40 million metric tons of nitrogen produced by the Haber-Bosch process are added to soil each year. However, this represents only about one-fifth the amount of nitrogen added to the world's soil by nitrogen-fixing bacteria (nitrogen is also added by thunderstorms and atmospheric deposition). Furthermore, the Haber-Bosch process is expensive in terms of energy; producing 2.5 kg of ammonia via the Haber-Bosch process requires the energy equivalent of 1,000 kg of coal. The costs of producing nitrogen account for about half of our $14 billion fertilizer bill. Consequently, manufacturing nitrogen-containing fertilizer requires more energy than any other aspect of crop production in the United States. To compound this problem, applications of nitrogen-containing fertilizers are inefficient, because crops absorb only about half of the nitrogen that is applied. The rest is absorbed by other organisms, leached from the soil in rainfall, or reconverted to gaseous nitrogen (N_2) by denitrifying bacteria such as *Micrococcus denitrificans*.

Chemical fertilizers are concentrated, easy to apply, and allow a grower to apply specific amounts of various nutrients. However, these fertilizers do not replenish humus in the soil. To maintain humus, growers usually plow under either the unharvested plants or a subsequent cover crop of barley or rye. The latter process is called green manuring and provides an excellent example of another kind of fertilizer: organic fertilizer.

ORGANIC FERTILIZERS

Organic fertilizers are essentially the same thing as humus. Although hardly new (planting a fish with corn seed is proverbial), the increased costs of chemical fertilizers have prompted a growing number of gardeners and farmers to rediscover organic fertilizers, which increase both the water retention and fertility of soil. Organic fertilizers include manure, dead animals and plants, fish scraps, and cottonseed meal. On a smaller scale, backyard gardeners often use compost, fish meal, lawn clippings, garbage, and a concoction called manure tea as organic fertilizers. We do not recommend fertilizing your houseplants with manure tea if guests are coming.

FOLIAR FERTILIZATION

Despite the presence of a thick cuticle, many plants can absorb nutrients through their leaves and stems. For example, iron is sprayed on azaleas and pineapples, and copper and zinc are sprayed on citrus to prevent mineral deficiencies. This type of fertilization is called foliar fertilization and is restricted primarily to micronutrients (that is, nutrients of which plants need small amounts).

property of clays would seem to make them ideal for plant growth. However, this is not entirely true, for along with the huge water-retention capacity of clay come some problems. For example, the small size of clay particles results in their packing tightly together—so tightly that the clay has low amounts of oxygen, due to small air spaces. This tight packing also retards the penetration of water into the soil (e.g., water penetrates clay about 20 times slower than it penetrates sand). As a result, much of the water that falls on clay soil runs off and is unavailable for plant growth. This runoff water also contributes to erosion (see reading 20.2, "Losing the Soil"). Tightly packed clay can impede root growth.

LOSING THE SOIL

Next time you're outside, pick up a handful of soil and imagine that it is the earth's surface. Then do this:

Drop three-fourths of the soil back to the ground. That's how much of the earth is covered by water. What's left in your hand represents the land.

Drop half of the remaining soil to account for deserts, glacial poles, and mountain peaks where crops won't grow.

Drop one-tenth to account for land covered by cities, houses, roads, and parking lots.

What's left in your hand represents all the soil that supports life on earth. This soil is trickling through our fingers at an alarming rate because of erosion. Indeed,

- Worldwide, more than 36 billion tons of soil are lost because of erosion each year. Most of this erosion is due to human activities. In the United States alone, more than 6 billion tons of soil are lost each year; that's enough soil to fill 320 million dump trucks, which, if parked end to end, would reach almost to the moon and back.

- Over the past 200 years, almost 100 million hectares of U.S. farmland have been abandoned because of erosion, salinization, and waterlogging.

reading figure 20.2

Barren fields are especially vulnerable to erosion.

- In one night, a heavy rain can wash away several tons of soil from an acre of land.

Erosion—the movement of soil by wind and water—often produces spectacular effects. For example, the breathtaking sculptures of the Grand Canyon result from millions of years of erosion. Equally spectacular, however, are other effects of erosion. For example, dust storms in the Great Plains have de-

posited topsoil on ships hundreds of miles out in the Atlantic Ocean, and the Mississippi River annually dumps more than a quarter of a billion tons of the Midwest's topsoil into a murky Gulf of Mexico. This loss of topsoil is costly: the eroded cropland that pollutes rivers and lakes costs Americans billions of dollars each year.

Why does all of this erosion occur? Although erosion is not caused only by humans—after all, the Grand Canyon formed long before humans began gathering at its edge with their cameras—human activities have greatly increased erosion. Many causes underlie this human-related increase in erosion, but most are related to economics. For example, the food demands of a growing population have opened new international markets for farmers, and massive grain sales have made crop prices soar. The increased profits gained from these exports resulted primarily from farmers growing crops on an additional 60 million acres of land, much of which was fragile grassland. For example, the grasslands of the Great Plains were stripped and used to grow wheat; when these unprotected soils dried, they were blown away by the winds.

Despite sodbuster legislation aimed at diminishing erosion, the problem persists. Today, exported crops use more than one-third of our cropland, most of which is affected by erosion (reading fig. 20.2).

c o n c e p t

There are three kinds of soil particles: sand, silt, and clay. Clay particles, which have a sheetlike structure, are the smallest of these particles. Because clays are negatively charged, they bind cations. Plants exchange H^+ for these cations in a process called cation exchange. Sand is not charged and does not bind cations.

Humus

Humus is the decomposing organic matter in soil. The amount of humus in soil varies: *mineral soils* contain only 1%–10% humus, while *organic soils* typically contain about 30%

humus. The most extreme examples of organic soils are the histosols of swamps and bogs, which may contain more than 90% humus. These soils are usually so acidic that decomposers can hardly grow in them. As a result, humus accumulates faster than it is broken down.

The amount of humus in soil affects the soil and its plants in several ways:

Its lightweight and spongy texture increases the water-retention capacity of the soil. Water absorption by humus decreases runoff, thereby slowing erosion.

Most humus is rich in organic acids and has a negative charge. As a result, humus increases the cation-exchange capacity of the soil.

Humus swells and shrinks as it absorbs water and later dries. This periodic swelling and shrinking aerates the soil.

Humus is a reservoir of nutrients for plants. Like time-release vitamins, humus gradually releases nutrients as it is degraded by decomposers. Humus also provides habitats for many organisms that mix and concentrate nutrients.

Most plants grow best in soil containing 10%–20% humus.

Since humus is constantly being decomposed by bacteria and fungi, it must be replenished. In natural ecosystems, humus is replenished by events such as leaf abscission and the addition of animal wastes. Many gardeners replenish humus in soil by *composting*, a method of converting organic wastes to fertilizer and a soil conditioner. The process is simple: organic wastes (e.g., lawn clippings) are gathered into a compost pile, where they are degraded by fungi and bacteria. During winter, this decomposition may take several months, while during summer it may be completed in as little as 3 weeks. The resulting compost costs less than commercial fertilizer, provides an efficient means of disposing of waste products, and contains nearly all of the nutrients needed by plants. Moreover, composting releases nutrients gradually as the materials decompose, thereby providing plants with a continuous balance of nutrients. On a larger scale, farmers often plow under the stubble of plants remaining after a harvest to maintain the humus in soil used to grow crops.

c o n c e p t

Humus is the decomposing organic matter in soil. Humus retains water and is a source of nutrients for plants.

Air

About 25%–50% of the volume of most soils is air. This "empty space" has a critical role: it is the conduit by which oxygen reaches the roots. When water replaces a soil's air for a long period, such as in flooded soil or overwatered houseplants, roots can become anaerobic and die.

Clay soils, with their tightly packed clay particles, contain less air than do sandy soils. Thus, aeration of a clay soil can be improved by adding silt or sand. The resulting loam often aggregates into macroscopic clumps separated by large air spaces, which have an effect similar to that of earthworms: they increase aeration and the penetration of water into the soil, thereby stimulating root growth.

Water

Plants use huge amounts of water. For example, a mature corn plant needs about 100 l of water per month for optimal growth. This water and its dissolved minerals are obtained from the soil by roots, which are exquisitely designed for gathering water: their continuous growth exploits new territory, and their root hairs greatly increase the surface area for water absorption. Indeed, 1 cm³ of soil under a patch of bluegrass (*Poa pratensis*)

contains more than 150,000 root hairs having almost 400 cm² of surface area. However, even this extensive collecting system does not ensure that a plant can obtain enough water from the soil, since the soil itself also influences the amount of water available to a plant. Recall, for example, that clays hold more water than sandy soils. Another factor that affects the ability of a plant to obtain water is the type of water in the soil.

Chemically Bound Water Some of the water in soil is bound chemically. This includes the water locked in the crystals of minerals and the water hydrated to the surface of clay micelles. Chemically bound water is unavailable to plants.

Unbound Water The remaining water in soil is unbound water and is held by adhesion (the attraction of water molecules for solid particles) and cohesion (the attraction of water molecules for each other). Based on the tension holding unbound water in soil, we can distinguish three types of unbound water: **gravitational water, capillary water,** and **hygroscopic water** (fig. 20.3). To illustrate the differences in these types of water, consider a soil sample to which you have just added an excess amount of water. Some of this water drains from the soil due to gravity, thereby leaving the soil's air spaces filled with air. The water that drains from the soil by gravity is called **gravitational water.** In areas having large amounts of rainfall, the excess gravitational water percolates to lower soil levels, carrying with it nutrients from the upper layers of the soil. This leaching helps form many soils.

When all gravitational water has drained from the soil, the soil is said to be at **field capacity.** Most of the remaining water is held by surface tension (i.e., capillary action) in small pores in the soil; this is called **capillary water.** Since it is held weakly in these pores, most capillary water is easily absorbed by plants. Thus, *capillary water is the main source of water and dissolved minerals for plants.* However, not all capillary water is available to plants; it is gradually lost by evaporation and absorption. When the tension at which the remaining water is held equals the ability of plants to extract it, the amount of water remaining in the soil is the **permanent wilting point.** Thus, the permanent wilting point is the lower limit of soil moisture that can support plant growth. A typical permanent wilting percentage for clay may be as high as 40%, while that of sand may be as low as 4%.

When all of the capillary water is gone, the only unbound water remaining in the soil is **hygroscopic water,** which is held tightly by small soil particles. Plants cannot extract hygroscopic water from soil.

c o n c e p t

Gravitational water is water drained by gravity from the soil, whereas capillary water is held by capillary action in pores of the soil. Hygroscopic water is held tightly by soil particles and is unavailable to plants. Almost all water absorbed by plants is capillary water.

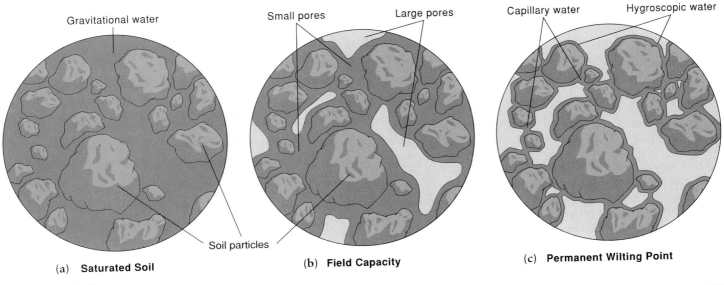

(a) **Saturated Soil**   (b) **Field Capacity**   (c) **Permanent Wilting Point**

Gravitational water   *Small pores*   *Large pores*   *Capillary water*   *Hygroscopic water*   *Soil particles*

figure 20.3

Soil under three conditions of hydration. (a) Water-saturated: all pores are filled with water. (b) Field capacity: large pores are filled with air, whereas small pores are filled with water. (c) Permanent wilting point: all pores are air-filled; films of hygroscopic water (which roots cannot absorb) surround soil particles, and small amounts of capillary water (which roots can absorb) occur at points of particle contact. The permanent wilting point (c) is the smallest amount of hydration that supports plant activity.

Living Organisms

Soil is more than a collection of soil particles, humus, air, and water. It teems with life: on average, about 0.1% of the weight of soil is living organisms. Although this percentage may not impress you, consider what it means: 1 kg of fertile soil contains about 2 trillion bacteria, 400 million fungi, 50 million algae, 30 million protozoa, nematodes, other worms, and insects. Together, these organisms exert a tremendous influence on soil and the plants it supports.

Organisms mix and refine the soil. One earthworm can digest more than a ton of soil in 1 year. This is why Aristotle called earthworms "the intestines of the earth": they aerate and refine the soil by processing it through their guts.

Organisms add humus to the soil. Plants, especially grasses, are the primary source of organic matter in soil. For example, a 4-month-old rye plant produces almost 11,000 kilometers of roots and root hairs, most of which eventually die and become humus. Because humus contains large amounts of nutrients, the biological activity of soil is usually an excellent indicator of its fertility.

Respiration by these organisms increases the amount of CO_2 in the soil. CO_2 in the soil dissolves in water to form carbonic acid, which decreases the soil pH. This, in turn, alters the availability of several nutrients.

Many organisms affect the availability of nutrients. For example, roots of lupine (*Lupinus albus*) growing in phosphorus-deficient soil secrete large amounts of H^+, which can decrease the pH of the soil and thus increase the availability of other nutrients to plants. Similarly, plants and microbes secrete phosphatases into the soil. These enzymes liberate phosphate from mineral particles, making more of it available to plants.

Some organisms make the soil inhospitable for other plants. Many plant pathogens, such as nematodes, live in the soil. Furthermore, some plants produce chemicals that inhibit the growth of other plants. For example, the soft-leaved purple sage (*Salvia leucophylla*) produces terpenoids that inhibit the growth of nearby plants. As a result, colonies of *Salvia* are typically surrounded by bare zones in which no other plants grow.

Finally, soils are a repository for numerous dormant organisms. Among the most influential of these dormant organisms are seeds. Indeed, a cubic meter of typical soil can contain as many as 5,000 viable seeds (see "Seed Banks" in Chapter 17). Although this number of seeds is impressive, it pales when compared to that of some fertile croplands, which contain as many as 80,000 weed seeds per cubic meter. This part of the soil ecosystem is often a problem for farmers.

c o n c e p t

Living organisms in soil refine and mix the soil, add organic matter, increase the amount of CO_2 in the soil, influence the availability of nutrients, and often make the soil uninhabitable for other organisms.

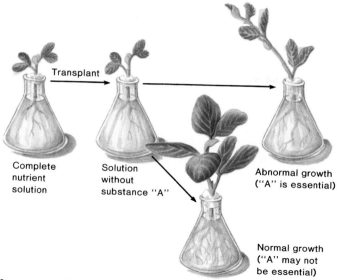

figure 20.4

Discovering essential plant nutrients. Plant physiologists transplant a seedling to a solution lacking only one of the ingredients thought to be essential for growth, substance "A" in this example. If the plant grows and reproduces normally after being transplanted, the missing ingredient is assumed to be nonessential.

Plant Nutrition

In Chapter 7 you were introduced to Jan-Baptista van Helmont, a Dutch physician who measured the growth of a willow tree for several years. That plant, you will recall, gained more than 70 kg yet reduced the weight of the soil by less than 60 g. Van Helmont attributed the growth of his willow tree to the water he had added to the soil. This conclusion was partially correct, since most of a plant is water. However, only later did botanists discover that plant growth also depends on several other nutrients that, like water, are absorbed from the soil and are essential for growth.

Essential Elements

More than 60 elements have been found in plant tissues. These elements range from those as common as carbon and hydrogen to those as exotic as platinum, uranium, and gold. Are all of these elements essential for growth? If so, what functions do they perform?

Answering these questions has proven to be a formidable task. We must first define what we mean by essential. An element is essential if

1. it is required for normal growth and reproduction,

2. no other element can replace it and correct the deficiency, and

3. it has a direct or indirect action in plant metabolism.

Thus, the mere presence of an element in a plant does not necessarily mean it is essential. Armed with this definition, botanists

figure 20.5

Hydroponic farming. In this apparatus, a chemically defined solution of nutrients flows over the roots of lettuce.

began studying the 92 naturally occurring elements to determine how each might influence plant growth. Their approach, at least in theory, was simple: an element would be deemed essential only if a plant could not grow and reproduce without it (fig. 20.4).

Many of the early experiments to determine the essentiality of an element involved *hydroponics*, a technique for growing plants without soil in a chemically defined liquid medium (fig. 20.5). By the middle of the 1800s, these experiments had demonstrated that plants require at least nine elements: carbon, hydrogen, oxygen, phosphorus, potassium, nitrogen, sulfur, calcium, and magnesium. Since these elements are required in relatively large amounts (i.e., usually more than 0.5% of the dry weight of the plant), they became known as **macronutrients** (fig. 20.6).

Subsequent studies of the essentiality of mineral nutrients encountered several problems, perhaps the largest of which was contamination. For example, "pure" water and chemicals used to prepare nutrient solutions often contained enough impurities to satisfy plants' requirements for several elements. Furthermore, the growth containers used in hydroponics experiments were often made of glass that contained large amounts of boron. When these containers were filled with nutrient solutions, enough boron dissolved out of the glass to satisfy the plant's requirement for boron. Such impurities, combined with insensitive methods for detecting many elements, made it impossible to determine the essentiality of many elements. To establish the essential nature of some micronutrients, plants had to be grown through several hydroponic generations, because there were sufficient amounts of the element in the original seed to permit normal growth and reproduction.

Progress in analytical chemistry has improved the purity of chemicals and increased the detectability of specific elements, thereby allowing botanists to discover seven other essential elements: iron, chlorine, copper, manganese, zinc, molybdenum,

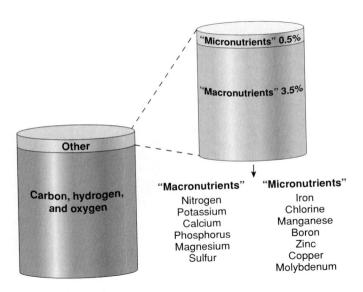

figure 20.6

The proportional weights of various elements in plants. Macronutrients and micronutrients together total only 4% of the total weight of the plant, but they are essential to the plant's life and growth.

and boron. Because these nutrients are required in relatively small amounts (i.e., usually only a few parts per million), they are called **micronutrients,** or trace elements (fig. 20.6). Although required in smaller amounts than macronutrients, micronutrients are equally essentially for growth. The 4 metric tons of carbon, hydrogen, and oxygen were no more important to our Idaho potato-farmer's crop than were its 2.5 g of copper.

c o n c e p t

Plants require at least 16 essential elements. Carbon, hydrogen, oxygen, phosphorus, potassium, nitrogen, sulfur, calcium, and magnesium are required in relatively large amounts and are called macronutrients. Iron, chlorine, copper, manganese, zinc, boron, and molybdenum are required in relatively small amounts and are called micronutrients.

There is no reason to think that the 16 essential elements listed are the only ones that plants require. Future improvements in analytical technology may reveal other essential elements. For now, we can only say that plants require *at least* 16 elements.

Functions of Essential Elements

The functions of the 16 elements essential for plant growth are summarized in table 20.2. A quick survey of this table leads to three important generalizations:

1. **Plants require different amounts of different elements.** Most plants require about 60 million times more hydrogen atoms than molybdenum atoms. These differing requirements reflect the differing uses of these elements: hydrogen is in almost all compounds in plants, whereas molybdenum occurs in only a few.

2. **Different elements are absorbed in different forms.** Calcium and iron are absorbed as cations, whereas phosphorus and sulfur are absorbed as anions.

3. **Most elements have several functions.** Potassium is involved in starch synthesis, affects protein conformation, and activates enzymes. Interestingly, the functions of most micronutrients in plants are similar to those in animals.

Although the functions of essential elements are diverse, they can be grouped into four general categories:

1. **Essential elements can be parts of structural units.** Carbon, hydrogen, and oxygen make up carbohydrates such as cellulose. Similarly, nitrogen is an integral part of proteins.

2. **Essential elements can be parts of compounds involved in metabolism.** Magnesium is part of chlorophyll, and phosphorus is part of ATP and nucleic acids.

3. **Essential elements can activate or inhibit enzymes.** Magnesium stimulates several respiratory enzymes, while calcium inhibits several enzymes. In some cases, these enzymes may be those responsible for synthesizing plant hormones.

4. **Essential elements can alter the osmotic potential of a cell.** For example, the movement of potassium into and out of guard cells is responsible for osmotic changes that cause stomata to open and close.

c o n c e p t

Essential elements are absorbed in differing amounts and forms. Once absorbed, these elements serve as parts of structural units and metabolites, activate and inhibit enzymes, and influence the osmotic potential of a cell.

The 16 elements listed in table 20.2 are essential for all plants studied so far. Some other elements appear to be essential for only a few plants. For example,

Silicon (Si) is essential for some grasses and horsetails. Silica-containing cells comprise up to 2%–16% of the dry weight of these kinds of plants.

Sodium (Na) is essential for CAM plants, C_4 plants, and several *halophytes*, which are plants that grow in salty soils.

Nickel (Ni) may be required by soybeans, and **selenium** (Se) is required by some halophytes such as *Stanleya* (fig. 20.7; also see reading 21.2, "Living in Salty Soil").

These observations indicate that the essentiality of an element depends on the genotype of a plant. Similarly, different cultivars of some crop plants have differing sensitivities to mineral deficiencies. Identifying and incorporating genes for this tolerance to mineral deficiency into other crops is currently a major area of research.

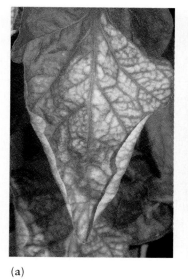

(a)   (b)

figure 20.7

The prince's plume (*Stanleya elata*), shown here growing in Death Valley, California, requires selenium. Its selenium level is so high that this plant is toxic to browsing mammals.

Deficiencies of Essential Elements

Plants require large amounts of nitrogen, phosphorus, and potassium. As a result, these are the elements most likely to be deficient in soil. Remedying deficiencies of these elements often requires adding large amounts of fertilizer. Deficiencies of micronutrients are corrected more easily; for example, molybdenum-deficient soil can be replenished by annually adding less than 4 g of molybdenum to a hectare of land, and deficiencies of zinc and iron have been corrected by driving zinc-coated tacks or iron nails into the trunks of trees. The deficiency symptoms of essential elements are listed in table 20.2 and are shown in figure 20.8.

(c)

figure 20.8

Nutrient deficiencies produce distinctive symptoms. (a) Potassium deficiency causes leaves to curl at their edges. (b) Phosphorus deficiency causes dark green or purple leaves in seedlings. (c) Iron deficiency causes chlorotic leaves (veins remain green). A deficiency of manganese produces similar symptoms.

table 20.2

Some Functions and Deficiency Symptoms of Essential Elements

Essential Elements	Chemical Symbol	Atomic Weight	Form Available to Plants	Percentage of Dry Tissue	Number of Atoms Relative to Molybdenum
MICRONUTRIENTS					
Molybdenum	Mo	95.95	MoO_4^{2-}	0.00001	1
Copper	Cu	63.54	Cu^+, Cu^{2+}	0.0006	100
Zinc	Zn	65.38	Zn^{2+}	0.002	300
Manganese	Mn	54.94	Mn^{2+}	0.005	1,000
Boron	B	10.83	H_3BO_3	0.002	2,000
Iron	Fe	55.85	Fe^{3+}, Fe^{2+}	0.01	2,000
Chlorine	Cl	35.46	Cl^-	0.01	3,000
MACRONUTRIENTS					
Sulfur	S	32.07	SO_4^{2-}	0.1	30,000
Phosphorus	P	30.98	$H_2PO_4^-$, HPO_4^{2-}	0.2	60,000
Magnesium	Mg	24.32	Mg^{2+}	0.2	80,000
Calcium	Ca	40.08	Ca^{2+}	0.5	125,000
Potassium	K	39.10	K^+	1.0	250,000
Nitrogen	N	14.01	NO_3^-, NH_4^+	1.5	1,000,000
Oxygen	O	16.00	O_2, H_2O	45	30,000,000
Carbon	C	12.01	CO_2	45	35,000,000
Hydrogen	H	1.01	H_2O	6	60,000,000

Functions	Deficiency Symptoms (see fig. 20.8)
Part of nitrate reductase Essential for N fixation	Chlorosis or twisting of young leaves
Component of plastocyanin, a plastid pigment Present in lignin of xylary elements Activates enzymes	Young leaves dark green, twisted, and wilted; tips of roots and shoots remain alive Rarely deficient
Necessary for formation of pollen Involved in auxin synthesis Maintenance of ribosome structure Activates enzymes	Chlorosis, smaller leaves, reduced internodes; distorted leaf margins; older leaves most affected
Photosynthetic O_2 evolution Enzyme activator Electron transfers	Interveinal chlorosis; appears first on older leaves; necrosis common; disorganization of lamellar membranes
Essential for growth of pollen tubes Regulation of enzyme function Possible role in sugar transport	Death of apical meristems; leaves twisted and pale at base; swollen, discolored root tips; young tissue most affected
Required for synthesis of chlorophyll Component of cytochromes and ferredoxin Cofactor of peroxidase and some other enzymes	Interveinal chlorosis; short and slender stems; buds remain alive; affects young leaves first
Activates photosynthetic elements Functions in water balance	Wilted leaves; chlorosis; necrosis; stunted, thickened roots
Part of coenzyme A and the amino acids and cysteine and methionine Can be absorbed through stomata as gaseous SO_2	Interveinal chlorosis; usually no necrosis; affects young leaves Rarely deficient
Part of nucleic acids, sugar phosphates, and ATP Component of phospholipids of membranes Coenzymes	Stunted growth, dark green pigmentation; accumulation of anthocyanin pigments, delayed maturity; affects entire plant Second to N, P is element most likely to be deficient
Part of chlorophyll Enzyme activator Protein synthesis	Chlorosis and reddening of leaves; leaf tips turn upward; older leaves most affected
Membrane integrity In middle lamella Functions as "second messenger" to coordinate plant's responses to many environmental stimuli Reversibly binds with calmodulin, which activates many enzymes	Death of root and shoot tips; young leaves and shoots most affected
Regulates osmotic pressure of guard cells, thereby controlling opening and closing of stomata Activates more than 60 enzymes Necessary for starch formation	Chlorosis and necrosis, weak stems and roots; roots more susceptible to disease; older leaves most affected
Part of nucleic acids, chlorophyll, amino acids, protein, nucleotides, and coenzymes	General chlorosis, stunted growth; purplish coloration due to accumulation of anthocyanin pigments N is element most likely to be deficient
Major component of plant's organic compounds	Rarely limiting enough as nutrient to cause specific symptoms
Major component of plant's organic compounds	Rarely limiting enough as nutrient to cause specific symptoms
Major component of plant's organic compounds	Rarely limiting enough as nutrient to cause specific symptoms

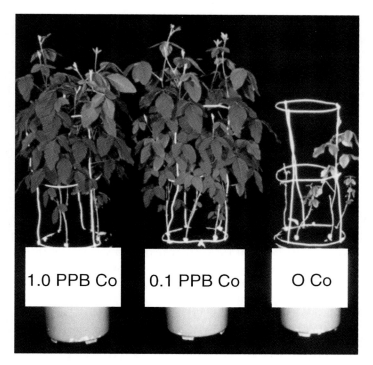

1.0 PPB Co   0.1 PPB Co   O Co

figure 20.9

Cobalt is required by nitrogen-fixing bacteria in root nodules. When cobalt is available (two plants on left, grown at cobalt concentrations of 1.0 and 0.1 parts per billion),[1] bacteria fix nitrogen and stimulate growth in soybeans; when cobalt is unavailable (plant on right), bacteria cannot fix nitrogen, so soybean growth is slow.

Beneficial Elements

Several elements in plants are not essential but will promote growth. For example, cobalt stimulates growth of legumes such as beans, clover, and alfalfa. This stimulation of growth by cobalt is due to its use not by the plant itself but by nitrogen-fixing bacteria that live in roots of the plants. In these plants, as little as 1 mg of cobalt per liter of nutrient solution stimulates growth dramatically (fig. 20.9).

Other Elements in Plant Tissues

Plants such as alpine pennycress (Thlaspi caerulescens) are hyperaccumulators, meaning that they concentrate trace elements, heavy metals, or radionuclides at levels 100-fold (or more) greater than normal. Some examples of metals concentrated by hyperaccumulators are:

Lead (Pb) has been until recently an antiknock component of most gasolines and is abundant along highways. Plants growing near these highways often absorb this lead, as do some grasses that grow near lead mines.

Cadmium (Cd), a contaminant in many chemical fertilizers, is common in plants growing in fertilized soils and near many military sites.

1. One part per billion (ppb) = 1 microliter in 1,000 liters = 1 second in 32 years.

Gold (Au) often occurs in plants such as *Phacelia sericea* that grow near gold mines. The gold content of these plants has been used by geologists to locate deposits of gold in soil.

Strontium (Sr) is sometimes present in plants and will partially substitute for calcium. Radioactive strontium (^{90}Sr) is abundant in plants growing near nuclear test sites and sites of nuclear accidents such as Chernobyl.

The use of hyperaccumulators to remove toxic substances from soil is part of a growing practice called *phytoremediation*. For example, at one site, a field of *Brassica juncea* (a member of the mustard family) decreased the concentration of selenium by more than half down to a depth of 1 m. *Sebertia acuminata* accumulates so much nickel (up to 20% of its dry weight) that its cell sap is jade green (in chromium accumulators it's blue). Selenium may account for more than 1% of the dry weight of locoweed (*Astragalus*). If locoweed is eaten by livestock, the large amounts of selenium in these plants affect the animals' nervous systems, causing them to stagger about as if intoxicated. This illness, called *"blind staggers,"* often kills livestock and has even caused ranchers to abandon their ranches. Similarly, bees that harvest pollen from locoweed often produce honey containing large amounts of selenium. Locoweed uses selenium to help combat the toxic levels of phosphorus characteristic of selenium-rich soils: the presence of selenium decreases the absorption of phosphorus, thereby preventing phosphorus toxicity. Hyperaccumulators tolerate high concentrations of metals by using *phytochelatins*, which are small polypeptides that bind metals and, in the process, reduce their toxicity. Metals are usually stored in vacuoles.

Phytoremediation is expected to save some of the $42 billion the United States has budgeted for soil cleanups over the next 5 years (costs range from $190,000 to $2,800,000 per hectare). The hyperaccumulation of metals also benefits the plants by helping them to evade weak-stomached predators, including caterpillars, fungi, bacteria, and humans. High-metal plants can poison unwelcome guests.

Obtaining Essential Elements

Obtaining enough essential elements is an ongoing problem for most plants. Many of the tropic and nastic movements discussed in the previous chapter help solve this problem. For example, positive gravitropism in roots increases their chances of encountering dissolved minerals in soil. However, plants' adaptations for absorbing nutrients are not limited to these growth responses. Rather, plant nutrition is an elaborately orchestrated process that depends on a variety of factors, ranging from the architecture of plant cells to the assistance of other organisms. Plant nutrition is based not only on the presence of nutrients but also on their *availability*.

Factors Determining the Availability of Essential Elements

The availability of an essential element is influenced by factors such as the gradual breakdown of rocks into mineral particles,

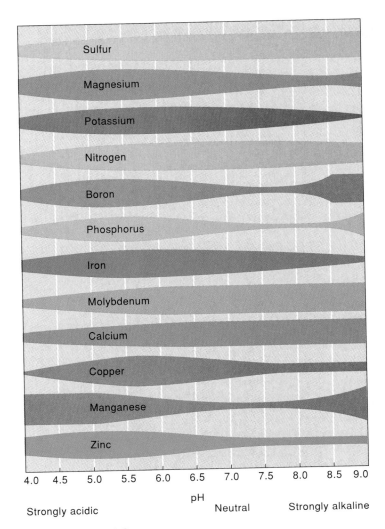

figure 20.10

How the pH of soil affects the availability of nutrients. The width of the shaded areas corresponds to the availability of the nutrient to roots.

the decay of humus, and the cation-exchange capacity of the soil. Four other factors that determine nutrient availability are soil pH, chelators, other elements, and the presence of other organisms.

Soil pH Neutral and acidic soils. Most soils have a pH between 6 and 7, and thus can support almost any plant. However, cation exchange, the decomposition of humus, and acidic precipitation can decrease the pH of soil to as low as 3. The pH of the soil affects the availability of many plant nutrients (fig. 20.10). For example, decreasing the pH of soil increases the availability of iron, manganese, and aluminum, often to toxic levels. Excess amounts of these elements often react with phosphate to produce insoluble precipitates. Since these precipitates are unavailable to plants, phosphorus deficiencies are common in acidic soils. Not surprisingly, acid rain has contributed significantly to the development of acidic soils in some areas. Acid rain has leached these soils of large amounts of the base minerals that neutralize acids and are essential to plant growth.

Thousands of species of plants (e.g., cranberry, blueberry) can grow well in acidic soils. The pH of acidic soil can be corrected by adding limestone ($CaCO_3$) or lime (CaO).

Alkaline soils. Alkaline soils having a pH above 8 usually form from limestone. Alkalinity decreases the availability of nutrients. For example, the excess amounts of Ca^{2+} in alkaline soils react with phosphate to produce insoluble $Ca_3(PO_4)_2$, which is unavailable to plants. The availability of iron and manganese also decreases in alkaline soil, primarily because they form insoluble precipitates.

Many plants, such as alfalfa and sweet clover, grow well in alkaline soils. The pH of alkaline soils can be neutralized by adding sulfur, which bacteria readily oxidize to sulfate. This sulfate then dissolves in water to form sulfuric acid, an acid that decreases the pH of the soil.

Chelators   A **chelator** is a compound that donates electrons to a metal cation such as Fe^{3+} and thereby increases its solubility. For example, versene is a chelator that maintains iron in soil as Fe^{2+}, which is more soluble than Fe^{3+}. Many organisms help their own cause by secreting chelators into the soil. By doing so, they increase the availability of many elements.

Other Elements   The requirement of a plant for a particular element often depends on the presence of other elements. For example, sodium can substitute for part of a plant's potassium requirement. Furthermore, a particular nutrient can affect different plants in different ways: the same amount of iron in soil that supports normal growth of alfalfa (*Medicago sativa*) can cause chlorosis in garden bean (*Phaseolus vulgaris*).

Other Organisms   The amounts of nutrients available to plants depend largely on the presence and activities of other organisms. Other organisms in soil compete with plants for nutrients and thereby decrease the amount of nutrients available to plants. Meanwhile, decomposers such as fungi degrade humus and thereby increase the supply of nutrients available to plants.

c   o   n   c   e   p   t

The availability of essential elements is influenced by mineral particles, cation exchange, decay of humus, pH, chelators, other elements, and other organisms.

The preceding discussion should convince you that nutrient availability is a complex phenomenon. Yet the mere availability of nutrients only begins the story, for to be useful to plants, these nutrients must also be located and absorbed.

Locating Essential Elements

Locating most of the essential elements is not a problem for plants. For example, CO_2 is always present in the air, and water is usually abundant enough to support at least minimal

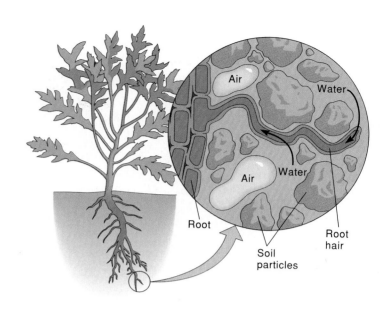

figure 20.11

Root hairs absorb water and nutrients from the soil.

growth. As a result, deficiencies of carbon, hydrogen, and oxygen rarely occur in plants. However, nitrogen in soils almost always limits plant growth, and concentrations of some micronutrients may be undetectable. Thus, growth and development largely depend on a plant's ability to locate these nutrients. One way that plants locate essential elements is by efficiently exploring the soil. For example, a mature oak tree can have several million root tips distributed throughout the soil. Each of these root tips has thousands of root hairs, each of which mines a different area for water and dissolved minerals (fig. 20.11). As a consequence of this extensive collecting system, few available sources of nutrients go untapped by plants.

Absorbing Essential Elements

Essential elements are useless to plants until they are absorbed. Absorption of nutrients depends on (1) exposing a large surface for absorption, and (2) having a mechanism for selectively absorbing and excluding elements.

Exposing a Large Surface Area for Absorption   Roots and root hairs not only permeate the soil but also greatly increase the surface area available for absorbing nutrients. For example, roots of a 4-month-old rye plant have a surface area exceeding 600 m^2, and the 150,000 root hairs in only 1 cm^3 of soil beneath a patch of bluegrass have a surface area equal to approximately two-thirds of this page. Exposing this much surface area to the environment greatly increases the absorptive capabilities of plants.

The structural design of plants also increases the surface area available for absorbing nutrients. To understand this, examine the root cross section shown in figure 20.12. When water and dissolved minerals move through the apoplast, they

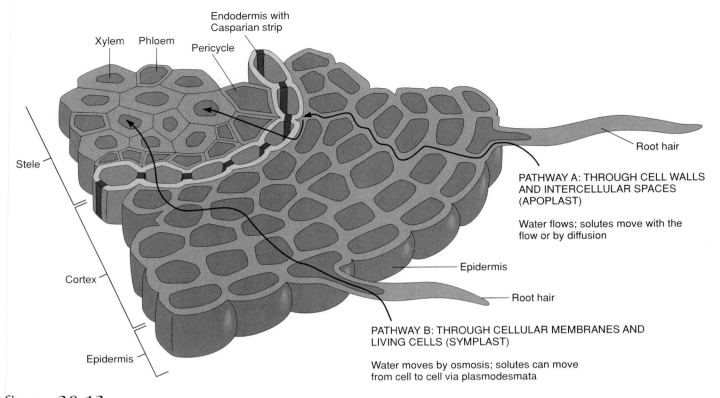

figure 20.12

Apoplastic versus symplastic pathways in roots. Casparian strips in cell walls of the endodermis block off the apoplastic pathway all around the stele of the root. As a result, water and dissolved minerals enter the stele via the symplastic pathway, ×100.

remain outside of the cell membrane. As a result of this apoplastic movement, the cell membranes of cortical cells become absorptive surfaces equivalent to those of root hairs. Thus, the absorptive surface area of most roots is not restricted to the epidermis and its root hairs; it also includes all of the cell membranes of the cortex. Cells of the cortex increase the absorptive surface area and thereby improve the nutrient uptake by a plant.

Absorption The concentrations of elements in plants differ from those in soil. For example, aluminum is usually abundant in soil but rare in plant tissues. Conversely, most micronutrients are usually more concentrated in plants than in soil. Therefore, plants must *selectively* absorb elements from the soil. This selectivity depends on the presence of **ion-specific carriers** in the membranes of plant cells. For example, the membranes of cortical cells contain a K^+-specific carrier that requires Mg^{2+} to function. Such carriers enable a plant to absorb and maintain adequate amounts of various essential elements in their tissues, even though the elements may be more or less concentrated in the soil. Absorption of these elements requires energy.

c o n c e p t

All essential elements except carbon are absorbed by roots. This absorption is aided by the large surface areas of roots and the presence of ion-specific carriers in the cell membrane.

The Fate of Absorbed Nutrients

Once absorbed into the cytoplasm, a mineral can be used immediately or transported to a different part of the plant. As you learned in Chapter 15, transport from the cortex requires that the nutrient move through the endodermis and pericycle before being loaded into the xylem for transport (fig. 20.12). This transport is critical, because different parts of the plant require different amounts of essential elements.

Modifying the Form of Essential Elements Most nutrients are used in the same form in which they are absorbed. For example, iron is incorporated directly into cytochromes, and phosphates are added directly to ADP to form ATP. Other elements, most notably nitrate and sulfate, are reduced before use.

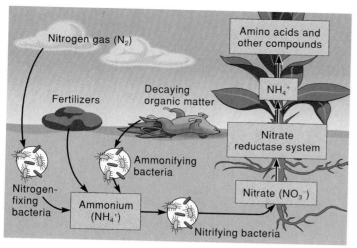

figure 20.13

How plants get nitrogen from the soil.

figure 20.14

Scanning electron micrograph of mycorrhizae: root of a lodgepole pine (*Pinus contorta*) associated with filaments of mycorrhizae.

Nitrate. Most nitrogen in fertilizer is in the form of ammonium salts. Ideally, this would ensure high availability of nitrogen to plants, since the NH_4^+ cation should bind to negatively charged clay micelles. However, NH_4^+ in soil is rapidly converted to NO_3^- by bacteria such as *Nitrosomonas* in a process called **nitrification** (fig. 20.13). As a result, most nitrogen in soil is in nitrate. Because nitrate is negatively charged, it does not bind to soil and is readily leached from the soil by rainfall. This is why nitrogen must be replenished by adding fertilizer.

Plants must reduce NO_3^- to ammonium before it can be incorporated into proteins. This conversion is catalyzed by a group of enzymes collectively called the **nitrate reductase system** (fig. 20.13). These enzymes produce amino acids such as glutamine, which are transported throughout the plant. This reaction is important because it provides a basic supply of nitrogen-containing compounds that can be used for protein synthesis.

Plants grown hydroponically in the absence of NO_3^- contain negligible levels of the nitrate reductase system. If given NO_3^-, plants rapidly synthesize the nitrate reductase system; that is, the nitrate reductase system is inducible. Although common in bacteria, inducible enzymes such as nitrate reductase are rare in plants.

Sulfate. Sulfur is absorbed almost exclusively as sulfate (SO_4^{2-}) and must therefore be reduced before being incorporated into proteins. The enzyme that reduces SO_4^{2-} to S^{2-} is ATP sulfurylase. Almost 90% of this reduced sulfur is incorporated into the amino acids cysteine and methionine.

c o n c e p t

Many essential elements are used in the form in which they are absorbed. Other nutrients, most notably nitrate and sulfate, are reduced before use.

Mobility of Essential Elements within Plants Utilization of an essential element often affects its movement in plants. For example, boron, iron, and calcium are *immobile elements:* they do not move once they are used in the shoot. As a result, deficiency symptoms of immobile elements always appear first at the shoot tip and in the youngest leaves. Conversely, potassium, nitrogen, phosphorus, magnesium, and sulfur can be transported after being used in the shoot. These elements are called *mobile elements;* their deficiency symptoms appear first in older leaves.

Associations with Other Organisms

Despite their structural and functional adaptations for obtaining nutrients, plants are not as independent as many people think they are. Indeed, their mineral nutrition often depends on associations with other organisms. These associations range from mutually beneficial symbioses to parasitisms, whereby plants rely on other organisms as nutrient collectors as well as prey.

Fungi When plants invaded the land some 400 million years ago, one of the biggest problems they faced was obtaining enough nutrients. Their aquatic habitat contained many decomposers that degraded large molecules, thereby increasing the availability of nutrients. As a result, aquatic organisms were literally bathed in nutrients. On land, however, the situation was different: most soils contained few fungi and therefore did not contain enough nutrients to sustain plant growth.

Plants adapted to life on land by forming mutually beneficial associations called **mycorrhizae** between their roots and fungi (fig. 20.14). In effect, plants took the fungi with them onto the land and thereby helped ensure the continued availability of nutrients for their growth. Plants benefited from the ability of

COMPETING FOR THE SOIL'S NITROGEN

Many plants produce large amounts of phenolic compounds that help ward off pathogens and herbivores (see Chapter 2). These compounds may also function in plant nutrition by making some nutrients available to the parent plant and worthless to others. To appreciate this, consider the Bishop pine (*Pinus muricata*). These pines produce phenolic compounds in their leaves: as the fertility of the soil decreases, the concentration of phenolics in leaves increases. When these leaves are shed from the plant, the phenolic compounds leach into the soil, where they bind with organic nitrogen and make it unusable by most other organisms. The only organisms that can release the nitrogen from the phenolic compounds are the fungi that live in the trees' mycorrhizae. These fungi supply the newly acquired nitrogen to the pines, thereby enabling them to outcompete other plants for this nutrient.

figure 20.15

Mycorrhizae stimulate the growth of plants. This photo shows how mycorrhizae affect the growth of lemon trees (*Citrus limon*): the plants on the right were grown with mycorrhizae, whereas those on the left were grown without mycorrhizae. These trees are 4.5 months old.

the fungi to efficiently absorb and make available nutrients from the environment. The fungi also benefited, since the plants provided them with carbohydrates for growth.

Mycorrhizae occur in more than 80% of all plants studied. From a functional point of view, mycorrhizae replace root hairs; that is, mycorrhizae increase nutrient absorption. This is especially true for the absorption of phosphorus: roots with mycorrhizae absorb as much as 4 times more phosphorus than do roots that lack them (also see reading 20.3, "Competing for the Soil's Nitrogen"). Two factors contributing to this efficient absorption are that (1) the fungal hyphae have a much greater surface area than roots, and (2) the fungal hyphae permeate a greater soil volume than do the roots. As a result of this increased absorption, mycorrhizae often dramatically increase plant growth; for example, they increase the growth of wheat by more than 200%, corn by 100%, and onions by more than 3,000%. Mycorrhizae also improve a plant's endurance of drought, disease, extreme temperatures, and high salinity, and are important for cycling nutrients, reclaiming strip-mined land, and establishing nursery stocks.

Mycorrhizae play critical roles in determining whether tree species can invade a disturbed area and are especially important for plants growing in nutrient-poor soils. For example, "pioneer" plants such as alder that grow on nutrient-deficient soil invariably have mycorrhizae, as did the first plants that invaded the land. Today, many plants even *require* mycorrhizae for vigorous growth. For example, orchid seeds germinate only in the presence of a mycorrhizal fungus, and many citrus trees and gymnosperms are difficult to grow unless mycorrhizae are present (fig. 20.15).

c o n c e p t

Mycorrhizae are mutually beneficial associations between roots and fungi. These fungi increase the absorption of nutrients and tolerance of stress.

Bacteria Nitrogen is usually deficient in soils; this deficiency often limits the growth of plants and, as a consequence, animals. The deficiency of nitrogen in soil is somewhat paradoxical, since the atmosphere is almost 80% N_2 ($N\equiv N$). The triple bond holding the two atoms of gaseous nitrogen together makes N_2 stable and therefore unusable by most organisms. However, a few bacteria contain an enzyme complex called **nitrogenase** that can reduce atmospheric N_2 to ammonia (fig. 20.16). These bacteria live alone as well as symbiotically with plants.

Free-living nitrogen fixers. Many bacteria capable of fixing atmospheric nitrogen are free-living; that is, they do not form any associations with other organisms. These prokaryotes, which include heterotrophic bacteria (e.g., *Azotobacter*), photosynthetic bacteria (e.g., *Heliobacillus*), and cyanobacteria (e.g., *Nostoc*), fix only the nitrogen that they need. On average, this amounts to about 7 kg of nitrogen per hectare per year, which is roughly equivalent to the amount of nitrogen added to soil by precipitation containing atmospheric pollutants. When these bacteria die, their nitrogen becomes available to plants.

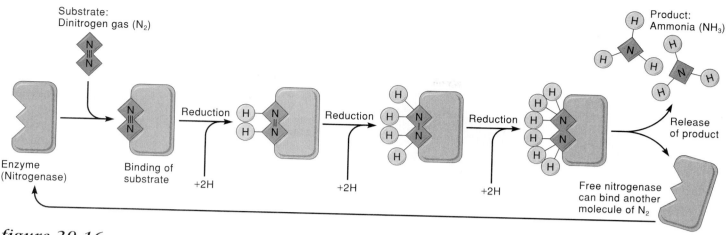

figure 20.16

Nitrogenase is a bacterial enzyme complex that converts atmospheric nitrogen (N_2) to ammonia (NH_3).

(a)

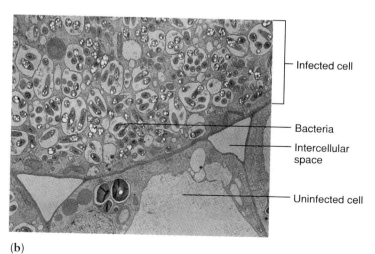

(b)

figure 20.17

Nitrogen fixation in nodules of roots. (a) Root nodules that were induced by *Rhizobium* bacteria. (b) Infected cells of nodules contain nitrogen-fixing bacteria, ×11,500.

Symbiotic nitrogen fixers. Some plants have established symbioses with nitrogen-fixing bacteria and therefore can obtain much of their nitrogen from the atmosphere. For example, the roots of many legumes such as beans and alfalfa are often infected with *Rhizobium* bacteria. These bacteria secrete carbohydrate-like molecules called lipo-chitooligosaccharides that trigger formation of nodules (figs. 20.17, 15.19a); the plant hormone ethylene is also involved in the formation of nodules in these plants. Unlike free-living bacteria, *Rhizobium* fixes almost 10 times more nitrogen than it uses. The excess nitrogen is released into the host plant, where it is used for growth. In return, the bacteria get carbohydrates from the host plant. Interestingly, the presence of NO_3^- decreases infection of the host by *Rhizobium*, the growth of nodules, and the activity of nitrogenase, thereby decreasing nitrogen fixation. Consequently, applying nitrogen-containing fertilizer to legumes seldom increases yield.

Nodules containing nitrogen-fixing bacteria are not restricted to roots. For example, nodules are common on leaves of members of certain families, such as the Myrsinaceae and Rubiaceae; removing these nodules often causes deformation, dwarfing, and chlorosis. However, nitrogen may not be the most important contribution of nodules to these plants, since applying nitrogen does not always relieve these symptoms.

Nitrogen-fixing prokaryotes also form symbioses with plants other than legumes. For example, alders (*Alnus*) and mountain lilac (*Ceanothus*) contain *Frankia*, a nitrogen-fixing bacterium, and nitrogen-fixing bacteria form symbioses with such diverse plants as cycads, mosses (*Sphagnum*), liverworts (*Blasia*), and tropical trees (*Parasponia*). These associations, like those with legumes, often have important agricultural implications. For example, the cyanobacterium *Anabaena* grows symbiotically in leaf cavities of water fern (*Azolla*) (fig. 20.18). When infected *Azolla* are grown in rice paddies, some of the nitrogen fixed by *Anabaena* dissolves in the water of the paddy and becomes available to rice plants. As a result, rice crops in China and southeast Asia can be grown in nitrogen-deficient soils. Botanists are now searching for new strains of *Anabaena* to increase rice production.

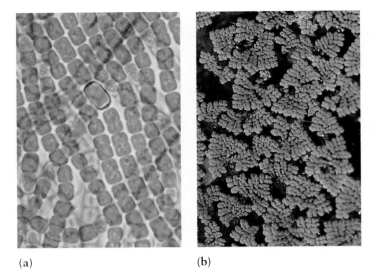

(a)   (b)

(c)

figure 20.18

Anabaena and rice production. (a) Cultivation of rice is aided by nitrogen fixation by the cyanobacterium *Anabaena azollae*, ×400. (b) *Anabaena* lives and fixes nitrogen in cavities of the water fern, *Azolla*. (c) *Azolla* is grown in rice paddies of the warmer parts of Asia. Later, the paddies are drained, and the *Azolla* is plowed into the soil, thereby enriching it with nitrogen for the rice crop. Nitrogen fertilizers do not have to be used in such paddies.

c o n c e p t

Many plants, especially legumes, establish mutually beneficial symbioses with nitrogen-fixing bacteria. The bacteria reduce atmospheric N_2 to ammonia, which the host uses for growth.

The Lore of Plants

Although it often seems that bacteria are the aggressors, nodule formation is an example of plants taking the first step in controlling the metabolism of another organism. Roots of alfalfa and other legumes secrete flavonoids that bind to and activate a bacterial nodulation gene called *nodD*. The *nodD* gene product activates other *nod* genes, whose products then turn on plant genes that control the growth and function of the nodule.

Not all symbioses between plants and nitrogen-fixing bacteria involve nodules. For example, you learned in Chapter 15 that roots secrete mucigel that helps them move through soil. Mucigel also houses nitrogen-fixing bacteria such as *Spirillum*, *Azospirillum*, and *Klebsiella*. These bacteria, which are common on grasses such as corn and sorghum, fix nitrogen that can be absorbed by the host plant. In return, the bacteria receive sugars, amino acids, and proteins that the plants secrete into their mucilage.

Although each of these symbioses between nitrogen-fixing bacteria and plants evolved independently, they all have a common function: they make nitrogen from the atmosphere available to plants. This recycling of nitrogen by bacteria is an important aspect of how plants obtain nitrogen (fig. 20.13). For example, factors such as harvesting crops remove more than 21 million metric tons of nitrogen per year from the soil. Nitrogen fixation by bacteria replenishes almost two-thirds of this loss. The remaining nitrogen is resupplied to the soil in fertilizers and other sources.

Nitrogen fixation has tremendous agricultural implications. For example, much of the nitrogen fixed in nodules is released into the soil. Many farmers exploit this by practicing *crop rotation*, by which leguminous crops are alternated with nonlegumes as a means of maintaining the nitrogen in the soil. Plowing under a leguminous crop has even more dramatic effects. For example, plowing under a crop of alfalfa adds humus and more than 300 kg of nitrogen per hectare—more than 45 times that contributed by free-living bacteria. Nitrogen fixation also reduces the need to apply expensive fertilizers (see reading 20.1, "Putting Things Back"). Molecular biologists are busy trying to incorporate nitrogen fixation into other crop plants such as corn and wheat.

Carnivorous Plants   Biologists have long been fascinated with carnivorous plants. Indeed, the botanists who discovered carnivorous plants referred to them as "the great wonder of the vegetable world" and "carnivorous vegetables." Today, the bizarre life-styles of these plants remain a favorite subject of B-movies. Although unusual, the carnivorous nature of

figure 20.19

Pitcher plants (*Sarracenia purpurea*) have leaves modified into pitchers that contain digestive enzymes (a). The downward-pointing hairs (b) help lure insects to their doom in the trap. There, the insects are digested and their nutrients are absorbed.

figure 20.20

Cobra lily (*Darlingtonia californica*) lures insects by sight and smell into its pitcher trap. Light shining through the "windows" atop the dome fosters a false sense of escape.

these plants serves an important purpose: it is an adaptation to the nutrient-poor histosols in which they grow. To obtain enough nutrients, especially nitrogen, carnivorous plants have evolved a life-style that resembles that of many heterotrophs—namely, extracting nutrients from the nitrogen-rich flesh of other organisms. The modifications for this life-style include sticky, flypaperlike surfaces, springlike trapdoors, and pitfalls leading to vats filled with digestive enzymes. You've already been introduced to sundew (fig. 19.12) and Venus's-flytrap (fig. 19.11), both of which utilize seismonasty to trap and digest their prey. However, the trapping mechanisms of carnivorous plants are not limited to these kinds of traps.

Passive traps involve no movement to trap prey. For example, leaves of pitcher plants, (*Sarracenia;* figs. 20.19, 14.24c) and cobra lily (*Darlingtonia;* fig. 20.20), are modified into tubular vats. Insects drawn into these pitfalls can initially walk across the downward-pointing hairs lining the pitcher (fig. 20.19). However, about halfway down the pitcher, the

hairs are replaced by a slick wax to which insects cannot adhere, and the unsuspecting prey slips into the digestive enzymes, where they are digested within a few hours or days. Pitchers of some carnivorous plants attain impressive sizes: pitchers of *Sarracenia* can hold more than a liter, while the lavatory-shaped pitchers of *Nepenthes rajah* can be large enough to hold a rugby football. Despite the large size of such traps, their prey are usually small insects.

Genlisia has unique and complex traps. The entrances of these traps contain numerous clawlike hairs that allow an animal to enter but not to leave. As a result, prey eventually die in the trap and are digested. These passive traps are designed like lobster traps, or, as Charles Darwin put it, like "the eel trap, though more complex."

Active traps move to trap prey. Sundew and Venus's-flytrap have active traps, as does butterwort (*Pinguicula*)—sticky leaves of this plant often fold over trapped prey to speed digestion in what Charles Darwin called a "temporary stomach."

figure 20.21

A bladderwort (*Utricularia minor*) trapping a mosquito larva, ×6.

A butterwort such as *P. grandiflora* grows a new "trap" (i.e., leaf) about every 5 days, so that more than 400 cm² of catching surface form in a single season. In Elizabethan times, *Pinguicula* supposedly protected cattle from the mischief of elf arrows and protected humans from the spells of witches. Despite such folklore and the claims of Hollywood movies, these plants seldom trap prey larger than houseflies.

One of the most ingenious traps belongs to bladderwort (*Utricularia*), an aquatic, carnivorous plant that supplements its diet with protozoa, mosquito larvae, and small arthropods. The traps of *Utricularia* range in diameter from 0.25–5 mm and are elastic (hence the name *bladderwort*). These traps involve a trapdoor mechanism set off by trigger hairs (fig. 20.21). Here's how they work:

Cells of the trap secrete ions out of the trap, thereby causing water to leave the trap via osmosis.

This loss of water creates a negative pressure in the trap, which establishes a balanced tension that closes the trapdoor.

When an unsuspecting prey brushes against the trap's trigger hairs, the tension is abolished and the door swings open.

Suction created by this release of negative pressure sweeps the victim into the trap.

Release of the negative pressure also causes the trapdoor to swing back to its normal closed position, leaving the victim a prisoner.

Soon thereafter, the trap secretes enzymes that kill and digest the prey.

Carnivorous plants seduce prey with a package of handsome advertisements that include bright colors and exotic scents, much as flowers attract their pollinators. For example, the traps of Venus's-flytrap and many pitcher plants produce attractants that lure animals to their demise. Like flowers, traps also lure prey with attractive patterns invisible to our eyes but exciting to an insect. The ultraviolet patterns of many pitchers appear to an insect as a map to the trap's interior, and regions near the trigger hairs of a Venus's-flytrap shine bright against the trap's dark margins. Some carnivorous plants, like flowers, may use ultraviolet patterns to select particular prey. Sundew, for example, has a penchant for a single species of springtail insects.

Once trapped, prey are digested by enzymes secreted by specialized glands lining the trap. The absorption of nutrients from traps is critical to survival: plants that trap prey are more vigorous and produce more flowers and seed than do plants that do not trap prey.

c o n c e p t

Carnivorous plants usually grow in nitrogen-poor soils of bogs. These plants get most of their nitrogen by trapping and digesting animals. Their traps are elaborate and range from sticky surfaces to elaborate trapdoors.

Parasitic Plants   Not all associations that improve the mineral nutrition of plants are mutually beneficial. Rather, several plants are **parasites**, meaning that they harm other plants as they obtain nutrients from them (figs. 20.22, 20.23). For example, cancer root (*Orobanche*) and mistletoe (*Phoradendron*) parasitize hardwood trees (fig. 20.22), whereas trees such as sandalwood (*Santalum*) obtain their nutrients from nearby grasses. Indian pipe (*Monotropa*) obtains its nutrients from mycorrhizae of trees (fig. 20.23b).

Many parasitic plants lack chlorophyll and therefore depend entirely on their host for nutrients. However, the presence of chlorophyll does not guarantee an independent life-style. For example, mistletoe and witchweed (*Striga*) are green yet grow only as parasites. The green portions of these parasites contain only small amounts of ribulose bisphosphate carboxylase and therefore are not photosynthetic (see reading 15.1, "Finding a Host").

The link between parasites and their hosts is called a **haustorium**. In many plants, the link is one or more xylem-to-xylem connections between the two plants (fig. 20.24). Thus, the parasite depends largely on evaporation of water from its leaves as a means of pulling nutrient-containing water from the xylem of its host. To accomplish this, the stomata of many parasitic plants always remain at least partially open, even at night or during a mild drought. Evaporation of water from these stomata brings a continuous supply of nutrients from the host.

(a)

(b)

figure 20.22

Mistletoe (*Phoradendron flavescens*), a parasitic plant. (a) This oak tree is heavily infected by mistletoe. (b) Despite being green, mistletoe is not photosynthetic.

(a)

(b)

figure 20.23

Parasitic plants. (a) Orange-brown twining stems of dodder (*Cuscuta* sp.) wrap around hosts; dodder is nonphotosynthetic but obtains water, sugars, and nutrients through specialized organs called haustoria that penetrate the host. (b) Indian pipe (*Monotropa uniflora*) also does not conduct photosynthesis: it gets its nutrients indirectly from other plants, by way of a fungal bridge in the soil.

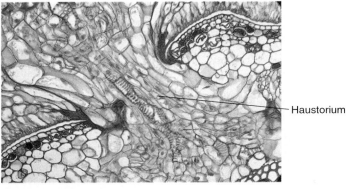

—Haustorium

figure 20.24

Haustorium of dodder (*Cuscuta campestris*; see fig. 20.23a) in the stem tissues of garden balsam (*Impatiens balsamina*). Note the vascular tissue in the haustorium, ×100.

figure 20.25

Epiphytic bromeliad growing on a tree.

Parasitic plants obtain their nutrients directly from other plants. The parasite and host are linked by a special organ called a haustorium.

Gathering Nutrients in a Rain Forest

The transfer of nutrients from plant to plant without letting them enter the soil is not restricted to parasitic plants. Indeed, such transfers are common in Amazon rain forests, where soils are notoriously deficient in most nutrients. The frequent deluges of rain in these forests leach large amounts of calcium, phosphorus, and magnesium from the forest canopy. Consequently, plants in a forest depend largely on their ability to gather the nutrient-rich rainwater that trickles down stems. Many plants do this by forming extensive, shallow root systems, whereas other plants intercept nutrients before they reach the soil. For example, the roots of trees such as *Eperua purpurea* grow upward in the fissures of the bark of nearby trees. These fissures are predictable pathways for flowing nutrients, which the roots absorb from the fissures without letting them enter the soil. Like their counterparts in the soil, these roots are extensively branched and occur throughout the canopy; they may extend more than 13 m into the air. Interestingly, the roots of *Eperua* grow up along its own stems as well as those of nearby trees, but roots of nearby trees do not grow onto *Eperua* stems. As a result, *Eperua* can exclusively recycle its own nutrients, as well as those of its neighbors.

Why do the roots of these tropical trees grow up, while most others do not? Do these roots respond to nutrients coming down the tree trunks, or do they fortuitously climb any vertical support that they happen to contact? To answer these questions, Robert Sanford of Stanford University sank drainpipes into the forest floor, attaching a reservoir atop each pipe. One-third of the reservoirs contained nutrient-rich cattle manure, one-third contained leaf litter, and the remaining third were empty. Within 4 months, climbing roots appeared on all of the "trees" containing manure, on 25% of those containing litter, and on none of the empty ones. Thus, the behavior of the roots of these plants is an adaptation to the low soil-nutrient availability of the Amazon rain forest.

Rain forests also contain many **epiphytes** (from the Greek words *epi*, meaning "upon," and *phyton*, meaning "plant"), which are plants that grow independently on other plants (fig. 20.25). Common epiphytes include bromeliads, orchids, staghorn ferns (*Platycerium bifurcatum*), and even some cacti. These plants grow slowly and must absorb their nutrients from sources other than the soil; thus, they are well adapted for aerial nutrition. Some epiphytes form their own "flowerpot" to trap nutrients. Similarly, bird's-nest ferns (*Asplenium nidus*) accumulate rainfall and litter between their closely packed leaves; the ferns then get their nutrients by growing roots among these leaves. Among the most specialized of epiphytes is Spanish moss (*Tillandsia usneoides*), which grows throughout the southeastern United States. These flowering plants lack roots; they absorb water and nutrients through hairs that coat their stems and leaves.

Chapter Summary

Soils form from rocks by a process called weathering, a slow process that produces soil layers called horizons. Each horizon has different properties that affect plant growth and development. There are many different kinds of soils, each of which consists of mineral particles, humus, air, water, and living organisms:

- The three main kinds of mineral particles are sand, silt, and clay. Clay particles have a sheetlike structure and are the smallest. Clay retains more water than does silt or sand. Also, the negative charges of clay particles retain cations in the soil. Plants can exchange H^+ for these cations via cation exchange. Because sand particles have no charge, they do not retain cations and have a low cation-exchange capacity.

- Humus is the decomposing organic matter in soil. Humus retains water and is a source of nutrients.

- Air makes up 25%–50% of soil volume. Air provides oxygen for roots and organisms living in the soil.

- Water in soil exists in two forms: chemically bound water and unbound water: Chemically bound water is unavailable to plants. Soil contains three types of unbound water: (1) Gravitational water is water drained from soil by gravity. (2) Capillary water is held by capillary action in pores in the soil. (3) Hygroscopic water, which is water held tightly by the soil particles, is unavailable to plants. Almost all water absorbed by plants is capillary water.

- Living organisms in soil refine and mix the soil, add organic matter, increase the amount of CO_2, affect the availability of nutrients, and often make the soil uninhabitable for other organisms.

Plants require at least 16 essential elements, which are necessary for normal growth and reproduction, cannot be substituted for, and directly or indirectly affect metabolism. Carbon, hydrogen, oxygen, phosphorus, potassium, nitrogen, sulfur, calcium, and magnesium are required in relatively large amounts and are called macronutrients. Iron, chlorine, copper, manganese, zinc, boron, and molybdenum are required in relatively small amounts and are called micronutrients.

Macronutrients and micronutrients function as parts of structural units, parts of compounds involved in metabolism, activators and inhibitors of enzymes, and controlling agents of a cell's osmotic potential. Beneficial elements are not essential but stimulate growth.

The availability of essential elements is influenced by the soil's mineral particles, cation-exchange capacity, decay of humus, pH, chelators, other elements, and organisms. Most essential elements are absorbed by roots. This absorption is facilitated by the large surface area of roots and the presence of ion-specific carriers.

Plants establish several kinds of associations with other organisms to obtain nutrients:

- The roots of almost all plants have mutually beneficial symbioses with fungi. These mutualisms, which are called mycorrhizae, increase the absorption of water and nutrients.

- Many plants, especially legumes, establish mutually beneficial symbioses with nitrogen-fixing bacteria. The bacteria reduce atmospheric N_2 to ammonia, which the host uses for growth. In return, the bacteria receive carbohydrates from the host plant.

- Carnivorous plants usually grow in nitrogen-poor soils of bogs. These plants get much of their nitrogen by trapping and digesting insects and other small organisms. Their traps are elaborate and range from sticky surfaces to elaborate trapdoors.

- Parasitic plants get their nutrients directly from other plants. The parasite and host are linked by special organs called haustoria.

Rain in tropical rain forests contains many nutrients leached from the forest's dense canopy. Some trees in the rain forest have roots that grow up along the trunks of adjacent trees and absorb the nutrients from rainfall trickling down the trees.

 ## What Are Botanists Doing?

The inhibition of one species of plant by chemicals from another species of plant is called **allelopathy**. Go to the library and find out why allelopathy is of concern in agriculture.

 ## Writing to Learn Botany

Crop plants absorb only about half of the nitrogen supplied in fertilizer. What happens to the rest? What are the consequences of this?

Questions for Further Thought and Study

1. What is the difference between an essential element and a beneficial element?

2. Criteria for establishing the essentiality of an element for plants have been around for several decades. Why, then, are essential elements still being discovered?

3. How are root nodules different from mycorrhizae?

4. How does the size of soil particles affect their ability to retain water?

5. What is cation exchange, and why is it important for plant nutrition?

6. What are the consequences of acid rain on nutrient availability to plants?

7. Chlorophyll contains nitrogen, but carotenoids do not. Therefore, could a nitrogen-deficient plant still have color? Why or why not?

8. Traps of Venus's-flytraps triggered by inanimate objects or sugars do not secrete digestive enzymes. What is the adaptive importance of this?

9. Serpentine soils are low in calcium. What would be the consequences of replenishing calcium in this soil by adding lime? Would these consequences be different if calcium were added as $CaCl_2$ instead of lime?

10. Describe how the pH of soil affects the availability of phosphorus.

11. Describe the major types of soils and how each is suited for plant growth.

12. Explain why removing living organisms from a soil can modify the properties of a soil.

13. Explain why the histosols of bogs are nutrient-poor yet contain large amounts of organic matter.

14. Discuss the adaptations of plants for obtaining sufficient amounts of nutrients.

15. *Rhizobium* produces cytokinin. How might this relate to its ability to produce nodules in roots?

16. How does the curling of tips of root hairs relate to the establishment of root nodules? Go to the library and find the answer to this question.

Web Sites

Review the "Doing Botany Yourself" essay and assignments for Chapter 20 on the *Botany Home Page*. What experiments would you do to test the hypotheses? What data can you gather on the Web to help you refine your experiments?

Here are some other sites that you may find interesting:

http://gnv.ifas.ufl.edu/-fairsweb/text/mg/11407.html

This site discusses the 16 essential elements needed by plants.

http://www.uoguelph.ca/CBS/Botany/roots.htm

The Root Biology and Mycorrhiza site focuses on the development of roots and their associations with soil organisms, especially mycorrhizae. At a related site *http://ozone.crle.uoguelph.ca/manure/soil.testing/tables.html* you'll learn about minerals in soil as well as about tests to measure the concentrations of those minerals.

http://www.hpl.hp.com/bpt/cp_home

This Carnivorous Plant Database includes over 3,000 descriptions of carnivorous plants. For some entries, photos and other information are available by clicking on highlighted descriptions in the entry.

http://www.tile.net/tile/news/bion29.html

At this site you'll find a newsgroup interested in nitrogen fixation.

http://meena.cc.uregina.ca/~luishus/bio/botany.html

Try some of your own surfing through this list of Web sites of interest to botanists.

http://hammock.ifas.ufl,edu/txt/fairs/16662

This site provides additional information about how pH affects the availability of plant nutrients.

Suggested Readings

Articles

Clarkson, D. T. 1985. Factors affecting mineral nutrient acquisition by plants. *Annual Review of Plant Physiology* 36:77–115.

del Rio, C. M. 1996. Murder by mistletoe. *Natural History* (February):64–68.

Gibbons, B. 1984. Do we treat our soil like dirt? *National Geographic* 166:350–388.

Kaiser, J. 1996. Acid rain's dirty business: Stealing minerals from soil. *Science* 272:198.

Nap, J-P., and T. Bisseling. 1990. Developmental biology of a plant prokaryote symbiosis: The legume root nodule. *Science* 250:948–954.

Penmetsa, R. V. and D. R. Cook. 1997. A legume ethylene-insensitive mutant hyperinfected by its rhizobial symbiont. *Science* 275:527–530.

Richter, D. D., and D. Markewitz. 1995. How deep is soil? *BioScience* 45:600–609.

Stewart, G. R., and M. C. Press. 1990. The physiology and biochemistry of parasitic angiosperms. *Annual Review of Plant Physiology and Plant Molecular Biology* 41:127–151.

Vijn, I. et al. 1993. Nod factors and nodulation in plants. *Science* 260:1764–1765.

Books

Galston, A. 1994. *Life Processes in Plants.* New York: W. H. Freeman and Co.

Haynes, R. J. 1986. *Mineral Nitrogen in the Plant-Soil System.* New York: Academic Press.

Slack, A. 1979. *Carnivorous Plants.* London: Ebury Press.

Big trees of California (*Sequoiadendron giganteum*). How does water reach the tops of these tall trees?

Movement of Water and Solutes

21

Chapter Outline

Chapter Overview

Plants did not colonize land until about 400 million years ago. The biggest challenges that plants faced as they invaded land were obtaining and transporting water in the dry terrestrial environment. These challenges to survival eventually led to the evolution of roots and a specialized vascular system to gather and transport water, and a waxy epidermis and stomata to conserve water. The evolution of multicellularity and tissue specialization also produced a system to move the products of photosynthesis to distant sites of use and storage.

This chapter describes the movement of water and solutes in plants. Specifically, it describes two tissues specialized for interorgan commerce: xylem, which carries water and dissolved minerals from roots to leaves, and phloem, which transports water and organic compounds throughout the plant. The evolution of these tissues was largely responsible for the widespread invasion of land by plants.

The survival of all plants depends on their ability to transport a variety of substances into, through, and out of their bodies. At the cellular level, diffusion is usually adequate for this movement: for example, small molecules can diffuse across a cell that is 50 μm wide in less than 1 second. Thus, primitive single-celled organisms such as algae typically relied on diffusion as a primary means of transport. However, the evolution of multicellularity rendered diffusion woefully inadequate for moving substances throughout a plant because the time required for diffusion is inversely proportional to the square of the distance involved. This means that a molecule diffuses across 1 μm 10^8 times faster than it diffuses across 1 cm, and the same molecule that diffuses across an algal cell in less than a second requires more than 8 years to diffuse only 1 m in the watery environment of a multicellular plant.

Transport between adjacent cells of multicellular organisms improved with the advent of cytoplasmic streaming, which speeded transport to about 5 cm per hour. Although cytoplasmic streaming is faster than diffusion, it too became inadequate to meet the needs of long-distance transport in ever-larger multicellular plants.

Meanwhile, multicellularity was not the only evolutionary trend placing increased demands on the transport systems of plants. Plants colonizing the land were in a "space race" to intercept light for photosynthesis. In large part, this meant growing taller and exposing large leaves to the dry, hostile environment. However, the algaelike photosynthetic cells in leaves could function only in a highly humid environment, which was far from the soil's precious water. Absorbing this water was accomplished by roots and root hairs in ferns and other plants. However, merely absorbing water was not enough, because leaves were located at ever-increasing distances from the roots. Thus, plants required a system for transporting water from the soil to their leaves. Multicellularity also required a system for transporting the sugars made in leaves to distant sites for storage and use. Thus, multicellularity and colonization of the land were largely responsible for the evolution of xylem and phloem, the two long-distance transport systems in plants. These systems have proven to be successful, for in few instances do deficiencies in internal transport limit plant growth.

Moving Water and Minerals in the Xylem

Water is the most abundant compound in plant cells: it accounts for 85%–95% of the weight of most plants and 5%–10% of the weight of seeds. Water is used to make organic compounds (e.g., sugars), support the plant (via turgor pressure), as a solvent in which important reactions occur, and as the medium in which other solvents move. Given the critical roles water plays in plants, it seems peculiar that more than 95% of the water gathered by a plant evaporates back into the atmosphere, often within only a few hours after being absorbed. This evaporation of water from the shoot of a plant is called **transpiration.** Most transpiration from leaves is through stomata and is a result of leaf architecture.

Leaf Architecture and Transpiration

Leaves are exquisitely adapted for photosynthesis (see discussion in Chapters 7 and 14). However, the adaptations that enhance photosynthesis also enhance a plant's greatest threat: dehydration. For example, the rate of gas exchange depends, among other things, on the amount of surface area available for exchange and evaporation. The loose internal arrangement of cells in leaves produces a large internal surface area for evaporation (fig. 21.1). Indeed, the internal surface area of a leaf may be more than 200 times greater than its external surface area. This amplifies transpiration by increasing the surface area for evaporation.

The internal surface area of a leaf is linked to the atmosphere by **intercellular spaces** that occupy as much as 70% of the volume of some leaves. The pores that link the internal surface area with the atmosphere are **stomata.** These pores are abundant; a squash leaf, for example, may have more than 80 million stomata. Leaves also have an efficient plumbing system of veins for distributing water to their internal evaporative surface; 1 cm^2 of leaf can have as many as 6,000 vein endings. As a result of this leaf architecture, a well-watered plant can lose tremendous amounts of water via transpiration; for example, a corn plant transpires almost 500 l of water during a 4-month growing season. If humans required the same amount of water, we would each have to drink about 40 l of water per day (and visit the bathroom often).

Water lost via transpiration is replaced by water absorbed from the soil by roots. This water moves to the leaves as fast as 75 cm min^{-1}, which is about the speed of the tip of a second hand sweeping around the face of a clock. Despite their huge requirements for water, however, plants have no active mechanism for acquiring water when they are stressed. Thus, losses of water via transpiration do not indicate an excess of water; on the contrary, water is the primary factor that limits plant growth in most areas.

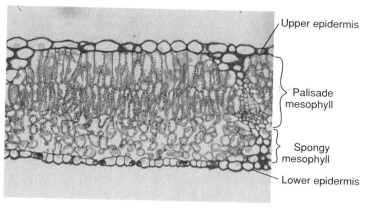

figure 21.1

Cross section of a lilac leaf (*Syringa* sp.), showing palisade and spongy regions of mesophyll. The loose arrangement of mesophyll cells exposes large amounts of surface area.

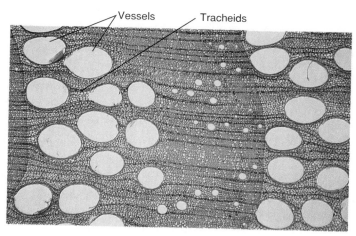

figure 21.2

Cross section of wood of oak (*Quercus* sp.), showing vessels and tracheids. Vessels and tracheids transport water and dissolved minerals from roots to shoots.

Structure of the Conducting Cells

Water must move rapidly through plants to replace water lost by transpiration. Through what tissue does this water move? To answer this question, consider the results of two simple experiments:

1. Removing the bark from a tree does not significantly alter transpiration.

2. When roots are exposed to a soluble dye, only xylary elements in the stem contain the dye.

These results suggest that water and dissolved minerals move from roots to leaves through xylary elements.

Recall from Chapter 13 that there are two kinds of xylary elements in plants: tracheids and vessel elements (fig. 21.2). Both of these cell types are well designed for conducting water: they are hollow and dead at maturity and therefore have no cellular organelles to retard water flow. Furthermore, both have thick cell walls and can therefore withstand changes in pressure associated with water flow caused by transpiration. However, tracheids are usually long (up to 10 mm) and thin (10–15 μm in diameter), and they overlap each other. Their walls contain many thin areas called *bordered pits*, which are valvelike structures that are especially abundant in portions of the walls where tracheids overlap (figs. 21.3, 13.24). Pits link tracheids into long, water-conducting chains and allow a slow flow of water through xylem.

Vessel elements are shorter and wider than tracheids: their diameters are usually 40–80 μm, and may be as large as 500 μm. Vessel elements are stacked end-to-end, and the end-walls separating adjacent vessel elements are often either wholly or partially dissolved (figs. 21.4, 13.26). As a result, vessel elements form cellulose pipes called *vessels* ranging in length from a centimeter to more than a meter. Water can move longer distances in vessels than in tracheids before having to traverse a pit. Moreover, the vessels' larger diameters and dissolved end-walls allow water to move faster than in tracheids. This

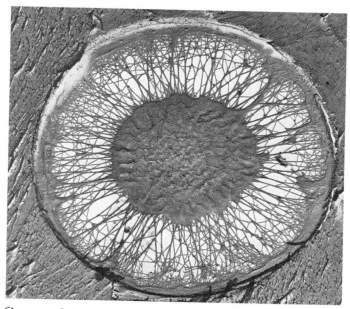

figure 21.3

Bordered pits are valvelike structures that are abundant in walls where tracheids overlap one another. Pits link tracheids into long, water-conducting chains and allow water to flow through xylem, ×6,050.

increased flow rate in vessels is one reason why angiosperms dominate today's landscapes. Most angiosperms contain both tracheids and vessels, while gymnosperms contain only tracheids (see reading 21.1, "The Risks of Having Vessels").

c o n c e p t

Water and dissolved minerals move from roots to leaves in dead conducting cells of the xylem. Most of this water evaporates from leaves via a process called transpiration.

THE RISKS OF HAVING VESSELS

Gymnosperms have only tracheids, while angiosperms have tracheids and vessels. This difference in xylary structure provides angiosperms with tremendous potential benefits as well as risks. For example, consider a tracheid with a diameter of 10 μm and a vessel with a diameter of 80 μm. Flow through cylindrical pipes such as these is proportional to the fourth power of their radius (i.e., radius^4), so that the flow rate through these cells would be as given in the table that accompanies this reading. As is shown in this table, water flows 4,096 times faster (i.e., 2,560,000/625) through the vessel than through the tracheid. Of what value, then, are tracheids?

The increased diameter that allows such rapid water transport in vessels of angiosperms also puts these plants at risk: the larger water columns of vessels have a lower tensile strength than do the thinner columns of tracheids, thereby making them more likely to break when stressed by freezing or by wind-induced bending.

An air bubble forms where a water column breaks, which stops the flow. Capillarity then shapes the air bubble into a sphere, which prevents it from moving through bordered pits into adjacent cells. Since the walls separating adjacent vessel elements are dissolved, an air bubble stops water flow in an entire vessel rather than in a single cell, as in tracheids. When air bubbles form in vessels, water flow is delegated to smaller vessels and tracheids of the wood. Thus, tracheids are the backup system of woody angiosperms: they ensure that water flow doesn't stop when air bubbles form in vessels.

Bubbles in vessel elements are common in many angiosperms. For example, the flow of water in virtually all of the large vessel elements of oak (*Quercus*) and ash (*Fraxinus*) is stopped by air bubbles by the end of a growing season. When this occurs, water moves in tracheids until the next spring, when the vascular cambium produces a new set of vessels.

Cell Type	Diameter	Radius	Flow Rate (radius^4)
Tracheid	10 μm	5 μm	$(5)^4 = 625$
Vessel element	80 μm	40 μm	$(40)^4 = 2,560,000$

Water Potential: The Force Responsible for Water Movement

The movement of water through plants is a physical process that requires no metabolic energy. Rather, water flows passively from one place to another because of differences in potential energy. The potential energy of water in a particular system compared to pure water at atmospheric pressure and at the same temperature is termed the **water potential** and is abbreviated by the Greek letter *psi*, Ψ. Lowering the potential energy of water lowers the water potential, and increasing the potential energy of water increases the water potential. Differences in water potential determine the direction that water moves: *water always flows passively from areas of high water potential to areas of lower water potential*. The movement of water into, through, and out of plants is regulated by water potential.

Components of Water Potential

Water potential is a pressure (energy per unit volume) measured in units called **pascals** (Pa) or, more conveniently, **megapascals** (MPa; 10^6 Pa). These metric units relate to the more common English units for measuring pressure as follows: 1 atmosphere = 14.7 pounds per square inch = 760 mm mercury (at sea level, 45° latitude) = 1.0 bar = 0.10 MPa = 1.0×10^5 Pa. A car tire is typically inflated to about 0.2 MPa; the water pressure in home plumbing is usually 0.2 to 0.3 MPa; and the water pressure under 5 m of water is about 0.05 MPa.

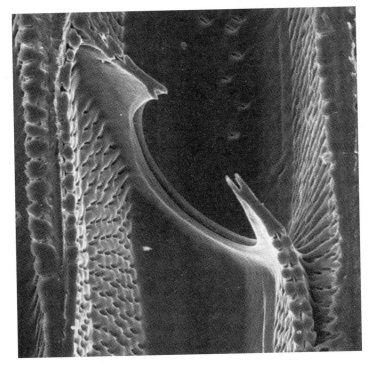

figure 21.4

Scanning electron micrograph of part of a vessel. Vessel elements are stacked end-to-end; the end-walls separating adjacent vessel elements are often either wholly or partially dissolved.

21–5

The potential energy of water in an open beaker is difficult to measure and is arbitrarily set equal to zero. The potential of this water can be modified by solutes, physical pressure, and wettable surfaces (matrices). Consequently, the water potential in a particular system (such as a cell) equals the sum of its pressure potential (Ψ_p), solute (or osmotic) potential (Ψ_π), and matric potential (Ψ_m):[1]

$$\Psi_{cell} = \Psi_p + \Psi_\pi + \Psi_m$$

Let's examine how each of these components affects the water status of plants.

Pressure Potential (Ψ_p) The **pressure potential** is a physical pressure and may be positive or negative. For example, water in dead xylary elements of transpiring plants is often under a negative pressure (i.e., tension) of less than -2 MPa; therefore, Ψ_p of water in xylem is usually negative. Conversely, water in living cells is usually under a positive pressure, much like the air in a balloon. In these cells, Ψ_p is greater than zero. Changes in the pressure potential can alter the volume of a cell by as much as 40%.

Solute Potential (Ψ_π) The **solute potential** is proportional to the number of dissolved solute particles and is independent of the type of particles. A particle may be a molecule of sugar, an ion of sodium, or any other dissolved chemical. These particles interact with and reduce the activity of water molecules, thereby decreasing the potential energy of the water. Thus, adding solutes always lowers the water potential, and Ψ_π is always ≤ 0. The solute potential of leaves of most crop plants typically ranges from -1 to -2 MPa.

Matric Potential (Ψ_m) The **matric potential** results from the adhesion of water to wettable surfaces such as cell walls and the cytoplasmic matrix and is the force required to remove water from these surfaces. This adhesion decreases the potential energy of the interacting molecules. As a result, Ψ_m is always ≤ 0.

The Ψ_m of vacuolated cells is trivial (usually between 0 and -0.01 MPa) and is therefore ignored in most calculations of water potential. However, Ψ_m is an important aspect of the water potential of densely cytoplasmic (e.g., meristematic) cells, desiccated tissues, dry seeds, and soils. For example, more than 4 MPa of pressure are required during imbibition to rupture the seed coat of a walnut; this force is generated by an uptake of water caused by the low matric potential of the seed.

---

1. Many botanists also include a gravitational potential in the calculation of water potential. Gravitational potential is of negligible importance in a root or leaf but becomes important when comparing water potentials at different heights in trees and soils. The upward movement of water in a tree must overcome a gravitational potential of approximately 0.01 MPa per meter of height.

Water Potential and Water Movement

To understand how Ψ_π and Ψ_p interact to account for the water potential of a cell (Ψ_{cell}), consider an algal cell with a water potential of -0.2 MPa and a solute potential (Ψ_π) of -0.5 MPa (for these calculations, we'll assume that the matric potential is negligible). The pressure potential (Ψ_p) of this algal cell is

$$\Psi_{cell} = \Psi_\pi + \Psi_p$$
$$-0.2 \text{ MPa} = -0.5 \text{ MPa} + \Psi_p$$
$$\Psi_p = +0.3 \text{ MPa}$$

Now suppose that this algal cell is put in a beaker of pure water ($\Psi = 0$). The water-potential gradient between the water and the cell causes water to enter the cell, inflating the cell with water and producing turgor pressure—the hydrostatic force that shapes plants and plant cells and is resisted by the rigid cell wall. Eventually, the pressure potential equals the solute potential, which prevents net uptake of water by the cell. At this point the cell is fully turgid, and its water potential equals zero. Thus, the increase in water potential from -0.2 to 0 MPa is due to a change in the Ψ_p component of water potential:

$$\Psi_{cell} = \Psi_\pi + \Psi_p$$
$$0 = -0.5 + \Psi_p$$
$$\Psi_p = +0.5 \text{ MPa}$$

If the cell is now placed in air, it will begin to dry. As this occurs, the pressure potential of the cell decreases. When $\Psi_p = 0$, the protoplast barely fills the cell, and there is no turgor. This condition is termed *incipient plasmolysis* and is the point at which $\Psi_{cell} = \Psi_p$ (remember, $\Psi_p = 0$ at incipient plasmolysis).

Now consider a root hair growing in soil. Suppose that the root hair has a water potential of -0.5 MPa, and the soil has a water potential of -0.2 MPa. The water-potential gradient between the soil and root hair causes water to enter the root hair. As the soil dries, however, its water potential decreases. When the water potential of the soil becomes less than that of the root hair, water moves from the root hair into the soil. Similarly, the presence of salts in soil decreases the water potential of the soil. If large amounts of solutes are present, as in salty or overfertilized soils, water moves into the soil from roots, and the plants appear "burned" (see reading 21.2, "Living in Salty Soil").

c o n c e p t

Water potential is the potential energy of water and indicates the tendency of water to move from one place to another. Water potential is influenced by pressure, solutes, and wettable surfaces. Water always moves passively from areas of high water potential to areas of lower water potential.

LIVING IN SALTY SOIL

Halophytes are plants that grow in salty soil. These soils occur in regions that range from hot, dry, salty deserts to moist, cool, salty marshes. One major problem faced by halophytes is obtaining enough water for growth. To do this, they must reduce their water potential below that of the soil so that water will flow into their roots. This is no small task, since the soil in which many halophytes grow often has a water potential less than −4 MPa. Plants such as shadescale (*Atriplex*) and greasewood (*Sarcobatus*) are halophytes typical of the moderately salty soils of the deserts of the southwestern United States. However, increasingly salty soil requires that plants be able to reduce their water potential even further to obtain water. These increased demands restrict the vegetation of more saline soils to only a few plants, such as pickleweed (*Salicornia*) and iodine bush (*Allenrolfea*). These plants are adept at reducing their water potential; pickleweed, for example, can reduce its water potential to almost −8 MPa. Although these adaptations are impressive, they are not limitless: no plants grow on the salt flats of Utah.

Another problem faced by halophytes is potentially toxic amounts of several ions, especially Na⁺ and Cl⁻. Unlike most other plants, halophytes not only tolerate large amounts of ions but often use them to their advantage. For example, halophytes such as *Spartina* and *Limonium* transport Na⁺ and Cl⁻ to their leaves, where these ions are

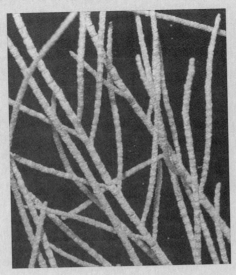

reading figure 21.2

Some halophytes excrete salt by means of specialized glands. Shown here are the salt-covered, scalelike leaves on the twigs of tamarisk (*Tamarix ramosissima*) from the Mojave Desert in California.

excreted into salt glands. Salt in the leaves reduces their water potential, which helps to pull water into them (reading fig. 21.2).

Understanding how halophytes tolerate such large amounts of salt is becoming increasingly important as fertilization and irrigation transform productive soils into salty, less fertile soils. Irrigation water seldom reaches the water table; rather, through evaporation and transpiration it leaves its solutes behind in the soil, where they form an encrusting layer of salt, much like the salt covering the edge of a margarita glass. In some areas, these salts are flushed from fields into rivers, where they become a problem for farmers living downstream who also use the river's water for irrigation. Arizona and other arid states have installed desalting facilities that can purify more than 100 million gallons of water per day. Another approach to reclaiming saline soil is growing and harvesting halophytes*; because these plants accumulate large amounts of salt, their harvest actually decreases the soil's salinity.

Many ancient civilizations arose by diverting rivers to irrigate arid lands to grow crops. The most productive use of irrigation occurred in the so-called Fertile Crescent, a broad valley formed by the Tigris and Euphrates rivers in what is now Iraq. At the peak of its productivity, this irrigated region supported more than a million people and was the point from which similar civilizations spread into present-day Afghanistan, Pakistan, and Iran. Slowly but surely, however, the salts washed from rocks and soil at higher elevations became concentrated in the irrigated fields as water evaporated and was transpired by plants. As a result, all of these civilizations ultimately collapsed for the same reason: their soils became too salty to grow crops.

---

*Growing *Salicornia* as a potential food crop is a pet project of Marlon Brando; he likes *Salicornia* cookies.

How Does Water Move in Plants?

Any hypothesis for water movement in plants must be based on water moving from areas of high water potential to areas of lower water potential. It must also account for an even more obvious requirement: water must reach the tops of tall trees such as the one pictured at the beginning of this chapter (p. 496). The forces involved in lifting water to treetops are considerable. For example, "Tallest Tree" in Redwood National Park, California, is over 110 m high; moving water to the top of this tree requires a water-potential gradient exceeding 2 MPa. Let's see how some of the hypotheses proposed for water movement in plants match up with these requirements.

HYPOTHESIS 1

Water moves up xylary elements via capillarity.
Capillarity results from the adhesion of water to the surfaces of small tubes. This adhesion pulls water up the tube and is visible as the curved meniscus atop the water column in a straw or glass tube. However, in tubes having the diameter of a xylary element, capillarity raises water less than 1 m. Therefore, capillarity alone cannot account for the movement of water to the tops of trees.

21–7

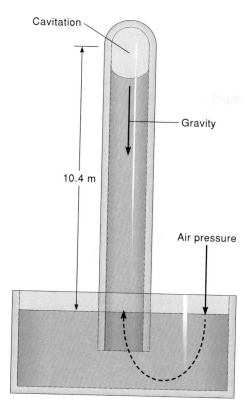

figure 21.5

The effect of atmospheric pressure on a column of water. The weight of the water in the tube pulls the water column down, while atmospheric pressure pushes water up the tube. The counteracting pressures equilibrate when the water column is about 10.4 meters high.

HYPOTHESIS 2

Water is pushed up xylary elements by atmospheric pressure.

To understand this hypothesis, imagine filling a long, hollow tube with water, closing it at one end, and placing the tube, open-end down, in a tub of water. Movement of the water column is balanced by two opposing forces: the weight of the water in the tube pulls the water column down, while atmospheric pressure pushes water up the tube. These counteracting pressures reach equilibrium when the water column is about 10.4 m high (fig. 21.5). When the length of the tube exceeds 10.4 m, the water column *cavitates,* meaning that it forms a partial vacuum filled with water vapor in the upper, closed end of the tube. Because atmospheric pressure raises a column of water only about 10.4 m, it cannot account for the movement of water to the tops of tall trees.

HYPOTHESIS 3

Water is pumped up the xylary elements.

Water in xylem moves in xylary elements, which are dead. Furthermore, there are no "water-pumping cells" in xylem. Therefore, water is not actively pumped through the xylem.

figure 21.6

Strawberry leaves with drops of guttation water at vein endings along the edges. Guttation is caused by root pressure.

HYPOTHESIS 4

Water is pushed up by root pressure.

On many mornings you have probably seen leaves like those shown in figure 21.6 that have water droplets at their edges. This loss of liquid water from leaves of intact plants is called *guttation* and is common in herbaceous plants growing in moist soil on cool, damp mornings. Guttation is caused by root pressure generated as follows:

1. Minerals actively absorbed at night are pumped into the apoplast surrounding xylary elements.

2. This influx of solutes decreases the water potential of the xylary element, thereby causing water to move into it from surrounding cells.

3. Since there is only negligible transpiration at night, the pressure in the xylem increases as high as +0.2 MPa.

4. Eventually, this pressure forces liquid water out of the leaves.

Guttation continues as long as the plant is kept under conditions favoring rapid absorption of minerals and minimum transpiration, such as in wet soils at night. The amount of water lost by guttation from most plants is relatively small. However, a notable exception is taro (*Colocasia esculenta*), a tropical plant from which Polynesians make poi. Individual leaves of taro plants can lose more than 300 ml of water per night by guttation. Although root pressure can push water several meters up a plant, it cannot push water to treetops.

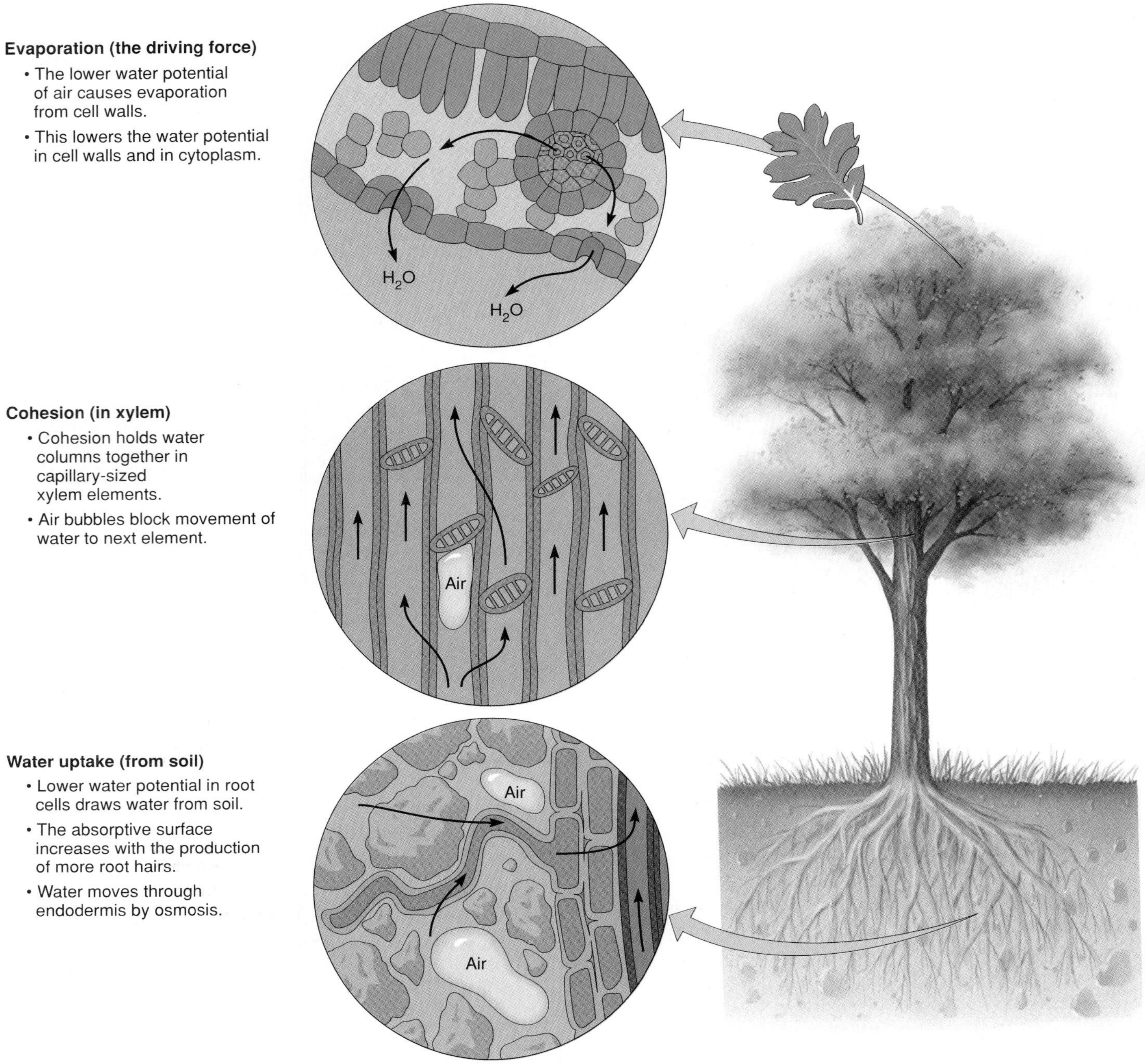

Evaporation (the driving force)
- The lower water potential of air causes evaporation from cell walls.
- This lowers the water potential in cell walls and in cytoplasm.

H_2O

H_2O

Cohesion (in xylem)
- Cohesion holds water columns together in capillary-sized xylem elements.
- Air bubbles block movement of water to next element.

Air

Water uptake (from soil)
- Lower water potential in root cells draws water from soil.
- The absorptive surface increases with the production of more root hairs.
- Water moves through endodermis by osmosis.

Air

Air

Air

figure 21.7

A summary of the transpiration-cohesion hypothesis for the ascent of water in plants.

HYPOTHESIS 5

Water is pulled up plants by evaporation.
This hypothesis was formulated more than a century ago and today is referred to as the *transpiration-cohesion hypothesis* for water movement. It is summarized in figure 21.7 and describes the process as follows:

1. Solar-powered transpiration of water dries the cell walls of mesophyll cells.

2. This loss of water from the cell walls lowers the water potential of the cell, thereby causing it to take up water from neighboring cells that have a higher water potential because they are farther away from the air spaces.

3. Cells farther from the site of evaporation have even larger water potentials, thus causing water to move from cell to cell toward the air spaces.

4. Cells bordering tracheids replace their water with water from the xylem. This loss of water from xylary elements creates a negative pressure (i.e., $\Psi_p < 0$), thereby lifting the water column up the plant.

5. The negative pressure in the xylem decreases the water potential all the way down to the tips of roots, even in the tallest trees. As a result, water flows passively from the soil, across the root cortex, and into the stele. Water in the stele is then pulled up the xylem to leaves to replace water lost via transpiration.

Although this hypothesis sounds logical, it must meet the criteria stated earlier in this section. Let's see if it does.

Can transpiration lift a column of water to the top of a tall tree? Stated another way, is there a water-potential gradient favoring movement of water from soil to plant to atmosphere? Yes. The following are typical values for water potentials at various points along the transpiration path for a small tree growing in moist soil:

$$\Psi_{soil} = -0.05 \text{ MPa}$$

$$\Psi_{root} = -0.2 \text{ MPa}$$

$$\Psi_{stem} = -0.5 \text{ MPa}$$

$$\Psi_{leaf} = -1.5 \text{ MPa}$$

$$\Psi_{atmosphere} = -100 \text{ MPa (at 50\% relative humidity at 22° C)}$$

The water potential of the atmosphere is highly variable, yet it is almost always less than that of leaves. For example, air with a relative humidity of 90% has a water potential less than -13 MPa.

Is water under a negative pressure in the xylem? Yes. The transpiration-cohesion hypothesis states that water is a continuous hydraulic system that is pulled up through plants and should therefore be under a negative pressure or tension. Tensions are greater at the tops of trees than at their bases (fig. 21.8), and the diameters of roots and tree trunks shrink when transpiration is greatest, as would be expected if the xylem were under a negative pressure. Finally, water in branches begins moving sooner than water in the trunk of the tree, as predicted if water were pulled rather than pushed up a plant. Therefore, water in xylem is under a negative pressure in transpiring plants.

Can columns of water withstand the tensions necessary to be pulled to the tops of tall trees? Yes. Water is a polar molecule, and therefore it coheres. This cohesion produces a high tensile strength—so high that it is easier to pull apart molecules in fine wires of some metals than to pull apart thin columns of water. A thin column of water in the xylem can withstand tensions of up to -30 MPa, which far exceed the approximately -2 MPa required to raise water to the tops of trees.

Is the adhesion between water and cell walls strong enough to support a column of water? Again, yes. These forces far exceed the -2 MPa needed to support a water column. Adhesion and cell-wall hydration prevent gravity from draining the water from xylary elements.

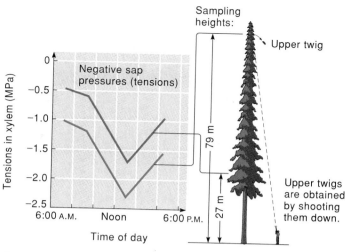

figure 21.8

Throughout the day, water tensions are greater (i.e., more negative) at the tops of trees than they are toward their bases.

c o n c e p t

The transpiration-cohesion hypothesis best explains how water moves in xylem. The driving force for the movement of water is a water-potential gradient generated by transpiration. Transpiration from leaves lifts water up plants. Cohesion of water molecules prevents the columns of water from breaking, and adhesion of water molecules to cell walls prevents gravity from draining the water column.

Factors Affecting Transpiration

Environmental Factors

Atmospheric Humidity Transpiration occurs as long as the water potential of the atmosphere is less (i.e., more negative) than the water potential of the leaf. Dry air increases this gradient and therefore increases transpiration. Similarly, transpiration typically slows in humid air.

Internal Concentration of CO_2 The concentration of CO_2 in the atmosphere rarely deviates much from 0.035%. However, the CO_2 concentration *in leaves* changes considerably, especially when stomata close and photosynthesis removes CO_2 from the intercellular spaces of the leaf. Low concentrations of CO_2 in leaves cause stomata to open, whereas high concentrations cause them to close. Thus, a reduced supply of CO_2 for photosynthesis (i.e., a low internal concentration of CO_2) opens stomata and, as a result, increases transpiration.

Wind The thin, moist layer of air adjacent to a transpiring leaf is called its *boundary layer*. A thick boundary layer decreases the diffusion gradient and therefore decreases transpiration.

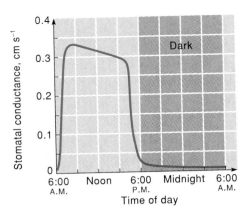

figure 21.9

The diurnal curve of stomatal opening. The data are expressed as stomatal conductance (cm s[-1]), an indication of the capacity for diffusion through stomata and an indirect measurement of stomatal opening. A stomatal conductance of zero indicates that stomata are closed (i.e., there is no transpiration). Higher conductance values indicate stomata are progressively more open. Stomata open rapidly in light and close at the end of the daylight period. Stomata remain closed throughout the dark period (except in CAM plants; see Chapter 7).

Wind usually replaces the boundary layer with drier air, thereby increasing the water-potential gradient and increasing transpiration.

The leaves of many grasses can temporarily increase the thickness of their boundary layer and thereby temporarily reduce transpiration. For example, you learned in Chapter 13 that the upper epidermis of the leaves of many grasses contains large, vacuolate cells called *bulliform cells,* which are sensitive to water loss. These cells shrink when they desiccate, thereby rolling the leaf into a cylinder. This shape increases the leaf's boundary layer and reduces the amount of light that reaches the leaf, thereby decreasing transpiration.

Air Temperature In direct sunlight the temperature of a leaf may exceed that of the air by as much as 10° C. Increasing the leaf temperature increases the water-vapor pressure in the leaf, which, in turn, increases its water potential and leads to faster rates of transpiration. Transpiration is most rapid at 20°–30° C.

Soil Any factor that affects the availability of water also affects transpiration; therefore, transpiration is affected by the water content of soil. Plants can absorb water from soil as long as their water potential is less than that of the soil.

Plants function as wicks that evaporate subsurface water from soil, which explains why soils covered by plants lose water faster than does bare soil. Indeed, almost all water lost from below 15 cm in the soil is lost via transpiration. Weeds therefore not only compete with crop plants for light and nutrients but also decrease the availability of water in soil.

Light Intensity Light usually causes stomata to open and therefore increases transpiration. Although stomata typically open at sunrise and close at sunset, these are not "all-or-none" effects; instead, stomata open gradually in the morning over a period of about 1 hour and gradually close throughout the afternoon (fig. 21.9). The effect of light on stomatal opening and closing is indirect: light promotes stomatal opening by stimulating photosynthesis, which decreases the internal concentration of CO_2 in the leaf. The stomatal responses of green and nongreen portions of variegated leaves support this conclusion: stomata in nongreen regions respond to light and dark more slowly than do those in green regions, presumably because light has less effect on the internal concentration of CO_2 there. The regulation of transpiration by light is important, since it prevents plants from needlessly losing water when there is not enough light for photosynthesis.

Transpiration is greatest in plants growing in moist soil on a sunny, dry, warm, and windy day. In these conditions, transpiration often exceeds the ability of plants to absorb water. As a result, many plants wilt at midday, even though the soil in which they are growing may be near field capacity.

Adaptations That Affect Transpiration

Physiological Adaptations Guard cells arbitrate the transpiration-photosynthesis compromise; that is, they control carbon fixation as well as transpiration. How efficiently do they do this?

Photosynthesis and water-use efficiency. The efficiency of plants in fixing CO_2 while losing water is termed the *water-use efficiency* and is defined as the amount of water lost divided by the amount of CO_2 fixed:

Water-use efficiency = (g H₂O lost)/(g CO₂ fixed)

The water-use efficiency of C_3 plants (e.g., tomato) is about 600. Since C_4 plants (e.g., corn) have no apparent photorespiration, they can fix large amounts of CO_2 with their stomata only partially open, thereby conserving water. Thus, C_4 plants need less water for growth than do C_3 plants: their water-use efficiency is about 300. CAM plants such as cacti open their stomata at night when temperatures are cooler and thereby reduce transpiration. Although their limited means of CO_2 fixation results in slow growth, it also conserves water: a typical CAM plant has a water-use efficiency of about 50. This increased efficiency allows CAM plants to grow where other plants cannot.

Leaf abscission and dormancy. Another effective means of decreasing transpiration is to reduce the evaporative surface area. One way that many plants accomplish this is by getting rid of their leaves when water or temperature becomes limiting, such as in the cold of winter or during prolonged drought (fig. 21.10).

(a)

(b)

figure 21.10

Opportune production of leaves. Plants in hot, dry environments lose great amounts of water through their leaves. One adaptation to reduce transpiration is to have leaves only when water is available. The ocotillo (*Fouquieria splendens*) plant shown here does this. (a) During dry periods, the spiny, leafless stems appear almost dead. (b) When water is available, leaves form rapidly and become photosynthetic.

figure 21.11

Leaves of *Eucalyptus* hang vertically; their flat surfaces are not presented directly to the midday sun, thereby minimizing heating and water loss.

Leaf position. Plants such as turkey oak (*Quercus laevis*) reorient their leaves as a means of altering transpiration. For example, *Eucalyptus* trees often reposition their vertically oriented leaves, thereby avoiding the midday sun (fig. 21.11). This decreases the leaf temperature and thereby reduces transpiration.

Circadian rhythms and stomatal cycling. The open-at-dawn/close-at-night cycle of stomatal opening and closing is also influenced by an internal biological clock (see Chapter 19). For example, plants grown in a photoperiod of 16 hr light:8 hr dark continue to open and close their stomata according to this photoperiod even when they are placed in the dark. This rhythmic opening and closing of stomata often occurs independently of the water status of a plant. We know little about the biological clock that controls this rhythmic opening and closing of stomata.

Abscisic acid. Mutants have played an important role in our understanding of plant physiology and plant-water relations. For example, the so-called wilty mutants of tomato helped identify the role of abscisic acid in transpiration (Chapter 18). Stomata of these mutants are always open, which causes the plants to desiccate and appear wilty. These mutants contain only about 10% as much abscisic acid as do normal plants. Moreover, adding abscisic acid to them causes their stomata to close and the plants to regain turgor, suggesting that abscisic acid prevents desiccation. Subsequent research

has supported this conclusion. For example, applying as little as 10^{-6} M abscisic acid causes the stomata of most plants to close, and desiccation stimulates the synthesis of abscisic acid in the mesophyll cells of leaves. Plants use abscisic acid as a messenger to conserve water during drought (see Chapter 18).

c o n c e p t

Leaf abscission, water-use efficiency, leaf position, circadian rhythms, and the production of abscisic acid are physiological adaptations of plants that affect transpiration.

Structural Adaptations *Cuticle.* The retention of water, and thus survival, would be almost impossible for plants without a cuticle. The cuticle is an effective means of conserving water: less than 5% of the water lost by a plant evaporates through the cuticle. In general, thicker cuticles provide more protection from desiccation than thin ones. Desert plants typically have thick cuticles, while those of aquatic plants are thin or absent.

Trichomes. Although trichomes increase the thickness of the boundary layer overlying a leaf, the primary means by which trichomes decrease transpiration is by reflecting light and thus decreasing the temperature of a leaf.

MAKING MAPLE SYRUP

From early March through early April of each year, many farms in western Massachusetts display the same sign: "Maple Syrup for Sale." Although the product is delicious, most people are no longer accustomed to maple syrup and prefer the commercial syrup sold in most grocery stores, which is cheaper. Commercial syrup contains only about 2% maple syrup (the rest is corn syrup and sugar) and tastes different from pure maple syrup.

Maple syrup is harvested from the trunks of sugar maple trees, *Acer saccharum* (reading fig. 21.3a). Unlike the production of other crops, the production of maple sugar requires widely fluctuating temperatures. From late autumn to early spring, a brief period of daily freezes and thaws triggers the flow of sap in sugar maple trees. This freezing and thawing is important because the flow stops if temperatures are constantly below or just above freezing. During cold nights, starch made during the previous summer and stored in wood is converted to sugar. The next day's warm temperatures create a positive pressure in the xylem's sapwood. When tubes called spiles are driven into the sapwood, this pressure pushes the sugary sap out of the trunk at a rate of 100–400 drops per minute (on a good day) (reading fig. 21.3b). Some trees produce as much as 150 l of sap per season; attaching vacuum pumps to the tubes can increase the seasonal yield by nearly threefold. The sugar content of maple sap ranges from 1%–10% but is usually about 3%. Once collected, the sap is boiled to remove the excess water; about 40 l of sap are required to make

reading figure 21.3a
This maple "rancher" is collecting sap from a sugar maple (*Acer saccharum*).

reading figure 21.3b
Sap drips from a metal tube driven into the sapwood of a sugar maple.

1 l of syrup. The characteristic taste of maple syrup is due to this heating and the presence of several amino acids in the sap.

The production of maple syrup has several other unusual characteristics. For example,

- Flow usually stops in the afternoon and does not start again until the temperature rises above freezing the next morning. Maple "ranchers" harvest most sap between about 9:00 A.M. and noon.

- Because it depends on the weather, the flow of sap is often sporadic. There may be 5–10 "runs" of sap-flow during a single spring.

- The concentration of sugar in sap is low early in the season, rises to a maximum, and then gradually decreases at season's end.

- Farmers use power drills with half-inch bits to drill about three holes into each tree. Each hole is 5–7 cm deep. Because this tapping removes less than 10% of a maple tree's sugar, trees can be tapped repeatedly without producing significant damage. Some of the trees tapped today were probably also tapped by pilgrims in the 1600s.

Production of maple syrup has been changed little by science and modern technology. We still do not understand precisely how the freeze-thaw cycle stimulates flow of sap, and experiments aimed at increasing sap-flow have, for the most part, failed. Despite remaining a primarily regional product, maple syrup is enjoying a renewed popularity and is now included in foods ranging from pumpkin soup to barbecued pork ribs.

Sunken stomata. Sunken stomata (i.e., stomata in crypts) increase the boundary layer surrounding guard cells. This is why plants with sunken stomata typically transpire less than do plants with raised stomata that are flush with the epidermal surface.

Reduced leaf area. Many desert plants (e.g., cacti) have greatly reduced leaves, thereby decreasing their evaporative surface area. In these plants, succulent stems that store large amounts of water often replace leaves as the primary photosynthetic organs.

c o n c e p t

Cuticles, trichomes, reduced leaves, and sunken stomata are structural adaptations that reduce transpiration.

Many of the factors that affect transpiration do so by influencing the opening or closing of stomata. For example, decreasing the internal concentration of CO_2 in a leaf opens stomata and therefore increases transpiration. Thus, understanding the movement of water through plants and into the atmosphere is largely a matter of understanding how stomata open and close.

Stomata

Stomata regulate gas exchange between the atmosphere and a plant, and are a key adaptation to life on land. That is, they moderate the compromise between photosynthesis and water loss—or, as one botanist aptly stated, "providing food while preventing thirst."

Structure and Distribution of Stomata   Leaves have anywhere from 1,000 to 100,000 stomata per square centimeter of leaf area. Each *stomatal apparatus* consists of two guard cells and adjacent epidermal cells called *subsidiary cells,* all of which surround a pore (see figs. 13.17 through 13.19). Guard cells and stomata have several distinguishing features:

- Unlike other epidermal cells, guard cells usually contain chloroplasts. The amount of starch in these chloroplasts increases at night and decreases during the day.

- Guard cells have distinctive shapes: in dicots they are crescent-shaped; in most grasses they are shaped like dumbbells.

- Guard cells typically lack functional plasmodesmata. Water and solutes enter guard cells from the apoplast.

Guard cells surround the only pores in an intact epidermis. When wide open, stomatal pores are usually only 3–12 μm wide and 10–40 μm long. Although these dimensions seem small, they are gigantic compared to the sizes of gas molecules that move through the pores. For example, a water molecule has a diameter of 0.00025 μm; even when a stoma is closed and has a diameter of only 1 μm, several thousand water molecules can pass through the closed stoma at the same instant.

Opening the Stomatal Pore   Guard cells control the size of a stomatal pore by changing shape; their unusually elastic walls buckle outward when stomata open and sag inward when stomata close. What causes these changes?

Hypothesis 1

Stomata close as a result of the desiccation of guard cells.
This hypothesis states that stomata close as a result of desiccation of the plant and its guard cells, thereby decreasing water loss during times of reduced availability. However, stomata often close when plants are fully turgid and well watered—at night, for example. Therefore, desiccation is not the primary means of closing stomata.

Hypothesis 2

Sugars produced by chloroplasts in guard cells reduce the solute potential of the cell, thereby drawing water into the guard cells and causing them to bow apart.
According to this hypothesis, the light-stimulated opening of guard cells results from synthesis of sugar in guard-cell chloroplasts. These sugars would lower the solute potential of the guard cells, thereby drawing water into them and making them bow apart.

This hypothesis was initially attractive because it explained the accumulation of starch in guard cells at night and the loss of starch during the day: incorporating sugars into insoluble starch at night would increase the solute potential of the guard cells and cause water to leave them, thereby explaining why stomata close at night. Although logical, this hypothesis does not explain several aspects of stomatal functioning. Specifically, chloroplasts in guard cells do not produce enough sugar to account for the decreases in solute potential necessary to open stomata. Furthermore, guard cells of plants such as onion (*Allium*) lack starch yet function normally. Therefore, stomatal opening and closing are not caused by the production of sugars in light and their conversion to starch in the dark.

The failure of these two hypotheses to explain stomatal behavior frustrated botanists for several years. However, botanists later discovered that stomatal opening correlates with a huge increase in the amount of K^+ in guard cells and that stomatal closing correlates with a corresponding decrease in the concentration of K^+ in guard cells. That is, the osmotic gradient responsible for drawing water into guard cells is an influx of K^+ from subsidiary cells. This discovery led to other hypotheses for stomatal function.

Hypothesis 3

Light provides energy to pump K^+ into guard cells.
This hypothesis was attractive because it correlated light-stimulated opening of stomata with a mechanism for pumping K^+ into guard cells. However, botanists soon discovered that this hypothesis could not explain stomatal opening in the dark in CO_2-free air.

Hypothesis 4

Guard cells pump K^+ in and H^+ out during stomatal opening.
Guard cells make large amounts of organic acids during stomatal opening. The most abundant of these is malic acid, which ionizes and releases H^+ into the cytoplasm of the guard cell. The H^+ is pumped out of the guard cell as K^+ is pumped in and stored in the vacuole. The accumulation of K^+ in guard cells may reach 0.5 M, which decreases the solute potential of the guard cell, despite a twofold dilution resulting from expansion of the guard cell. Similarly, stomata close when guard cells pump K^+ into the apoplast: this causes water to leave the guard cells via osmosis and

the pore to close. Subsidiary cells and other epidermal cells surrounding the guard cells function as reservoirs where K^+ is stored when stomata are closed.

The active pumping of K^+ into guard cells remains the most widely accepted mechanism for stomatal opening. Moreover, it accounts for other aspects of stomatal opening and closing. For example, flushing leaves with CO_2-free air triggers accumulation of K^+ by guard cells, thereby causing stomata to open. Furthermore, isolated pairs of guard cells open when K^+ is present.

c o n c e p t

Stomata open as a result of the active uptake of K^+ by guard cells. This K^+ decreases the solute potential of guard cells, thereby causing water to enter the guard cells and turgor pressure to increase. This increase in turgor pressure causes guard cells to bow apart and form the stomatal pore.

Early studies indicated that in some plants, the walls of guard cells adjacent to the stomatal pore were thicker and less elastic than walls along other sides of the cell. This difference in wall elasticity and thickness was presumed to be the basis for guard cells bowing apart when turgid: the cell walls adjacent to the pore stretched less than did other walls, thereby forming a pore between the cells. However, more recent studies have shown that the guard cells of most plants lack a thick, inelastic wall bordering the pore. Therefore, a differential thickness in cell walls of guard cells cannot account for stomatal opening. Rather, stomatal opening is due to radial micellation of guard cells by cellulose microfibrils arranged much like the belts in a belted tire. These microfibrils are inelastic and restrict radial expansion of guard cells while allowing increases in length. When the guard cells lengthen due to the K^+-driven influx of water, they bow apart and form the stomatal pore (fig. 21.12).

A Model to Regulate the Photosynthesis-Transpiration Compromise

Botanists have spent decades trying to understand the interactions that affect transpiration. The most popular model to account for the compromise between photosynthesis and transpiration is shown in figure 21.13 and involves two feedback loops. One loop monitors the needs of photosynthesis by measuring the internal concentration of CO_2:

1. Photosynthesis decreases the internal concentration of CO_2 in a leaf.
2. This reduction in internal CO_2 causes guard cells to take up K^+.
3. The accumulation of K^+ in guard cells causes stomata to open.
4. CO_2 diffuses through open stomata to mesophyll cells, where it is used in photosynthesis.

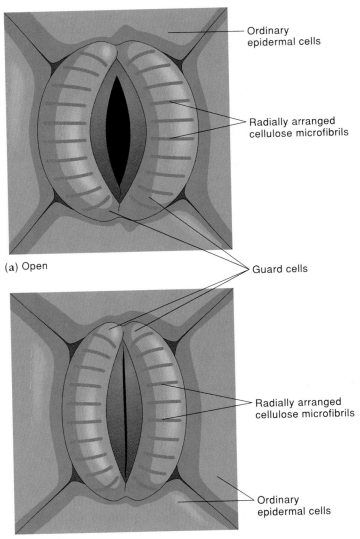

(a) Open

(b) Closed

figure 21.12

Stomatal opening is due to the movement of water into guard cells, which have radially arranged, inelastic microfibrils in their cell walls. (a) When water moves into guard cells, the radial reinforcements of the cells cause the cells to bow apart, thereby forming a pore. Movement of water into the guard cell occurs via osmosis triggered by the pumping of K^+ into the cells. (b) When water leaves guard cells, the guard cells collapse, and the stomatal pore closes. The movement of water out of guard cells occurs via osmosis triggered by the pumping of K^+ out of the cells.

The second loop protects against desiccation:

1. Desiccation stimulates synthesis of abscisic acid.
2. Abscisic acid moves to guard cells in the transpiration stream.
3. Abscisic acid stimulates transport of K^+ out of guard cells.
4. This loss of K^+ causes stomata to close, thereby conserving water.

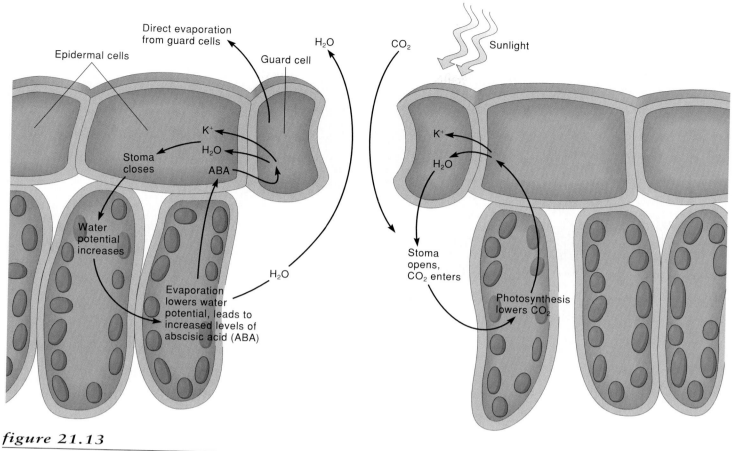

figure 21.13

Model of the transpiration-photosynthesis compromise. The right side shows the feedback loop that monitors photosynthesis; the left side shows the feedback loop that protects against desiccation.

These two loops cause stomata to open when light is sufficient for photosynthesis and to close when the risks of dehydration exceed the potential gains of photosynthesis.

What Is the Adaptive Value of Transpiration?

Transpiration has several beneficial effects for plants. It can move water from roots to shoots, it can cool a leaf as much as 10–15° C below air temperature, and it can move minerals in the transpiration stream. These benefits, however, do not seem to be *essential* for plant growth. For example, most minerals can move independently of transpiration, and leaves in full sunlight are seldom damaged by increases in temperature that occur when stomata close and transpiration is reduced by wilting at midday. This brings us back to our original question: What is the adaptive value of transpiration? Although in the future we may discover an essential role for transpiration, today most botanists regard transpiration as an unavoidable "evil"— unavoidable because of leaf structure and "evil" because it desiccates and often injures leaves. Except in dry areas, the evolution of leaves capable of high rates of photosynthesis was of greater survival value than that of more water-efficient leaves.

The Lore of Plants

The interior of a forest can be 10–15° C cooler than the surrounding countryside because of transpiration. Indeed, each large tree of a forest has the cooling capacity of about five air conditioners.

Coping with Extremes: Too Much or Too Little Water

Plants can be grouped into three categories based on their adaptations to the amount of water in the environment in which they live: mesophytes, hydrophytes, and xerophytes. We have already discussed many of the features of **mesophytes**, which grow in environments that have moderate amounts of water. A few plants have evolved ways of coping with extreme environments, such as those containing much or little water.

figure 21.14

In mangroves, gases diffuse to submerged roots through pneumatophores.

figure 21.15

Aeration in rice. Submerged leaves of rice are never wetted. Continuous air layers, which appear silvery under water, are trapped between hydrophobic leaf surfaces and the surrounding water. These layers of air are open to the atmosphere and form a low-resistance pathway for gas exchange.

Hydrophytes

Plants growing in water or excessively wet soil are called **hydrophytes**. These plants have few characteristics for conserving water; for example, they usually lack a cuticle and have reduced amounts of xylem.

The presence of too much water makes gas exchange a serious problem for hydrophytes; for example, the diffusion of O_2 is 10,000 times slower in water than in air, and the concentration of O_2 is 30 times less in water than in air. Also, the water in which hydrophytes grow often absorbs much light, thereby shading the leaves. Hydrophytes have evolved several adaptations that minimize these problems:

- Hydrophytes usually contain large amounts of aerenchyma tissue, which, you'll recall, has large amounts of air-containing intercellular spaces. In plants such as mangroves, aerenchyma-filled roots called pneumatophores project upward from the soil into the air and function as snorkels through which gases can diffuse to submerged roots (fig. 21.14).

- The underwater foliage of many hydrophytes is thin or dissected, so that no cell is far from water and its circulating currents.

- Hydrophytes use light efficiently; almost all their cells are photosynthetic, including epidermal cells.

Partially submerged plants such as rice use trapped layers of air as routes of gas exchange (fig. 21.15). In these plants, layers of air continuous with the atmosphere are trapped between the hydrophobic surface of the leaf and the surrounding water. Gases diffuse from the atmosphere to submerged parts of the plant through these trapped layers of air.

Xerophytes

The problem of gathering enough water for growth is intensified by excessively dry habitats such as deserts. Plants that live in these dry environments are called **xerophytes** and have adaptations such as reduced surface area, succulent leaves and stems, a thick cuticle, sunken stomata, and Crassulacean acid metabolism.

Plants such as mesquite and many prairie grasses *avoid* drought by growing deep roots, whereas succulents such as cacti *endure* drought by storing water in succulent tissues. For example, a large saguaro cactus (*Carnegiea gigantea*) may contain several tons of water in the parenchyma cells of its stem. Plants gather this water by producing hydrophilic colloids that decrease their water potential. However, not all plants that live in deserts are xerophytes. For example, desert star (*Monoptilon bellidiforme*) escapes drought by completing its life cycle during the brief rainy season in the desert. These plants are mesophytes that escape the droughts typical of xeric environments.

Transporting Organic Solutes in the Phloem

The first serious studies of how organic solutes move in plants were done in the 1800s by Theodor Hartig, a German botanist. Hartig was a forester interested in determining how the products of photosynthesis move in trees. In 1837 Hartig discovered a new cell type in the bark of trees. He called these cells *sieve tube members* and suspected that they were the conduits for moving sugars

figure 21.16

This girdle on a black cherry tree (*Prunus serotina*) blocked the flow of nutrients through phloem from above, stopping growth below.

from leaves to roots. To test his hypothesis, Hartig reasoned that if sieve tubes of bark were the cells through which nutrients moved, then removing a ring of bark from the tree trunk should cause nutrients to accumulate above this so-called girdle. This is exactly what happened (fig. 21.16). Because the wood was intact, Hartig concluded that nutrients move through the bark and not through the secondary xylem of the trunk. In the late 1850s Hartig began other experiments that eventually linked translocation with sieve tubes. He made a series of shallow cuts into the sieve tubes and, to his surprise, discovered that sap oozed from these incisions. Hartig concluded that organic solutes moved in sieve tubes. More recent studies with radioactive tracers have confirmed Hartig's conclusion: sugars and other organic substances move almost exclusively in sieve tubes of the phloem.

Structure of Conducting Cells

Hartig's experiments stimulated an enormous amount of research and eventually a storm of controversy. The reason for this research and controversy was relatively simple: structure and function are inseparable, and one cannot understand phloem transport without understanding the structure of the conducting cells.

The early studies of sieve tube structure provided valuable information; namely, that sieve tube members are arranged end-to-end and are associated with files of parenchyma-like cells called *companion cells* (see Chapter 13). Companion cells and sieve tube members function as a single unit.

Sieve tube members are cylinders connected by sievelike areas called *sieve plates,* each of which has numerous *sieve*

pores 1–5 μm in diameter. Sieve pores may occupy more than 50% of the area of a sieve plate. The nature of these pores has generated much controversy. Open sieve pores offer much less resistance to solute flow than do clogged pores. Since solutes traverse as many as 12,000 sieve plates per meter, knowing whether sieve pores are open or occluded is fundamental to understanding phloem transport.

Most of the early studies of sieve tube structure indicated that sieve pores were clogged with a proteinaceous material called *P-protein,* or (less elegantly) *slime.* Some of these studies also showed that sieve pores were frequently clogged with *callose,* an amorphous polymer of glucose. These occlusions made it difficult to envision how solutes could move through sieve tubes. Nevertheless, many botanists accepted this structure and began devising models for phloem transport based on clogged sieve pores. Other botanists, however, suspected that this clogging was an artifact; that is, that sieve pores are open in functional sieve tubes. If their suspicions were correct, these botanists not only would have to explain why the clogged pores were artifacts but also provide evidence that the sieve pores were, in fact, open in functioning sieve tube members.

Botanists explained the clogged sieve pores as follows. Sieve elements are extremely sensitive to manipulation and preparation for microscopic examination. Because sieve tubes contain large amounts of sugars, water is drawn into them by osmosis and generates a high hydrostatic pressure. When the sieve tubes are severed during sampling or when their membranes are altered by the chemical fixatives used for microscopy, this pressure is released, and their contents surge toward the site of pressure release. Surging P-protein becomes trapped in the sieve pores and gives the appearance that the pores are normally clogged, when in fact the clogging is due only to preparation of the tissue for study.

Although this explanation for clogged sieve pores seemed logical, a more formidable task remained: it had to be *shown* that the pores were open in functioning sieve tubes. Botanists attacked this challenge by using a new means of fixation: instead of fixing the severed sieve tubes with slow-acting chemicals, the tissues were fixed with rapid freezing. These fixations showed that sieve pores are not clogged by P-protein, callose, or any other material. That is, sieve pores are open in functional sieve tubes.

Sieve tubes in most plants are short-lived; they usually function only during the season in which they are formed. In these plants, sieve tubes are eventually replaced by cells derived from the vascular cambium. Plants such as basswood (*Tilia*), however, retain functional sieve elements for several years. During dormancy, sieve pores become clogged with callose and nonfunctional. When growth resumes in the spring, this callose is hydrolyzed, and the sieve tubes again transport sugars; the products of callose hydrolysis are used as substrates for the renewal of growth.

Callose and P-protein are located along the periphery of functioning sieve tube members. Moreover, callose is rapidly synthesized when sieve tubes are wounded. Callose plugs the pores of wounded sieve tubes and prevents the loss of assimilated nutrients through the wound. Similarly, P-protein rapidly clogs the pores of wounded sieve tubes and minimizes the loss of sugars.

Sugars and other organic compounds move throughout plants in the phloem. These solutes move through living cells and under a positive pressure.

How Substances Move in the Phloem

Several models for phloem transport have been proposed since Hartig reported his observations in the mid-1800s. The validity of these models depends on their ability to explain and accurately predict phloem transport. We have already established the *path* of transport: solutes move in sieve tubes that have open sieve pores. Another important fact that any model for phloem transport must explain is the *rate* of phloem transport.

Rates of Phloem Transport

Solutes move fast in the phloem: peak rates of transport may exceed 2 m h⁻¹. The consequences of this rate of transport are impressive; as much as 20 l of sugary sap can be collected per day from the severed stems of sugar palm. At the cellular level, solute transport is turbulent. A sieve element 0.5 mm long empties and fills every 2 seconds, thereby delivering about 5–10 g sugar h⁻¹ cm⁻² of phloem to sites of sugar storage or utilization. The largest estimate for sugar transport through sieve tubes is 180 g sugar h⁻¹ cm⁻² of phloem area. Picture what this means: a gram of solid sugar passes through a square centimeter of sieve tube area every 20 seconds.

Models for Phloem Transport

The validity of a biological model depends on how well it accounts for and predicts a biological activity. Let's see how some of the proposed models for phloem transport compare with the experimental evidence.

HYPOTHESIS 1

Solutes move through the phloem by diffusion.
Diffusion is much too slow to account for phloem transport in plants. If a 10% sucrose solution were connected to a pan of pure water by a 1-m tube with a cross-sectional area of 1 cm²: the diffusion of only 1 mg of sucrose to the pan of pure water would require almost 3 years. Therefore, mechanisms more rapid than diffusion must be involved in solute transport in phloem.

HYPOTHESIS 2

Solutes move through the phloem by cytoplasmic streaming.
Cytoplasmic streaming is also too slow to account for phloem transport. Moreover, streaming apparently does not occur in mature sieve tubes. Therefore, cytoplasmic streaming alone cannot account for the movement of solutes in phloem.

HYPOTHESIS 3

Solutes move through the phloem via pressure flow.
In 1926 German plant physiologist Ernst Münch proposed the pressure-flow model, which states that a turgor-pressure gradient drives the unidirectional flow of solutes and water through sieve tubes. According to this model, solutes move through sieve tubes along a pressure gradient in a manner similar to the movement of water through a garden hose.

Like all good models, Münch's model is testable. Consider the experimental setup shown in figure 21.17, in which (1) A and B are osmometers permeable only to water, (2) osmometer A contains more solutes than osmometer B, (3) both osmometers are submerged in water, and (4) the osmometers and water container have open connections that offer little resistance to the flow of water and solutes. Here's what happens:

1. Because the osmometers contain solutes, their water potential is less than that of the surrounding water. Consequently, water enters A and B by osmosis, which produces turgor pressure.

2. Because osmometer A contains more solutes than osmometer B, the turgor pressure in A will exceed that in B. Thus, the differing solute concentrations in the two osmometers produce a pressure gradient.

3. Because the osmometers are connected, the pressure gradient is transmitted throughout the entire system. This pressure moves fluid in A through the connecting tube to B. Solutes in the flowing fluid are carried passively from A to B via mass flow.

4. The movement of fluid from A to B builds pressure in B, which causes water to leave osmometer B, thereby creating a circulating system.

Four requirements must be satisfied for Münch's model to work:

1. There must be an osmotic gradient between the two osmometers.

2. Selectively permeable membranes must be present to establish a pressure gradient.

3. There must be an open channel between the two osmometers to allow flow.

4. The surrounding medium must have a water potential that exceeds (i.e., is less negative than) that of the most negative osmometer.

Let's now use these criteria to determine if Münch's model can explain the movement of solutes in phloem.

Does an osmotic gradient occur in the phloem? Yes. This gradient occurs between sources and sinks. *Sources* are the sites where sugars are made by photosynthesis or hydrolysis of starch. Sources are analogous to osmometer A in figure 21.17: they contain large amounts of sugar, and their solute concentration is high. Chlorenchyma cells are examples of sources. *Sinks* are sites where sugars are used or

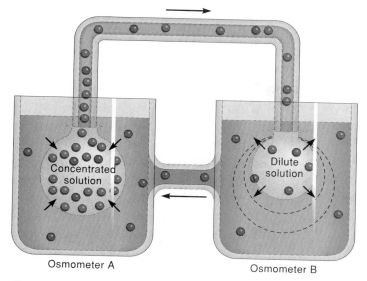

figure 21.17

Model of the pressure-flow hypothesis for solute movement. Solutes move in a mass flow generated by differences in pressure in osmometers A and B.

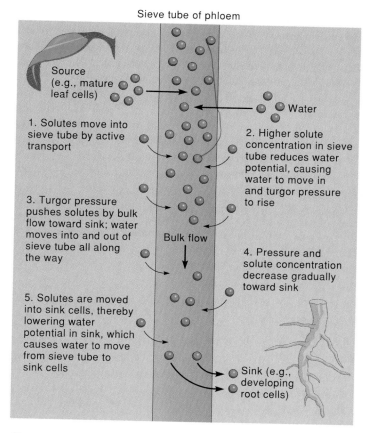

Sieve tube of phloem

Source (e.g., mature leaf cells)

1. Solutes move into sieve tube by active transport

Water

2. Higher solute concentration in sieve tube reduces water potential, causing water to move in and turgor pressure to rise

3. Turgor pressure pushes solutes by bulk flow toward sink; water moves into and out of sieve tube all along the way

Bulk flow

4. Pressure and solute concentration decrease gradually toward sink

5. Solutes are moved into sink cells, thereby lowering water potential in sink, which causes water to move from sieve tube to sink cells

Sink (e.g., developing root cells)

figure 21.18

Proposed mechanism for pressure flow in the phloem of flowering plants. Solutes move via mass flow from sources to sinks.

stored as insoluble starch. Sinks are analogous to osmometer B: they contain less sugar than do the sources, and therefore their solute concentration is relatively low. Roots, active meristems, and developing fruits are examples of sinks.

The osmotic gradients between sources and sinks are large enough to account for phloem transport. For example, the solute concentration decreases significantly in sieve tubes linking sources and sinks of white ash (*Fraxinus americana*). Furthermore, solute transport is directly proportional to the magnitude of this gradient: steeper gradients correlate with faster transport. Finally, the contents of sieve tubes are under a positive pressure, thereby explaining Hartig's observation that sap oozes from punctured sieve tubes. Taken together, these observations indicate that osmotic gradients occur in phloem and that these gradients are important for phloem transport.

Are semipermeable membranes present to form and maintain a pressure gradient? Yes. Conducting elements in the phloem must be alive and have intact membranes to function.

Does a channel exist between sources and sinks that allows for the flow of solutes? Yes. The sieve pores of sieve elements are open and form a conduit through which solutes can move.

Is the water potential of the surrounding medium more negative than that of the sink? Yes. The solute concentration in sieve tubes is 2–3 times greater than that of surrounding cells.

In summary, the pressure-flow model is attractive because it explains source-to-sink movement in plants (fig. 21.18):

1. Sucrose produced at a source is loaded into a sieve tube.

2. This loading decreases the water potential, which causes water to enter the sieve tube by osmosis.

3. The influx of water into sieve tubes creates a pressure gradient that carries sucrose to a sink, where it is unloaded.

4. Removing sucrose at the sink increases the water potential there, causing water to move out of the sieve tube at the sink.

5. Sucrose is either used or stored at the sink. Water exiting the sieve tube at the sink returns to the xylem and is recirculated.

The ability of the pressure-flow model to account for and accurately predict the characteristics of phloem transport makes it the most widely accepted model for phloem transport.

c o n c e p t

Transport of organic solutes in phloem is best explained by the pressure-flow model. This model states that sugars are carried by water along a gradient of turgor pressure generated by an osmotically driven influx of water at sources and an exit at sinks.

Unlike the unidirectional upward flow of water in xylem, phloem solutes move in all directions; these directions are determined by the presence of sources and sinks. However, the overall scheme of sugar transport in plants is not as simple as moving solutes from one source to one sink. Rather, sucrose may be required simultaneously by *several* sinks. The presence of numerous sinks produces two interesting questions:

Do all sinks receive equal amounts of sugars? No. Larger sinks receive more sugars than other sinks. Developing fruits and flowers are strong sinks that often take priority over other sinks, thereby explaining why vegetative growth often slows or stops until fruit formation is complete.

Do all sources contribute equally to sinks, or is a sink preferentially supplied with sugars by only a few sources? Sources contribute primarily to the nearest sink rather than equally to all sinks. For example, leaves near roots usually transport their sugars to roots, while leaves near tips of stems transport sugars to the shoot apex. Leaves midway between roots and shoot tips transport their sugars to both of these sinks.

Loading and Unloading the Phloem

Phloem Loading The Münch pressure-flow model accounts for the movement of solutes once they are loaded into sieve tubes. However, sieve tubes themselves are not the sources or sinks; that is, sugars are neither produced, needed, nor stored in sieve tubes. Rather, sieve tubes *link* sources and sinks. How, then, do solutes get from sources into sieve tubes and from sieve tubes to sinks?

Consider a chlorenchyma cell (i.e., a source) in a leaf and a cortical cell (i.e., a sink) in a root. Most chlorenchyma cells are 2–4 cell layers from a vein. Thus, sugars made in chlorenchyma cells must be transported across several other chlorenchyma cells before they can be loaded into a sieve tube. The movement of solutes between adjacent chlorenchyma cells is symplastic and is enhanced by the many plasmodesmata that link these cells. That is, symplastic transport accounts for the movement of solutes to chlorenchyma cells bordering the vein. However, plasmodesmata rarely occur between mesophyll cells and companion or sieve cells. This absence of symplastic links between chlorenchyma cells and sieve tubes puzzled botanists because it suggested that sugars had to move through the cell wall (i.e., the apoplast) before being loaded into sieve tubes. Experiments with radioactive tracers confirmed that this is what often occurs. For example, radioactive $^{14}CO_2$ is rapidly incorporated into sugars by chlorenchyma cells of leaves. Soon thereafter, the resulting radioactive sugars can be detected in the cell walls between chlorenchyma cells and veins. Radioactive sugars cannot be detected between adjacent chlorenchyma cells, providing further evidence that solute transport between chlorenchyma cells is symplastic.

Sugars in the cell wall are loaded into sieve tubes by companion cells, which often have many plasmodesmata and include structures similar to those of transfer cells (fig. 21.19). The elaborate ingrowths of the cell wall and cell membrane of transfer cells provide a large surface area for transporting sugars from the cell wall into the sieve tube. The loading of sieve

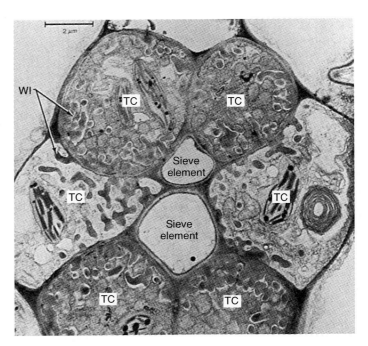

figure 21.19

Cross section of a leaf vein from common groundsel (*Senecio vulgaris*), showing sieve elements and transfer cells. The wall ingrowths (WI) of transfer cells (TC) produce large surface areas that are used to enhance loading and unloading of sieve elements.

tubes from the apoplast requires metabolic energy and is driven indirectly by a proton gradient generated at the expense of ATP (fig. 21.20). Here's how it occurs:

1. ATP and an H+-carrier in the cell membrane are used to pump protons out of the sieve tube.

2. This pumping forms a proton gradient across the membrane, with the largest concentration of H+ being on the outside of the membrane. Electrical neutrality is maintained by K+ entering the sieve tube, which accounts for the large amount (1–2 mg ml^{-1}) of K+ in phloem sap.

3. Diffusion of protons back into the sieve tube is coupled to a carrier and powers the transport of sugar into the sieve tube.

How do we know that phloem loading occurs like this? The answer is experimental evidence:

- *Requirement for ATP:* ATP is common in phloem sap, and phloem loading stops when a plant is exposed to respiratory poisons such as cyanide. Furthermore, supplying sieve tubes with ATP significantly increases phloem loading. Finally, sugars are pumped into sieve tubes against a concentration gradient: the concentration of sucrose in a chlorenchyma cell is typically 10–50 mM, while that in a sieve tube of a minor vein of a leaf may be as high as 1 M. Taken together, these results indicate that energy from ATP is required for phloem loading.

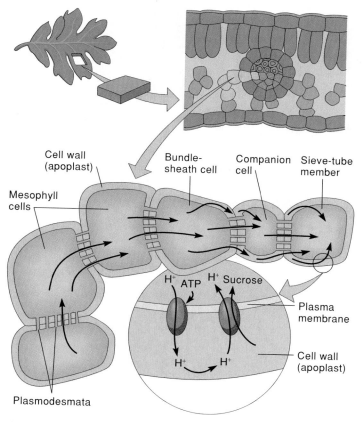

Cell wall (apoplast)

Mesophyll cells

Bundle-sheath cell

Companion cell

Sieve-tube member

H⁺ ATP H⁺ Sucrose

Plasma membrane

H⁺ H⁺

Cell wall (apoplast)

Plasmodesmata

figure 21.20

Model of sucrose loading into phloem. According to this model, H^+ is pumped out of the sieve tube, using energy from ATP. Sucrose is transported into the sieve tube.

- *Proton pumping:* Phloem loading coincides with an increase in the pH of sieve tubes to about 8 and a simultaneous decrease of the apoplast pH to about 5.5. These changes are consistent with secretion of H^+ from the sieve tube into the cell wall. The loading of sieve tubes also coincides with an increase in the concentration of K^+ in the sieve tube, an observation consistent with the idea that K^+ is absorbed by sieve tubes to maintain electrical neutrality.

- *H^+ and phloem loading:* Bathing the apoplast with neutral buffers inhibits phloem loading, and acidifying the apoplast enhances phloem loading. This is consistent with a proton gradient being necessary for the loading of sugars into sieve tubes.

Sugars and other solutes are loaded selectively into sieve tubes: only those solutes that are transported will be loaded. For example, sucrose is always present in sieve tubes, while glucose is rarely present. If veins are bathed in a solution of sucrose and glucose, only sucrose will be loaded into sieve tubes. This selectivity depends on specialized membrane-carriers in the cell membrane of sieve tubes and companion cells.

Phloem Unloading Unloading of solutes from sieve tube members can occur symplastically or apoplastically. In vegetative sinks that are growing, such as roots and young leaves, unloading is usually symplastic. In other sinks, unloading is usually apoplastic. The mechanisms underlying phloem unloading may vary in different species.

Influence of the Environment on Phloem Transport

Several environmental factors affect phloem transport. One of them is light, which promotes photosynthesis and increases the production of sugars. As a result, increased light intensity generally promotes transport to roots. Similarly, darkness stimulates translocation from roots to shoots.

Mineral deficiencies are also important in phloem transport. For example, phloem transport is slow in boron-deficient plants; transport increases dramatically when plants are supplied with boron. Potassium deficiencies also decrease phloem transport, presumably because of the dependence of phloem loading on K^+ uptake.

Finally, there is some evidence that hormones affect transport in phloem. These effects are presumably due to altered growth and metabolism rather than to direct stimulation of phloem loading or unloading.

What Moves in the Phloem?

You have already learned that sieve tubes transport sugars and some other organic solutes. But exactly which and how much of these substances move in sieve tubes?

Methods of Examining the Contents of Sieve Tubes

One seemingly easy way to determine the content of a sieve tube would be to collect phloem sap as it oozes from a wounded stem. However, there are problems with this approach, because wounds rupture and kill other cells, and their contents mix with and contaminate the exudate from phloem. Cutting also releases the pressure in sieve tubes, which decreases their water potential and causes water to move into the cells by osmosis. Ideally, a botanist could delicately insert a microneedle into a single sieve tube without a sudden release of pressure.

Such a microneedle was discovered in 1953. It was provided by two insect physiologists, J. S. Kennedy and T. E. Miller. These scientists were studying the feeding habits of aphids, insects that obtain food from phloem. Kennedy and Miller noted that aphids have natural phloem-probes that tap into individual sieve tubes: each insect has a needlelike mouthpart called a stylet (fig. 21.21). Because the contents of sieve tubes are under positive pressure, their contents surge into the aphid as it sits calmly and enjoys its sugary meal. This rush of phloem sap is often overwhelming; when this occurs, the aphid secretes excess sap as a drop of sugary *honeydew*, which

(a)

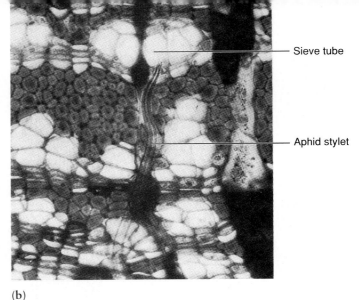

Sieve tube

Aphid stylet

(b)

figure 21.21

Aphids are phloem-feeders. (a) An aphid feeding on a plant stem; excess phloem sap (droplet of honeydew) passes through the aphid. (b) Micrograph showing an aphid stylet in a phloem cell.

eventually drops onto underlying plants, sidewalks, and cars. But how can one be sure that the aphid does not alter the contents of the sap before secreting it as honeydew? Some clever microsurgery solved the problem:

1. An aphid was allowed to insert its stylus into a sieve tube member.

2. When the feeding started, the aphid was anesthetized with a gentle stream of CO_2.

3. The stylet was then severed from the aphid's body, leaving only the open-ended stylet inserted into a sieve tube member.

Positive pressure in the sieve tube pushed sugary sap through the severed stylet for several days: collection rates averaged about 1 mm³ of phloem exudate per hour. Kennedy and Miller wrote that the technique "might also be of use to plant physiologists." They were right. Botanists soon took Kennedy and Miller's suggestion, transforming the lowly aphid from a garden pest to an important research tool and technician. Thanks to a timely assist from the insect world, botanists could finally determine exactly what is in a sieve tube.

Contents of Sieve Tubes

The most abundant compound in a sieve tube is water. More than 90% of the solutes in sieve tubes are carbohydrates. In most plants, these carbohydrates move largely or entirely as sucrose. The concentration of sucrose may be as high as 30%, thereby giving phloem sap a syrupy thickness. However, not all plants transport only sucrose; a few plant families also transport other sugars, such as raffinose, stachyose, and verbascose. These sugars are similar and consist of sucrose attached to one or more D-galactose units. Like sucrose, they are

all nonreducing sugars, which are less reactive and less labile to enzymatic breakdown than are reducing sugars such as glucose and fructose.

Some plants also transport sugar alcohols in their phloem. For example, apple and cherry trees transport sorbitol, and mannitol moves in the phloem of ash (*Fraxinus*). While the biblical manna that was miraculously supplied to the Israelites may have come from various sources, commercial manna (the source of mannitol) is the dried phloem-exudate of manna ash (*Fraxinus ornus*) and related plants. Sieve tubes also contain ATP and nitrogen-containing compounds such as amino acids, especially during senescence of leaves and flowers. Sieve tubes also transport hormones, alkaloids, viruses, and inorganic ions, especially K^+.

Exchange between the Phloem and Xylem

The contents of xylem and phloem are in aqueous equilibrium; that is, they have a similar water potential. For example, water entering loaded sieve tubes comes from the xylem, and water leaving unloaded sieve tubes returns to the xylem and is recirculated. The movement of water between xylem and phloem is occasionally accompanied by an exchange of the solutes of these tissues.

Chapter Summary

Multicellularity of plants and their colonization of the land corresponded with the evolution of systems for long-distance transport. The vascular systems responsible for this transport are xylem, which moves water and dissolved minerals from the soil to leaves, and phloem, which moves sugars and other organic compounds throughout a plant. Water movement in both of these systems occurs because of differences in water potential, which is a measure of the potential energy of the water in the system. Water potential is the sum of the energy attributable to pressure, solutes, and wettable surfaces. Water always moves from areas of high water potential to areas of lower water potential.

Water moves from roots to leaves in tracheids and vessels of xylem. These conducting cells are dead and hollow at maturity, and have thick walls that can withstand the negative pressures characteristic of xylem transport. Leaves expose large, evaporative surface areas to the dry atmosphere. The evaporation of water from shoots is called transpiration.

Movement of water through the xylem under negative pressure is best explained by the transpiration-cohesion hypothesis. The driving force for water movement is the evaporation of water from the walls of leaf cells, which decreases the water potential of the leaf and thus pulls replacement water from the xylem. The water-potential gradient that lifts water through the xylem extends the length of the plant and into the soil, thereby pulling water into the plant. The movement of water in plants is also promoted by the strong cohesion of water molecules. The adhesion of water to the cell walls of tracheids and vessels helps to prevent gravity from draining the water columns.

Many factors affect transpiration:

- *Environmental factors* that affect transpiration include atmospheric humidity, internal concentration of CO_2 in leaves, air movement, air temperature, availability of water in the soil, and light.

- *Physiological adaptations* that affect transpiration include water-use efficiency, leaf abscission, dormancy, leaf position, stomatal cycling, circadian rhythms, and the synthesis of abscisic acid.

- *Structural adaptations* that reduce transpiration include the presence of a cuticle, trichomes, sunken stomata, and decreased leaf area.

All of the factors that influence transpiration do so primarily by affecting the opening and closing of stomata. A stoma consists of two guard cells surrounded by specialized epidermal cells called subsidiary cells. Guard cells open and close a small pore; this opening and closing is controlled by the uptake and loss of K^+ from subsidiary cells. The uptake of K^+ decreases the water potential of the guard cells, which causes water to enter them and increases their turgor pressure. Radial micellations of cellulose microfibrils prevent guard cells from expanding radially. Thus, the increase in turgor pressure causes them to lengthen and bow apart, which forms the stomatal pore.

The pattern of stomatal opening and closing is a compromise between obtaining enough CO_2 for photosynthesis and conserving water to prevent desiccation. This compromise is mediated by two feedback loops. One loop meets the needs of photosynthesis by monitoring the internal concentration of CO_2; low concentrations of CO_2 in a leaf cause stomata to open. The other loop protects the plant from desiccation by producing abscisic acid during dry periods, which causes stomata to close.

Several plants are adapted to growing in the presence of too much or too little water:

- *Hydrophytes* are plants that grow in water or in excessively wet soil. These plants have few characteristics for conserving water and are specialized for gas exchange. Hydrophytes usually contain large amounts of aerenchyma tissue.

- *Xerophytes* are plants that live in unusually dry environments. These plants conserve water with thick cuticles, sunken stomata, Crassulacean acid metabolism, decreased leaf area, and succulent tissue.

Organic solutes move through phloem in sieve tube members, which are living cells arranged into pipelike structures called sieve tubes. Solutes move under positive pressure in sieve tubes. The primary solute transported in sieve tubes is sucrose.

Phloem transport is best explained by the pressure-flow model. According to this model, companion cells load sugars into sieve tubes at sites called sources. These sugars decrease the water potential of the sieve tube, so that water enters the cell and increases its turgor pressure. Meanwhile, the turgor pressure in sieve tubes decreases at sinks, where sugars are unloaded. As a result, sugars in sieve tubes are carried along a gradient of turgor pressure generated by an osmotically driven influx of water at sources and an exit at sinks. Phloem transport is affected by hormones, light, and mineral deficiencies.

 What Are Botanists Doing?

Phloem loading and unloading are hot topics in botanical research. Using these as keywords, do a library search to find out what studies have been done on these topics in the past five years.

 Writing to Learn Botany

The contents of sieve tubes are under a positive pressure, while those of nearby xylary elements are under a negative pressure. How, then, could these cells have the same water potential?

Questions for Further Thought and Study

1. Where do the ions that move during stomatal opening and closing come from?

2. Discuss how epidermal cells control photosynthesis and transpiration.

3. Define water-use efficiency. Which plants use water most efficiently: C_3, C_4, or CAM plants? What is the basis for this increased efficiency?

4. Is it a good idea to fertilize plants during a drought? Why or why not?

5. Trace the path of water from soil to plant to atmosphere. Describe the structure and function of each cell that the water molecule passes through.

6. Discuss how and why water potential changes during a day.

7. What is the selective advantage of producing thousands of vessels and tracheids instead of one giant conduit for water transport?

8. Suppose you collected the water droplets formed by guttation. What would you expect to find dissolved in these droplets, if anything? Why?

9. Describe what happens to the water potential of guard cells and adjacent epidermal cells when stomata open and close.

10. During periods of reduced transpiration, almost all water moves through large vessel elements and large tracheids. Smaller tracheids are used to move water only when water flow increases. Why?

11. Leaves of redwood (*Sequoia sempervirens*) can absorb water from fog that bathes them along the California coast. Explain, in terms of water potential, how this absorption could occur.

12. Adding salt to roads is a common way of minimizing ice formation during winter. What are the consequences for plants living along the side of the road?

13. Describe how leaf architecture affects transpiration.

14. Plants have been likened to an open-ended tube stuck in soil. In what ways is this analogy correct? In what ways is it incorrect?

15. Discuss the adaptations of plants for living in excessively wet and dry environments.

16. Compare and contrast the movement of substances in phloem and in xylem. Include comparisons of the conducting cells, substances transported, directions of transport, and the driving force for each.

17. Discuss how sugars made in a mesophyll cell of a leaf are transported into a sieve tube.

18. Summarize the evidence supporting the pressure-flow hypothesis for phloem transport.

19. Could a cell have a water potential greater than zero? Why or why not?

20. Stomata, when open, occupy less than 1%–2% of the area of a leaf. However, diffusion of water through stomata often exceeds the amount that evaporates from a freestanding water surface having the same area as the stomatal pores. Explain this "paradox of the pores."

21. Vessels at the base of a tree are usually longer and wider than those near the shoot tips. For example, less than 10% of the vessels that occur 11 m high in a tree are longer than 4 cm, while more than half of the vessels at the base of the trunk exceed 4 cm. What is the adaptive significance of this?

Web Sites

Review the "Doing Botany Yourself" essay and assignments for Chapter 21 on the *Botany Home Page*. What experiments would you do to test the hypotheses? What data can you gather on the Web to help you refine your experiments? Here are some other sites that you may find interesting:

http://sol.uvic.ca/treephys
This site lets you preview *Tree Physiology,* a journal that publishes articles about all aspects of trees.

http://www.cco.caltech.edu/~aquaria/Krib/Plants/
This site presents descriptions, photos, and information about hydrophytes. You'll also learn how to grow these hydrohytes in an aquarium.

http://www.graylab.ac.uk/usr/hodgkiss/succule.html
This site contains information about xerophytes. You'll be especially interested in this site if you grow, conserve, or just like cacti.

http://www.clc.uc.edu/~ladybug/maple/waffle.html
This site includes photos and descriptions of a harvest of sap used to make maple syrup (see reading 21.3, "Making Maple Syrup").

http://www.graylab.ac.uk/usr/hodgkiss/americ.html
This site provides selected images of succulent plants from North America.

Suggested Readings

Articles

Clifford, P. E., C. E. Offler, and J. W. Patrick. 1989. How are sugars unloaded from the phloem? Some answers from a novel experimental system. *Journal of Biological Education* 23:147–151.

Evert, R. F. 1982. Sieve tube structure in relation to function. *BioScience* 32:789–795.

Van Bel, A. J. E. 1993. Strategies of phloem loading. *Annual Review of Plant Physiology and Plant Molecular Biology* 44:253–282.

Zimmermann, M. H. 1982. Piping water to the treetops. *Natural History* 91:6–13.

Books

Flowers, T. J., and A. R. Yeo. 1992. *Solute Transport in Plants.* London: Blackie Academic and Professional Publishing.

Moorby, J. 1981. *Transport Systems in Plants.* New York: Longman.

Steward, F. C. 1986. *Water and Solutes in Plants, Vol. 9: Plant Physiology: A Treatise.* New York: Academic Press.

Evolution

Throughout recorded history, people have wondered how the great variety of earth's organisms came to be. Among the earliest thinkers was Aristotle, who thought that each species arose independently from inorganic matter and did not change. His *scala naturae* (scale of nature) stretched from nonliving matter through lower forms of life to humans at the top. Aristotle believed that each species held a permanent place on the scale but that all species were imperfect. The work of a perfect God was found only in the transcendent world of ideas, not in the variation of organisms around us.

Today our ideas about the diversity of life differ markedly from those of Aristotle and his contemporaries. We view populations of organisms as entities that change as they adapt to changing environments. The foundation for understanding how biological change occurs and how such change forms new species rests on the theories of organic evolution. The earliest of these theories is most closely identified with Charles Darwin, the greatest naturalist-philosopher of the nineteenth century. His ideas about evolution have impacted a wider array of human endeavors than any other scientific advancement of the past 150 years.

How Darwin arrived at his ideas and how they have affected science, philosophy, religion, and attitudes is one of the most fascinating stories of human achievement. This subject occupies a significant portion of hundreds of books spanning the past century. For this textbook, the most important aspect of the story is how Darwin's work provided the first scientific explanation for the diversity of life and a mechanism for evolutionary change.

Evolution may be the grandest puzzle in all of natural science. Our clues about this puzzle are sometimes ancient and sometimes modern. This fossilized *Ginkgo* leaf dates from the Paleocene epoch over 65 million years ago. A comparison with its modern counterpart raises tantalizing questions about change. How much? How little? What happened during those 65 million years?

Evolution

22

Chapter Outline

Chapter Overview

Evolution is the most comprehensive theme in biology because it explains how the first forms of life diversified into the organisms of today. Evolution implies change, but it specifically refers to the modification and descent of successful forms of life. The foundation for modern evolutionary thought was described by Charles Darwin more than a century ago; Darwin also described natural selection as the main force that makes evolution work. Although Darwin provided evidence for evolution by the mechanism of natural selection, he and his contemporaries did not know how parental traits could be changed and passed to offspring. Beginning with the work of Gregor Mendel, however, genetics provided some clues as to how genetic variation arose and how it was inherited. After the turn of the century, evolutionary biologists began to learn how genes behave at the population level. More recently, explanations of how genes evolve have come from information about their molecular biology.

Organic evolution implies change in living things, and its definition in texts and dictionaries is usually something like this: the development of complex forms of life from simple ancestors via genetic change. Unfortunately, this definition is oversimplified to the point of being only partially correct. There is no simple definition of evolution; a full description would include a number of components, or postulates, central to the theory of evolution. Such a theory is popularly attributed to Charles Darwin, even though the word *evolution* did not appear in his famous book, *On the Origin of Species*.[1] It is more accurate to think of Darwin's work as a set of three theories. Two of his theories are still accepted: the theory of *descent with modification,* which explains the pattern of biological diversity, and the theory of *natural selection,* which explains the primary mechanism by which evolution works. The third theory proposed by Darwin, the theory of *pangenesis,* was an incorrect explanation of inheritance. This theory was soon discarded and later replaced by Gregor Mendel's theory of *inheritance* and the *chromosome theory of heredity* (see Chapter 8).

A set of theories to define evolution seems cumbersome, so evolutionary biologists have spent a lot of time trying to refine, reinterpret, and update Darwin's work into a more modern and streamlined form. As a result,

we now have the genetic or synthetic theory of evolution.[2] The postulates of this theory include many of Darwin's views on evolution, as well as modern ideas about inheritance and a theoretical foundation for genetic change within populations. Ideas about population genetics have been especially important to understanding how populations evolve.

As you read this chapter, keep table 22.1 handy as a reference for the major postulates of currently accepted theories of evolution. It is especially important to note that Darwin proposed separate explanations for the *pattern* of evolution (descent with modification) versus the main *process* of evolution (natural selection). Also note that the synthetic theory incorporates ideas from Darwin, Mendel, modern genetics, and population genetics.

Like all important ideas, evolution attracts controversy; it has attracted more controversy than other scientific ideas because it has affected not only science but also philosophy, religion, and attitudes. Unfortunately, evolution is often seen as competing with religious ideas as a way of explaining the origin and diversity of life. This competition is inappropriate, since evolutionary theory is scientific and religions are not. Evolution is evaluated according to the scientific method; the existence of a divine creator is a product of faith and cannot be evaluated scientifically.[3]

Evolution before Darwin

Greek Philosophy

Several people, including some classical Greek philosophers, recognized the gradual evolution of life as early as 2,000 years before Darwin. However, the influential teachings of Plato, Aristotle, and others denied the concept of an evolving world and profoundly influenced Western culture. Plato's philosophy of idealism stated that living organisms were merely variations of ideal forms. Each species was created by God, with modern individuals tracing their ancestry to the individual created by God. Aristotle, the natural historian, also wrote of the divine creation of eternal forms. He also recognized that organisms ranged from relatively simple to complex, and he organized them on a scale of increasing complexity called *scala naturae* (scale of nature). That scale implied a linear increase in complexity and was completely uniform: there were no missing

rungs and no movement up or down the ladder. However, the ladder implied nothing about the origin of various groups. This organization of static, ideal forms endured for 2,000 years. According to Aristotle, organisms did not evolve.

The Judeo-Christian culture also promoted prejudice against evolution. Most people believed in the special creation of each species, as described in the Bible's book of Genesis. At least through the 1600s, people believed that plants and animals were created in their current form during the 6 days of creation. This doctrine of fixed species was never convincingly challenged before Darwin. Even during Darwin's time, biology in the Western world was dominated by natural theology, the attempt to discover God's plan by studying God's works. To natural theologians, variation and adaptation in organisms "proved" that each species was designed by God for a particular purpose. Besides, these theologians believed that the earth was only a few thousand years old. Clearly, this wasn't long enough for significant evolutionary change.

---

1. The full title of Darwin's book was *On the Origin of Species by Means of Natural Selection, or the Preservation of Favoured Races in the Struggle for Life.*

2. *Synthetic* refers to the synthesis of information and ideas from different fields of biology into a cohesive framework of evolutionary thought.

3. This distinction has not stopped some religious groups from inventing a discipline called *scientific creationism,* which has nothing to do with scientific thinking.

table 22.1

Major Postulates of Theories That Encompass Evolution

Theory of Descent with Modification	Theory of Natural Selection	Genetic (Synthetic) Theory of Evolution
1. All life evolved from one simple kind of organism or from a few simple kinds.	1. A population of organisms has the tendency and the potential to increase at a geometric rate.	1. Evolution is the change of gene (allele) frequencies in a gene pool over many generations.
2. Each species, fossilized or living, arose from another species that preceded it in time.	2. The number of individuals in a population remains fairly constant.	2. Each species is an isolated pool of genes possessing regional gene complexes that are connected by gene flow.
3. Accumulation of evolutionary adaptations was gradual and of long duration.	3. The conditions supporting life are limited.	3. An individual contains only a portion of the genes in the gene pool, and the portions are different for each individual.
4. Over long periods of time, new genera, new families, new orders, new classes, and new phyla (divisions) arose by a continuation of the kinds of evolution that produced new species.	4. The environments of most organisms have been in constant change throughout geologic time.	4. The kinds of genes and gene combinations in an individual that reproduces sexually came from the transmissible halves of the parents, from recombination, and from mutation.
5. Each species originated in a single geographic location.	5. Only a fraction of the offspring in a population will live to produce offspring.	5. An individual with a phenotype that favors the production of more offspring will contribute a larger proportion of genes and gene combinations to the gene pool.
6. The greater the similarity between two groups of organisms, the closer is their relationship and the closer in geologic time is their common ancestral group.	6. Individuals in a population are not all the same; they have heritable variations.	6. Isolation that restricts gene flow between a subpopulation and its parent population is essential if the subpopulation is to evolve into a new species.
7. Extinction of old forms (species, etc.) is a consequence of the production of new forms or environmental change.	7. The struggle for existence determines which traits are favorable or unfavorable by determining the success of the individuals who possess the traits.	7. Changes of gene (allele) frequencies come about by natural selection, migration, gene flow, mutation, and random genetic changes. Natural selection is the most important cause of changes in gene (allele) frequency.
8. Once a species or other group has become extinct, it never reappears.	8. Individuals having favorable traits will, on average, produce more offspring, and those with unfavorable traits will produce fewer offspring.	8. Evolution of a species may result in a chronological sequence of species without an increase in the number of species, in a group of new species, or in combinations of these two possibilities.
9. Evolution continues today in generally the same manner as during preceding geologic eras.	9. Natural selection causes the accumulation of favorable traits and the loss of unfavorable traits such that a new species may arise.	9. Speciation is completed when variations have accumulated in a species subpopulation to the extent that genetic exchanges with the parent population, or with "sister" populations, do not occur, even though the two populations meet.
10. The geologic record is incomplete.		10. Mutations are the ultimate source of new genes in a gene pool.

Catastrophism

The relevance of fossils to organic evolution was first recognized by Georges Cuvier, a French anatomist working in the Paris Basin of France. He realized that the history of living organisms was recorded in a succession of fossil species trapped and preserved in chronologically ordered layers of rock (fig. 22.1). Cuvier noted that each layer contained fossils from a different set of ancient species. Furthermore, deeper layers revealed plants and animals that were increasingly different from today's organisms. This evidence implied change, but Cuvier still opposed the concept of evolving species.

As an anatomist, Cuvier noted similarities among fossils as well as differences and believed that similar structures reflected a grand design by a Creator. He also believed that boundaries between geological strata of fossils represented catastrophic events such as floods or drought. These events destroyed some of the resident species, which therefore would be absent in the next stratum and would not be living today. This view of natural history is called **catastrophism.** To Cuvier, fossils were organisms that had died in a series of catastrophes, after which the extinct plants and animals were re-placed by the immigration of distant species to the devastated region. Through the eighteenth century, most naturalists invoked catastrophism rather than organic evolution to explain fossils. However, evidence that species changed over time was steadily accumulating.

Gradualism

In 1795 James Hutton argued that the earth was much older than a few thousand years. Hutton explained that profound geological change occurred slowly but continuously by the process of **gradualism** (sometimes called uniformitarianism). The past history of the earth can be interpreted and deciphered in view of present natural laws. Sedimentary rock that encased fossils formed by the gradual accumulation of sediments in lakes, rivers, and oceans (fig. 22.2). Hutton's explanation was widely debated by geologists, including Scottish lawyer-turned-geologist Charles Lyell. Lyell believed that Hutton's evidence for gradualism indicated that the earth was millions rather than thousands of years old. Furthermore, Lyell believed that even slow and subtle processes could cause substantial change over a long time.

figure 22.1

Geological processes have exposed layers of ancient soils and fossils in this canyon. These layers of rock hold "snapshots" of plants and plant communities of the past and record the sequence of ancient organisms. The deeper the layer, the older the fossils it contains. These preserved impressions of ancient plants hold the most tangible clues to evolutionary history.

Catastrophists and gradualists such as Cuvier, Hutton, and Lyell were good geologists convinced of an ancient earth, but they were at odds over how to explain the appearance and disappearance of species in the fossil record. They were all creationists. Lyell, in particular, invoked an unspecified Creator as the source of new species that were added gradually to the earth's flora and fauna. Later, he only grudgingly accepted Darwin's ideas about evolution.

c o n c e p t

Before Darwin's development of evolutionary theory, species were believed to be divine and unchanging creations that appeared on earth a few thousand years ago. Studies by Cuvier, Hutton, and Lyell of fossils and geology indicated that the earth was old and slowly changing, and that life existed millions of years ago.

figure 22.2

This fossil of a Triassic seed plant, *Dicroidium* sp., is about 200 million years old. The gradual accumulation of sediments that surrounded and fossilized such plants indicated to Charles Lyell in the early 1800s that the earth was millions rather than thousands of years old.

Lamarckism

In 1809, the year in which Charles Darwin was born, Jean Baptiste Lamarck (1744–1829) proposed that modern species had *descended* from other species (fig. 22.3). Lamarck proposed some of the ideas found later in Darwin's work, such as the origin of species from preexisting species and the ability of organisms to adapt to changing environments. However, he is most remembered for his theory of the inheritance of acquired characteristics. According to this theory, traits acquired by an individual during its life were passed to its offspring, a theory caricatured by many detractors in drawings showing giraffes obtaining their long necks from previous giraffes who stretched to eat the leaves of high tree branches. According to Lamarck, stretching increased the length of their necks, and this *acquired* characteristic was passed to the next generation. Their necks were stretched further in each successive generation. Using similar reasoning, Lamarck also proposed that the use and disuse of a feature governed the fate of that feature in successive generations.[4] Organs of the body that were used extensively to cope with the environment become larger and stronger, while organs that were not used deteriorated. Instead of the static scheme of Linnaeus, Lamarck provided a dynamic one.

Lamarck was the first naturalist to present a unified theory that attempted to explain the changes in organisms from one generation to the next. Lamarck was a dedicated and observant biologist, and his theories stimulated inquiry into the history and diversity of organisms. However, he had no insight into the laws of heredity, and the mechanisms he proposed for change were wrong. Individuals do not inherit acquired characteristics. Inheriting acquired characteristics would mean that a feature acquired during an organism's lifetime would somehow have to be transmitted to the genes of that organism's gametes. There is no evidence that this happens.

c o n c e p t

Lamarck recognized change in organisms but incorrectly explained it as the inheritance of acquired characteristics. Acquired characteristics are not heritable.

Darwin

As a boy in Shrewsbury, England, Charles Darwin was fascinated with nature. His well-to-do father pressured Charles to pursue the family tradition of a medical career at the University of Edinburgh, but Charles was soon bored and discouraged by medicine and the ghastly nature of surgery, so he quit school. He then enrolled at Christ College of Cambridge University to

---

4. Lamarck either ignored or was unaware of the fact that hundreds of generations of Jewish males had been circumcised without giving rise to a single line of already-circumcised males.

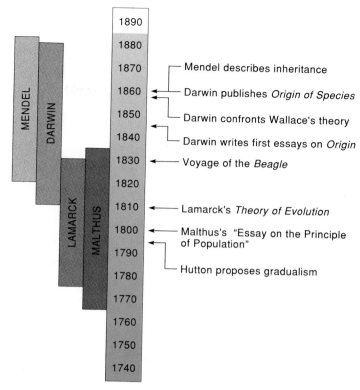

figure 22.3

This historical time line of major events and people who developed evolutionary theory shows that Darwin and Mendel were contemporaries. Notice that the work and publications of Lamarck and Malthus preceded those of Darwin and Mendel by more than 30 years.

study theology, which appealed to him. Indeed, most naturalists at that time were also clergymen. As a clergyman, Charles's pursuit of natural history was accepted by his peers, and he graduated from Cambridge in 1831.

One of Darwin's botany professors at Cambridge was John Stevens Henslow, who became a lifelong friend. Henslow recommended Darwin as a keen observer, naturalist, and companion to Captain Robert FitzRoy for his voyage aboard the HMS *Beagle* to survey lands around the world. Darwin's father initially refused to let Charles go on the trip. Fortunately, Charles's father was swayed by the supportive appeal of his uncle Josiah Wedgewood, son of the famous porcelain maker and father of Darwin's future wife, Emma. Darwin's father finally allowed Charles to join the voyage of the *Beagle*.

The Voyage of the *Beagle*

The *Beagle* sailed from England in 1831 to chart some of the remote islands and coastline of South America (fig. 22.4). The 22-year-old Darwin spent much of his time on shore collecting plants and animals and making geological and biological observations. The flora and fauna of South America were remarkably more diverse and exotic than those of Europe, and

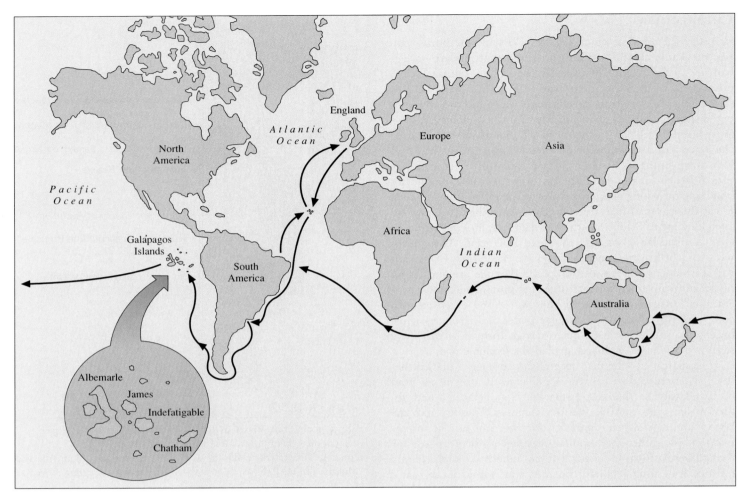

figure 22.4

The *Beagle* voyaged around the world and covered 60,000 km between 1831 and 1836. At each of many stops, Darwin observed and chronicled variation among and within species. Later he used this evidence to support his theories of speciation and natural selection.

they gave Darwin examples of the varied adaptations of organisms to life in mountains, jungles, grasslands, and desolate islands such as the Galápagos Islands and Tierra del Fuego of South America.

The geographic distribution of South American organisms both fascinated and puzzled Darwin, and he pondered why they were so diverse. During the voyage, Darwin also read Lyell's *Principles of Geology* and realized that the earth was very old and constantly changing. This idea, along with his collections and observations, swayed Darwin from his previously staunch, literal interpretation of the Bible. Slowly, he embraced the idea that the organisms that fascinated him may have changed or evolved along with a slowly changing environment.

Charles Darwin had extraordinary powers of observation, and he made some of his most profound observations on the Galápagos Islands. These islands are near the equator and about 900 km west of South America (fig. 22.4). Most of the plants and animals found there are endemic to the islands (i.e., they live nowhere else), but they strongly resemble species from the closest mainland. Among the more famous examples of unusual

organisms on these islands were finches (fig. 22.5). Darwin collected 14 types of finches that were similar but appeared to be different species. Furthermore, these finches were curiously similar to the mainland species yet slightly different. If species were unique and unchanging creations, then why, Darwin wondered, did so many of the Galápagos finches resemble nearby South American species rather than European or African organisms? If finches on the islands had been specially created, why didn't they all look the same? Darwin later concluded that the island finches had descended from a population of ancestral finches from the mainland. The population of mainland finches changed and diversified after becoming isolated on the islands, giving rise to the 14 species endemic to the islands today.

Since the time of Darwin, many other groups of organisms have been discovered to be descended from a single ancestral type. One of the best-known examples of such diversification in plants is the genus *Phlox*, which evolved from an ancestral bee-pollinated species to a variety of species that undergo pollination by beetles, flies, bats, hummingbirds, moths, or butterflies.

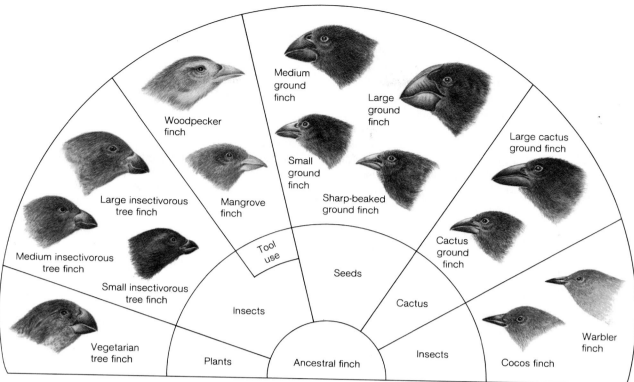

figure 22.5

Darwin theorized that the variety of finches on the Galápagos Islands arose from mainland species. He believed that many generations after their ancestors' arrival, they had adapted to the islands' various environments and food sources and had become unique, reproductively isolated species. Similar examples of evolutionary diversification after separation from ancestral populations occur among plants (see fig. 22.12).

The Formulation of Darwin's Theories

After returning to England in 1836, Darwin did not immediately begin writing his book on the origin of species. Instead, he methodically organized his observations and extensively analyzed a tremendous number of specimens and data. During his voyage, Darwin shipped many crates of carefully packed specimens back to England.

In 1838, 2 years after returning to England, Darwin encountered an inspiration for his theory of natural selection: *An Essay on the Principle of Population,* by Thomas Malthus. Malthus was an economist and clergyman who wrote that populations had an inherent tendency to increase in geometric proportions. Malthus also claimed that resources to support this growth may increase slowly or not at all. He therefore reasoned that because continued growth of a species would outstrip needed resources (especially the food supply), growth would be limited (fig. 22.6). Specifically, Malthus warned of the explosive growth of the human population. Malthus pessimistically warned that humans were reproducing so fast that war, famine, disease, and other human sufferings must eventually limit growth. He believed that humans were doomed to suffering, and he foresaw the need for birth control and social change, such as delayed marriage.

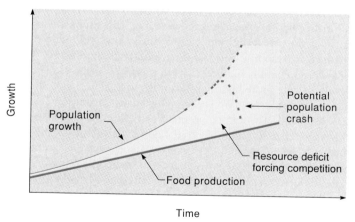

figure 22.6

The predictions of Malthus guided Darwin's formulation of natural selection theory. Malthus believed that the deficit between population growth and food production forced competition and would later crash the human population. Darwin examined these predictions closely and concluded that expanding competition creates strong selection pressures that shape future generations.

Darwin disagreed with Malthus's social views but adopted his ideas that populations tend to increase geometrically and outstrip their resources. Logically, some newborn organisms will die in competitive environments. Darwin observed that wild organisms have variable traits, and he reasoned that in a resource-limited environment, the hardiest or best-suited individuals have a competitive and reproductive advantage over weaker individuals. As a result, traits of well-adapted individuals would increase in successive generations, and the traits of poorly adapted individuals would decrease. Darwin called this process **natural selection,** and he saw a similar mechanism in the modern world among animal breeders who practiced **artificial selection.** These breeders changed the population by selecting organisms with desirable traits and destroying those with undesirable traits.

In Darwin's view, the history of changes of organisms could be represented as a tree with multiple branches (fig. 22.7). The trunk and the bases of the branches represented common ancestors that gave rise to all the subsequent species represented by the more distal branches, some of which were common ancestors of other diverging branches and species. The diversity we see today is represented by all the tips of the branches, which may become extinct or may produce new branches through speciation. Most branches of the evolutionary tree (possibly as many as 99%) have been dead ends; that is, they did not produce new species but became extinct. Darwin often referred to "descent with modification"; the pattern of this descent included many branches.

Darwin was convinced that a population of a common ancestral species could have given rise to a new species by accumulating adaptations to a new environment. New and different adaptations were especially likely when an ancestral species became divided into populations isolated by geographical barriers. Each of the separated populations might have reproduced and changed in response to local environmental conditions. As an isolated population adapted to its new environment, its members would deviate more and more from the appearance and adaptations of the ancestral population. After many generations, an isolated population would be different enough to be a new species. Darwin believed that this had happened to the Galápagos finches. The populations on the Galápagos Islands gradually adapted to the island environment and its food supply and became different from their mainland ancestors. In addition, the populations on different islands diverged from one another.

Within 5 years Darwin had established the major postulates of his theory of natural selection. But he hesitated to publish his idea that species arose by adaptations of organisms to their environment. His idea strongly contradicted public opinion, even though the concept of a changing earth and changes in its inhabitants was gaining credibility among naturalists. Darwin overcame his hesitation in 1858 after receiving a letter from Alfred Russel Wallace containing a manuscript about a theory of natural selection that was remarkably similar to his own. Darwin was shocked and discouraged. He wrote to Lyell, "I never saw a more striking coincidence; . . . all my originality, whatever it may amount to, will be smashed." However, Darwin and Wallace were cooperative scientists interested in the truth. Their theory was presented jointly at a scientific conference (fig. 22.8), after which Darwin quickly finished and published his *On the Origin of Species* on Thanksgiving Day in 1859. The first edition sold out on the first day it appeared.

Darwin explained and documented his ideas so much more extensively than Wallace that Darwin is known as the leading author of the theory of natural selection. The theory was revolutionary not only because it was unique but also because Darwin presented a strong argument based on overwhelming evidence. Interestingly, the first edition of *On the Origin of Species* did not contain the word *evolution*.

Darwin's *On the Origin of Species* shook the foundations of Western culture because his explanation for the origin of and change in species did not invoke supernatural intervention. Notably, Darwin did not confirm or deny the underlying "divinity" of species, but he did propose a tangible, causal mechanism for gradual evolution of modern forms from ancestral species. Darwin did not speculate on the role of God in evolution; but he proposed a plausible explanation for evolution (i.e., natural selection), which he deduced in an organized fashion from abundant evidence. To Darwin, evolution was the gradual and selective accumulation of adaptations that were beneficial to a species in its environment—no more, no less.

c o n c e p t

According to Darwin, evolution includes gradual and selective accumulation of the beneficial adaptations of a population to its environment and a reduction of detrimental characteristics. Natural selection is a mechanism that accounts for these heritable changes in populations.

Evidence for Descent with Modification

Darwin meant to publish a full explanation of evolution based on a thorough analysis of the evidence he had gathered, but he never did. *On the Origin of Species* was an abstract of his work, which he rushed into publication when Wallace's work appeared. Although not complete, Darwin's work provided several kinds of evidence for the theory of descent with modification. Work since Darwin's has added much new information, some of which is summarized next. Note that evidence for evolution by descent with modification is observational and indirect. Nevertheless, taken together, different kinds of indirect evidence provide the basis for strong inference of the process of evolution.

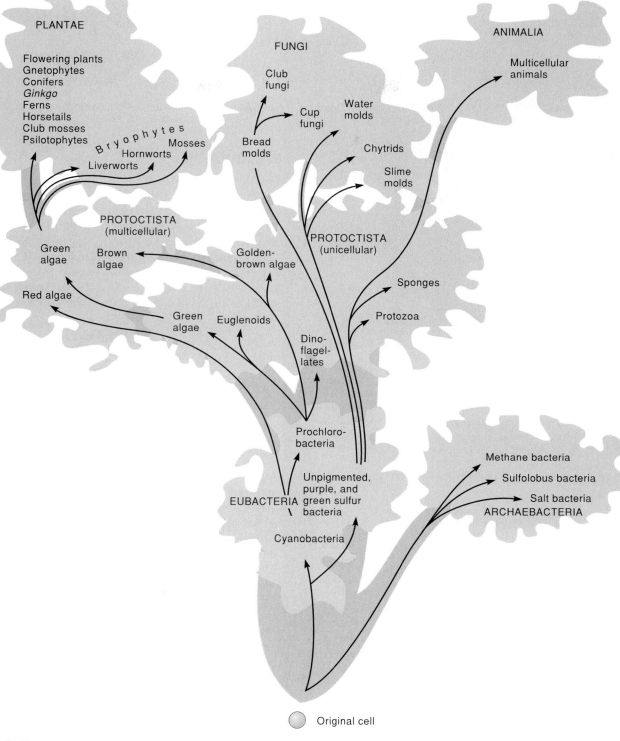

PLANTAE

Flowering plants
Gnetophytes
Conifers
Ginkgo
Ferns
Horsetails
Club mosses
Psilotophytes

B r y o p h y t e s

Mosses
Hornworts
Liverworts

FUNGI

Club
fungi

Cup
fungi

Water
molds

Bread
molds

Chytrids

Slime
molds

ANIMALIA

Multicellular
animals

PROTOCTISTA
(multicellular)

Green
algae

Brown
algae

Red algae

Golden-
brown algae

Green
algae

Euglenoids

PROTOCTISTA
(unicellular)

Dino-
flagel-
lates

Sponges

Protozoa

Prochloro-
bacteria

Unpigmented,
purple, and
green sulfur
bacteria

EUBACTERIA

Cyanobacteria

Methane bacteria

Sulfolobus bacteria

Salt bacteria
ARCHAEBACTERIA

Original cell

figure 22.7

The evolutionary history of organisms can be portrayed as a tree growing through time. Ancient forms are low on the tree. At the tips of the branches are modern forms, and the base of diverging branches represents the common ancestry of the newly evolved kinds of organisms. Prokaryotes (bacteria) are the most ancient organisms and probably gave rise to eukaryotic organisms that were ancestors to plants, animals, and fungi.

figure 22.8

Alfred Wallace presented his paper on natural selection (coauthored with Darwin) to the Linnaean Society on July 1, 1858. Speculation was then replaced with written publication that demanded scrutiny. It marked the beginning of an intense emotional, political, and religious controversy over the history of life and the ancestry of humans in particular.

The Age of the Earth

If evolution occurs gradually, then the earth must be old enough to allow it to happen. Many physicists in Darwin's day believed that the earth was only a few thousand years old—a period too brief for all living things to have evolved from a single common ancestor.

We now estimate the age of the earth by comparing ratios of radioactive to nonradioactive elements in rocks. For example, physicists have evidence that the ratio of uranium-238 (U^{238}) to lead when the solar system was formed was about double the current ratio. We know that the **half-life** of U^{238}— that is, the time in which half of the U^{238} in a sample decays to something other than U^{238}—is about 4.5 billion years. This indicates that the earth is between 4.5 and 5.0 billion years old, since the ratio of U^{238} to lead has decreased to about half of what it was when the earth was formed. Biologists have examined evidence and calculated that this is long enough for the evolution of all of the organisms of today to have occurred from a single ancestor. Incidentally, meteorites are also estimated to be about 4.5 billion years old, which supports the idea that the earth was formed simultaneously with other parts of the solar system.

Fossil Evidence

The oldest organisms are probably either absent or we can't recognize their preservation in rocks. The oldest recognizable fossils are of bacteria, some of which were photosynthetic, that lived at least 3.5 billion years ago (fig. 22.9). This age establishes a minimum time span for organic evolution. Interestingly, recent observations and research by J. W. Schopf at the University of California have indicated that many fossils of bacteria look the same as modern species. For these species, he concludes that evolutionary change seems to be arrested.

5:20 A.M.: First life? Bacteria appear in the fossil record.

Darwin predicted that the fossil record would contain links among related organisms, from ancient to progressively more recent species. Much of the information about fossils that has subsequently been gathered agrees with Darwin's prediction. Some of the best fossils are from bones, especially teeth, so zoologists can construct nearly complete series for some groups, such as horses, humans, other mammals, and reptiles. The most complete record of plants is for those with vascular tissue; however, there is not a clear lineage of plants, one giving rise to the other, as there seems to be for other organisms. Nevertheless, 400-million-year-old fossils of the first vascular plants, if not directly ancestral to present-day groups, look like they share a common ancestry with certain groups of living vascular plants (fig. 22.10). More details about the relationships of fossilized and living vascular plants are presented in Chapters 28–30.

9:52 P.M.: The first vascular plants appear only 2 hours from the end of the "day."

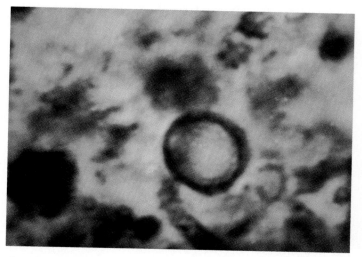

figure 22.9

Fossilized bacterial cells such as these have been found in ancient rocks of Australia and South Africa. These rocks are about 3.5 billion years old and predate the earliest known fossils of eukaryotes by at least 2 billion years. Chemical evidence within these fossils suggests that photosynthesis was occurring 3.5 billion years ago.

figure 22.10

This modern (*Psilotum*) plant resembles the fossils of ancient ancestors. Although the fossil lineage of vascular plants is incomplete, the earliest fossils show common ancestry with some simple plants such as *Psilotum*. This modern *Psilotum* has dichotomous branching, terminal sporangia, simple vascular tissue, and no roots or leaves.

Homology

Features that are similar because of common descent show **homology**. Many organisms have features that appear to have come from the same ancestor. In flowering plants, for example, the ovarian tissue surrounding the seeds of magnolia resembles that of lily, poppy, oak, and orange, because all of these plants inherited it from a common ancestor. Likewise, the red pigments of cactus flowers, beet roots, *Portulaca* stems and flowers, and spinach petioles are homologous. These pigments are of the same chemical type, called *betalains* (see Chapter 2), which is unlike other types of red pigments. Most of the major features of plants that we've discussed, such as their vascular structure and their hormonal responses to stimuli, probably descended from a common ancestor of the plant kingdom. Plants also have features that are homologous with features of other kinds of organisms. Cells with nuclei, chromosomes with histones, and genes with introns were inherited from the common ancestor of plants, animals, and fungi. Similarly, photosystem I in plants and in certain photosynthetic bacteria (see Chapter 7) probably came from a common ancestor.

Convergence

Whether similar features are homologous is often unclear because not all similarities are due to homology. The issue of homology is clouded because similar features may arise in distantly related plants that live in similar environments. One of the most common examples, which often fools all but the most astute observers, is the stem succulence of cacti and euphorbias (fig. 22.11). Instead of homology, this feature shows **convergence**, which is one explanation for a similarity that does not

come from a common ancestor. In this case, descendants of a pre-cactus ancestor became adapted to arid environments in the Americas by developing succulent stems. Likewise, descendants of a pre-euphorbia ancestor evolved succulent stems in response to arid environments in Africa. Entire plant communities in different geographical areas also appear similar because they are adapted to similar environments. Shrublands around the Mediterranean Sea, the coast of southern California, and the coast of central Chile are similar, even though the plants that comprise them are not closely related to one another.

Similarities among distantly related organisms are not likely to result from coincidence. Instead, nonhomologous features of plants and plant communities indicate that different organisms have evolved in similar ways to similar environments in different areas of the world.

Biogeography

Biogeography is the study of the geographical distribution of organisms. The distribution of today's organisms provides strong evidence for evolution because biogeography reflects the history of living species. That is, it reveals their movement and change. From this information we can determine the ecological factors that control a species' distribution and the events that led to this distribution. In other words, the biogeography of a species partially records its evolution.

Darwin and Wallace both questioned the apparent discrepancy between divine creationism and the biogeography of species that they observed. Why wouldn't similar environments in distant parts of the world have the same divinely created species? What explained the oddly disjunct distributions of organisms such as *Sanicula crassicaulis* (a member of the carrot

(a)

(b)

figure 22.11

Stem succulents: (a) cactus and (b) *Euphorbia*. These plants are strikingly similar not because they share a common ancestry, as you might suspect, but because they have similar adaptations to arid environments. Their similarity illustrates convergence; unrelated plants may accrue similar adaptations due to similar environmental pressures.

family), found only in California and southern Chile; or creosote bushes (*Larrea*), which live in the deserts of western North America and southern South America; or skunk cabbage (*Symplocarpus foetidus*) and the tulip tree (*Liriodendron tulipifera*), which occur in eastern Asia and eastern North America; or cacti, which are all American except for one species in Madagascar? How have these patterns of distribution come about? The answers lie in the study of biogeography.

Island biogeography has been particularly informative about the movement and evolution of organisms. *Island* refers not only to a habitat surrounded by water but also to any area surrounded by an aberrant habitat, such as a mountain surrounded by desert or a clump of trees surrounded by prairie. Islands often have species found nowhere else in the world, and the closest relatives of these species may inhabit a different environment on the nearby mainland. This is best explained by migration, adaptation to a new environment, and subsequent speciation. Populations from an ancestral group living in a particular place (e.g., the mainland) often spread or radiate to other habitats (e.g., islands). These new habitats include new environmental conditions that promote the evolution of new adaptations. Migration to new environments followed by adaptation and speciation is called **adaptive radiation** (fig. 22.12).

The finches that Darwin found on the Galápagos Islands are an example of adaptive radiation. Botanists have also found many examples of adaptive radiation in plants, most notably on the islands of Hawaii. A well-studied example involves the stick-

tights (*Bidens* sp.), so named because their fruits stick to socks or other clothing when you walk among the plants. This stickiness explains how they may have attached and been transported by seabirds from the mainland to the Hawaiian Islands. The sticktights in Hawaii apparently radiated from a single colonizer into the more than 40 species that now occur only on the islands. This radiation was accompanied by a loss of dispersal and colonizing ability; the first weedy sticktights produced progressively larger and less weedy descendants. The most recently appearing species may be *Bidens ctenophylla*, which is a long-lived, woody tree whose fruits are larger and fewer than those of its weedy ancestor and do not stick to anything (fig. 22.13). It no longer has the weedy life-style that characterized its earliest island ancestor.

The Lore of Plants

Islands can be virtual "hot spots" of evolution. Environmental conditions on New Caledonia, an island distant from the east coast of Australia, have produced one of the world's unique floras. The forests of New Caledonia contain more than 1,575 species of plants, of which an astonishing 89% are found nowhere else in the world. Unfortunately, disruption of the environment by logging, mining, brush fires, and so on has reduced the amount of undisturbed forest to less than 1,500 km² covering less than 9% of the island.

(a)

(b)

(c)

figure 22.12

Isolated and diverse environments such as the Hawaiian Islands fuel adaptive radiation and divergence. Radiation has been striking and rapid for the three species shown here, (a) *Argyroxiphium sandwicense*, (b) *Wilkesia gymnoxiphium*, and (c) *Dubautia ciliolata*. These and twenty-five other closely related species probably descended from a single species arriving from the Pacific Coast of North America. Adaptive radiation on the islands has produced diverse morphologies even though the species differ in only a few critical genes.

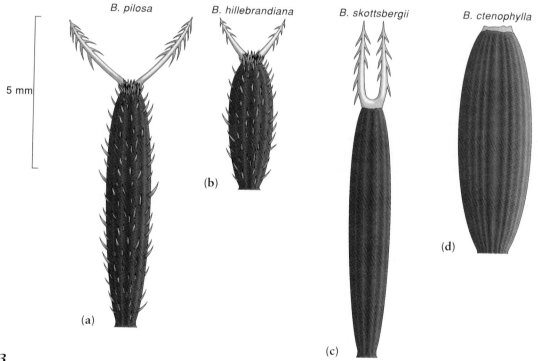

B. pilosa

B. hillebrandiana

B. skottsbergii

B. ctenophylla

5 mm

(a)

(b)

(c)

(d)

figure 22.13

Variations in fruit morphology show adaptations to new environments and the pattern of colonization and inland invasion by Hawaiian species of *Bidens*. (a) *B. pilosa* was apparently the ancestral colonizer. It was easily dispersed from the American mainland by birds carrying its sticky fruits with upward-pointing bristles and spreading, barbed awns. (b) Today, a resident species on the coast of Hawaii, *B. hillebrandiana*, has ornamented fruit similar to *B. pilosa*. (c) More inland, the low-elevation species *B. skottsbergii* has fruit with reduced dispersal ability due to loss of bristles and closing of awns. (d) Still farther inland, the central mountain habitats have *B. ctenophylla* with no bristles or awns. Apparently, habitats progressively inland do not favor fruit with strong features for animal dispersal.

Molecular Evidence

Since the late 1960s, evolutionary biologists have been comparing proteins and genes of various species, much as they have done with morphology and anatomy for more than a century. The most widely used comparisons have been based on cytochrome *c* oxidase, an enzyme of oxidative electron transport (see Chapter 6). The universal occurrence of this enzyme indicates that it evolved early in the origin of life and has been inherited by all of the descendants from a single ancestor; thus, it is evidence of a single origin of life. In addition, its amino acid sequence has slowly changed as organisms have descended with modification during evolution, and it now contains information about the pattern of descent among organisms. Few differences in amino acids between organisms indicate a more recent common ancestry and a close evolutionary relationship. The oldest ancestor gave rise to fungi, plants, and animals, and the most recent ancestors are the one shared by donkeys and horses and another shared by humans and monkeys. Recent examination of amino acid sequences for wheat, for example, shows it to be more closely related to animals than to fungi.

More recently, evolutionary biologists have also begun to compare sequences of DNA and RNA, and many other kinds of molecular information to deduce patterns of evolutionary descent. Recent research by Svante Pääbo and others at the University of Munich has shown that genetic information is retrievable from DNA in fossils many thousands of years old. The results of these comparisons are generally consistent with the patterns of other homologous features. Specific examples of molecular evolutionary comparisons are discussed for each group of organisms presented in Chapters 25–31.

c o n c e p t

Modern evidence for evolutionary descent includes the age of the earth, the fossil record, homology, the convergence of adaptations of different species in similar environments, the similarity of species in different parts of the world, and similarities of nucleotide and amino acid sequences among species descended from a common ancestor.

Evidence for Natural Selection

Although the theory of descent with modification was revolutionary, even during Darwin's time it was supported by convincing evidence. Darwin's greatest contribution to evolutionary biology, however, was not the concept of change but his explanation of a *mechanism* for change. To Darwin, observable evidence verified the results of evolution, but the specific *process* was apparently so subtle that it had eluded other naturalists. That process was natural selection. Other processes change populations, but natural selection is the most important.

Even if the advantages of some characteristics are only slightly greater than those of others, the favorable characteristics will inevitably accumulate *after many generations* of being "selected"—that is, disproportionately reproduced. This accumulation of characteristics is synonymous with long-term genetic change, which is the heart of evolution. Of course, the same mechanism decreases the occurrence of unfavorable characteristics. *Remember that natural selection and evolution are not identical: natural selection is a mechanism that results in evolution. The important result of natural selection is that it adapts a population to its environment.*

c o n c e p t

Natural selection entails the differential reproduction of different genotypes. Genetic variation provides traits that determine an organism's success in the environment. Natural selection is the mechanism that fosters the reproduction of advantageous traits in the next generation.

Comparison with Artificial Selection

Darwin observed that farmers often selected certain animals or plants for reproduction based on their desirable traits. Humans have practiced the selection and controlled mating of plants and animals for centuries, a practice we now call *artificial selection*. Animal breeders know that to produce fatter hogs, they must allow only the fattest hogs to mate. Plant breeders know that to grow the largest tomatoes, they must allow cross-pollination only between plants that produce large tomatoes. These ideas are not new: most of our fruits and vegetables have been artificially selected for size, yield, pest-resistance, and so on since agriculture began more than 10,000 years ago (see reading 22.1, "Selection for Monsters").

The practice of artificial selection led Darwin to conclude that a similar process occurs in nature, which he called *natural selection*. The main difference is that humans are the agents of artificial selection, whereas the forces of nature are the agents of natural selection. Many of the postulates of natural selection, therefore, can be deduced from the process of artificial selection. Furthermore, unlike the evidence for descent with modification, the evidence for several of the postulates of natural selection is direct or experimental. Some of the postulates of natural selection can be compared with their counterparts in artificial selection as follows (postulate numbers come from table 22.1).

POSTULATE 1

A population of organisms has the tendency and the potential to increase at a geometric rate.

How many seeds does a tomato produce? How many does a tomato plant or a whole field of tomatoes produce? If the field of tomatoes is analogous to a population in nature, then many more seeds are produced than ever grow into new plants. If all seeds produced new plants, the earth would be knee-deep in tomato plants by the end of one growing season. The overabundance of tomato seeds is evidence that a population has the

SELECTION FOR MONSTERS

Monstrous plants are plants that differ markedly from the norm. We have bred and selected for giant seeds, fruits, flowers, and vegetative organs. For example, we have selected and perpetuated cultivars of banana, grape, sugarcane, and other plants for increased bulk, sugar content, and visual appeal. Most of these varieties can no longer produce viable seeds, but our production of tasty tissue has been remarkable.

Nowhere is our selection for tasty tissue more intense than among pumpkin growers. These growers have spawned two groups—the World Pumpkin Confederation and the rival Great Pumpkin Commonwealth—that sponsor Weigh-Offs to crown world champion pumpkins. With tens of thousands of dollars of prize money at stake, the selection pressures are intense: farmers work in their patches as much as 5 hours per day on a pumpkin, caring for the soil, eliminating weeds, and covering the pumpkins with plastic and blankets at night. On a good day, a pumpkin can gain 30 lb. How big will the winner be? Almost 1,000 lb. Or, as one grower put it, "Envision a small Volkswagen; an orange one."

Of course, Volkswagen-size pumpkins won't spring from the kind of seeds that you pick up at the local grocery store; the winning pumpkins almost always come from winning parents. For example, one year, five of the top 20 pumpkins came from a pumpkin that weighed 600 lb; seeds from that pumpkin produced seven pumpkins that each weighed over 800 lb. These winners become stud pumpkins, each producing about 600 seeds. Thirty seeds are returned to the grower, while the rest are sold for as much as $25 per seed.

Our rampage through the genome of plants is also apparent in the genus *Brassica* of the mustard (Brassicaceae; also called Cruciferae) family. This genus contains perennial, biennial, and annual herbs, with 100 wild species found in north temperate parts of the world. You can easily recognize the family by its characteristic flower having four petals that form a cross (hence the

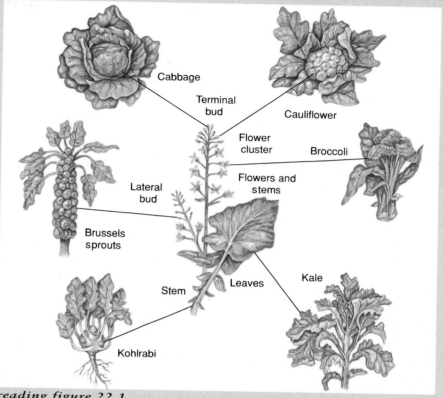

reading figure 22.1

Artificial selection has produced this variety of common vegetables from a single common ancestor. All of these vegetables are the same genus, *Brassica* spp., but various morphological features have been artificially selected to enhance different textures and flavors.

family name); six stamens, of which two are short; and its special fruit, the silique. Although the cultivated brassicas are probably derived from several wild species, a reasonable ancestor is colewort (*B. oleracea*), a scrubby perennial native to Europe and Asia. From colewort, we have selected and developed a wide array of common vegetables (reading fig. 22.1). One line produced kale, cauliflower, and broccoli.* A second line gave us kohlrabi, with a short stem enlarged into an aboveground tuberous vegetable, and rutabaga, whose tuberous storage stem develops belowground. The turnip's (*B. rapa*) storage organ is also a tuber, and its true root extends down from the swollen stem. Although all the leafy vegetable members of the genus may not be derived from colewort, they are all closely related. Several species of mustard grow

wild, and we exploit their edible leaves (*B. juncea*) and particularly their seeds (*B. nigra* and *B. alba*), which are used as mustard. Pakchoi (*B. chinensis*) is an important leafy vegetable in Asia, as are the closely related petsai (*B. pekinensis*) and false pakchoi (*B. parachinensis*). Rape or colza (*B. napus*) is grown primarily for its seeds, which yield an edible oil.

Text on "Monstrous Plants" from *The Green World: An Introduction to Plants and People*, Second Edition by Richard M. Klein. Copyright © 1987 by Harper & Row, Publishers, Inc. Reprinted by permission of Addison Wesley Educational Publishers, Inc.

*Broccoli was created by Italian horticulturists. Among the descendants of these horticulturists was Albert Broccoli, who produced 17 James Bond movies.

potential to increase. Imagine the potential for certain orchid populations to grow when each fruit produces more than 20,000 seeds!

POSTULATE 5

Only a fraction of the offspring in a population will live to produce offspring.

In agriculture, most plants and animals are harvested before they reproduce. Humans choose which individuals to harvest; that is, we are the agents of selection that limit reproduction. In nature, most plants die as seeds or seedlings. Conditions for germination or growth may not be good enough, pests may destroy the seeds or seedlings, or a freeze or drought may wipe them out. Saguaro cacti have been extensively studied in this regard. Fruits of the saguaro produce dozens of tiny black seeds, most of which will germinate and grow under cultivation. A population of saguaros probably produces millions of seeds, yet in most populations very few seedlings appear during any one year. Moreover, only a few new seedlings live more than a few years, and a saguaro must be 10–20 years old before it can form flowers and fruit. Some populations have no seedlings for several years in a row.

POSTULATE 6

Individuals in a population are not all the same; some have heritable variations.

Even in a field of corn, which may seem uniform, certain features vary from one plant to another. This was shown in a long-term experiment in artificial selection that began in 1896 at the Illinois Agricultural Experiment Station. At the beginning of the experiment, two types of kernels were selected for reproduction, one of slightly higher protein content than the other. Every year, the plants producing the highest amount of protein were hybridized with one another, and plants producing the lowest amount of protein were hybridized with one another. At the beginning of the experiment, the average protein content per kernel was 10.9%. After fifty generations, the low-protein line was down to 4.9%, and the high-protein line was up to 19.4%. In this case, heritable variation occurred in protein content, and humans selected only certain individuals for breeding.

The evidence for heritable variation in corn is that the protein content changed over time because of strong selection pressure by humans. What evidence do we have that heritable variation occurs in nature? In other words, is natural variation due to genetic differences (heritable), or is it due to different responses of individuals to their environment (nonheritable)? We know that the sizes of leaves and fruits, the number of seeds, and other features of cultivated plants can be influenced by how much fertilizer and water we give them. If water or nutrient availability is uneven in a certain habitat, then certain plants could be bigger merely because they get more water or more nitrogen from the soil.

Transplant experiments reveal evidence for heritable variation. Genetically controlled features do not change after individuals are transplanted from one environment to another. A classic example of this kind of experiment was done with yarrow (*Achillea lanulosa*), a highly variable species that grows along the Pacific Coast and in nearby mountains. When transplanted to a garden at Stanford University near the coast and to another at timberline in the Sierra Nevada, plants from different habitats maintained their basic features, sometimes even to their demise. Plants from the coast, for example, normally bloom in the fall. In the timberline garden they still flowered in the fall, but fall frost caught them before they could set seed. This experiment shows that some adaptive features of yarrow in different habitats are controlled genetically.

Not all features are heritable, however. In another transplant experiment, clones of the same individual of cinquefoil (*Potentilla glandulosa*) were grown in different environments and showed variation among environments. Since all plants were identical genetically, this subsequent variability was attributed to individual responses to different environments, not to heritable variation.

POSTULATE 8

Individuals having favorable traits will, on average, produce more offspring, and those with unfavorable traits will produce fewer offspring.

High-protein corn, fat hogs, and large tomatoes are favorable traits for artificial selection. Organisms with these traits produce more offspring because people select them for breeding. In nature, if a plant is rugged, disease-resistant, a good competitor for resources, and an efficient photosynthesizer, then we also expect it to reproduce more frequently than plants that are not so rugged or disease-resistant. For example, a plant having a larger leaf-surface area is adapted to a low-light environment and will have greater reproductive success than small-leaved plants in the same environment. Random events such as storms and floods may cause the unexpected, but under normal circumstances the organisms that are most attuned to their environment will be most successful.

A well-documented example of natural selection involves the work of Janis Antonovics with plants growing in soil containing high concentrations of copper, zinc, or lead (i.e., the waste products from mines). Most plants can't grow in soils contaminated with these metals; however, some species of grasses and weeds do grow around these mines. Surprisingly, these plants are the same species that grow in nearby pastures; apparently, some members of the species can tolerate toxic metals. In one experiment, Antonovics planted seeds from the pasture population of grasses in soil that was rich in copper. Only 1 in 7,000 survived. Thus, copper selected intensely against the intolerant phenotype.

In a related experiment, Antonovics gathered seeds from metal-tolerant adults growing on contaminated soil and planted them in the same contaminated soil. Not all the offspring survived. Moreover, the offspring varied more in their tolerance to metals than did their parents. This limited generation of seedlings revealed that the parents still housed significant genetic variation for metal tolerance. Alleles for intolerance were still present, even though metal in the soil continually selected against intolerant individuals. Natural selection was operating in this population, and genetic variation was still present.

c o n c e p t

Direct evidence exists for several of the postulates of natural selection: the overproduction of seeds indicates that populations can increase at a geometric rate; a small fraction of seeds grows into reproductively mature plants; different individuals in a population have variable, genetically controlled features; plants that are better adapted to a habitat have more reproductive success in that habitat than plants that are not so well adapted to it.

Subtle Features of Natural Selection

Populations

Evolution is a genetic change in populations over many generations. Thus, the concept of a **population** is critical to understanding natural selection and evolution because a population is the smallest unit that can evolve.

A population is a group of interbreeding individuals of the same species sharing the same territory at the same time. Members of the same species may be widely distributed and divided into many populations, but a population is specific to a location, and members of a population must be close enough to interbreed. Because environmental conditions are also location-specific, natural selection of traits adaptive to an environment occurs in local populations.

The concept of a population was alien to many of the original readers of Darwin's work. Indeed, the idea of a population that included unique and varied individuals was in stark contrast to the constancy associated with divinely created species. The population concept implied that variation occurred within a species and that each individual was unique; there was no average or ideal form of a species. In natural populations, an average individual or the average for a characteristic (e.g., that the average American family has 2.2 children) is merely a statistical abstraction.

Survival of the Fittest or Reproduction of the Fittest?

The phrase *survival of the fittest* is not entirely appropriate to natural selection. It implies that the singular direction of natural selection is to increase survival, but this is inaccurate. More precisely, natural selection promotes traits that increase the passage of genes to the next generation. That is, it promotes successful *reproduction*. Of course, survival is a prerequisite to reproduction, but reproductive success can be enhanced by traits not associated with long survival times. For example, annual flowering plants are diverse and abundant, even though individuals do not live as long as perennial plants. Their survivorships are different, but they both have successful reproductive strategies. The term *survival*, however, may be appropriate if one thinks in terms of the survival, over generations, of species or genetic traits rather than of individuals.

Successful reproductive strategies are highly varied and strongly selected. For example, mushrooms and cottonwood trees produce thousands or millions of spores and seeds, respectively, whereas an avocado tree produces only a few. The seeds of an orchid may be the size of a pinhead, while those of a coconut tree may be as large as a grapefruit and loaded with food reserves for the developing embryo. Early reproduction, abundant reserves for the embryo, easily dispersed seeds, noxious-tasting seeds, large numbers of seeds, and conspicuous flowers all promote reproduction *in appropriate environments*. Such strategies are selected if they ensure production of the next generation. Reproductive traits are more critical than survival traits to natural selection.

Natural Selection in the Context of the Environment

Successful adaptations and natural selection are inseparable from the environment. For example, an adaptation of succulent leaves for water storage is neither positive nor negative unless put in the context of an environment. Succulent leaves may promote reproductive success in a desert but not in a rain forest. That is, the same trait can be positively selected in one environment and negatively selected in another environment. Thus, environmental conditions are the forces that determine the nature and outcome of natural selection. These environmental conditions are called **selection pressures**. For example, fungal disease is a selective pressure. A dispersed seed may have a tough, protective seed coat that prevents fungal invasion. However, the seed coat may also prevent the penetration of water needed for germination. Thus, thicker seed coats may be advantageous in moist, fungus-rich environments, while thinner seed coats are selected in drier, less fungus-rich environments. If the embryo is disease-resistant, however, then a thin seed coat may be best in moist as well as dry habitats.

Organisms must deal with various and often contradictory environmental factors. For example, warmth may increase metabolism for rapid growth, but it may also dry the soil rapidly. Different species solve problems with a variety of strategies promoted by natural selection in the context of the environment.

Selection as a Noncreative Force

Natural selection does not create new alleles, traits, or adaptations. Instead, it channels genetic variation that is already present in a population by differentially reproducing traits with negative or positive adaptive value. If the alleles and

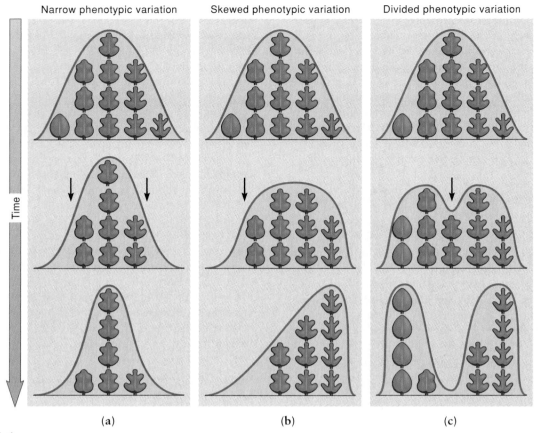

Narrow phenotypic variation   Skewed phenotypic variation   Divided phenotypic variation

Time

(a)   (b)   (c)

figure 22.14

Stabilizing, directional, and diversifying selection for leaf margination will change phenotypic frequencies over time. The height along the curve represents the number of individuals with that phenotype, and the arrows indicate negative selective pressure. (a) Stabilizing selection reduces variation and preserves a narrow phenotype best adapted to the environment. (b) Directional selection promotes phenotypes at one end of the range of variation, and the distribution of phenotypes moves toward that end. (c) Diversifying selection promotes extremes of a characteristic and may produce two distinct populations.

genetic combinations for traits such as color, competitiveness, or starch storage are absent in the population, then these adaptations cannot develop, regardless of the intensity of natural selection.

A common misconception is that natural selection brings organisms toward an ideal, maximum fitness. Natural selection does not lead to perfection; rather, it promotes genetic combinations that "work" in the local environment. If environments are different, then naturally selected organisms will be different. Variation is an expected consequence of natural selection among populations.

c o n c e p t

Natural selection operates on populations and their interactions with the local environment. Rather than create new adaptations, natural selection promotes the most adaptive, available variants through increased reproduction and retards nonadaptive variants through decreased reproduction.

Kinds of Natural Selection

Selection can alter the occurrence of a trait in a population in various ways. In a simple case, positive selection for tolerance to toxic metals can increase the number of tolerant versus intolerant plants in a population over many generations. But genetic traits are usually not in clear categories such as tolerant and intolerant. Instead, populations contain a continuum of values for a trait; that is, some individuals may be slightly more or slightly less tolerant than others. Thus, natural selection can move the distribution of values for a phenotype along a continuous scale (fig. 22.14). We can graphically represent this scale as the x-axis—a continuum of values for the alleles and the trait they affect. The y-axis is the frequency (i.e., number) of individuals expressing each value. This simple representation is more ideal than realistic, but it illustrates the different ways that natural selection can alter the genetic composition of a population. The peak of the curve represents the most common value for the trait, and the breadth of the curve represents the degree of variation in the population. Different kinds of selection alter the shape of the curve.

Stabilizing Selection

Stabilizing selection promotes the norm and reduces both extremes (fig. 22.14a). This kind of selection operates in most populations, and individuals with traits near the peak are positively selected. Organisms near the extremes of variation are selected against and rarely pass their extreme traits to the next generation. As a result, the frequency of alleles producing the optimal phenotype continues to increase, while alleles interacting with the environment to produce extreme phenotypes decrease. Stabilizing selection in most populations eliminates extreme individuals, including mutant forms. The resulting stable or predictable distribution of traits is most common in stable, unchanging environments.

Directional Selection

In changing environments, **directional selection** is a primary mechanism of change. In directional selection, individuals at one extreme of population variation are negatively selected, whereas those at the other extreme are positively selected. Over many generations, the average condition moves toward the positively selected extreme (fig. 22.14b).

Directional selection can cause change over a relatively short period of intense selection. For example, plants studied in a grazed pasture in Maryland were shorter than the same species in ungrazed pastures. Were they short due to grazing or due to genetic differences? To separate the effects of grazing from genetics, some of the short plants were planted away from the selection pressure of grazing herbivores. Surprisingly, some of the transplants remained short even without grazing. Apparently, directional selection pressure by grazers had genetically altered the grazed population in the direction of short plants less susceptible to herbivores. Directional selection is the main type of selection used in plant breeding.

Diversifying Selection

Diversifying selection (also called *disruptive* or *catastrophic* selection) is common in environments that are suddenly or drastically changing. These environments become unfavorable to the previously successful and adaptive peak of a trait. As a result, peak values for a phenotype are selected against because extreme values in an altered environment are just as likely to enhance successful reproduction (fig. 22.14c). Sometimes organisms with extreme values escape catastrophic negative selection because they live at the periphery of the population. They survive to reproduce, while normal, centrally located organisms are negatively selected. As you might suspect, drastic environmental changes are usually short-lived, but they significantly increase genetic diversity and temporarily decrease population size.

Genetic diversity of sugar maples in Ohio is maintained by opposing or diversifying selective pressures. The southern strain of sugar maples can survive occasional droughts and cool winters, while the northern strain is better adapted to the coldest winters. In Ohio a cool, dry winter may promote the southern strain and its alleles, but the following year a severe winter may select for the northern strain. Diversity is maintained as both types intermingle in the Ohio population. Sugar maples with intermediate tolerances are not favored in either cold or warm climates.

The Genetic Basis of Evolution

Among the most important developments in evolutionary biology since Darwin is the application of genetics to the theory of natural selection. A major problem with the initial understanding and acceptance of evolution by natural selection was a lack of knowledge of the genetic causes of variation (see reading 22.2, "Darwin's Big Mistake"). Mendel's work was unknown. Most biologists believed that parental traits "blended" to produce intermediate traits in offspring.

Evolutionary theory came of age when biologists stopped thinking in terms of individual organisms and began thinking in terms of populations and frequencies of genes and alleles. The integration of Mendelian genetics and evolution by natural selection is called **population genetics**. For example, dandelions growing in a valley or water hyacinths growing in a lake constitute populations. The unifying property of a population is its *gene pool*. The gene pool is the total genes in a population at any one time; it consists of all genes at all loci in all individuals of the population. Members of the next generation get their genes from this pool, and individual organisms temporarily contain a small portion of the gene pool. To population geneticists and to evolutionary theory, the gene pool, its variations, and the mechanisms causing change are more important than the individual. Indeed, because populations often include thousands or millions of organisms, the fate of one individual usually has little influence on the fate of the whole population (fig. 22.15).

Evolutionary theory presumes that some alleles of natural populations increase in frequency from generation to generation, while others decrease. Favorable combinations of alleles in a genotype are more likely to occur in greater proportion in the next generation. In contrast, unfavorable alleles are less likely to be passed to the next generation, and their frequency will be reduced. The change in the frequency of alleles within a gene pool over many generations is a geneticist's definition of evolution.

DARWIN'S BIG MISTAKE

In seeking a genetic mechanism for traits passing from one generation to another, Darwin addressed the previously unquestioned assumption that the parent's body somehow influenced the form of the offspring. In this regard, Darwin subscribed to Lamarck's theory of the inheritance of acquired characteristics. But Darwin took the idea of somatic influence on heritable traits a step further. He knew that there must be some physical connection between parent and offspring. He explained this connection in his theory of pangenesis, which was the greatest mistake of his scientific career. According to this theory, every part of a mature individual produces tiny packets of heritable information, which Darwin called **gemmules.** To be transmitted to offspring, gemmules from all over the body were somehow transported to the reproductive organs and packed into the gametes before fertilization. The gemmules of the two parents would blend together in the offspring, where they would sprout wherever appropriate to determine the features of each part of the new organism. Darwin's ideas about pangenesis probably arose from his belief that the use and disuse of organs influenced heredity, which was a widely held view at the time. In that context, pangenesis was a perfectly logical theory, although it was inaccurate.

figure 22.15

Large populations such as these white prickly poppies (*Argemone albiflora*) share a common gene pool and frequently cross-pollinate. Population geneticists study the genetics of populations rather than the genetics of an individual.

Fitness is a measure of an organism's evolutionary success but does not strictly refer to its health, survival, or adaptation to the environment. Rather, an individual's fitness is measured by its number of surviving offspring relative to the number surviving from other individuals in the population. Fitness represents the extent to which an individual's alleles will occur in succeeding generations. Highly fit individuals produce more surviving offspring than do less fit individuals.

The Hardy-Weinberg Equilibrium

Population genetics is the mathematical study of the events occurring in gene pools that modify gene frequencies. Among the most significant models of gene frequencies is the **Hardy-Weinberg equilibrium,** named for G. H. Hardy and W. Weinberg, who independently developed and proposed this concept in 1908.[5] This model describes a gene pool whose various allelic frequencies are at equilibrium under certain conditions. Hardy-Weinberg equilibrium is not concerned with the genotypes or phenotypes of individuals but with the composition of the entire gene pool. It predicts genetic equilibrium (nonevolution). Such equilibrium is rarely attained in natural populations, but it is a good starting point for explaining how gene pools change.

To illustrate Hardy-Weinberg equilibrium, let's consider a simple, hypothetical population of diploid wildflowers carrying only two alleles, A and a, for flower color. The allele A is a dominant allele for blue flowers, and its frequency in the gene pool is designated as p. The allele a for white flowers is recessive, and its frequency is q. Since A and a are the only alleles occurring at this gene locus, the sum of their frequencies (i.e., their proportion of the total) must equal 1.00 (i.e., 100%). For example, if 60% of the alleles are A, then 40% of the alleles must be a. That is,

$$p + q = 1.00$$

$$q = 1.00 - p = 1.00 - 0.60 = 0.40$$

In this example, 60% of all gametes (sperm cells and egg cells) produced by the population carry A, while 40% of all gametes carry a. For a randomly mating population, we can portray the genotypes of the offspring and their relative frequencies from the fusion of eggs and sperm as follows:

Genotypes of the Next Generation

	A	Eggs	a
A	AA		Aa
Sperm			
a	Aa		aa

5. Others had proposed the model earlier, including W. E. Castle and S. S. Tschetverikov, but the tradition of naming it for Hardy and Weinberg is too deeply ingrained to overcome. It is often inappropriately referred to as the Hardy-Weinberg law.

Genotypic Frequencies of the Next Generation

	$A = 0.60$	Eggs	$a = 0.40$
$A = 0.60$	0.36		0.24
Sperm			
$a = 0.40$	0.24		0.16

This table is a Punnett square (see Chapter 8), which helps biologists visualize all mating possibilities. The genotypes resulting from random mating within this population are 36% AA, 48% Aa (i.e., 24% + 24%), and 16% aa. These same results can be calculated using the following Hardy-Weinberg notation:

$$p + q = 1.00$$

$$\text{Therefore } (p + q)^2 = (1.00)^2$$

$$\text{Therefore } p^2 + 2pq + q^2 = 1.00$$

This equation is an expansion of the binomial $(p + q)^2$, which is a model of the frequencies of genotypes. Rather than using the Punnett square, we can directly calculate genotypic frequencies for the next generation from the allelic frequencies p and q:

$$\text{Frequency of } AA = p^2 = (0.6)(0.6) = 0.36$$

$$\text{(blue flowers)}$$

$$\text{Frequency of } Aa = 2pq = (2)(0.6)(0.4) = 0.48 \text{ (blue flowers)}$$

$$\text{Frequency of } aa = q^2 = (0.4)(0.4) = 0.16$$

$$\text{(white flowers)}$$

The frequencies of the two possible phenotypes (A ___ and aa) can also be calculated for the next generation:

$$\text{Blue flowers} = AA + Aa = p^2 + 2pq$$

$$= 0.36 + 0.48 = 0.84$$

$$\text{White flowers} = aa = q^2 = 0.16$$

This raises an important question: How have the frequencies of the two alleles in this gene pool changed after production of the next generation? To answer this, we must solve for p and q from the new genotypic frequencies and compare our results with the original values of p and q. If 36% of the offspring are AA (i.e., p^2) and half of the alleles of the Aa offspring (i.e., $2pq$ = 48%) are A, then the total frequency of A (i.e., p) is $0.36 + [(1/2)(0.48)] = 0.60$. In a similar calculation the new frequency of a (=q) = 0.40. Thus, *the allelic frequencies have not changed.* In fact, even after a second, third, or fourth generation of random mating, the allelic frequencies do not change. Thus, *the Hardy-Weinberg equilibrium predicts that the proportional gene frequencies of dominant and recessive alleles are retained from generation to generation in a randomly mating population.* The genotypic frequencies will also be in equilibrium after at least one generation. There will be just as many white flowers after multiple generations as in the previous generation, even though white flowers are a recessive trait. However, there are constraints and assumptions associated with maintaining this equilibrium. Specifically, Hardy-Weinberg equilibrium makes the following assumptions:

1. The population is large. Changes in the gene frequencies of large populations are less likely to occur by chance. In a small population, the random loss of one or more individual genotypes (such as by failure to breed) can eliminate one or more alleles from the population.

2. Individuals do not migrate in or out of the population. The net movement of individuals between populations (which occurs via seed dispersal in plants) must not be extensive enough to change gene frequencies. In natural populations the degree of migration varies considerably.

3. Mutations do not occur. In natural populations the mutation rate varies considerably. Mutations are often spontaneous.

4. Reproduction is random. That is, every member of a population must have an equal opportunity to reproduce with any other member. In the terms of Hardy-Weinberg population genetics, genotypes mate in proportion to their frequency.

5. There is no natural selection; that is, all alleles and combinations of alleles have equal fitness. The number of offspring must be independent of genotype.

c o n c e p t

Population genetics deals with entire gene pools rather than individuals. Hardy-Weinberg equilibrium predicts that allelic frequencies will not change over multiple generations in a large, isolated, randomly mating population that is not subject to selection or mutation.

The Significance of Hardy-Weinberg Equilibrium

Hardy-Weinberg equations are valuable models, even though no natural populations are in Hardy-Weinberg equilibrium for all of their gene loci. (Populations may maintain Hardy-Weinberg equilibrium for specific genes that are not adaptive.) Nevertheless, by understanding equilibrium, we can study how natural populations deviate from this condition. Although the Hardy-Weinberg model is of an unchanging population, it allows us to measure change and examine the forces causing that change.

We can investigate basic questions about simple population genetics and Hardy-Weinberg equilibrium. For example, do dominant alleles eventually drive recessive alleles from the population? Is a lethal recessive allele quickly eliminated from a population? Are less frequently occurring alleles lost from the gene pool after many generations?

The answer to all these questions is no. Hardy-Weinberg equations predict that the relative frequencies of alleles do not change; even rare alleles will not disappear just because they are rare. Natural selection alters their frequencies regardless of whether they confer an advantage or a disadvantage. Nevertheless, selection rarely changes the frequency of a recessive allele because such alleles are so infrequently expressed in the phenotype. For example, if the frequency of a recessive allele is 0.01 (1%), then the probability of homozygous expression is 1 in 10,000 (i.e., 0.01 × 0.01). Well-known examples of such rare alleles include those that cause simple metabolic defects that are easily observable, such as deficiency of the amino acid cysteine in the mouse-ear cress (*Arabidopsis thaliana*), deficiency of the vitamin thiamine in tomato (*Lycopersicon esculentum*), reduced ability to use magnesium in celery (*Apium graveolens*), and deficiency of the hormone gibberellin in maize (*Zea mays*). Such recessive alleles do not occur in the homozygous state, and natural selection only operates on phenotypes. In summary, Hardy-Weinberg equilibrium shows that genetic diversity tends to be maintained, thereby making evolution possible.

Hardy-Weinberg equilibrium also demonstrates that the recombination of genes during sexual reproduction does not alter the total gene pool of a population. Rather, sexual reproduction provides progeny with different genetic combinations of the alleles *already present*. Segregation by meiosis and recombination by fertilization neither add alleles to nor remove alleles from the population. Changes in allelic frequencies occur through agents such as migration, mutation, and selection.

c o n c e p t

Hardy-Weinberg equilibrium offers population geneticists a model of stability for comparison with more natural, changing populations. Understanding deviations from equilibrium promotes investigation of the causes of change.

Sources of Genetic Variation

Selection is probably the most important cause of deviation from Hardy-Weinberg equilibrium because it promotes genetic change in the context of the environment and requires genetic variation. Natural selection can change the frequency of phenotypes, but it must have genetic variation as a raw material to effect change. Where does genetic variation originate? Where do new alleles come from?

You've already learned about a major source of genetic variation: mutation (see Chapter 8). In this section we describe other sources of variation upon which natural selection can operate. Understanding these processes is essential to understanding evolutionary theory, because natural selection does not affect all genes in the same way. Changes that occur at the level of genes and chromosomes are mutations, which were discussed in Chapter 8. Genetic variation within populations results largely from gene flow.

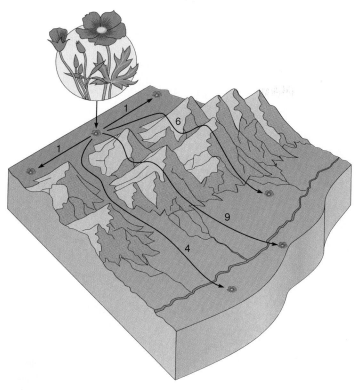

figure 22.16

Gene flow is inhibited by geographical barriers between populations. Dots represent distance of populations. Arrows show directions alleles may flow to these populations. Numbers are the generations (time) needed for gene flow across barriers. For example, we may expect nine generations for an allele to flow to compatible plants across mountains and rivers. Although distance and barriers slow the flow of alleles to a distant population, even occasional foreign alleles may maintain significant genetic variation within that population and increase its similarity to neighboring populations.

Gene Flow

Gene flow is the transfer of genetic material from one population to another (and less frequently from one species to another) (fig. 22.16). This typically occurs through migration of individuals or movement of seeds or pollen to neighboring populations or, in some cases, to distant populations. For example, the grass *Bothriochloa intermedia* seems to have incorporated genes from many other grasses, including *B. ischaemum* in Pakistan, *B. insculpta* in eastern Africa, and *Capillipedium parviflorum* in northern Australia.

Gene flow boosts the variation within a population and decreases genetic differences between populations; it minimizes geographic variation in a species gene pool. Genes flow frequently between neighboring populations; gene flow between distant or isolated populations is less frequent, which allows their gene pools to diverge over time. Significant reduction in gene flow partially explains why islands isolated by water and geographically more varied are more likely to produce new species than are vast expanses of grassland and why lakes and streams contain more geographic variation among populations than do oceans. Separated populations of a species are seldom genetically identical, and the differences coincide with the distance between populations. As you'll learn in the next chapter, geographic barriers that prevent gene flow are important in the formation of new species.

Gene flow in plant populations is difficult to measure, but it can be experimentally estimated by planting recessive homozygotes at various distances from a strain marked with a dominant allele and then examining the distribution of heterozygous progeny. Using this technique, A. J. Bateman measured pollen dispersal in wind-pollinated (e.g., corn) and insect-pollinated (e.g., radish) crops. The proportion of corn plants receiving the dominant allele by gene flow decreased exponentially with distance and was reduced to 1% at only 13–16 m from the pollen source. Similarly, most pollen of insect-pollinated plants is carried only a short distance. However, the small proportion of pollen that is carried farther may contribute importantly to gene flow.

Genetic Drift

Genetic drift refers to changes due to chance in the gene pool of a small population; that is, genetic drift reduces variability (decreases genetic variation) by the loss of alleles. In small populations, chance events such as mutation, mating, or pollination may significantly affect the gene pool and change gene frequencies independently of natural selection (fig. 22.17). If, for example, one individual in a small population carries the only copy of an allele, then the passage of that allele to the next generation may depend largely on the vagaries of insect pollination or random, lethal storms rather than natural selection. Favorable alleles in a small population can be eliminated by chance alone. Similarly, catastrophic damage to or death of well-adapted individuals may increase the frequency of the alleles of less fit but surviving individuals. Current research indicates that genetic drift may be a more significant force for changing gene frequencies than previously assumed. This would be especially true for the frequencies of genes that are not subjected to heavy selection pressure.

c o n c e p t

Gene flow homogenizes the gene pools of neighboring populations. Genetic variation within a population is increased by gene flow. Genetic drift reduces variability by the loss of alleles.

Maintaining Genetic Variation

If natural selection continually promotes genes that promote fitness and adaptation to the environment, we might expect variation to slowly decrease. For example, during stabilizing and directional selection, a population's gene pool may become increasingly homogeneous or consistent. This doesn't usually happen, however, and the reasons are not well understood. Apparently, there are forces that promote mixing and maintain genetic variation.

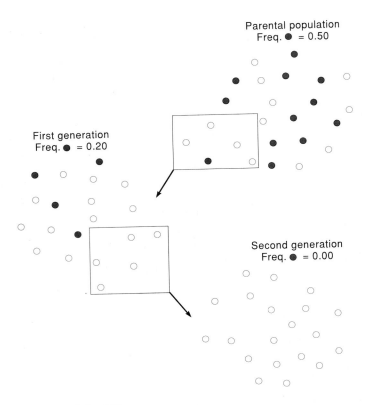

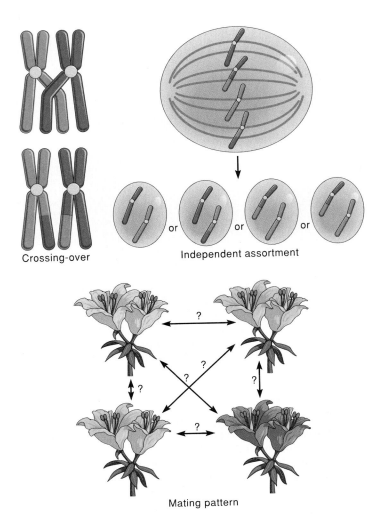

figure 22.17

Genetic drift over three generations may alter gene frequencies. Open and closed circles represent individuals with light and dark alleles, respectively. When a new generation is derived from only a small portion of the parent population, there may be sampling error—that is, the parent population is poorly "sampled." Sampling error between these generations has caused the frequency of dark alleles to drift from 0.50 to 0.00.

figure 22.18

Three mechanisms of sexual reproduction enhance genetic variation. Crossing-over recombines alleles. Independent assortment varies combinations of chromosomes in each gamete. Different mating patterns (possible crosses) vary combinations of parents. Mating pattern is subject to selection pressure, chance, and spatial distribution, and increases exponentially with the number of individuals in the population.

Sexual reproduction is probably the most important factor that promotes genetic variation in plants and animals (fig. 22.18). Specifically, sexual reproduction (1) includes crossing-over and genetic recombination after chromosome synapsis, (2) assorts chromosomes during meiosis, and (3) combines genetic material from two different parents. All three of these mechanisms produce new combinations of genes. In contrast, asexual organisms acquire genetic variation only through mutation. Their offspring are clones of the parents.

Outbreeding, meaning reproduction with individuals other than the self, promotes new genetic combinations. Although many plants are self-compatible, several mechanisms have evolved to ensure sexual reproduction with other individuals (fig. 22.19). For outbreeding to occur, pollen must land on the stigma of a flower of another individual. Holly trees and date palms, for example, have male flowers on one tree and female flowers on another. In other plants, such as avocado, gamete production is separated by time: when a flower is producing pollen, the stigma of the same flower is not receptive.

Stigmas are commonly receptive to pollen only after the flower, or all flowers on the same plant, have shed their pollen. Other plants have anthers and nectaries arranged in variable patterns on the flowers of different individuals. Flowers of the primrose (*Primula vulgaris*) and the bogbean (*Menyanthes trifoliata*) include two arrangements of styles and anthers (fig. 22.20). The "pin" flowers have a long style and low-set anthers, whereas the "thrum" flowers have a short style and elevated anthers. The contrasting positions of style and anthers ensure that the insects collecting pollen on their abdomens while drinking nectar from a plant with pin flowers will be more likely to pollinate a plant with thrum flowers.

figure 22.19

The bizarre flower of the bird-of-paradise (*Strelitzia reginae*) is decorated with colors and shapes to attract bird pollinators. Successful attraction of flying pollinators promotes outbreeding, which allows for new genetic combinations.

figure 22.20

The flowers of bogbean (*Menyanthes trifoliata*) may be either long-styled or short-styled. This difference increases the chances of cross-pollination because a pollinator that is dusted with pollen from a short-styled flower is more likely to successfully pollinate a long-styled flower.

Genetic systems for self-incompatibility also promote outbreeding. For example, plants such as evening primrose (*Oenothera*) cannot fertilize themselves because the ovule, stigma, style, and pollen have the same allele for self-sterility. They can only fertilize another individual if it has a different allele at the locus for self-sterility. This system works well if the population has many different alleles, so that many different genotypes can fertilize many other genotypes. A population of red clover may have over 200 such alleles.

Diploidy also preserves variation. In haploid individuals all alleles are immediately expressed in the phenotype, but in diploid organisms genetic variation in the form of "hidden" recessive alleles can be carried and reproduced but not expressed or subjected to intense selection. For example, if recessive allele *a* has a frequency of 0.01 in the gene pool, it will be expressed as *aa* in the phenotype in only one of every 10,000 individuals ($q^2 = 0.01 \times 0.01 = 0.0001$). Furthermore, it would take 100 generations of complete negative selection to diminish the frequency by half. Because removal or promotion of this allele by selection of phenotypes is very slow, it will remain in the population for a long time.

Also associated with diploidy and the maintenance of genetic variation is **heterozygote superiority.** As stated earlier, recessive alleles can be hidden in the heterozygous genotype. For reasons beyond the scope of this chapter, the heterozygous condition is often more fit than either the homozygous dominant or the homozygous recessive condition. This phenomenon is called heterozygote superiority and may increase the frequency of a recessive allele carried in the heterozygote, even if the recessive allele in the homozygous condition is harmful.

Heterozygote superiority is also a benefit of outbreeding and is responsible for **heterosis** (sometimes called *hybrid vigor;* fig. 22.21). Hybrid corn, for example, bred from two different strains, is extremely vigorous and hardy because it is more heterozygous at many of its loci than either of the parents. The controlled process of corn improvement initially involves 5–7 years of inbreeding and selection of desired features. This inbreeding produces purebred and strongly homozygous strains. After repeated inbreeding, two purebred strains are crossed to produce a hybrid. Frequently, another hybrid is produced from two other purebred, homozygous strains, and the two hybrids are then crossed. The result is a double-hybrid strain of seeds with strong hybrid vigor.

c o n c e p t

Genetic variation is maintained by sexual reproduction, outbreeding, diploidy, heterozygote superiority, and some forms of selection.

Macroevolution versus Microevolution

The foregoing discussion is a brief synopsis of some of the best-supported ideas about evolution. It is not the final word, however, because evolutionary biology itself is evolving. More recent ideas about evolution involve such concepts as

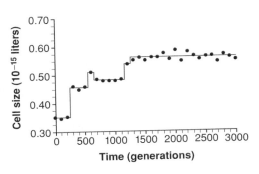

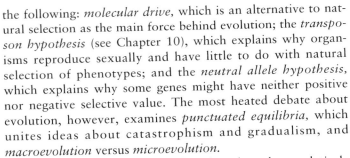

figure 22.21

Hybrid corn may be superior to inbred strains (hybrid superiority) because outbreeding promotes heterosis. Hybrids shown in the background are taller and more robust than the inbred strains shown in the foreground. Corn ears (inset) produced by the hybrids are much larger.

figure 22.22

Changes in the average size of cells in a population of *E. coli* during 3,000 generations. For this population, the only source of genetic changes was spontaneous mutations. The pattern shown here—that is, stasis followed by rapid change—is consistent with the pattern suggested by punctuated equilibria.

the following: *molecular drive*, which is an alternative to natural selection as the main force behind evolution; the *transposon hypothesis* (see Chapter 10), which explains why organisms reproduce sexually and have little to do with natural selection of phenotypes; and the *neutral allele hypothesis*, which explains why some genes might have neither positive nor negative selective value. The most heated debate about evolution, however, examines *punctuated equilibria*, which unites ideas about catastrophism and gradualism, and *macroevolution* versus *microevolution*.

The fossil record shows that there have been relatively short periods of explosive diversification in large groups of organisms, followed by long periods of little or no change. Some taxa remain relatively unchanged for millions of years, then almost disappear, only to be replaced over tens of thousands of years by other groups. Cone-bearing plants and dinosaurs, for example, existed for tens of millions of years and dominated the earth, but these plants dwindled rapidly, and the dinosaurs disappeared completely about 65 million years ago. A large diversity of flowering plants and mammals quickly took their place. This pattern of evolutionary change is one of stasis (no large-scale change) punctuated with short periods of rapid change—that is, gradualism followed by catastrophism. Such a pattern is referred to as *punctuated equilibria*.

Some of the most interesting evidence for the claim that minor genetic changes can produce large evolutionary leaps was reported in 1996 by Jane Dorweiler, a graduate student at the University of Minnesota. Dorweiler studied teosinte, the presumed ancestor of domesticated corn. Teosinte is a bushy weed whose kernels are encased in a lignified armor that is hard enough to crack the teeth of even the most ardent corn lover. Dorweiler found that giving teosinte a single portion of chromosome 4 alters flower development and results in kernels that are exposed like those of domesticated corn. That hybrid, according to Dorweiler, is what teosinte may have looked like 7,000 to 10,000 years ago, when archaeologists believe teosinte was domesticated in what is now Mexico.

Other interesting evidence for punctuated equilibria was also reported in 1996 by Richard Lenski and his colleagues at Michigan State University. Lenski's group studied the bacterial workhorse, *Escherischia coli*. Their experiment started in 1988 when Lenski placed one *E. coli* in a flask containing a low-sugar broth; the scarcity of food in the broth provided a force for selection. Unlike many bacteria, the strain of *E. coli* that Lenski and his colleagues used could not exchange DNA; hence, spontaneous mutations—which occurred at the rate of 10^6 per day in each population—were the only source of genetic changes in the population. After starting their culture, Lenski or a coworker transferred a bit of the broth into a fresh flask every day to keep the cells growing and dividing; every 2 weeks they froze a sample for later analysis. After 4 years, they'd gathered a freezer full of information—about 10,000 generations of bacterial evolution. They found that cells tend to have one average size, then suddenly go through spurts of growth (fig. 22.22). Larger cells were thought to be an advantage, because they could store more materials and produce more energy reserves.

22–28

Lenski and his colleagues also studied the fitness of different generations of the bacteria by pitting ancestral cells versus descendants in the same flask. Later (i.e., larger) generations had higher fitness, producing more offspring than their ancestors. Clearly, the availability of beneficial mutations limited the rate of evolution. When such mutations appeared, they swept through the population rapidly.

Lenski's data suggest that a simple genetic mechanism—natural selection of rare, beneficial mutations—can lead to punctuated patterns of evolution. Some biologists question how Lenski's data apply to sexually reproducing organisms, which have more complicated patterns of evolution. Other biologists discount this concern by arguing that the same evolutionary forces are at work, no matter what the organism or the scale.

Macroevolution is the origin or disappearance (extinction) of taxonomic categories above the species level, and the results are new genera, families, and phyla. These events are on a grand scale and include novel, adaptive designs (e.g., tracheids and vessels), broad diversification of organisms (e.g., the explosive radiation of flowering plants), and sometimes massive extinctions. Punctuated equilibria is considered a pattern of macroevolution. **Microevolution** is the progressive change in gene frequencies within populations over many generations. Natural selection, mutation, genetic drift, molecular events, and so on are primary forces of microevolution and can culminate in **speciation,** the formation of new species. Notably, speciation is interpreted by some botanists to be a culmination of microevolution, while others see speciation as the most fundamental event of macroevolution.

These two concepts are sometimes contrasted because of the suggestion that the processes of macroevolution are somehow fundamentally different from those of microevolution. However, direct evidence is lacking for such ideas about macroevolution. Conversely, microevolution, which occurs by natural selection and other mechanisms described in this chapter, is a robust experimental discipline that draws on an abundance of evidence. We therefore offer no further discussion of macroevolution; the next chapter explains the processes of microevolution.

Chapter Summary

Evolution refers to all the changes of life on earth from its earliest beginnings to the diversity of today. Today's organisms arose through the modification of ancient forms of life that was fueled by accumulations of genetic changes through many generations. Evolution involves a change in the frequencies of alleles in a population's gene pool from one generation to the next.

Evolution was an intimidating idea; it violated preconceived notions that species were divine and unchanging creations. Georges Cuvier examined fossils as evidence of change, but he proposed that catastrophes accounted for changes in ancient assemblies of species. James Hutton and Charles Lyell argued that the earth was millions rather than thousands of years old and that organisms had changed through time. Lamarck tried to explain changes in species by inheritance of acquired characteristics; however, such characteristics are not heritable from generation to generation.

Charles Darwin voyaged to South America for 5 years and observed remarkably diverse adaptations and distributions of species that eventually caused him to question divine creation. He adopted Lyell's contention that the earth was old and gradually changing and subscribed to Malthus's observation that species increase geometrically and face limited resources. From these observations and ideas, Darwin conceived of a mechanism called *natural selection* to explain the development of diverse species through time. At the same time, Alfred Russel Wallace also developed a theory of natural selection, and both naturalists presented their work in 1858. The next year Darwin published his extensive documentation and explanation for natural selection in *On the Origin of Species.*

Support for evolutionary theory includes both indirect and direct evidence. Indirect evidence for descent with modification comes from the age of the earth, the fossil record, homology, convergence, biogeography, and molecular biology.

The essence of natural selection is the differential reproduction of genetic variants. Organisms can produce more offspring than needed to replace themselves and more than environmental resources can support. Some offspring die before maturity, and those with genes that confer the best adaptation to their environment will reproduce the most. The frequency of reproduced genotypes increases with each generation and that of unfavorable genotypes will be reduced. Natural selection functions at the population level, and a population, not an individual, is the smallest unit that can evolve. Populations evolve in the context of their environment, which confers success or failure to phenotypes. Selection does not create new adaptations but promotes the most adaptive genetic variation available through increased reproduction. Direct evidence for natural selection comes from experimentation and from comparisons with artificial selection.

Stabilizing selection in constant environments promotes the normal, most prevalent phenotype and reduces variation. Directional selection is common in changing environments and fosters an extreme phenotype while reducing the occurrence of the contrasting phenotype. This shifts the population's distribution of phenotypes for a trait toward an extreme. Diversifying selection occurs in drastically changing environments; it reduces the normal, most frequent phenotype and promotes the extremes, thereby maintaining overall variation.

Evolution and natural selection are genetic processes. Population genetics deals with entire gene pools rather than individuals. The Hardy-Weinberg equilibrium predicts that allelic frequencies will not change over multiple generations in a large, isolated, randomly mating population that is not subject to selection or mutation. The Hardy-Weinberg equilibrium is important because it provides population geneticists with a model of stability for comparison to more natural, changing populations. A measure of deviation from equilibrium is needed to investigate causes of change.

Genetic variation as the raw material for natural selection is introduced into populations by mutation, gene flow, and genetic drift. Mutations are sudden, heritable changes in genes or chromosomes. Gene flow is the transfer of genetic material from one population to another by migration of individuals or reproductive cells. Genetic drift is random, but it may produce significant changes in allelic frequencies in small populations that are subject to chance events such as natural catastrophes, mutation, and nonrandom mating.

Genetic variation is maintained by sexual reproduction, which assorts chromosomes during meiosis and recombines genetic information during crossing-over and the combining of genes from different parents. Outbreeding promotes new genetic combinations by ensuring sexual reproduction with other individuals rather than within the same individual. Diploidy preserves variation by effectively hiding recessive alleles in the heterozygous condition and isolating these alleles from selection, which operates only on phenotypes. Heterozygote superiority also maintains diversity by promoting the heterozygous condition over either dominant or recessive homozygotes.

More recent ideas about evolution include alternatives to natural selection (e.g., molecular drive and the transposon hypothesis) as the main driving force of evolution. Other ideas involve neutral alleles and macroevolution versus microevolution. Microevolution explains the origin of species, while macroevolution explains the origin of higher taxonomic categories.

 ## What Are Botanists Doing?

Piecing together the phylogenetic tree that maps the evolutionary path of organisms can sometimes be tedious and far removed from everyday events. Interview a taxonomist and ask specifically what relevance his or her work has to current topics such as biodiversity or pharmacology.

 ## Writing to Learn Botany

Thomas Malthus was pessimistic about the future of the human race. Consider information presented in the Epilogue of this book and speculate on what factors have diverted humanity from the fate predicted by Malthus. Or have we actually been diverted?

Questions for Further Thought and Study

1. In earlier chapters of this book you've seen examples of evolution—for example, how plants form symbioses with other organisms (e.g., fungi, ants, and bacteria). You've also seen more subtle examples of biochemical adaptations, such as how plants make compounds that deter herbivores, and even how other organisms such as grasshoppers have evolved ways of using plants' defenses for their own survival. What would be the selection pressures for these examples of biochemical evolution?

2. Lamarck was an insightful evolutionist. What was the major problem with his theory?

3. How are gametes, eggs, and embryos subject to environmental selection pressures?

4. Assume that a randomly mating population of blue-flowered and white-flowered plants produces offspring consisting of thirty-one blue-flowered plants and seventy-six white-flowered plants. Explain these results in terms of Hardy-Weinberg equilibrium and its assumptions.

5. Describe how stabilizing selection changes the gene pool.

6. What would be the consequences if a population reproduced without the influence of natural selection?

7. Assume heterozygote superiority operates for a gene locus of two alleles. How does selection for the heterozygote affect the relative frequencies of the two alleles?

8. Under what circumstances might a single mutation significantly affect a population?

9. How would you explain positive selection for survival beyond the oldest reproductive age?

10. How is evolution different from natural selection?

11. Why can't individuals evolve?

12. What was the importance of artificial selection to the formulation of Darwin's theory of natural selection?

13. If natural populations are rarely, if ever, in Hardy-Weinberg equilibrium, why is this model important to population geneticists?

14. How would you determine if most mutations were harmful?

15. Herbert Spencer was an eighteenth-century English sociologist who believed in applying Darwin's concepts to human society. He popularized the phrase *survival of the fittest* and suggested that the unemployed, the sick, and other "burdens on society" be allowed to die rather than be objects of public assistance and charity. What do you think of this idea and of the validity of his interpretation of the concept of natural selection?

16. Hypotheses to account for evolutionary events such as the pace of evolution and the cause of mass extinctions are often conflicting. Those who do not believe that evolution has taken place sometimes perceive these conflicts among scientists as signs of doubt that evolution occurs. Scientists, however, maintain that conflicting hypotheses do not argue against evolution at all but demonstrate the process of scientific thinking. What is your opinion? Why are hypotheses particularly important in understanding evolutionary processes compared to other fields of biology?

17. Mimicry is a common biological phenomenon. For example, the Notodontidae moth (*Antaea jaraguana*) from Brazil has a wing pattern that resembles a dead leaf and its venation. How would an insect benefit by resembling a plant?

18. James Watson, who along with Francis Crick discovered the structure of DNA (see Chapter 9), said: "Today, the theory of evolution is an accepted fact for everyone but a fundamentalist minority, whose objections are based not on reasoning but on doctrinaire adherance [sic] to religious principles." Do you agree with Watson? Why or why not?

Web Sites

Review the "Doing Botany Yourself" essay and assignments for Chapter 22 on the *Botany Home Page*. What experiments would you do to test the hypotheses? What data can you gather on the Web to help you refine your experiments?

Here are some other sites that you may find interesting:

http://rumba.ics.uci.edu:8080/faqs/

At this site you can read an essay about evolution, discussions of common misconceptions, and an explanation of how Darwin's theory has gradually given way to the currently accepted view of evolution.

http://sunrae.uel.ac.uk/palaeo/index.html

The International Organization of Palaeobotany (IOP) manages this Plant Fossil Record database (PFR) including descriptive details of most plant fossil genera and of those modern genera which have fossil species.

http://www.literature.org/Works/Charles-Darwin/

Even Charles Darwin would be impressed to see his works online! Browse a full copy of *On the Origin of Species* or *Voyage of the Beagle*.

http://www.ucmp.berkeley.edu/

The Museum of Paleontology at the University of California at Berkeley provides information and pictures of fossil remains that offer clues to the phylogeny of today's plants.

Suggested Readings

Articles

Ayala, F. J. 1978. The mechanisms of evolution. *Scientific American* 239(September):56–69.

Hartman, H. 1990. The evolution of natural selection: Darwin versus Wallace. *Perspectives in Biology and Medicine* 34:78–88.

Mlot, C. 1996. Microbes hint at a mechanism behind punctuated equilibrium. *Science* 272: 1741.

Niklas, K. J. 1994. One giant step for life: Simple law-abiding plants led the invasion of hostile lands. *Natural History* 103:22–25.

Pääbo, S. 1993. Ancient DNA. *Scientific American* 269(November):86–92.

Pennisi, E. 1994. Static evolution: Is pond scum the same now as billions of years ago? *Science News* 145: 168–169.

Stebbins, G. L., and F. J. Ayala. 1985. The evolution of Darwinism. *Scientific American* 253(July):72–82.

Wilson, A. C. 1985. The molecular basis of evolution. *Scientific American* 253(October):164–173.

Books

Brown, J. H., and A. Gibson. 1983. *Biogeography*. St. Louis: C. V. Mosby.

Browne, J. 1995. *Charles Darwin. Vol. I. Voyaging*. New York: Knopf.

Darwin, C. 1967. *On the Origin of Species*. Facsimile 1st ed. of 1859. New York: Atheneum.

Dennett, D. C. 1995. *Darwin's Dangerous Idea: Evolution and the Meanings of Life*. New York: Simon & Schuster.

Endler, J. A. 1986. *Natural Selection in the Wild*. Princeton, NJ: Princeton University Press.

Hancock, J. F. 1992. *Plant Evolution and the Origin of Crop Species*. Englewood Cliffs, NJ: Prentice-Hall.

Stebbins, G. L. 1982. *Darwin to DNA, Molecules to Humanity*. San Francisco: W. H. Freeman.

Stewart, W. N., and G. W. Rothwell. 1993. *Paleobotany and the Evolution of Plants*. New York, NY: Cambridge University Press.

A sample of the diversity of flowering plants.

Speciation

23

Chapter Outline

Chapter Overview

The central phenomenon of evolution is the origin of new species—a group of organisms capable of breeding successfully with each other but not with members of other groups. Speciation is distinguished from broad evolutionary change by the formation of distinctive, interbreeding groups of organisms called species. Formation of a new species is a genetic event involving isolation of the gene pool of a population. Reproductive isolation, the most critical characteristic of a species, occurs via geographic isolation, temporal isolation, sterile or poorly developed hybrid offspring, or any other means of preventing fertilization by gametes from different species. Despite reproductive barriers, sympatric plants (i.e., plants that overlap geographically) hybridize frequently, and these hybrids may occasionally produce new species.

Allopatric speciation is the genetic divergence of two populations that are geographically separated. Models of allopatric speciation rely on the likelihood of gradual genetic divergence caused by mutation, genetic drift, and natural selection occurring in populations separated in space. Sympatric speciation frequently involves cellular nonseparation of chromosomes resulting in polyploidy (i.e., increased number of chromosome sets). Even though such polyploids and some hybrids may be sterile, they may represent a new, reproductively isolated, asexually reproducing species. In some cases, polyploidy may restore fertility of hybrids.

Evolutionary theory focuses on speciation—that is, the origin of new species—to explain the diversity of life. New species are the foundation of biological diversity. Today, 2–3 million species of living organisms have been named, and 2–10 times that many remain to be discovered. As vast as this diversity may appear, extrapolations from the fossil record indicate that today's species probably represent less than 1% of all those that have ever existed; more than 99% of all species have become extinct. Obviously, speciation has occurred countless times in all sorts of environments; it is as much a property of life as growth, reproduction, and death.

New species form from preexisting species as populations adapt to local conditions. This idea, first proposed by Charles Darwin, is now uniformly accepted in biology. However, the widespread agreement that speciation has occurred and continues to occur is spiced with plenty of arguments about *how* speciation works. The theory of natural selection (Chapter 22) provides some of the answers, at least in a general way. Many possible mechanisms of speciation have been described, some of which are based on scientific evidence and some of which are rational explanations but are based on little or no evidence.

Biologists are alternately excited or frustrated by the *species problem*—that is, the difficulty of defining a species. No single definition fits all species. But this does not keep biologists from using the term *species* or from using species names, as we have done throughout this text. At this point, however, we confess that our frequent use of species names in earlier chapters hides one of the biggest problems in biology: many different kinds of species have been described, but we don't know enough about most species to know how they fit different definitions of species. Nevertheless, knowing what kinds of species exist is important to understand how they might arise. The first objective of this chapter, therefore, is to define different kinds of species.

The second objective of this chapter is to explain how species are formed. If you suspect that our understanding of the mechanisms of speciation is muddied by the species problem, you are right. Ideas about how species arise have produced even more heated arguments than ideas about how to define a species. Many of these arguments are about the relative importance of the different mechanisms in nature. Furthermore, several different mechanisms for the origin of species are strongly supported by evidence; some of these mechanisms have been used by artificial selection to produce new species. As noted by Charles Darwin in *On the Origin of Species* the creation of new species by human activities is indirect evidence that the same mechanisms occur in nature. However, most of our evidence for natural mechanisms does not reveal how significant each mechanism might be.

What Is a Species?

The species is the fundamental category of biological classification; species are the building blocks of the classifications presented in subsequent chapters on diversity. Beginning with Darwin, species recognition has commanded a great deal of attention from biologists because of its importance in classifying and understanding speciation. *Species concepts* (that is, perceptions and definitions of species) remain a hot topic today: spirited arguments on this topic are heard every year at the annual meeting of the Botanical Society of America. At least a dozen species concepts have been proposed during the past two centuries. The most widely used are morphological species, biological species, genetic species, paleontological species, and evolutionary species concepts. We focus on these five species concepts in the following discussion.

The Morphological Species Concept

The **morphological species concept** holds that species are the smallest groups of organisms that can be consistently distinguished by their morphology. This is the most practical and widely used species concept among taxonomists, who deal with the identification and classification of organisms. Taxonomists use this concept mostly by default, since they know most of the species on earth only from their morphology. This concept is often called the *classical species concept,* since it generally has been used to describe species since people first began to classify organisms.

New species are usually named on the basis of the morphological species concept; morphological descriptions are all that we need to distinguish new species from previously known ones. Virtually all of the names of plant species used in this text are of morphological species. Likewise, species that are listed in such books as the *Manual of the Vascular Plants of Texas,* the *Manual of the Vascular Plants of the Northeastern United States and Adjacent Canada,* and *An Illustrated Flora of the Pacific States* are morphological species.

Most of the morphological species that you are familiar with are intuitively recognizable as distinct from one another. For example, you can distinguish a species of lily from one of pine or oak. With a little study, you could also learn to distinguish some of the 100 species of *Lilium* (lilies), the 93 species of *Pinus* (pines), and the 600 species of *Quercus* (oaks) from one another (fig. 23.1). In so doing, you would be learning these plants as morphological species.

(a)   (b)   (c)

figure 23.1

These three lilies are easily recognized by color and shape; that is, the morphological species concept clearly distinguishes them as separate species based on their general morphology. However, this concept gives no insight into their degree of reproductive isolation or their genetic similarity. Are they truly different species? (a) *Lilium regale.* (b) *Lilium auratum.* (c) *Lilium martagon.*

At this point it might seem that the morphological species concept allows you to recognize all the species that you need or want to know. So why go any further? The answer is that speciation has not always formed well-defined species; the lines between species are often blurred. In such cases, taxonomists often disagree on which plants are the same or different species. For example, the morphological species concept of the cactus genus *Opuntia* includes anywhere from 400 to 1,000 species, depending on whose judgment is followed.

c o n c e p t

Ideas about speciation are based on what kind of species exist and on the different ways that species arise. Biologists argue about how to define species and about which mechanisms of speciation might best explain biological diversity. The morphological species concept defines species by their observable morphological features. It is the most widely used species concept, because most species are known only by their morphology.

The Biological Species Concept

In contrast to a morphology-based concept of species, the **biological species concept** uses reproductive biology to define species. According to this concept, a species is a group of interfertile populations that are reproductively isolated from other such groups. The biological species concept was first proposed by ornithologists (bird specialists), because birds commonly occur in interfertile, reproductively isolated populations that are hard to distinguish morphologically. Most animal species probably fit this concept. However, many plants do not fit the biological species concept, so most botanical discussions of this concept focus on how poorly it applies to plants. For example, morphologically

different species that are geographically isolated may hybridize readily when they are grown together (fig. 23.2). Are they really separate species?

Exceptions to the biological species concept are common among morphological species that frequently hybridize where the two parent species overlap. Furthermore, plants often reproduce asexually or by self-pollination, which means that interfertility, or at least interbreeding, may be absent. Dandelions (*Taraxacum officinale*) and blackberries (*Rubus* sp.), for example, have played particular havoc with the biological species concept because they produce seed without fertilization. The biological species concept is useless in defining certain crop species, such as the commercial banana or the navel orange, because many of these species are exclusively asexual. The biological species concept is also useless with fossils, because information about the reproductive biology of extinct organisms is usually missing. The biological species concept is, however, evolutionarily important because it requires reproductive isolation.

The biological species concept emphasizes a fundamental assumption about speciation: a population is a distinct species only if it does not mix genes with other populations. Speciation can begin only when reproductive isolation occurs, and it may occur in several ways, ranging from geographical isolation to genetic or behavioral isolation (mostly in animals). Although plants often hybridize in nature, their hybrids are frequently weak and less fit. The reduced fitness of some hybrids provides some reproductive isolation, although it is incomplete.

c o n c e p t

A biological species is a group of interfertile populations that are reproductively isolated from other populations. Plants often do not fit this species concept because plant species may be interfertile, self-pollinating, or predominantly asexual.

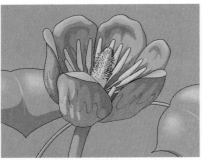
(a) *Liriodendron tulipifera*

(b) *Liriodendron chinensis*

(c) Hybrid, *L. tulipifera* × *L. chinensis*

figure 23.2

These two species of *Liriodendron* are geographically isolated from one another, but they readily hybridize when they are grown together. (a) *L. tulipifera* is a common tree of the eastern United States. (b) *L. chinensis* occurs in small populations in eastern China and Vietnam. (c) The hybrid occurs in botanical gardens where trees of both species are cultivated near one another.

The Genetic Species Concept

The **genetic species concept** defines a species by its genetic uniqueness, that is, by how different it is genetically from its nearest relative. Thus, the genetic species concept is like the morphological species concept, except that the recognition of species is based on genetic data instead of morphological data.

The uniqueness of a genetic species is typically expressed as its **genetic distance** from other species. The main drawback to measuring genetic distance is that the distance is based on only a small part of the genome. For example, allele frequencies can be obtained by electrophoresis for only about 30 different enzymes; several thousand other enzymes cannot be used to measure genetic distance because we do not know how to detect them. Recently, genetic distance has been calculated by comparing DNA fragments (RFLPs; see Chapter 11) and gene sequences. Like allele frequencies, however, these measures are limited to a small fraction of the genome.

Because the genetic species concept quantifies the definition of species, it is important for estimating the difference between a species and its nearest relative. This information is becoming increasingly useful for evaluating endangered species—that is, for determining whether an endangered species is either very different or not so different from its closest relative (see reading 23.1, "Genetic Distance: A Tool for Conservation").

c o n c e p t

The genetic species concept is based on a quantifiable comparison of genetic data. The uniqueness of a species is determined by its genetic distance from its nearest relatives.

The Paleontological Species Concept

Paleobotanists cannot deal directly with species concepts that involve genetic distance or reproductive isolation. Species in the fossil record are often fragmentary and are represented from only a few localities. Fossils best fit the morphological species concept but with a time dimension, because they appear and disappear in the fossil record. Such species are sometimes called **paleospecies**. A paleospecies has a unique morphology and exists during a particular geological time. Sometimes paleospecies seem to precede existing species, which indicates a gradual change from one species to another (see "Gradualism" in Chapter 22).

The Evolutionary Species Concept

None of the species concepts mentioned so far refers directly to evolution. For this reason, evolutionary biologists have proposed that an ancestral-descendant sequence of populations be called an **evolutionary species.** For example, a paleospecies that precedes a living species represents an ancestor-descendant sequence. They are, therefore, a single evolutionary species.

c o n c e p t

The paleontological species concept is based on distinguishing morphological species over geological time. A sequence of paleontological species along one evolutionary line is an evolutionary species.

Mechanisms of Reproductive Isolation

Reproductive isolation is the foundation of the species concept. Reproductive isolation protects the integrity of a species by preventing the gene pool from being contaminated by genes from other species. Mechanisms that prevent the production of fertile offspring from different species (hybrids) are **reproductive barriers** that promote reproductive isolation. The most obvious barrier is geographic separation: if two species are separated by a great distance, they will not interbreed. However, most barriers to hybridization are not absolute—other biological barriers operate to varying degrees in different species, especially if the species' ranges overlap.

GENETIC DISTANCE: A TOOL FOR CONSERVATION

Hundreds of plant and animal species are endangered in the United States: their habitat is disappearing, and they are in danger of extinction. Many of the species are endangered because of human activities such as land development, logging, and pollution. The cost of protecting endangered species involves restricting such activities, but such restrictions often slow economic development. Extremists on one side of the argument say that humans are the most important species, so let the bulldozers loose. Extremists on the other side say that species should be protected at all costs. Although it seems arrogant for one species, humans, to judge the value of another species, this is necessary when too many people demand too few resources. Which species should we value the most and spend the most to protect? One suggestion is to use genetic distance to determine how distinctive an endangered species is from its nearest neighbor (reading fig. 23.1). The more distinctive it is, the more protection it deserves.

More recently, conservation biologists have discovered that protecting endangered species is an inadequate, stopgap strategy. Species are endangered because of habitat destruction, and we should be evaluating the uniqueness of habitats as well as the value of species. Fortunately, measurement of genetic distance can indirectly compare habitats just as it compares endangered species. The uniqueness of a habitat is measured as the sum of genetic distances between each occupant species and its nearest relative. If a habitat is filled with populations that are highly distant from their nearest relatives, then the habitat is probably unique. This idea is still relatively new, however, and it has been used mostly for comparing animal species. Most plant conservationists have yet to adopt the idea in any significant way.

(a)

reading figure 23.1

(a) This autumn buttercup (*Ranunculus aestivalis*) is an endangered species that is known only from the Sevier River Valley in Utah. (b) Comparisons of DNA fingerprints among populations of the autumn buttercup and other kinds of buttercups show how genetically distinctive this endangered species is.

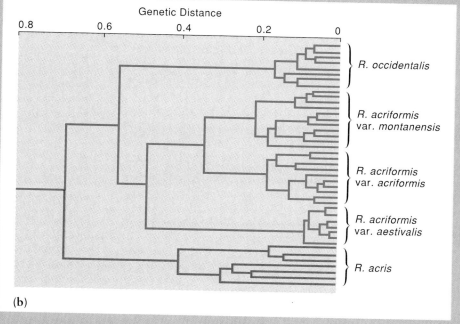

(b)

Reproductive barriers that isolate the gene pools of species may be prezygotic or postzygotic. **Prezygotic mechanisms** function before the formation of a zygote (i.e., they prevent successful fertilization). For example, a prezygotic barrier might block pollination or chemically impede fertilization of the egg. **Postzygotic mechanisms** operate after zygote formation by obstructing development of the hybrid zygote into a fertile adult.

Development of reproductive barriers is critical for speciation, and natural selection promotes pre- and postzygotic barriers. Parents that circumvent barriers often produce unfit offspring. Indeed, hybrid offspring resulting from parents that somehow circumvent reproductive barriers have incompatible genes from both parents. These genes provide the growing embryo with mixed signals, confuse development, and may kill

table 23.1

Mechanisms of Reproductive Isolation

1. Prezygotic mechanisms prevent mating or successful fertilization.

    a. Geographical isolation: Populations live in distant geographical areas.

    b. Microhabitat isolation: Populations live in different microhabitats and do not meet.

    c. Temporal isolation: Individuals are receptive to mating or flowering at different times.

    d. Mechanical isolation: Flowers are structurally different, and pollination is prevented.

    e. Gametic isolation: Female and male gametes from two species are incompatible.

2. Postzygotic mechanisms prevent production of fertile adults.

    a. Hybrid inviability: Hybrid zygotes do not develop to sexual maturity.

    b. Hybrid sterility: Hybrids do not produce viable gametes.

    c. Hybrid breakdown: The progeny of fertile hybrids are weak or infertile.

figure 23.3

This night-blooming *Cereus* (*Cereus* sp.) is reproductively isolated because its flowers open only at night for a limited time. This temporal isolation is a prezygotic barrier to reproduction.

the embryo. Even if the embryo survives, the mature organism is often weak and sterile. In contrast, parents that do not overcome reproductive barriers produce more viable, healthy offspring and pass their genes to the next generation. In this way, natural selection favors organisms that neither mate nor reproduce outside their species. The prezygotic and postzygotic barriers are summarized in table 23.1.

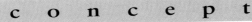

c o n c e p t

Reproductive isolation is critical to speciation and is maintained by a variety of reproductive barriers. Prezygotic barriers prevent successful fertilization. Postzygotic barriers prevent successful development of a zygote.

Prezygotic Barriers

Geographic Isolation

The gametes of individuals will not meet if the organisms are separated by significant distance. For example, two closely related species of small trees called redbuds look much alike and are compatible if artificially hybridized. However, the eastern redbud grows on the borders of streams east of the Mississippi River and is geographically distant from the western species in California, Nevada, and Arizona. This distance isolates the two species.

Microhabitat Isolation

Species living in different microhabitats within the same area may not encounter each other. Although they are not geographically isolated, they are significantly separated in space. For example, scarlet oaks (*Quercus coccinea*) of the eastern United States grow in the same broad range as black oaks

(*Q. velutina*). Scarlet oaks grow in moist, slowly draining, acidic microhabitats, whereas black oaks grow in dry, well-drained microhabitats. The separation of these two habitats within the same broad area sufficiently discourages hybrid formation. Although both species are wind-pollinated and will hybridize if brought together artificially, natural hybrids are rare. Such hybrids often are successful only in disturbed areas.

Temporal Isolation

Two species will not interbreed if they reproduce during different times of the day, season, or year (fig. 23.3). For example, three orchid species of *Dendrobium* live in the same tropical forest but produce flowers at different times. Flowering in all three species is triggered by the same event—a thunderstorm; however, one species opens its flowers 8 days after the storm, a second species flowers 9 days after the storm, and the third flowers 10 days after the storm. Pollination occurs only on a single day because the flowers open in the morning and close the following evening. Consequently, all three species are reproductively isolated in time.

Mechanical Isolation

Mating or pollination sometimes fails because plants are anatomically incompatible (fig. 23.4). For example, four closely related species of Peruvian *Catasetum* orchids with overlapping distributions have flowers with mechanical barriers adapted to the anatomy of a single species of bee. These orchids can be artificially hybridized, but no natural hybrids occur. Close examination of the flowers and bees has shown that the flower of one species will attach and receive pollen only from the insect's head, the flower of another species attaches pollen to the insect's back,

23–7

figure 23.4

Mechanical isolation. The flower morphology of this zebra orchid makes its pollen accessible to only a few kinds of pollinators such as this wasp. The flower releases sex pheromones that prompt the pseudocopulation by this wasp.

figure 23.5

Many hybrids such as this orchid, *Dendrobium crepidatum,* are sterile because they cannot produce normal gametes. This sterility is a postzygotic barrier that reinforces reproductive isolation of the parent species.

a third to the abdomen, and the fourth only to the left front leg. Pollen is transferred only between plants of the same species with the same floral anatomy.

Gametic Isolation

Flowers can distinguish between pollen of their species and pollen of a different species. Even if pollination should occur, germination of the pollen grain may be retarded by components in the stigma and style. Surfaces of the egg and sperm may have compatibility factors that promote or prevent fertilization.

Another type of gametic isolation involves malfunctions of the guidance system that directs a pollen tube to an ovule. Two genes, *pop2* and *pop3,* are critical for this system to function. Indeed, the system malfunctions only when both genes are defective in both the pollen and pistil.

Postzygotic Barriers

Hybrid Inviability

If a hybrid zygote forms, the embryo's development may be prevented because the combined genetic instructions are unworkable (e.g., the wrong gene products are produced) or because the control mechanisms are inoperable.

Hybrid Sterility

If two species produce healthy hybrid offspring, the offspring are often sterile (fig. 23.5). In most sterile hybrids, meiosis produces abnormal gametes because the sets of chromosomes contributed by the parents are different in number or structure.

Hybrid Breakdown

Sometimes the first generation of hybrids is healthy and fertile, but when they, in turn, mate with each other or with members of parent populations, their offspring are frail or sterile. For example, some species of cotton produce fertile hybrids, but the progeny of the next generation die as embryos or as weak, vulnerable plants.

Population Divergence and Speciation

To form a new species, the gene pool of a population must diverge from that of the parent (i.e., ancestral) population. Individuals in a diverging population become different—so

table 23.2

Forces of Genetic Deviation within a Population

Promoters of Genetic Variation	Promoters of Genetic Homogeneity
Mutation	Random sexual mating
Abnormal meiosis (polyploidy)	Stabilizing selection
Sexual recombination	Asexual reproduction
Genetic drift	
Disruptive selection	
Environmental gradients	
Geographical separation	
Geological events	

different that they usually look different from the ancestral population, exploit different habitats, and respond to the environment differently. The mechanisms controlling this divergence are not predictable or easily observed, because some forces homogenize, mix, and minimize variation, while other forces enhance and promote divergence (table 23.2). The first step in speciation, however, is partial reproductive isolation, which initiates genetic divergence. Once the gene pool of a population is isolated, it inevitably evolves further, and its gene frequencies change because of selection, genetic drift, and mutations in subsequent generations. For example, mutations occur randomly, and isolated populations of the same species will randomly accumulate different mutations, even if the environments are identical. Enough mutations may accumulate that the two populations diverge and no longer interbreed. Similarly, genetic drift and environmental variation can enhance the divergence of recently isolated populations. Sexual reproduction does not usually promote divergence directly, because it mixes genetic information within a population and can slow isolation; however, it does supply new combinations of genes. Such a new genetic combination arising in a splintered population may allow some individuals to exploit a new or different patch of environment. Once established in a new environment, a population's morphology, competitiveness, or reproductive life history may change and enhance isolation from the parent population. Separation of a gene pool of a population from other populations of the parent species is a crucial event in the origin of a species.

c o n c e p t

Speciation involves reproductive isolation of a gene pool and results from genetic divergence of populations. Reproductively isolated populations continue to evolve through mutation, genetic drift, and selection. Therefore, physical or temporal separation of populations allows genetic divergence to increase.

There are three modes of speciation, each based on the geographical relationship of a new species to its ancestral species: allopatric speciation, sympatric speciation, and parapatric speciation (fig. 23.6).

Allopatric Speciation

One obstacle to interbreeding among organisms is geographic isolation. Populations that are separated geographically are **allopatric** (meaning "different homeland"); similarly, **allopatric speciation** is the genetic divergence of two populations that are geographically separated. Separated populations occupying different environments are likely to accumulate genetic differences, and their phenotypes will diverge. This divergence initiates speciation and is promoted by different selective pressures in different environments. Isolated populations diverge as they accumulate different genetic characteristics and may become different species.

Geological processes can segment a population. For example, mountain ranges, canyons, lakes, and rivers can separate organisms. The formation of islands and deserts creates geographic barriers (fig. 23.7). The effectiveness of a barrier depends on the ability of an organism or its spores, pollen, or seeds to cross the barrier.

Small isolated populations often become extinct, but those that survive diverge from their ancestors faster than large populations. Recall that genetic drift and mutations are more significant in small populations; large populations have great genetic momentum and change more slowly. Their genetic drift is insignificant, and natural selection may require many generations to replace alleles and change the phenotype of a large population. For example, large populations of North American and European sycamore trees have been allopatric for 30 million years, but they can still interbreed; that is, they have not evolved into separate species. In contrast, the gene pool of a small population can change in relatively few generations. A few reproductively successful individuals with favorable gene combinations can greatly affect the evolution of a small population and may lead to speciation in as few as hundreds or thousands of generations. Small populations become a new species faster because they are genetically more transient and erratic.

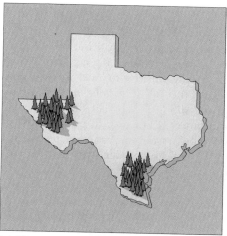

(a) Allopatric populations

(b) Parapatric populations

(c) Sympatric populations

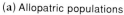

figure 23.6

Three modes of speciation. (a) Allopatric speciation occurs when populations are geographically isolated. (b) During parapatric speciation, populations share a common border and hybridize somewhat. But subtle differences in the adjacent environments are distinctive and lead to speciation. (c) During sympatric speciation, reproductive isolation occurs in the midst of intermingled populations. Isolation is usually achieved by abrupt genetic changes of an individual within the population.

Allopatric speciation is not easy to detect. In nature, reproductive isolation is not tested because separated organisms do not have the opportunity to sexually reproduce. An experiment using artificial conditions to test compatibility is not a definitive measure of speciation. Nevertheless, if successful hybrids are not produced, biologists usually designate the allopatric populations as true species. The best measure of genetic divergence in allopatric populations occurs when geographical barriers are naturally removed and these populations come together. Interbreeding will occur readily if divergence is only slight.

Surprisingly, the production of a few hybrids may reinforce genetic divergence because hybrid fertility is usually low. In this case, selection is against parents that are likely to hybridize and to produce weak, unfit offspring. Selection favors parents that are less likely to interbreed. In other words, parents with characteristics that prevent hybridization (usually prezygotic barriers) typically have more viable offspring than do parents that occasionally produce hybrids. This reinforces prezygotic barriers and enhances divergence. The populations diverge as characteristics that promoted hybridization are slowly eliminated. Increased reproductive isolation by selection against hybrids is difficult to demonstrate in natural populations.

Parapatric Speciation

Parapatric (meaning "parallel native land") populations have separate ranges that meet along a common border. Often this common border follows a discontinuity in some important environmental feature (fig. 23.6b). For example, a zone where the soil changes may mark the common border for two parapatric populations that are adapted to different soils. There, the gene pools of the populations would differ somewhat because of the plants' adaptations to different soils. Because most plants breed only with their neighbors, gene flow between populations would seldom go far beyond the area where the populations of plants meet. Nevertheless, there would be a limited flow of genes between the populations via interbreeding at the contact zone. **Parapatric speciation** could occur if, despite the contact, strong reproductive isolating mechanisms develop as the populations slowly split into ecological units.

Sympatric Speciation

Populations that overlap geographically are **sympatric** (meaning "same homeland"). **Sympatric speciation** is the production of new species within a single population or overlapping populations (fig. 23.6c). Speciation of sympatric populations may seem unlikely, but there are mechanisms that reproductively isolate groups within the same environment. For example, a slight change in color, shape, or chemical attractants in a mutated flower may make it unattractive to a pollinator. Such a mutation prevents the individuals from being pollinated by the same pollinator that visits the surrounding plants and may effectively isolate them reproductively from the parent population. The isolated plants may survive by self-fertilization, reproduction with limited members of their own kind, or hybridization with a nearby population of a different species. Once the small subpopulation is isolated, it may diverge genetically and form a new species. In addition to mutation, sympatric speciation may also result from chromosomal misalignment during meiosis, forming polyploid offspring, and from successful hybridization by an individual within the sympatric population with another species.

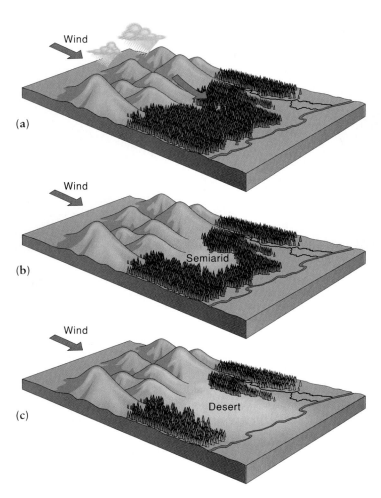

figure 23.7

The formation of a desert can separate populations and promote their genetic divergence. (a) Heavy rainfall on the left of the mountains leaves a dry wind on the leeward side. (b) A desert can form and expand to divide a population into reproductively isolated subpopulations. (c) Accumulations of adaptations and incompatibilities by the two allopatric populations can lead to speciation.

c o n c e p t

Allopatric populations are separated geographically (e.g., by mountains or rivers) and will speciate as genetic variations accumulate and reproductive barriers form. When successful hybridization can no longer occur, the allopatric populations represent different species. Parapatric speciation can occur at a boundary between two populations where some gene flow occurs but at a rate too slow to overcome divergence of the gene pools of the two neighboring populations. Sympatric speciation occurs within a single population or among overlapping populations. Mechanisms for sympatric speciation include mutations, hybridization, and polyploidy.

Plant Responses to Environmental Gradients

Complex environments and sharp boundaries such as lakes, streams, and soil types produce patchy distributions of plant species and populations within species. In these patches plants are immobile and are especially sensitive to local selective pressures. Limited movement of their alleles emphasizes the importance of selection and the subtle differences of the environment even over short distances. Even though pollen dispersal varies greatly by vector (wind, insect, etc.) and by species, as a general rule less than 1% of released pollen will reach individuals more than 300 m away (fig. 23.8). Considering all the other constraints on gene flow, isolation is often ineffective beyond 100 m. This means that pressures of the local environment may overshadow gene flow from dispersed pollen.

One response to local conditions is modification of morphology. **Morphological plasticity** means that plants display different growth forms in response to different environmental conditions even without genetic change. As a result, organisms with the same genes often look different according to the environment. For example, the height of plants or the size of their leaves may reflect the availability of water. The genomes of most plants allow enough morphological variability for the environment to promote considerable variation.

c o n c e p t

Plants are morphologically plastic, meaning that their genome allows for great morphological variation according to the local environmental conditions.

Some populations or species range over a large geographic area with a gradient of climatic features that cause organisms at extremes to change morphologically and to diverge genetically. A north-to-south gradient may include a range of temperatures and possibly a gradient of plants—taller ones at the southern end grade into shorter plants toward the north. This continuum of a characteristic is a **cline** of morphology. Members of a true cline are reproductively compatible, and gene flow can occur throughout the cline (fig. 23.9).

Different morphologies along a cline may be due to genetics as well as morphological plasticity. That is, short northern plants transplanted to the south may not grow as tall as the southern plants, indicating a genetic difference. Cline formation is a simple form of genetic divergence, and the degree of divergence is proportional to the length of the cline. Few species form smooth clines.

c o n c e p t

A cline consists of plants that diverge morphologically along an environmental gradient, but gene flow occurs throughout the population. Plants at the extremes of a cline are morphologically plastic and usually appear different.

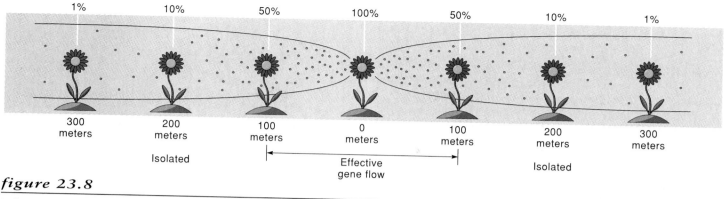

figure 23.8

Pollen dispersal is limited; less than 1% will travel as far as 300 m. Therefore, effective gene flow only occurs between individuals less than 100 m apart. Limited movement of pollen and its alleles allows plants to genetically diverge in local habitats.

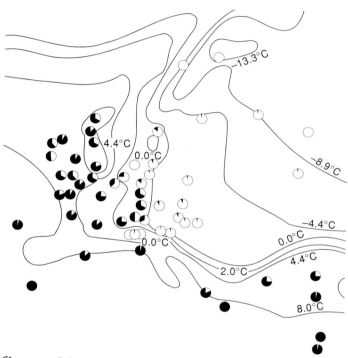

figure 23.9

Clines. Darker circles indicate increased cyanide production by a European species of clover; this cyanide discourages herbivores. A cline of the cyanogenic phenotype occurs along a temperature gradient: the production of cyanide occurs only when temperatures are above freezing (0° C). This cline is apparently determined by a balance between the advantage of being unpalatable to herbivores and the disadvantage of frost rupturing cellular membranes, releasing cyanide in the plant's tissues.

Distinct environmental patches (microhabitats) along a cline can isolate segments of a cline into subpopulations. These discontinuous subpopulations may diverge further because they are subject to local selection pressures and develop combinations of alleles (adaptations) that are best suited to the local environment. The differences among subpopulations will be genetically based. Fragmented subpopulations that have genetically diverged in association with their environments are called **ecotypes**. Ecotypes may look markedly different, even though they are all reproductively compatible and produce viable offspring. Ecotypes that differ physiologically (e.g., frost tolerances, time of flowering, photosynthetic rates) may not look different, but they react differently to their environments. Well-defined ecotypes occur where microhabitats have well-defined boundaries.

A well-known description of an ecotypic variation was presented by Jens Clausen, David Keck, and William Hiesey, who studied the genetic adaptation of isolated populations along a variety of Californian climatic zones. These zones included a sea-level station near the Pacific coast, a midaltitude station in the Sierra Nevada Mountains, and a timberline station at 3,000 m elevation. One of their experimental plants, *Potentilla glandulosa*, a member of the rose family, grew at all three locations and showed genetic adaptations to each habitat. Because this plant reproduces asexually by runners (as well as sexually), clones could be used to control for genetic variation and emphasize habitat-induced variation. Individual plants were gathered from a variety of habitats along the gradient and grown side by side at a field station in each climatic zone. At least three distinct ecotypes became apparent. Each had distinct morphological traits that were adaptive to its original environment; that is, these subpopulations (ecotypes) had diverged genetically. Besides different morphological traits, each of the ecotypes flowered at a different time of the year in response to the length of the growing season. The coastal ecotype flowered in mid-April, the foothills ecotype flowered in May–June, and the alpine ecotype flowered in late August.

Populations of aspen (*Populus tremuloides*) high in the mountains of Utah have become a genetically based ecotype, producing leaves early in summer and reproducing only vegetatively (fig. 23.10). Most aspens can produce viable seeds, but they also reproduce asexually by producing suckers from roots. In the Utah mountains aspen seeds rarely germinate, because the climate has rainfall early in the summer; there, aspens reproduce only vegetatively. Under those conditions, an ecotype producing leaves earlier in the growing season and reproducing asexually

figure 23.10

This stand of aspen is probably a clone of asexually reproducing organisms. This ecotype is adapted to its high-mountain, summer-drought environment. It produces leaves early in spring and reproduces only asexually.

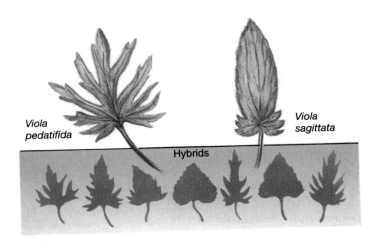

figure 23.11

Hybridization can produce individuals with phenotypes that do not occur in either parent and represent a variety of intermediate forms.

has circumvented the dry season and successfully colonized higher elevations of the Utah mountains. Considering that the climate of summer drought began about 8,000 years ago in Utah, it is possible that some clones of aspen are 8,000 years old.

c o n c e p t

Genetic divergence may occur along an environmental gradient across the range of a population. In a patchy or graded environment, fragmented subpopulations with significant genetic divergence, called *ecotypes*, may form. Gene flow between ecotypes is limited but greater than that along a cline.

Hybridization

Hybridization is the production of offspring from parents of different species or genetically distant strains within a species. Hybridization doesn't occur often, but it can produce individuals with new genetic combinations (fig. 23.11). If these combinations are successful, they represent evolutionary change and possibly a new species.

In 1760 Josef Kölreuter successfully conducted the first controlled hybridization of two plant species that resulted in fertile offspring. He crossed two species of tobacco and produced offspring that were considerably different from the parents. More crossings among the hybrids produced highly variable offspring: some resembled one grandparent, some resembled the other grandparent, and some looked different from either of the original species. He concluded that hybridization mixed the genome from the original parents and provided considerable genetic variation.

Hybridization occurs in animals but is far more common in plants. Many species of oaks (*Quercus*), cottonwoods (*Populus*), aspens (*Populus*), and other trees and shrubs readily hybridize, but their offspring are often sterile (fig. 23.12). Plants such as Kentucky bluegrass (*Poa pratensis*) have frequently hybridized with a variety of species within the genus and produced hundreds of genetic strains that are well-adapted to their local environments.

Plant hybrids are often sterile because their chromosomes cannot synapse during meiosis (fig. 23.12). Since the two haploid sets of chromosomes come from different parents, they have no homologues, and meiosis without synapsis will not produce viable gametes. (Later in this chapter we discuss how polyploidy may restore fertility to sterile hybrids.) Many sterile hybrids are **apomictic**, meaning that they reproduce vegetatively rather than sexually. However, sterility or infrequent outcrossing does not mean lack of success. Vegetative reproduction may be particularly appropriate in disturbed areas or in environments such as the Arctic that have a poor climate for pollination. For example, apomictic hybrids of blackberry (*Rubus*) are particularly well adapted to thrive in the ever-increasing habitats disturbed by humans.

Hybridization, even producing sterile offspring, may be a major mechanism for the production of new species suited to different habitats. However, sterile individuals do not interbreed and therefore do not fit the biological definition of a species. Some hybrid varieties of dandelions produce seeds, but the embryo is produced asexually. All the individuals in an immediate area may be clones. Nevertheless, hybrids may be considered true species because they are reproductively isolated from their parents.

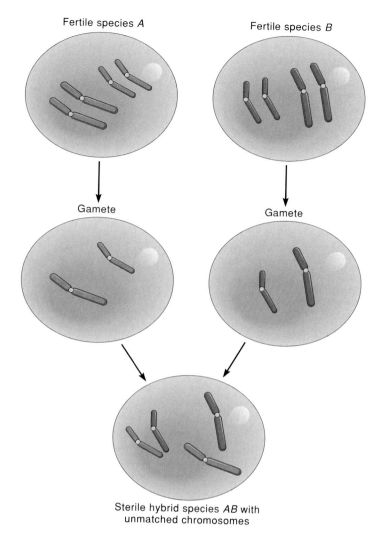

Fertile species *A*

Fertile species *B*

Gamete

Gamete

Sterile hybrid species *AB* with
unmatched chromosomes

figure 23.12

Hybrids are often sterile because their chromosomes are unpaired and
cannot synapse during meiosis.

Hybrid offspring may be superior and outcompete their
parents in a variety of ways. For example, frequent disturbance
or environmental change such as cooler temperature or less
rainfall may favor new hybrid genotypes and select against parent genotypes. Some hybrids may be genetically varied enough
to inhabit environments that are unsuitable for the parents.
Even in stable environments, hybrids may have an advantage
due to heterozygote superiority (see Chapter 22). Multiple alleles may be favorable; for example, an organism may be better
off with alleles coding for two different enzymes instead of one.
Hybrids often have **allozymes,** which are enzymes controlled by
alternate alleles. Even if these enzymes catalyze the same reaction, they may have different tolerances, optimal reaction temperatures, or substrate affinities. They can be useful at different
stages of an organism's life, or they may broaden the range of
potential environments.

Hybrids are especially superior to inbred parents that are
usually homozygous at many loci, especially homozygous recessive loci. In contrast, the more heterozygous hybrids will
have a dominant and often advantageous allele at many of the
previously recessive loci. This contributes to the vigor of many
first-generation hybrids.

Hybridization may also move genes between species or
populations. For example, a hybrid will occasionally reproduce with one of its parents. In this way alleles from one
species enter the gene pool of another species (fig. 23.13). This
method of gene movement is called **introgression.** Introgression is the flow of genes that bypass reproductive barriers and
move between populations via the mating of fertile hybrids
with parent populations. Hybrids of *Poa* frequently cross with
other grasses and introduce new alleles to these grasses. Some
genes of corn (*Zea mays*) can be traced to teosinte (*Zea mexicana*), a related grass; this gene flow probably occurred
through a hybrid. Introgression provides a tool for crop geneticists to obtain additional genetic variation for artificial selection and crop improvement.

c o n c e p t

Hybridization between two species is more common in plants
than animals and typically produces sterile offspring. Hybrids
may be successful in disturbed or new environments and may initiate a new species. Hybrids that can reproduce with other species
can cause introgression, a type of gene flow between two species.

Polyploidy

The most widespread cellular process affecting plant evolution
is polyploidy. Organisms with more than two sets of chromosomes are **polyploids** (poly = many, ploid = sets) and result from
aberrant chromosomal separation during cell division (fig.
23.14). Domesticated grains such as durum wheat (tetraploid),
oats (hexaploid), and rye (hexaploid) are polyploids, as are cotton, tobacco, and potato. Polyploids also include garden flowers
such as chrysanthemums, pansies, and daylilies. Polyploidy is
rare among animals, fungi, and most gymnosperms but occurs
regularly among angiosperms and ferns. The rarity of polyploidy among animals may be related to their having distinct
sex chromosomes. Polyploid sets of sex chromosomes in animals may fatally disrupt hormonal and sexual development.

Estimates of polyploidy in angiosperms have ranged
from 30% to 80%. In 1994, Jane Masterson of the University
of Chicago tested these estimates by studying a feature affected by polyploidy: cellular size. Because cellular size correlates with DNA content and thus with chromosome number,
Masterson could estimate ploidy levels by studying cellular
size in fossils of extinct plants. Masterson's clever research enabled her to infer that the haploid chromosome number in
many angiosperms is 7 to 9 and that approximately 70% of
angiosperms have polyploidy in their history.

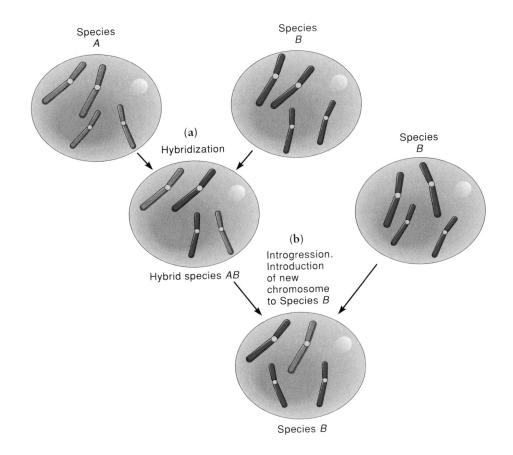

Species
A

Species
B

(a)
Hybridization

Species
B

Hybrid species AB

(b)
Introgression.
Introduction
of new
chromosome
to Species B

Species B

figure 23.13

Introgression. A hybrid will occasionally cross with one of its parent species. (a) In this schematic example, Species A hybridized with Species B. (b) This hybrid crossed with an individual of Species B. Their offspring was compatible with Species B and therefore carried a chromosome from Species A to the Species B population as it continued to reproduce with Species B individuals.

The most common polyploids are tetraploids, with four sets of chromosomes. Different species in the same genus often have different numbers of chromosome sets. For example, various species of wheat (*Triticum*) are diploid, tetraploid, and hexaploid (see reading 10.1 "Polyploidy in Plants"; also see Epilogue).

Sexually reproducing polyploids usually have an even number of sets of chromosomes that allows successful separation during meiosis. Successive generations remain fertile and maintain a constant number of chromosomes. Triploids and pentaploids are usually sterile due to unsuccessful meiosis. Evolutionarily, they are rarely successful, but they may be sought after as novelties. For example, some orchids are sterile triploids or pentaploids that flower more profusely without the demands of sexual production of seeds and fruit (fig. 23.15).

Polyploids with many copies of the same chromosome set derived within a single species are **autopolyploids.** These plants usually form within a single species when chromosomes are duplicated but not separated. When spindle fibers are disrupted, the cell may not divide cleanly (fig. 23.14a); therefore, the resulting cell has twice the original number of chromosomes. This

nondisjunction of chromosomes can be induced chemically by colchicine, a drug that is derived from the autumn crocus (*Colchicum autumnale*). Colchicine disrupts spindle formation and prevents chromosome separation during mitosis.

Autopolyploidy occurs naturally in many genera, including *Sedum* and *Galax;* some domestic plants have been induced to become autopolyploids to promote desired characteristics. For example, polyploids often have larger cells, thicker leaves, increased water retention, slower growth, delayed flowering, and flowering over a longer season. Cold tolerance is also enhanced, and the number of polyploid species increases with increasing latitude. For these reasons, natural autopolyploids are most common in harsh environments where selection pressures are intense. Enhanced tolerance to harsh environmental factors is often the competitive edge that allows a polyploid species to establish a permanent population.

Another common type of polyploidy is **allopolyploidy.** Allopolyploids have two or more distinct sets of chromosomes typically derived by duplication and nondisjunction in hybrids already having mixed chromosome sets. Note that the

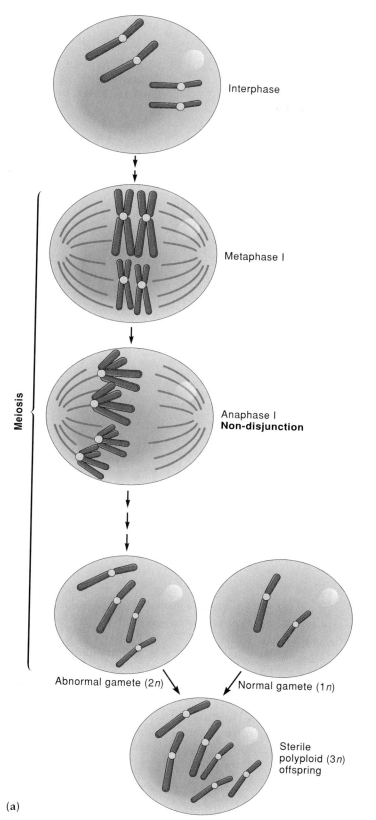

Interphase

Metaphase I

Anaphase I
Non-disjunction

Meiosis

Abnormal gamete (2*n*)

Normal gamete (1*n*)

Sterile polyploid (3*n*) offspring

(a)

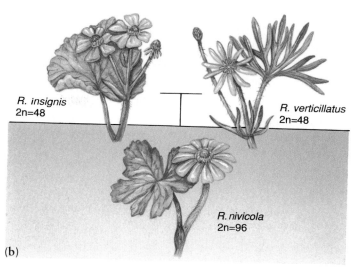

R. insignis
2n=48

R. verticillatus
2n=48

R. nivicola
2n=96

(b)

figure 23.14

Polyploidy by nondisjunction of chromosomes. (a) Nondisjunction during meiosis (anaphase I) produces an abnormal gamete that may lead to sterile polyploid offspring. If chromosome numbers are multiplied within a species, it is autopolyploidy. If multiple sets of chromosomes form by polyploidy of a hybrid or by fusion of gametes from different species, it is allopolyploidy. (b) Three species of *Ranunculus* live on the North Island of New Zealand. Two of these species have 24 pairs of chromosomes (2*n*=48). Allopolyploidy of the hybrid from the two parent species probably produced the third species with twice the chromosome number (2*n*=96) as the hybridizing parents.

figure 23.15

Hybridization of orchids to produce new color patterns often results in sterile polyploid offspring. Fortunately, orchids such as this hybrid *Dendrobium* sp. can be cloned asexually, so the infertility of a triploid or pentaploid hybrid is not of great concern to horticulturists.

POLYPLOIDS IN THE KITCHEN

In 1924, a Russian geneticist had a tasty idea: crossing a cabbage and a radish to produce a new kind of vegetable called a cabbish, having the crispy leaves of cabbage and the red, sharp-tasting root of radish. The geneticist knew that he could probably produce a hybrid, because cabbages and radishes each have 18 pairs of chromosomes. Unfortunately, the hybrid that he produced was sterile because the chromosomes of radishes differed so much from those of cabbage that they did not pair during meiosis. Consequently, the hybrid could not produce viable gametes.

However, the hybrid spontaneously produced a tetraploid having a total of 72 chromosomes: two sets from radish and two from cabbage. In this tetraploid, each chromosome of cabbage and radish had a homologous partner, which allowed pairing to occur during meiosis. The gametes produced by the tetraploid each had a complete haploid set of chromosomes from radish and cabbage. Therefore, the hybrid was an allopolyploid, a polyploid containing the chromosomes of different species. Although the allopolyploid was fertile, it never reached our dinner tables because it

was a "radbage," not a "cabbish": it had distasteful, radishlike leaves and uninteresting, cabbagelike roots.

Although the cabbish experiment was a culinary failure, other allopolyploids have been successful. For example, the development of Western civilization depended largely on wheat, an allopolyploid that is a combination of three species that each contributed two sets of chromosomes (see reading 10.1; also see Epilogue). Thus, while you may never sink your teeth into a cabbish, you probably eat an allopolyploid every day.

figure 23.16

The daylily on the left is diploid, whereas the one on the right is tetraploid. Such polyploids often have exceptionally robust petals, leaves, and stems.

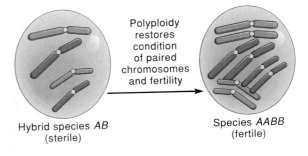

Polyploidy restores condition of paired chromosomes and fertility

Hybrid species *AB* (sterile) → Species *AABB* (fertile)

figure 23.17

Polyploidy can restore fertility. Hybrids are often sterile because their chromosomes are unpaired. Polyploidy can restore the condition of paired chromosomes necessary for successful meiosis and sexual fertility.

distinction between auto- and allopolyploidy is complex and not always rigid. However, most plant hybrids are considered allopolyploids, including cultured hybrids that have been artificially selected for their large cells, plump plant parts, high water content, and drought resistance (fig. 23.16).

Polyploidy, especially allopolyploidy, can restore sexual viability to an otherwise sterile hybrid. Recall that hybrids without homologues cannot complete meiosis. Allopolyploidy solves this problem by providing duplicate sets of chromosomes that synapse during meiosis. The diploid hybrid between

the primroses *Primula verticillata* and *P. floribunda* is sterile, but tetraploid hybrids called *P. kewensis* have been produced artificially and are fertile.

Polyploidy, especially hybridization followed by allopolyploidy, can be an immediate mechanism of sympatric speciation in higher plants (recall that a hybrid between two species is often sterile because chromosomes are not paired). Polyploidy restores homologues and fertility (fig. 23.17). The resulting allopolyploid cannot readily reproduce with either parent and immediately becomes a new species. Bread wheat, *Triticum aestivum*, is an allopolyploid species with 42 chromosomes that probably arose about 8,000 years ago by hybridization between a wheat species with 28 chromosomes and a grass with 14 chromosomes. The sterile hybrid had 21 chromosomes, but doubling of the chromosomes following hybridization produced a fertile allopolyploid wheat (see reading 10.1, "Polyploidy in Plants"; also see Epilogue). Rutabaga (*Brassica napus*, 38 chromosomes, $n = 19$) arose as an allopolyploid of a hybrid between cabbage (*Brassica oleracea*, 18 chromosomes, *n* = 9) and turnip (*Brassica rapa*, 20 chromosomes, *n* = 10).

Similar sympatric speciation by backcrossing of autopolyploids is apparently rare. Recall that autopolyploidy produces multiple sets of chromosomes that are often tetraploid. If a tetraploid plant backcrosses with a diploid individual, the hybrid will probably be sterile due to unbalanced sets of chromosomes. If the hybrid can successfully reproduce asexually, it may become a new species as it diverges from the parent population.

c o n c e p t

Polyploids have more than the typical diploid number of chromosomes. Polyploidy is a common evolutionary mechanism that can produce immediate speciation and restore fertility to hybrids.

Chapter Summary

Speciation—the origin of new species—is the central concept of evolutionary theory and is the mechanism that explains the vast diversity of organisms. Speciation is difficult to describe because it includes a variety of mechanisms; moreover, a species is difficult to define clearly. The most general definition of a species is a group of individuals that can successfully breed with one another but not with members of other groups. This emphasizes reproductive isolation. Distinguishing between species and between the mechanisms that produce new species is further confounded because all members of a species are not identical.

During speciation, segments of the gene pool of a plant population diverge—sometimes rapidly through processes such as polyploidy and hybridization, and sometimes slowly through responses to natural selection pressures. As parts of the population become different, they lose some features and accumulate others that prevent interbreeding. This genetic accumulation and divergence is promoted by environmental variation, different mutations, and different rates of gene flow and genetic drift. Ultimately, members of two different populations acquire different traits and are reproductively isolated. They no longer produce fertile offspring (hybrids), and the two populations represent two species.

Biologists need a working definition of a species as a framework for questions and research. Morphologically, a species is a group of organisms that look the same and appear significantly different from other organisms. This definition is easily applied and communicated. Biologically, a species is one or more populations whose members can reproduce with each other in nature to produce fertile progeny but cannot successfully reproduce with members of other populations and species. This reproductive isolation is frequently violated by hybridization. Genetically, a species is a group of organisms that exchange genes.

Reproductive isolation is maintained by prezygotic barriers that prevent fertilization of gametes from individuals of two different species or by postzygotic barriers that result in sterile or unfit hybrid offspring. Prezygotic barriers include (1) geographical isolation of populations in distant areas, (2) microhabitat isolation of populations in the same general area, (3) temporal isolation of organisms receptive to mating at different times, (4) mechanical isolation of organisms that are structurally incompatible for reproduction, and (5) gametic isolation from incompatible gametes. Postzygotic barriers include (1) inviability of hybrid zygotes that cannot develop to sexual maturity, (2) sterile hybrids, and (3) the breakdown of successive hybrid generations that become weak or infertile.

For a subpopulation to form a new species, the gene pool must diverge from that of the parent population. Divergence during allopatric speciation of geographically separated populations is usually gradual and results from genetic differences accumulated through mutation, genetic drift, and selection. When divergence is great enough to eliminate interbreeding, then the two populations are different species. However, it is difficult to test the reproductive isolation of naturally separated populations, because the bringing together of potential parents by researchers introduces artificial conditions.

Populations that overlap geographically are sympatric. Speciation in these populations may begin with the formation of a cline, which is the pattern of variation among individuals of the same species distributed along an environmental gradient. Demands of the local environment along a cline may overshadow gene flow and lead to variation. Plants are morphologically plastic and display different growth forms in response to different environmental conditions, even without significant genetic change. If the environment is distinctly patchy, then the cline may also develop ecotypes, which are subpopulations subject to local selection and which diverge both genetically and morphologically. Significant divergence can lead to speciation.

Hybridization is the production of offspring from parents of different species, which produces new genetic combinations. New combinations may lead to speciation if the hybrids are fit and reproductively isolated from the parent populations. Plant hybrids are often sterile because sets of chromosomes are typically unbalanced and cannot pair during meiosis. However, many hybrids reproduce vegetatively and are especially successful in stressful or disturbed environments. Successful hybrids may be ecologically superior to their parent populations and become a new species. Hybrids that backcross with parent populations may introduce genes from one species to another by introgression.

Polyploids have more than two sets of chromosomes; they result from the nondisjunction of chromosomes during cell division. Polyploidy, which is common in higher plants, is an instantaneous mechanism of speciation because polyploids are reproductively incompatible with the parent population. Polyploids with multiple copies of the same sets of chromosomes are autopolyploids. Polyploids with two or more distinct sets of chromosomes are allopolyploids. Polyploidy may restore fertility to a hybrid by providing paired chromosomes.

Writing to Learn Botany

What are the comparative advantages and disadvantages of each of the various species concepts?

Questions for Further Thought and Study

1. Describe how reproductive contact between two populations may reinforce their genetic divergence.

2. When would the production of an offspring not necessarily indicate that the two parents were of the same species?

3. What role could a mountain range play in speciation?

4. What are the basic differences between sympatric and allopatric speciation?

5. Why are many hybrids sterile?

6. In what ways does the environment affect speciation?

7. Why are hybrids often more successful in disturbed environments?

Web Sites

Review the "Doing Botany Yourself" essay and assignments for Chapter 23 on the *Botany Home Page*. What experiments would you do to test the hypotheses? What data can you gather on the Web to help you refine your experiments?

Here are some other sites that you may find interesting:

http://phylogeny.arizona.edu/tree/life.html
This site shows you the Tree of Life, which traces the roots of life through all of its branches. The site also includes links to other web sites at each branch.

http://www.book.uci.edu/Books/Moon/flora_plants.html
Hawaii's indigenous and endemic plants, flowers, and trees are both fascinating and beautiful, but unfortunately, like everything else that was native, they are quickly disappearing.

http://www.uel.ac.uk/palaeo/pfr2/pfr.htm
Names, places and ages can be searched in the Plant Fossil Record, and the occurrences are instantly plotted on a global map. Patterns of migration and evolution through geological time can be clearly examined to help better understand the history of climatic and environmental change.

Suggested Readings

Articles

Elisens, W. J. 1989. Genetic variation and evolution of the Galápagos shrub snapdragon. *National Geographic Research* 5:98–110.

Gibbons, A. 1996. On the many origins of species. *Science* 273:1496–1499.

Masterson, J. 1994. Stomatal size in fossil plants: Evidence for polyploidy in majority of angiosperms. *Science* 264:421–423.

May, R. M. 1992. How many species inhabit the earth? *Scientific American* 267 (October):42–49.

Mayr, E. 1989. Speciational evolution or punctuated equilibria. *Journal of Social and Biological Structures* 12:137–159.

————. 1992. A local flora and the biological species concept. *American Journal of Botany* 79:222–238.

Moffat, A. S. 1995. Plants proving their worth in toxic metal cleanup. *Science* 269:302–303.

Rose, M. R., and W. F. Doolittle. 1983. Molecular biological mechanisms of speciation. *Science* 220:157–162.

Slatkin, M. 1987. Gene flow and the geographic structure of natural populations. *Science* 236:787–792.

Templeton, A. R. 1981. Mechanisms of speciation—A population genetics approach. *Annual Review of Ecology and Systematics* 12:23–41.

Books

Grant, V. 1981. *Plant Speciation*. 2d ed. New York: Columbia University Press.

Hancock, J. F. 1992. *Plant Evolution and the Origin of Crop Species*. Englewood Cliffs, NJ: Prentice-Hall.

Margulis, L., and R. Fester, eds. 1991. *Symbiosis as a Source of Evolutionary Innovation: Speciation and Morphogenesis*. Cambridge, MA: MIT Press.

Otte, D., and J. A. Endler, eds. 1989. *Speciation and Its Consequences*. Sunderland, MA: Sinauer Associates.

Proctor, M., P. Yeo, and A. Lack. 1996. *The Natural History of Pollination*. Portland, OR: Timber Press.

Diversity . . .

P lants dominate our planet and our lives, thanks largely to their diversity and remarkable adaptations. In the previous unit we discussed how evolution produces this diversity. In this unit we discuss the nature and extent of the diversity of plants. Specifically, we discuss the major groups of plants: bryophytes, ferns and their "allies," gymnosperms, and angiosperms. We also discuss algae, because of their presumed relationship to plants, and certain other organisms—bacteria and fungi—that are traditionally included in botany textbooks. Although bacteria, fungi, and most algae are no longer classified as plants, botanists once thought they were.

Our treatment of diversity includes discussions of the distinguishing features of each group of organisms. You will also see how botanists use some of the techniques described in the next chapter to construct evolutionary trees to understand how the diversity of different groups of organisms might have arisen. Moreover, you will learn about some of the unresolved problems concerning each group. We include these discussions to emphasize the changing nature of our understanding of plant diversity.

CAROLI LINNÆI

S:Æ R:GIÆ M:TIS SVECIÆ ARCHIATRI; MEDIC. & BOTAN.
PROFESS. UPSAL; EQUITIS AUR. DE STELLA POLARI;
nec non ACAD. IMPER. MONSPEL. BEROL. TOLOS.
UPSAL. STOCKH. SOC. & PARIS. CORESP.

SPECIES PLANTARUM,

EXHIBENTES

PLANTAS RITE COGNITAS,

AD

GENERA RELATAS,

CUM

DIFFERENTIIS SPECIFICIS,
NOMINIBUS TRIVIALIBUS,
SYNONYMIS SELECTIS,
LOCIS NATALIBUS,
SECUNDUM

SYSTEMA SEXUALE

DIGESTAS.

TOMUS I.

Cum Privilegio S. R. M:tis Sveciæ & S. R. M:tis Polonicæ ac Electoris Saxon.

HOLMIÆ,
IMPENSIS LAURENTII SALVII.
1753.

—Title page of Linnaeus's *Species Plantarum*. In this two-volume work, Carolus Linnaeus established the rules of botanical nomenclature that are still followed today.

SVECIÆ ARCHIATRI; MEDIC. &
AL; EQUITIS AUR. DE STELLA PO
AD. IMPER. MONSPEL. BEROL. TOI
STOCKH. SOC. & PARIS. CORESP.

Systems of Classification

24

Chapter Outline

Chapter Overview

Classification is often viewed by nonspecialists as the discipline of biology that places organisms in their proper categories, where they remain permanently. Nothing could be further from the truth. Taxonomists, whose specialty is the discovery and classification of organisms, throughout history have had to revise and modernize classification systems as new species were discovered and as new knowledge about plants and other organisms was obtained in different disciplines of biology. The first classifications of plants, for example, made more than 2,000 years ago, dealt with hundreds of species and were based on features that could be seen with the naked eye. The number of known plant species is now in the hundreds of thousands, and taxonomists must integrate knowledge from genetics, ecology, anatomy, chemistry, physiology, and other areas into their classifications.

For many centuries, the first major dividing line among all living things was between plants and animals. Now we also recognize bacteria, fungi, and other kinds of organisms as equals to plants and animals at the highest levels of classification. The history of classification, therefore, is one of continual change. The subject of classification is more dynamic now than at any time in the past.

Perhaps the first requirement for classifying plants—putting them into groups—is to name them. Once a plant has a name, anyone using that name can expect other people to associate it with a specific plant. Although there are few records of plant names from more than about 4,000 years ago, you can nevertheless imagine that plant names were important to early humans for talking about plants that were used for medicines and foods. Plant names are still important for these reasons, but they are also important for tracking the diversity of the plant kingdom, monitoring the effects of environmental change, reporting botanical research, cultivating crops and ornamental plants, and all other aspects of the uses of plants.

Taxonomy is the science of discovering, describing, naming, and classifying organisms. If naming plants is a natural human endeavor, then everyone is to some degree or another a plant taxonomist. Many people also share a tendency to organize plants into groups. Edible plants, medicinal plants, and roses, for example, are recognized as groups. Rationales for grouping plants have varied over time, with some systems of classification having been influenced by religious or ethnic considerations. In northern Mexico, for example, the Tarahumara tribe classifies some plants on the basis of ceremonial uses involving hallucinogens. Tarahumarans group botanically unrelated plants such as *hikuli*—known to us as peyote cactus (*Lophophora williamsii*)—and *dowaka* (*Tillandsia benthamiana*, a member of the bromeliad family) because these plants are used together in ceremonies in the belief that their combined psychoactivity is greater than that of either plant by itself.

Although informal classifications are widespread, they are generally not recorded, nor are they used outside of the tribes or small groups of people who invented them. The focus of this chapter is on classifications of plants that have been recorded and can be used to follow the historical development of taxonomic thought up to the present.

The history of classification is one of continual change. Change, which still occurs today, has come from the discovery of new species, the acquisition of new knowledge about the features of plants and other organisms, and the development of new philosophies and methods of classification. To a large extent, new methods are due to the development of new technologies that are applicable to classification, especially technologies involving computers and molecular genetics. After reading this chapter, you will probably realize that classification has been, and continues to be, one of the most dynamic aspects of biology. You will also see that, partly because of its dynamic nature, the subject of classification tolerates a large diversity of ideas and opinions on how organisms should be classified.

The Early History of Classification

Any records of biological discoveries before the fourth century B.C. have either been lost or are too fragmentary to be useful. During the fourth century B.C., however, the famous Greek philosopher Aristotle wrote several scientific essays, primarily about animals. Unfortunately, all of Aristotle's writings about plants have been lost, and it was his brilliant student, Theophrastus of Eresus, who is credited with giving us the first known classification system for living organisms. In his *Historia Plantarum*,[1] Theophrastus described the parts, uses, and habitats of plants and sorted them primarily by the different forms of their leaves and whether they were trees, shrubs, or herbs. In so doing, he classified nearly 500 plants. For several centuries after Theophrastus, many new classifications were written by Greek and Roman scholars. Probably the most important of these was *De Materia Medica*, meaning "about materials medicinal," written in the first century A.D. by the Greek physician Dioscorides. The purpose of that book, which included all 600 medicinally useful plants known at the time, was to improve medical service in the Roman Empire. Nevertheless, it became the principal book of plant classification in Western civilization for nearly 1,500 years. Several European botanists and physicians used the work of Dioscorides during the fifteenth and sixteenth centuries as a basis for their *herbals*, which were illustrated books on the presumed medicinal uses of plants. Although some of the herbals had excellent drawings, they contained much folklore and many stories that became legends. The herbals also led to the development of the *Doctrine of Signatures*, which held that if a plant part resembled a part of the human body, it would be useful in treating ailments of that part. For example, walnut meats, which resemble tiny brains, were used to treat brain disorders, and *Hepatica* leaves, whose lobes resemble those of the liver, were used for liver ailments.

Beginning in about the fifteenth century, classification became more complicated as European explorers returned to Europe with new plants from other continents. As the plants

1. *Historia Plantarum* is Latin for "History of Plants." The works of Theophrastus and other early scientists of Western civilization were translated into Latin once it became the language of scholars. Major essays and books were published in Latin as recently as the late nineteenth century. Today, descriptions of new species of organisms are still published in Latin.

accumulated, botanists began to turn their attention from medicinal plants alone to classifying and cataloging all types of plants. This effort was aided by the practice of categorizing each plant by a few unique features. By 1623 the Swiss botanist Gaspard Bauhin had used this method to classify about 6,000 kinds of plants. Some of the plants he classified had two-part Latin names. All of this occurred more than 100 years before the great Swedish botanist Carolus Linnaeus introduced the binomial system of nomenclature, which we still use today.

Many botanists of the late seventeenth and early eighteenth centuries focused nearly all of their work on classification. To these scientists, to name a plant was to know a plant. Among these botanists was Protestant cleric John Ray, whose three-volume *Historia Plantarum Generalis* was a detailed classification of more than 18,000 kinds of plants. In that book, Ray divided the flowering plants into dicotyledons and monocotyledons. Ray's classification was significant because it grouped plants on the basis of multiple similarities rather than on just a few key features. This was a notable advance in thinking, because it started to show the natural relationships of plants—what we would now call their evolutionary relationships—even though Ray still believed species to be unchanging. Ray's natural classification, coming more than a century before Charles Darwin's ideas about evolution, was prophetic; many of the plant families defined by Ray are still recognized today. The most highly developed system of plant classification before Linnaeus was that of the French botanist J. P. de Tournefort. He was the first to clearly characterize **genera** (singular: **genus**) as a taxonomic rank between family and species. De Tournefort's system included 9,000 species in 700 genera.

c o n c e p t

The earliest classifications of plants were primarily utilitarian classifications; that is, they included mostly medicinal or other useful plants. Beginning in about the fifteenth century, classifications became more complicated as thousands of new plants were discovered during worldwide exploration. These plants were classified on the basis of a few key features until the seventeenth century, when John Ray developed the first classification based on multiple features. Ray's classification showed natural relationships among plants. The system of J. P. de Tournefort was the first to characterize genera as an important taxonomic rank.

Carolus Linnaeus

Swedish physician and naturalist Carolus Linnaeus (1707–1778) merits special attention in any discussion of the history of classification because he is credited with giving two-part scientific names to organisms, which taxonomists still do today (Linnaeus even converted his Swedish name, Carl von Linné, to the two-part Latin version we use today). His system is called the **binomial system of nomenclature** because all of the scientific names of organisms have two parts; that is, they are *binomials*

("two names"). Although some botanists want to replace Linnaeus's system with one based on evolutionary relationships, Linnaeus's system remains the most widely used system for classifying plants and other organisms (see reading 24.2, "Is It Time to Change the System?").

Linnaeus's classification was published in 1753 in a two-volume set called *Species Plantarum* ("Species of Plants"), which included about 7,300 kinds of plants. Linnaeus tried to do more than simply publish long lists of plants with Latin binomials. He also organized plants into twenty-four classes. These classes were based primarily on the features of stamens, including the number of stamens per flower, whether or not they were fused together, and whether or not they occurred on the same flower as the pistils. For example, plants in the class Triandria had three stamens per flower, and those of the class Monadelphia had fused stamen filaments. In addition, within each class of plants with flowers, Linnaeus recognized subunits based on the number of pistils. Organisms such as fungi and algae, which lack flowers, were placed in a class of their own.

Since Linnaeus's classification was based on a few reproductive features, it often did not reflect natural relationships and is therefore referred to as an artificial classification. Nevertheless, the system was more convenient and more comprehensive than any other system available at that time.

When Linnaeus began his work, it was customary to use descriptive Latin phrase names for both plants and animals. The first word of the phrase constituted the genus to which the organism belonged. For example, all known poplars were given phrase names beginning with the word *Populus*. Similarly, the phrases for willows began with *Salix*, those for roses with *Rosa*, and those for mints with *Mentha*. The complete phrase name for peppermint was *Mentha floribus capitatus, foliis lanceolatis serratis subpetiolatis*, or "Mint with flowers in a head, leaves lance-shaped, saw-toothed, with very short petioles" (fig. 24.1). Although such more-than-a-mouthful names were specific, they were far too cumbersome to be useful.

In addition to including a referenced list of all the Latin phrase names previously given to plants, Linnaeus also changed some of the phrases to emphasize similarities among groups of plants, and he limited the phrases to a maximum of twelve words. Furthermore, he used such similarities for placing groups of plants in the same genus; each member of a genus was called a **species**.[2]

In the margin next to the phrase, Linnaeus listed a word that, when combined with the genus name, formed a convenient abbreviation for a species. For example, Linnaeus adopted *Vitis*, the first word in the Latin phrases for grapes, as the genus name for grapes. The word *vinifera* was placed in the margin next to the phrase describing the common wine grape, and the word *vulpina* next to the phrase for the winter grape. In doing so, Linnaeus designated the abbreviated names for two species of grapes as *Vitis vinifera* and *Vitis vulpina*. Similarly, peppermint was designated *Mentha piperita*.

---

2. Note that *species* is like the word *sheep* in that it is spelled and pronounced the same way in either singular or plural usage. There is no such thing as a plant or animal specie.

figure 24.1

Peppermint plants (*Mentha piperita*). *Mentha* is the genus name that means this plant is a mint; the full name, *Mentha piperita*, indicates what kind of mint.

Although Linnaeus originally considered the phrase names to be the scientific names of plant species, he and those who followed him eventually replaced all the phrase names with abbreviated ones; that is, with binomials. Today all scientific names of plants are binomials. In addition, the complete scientific name also includes the initials or name of the person or persons who first described the species.[3] Accordingly, the scientific names of plants that were first described and named by Linnaeus still bear an *L.* after the binomial. Thus *Mentha piperita* L., *Plantago major* L. (common plantain), *Populus alba* L. (silver poplar), and *Hedera helix* L. (English ivy) are current scientific names that came from Linnaeus. Plant species that were discovered

figure 24.2

Hoary manzanita (*Arctostaphylos canescens* Eastw.) of California. A person's full or abbreviated name after a plant species name indicates who named the species. In this case, the namer was Alice Eastwood.

after Linnaeus, such as *Arctostaphylos canescens* Eastw. (hoary manzanita; fig. 24.2), bear reference to more recent taxonomists.[4] In this case, *Eastw.* is a standard abbreviation for Alice Eastwood, who was curator of botany at the California Academy of Sciences from 1892 through the first half of the twentieth century. For more information about plants' names, see reading 24.1, "How to Name a New Plant Species."

Many plants have two authorities listed after the Latin binomial. The full scientific name of the giant sequoia, for example, is *Sequoiadendron giganteum* (Lindl.) Buchh. In this instance, John Lindley, a nineteenth-century English botanist, first described the giant sequoia and gave it the binomial *Wellingtonia gigantea*. However, the name *Wellingtonia* had already been used for a genus of dicots, and so Illinois botanist John Buchholz later transferred the giant sequoia to the genus *Sequoiadendron*. Since both botanists helped form the scientific name by which the giant sequoia is currently known, both are cited after the Latin binomial. The author of the original description is given in parentheses.

---

3. In contrast to plant names, the full scientific names for animals do not include references to authors of species names.

4. In scientific publications the author of a plant species is supposed to be included with the scientific name the first time the binomial is mentioned. Thereafter, it is not necessary to include the author. However, the editors of some journals prefer omission of the name's authority altogether. This usually does not cause confusion unless the same scientific name has been used by two different authors for two different plants or unless two authors have given different names to the same plant.

HOW TO NAME A NEW PLANT SPECIES

New species are discovered every day, usually in little-explored regions of the world such as tropical rain forests. New plant species, however, may be found almost anywhere that wild plants grow. When a botanist finds a plant that does not match the description of any known plant species, a new species may have been discovered. Because a new species usually fits into an existing genus, the botanist needs to show how the new plant differs from other species in the genus and give it a name. Botanists are guided in such a task by the *International Code of Botanical Nomenclature*, which is the rulebook for naming new plants. The new name must be one that has not been used for another plant and must be accompanied by a description of the species or an explanation (diagnosis) of how the new species is different from other, similar species in Latin. The name and description or diagnosis must be published to establish the validity of the new name. Editors of journals or books for this purpose send the species proposal to specialists to verify that the species is a new one and that the name is unique.

The name of each new species must also be represented by a single specimen, called the type specimen, that was used to describe the species (reading fig. 24.1). Specimens (including type specimens) are generally made by pressing the freshly collected plants as flat as possible between sheets of newspaper and cardboard, drying them, and then mounting them on large sheets of paper for safekeeping in a herbarium. A herbarium is like a library of preserved plant specimens. Large herbaria, such as that of the New York Botanical Garden or the Gray Herbarium of Harvard University, maintain millions of specimens, thousands of which are type specimens.

Herbarium collections are vital for scientific research. They are especially useful for comparing new species with previously known ones, keeping records of the discovery of new species, keeping track of species migration and extinction, and knowing which species are becoming rare or endangered due to human activities.

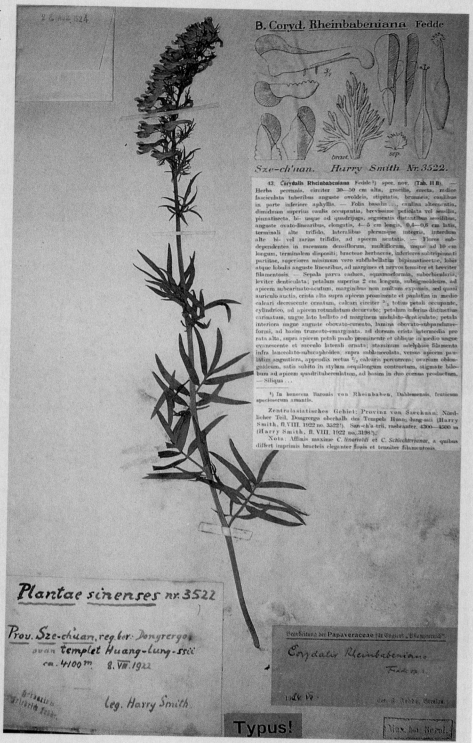

reading figure 24.1

A type specimen and a description of a new species. Each species name must be represented by such a type specimen.

Although Bauhin was the first to use two-part names for some plants, Linnaeus is credited with the first use of the binomial system of nomenclature for all species. His classification of plants was based primarily on features of stamens, which he used to organize plants into twenty-four classes. Because Linnaeus's system was based on just a few reproductive features, it was considered to be an artificial system of classification.

Early Post-Linnaean Classifications

By the end of the eighteenth century, botanists had begun to oppose the artificial system of Linnaeus because his system often placed unrelated plants together. For example, cherries and cacti were placed in the same class because they both have many stamens per flower. Instead, the development of a natural system became a major goal of classification. As in Ray's earlier classification, however, the idea of a natural system had no evolutionary foundation. Rather, natural systems of classification were meant to reflect the divine creation of related groups.

The most significant of the natural systems of classification after Linnaeus included *Genera Plantarum*, "Genera of Plants," by French botanist Antoine Laurent de Jussieu; *Prodromus Systematis Naturalis Regni Vegetabilis*, "Forerunner of the Natural System of the Vegetable Kingdom," which was started by Swiss botanist Augustin Pyrame de Candolle in 1824 and finished by others 50 years later; and another *Genera Plantarum*, this one published from 1862 to 1883 by two Englishmen, George Bentham and Sir Joseph Dalton Hooker. De Jussieu's *Genera Plantarum* is credited with creating wide acceptance of the idea of natural systems. De Candolle's *Prodromus* was essentially a large-scale expansion of Linnaeus's *Species Plantarum*. Nevertheless, *Prodromus* included many new families and original descriptions of plants, several of which remain the only worldwide treatment of certain groups. Bentham and Hooker's classification was based on the natural systems of de Jussieu and de Candolle, but Bentham and Hooker used their own descriptions of plants rather than relying on previously published ones. Partly for this reason, Bentham and Hooker's treatment remains an important resource for plant taxonomists today.

Honoring a "Disregarded" Botanist

When he was 25 years old, Swedish botanist and explorer Carolus Linnaeus (1707–1778) was paid by the Swedish Royal Society of Science to make the first scientific survey of the plants, animals, and geology of Lapland, a vast region of northern Europe almost entirely within the Arctic Circle. North of Gävle on the central east coast of Sweden, Linnaeus came upon the twinflower, a woodland plant common in the cold, swampy, coniferous forests of that area. The twinflower (see photo at right) is easily recognized as a creeping, dainty, shade-loving plant that, during May and June, produces pairs of pink to purplish, nodding flowers.

Although twinflower had already been named by another botanist, it was later renamed after Linnaeus. Today twinflower remains known as *Linnaea borealis* and is the only species of the genus (in North America two varieties of the species are recognized). The Latin binomial of twinflower denotes its taxonomic history: *Linnaea* for Linnaeus and *borealis* for its boreal (i.e., northern) location. Although Linnaeus is now regarded as a giant of biology's past, with his work having been the foundation for the science of modern taxonomy, he may not have had great self-esteem. As he later wrote, "*Linnaea* . . . a plant of Lapland, lowly, insignificant, disregarded, flowering for but a brief space—from Linnaeus who resembles it."

Twinflower (*Linnaea borealis*).

Is It Time to Change the System?

Carolus Linnaeus classified plants according to their reproductive parts, thereby endowing them with sex lives. For example, Linnaeus's system was based on *nuptiae plantarum*—the marriages of plants—that involved plants functioning as "husbands" and "wives" and enjoying sexual relations in "bridal beds." Linnaeus's system of classification also reflected eighteenth-century values; for example, male parts of flowers were given more taxonomic significance than were female parts. This judgment was based not on empirical evidence, but rather on traditional tenets of gender-related bias. Indeed, Linnaeus's system was artificial as well as explicit (one opponent called it "loathsome harlotry"); that's why most of the details of Linnaeus's system have largely been abandoned. What survives of Linnaeus's system is its method of hierarchical classification and its custom of binomial nomenclature.

Although Linnaeus's classification system has been a bedrock of biology for more than two centuries, a growing number of biologists now want to scrap Linnaeus's system. These biologists argue that Linnaeus's system is not only outdated, but also misleading. Whereas Linnaeus classified organisms according to appearances, biologists today view organisms as related by evolutionary history. Because Linnaeus knew nothing of evolution, his rules and taxonomic categories have no biological meaning. As a result, his system of classification can confuse the evolutionary relationships of organisms.

Will biologists abandon Linnaeus's system? Although most biologists acknowledge the limitations of Linnaeus's system, they aren't ready to try something new. Indeed, abandoning Linnaeus's system of classification would involve a confusing transition period; moreover, many taxonomists wonder if any new system—regardless of its strengths—could displace the entrenched system of Linnaeus.

The debate continues . . .

Benefits of a Phylogenetic Classification of Life

Systematics is the scientific study of organisms and their phylogenetic relationships. Scientists who specialize in systematics are called *systematists*. Recently, systematists worldwide have begun to organize a campaign to classify all organisms on earth in a comprehensive phylogenetic system before the year 2020. This monumental effort has at least six goals that should have major benefits to science and human society:

1. To guide the search for new crop species, novel genes, more biological control agents, and new products from organisms.

2. To help resource managers, conservation biologists, and policy makers set environmental priorities.

3. To show how all fields of biology can be linked through comparative, interdisciplinary studies encompassing all groups of organisms.

4. To provide a basis for managing biological knowledge and for communicating it across science.

5. To establish a framework for estimating rates of extinction and of global environmental change.

6. To provide the evolutionary context for understanding the history of life.*

The goals of this multinational project are especially ambitious because there may be several million species yet to be discovered, hundreds of thousands of which may be plants.

Systematists estimate that the research needed to obtain a phylogenetic classification of life within the next 25 years will cost about $3 billion per year. Currently about $500 million is spent annually on global systematics. At the present rate of research funding and species extinction, systematists estimate that more than half of the species or organisms now living will be extinct before the project can be completed. Moreover, they estimate that most extinctions will be of undiscovered species.

---

*Adapted from "Systematics Agenda 2000: Charting the Biosphere," a pamphlet written by internationally renowned systematists to publicize their concerns about the rapid worldwide extinction of species due to human activities.

Modern Classifications

Until the latter half of the nineteenth century, virtually all classification systems were based on the belief that living organisms had not changed since their creation and would not change in the future. As the ideas of Charles Darwin and Alfred Wallace spread and became accepted, however, these systems fell into disfavor, and classifications based on evolutionary relationships began to be developed.

Classifications that try to reflect evolution are said to be **phylogenetic.** Taxonomists consider phylogenetic systems to be superior for several reasons (see reading 24.3, "Benefits of a Phylogenetic Classification of Life"). However, taxonomists do not agree on which phylogenetic system is the best. Accordingly, various phylogenetic classifications have been proposed, each differing in its basic assumptions about the features that best reflect the phylogeny of plants.

The first major phylogenetic system of plant classification was proposed around the turn of the nineteenth century.

figure 24.3

Male flowers of oak (*Quercus*). These flowers have no petals. The inset is of a single oak flower.

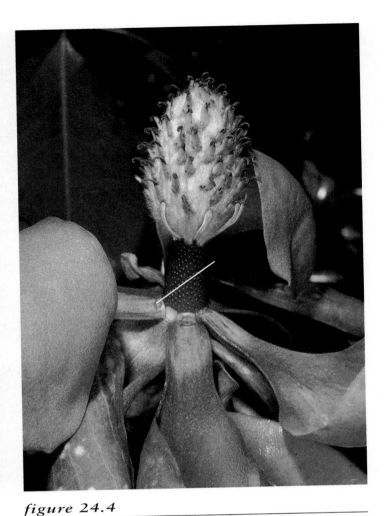

figure 24.4

Magnolia flower after many of the flower parts have fallen off, showing how they were attached in a spiral. One of the spiral lines is indicated by the line drawn on the floral axis.

Subsequent modifications and proposals for new systems by many botanists led to one of the most complete phylogenetic treatments that is still in use today. This is *Die natürlichen Pflanzenfamilien* ("The Natural Plant Families"), published from 1887 to 1915 by the German botanists Adolf Engler and Karl Prantl. Engler and Prantl recognized about 100,000 species of plants, which were organized in their presumed evolutionary sequence.[5] To begin such a sequence, Engler and Prantl first assumed that flowers lacking petals, such as those of walnuts, willows, and oaks, were primitive and that monocots were more primitive than dicots (fig. 24.3). Engler and Prantl's system is now in its twelfth edition, which was published in 1964. However, current ideas about the nature of primitive flowers differ considerably from those of Engler and Prantl (see Chapter 31).

Classifications produced during the twentieth century marked the beginning of another trend in major treatments of plants: the taxonomic focus in major works changed from being comprehensive for all plants to being specialized for different groups of plants. The flowering plants have received the most attention because they are the largest and most diverse of all plant groups. For this reason, the most significant new ideas about plant classification in the twentieth century have come from studies of flowering plants.

The first American to contribute significantly to the development of systems of classification was Charles Bessey, a botanist at the University of Nebraska. Bessey rejected some of Engler and Prantl's basic assumptions about such matters as the nature of primitive flowers. In following the ideas of J. D. Hooker, Bessey assumed, for example, that flowers such as magnolias, which have many separate spirally arranged parts, were the primitive ancestors of all other flowers (fig. 24.4).

Many of the ideas in Bessey's 1915 publication are still followed today. Nevertheless, many new phylogenetic systems

---

5. Note that by the beginning of the twentieth century, major scientific papers were no longer written exclusively in Latin. Engler and Prantl's work, for example, was published in German.

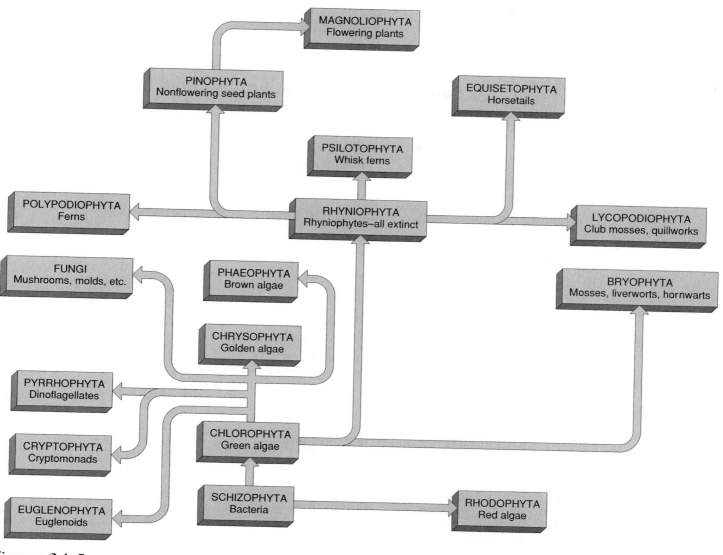

figure 24.5

Recent example of a traditional classification and phylogeny of plants (by the late Arthur Cronquist, formerly of the New York Botanical Garden). Not all of the names in this classification are still generally accepted.

of classification have been published since Bessey's system was first proposed. All of these systems stress what their authors believe are the primitive features of plants versus those that have undergone evolutionary change; that is, they emphasize primitive features versus derived features. These more recent classifications also draw lines of descent between related groups of plants (fig. 24.5).

At this point you may be wondering why there are so many phylogenetic classifications of plants. Why, in other words, if there is general agreement about the desirability of phylogenetic systems, is there no such general agreement about which system is best? There are at least two good answers to this question. One is that classification involves a subjective selection of characters (features) and a subjective evaluation of their importance for designating species, genera,

families, or any other taxonomic grouping. A good example of this subjectivity is the classification of flowering plants whose fruits are *legumes* (see Chapter 17). Legume-producing plants can be divided into three groups on the basis of different flower types. One type is exemplified by the common pea (*Pisum*) flower, which is bilaterally symmetrical and has five petals, the middle petal being larger and exterior in the bud to the other four. A second type, represented by honey locust (*Gleditsia*), also has a bilaterally symmetrical flower with five petals, but the middle petal is not larger and is interior in the bud to the other four. Flowers of the third type, represented by *Acacia* and *Mimosa*, are radially symmetrical. Taxonomists who consider the fruit (i.e., legume) to be the unifying feature recognize a single family, the Fabaceae, with three subfamilies based on flower type. Other taxonomists, who consider differences

figure 24.6

This diagram shows how Dr. Lumper and Dr. Splitter would differ in their classifications of the legume-bearing plants *Pisum*-type, *Gleditsia*-type, and *Mimosa*-type.

whose underlying assumptions about the evolution of characters often are not obvious and cannot easily be scientifically evaluated. Many taxonomists, therefore, consider traditional classifications to be more artistic than scientific, despite the fact that most of the recent classifications of flowering plants as a whole have been traditional. Currently, four of these traditional classifications are widely accepted: those of Americans Robert Thorne and Arthur Cronquist; that of Russian botanist Armen Takhtajan; and that of Swedish botanist Rolf Dahlgren. These four systems differ from earlier ones primarily in their use of sophisticated types of data, which were unavailable to earlier taxonomists. The newer types of data, for example, include information from genetics, ecology, anatomy, chemistry, and physiology of plants. Despite the sophistication of such data, however, intuitive assumptions about the significance of characters have been made in the same way that assumptions were previously made.

c o n c e p t

Post-Linnaean classification systems were said to be natural because they were based on nonevolutionary views of relationships among plants. Conversely, modern systems have attempted to be phylogenetic. Several different phylogenetic classifications of flowering plants are currently used widely by different taxonomists.

Cladistics

Systematists are now developing new approaches to phylogenetic classifications. Keep in mind, though, that ultimately all classifications are human artifacts and that there is no "universal truth" to be found. Nevertheless, these newer approaches are more explicit in their assumptions and are more testable scientifically. Perhaps the most widely adopted of the new approaches of the past two decades is one called **cladistics**. Cladistics is generally defined as a set of concepts and methods for determining **cladograms**, which depict branching patterns of evolution (fig. 24.7).

For cladistics, a *character state* is the form or value of a character. For example, the character could be flower color, and the state could be red or blue. Or the character could be the pigment chemical-type for red pigments, whose states are either a *betalain-type* red chemical or an *anthocyanin-type* red chemical (fig. 24.8). The states, or values, of each character must be explicit. Furthermore, once characters and their states are declared, much emphasis in cladistics is placed on assumptions about which states are primitive (ancestral) and which are derived (evolved). If the anthocyanin type of pigment chemical is assumed to be primitive, then the betalain type is derived, and vice versa. By assuming anthocyanin type to be primitive, we can see that plants with betalain-type pigments have a recent common ancestor in which betalain pigments first appeared (fig. 24.9a). In this case, the occurrence of betalain pigments indicates a shared evolutionary change (*shared derivation*) that was passed from an ancestral betalain producer to its descendant betalain producers.

in flower types more significant, divide the group into three families: the Fabaceae (*Pisum*-type), the Caesalpiniaceae (*Gleditsia*-type), and the Mimosaceae (*Mimosa*-type). Taxonomists who habitually take the more conservative view, such as considering all leguminous plants in one family, are often called *lumpers*. Those who divide families or other groups more liberally are often called *splitters* (fig. 24.6). Both views are widely accepted, although heated debates over lumping or splitting specific groups have been common.

A second explanation for the number of phylogenetic classifications of plants comes from differing views of taxonomists on how to derive evolutionary relationships from the features, or characters, of plants. Some make intuitive judgments

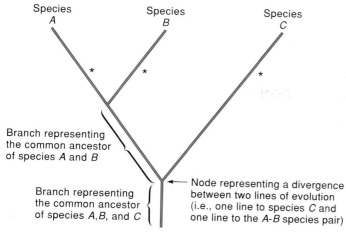

Species *A*   Species *B*   Species *C*

Branch representing
the common ancestor
of species *A* and *B*

Branch representing
the common ancestor
of species *A*, *B*, and *C*

Node representing a divergence
between two lines of evolution
(i.e., one line to species *C* and
one line to the *A-B* species pair)

* Terminal branches representing the evolution of individual species
after divergence from ancestors shared with other species

figure 24.7

This cladogram shows the phylogenetic relationships of three species,
all of which arose from a single ancestor. As shown here, Species *A*
and Species *B* are more closely related to each other than either is to
Species *C*.

What if, instead, we assume the betalain type to be primitive? The same cladogram would then show two evolutionary
origins of the anthocyanin type, one in the flowering plants
and one in the nonflowering seed plants (fig. 24.9b). Alternatively, another cladogram based on the evolution of pigment
chemical-types could be made (fig. 24.9c). The importance of
this alternative cladogram is that it represents an alternative
hypothesis about phylogenetic relationships: it minimizes the
number of times the anthocyanin type arose and shows instead
that flowering plants had two evolutionary origins. As an alternative hypothesis, it can be tested by evaluating additional
characters. In cladistics, testing hypotheses entails finding additional characters whose states may corroborate some cladograms but not others. Those that are not corroborated by additional data can be rejected, and those that are corroborated
can be maintained for further testing.

c o n c e p t

Cladistics is a set of concepts and methods that are represented
in evolutionary branching diagrams called *cladograms*, which
express hypotheses about evolutionary relationships among organisms. The evolutionary relationships shown in cladograms
are based on shared-derived character states.

How Do We Choose the Best Cladogram?

The alternative cladograms based on pigment chemical-type
show that betalain biosynthesis arose once in plant evolution
(fig. 24.9a), that anthocyanin biosynthesis arose twice in

(a)

(b)

figure 24.8

Two possible states of the character, *chemical type of red pigment.*
(a) The flowers of this rose (*Rosa* sp.) contain an anthocyanin-type
red pigment. (b) The red flowers of this hedgehog cactus
(*Echinocereus triglochidiatus*) contain a betalain-type red pigment.

evolution (fig. 24.9b), or that anthocyanin biosynthesis arose
once, but flowering arose twice (fig. 24.9c). Anthocyanin
biosynthesis involves a complex metabolic pathway, and it is
intuitively hard to believe that it could have two independent
evolutionary origins or that flowers and all their associated
features arose twice. The simpler scenario, that each of these
arose only once, is somehow more satisfactory. Such intuition
is consistent with a principle of logic called Occam's razor,
which is attributed to a fourteenth-century English philosopher named William of Occam. This principle states that a

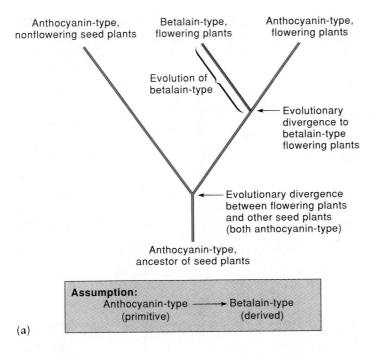

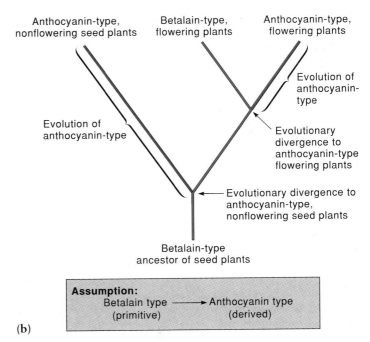

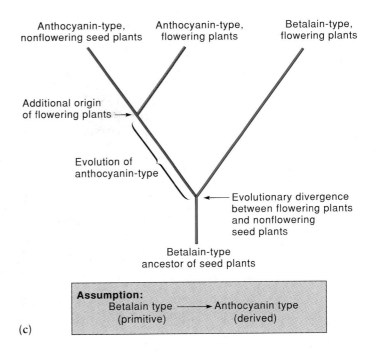

figure 24.9

Alternative cladograms for plants with anthocyanins or betalains.
(a) Cladogram built by assuming that anthocyanins were primitive
and betalains were derived. (b) Cladogram showing how character
evolution would look if betalains were primitive and anthocyanins
were derived twice. (c) Alternative cladogram showing that betalains
were primitive and anthocyanins were derived once.

person should not make more assumptions than the minimum
needed to explain anything. The principle is often called the
principle of **parsimony,** which is what systematists call it as it
is applied to cladistics. Nevertheless, understanding nature is
tricky because there are often exceptions to the principles that
we discover (see fig. 24.10).

In cladistic terms, the principle of parsimony can be re-
stated to say that the best cladogram requires us to make the
fewest assumptions about evolutionary changes. Parsimony is
especially powerful when there can be many possible cladograms for a group of organisms. The number of possible
cladograms would be easy to evaluate for three kinds of organisms (3 possibilities), for four kinds (26 possibilities), or
maybe even for five kinds (125 possibilities). However, the
number of imaginable cladograms increases more than exponentially as the number of different organisms increases. Indeed, there are more than 7 *trillion* possible cladograms that
can be made for 15 kinds of organisms.

Cladistics and Classification

The goal of cladistics or any other approach to making phylogenies is to evaluate and improve classifications. Opinions

vary, however, as to how the branching points on a cladogram
should be used to designate taxonomic levels and usually will
not convince splitters to lump or lumpers to split. In spite of
its newness and the competing ideas about how it should be
applied to classification, however, cladistics has been used to
infer the phylogenetic relationships of many different plant
groups. Several of these phylogenies will be helpful in learning
about different groups as they are discussed in the following
chapters in this unit.

figure 24.10

The red-capped *Amanita* (*A. muscaria*) mushroom. Although the red pigment in this fungus is a betalain that resembles the pigment that occurs in the plants shown in figure 24.8, this feature alone does not mean that fungi are closely related to plants.

A Theoretical Foundation of Biological Classification

We can make several basic assumptions, or *postulates*, about how we classify organisms. Some of these postulates are more obvious than others, but the main ones so far are:

1. Organisms exhibit different degrees of similarities and differences among individuals and groups.

2. Those organisms that are similar in nearly all respects are a species.

3. Species that share some of their features comprise a genus.

4. On the basis of their shared features, similar genera can be organized into a family; likewise, families and larger groups can be organized into successively higher levels of a taxonomic hierarchy.

5. The greater the similarity among organisms and among groups, the closer is their evolutionary relationship.

OR

The greater the number of shared derivations among organisms, the closer is their relationship.

Taken together these postulates function as the theoretical foundation of biological classification. As such, they can be called the *theory of biological classification*. Like all modern theories of biology, the theory of biological classification is constantly being tested and modified. Systematists, for example, are especially interested in what a species is (postulate 2; see Chapter 23). There is no general agreement about how many similarities are enough to call a group of organisms a species. Postulate 2 is often modified to include a reproductive component: Species are an interbreeding or potentially interbreeding population. This addition to the postulate is more applicable to animal species, however, because plant species often hybridize with one another but are still considered separate species (see Chapter 23).

Postulate 4 is most often applied to form a hierarchy that has species at the lowest level and kingdom at the highest level. According to such a hierarchy, species are organized into genera, genera into families, families into orders, orders into **classes**, classes into **divisions**,[6] and divisions into kingdoms.

Each category is a **taxon** (plural, **taxa**), which is a general term for any level of classification, such as species, genus, or family. A good way to envision the postulate of a taxonomic hierarchy is to see how it looks for a few examples, as shown in table 24.1.

Classification of Kingdoms

Botany does not exist in a vacuum; plant biology is best understood as it relates to the biology of other kinds of organisms. With regard to diversity, this means that the classification of plants as a group is best understood as it relates to the classification of animals, bacteria, and other groups of organisms. In finding out what plants are and where they came from—that is, in defining the plant kingdom—it is helpful, therefore, to consider relationships of members of the plant kingdom with members of other kingdoms.

A good starting place for evaluating relationships of plants to organisms of other kingdoms might be to determine what kingdoms there are and how the organisms are similar to or different from plants. Before the middle of the twentieth century almost everyone viewed living organisms as either plants or animals. Those who didn't and classified living organisms in three or four kingdoms were ahead of their time and generally not followed by their colleagues. In the 1950s, however, biologists began to consider features of the biochemistry and ultrastructure of cells as fundamental to classification and finally started accepting the idea that living things comprised more than two kingdoms. Some even went so far as to divide all living things into thirteen kingdoms.

---

6. Division is equivalent to the term **phylum**, which is used in classifying animals. Although the use of the word *phylum* for all living organisms is supported by many taxonomists, the *International Code of Botanical Nomenclature*, which governs the use of plant names, permits only the use of the word *division* for that taxonomic level of plants.

table 24.1

The Taxonomic Hierarchy for Four Species of Plants

Kingdom:	Plantae	Plantae	Plantae	Plantae
Division:	Magnoliophyta	Magnoliophyta	Magnoliophyta	Pinophyta
Class:	Liliopsida	Liliopsida	Magnoliopsida	Pinopsida
Order:	Zingiberales	Commelinales	Fagales	Pinales
Family:	Musaceae	Poaceae	Fagaceae	Pinaceae
Genus:	*Musa*	*Hordeum*	*Quercus*	*Pinus*
Species:	*Musa acuminata*	*Hordeum vulgare*	*Quercus alba*	*Pinus ponderosa*

Note: The common names of these species are, left to right, banana, barley, white oak, and ponderosa pine. The Magnoliophyta are the flowering plants; the Pinophyta are the conifers. Liliopsida is the current name for the monocots; Magnoliopsida is the name for the dicots.

table 24.2

Major Characteristics of the Five Kingdoms

	Monera	Protoctista	Fungi	Plantae	Animalia
Cell type	Prokaryotic	Eukaryotic	Eukaryotic	Eukaryotic	Eukaryotic
Genetic material	DNA not associated with protein in chromosomes	DNA associated with protein in chromosomes	DNA associated with protein in chromosomes	DNA associated with protein in chromosomes	DNA associated with protein in chromosomes
Gene structure	Introns absent in most groups, present in some	Introns present	Introns present	Introns present	Introns present
Nuclear envelope	Absent	Double or single	Double	Double	Double
Membrane-bound organelles	Absent	Present	Present	Present	Present
Chloroplasts	Absent	Present or absent	Absent	Present	Absent
Cell wall	Noncellulosic	Absent or present; cellulosic or various types	Chitinous or cellulosic	Cellulosic	Absent
Means of genetic recombination	Conjugation, transduction, or plasmid transfer	Fertilization and meiosis in most groups	Fertilization and meiosis	Fertilization and meiosis	Fertilization and meiosis
Mode of nutrition	Autotrophic or heterotrophic by absorption	Autotrophic or heterotrophic by absorption or phagocytosis	Heterotrophic by absorption	Autotrophic	Heterotrophic by ingestion
Flagella/cilia	Solid, rotating	9+2 microfibrils	Absent	Absent or 9+2 microfibrils in gametes of some groups	9+2 microfibrils
Multicellularity/cell specialization	Absent	Absent in most, present in algae and water molds	Present	Present	Present
Nervous system	Absent	Absent or simple	Absent	Absent	Present, often complex
Respiration	Aerobic or anaerobic	Aerobic	Aerobic or anaerobic	Aerobic	Aerobic
Life cycle	Haploid; fission	Mostly haploid; some forms diploid	Mostly haploid, often two nuclei per cell; some alternation of generations	Alternation of generations	Diploid except for gametes
Unicellular spore formation	Present in some groups	Present in some groups	Present in all groups	Present in all groups	Absent

A Modern System of Classification

The most popular classification at the kingdom level generally includes plants, animals, fungi, and a bacterial hodgepodge in four separate kingdoms, with a wide variety of other kinds of organisms thrown together into a fifth kingdom. The first four kingdoms are referred to as the **Kingdom Plantae**, the **Kingdom Animalia**, the **Kingdom Fungi**, and the **Kingdom Monera**, respectively. Organisms in the fifth group are in the **Kingdom Protoctista**. Robert H. Whittaker is credited with the suggestion, made in 1969, that all living things could be divided into five kingdoms. His classification, based on numerous features, is summarized in table 24.2. A more recent and more comprehensive five-kingdom classification, proposed by Lynn Margulis and Karlene Schwartz, is shown in figure 24.11; it provides a phylogeny that was based on traditional methods of systematics.

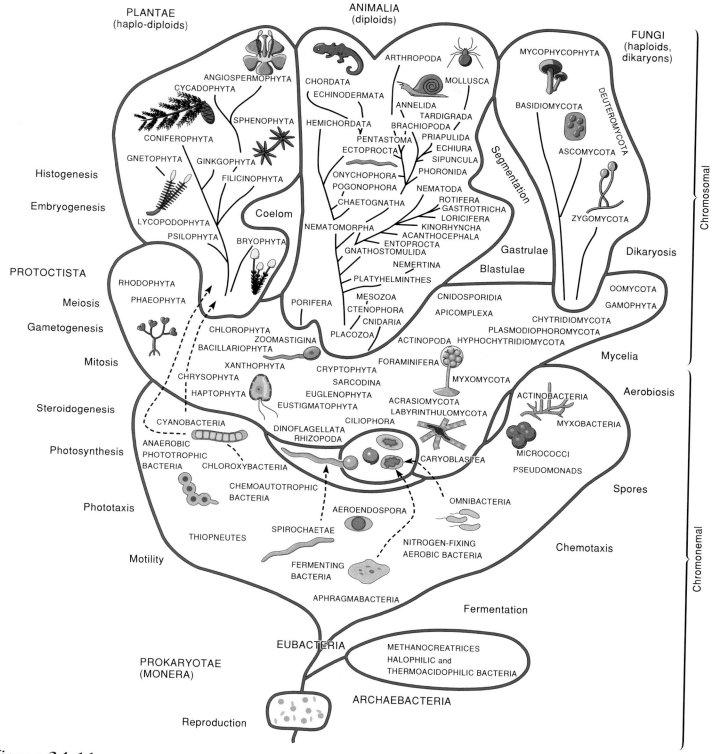

figure 24.11

Classification and phylogeny of five kingdoms by Margulis and Schwartz. Some of the names in this phylogeny are no longer accepted and are therefore not used in this textbook.

From Five Kingdoms *2/e by Margulis and Schwartz. Copyright © 1988 by W. H. Freeman and Company. Reprinted with permission.*

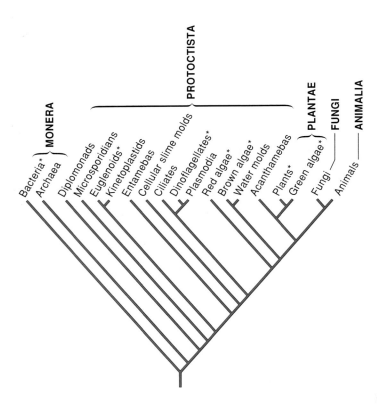

* Some or all members of these groups photosynthesize with chlorophyll *a*.

figure 24.12

RNA-based molecular phylogeny of life. Kingdoms are indicated according to the five-kingdom system. Organisms with chlorophyll *a* were classified as plants until recently; some botanists still classify them in the plant kingdom, with the exception of bacteria that contain chlorophyll.

Other Hypotheses for the Classification of Kingdoms

Phylogenetic hypotheses like that of Margulis and Schwartz are continually being scrutinized, especially by systematists who use molecular data for inferring evolutionary relationships. Molecular data, such as nucleotide sequences of genes or amino acid sequences of proteins, can be compared in the same way that morphological data are compared, either by overall similarity or by cladistic methods. Nucleotide or amino acid sequences are obtained for the same gene or the same protein from each of the organisms in the group. The characters in nucleotide sequences are the positions of the nucleotides along the sequence; the character states are whatever the nucleotide is at each position (i.e., adenine, guanine, cytosine, or thymine). For proteins, amino acid position is the character, and the character state can be any one of twenty amino acids.

A phylogeny based on molecular data is called a *molecular phylogeny*. Several molecular phylogenies have been proposed for the kingdoms of organisms, each differing in the gene or protein of choice in the method of data analysis (fig. 24.12).

Although the details vary, the different phylogenies generally show that the Kingdom Monera consists of two main groups and that the protoctists are scattered among several branches of the phylogeny of life. A recent system based on molecular phylogeny classifies all of life into one of three domains: Eubacteria (true bacteria), Archaebacteria (primitive bacteria), and Eukaryota (eukaryotes) (see Chapter 25). There is no such consensus regarding the classification of protoctists, although the results shown in figure 24.12 indicate that these organisms may comprise as many as seven to ten kingdoms.

What Kinds of Organisms Should a Botany Text Include?

Traditionally, botany texts have presented overviews of all organisms containing chlorophyll *a*: green plants, all groups of algae, dinoflagellates, euglenoids, and certain kinds of bacteria. Furthermore, for comparison with bacteria containing chlorophyll *a*, all other groups of bacteria have often been included as well. Fungi also have received significant coverage in botany texts, mostly because fungi were previously classified as plants. Finally, for good measure, botany texts have usually provided brief surveys of the diversity of viruses. If we stand back and take a look at this odd assortment of diversity, we see that it encompasses all but the animals, with at least four kingdoms plus a dash of the nonliving (i.e., viruses) represented.

The traditional inclusion of nonplants in botany texts also gets indirect approval from the Botanical Society of America. This organization maintains special sections for mycology, microbiology, and phycology (algae) and uses space in its journal, the *American Journal of Botany*, for research papers on these subjects. Nevertheless, a botany text is supposed to be about plants, which implies that a unit on diversity such as this should exclude nonplants. Such treatment would delight some instructors and students but dismay others. We have therefore chosen a compromise. The chapters on diversity that follow minimize the viruses, dinoflagellates, euglenoids, and flagellated molds and include briefer-than-usual overviews of bacteria, fungi, and protoctists. This enables us to provide a more expansive coverage of the plant kingdom.

Chapter Summary

The first record of a system of classification is that of Theophrastus, who is credited with classifying nearly 500 plants in the fourth century B.C. In the first century A.D., Dioscorides wrote a book that listed all known medicinally useful plants; this was the principal book of plant classification for nearly 1,500 years. Worldwide exploration and the discovery of new plants inspired many revisions of plant classification, beginning in about the fifteenth century. Subsequent major works included those of Gaspard Bauhin (1623), John Ray (ca. 1700), J. P. de Tournefort (ca. 1700), and Carolus Linnaeus (1753). Linnaeus's system is especially important because it divided all plants into twenty-four classes, and it designated each species by a Latin binomial. The binomial system of nomenclature is still used today for all species of organisms.

Classifications after Linnaeus continued to use the binomial system but adopted Ray's view that taxonomic categories should be natural—that is, that they should be based on natural relationships among plant groups. Natural classifications before Charles Darwin and Alfred Wallace, however, were meant to reflect the divine creation of related groups.

Once the ideas of Darwin and Wallace became widely accepted, classifications were revised to show the evolutionary relatedness among plant groups. These classifications are called *phylogenetic classifications*. Although these systems are still important, they have been modified and revised repeatedly throughout the twentieth century. These modern classifications are based largely on the intuition of their authors, whose assumptions about character evolution are usually not obvious and cannot be easily evaluated scientifically.

The abundance of different phylogenetic classifications of plants has prompted taxonomists to develop more explicit methods of building phylogenies. The most widely adopted of these newer methods comes from cladistics, whose goal is to determine evolutionary branching patterns among hierarchies of organisms. Cladistics is based on determining the evolutionary direction of character-state changes and showing phylogenetic relatedness based on shared derivations.

After the 1950s, as classification systems were being further developed for plants, it became apparent that many kinds of organisms traditionally included in the plant kingdom were not plants. Defining the plant kingdom, therefore, has involved determining the phylogenetic relatedness of plants to organisms of other kingdoms. At present, biologists classify organisms into five kingdoms by traditional phylogenetic methods. In contrast, the methods of cladistics, especially as they are applied to molecular data (e.g., DNA, RNA, proteins), provide phylogenies that show the five-kingdom system to be inadequate for making phylogenetic classification of all living things.

What Are Botanists Doing?

Several systems of plant classification now depend on techniques of molecular biology (e.g., gene sequencing). Go to the library and read an article about how molecular biology is changing our ideas about plant classification. Are molecular data more important than traditional data such as leaf shape and flower morphology? Why or why not?

Writing to Learn Botany

Paleontologist Stephen Jay Gould said the following about taxonomy:
"Taxonomy is often regarded as the dullest of subjects, fit only for mindless ordering and sometimes denigrated within science as mere 'stamp collecting.' . . . If systems of classification were mere hat racks for handling the facts of the world, this disdain might be justified. But classifications both reflect and direct our thinking. The way we order represents the way we think. Historical changes in classification are fossilized indicators of conceptual revolutions."
What was Gould saying? Do you agree or disagree? Why?

Questions for Further Thought and Study

1. Few, if any, intelligent persons would mistake a giraffe for an elephant, or a tomato for a potato, nor would most know or care about the scientific names of organisms. Why, then, should we be concerned about all organisms having scientific names?

2. Contrast and explain the similarities and differences among artificial, natural, and phylogenetic systems of classification.

3. How might the arrangement of plants in Linnaeus's *Species Plantarum* have shown or not have shown relationships among plants at different taxonomic levels?

4. Most early taxonomists used the form and structure of major plant parts in their systems of classification. What additional features of plants can modern taxonomists use? How might they be used for classification?

5. Linnaeus recognized 3 kingdoms of organisms; others have recognized up to 13 kingdoms. Is there a "correct" system of classification? Why or why not? How might you explain the abundance of classifications of plants that exists today?

6. What explanations might there be for the current absence of a cladistic classification of all plants?

7. Isaac Asimov wrote, "The card-player begins by arranging his hand for maximum sense. Scientists do the same with the facts they gather." How does Asimov's analogy relate to plant classification? Explain your thinking.

Web Sites

Review the "Doing Botany Yourself" essay and assignments for Chapter 24 on the *Botany Home Page*. What experiments would you do to test the hypotheses? What data can you gather on the Web to help you refine your experiments?

Here are some other sites that you may find interesting:

http://www.ucmp.berkeley.edu/history/linnaeus.html
This site describes the life and impact of Carolus Linnaeus on modern biology. Visit this site to learn about Linnaeus's many accomplishments (he was the personal physician to the Swedish royal family) and to see his botanical garden.

http://ucmp1.berkeley.edu/exhibits.html
This page from the California Museum of Paleontology teaches you about systematics and the scientists who contributed to our understanding of evolution.

http://www.herbaria.harvard.edu/

The Harvard University Herbaria contain a wealth of information ranging from libraries, collections, publications, and people, to education, databases, and research. Use its search facility for information about your favorite plants.

http://phylogeny.arizona.edu/tree/phylogeny.html

The Tree of Life is a project designed to contain information about the phylogenetic relationships of organisms, to link biological information available on the Internet in the form of a phylogenetic navigator, and to illustrate the diversity and unity of living organisms.

Suggested Readings

Articles

Doyle, J. J. 1993. DNA, phylogeny, and the flowering of plant systematics. *Bioscience* 43:380–389.

Funk, V. A., and T. F. Stuessy. 1978. Cladistics for the practicing taxonomist. *Systematic Botany* 3:159–178.

Gaffney, E. S., L. Dingus, and M. K. Smith. 1995. Why cladistics? *Natural History* 104:33–35.

Hillis, D. M., J. P. Huelsenbeck, and C. W. Cunningham. 1994. Application and accuracy of molecular phylogenies. *Science* 264:671–677.

Leedale, G. F. 1974. How many are the kingdoms of organisms? *Taxon* 23:261–270.

Schiebinger, L. 1996. The loves of the plants. *Scientific American* 274: (February) 110–115.

Wainwright, P. O., G. Hinkle, M. L. Sogin, and S. K. Stickel. 1993. Monophyletic origins of the Metazoa: An evolutionary link with fungi. *Science* 260:340–342.

Books

Crawford, D. J. 1990. *Plant Molecular Systematics: Macromolecular Approaches.* New York: John Wiley.

Gledhill, D. 1989. *The Names of Plants.* 2d ed. New York: Cambridge University Press.

Hillis, D. M., and C. Moritz, eds. 1990. *Molecular Systematics.* Sunderland, MA: Sinauer.

Hoch, D. C., and A. G. Stephenson, eds. 1995. *Experimental and Molecular Approaches to Plant Biosystematics.* St. Louis: Missouri Botanical Garden.

Jones, S. B., Jr., and A. E. Luchsinger. 1986. *Plant Systematics.* 2d ed. New York: McGraw-Hill.

Scagel, R. F., R. J. Bandoni, J. R. Maze, G. E. Rouse, W. B. Scholfield, and J. R. Stein. 1984. *Plants: An Evolutionary Survey.* Belmont, CA: Wadsworth.

Stuessy, T. F. 1990. *Plant Taxonomy: The Systematic Evaluation of Comparative Data.* New York: Columbia University Press.

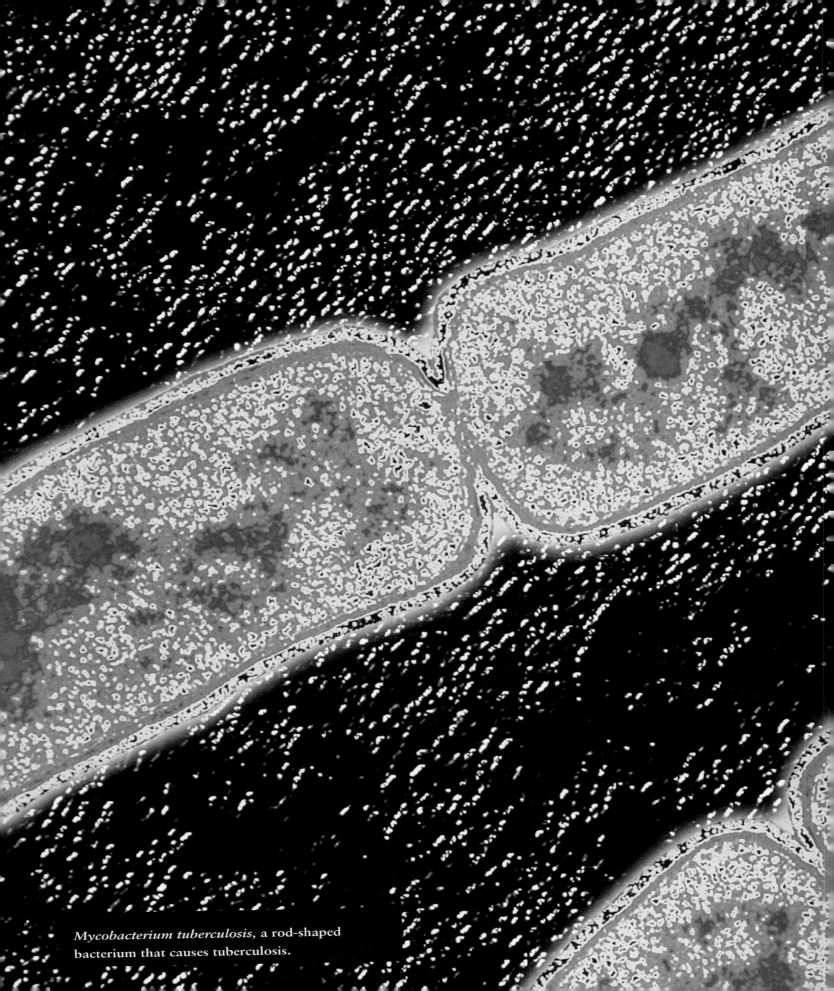

Mycobacterium tuberculosis, a rod-shaped
bacterium that causes tuberculosis.

Bacteria

25

Chapter Outline

Chapter Overview

Many kinds of organisms have posed problems for the classification and biology of plants, and, at the same time, offered insight into botanical processes. In the past bacteria and fungi have even been variously classified as members of the plant kingdom. Most scientists agree now, however, that none of these organisms share a close phylogenetic relationship with plants and should not be called plants. Nevertheless, any presentation about the diversity of plants is not complete without some mention of these organisms and their influence on plant biology. Accordingly, the next two chapters present a brief overview of bacteria and fungi.

Ideas about the biology of plants often spring from information about the biology of other organisms, especially the chlorophyll-containing bacteria, unicellular fungi, and algae. Knowledge about plants gained from other organisms includes understanding how the mitotic spindle works in a diatom, how genes work in bacteria and in yeasts, and how carbon is fixed during photosynthesis in unicellular algae. Botanists extrapolate this information to plants and, whenever possible, seek evidence that plants are similar to or different from other organisms. Accordingly, botanists think that the push and pull of spindle fibers is the same in plants as in isolated spindles of diatoms, that well-known mechanisms of gene action in fungi are the same for plants, and that the Calvin cycle for carbon fixation in green algae also occurs in plants.

Besides their importance in our study of basic plant science, other kinds of organisms are also directly involved in the growth and success of plants. The ecology, physiology, chemistry, reproductive biology, and almost all other aspects of the biology of plants are heavily influenced by symbiotic and disease-causing bacteria, fungi, and other kinds of organisms.

Bacteria

Bacteria are the most abundant organisms on earth. They live almost everywhere: at the bottom of oceans, in our mouths, on the surface of hot coals, and in puddles on glaciers. Most people know something about bacteria because some bacteria make people sick. This limited awareness of bacteria gives them an undeserved bad reputation, because most bacteria do not cause diseases. For example, Robert J. Price, a seafood technology specialist at the University of California at Davis, told the press in 1990 that about once a month he receives a report of seafood that glows in the dark. The glow usually comes from *Photobacterium phosphoreum*, a bacterium that produces light instead of heat as it respires. Such bacteria occur on the skin and in the intestines of many fish and shellfish, and they glow when salt is added to cooked seafood. These luminescent bacteria are harmless; no one has ever reported getting sick from eating seafood on which the bacteria were growing.

Harmless bacteria also live in our intestines, and we depend upon them for normal digestion. There may be two or three dozen species of bacteria in the human digestive tract. They are usually so abundant that the dry weight of normal feces can consist of up to 80% bacterial cells.

Features of Bacteria

The main features of bacteria were mentioned briefly in Chapter 24 (see table 24.2). The distinguishing feature of these organisms is that they are prokaryotic, which means that they have no membrane-bound organelles—that is, no nucleus, no mitochondria, and no chloroplasts. Metabolic reactions in the organelles of eukaryotes occur instead in folds of the plasma membrane in bacteria. Also, because bacterial DNA is not associated with histone proteins in chromosomes, the "chromosomes" of bacteria are sometimes referred to as *genophores* to distinguish them from true chromosomes. Multicellular forms of bacteria are rare except among the cyanobacteria. Cell walls, which occur in most bacteria, differ markedly from those of eukaryotes. Indeed, the cell walls of bacteria are usually made of **peptidoglycans,** in which carbohydrate polymers are interconnected with short chains of amino acids. Contrary to those of eukaryotes, most bacterial cell walls contain muramic acid. Different bacteria may also have various layers outside their walls and may have flagella or shorter, thinner filaments called *pili*. Genetic material can move from donor to recipient through pili, but reproduction is asexual by means of fission or budding. Many bacteria are strictly anaerobic and are killed by oxygen. Others can live anaerobically or aerobically, and still others require oxygen for respiration.

Although most bacteria are 1–5 μm in diameter (much smaller than plant cells), bacteria known as mycoplasmas are often only 0.2 μm across (fig. 25.1a).[1] At the other extreme, some cyanobacteria are up to 3 mm long (fig. 25.1b), and the recently discovered *Epulopiscium fishelsonii*—the largest known nonphotosynthetic bacterium—is as large as a small hyphen (0.3 mm long) (fig. 25.1c). This should help you appreciate the size of a typical bacterium: it would take about 9 trillion average-sized bacteria to fill a box the size of a 5-stick package of chewing gum.

Bacteria are mostly simple unicells, but there are some exceptions to such simple organization. For example, some species have cells that cohere in chains; similarly, filamentous forms occur among the cyanobacteria, others have flagella and live as motile individuals or colonies, and a group called the **myxobacteria** often form upright, multicellular reproductive bodies (fig. 25.1d). These are thought to be the most complex forms among bacteria.

Several groups of bacteria contain chlorophyll or other photosynthetic pigments and are capable of either plantlike photosynthesis or photosynthesis that is unique to bacteria.

---

1. Mycoplasmas are the smallest free-living microorganisms. Unlike other bacteria, mycoplasmas lack a cell wall and are too small to be seen with a light microscope. They cause a variety of conditions ranging from kidney stones to premature labor. Mycoplasmas also cause several plant diseases that are associated with symptoms of yellowing and stunting (e.g., citrus stubborn disease).

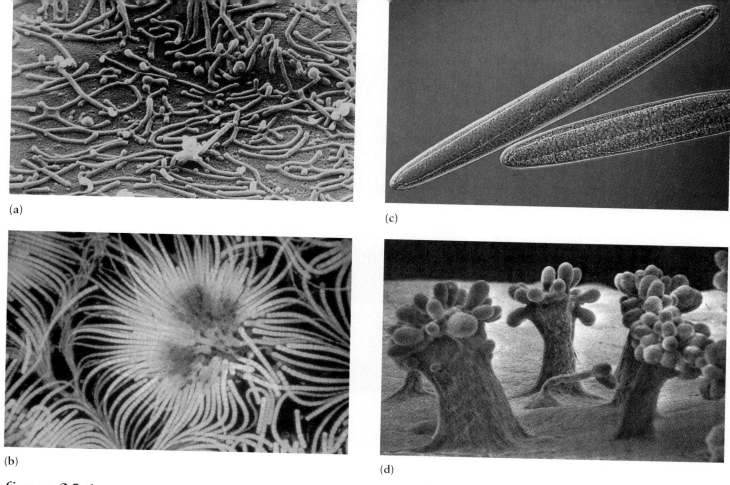

(a)

(c)

(b)

(d)

figure 25.1

Mycoplasmas and bacteria have various shapes, but their biochemical and metabolic diversity far surpasses their morphological variation. (a) *Pneumonia mycoplasma* bacteria, ×62,000. (b) Strands of the cyanobacterium *Gloeotrichia echinulata*, ×63. (c) *Epulopiscium fishelsonii*, a very large bacterium discovered in the guts of surgeonfish. Each cell is about 0.3 mm long. (d) *Chondromyces crocatus*, a myxobacterium.

Many aquatic bacteria have a membranous structure, called a **gas vacuole,** in each of their cells. As the name suggests, these vacuoles are filled with gas and enable the bacteria to float.

The distinguishing features of different bacteria are their shapes, metabolism, and chemical composition. Bacterial cells typically have one of three shapes (fig. 25.2): spherical (also called cocci), rod-shaped (bacilli), or corkscrew-shaped (spirilla). Bacteria can be further distinguished by a diagnostic test called the **Gram stain,** named after Danish physician H. C. Gram. This stain distinguishes two bacterial groups: Gram-positive bacteria and Gram-negative bacteria (fig. 25.3). **Gram-positive** bacteria take up a purple stain in their thick, peptidoglycan-containing cell walls. **Gram-negative** bacteria cannot retain the stain, either because they have thin cell walls surrounded by lipids or because they have no cell walls at all. Notably, Gram-negative pathogens are generally more virulent because their lipid coating is often toxic, protects the bacterium from host defenses, and resists penetration by antibiotics.

Bacteria are far more diverse metabolically than all of the eukaryotes. This diversity is reflected in the range of energy sources that bacteria can use. Different bacteria can use sulfide, iron, methane, or carbon dioxide in energy metabolism. Moreover, many bacteria can fix nitrogen, which means they can convert atmospheric nitrogen (N_2) into metabolically useful forms (e.g., ammonia). Some bacteria such as *Nitrosomonas* are nitrifying; they convert ammonia into nitrite (NO_2^-). Still other genera such as *Nitrobacter* oxidize nitrite ions, which are toxic to plants, to nitrate (NO_3^-), an important nutrient. Cyanobacteria have both photosystems I and II, although their only chlorophyll is chlorophyll *a*. Purple sulfur bacteria and green sulfur bacteria have a different type of chlorophyll, called **bacteriochlorophyll,** and a photosynthetic pathway that is analogous but not identical to photosystem I. Bacteria can also ferment a variety of sugars and organic acids, thereby producing important commercial products such as ethanol, methanol, acetone, lactic acid, acetic acid, and propionic acid. Bacteria have also been used to clean up oil spills because they can digest many of the chemicals in petroleum oil. Still other bacteria can detoxify polychlorinated biphenyls (PCBs), a class of human-made pollutants that has become a worldwide environmental problem.

In a rather extreme example of metabolic diversity, two researchers from the Pacific Northwest Laboratory in

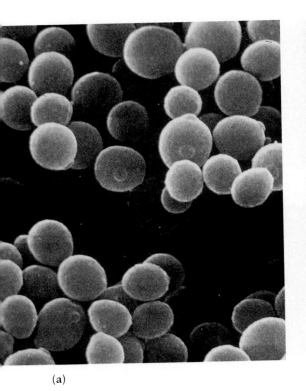

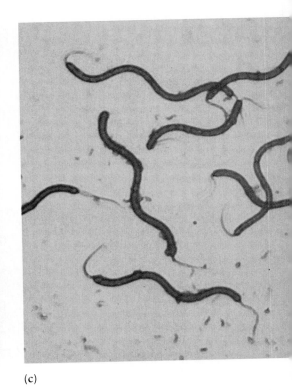

(a)   (b)   (c)

figure 25.2

The three basic shapes of bacteria. (a) Spherical (*Micrococcus*), ×40,000. (b) Rod (*Bacillus*), ×60,000. (c) Corkscrew (*Spirilla*), ×450.

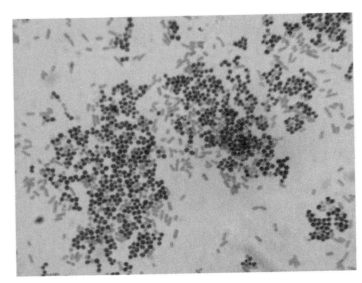

figure 25.3

The Gram stain. Photomicrograph of bacteria that are Gram positive (purple) and Gram negative (red), ×700. The species are *Staphylococcus aureus* (+) and *Escherichia coli* (−).

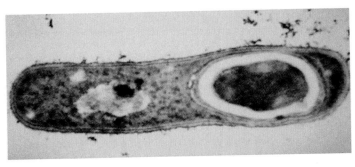

figure 25.4

Electron micrograph of *Bacillus subtilis* bacterium showing an endospore, ×49,000.

Richland, Washington, reported in 1995 that they've dipped more than a thousand meters into basalt rocks near the Columbia River and found bacteria that seem to get along fine without either oxygen or organic molecules. These bacteria apparently live on rocks alone! The microbes appear to get energy from hydrogen generated in a reaction between iron-rich minerals in basalt, a volcanic rock, and groundwater.

Some bacteria form special, durable **endospores** (fig. 25.4). These spores, which are relatively thick-walled, enable bacteria to survive adverse conditions. Endospores, including those produced by the notorious botulism bacteria (*Clostridium botulinum*), are exceptionally resistant to heat: some can survive boiling water for over an hour, while others can survive desiccation, sunlight, a vacuum, and various chemicals (see reading 25.1). Many may remain dormant for hundreds of years.

DEADLY FOOD! BEWARE!

The bacterium *Clostridium botulinum* can grow in food products and produces a toxin called botulinum, the most toxic substance known. Microbiologists estimate that 1 g of this toxin can kill 14 million adults! The good news is that *C. botulinum* requires anaerobic conditions for growth, which limits its prevalence. The bad news is that *C. botulinum* is extremely tolerant to stress; it can withstand boiling water (100° C) for short periods but is killed at 120° C in 5 min. This tolerance makes *C. botulinum* a serious concern for people who can vegetables. If home canning is not done properly, this bacterium will grow in the anaerobic conditions of the sealed container and be extremely poisonous. Several adults and infants die every year from botulism in the United States.

Tolerance to stress is enhanced in *C. botulinum* and many other bacteria by the formation of thick-walled endospores that surround their chromosome and a small portion of the surrounding cytoplasm (reading

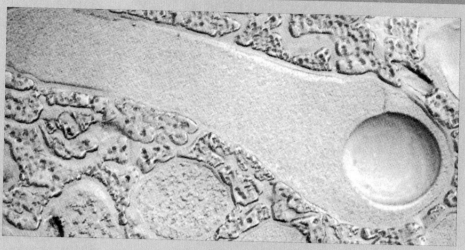

reading figure 25.1

Endospores. The round circle at the lower right is an endospore forming within a cell of *Clostridium botulinum*, the bacterium that causes botulism. These resistant endospores enable the bacterium to survive in improperly sterilized canned and bottled foods.

fig. 25.1). These highly resistant endospores may later germinate and grow after decades or even centuries of inactivity. The endospores of *C. botulinum* can germinate in poorly prepared canned goods, so never eat food from a swollen (gas-filled) can of food; you risk contracting botulism leading to nerve paralysis, severe vomiting, and death.

c o n c e p t

Bacteria are metabolically the most diverse of all groups of organisms. They use a variety of chemicals as sources of metabolic energy, and they have different kinds of photosynthesis.

The Generalized Life Cycle of Bacteria

Bacteria reproduce mainly by **fission,** which occurs when a cell reaches a certain size and then pinches in half to form two cells (fig. 25.5). During fission, the genophore attaches to the plasma membrane, where it replicates. Cell division occurs between the two genophores, thereby ensuring that each new cell gets an identical molecule of DNA.

Because bacteria have no nuclei, they cannot undergo meiosis and fertilization. However, genetic material can move from one cell to another through a **conjugation pilus** (fig. 25.6). Genetic material that moves in this fashion usually involves plasmid DNA (see Chapter 11) but not the genophore.

Two other types of genetic transfer also occur in bacteria. One is **transformation,** which occurs when free DNA is taken up by cells, after which the genes of the absorbed DNA are expressed in their new host. The second type is **transduction,**

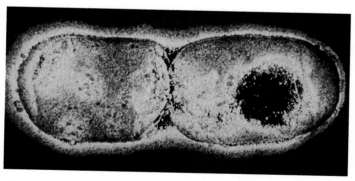

figure 25.5

Electron micrograph showing fission of *Escherichia coli*, a common bacterium. Fission is a form of asexual reproduction.

in which viruses enter a bacterial cell, and any DNA from the previous bacterial host of the virus can replace the DNA of the new host cell. Both of these processes have immense potential for genetic engineering because they can be manipulated in the laboratory (see Chapter 11).

In spite of the absence of sexuality and a consequent lack of genetic recombination from sexual reproduction, bacteria can evolve quickly. About 5,000 genes are present in the DNA of a typical bacterium. Mutations occur randomly approximately

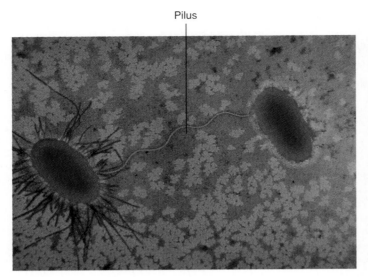

Pilus

figure 25.6

Electron micrograph of bacterial cells (*E. coli*) with a conjugation pilus. Genetic material, especially plasmids, moves through the pilus.

once in every 200,000 genes. This means that about one out of every 40 bacteria in a given population may spontaneously develop a mutant characteristic. At this rate, as many as 50 million mutant bacteria could be present among the more than 2 billion bacteria inhabiting a single gram of garden soil. This high rate of mutation in bacteria is important to the medical industry, because random mutations often enable bacteria to become resistant to antibiotics and disinfectants. Such drug-resistant bacteria often spread when susceptible strains are eliminated by the overuse of antibacterial agents.

c o n c e p t

Bacteria reproduce only asexually. However, genetic exchange can occur between a donor and a recipient. New genotypes of bacteria arise by absorbing DNA into their plasmid genomes, genetic transfer from viruses, and high rates of mutation.

Where Bacteria Grow

Bacteria are everywhere. They grow in such extreme environmental conditions that they must be considered the most hardy of all organisms. They live in the intestines of all kinds of animals, in soil, in clouds, in water, on airborne dust particles, and in smog. Many can tolerate hot acids, others can survive temperatures below freezing for years, and still others thrive in boiling hot springs. Their tolerance of high temperature requires methods of sterilization that include a combination of high heat (120° C) and high pressure. However, there are deep-sea bacteria that live near volcanic vents, where the temperature approaches 360° C and the pressure can reach

figure 25.7

Root nodules containing nitrogen-fixing bacteria are common in legumes such as peas and soybeans.

more than 26 megapascals (ca. 260 atmospheres). For comparison, the water pressure in home plumbing is usually 0.2 to 0.3 megapascals.

Bacterial Ecology

Bacteria play an important role in every habitat on earth. Photosynthetic bacteria help maintain the global carbon balance, and nitrogen-fixing bacteria account for about a quarter of the total nitrogen fixed in oceans. Some of the most important bacteria in agriculture live and fix nitrogen in the root nodules of crops such as alfalfa, soybean, and pea (fig. 25.7; also see Chapter 20).

Saprobic bacteria—those that obtain nourishment from dead organic matter—are partially responsible for decomposing and recycling organic material in the soil. Decomposition by

table 25.1

Proposed Higher Taxa for Bacteria, According to *Bergey's Manual* of Systematic Bacteriology

Kingdom Prokaryotae [Monera]

    Division I. *Gracilicutes* (Gram-negative, typical bacteria)

        Class I. *Scotobacteria* (nonphotosynthetic bacteria)

        Class II. *Anoxyphotobacteria* (nonoxygenic photosynthetic bacteria)

        Class III. *Oxyphotobacteria* (oxygenic photosynthetic bacteria)

    Division II. *Firmicutes* (Gram-positive, typical bacteria)

        Class I. *Firmibacteria* (simple, Gram-positive bacteria)

        Class II. *Thallobacteria* (branching, filamentous bacteria)

    Division III. *Tenericutes* (bacteria lacking a cell wall)

        Class I. *Mollicutes*

    Division IV. *Mendosicutes* (bacteria with defective cell walls or lacking peptidoglycan)

        Class I. *Archaebacteria*

From *Bergey's Manual of Systematic Bacteriology*. Copyright © 1989 Williams & Wilkins. Reprinted by permission.

saprobic bacteria and their cohorts, saprobic fungi, produces as much as 90% of the biologically made CO_2 in the atmosphere. Without these organisms, the earth's surface would be covered by dead plants and animals in a matter of weeks. So much carbon would be tied up in organic matter that less atmospheric CO_2 would be available, which would limit photosynthesis by plants.

The Diversity and Relationships of Bacteria

Taxonomic disputes abound regarding Kingdom Monera. Different biologists estimate the number of bacterial species to be anywhere from 2,500 to 10,000. The most widely used manual for identifying bacteria, *Bergey's Manual of Systematic Bacteriology*, divides them into four divisions and seven classes (table 25.1), but this arrangement is artificial because it classifies groups based on a few similarities that may not be derived from a unique common ancestor. For example, Class Anoxyphotobacteria includes all bacteria that photosynthesize without producing oxygen. The likelihood that nonoxygenic photosynthesis fails to indicate relatedness is supported by a phylogeny of bacteria based on sequence analyses of ribosomal RNA (fig. 25.8a). That analysis suggests that nonoxygenic photosynthetic bacteria have widely scattered phylogenetic relationships among nonphotosynthetic bacteria. This means that groups such as the purple bacteria are more closely related to nonphotosynthetic bacteria such as *Escherichia coli* (colon bacterium) and *Rickettsia typhi* (typhus pathogen) than they are to other photosynthetic bacteria.

According to Carl Woese at the University of Illinois, who is one of the leading proponents of using molecular

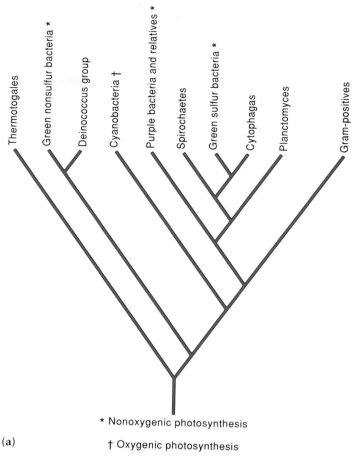

* Nonoxygenic photosynthesis

† Oxygenic photosynthesis

(a)

figure 25.8

(a) A phylogeny for Eubacteria based on similarities of ribosomal RNA sequences. Such similarities of sequences reliably indicate common ancestry.

phylogenetics for classifying prokaryotes, traditional classifications of bacteria like that in *Bergey's Manual* rely heavily on characters that are almost useless for determining phylogenetic relatedness. Nevertheless, traditional views and molecular phylogenies do agree on the classification of prokaryotes into two main groups, the **Eubacteria** (*eu* = true) and the **Archaebacteria** (*archae* = ancient). (Molecular systematists prefer to call them **Bacteria** and **Archaea**, respectively.) These two groups are often classified as subkingdoms of the Monera, although there is growing sentiment to separate the Eubacteria and Archaebacteria into their own kingdoms. Microbiologists in particular favor recognition of three domains for all of life: Archaebacteria, Eubacteria, and Eukaryota (fig. 25.8b).

The most convincing evidence for Archaebacteria being a separate taxonomic group comes from an analysis of the entire genome of *Methanococcus jannaschii*, a methane-producing archaebacterium that lives in near-boiling water near thermal vents at the bottom of the Pacific Ocean. The results of that analysis,

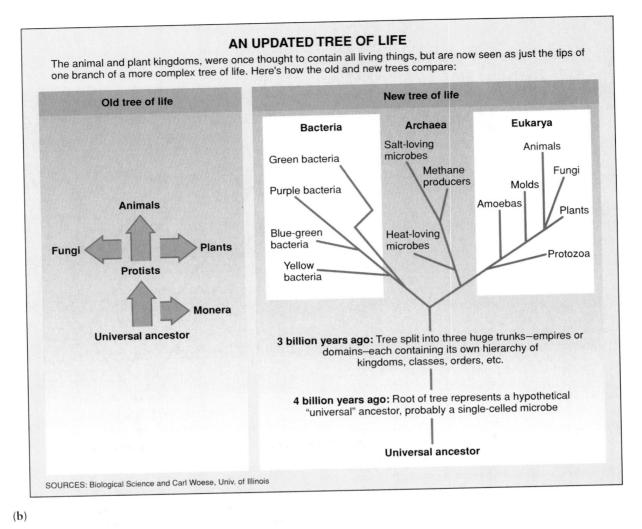

(b)

figure 25.8—(continued)

(b) A "tree of life" favored by a growing number of biologists. This classification is based primarily on analyses of the organisms' DNA.

published in 1996, were surprising: 56% of the archaebacterium's 1,738 genes are new to science, unlike any found in Bacteria or Eukaryotes. Genes related to energy production, cell division, and metabolism resemble those of Bacteria, whereas those for transcription, translation, and replication resemble those of Eukaryotes. Taken together, these data suggest that Archaebacteria and Eukaryotes share a common heritage that is independent of that of Bacteria. That is, Archaebacteria, which diverged from bacteria more than 3 billion years ago, may be more phylogenetically distant from Eubacteria than from eukaryotes.

Archaebacteria live in the harshest environments, including hot springs, salty lakes, and oxygen-starved swamps. However, they are also abundant in normal habitats. Interestingly, the waters of Antarctica are a surprisingly rich source of archaebacteria. Edward F. DeLong and his colleagues at the University of California reported in a 1994 issue of *Nature* that archaebacteria constitute 34% of the prokaryotic biomass in coastal Antarctic surface water. Archaebacteria are not just biological oddities; rather, they represent a major component of the ocean's biota.

Distinguishing features of the Eubacteria and Archaebacteria are still being discovered, but at present the Archaebacteria differ from the Eubacteria in the following ways:

1. The major lipids of Archaebacteria are ether-linked instead of ester-linked, as in the triglycerides of Eubacteria and eukaryotes.

2. Archaebacteria have cell walls of glycoproteins and polysaccharides but lack peptidoglycans.

Archaebacteria   Eubacteria   Eukaryotes

figure 25.9

The different shapes of the 30S subunit of ribosomes in Eubacteria, Archaebacteria, and eukaryotes provide evidence of the fundamental differences among these groups of organisms.

3. Archaebacteria have a single type of DNA-dependent RNA polymerase, which is more similar to the RNA polymerases of eukaryotes than to those of Eubacteria.

4. Archaebacterial ribosomes have a distinctive shape (fig. 25.9).

5. Archaebacteria can metabolize methane, grow at high temperatures, or grow in high concentrations of salt.

6. Like eukaryotic genes, archaebacterial genes may contain introns, which are absent in eubacterial genes.

Even a brief summary of examples from all groups of bacteria is beyond the scope of this textbook, but two groups are of special interest here because they photosynthesize much as plants do. These groups are the Cyanobacteria and the Chloroxybacteria. They are classified together in Class Oxyphotobacteria in *Bergey's Manual*, but many botanists view them as separate classes or divisions.

Cyanobacteria

The cyanobacteria are named for the blue-green pigment **phycocyanin**. Phycocyanin, along with chlorophyll *a*, gives some cyanobacteria a blue-green appearance, which led to their being called *blue-green algae* before the Kingdom Monera was recognized. Phycocyanin and its red counterpart, **phycoerythrin**, are proteins called **phycobilins**. These proteins, which occur only in the cyanobacteria and red algae, function as photosynthetic pigments in photosystem II. This contrasts with photosystem II in plants, which relies on chlorophyll *b*.

Cyanobacteria live in a wide variety of habitats, including snowfields; frozen lakes of Antarctica; extremely acidic, basic, salty, or pure (i.e., nutrient-poor) water; deserts; inside rocks; and hot springs where water temperatures approach 85° C. In Yellowstone National Park, a different species of cyanobacterium lives in each temperature range of hot springs. In such habitats the cyanobacteria precipitate chalky,

carbonate deposits, which become a rocklike substance called travertine. The travertine is often marked with brilliantly colored streaks of cyanobacteria (fig. 25.10). Cyanobacteria are usually the first photosynthetic organisms to appear on bare lava after a volcanic eruption. They also live on the shells of turtles and snails, as symbionts in lichens (which are discussed later in this chapter), and in protozoans, amoebas, aquatic ferns, the roots of tropical plants, sea anemones, and a variety of other hosts. Cyanobacteria that live symbiotically in other organisms often lack a cell wall and function essentially as chloroplasts inside their hosts.

The more than 1,500 known species of cyanobacteria are diverse in form (fig. 25.11); they range from single-celled spheres to filaments and colonies. Many of the cyanobacteria have a gelatinous sheath around their cells, which protects them and binds trace elements for metabolism. In some parts of the open ocean, cyanobacteria sometimes reach densities of 10 million cells per liter. By themselves, cyanobacteria account for 20% of the primary production in the seas and provide fodder for the rest of the marine food chain.

If the substrate in which they are growing is low in nitrogen, most cyanobacteria produce **heterocysts,** which are larger, thicker-walled cells that can fix nitrogen (fig. 25.12). Cyanobacteria may also produce thick-walled **akinetes** that resist desiccation and freezing. Akinetes can lie dormant during long, dry periods and then germinate to form a new chain of cells when water becomes available.

Cyanobacteria reproduce by fission, but they can also reproduce by fragmentation of filaments, budding, or multiple fission. Budding begins when a large swelling forms on a cell and enlarges until it becomes a mature cell. Multiple fission occurs when a cell enlarges and its contents divide several times. The wall of the original cell then breaks down and the smaller cells are released.

Cyanobacteria lack flagella, but filamentous forms like *Oscillatoria* exhibit curious gliding movements. Adjacent filaments glide up and down while touching each other, or they appear to rotate on their axes. These movements are apparently caused by the twisting of fibrils in the cell walls. *Synechococcus*, a nonfilamentous genus, includes marine forms whose cells can move at speeds of up to 25 μm (about the diameter of a pollen grain) per second. The mechanism responsible for this movement is unknown.

Chloroxybacteria

In 1976 Ralph A. Lewin of the Scripps Institute of Oceanography announced the discovery of a unicellular, prokaryotic organism living on or in sea squirts in the intertidal region of Baja California, Mexico. This prokaryote, although similar to cyanobacteria in structure and chemistry, possesses chlorophylls *a* and *b*, like higher plants, yet lacks phycobilins. Lewin named the new organism *Prochloron*. In 1984 a group of

figure 25.10

A hot spring at Yellowstone National Park is brilliantly colored by cyanobacteria. The smooth lining of the spring is a travertine layer (carbonate) precipitated by cyanobacteria that tolerate high temperatures.

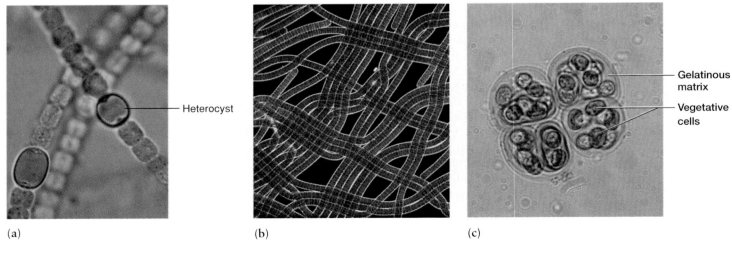

(a) (b) (c)

figure 25.11

Light micrographs of cyanobacteria. (a) *Anabaena*, ×850. (b) *Oscillatoria*, ×400. (c) *Gleocapsa*, ×300.

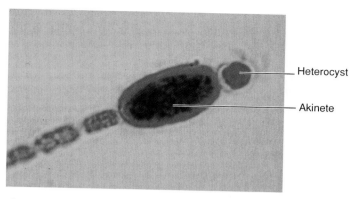

Heterocyst

Akinete

figure 25.12

Light micrograph of *Cylindrospermum* with a terminal heterocyst that fixes nitrogen and subterminal akinete that resists drying and freezing, ×500.

Dutch scientists discovered a similar organism in the shallow Loosdrecht lakes in the Netherlands. This organism, which they named *Prochlorothrix*, also has chlorophylls *a* and *b* and lacks phycobilins but differs from *Prochloron* in being filamentous and free-living. *Prochloron* and *Prochlorothrix* are now classified together, sometimes as cyanobacteria and sometimes as the only two members of their own class or division, the Chloroxybacteria.

The Lore of Plants

Some strains of *Pseudomonas aeruginosa* infect plants, causing "soft rot," a disease that turns leaves slimy and necrotic. The same bacteria cause fatal hospital-acquired infections, especially in burn patients with depressed immune systems. Surprisingly, humans and plants may suffer not just from the same organisms but also from products of the same virulence genes in bacteria that trigger disease. This discovery is particularly promising because it presents a new way to study the virulence genes of human pathogens. Plants are an easily cultured, low-cost model to replace the huge numbers of animals needed to screen for such genes and to reveal new virulence factors in animal as well as plant infections. Work with these plant models will provide a better understanding of how the bacteria cause disease and could lead to the discovery of mechanisms for new antibacterial drugs.

The chloroxybacteria have paired thylakoid membranes, unlike the unstacked membrane of cyanobacteria but like the stacked thylakoids of green algae and plants. Their stacked thylakoids, their use of chlorophyll *b* in photosystem II, and their lack of phycobilins have led scientists to speculate that chloroxybacteria descended from the same ancestral prokaryotes as the chloroplasts of plants.

concept

Biologists classify bacteria into two major groups, the Eubacteria and the Archaebacteria. The Eubacteria include the Cyanobacteria, which use chlorophyll *a* and phycobilins for photosynthesis, and the Chloroxybacteria, which resemble plants in their use of chlorophylls *a* and *b* for photosynthesis. The Archaebacteria are so distinctive that some biologists classify them as a separate kingdom. Archaebacteria seem to be more closely related to the eukaryotes than to the Eubacteria.

The Origin and Evolution of Bacteria

Bacteria-like cells were probably the first living organisms. Fossil evidence for bacteria goes back about 3.4 billion years, which is much older than the oldest fossil evidence for eukaryotes (1.5 billion years). Moreover, cyanobacteria-like fossils lived at least 2 billion years ago, which is evidence that oxygen-producing photosynthesis might be that old. Scientists believe that before the appearance of cyanobacteria, prokaryotes lived in an atmosphere that was probably no more than 1% oxygen. Once the cyanobacteria appeared, they began to transform the air into an oxygen-rich atmosphere that may have reached as much as 30% oxygen (the present level is about 21% oxygen). This transformation set the evolutionary stage for the diversification of aerobic bacteria and the origin of oxygen-respiring eukaryotes.

Ideas about the origin of bacteria are essentially ideas about the origin of life, since bacteria seem to be the oldest form of life that is still living. Scientists have long speculated that prebacteria arose by the spontaneous aggregation of simple molecules into polymers. Recent discoveries about the catalytic properties of RNA have led to the idea that prebacteria were probably regulated by small RNA molecules (see Chapter 11). According to this idea, the first bacteria evolved from RNA-based ancestors. All extant organisms, however, are DNA-based, so molecular phylogenies comparing the RNA-based relatives of modern-day bacteria cannot be made.

The Economic Importance of Bacteria

Bacteria affect every aspect of our lives. They are the pivotal organisms in many diseases of plants and animals, including humans, and they are the microscopic laborers in several multimillion-dollar industries. Thus, we can think of the economic importance of bacteria as both positive and negative. The following paragraphs discuss a few examples of each.

The Positive Aspects

The cyanobacterium *Spirulina* is cultivated for human consumption. This organism, which is common in saline lakes, has a dry-weight protein content of about 70%. *Spirulina* is harvested in Mexico and is commercially cultivated for human consumption in both Mexico and Israel. In the United States, *Spirulina* is sold as a nutritional food supplement in most health food stores.

The genus *Bacillus* includes many species important to humans. For example, three species have been approved by the U.S. Department of Agriculture for biological control of pests. *Bacillus thuringiensis* (BT) reproduces only in the intestinal tracts of caterpillars, which are killed by a toxin from the bacterium. This toxin is harmless to all other living organisms. BT is being used to control more than 100 species of plant pests, including tomato hornworms and corn borers. A variety of this species (var. *israelensis*) is being used to control mosquito larvae, and another variety (var. *nigeriensis*), which is even more effective as a mosquito control, may soon be approved for use in the United States.

The discovery in 1994 of a mutant of *Bacillus stearothermophilus* may eventually enable us to ferment ethanol from agricultural wastes as cheaply as we refine gasoline from crude oil. The mutant was discovered accidentally (the researchers suspect that a spore floated into the lab through a window) during a study of rapid anaerobic growth by bacteria at high temperatures. To everyone's surprise, the bacterium fermented wheat, corn, and sugar beet wastes at 75° C, producing ethanol in the process. The mutant has a big advantage over yeast (the most commonly used fermenting agent) because it can digest hemicelluloses that yeast ignores. This could dramatically increase the proportion of plant waste that can be made into fuel. In Brazil, where more than two-thirds of cars are powered by ethanol from sugarcane, only the juice of the plants is converted into fuel. If the leftover hemicelluloses from sugarcane were also fermented, the yield of ethanol would be more than doubled.

Several species of *Lactobacillus* are mainstays of the dairy, beverage, and baking industries. Lactobacilli are used to make acidophilus milk, kefir, cheeses, yogurt, and related foods. These bacteria are also used in the production of wine, beer, sourdough bread, sauerkraut, and many other commercial products.

Antibiotic chemicals from actinomycete bacteria (i.e., nonphotosynthetic bacteria with branched filaments) account for about two-thirds of the more than 4,000 antibiotics that have been discovered. Actinomycetes were the original sources of such well-known antibiotics as tetracycline, neomycin, Aureomycin, erythromycin, and streptomycin. Streptomycin is produced by *Streptomyces*, which is common in soil and is the largest genus of actinomycetes. Also, in the continuing struggle to produce effective antibiotics, research in the laboratory of Gary Strobel at Montana State University has lead from the study of Dutch elm disease to emphasize pseudomycins, a fungal-killing compound produced by bacteria. Development of antimycotics has not kept up with recent increases in fungal diseases, especially in immuno-compromised humans.

The various species of *Streptomyces* play a significant role in breaking down and recycling plant and animal products such as lignin, latex, and chitin. They also produce the characteristic odor of damp soil. *Frankia* species are nitrogen-fixing actinomycetes that, like other kinds of nitrogen-fixers, form nodules on the roots of legumes and other higher plants. *Thermoactinomyces* is common and active in compost piles, haystacks, silos, and sod, where it thrives at temperatures between 45° C and 60° C.

The Negative Aspects

In water supplies, cyanobacteria frequently clog filters, corrode steel and concrete, soften water, and produce undesirable odors or coloration in the water. Cyanobacteria also produce toxins. Fish that are immune to the toxins produced by certain cyanobacteria become toxic to their predators after eating these bacteria. In summer months, various cyanobacteria and algae often become abundant in bodies of freshwater, especially if the water is polluted. When this happens, a floating mat called a *bloom* may form. Such blooms can cover more than 2,000 km². Domestic animals and fowl have occasionally died as a result of drinking water that contains such a bloom. While the cyanobacteria of a bloom are alive, the oxygen content of the surrounding water may temporarily increase. When they die, however, the other bacteria that decompose the remains may so deplete the oxygen that fish and other aquatic organisms in the vicinity die from lack of oxygen. Many communities control cyanobacteria in reservoirs with dilute concentrations of copper sulfate.

Some bacilli cause diseases such as diphtheria, periodontal disease in humans, and anthrax in cattle. The most dangerous of the bacilluslike bacteria may be the clostridia: species of *Clostridium* cause tetanus, gas gangrene, and botulism (see reading 25.1, "Deadly Food! Beware!").

Viruses

Viruses are parasites that cannot replicate on their own outside of their host, and they completely lack cellular structure. Viruses, therefore, are often referred to as *particles* instead of organisms. In addition, viruses do not grow by increasing in size or by dividing, nor do they respond to external stimuli. They have few (if any) enzymes, they cannot carry on independent metabolism, and they cannot move on their own. Nevertheless, viruses are incredibly numerous. In 1989 marine biologists at the University of Bergen in Norway discovered that a single teaspoon of seawater typically contains more than 1 billion viruses. Viruses have varying sizes. A small virus is about 17 nm long; that's about 1/1,000th the size of a pore in a hen's egg.

A ROGUES' GALLERY OF BACTERIA

Yersinia pestis—causes bubonic plague (The Black Death), a disease that killed about 70% of Europe's population during the years 1347–1425. In its heyday, *Yersinia pestis* more than earned its place as one of the Four Horsemen of the Apocalypse.

Vibrio cholerae—causes cholera, a disease that has killed millions of people throughout history. Florence Nightingale earned her reputation while treating cholera victims.

Helicobacter pylori—infects the stomach, where it causes ulcers and cancer. About 40% of the world's population harbors *Helicobacter pylori*. Barry Marshall, who isolated the bacterium in 1982, drank a culture of *Helicobacter pylori;* he got an ulcer.

Corynebacterium diphtheria—causes diphtheria, the disease that killed George Washington and thousands of others.

Salmonella typhi—causes typhoid ("stuporlike") fever.

Bacillus anthracis—causes anthrax, the "very severe plague" on the Pharaoh's cattle (Exodus 9). This bacterium was used by Pasteur to develop the first successful vaccine, and remains a top candidate for use in biological weapons. Soldiers who fought in the Gulf War were immunized against this bacterium.

Bordetella pertussis—causes whooping cough, a disease that affects 4 million and kills 400,000 people per year.

Borrelia burgdorferi—causes Lyme disease, a common ailment caused by insect bites in the United States and Europe.

Neisseria gonorrhoeae—causes gonorrhea.

Treponema pallidum—causes syphilis.

Campylobacter—causes 10% of all cases of diarrhea worldwide.

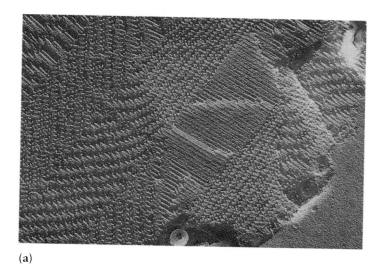

(a)

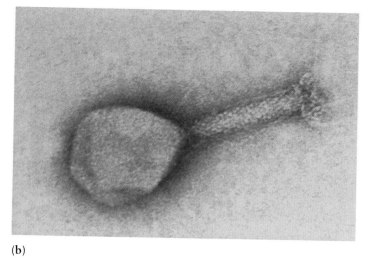

(b)

figure 25.13

Examples of viruses. (a) Tobacco mosaic viruses, ×95,000. (b) A bacteriophage, ×85,000.

Viruses contain DNA or RNA (but not both), which is either single-stranded or double-stranded. The DNA or RNA is surrounded by a protein coat that may be attached to more complicated structures (fig. 25.13). With few exceptions, plant viruses have only RNA and a simple coat made of one or a few kinds of proteins. Tobacco mosaic virus, for example, has a core of RNA that is surrounded by 2,200 copies of the same protein. The genome has four genes, one for the coat protein, two for replicase enzymes, and the fourth for a protein that probably enables the virus to spread from cell to cell in the plant.

(a)

(b)   (c)

figure 25.14

(a) A rice plant with symptoms of infection by rice dwarf virus (left). The plant on the right is not infected with the virus. (b) An ordinary tobacco leaf that has been ravaged by bollworms; however, a virus can protect a plant. (c) This plant leaf, armed with the HaSV virus—a virus that kills bollworms—fares better. A dead bollworm is circled.

Viruses are nonliving, parasitic particles that consist of nucleic acids surrounded by a protein coat.

Viruses have been classified either according to their hosts and the types of tissues or organs they affect, or according to their type of nucleic acid (DNA or RNA), their size and shape, the nature of their protein coats, the number of nucleic acid molecules in their cores, and the type of host they can invade. In spite of the widely held views that viruses are not organisms, animal and bacterial virologists have agreed to use a classification system that is modeled after that of Linnaeus. In this system, viruses are grouped into families and genera, which are comprised of serotypes instead of species. A serotype is a protein that is a unique antigen; it induces and binds to antibodies that are specific to it alone. For example, the common cold is caused by as many as 113 serotypes of the genus *Rhinovirus,* which has an additional two serotypes that invade cows. The genus *Rhinovirus* is classified in the family Picornaviridae, which also includes polio and hepatitis-A viruses (serotypes of *Enterovirus*) and foot-and-mouth disease viruses (serotypes of *Aphthovirus*).

In opposition to the preceding system, plant virologists have so far refused to classify plant viruses in a Linnaean-type system. Plant viruses are classified in groups based on the same kinds of features that are used to classify other viruses, but the names of plant viruses are usually derived from a prototype for the group. For example, viruses that are similar to the tobacco mosaic virus are members of the **tobamo** group. As you might expect, the opposing taxonomic philosophies of plant virologists and other virologists can lead to confusion. Many plant viruses, for example, are similar enough to animal viruses to be classified side by side in the same families or genera. Rice dwarf virus, for instance, resembles animal viruses in the family Reoviridae (fig. 25.14a). Animal virologists think it should be called *Phytoreovirus* (*phyto* to indicate a plant virus), which is how it appears in some virology textbooks. To confuse things further, alternate hosts for *Phytoreovirus* include leafhoppers, which are insects.

The classification of viruses is based mostly on composition, size, and shape. Viral taxonomy differs among plant virologists and other virologists.

Although viruses have a deservedly negative reputation, they can also be put to good use in controlling pests and in making vaccines and other medically important products. For example, in 1995 a group of Australian botanists armed a tobacco plant with the HaSV virus; that virus infects and kills the cotton bollworm, one of the five most economically destructive insect pests in the United States. The bioengineered plants produce HaSV particles that are harmless to the plant, but deadly to munching bollworms. In tests, the plants armed with the virus remained unharmed by bollworms while normal plants were ravaged (fig. 25.14b, c).

Botanists have used viruses to control other pests. Pest-control viruses, for example, include **Abby,** which is found only in the caterpillars of gypsy moths. Abby is one of 19 strains of a virus that forms particles in polyhedral crystals. These viruses are characterized by double-stranded DNA cores surrounded by protein coats that protect the viruses against chemicals, heat, and low pH. Gypsy moth caterpillars have been particularly destructive in forests of both North America and Europe, and it is hoped that this destruction may be greatly reduced by the dissemination of Abby in the forests under attack. Related viruses are receiving considerable attention as potential biological control agents for several other insect pests. For example, the U.S. Environmental Protection Agency has approved the use of viruses to control alfalfa looper, cotton bollworm, European pine sawfly, Douglas fir tussock moth, and other pests.

The Origin and Evolution of Viruses

There are no fossil remains of the oldest viruses. However, clues about the possible origin of viruses come from the similarity between viral genomes and their host genomes. For example, DNA sequences of several hundred nucleotides from cancer-causing viruses are almost the same as host DNA sequences. This could mean that viruses arose from fragments of DNA (or RNA) that became self-replicating in the host cell. Viruses that insert their DNA into the host genome, then recover it for making new particles after replication, may mimic this origin repeatedly.

Multiple origins for viruses from different hosts mean that viruses are more closely related to their hosts than they are to each other. Therefore, any phylogenetic classification should account for the different origins of plant and animal viruses. Accordingly, the view by plant virologists that rice dwarf virus, for example, should not be classified with animal viruses in the family Reoviridae has merit. This also means that classifications that rely on physical features such as size, shape, and number of nucleic acid molecules are artificial, because these features do not indicate phylogenetic relationships. Instead, because of genetic similarities to their hosts, plant, reoviruses are more closely related to plants than they are to animal reoviruses.

---

c o n c e p t

Viruses are more closely related to their hosts than they are to each other. Plant viruses, therefore, have more nucleotide-sequence similarity with plant hosts than they do with other viruses.

Can viruses evolve? They can in the sense that viruses are like genetic programs that are subject to change. As such, they can evolve faster than any organism. Evidence for such quick change comes from comparative studies of influenza viruses. Samples of these viruses were put into storage in the 1940s and used for genetic comparisons with the same kinds of viruses that were isolated every 10 years through the 1980s. These comparisons showed that the genetic difference between influenza viruses in the 1940s and 1980s was equivalent to the difference between species. This means that "speciation" in viruses can occur in 40 years or less. Different "species" of viruses may even arise from year to year, which may explain why new vaccines have to be developed almost constantly to combat new strains of viruses as they appear. This is far faster than evolution in eukaryotic organisms, which may require hundreds of thousands or, in some cases, millions of years for the formation of a new species.

Viroids and Virusoids

If it seems that nature has innumerable agents of disease, it may be because viruses are only one type of infectious particle that plagues living organisms. In plants (but not animals), such additional particles include **viroids** and **virusoids,** so named because they have some viruslike features. Like some viruses, for example, viroids have a single molecule of single-stranded RNA. Unlike viruses, the RNA molecule of a viroid is circular and is not surrounded by a protein coat. Viroids are known primarily from cultivated plants.

Virusoids are like viroids but are located inside the protein coat of a true virus. Virusoid RNA can be either circular or linear. Unlike viroids, virusoids are not infectious by themselves because they are replicated only in the presence of their host viruses. In some cases, the dependency is mutual; the virus cannot replicate in the absence of its virusoid.

Chapter Summary

Bacteria are prokaryotic, cannot undergo true sexual reproduction, and are the most metabolically diverse group of organisms. Biologists divide bacteria into two groups, the Eubacteria and the Archaebacteria, which some systematists now classify as two separate kingdoms. The Eubacteria include organisms that have photosystems like those of plants. These are the cyanobacteria, formerly called the *blue-green algae*, and the chloroxybacteria. The cyanobacteria have chlorophyll *a* and phycobilins as their main photosynthetic pigments. In contrast, like plants, the chloroxybacteria have chlorophylls *a* and *b*. The Archaebacteria and Eubacteria differ in their lipid linkages, cell-wall components, RNA polymerase types, ribosome shapes, energy sources, habitats, and gene structures.

The ability of bacteria to incorporate foreign DNA into their plasmid genomes and then replicate and express it in new cells is an important feature of bacteria for genetic engineering. This feature allows scientists to modify bacteria to control pests and diseases and to produce medically or agriculturally important biochemicals.

Bacteria are probably more closely related to the ancestor of life than any other group of organisms. Like all other organisms, however, bacteria are based on DNA as a genetic code that is transcribed into RNA and translated into proteins. In contrast, the first life-forms probably originated as RNA-based organisms.

Viruses are on the edge of life, since they are based on proteins and either DNA or RNA. However, because viruses lack many features associated with life, they are instead referred to as particles.

The structures of viruses provide the main characters for their classification. Viruses vary in whether they are based on DNA or RNA, how many kinds of proteins make up their coats, and their size. However, classification based on nucleic acid sequences is probably a more accurate reflection of the relationships of viruses. According to such comparisons, viruses are more closely related to their hosts than they are to each other. This suggests that viruses originated as genetic fragments from a host genome and that their origin has occurred repeatedly.

 ### What Are Botanists Doing?

Botanists have long known that some bacteria, called thermophiles, such as those at Yellowstone National Park, thrive in water as hot as 190° C. However, in 1982 a German microbiologist named Karl Stetter discovered the first organisms that thrive above 212° C in shallow hot springs off the coast of Sicily. The chemistry inside these microbes may point toward the original chemistry of life. What structure of molecules can withstand such brutal temperatures? To learn of the ongoing research on these microbes, go to the library and review the work of Mike Adams at the University of Georgia, where they grow microbes in a pressure cooker at 220° C. Your search will lead you to information on hyperthermophiles.

 ### Writing to Learn Botany

Some biologists refer to bacteria as the most primitive organisms on earth, while others believe that the persistence of prokaryotes in the environment for millions of years makes them the most advanced forms of life. How would you characterize bacteria and their complexity?

Questions for Further Thought and Study

1. If bacteria can only reproduce asexually, how can new forms arise?

2. Since many of the diseases of plants and animals are caused by bacteria, explain how it would or would not benefit humans if a virus that would eliminate all bacteria could be developed through genetic engineering.

3. If you could develop the most useful bacterium known to humans from various existing bacteria, what features would you combine in your new bacterium?

4. If viruses do not respond to external stimuli, lack enzymes, have no means of locomotion, and cannot reproduce outside of a living cell, should they be classified as lifeless, inanimate objects? Explain.

5. The morphology of cyanobacteria has changed relatively little during the past billion or so years; indeed, cyanobacteria have persisted 2–10 times longer than other "living fossils" (e.g., crocodiles). How would you explain this observation?

6. Recent evidence suggests that Archaebacteria have diverged less from the common ancestor they share with the eukaryotes than the eukaryotes have. How would you test this hypothesis?

7. Paleontologist Stephen Gould claims that bacteria, not humans, are evolution's most important product. Do you agree? Write a short essay to explain your answer.

8. Go to the library and read how we now use the following bacteria for economically important purposes: *Rhodococcus chlorophenolicus, Alcaligenes entrophus, Enterobacter agglomerans, Photobacterium phosphoreum, Arthrobacter globiformis.*

Web Sites

Review the "Doing Botany Yourself" essay and assignments for Chapter 25 on the *Botany Home Page*. What experiments would you do to test the hypotheses? What data can you gather on the Web to help you refine your experiments?

Here are some other sites that you may find interesting:

http://www.ucmp.berkeley.edu/bacteria/bacteria.html

This site includes an introduction to the bacteria, as well as links to *The Microbial Underground, Digital Learning Center for Microbial Ecology, Microbe Zoo,* and *Bugs in the News,* a set of interesting articles about bacteria and other microbes.

http://www.hhmi.org/beyondBio101

Cells Alive includes a variety of biology-related topics, including a Feature of the Month. Visit this site to learn about HIV infection, bacterial motility, bacterial division, penicillin, *Helicobacter pylori* (the bacterium that causes ulcers), and the sizes of viruses and bacteria.

http://golgi.harvard.edu/biopages/all.html

Are virtual libraries the wave of the future? What is a virtual library anyway? It'll take you a while to browse in this library of bioscience and related subdisciplines. A remarkable number of journals and databases are at your fingertips.

http://ag.arizona.edu/~zxiong

Xiong's Virology Page presents an excellent taxonomy and description of plant viruses.

Suggested Readings

Articles

Barinaga, M. 1994. Molecular evolution: Archae and eukaryotes grow closer. *Science* 264:1251.

Butler, P. J. G., and A. Klug. 1978. The assembly of a virus. *Scientific American* 239 (May):62–69.

Carmichael, W. W. 1994. The toxins of cyanobacteria. *Scientific American* 270:64–72.

Zimmer, Carl. 1995. Triumph of the Archaea. *Discover* (February):30–31.

Books

Goto, M. 1992. *Fundamentals of Bacterial Plant Pathology.* San Diego: Academic Press.

Matthews, R. E. F. 1992. *Fundamentals of Plant Virology.* San Diego: Academic Press.

Turkey tail bracket fungi.

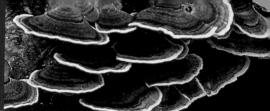

Fungi

26

Chapter Outline

Chapter Overview

Fungi occupy an odd place in our perspective on living organisms. We smile at the aroma of yeast in fresh-baked bread, yet are quite irritated by the itching of athlete's foot and disgusted by the welts of a ringworm infection. Rusts and smuts ruin some of our valuable crops, but we rarely consider the vital role of fungi as decomposers, allowing new plants to replace the old. Probably most confusing of all is that some fungi look like plants, while others look more like used motor oil.

The only cure for our naive confusion is to examine the characteristics that define the Kingdom Fungi and distinguish all its divisions and subgroups, and to relate their characteristics to their role in the biosphere. At the heart of this chapter is the relationship between an absorptive style of nutrition and fungi's unique roles as decomposers, parasites, food, and pathogens.

Fungi

All fungi are eukaryotic, and most are nonmotile, spore-producing filaments. Cell walls surround their cells during some or most stages in their life cycle, a characteristic that was partly responsible for their original classification in the plant kingdom. But fungal cells are heterotrophic and lack chlorophyll, a basic characteristic of plants. Their cell walls are made of **chitin** combined with other complex carbohydrates (chitin is also the main component of the exoskeletons of insects, spiders, and crustaceans).

Fungi are absorptive heterotrophs; that is, they absorb their food through the cell wall and cell membrane. To do this, they secrete digestive enzymes into their immediate environment and absorb the liquified products. Fungi are particularly important decomposers, aiding the breakdown of dead matter and the recycling of inorganic as well as organic molecules in an ecosystem.

Features of Fungi

The main vegetative feature of most fungi is a tubular, threadlike, whitish or colorless filament called a **hypha** (plural, **hyphae**). Hyphae, which vary in thickness between 0.5 and 100 μm, grow only at their tips and can grow indefinitely in favorable conditions. The hyphae of most fungi branch repeatedly, intertwine, or fuse with other hyphae, forming a mass known as a **mycelium** (plural, **mycelia**). Mycelia can become quite compact and may take on the form of parenchymalike tissue, in which individual hyphae are indistinguishable from one another. Such compact mycelia form mushrooms and similar spore-bearing structures that have traditionally been referred to as *fruiting bodies* (fig. 26.1). When growth is undisturbed, a mycelium tends to grow more or less equally in all directions from its point of origin, sometimes for hundreds of years. Recent size estimates of single-mycelium mats in the states of Washington and Michigan, each of whose mycologists claim world records for the largest fungus, put them at several metric tons for a single mycelium.

The hyphae of most fungi are partitioned into cells at regular intervals by cross walls, or **septa** (singular, **septum**), but some groups are nonseptate. The cells of septate hyphae may be uninucleate, **dikaryotic** (carrying a nucleus from each of two mating strains), or multinucleate (coenocytic). Nonseptate hyphae are all coenocytic. Unlike mitosis in plants, mitosis in fungi occurs almost completely within a persistent nuclear envelope.

The main storage carbohydrate of fungi is *glycogen*, which is also the main storage carbohydrate of animals (but not plants). All species of fungi are either saprobes or *symbionts* (i.e., live with other organisms). As symbionts, they may be parasitic, they may benefit their host, or they may be parasitized by their symbiont.

concept

Fungi have plantlike features, but they lack chlorophyll and their cell walls usually contain chitin. The main body of a fungus is a hypha or an aggregate of hyphae called a *mycelium*. Hyphae are either uninucleate, dikaryotic, or coenocytic. Mitosis in fungi is unique because it occurs within a persistent nuclear envelope.

The Generalized Life Cycle of Fungi

Reproduction in fungi can be either asexual or sexual; most fungi exhibit both forms. Asexual reproduction usually occurs by mitosis and cell divisions or by budding. Budding is typical of yeasts and other unicellular fungi; it produces new individuals by small outgrowths that pinch off from parent cells. Other forms of asexual reproduction include mycelial fragmentation and the mitotic production of spores. Such spores form either within a specialized container (**sporangium**) or in rows at the tips of specialized hyphae (fig. 26.2).

The sexual life cycles of many fungi involve a dominant haploid phase and a diploid phase consisting of just the zygote. The haploid phase consists of two parts in some fungi: one part derives from spore germination, and one part forms after the cells, but not the nuclei, of different mating strains fuse. The resulting dikaryotic cells can then grow into a dikaryotic mycelium. Fertilization eventually occurs in some of the dikaryotic cells, which form zygotes that undergo meiosis. These meiotic products develop into different kinds of spores, depending on what type of fruiting body the fungus produces. However, at no time in the life cycle are there any flagellated cells. Further details of fungal reproduction are presented in the discussions of particular groups of fungi later in this chapter.

figure 26.1

The diversity of fruiting bodies in fungi. (a) Oyster mushrooms (*Pleurotus ostreatus*). (b) A jelly fungus (*Tremella mesenterica*). (c) A bird's nest fungus (*Cyathus striatus*). (d) A shelf or bracket fungus (*Grifola sulphurea*). (e) Puffballs (*Lycoperdon* sp.). Note the spores being released. (f) A common stinkhorn fungus (*Phallus impudicus*).

Where Fungi Grow

The distribution and ecology of fungi resemble those of bacteria. Saprobic fungi and bacteria both decompose and recycle plant and animal remains, and have probably done so for as long as 2 billion years. Fungi attack virtually all organic materials. They can also etch the lenses of cameras, telescopes, and binoculars. In lichens, fungi can even degrade rocks.

Most diseases of living plants are caused by fungi. Rusts and other fungi frequently attack crops and stored foods, causing millions of dollars of losses annually. Fungi also cause several diseases in humans, especially diseases of the skin and lungs. Although most fungi are not poisonous, the few exceptions include mushrooms that are relatively common and widespread. Consumption of such fungi has often proven fatal; no effective antidote for human poisoning by mushrooms has been found.

The Lore of Plants

Ancient Fungal Farmers

Ants have grown crops of fungi for about 50 million years. When the ants moved into new areas, the fungus was brought by the queen ant. The ants then used this "starter packet" to clone a new crop, thereby explaining why many species of ants are nourished by the same fungus as were their ancestors millions of years ago. The dependence of the ants and fungus is mutual: the ants can't survive without the fungus, and the fungus can't propagate on its own.

(a)

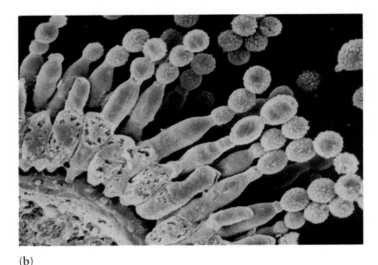

(b)

figure 26.2

Fungal spores and sporangia for asexual reproduction. (a) A sporangium of *Rhizopus stolonifer*, ×500, releases thousands of spores when it matures and breaks open. (b) In some fungi such as *Aspergillus*, asexual spores called conidia may be pinched off in chains from hyphal tips, ×1,600.

The Diversity of Fungi

Nearly 70,000 species of fungi are known, and descriptions of more than 1,000 new species are published each year. Moreover, biologists believe that more than half of all existing fungi have yet to be described. Mycologists (scientists who study fungi*) also suspect that unknown numbers of undiscovered species have already become extinct, due primarily to human activities such as the clearing of tropical rain forests and other natural habitats. In addition, fungal diversity is decreasing in industrialized countries because of environmental pollution (see reading 26.1, "Disappearing Fungi").

Fungi are most often classified into three sexually reproducing groups, variously treated as divisions, subdivisions, or classes. Regardless of their taxonomic level, the common names of these three groups persist as the **zygomycetes,** the **ascomycetes,** and the **basidiomycetes.** Furthermore, fungi that apparently cannot reproduce sexually, which are referred to as imperfect fungi or deuteromycetes, are often put into their own division. Most deuteromycetes, however, are believed to be ascomycetes. The distinguishing features of each of these groups primarily involve differences in reproduction. These differences are described in the next few sections, along with general information about each group.

The Zygomycetes

Zygomycetes have mostly coenocytic mycelia, with septa occurring in the hyphae of some species. The name *zygomycete* refers to the thick-walled, sometimes elaborately ornamented spore container—the **zygosporangium** (plural, **zygosporangia**)—that zygomycetes form during sexual reproduction.

More than 750 species of zygomycetes are known, examples of which are the hat-throwing fungi (*Pilobolus;* see the discussion in the section on "The Physiology of Fungi"), which grow on horse dung; the fly fungi (*Entomophthora*), which are most commonly seen growing from dead flies on window panes; and most of the fungi that are symbiotic with plant roots (for example, *Gigaspora*) (fig. 26.3). One of the best known and most widespread of the zygomycetes is *Rhizopus stolonifer,* a common bread mold that is also the chief cause of "leak," a disease of strawberries that appears while they are being transported to market. Like all zygomycetes, *Rhizopus* forms asexual spores in sporangia and sexual reproduction produces a zygosporangium. Because *Rhizopus* is so common, the following discussion of zygomycete reproduction uses this genus as representative of the entire group.

The mycelia of *Rhizopus* occur in two morphologically identical but reproductively different mating strains, usually designated as plus (+) or minus (−). Sexual reproduction begins when mature hyphae of two different mating strains come close to each other and start to develop ovoid swellings (fig. 26.4). Once these swellings touch, a septum forms in each swelling a short distance behind the point of contact.

---

*The term mycology (the study of fungi) is derived from *myketos,* the Greek word for *fungus.*

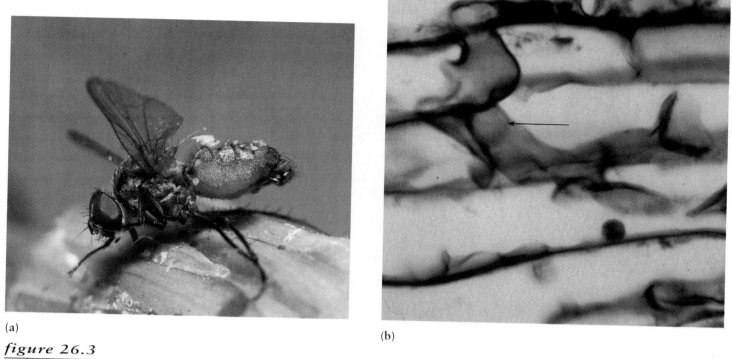

(a)

(b)

figure 26.3

Zygomycete fungi. (a) *Entomophthora* growing on a housefly. (b) *Gigaspora*, a mycorrhizal fungus penetrating a root of a cotton plant. The arrow indicates the penetrating fungus.

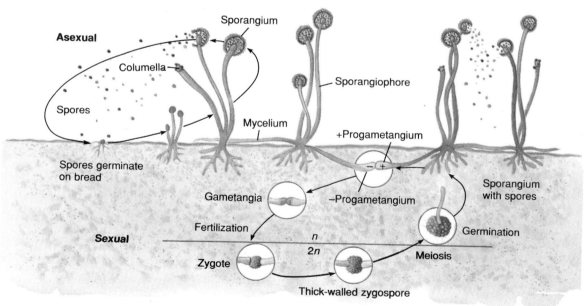

figure 26.4

During asexual reproduction of *Rhizopus stolonifer*, haploid nuclei packaged as spores germinate into asexually reproduced clones of the parent mycelium. During sexual reproduction, one or more pairs of haploid nuclei fuse in the zygote formed at the point of contact of two parent strains. The diploid nuclei soon undergo meiosis, and the resulting cells germinate as a new haploid strain.

European gourmets with a taste for the subtle flavors of fresh wild mushrooms are discovering that these delicacies are increasingly harder to find. A few years ago it was easy to pick a basket of the most prized fungus of all, the apricot-scented chanterelle (reading fig. 26.1). However, not only are these mushrooms becoming scarce, they are also getting smaller: it took 50 times as many chanterelles to make up a kilogram in 1975 as it did in 1958. Other fungi are also becoming rare. For example, the average number of fungal species in Holland has dropped from 37 to 12 per 1,000 m^2. These and many similar observations by other mycologists suggest that a mass extinction of mushrooms is happening all over Europe. But why?

Ecologists have recently begun to notice a negative correlation between the abundance and diversity of fungi and the amount of pollution. This negative correlation is stronger for fungi than for plants or other kinds of organisms. Apparently, fungi are more sensitive to air pollution than are plants because fungi have no protective covering, whereas the aerial parts of plants are protected by cuticles and bark. This distinction is a functional one: some fungi absorb water directly from air, along with whatever else is in the air, but plants get water from soil through their roots. It seems, therefore, that fungi are being driven to extinction by bad air.*

reading figure 26.1
A prized chanterelle mushroom.

Although the loss of a few gourmet treats does not seem to be important, the loss of fungi is harmful to forests. Hyphae associated with roots absorb some minerals more readily than do plants and then transport them into the plants. This association normally changes as a tree gets older; one species of fungus gives way to another in a steady progression. However, this progression is speeding up in European forests; that is, "old-age" fungi are more frequently associated with "middle-age" trees. Unfortunately, trees associated with the right fungi but at the wrong age tend to die early. Changes in the patterns of fungus-plant associations and the loss of fungal diversity are early warning signals of problems for trees. This may explain why forests are dying faster where the loss of fungal diversity is heaviest.

Similar disappearances of fungi have not been noticed in the United States, but they have probably occurred. Mycologists hypothesize that many species of fungi in the United States have not yet been described, and since people in the United States do not have a long history of collecting and eating wild fungi, there are few historical data from mushroom collectors about fungal diversity. This means that there is simply little or no information on the decline of edible wild fungi in the United States; it does not mean that such decline has not occurred. Suggestions that fungi are on the decline in the United States are nevertheless supported by collection records for lichens. These records show that most species that were native to areas such as the Los Angeles basin can no longer be found there.

---

*Colonies of luminescent fungi are sometimes maintained aboard spaceflights. Sensitive to escaped fuel or other noxious fumes, the fungi stop luminescing when exposed to as little as 0.02 parts per million of fuel. Thus, the fungi serve as an early warning system for noxious fumes, much as the death of canaries warned miners of the lack of oxygen or the presence of dangerous gases such as methane.

The two multinucleate cells that form by this process are the **gametangia** (gamete containers). These gametangia (singular, **gametangium**) soon fuse, the wall separating them disintegrates, and each + nucleus unites with a – nucleus to form diploid nuclei. Any unfertilized nuclei disintegrate. The cell containing the diploid nuclei is a multinucleate zygote, which enlarges into the zygosporangium. After a resting period of a few months, the zygosporangium splits open, thereby allowing one or more filaments to grow out of it. Although meiosis has not been observed in *Rhizopus*, mycologists believe that it occurs just before these filaments germinate. As the filaments grow, the tip of each one swells, and the haploid nuclei from the zygosporangium migrate into it. A septum then forms between the swelling and the filament, and a cell wall forms around the cytoplasm that immediately surrounds each nucleus. At this point, each walled cell is a **spore**, and the swelling that contains the spores is the sporangium. Eventually, when the sporangium wall disintegrates, the spores are released into the environment, where they germinate and form new hyphae.

In addition to sexual reproduction by zygosporangia, each hypha can also sprout branches that form swellings at their tips. These swellings become sporangia filled with mitotically derived nuclei. When a cell wall forms around the cytoplasm that immediately surrounds each nucleus, it becomes a spore. These are asexual spores because neither fertilization nor meiosis is involved in their production. Because they are produced asexually, these spores form clones of the parent hypha. Most reproduction in *Rhizopus* and other zygomycetes is from asexual spores.

The Ascomycetes

Asco means "sac," which refers to the saclike structures where spores form in ascomycetes. The spore sacs are called **asci** (singular, **ascus**). There are about 30,000 known species of ascomycetes, the most famous of which are brewer's or baker's yeast (*Saccharomyces cerevisiae*), bread mold (*Neurospora crassa*), truffles (*Tuber melanosporum*), and morels (*Morchella esculenta*) (fig. 26.5). Yeast is probably the most commercially important ascomycete; bread mold is famous primarily as a research organism. People have used yeast to make beer since the fourth millenium B.C. Truffles and morels are gourmet delicacies, primarily in North America and Europe.[1] Ascomycetes are also of interest because they cause plant diseases such as chestnut blight, Dutch elm disease, apple scab rot, apple bitter rot, stem rot of strawberries, powdery mildew, and brown rot of peaches, plums, and apricots. Ascomycetes are the most common type of fungi that occur in lichens.

(a)

(b)

figure 26.5

Fruiting bodies of ascomycetes. (a) Truffles. (b) Morel. Such ascomata occur in an impressive variety.

---

1. Morels, which sell for as much as $40 per lb ($88 per kg), are an important part of the $50-million-per-year wild mushroom industry. Commercial mushroom pickers in the Northwest earn $70 to $500 per day.

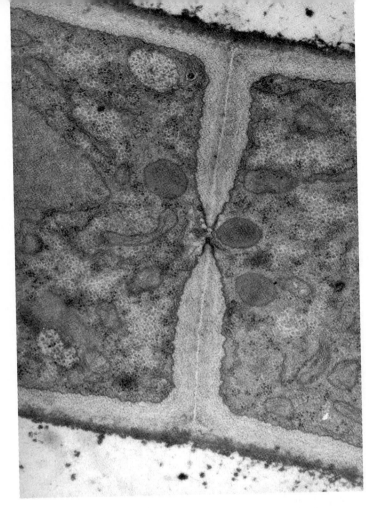

figure 26.6

Electron micrograph of a septal pore of an ascomycete. Nuclei and cytoplasm can move easily from one cell to the next through this pore.

The simplest ascomycetes are yeasts, which are unicellular. Other ascomycetes are filamentous with septate hyphae. Hyphal cells can be either uninucleate or multinucleate, but nuclei and cytoplasm are freely exchanged between adjacent cells through septal pores (fig. 26.6).

Sexual reproduction in the ascomycetes begins when uninucleate hyphae of opposite mating strains touch each other (fig. 26.7). Before contact, however, a hypha will form either male or female gametangia, depending on which mating strain it is; each gametangium forms around several nuclei at the tip of a short hyphal branch. Each female gametangium, called the **ascogonium** (plural, **ascogonia**), then sprouts slender outgrowths, called **trichogynes**, that grow toward the male gametangia, which are called **antheridia** (singular, **antheridium**).[2] Once the trichogyne touches the antheridium, nuclei migrate through it from the antheridium to the ascogonium.

2. Maleness and femaleness are artificial here. Males and females are so designated simply because their gametes either move, as male gametes do, or are stationary, as female gametes are.

The intermingling of nuclei from different mating strains is not followed immediately by fertilization. Instead, the ascogonium sprouts new hyphae, **ascogenous hyphae,** whose cells are dikaryotic, with one nuclear type from each of the mating strains in each cell. Cell division occurs in such a way that every cell contains one of each type of nucleus. These dikaryotic hyphae may grow extensively among the uninucleate hyphae of the parent mating strains, often becoming compacted into different kinds of fruiting bodies. As the fruiting bodies begin to form, the terminal cell of each dikaryotic hypha forms a hook. Within this cell the two nuclei divide with parallel spindles. Septa form to distribute the four nuclei into three cells (fig. 26.7). Nuclei in the enlarging terminal cell fuse to form a zygote nucleus, the only diploid nucleus in the life cycle. This zygote nucleus immediately undergoes meiosis. Mitosis follows meiosis in such a way that the resulting eight nuclei are arranged in a row. Each nucleus becomes a spore when a cell wall forms around it and the immediately surrounding cytoplasm. By this time, the cell in which fertilization occurred has enlarged into an ascus that contains eight spores.

The fruiting body of ascomycetes is called an **ascoma** (plural: ascomata) (sometimes called an ascocarp). The spores released from it are the **ascospores,** and the layer of asci is called the **hymenium.** There are three main types of ascomata: (1) a **cleistothecium** is a completely closed, almost spherical ascoma; (2) a **perithecium** is a usually flask-shaped ascoma that opens by a single pore; (3) an **apothecium** is shaped like an open cup.

Unlike the zygomycetes, the ascomycetes do not form spores within sporangia during asexual reproduction. Instead, the spores are pinched off in neat rows at the tips of exposed hyphae. Asexual spores that are formed by pinching off hyphal tips are called **conidia,** and the hyphae that bear them are called **conidiophores** (fig. 26.8).

The Deuteromycetes

The deuteromycetes, also called the "imperfect fungi," are defined by a single feature: the absence of sexual reproduction. The approximately 17,000 species of imperfect fungi reproduce almost exclusively by conidia. Because their asexual reproduction usually resembles that of the ascomycetes, most of the deuteromycetes probably descended from an ascomycete ancestor that lost the ability to reproduce sexually. This suggestion is supported by the observation that whenever sexual reproduction is discovered in a deuteromycete, it is usually of the ascomycete type. However, the sexual reproduction discovered in a few deuteromycetes resembles that of the basidiomycetes (discussed later in this chapter), which means that a small proportion of deuteromycetes are of basidiomycete origin.

When sexual reproduction is found in a deuteromycete, the species is then reassigned to an appropriate genus of sexual fungi. The taxonomic reassignment involves replacing the old deuteromycete genus name but keeping the second part of the binomial. The new genus name can be either new or one that already exists, depending on the similarity of the species in

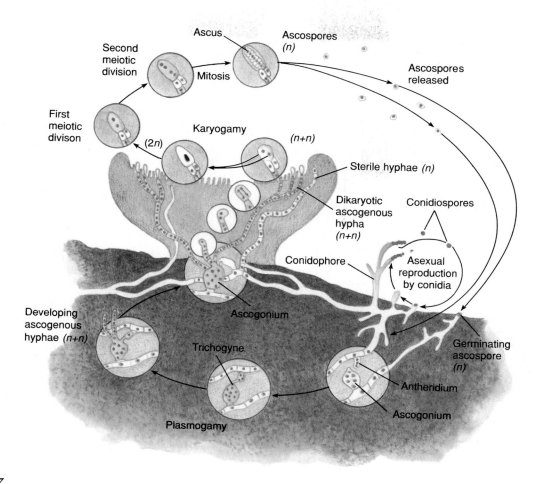

figure 26.7

The sexual life cycle of an ascomycete. Haploid nuclei housed in special reproductive swellings from each strain intermingle and grow as dikaryotic hyphae. On the surface of the fruiting body the nuclei fuse (karyogamy), undergo meiosis, and produce reproductive ascospores.

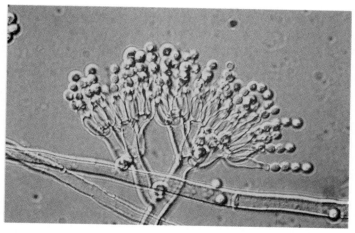

figure 26.8

Light micrograph of the conidiophores of *Penicillium*, ×790. Chains of conidia form at the tips of the conidiophores, and each conidium breaks off as a spore for asexual reproduction.

question to other fungi. For example, when the deuteromycete species *Aspergillus chevalieri* was discovered to produce ascomata, it was reclassified as *Eurotium chevalieri*, an ascomycete. Nevertheless, most mycologists still use both names: *A. chevalieri* for the conidial stage and *E. chevalieri* for the ascoma stage. Classification of sexual and asexual fungi into separate groups results in the same species being classified in two genera in two divisions. Furthermore, species of *Aspergillus* that have no known sexual counterpart are probably more closely related to *Eurotium* than to other deuteromycetes.

Deuteromycetes are mostly free-living and terrestrial, but some are pathogenic. The best known of the pathogenic deuteromycetes include the causal agents of a respiratory disease called aspergillosus (*Aspergillus niger*), athlete's foot (*Epidermophyton floccosum*), ringworm (*Microsporum canis*), and candida "yeast" infections (*Candida albicans*). The most famous deuteromycetes are species in the genus *Penicillium*: *P. notatum* for its role in the discovery of penicillin, *P. chrysogenum* for the commercial production of penicillin, *P. griseofulvum* for the

THE YEAST GENOME

Yeasts have long been a favorite experimental organism of biologists. These ascomycetes are single-celled and simple, and their genome—which is only one-fiftieth the size of the human genome—is considered to be a basic, no-frills set of instructions for maintaining the cells of higher organisms, including those of humans. Baker's yeast has only sixteen chromosomes, number three of which is the third smallest. Although this chromosome is relatively small and simple, it took a huge amount of work—147 biologists working in 35 labs in 17 countries—to spell out its DNA sequence. The chromosome consists of 315,356 base-pairs, making it the longest stretch of DNA ever sequenced and the first chromosome of any organism to be sequenced end-to-end (reading fig. 26.2).

When the sequencing project began in 1989, biologists had mapped just 34 genes on chromosome 3. However, the completion of its sequencing in 1992 showed that the chromosome has 182 genes. This huge number surprised biologists, who are now trying to figure out the function of each of the genes. The sequencing has produced other surprises: for example, only 10% of the new genes

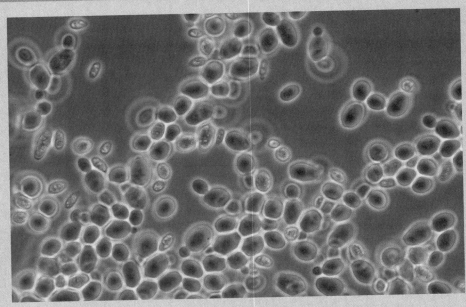

reading figure 26.2

Light micrograph of yeast. The third of its sixteen chromosomes was the first chromosome of any organism to be sequenced from end to end.

bear any resemblance to known genes of other organisms. One of the genes, strangely enough, codes for a protein normally used by some bacteria to fix nitrogen (see Chapter 20). Although yeasts do not fix nitrogen, they still need the gene. Indeed, when the gene was deleted, the yeast died, suggesting that the gene may be needed for other aspects of general metabolism.

---

production of griseofulvin (the only effective antibiotic against ringworm and athlete's foot), and *P. roquefortii* and *P. camembertii*, which are used to make Roquefort and Camembert cheeses, respectively (fig. 26.9).

Lichens

Lichens are symbiotic relationships consisting of a fungus and a green alga, a fungus and a cyanobacterium, or a fungus with both, in a body called a **thallus** (plural, **thalli**). Within this symbiosis the fungus gets carbohydrates from algae and cyanobacteria and fixed nitrogen from cyanobacteria; the photosynthetic organisms apparently receive nutrients, protection, and a receptive substrate for growth from the surrounding fungus.

The lichen thallus generally consists of several layers of cells or hyphae of an ascomycete (fig. 26.10a), although about twenty lichenized basidiomycetes are known. At the surface of the thallus is a protective layer where hyphae are so compressed that they resemble parenchyma cells. Below this upper layer is the algal or cyanobacterial layer, where the photosynthetic organisms are scattered among hyphae. Next is a layer of loosely packed hyphae that occupies at least half the volume of the thallus. Finally, a bottom layer of tightly packed hyphae is frequently but not always present. This layer is often accompanied by anchoring strands of hyphae.

Each of the approximately 20,000 known species of lichens is assumed to have its own unique fungal component, but the photosynthetic symbionts in about 90% of lichens come from five genera: the green algae *Trebouxia*, *Pseudotrebouxia*, and *Trentepohlia*, and the cyanobacteria *Nostoc* and *Anabaena*. Although a green alga and a cyanobacterium in a lichen might have their own Latin binomials, the names of lichens are derived from the name of the fungal symbiont. Their classifications are based on the features of fungi, and lichen species are separated into families, orders, and classes apart from nonlichenized fungi.

Individual symbionts of lichens can be isolated and cultured separately, but in pure culture the fungi often assume indefinite, compact shapes, and the algae and cyanobacteria

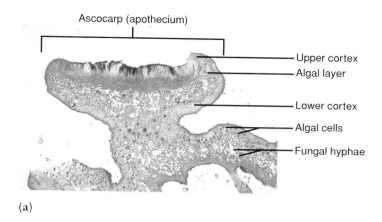

(a)

figure 26.9

Commercial products from fungi of the genus *Penicillium*. (a) Bottle of penicillin. (b) Roquefort cheese.

grow faster than they do when they are part of a lichen. The lichen-forming fungi rarely grow by themselves in nature, but the algal and cyanobacterial components of lichens often do. These observations indicate that the symbiosis between the fungus and the algae within the lichen may have a parasitic nature, since the fungus retards the growth of the algae.

Lichen thalli grow slowly, sometimes up to 1 cm or as little as 0.1 mm per year. On the basis of such growth rates, lichenologists estimate that some lichens have lived at least 4,500 years. Furthermore, lichens tolerate environmental conditions that are too extreme for most other forms of life. Indeed, lichens live on bare rocks in the blazing sun or bitter cold in deserts, in both arctic and antarctic regions, on trees, and just below the permanent snow line of high mountains where nothing else will grow. One species grows completely submerged on ocean rocks. They even attach themselves to artificial substances such as glass, concrete, and asbestos.

Although lichens can withstand many environmental extremes, they are, like other fungi, sensitive to air pollution. Indeed, lichens are effective monitors of air pollution: lichens in or near large cities (and around polluting industries such as power plants) throughout the world have disappeared rapidly during the twentieth century. Similarly, they have recolonized areas in which air quality has improved.

(b)

figure 26.10

Lichens. (a) Section through part of a foliose lichen thallus (*Physcia* sp.). (b) A fruticose lichen (*Cladonia deformis*). The red structures are immature apothecia (ascocarps).

Lichens are dispersed in nature primarily by asexual means. Rain, wind, running water, and animals disperse lichen fragments that grow into new thalli. The fungi reproduce sexually, as evidenced by the brightly colored apothecia often produced by lichens (figs. 26.10b, 26.11). Sexual reproduction is a bit of a problem though, because spores, or hyphae from them, must contact algae or cyanobacteria before the lichen can develop. No one has ever observed such an initiation of a new thallus in nature.

concept

Lichens are symbiotic relationships consisting of a fungus and a green alga, a fungus and a cyanobacterium, or a fungus with both.

The Lore of Plants

The stinkhorn fungus *Dictyophora* is one of the world's fastest-growing organisms: it pushes out of the ground at a rate of about 0.5 cm per minute. The growth is so fast that a crackling can be heard as the tissues of the fungus swell and stretch. During growth, a delicate, netlike veil forms around the fungus (this is the basis for the other common name of the fungus, "the lady of the veil"). The fungus then decomposes and, in the process, produces a strong odor of decaying flesh. This odor attracts flies, which crawl over the fungus and collect spores on their feet, thereby ensuring that the spores are carried to new areas.

The Basidiomycetes

Like ascomycetes, basidiomycetes are filamentous with uninucleate or multinucleate (i.e., dikaryotic) hyphae. Basidiomycetes are so named because they form spores on structures called **basidia** (singular, **basidium**), which are hyphal swellings that bear spores on tiny pegs. The fruiting body is a **basidioma** (sometimes called a basidiocarp), which is best known in the form of mushrooms, bracket fungi, jelly fungi, puffballs, earth stars, coral fungi, bird's nest fungi, and stinkhorns (see fig. 26.1). Basidiomycetes also include rusts and smuts, which cause plant diseases and do not form basidiomata (fig. 26.12).

Except for the absence of basidiomata in some species, the basidiomycetes have a common reproductive cycle (fig. 26.13). The discussion that follows regarding reproduction in such basidiomycetes as the common edible mushroom, *Agaricus brunnescens*, represents the entire basidiomycete group.

Reproduction by *A. brunnescens* begins when uninucleate hyphae of different mating types touch each other. Unlike the ascomycetes, the basidiomycetes have no specialized gametangia.

(a)

(b)

figure 26.11

Types of lichens. (a) Crustose lichens. (b) A fruticose lichen. (c) A foliose lichen.

figure 26.12

Wheat infected with wheat rust (*Puccinia graminis*).

Asexual reproduction in the basidiomycetes is primarily by means of conidia, budding, or fragmentation. Fragmentation involves the breakup of hyphae into single cells that develop into mycelia without the formation of thick walls or of internally produced spores.

The Physiology of Fungi

Fungal metabolism is similar to plant metabolism, except that fungi are never photosynthetic, and they can metabolize a wider variety of carbon sources, including sugar, plastic, and jet fuel. Fungi may also show some of the same responses to light as do plants. For example, many fungi cannot reproduce without light, and others show positive phototropic responses. Fungal phototropism has been studied most thoroughly in the hat-throwing fungus (*Pilobolus*) (fig. 26.15). This zygomycete produces upright hyphae up to about 10 mm tall, primarily on horse dung. The sporangium-bearing hyphae grow toward light so that the light enters a lenslike swelling below the sporangium. Although the mechanism of this response to light is not known, some kind of photoreceptor probably influences the swelling to split, enabling the explosive release of turgor pressure in the cells of the swelling. The explosion of the swelling blasts its sporangium up to 2 m away, at a speed of almost 60 km per hour, which is faster than the speed limit on most city streets. The sporangia stick to grass that may be eaten by a horse, through which the spores pass unharmed and germinate on the next round of dung.

The Ecology of Fungi

The ecology of fungi is especially important for plants. For example, tiny orchid seeds cannot germinate until they are invaded by hyphae of the soil fungus *Rhizoctonia*, and plants of all kinds are healthier when their underground parts associate with soil fungi. Members of all 400 or so families of flowering plants, with the exception of possibly fewer than a dozen, form mycorrhizal association (see Chapter 20). A mycorrhiza, meaning "fungus-root," is an association between a fungus and the underground parts of a plant. The association is a mutually beneficial symbiosis: the plants provide a source of carbon for the fungus, and the fungus absorbs phosphorus or other minerals that the plant cannot otherwise get easily from the soil.

Mycorrhizae are divided into two groups, depending on whether the fungal component penetrates the plant (**endomycorrhizae**) or forms only an external mantle around the plant's

Instead, a dikaryotic hypha grows from the fusion point of the two parent hyphae. As in the ascomycetes, the dikaryotic hyphae of basidiomycetes divide in such a way that each cell has one nucleus from each type of parent nucleus. The basidiomycetes are unique, however, because their dikaryotic hyphal growth involves the formation of a small bypass loop between cells. This loop is called a **clamp connection** because it looks like it clamps adjacent cells together (fig. 26.14).

The dikaryotic hyphae constitute the vegetative stage of basidiomycetes. As these dikaryotic hyphae grow, they form a compact mycelium that eventually develops into a small, closed basidioma. As these mushroom-type basidiomata develop, basidia form on fleshy plates called **gills,** in what will become the mushroom's cap (fig. 26.13). As basidia form, the nuclei in cells at the hyphal tips unite into diploid zygotes and then immediately undergo meiosis. The haploid nuclei from meiosis migrate into the pegs that extend from the basidia, and the tips of the pegs expand into spores.

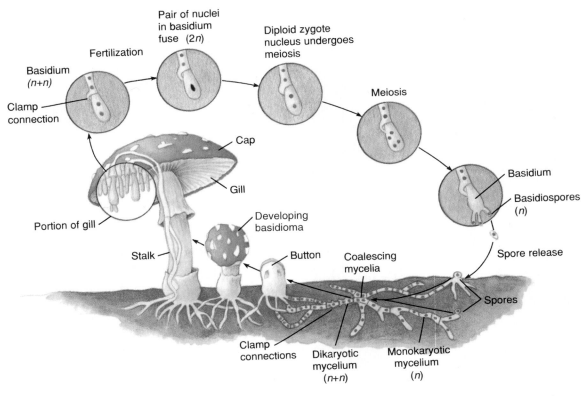

figure 26.13

The sexual life cycle of a mushroom. Coalescing mycelia from compatible strains form dikaryotic hyphae, which grow into the familiar fruiting body of mushrooms. On the gills of the mushroom cap the nuclei in dikaryotic cells fuse, undergo meiosis, and produce basidiospores to germinate into a new mycelium.

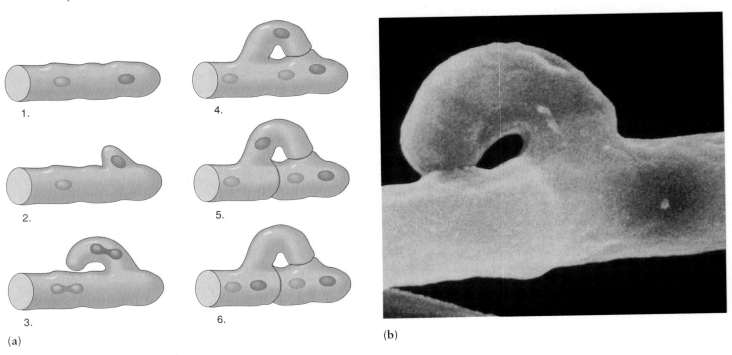

figure 26.14

Clamp connections common in basidiomycetes. (a) Development of a clamp connection: (1) adjacent nuclei in dikaryotic cell; (2) clamp connection develops; (3) each nucleus divides; (4, 5) septa form between cells of each pair of daughter nuclei; (6) the isolated pair in the end cell prepares for fusion. (b) Scanning electron micrograph of a hypha showing a clamp connection, ×20,000.

FUNGAL FOOLERY

Rock cress (*Arabis holboellii*) usually produces delicate, pale purplish-pink flowers. However, plants infected by the rust fungus *Puccinia monoica* (a pathogen of several species of mustard) produce twice as many leaves and, in one species, a dense cluster of yellow leaves that make the plant look like a buttercup, both to many insects and botanists alike. The leaves appear yellow because they are covered with spermagonia that produce a sugary fluid that attracts pollinators such as flies, bees, and butterflies. This fluid contains 10–100 times more sugar than does the nectar of nearby flowers, and the pollinators stay up to 5 times longer at the pseudoflowers than they do at real flowers. As the pollinators shuttle from one counterfeit flower to another, they carry fungal gametes between mating types, thereby increasing the reproductive success of the fungus (reading fig. 26.3).

(a)

(b)

reading figure 26.3

(a) A typical rock cress (*Arabis* sp.) in flower. (b) The yellow, flowerlike leaves of a rock cress that has been invaded by a rust fungus. The fake flower exudes a fragrant, sugary fluid attractive to insects, which inadvertently redistribute fungal sex cells that adhere to their bodies.

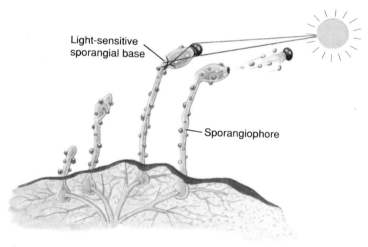

Light-sensitive sporangial base

Sporangiophore

figure 26.15

How light influences the release of sporangia of *Pilobolus*. The sporangial base receives light and enhances turgor pressure such that much of the sporangium is shot from the end when it ruptures.

roots (**ectomycorrhizae**) (fig. 26.16). Endomycorrhizal fungi are mostly zygomycetes. They occur in about 80% of all vascular plants, usually forming balloonlike structures (vesicles) or treelike structures (arbuscles). These structures have led to the name **vesicular-arbuscular (V-A)** mycorrhizae for associations that have these fungal components (fig. 26.16a). Endomycorrhizae are especially important to tropical plants, because the fungi can help plants obtain phosphorus from the phosphate-poor soils that characterize tropical habitats. Ectomycorrhizal fungi are primarily associated with the roots of trees and shrubs in temperate regions. Ectomycorrhizal fungi apparently replace root hairs, which may be absent in ectomycorrhizal roots (fig. 26.16b). Most ectomycorrhizal fungi are basidiomycetes, but some are ascomycetes.

c o n c e p t

The most significant association of fungi with plants is the formation of mycorrhizae. The fungal symbiont in a mycorrhiza gets carbohydrates from the plant host, and the plant gets minerals from the fungus. Most plants depend on this association.

The Economic Importance of Fungi

Fungi have greatly benefited human societies as sources of industrial chemicals, antibiotics, medicines, and vitamins. They are the mainstay of the brewing and baking industries, and are also important for making certain dairy foods, including gourmet cheeses. Fungi also cause many plant and animal diseases.

Fungi produce gallic acid, which is used in photographic developers, dyes, and indelible black ink, and in the production

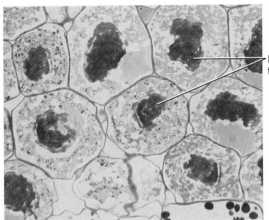

(a)

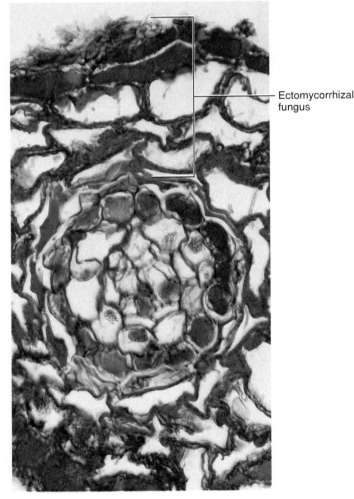

(b)

figure 26.16

Mycorrhizae. (a) Light micrograph of V-A endomycorrhizal fungus in plant root cells, ×400. (b) Transverse section of a plant root surrounded by ectomycorrhizal fungus.

figure 26.17

Plant infected with ergot (*Claviceps purpurea*). Flour made from grain crops infected with ergot can cause hysteria and convulsions if eaten.

of artificial flavoring and perfumes, chlorine, alcohols, and several acids. Fungi are also used to make plastics, toothpaste, and soap, and in the silvering of mirrors. In Japan, almost 500,000 metric tons of fungus-fermented soybean curd (tofu, miso) are consumed annually.

Different strains of the rust fungus, *Puccinia graminis*, cause billions of dollars of damage annually to food and timber crops throughout the world (fig. 26.12). Plant breeders are constantly faced with the challenge of developing rust-resistant varieties of crops. As a result of this effort, for example, Donald Winkelmann of the International Maize and Wheat Improvement Center in Mexico City recently announced that the defeat of wheat-rust is imminent. Scientists at the center have found a Brazilian-grown wheat plant that somehow controls the disease by slowing its growth. The "slow-rust" genes of this plant have been incorporated into cultivated wheat by hybridization experiments, and the new rust-resistant wheat is now being grown in more than 100 countries.

Another plant disease that has had a significant impact on human society is caused by the ergot fungus, *Claviceps purpurea* (fig. 26.17). This fungus infects the inflorescences of rye and other grain crops. Ergot seldom damages a crop significantly, but it produces several powerful drugs in the maturing grain. If infected rye is harvested and milled, a disease known as ergotism may occur in those who eat the bread made from the contaminated flour. The disease can affect the central nervous system, often causing hysteria, convulsions, and even death. Another form of ergotism causes gangrene of the limbs, and cattle that eat infected grass often abort their calves. Regardless, ergot-derived drugs have been used since the sixteenth century to hasten childbirth by stimulating uterine contraction. Because other ergot drugs constrict blood vessels, they are used to stop the bleeding often associated with childbirth. Ergot drugs have also been used to treat migraine headaches, heart palpitations, nervous stomach, menopausal disorders, and several other medicinal problems.

The Lore of Plants

When Fungi Dominated the Earth

At the end of the Permian period (about 250 million years ago), there was a mass extinction (e.g., about 90% of marine animal species died). Biologists originally thought that land plants weathered the late Permian crisis without much loss. However, research published in 1996 showed that this was probably incorrect. Indeed, so many trees died at the end of the Permian period that fungi dominated the land for a brief geologic period, living off of the huge amount of dead vegetation that covered the planet.

The Relationships of Fungi

The fossil record of fungi is more fragmentary than that of any other group of multicellular eukaryotes (see reading 26.4, "Fossil Mushrooms"). Nevertheless, mycologists have used phylogenetic systematics to infer the evolutionary relationships of fungi. The main questions about fungal relationships involve the zygomycetes, ascomycetes, and basidiomycetes. Traditionally, the zygomycetes have been thought of as "lower" fungi and the ascomycetes and basidiomycetes as "higher" fungi. This terminology reveals the traditional view of mycologists that the zygomycetes are more primitive, and the ascomycetes and basidiomycetes are more advanced and closely related to each other. Based on a cladistic analysis of fungi, Anders Tehler of the University of Stockholm has proposed a phylogeny that bears out this traditional view (fig. 26.18). He also supports the traditional view that some flagellated unicells, which are discussed in the next section of this chapter, should be included in Kingdom Fungi. These organisms (slime molds and water molds) are now more often classified as protoctists, a view supported by molecular phylogenies that show little relationship between these flagellated molds and fungi. In addition to suggesting evolutionary relatedness, Tehler took the next step for phylogenetic classification by designating taxonomic categories based on relationships among monophyletic groups. These categories, as well as some of the character state changes that support the relationships suggested by Tehler, are also shown in figure 26.18. Note that one of the surprises in this cladogram is the close relationship between the basidiomycetes and *Saccharomyces*, which is

traditionally classified as an ascomycete. One of the features that supports this idea is the shared origin of **meio-blastospores.** This is the name for spores that arise by budding from haploid, meiotically produced spores.

c o n c e p t

The zygomycetes are probably the closest living representatives of the ancestor of true fungi. The ascomycetes and basidiomycetes are more closely related to each other than either group is to the zygomycetes.

Slime Molds and Water Molds

Two groups of organisms that have been grouped with fungi are informally referred to as *slime molds* and *water molds*. Slime molds include the plasmodial slime molds and the cellular slime molds (fig. 26.19a–c); water molds are grouped into those with uniflagellate motile cells and those with biflagellate motile cells (fig. 26.19d). Some classifications still treat slime molds and water molds as fungi, but they are not closely related to fungi. Most biologists now regard those molds as members of Kingdom Protoctista. As protoctists, they are usually classified into four divisions, whose main characteristics are summarized in table 26.1.

Slime molds and water molds resemble fungi in that they are heterotrophic, store glycogen, and have cell walls that contain cellulose and chitin. They differ from fungi in that they form swimming cells that have flagella or that are amoeboid. The slime molds also differ from true fungi by being **phagotrophic;** that is, they ingest solid food particles. In contrast, fungi obtain their nutrition by absorbing dissolved nutrients. Furthermore,

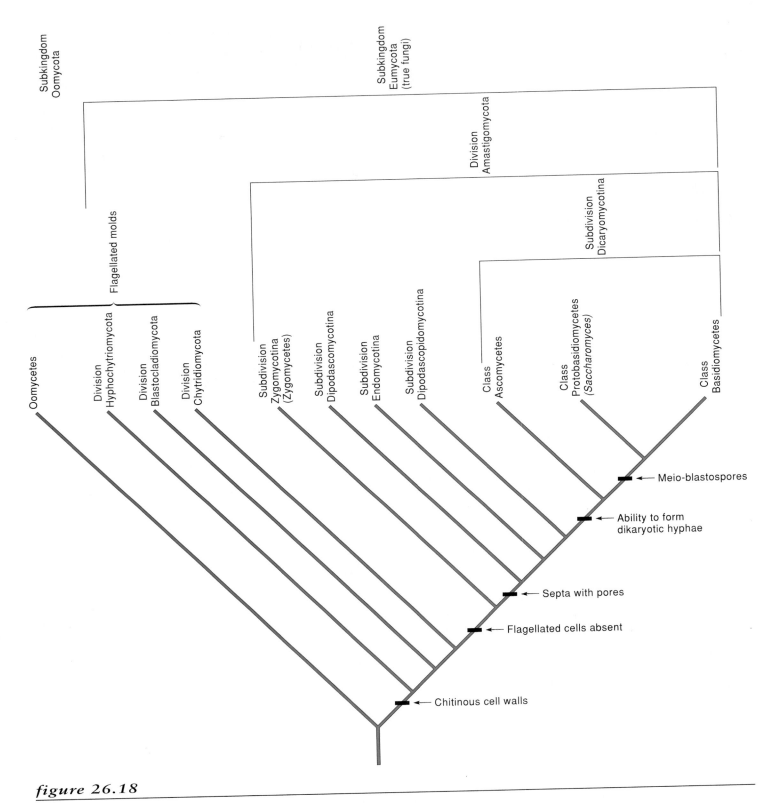

figure 26.18

A cladistic analysis of fungi.

(a)

(b)

(c)

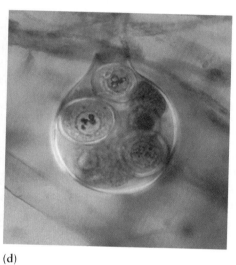

(d)

figure 26.19

Slime molds and water molds. (a) Plasmodium of the common slime mold *Physarum polycephalum*. (b) Sporangia of the slime mold *Stemonitis splendens*. (c) An immature sporangium of *Dictyostelium*, a cellular slime mold. (d) An oogonium of the water mold *Saprolegnia*.

table 26.1

The Main Characteristics of Slime Molds and Water Molds

Division	No. of Species	Mode of Nutrition	Storage Carbohydrate	Motility	Cell-Wall Composition	Main Habitat
Acrasiomycota (cellular slime molds)	70	Phagotrophic	Glycogen	Amoeboid	Cellulose	Terrestrial
Myxomycota (plasmodial slime molds)	500	Phagotrophic	Glycogen	2 flagella (whiplash)	None	Terrestrial
Chytridiomycota (uniflagellate water molds)	575	Absorptive	Glycogen	1 flagellum (whiplash)	Chitin, glucan*	Freshwater or marine
Oomycota (biflagellate water molds)	580	Absorptive	Glycogen or mycolaminarin	2 flagella (1 whiplash, 1 tinsel)	Cellulose, chitin, glucan*	Freshwater

*Glucan refers to structural polymers of glucose other than cellulose.

some of the water molds store a carbohydrate, **mycolaminarin,** that is more similar to that of brown algae (see Chapter 27) than it is to the storage carbohydrates of fungi, plants, or animals.

c o n c e p t

Slime molds and water molds that were traditionally considered to be fungi are more appropriately classified in Kingdom Protoctista.

Considerable scientific attention has been paid to the cellular slime mold *Dictyostelium discoideum* (fig. 26.19c). Normally, this organism lives as single, amoeboid cells. When food gets scarce, however, the cells stream together into a moving slug, which ultimately settles down and differentiates into a stalk with a spore-bearing fruiting body at its top (fig. 26.19c). The development from identical, free-living cells to a multicellular organism in *Dictyostelium* simulates many of the properties of cells that form embryos in much more complicated organisms, including mammals. For this reason, *Dictyostelium* has been studied for decades as a model for the developmental biology of complex organisms (see p. 68). More recently, a new technique has been devised for tagging and isolating the estimated 300 developmental genes in this slime mold. Using this technique, scientists expect to obtain the DNA sequences of almost all of these genes by late in 1997. In addition to its significance for the genetics of slime molds, this work will also be important for the Human Genome Project: matches between developmental genes of the slime mold and unknown genes in the human genome will help identify developmental genes in humans.

Chapter Summary

Fungi have many plantlike features. They have multicellular or coenocytic thalli or filaments that look plantlike, and they often have cellulose in their cell walls. Unlike plants, though, fungi have only one diploid cell in their life cycle, they can have chitin in their cell walls, they store glycogen, and they can reproduce asexually by spores. All fungi are heterotrophic, including saprobic and parasitic forms. Most plant diseases are caused by fungi. The two most complicated kinds of fungi, the ascomycetes and the basidiomycetes, are useful for making many commercial products. Mushrooms and truffles, which are basidiomycetes and ascomycetes, respectively, include species that are eaten as delicacies.

Fungi have a fragmentary fossil history. The two groups of "higher" fungi (ascomycetes and basidiomycetes) are believed to be more recent and closely related to each other than either one is to the zygomycetes. The zygomycetes are the group that is most representative of the ancestor of true fungi.

Slime molds and water molds were classified as fungi by early botanists. These molds have funguslike, animal-like, and plantlike features, but they are now classified in Kingdom Protoctista.

 What Are Botanists Doing?

Although most woodland mushrooms are either saprobic or mycorrhizal, the saprobic oyster fungus shown in figure 26.1a is also predatory. Predatory fungi lure, trap, lasso, paralyze, colonize, or enzymatically dissolve their prey. Go to the library and read a recent article about this unusual life-style. Summarize the main points of the paper. What are some questions you have about these fungi? What experiments could you do to answer these questions?

 Writing to Learn Botany

What features of fungi prompted early botanists to classify them as plants? Why are fungi no longer considered to be plants?

Questions for Further Thought and Study

1. If a single mushroom can produce a trillion spores, each of which can germinate and develop, why are we not overrun with mushrooms and other fungi that also produce prodigious numbers of spores?

2. Pollution, loss of habitat, and other factors contribute to the extinction of many species of animals and plants. Fungi are particularly sensitive to pollution, and many species are threatened or have disappeared. Should we be as concerned about the loss of fungal species as we are about the loss of, for example, Africa's big game animals? Explain.

3. For many years the U.S. Forest Service had a program to eradicate gooseberry bushes, which are the host for one phase of the life cycle of the fungus that produces white pine blister rust. The idea was that if gooseberry bushes could be eliminated, then the fungus could not reproduce and form spores that infect white pines. This program was eventually abandoned in favor of growing rust-resistant, nonnative pines instead. Can you think of any possible alternative measures to control the fungus?

4. Conifer seeds planted in fertilized, sterilized soil may germinate when watered but do not grow nearly as well at first as their counterparts in the forest. What might explain this phenomenon?

5. In 1994 researchers discovered that spores of some fungi germinate when exposed to ethylene. Of what significance is this?

Web Sites

Review the "Doing Botany Yourself" essay and assignments for Chapter 26 on the *Botany Home Page*. What experiments would you do to test the hypothesis? What data can you gather on the Web to help you refine your experiments?

http://worms.cmb.nwu.edu/dicty.html
http://www.ucl.ac.uk/~dmcbrob/dicty.html
Cellular Slime Mold WWW Server provides all you ever wanted to know about cellular slime molds.

http://muse.bio.cornell.edu/~fungi/
Cornell University maintains an excellent resource for information on fungi and all aspects of mycology.

http://www.igc.apc.org/mushroom/welco.html
Mycelium is your WWW connection to the fascinating world of mushrooms and other fungi. Like the World Wide Web, *Mycelium* is also a web—one that weaves its way underneath a mushroom.

http://www.halcyon.com/mycomed/fppage.html
Fungi Perfecti has made the jump to the World Wide Web! You'll find this page to be both entertaining and educational.

http://ucmp1.berkeley.edu/fungi/fungi.html
This site, maintained by the University of California Museum of Paleontology, includes interesting information about the fossil record, ecology, classification, and physiology of fungi.

Suggested Readings

Articles

Batra, L. R., ed. 1967. Insect-fungus symbiosis: Nutrition, mutualism and commensalism. *Scientific American* 217 (5):112–120.

Chang, S. T., and P. G. Miles. 1984. A new look at cultivated mushrooms. *BioScience* 34:358–362.

Lewis, R. 1994. A new place for fungi? Molecular evolution studies suggest fungi should be taxonomically transposed. *Bioscience* 44:389–391.

Litten, W. 1975. The most poisonous mushrooms. *Scientific American* 232:14–22.

Marx, J. 1992. *Dictyostelium* researchers expect gene bonanza. *Science* 258:402–403.

Seaward, M. R. D. 1989. Lichens as monitors of recent changes in air pollution. *Plants Today* (March–April): 64–69.

Books

Carlile, M., and S. Watkinson. 1994. *The Fungi*. San Diego: Academic Press.

Harley, J. L., and S. E. Smith. 1983. *Mycorrhizal Symbiosis*. New York: Academic Press.

Ingram, D. S., and A. Hudson. 1994. *Shape and Form in Plants and Fungi*. San Diego: Academic Press.

Nash, T. H. 1995. *Lichen Biology*. New York: Cambridge University Press.

Pirozynski, K. A., and D. L. Hawksworth, eds. 1988. *Coevolution of Fungi with Plants and Animals*. San Diego: Academic Press.

Smith, A. H., and N. Weber. 1980. *The Mushroom Hunter's Field Guide*. En. ed. Ann Arbor: University of Michigan Press.

A kelp from along the coast of California. Kelps are brown algae; giant kelps can grow to more than 60 m long.

Algae

27

Chapter Outline

Chapter Overview

Excluding blue-green algae (see "Cyanobacteria" in Chapter 25), algae include some of the smallest and largest eukaryotes; they can be unicellular, colonial, filamentous, coenocytic, or multicellular. Sexual life cycles of algae can be dominated by a haploid phase or a diploid phase, or they can be split evenly between the two. Although algae live in a diversity of habitats that is rivaled only by that of bacteria and fungi, most algae live in water. The great diversity of algae has led to many conflicting ideas about how they should be classified, but three main groups are consistently recognized as their own divisions: the green algae, the brown algae, and the red algae. There may be anywhere from seven to fifteen divisions of algae, depending on the views of different taxonomists. Most divisions, consisting mostly of unicellular forms, are variously aligned with the "big three" or are classified with protozoans or other single-celled protoctists.

The unifying feature of algae that has traditionally led botanists to classify them with plants is their ability to photosynthesize using chlorophyll *a*. Certain groups of algae also share with plants the presence of chlorophyll *b*, starch for energy storage, and cellulose in their cell walls.

People often think of algae only as pond scum or seaweeds, which they are, but algae are more diverse in their growth forms and habitats than is commonly assumed. Many species also grow in soil, snow, or clouds, or symbiotically with plants, animals, and fungi. Some algae live as epiphytes on aquatic plants, tropical plants, or other algae, and still others live on the fur of animals (fig. 27.1).

The wide variety of habitats occupied by algae is matched by their morphological diversity. Algae range in size from single-celled green algae, which may reach a maximum of 50 μm in diameter, to giant kelps whose length may exceed 60 m (see the photo that opens this chapter). Algae include species that either swim by flagella or produce certain cells that do so. Because some marine and aquatic algae have no flagellated cells, their gametes must float or glide to their targets.

Many botanists still consider all or some algae to be plants; others exclude them all. This creates many interesting (and often heated) arguments about the best classification of algae. For example, all groups of algae—like plants—have chlorophyll *a* and plantlike photosynthesis. A few, such as the green algae, also have chlorophyll *b* and photosynthesize exactly as plants do; indeed, the pathway for carbon fixation in plants was discovered in *Chlorella,* a green alga (see Chapter 7).

Conversely, algae exhibit a wide variety of life cycles that are unknown in plants, and different groups of algae have cells walls, carbohydrate reserves, and pigments that are unlike those of plants.

In this overview of algal diversity, we emphasize the three groups that have the greatest variety of forms: the green algae, the brown algae, and the red algae. Other groups, such as the diatoms, may include more species and be distributed more widely, but the green, brown, and red algae include representatives of almost all morphological and reproductive types that occur in algae. We present the key features of some of the other groups of algae later in this chapter, but in less detail.

General Features of Algae

Algae are an informally defined group of eukaryotes that are usually classified in seven divisions. The names and main features of each of these divisions are given in table 27.1. These divisions are at least partially distinguished by their pigments, their energy-storage polymers, their cell-wall components, and the number and types of their flagella. In addition, the organisms within each division have a variety of life cycles; representative life cycles are presented in this chapter.

Overview of Vegetative Organization

Although the simplest algae are unicellular (fig. 27.2), the most complex algae rival the giant redwoods as the largest of all photosynthetic organisms. Most algae are somewhere between these two extremes. Colonial algae are those with groups of cells that are loosely attached to each other and sometimes surrounded by a slimy sheath. Filamentous algae are either branched or unbranched and have either uninucleate or multinucleate cells. Some filamentous algae are coenocytic because they have no cross walls (fig. 27.3).

The kelps and other macroscopic algae have organs that resemble leaves, stems, and roots. The blades of the leaflike structures consist of parenchyma cells. The stemlike organs, called **stipes,** have many cell types, including sieve cells (fig. 27.4). Sieve cells of kelps occur in sieve tubes, like the phloem of vascular plants. However, botanists hypothesize that the sieve cells of algae arose separately from plant cells (that is, that algal and plant sieve cells arose by evolutionary convergence).

The divisions of algae are sufficiently different from one another that botanists hypothesize that the algae evolved from several different ancestors. If this is true, then algae are not a monophyletic group. This conclusion is currently supported by molecular phylogenies based on comparisons of ribosomal RNA sequences (see fig. 24.12).

Generalized Life Cycles

As in plants and fungi, sexual reproduction in algae entails an **alternation of generations** between diploid and haploid phases, which alternate between fertilization and meiosis. The diploid phase produces the haploid phase by meiosis; the haploid gametes then fuse to make a zygote that starts another diploid phase. Unlike those of plants, however, the diploid and haploid phases of algae are usually free-living; that is, neither phase is attached to the other phase at maturity. Moreover, unlike that of fungi, the diploid phase in many algae is plantlike in being multicellular.

(a)

(b)

figure 27.1

Diverse habitats of algae. (a) *Pleurococcus* growing on a yew tree and stone. (b) Algae growing on the fur of a three-toed sloth in Costa Rica.

table 27.1

Comparison of the Main Features of Algal Divisions*

Division	Habitat	Photosynthetic Pigments	Cell-Wall Components	Carbohydrate Storage	Flagella
Chlorophyta (green algae)	Mostly freshwater, some marine, terrestrial, or airborne	Chlorophylls *a* and *b*, carotenoids	Polysaccharides, including cellulose	Starch	None, 1–8, or dozens; whiplash
Phaeophyta (brown algae)	Almost all marine, rarely freshwater	Chlorophylls *a* and *c*, fucoxanthin and other carotenoids	Cellulose, alginic acid, and sulfated polysaccharides	Laminarin, mannitol	2, lateral; forward tinsel, rearward whiplash
Rhodophyta (red algae)	Mostly marine, some freshwater	Chlorophyll *a*, carotenoids, phycobilins	Cellulose, pectin, calcium salts	Floridean starch	None
Chrysophyta (diatoms, yellow-green and golden-brown algae)	Marine and freshwater, some terrestrial or airborne	Chlorophylls *a* and *c*, fucoxanthin and other carotenoids	Cellulose wall or silica shell; sometimes absent	Chrysolaminarin	None, 1, or 2; whiplash or tinsel
Euglenophyta (euglenoids)	Marine or freshwater, some airborne	Chlorophylls *a* and *b*, carotenoids	Absent	Paramylon	1–3; tinsel
Pyrrhophyta (dinoflagellates)	Marine and freshwater, some airborne	Chlorophylls *a* and *c*, peridinin and other carotenoids	Armorlike plates that may be cellulosic	Starch	None or 2; tinsel
Cryptophyta (cryptomonads)	Mostly freshwater, some marine	Chlorophylls *a* and *c*, carotenoids, phycobilins	Absent	Starch	2; tinsel

*The seven-division system represents the simplest traditional classification of algae still in use today; some treatments divide the algae into as many as fifteen divisions.

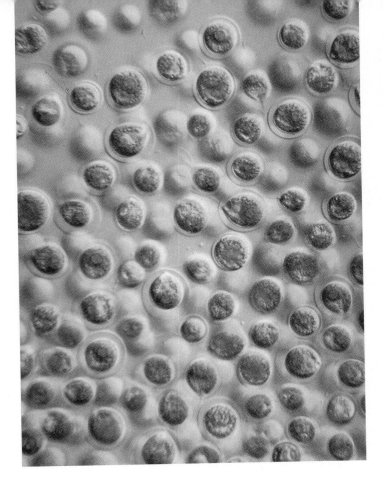

figure 27.2

Algae vary greatly in size. *Chlamydomonas*, shown here, is a single-celled green alga that is less than 100 µm long, ×150. Compare the microscopic size of *Chlamydomonas* with the kelp shown in the opening photograph of this chapter.

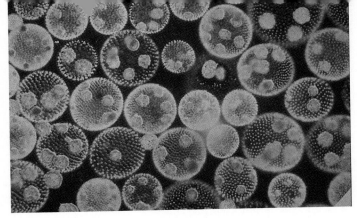

(a)

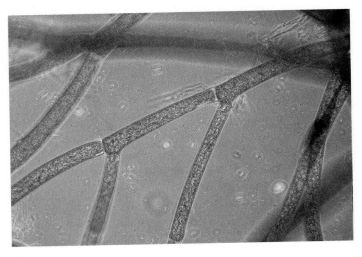

(b)

figure 27.3

Examples of other common types of vegetative organization of algae. (a) *Volvox* is a colonial green alga, ×50. (b) *Cladophora* forms branched filaments that consist of multinucleate cells, ×100.

Three distinct versions of a generalized life cycle occur among algae. One version resembles the life cycle of plants (see Chapters 28–31) because certain cells of a multicellular diploid phase undergo meiosis to make spores (fig. 27.5a). The diploid phase is therefore a sporophyte. Diploid sporangia produce haploid spores that can be either motile (**zoospores**) or nonmotile (**aplanospores**). Spores grow by mitosis, which may or may not be followed by cytokinesis, into the haploid phase of the life cycle. This haploid phase produces gametes and is therefore a gametophyte. The gametes can be either motile or nonmotile. Fertilization restores the diploid phase. This type of life cycle is based on **sporic meiosis**, because meiosis produces spores. Sporic meiosis is common among all divisions of algae that have multicellular forms.

The second type of life cycle resembles that of animals. Certain cells of a multicellular diploid phase undergo meiosis to make gametes, not spores (fig. 27.5b). Because meiosis produces gametes directly, it is called **gametic meiosis.** Gametic meiosis is rare in the algae.

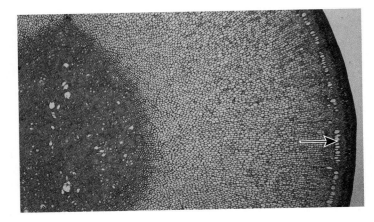

figure 27.4

Cross section of a stipe of *Laminaria cloustonii*. The arrow points to a region of the outer cortex that has many sieve cells, which are common in kelps.

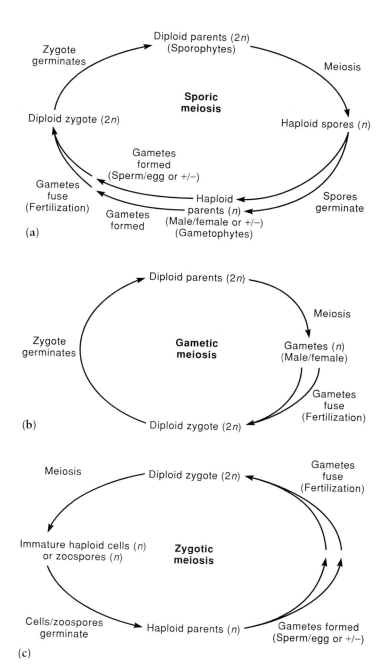

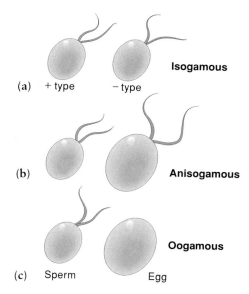

figure 27.6

Gametes of algae. (a) Isogametes are gametes of opposite mating types that look alike; they are generally designated as + and − mating strains. (b) Flagellated gametes of different sizes are called anisogametes. (c) In oogamy, one gamete is an egg that is large and nonmotile, while the other is usually a much smaller gamete, called a sperm, that is motile or nonmotile.

figure 27.5

The three types of life cycles in algae. (a) Diploid parents form spores by sporic meiosis; spores produce haploid parents, which produce gametes. (b) Diploid parents produce gametes directly by gametic meiosis, thereby bypassing a haploid parent phase. (c) Haploid parents produce gametes; gametes fuse into a zygote, which is the only diploid cell in this type of life cycle.

The third type of life cycle resembles that of fungi, in that the only diploid cells are the zygotes (fig. 27.5c). This means that the dominant phase of the life cycle is haploid, regardless of whether the organism is unicellular or multicellular. The haploid phase forms when the zygote undergoes meiosis. Although the zygote may produce spores when it divides, this type of meiosis is called **zygotic meiosis** to distinguish it

from the sporic meiosis of a multicellular diploid phase. Most of the green algae, including almost all of the unicellular forms, reproduce by zygotic meiosis.

The gametes of algae show more morphological diversity than any other group of organisms. With the exception of the red algae, at least some of the members of every algal group produce gametes that swim by one or more flagella. In addition, in some species the gametes are neither male nor female; all the gametes look alike (fig. 27.6a). Such gametes are called **isogametes,** and organisms that reproduce by isogametes are referred to as **isogamous.** Isogametes are not genetically identical, however; they belong to one of two mating strains and therefore are different genetically. Mating strains are arbitrarily designated as + or − strains. Fertilization can occur only between gametes of different strains.

Algae whose gametes are flagellated and of two different sizes are **anisogamous** (fig. 27.6b); the gametes are **anisogametes.** The smaller anisogamete is traditionally referred to as male and the larger as female.

The most pronounced differences between gametes occur in algae that are **oogamous.** In oogamy, one gamete is large and nonmotile, and the other is small and either motile or nonmotile (fig. 27.6c). The larger gamete is designated the egg, and the smaller one the sperm.

Asexual Reproduction

Unicellular algae reproduce asexually by mitosis and cell division. Multicellular algae also reproduce asexually, by either vegetative fragments, vegetative propagation (i.e., growth of new individuals from rootlike structures), or mitotically produced spores that form clones of the parent. Like spores produced by meiosis, mitotically derived spores are either motile or nonmotile. Mitotically derived spores are sometimes indistinguishable from meiotically derived spores.

Vegetative propagation and mitotic spore production occur almost continuously during a growing season. Because asexual reproduction is generally much faster than sexual reproduction, most algal populations consist of several clones.

c o n c e p t

Algae include organisms that are unicellular, colonial, filamentous, or large and multicellular. They reproduce sexually by fertilization and meiosis that is either sporic, gametic, or zygotic. Algae also reproduce asexually by different kinds of motile or nonmotile spores, or by different vegetative means.

Division Chlorophyta: The Green Algae

The Division Chlorophyta, which includes about 7,500 species, has a higher diversity of vegetative organization, life cycles, and habitats than any other group of algae. Most green algae live in fresh water, but different species also occur in marine habitats, clouds, snowbanks, soil, or on the shady moist sides of trees, buildings, and fences. Green algae also live symbiotically with several different kinds of animals and with the fungi of lichens (see Chapter 26). Green algae and plants share many important characteristics that support the hypothesis that plants arose from a green-algalike ancestor (see reading 27.1, "Evolutionary Relationships of Green Algae and Plants").

Unicellular Green Algae

There are two types of unicellular algae: motile and nonmotile. The most well-known example of a motile green alga is *Chlamydomonas* (see fig. 27.2). Cells of *Chlamydomonas* swim by means of two equal flagella at the anterior end of each cell. The dominant feature of each cell is a single, large chloroplast that encloses at least one **pyrenoid** and usually a **stigma** (fig. 27.7). Each pyrenoid contains the enzyme RuBP carboxylase-oxygenase (see Chapter 7) and is surrounded by starch granules. The red stigma contains the pigment rhodopsin, the same pigment that vertebrates use in vision. The stigma functions as a light-receptor for phototaxis.

Most cells in a population of *Chlamydomonas* are products of asexual reproduction by mitosis and cell division. However, environmental conditions such as nitrogen availability or daylength stimulate sexual reproduction (fig. 27.8). Sexual reproduction begins when flagella of cells from compatible mating strains touch. At first, cells aggregate into clusters;

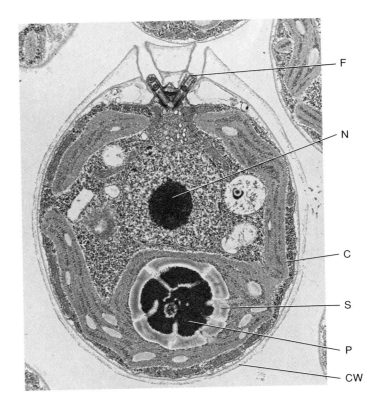

figure 27.7

Transmission electron micrograph of a *Chlamydomonas* cell in median longitudinal section, ×10,500. C = chloroplast, CW = cell wall, F = flagellum, N = nucleus, P = pyrenoid, S = starch granule. A stigma was not preserved in this cell.

then a delicate cytoplasmic thread develops between pairs of cells. The cell walls dissolve in the region of this thread, and the cytoplasm of the two cells moves together and fuses. Fertilization occurs when the two nuclei fuse into a large, diploid nucleus. This diploid zygote is surrounded by a thick, spiny wall that resists heat and desiccation. The zygote then becomes dormant, after which it divides by meiosis and releases four haploid zoospores, two of each mating strain.

In stressful conditions, *Chlamydomonas* retracts its flagella and becomes dormant. When water becomes available again, *Chlamydomonas* regrows its flagella, enlarges, and reproduces. After a rainfall, *Chlamydomonas* cells that have been dormant grow quickly in puddles and drainage ditches.

Most unicellular green algae that are nonmotile can produce zoospores. One example is *Tetracystis*, a soil-inhabiting alga that releases zoospores from mitotic cell division whenever there is enough moisture for swimming. Each zoospore looks like a cell of *Chlamydomonas* (fig. 27.9a). Zoospores lose their flagella quickly, and the cells enlarge and divide by mitotic cell division. New vegetative cells often stick together in two-celled or four-celled complexes (fig. 27.9b). In some species of *Tetracystis*, the zoospores may function as gametes. By analogy with *Chlamydomonas*, meiosis in *Tetracystis* is thought to be zygotic, but this has not been observed.

EVOLUTIONARY RELATIONSHIPS OF GREEN ALGAE AND PLANTS

Botanists have long believed that green algae, specifically the charophycean green algae, are the closest relatives of plants, and recent molecular phylogenies support this idea (see Chapter 24 and reading fig. 27.1). However, botanists are not satisfied with general agreement on such an apparently easy question; the question now asked is, "Which charophycean is most closely related to plants?" This is a more interesting question scientifically, because there is more than one possible answer, depending on the significance of different characters.

One possibility is presented in cladogram A, which shows the relationships between the land plants and the charophycean green algae that have been traditionally thought to be their closest relatives. Some of the characters that support these relationships are noted on the cladogram. The cladogram shows that the closest green alga to the land plants is *Coleochaete*; this relationship is supported by the shared derivation of the feature that zygotes are retained by the gametophyte. However, an equally likely possibility is presented in cladogram B, in which the closest green alga to the land plants is *Chara*. This relationship is supported by the evolution of flavonoid biosynthesis. Note that in both cladograms, however, one or the other derived state (i.e., zygote retention or flavonoid biosynthesis) is a parallelism: flavonoid biosynthesis in A and zygote retention in B. This means that flavonoid biosynthesis arose twice according to cladogram A or zygote retention arose twice according to cladogram B. Botanists who regard zygote retention as more significant than flavonoid biosynthesis accept *Coleochaete* as the green alga that is most closely related to land plants. Others consider flavonoid biosynthesis to be too complicated to have arisen twice in evolution, which leads them to choose *Chara* as the closest relative to land plants. Molecular phylogenies may help resolve this issue. At the moment, overall similarities of ribosomal RNA sequences support cladogram B.

Perhaps the most significant and generally agreed upon result of the cladistic analysis of charophycean green algae and land plants is that the charophyceans are apparently a paraphyletic group. This has led plant systematists to consider classifying all of the organisms shown in these cladograms in one supergroup, the Streptophyta, and redefining the Chlorophyta to exclude the streptophytes. No decision has been made, however, about whether the streptophytes should be classified as a division, a superdivision, or any other taxonomic category.

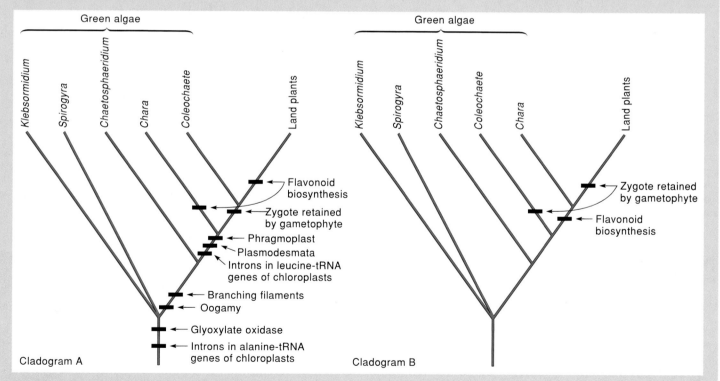

reading figure 27.1

Cladograms A and B represent two interpretations of the relationship of green algae and plants. A closer relationship between plants and *Coleochaete* (cladogram A) is supported by the shared feature of zygote retention by the gametophyte. In contrast, a closer relationship between plants and *Chara* (cladogram B) is supported by their shared ability to synthesize flavonoids.

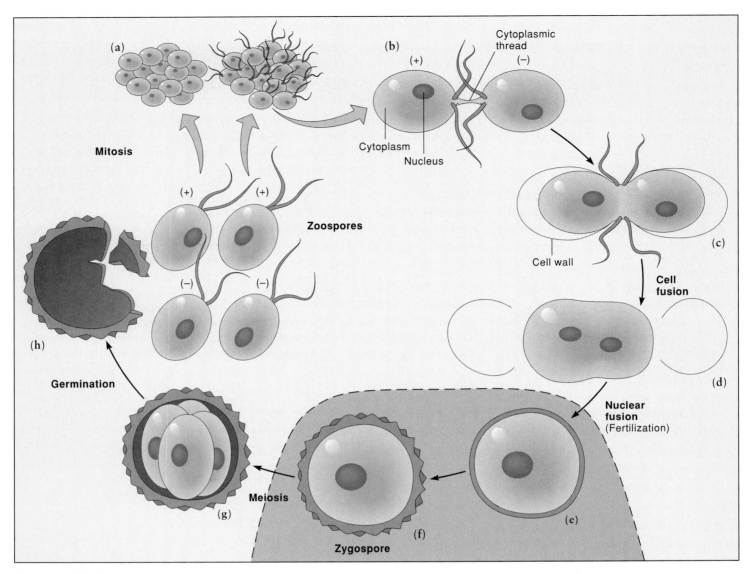

figure 27.8

Stages in sexual reproduction in *Chlamydomonas*. (a) Cells from compatible mating strains aggregate into clumps. (b) A cytoplasmic thread forms between cells of opposite mating strains, indicated by "+" and "−". (c–d) Cytoplasm of both cells fuses; the old cell walls are discarded. (e) Nuclei fuse (fertilization), and a new cell wall forms around the zygote. (f) The cell wall surrounding the zygote thickens and becomes spiny, becoming a zygospore. (g) Four haploid cells are produced by meiosis and cytokinesis of the zygote. (h) Two of the new cells after meiosis are of one mating strain, and two are of the other mating strain.

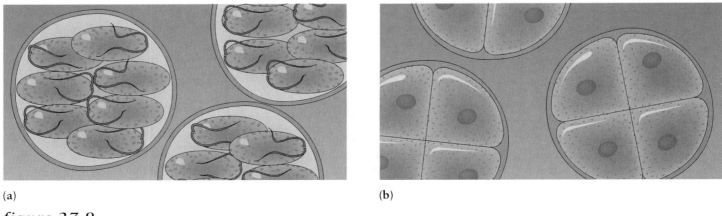

(a)   (b)

figure 27.9

Tetracystis, a nonmotile green alga. (a) *Tetracystis* forms biflagellate zoospores whenever there is enough moisture for swimming. (b) When water is scarce, these algae often stick together in two-celled or four-celled complexes.

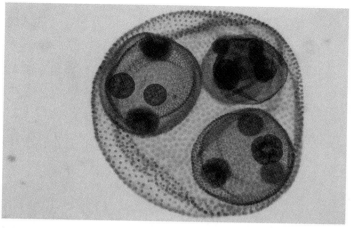

(a)

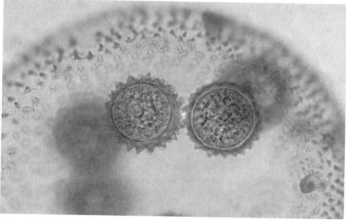

(b)

figure 27.10

Volvox. (a) *Volvox,* a motile, colonial green alga containing daughter colonies, ×250. (b) Portion of a colony with spiny-coated zygotes, ×500.

Some unicellular green algae such as *Chlorella* seem to reproduce exclusively by asexual means. Zoospores and sexual reproduction have not been observed in *Chlorella,* so the rapid growth of a *Chlorella* population is apparently due to asexual reproduction alone.

Colonial Green Algae

Like the unicellular green algae, the colonial green algae include both motile and nonmotile organisms. An example of colonial green algae is shown in figure 27.3a. The main features of representative genera of colonial green algae are discussed next.

The largest and most spectacular motile colonies of green algae occur in the genus *Volvox* (fig. 27.10). Each *Volvox* colony may contain thousands of *Chlamydomonas*-like cells arranged at the periphery of a hollow sphere. The cells are attached to each other by delicate cytoplasmic

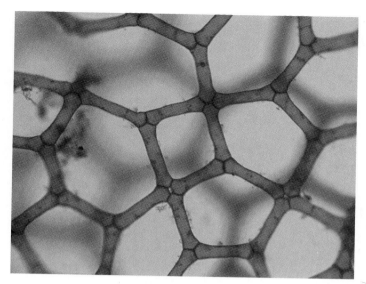

figure 27.11

Hydrodictyon, a nonmotile, colonial green alga, with coenocytic cells, ×100. This alga is called the "water net" because of the netlike arrangement of the cells.

threads. Their flagella beat in a coordinated motion that rolls the colony like a ball. During asexual reproduction, certain cells enlarge and grow inward, where they pinch off and form **daughter colonies** (fig. 27.10a). These colonies are released when the parent colony disintegrates. During sexual reproduction, *Volvox* is oogamous; the same enlarged cells that grow into daughter colonies may instead become fertile eggs (fig. 27.10a). Packets of sperm are produced elsewhere in the same colony or in a different colony; that is, a colony is either **bisexual** or **unisexual**. The sperm cells swim to the eggs and fertilize them, thereby forming zygotes (fig. 27.10b). The zygotes go through a dormant period in the parent colony before they undergo meiosis and release haploid zoospores.

Mature colonies of *Hydrodictyon,* the "water net," consist of elongated, coenocytic cells joined at their ends to make polygonal shapes (fig. 27.11). The whole colony is a hollow cylinder that can grow up to 75 cm in length if it is not crowded by other colonies. Asexual zoospores develop when mature coenocytic cells cleave into numerous uninucleate segments, each of which produces a biflagellated zoospore. Miniature nets grow within the parent cell from zoospores that come together into preliminary polygonal shapes and lose their flagella. Young nets are released as their cells expand and rupture the parent cell. If biflagellated cells are discharged from the parent cell before they assemble into young nets, they function as isogametes. Zygotic meiosis produces four biflagellated zoospores, each of which enlarges and divides mitotically into a coenocytic cell in a polyhedral gelatinous matrix. This large coenocytic cell cleaves to another set of zoospores, which make small nets in the matrix, just as asexual zoospores do in cells of mature colonies.

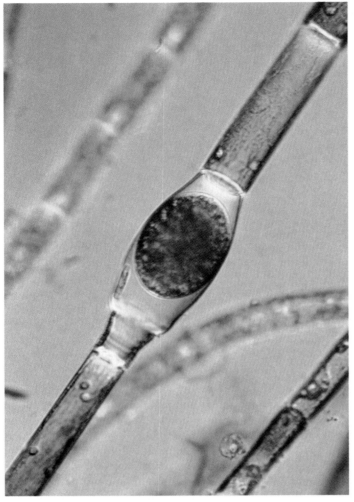

figure 27.12

Light micrograph of an unbranched filament of *Oedogonium*, ×780. The enlarged cell is an oogonium.

Filamentous Green Algae

Filamentous green algae are microscopic and are either branched or unbranched (figs. 27.3b and 27.12). Filamentous green algae often grow epiphytically on aquatic flowering plants; they also attach to rocks or other objects underwater. Some filaments are free-floating.

Asexual reproduction in filamentous green algae is usually by quadriflagellate or biflagellate zoospores, but some organisms produce zoospores with up to 120 flagella (fig. 27.13). In sexual reproduction, meiosis is usually zygotic or sporic, while gametic meiosis is rare. In most filamentous green algae, gametes are either oogamous or isogamous. In oogamous filaments, such as those of *Oedogonium*, the egg forms in an enlarged container called the **oogonium** (fig. 27.12). Sperm arise in pairs in each of several boxlike cells, together called the **antheridium**. Like zoospores, sperm have many flagella (approximately 30) in *Oedogonium*, but they

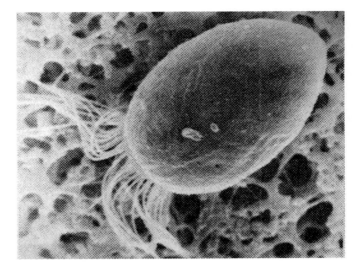

figure 27.13

Zoospores of *Oedogonium* have up to 120 flagella, ×1,000.

are biflagellate or quadriflagellate in most other genera. Although meiosis is zygotic in oogamous organisms, many isogamous genera have sporic meiosis. In the genus *Cladophora*, for example, the zygote germinates into a diploid branched filament that looks like the haploid branched filaments (see fig. 27.3b). Although the sporophyte and the gametophyte are **isomorphic**—that is, vegetatively they look alike—they differ reproductively. After meiosis, the sporophyte releases quadriflagellate zoospores that grow into filamentous gametophytes. At maturity, these gametophytes produce biflagellate isogametes. Sexual reproduction in *Cladophora* is summarized in figure 27.14. The life cycle of *Cladophora* resembles that of plants because meiosis is sporic.

Some filamentous green algae reproduce differently than do *Oedogonium* and *Cladophora*. The most common representative of other types of reproduction occurs in the genus *Spirogyra*, which is named for the spiral chloroplasts in each cell (fig. 27.15a). Species of this genus grow as frothy or slimy green masses of unbranched filaments that float in the water of small ponds. Flagellated cells are absent in all species of *Spirogyra*, and asexual reproduction is restricted to fragments of filaments that form new filaments. Sexual reproduction in *Spirogyra* is called **conjugation.** During conjugation, filaments growing side by side in a dense mass form small protuberances, or **papillae,** that grow toward each other. Upon contact, the end-walls of the papillae dissolve to create a tube. The contents of the cell of one filament then move through the tube, and the cytoplasm and nuclei of both cells fuse. Although each cell is apparently an isogamete, the migrant cell is considered to be the male. Conjugation occurs between almost every pair of cells along the length of the two filaments (fig. 27.15b). The diploid nucleus in each cell undergoes meiosis, but only one of the four haploid cells grows into a new filament. The other three nuclei disintegrate.

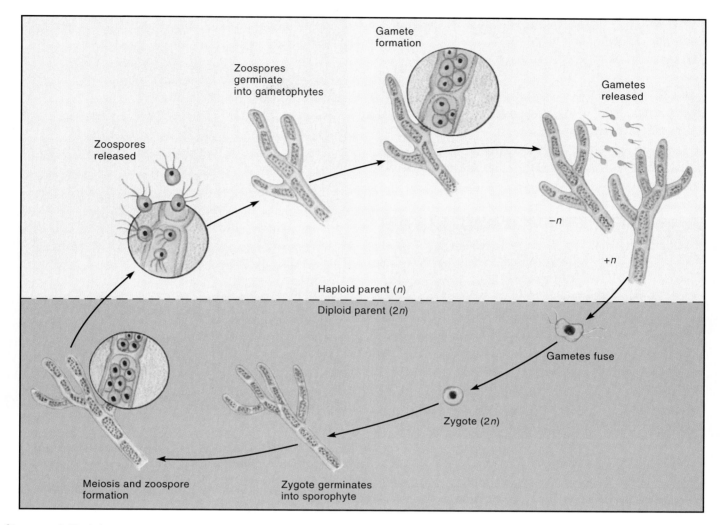

Zoospores
germinate
into gametophytes

Gamete
formation

Gametes
released

Zoospores
released

Zoospores
germinate
into gametophytes

−n

+n

Haploid parent (n)

Diploid parent (2n)

Gametes fuse

Zygote (2n)

Meiosis and zoospore
formation

Zygote germinates
into sporophyte

figure 27.14

Alternation of isomorphic phases in the life cycle of *Cladophora*. As in plants, meiosis in *Cladophora* is sporic, meaning that its spores are produced by meiosis.

The most complex and plantlike green algae in overall appearance are the stoneworts (also known as tank mosses and brittleworts). The branched filaments of stoneworts, such as *Chara,* are whorled like those of some plants, and they have leaflike structures at their nodes (fig. 27.16). A single apical cell produces cells that form nodes, internodes, and leaflike structures. Indeed, *Chara* is often identified incorrectly as an aquatic vascular plant because of its cylindrical branches attached to nodes, its large size (up to a meter long), and its plantlike body. However, when *Chara* is collected with its distinctive reproductive structures present, there is no doubt that it is a stonewort.

The sexual organs of stoneworts are unique in the algae. Antheridia and oogonia are multicellular, and each is surrounded by a layer of sterile cells (fig. 27.17). Herein lies a botanical and taxonomic dilemma: in other algae, sex organs are either unicellular or, if multicellular, entirely fertile (i.e., not covered by sterile cells). If *Chara* is considered ancestral to plants, the sterile jacket of cells that surrounds their reproductive organs is yet another shared characteristic.

Stoneworts, like plants, can also make flavonoids (see reading 27.1, "Evolutionary Relationships of Green Algae and Plants"). No other algae have yet been found to do this.

Other Multicellular Forms

Seaweeds in the genus *Ulva,* the sea lettuce, are leafy green algae. These seaweeds are widely distributed on rocks, wood, and larger algae in shallow marine habitats. As in some filamentous green algae, the sporophytes and gametophytes of *Ulva* are isomorphic (fig. 27.18).

A population of sea lettuces includes three kinds of organisms: the sporophyte, the male gametophytes, and the female gametophytes. Sporophytes release large quadriflagellate

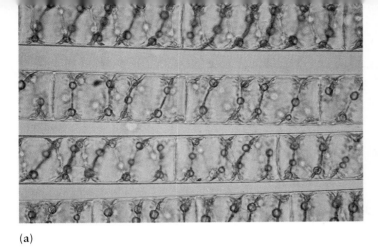

(a)

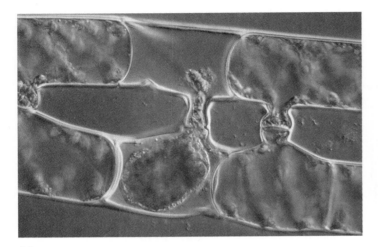

(b)

figure 27.15

Spirogyra. (a) Light micrograph of *Spirogyra* filaments showing spiral chloroplasts, ×150. (b) Phase-contrast light micrograph of *Spirogyra* filaments in a late stage of conjugation, ×250.

figure 27.16

Chara is a stonewort that has branched filaments and whorls of leaflike organs. The brown structures are antheridia and oogonia.

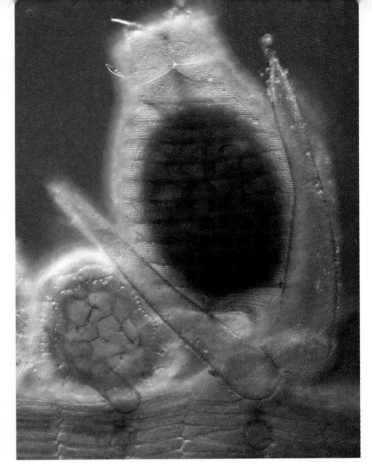

figure 27.17

Portion of a filament of *Chara* having mature antheridium (left) and an apparently fertilized oogonium containing a darkened, starch-filled zygote, ×1,500. *Chara* and other stoneworts are unique among algae due to the layer of sterile cells surrounding their reproductive organs.

figure 27.18

Ulva, a green alga that is commonly referred to as sea lettuce. Populations of *Ulva* consist of isomorphic sporophytes and gametophytes.

zoospores produced by meiosis. Half of these zoospores grow into male gametophytes, and half grow into female gametophytes. The male gametophytes produce small biflagellate gametes, and the female gametophytes produce large biflagellate gametes. Sexual reproduction is bypassed when gametes that fail to unite grow directly into new haploid organisms.

Classification of the Green Algae

Less than two decades ago, the classification of green algae was based primarily on features such as vegetative organization (unicellular, filamentous, colonial), the presence or absence of flagellated cells, and the type of life cycle. Current classifications also use features determined from electron microscopy and comparative biochemistry. Because these subcellular characteristics are time-consuming to determine, most green algae have not been surveyed. Consequently, the classification of green algae is changing rapidly as more organisms are examined.

One recent classification divides green algae into five classes. Representative organisms from each of the three largest of these classes were discussed earlier in this chapter: the Chlorophyceae (*Chlamydomonas, Tetracystis, Chlorella, Volvox, Hydrodictyon,* and *Oedogonium*), the Charophyceae (*Spirogyra* and *Chara*), and the Ulvophyceae (*Ulva* and *Cladophora*). The two smaller classes are the Pleurastrophyceae and the Micromonadophyceae. Several relatively new kinds of features are used to define groups of green algae in current classifications, including types of cell division, details of flagellar ultrastructure, enzymes used in photorespiration, and biosynthetic capabilities.

A phylogenetic study in 1994 showed how these and many other characters could be combined with sequence data from ribosomal RNA genes to evaluate the classification and relationships of green algae. The results confirmed the long-held view that the Charophyceae are the closest green algal relatives to the land plants (fig. 27.19; also see reading 27.1, "Evolutionary Relationships of Green Algae and Plants"). This relationship is supported by the occurrence of glyoxylate oxidase and introns in the alanine-tRNA genes of chloroplasts in the Charophyceae and in land plants. The results also showed that the Chlorophyceae, Pleurastrophyceae, and Ulvophyceae clustered together into a single branch, but the position of the Micromonadophyceae could not be resolved.

c o n c e p t

The green algae are the most diverse of algae. Green algae include examples of all the major types of vegetative organization, including organization into colonies, which is absent or rare in other divisions. Although the most common type of meiosis in green algae is zygotic, both sporic meiosis and gametic meiosis also occur. Gametes may be isogamous, anisogamous, or oogamous. Asexual reproduction is common and involves motile or nonmotile spores, or growth by vegetative means.

Division Phaeophyta: The Brown Algae

Almost all of the approximately 1,500 species of brown algae are marine organisms, but a few species live in freshwater. The Division Phaeophyta includes a wide range of morphological types, from the most complex of all the algae, the kelps (see the photo at the beginning of this chapter), to microscopic, branched filaments (fig. 27.20). There are no unicellular, colonial, or unbranched filamentous organisms among the Phaeophyta.

The flagella of brown algae are characteristic of this division. Most species of brown algae produce motile cells having two flagella, which are usually on the side of the cell. The shorter of the two flagella is a smooth **whiplash** type that points toward the back of the cell. The longer flagellum points toward the front of the cell and is a **tinsel** type, meaning that it has several rows of small appendages projecting from it (fig. 27.21). Except for the Order Dictyotales, all of whose members produce motile cells having a single tinsel flagellum, all brown algae release biflagellate motile cells at some time in their life history, either as gametes or as zoospores.

Filamentous Brown Algae

Ectocarpus, perhaps the best known of the filamentous brown algae, grows on rocks or on larger marine algae along ocean shores worldwide. The outward appearance and reproductive cycle of *Ectocarpus* resemble those of some green algae. Small, branched filaments are either haploid or diploid. The haploid gametophyte and the diploid sporophyte are isomorphic, as they are in the green algae *Cladophora* and *Ulva*. In the sporophyte, cells at the ends of lateral branches enlarge into **unilocular** (i.e., having one chamber) sporangia. After the diploid nucleus undergoes meiosis, the new haploid nuclei divide mitotically, forming 32–64 zoospores (fig. 27.22). Each zoospore germinates into a haploid gametophyte. When the gametophyte is mature, isogametes form in elongated multicellular organs called **plurilocular** (i.e., having many chambers) gametangia on lateral branches. Although they are isogamous, the cells of one mating strain settle on the ocean floor and secrete **ectocarpene**, which attracts cells of the other mating strain. The ectocarpene-secreting gametes are designated as female, and the gametes that are attracted to them are designated as male. The gametophytes of *Ectocarpus* are unisexual; that is, each is either male or female.

Asexual reproduction in *Ectocarpus* is versatile, because mitotic zoospores can be produced by diploid and haploid phases. Plurilocular organs on diploid individuals release diploid zoospores, which germinate into new sporophytes. Plurilocular organs of the diploid phase are morphologically indistinguishable from those of the haploid phase. Moreover, gametes that fail to unite may become zoospores and germinate into new haploid organisms.

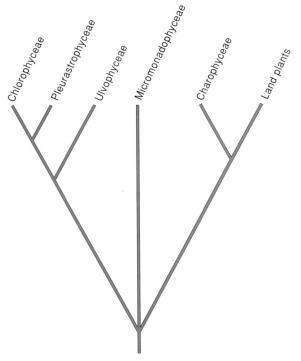

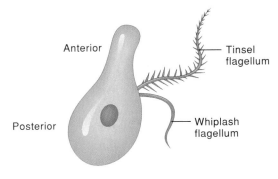

figure 27.21

figure 27.21

Motile cells of brown algae typically have two lateral flagella, one whiplash and one tinsel.

figure 27.19

This cladogram shows our current understanding of the phylogenetic relationships of the five classes of green algae. It is derived from a combined analysis of biochemical, ultrastructural, morphological, and molecular (DNA) data.

figure 27.20

Light micrograph of *Ectocarpus* filaments, which are among the smallest of the brown algae.

The full range of reproductive alternatives in a single species of *Ectocarpus* is known only from a single population of *E. siliculosus* near Naples, Italy. Other populations of this species, from the coasts of England and the northeastern United States, are not as versatile. No other species have been studied enough to know whether the *E. siliculosus* population near Naples is typical of the genus. However, reproduction in the Naples population shows that other populations and species of *Ectocarpus* may have the latent genetic capability to produce zoospores regardless of ploidy level.

The production of zoospores by haploid and diploid phases blurs the distinction between sporophytes and gametophytes. This is apparently a general phenomenon, since spore production by "gametophytes" also occurs in some plants. Such variations in basic life histories are evidence that chromosome number does not determine whether a "sporophyte" or a "gametophyte" develops when a specific cell germinates.

Kelps and Rockweeds

Kelps and rockweeds dominate shorelines and nearby offshore habitats in cool climates worldwide. Some kelps are free-floating; they can grow rapidly by vegetative propagation into massive populations in the open ocean. Kelps and rockweeds are of interest because of their vegetative organization and the distinctiveness of their life cycle in comparison with those of green algae.

The life cycle of kelps is dominated by a large sporophyte. The huge sporophytes of *Macrocystis* (giant kelp), for example, consist of stipes, blades, and branching **holdfasts** that anchor them to the substrate. The leafy blades often have air bladders that keep the kelps afloat (fig. 27.23). Intercalary growth occurs in the stipe, so the oldest tissues are at the tips of the blades.

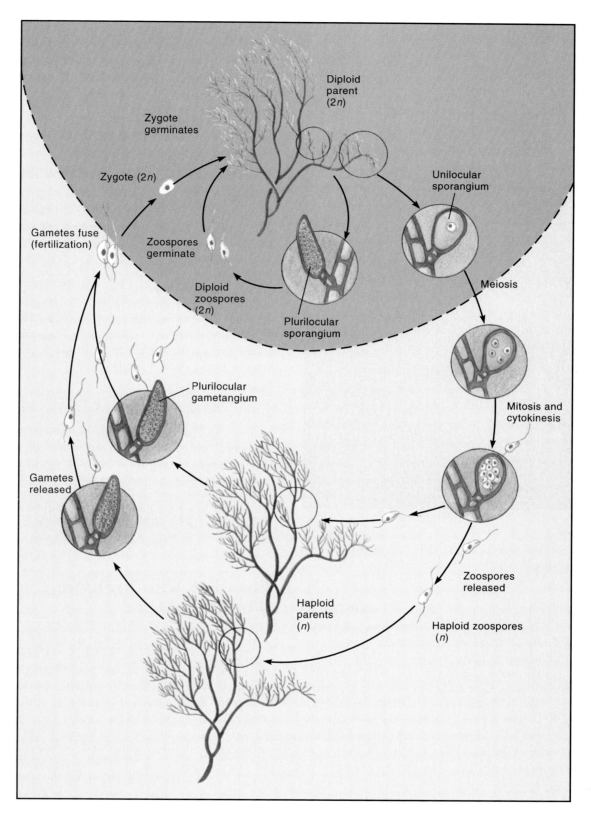

Labels within figure:

- Diploid parent (2n)
- Zygote germinates
- Zygote (2n)
- Unilocular sporangium
- Gametes fuse (fertilization)
- Zoospores germinate
- Meiosis
- Diploid zoospores (2n)
- Plurilocular sporangium
- Mitosis and cytokinesis
- Plurilocular gametangium
- Gametes released
- Zoospores released
- Haploid parents (n)
- Haploid zoospores (n)

figure 27.22

The life cycle of *Ectocarpus* alternates between isomorphic gametophytes and sporophytes, as in some green algae.

figure 27.23

Giant kelp (*Macrocystis*), showing stipes, blades, and air bladders. Giant kelps are anchored to the ocean floor by holdfasts.

During sexual reproduction in kelps, certain cells on the surfaces of some blades enlarge into unilocular sporangia, each of which releases 32–64 biflagellate zoospores. Sporangia occur in large groups called **sori** (singular, **sorus**). The spores of some kelps have sex chromosomes of the X-Y type, which are common in animals but rare in algae and plants. Each spore germinates into a microscopic filament that is either a male or a female gametophyte. The female gametophytes have oogonia that extrude eggs, and the male gametophytes produce biflagellate sperm that swim to the eggs. After fertilization, the zygote germinates into a new sporophyte. The life cycle of kelps resembles those of *Ulva*, *Cladophora*, and *Ectocarpus* because they all have sporic meiosis. However, the kelps differ in the dominance of the sporophytic phase of the life cycle.

Rockweeds are similar to the kelps. However, meiosis in rockweeds is gametic, as it is in animals. In the rockweed genus *Fucus*, the fertile swollen tips of branches are embedded with **conceptacles,** which are containers that bear oogonia and antheridia (fig. 27.24). Eggs and sperm are forced out of the conceptacles by a slimy substance that absorbs water and sets the gametes free. Immediately after fertilization, the zygote germinates into another diploid rockweed.

Classification of the Brown Algae

The Division Phaeophyta has just one class, the Phaeophyceae, which includes about thirteen orders. Classification of brown algae is based on the complexity of vegetative organization and the type of life cycle. There are three main orders of brown algae based on these kinds of features; these groups are represented by *Ectocarpus* (Order Ectocarpales), *Macrocystis* (Order Laminariales), and *Fucus* (Order Fucales). The filamentous brown algae are assumed to be primitive because of the simplicity of their vegetative organization. These and other assumptions about the evolutionary relationships within the Phaeophyta have yet to be examined in the framework of testable phylogenetic hypotheses.

c o n c e p t

Brown algae are exclusively filamentous or multicellular and complex. They include the largest of the algae, the giant kelps. Although sexual reproduction generally involves sporic meiosis and isogamous or anisogamous gametes, some brown algae are oogamous and have gametic meiosis. Spores and motile gametes generally have two lateral flagella. Asexual reproduction is mostly by zoospores.

Division Rhodophyta: The Red Algae

Like the brown algae, the red algae are mostly marine organisms that are either microscopic filaments or macroscopic leafy branches. One group, the coralline algae, have cell walls impregnated with carbonates of calcium and magnesium, which makes the walls hard and crusty (fig. 27.25). The coralline algae are an important part of coral reefs. The Division Rhodophyta also includes unicellular species and species with coenocytic filaments, features that are common in the Chlorophyta but unknown in the Phaeophyta. The most significant features of the approximately 3,900 species of red algae are their phycobilins, their single thylakoid lamella per band, and their lack of flagellated cells.

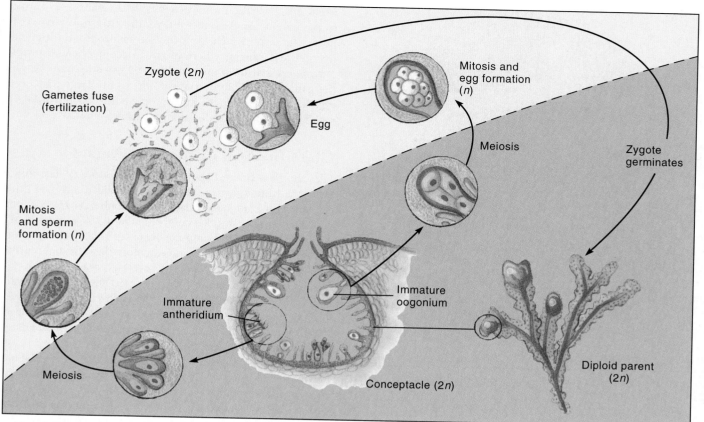

figure 27.24

Conceptacles of *Fucus* are embedded containers that bear oogonia and antheridia. Meiosis in oogonia and antheridia produces eggs and sperm, respectively, which are the only haploid cells in the life cycle of *Fucus*. Meiosis is therefore gametic.

figure 27.25

Corallina, a macroscopic red alga. The hard, crusty cell walls of *Corallina* contain carbonates of calcium and magnesium.

Red algae are red because they have an abundance of phycoerythrin, a red phycobilin. Phycoerythrin absorbs blue light, which penetrates more deeply into water than do other colors of light. This means that red algae can photosynthesize at greater depths than other algae and explains why some red algae can grow at depths of more than 200 m.

The gametes of red algae are all nonmotile. The female gametes are therefore eggs, so sexual reproduction in this division is oogamous. The nonmotile male gametes are called **spermatia** (singular, **spermatium**) to distinguish them from the motile sperm of other algae.

Unicellular Red Algae

The simplest red algae are represented by unicellular organisms in the genus *Porphyridium*. Each single-celled individual lacks a cell wall and has a prominent, star-shaped chloroplast (fig. 27.26). *Porphyridium* often grows on the surfaces of moist soils or pots in greenhouses; populations form shiny, blood-red or bluish-green patches, depending on the relative

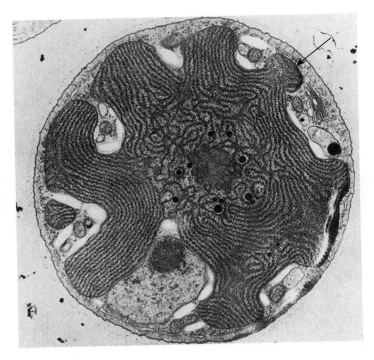

figure 27.26

Electron micrograph of a star-shaped chloroplast (arrow) of *Porphyridium,* ×15,000.

amounts of red or blue phycobilins present. The cells move by secreting a slimy substance from one end. Reproduction is entirely asexual by mitotic cell division.

Multicellular Red Algae

Red algae are mostly macroscopic and leafy or filamentous. Most of these larger forms have an unusual three-phase life cycle. In *Polysiphonia,* for example, the haploid phase consists of male and female gametophytes that look alike (fig. 27.27). These filamentous gametophytes are coenocytic. The males release free-floating spermatia that are carried by currents to the female gametophytes. The egg container has a special name, the **carpogonium,** because of its unique appearance and function in sexual reproduction. As it is formed, the carpogonium sprouts a sticky, hairlike projection, called a **trichogyne,** that spermatia stick to as they float around the female gametophyte. Once it gets stuck, each spermatium releases its nucleus into the trichogyne, after which the nucleus migrates toward the egg nucleus in the carpogonium. After the zygote is formed by fertilization between a spermatium and an egg, it moves to an **auxiliary cell** and divides mitotically into a multicellular sporophyte. This phase of the life cycle is the **carposporophyte,** and it releases diploid **carpospores** that are produced mitotically. Carpospores settle onto rocks or other substrates and germinate into diploid sporophytes, called **tetrasporophytes.** Because the tetrasporophyte looks like the gametophytes, these two phases are

isomorphic. Unlike the carposporophyte, which stays attached to the parent female gametophyte, the tetrasporophyte is free-living. Lateral branches on the tetrasporophyte differentiate into **tetrasporangia,** where meiosis occurs. Meiosis produces **tetraspores** that form male and female gametophytes. Although the gametophytes and tetrasporophytes are large and isomorphic in *Polysiphonia,* the microscopic tetrasporophytes of other red algae are much smaller than the gametophytes.

Classification of the Red Algae

Like the Phaeophyta, the Division Rhodophyta has only one class, the Rhodophyceae. This class is divided into two groups, the subclass Bangiophycidae and the subclass Florideophycidae. The Bangiophycidae comprise about 1% of the red algae. They are unicellular or multicellular, and each cell has a single nucleus and a single, star-shaped chloroplast. Sexual reproduction in the Bangiophycidae is rare (see reading 27.2, "The Strange Case of *Bangia*"). In contrast, the Florideophycidae are exclusively multicellular, and each cell has many disc-shaped chloroplasts. Sexual reproduction is well developed in the Florideophycidae; most species have a three-phase life cycle like that of *Polysiphonia.*

Several phylogenies have been proposed for the red algae, but all are plagued by apparent parallelisms. Nevertheless, all eleven orders of the subclass Florideophycidae can be distinguished by two shared-derived character states: the presence of tetrasporangia and the formation of **gonimoblasts,** which are unbranched filaments that bear sporangia in carposporophytes. Earlier suspicions that the Bangiophycidae is an artificial subclass have been borne out; it appears as a paraphyletic group in every cladistic analysis that has been done so far.

c o n c e p t

Red algae are unicellular, filamentous, or multicellular and complex. Unicellular and small filamentous members of this division rarely reproduce sexually. Multicellular red algae have a more complex life history than any other group of algae because their reproductive cycles have three phases: the gametophyte, the carposporophyte, and the tetrasporophyte. Red algae do not produce motile cells.

Other Divisions of Algae

In addition to the three divisions already discussed, algae also include the Division Chrysophyta and three divisions that have only unicellular species. Each of these four divisions has one or more features that distinguish it from other algae (see table 27.1). Each division also shares important biochemical and ultrastructural characteristics with the green, brown, or red algae. Some of the unicellular species resemble protozoans because

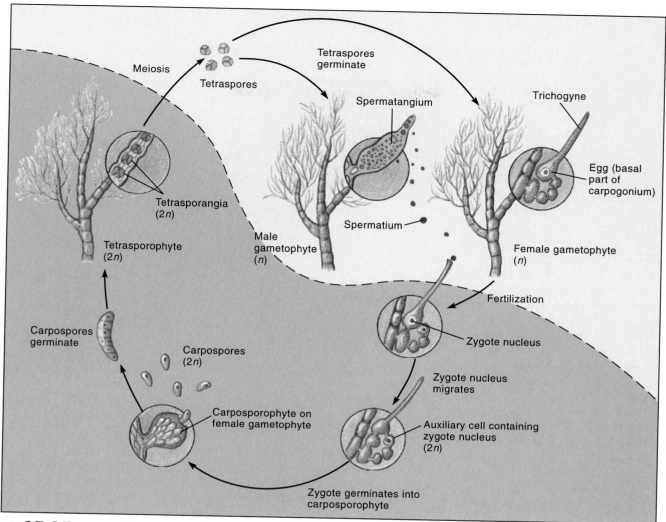

figure 27.27

Life cycle of *Polysiphonia*. This genus represents a three-phase life cycle that characterizes most red algae; there are two sporophytic phases and one gametophytic phase.

they lack chlorophyll and engulf their food. Although photosynthetic pigments and chloroplast structures probably reflect a common ancestry, the origin of other features is not clear. The possible origins of algae are discussed later in this chapter.

Division Chrysophyta: Diatoms and the Golden-Brown and Yellow-Green Algae

The Chrysophyta form the largest division of algae, with more than 11,000 species. About 10,000 species are diatoms in the Class Bacillariophyceae, almost all of which are unicellular with cells that are either bilaterally symmetrical (**pennate**, fig. 27.28a) or radially symmetrical (**centric**, fig. 27.28b). They lack flagella, except for some male gametes, but some diatoms can be motile by secreting a sticky substance through pores that are next to a fine groove called a **raphe**. This substance sticks to other objects, enabling the diatom to pull itself toward them. Diatoms are unique because of their exquisitely ornamented glass shells.

The golden-brown algae (Chrysophyceae) and yellow-green algae (Xanthophyceae) are also mostly unicellular, but some are colonial or filamentous and coenocytic (fig. 27.29). Chrysophytes occur in both freshwater and saltwater.

Sexual reproduction in chrysophytes includes isogamy, anisogamy, and oogamy. Meiosis is gametic in the diatoms but zygotic in the other classes. Reproduction has been more thoroughly studied in diatoms, probably because of their large number and their great economic importance (see "The Economic Importance of Algae" in this chapter). In asexual

THE STRANGE CASE OF *BANGIA*

Bangia atropurpurea is a macroscopic, filamentous red alga that grows attached to rocks or wood in marine and freshwater (reading fig. 27.2). It rarely reproduces sexually. Algae like *Bangia* are often brought into the laboratory to be studied in culture because it is difficult to observe all of the components of their life cycles in nature. Botanists who collect these algae often do not know what phase of the algal life cycle they have collected. For example, the red alga *Bangia atropurpurea* had only one known phase for a long time, until it was brought into laboratory culture. At first, when it was grown in no more than 12 hours of light per day, *Bangia* produced spores that formed more *Bangia*. However, when the daylength was increased to more than 12 hours, *Bangia* produced spores that grew into another genus, *Conchocelis*, whose alternate phase had also been previously unknown. Furthermore, when cultures were switched back to less than 12 hours of light, they produced spores that grew back into *Bangia*. These two genera—that is, *Bangia atropurpurea* and *Conchocelis*—turned out to be alternate phases of the same organism. Because *Bangia* was named first, the genus name for the two phases is *Bangia*. Nevertheless,

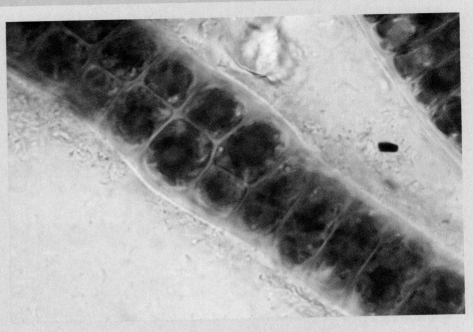

reading figure 27.2

Bangia atropurpurea.

biologists still refer to this organism as having a *Bangia* phase alternating with a *Conchocelis* phase in its life cycle.

The puzzle about the relationship between *Bangia* and *Conchocelis* is still not fully resolved, because the sexual cycle is poorly understood. Both phases have the same chromosome number, either $n = 3$ or $n = 10$; therefore, if fertilization occurs at all, then meiosis must be zygotic.

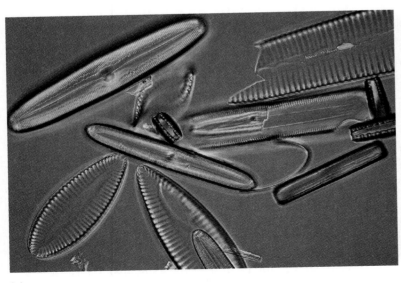

(a)

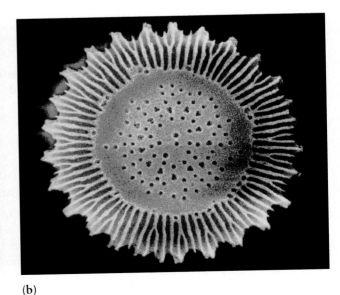

(b)

figure 27.28

Diatoms. (a) Scanning electron micrograph of a pennate diatom, ×1,000. (b) Scanning electron micrograph of a centric diatom, ×1,000.

(a)

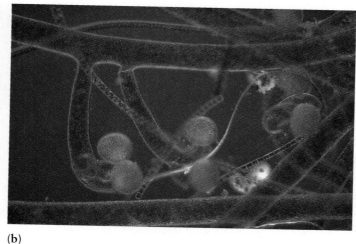

(b)

figure 27.29

Other chrysophytes. (a) *Dinobryon* is an example of a colonial chrysophyte whose cells live in urn-shaped shells that are attached in a branched series, ×150. (b) *Vaucheria* is a coenocytic, filamentous chrysophyte, ×100. The spherical structures are cells that were walled off at the tip of a filament; they will form flagella, swim away, and eventually settle and grow into new individuals.

reproduction, diatom cells divide mitotically while they are still encased in their rigid glass (SiO_2) walls. The walls consist of two parts, called **valves,** one of which fits tightly over the other like the lid of a petri dish (fig. 27.30). The new cells inside the rigid wall expand and force the valves to separate. New inner valves form inside the old valves so that one of the new cells is smaller than the parent cell and the other new cell is the same size. Division of the smaller cell again produces two new cells, one of which will be even smaller. Thus, in a population of diatoms, some cells are about half the size of others. When the lower size limit is reached, the smallest cells become sexual and produce new large cells.

Vegetative cells of diatoms are diploid and undergo meiosis to produce gametes. In certain diatoms, three of the four meiotic products disintegrate, and the remaining cell becomes an egg. In others, all four products become uniflagellate sperms, which are attracted to the egg for fertilization.

In some other diatoms, two of the meiotic products disintegrate, and the remaining two cells function as isogametes, which fertilize another pair of isogametes to make two zygotes. The zygotes of diatoms, called **auxospores,** lack cell walls and expand considerably before making new glass walls.

The Chrysophyta share many features with the Phaeophyta. For example, except for the yellow-green algae, the Chrysophyta have chlorophylls *a* and *c* and the golden-brown pigment fucoxanthin. The color of the yellow-green algae comes from the dominance of yellow beta-carotene over green chlorophyll. Motile cells can have one to three whiplash or tinsel flagella. Some motile cells have lateral flagella and look like the motile cells of the brown algae. The photosynthetic storage carbohydrate of the Chrysophyta is **chrysolaminarin,** which is similar to the brown algal laminarin.

Classification of the Chrysophyta is controversial. Some botanists argue that the similarities between the Chrysophyta and Phaeophyta are significant enough to merge both groups into the same division. Others argue that the reproductive and vegetative features of *Vaucheria* and other Xanthophyceae are sufficiently distinctive to elevate this group to a division, the Xanthophyta. Regardless of their taxonomic status, the Chrysophyta have chloroplasts that are apparently derived from the same prokaryotic endosymbiont that gave rise to chloroplasts in the brown algae (see reading 3.2, "Cellular Invasion: Origin of Chloroplasts and Mitochondria").

Divisions of Unicellular Algae

Three divisions of algae consist exclusively of flagellated unicellular organisms: the euglenoids (**Euglenophyta,** 800 species), the dinoflagellates (**Pyrrhophyta,** 3,000 species), and the cryptomonads (**Cryptophyta,** 100 species). These three divisions are traditionally studied by botanists and zoologists alike because of the plantlike and animal-like features of different species. Each division includes species that are nonphotosynthetic and species that have chlorophyll *a* and beta-carotene. In addition, the euglenoids have chlorophyll *b*, and the dinoflagellates and cryptomonads have chlorophyll *c*; cryptomonads also have phycobilins. The chloroplasts in all three divisions are surrounded by an extra envelope of endoplasmic reticulum. The diversity of pigments and structural features among the chloroplasts of these groups is indirect evidence that their plastids came from endosymbiotic eukaryotes.

Euglenoids and cryptomonads have no cell walls. Instead, their cells are bounded by a flexible **pellicle,** which is a plasma membrane that has extra inner layers of proteins and a

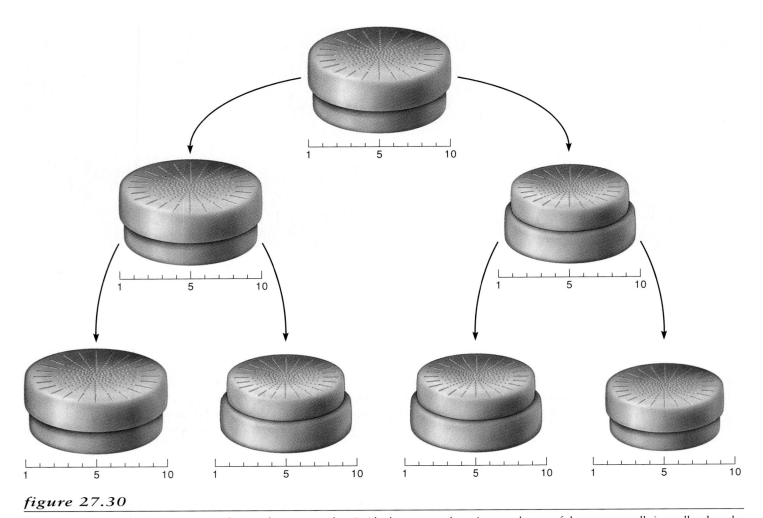

figure 27.30

Cell division in diatoms. Each successive division forms new valves inside the parent valves. As a result, one of the two new cells is smaller than the parent cell, and one is the same size.

grainy outer surface (fig. 27.31). Dinoflagellates also lack cell walls, but most species have cellulose plates interior to the plasma membrane. These plates are grouped into armorlike arrangements that are useful for identifying different genera (fig. 27.32). The dinoflagellates have the most distinctive flagellar pattern among the unicellular algae. Both flagella are lateral, but one coils around the cell like a belt, while the other trails the cell as a rudder. The rhythmic beating of these flagella propels the dinoflagellate through the water like a spinning top. Most dinoflagellates live in saltwater habitats.

Sexual reproduction is unknown in the Euglenophyta and Cryptophyta; reproduction occurs only by mitotic cell division. In euglenoids, the nuclear envelope persists during mitosis. The cryptomonads have a nucleus-like organelle between the outer chloroplast membrane and the chloroplast ER. This organelle, called a **nucleomorph**, is not present in any other algae.

A few species of the Pyrrhophyta reproduce sexually, but most species are exclusively asexual. Sexual dinoflagellates are either isogamous or anisogamous, and meiosis is usually zygotic. Some dinoflagellates are apparently diploid, but their sexual cycles are not well understood. Nuclear division in the Pyrrhophyta is unusual, because dinoflagellate chromosomes are permanently condensed and lack histones. Like the euglenoids, the dinoflagellates have a persistent nuclear envelope during mitosis.

c o n c e p t

The largest group of algae are the diatoms, which are unicellular organisms. Regardless of their life histories, diatoms are usually classified with the golden-brown and yellow-green algae in the Division Chrysophyta because of their similar photosynthetic accessory pigments. The relationships of the other divisions of unicellular algae are not so clear. Many organisms in these other divisions are nonphotosynthetic and are often classified as protozoans.

figure 27.31

Colored scanning electron micrograph of *Euglena*, ×2,800. *Euglena* cells are bounded by a flexible periplast instead of a cell wall.

figure 27.32

Many dinoflagellates have cellulose-containing armor plates that give these organisms a highly sculptured appearance.

The Fossil History of Algae

The earliest photosynthetic organisms were undoubtedly prokaryotic; their fossils are more than 3 billion years old. The first photosynthetic eukaryotes may be represented by fossils of the late Precambrian era (1.2 to 1.4 billion years old) from Bitter Springs, Australia. These fossils have structures that resemble nuclei and pyrenoids, but not all paleobotanists agree that they are truly eukaryotic organelles. Nevertheless, green algae certainly existed in the Paleozoic era, approximately 500 million years ago. They are common in the fossil record because they are so well preserved as imprints in calcium carbonate that precipitated around them when they were alive. Most of the earliest green algae resemble living species that have branched, coenocytic filaments. Oogonia from *Chara*-like green algae also occur as calcium carbonate imprints; such representatives of the most complex green algae are more recent, about 360 million years old. Complex red algae probably lived slightly earlier than complex green algae. Ancestors of the coralline red algae are represented in calcium carbonate deposits that are about 400 million years old.

Algae that are commonly found in the fossil record include dinoflagellates, golden algae, and diatoms. Dinoflagellates are preserved as zygotes that formed thick, protective plates around themselves. Golden algae that secrete shells of calcium carbonate also make up significant fossil deposits. The best-known and perhaps the most widespread fossil algae are diatoms. The shells of these algae become "instant fossils" when a cell dies, because silica does not decompose. Diatoms have been identified in sediments deposited about 120 million years ago. Organisms that seem to be diatoms lived almost 200 million years ago. The fine, sculptured patterns of diatom cell walls make them easily distinguishable from one another. Approximately 40,000 species of fossil diatoms, all of which are probably extinct, have been named from different cell-wall

figure 27.33

Deposits of diatomaceous earth such as this one at Lompoc, California, are mined for commercial applications of diatoms.

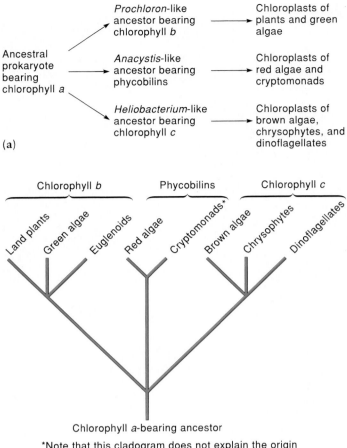

(a)

(b) *Note that this cladogram does not explain the origin of chlorophyll *c* in the cryptomonads (see Table 27.1).

figure 27.34

Evolution of the algae. (a) One possible evolutionary pathway for the origin of chloroplasts from prokaryotic ancestors based on the evolution of their main photosynthetic pigments. (b) Example of a cladogram of the main groups of algae, showing an unresolved trichotomy at the base and later divergence of different divisions on each of the three evolutionary branches.

patterns. Diatom cell walls sometimes settle in large deposits, the largest of which is about 900 m deep and 2 km long, in Lompoc, California (fig. 27.33). Diatoms in this deposit are harvested commercially and used in many ways (see p. 658).

The most complex fossil algae are in the genus *Prototaxites,* which is about 400 million years old. Specimens of this genus have massive, trunklike stipes that are often misidentified as tree stumps. They also contain tubes of two kinds of cells laid end-to-end, which seem to resemble vascular tissue. The larger tubes are hollow, however, and the smaller ones have cross walls with pores that look like those of some algae and fungi. *Prototaxites* is probably related to the Phaeophyta, because no other algae have such large, complex stipes.

Unicellular organisms that resemble dinoflagellates also occur in the Bitter Springs deposit. Like the other organisms in that deposit, however, their resemblance to extant groups may be superficial. The oldest fossils that are confirmed as dinoflagellates are about 430 million years old.

The Origins of Algae

Botanists hypothesize that the chloroplasts of algae came from at least three kinds of photosynthetic prokaryotes. Chloroplasts of the green algae and euglenoids came from a *Prochloron*-like ancestor of the Chloroxybacteria that had chlorophylls *a* and *b*. Chloroplasts of the red algae and cryptomonads came from an *Anacystis*-like ancestor of the Cyanobacteria that had chlorophyll *a* and phycobilins. The genus *Heliobacterium* of the Eubacteria represents the most likely ancestor of the chloroplasts of the Phaeophyta, Chrysophyta, and Pyrrhophyta, which all have chlorophylls *a* and *c*.

One hypothesis about the evolutionary relationships of the three main chloroplast ancestors is shown in figure 27.34a. This hypothesis shows that chlorophyll *a* evolved once from a porphyrin-bearing ancestor. However, there is no clear candidate for the porphyrin-bearing ancestor of the photosynthetic prokaryotes, since porphyrins are also common in all organisms as components of cytochromes.

Aside from the evolution of their chloroplasts from prokaryotes, there are no good characters for evaluating the evolutionary relationships of the three main groups of algae. This means that a cladogram of these three groups must be three-parted at its base, as shown in figure 27.34b. Assuming that cellulosic cell walls evolved only once, then this character is a shared derivation for all divisions of algae. If so, we must assume that groups without cell walls lost the ability to make cellulose, even though their ancestors had it. The cladogram must also include other protoctists, such as water molds, because they have cellulosic cell walls, too.

The red algae and cryptomonads appear to be closely related, probably because of an additional endosymbiosis. According to this hypothesis, the cryptomonads evolved by acquiring a eukaryotic red algal cell; the nucleus of the endosymbiont has been reduced to what we now see as a nucleomorph.

The relationship between the dinoflagellates and the other algae that have chlorophyll *c* is more puzzling. The prokaryote-like chromosomes of dinoflagellates are so different from those of other eukaryotes that the dinoflagellates are sometimes called **mesokaryotes.** The mesokaryotic nucleus in dinoflagellates evolved independently of the eukaryotic nucleus in other algae. However, the dinoflagellate chloroplast apparently shares its ancestry with chloroplasts of the Phaeophyta and Chrysophyta.

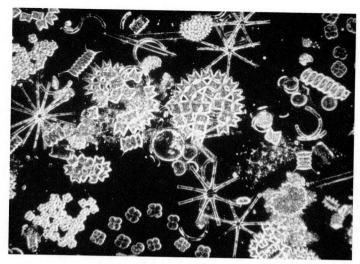

figure 27.35

Light micrograph of a drop of pond water, ×40. Microscopic organisms, including algae, are the plankton that are eaten by many aquatic and marine animals.

c o n c e p t

The fossil history and evolutionary origins of algae show that there are three main evolutionary lines. These are best explained by the character of their chloroplasts. One line has chlorophylls *a* and *b* from a *Prochloron*-like ancestor, one has chlorophyll *a* and phycobilins from an *Anacystis*-like ancestor, and the third line has chlorophylls *a* and *c* from a *Heliobacterium*-like ancestor. The divergence of these three groups probably began at least 500 million years ago, but their origins may be more than 1 billion years old.

The Ecology of Algae

Algae are nearly ubiquitous and occupy a wide variety of habitats. They live primarily in marine and freshwater habitats, either free-floating or attached to rocks, wood pilings, shells of shellfish, or other algae. Many species are terrestrial on moist soil, rocks, stone roofs and walls, or tree bark. Some of their more unusual habitats include clouds and airborne dust, snow, and the fur of certain animals. Some unicellular species are the food-producing endosymbionts inside the cells of protozoans, sponges, sea slugs, sea anemones, and saltwater fish. Green algae live symbiotically with fungi and cyanobacteria in lichens.

The most significant ecological role of algae is as plankton (fig. 27.35). Algae and other unicellular organisms are eaten by small animals, which, in turn, are eaten by larger animals. Thus, algae are the primary producers that support life in marine and freshwater habitats. Commercial fish farmers often exploit this relationship by fertilizing tanks and ponds to enhance the growth of plankton. The oxygen that planktonic algae produce is equally important to life; perhaps 50%–70% of the earth's atmospheric oxygen comes from unicellular marine algae.

Brown and red algae form "forests" in intertidal zones and shallow coastal waters. These marine forests are habitats for teeming populations of many kinds of animals. Coral reefs are special habitats whose primary production comes from coralline red algae. Some red algae precipitate calcium carbonate around them or bond the calcium carbonate from marine sponges. These interactions of red algae, sponges, and other organisms form coral reefs. Corals and many other kinds of nonphotosynthetic organisms also harbor green algae, yellow-green algae, or dinoflagellates as food-producing endosymbionts. Coral reefs can form extensive, highly complex ecosystems that support many species of marine animals. The most spectacular example of a coral reef is the Great Barrier Reef off the east coast of Queensland, Australia.

Planktonic and other microscopic algae are especially conspicuous when their populations expand rapidly into large **blooms.** These blooms may be induced by warm temperatures, phosphate pollutants, or other factors that enhance algal growth. In freshwater, algal blooms sometimes get so large that when they ultimately begin to decay, oxygen is used by the respiration of decomposing bacteria faster than it can be replenished by algal photosynthesis. When this happens, fish and other animals die from lack of oxygen, and their decaying bodies make the problems worse.

Blooms of some dinoflagellates are called *red tides* because of the abundance of these red-pigmented organisms. Unlike other algal blooms, the cause of red tides is unknown. Red tides usually occur at least 10 miles offshore, and are especially common along the Gulf coast of Florida and the coast of central California. One of the most interesting features of many red tides is that they are often bioluminescent—they glow in the dark just like fireflies. Red tides kill millions of fish each year because of a nerve toxin called brevetoxin secreted by dinoflagellates. Brevetoxin is responsible for the red or brown discoloration of water during a red tide. Shellfish that eat dinoflagellates appear to be immune to these toxins, but humans or other vertebrates that eat these shellfish may suffer poisoning and death. In 1996 red tides killed hundreds of endangered manatees off southwest Florida.

The Lore of Plants

During a bloom, algae can be so abundant that only 0.03 m^3 (about 1 ft^3) of water can contain more than 13 million organisms.

The Physiology of Algae

The Influence of Light and Temperature on Reproduction

Reproduction in many multicellular algae is influenced by daylength and temperature (see reading 27.2, "The Strange Case of *Bangia*"). For example, gametophytes of the brown alga *Sphacelaria* produce gametes at 4° C and 12° C under a 12-hour photoperiod. At 20° C, they reproduce only asexually. Reproduction is also asexual when temperatures are lower but the photoperiod is longer.

Euglena cells point their eyespots toward light, so that the cells are parallel to the direction of the light. In this way, *Euglena* cells are oriented for swimming toward the light. Conversely, in multicellular algae, zoospores with eyespots often swim away from the light toward the seabed, where they can germinate. Regardless of the response, cells with eyespots have light receptors. Light receptors are usually associated with the absorption of blue light by carotenoids, but in *Euglena* the receptor pigment may be riboflavin.

Osmotic Regulation

The unicellular, wall-less green alga *Dunaliella* has a remarkable range of tolerance for different salt concentrations: it can live in water with NaCl concentrations between 0.05 M and 5 M (seawater is about 0.03 M). *Dunaliella* maintains cytoplasmic concentrations of Na$^+$ at 50–100 times lower than those of the external environment. Because it has no cell wall and it is easy to grow in laboratory culture, *Dunaliella* is used as a model for studying membrane functions. Much of our information on osmotic regulation and control of membrane potential comes from this organism.

Silicon Metabolism

Silicon is an essential element for the development of diatoms, certain seedless vascular plants, and many flowering plants. In living organisms, silicon is only known in the form of silica (SiO_2). The physiology of silica metabolism is best known in diatoms, because they can be easily manipulated in laboratory cultures. Silicon is absorbed by diatoms as silicic acid, $Si(OH)_4$. In general, growing cells absorb silicic acid much faster in the light than in the dark.

Diatoms seem to have an ecological and evolutionary advantage because they can make cell walls out of silica. Silica cell walls are more economical to make, since their synthesis requires up to 10 times less metabolic energy than does the synthesis of cellulose or other cell-wall materials. Moreover, silica is highly adsorptive for cationic minerals (e.g., Ca^{2+}, Mg^{2+}) because of the density of electronegative oxygen atoms. Such adsorption, along with the intricately sculpted valves that have a large surface area, may help diatoms to compete for nutrients, especially when nutrients are at low concentrations.

c o n c e p t

Algae are the major primary producers in marine and freshwater habitats. In the form of plankton, algae are the foundation of food webs that support animal life in water. The rapid vegetative growth of kelps and other macroscopic algae is also important in "marine forest" habitats that occur along some coastlines. Algal reproduction and growth are sensitive to light and temperature.

The Economic Importance of Algae

Diatoms

As a group, the diatoms are perhaps the most economically important algae. Indeed, they have a substantial role for the fishing industry because they are food for freshwater and marine animals. Diatom cell walls are also harvested from large deposits as *diatomaceous earth* (see fig. 27.33), which is used in many industries. For example, diatomaceous earth is an abrasive in metal polishes and a few brands of toothpaste; it is also used in filters for cleaning swimming pools and clarifying beer and wine. Reflective paint on highways, road signs, and license plates also contains cell walls of diatoms. Diatomaceous earth is an absorbent in kitty litter, and one of the recipes for dynamite calls for diatomaceous earth as an inert absorbent for mixing with nitroglycerin. Because much of the earth's fossil oil originates from or is absorbed by diatoms, the presence of diatoms in sample cores of the earth is often a useful indicator of oil deposits.

Industrial Polymers

Red and brown algae produce cell-wall polysaccharides that have many industrial uses. The main red algal polysaccharides are **carrageenan** from seaweeds such as Irish moss (*Chondrus crispus*) and **agar** from several other seaweeds, including species of *Gracilaria* (fig. 27.36). Carrageenan is used to stabilize or emulsify paints, cosmetics, cream-containing foods, and

figure 27.36

Irish moss (*Chondrus crispus*) is a red alga that is commercially important as a source of carrageenan.

chocolate. The main use of agar is in preparing culture media to grow bacteria and other microorganisms for laboratory research. Agar is also used as a gel for canning fish and meat and for making desserts. Some medicines contain agar as an inert carrier. The most useful polysaccharide from brown algae is **algin,** which is obtained from species of *Macrocystis* that are cultivated in large kelp beds. The uses of algin are similar to those of carrageenan and agar.

One of the more promising uses of algal polymers is in medicine. Polysaccharides extracted from certain unicellular and colonial green algae stimulate the immune systems of animals, which may enhance resistance to disease. Although several patents for these potential medicines have been awarded in Japan, algal polysaccharides are not yet approved for human medicinal use.

Algae as Food

Many algae have been used traditionally as food, although most algae are not particularly nutritious. Some seaweeds, however, contain a lot of iodine, which is an important mineral in the thyroid gland. Furthermore, although much of the cell-wall material of algae cannot be digested by humans, some seaweeds contain useful amounts of protein. These include nori (*Porphyra tenera*), a red alga that is used mostly as a flavoring in soups and salads and as a wrapping around small rolls of rice (*maki-sushi*). Species of *Ulva*, *Laminaria*, and other seaweeds are also eaten fresh or pickled.

Other Uses of Algae

In recent years, algae such as *Spirulina* and *Dunaliella* have become increasingly important as sources of vitamins, industrial chemicals, food additives, and fertilizer. Algae are also having increasingly important roles in treating sewage and industrial wastes. These uses of algae as sewage treatment "plants" are especially important in areas that cannot afford basic sanitation services.

Chapter Summary

The vegetative organization of algae includes microscopic unicellular, colonial, and filamentous organisms and complex macroscopic seaweeds. Of the seven divisions of algae, three include macroscopic forms, and four consist almost entirely of unicellular organisms. Most algae reproduce both sexually and asexually. In sexual reproduction, gametes may look alike (isogamy), be of two different sizes (anisogamy), or consist of sperm and eggs (oogamy). In most green algae, zygotes undergo meiosis, so the only diploid cell in the life cycle is the zygote itself. In other green algae and other multicellular algae, meiosis is either sporic or gametic. Only the red algae lack motile cells in all phases of the life cycle. In filamentous algae, the sexual alternation between two phases may occur between dominant diploid and reduced haploid phases, or the reverse. Some algae alternate between isomorphic sporophytes and gametophytes; larger seaweeds, such as kelps and rockweeds, have dominant diploid phases that alternate with microscopic haploid filaments or with gametes as the only haploid cells in the life cycle.

The evolutionary relationships of algae show that green algae are the ancestors of plants. Within the green algae, genera of the Charophyceae are apparently the closest relatives of land plants. Some of the most significant features shared by green algae and plants include the development of a phragmoplast during cytokinesis, the retention of the zygote by the gametophyte, and the ability to make flavonoids.

The fossil record of algae contains representatives that are at least 500 million years old. Some fossils that may be algae are up to 1.4 billion years old, but there is no general agreement that these fossils are truly eukaryotic. The most common fossil algae are diatoms, because their glass cell walls do not decompose in most sediments. The first likely brown algae are represented by giant stipes that are about 400 million years old.

Relationships among algae are shown by the features of their chloroplasts; however, these features do not show which two of the three main evolutionary lines of algae are more closely related to each other. The cryptomonads (Cryptophyta) are closest to the red algae; the diatoms and other Chrysophyta are closest to the brown algae. Dinoflagellates (Pyrrhophyta) may also be closely related to the brown algae, but dinoflagellate nuclei are mesokaryotic and apparently very primitive. The euglenoids (Euglenophyta) share their main photosynthetic features with the green algae.

The most important ecological role of algae is as plankton. In this role, algae are the primary producers in the food webs of marine and freshwater habitats; plankton feeders are then eaten by other animals. Certain red algae cause calcium carbonate to precipitate around them, or they cement the calcium carbonate from marine sponges. The interaction of red algae with sponges, corals, and other sea life forms coral reefs.

The growth and reproduction of algae are sensitive to photoperiod and temperature. Some kelps reproduce sexually only at lower temperatures, whereas certain red algae alternate phases only when daylength is increased or decreased. Unicellular marine algae, especially those without cell walls, have evolved effective mechanisms for maintaining osmotic potential in spite of high salt concentrations in the external environment. Perhaps the most unusual aspect of algal physiology is the silicon metabolism of diatoms. These algae absorb silicic acid and convert it into silica-based cell-wall materials.

Diatoms are probably the most economically important group of algae because they produce large deposits of discarded cell walls. These fine glass particles are used in many commercial, nonfood products. Polysaccharides from kelps and red algae are used to stabilize and emulsify many foods and nonfood products. Some algae are used directly as human food.

 ### What Are Botanists Doing?

Go to the library and read an article about the classification of green algae. Summarize the main points of the article. Considering what you know about algae, were you convinced by the arguments presented by the author(s)? Explain your answer.

 ### Writing to Learn Botany

Although plants probably evolved from green algae, they are classified in different kingdoms by some botanists. Do you agree with this classification? Explain your answer.

Questions for Further Thought and Study

1. What is the evidence that green algae and plants had a common ancestor that is not shared with either the brown algae or the red algae?

2. How can you tell if isogamy, anisogamy, or oogamy is each the result of parallel evolution?

3. Why can red algae grow at greater depths than other algae?

4. Why are only some algae well preserved in the fossil record?

5. Why are algal polysaccharides so important as emulsifiers in many commercial products?

6. What features of algae are plantlike? What features of algae are not plantlike?

7. Botanists generally agree that the chloroplasts of green algae, brown algae, and red algae are derived from three different prokaryotic endosymbionts. How do the other divisions of algae fit into this three-part pattern of evolution? What features of these other algal divisions do not seem to fit this pattern?

8. Both *Ectocarpus* and *Ulva* have life cycles that alternate between isomorphic phases. *Ectocarpus* populations have two kinds of individuals, and *Ulva* populations have three kinds of individuals. Why are they different? What might be the adaptive significance of this difference?

9. What features of coralline red algae are important for their role in making coral reefs? Why is the growth pattern of coral reefs determined by algae, even though the main components of reefs are corals?

10. What features of charophytes distinguish them from other algae? What is the potential evolutionary significance of these distinguishing features?

11. Figure 27.28b shows pores in the glass shells of diatoms. What function might these pores have?

12. When grown in darkness, euglenoids such as *Euglena* lose their chloroplasts and live as heterotrophs. Of what selective advantage is this?

Web Sites

Review the "Doing Botany Yourself" essay and assignments for Chapter 27 on the *Botany Home Page*. What experiments would you do to test the hypotheses? What data can you gather on the Web to help you refine your experiments?

Here are some other sites that you may find interesting:

http://ucmp1.berkeley.edu/greenalgae/greenalgae.html
At this site you'll find interesting information about the green algae (Chlorophyta).

http://www.indiana.edu/~diatom/diatom.html
The DIATOM-L is an electronic distribution list of researchers involved with diatom ecology, taxonomy, paleoecology, and other Bacillariophycean pursuits.

http://chrs1.chem.lsu.edu/~wwwpb/Chapman/algae.html
Review the Algae Gallery of the Chapman Lab, Department of Biology, Louisiana State University, USA for high-resolution photographs and interesting information about these protists.

http://seaweed.ucg.ie/seaweed.html
Check out the "seaweed page" sponsored through the Department of Botany, University College, Galway.

Suggested Readings

Articles

Anderson, D. M. 1994. Red tides. *Scientific American* 271 (August):62–68.

Anderson, R. A. 1992. Diversity of eukaryotic algae. *Biodiversity and Conservation* 1:267–292.

Bremer, K. 1985. A summary of green plant phylogeny and classification. *Cladistics* 1:369–385.

Corliss, J. O. 1984. The Protista kingdom and its 45 phyla. *BioSystems* 17:87–126.

Graham, L. E. 1984. *Coleochaete* and the origin of land plants. *American Journal of Botany* 71:603–608.

Manhart, J. R., and J. D. Palmer. 1990. The gain of two chloroplast tRNA introns marks the green algal ancestors of land plants. *Nature* 345:268–270.

Maxwell, Christine D. 1991. A seaweed buffet. *The American Biology Teacher* 53:159–161.

Mishler, B. D., L. A. Lewis, M. A. Buchheim, K. S. Renzaglia, D. J. Garbary, C. F. Delwiche, F. W. Zechman, T. S. Kantz, and R. L. Chapman. 1994. Phylogenetic relationships of the "green algae" and "bryophytes." *Annals of the Missouri Botanical Garden* 81:451–483.

Saffo, M. B. 1987. New light on seaweeds. *BioScience* 37:654–664.

Books

Bold, H. C., and M. J. Wynne. 1985. *Introduction to the Algae.* 2d ed. Englewood Cliffs, NJ: Prentice-Hall.

Graham, L. E. 1993. *Origin of Land Plants.* New York: Wiley.

Lee, R. E. 1989. *Phycology.* 2d ed. New York: Cambridge University Press.

Lembi, C. A., and J. R. Waaland, eds. 1988. *Algae and Human Affairs.* New York: Cambridge University Press.

Round, F. E., ed. 1988. *Algae and the Aquatic Environment.* Bristol: Biopress Ltd.

Van den Hoek, C., D. Mann, and H. M. Johns. 1996. *Algae.* New York: Cambridge University Press.

The discharge of spores from
the moss *Polytrichum*.

Bryophytes

28

Chapter Outline

Chapter Overview

Plants are multicellular eukaryotes with cellulose-rich cell walls, chloroplasts containing chlorophylls *a* and *b* and carotenoids, and starch as their primary food reserve. Plants evolved from green algae and are currently divided into two main groups: bryophytes and vascular plants. Vascular plants include ferns, conifers, flowering plants, and related groups, and will be considered in subsequent chapters.

Bryophytes include mosses, liverworts, and hornworts. Although bryophytes are most noticeable when they grow in dense mats, they can grow just about everywhere—including on bark and exposed rocks, where other plants cannot grow. Bryophytes are especially common in moist areas and lack the specialized vascular tissues that characterize other groups of plants. Gametophytes dominate the life cycle of bryophytes. Sporophytes are short-lived and obtain their food from the gametophytes to which they are attached.

Although most biologists agree that early photosynthetic organisms evolved in the water, true plants are virtually absent in the oceans, and those that occupy freshwater habitats seem to be highly specialized. Aquatic habitats offer tremendous stability when compared to land; there is always plenty of water, and the temperature range is small. Yet this watery medium has its drawbacks. Light does not penetrate far into the water, and most stationary plants cannot grow at depths greater than a few meters. Furthermore, limited amounts of CO_2 (for photosynthesis), nutrients, and O_2 for their submerged parts (especially their roots) can sometimes make freshwater a very inhospitable environment. Clearly, a movement to land would open vast new opportunities to obtain gases and light, if the adventurous plant could solve its water problems.

If we speculate about those first organisms to survive on the land, we would probably consider them to be simple organisms with no organized vascular systems, perhaps like the **bryophytes** (mosses and liverworts). After all, there was no selection pressure for any wasteful vascular tissue while these organisms were living in the water. When we consider these organisms, we are tempted to think about wet habitats where the mosses are close to water, basking in the sun of a bog or cooling off in the spray of a waterfall. Certainly, these are habitats where bryophytes are common. But keep thinking. What about the rocks on a cliff or the sand of the dunes? In fact, can you think of a habitat that has plants where it is impossible to find mosses? There aren't many, and if you visualize some of the rocky habitats in your mind, certainly these organisms undergo tremendous changes in moisture and temperature, even within a single day, occupying habitats where no vascular plants can survive.

Even with so many diverse habitats occupied by plants today, we still consider the move from water to land to have been a major one. The greatest challenge to plants colonizing land was to keep their cells wet. Land plants responded to this challenge in two ways. Some, the ones we call **vascular plants,** acquired lignin, developed a complex water-transport system, and encased themselves in a waxy, waterproof **cuticle.** These adaptations enabled land plants to maintain a watery environment in their leaves, despite the fact that they were suspended in desertlike air. Other plants, the bryophytes, developed other strategies that we are only beginning to understand. They lack lignin, and as a result they lack the complex transport system of other plants. They have only a thin cuticle, if any, and their leaves are only one cell thick, making it easy to lose water. Yet, there are about 16,000 species of bryophytes, more than any other group of plants besides flowering plants.

Add to these simple survival problems the problem of transferring gametes from a male organ to a female organ when the male gamete, the sperm, requires free water to swim! It seems that one of the best "solutions" was to produce gametes only when water was available, but that required developing the gametangia well in advance to be ready on time. Something had to trigger the plants to stop using all their energy for growth and put some of it into making gametangia. That is, a method of receiving and responding to environmental signals was necessary. Nevertheless, even the bryophytes have been successful at organizing their life cycles in a way best suited to their individual environments.

Features of Bryophytes

Bryophytes have the following general features:

1. Most bryophytes are small, compact, green plants (fig. 28.1). Like green algae, they produce chlorophylls *a* and *b*, starch, cellulose cell walls, and motile sperm. Bryophytes usually grow very slowly.

2. Bryophytes lack well-developed vascular tissues and lignified tissues. As a result, they grow low to the ground and absorb water by capillarity. However, some mosses have a central strand of conducting cells that are functionally equivalent to xylem and phloem.

3. Organs such as leaves and roots of vascular plants are defined by the arrangement of their vascular tissues. Since bryophytes lack true vascular tissues, they also lack true leaves and roots. However, many bryophytes have structures that are similar and functionally equivalent to leaves, so they are often referred to as being "leafy."

4. Bryophytes get their nutrients from dust, rainwater, and substances dissolved in water at the soil's surface. Tiny **rhizoids** (hairlike extensions of epidermal cells) along their lower surface anchor the plants but absorb only small amounts of water and minerals. Water and dissolved minerals move by capillarity over the surface of bryophytes.

5. The gametophyte dominates the life cycle; it is often associated with mycorrhizal fungi and is usually perennial. Gametes form by mitosis in multicellular gametangia called **antheridia** (male) and **archegonia** (female) (fig. 28.2; to understand where these structures

28–3

(a)

(b)

(c)

(d)

figure 28.1

The diversity of bryophytes. (a) A hornwort (*Anthoceros*) growing in a highland rain forest. (b) A liverwort (*Conocephalum*) growing near the entrance of a cave. (c) A moss (*Dawsonia*) growing on Mount Kinabalu, Borneo. (d) Another moss (*Hylocomium splendens*) growing in a rain forest in Olympic National Park, Washington.

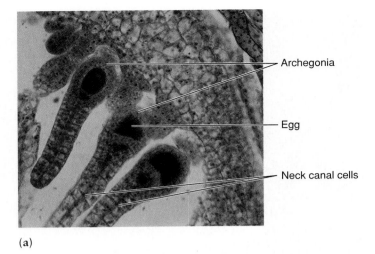

(a)

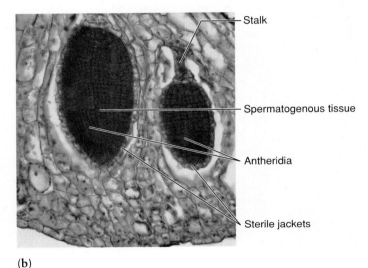

(b)

Labels for (a): Archegonia, Egg, Neck canal cells

Labels for (b): Stalk, Spermatogenous tissue, Antheridia, Sterile jackets

figure 28.2

Gametangia of the liverwort *Marchantia*, a type of bryophyte.
(a) Archegonia, each containing an egg, ×100. (b) Antheridia, each containing spermatogenous tissue, ×100. The locations of these gametangia in bryophytes are shown in figure 28.16.

occur in bryophytes, see fig. 28.3). Each flask-shaped archegonium produces one egg, and each saclike antheridium produces many sperm. Gametangia are protected by a sterile sheath of cells.

6. The sporophyte is often short-lived and unbranched, and produces a terminal sporangium. Although photosynthetic for most of its life, the sporophyte is permanently attached to and partially dependent on the gametophyte. This contrasts with the situation in green algae such as *Ulva* and *Ulothrix*, in which the sporophyte and gametophyte live independently of each other. The sporophytes of bryophytes have no direct connection to the ground. Spores have a cutinized coat and are usually dispersed by wind.

7. Biflagellate sperm swim through water to eggs. Thus, bryophytes need free water for sexual reproduction (in this regard they are like amphibians). Unlike those of nonplants, fertilized eggs develop in (and are nourished by) protective organs. Most reproduction in bryophytes is asexual.

Although small and inconspicuous, bryophytes are studied intensely. For example, much of the early work on sex determination in plants involved bryophytes. In unisexual species, spores that produce female gametophytes contain a large X chromosome, while those that produce male gametophytes contain a corresponding and smaller Y chromosome. This topic remains actively studied, as are the evolutionary relationships of bryophytes.

A Generalized Life Cycle of Bryophytes

Unlike many green algae, bryophytes have heteromorphic alternation of generations; that is, the sporophyte and gametophyte are distinctly different (fig. 28.3). Haploid spores formed by meiosis begin the gametophyte generation. In most bryophytes, the spore germinates to form a **protonema** (plural, **protonemata**), which becomes a leafy gametophyte. These gametophytes are more diverse in bryophytes than in any other group of plants. Moreover, they are the dominant (and most visible) structure of bryophytes that you are likely to encounter. Rhizoids attach the gametophyte to its substrate. The gametophyte grows from an apical cell and produces antheridia and archegonia. Antheridia (singular, antheridium) produce sperm, and archegonia (singular, archegonium) produce eggs. Sexual reproduction in bryophytes requires free water, because the sperm must swim to the egg. Since sperm cannot swim far, antheridia and archegonia must be close to each other for sexual reproduction to occur. Sperm released from an antheridium do not swim randomly; rather, they are attracted by a gradient of a still-unidentified substance(s) produced by an archegonium. A sperm fertilizes an egg in the archegonium, thereby forming a diploid zygote and beginning the sporophyte generation of the life cycle.

The developing sporophyte depends on the gametophyte for nutrition and water. Although sporophytes develop differently in mosses, liverworts, and hornworts, they all ultimately produce sporangia containing sporogenous (spore-producing) tissue. This tissue undergoes meiosis to produce spores, which are released to the environment. If a spore lands in a dry area, it can remain dormant, often for several decades. When water becomes available, the spore germinates (sometimes within a few hours) and forms the gametophyte, thus completing the sexual life cycle.

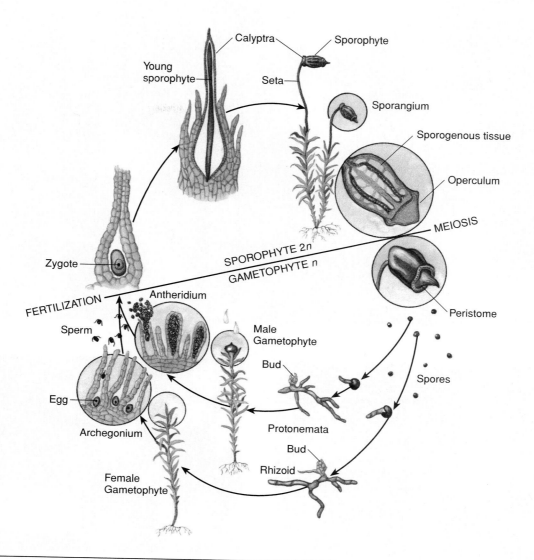

Calyptra
Sporophyte
Young sporophyte
Seta
Sporangium
Sporogenous tissue
Operculum
MEIOSIS
SPOROPHYTE 2n
GAMETOPHYTE n
Zygote
Peristome
FERTILIZATION
Antheridium
Sperm
Male Gametophyte
Bud
Spores
Egg
Archegonium
Protonemata
Bud
Female Gametophyte
Rhizoid

figure 28.3

The general life cycle of a moss, a type of bryophyte. Spores germinate to produce a protonema, which becomes a leafy gametophyte. This gametophyte is the dominant visual structure that you are most likely to see.

c o n c e p t

Bryophytes are small plants lacking complex vascular tissues and supporting tissues. Their life cycle is dominated by free-living, photosynthetic gametophytes that produce archegonia and antheridia, each surrounded by a sterile, protective sheath of cells. Bryophytes require free water for sexual reproduction. Egg and sperm fuse to form a zygote that develops in an archegonium and begins the sporophyte generation. The sporophyte, which is attached to and dependent on the gametophyte for nourishment, produces spores that form gametophytes, thus completing the life cycle.

Some bryophytes produce only antheridia or archegonia. Furthermore, the sporophytes of some mosses have never been found. These bryophytes apparently reproduce only asexually.

The Diversity of Bryophytes

The three major groups of bryophytes are variously treated as classes of a single division, Bryophyta, or as three separate divisions. In the latter case, they are Division Bryophyta (mosses), Division Hepatophyta (liverworts), and Division Anthocerotophyta (hornworts). The classification of all bryophytes in a single division implies a monophyletic origin for mosses, liverworts, and hornworts. As you will see later in this chapter, this idea is in dispute (see "The History and Relationships of Bryophytes").

Mosses

Mosses are remarkably successful plants that thrive alongside more conspicuous vascular plants. The approximately 12,000 species of mosses make up the largest and most familiar group of bryophytes.

(a)

(b)

(c)

(d)

(e)

figure 28.4

The diversity of mosses. (a) Mosses often dominate wet places, such as this waterfall. (b) Moss (*Rhacomitrium*) colonizing a lava bed in Iceland. (c) Cushion moss growing in Antarctica, where daily summer temperatures range from –10° to –30° C. (d) *Splachnum luteum*, a moss that resembles a small flowering plant. The colors of the moss and the chemicals it produces attract insects that disseminate its spores. (e) *Grimmia*, a rock moss that survives on bare rocks, often in scorching sun. In this photo, *Grimmia* is growing among larger, more conspicuous lichens.

figure 28.5

Gametophytes of *Polytrichum*, the hairy-cap moss, showing antheridial heads. Note the radial symmetry and differentiation into "leaves" and "stems."

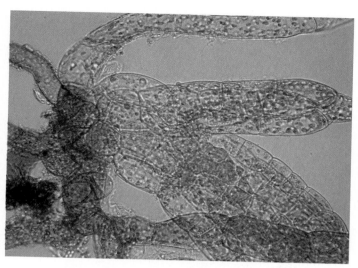

figure 28.6

Moss protonema with a developing bud, ×450.

Moss morphology is very diverse (fig. 28.4). The smallest moss may be the pygmy moss, which is only 1–2 mm tall and completes its life cycle within a few weeks. Luminous mosses such as *Schistostega* (cave moss) and *Mittenia* often grow near entrances to caves and glow an eerie golden green. The upper surface of these mosses is made of curved, lenslike cells that concentrate the cave's dim light onto chloroplasts for photosynthesis.

The Lore of Plants

Cave moss (*Schistostega*) has long fascinated the Japanese. Its eerie glow has been the subject of countless books, newspaper and magazine articles, television shows, and even an opera. There is a national monument to *Schistostega* near the coast of Hokkaido, where it grows near a small cave.

Contrary to what you may think, not all organisms referred to as *mosses* are true mosses. For example, Irish moss and other sea mosses are red algae, Iceland moss and reindeer moss are lichens, club mosses (i.e., a type of lycopod) are seedless vascular plants, and Spanish moss is a flowering plant in the pineapple family.

Mosses have several features that distinguish them from other bryophytes:

1. Gametophytes of most mosses are radially symmetrical (fig. 28.5). Their rhizoids are multicellular and lack chlorophyll.

2. Their spores form a filamentous **protonema** (meaning "first thread") that is an extensive, branched phase of the life cycle (fig. 28.6). Unlike the cross walls of algae, the cross walls of protonemata are oblique.

3. Their gametophytes are always leafy. Leaves are not lobed, are arranged spirally, usually have a midrib, are usually only one cell thick, and have no mesophyll, stomata, or petiole. Chloroplasts are lens-shaped.

4. Intermixed with archegonia and antheridia are sterile, absorptive filaments called **paraphyses** (singular, **paraphysis**).

5. The sterile jacket surrounding the sporangium has stomata and opens by means of an apical lid called an **operculum** that dries and falls off of the capsule. This exposes the tooth-shaped segments of the peristome (fig. 28.7).

6. Their spores can be dispersed over relatively large distances.

Keep these characteristics in mind as we examine the life cycle of a typical moss.

Gametophyte

A moss spore germinates to form a green protonema, which is strikingly similar to a filamentous green alga (fig. 28.6). The protonema grows rapidly from its tip—often more than 40 cm in only a few months—and branches into a tangled mass over the substrate. As the protonema grows, it forms specialized cells having fewer chloroplasts and oblique cross walls. At this stage, the protonema accumulates hormones (e.g., cytokinins) and forms buds that become leafy gametophytes with stemlike axes.

Protonemata can form from many kinds of cells. For example, fragments of gametophytes can usually form a protonema (fig. 28.8). This type of fragmentation is an important means of asexual reproduction in mosses. The Japanese

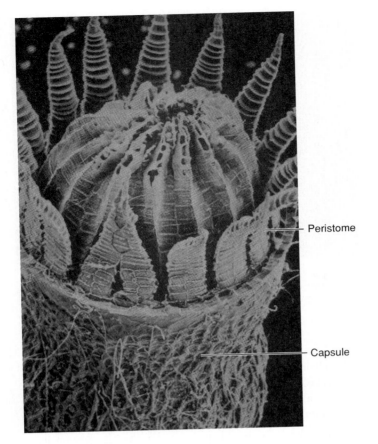

figure 28.7

Scanning electron micrograph of the spore capsule of a moss, ×125.

Peristome

Capsule

figure 28.8

A leaf tip of the moss *Bryum capillare*. Protonemata are growing from a cell on the lower surface of the leaf, while a new shoot is developing from the upper surface.

use tiny fragments of *Polytrichum* to create their lavish moss gardens. In nature, the fragmentation of mosses is so extensive that a cubic meter of arctic snow can contain more than 500 fragments of moss gametophytes, each of which can form a protonema.

In most mosses, archegonia and antheridia form on separate gametophytes, which means the gametophytes are **unisexual.** Other mosses are **bisexual,** meaning that antheridia and archegonia form on the same gametophyte. Gametangia form at the tips of shoots or on separate branches. Archegonia terminate the main shoot or form on reduced lateral branches (fig. 28.9a). The swollen base of the archegonium is called the **venter** and contains a single egg. The slender region above the venter is the **neck.** Cells in the center of the neck disintegrate and produce a liquid that attracts sperm and provides a fluid-filled canal through which sperm swim to reach the egg. Archegonia are often separated by rows of paraphyses that absorb water.

Round or sausage-shaped antheridia form atop short stalks (fig. 28.9b). Inside each antheridium is a mass of tissue that forms coiled sperm. Water-absorbing paraphyses swell the antheridium and help expel the sperm, often in a cottony, mucilaginous mass. Once released, the mass of sperm disperses as individual cells. These sperm are attracted to archegonia by compounds such as sugars, proteins, and acids released by the

archegonia. Sperm swim through the fluid-filled canal to the venter, where one sperm fertilizes the egg. The resulting zygote begins the sporophyte generation.

The gametophytes of some mosses (e.g., *Dawsonia, Polytrichum, Atrichum, Mnium*) have a central strand of conducting tissues made of specialized cells called **leptoids** and **hydroids.** Leptoids resemble sieve tubes in vascular plants because they transport sugars, lack nuclei, have many plasmodesmata, contain callose, and are associated with parenchyma. Hydroids are tracheid-like cells that are dead and empty at maturity and transport water and dissolved minerals. Unlike tracheids, however, hydroids lack lignified secondary cell walls and therefore contribute little support to the gametophyte.

Sporophyte

The sporophyte generation in a moss life cycle begins when an egg and sperm fuse to form a diploid zygote. The spindle-shaped embryo begins developing in the archegonium; cells of the venter divide to accommodate its growth. Cells of the upper part of the archegonium also divide and produce a protective sheath called the **calyptra,** which remains atop the capsule like a tiny pixie cap (fig. 28.10). The calyptra has different shapes in different mosses; for example, it looks like a tiny candlesnuffer in extinguisher mosses. In hairy-cap mosses, such as *Polytrichum*, the calyptra prevents desiccation of immature cells of the sporophyte. If the calyptra is removed too early, the sporophyte stops growing, and no sporangium forms. Stomata on the sporangium have doughnut-shaped

Unit Eight Diversity

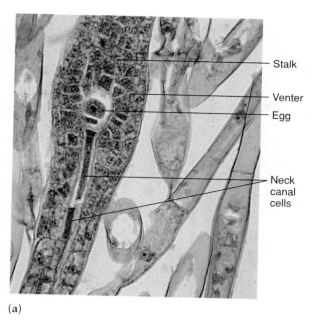

- Stalk
- Venter
- Egg
- Neck canal cells

(a)

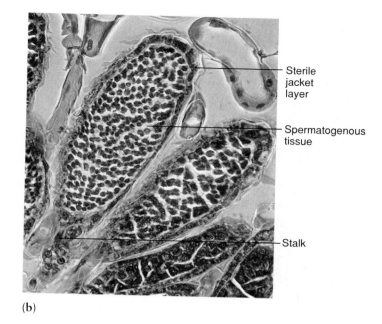

- Sterile jacket layer
- Spermatogenous tissue
- Stalk

(b)

figure 28.9

Gametangia in mosses. (a) Archegonia each contain an egg. Sperm reach the egg by swimming through the canal in the center of the neck. (b) Antheridia contain spermatogenous tissue that gives rise to sperm.

figure 28.10

Capsules of hairy-cap moss (*Polytrichum commune*).

guard cells that remain open unless the surrounding air becomes excessively dry. Once the sporangium is mature, the stomata close permanently.

The mature sporophyte has three parts: a **foot**, a **seta**, and a **capsule** (fig. 28.11a). The foot, which penetrates the base of the venter and grows into the gametophyte, absorbs water, minerals, and nutrients from the gametophyte. The wiry seta elongates and raises the capsule as much as 15 cm above the gametophyte. In the process, the top of the exposed archegonium is torn from the gametophyte and forms the calyptra.

The **capsule** is a sporangium that begins developing while still within the protective calyptra. The outer cells form a sterile, protective jacket, and the innermost cells elongate to form another sterile tissue called the **columella**. Sporogenous cells form between the jacket and the columella and undergo meiosis to form as many as 50 million haploid spores per capsule. When these spores are mature, the calyptra falls off, thereby exposing the operculum (i.e., lid of the capsule). Contraction forces associated with drying often then burst off the operculum, exposing tooth-shaped segments of the **peristome** (figs. 28.7, 28.11b). In some mosses the peristome is cone-shaped and has pores through which spores are released. The mature sporophyte is usually yellow or brown (fig. 28.11c), and its spores are dispersed by wind (however, see reading 28.1, "Shooting Spores"). Spores that land in suitable environments form protonemata, thus completing the life cycle.

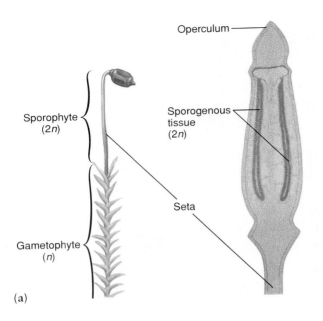

(a)

figure 28.11

The structure of mosses. (a) Diagram of parts of a mature moss sporophyte. (b) Photograph of a moss capsule showing the tooth-shaped segments of the peristome. (c) Moss sporophytes.

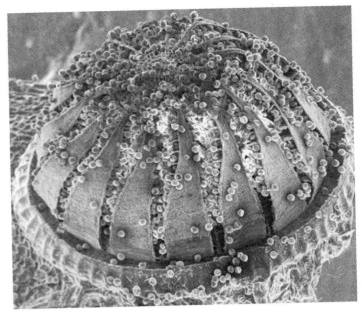

(b)

(c)

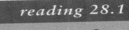

reading 28.1

SHOOTING SPORES

The *Sphagnum* growing in peat bogs uses a battery of natural air guns to spread its tiny spores. As they mature, the spore capsules shrink to about a fourth of their original size. This shrinkage, which compresses the air in the capsule, also shapes the capsule into a tiny gun barrel, with its own airtight cap (i.e., operculum). Eventually the cap is blown off the capsule with an audible pop, firing the spores as far as 2 m.

c o n c e p t

Mosses are the largest and most familiar group of bryophytes. Their gametophytes are leafy and usually radially symmetrical. Sperm swim through the neck of an archegonium to reach an egg held in the venter. Spores are released from sporangia through the peristome. Spores germinate and form a filamentous protonema, which grows into a gametophyte. The gametophytes of some mosses contain special conducting tissues that resemble sieve tubes and tracheids.

Liverworts

Liverworts were named during medieval times, when herbalists followed the *Doctrine of Signatures* (see Chapter 24). A few liverworts are lobed and thus have a fanciful resemblance to the human liver, so the word *liver* was combined with *wort* (herb) to form the name *liverwort*. Although we now know that the *Doctrine of Signatures* is invalid and that eating liverworts will not help an ailing liver, the name *liverwort* has endured. Almost 8,500 species of liverworts have been named.

Liverworts range in size from tiny, leafy filaments less than 0.5 mm in diameter to a thallus more than 20 cm wide (fig. 28.12). All liverworts have a prominent gametophyte,

figure 28.12

The diversity of liverworts. (a) Leafy liverworts growing on a leaf of an evergreen tree in the Amazon Basin in Brazil. (b) *Calypogeia muelleriana*, a leafy liverwort having unlobed leaves. (c) *Bazzania trilobata*, a leafy liverwort having dichotomous forking. (d) *Fossombronia cristula*, a thallose liverwort. (e) An aquatic thallose liverwort (*Ricciocarpus natans*) floating on the shallows of a pond. (f) *Conocephalum conicum*, a thallose liverwort.

(a)

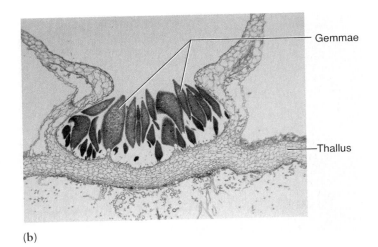

Gemmae

Thallus

(b)

figure 28.13

Gemmae cups of liverworts. (a) Gemmae cups ("splash cups") containing gemmae on the gametophytes of *Lunularia*. Gemmae are splashed out of the cups by raindrops, after which the gemmae can then grow into new gametophytes, each identical to the parent plant that produced it by mitosis. (b) Cross section of a thallus showing a gemmae cup, ×10.

which sometimes has a cuticle. Also, the spores of liverworts have thick walls. These features are believed to be important adaptations that enable liverworts to live on land. Liverworts also have the following distinguishing features:

1. Their rhizoids are always unicellular.

2. Their gametophytes are leafy or thallose and are often lobed and bilaterally symmetrical. They lack a midrib. The upper side of the thallus is photosynthetic; the lower side lacks chlorophyll and is used for storage.

3. Their sporangia are often unstalked.

4. They shed spores from sporangia for a relatively short time.

Liverworts frequently reproduce asexually. One way they do this involves the death of older parts of the plant. When this occurs, the growing areas are left isolated from the parent plant. A second means of asexual reproduction involves the production of ovoid, star-shaped, or lens-shaped pieces of tissue called **gemmae** (singular, **gemma**). Gemmae form in small "splash cups" on the upper surface of the gametophyte (fig. 28.13). Gemmae become detached from their cups by falling raindrops and are often splashed up to a meter away from the parent plant. Gemmae grow into gametophytes. Pieces of gametophytes that are broken or torn from the parent plant can also regenerate entire plants.

Gametophyte: Leafy Liverworts

The gametophytes of liverworts have two shapes: leafy and thallose (see fig. 28.12). Approximately 80% of liverworts are leafy; they usually grow in wetter areas than mosses and are abundant in tropical jungles and fog belts. Their gametophytes have a leafy stem and resemble mosses, except that leafy liverworts appear bilaterally symmetrical while most mosses appear radially symmetrical. Leafy liverworts have three ranks of sessile leaves, two that project laterally and one (often colorless) along their lower side. These leaves are usually one cell thick and lack a midrib. The overall appearance of the gametophyte is flattened.

Gametophyte: Thallose Liverworts

Thallose liverworts have a flat, ribbonlike gametophyte. *Marchantia* is a thallose liverwort and is the most intensively studied liverwort. As a thallose type, it is most specialized and, consequently, is an atypical representative.

The gametophytes of *Marchantia* are perennial and branch dichotomously. Each branch grows from an apical cell in the notch at the tip of a lengthwise groove. The flat gametophyte is 2–20 mm wide, and its surface consists of small, diamond-shaped plates (fig. 28.14). Each plate is covered by cutin and delimits an underlying chamber. These chambers contain filaments of photosynthetic cells arranged like plants in a cactus garden. Each chamber is connected to the atmosphere by a chimneylike pore surrounded by barrel-shaped cells. This pore is analogous to a stoma. Colorless rhizoids and scales form along the lower side of the gametophyte.

Many liverworts such as *Marchantia* are unisexual (fig. 28.15). In *Marchantia*, archegonia form on the underside of rays protruding from the umbrella-shaped caps of **archegoniophores** (archegonium bearers), and antheridia form in chambers along the upper side of disk-shaped caps of

28–13

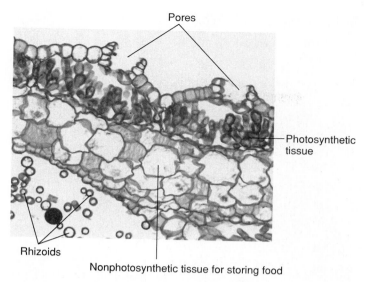

figure 28.14

Marchantia gametophyte, showing the location of photosynthetic tissue, storage tissue, and rhizoids.

antheridiophores (antheridium bearers) (fig. 28.16). Other thallose liverworts are structurally simpler than *Marchantia*. For example, the gametophytes of *Pellia* lack chambers and pores, and tiny *Riccia* can produce archegonia and antheridia on the same plant.

Sporophyte

The sporophyte of liverworts lacks stomata and is anchored in the gametophyte by a knoblike foot. The sporophytes of many liverworts are spherical, unstalked, and held within the gametophyte until they shed their spores. In other liverworts, a stalklike seta grows upward from a foot. The capsule forms atop the seta and is covered by a calyptra. Inside the capsule are sporogenous cells that undergo meiosis to form spores. Among these spores are long, dead cells called **elaters**. Because elaters are hygroscopic, they twist violently when they dry, thus dispersing spores.

c o n c e p t

The gametophytes of liverworts are either leafy or thallose, and they often reproduce asexually by pieces of tissue called gemmae. Antheridia and archegonia in some liverworts are borne on antheridiophores and archegoniophores, respectively. Hygroscopic elaters help disperse spores.

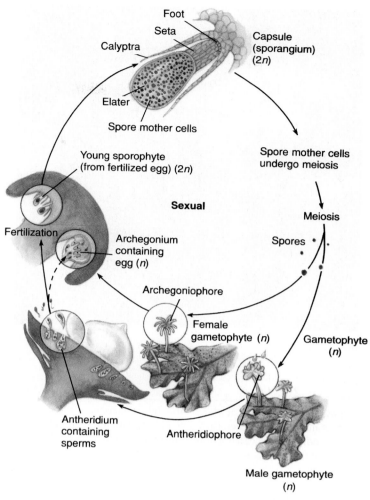

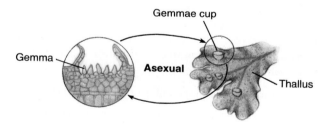

figure 28.15

Life cycle of *Marchantia*, a thallose liverwort. During sexual reproduction, spores produced in the capsule germinate to form independent male and female gametophytes. The archegoniophore produces archegonia, each of which contains an egg; the antheridiophore produces antheridia, each of which produces many sperm. After fertilization, the sporophyte develops within the archegonium and produces a capsule with spores. *Marchantia* reproduces asexually by fragmentation and gemmae.

(a)

(b)

figure 28.16

Reproductive structures of *Marchantia*. (a) Gametophyte with archegoniophores (female). (b) Gametophyte with antheridiophores (male); note the archegoniophore in the upper right. The cellular details of these structures are shown in figure 28.2.

Hornworts

The hornworts are the smallest group of bryophytes; there are only about 100 species in six genera. The most familiar hornwort is *Anthoceros*, a temperate genus (fig. 28.17). Hornworts have several features that distinguish them from most other bryophytes:

1. The sporophyte is shaped like a tapered horn (fig. 28.18).

2. Each photosynthetic cell contains only one chloroplast. Each chloroplast is associated with a starch-storing pyrenoid, as are the chloroplasts of green algae and *Isoetes*, a vascular plant.

3. The sporophyte has an intercalary meristem at its base above the foot. As a result, the sporophyte grows indeterminately.

4. Archegonia are not discrete organs. Rather, they are embedded in the thallus and are in contact with the surrounding vegetative cells.

5. The thallus of the gametophyte has stomalike structures. Stomata (even nonfunctional ones) are absent from the gametophytes of all other plants.

6. The cavities of hornwort gametophytes are usually filled with mucilage.

figure 28.17

Hornworts, a type of bryophyte.

Gametophyte

The flat, dark green gametophytes of hornworts are structurally simpler than those of other bryophytes. They are dorsiventrally flattened and superficially resemble thallose liverworts such as *Marchantia*. The gametophyte is round or

(a)

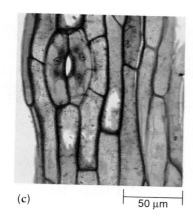

(c)

50 μm

Mature
sporangium
splits open
to release spores

(b)

Gametophyte

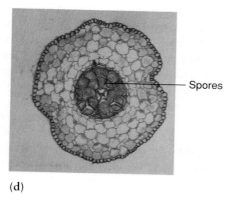

Spores

(d)

figure 28.18

Anthoceros, a hornwort. (a, b) *Anthoceros* with mature sporangia. (c) Stomalike structure on the gametophyte of *Anthoceros*; such structures are absent from the gametophytes of all other plants. (d) Spores form along the central axis of the sporophyte.

slightly oblong, and there is little internal differentiation. Hornwort gametophytes are annual or perennial and are anchored to the substrate by rhizoids.

Most hornworts are unisexual. Sex organs form on the upper surface of the thallus. One or more antheridia resembling those of liverworts form in roofed chambers in the upper portion of the thallus, and archegonia form in rows beneath the surface. Hornworts reproduce asexually by fragmentation.

The pores and cavities of hornwort gametophytes are filled with mucilage instead of air. Nitrogen-fixing cyanobacteria such as *Nostoc* live in this mucilage and release nitrogenous compounds to the hornwort.

Sporophyte

The sporophytes of hornworts differ remarkably from those of other bryophytes. Hornwort sporophytes are long, green spindles (1–4 cm long) having tapered tips (fig. 28.17). They have a distinct epidermis and stomalike structures but lack setae. Sporophytes begin forming beneath the surface of the thallus. During development, their tips pierce the upper surface of the thallus and appear as miniature horns. The base of the green capsule remains embedded in the gametophyte. Spores form along the central axis of the capsule from sporogenous cells. The sporophyte remains photosynthetic and can live for several months until spores are released. This semi-independence is viewed by many botanists as an evolutionary step toward the independent sporophytes that characterize vascular plants. Indeed, many botanists no longer consider hornworts to be bryophytes; they consider them to be more closely related to ferns.

Hornworts comprise the smallest group of bryophytes. Their sporophytes are shaped like tapered horns, and their photosynthetic cells each contain one chloroplast associated with a pyrenoid. The sporophyte has an intercalary meristem, and archegonia are not discrete organs.

figure 28.19

Andreaea rothii, a rock moss.

The Ecology and Distribution of Bryophytes

Bryophytes live in almost all places that plants can grow and in many places where vascular plants cannot grow (see fig. 28.4). This is one of the advantages of being small and nonvascular: bryophytes can live on impermeable substrates, thereby avoiding competition with vascular plants. Indeed, bryophytes grow in cracks of sidewalks, on moist soil, rooftops, the faces of cliffs, tombstones, and birds' nests; they carpet forest floors, dangle like drapery from branches, and sheathe the trunks of trees in rain forests. Examples of the extremes of bryophyte habitats include exposed rocks, volcanically heated soil (up to 55° C), and Antarctica, where summer temperatures seldom exceed −10° C. However, the most unusual habitat for bryophytes is reserved for *Splachnum*, the mammal dung moss (see fig. 28.4d). This moss produces a colored stalk and releases a putrid odor that attracts flies. Unlike the spores of other mosses, those of *Splachnum* are sticky and adhere to the visiting flies. The spores are disseminated when the flies move from the moss to piles of dung.

Bryophytes, which are often the first plants to invade an area after a fire, grow at elevations ranging from sea level to 5,500 m. There are no marine bryophytes, but some, such as dune mosses, grow near the seashore. Each group of bryophytes has aquatic species. Bryophytes dominate the vegetation in peatlands. Mosses are especially abundant in the arctic and antarctic, where they far outnumber vascular plants.

Some bryophytes can withstand many years of dehydration and often grow in deserts. Other bryophytes can withstand prolonged periods of dark and freezing, which explains why they are the most abundant plants in Antarctica. Although generally widespread, some bryophytes grow only in specific habitats, such as on the bones and antlers of dead reindeer or on animal dung.

The ecology of bryophytes is determined by their gametophytes. Bryophytes, along with lichens, are among the first organisms to colonize bare rocks and volcanic upheavals. For example, *Andreaea* is a black to reddish brown moss that grows on exposed rocks (fig. 28.19). These organisms convert rock to soil, thus paving the way for colonization by other organisms.

Many bryophytes, such as the genus *Hypnum*, are notoriously sensitive to pollution, especially sulfur dioxide. As a result,

figure 28.20

The moss *Scopelophila cataractae* growing on copper-rich soil.

most bryophytes are rare in cities and industrialized areas; for example, twenty-three species of bryophytes that grew in Amsterdam in 1900 no longer grow in that area. Other mosses, such as *Ceratodon* and *Bryum*, thrive in urban and polluted areas. Unlike pollution-sensitive mosses, *Ceratodon* and *Bryum* have short-lived protonemata and produce gametophytes rapidly.

Bryophytes increase the humus in soil and often indicate the presence or absence of particular salts, acids, and minerals. For example, the liverwort *Carrpos* grows only on gypsum-rich "salt pans," and *Mielichhoferia* and *Scopelophila* are mosses that grow only on copper-rich substrates (fig. 28.20). Other bryophytes concentrate elements such as barium, lead, strontium, and zinc—sometimes at more than 200 times their levels in the soil. Bryophytes typically accumulate twice as much radioactivity as flowering plants on a weight-for-weight basis.

Bryophytes usually grow very slowly. However, peat moss (*Sphagnum*) accumulates at rates exceeding 12 metric tons per hectare annually (fig. 28.21); for comparison, yields of corn and rice average about 6 and 5.5 metric tons per hectare, respectively. Some mosses in Antarctica also grow rapidly; indeed, many mosses of the Northern Hemisphere grow more slowly than do those in Antarctica.

Carpets of bryophytes are excellent seedbeds for vascular plants (fig. 28.22). Peat moss can absorb 20–25 times its weight in water; this is why peat moss is an excellent soil conditioner that helps to prevent flooding and minimizes erosion. Peat moss produces large amounts of acids and antiseptics that kill decomposers. As a result, dead peat moss accumulates and forms large deposits called **peat bogs.** The pH of these bogs often approaches 3.0 (i.e., about the same pH as vinegar) and thus eliminates all but the most acid-tolerant plants, such as cranberry, a few carnivorous plants, blueberry, and black spruce, which is often a climax species in acidic communities. The acidity of peat bogs also makes these communities very stable; some peat bogs are estimated to be more than 50,000 years old. Peat bogs cover approximately 1% of the earth's surface, an area that is equivalent to half of the United States.

figure 28.21

Sphagnum. (a) Carpet of peat moss. (b) Close-up of photosynthetic (i.e., chloroplast-containing) and water-retaining cells of *Sphagnum*.

(a)

(c)

(b)

figure 28.22

Peat and peat bogs. (a) A peat bog showing the diversity of plants that grow on peat moss. (b) A family in Ireland stacking peat sods to dry. This peat will later be used as fuel to heat their home. (c) The bucolic image of peat fires smoldering on small hearths misrepresents peat's use. Indeed, more than 95% of the world's peat harvest is burned to generate electricity. Peat-fired power plants are preferred over coal-fired power plants because of peat's low sulfur and ash content.

SECRETS OF THE BOG

In the 1950s, several countries in northern Europe began mining peat as fuel. This peat from bogs not only provided valuable fuel but also yielded some 2,000-year-old secrets that were as grotesque as they were valuable. Horrified workers uncovered several hundred human bodies. These bodies had been tanned by the bog's acids (i.e., much as we use acids to tan leather), and many were in such good condition that the workers suspected that they had been recently dumped into the bogs. Furthermore, many of these bodies were mutilated; they had slit throats, severed vertebrae, and nooses around their necks.

Further study showed that the bodies were actually 2,000–3,000 years old. The acidity and anaerobic environment of the bogs had preserved the bodies by inhibiting the growth of bacteria and fungi that normally decompose dead organic material. Archaeologists gave the bodies that were pulled from the peat bogs names such as Lindow man (reading fig. 28.2) and Tollund man, referring to the geographical locations of the bogs that became graves. The most intensively studied of these bog people is Lindow man, a fellow who lived about the time of Aristotle. The

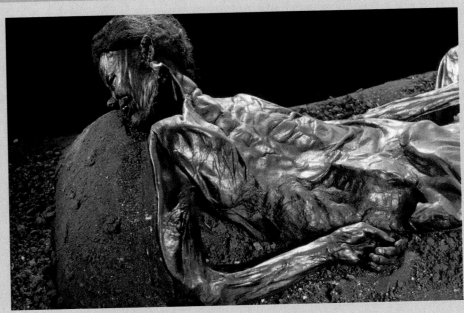

reading figure 28.2
Lindow man.

preserved stomach-contents of Lindow man showed that his last meal was barley and linseed gruel. Thanks to the preservative effects of the bog, archaeologists have also learned about musical instruments and household items of the 2,000-year-old civilizations and determined that bogs were the sites of human sacrifice in religious ceremonies.

The acidity and anaerobic environment of peat bogs also preserve dead plants and animals. Paleobotanists have used pollen preserved in peat to determine which plants grew in a particular area and how they have changed over time. The preservative effects of peat bogs have also yielded horrifying secrets about past civilizations (see reading 28.2, "Secrets of the Bog").

The Physiology of Bryophytes

The Influence of Light

Several bryophytes produce gametangia in response to daylength. For example, most leafy liverworts, such as *Porella*, produce antheridia and archegonia only during long days; other bryophytes, such as *Riccia*, *Anthoceros*, and *Sphagnum*, produce gametangia only on short days. Bryophytes such as *Pogonatum* are day-neutral.

Light also affects how protonemata develop. For example, consider these experimental observations:

1. The protonemata of *Funaria* grow in darkness if supplied with nutrients but produce no leafy, upright shoots. Similar growth occurs in blue light.

2. Changing the light from blue to red induces the formation of buds.

3. Buds form in the light or dark within 2 days if the protonema is supplied with the hormone cytokinin.

These results suggest that light may affect growth and development by altering the production of cytokinins. Budding is inhibited by auxin, another hormone (for more information about plant hormones, see Chapter 18).

Hormones and Apical Dominance

The growth of protonemata requires auxin, a hormone that is made in the apex of the thallus. The protonemata of many mosses exhibit a strong apical dominance, meaning that buds farthest from the apex develop earlier than buds closer to the apex. Removing the apex releases this apical dominance.

Applying auxin to a detipped protonema restores apical dominance, suggesting that auxin or a similar compound may control apical dominance in the plant. Auxin also influences the formation of gemmae and rhizoids.

Interactions with Other Organisms

Mats of bryophytes house a rich diversity of other organisms, especially invertebrates. Interestingly, soil bacteria such as *Agrobacterium* can induce formation of buds on protonemata of mosses such as *Pylaisiella*. Species of *Agrobacterium* that induce formation of roots in vascular plants also induce formation of rhizoids in many bryophytes.

The Formation of Gametangia

In bryophytes such as *Riccia*, cytokinins and auxin induce the formation of archegonia, whereas in mosses such as *Barbula* and *Bryum*, auxin induces the formation of antheridia. In *Ricciocarpus* and other liverworts, the formation of archegonia is delayed, and young plants have only antheridia.

Water Relations

Bryophytes absorb most of their water from the atmosphere and the surface of their substrate. Their water content varies from less than 50% to more than 2,500% of their dry weight. Many bryophytes are extremely tolerant of desiccation. For example, star moss (*Tortula ruralis*; also called *dune moss*) becomes photosynthetic within a few hours of being moistened, even after being air-dried for almost a year (see reading 28.3 "Quick Recovery"). When dry, *Tortula* can tolerate temperatures ranging from over 100° C down to that of liquid nitrogen (–196° C), a temperature range of about 300° C. Some bryophytes are photosynthetic at temperatures below freezing.

Sporophyte Physiology

We know relatively little about the physiology of bryophyte sporophytes. The sporophytes of most bryophytes (e.g., *Mnium* and *Pleuridium*) receive 80%–90% of their water and carbohydrates from their gametophytes, and it typically takes 1–4 days for solutes to move from the gametophyte to the sporophyte.

c o n c e p t

Light and hormones influence several aspects of bryophyte growth and development, including the production of gametangia, the branching of protonemata, and apical dominance. Bryophytes absorb water and dissolved minerals from the atmosphere or from the surface of their substrate. Some bryophytes are extremely tolerant of desiccation.

The History and Relationships of Bryophytes

The fossil record of bryophytes is poor and shows relatively little diversity. Fossils dating from the Carboniferous period (286–360 million years ago) have been interpreted as being mosses. The most ancient liverworts (*Pallaviciniites devonicus*) date to the Devonian period (360–408 million years ago), and other fossils that resemble liverworts appear in the Carboniferous and succeeding periods. Many of these fossil bryophytes have vague relationships and are classified as *Muscites*, a catchall genus for problematic mosslike fossils, or *Thallites*, an equivalent all-purpose genus for liverwort-like fossils. The gametophytes of most fossilized bryophytes are rarely associated with sporophytes, which contributes to the uncertainty of their identity. Although fossils of genera such as *Sporogonites* and *Torticaulis* dating from the Devonian and late Silurian periods (408–438 million years ago) may be mosses, these plants are preserved so poorly that clear interpretation is difficult. The problem is further complicated because the sporangia of some bryophytes are almost indistinguishable from those of ancient lignified plants.

In spite of the meager fossil record, botanists suspect that bryophytes diverged from an ancestor common to vascular plants more than 430 million years ago (i.e., some time during the Silurian period). There is no general agreement as to what the common ancestor was or even whether

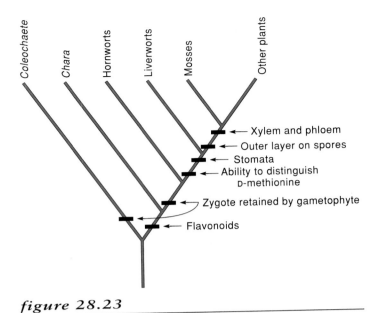

figure 28.23

Cladogram comparing characters that indicate a possible common ancestry of green algae, bryophytes, and vascular plants.

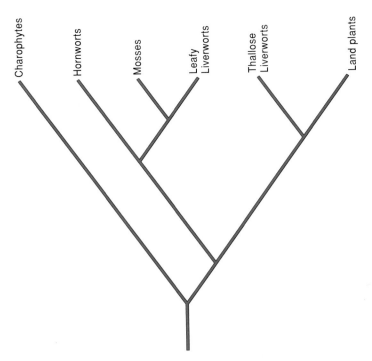

figure 28.24

The evolutionary history of bryophytes, based on a cladistic analysis of ribosomal RNA sequences.

bryophytes diverged as a monophyletic group. Key features in identifying a possible ancestor may be the flagellated sperm, photosynthetic pigments, similar cell walls, and storage of starch that are common among bryophytes, vascular plants, and green algae. Among the green algae, the most likely ancestral group may be the charophytes (e.g., *Chara, Coleochaete*), because they are the only group of algae with a flavonoid biosynthetic pathway, a feature of plants (see reading 27.1, "Evolutionary Relationships of Green Algae and Plants"). Another possibility is that the ancestor is represented by the green alga *Coleochaete*, which retains the zygote on the gametophyte, as do plants, but does not produce flavonoids (fig. 28.23). *Coleochaete* also has a thalluslike body that resembles a thalloid liverwort.

The bryophytes as a group seem to be paraphyletic, based on two shared-derived characters that are present in liverworts, mosses, and plants but not hornworts: the occurrence of stomata and the enzymatic ability to distinguish D-methionine from L-methionine. Also, if leptoids and hydroids are interpreted as vascular tissue like that of plants, then mosses are closer to land plants than to other bryophytes. These comparisons confirm the traditional view that bryophytes should not be classified as a single division.

Disputes and unanswered questions about bryophyte evolution are also being approached by molecular phylogenetic studies. This work shows promise but also raises some new questions. For example, the evolutionary relationships shown in figure 28.24, which are based on a cladistic analysis of ribosomal RNA sequences, confirm most of the suggestions based on other features. However, the cladogram also indicates that thallose liverworts are more closely related to vascular plants than they are to leafy liverworts. Keep in mind, though, that this evolutionary tree represents just one of several hypotheses about the relationships of bryophytes; it will undoubtedly be tested, modified, and improved as botanists pursue more kinds of molecular and morphological data.

The Economic Importance of Bryophytes

Bryophytes are generally not edible, although liverworts were often soaked in wine and eaten in the 1500s. Today, the earthy aroma of Scotch whiskey is partly due to the fact that the malted barley is dried over peat fires.

Mosses are used as stuffing in furniture, as a soil conditioner, as an absorbent in oil spills, and for cushioning. Florists use peat moss as a damp cushion when shipping plants. *Sphagnum* has also been used by aboriginal people for diapers and as a disinfectant. Because of its acidity, peat moss is an ideal dressing for wounds. Indeed, the British used more than 1 million such dressings per month during World War I, and the Red Cross refers to *Sphagnum* as a wound dressing in its publications. North American Indians used *Mnium* and *Bryum* to treat burns. *Dicranoweisia* has been used to waterproof roofs in Europe.

The rapid growth of peat moss suggests that peat bogs may be an important source of renewable energy. Peat and lignite (a soft, brown type of coal) have several properties that make them good sources of fuel. For example, peat has a

28–21

Unit Eight Diversity

caloric value of 3,300 calories per gram, a value that is greater than that of wood (but only half that of coal). Furthermore, peat is abundant. The United States (excluding Alaska) has more than 60 billion tons of peat, an amount of fuel equivalent to approximately 240 billion barrels of oil. Countries of the former Soviet Union have an annual harvest of more than 200 million tons of peat, which is used as fuel for nearly eighty power plants. Ireland obtains more than 20% of its energy from peat, and the United States annually harvests more than a million tons of peat in twenty-two states.

Chapter Summary

Bryophytes, which include mosses, liverworts, and hornworts, are small plants that produce chlorophylls *a* and *b*, starch, cellulosic cell walls, and motile sperm. They lack complex vascular tissues and supporting tissue, and absorb water by capillarity. Bryophytes have a heteromorphic alternation of generations dominated by a photosynthetic and independent gametophyte. Eggs and sperm form in gametangia called archegonia and antheridia, respectively, both of which are surrounded by a protective jacket of sterile cells.

Bryophytes require free water for sexual reproduction, so that sperm can swim to eggs. The fertilized egg begins the diploid sporophyte generation, which is attached to and nutritionally dependent on the gametophyte. Haploid spores formed by meiosis are released from sporangia and are usually dispersed by wind. Spores germinate and form gametophytes, thus completing the life cycle. Bryophytes have modest growth requirements and grow almost everywhere that other plants grow. Most reproduction in bryophytes is asexual.

Mosses are the largest and most familiar group of bryophytes. They produce multicellular rhizoids and have leafy gametophytes that are radially symmetrical. Spores form in sporangia that open by means of a lidlike operculum. A spore produces a filamentous and branched protonema, which develops into the gametophyte. Most mosses are unisexual. Sperm released from antheridia swim to archegonia and through the neck canal to the egg located in the venter. The zygote, which begins the sporophyte generation, is covered by a protective calyptra. The sporophyte consists of a foot, seta, and capsule (sporangium). The foot penetrates the gametophyte and absorbs food and water for the sporophyte. The stalklike seta raises the spore-producing capsule above the gametophyte.

Liverworts have leafy or thallose gametophytes that are either lobed or bilaterally symmetrical. They have unicellular rhizoids and frequently reproduce asexually by pieces of tissue called gemmae. Most liverworts are leafy and unisexual. The gametangia of some species form on upright, stalked structures called antheridiophores and archegoniophores. Hygroscopic elaters help to disperse spores from sporangia.

Hornworts are the smallest group of bryophytes. Their sporophytes are shaped like tapered horns and have an intercalary meristem at their base. The thallus of the gametophyte has stomatalike structures, and each photosynthetic cell has a chloroplast associated with a pyrenoid. Archegonia in hornworts are not discrete organs. Some botanists do not consider hornworts to be bryophytes.

The ecology of bryophytes is determined by their gametophytes. Bryophytes reduce erosion, condition soil, and are often among the first organisms to invade disturbed areas. Many bryophytes grow in specific habitats and are sensitive to pollution.

Light and hormones strongly affect several aspects of bryophyte growth and development, including the branching of protonemata, apical dominance, and the formation of gametangia. Some bryophytes are extremely tolerant of desiccation.

The evolutionary history of bryophytes is vague. They probably diverged after plants invaded land. Botanists do not agree on the relationships of bryophytes to other groups of plants.

 ## What Are Botanists Doing?

Go to the library and read a recent article in a popular magazine (e.g., *Natural History*) about bryophytes. Summarize the main points of the article. How does the information in the article relate to what you have learned from this chapter?

 ## Writing to Learn Botany

Consider the results of the following experiment:
- If the zygote is removed from an archegonium of a moss, it forms a protonema and a leafy gametophyte.
- If a piece of the gametophyte is transplanted into an archegonium, it will form a sporophyte.

What do you conclude from these results?

Questions for Further Thought and Study

1. Mosses and liverworts have been used extensively to monitor radioactive fallout from the Chernobyl reactor accident that occurred in April 1986. What features of these organisms make them ideal for such a use?

2. Much of the early work on sex chromosomes in plants was done with bryophytes. The spores of unisexual species contain either a large X chromosome or a smaller Y chromosome, depending on whether they will form female or male gametophytes, respectively. In light of these findings, consider the following experiment: The lower portion of the capsule of many mosses can be removed and cultured in the laboratory. Given the proper nutrients, this diploid tissue does not form a sporophyte but forms a diploid gametophyte by a process called **apospory** (without spore formation; i.e., without meiosis). If apospory occurs in a species that normally forms unisexual gametophytes, the diploid gametophytes are bisexual. How can you explain this phenomenon?

3. What is the evidence that bryophytes and vascular plants had a common ancestry? What does this evidence suggest about the common ancestor of bryophytes and other plants?

4. Although bryophytes require free water for sexual reproduction, several bryophytes grow in deserts. How do you think these bryophytes reproduce in spite of their mostly dry environment?

5. What might be the adaptive significance of unisexual versus bisexual gametophytes?

6. Why are there so few fossils of bryophytes?

7. What does the fossil record show regarding the relationships of bryophytes to other plants?

8. Why is it important that pollution-tolerant bryophytes have short-lived protonemata?

9. When are bryophytes most vulnerable?

Web Sites

Review the "Doing Botany Yourself" essay and assignments for Chapter 28 on the *Botany Home Page*. What experiments would you do to test the hypotheses? What data can you gather on the Web to help you refine your experiments? Here are some other sites that you may find interesting:

http://128.146.143.171/hvp/tmi/hort300/liver2.htm

This site reviews the basic features of mosses and liverworts. You'll also learn about the "invisible industry" of Wisconsin.

http://www.science.siu.edu/bryophytes/

At this site you'll see photographs of bryophytes, as well as general information about these interesting plants.

http://ucjeps.berkeley.edu/bryolab/abls.html

This site is maintained by the American Bryological and Lichenological Society. Here you'll find Internet resources about bryophytes, as well as links to other flora.

Suggested Readings

Articles

Levanthes, L. E. 1987. Mysteries of the bog. *National Geographic* 171(3):396–420.

Miller, N. G. 1980. Bogs, bales, and BTUs: A primer on peat. *Horticulture* 59:38–45.

Stebbins, G. L., and G. J. C. Hill. 1980. Did multicellular plants invade land? *American Naturalist* 115:342–353.

Thieret, J. W. 1955. Bryophytes as economic plants. *Economic Botany* 10:75–91.

Wang, T. L., and D. J. Cove. 1989. Mosses—lower plants with high potential. *Plants Today* (March–April): 44–50.

Books

Bates, J. W., and A. M. Farmer, eds. 1992. *Bryophytes and Lichens in a Changing Environment.* Oxford: Oxford University Press.

Chopra, R. N., and P. K. Kumra. 1988. *Biology of Bryophytes.* New York: John Wiley.

Conard, H. S., and P. L. Redfearn, Jr. 1979. *How to Know the Mosses and Liverworts.* New York: Academic Press.

Dyer, A. F., and J. G. Duckett. 1984. *The Experimental Biology of Bryophytes.* London: Academic Press.

Smith, A. J. E. 1982. *Bryophyte Ecology.* London: Chapman and Hall.

Certain species of *Selaginella*, a seedless vascular plant, have iridescent blue leaves, as shown here.

Seedless Vascular Plants

29

Chapter Outline

Chapter Overview

Seedless vascular plants share features with bryophytes, including the same types of pigments, the basic life cycle, and the storage of starch as their primary food reserve. However, the evolution of vascular tissue enabled vascular plants to invade and dominate the drier habitats on land more effectively than could nonvascular plants. The sporophyte of seedless vascular plants dominates the life cycle; the gametophyte is always smaller, sometimes microscopic, and nutritionally independent of the parent sporophyte. The ancestors of modern seedless vascular plants dominated the earth's vegetation for more than 250 million years, eventually giving way to seed plants. Most species of seedless vascular plants are true ferns, but they also include horsetails, whisk ferns, and club mosses. Most varieties of these plants live in tropical areas.

Vascular plants, like bryophytes, evolved from aquatic, nonvascular ancestors. In contrast to the bryophytes, however, vascular plants become the dominant organisms on land, primarily because of their diversity and large size. Early vascular plants developed upright stems with large leaves to absorb sunlight for photosynthesis. Roots anchored and supported the stems and absorbed water and minerals from the soil.

Structural support for large vascular plants came primarily from the lignin that strengthened their secondary cell walls. Conducting tissues enabled them to transport water and nutrients rapidly and efficiently throughout the larger plant body. A waterproof cuticle and functional stomata controlled gas exchange and minimized water loss.

Although all of the earliest vascular plants are extinct, their successful strategies for adapting to land habitats persist in their descendants. There are more than 250,000 species of vascular plants living today. These plants are conveniently divided into two groups: those that produce seeds and those that do not. In this chapter, we discuss the seedless vascular plants, which are divided into four divisions of extant plants.

The seedless vascular plants include about 13,000 species, most of which are ferns (Division Pteridophyta, with about 12,000 species) (fig. 29.1). The other three divisions consist of the whisk ferns (Division Psilotophyta), the club mosses (Division Lycopodophyta), and the horsetails (Division Equisetophyta). The seedless vascular plants are distinguished from seed plants by their vascular organization and their lack of seeds; they reproduce by spores in much the same way that bryophytes do.

Features of Seedless Vascular Plants

Vascular plants are not watertight, nor do they use or conserve water efficiently. Water escapes when stomata open and exchange gases for photosynthesis. However, vascular plants can continuously replace water that is lost and keep the entire plant body moist; water lost by transpiration is replaced by absorption into the roots. When water loss increases, water is absorbed and transported faster to compensate for the loss. When the demand for water uptake exceeds the ability of the roots to absorb it, then stomata close and prevent further loss. Bryophytes lack such control over their use of water.

Although many features of seedless vascular plants are not found in the bryophytes, these two groups do have many characters in common. Some of these features are shared with a few algae as well. For comparison with the algae and bryophytes, the general features of seedless vascular plants are summarized as follows:

1. The life cycle of seedless vascular plants is similar to those of bryophytes and algae that exhibit sporic meiosis (fig. 29.2). The diploid sporophyte produces haploid spores by meiosis. Each spore germinates and grows into a gametophyte that produces gametes by mitosis. The gametes (eggs and sperm) fuse to form diploid zygotes.

2. Eggs are produced in archegonia, and sperm are produced in antheridia.

3. The zygote germinates to produce a multicellular embryo that depends on the gametophyte for its nutrition. To complete the life cycle, the embryo grows into a mature sporophyte. A multicellular embryo also characterizes the bryophyte life cycle, but it is absent in algae.

4. Seedless vascular plants produce chlorophylls a and b, carotenoids, starch, cellulose cell walls, and motile sperm. These features are also shared with bryophytes and many algae.

5. As in bryophytes and a few green algae, a cell plate separates new cells during cytokinesis in seedless vascular plants.

6. Each sporangium is protected by a multicellular jacket of nonreproductive cells. Spores are dispersed from sporangia by wind and are cutinized to resist desiccation.

7. Seedless vascular plants have a well-developed cuticle to minimize water loss. They also have stomata to allow gas exchange for photosynthesis. Although hornworts and mosses have stomata, they are not well developed and do not function efficiently to prevent water loss. Most bryophytes lack a cuticle.

8. Flagellated sperm swim through water to eggs. Like bryophytes, seedless vascular plants require free water for sexual reproduction.

9. Seedless vascular plants have well-developed vascular tissues. Xylem transports water and dissolved minerals great distances from the soil. Carbohydrates are transported up and down throughout the plant.

10. Many seedless vascular plants have lignin and cellulose in their secondary cell walls. Lignin strengthens cellulosic microfibrils, thereby enabling plants to stand upright to much greater heights than bryophytes.

figure 29.1

The ferns, represented here by bracken fern (*Pteridium aquilinum*), are the largest group of seedless vascular plants.

11. Sporophytes and gametophytes of seedless vascular plants are nutritionally independent of each other. Sporophytes are photosynthetic, and gametophytes are either photosynthetic or saprophytic (i.e., they obtain nutrition from dead and decaying organic matter).

12. The sporophyte of seedless vascular plants, which dominates the life cycle, is long-lived and often highly branched.

c o n c e p t

Seedless vascular plants share many of their reproductive and vegetative features with bryophytes. However, the evolution of a resistant cuticle, complex stomata, vascular tissue, and lignin enabled these plants to dominate habitats on land.

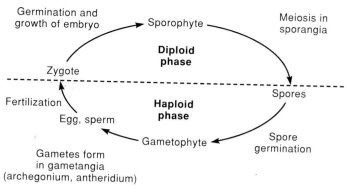

figure 29.2

The generalized life cycle of seedless vascular plants. Sexual reproduction in these plants entails sporic meiosis.

A Generalized Life Cycle of Seedless Vascular Plants

As in bryophytes, sexual reproduction in seedless vascular plants entails alternation of generations that consist of heteromorphic diploid and haploid phases. Meiosis produces haploid spores in the sporangia of diploid sporophytes. Spores then germinate and grow into gametophytes that produce eggs in archegonia and sperm in antheridia. Sperm must have water to swim to archegonia, where fertilization occurs to produce a diploid zygote. The zygote starts another diploid phase (fig. 29.2).

The Sporophyte

The life cycle of seedless vascular plants is dominated by the sporophyte, which is the "plant" that everyone thinks of when they think of plants. Sporophytes become nutritionally independent soon after the zygote grows out of the archegonium. In many genera, the sporophytes are perennial; new growth sprouts from underground rhizomes year after year. For example, the branching rhizomes of horsetails form extensive underground networks, and in ferns new leaves grow from the same rhizome every year. Sporophytes commonly reproduce asexually, producing populations consisting entirely of clones.

The Gametophyte

The gametophyte of seedless vascular plants is often short-lived and small, sometimes microscopic. With few exceptions, it is nutritionally independent of the sporophyte as soon as the spore leaves the sporangium. The gametophytes of whisk ferns, club mosses, and many ferns are saprophytic; they obtain their nutrition from decaying organic matter. In horsetails and most ferns, the gametophytes are photosynthetic. The gametophytes of many ferns look somewhat like thallose liverworts; they sometimes reproduce asexually and, like many sporophytes, can produce extensive clonal populations where the sporophytes are absent or rare. One example that has fascinated

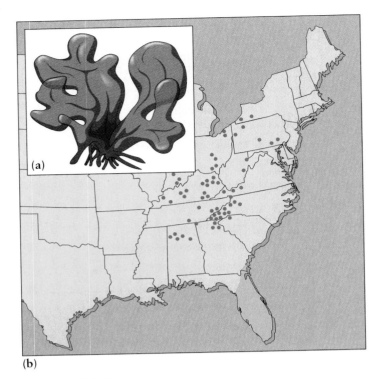

(b)

figure 29.3

(a) The Appalachian gametophyte of the fern *Vittaria*. (b) Map showing the distribution of the Appalachian gametophyte (dots) and of the sporophyte (shading) of its nearest relative, *Vittaria lineata*.

botanists for many years is the "Appalachian gametophyte" of the fern genus *Vittaria*, whose sporophyte is thought to be extinct. The gametophyte is distributed over several thousand square miles in the eastern half of the United States, but it is not known to produce a sporophyte (fig. 29.3).

Organization of Reproductive Structures

Organs that produce gametes and spores in seedless vascular plants are similar in several ways to those of bryophytes (fig. 29.4). Antheridia and archegonia are multicellular and protected by a layer of sterile cells. Each antheridium produces many sperm cells, but each archegonium makes only one egg.

The sporangia of many vascular plants are aggregated into cones, called **strobili** (singular, **strobilus**). A strobilus is essentially a stem tip with several closely packed leaves or branches that bear sporangia (fig. 29.5). The strobilus does not consistently distinguish vascular from nonvascular plants, however, since it occurs primarily in club mosses, spike mosses, and horsetails but not in ferns. Nevertheless, the strobilus is one of the most significant developments in reproductive organization among plants because it foreshadows the

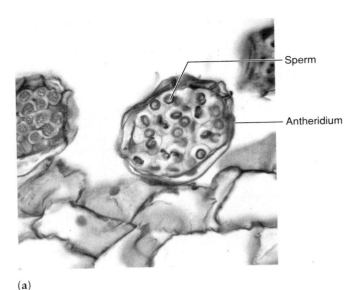

(a)

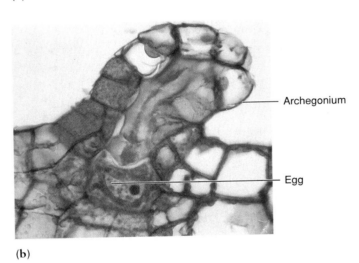

(b)

figure 29.4

The sexual organs of ferns are similar to those of bryophytes: antheridia and archegonia are multicellular and include a protective layer of sterile cells. (a) Antheridia are nearly spherical, with each antheridium producing many flagellated sperm cells (×125). (b) Archegonia are flask-shaped, with a narrow neck on top of an expanded region containing a single nonmotile egg (×150).

flower. The flowering plants, which are the most highly developed group of plants, bear sporangia exclusively in flowers that are highly modified strobili.

c o n c e p t

Seedless vascular plants reproduce by means of spores; the sporophyte, unlike those of the bryophytes, is the dominant phase of the life cycle. Gametes form in archegonia and antheridia. Spores are produced in sporangia that in some plants are aggregated into cones (strobili). Spores are usually dispersed by wind.

figure 29.5

Strobili, or cones, are aggregations of closely packed sporangium-bearing branches or leaves. Shown here are strobili of *Lycopodium obscurum*, a club moss.

Organization of Vascular Tissues

Vascular tissues in stems and roots have two main functions: conduction and support. Nevertheless, the organization of vascular tissues is variable. For example, not all stems have a central pith, and phloem is not always outside of xylem but may be on the inside as well. Regardless of the organization, each arrangement of vascular tissues is referred to as a **stele.** A stele is the central cylinder of tissue inside the cortex of stems and roots. It includes xylem, phloem, and pith, if any. The basic terminology of stem and root structure was presented in Chapters 14 and 15. Those chapters focused on seed plants, but the basic cell and tissue types are the same in the seedless vascular plants.

Simple Steles

The simplest type of stele is a solid or nearly solid core of xylem and phloem. This basic stele type is called a **protostele,** which occurred in the earliest vascular plants. Among the extant seedless vascular plants, protostelic stems occur only in the psilotophytes and a few lycopods and pteridophytes. However, with the exception of one class of flowering plants (monocots), protostelic roots are common in all groups of vascular plants.

Complex Steles

Vascular tissues in complex steles are interspersed with nonvascular, parenchymatous tissues that comprise the pith. Like the tissues in the cortex, the pith is derived from the ground meristem during primary growth (see Chapter 14). Among seedless vascular plants, most ferns have a pith that is surrounded by a tube of vascular tissues. This arrangement of pith and a vascular cylinder is referred to as a **siphonostele** (fig. 29.6a). The xylem is often bounded on both sides by phloem in fern siphonosteles, but in many species the phloem is restricted to the outside of the xylem.

The configuration of siphonosteles, in contrast to that of protosteles, is often influenced by leaf traces. As a trace branches from the siphonostele, it interrupts the vascular tissue and creates what is called a **leaf gap** (fig. 29.7). A single leaf trace can form a wide leaf gap, or several nearby leaf traces can cause the siphonostele to appear highly dissected (fig. 29.6b). Such a siphonostele, which is called a **dictyostele,** is common in ferns.

The second major type of complex stele is a **eustele.** A eustele has vascular bundles that, like those of a siphonostele, are arranged in a circle around the pith (fig. 29.8). The bundles appear discrete in cross section, but they are actually interconnected longitudinally. Among seedless vascular plants, eusteles are common among horsetails and ferns.

Microphylls and Megaphylls

Two kinds of leaves occur in the sporophytes of seedless vascular plants: **megaphylls** and **microphylls.** Megaphylls are what you usually think of as leaves and are defined by their inferred evolutionary origin: they are believed to have arisen from modified branches. The most widely accepted suggestion for how this happened is that branching patterns gradually changed by the twisting of smaller branches into one plane (**planation**), the faster growth of some branches over others (**overtopping**), and the formation of parenchyma tissue between the lateral branches (**webbing**) (fig. 29.9). Conversely, microphylls probably first appeared from **enations,** which are flaps of tissue that protrude from stems. Megaphylls are identifiable by their multiple, branched veins, whereas microphylls have a single, unbranched vein.

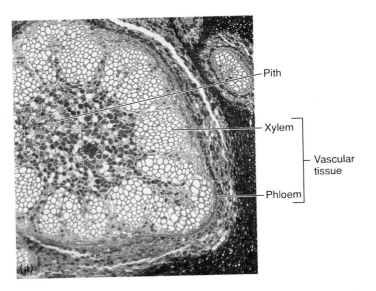

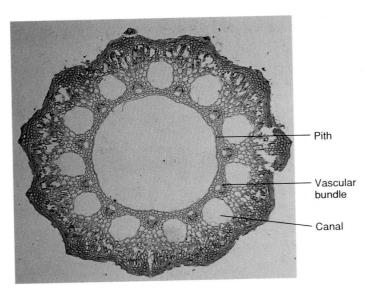

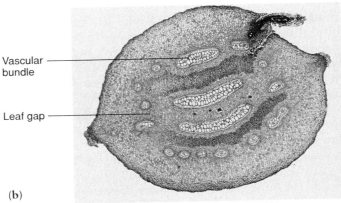

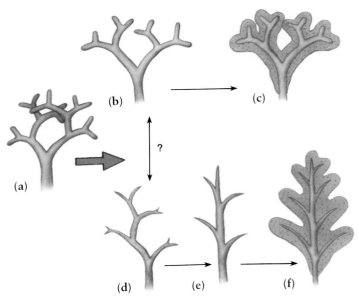

figure 29.6

Steles in seedless vascular plants. (a) The stems of most ferns are characterized by a siphonostele, which is a cylinder of vascular tissue that surrounds a central pith. The example here is from a species of *Osmunda*. (b) In other ferns, such as *Pteridium*, the vascular cylinder is interrupted by leaf gaps, thereby forming a dictyostele.

figure 29.8

This cross section of an *Equisetum* stem shows an example of a eustele, which is the arrangement of discrete vascular bundles in a circle around the pith. *Equisetum* stems also have different kinds of canals, which appear in cross section as circular clear areas.

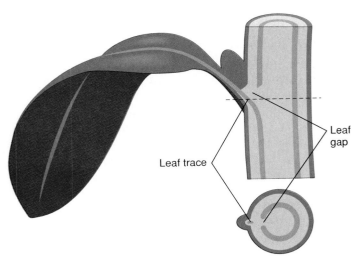

figure 29.7

Longitudinal and cross-sectional views of a leaf gap.

figure 29.9

Possible pathways for the evolution of leaves from branches. (a) The earliest branching patterns consisted of equal, dichotomously branched (i.e., branches in pairs) systems that were three-dimensional. This pattern changed by planation to flattened branching systems that either remained equally dichotomous (b) or that became unequally branched (d) and then uniaxial with side branches (e) by overtopping. Finally, parenchyma tissue filled in the spaces between branches by webbing (c, f). (The two-headed arrow and question mark between (b) and (d) mean that the direction of evolution may have been in either or both directions.)

Arrangements of vascular tissues vary from simple to complex among seedless vascular plants. Whisk ferns have the simplest steles, and ferns have the most complex. Seedless vascular plants have either microphylls, which first arose from enations, or megaphylls, which evolved from modified branches.

The Diversity of Seedless Vascular Plants

The four divisions of living seedless vascular plants are distinguished by both sporophytic and gametophytic features; significant variation exists among their branching patterns, leaf morphology, vascular organization, and underground absorptive organs. The types and arrangements of sporangia on the sporophyte can also be used in defining divisions. Gametophytes vary in their origin from different spore types and in their sources of nutrition. No single characteristic defines each division; rather, classification depends on sets of features.

Division Psilotophyta: Whisk Ferns

The Psilotophyta are the simplest vascular plants, primarily because they have no roots and most species have no obvious leaves. Their traditional classification as a separate division has become controversial because some botanists believe that they are simplified ferns. This controversy is discussed at the end of this chapter (see "The Puzzle of the Psilotophyta").

Instead of roots with root hairs, the psilotophytes have rhizomes with absorptive rhizoids. The larger of the two genera in the division, *Psilotum*, has enations instead of leaves (fig. 29.10b). Botanists hypothesize that the enations of *Psilotum* may be the reduced remnants of leaves, which were larger and more leaflike in the ancestors of the psilotophytes. The stems of *Psilotum* are green and photosynthetic.

The name *whisk ferns* for this group comes from the highly branched stems of *Psilotum*, which give the plant the appearance of a whisk broom (fig. 29.10a). *Psilotum* is widespread in subtropical regions of the southern United States and Asia, and it is a popular and easily cultivated plant that is grown in greenhouses worldwide.

The other genus in the division, *Tmesipteris,* has leaflike structures, each with a single vascular strand. Botanists are not sure whether these structures are flattened branchlets or leaves. *Tmesipteris* is restricted to islands in the South Pacific, where it often grows as an epiphyte on the trunks of tree ferns. This genus is rarely cultivated.

(a)

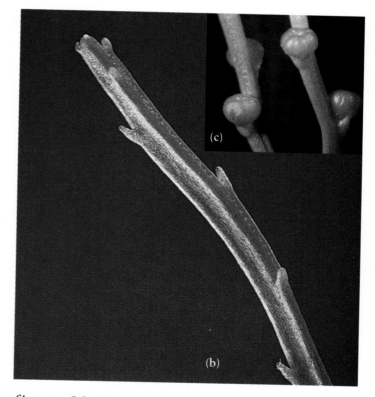

(c)

(b)

figure 29.10

(a) Whisk fern (*Psilotum nudum*). (b) The stems of *Psilotum* are leafless but bear lobed sporangia (c) that are called synangia.

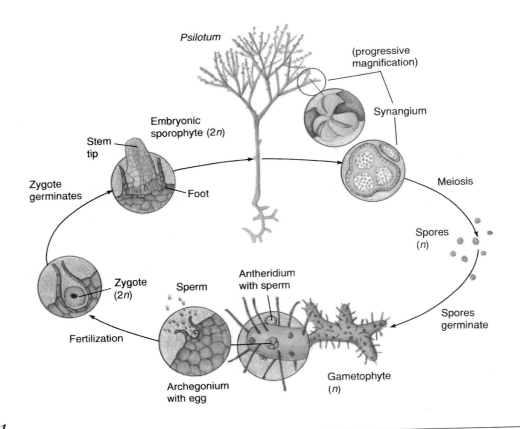

figure 29.11

The life cycle of *Psilotum* alternates between a dominant sporophyte and a microscopic, bisexual gametophyte.

The stems of *Psilotum* are protostelic and usually dichotomously branched. Lobed sporangia form at the tips of short lateral branches, each of which grows in the axil of a forked projection on the stem. The occurrence of separate vascular strands to each sporangial lobe suggests that each lobe evolved from a separate sporangium. This type of fused sporangium is called a **synangium** (fig. 29.10c). Meiosis occurs in the synangium, after which it splits open and releases its spores.

The spores of *Psilotum* germinate into microscopic, brownish gametophytes that live just below the soil surface. These gametophytes are usually cylindrical but sometimes have one dichotomous branch (fig. 29.11). Their cells are inhabited by a mycorrhizal fungus that absorbs nitrates, phosphates, and organic compounds used by the gametophyte. Sex organs develop from surface cells of monoecious (i.e., bisexual) gametophytes. Antheridia protrude slightly above the surface and contain a small numbers of coiled, multiflagellate sperm; archegonia are partially sunken and have necks that are shorter than those in bryophytes. The sperm cells, each having eight to ten flagella, must swim to the archegonium, often through thin layers of water on the surface of the gametophyte. Because the gametophytes are monoecious, fertilization can occur between eggs and sperm from either different gametophytes (cross-fertilization) or

the same gametophyte (self-fertilization). Following fertilization, the zygote forms a foot and an embryonic stem. The primary stem is a branched rhizome that separates from the foot. The sporophyte becomes infected with a mycorrhizal fungus as the rhizome emerges from the gametophyte.

Division Lycopodophyta: Club Mosses

The Division Lycopodophyta consists of 10 to 15 genera and more than 1,100 species that live in various habitats worldwide. They are primarily tropical but also form a conspicuous part of the flora in temperate regions. Most of the species are included in two genera, the club mosses (*Lycopodium*, about 400 species) and the spike mosses (*Selaginella*, about 700 species), both of which get their common names from their club-shaped or spike-shaped strobili (see fig. 29.5). Most species are terrestrial, but many are epiphytic. One species of *Selaginella*, called the resurrection plant (*S. lepidophylla*) because of its ability to defy drought conditions, occurs in the deserts of southwestern United States and Mexico. During periods of drought, this plant forms a tight, dried-up ball (fig. 29.12a); when rain comes, its branches expand and come alive again with photosynthetic activity (fig. 29.12b).

(a)

(b)

figure 29.12

Resurrection plant *Selaginella lepidophylla*, found in the southwestern United States and Mexico, (a) shown in its dried state, (b) but when rain comes, its branches expand and once again become green and photosynthetic.

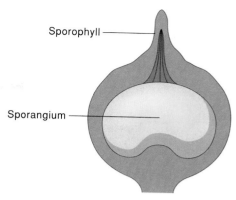

Sporophyll

Sporangium

figure 29.13

A sporophyll of *Lycopodium*, a club moss.

The sporophytes of club mosses are differentiated into leaves, stems, and roots. The roots branch from perennial rhizomes that sometimes grow outward from a central point to form "fairy rings" (as many mushrooms do). One such fairy ring of a *Lycopodium*, when measured for its size and annual growth rate, was calculated to have started growing in 1839.

In mature club mosses, some leaves bear a single sporangium on their upper surfaces. These fertile leaves are called **sporophylls** to distinguish them from sterile vegetative leaves (fig. 29.13). In *Selaginella* and many species of *Lycopodium*, sporophylls are nonphotosynthetic and are packed together at the tops of stems into club-shaped or spike-shaped strobili. In other species of *Lycopodium*, the sporophylls are scattered among the sterile leaves; the two types of leaves are indistinguishable except for the presence of sporangia on the former.

There are two types of spore production in the Lycopodophyta. In *Lycopodium*, spore production is homosporous (*homo* = same), meaning that all spores are morphologically indistinguishable (fig. 29.14). Whisk ferns and all bryophytes are also homosporous. The life cycle of each of these plants alternates between a sporophyte that produces one kind of spore and a gametophyte that may be unisexual or bisexual. Recall from Chapter 28 that the liverwort *Marchantia* has unisexual gametophytes, either male or female. On the other hand, other bryophytes, as well as *Psilotum*, produce bisexual gametophytes.

The gametophytes of *Lycopodium*, which are either photosynthetic and short-lived or saprophytic and long-lived (sometimes up to 10 years), are apparently bisexual because antheridia and archegonia occur on the same plant. However, studies of isozyme inheritance (see p. 177) show that they undergo little self-fertilization. It is not known how cross-fertilization is promoted, but the gametophytes of some *Lycopodium* species are protandrous, meaning that they produce antheridia first, then archegonia. In a population of gametophytes of different ages, the young ones would all be male, thereby promoting outcrossing, if only temporarily. Ultimately, there can be many archegonia on a sexually mature individual, but generally only one zygote forms on each short-lived gametophyte. In contrast, longlived gametophytes can support more than one young sporophyte in various stages of development.

In *Selaginella*, spore production is **heterosporous** (*hetero* = different) because there are two kinds of spores, a small type and a large type (fig. 29.15). The small spores are called **microspores** and produce male gametophytes; the large spores

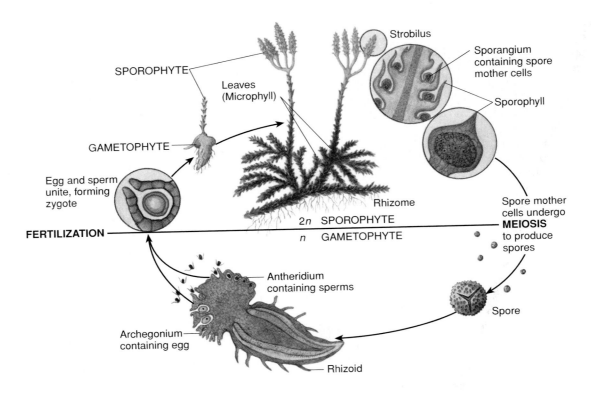

figure 29.14

The life cycle of *Lycopodium* alternates between a homosporous sporophyte and a gametophyte that is morphologically bisexual but may be functionally unisexual.

are called **megaspores** and produce female gametophytes. Microspores and megaspores are produced in **microsporangia** and **megasporangia,** respectively; microsporangia are borne on **microsporophylls,** and megasporangia on **megasporophylls.** Both types of sporophylls usually occur in the same strobilus in *Selaginella*. In the megasporangium, usually one cell undergoes meiosis to produce four megaspores. In contrast, many cells divide meiotically in the microsporangium, producing hundreds or thousands of spores.

The megaspores of *Selaginella* divide numerous times inside the spore walls, producing a multicellular female gametophyte within each spore. As the gametophyte grows and matures, the spore wall breaks open to expose the archegonia (fig. 29.15). Likewise, a multicellular male gametophyte forms in each microspore. The male gametophyte consists almost entirely of a single antheridium, which protrudes from the spore just before releasing the biflagellate sperm (fig. 29.15). The sperm must swim from the tiny male gametophyte to the large female gametophyte to fertilize the eggs. Once the zygote germinates, the resulting embryo develops a foot and an embryonic root, stem, and leaf.

The Lore of Plants

According to Druid mythology, Druid nuns living on the Isle of Sain in the Loire River valley of France had elaborate rituals to gather *Lycopodium*, which they displayed on their altars. The plants were considered to be sacred, and their display was believed to bring good fortune to the Druids.

The Lycopodophyta also includes the quillworts (*Isoetes*), so named because of their narrow, quill-like leaves (fig. 29.16). Quillworts, which are almost exclusively aquatic, live in freshwater habitats on almost every continent. Most of the leaves of quillworts are fertile and do not aggregate into strobili; some leaves produce sporangia that abort before they mature. Leaves grow from a corm instead of a rhizome. The corm has a peculiar cambium that adds both secondary phloem and secondary xylem to its interior, rather than to both sides, as in woody plants. This secondary growth in quillworts is unique among extant Lycopodophyta, although some extinct members of this division also had secondary tissue. Quillworts are also distinctive in the

figure 29.15

The life cycle of *Selaginella* alternates between heterosporous sporophytes and unisexual gametophytes. Microspores produce male gametophytes, and megaspores produce female gametophytes.

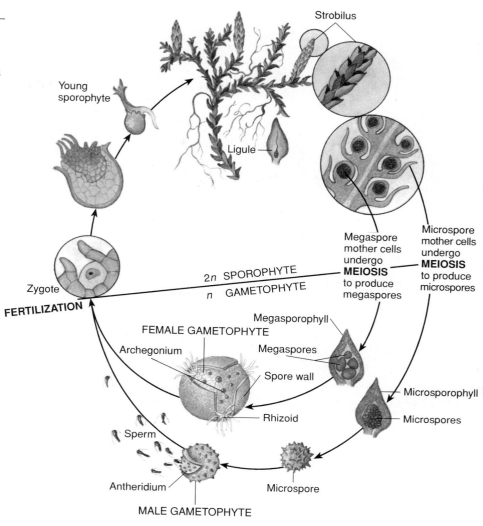

figure 29.16

Quillworts (*Isoetes* sp.) are so named because of their narrow, quill-like leaves.

division because they have CAM photosynthesis (see Chapter 7). Some lake-dwelling species in the Peruvian Andes (an extreme form) have no stomata in their leaves and obtain carbon dioxide for photosynthesis from the muddy substrate where they grow.

Division Equisetophyta: Horsetails

Equisetum is the only living genus of the Equisetophyta. Some of the fifteen species of *Equisetum* have branched stems (fig. 29.17) and some have unbranched stems (see fig. 29.21a). *Equisetum* species are called horsetails or scouring rushes, the latter because their epidermal tissue contains abrasive particles of silica. Scouring rushes were used by Native Americans to polish bows and arrows and by early colonists and pioneers to scrub their pots and pans. *Equisetum* occurs worldwide in moist habitats along streams or forest edges. Its rhizomes are highly branched and perennial. Because its rhizomes can grow rapidly and its aerial stems are poisonous to livestock, *Equisetum* can be a serious problem for farmers and ranchers. Gardeners have often chopped up the rhizome while trying to remove it from the soil, only to have new plants arise from each of the fragments left behind.

figure 29.17

A species of horsetail with photosynthetic stems and whorled branches (in background), and nonphotosynthetic reproductive stems.

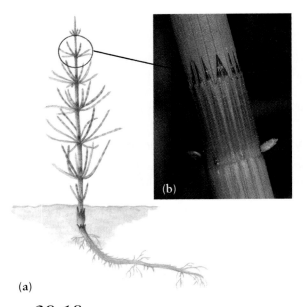

figure 29.18

Horsetails. (a) Diagram of a horsetail. (b) The conspicuous "joints" of *Equisetum* stems consist of whorls of small leaves. These leaves, whose brown tips give the appearance of a collar just above the node, are fused into a sheath around the stem.

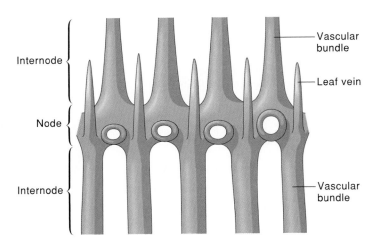

figure 29.19

As shown in this longitudinal view of the vascular tissue of *Equisetum*, the discrete vascular bundles of eustelic internodes meet to form a continuous, siphonostelic pattern at the node.

Although *Equisetum* has true leaves, the stem is the dominant photosynthetic organ of the plant body. The most conspicuous feature of the stem is the series of "joints" made by whorls of small leaves (fig. 29.18). The leaves are fused along most of their length, but their brown tips give the appearance of a collar around the stem just above each node. When the stems are pulled apart, they break easily at the nodes to yield pipelike internodal pieces. In addition, *Equisetum* stems are notable for a branching pattern that is unique among vascular plants. Instead of growing from axillary buds opposite the leaves, the lateral branches of horsetails sprout from between the leaf bases.

The internal structure of a horsetail stem fascinates botanists because of its complexity. As seen in cross section at an internode, the most prominent features of the stem are the epidermal ridges and the internal canal system (see fig. 29.8). The internodal stele is a eustele because of its regular, discrete vascular bundles; however, the vascular strands meet at the nodes in such a way that they form a siphonostele (fig. 29.19). Furthermore, at the node the central canal is interrupted with a short cylinder of pith.

Equisetum is homosporous, and sporangia occur in terminal strobili, which are either on vegetative stems or on nongreen fertile branches that develop from the rhizome (fig. 29.20). Rather than being on sporophylls, the sporangia of *Equisetum* are borne on branches called *sporangiophores* (fig. 29.21). Several fingerlike sporangia are attached to each sporangiophore, giving the overall appearance of tiny sausages hanging from the underside of a small umbrella. When sporangia are mature, the strobili elongate to separate the sporangiophores from each other, so that spores can be released into the air. The spores have an extra layer of wall material that peels off in four flattened strips called **elaters,** which entangle the spores and

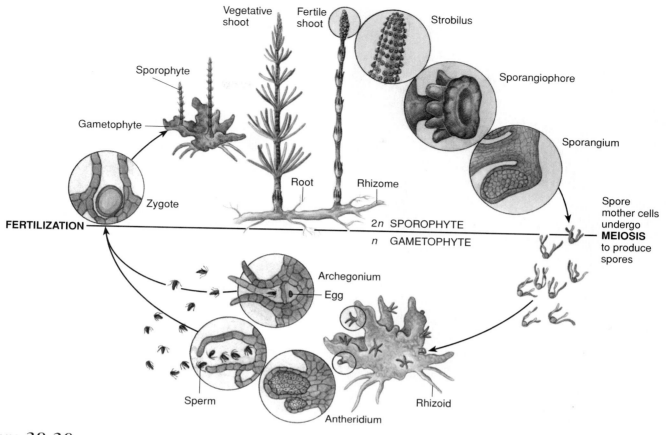

Vegetative shoot

Fertile shoot

Strobilus

Sporophyte

Sporangiophore

Gametophyte

Sporangium

Zygote

Root

Rhizome

2n SPOROPHYTE

Spore mother cells undergo **MEIOSIS** to produce spores

FERTILIZATION

n GAMETOPHYTE

Archegonium

Egg

Sperm

Rhizoid

Antheridium

figure 29.20

The life cycle of *Equisetum* includes a homosporous sporophyte and a photosynthetic, apparently bisexual, gametophyte.

keep gametophytes closer together upon spore germination (fig 29.22). Gametes from closely packed gametophytes have a better chance of being fertilized. Although the term *elater* is also used in relation to liverworts, the elaters of *Equisetum* are not cellular like the elaters of *Marchantia*.

The gametophytes of *Equisetum* are photosynthetic, pincushion-shaped plants that can grow up to 1 cm in diameter. They are differentiated into a basal region with rhizoids and a branched region with bright green, platelike lobes (fig. 29.23). The sexuality of *Equisetum* gametophytes is not well understood because it is variable and appears to be related to environmental conditions (see "The Physiology of Seedless Vascular Plants" in this chapter).

All gametophytes are potentially monoecious, with antheridia borne on the upright lobes and archegonia embedded in cushions at the bases of the lobes. The antheridia release multiflagellate sperm that swim to the archegonia to fertilize the eggs.

As it grows, the zygote produces two regions of differentiation: (1) the stem and leaf sheath and (2) the foot and root (fig. 29.24). The root grows through the gametophyte and establishes the nutritional independence of the emergent sporophyte. Each gametophyte can bear many young sporophytes.

Division Pteridophyta: Ferns

Ferns include approximately 12,000 living species, making this division by far the largest among the seedless vascular plants. Ferns are primarily tropical plants, but some species inhabit temperate regions, and some even live in deserts. North or south of the tropics, the abundance of ferns decreases because of decreasing moisture. In Guam, for example, about one-eighth of the species of vascular plants are ferns, but ferns constitute only about one-fiftieth of the total species in California.

Some genera of ferns have leaves that are the largest and most complex in the plant kingdom. For example, one species of tree fern in the genus *Marattia* has leaves that are up to 9 m long and 4.5 m wide, which is nearly the size of a two-car garage (fig. 29.25a). At the other extreme, the aquatic ferns *Salvinia* and *Azolla* have relatively tiny leaves (fig. 29.25b). For most people, however, a typical example of a fern is the bracken fern, *Pteridium aquilinum* (see fig. 29.1). The bracken fern, like most ferns, is classified in the Order Pteridales, which includes about 10,000 species. Because the bracken fern and its relatives are typical examples of the largest order of ferns, their characteristics are used in the following discussion to illustrate the general features of ferns.

(a)

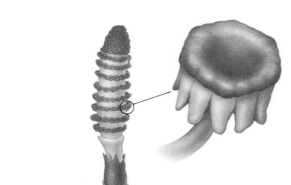

(b)

figure 29.21

(a) *Equisetum* stem with a strobilus. Segments that appear to be separating from one another are sporangiophores. (b) Each sporangiophore consists of a stalk bearing a cap to which several sporangia are attached.

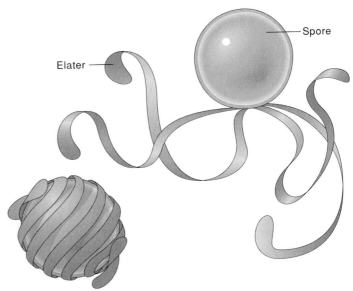

Spore

Elater

figure 29.22

The spores of *Equisetum* have an extra layer of wall material that forms four strips called elaters. Elaters entangle the spores together, thereby promoting outcrossing when gametophytes germinate close to one another.

Rhizoid

figure 29.23

Gametophytes of *Equisetum* consist of a basal region with rhizoids and a branched region with bright green lobes.

The most conspicuous parts of a fern are its leaves, or **fronds**. New leaves grow from a fleshy, siphonostelic or eustelic rhizome each year. Early leaves are lopsided because they grow faster on their lower surface than on their upper surface. This growth pattern, which is called **circinate vernation**, produces young leaves that are coiled into "fiddleheads" (fig. 29.26). New fiddleheads arise close to the growing tip of the rhizome at the beginning of each growing season. The leaves of most ferns die back each year, but the leaves of the

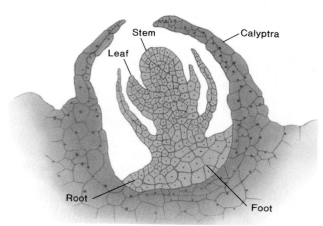

figure 29.24

Emergent sporophyte of *Equisetum* showing regions of differentiation.

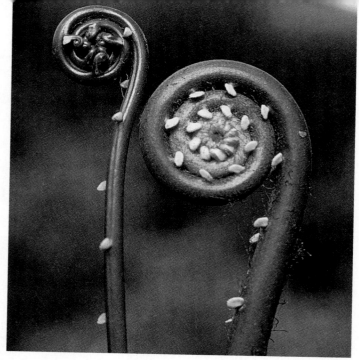

figure 29.26

The coiled "fiddleheads" of fern leaves are rolled-up leaf buds. Fiddleheads are formed by a pattern of growth called circinate vernation, as shown in this *Blechnum* fern.

figure 29.25

The diversity of ferns. Ferns range from (a) *Marattia* (×1/10) and other tree ferns, some of whose leaves are the largest known among plants, to (b) tiny-leaved aquatic ferns such as *Azolla* (×3/4).

figure 29.27

The tips of leaves of the walking fern (*Asplenium* sp.) can become meristematic and form new plants. In this photograph, the tip of the long leaf on the plant to the left has formed the new plant (actually a clone) to the right.

walking fern (*Asplenium rhizophyllum*) can form new plants. Near the tip of each leaf, certain cells revert to meristems and grow into new roots, leaves and a rhizome (fig. 29.27).

Fern leaves are usually fertile but do not form strobili. The leaves have dark spots on their lower surfaces, each of which is a collection of sporangia, together called a **sorus** (plural, **sori**). The sori of some species are covered by an outgrowth from the leaf surface called an **indusium,** while the sori of other species either are not covered or are enfolded by the edge of the leaf (fig. 29.28). With few exceptions, ferns are homosporous.

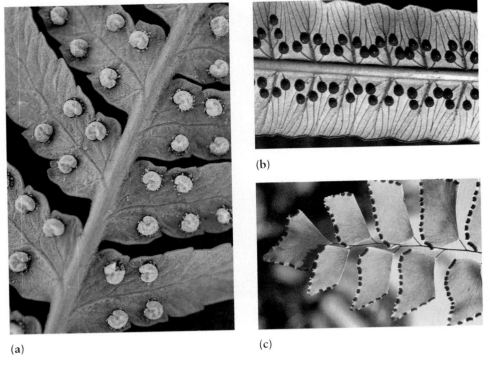

(a)

(b)

(c)

figure 29.28

Fern sporangia. Most ferns have sporangia aggregated into clusters, called sori, on the undersides of the leaves. (a) In some ferns, such as the marginal woodfern (*Dryopteris marginalis*), each sorus is covered by a flap of leaf tissue called an indusium. (b) Other ferns bear uncovered sori, as shown here in *Alsophila sinuata*. (c) In still other ferns, as in the giant maidenhair fern (*Adiantum trapeziforme*), sori occur along leaf edges and are enfolded by the edge of the leaf itself.

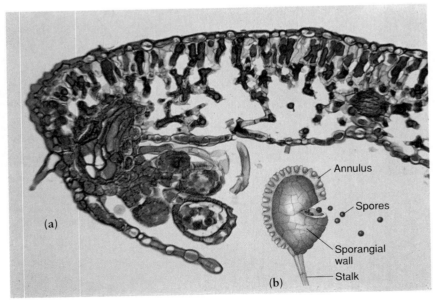

(a)

(b)

Annulus

Spores

Sporangial wall

Stalk

figure 29.29

Leptosporangiate ferns have a unique type of sporangium. (a) Cross section of a leaf of *Pteris* showing a sorus of leptosporangiate sporangia. (b) Leptosporangiate sporangia are characterized by thin walls, a delicate stalk, relatively few spores, and usually an annulus.

Two kinds of sporangium development occur in the ferns. Some ferns are **eusporangiate**, like most vascular plants. This means that their sporangia arise from multiple cells on the leaf surface, are relatively large, with either massive stalks or no stalks at all, and contain many spores surrounded by a multilayered sporangial wall. Tree ferns are examples of eusporangiate ferns (see fig. 29.25a). In contrast, most ferns are **leptosporangiate.** Their sporangia arise from a single surface cell, are relatively small, with a delicate stalk and a thin sporangial wall (fig. 29.29). The small number of spores per leptosporangium is a multiple of 4, varying between 16 and 512 (often 16 or 32) in homosporous species. Nevertheless, each plant can produce millions of spores because of the large number of sporangia per sorus and the enormous number of sori per leaf. For example, one mature plant of *Thelypteris dentata* can produce more than 50 million spores each season. The most common ferns in North America are leptosporangiate. Bracken fern (see fig. 29.1) and all houseplant ferns are included in this group.

Spores are catapulted from their sporangia by an ingenious method that has attracted much attention from botanists. The flinging action comes from the behavior of an incomplete ring of cells, called the **annulus,** that encircles the sporangium (fig. 29.29b). The thin outer cell walls of the annulus slowly contract as they dry, creating a pulling force that ruptures the thin-walled lip cells and the outer walls of the sporangium. As drying continues, the water tension increases in the annulus, sometimes exceeding 300 atmospheres of pressure. Ultimately, the water evaporates completely, and the tension is broken. The sporangium then snaps back into its original position, ejecting the spores forcefully to a distance of about 1 cm.

In most homosporous ferns, spores germinate at first into protonemata (fig. 29.30). These protonemata eventually differentiate and form green, heart-shaped gametophytes that are anchored to the soil by rhizoids (fig. 29.31). The gametophytes are usually bisexual with the sex organs on the lower surface. Archegonia are sunken in the gametophyte tissue near

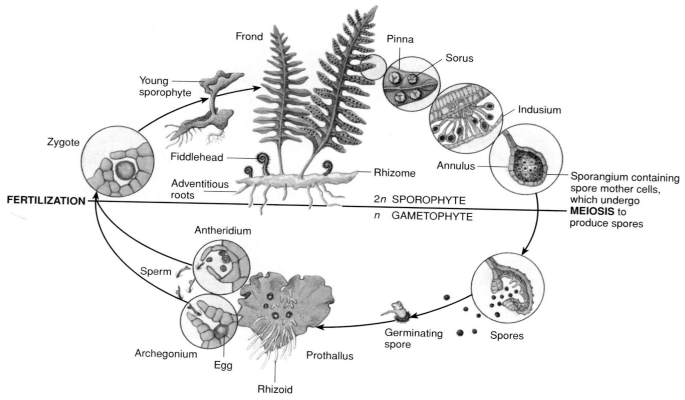

figure 29.30

Life cycle of the fern *Polypodium*. Most ferns are like *Polypodium* in that they are leptosporangiate, homosporous, and gametophytically bisexual.

(a)

(b)

figure 29.31

Fern gametophyte. (a) A heart-shaped fern gametophyte with rhizoids and reproductive organs (×10), which are the dark spots (b) near the notch of the heart (×50).

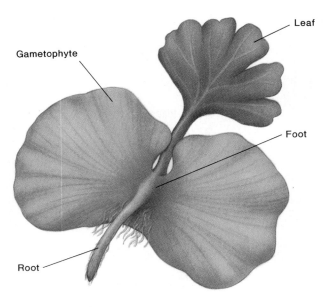

figure 29.32

Young fern sporophyte growing out of its gametophyte parent. Shortly after this stage, the gametophyte dies and begins to decompose.

Labels on figure: Leaf, Gametophyte, Foot, Root

figure 29.33

The clover fern (*Marsilea* sp.) is an example of a heterosporous fern. Like many heterosporous ferns, *Marsilea* is aquatic; the roots of these plants are anchored in the muddy bottoms of ponds or ditches.

the notch, their necks sticking out slightly; antheridia protrude from the surface near the tip and are intermingled with rhizoids. The multiflagellate sperm released by the antheridia swim into the neck of the archegonium to reach the egg. Once fertilization occurs, the zygote germinates into a young sporophyte that quickly becomes independent of the host gametophyte (fig. 29.32). Shortly thereafter, the short-lived gametophyte dies and begins to decompose.

Heterosporous ferns are decidedly unfernlike in their overall appearance. *Azolla*, for example, is a floating aquatic fern having tiny leaves that are crowded onto slender stems (see fig. 29.25b). Other genera, such as *Marsilea*, grow from roots buried in the muddy bottoms of ponds or ditches; only their leaves reach the water surface and float upon it (fig. 29.33). The heterosporous ferns are leptosporangiate, but their sporangia form in specialized, hardened structures called **sporocarps.** Each sporocarp bears several sori consisting of many sporangia. When the spores are mature, they are either released into the surrounding water or pushed out of the sporocarp by the swelling of a gelatinous, hygroscopic filament.

c o n c e p t

More than 12,000 of the approximately 13,000 species of seedless vascular plants are ferns, Division Pteridophyta. The remainder are classified in three divisions: the Psilotophyta (whisk ferns), Lycopodophyta (club mosses), and Equisetophyta (horsetails). Some of the ferns and club mosses are heterosporous; that is, they have one type of spore that grows exclusively into a male gametophyte and one type that becomes exclusively female. All seedless vascular plants produce flagellated sperm and therefore require free water to reproduce sexually.

The Physiology of Seedless Vascular Plants

The physiology of seedless vascular plants is most conveniently studied in the gametophytic phase. Spores can be germinated and gametophytes can be easily grown in laboratory cultures. Because of their independence from the sporophyte phase, these haploid forms are more easily studied in seedless vascular plants than in seed plants. Most of our knowledge of the physiology of seedless vascular plants comes from the largest group, the ferns.

The Influence of Light

Light is especially influential in controlling the growth of fern gametophytes. For example, spore germination in ferns is controlled by light: red light, through the action of phytochrome (see Chapter 19), induces spore germination, and blue light prevents spore germination. Red light also affects protonemata by inducing apical growth and positive phototropism, increasing the gap phase (G-2) in mitosis, and delaying cell-plate formation in cytokinesis. All of these phenomena are inhibited by blue light, but the pigment that absorbs in that region is unknown.

The Environment and Gametophyte Sexuality

Homosporous plants often produce bisexual gametophytes, but some species of *Equisetum* and some ferns instead produce unisexual gametophytes or mixtures of unisexual and bisexual gametophytes. In the latter, sexuality can be affected

Belly-Crawling Through Cemeteries: Documenting the Distribution of the Winter Grapefern in Texas

Jeff Stevens   **Robin Gooch**
Baylor University *Baylor University*

What would you think of two students who spend much of their time crawling around on their bellies in old country cemeteries? For Robin Gooch and Jeff Stevens, both seniors majoring in biology at Baylor University, it's just another episode of searching for the winter grapefern (*Botrychium lunarioides*) (Michx.) Sw. While this behavior seems a little odd, the best way to find the fern is to get down to its level. The fern sporulates in late winter to early spring and is difficult to locate because it grows at ground level and is often camouflaged among other vegetation. The fern is often found elsewhere, but cemeteries provide a favorable habitat; in Texas, with its stringent trespassing laws, permission to gain access to cemeteries is not required.

Robin and Jeff's research is a continuation of a project initiated the previous year (see reference) in which the winter grapefern was documented as occurring in ten counties in central Texas. At that time the fern was known from two localities in eastern Texas, each approximately 32 km from Louisiana or Arkansas. The new records extended the known distribution of the fern 161 to 273 km to the west and suggested a need for further searches. It also showed that students can do important botanical research.

The thrust of the research involves finding additional records for the fern and, once found, determining the population by use of a belt transect. Cemeteries are first located on a surface geological map, then visited to determine if conditions are suitable for the fern. Since the ferns are difficult to find, Robin and Jeff normally search for indicator species—that is, conspicuous plants that occur in association with the winter grapefern. The best indicator plants are *Ophioglossum crotalophoroides* Walter and *Hedyotis boscii* DC. If these plants are present, Robin and Jeff search the cemetery. A voucher specimen, for deposit in the Baylor University Herbarium, is collected to document the occurrence. Currently, search efforts are concentrated in north Texas near Oklahoma, the southern part of the Post Oak Savannah just east of Austin and San Antonio, and the Cross Timbers about 100 miles west of Waco, an area well-known for its peanut farming—hence sandy soil. The field studies are under the direction of Walter C. Holmes.

Do, L. H., R. D. Gooch, J. R. Stevens, W. C. Holmes, & J. R. Singhurst. 1996. New county records of *Botrychium lunarioides* in Texas. *American Fern Journal* 86:28–31.

---

by variations in temperature or the amount of light. In *Equisetum*, for example, spores germinating at a high temperature (32° C) produce only male gametophytes, while at a lower temperature (15° C) only half of them produce males, and the other half produce gametophytes that are female at first, then bisexual. Likewise, a greater proportion of males grows from spores at high light intensity or from spores germinating in red light.

Reproductive Hormones

When the spores of many ferns germinate, the first heart-shaped gametophytes secrete a substance that induces younger gametophytes to develop antheridia. These substances, called **antheridiogens,** can also break dormancy in darkness, thereby eliminating the usual light requirement for spore germination. Some antheridiogens act like gibberellins in gametophytes, and an antheridiogen from at least one fern (*Anemia*) has a gibberellin-type chemical structure (fig. 29.34). In the genus *Cerotopteris*, however, abscisic acid blocks the formation of antheridia that are induced by antheridiogens.

concept

Many physiological studies of seedless vascular plants have focused on the gametophyte. Growth and development are often controlled by light and temperature. Sexuality in some ferns may be controlled by hormones, some of which are chemically and functionally similar to gibberellins.

The Ecology of Seedless Vascular Plants

One of the most interesting ecological aspects of any group of plants involves their interactions with other organisms. For example, the occurrence of mycorrhizal fungi in the gametophytes of *Psilotum* and *Lycopodium* and in the rhizomes and roots of many plants suggests a general symbiotic relationship between fungi and plants. Fungal filaments extract certain minerals from the soil better than do rhizoids or root hairs. Plants obtain such minerals from the fungi associated with them, while the fungi, in return, receive organic nutrients from

(a)

(a)

figure 29.34

(a) The gametophytes of this fern, *Anemia rotundifolia*, secrete gibberellin-like substances, called antheridiogens, that induce other, younger gametophytes nearby to develop antheridia; older gametophytes have antheridia and archegonia. (b) The chemical structure of an antheridiogen.

their hosts. Similar fungi found in fossil plants probably had the same relationship with their hosts more than 400 million years ago. Indeed, this association with fungi is thought to have hastened the invasion of land by plants.

Several tropical species of ferns harbor ant nests in their rhizomes (fig. 29.35). Roots grow into the ant-carved tunnels and absorb nutrients from the decaying organic matter brought in by the ants, and ants use the sporangia as food. Spores are often shaken loose and dispersed from the parent plant as the ants handle the sporangia. Botanists speculate that the loss of spores to ants is sufficiently offset by the advantage of having at least some of the spores dispersed.

The gametophytes of some homosporous vascular plants secrete chemicals that inhibit the growth and reproduction of other plants (see reading 29.1, "Killer Plants"). In many plants,

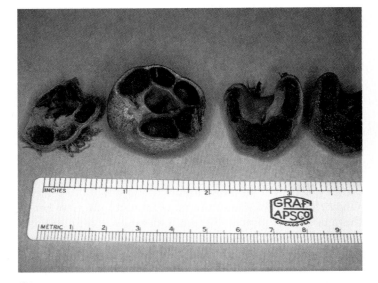

(b)

figure 29.35

(a) *Solanopteris brunei*, a fern native to Costa Rica, harbors nests of ants. (b) Ants live in podlike chambers that are formed by the plant.

KILLER PLANTS

Despite the enormous number of spores produced by each seedless vascular plant, relatively few develop into mature gametophytes in nature. Some of this loss can be attributed to bad luck; most spores are carried by air currents to places that are too dry, too nutrient-poor, or in some other way unfavorable for germination. Spore germination can also be inhibited by substances secreted by other plants, either sporophytes or other gametophytes. In *Thelypteris normalis*, for example, the growth of gametophytes is inhibited by secretions diffusing from the roots of the sporophyte. The two active chemicals, named **thelypterin A** and **thelypterin B,** are indole derivatives that inhibit cell division in the gametophytes of *Thelypteris* and other fern genera. They have no effect on the growth of young sporophytes.

Thelypterins are similar to auxin, both in structure and in function (see Chapter 18). The apparent function of these chemicals is to prevent the growth of gametophytes. The immediate advantage is that competition for potentially limited resources (minerals, water, etc.) is stopped before it even begins.

The secretion by one organism of chemicals that inhibit the growth of another organism is called **allelopathy** (allelopathy also occurs in seed plants and fungi). Penicillin, for example, is a classic example of an allelopathic substance. In most instances, however, one species inhibits another species; yet sporophyte-derived thelypterins inhibit gametophytes of the same species (reading fig. 29.1). This is a somewhat troubling interaction to decipher, because the sporophyte kills its own offspring. How such an interaction could evolve and be maintained has not been clearly explained. Our understanding of the long-term advantages, if any, of this intraspecific allelopathy remains incomplete.

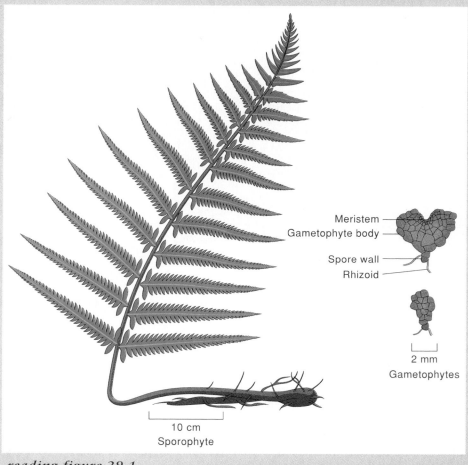

reading figure 29.1

The upper gametophyte of *Thelypteris normalis* shown at the right was grown for 14 days without a nearby sporophyte, whereas the lower gametophyte was grown for 131 days in the presence of a sporophyte. Note the absence of a meristem in the lower gametophyte.

high levels of polyploidy are associated with greater production of various chemicals that can protect plants from insects and fungi. Approximately 95% of seedless vascular plants are polyploid, which may explain why ferns, for example, are remarkably resistant to herbivores and fungal parasites.

The ecology of many ferns is well understood because these plants can be such noxious weeds. For example, when people burn native forests to establish pastures, bracken ferns can quickly invade the newly available habitats (see fig. 29.1). Populations of this plant spread rapidly from an extensive network of fast-growing rhizomes. The problem of bracken infestation is worsened by the toxicity of this plant to the cattle raised in such pastures. Herbicides can alleviate the problem somewhat, but this damages other plants and leaves a toxic residue in the soil. Weed scientists are now studying how to control bracken fern by introducing pathogenic fungi into invasive populations.

Kariba weed (*Salvinia molesta*), an aquatic fern, can be a more serious problem than bracken fern, even though it is not poisonous (fig. 29.36). Trouble with this plant began after it became established in Papua, New Guinea, north of Australia. The invasion of kariba weed overtook waterways and eliminated native species of plants, eventually threatening

(b)

(a)

figure 29.36

The kariba weed (a) reproduces so fast that it can (a) crowd out other kinds of plants and (b) overtake waterways.

the availability of water to 80,000 people. Kariba weed is sterile but reproduces asexually very rapidly, doubling its mass approximately every 2 days. From a genetic viewpoint, millions of tons of this weed may represent a single individual, making it possibly the largest organism on earth. In the 1980s, botanists made a significant breakthrough in understanding the ecology of this organism: its native range was discovered in southeastern Brazil, where a new beetle species was found that feeds exclusively on kariba weed. This beetle is now being used successfully to control invasions of kariba weed. In Papua, New Guinea, the beetles have eaten about 2 million metric tons (2 billion kg) of the plant, reducing the water surface it covered by more than 90%.

c o n c e p t

Seedless vascular plants, especially ferns, are often weedy because they can reproduce rapidly by asexual means. Some species cause problems when they invade pastures (bracken fern) or waterways (kariba weed). Effective biological control of these plants entails introducing their natural enemies, such as pathogenic fungi or herbivorous insects, into areas of rampant growth.

The Economic Importance of Seedless Vascular Plants

Seedless vascular plants are of little economic importance today. Perhaps their greatest economic impact comes from their aid in discovering fossil fuel deposits (see reading 29.2, "Fossil Spores"). Many seedless vascular plants, especially ferns, are often found in greenhouses or are grown as houseplants and ground covers. Western pioneers used to use *Equisetum* to scrub their dishes, and before the invention of flashbulbs, photographers used flash powder that consisted almost entirely of dried *Lycopodium* spores. A pound of spores can still be purchased for only a few dollars from scientific supply companies. In China, where petroleum-based fertilizers are not affordable, *Azolla* is substituted as a rotated crop in rice paddies. This aquatic fern hosts a cyanobacterium, *Anabaena azollae,* that fixes nitrogen from the air, thereby acting as a fertilizer to replenish the nitrates removed from the soil by other crop plants (fig. 29.37).

The evolution of biochemical complexity in land plants coincided with their morphological advancements. Besides structural compounds like lignin, chemicals also evolved to

FOSSIL SPORES

During the early stages of spore formation, a tough polymer called **sporopollenin** is built into spore walls. Because of the stability and resistance to decay of this substance, spores are essentially "instant" fossils (reading fig. 29.2). Indeed, the recovery of spores from sediments entails boiling samples in concentrated acids to dissolve the rocks that contain them. The spores themselves, regardless of whether they are fresh or millions of years old, are unharmed by this process. Their remarkable toughness makes spores, as well as pollen from seed plants, ideal records of plant communities, both past and present.

Exploration for fossil fuels, particularly oil, has been made much easier by the finding that plants are abundantly repre-

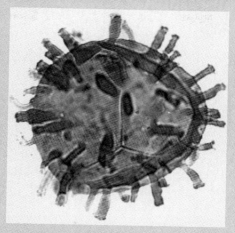

reading figure 29.2
The fossil spore *Raistrickia crocea*.

sented by spore and pollen "communities," the remnants of extinct floras that thrived along ancient shorelines. Specific kinds of spore communities are associated with oil deposits. Core samples are taken by special drills that often penetrate more than several thousand meters below the earth's surface. Oil is usually discovered when the oil-indicator spores and pollen are found in the cores. Much of the analytical success of such efforts depends on **palynologists** (specialists who study pollen and spores); accurate identification of the microfossils in the core samples provides a reliable age for each layer of rocks relatively quickly and inexpensively. These methods have been effective, for example, in explorations of Alaskan oil fields.

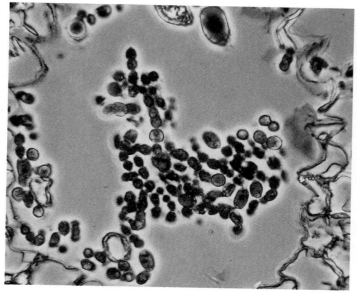

figure 29.37

Leaf lobes of *Azolla* contain cavities, as shown here, inhabited by the cyanobacterium, *Anabaena azollae*.

provide protection from ultraviolet light, parasitic fungi, and protozoa and other predators. The continued biochemical diversification of land plants has resulted in a wealth of chemicals that have been useful to humans. For example, Native Americans treated wounds and nosebleeds with spores from one species of club moss, *Lycopodium clavatum*, which have

antibiotic and blood-coagulant properties. Resin from the rhizomes of the marginal fern *Dryopteris marginalis* was once taken internally to get rid of intestinal tapeworms. As is true for most medicinal plants, the exact identity of the active ingredient from these plants has not been determined. We do know, however, that many *Lycopodium* species synthesize complex, nitrogen-containing chemicals called **alkaloids** that are potent animal poisons. The dried and powdered leaves containing these chemicals are used directly as pesticides in parts of Eastern Europe. More recently, extracts from the fiddleheads of bracken fern, which is considered to be a delicacy by some people, were discovered to cause intestinal cancer.

The Origin and History of Seedless Vascular Plants

Fossils of vascular plants are much more abundant than those of bryophytes. Vascular tissues, cuticles, and spores are particularly well preserved because of their resistance to decomposition. Although the fossil record is still far from complete, several thousand extinct species have been discovered. Most of these are from species or genera; that is, they are known only from spores or fragments of the plant body and cannot be matched with an entire plant (see reading 29.3, "How Does a Plant Become a Fossil?"). The oldest from genera, consisting of cuticles and spore tetrads, were discovered in late Ordovician rocks in Libya. Many botanists are not convinced that these plants had vascular tissue, however, since tracheids or other evidence of vascularization has not been found with them.

How Does a Plant Become a Fossil?

Archaeologists trying to piece together evidence of past civilizations search for telltale shards of pottery and other artifacts. Paleobotanists use similar direct evidence, in the form of fossils.

The public perception that fossils are just old plants or animals in rocks doesn't do justice to the wide variety of things that are called fossils. Different types of fossils form, depending on where the organisms grew and how fast they were buried in sediment. The preservation of most organisms has occurred in water as sedimentary particles were deposited on plant parts that fell into the water. Heavy sediment flattened leaves or other plant parts, squeezing out water and leaving only a thin film of tissue, called a **compression** fossil (reading fig. 29.3a). Cellular structure rarely survived this process, but well-prepared cuticles have occasionally been found in deposits of compression fossils. Intact cells (and even DNA) have also been found in compressions, as in the Miocene deposit (ca. 20 million years old) at a fossil site named Clarkia Lake in Idaho.

Other types of fossils usually lack plant tissue. An **impression,** for example, is an imprint of an organism that is left behind when the organic remains have been completely destroyed. Only the contour of the plant remains (reading fig. 29.3b).

In a third type of fossilization, tissues became surrounded by hardened sediment and then decayed. The hollow negative of the original tissue is called a **mold.** In conditions that allowed the mold to become filled with other sediment that conformed to the contours of the mold, the resultant fossil is called a **cast.** Fossil molds and casts, although formed on a geological time scale, are analogous to those that modern artists make of their subjects.

An interesting but poorly understood process of fossilization involves the replacement of cell contents with minerals. The compaction of such mineralized tissues essentially transforms the organic material into rock. Such fossils, called **petrifactions,** make areas like the Petrified Forest of Arizona famous among botanists and tourists alike (reading fig. 29.3c).

Plant fossils are rarely found as whole plants. Different plant parts in the same fossil bed are given separate species and genus names, referred to as *form* species or genera. Sometimes two or more parts are connected to each other, and the plant they came from can be drawn as a reconstruction of the original. One such example, the genus *Lepidodendron,* is shown in reading figure 29.3d. *Lepidodendron* is actually the name given to stem fragments; the rest of the plant is derived from several unconnected organs and other fragments, each of which was given a separate name when it was discovered.

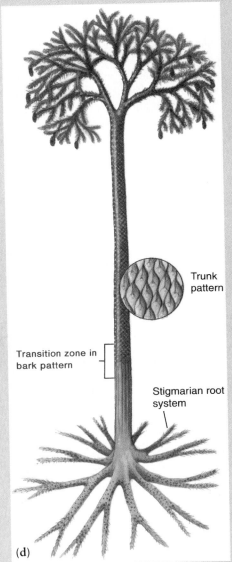

reading figure 29.3

Types of fossils include: (a) compressions, which are made when heavy sediment squeezes out water and leaves only a thin film of tissue; (b) impressions, which are the imprints that are left in rocks after the organic remains of an organism have been completely destroyed; and (c) petrifactions, which form when cell contents are replaced with minerals, thereby transforming the organic material into rock. Petrifactions are so common in some areas that they form "petrified" forests, such as the Petrified Forest of Arizona. (d) Reconstruction of the genus *Lepidodendron*.

Trunk pattern

Transition zone in bark pattern

Stigmarian root system

(a)

(b)

figure 29.38

(a) Reconstruction of a Devonian habitat. The low-growing seedless vascular plants of this period included *Cooksonia, Zosterophyllum,* and *Psilophyton.* (b) Swamp forest of the Carboniferous period. Most of the trees in this reconstruction are of giant club mosses. Plants with frondlike leaves (e.g., left foreground) are seed ferns. The tree on the right that has bottlebrush-appearing branches is *Calamites,* a giant horsetail.

The oldest fossils that are unquestionably vascular plants consist of well-preserved vegetative and reproductive structures. These plants, named *Cooksonia,* were rootless and leafless, with slender, dichotomously branched stems (fig. 29.38). They produced homosporous spores in sporangia that developed from the expanded tips of their branches. *Cooksonia* first appeared in the late Silurian period, about 420 million years ago. These first known invaders of land were abundant in the tidal mudflats of New York State, South Wales, the Czech Republic, and Podolia (Ukraine). Their distribution probably included many other areas that experienced periodic flooding, but fossils in these areas were either not preserved or have not yet been discovered.

Cooksonia represents the opening of the floodgates of diversification for vascular plants. There was little competition for colonizing immense areas of land. Although *Cooksonia* was the most successful of the land invaders, it gave way to larger and more complex kinds of plants during the Devonian period. Plants grew taller and thicker, and they began to develop leaves that had a larger surface area for photosynthesis. The development of roots and stem cuticles allowed plants to move even farther from their swampy origins into increasingly drier habitats. The development of mycorrhizae probably

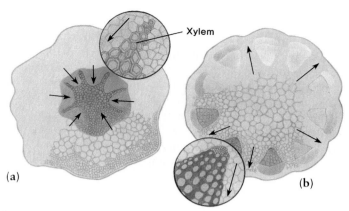

(a)

(b)

figure 29.39

Examples of exarch (a) and endarch (b) xylem development. Exarch xylem matures toward the center of the stem (i.e., centripetally), whereas endarch xylem matures toward the perimeter of the stem (i.e., centrifugally).

played an important role in enabling plants to get nutrients from soil. Spores became more resistant to desiccation, which enabled them to tolerate dispersal by wind without drying out.

The First Vascular Plants

Depending on their quality and the extent of preservation, fragments of fossil plants can occasionally be matched so that a significant portion of the original plant can be reconstructed. Perhaps 200 or so fossil species fit this category. Of these, not more than two dozen are known in superb detail from a large number of specimens. Our picture of the earliest vascular plants relies heavily on this small number of well-known plants.

Devonian plants reveal two distinct evolutionary lines. Plants in one line, referred to as the **zosterophyllophytes** (Division **Zosterophyllophyta**), superficially resemble the living genus *Zostera,* the marine eelgrasses. Zosterophyllophytes grew in clusters of upright, leafless branches that arose from horizontal stems, much as *Cooksonia* did (see fig. 29.38a). However, their sporangia were attached laterally near the ends of branches, which gave them the appearance of loosely aggregated strobili.

The xylem in zosterophyllophytes was **exarch,** meaning that it developed centripetally; that is, the youngest xylem cells were closer to the periphery of the stem than were the older xylem cells (fig. 29.39a). Even though *Zosterophyllum* and its relatives were extinct by the late Devonian period, the zosterophyllophytes probably gave rise to the ancestors of the Lycopodophyta, including now-extinct trees and present-day club mosses. Botanists believe that the club mosses and their extinct relatives inherited the features of lateral sporangial attachment and exarch protosteles from zosterophyllophyte progenitors.

The second evolutionary line includes *Cooksonia* and its relatives in Division **Rhyniophyta.** This division is named for the remarkably preserved assemblage of Devonian fossils near the village of Rhynie, Scotland. **Rhyniophytes** looked like zosterophyllophytes, and members of the two divisions often occur in the same fossil beds. The main difference between them is that the rhyniophytes had terminal sporangia, whereas the zosterophyllophytes had lateral sporangia. Also, the xylem of rhyniophytes developed centrifugally (**endarch**) from the interior of the protostele (fig. 29.39b), not from the periphery, as in zosterophyllophytes. Although *Cooksonia*, a rhyniophyte, is the oldest known vascular plant, it is unlikely that the rhyniophytes gave rise to the zosterophyllophytes. These two divisions must have shared a common ancestor at some point in their history, but their exact relationship to one another is unknown.

Rhyniophytes occupy an important position in the history of modern plants because they were the ancestral stock for the Division **Trimerophytophyta**. Furthermore, because of their large size and structural complexity, the **trimerophytes** are believed to be the most likely precursors of ferns, horsetails, and seed plants. Before these modern groups arose, however, plants had to evolve leaves and roots.

The possible origin of roots is even less clear than the origin of leaves, but botanists assume that roots came from primitive rhizome branches that developed protective tissue (i.e., root caps) to cover their growing tips. Indirect evidence for such an origin is that roots are generally protostelic, as were primitive rhizomes. In addition, the roots of seedless vascular plants are almost exclusively adventitious, which may also be a holdover from their origins as rhizome branches. However, the xylem of roots is exarch, whereas the rhyniophytes and trimerophytes, the oldest ancestors of modern plants, had endarch protosteles.

Giant Club Mosses, Giant Horsetails, and Tree Ferns: Coal-Age Trees

The first land plants were relatively small and herbaceous. With the advent of leaves, thicker cuticles, roots, and perhaps most important, lignin, plants could grow larger and could more effectively exploit the year-round growth conditions during the Carboniferous period. By the middle of the Carboniferous period (about 300 million years ago), some plant groups had evolved into large trees, and their herbaceous progenitors were well on their way to becoming extinct. These first trees included the earliest seed plants (discussed in the next chapter) as well as several groups of seedless vascular plants. The fossils of these huge plants formed the extensive coal deposits that characterize the Coal Age.

The swamp-dwelling lycopodophyte trees, or giant club mosses, grew nearly 40 m tall and dominated the Carboniferous period. Like their zosterophyllophyte progenitors, *Lepidodendron* and related trees were protostelic, although *Lepidodendron* was protostelic only at the base of the stem and the tips of branches; in between, the stem and branches were siphonostelic. The massive stems were supported by extensive periderm tissue and a small amount of secondary xylem. Underground they had a modified branch system with rootlike appendages whose internal structure was like the roots of *Isoetes*. These giant club mosses also had microphyllous leaves, and their sporophylls were packed

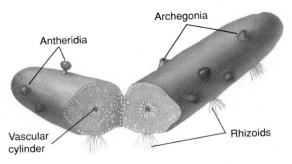

figure 29.40

Schematic drawing of the "gametophyte" of *Rhynia gwynne-vaughanii*.

together into strobili. Trees in the Lycopodophyta were mostly heterosporous. A single strobilus from one of these plants, up to 0.5 m long, may have produced as many as 8 billion microspores or as many as several hundred megaspores. Although the wastage rate of spores must have been enormous, spore dispersal from tall trees enabled the trees to colonize large areas of land successfully.

Giant horsetails such as *Calamites* lived in swampy forests alongside the giant club mosses. Secondary xylem enabled *Calamites* to grow to almost 20 m tall. With few exceptions, the plants were homosporous, and all the sporangia were borne on sporangiophores in terminal strobili. Unlike the giant club mosses, the giant horsetails had extensive networks of underground stems that allowed them to spread rapidly by vegetative reproduction. Because of such rapid reproduction by rhizomes, aboveground stems often occurred in dense, bamboolike thickets.

The Carboniferous period also saw the origin and diversification of tree ferns. Although all of the genera of tree ferns from the Carboniferous are extinct, as a group these primitive plants are still well represented in tropical regions of the world today.

At the end of the Carboniferous period, the giant club mosses began to disappear rapidly and were gone completely after the mass extinction that characterized the end of the Permian period, approximately 250 million years ago. In contrast, the giant horsetails continued to dominate the landscape for another 100 million years or so, until the mid-Cretaceous period, when they also died out. Nevertheless, many of the primitive features of these extinct trees are maintained in the smaller, modern members of these groups.

Ancient Gametophytes

Our knowledge about the structure of extinct plants is based primarily upon features of the sporophytes. Preservation of gametophyte tissue from vascular plants was probably even poorer than for the gametophytes of bryophytes. One intriguing exception to this general pattern may be a controversial Devonian plant named *Rhynia gwynne-vaughanii* (fig. 29.40). This species is known only from vascularized, rhizomelike fragments found in the Rhynie fossil beds. Some botanists interpret the small bumps on the surface of these fragments to be archegonia and antheridia, which means that the fossil fragments are gametophytic. If these structures are indeed sex organs, the gametophytes bearing them resemble those of *Psilotum.* However, the microscopic features of *Rhynia gwynne-vaughanii* are not interpreted as sex organs by all botanists; some botanists believe they are unusual secretory structures on an otherwise ordinary sporophyte rhizome. The argument about whether this plant is a gametophyte or a sporophyte remains unresolved.

c o n c e p t

The first vascular land plants are known from fossil sporophytes of the Silurian period, approximately 420 million years ago. Early vascular plants were small and herbaceous, and they grew in swampy, tropical areas. Like present-day *Psilotum,* they had no leaves or roots but instead had photosynthetic stems and underground rhizomes with absorptive rhizoids. Descendants of these plants evolved leaves, roots, and strong structural support tissues. All groups developed tree forms that dominated the earth for millions of years, and then most died out. The closest living descendants of these early trees are the club mosses, horsetails, and tree ferns. Other evolutionary lines produced the seed plants.

The Puzzle of the Psilotophyta

Progress toward a phylogenetic classification of seedless vascular plants has been enlivened considerably by arguments about how to classify the psilotophytes. The traditional view is that *Psilotum* and *Tmesipteris* comprise a separate division that, with the exception of mosses, may be the most primitive of the vascular plants. This view is based on such assumed primitive features as lack of roots, enations instead of leaves, protosteles, an inconspicuous bisexual gametophyte, and dichotomous branching. Starting in 1969, however, David Bierhorst did a series of studies that led him to propose that *Psilotum* and *Tmesipteris* should be classified as ferns. He noted that many of the features of psilotophytes are shared with ferns, especially the primitive genera *Stromatopteris* and *Actinostachys.* Furthermore, by comparison with these ferns, the shoots of the psilotophytes can be interpreted as fronds. At that time, some botanists saw Bierhorst's proposal as an unacceptable departure from traditional classification. Some botanists still do.

Three new kinds of information are now available for further evaluation of the traditional hypothesis versus Bierhorst's alternative hypothesis about the relationships of the psilotophytes to ferns. One kind involves flavonoids: the psilotophytes synthesize a different class of flavonoids than is found in the ferns. Some botanists interpret this as a significant difference that does not support a close relationship between psilotophytes and ferns. The second kind of information comes from a cladistic analysis of all plants, which places the psilotophytes near the ancestral base of seedless vascular plants, just above

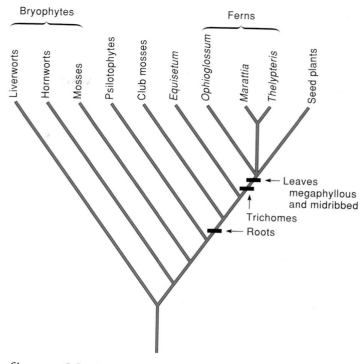

figure 29.41

Cladogram of land plants. The presence of trichomes and megaphyllous leaves is evidence for a single evolutionary line to ferns and seed plants, but the relationship of different kinds of ferns to one another or to seed plants is unclear.

the mosses (fig. 29.41). Although the cladogram was derived from the same kinds of characters that Bierhorst used for his hypothesis, the conflicting results of the two analyses are probably caused by different assumptions about the importance and evolutionary direction of the characters used.

This conflict may be resolved by yet a third kind of new information, which involves the structure of the chloroplast genome in plants. In a certain region of about 30,000 base-pairs of the genome, all genes point in one direction in some plants and in the opposite direction in others (see Chapter 11). Botanists believe that this structural difference in the chloroplast genome is highly significant and that it represents a mutation that probably occurred just once in the evolutionary history of plants. A survey of plants in 1992 found that one orientation occurs in bryophytes and club mosses, while the other orientation occurs in psilotophytes, *Equisetum,* ferns, and seed plants. Assuming that the primitive orientation occurs in the bryophytes and club mosses, then all other plants share a common ancestor that mutated to the other orientation. Although this single feature does not show how close the relationship might be between psilotophytes and ferns, it does provide evidence for rejecting the traditional hypothesis about the primitive position of the Psilotophyta. Furthermore, it also supports the widely held hypothesis among paleobotanists that the club mosses represent the oldest vascular plants.

Chapter Summary

Seedless vascular plants are primarily ferns, but they also include whisk ferns (which are not true ferns), club mosses, and horsetails. Features of seedless vascular plants that enable them to thrive on land include a resistant cuticle, complex stomata, vascular tissue, absorptive root hairs, and desiccation-resistant spores. Vascular tissue in ferns is more complex than in plants of the other three divisions.

Sexual reproduction in seedless vascular plants is similar to that of bryophytes; that is, spores germinate into gametophytes. The life cycle of seedless vascular plants is dominated by the sporophyte phase. Spores are generally dispersed by wind, although the spores of some ferns are dispersed by ants. All horsetails (*Equisetum*) and some club mosses (all species of *Selaginella* but not all species of *Lycopodium*) bear sporangia in densely packed cones, or strobili. *Selaginella*, *Isoetes*, and some ferns produce dioecious gametophytes from two kinds of spores, a phenomenon called heterospory. The gametophytes of seedless vascular plants are nutritionally independent of their parent sporophytes, but not all are photosynthetic; the gametophytes of whisk ferns, some club mosses, spike mosses, and several ferns are saprophytic. Many seedless vascular plants can reproduce vegetatively, either by the sporophyte or the gametophyte or both.

The seedless vascular plants are of little direct economic value. Some species were used as folk medicines, pesticides, or photographic flash powder. Fossil plants have some indirect economic value because geologists use the location of spores in the earth's strata to judge where oil deposits might occur. Some weedy plants cause problems because they diminish the usefulness of pastureland, clog waterways, and eliminate native plant species.

The fossil history of vascular plants begins with *Cooksonia*, a leafless and rootless progenitor of most modern plant groups. The first plants with leaves and roots appeared more than 150 million years after plants first invaded land. Giant horsetails, giant club mosses, and tree ferns dominated the vegetation of the Carboniferous period. Although these seedless trees ultimately gave way to seed plants, their descendants live on as herbaceous lycopodophytes and as modern tree ferns.

There are conflicting ideas about the evolutionary relationships of the Division Psilotophyta. One view is that it is a division of primitive plants, and the other is that psilotophytes are modified ferns.

 What Are Botanists Doing?

Go to the library and read a recent article about the use of *Azolla* in rice production in China. Is *Azolla* used in rice production in the United States? Why or why not?

Writing to Learn Botany

What is the potential evolutionary significance of heterospory? What are the disadvantages of heterospory, and how have seedless vascular plants overcome these disadvantages?

Questions for Further Thought and Study

1. How are the gametophytes of seedless vascular plants similar to the gametophytes of bryophytes? How are they different?

2. Why is *Cooksonia*, which is the oldest confirmed plant fossil, not considered to be the oldest ancestor of all modern plant groups?

3. What are the major distinguishing features of the divisions of extant seedless vascular plants?

4. Most seedless vascular plants are polyploid. What features of their reproductive biology or ecology might explain this phenomenon?

5. *Lycopodium* means "foot of the wolf." What feature of the plant is probably responsible for this name?

6. A tracheid-based vascular system is supposedly less efficient than one based on vessels. Why? How does the geographical distribution of seedless vascular plants support this hypothesis? How does it conflict with this hypothesis?

7. Why are the fertile appendages of *Equisetum* called sporangiophores instead of sporophylls?

8. What are the distinguishing features of leptosporangiate and eusporangiate development?

Web Sites

Review the "Doing Botany Yourself" essay and assignments for Chapter 29 on the *Botany Home Page*. What experiments would you do to test the hypotheses? What data can you gather on the Web to help you refine your experiments?

Here are some other sites that you may find interesting:

http://osprey.anbg.gov.au/ferns/ferns.html

This site, maintained by the Australian National Botanical Gardens, describes how to grow ferns, including how to prepare soil, fertilize, and propagate ferns from spores.

http://www.usc.edu:80/dept/materials_science/MDA125/ plants/sld028.htm

http://www.marceline.k12.mo.us/~student/hainds/ferns/ferns. html

At these sites you'll find photographs and information about the evolution, reproduction, and economic importance of ferns, as well as links to other sites.

http:/www.cc.mancol.edu:80/science/biology/plants_new/ vascular/spheno.html

Here you'll find descriptions of the morophology and life cycle of *Equisetum*.

http://herb.biol.uregina.ca/liu/bio/idb.shtml

Welcome to the *Internet Directory for Botany*. This site includes a searchable list of botanical gardens, botanical journals, etc.

http://www.mobot.org/CPC/welcome.html

The Center for Plant Conservation (CPC) is dedicated to conserving rare plants native to the United States. The Center also complements and supports in situ conservation efforts by implementing rare plant research and education programs.

http://www.anbg.gov.au/jrc/ferns-man-ng.html

Welcome to Ferns and Man in New Guinea. Over 90 species in 42 genera of ferns that are used by people, or affect people in some way or other, are found in New Guinea and the Solomon Islands. This site discusses their use and importance as food, medicine, structural and aesthetic material, weeds, pests, and items of traditional ritual significance.

Suggested Readings

Articles

Banks, H. P. 1975. Early vascular plants: Proof and conjecture. *BioScience* 25:730–737.

Barrett, S. C. H. 1989. Waterweed invasions. *Scientific American* 261 (October):90–97.

Graham, L. E. 1985. The origin of the life cycle of land plants. *American Scientist* 73:178–186.

Niklas, K. J. 1981. The chemistry of fossil plants. *BioScience* 31:820–825.

Raghavan, V. 1992. Germination of fern spores. *American Scientist* 80:176–185.

Rothwell, G. W., and R. A. Stockey. 1994. Pteridophytes and gymnosperms: Current concepts of structure, evolutionary history, and phylogeny. *Journal of Plant Research* 107: 409–411.

Voeller, B. 1971. Developmental physiology of fern gametophytes: Relevance for biology. *BioScience* 21:266–270.

Books

Foster, F. G. 1984. *Ferns to Know and Grow*. Portland: Timber Press.

Gifford, E. M., and A. S. Foster. 1989. *Morphology and Evolution of Vascular Plants*. New York: W. H. Freeman.

Jones, D. L. 1987. *Encyclopedia of Ferns*. Portland: Timber Press.

Stewart, W. N. 1983. *Paleobotany and the Evolution of Plants*. Cambridge: Cambridge University Press.

Thomas, B. 1981. *The Evolution of Plants and Flowers*. New York: St. Martin's Press.

Tryon, R. M., and A. F. Tryon. 1982. *Ferns and Allied Plants*. New York: Springer-Verlag.

West, R. G. 1977. *Studying the Past by Pollen Analysis*. Oxford Biology Reader No. 10. Oxford: Oxford University Press.

Pollen being released into the wind from a cluster of male pinecones. Gymnosperms such as pine depend on the wind to carry pollen from the male cones to the female cones.

Gymnosperms

30

Chapter Outline

Chapter Overview

A major change occurred in plant evolution when seeds appeared about 360 million years ago, a time when most continents were flooded by tropical swamps. This change signified the increasing protection of the female gametophyte, which was followed by the evolution of pollen. Later, when the earth became cooler and drier, these reproductive changes were accompanied by changes in vegetative features, such as sunken stomata, thicker cuticles, and tougher tissues, that enabled land plants to minimize water loss. Changing environments fostered the diversification and extinction of several groups of seed plants; some groups were more successful than others in surviving to modern times. The most dominant of the temporary successes may have been the seed ferns, which lasted about 70 million years, until the environment got too cool and dry for them during the early Permian period. The cycads, conifers, and cycadeoids thrived after the seed ferns disappeared. Of these, only a few descendants of the cycads and conifers remain.

The cycadeoids and cycads were so abundant during the Jurassic that this period is known as the Age of Cycads, at least among botanists (other biologists usually refer to it as the Age of Dinosaurs). The cycadeoids went extinct about the same time as the dinosaurs, about 65 million years ago. The fossil record of gymnosperms provides much information on the earliest of the seed plants—information that helps us understand how the modern gymnosperms evolved.

One of the most significant events in the evolution of vascular plants was the origin of the seed. Ideas about how seeds arose are based on interpretations of the earliest fossils that look like seeds. The oldest such fossil is from the late Devonian period (about 360 million years ago), which was more than 40 million years after the first vascular plants appeared. This fossil has been interpreted by some botanists as a loose cluster of seeds, each consisting of a megasporangium surrounded by a protective, integument-like layer with several fingerlike projections (fig. 30.1). Other fossils, from a little later in the Devonian and from the early Carboniferous period, are more seedlike; they show the megasporangium becoming progressively more enclosed by the integumentary layer (fig. 30.2). Paleobotanists believe that this progression represents the evolutionary development of seeds.

The counterpart of the seed is the pollen grain, which is an immature male gametophyte combined with and contained within a microspore wall (see Chapter 17). Pollen grains evolved by a continued reduction of the male gametophyte and its more secure retention inside the protective spore wall. This description of a pollen grain would also seem to apply to the microspore of *Selaginella* and other seedless vascular plants whose male gametophytes are mostly enclosed in the spore wall, but botanists distinguish true pollen from microspores.

Morphologically, the first pollen differed from microspores in the location of the gametophyte's germination from the spore wall. Microspores germinated from the **tetrad scar,** which is the place on a spore where it was attached to three other spores in a meiotic group of four. In contrast, the first true pollen grains germinated from the spore wall opposite the scar. The development of pollen was also accompanied by the disappearance of the tetrad scar entirely (fig. 30.3). The first true pollen did not appear in the fossil record until the Mesozoic era, more than 150 million years after the origin of seeds.

Features of Gymnosperms

The term *gymnosperm* derives from the Greek word roots *gymnos,* meaning "naked," and *sperma,* meaning "seed."[1] Gymnosperms are plants whose pollen goes directly to ovules (unfertilized seeds) instead of to a stigma (as in the flowering plants) and whose seeds are naked (i.e., are not enclosed in fruits). Thus, by definition, gymnosperms are all fruitless seed plants. Examples of gymnosperms include maidenhair tree (*Ginkgo*), cycads, conifers, and members of the Gnetophyta (e.g., *Ephedra, Gnetum*) (fig. 30.4).

Gymnosperms are characterized by secondary growth that usually forms woody trees or shrubs, but some species are more vinelike. Most gymnosperms lack vessels in their xylem, with the exception of the gnetophytes. In this regard, the gymnosperms are like most of the seedless vascular plants and some primitive angiosperms.

Considering the relatively small number of living gymnosperms (about 720 species in 65 genera), they are remarkably diverse in their reproductive structures and leaf types. Microsporangiate strobili may be either loosely arranged, laden with up to a thousand microsporangia, packed into multiple clusters, or flowerlike (fig. 30.4). Leaf types range from simple, flat blades to needles, compound frondlike leaves, and highly reduced *Equisetum*-like leaves.

Gymnosperms, like angiosperms, differ from the seedless plants in not requiring water for sperm to swim in to reach the egg. Only the cycads and the maidenhair tree have flagellated sperm, but these and other seed plants must be pollinated by wind, animals, or water. This means that the movement of male gametes to female gametes in seed plants relies on airborne transport, not on water transport. Consequently, most gymnosperms produce huge amounts of pollen. For example, each male of a pine tree cone annually releases an estimated 1–2 million pollen grains. Similarly, the immense conifer forests of Sweden have been estimated to release 75,000 tons of pollen every spring.

The most dramatic differences between gymnosperms and other plants involve pollen and seeds and the organs that bear them. These features often differ significantly from those of comparable organs of flowering plants.

The Gymnosperm Pollen Grain

Although pollen evolved by the reduction of the male gametophyte, the pollen grains of gymnosperms still retain vegetative remnants of the ancestral gametophyte thallus. These remnants usually consist of one or two cells, called **prothallial cells,** which often disintegrate before fertilization. The exceptions include *Gnetum* and a few of the Pinophyta, whose pollen grains, like those of angiosperms, have no prothallial cells.

The pollen grains of gymnosperms also apparently contain vestiges of ancestral antheridia. The generative cell divides to form a **stalk cell,** which seems to be all that remains of an

---

1. The ancient Greeks went to a *gymnasium* for physical exercise, for which they removed their clothing. We still have gymnasiums, but we don't use them in quite the same way as the ancient Greeks did.

figure 30.1

Reconstruction of ovules (with fingerlike projections) of a plant that grew in the late Devonian period (about 360 million years ago). Some botanists interpret this structure as a loose cluster of seeds.

Source: After G. W. Rothwell.

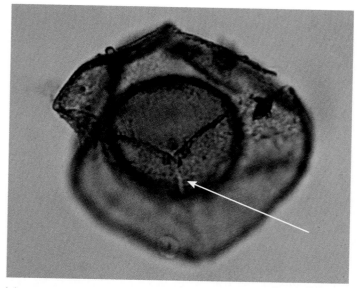

(a)

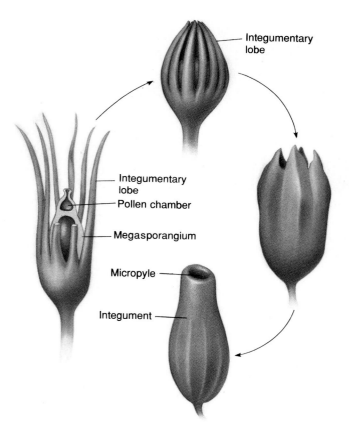

Integumentary lobe

Integumentary lobe

Pollen chamber

Megasporangium

Micropyle

Integument

figure 30.2

The evolution of seeds. Reconstructed ovules from the early Carboniferous period show how they may have become progressively more enclosed by the integuments.

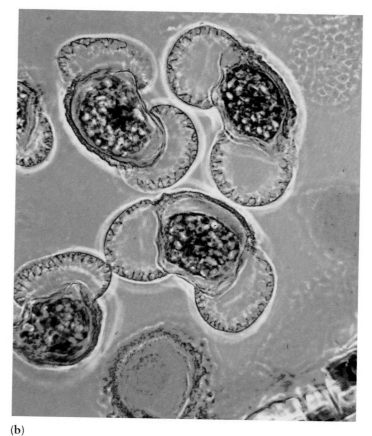

(b)

figure 30.3

(a) *Endosporites*, a lycopodophyte spore from the Carboniferous period, has a three-ridged tetrad scar (arrow) that characterizes microspores. (b) In contrast to spores, pollen grains of *Pinus* and other seed plants have no scar.

(a)

(c)

(d)

(b)

figure 30.4

Pollen-bearing cones of gymnosperms. (a) *Ginkgo.* (b) *Zamia.* (c) *Pinus.* (d) *Ephedra.*

antheridial stalk, and a **body cell,** which is the only sperm-producing cell in the male gametophyte (see fig. 30.8). The body cell is essentially a single-celled antheridium. Immediately before fertilization, the body cell divides to form two sperm cells. Among gymnosperms, exceptions to this type of male gametophyte development occur in *Gnetum* and *Welwitschia,* whose generative cells divide directly into sperm cells. The formation of sperm cells directly from generative cells is also a feature of the angiosperms.

The Gymnosperm Seed

The most consistent features shared by all living gymnosperms—features that distinguish this group from seedless plants or other seed plants—involve the seed. Most of the mass of a mature gymnosperm seed consists of an integumentary layer, a multicellular female gametophyte, and one or

30–5

more embryos; other cells in the seed come from the megasporangium and the megaspore wall. Ovule development begins when a single cell in the megasporangium undergoes meiosis, forming a linear tetrad of haploid megaspores. Three of the spores usually disintegrate, and the remaining **functional megaspore** undergoes repeated mitotic divisions that are not followed immediately by cytokinesis. In *Gnetum*, however, all four megaspore nuclei divide repeatedly inside a single spore wall, as in the tetrasporic embryo-sac development of *Lilium* (see Chapter 17). Either way, the result is a coenocytic stage that is called the **free-nuclear female gametophyte**. The number of free nuclei can be as low as 256 in some species of *Ephedra* to as high as about 8,000 in the maidenhair tree, which is much more than the 8–16 free nuclei in the female gametophytes of angiosperms. Cell walls later form around each nucleus, after which 2 or more (up to 200) archegonia develop at the micropylar end of the ovule. When mature, each archegonium contains a single egg, and all eggs may be fertilized. Archegonia are absent in some of the gnetophytes.

Multiple embryos in the same seed, which occasionally occur in some of the divisions of gymnosperms, form in one of two ways. The less common way, called **simple polyembryony**, occurs when two or more zygotes develop into embryos. Most of the time, all but one of the zygotes fail to develop. The more common origin of multiple embryos occurs when certain cells of a single embryo differentiate into more than one embryo. This mechanism is called **cleavage polyembryony** and produces clonal embryos.

The early stages of embryo development in gymnosperms are characterized by free-nuclear divisions of the zygote nucleus (except perhaps in some gnetophytes). This **free-nuclear embryo** may consist of as few as 4 nuclei in pines to as many as 256 nuclei in cycads. Once the gametophyte becomes cellular, cells near the micropylar end of the embryo elongate into **suspensor cells** in all divisions except the Ginkgophyta.

c o n c e p t

The most significant development leading to the evolution of gymnosperms was the origin of the seed. Seeds probably arose from a megasporangium that became surrounded by an integument-like layer of tissue with fingerlike projections. Pollen arose about 150 million years after the first seeds appeared. Pollen grains evolved by reducing the male gametophyte to fewer cells and retaining it within the microspore wall.

Seed-Bearing and Microsporangium-Bearing Structures

The simplest seed-bearing structures among gymnosperms are those of the maidenhair tree (*Ginkgo biloba*) and the yew family (*Taxus* and *Torreya*). Seeds of these groups are borne singly at the ends of stalks; *Ginkgo* ovules start out in pairs, but one aborts early in development (fig. 30.5a). The mature ovules of *Taxus* (yews) and *Torreya* (e.g., California nutmeg, *T. californica*) are surrounded by a fleshy, cuplike **aril** that is often bright red (fig. 30.5b). In contrast, most other gymnosperms produce seeds in complex strobili, structures similar to those discussed in Chapter 29. The smallest strobili include those of the junipers (*Juniperus*), which have fleshy scales that are fused into a berrylike structure (fig. 30.5c). The largest seed cones, which may be up to 1 m in length and weigh more than 15 kg, are produced by cycads (fig. 30.5d).

Megastrobili show a wide range of complexity, but botanists generally categorize them as either **simple strobili** or **compound strobili**. A simple strobilus consists of an unbranched axis that bears sporophylls, as in *Selaginella* and *Lycopodium* (see Chapter 29). The microstrobili of pines and the megastrobili and microstrobili of cycads are simple (fig. 30.6a). Based on comparative morphology of fossils, current thinking regards the megastrobilus of pines as a reduced branching system. Ovules are borne on reduced shoots that occur in the axils of spirally arranged bracts, which represent branches from the main axis of the cone (fig. 30.6b). The entire structure, known as a seed-scale complex, becomes woody at maturity. The compound strobili of *Ephedra* are more reduced (fig. 30.7); the seed cone (megastrobilus) has four to seven pairs of small bracts, but only the uppermost pair bears ovules. Likewise, the compound microstrobilus of *Ephedra* has up to seven pairs of bracts. In addition, microsporangia often occur in clusters on a single stalk (see fig. 30.4d). Botanists view the compound strobili of *Ephedra* as being more like flowers than the strobili of any other gymnosperms. Moreover, the stalked cluster of microsporangia is often referred to as a *stamen*.

Microsporangia generally occur in conelike strobili in gymnosperms. An exception is the microstrobilus of *Ginkgo*, which consists of many paired, antherlike sporangia on short stalks that are loosely arranged in a spiral along a main axis (see fig. 30.4a). Pine microstrobili are distinguished by having microsporangia borne on the lower (**abaxial**) surface of the sporophylls, instead of on the upper (**adaxial**) surface, as in cycads and in seedless vascular plants (fig. 30.6a). Pollen cones are usually much smaller than seed cones, except in cycads, where the pollen and seed cones may be the same size.

c o n c e p t

Most groups of seed plants have simple strobili, but some have compound strobili, and others bear seeds singly on individual stalks.

(a)

(c)

(b)

(d)

figure 30.5

(a) *Ginkgo* seeds attached to tree. (b) Yew (*Taxus cuspidata*) seeds. The fleshy, bright red coverings are arils. (c) Juniper (*Juniperus osteosperma*) "berries." (d) Seed-bearing cone of a cycad (*Zamia* sp.).

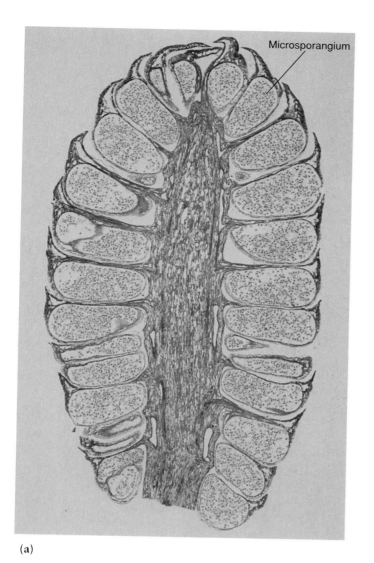

Microsporangium

(a)

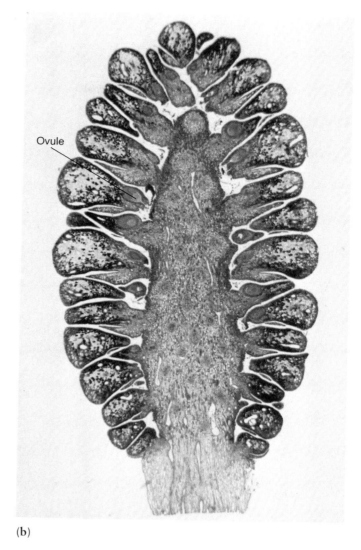

Ovule

(b)

figure 30.6

Pinecones. (a) Longitudinal section through a microstrobilus (pollen-bearing cone). (b) Longitudinal section through a megastrobilus (ovule-bearing cone). An intact pollen-bearing cone of *Pinus* is shown in figure 30.4c.

figure 30.7

Seed cones of *Ephedra*. The pollen-bearing cones of *Ephedra* are shown in figure 30.4d.

A Generalized Life Cycle of Gymnosperms

Most of the variation in life cycles among gymnosperms involves the timing of different reproductive events and how long different reproductive stages last. These aspects of gymnosperm life cycles are presented later in this chapter.

The alternation between sporophytic and gametophytic phases in gymnosperms is the same as that in other plants (fig. 30.8). Like the angiosperms, gymnosperms have heterosporous sporophytes, which means that the gametophytes are unisexual. Unlike lycopodophytes and angiosperms, which include species with bisporangiate strobili (e.g., perfect flowers in angiosperms), gymnosperms have only microstrobili

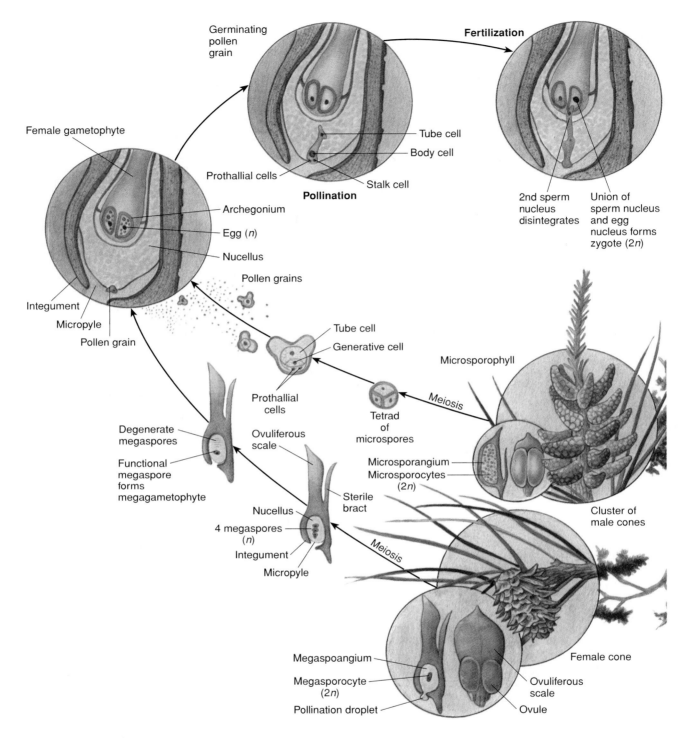

Germinating
pollen
grain

Fertilization

Female gametophyte

Tube cell

Body cell

Prothallial cells

Stalk cell

Pollination

2nd sperm
nucleus
disintegrates

Union of
sperm nucleus
and egg
nucleus forms
zygote (2n)

Archegonium

Egg (n)

Nucellus

Integument

Micropyle

Pollen grain

Pollen grains

Tube cell

Generative cell

Microsporophyll

Prothallial
cells

Meiosis

Tetrad
of
microspores

Microsporangium

Microsporocytes
(2n)

Cluster of
male cones

Degenerate
megaspores

Ovuliferous
scale

Functional
megaspore
forms
megagametophyte

Sterile
bract

Nucellus

4 megaspores
(n)

Integument

Micropyle

Meiosis

Megaspoangium

Megasporocyte
(2n)

Pollination droplet

Female cone

Ovuliferous
scale

Ovule

figure 30.8

Life cycle of *Pinus* (pine), a representative gymnosperm. In pine, pollination occurs more than a year before the ovule produces a mature female gametophyte. Seeds may mature more than 6 months after fertilization, which would be more than 18 months after pollination.

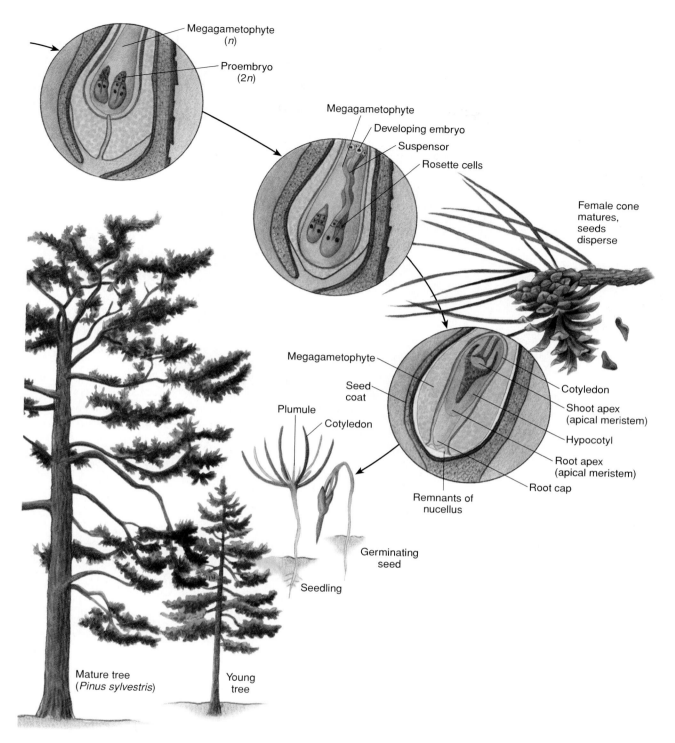

Megagametophyte
(n)

Proembryo
(2n)

Megagametophyte

Developing embryo

Suspensor

Rosette cells

Female cone
matures,
seeds
disperse

Megagametophyte

Seed
coat

Cotyledon

Shoot apex
(apical meristem)

Hypocotyl

Root apex
(apical meristem)

Root cap

Remnants of
nucellus

Plumule

Cotyledon

Germinating
seed

Seedling

Mature tree
(*Pinus sylvestris*)

Young
tree

figure 30.8 (continued)

A Fungus That May Save the Yews

In 1989 researchers at the Johns Hopkins University Oncology Center reported that tumors caused by ovarian cancer, which had not responded to traditional therapies (including, in some cases, surgery), had shrunk or disappeared in several patients after treatment with **taxol**. Taxol is a drug obtained from the bark of the Pacific yew (*Taxus brevifolia*), a gymnosperm that grows only in a few areas of the Pacific Northwest and is included among the trees that give shelter to the rare northern spotted owl. Unlike most other cancer drugs, which keep cancer cells from reproducing by damaging their DNA, taxol "freezes" the cancer cells early in the process of cell division. Unable to divide, the cells eventually die.

Unfortunately, there is far less taxol available than is required to meet the expected need. Pacific yew trees are small and do not occur in extensive strands, and they grow so slowly that they take more than 70 years to attain their full size. Three whole trees produce only enough taxol for a single treatment; a pound of the drug costs $250,000. Furthermore, the extraction of taxol requires removal of the bark, which kills the tree.

Despite its limited availability, scientists are excited about the potential of taxol in the fight against ovarian cancer. Medical science has launched several studies to find alternative sources of taxol. These include searching for relatives of the yew that may

reading figure 30.1

A scanning electron micrograph of *Taxomyces andreanae*, a recently discovered fungus occurring on bark of the Pacific yew. The fungus produces taxol, a cancer-fighting agent, in culture, ×2600.

also produce taxol, for economical methods of synthesizing taxol in the laboratory, for ways to grow taxol-producing cell cultures from the plant, and for the genes of taxol biosynthesis, which could be transferred

into bacteria for large-scale production of taxol in bacterial culture. Slow but steady progress is being made in all of these efforts, and another surprising possibility has recently been found. Chemist Andrea Stierle and plant pathologist Gary Strobel of Montana State University have discovered that a fungus that grows in the bark of Pacific yew also contains taxol. More important, the fungus produces taxol in culture, which means that it has its own genetic machinery for making the drug. The fungus, which turned out to be a new genus, was named *Taxomyces andreanae* (reading fig. 30.1). In March 1992 the research team at Montana State University filed a patent application for taxol production by *Taxomyces*, and in January 1993 the U.S. Food and Drug Administration approved taxol—in a record 5 months—for treating ovarian tumors. Bids for licensing the patent rights to pharmaceutical companies were being solicited by 1993.

The potential for large-scale production of taxol by *Taxomyces* is expected to satisfy additional demand for the drug in treating cancers of the breast, neck, and head. Worldwide, hundreds of thousands of cancer cases may be treatable by taxol. If the Pacific yew were the only source for this drug, the species would be destroyed long before the demand for taxol could be satisfied.

and megastrobili. This means that sexual reproduction in gymnosperms always requires the transfer of gametophyte material from one strobilus to another.

Pollination in gymnosperms involves a **pollination droplet** that protrudes from the micropyle when pollen grains are being shed (fig. 30.9). This droplet provides a large, sticky surface that catches the normally wind-borne pollen grains of gymnosperms so that the ovule is more likely to be fertilized. During pollination dozens of pollen grains may stick to each droplet. After pollination the droplet evaporates and contracts, carrying the pollen grains into the pollen chamber and into contact with the ovule.

The Diversity of Gymnosperms

There are considerably fewer gymnosperms than there are angiosperms. Most classifications of gymnosperms include about 65 genera and 720 species. For centuries they were lumped into a single class of seed plants, but botanists now consider them sufficiently diverse to be separated into four extant divisions: Ginkgophyta (maidenhair tree), Cycadophyta (cycads), Pinophyta (conifers), and Gnetophyta (gnetophytes).

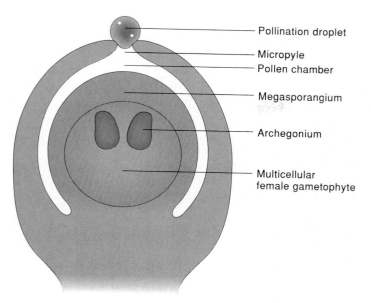

- Pollination droplet
- Micropyle
- Pollen chamber
- Megasporangium
- Archegonium
- Multicellular female gametophyte

figure 30.9

Diagrammatic representation of a longitudinal section through an ovule of pine (*Pinus*).

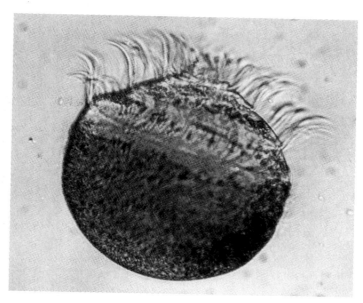

figure 30.10

Light micrograph of *Ginkgo* sperm with flagella, ×400.

Division Ginkgophyta: Maidenhair Tree

Ginkgo, the maidenhair tree, derives its Latin name from two Chinese words meaning "silver apricot." *Ginkgo biloba*, which has remained virtually unchanged for 80 million years, is the only living representative of the division. Its distinctive, fan-shaped leaves with dichotomous venation are produced on two types of shoots: relatively fast-growing long shoots and seedlings produce leaves with a distinct apical notch (hence, *biloba*), while slow-growing spur shoots produce leaves without a notch.

    Ginkgo is exclusively dioecious. After the pollination droplet withdraws with its load of pollen, pollen tubes begin to grow by digesting the tissue of the megasporangium. Once the female gametophyte is mature, the egg cells protrude toward the male gametophyte and engulf the sperm cells. Thus, although the sperm cells of *Ginkgo* are flagellated, the flagella are not needed for carrying the sperm to the egg (fig. 30.10). Furthermore, the pollen tubes, which function as digestive filaments, do not convey sperm to egg. This type of pollen tube is referred to as **haustorial**, because it is analogous to the haustorial filaments of parasitic fungi.

    The seeds of *Ginkgo* include a massive integument that consists of a fleshy outer layer, a hard and stony middle layer, and an inner layer that is dry and papery. Mature seeds have the size and appearance of small plums, but the fleshy integument has a nauseating odor and irritates the skin of some people. Nevertheless, pickled *Ginkgo* seeds are a delicacy in some parts of Asia.

    Ginkgo trees are deciduous, a feature that otherwise occurs only in a few members of the Pinophyta. *Ginkgo* is a popular cultivated tree, but it is apparently extinct in nature. All living *Ginkgo* trees are descendants of plants that were grown in temple gardens of China and Japan, although in the 1950s there were reports of some individuals still living in the wild in eastern China. Much of the genetic diversity of *Ginkgo* has probably been lost in cultivation, because most nurseries propagate only cuttings of microsporangiate trees to avoid the stinky and messy seeds.

Division Cycadophyta: Cycads

There are about 10 genera and 100 species of cycads, distributed primarily in the tropical and subtropical regions of the world. All species of cycads are dioecious. Although they are planted as ornamentals throughout the milder climatic areas of the world, only two species are native to the United States. Both are in the genus *Zamia*, and neither lives in the wild outside of Florida. Many species of cycads are threatened with extinction, and some may soon remain only in cultivation.

    Cycads have palmlike leaves that bear no resemblance to the leaves of other living gymnosperms. Under favorable conditions, cycads usually produce one crown of leaves each year. In some cycads, the roots grow at the surface of the soil and develop nodules containing nitrogen-fixing cyanobacteria.

    Cycads have simple, often shield-shaped or bractlike sporophylls that may be covered with thick hairs (see fig. 30.5d), which usually occur in large strobili. The seeds of cycads are more like those of *Ginkgo* than those of any other gymnosperm; they have a three-layered integument, but the inner layer is soft instead of papery.

BIG TREES

Throughout this book, you've read about redwoods (*Sequoia sempervirens*) and giant sequoias (*Sequoiadendron giganteum*). Before the 1840s the chief admirer of these trees was the chickaree, or Douglas squirrel—a clever-pawed animal that eats the fleshy green scales of sequoia cones. However, the enormous size of these trees soon made them wildly popular after their discovery in 1833. How big are these trees?

- The General Sherman Tree weighs an estimated 5.5 million kg.

- The stump of one felled tree, used for a Fourth of July cotillion in 1854, accommodated thirty-two waltzers (half of them in hooped skirts), plus musicians.

- William Waldorf Astor—on a bet—had a dinner table capable of seating forty dinner guests made from a cross section of a redwood tree.

- The Columbian Exposition of 1893 in Chicago displayed "the biggest plank ever sawed"—a piece of coastal redwood 5 m wide.

These days, such frivolous waste of natural resources would be met by a public outcry. But the greatest loss of redwoods came at the hands of the lumber industry. Logging companies, lured by returns of $1,350 for each $1.25 invested, cleared more than one-third of California's original 2 million acres of redwood forest between 1850 and 1925. Most of the remaining trees have been cut since then. To help preserve the trees, John Muir—the founder of the Sierra Club—ensured federal protection for Sequoia National Park in 1890.* As he said, "Any fool can destroy trees."

---

*Muir also helped secure federal protection for the Grand Canyon, the Petrified Forest of Arizona, and Yosemite. The 503-acre Muir Woods National Monument (just north of San Francisco) was established in 1907 in his honor. To learn more about Muir, visit the John Muir Exhibit home page at: http://www.sierraclub.org/john_muir_exhibit/

figure 30.11

Bristlecone pines (*Pinus longaeva*) are the oldest known plants that are not clones.

Also like *Ginkgo*, cycads have flagellated sperm. The sperm cells of cycads are the largest among plants (up to 400 µm in diameter) and can have about 10,000–70,000 spirally arranged flagella. The pollen tubes in cycads are chiefly haustorial, however; the sperm cells are released from the pollen grain and not through the pollen tubes.

c o n c e p t

Cycads and *Ginkgo* are dioecious and have flagellated sperm, haustorial pollen tubes, and fleshy integuments. Cycads, however, usually have large strobili and persistent, palmlike leaves, whereas *Ginkgo* has loose microstrobili, single-seeded stalks, and deciduous simple leaves.

Division Pinophyta: The Conifers

The informal name of this group, *conifers*, signifies plants that bear cones, even though other divisions of gymnosperms also include cone-bearing species. Members of the genus *Pinus*, considered typical for the Pinophyta, are the most abundant trees in the Northern Hemisphere; many of the species in conifer forests are pines. They have also been widely planted in parts of the Southern Hemisphere, but only the Merkus pine (*Pinus merkusii*), whose distribution barely extends south of the equator in Sumatra, is native south of the equator. Also included in this genus is the bristlecone pine (*P. longaeva*) of the western United States, some of which are the oldest known plants that are not clones (fig. 30.11). One specimen—a tree named Methuselah—is about 4,725 years old; this means that its first sprouts poked out of the stony soil when the Egyptians were building the Great Sphinx at Gizeh. It is 1,000 years older than Stonehenge, 2,000 years older than the Acropolis, and 3,700 years older than Westminster Abbey.

Like *Ginkgo*, pines have short shoots, long shoots, and two kinds of leaves. The more obvious type of leaf is the pine needle, which occurs in groups, called **fascicles,** of generally two to five needles. A few species have as many as eight needles per fascicle, and others have only one. Regardless of the number of needles, a fascicle always forms a cylinder when the leaves are

(a)

(a)

(b)

(b)

figure 30.12

Leaves of gymnosperms. (a) Nonfascicled leaves of coastal redwood (*Sequoia sempervirens*). (b) Scalelike leaves of juniper (*Juniperus*). Each branchlet is surrounded by the tiny green leaves.

figure 30.13

Pinecones. (a) First-year seed cones open for pollination. (b) Second-year pinecones at time of fertilization.

held together. Fascicles are actually short shoots that are surrounded at their base by small, nonphotosynthetic, scalelike leaves that usually fall off after 1 year of growth. The needle-bearing fascicles are also shed a few at a time, usually every 2–5 years, so that any pine tree, while appearing evergreen, has a complete change of leaves every 5 years or less. Bristlecone pines are an exception; their needles last an average of 25–30 years—about as long as the life span of a horse.

Other members of the Pinophyta in the Northern Hemisphere also have narrow leaves that often have a small point at the tip, but they do not occur in fascicles. These include yews, firs (*Abies*), larches (*Larix*), spruces (*Picea*), and the coastal redwood (*Sequoia sempervirens;* fig. 30.12a). In contrast, the leaves of cypresses (*Cupressus*) and juniper (*Juniperus*) are scalelike at maturity (fig. 30.12b). In addition, the leaves of some podocarps (*Podocarpus*), araucarias (*Araucaria*), and other Pinophyta of

the Southern Hemisphere have flat blades. Like pines, most of these other members of the Pinophyta are evergreen; however, larches, the bald cypress (*Taxodium distichum*), and the dawn redwood (*Metasequoia glyptostroboides*) are deciduous.

Diversity within the Pinophyta is reflected in a wide variety of reproductive structures and variations of the reproductive cycle. As already noted, seed cones may be absent, as in yews, or have a berrylike appearance, as in junipers. One of the most distinctive variations, however, involves the timing of different reproductive stages in pines. Unlike other gymnosperms, in which pollination, fertilization, and seed maturation occur within the same year, the pines have an extended reproductive cycle. The exact timing varies among species and in different localities.

At the time of pollination, the seed cones are small, and the ovule-bearing scales are slightly separated from one another, enabling pollen grains to reach the pollination droplet (fig. 30.13).

After the pollen grains are drawn into the ovule, the micropyle closes, and the seed cones seal up.[2] Meiosis begins in the megasporangium about a month later. The development of the female gametophyte from the functional megaspore is extremely slow, going through an extensive free-nuclear stage and finally becoming cellular about 13 months after pollination. During this time, pollen tubes digest most of the megasporangial tissue (i.e., as in the cycads and the maidenhair tree). About a month before the female gametophyte becomes fully cellular, the generative cell in each male gametophyte divides and forms a stalk cell and a body cell. Several days before the pollen tube reaches the female gametophyte, the body cell divides into two sperm cells, and each archegonium forms an egg.

Although the pollen tube is haustorial in pines, it also transports sperm cells to the female gametophyte. The seed cone is larger and woodier than it was the previous year, but it remains closed while the embryos develop. Following fertilization, the embryo undergoes free-nuclear division and then becomes cellular and matures before the following winter. By this time, the seed cones have grown into the mature, woody structures that we recognize as pinecones. The time between pollination and seed maturation is usually about 18 months, but it can range from 14–20 months.

When mature, the cones of most pines become dry, the scales are woody, and they open to release the winged seeds. The cones of some pines do not open so gracefully; these cones often require a fire to cause them to open and release their seeds. The cones of some pines explode like popcorn when heated.

c o n c e p t

Pines are the most abundant species of northern conifer forests. Many conifers have needlelike leaves that occur in fascicles, while others have broad, flat leaves. Reproduction in pines can require 18 or more months between pollination and the formation of mature seeds.

Division Gnetophyta: Gnetophytes

The gnetophytes include some of the most distinctive (if not bizarre) of all seed plants. There are three clearly defined genera, each in its own order, and seventy-one woody species. These genera are *Ephedra* (forty species), *Gnetum* (thirty species), and *Welwitschia* (one species).

Ephedra, whose species are either monoecious or dioecious, is known as *Mormon tea, ma huang,* or *joint fir.* The first two names come from its use as a stimulating or medicinal tea. Ma huang is an Asian species, *E. sinica,* which contains chemicals that are similar to those of human neurotransmitters. The name *joint fir* refers to the leafless appearance of *Ephedra* stems, which resemble the jointed stems of *Equisetum. Ephedra*

(a)

(b)

figure 30.14

Gnetophytes. (a) Leaves and immature seeds of *Gnetum*.
(b) *Welwitschia* plants in the Namib Desert of southwest Africa.

is not leafless, though; its leaves are small and lose their photosynthetic capability as they mature. Therefore, most photosynthesis in *Ephedra* occurs in green stems.

Members of the genus *Gnetum* inhabit tropical forests. These plants, which are dioecious, are either climbing vines or trees, all with broad, simple leaves similar to those of woody dicots (fig. 30.14a).

Welwitschia mirabilis is the sole living representative of its genus and looks more like something out of science fiction than a real plant (fig. 30.14b). This slow-growing species is confined to the Namib and Mossamedes Deserts of southwestern Africa, where most of its moisture is derived from fog that rolls in from the ocean at night. The woody stem, which is concave and bark-encrusted, may be as much as 1.5 m in diameter and is connected to a large taproot. Mature plants have a pair of large, strap-shaped leaves,

---

2. Romans considered these unopened cones to be symbols of virginity. Virgins, or those who wanted to appear as such, wore garlands of closed evergreen cones at ceremonies.

which persist throughout the life of the plant. Each leaf has a meristem at its base, which constantly replaces tissue that is lost at its drier, aging tip.

Like many other gnetophytes, *Welwitschia* is dioecious, producing male and female strobili on different plants. Fertilization in *Welwitschia* is unique, in that tubular growths from the eggs grow toward and unite with the pollen tubes; fertilization occurs within these united structures. Reproduction is otherwise similar to that of other gnetophytes.

Gnetophytes are unique among gymnosperms because after one of the sperm cells from a male gametophyte fertilizes an egg, the second sperm cell fuses with another cell in the same female gametophyte. Thus, gnetophytes undergo double fertilization, a process otherwise known only in the angiosperms. Unlike double fertilization in angiosperms, however, double fertilization in gnetophytes is not followed by the formation of endosperm. Instead, the diploid cell from fertilization by the second sperm disintegrates.

c o n c e p t

Gnetophytes are the most distinctive group of gymnosperms because of their many similarities with the angiosperms. In various gnetophytes, these include flowerlike compound strobili, vessels in the secondary xylem, loss of archegonia, loss of prothallial cells in the pollen, and double fertilization.

The Physiology and Ecology of Gymnosperms

The physiology and ecology of gymnosperms are like those of the flowering plants (see Chapters 18–21 and 31), but as a smaller group, the gymnosperms have a narrower range of features and habitats. Division Pinophyta is dominated by trees and shrubs that are well adapted for temperate or cold climates, especially where free water is scarce for part of the year. Accordingly, northern forests in some areas consist mostly of pines and their relatives, and southern forests consist mostly of araucarias and their relatives. These gymnosperms are adapted to such regions by their sunken stomata, thick cuticle, and tough hypodermis. Some gymnosperms, such as *Ephedra*, *Welwitschia*, and piñon pines, live in deserts. Only a few gymnosperms are tropical; the cycads and *Gnetum* are the main examples.

The Economic Importance of Gymnosperms

The gymnosperms are second only to the angiosperms in their daily impact on human activities and welfare. A detailed account of all they contribute would occupy many volumes, but a few of their main uses are described here.

The greatest economic impact of gymnosperms in the Northern Hemisphere comes from the use of their wood for making paper and lumber. Indeed, conifers overall produce about 75% of the world's timber and much of the pulp used to make paper. The chief source of pulpwood for newsprint and other paper in North America is the white spruce (*Picea glauca*). A single midweek issue of a large metropolitan newspaper may use an entire year's growth of 50 hectares of these trees, and that amount may double for weekend editions (see reading 30.3, "How Much Paper Do We Use?").

Before it was heavily logged, Douglas fir (*Pseudotsuga menziesii*) was widespread in the foothills and mountains of the West. In the Pacific Northwest, it grew into giant trees, which were second in size only to the redwoods (see reading 30.2, "Big Trees"). It is probably the most desired timber tree in the world today. The wood is strong and relatively free of knots as a result of rapid growth with less branching than in most other conifers; it is heavily used in plywoods and is a major source of large beams. The lumber industry has nearly eliminated native old-growth stands, but large numbers of new trees are being grown in managed forests, where they are usually logged before they reach 50 years of age.

Redwoods (*Sequoia sempervirens*) are also prized for their wood, which contains substances that inhibit the growth of fungi and bacteria.[3] The wood is light, strong, and soft, but it splits easily. It is used for some types of construction, furniture, posts, greenhouse benches, and for many other purposes.

Spruce wood is especially important to the music industry. The tracheids of spruces have spiral thickenings on the inner walls, which apparently give the wood a resonance that makes it ideal for use as soundboards in violins and related musical instruments.

Important wood products besides lumber and pulp include **resin,** which is the sticky, aromatic substance in the resin canals of conifers. It is a combination of a liquid solvent called **turpentine** and a waxy substance called **rosin.** Both turpentine and rosin are useful products, and a large industry centered in the southern United States and the south of France is devoted to their extraction and refinement. Turpentine and rosin are often referred to as *naval stores,* a term that originated when the British Royal Navy used large amounts of resin for caulking and sealing its sailing ships and for waterproofing wood, rope, and canvas (most naval stores and one-third or more of the lumber used in the United States today come from a group of southern yellow pines, particularly slash pine, *P. caribea* and *P. elliottii*). Pine resin was used by sailors in ancient Greece, Egypt, and Rome. Egyptians sealed their mummy wrappings with pine resin, and the Greeks lined their clay wine vessels with pine resin to

---

3. The word *Sequoia* was proposed by Austrian botanist Stephen Erdlicher to commemorate the eighteenth-century Cherokee leader Sequoyah, famed for devising an eighty-three-letter alphabet for the Cherokee language.

HOW MUCH PAPER DO WE USE?

Before the 1860s, most newspaper was made from linen or cotton rags (states such as Massachusetts appointed officials to help get rags to mills). Although many people suggested unusual sources for paper (including cabbage, dandelions, and asbestos), the most unusual suggestion was to make paper from rags used to wrap Egyptian mummies. Indeed, because each mummy was wrapped in about 30 lb of rags, investors calculated that they'd need only 13.5 million mummies per year to provide rags for American paper mills. In the 1860s Augustus Stanwood of Gardiner, Maine—to the dismay of archaeologists—started importing mummies. People in Maine were soon taking their lamb chops home in paper made from mummy rags.

When mummies became hard to obtain, investors began looking elsewhere for sources of paper. They quickly settled on wood, especially spruce. Wood pulp was first used to make newspapers in the 1860s, and the first all-wood issue of the *New York Times* was published on August 23, 1873. Today Americans use almost 200,000 tons of paper each day, enough to cover 1,350 m^2 (an area the size of Long Island). Our uses of paper are many: food packaging, newspapers, cardboard, toilet paper, and paperback novels are just some of the throw-away products that require a constant supply of wood pulp. Millions of trees are harvested every year to meet this demand. It is only a matter of time before the increasing demand for these products outstrips the supply of wood pulp. How can we help? The management of a large American publishing company, in an attempt to find ways of reducing paper consumption in the United States, tried trimming 2.5 cm from the width of all rolls of toilet paper in its building. It found that the employees still used the same number of rolls per month as they had previously. From this, it calculated that if all rolls of toilet paper in the United States were similarly trimmed, 1 million trees would be saved each year. This is just one example of a simple and painless way to reduce our huge demand for paper goods made from woody trees.

CHRISTMAS TREES

Although the origin of the Christmas tree is uncertain (suggestions range from Germans to ancient Druids), today there is no question that Americans consider decorated coniferous Christmas trees to be an important part of the holiday season. Our use of Christmas trees has an interesting history:

- Although the oldest mention of an American Christmas tree dates to 1747, the first Christmas tree didn't appear in the White House until 1856. That tree was decorated by the family of Franklin Pierce, a politician otherwise noted for putting stickum on postage stamps.

- The first commercial ornaments for Christmas trees (glass icicles and balls) came from Germany. Fake icicles were introduced in 1878.

- The first electric ornaments for Christmas trees were handblown lightbulbs made in 1882 by Thomas Edison's lab assistants.

Today, people in different parts of the United States have differing preferences for their Christmas trees. Scotch pine is the favorite in the North-Central United States; the West prefers Douglas fir; most Californians prefer Monterey pine; Southerners like white pine; and New Englanders prefer balsam fir. There's also a big market for fake trees, the first of which were sold by Sears, Roebuck, and Company in the 1880s. According to its catalog, you could buy a "tree" having 33 "limbs" for only 50 cents.

prevent leakage. Pine flavoring is still added to Greek wines, giving retsina its distinctive flavor. The unappreciative liken the taste to turpentine.

Turpentine is the premier paint and varnish solvent and is used to make deodorants, shaving lotions, drugs, and limonene—the lemon flavoring in commercial lemonade, lemon pudding, and lemon meringue pie. Ballerinas dip their shoes in resin to improve their grip on the stage; violinists drag their bows across blocks of resin to increase friction with the strings. Baseball pitchers use resin to improve their grip on the ball, and batters apply pine tar to the handles of bats to improve their grip (see reading 16.2, "The Bats of Summer: Botany and Our National Pastime").

The huge kauri pines (*Agathis australis* and *A. robusta*) of New Zealand, which are genetically different from true pines, are the source of a mixture of resins called *dammar* (fig. 30.15). Dammar is used in high-quality, colorless varnishes and was the resin originally used to make linoleum. Dammar, also called *amber*, is the only jewel of plant origin. It comes primarily in fossil form from former or present kauri pine forests and occurs as lumps of translucent material with a deep orange-yellow tint. These lumps, which weigh up to 45 kg, were believed in ancient times to protect the wearer from asthma, rheumatism, and witchcraft. The best amber comes from Russia; however, that supply, which was at its peak at the turn of the century, is now nearing exhaustion. Remarkably lifelike preservations of prehistoric insects in amber still have intact DNA. Other than amber, the most unusual resin product is the nest of the *Dianthidium* bee, which consists of pebbles and pine resin.

(a)

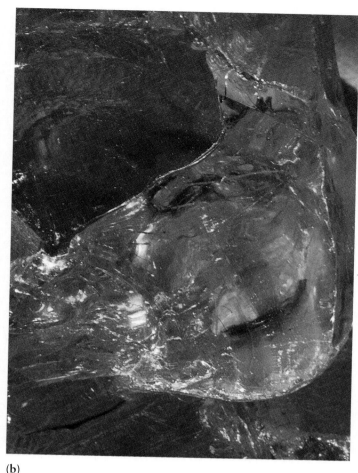

(b)

figure 30.15

(a) Kauri pine (*Agathis australis*) of New Zealand. (b) Kauri pine is the source of dammar.

---

c o n c e p t

In the Northern Hemisphere, gymnosperms account for most of the wood that is needed for lumber and paper. Worldwide, conifers are the main source of resin and amber.

 **The Lore of Plants**

"What the apple is among fruits . . . ," wrote horticulturist Liberty Hyde Bailey, "the pines represent among conifers." This may explain why ten states have chosen the pine as their state tree. Some states (Arkansas and Minnesota, for example) boast a generic pine. The state tree of Alabama is the longleaf pine, Michigan chose the eastern white pine, Nevada and New Mexico claim the piñon pine, Idaho the western white pine, and Montana the ponderosa pine. The state tree of Maine, the Pine Tree State, is also the eastern white pine—and its state "flower" is the pinecone.

The Origin and History of Gymnosperms

Ideas about the ancestry of modern gymnosperms are suggested by the fossil record. The most likely ancestors are the **progymnosperms,** which are classified in Division Progymnospermophyta. The prefix *pro* meaning "before" implies that these plants were not yet gymnosperms, but this depends on how the fossil structure shown in figure 30.1 is interpreted. If it is seen as a pre-seed or is merely said to be seedlike, then the progymnosperms were not seed plants; however, if it is seen as a primitive seed, then the Progymnospermophyta were the first gymnosperms. Either way, they are the most likely ancestors of the first plants that had seeds.

Some fossils are sufficiently distinctive to be classified in their own divisions, whereas others resemble modern forms closely enough to be classified in one of the extant divisions. All divisions of living gymnosperms have extinct relatives, and two divisions of gymnosperms are entirely extinct: the Pteridospermophyta (seed ferns) and the Cycadeoidophyta (cycadeoids).

Sporangia

figure 30.16

Progymnosperms. (a) *Archaeopteris* was a woody progymnosperm that grew to more than 20 m tall. (b) Wedge-shaped leaves, some bearing mature sporangia, occurred on frondlike primary branches.

Several phylogenetic relationships between extinct and extant gymnosperms have been proposed, based on traditional and modern phylogenetic methods. The following discussion presents an overview of the main features of the most important groups of extinct plants; it is followed by explanations of how past and present gymnosperms might be related.

Division Progymnospermophyta

The progymnosperms may be the oldest seed plants, depending on whether their reproductive structures were seeds (some botanists think they were and others think they were not). Progymnosperms lived from about the middle Devonian to the early Carboniferous periods (ca. 360–310 million years ago) (see fig. 30.16). These treelike or shrublike plants had a combination of features that resembled those of the seed plants as well as the seedless vascular plants. Some were heterosporous; they had simple leaves derived from flattened branch systems, and they formed seeds or seedlike reproductive structures. They also had a vascular cambium that could undergo radial divisions. The cambium produced large amounts of secondary xylem and phloem similar to that of modern gymnosperms, and the wood contained long tracheids and bordered pits.

Division Pteridospermophyta

The seed ferns appeared in the late Devonian period. Their leaves were so fernlike that they were originally grouped with the ferns; indeed, it is still not possible to distinguish pteridosperm leaves from fern fronds in the absence of reproductive structures (fig. 30.17). Seed ferns became so abundant during the late Carboniferous that this period became known as the Age of Ferns. In the early part of the twentieth century, however, botanists discovered that the fronds of these fernlike plants bore seeds, which means the Age of Ferns was really the Age of Seed Ferns (fig. 30.18; see also fig. 30.1). Although they are called seed ferns, these plants were not just ferns with seeds. Moreover, they did not evolve from the ferns. Rather, the seed ferns probably evolved from the progymnosperms along an evolutionary line that has no modern descendants.

Like most fossils, the seed ferns are known only from fragments, not from intact plants. In many cases, reproductive structures and vegetative bodies occur in the same strata but not attached to each other. Being appropriately cautious, paleobotanists cannot say which microsporangia or which seeds belonged to which plants. Each organ maintains a separate scientific name, often at the genus level, until it is found attached to an already named plant in the fossil record.

30–19

figure 30.17

Seed ferns, represented here by this reconstructed *Medullosa*, were abundant in the late Carboniferous period.

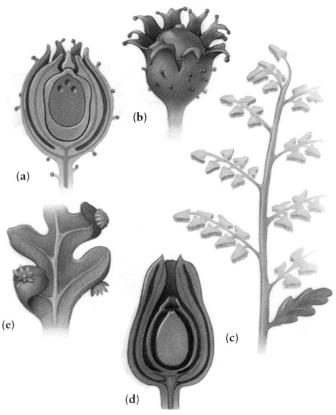

figure 30.19

Reproductive structures of representative seed ferns. (a, b) Seed-bearing cupules of *Lyginopteris*. (c) The probable pollen-bearing organ of *Lyginopteris*. (d) An ovule of *Callistophyton*. (e) The probable pollen-bearing organ of *Callistophyton*.

figure 30.18

Seeds were borne on the frondlike leaves of seed ferns, as shown in this drawing of *Emplectopteris*.

Nevertheless, frondlike fossils are associated with a variety of seed types and reproductive organs that are from plants that would probably be classified into several divisions if they were better known (e.g., fig. 30.19). It is likely, therefore, that the relationships of the seed ferns are more complicated than their classification in a single division would indicate.

Division Cycadeoidophyta

Cycadeoids are so named because of their superficial resemblance to cycads (fig. 30.20). Cycadeoids and cycads flourished together in the Jurassic period, which botanists call the Age of Cycads (and zoologists call the Age of Dinosaurs). Although the cycads are still living, the cycadeoids went extinct at the end of the Cretaceous period (ca. 65 million years ago), at about the same time the dinosaurs did. Cycadeoids differ from other gymnosperms, living or extinct, by having bisporangiate strobili—that is, megasporangia and microsporangia on the same strobilus (fig. 30.21). This situation occurs otherwise only in flowers. The sporangial and vegetative features of cycadeoids traditionally have led botanists to believe that cycadeoids arose from an ancestor among the seed ferns, a lineage generally thought to be unrelated to the flowering plants.

figure 30.20

Vegetatively, the cycadeoids such as *Cycadeoidea* were similar to cycads. Internally, like cycads, they had a broad pith, a small amount of xylem, a broad cortex, and a tough outer protective layer formed by persistent leaf bases.

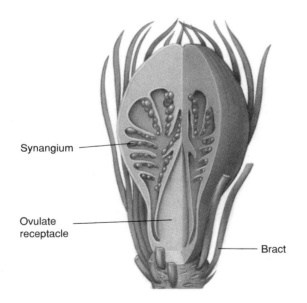

figure 30.21

Cycadeoid strobilus showing two kinds of sporangia.

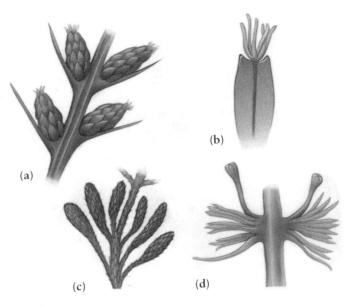

figure 30.22

Reproductive structures of the Cordaitales and Voltziales. (a) Portion of a branch with microsporophylls. (b) A microsporophyll with several microsporangia of *Cordaianthus* (Cordaitales). (c) Portion of a branch with megastrobili. (d) Two ovules of *Lebachia* (Voltziales).

Extinct Plants among Divisions of Living Gymnosperms

The fossil record contains many extinct species and genera of Pinophyta, Cycadophyta, Gnetophyta, and Ginkgophyta. Probably the most diverse of these divisions among the extinct gymnosperms is the Pinophyta. This division includes two extinct orders, whereas extinct members of the other divisions all fit into modern orders.

The most abundant plants of the extinct Pinophyta were the cordaites (Order Cordaitales). Cordaites were freely branching trees and shrubs with simple, strap-shaped leaves and compound strobili that bore either seeds or microsporangia (fig. 30.22a, b). Cordaites were commonly associated with seed ferns in the swamps and lush forests of the Carboniferous period, but they disappeared in the late Permian period (ca. 250 million years ago). At about the time the cordaites were waning, a second order of extinct Pinophyta, the Voltziales, began to flourish. The leaves and branching patterns of Voltziales resembled those of modern araucarias (fig. 30.22c, d). The Voltziales also had compound seed-bearing strobili, but their microstrobili were simple, like those of modern pines.

concept

As with all kinds of organisms, most groups of gymnosperms are extinct. They included the seed ferns, cycadeoids, at least two orders of the Pinophyta, and probably the progymnosperms. Most ancient cycads and ginkgophytes have also gone extinct.

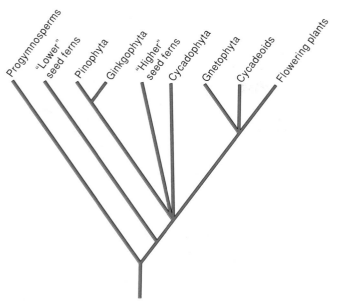

figure 30.23

Phylogenetic relationships among seed plants. Dichotomous branch-points show where botanists generally agree about such relationships, whereas multiple branch-points represent relationships that are not well understood.

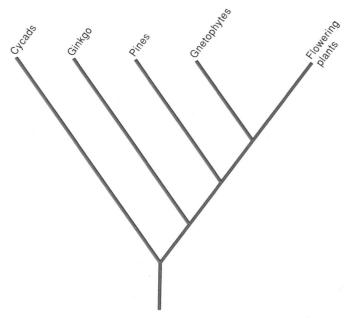

figure 30.24

Cladogram of living seed plants based on *rbc*L/DNA-sequence comparisons.

The Phylogenetic Relationships of Gymnosperms

The phylogeny of gymnosperms interests botanists because of its potential for revealing how reproductive and vegetative structures evolved, as a framework for the interpretation of ambiguous fossil structures, and for answering questions about the origin and evolution of flowering plants. Some of the traditional views about evolutionary relationships of gymnosperms have already been mentioned in this chapter, such as the progymnosperms being ancestral to all seed plants. Many conflicting suggestions have been made about the phylogenetic relationships of the gymnosperms. These conflicts derive mostly from different interpretations of fossil structures and from the assumed significance of various characters as indicators of ancestor-descendant relationships. Nevertheless, botanists generally agree on the ancestral position of the progymnosperms, the unique common ancestry of the Pinophyta and the Ginkgophyta, and the close relationship among the cycadeoids, the Gnetophyta, and the flowering plants. Most disagreements center around the scattered positions of different groups of seed ferns, the best phylogenetic fit for the cycads, and the relationships of the Gnetophyta and the cycadeoids to the flowering plants. Such unresolved relationships are depicted in a cladogram as branch-points with multiple branches (fig. 30.23).

At this point it would be nice to have a molecular phylogeny come to the rescue and help resolve the differences among competing cladograms. Unfortunately, DNA from plants has not been well preserved in fossils, except for some relatively recent material from about 20 million years ago. Nevertheless, a molecular phylogeny of extant seed plants has been proposed from a comparison of DNA sequences of the chloroplast gene *rbc*L (fig. 30.24). This phylogeny supports a close relationship between gnetophytes and angiosperms. However, it shows the cycads to be near the ancestral root of seed plants, and it fails to support the previously suggested relationship between the Pinophyta and Ginkgophyta. These results may be influenced by the absence of information from extinct species. Cycads, for example, probably do not represent the group closest to the ancestors of seed plants.

c o n c e p t

Various relationships among gymnosperms have been suggested, based on comparisons between living and extinct seed plants. In spite of their differences, suggested relationships consistently show a close tie between the gnetophytes and the angiosperms and a variety of relationships among the seed ferns.

Chapter Summary

Seedlike structures first appeared about 360 million years ago in progymnosperms. These structures consisted of megasporangia surrounded, but not enclosed, by an integumentary layer. Pollen arose more than 150 million years after the origin of seeds. Seed plants diverged and flourished throughout the Carboniferous period as seed ferns; diversification of ancestral Pinophyta and Cycadophyta followed. The Jurassic period was dominated by cycads, cycadeoids, and early members of the Pinophyta. This period is called the Age of Cycads.

Most of the features used for classifying living gymnosperms involve reproductive structures. Most gymnosperms bear seeds and microsporangia on strobili that are either simple or compound. Pollination requires that pollen be transferred from microsporangia to ovules. Ovules are generally exposed (i.e., naked) or temporarily enclosed by sporophylls or other branches from the main axis of the strobilus.

The Pinophyta, Cycadophyta, Ginkgophyta, and Gnetophyta are four divisions of gymnosperms with living representatives. Pines are abundant in northern conifer forests and are planted throughout the Southern Hemisphere. Bristlecone pines are long-lived. Other members of the Pinophyta dominate similar habitats in the Southern Hemisphere. Pine leaves form on short shoots called fascicles, which usually consist of two to five needles with deciduous, nonphotosynthetic, scalelike leaves at the base. The fascicles themselves are intermittently deciduous, usually lasting for less than 5 years.

Cycads are slow-growing gymnosperms of warmer climates. Their reproduction is similar to that of the pines, except that their sperm are motile by means of many flagella and their female strobili can be massive. All cycads are dioecious.

There is only one living species of *Ginkgo*, a tree with fan-shaped, dichotomously veined leaves. *Ginkgo* is dioecious and produces seeds that have a nauseating odor.

There are three distinct genera of gnetophytes. *Ephedra* species are monoecious or dioecious and native to northern drier areas. They are mostly shrubby and otherwise resemble horsetails in having whorled branches and small, essentially functionless leaves. Their strobili consist of paired bracts, some of which subtend ovules or contain microsporangia. Their life cycle is similar to that of conifers. *Gnetum* species are tropical vines or trees with broad leaves; their reproduction resembles that of *Ephedra*. *Welwitschia* is a bizarre, dioecious plant of southwest African deserts. It produces two large, strap-shaped leaves with basal meristems. The leaves arise from a concave, bark-encrusted, trunkless stem from which a taproot extends into the ground. Strobili form in the leaf axils.

Gymnosperms, especially pines and their relatives, are important as a source of lumber, wood pulp, and resin. Resin is used for making turpentine and rosin. Wood pulp is used as raw material for making paper.

Progymnosperms of the late Paleozoic era appear to have been the progenitors of gymnosperms. Progymnosperms reproduced by means of spores and seedlike organs. Progymnosperms gave rise to seed ferns, which had frondlike leaves but bore seeds. Cycadeoids, which had palmlike leaves like the cycads, were the only gymnosperms to have bisporangiate strobili.

The Cordaitales, the earliest conifers, and pteridosperms were abundant during the Carboniferous period. Cordaitalean stems and roots resembled those of modern gymnosperms, and their strobili were simple. They had strap-shaped leaves adapted to the drier conditions that developed in the Permian period.

Several hypotheses have been proposed for the phylogenetic relationships of gymnosperms. These hypotheses differ in the selection of characters and the significance placed on them for inferring ancestor-descendant relationships. Nevertheless, they all show the flowering plants to be closest to the gnetophytes among extant gymnosperms.

 What Are Botanists Doing?

Go to the library and read a recent article that describes how botanists are trying to increase production of taxol. Summarize the important points of the article.

 Writing to Learn Botany

What might be the adaptive advantage of pinecones that remain closed until heated by fire?

Questions for Further Thought and Study

1. The forms of gymnosperm leaves include tiny scales, huge strap-shaped leaves with basal meristems, needles, palmlike leaves, fan-shaped leaves, and simple, broad leaves. Can you think of any functional or ecological significance of the various forms?

2. Cycads, yews, and several other gymnosperms are dioecious; pines are monoecious. Can you think of any survival or other value to the plants in having their reproductive structures on separate plants?

3. In the early evolution of land plants, of what advantage was the development of seeds?

4. Discuss the economic importance of the gymnosperms.

5. How are gymnosperms similar to seedless vascular plants?

Web Sites

Review the "Doing Botany Yourself" essay and assignments for Chapter 30 on the *Botany Home Page*. What experiments would you do to test the hypotheses? What data can you gather on the Web to help you refine your experiments? Here are some other sites that you may find interesting:

http://fsias_s1.for.gov.bc.ca:80/pab/educate/treebook/ 32.htm#where

This site is part of *Tree Book,* a directory of trees in British Columbia. You'll find descriptions of many pines, including where they grow, their features, and other interesting information.

http://www.sonic.net/bristlecone/intro.html

At this award-winning site you'll find all sorts of fascinating information about one of earth's oldest living creatures: the bristlecone pine (*Pinus longaeva*). Check out this site to learn about the discovery, growth, and habitat of these fascinating organisms. You'll also find links to other sites containing information about bristlecone pines.

http://www.mq.edu.au:80/school/museum/garden/cycad.html

Here you'll find photographs and brief descriptions of cycads.

http://128.146.143.171/hvp/tmi/hort300/gymno.htm

At this site you'll find an excellent summary of the features of the various groups of gymnosperms.

http://www.next.com/~jmh/SeedsOfLife/home.html

Visit *The Seeds of Life*. "Though I do not believe that a plant will spring up where no seed has been, I have great faith in a seed. Convince me that you have a seed there, and I am prepared to expect wonders." *Faith in a Seed* - Henry D. Thoreau

Suggested Readings

Articles

Baum, D. 1994. *rbc*L and seed plant phylogeny. *Trends in Ecology and Evolution* 9:39–41.

Donoghue, M. J. 1994. Progress and prospects in reconstructing plant phylogeny. *Annals of the Missouri Botanical Garden* 81:405–418.

Owens, J. N, and V. Hardev. 1990. Sex expression in gymnosperms. *Critical Reviews in Plant Sciences* 9(4):281–294.

Poinar, G. O., Jr. 1993. Still life in amber. *The Sciences* 34(2):34–39.

Stone, R. 1993. Surprise! A fungus factory for taxol? *Science* 260:154–155.

Taylor, T. N. 1982. Reproductive biology in early seed plants. *BioScience* 32:23–28.

Books

Beck, C. B. 1988. *Origin and Evolution of Gymnosperms.* New York: Columbia University Press.

Jones, D. 1993. *Cycads of the World.* Washington, DC: Smithsonian.

Simpson, B. B., and M. Conner-Ogorzaly. 1995. *Economic Botany: Plants in Our World.* 2d ed. New York: McGraw-Hill.

van Geldenen, D. M., and J. R. P. van Hoey Smith. 1996. *Conifers.* Portland: Timber Press.

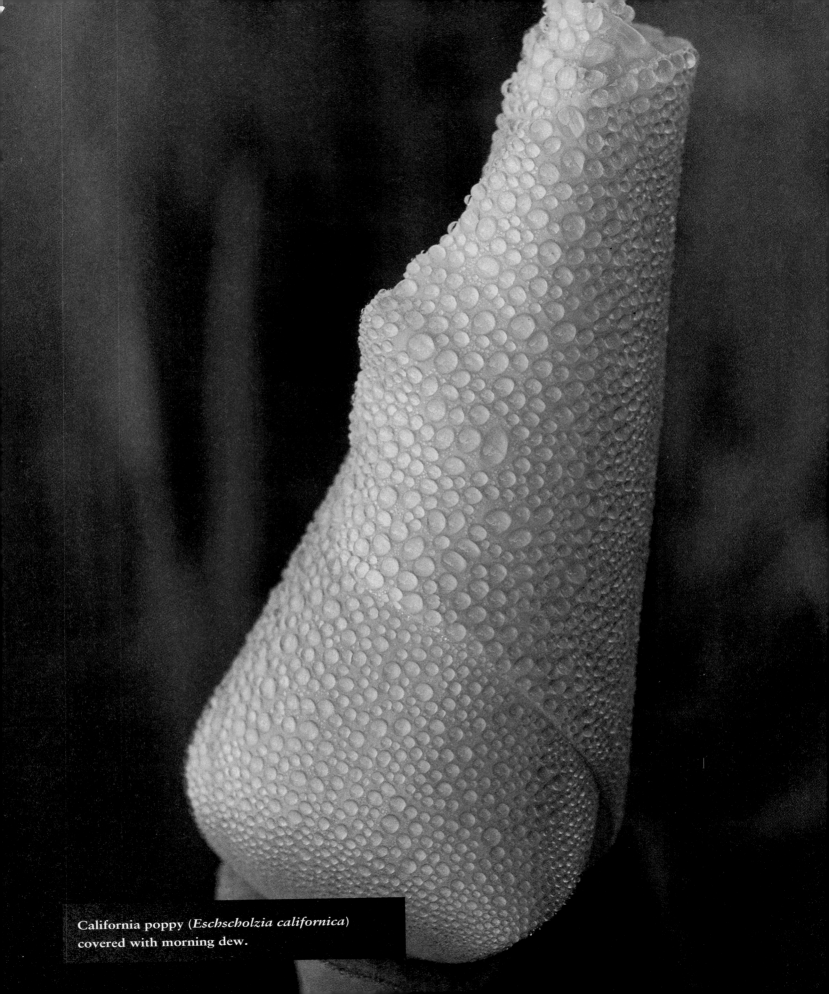

California poppy (*Eschscholzia californica*)
covered with morning dew.

Angiosperms

31

Chapter Outline

Chapter Overview

Flowering plants have been the dominant group of plants since the late Cretaceous and early Tertiary periods. Their sudden and abundant appearance in the fossil record has been a mystery since the time of Charles Darwin. To unravel this mystery, botanists have used evidence such as fossils, new ways of analyzing evolutionary relationships, and comparisons of modern molecular data from nucleic acid sequences. Although answers to the questions about when, where, and how the flowering plants arose have not yet been found, botanists have learned much about the evolution of angiosperms in the nearly century and a half since Darwin first pointed out the mystery of angiosperm origins.

Flowering plants (Division Magnoliophyta) are the most successful of all plant groups in terms of their diversity. The group includes more than 250,000 species and at least 12,000 genera. This group is often referred to as the angiosperms because their seeds are enclosed in a carpel. The carpel is the primary feature that distinguishes angiosperms from gymnosperms. Otherwise, the generalized life cycle of the angiosperms is similar to the life cycles of other groups of seed plants (fig. 31.1). The flower is similar to the bisporangiate strobili of cycadeoids, and double fertilization is a feature that angiosperms share with gnetophytes. The seeds of angiosperms have one or two cotyledons, which usually derive their nutrition from the endosperm and not the female gametophyte, as in gymnosperms. Although even a brief overview of angiosperm diversity would be too voluminous for this text, a few of the main ideas about the origin and the nature of the variation of reproductive morphology in this large plant group were provided in Chapter 17.

Angiosperms live in almost all terrestrial and aquatic habitats on earth. Except for conifer forests and moss-lichen tundras, angiosperms dominate all of the major terrestrial zones of vegetation. Moreover, angiosperms include some of the largest and some of the smallest plants (fig. 31.2); lilies, oak tress, lawn grasses, cacti, broccoli, and magnolias are all flowering plants. Such plants surround us and affect virtually all aspects of our daily lives. It is not surprising that this group of plants attracts the most attention from scientists and the public alike.

Most of the discussions about the genetics, physiology, ecology, structure, and economic importance of plants in this text focus on angiosperms, simply because most of our knowledge of plants is based on the angiosperms. Therefore, this chapter does not include information about these aspects of plant biology. Instead, this chapter presents some of the most interesting and puzzling questions about plants that are not explained in other chapters, including where the angiosperms came from and how they got to be what they are today.

More than a century ago, Charles Darwin referred to the sudden and abundant appearance of angiosperms in the fossil record as "an abominable mystery." In spite of the intense study of angiosperms since that time, we still do not have a completely satisfactory explanation for Darwin's abominable mystery. We do not know where they came from or how they evolved so quickly into such a diverse group. Botanists have used several different kinds of evidence to try to solve some of the mysteries associated with the evolution of flowering plants, including evidence from the fossil record, comparative morphology of extant plants, and, more recently, gene sequencing. The contributions of such evidence to studies of the origin and evolution of flowering plants are the subject of this chapter.

The Fossil Record of Angiosperms

Most plants from the past decomposed without leaving a trace of their existence. Indeed, botanists estimate that the fossil record of plant species may be only 1% complete and that at least 90% of the species that ever existed are extinct. Nevertheless, the fossil record of plants does provide a basis for some general ideas about where flowering plants came from and how they might have evolved.

The first fossils of vascular plants are more than 420 million years old (see Chapter 29), and the first seeds appeared as long as 360 million years ago (see Chapter 30). However, fossils of plant fragments that probably came from angiosperms are not known before the early Cretaceous period, about 135 million years ago. Unfortunately, most of the oldest of these fossils are so fragmented and incomplete that paleobotanists are not certain that they are angiosperms at all. Nevertheless, one particular fossil stands out because it consists of all the parts of a flower attached to a reasonably intact plant. This flowering plant is from a 120-million-year-old fossil deposit near Koonwarra, Australia.[1] Paleobotanists believe that this plant represents the ancestral type of flower. If this is true, then the features shared by the Koonwarra angiosperm and certain modern angiosperms may show which living plants are closest to the ancestral origin of the group.

---

1. In February 1996, paleobotanist Chris Hill reported a fossilized flower in 130-million-year-old clay rocks in the south of England. That plant, named *Bevhalstia pebja*, was a small herb about 25 cm high and had a primitive fernlike anatomy. Now that Hill has found this flower, botanists are hunting for pollen that will clinch its identification.

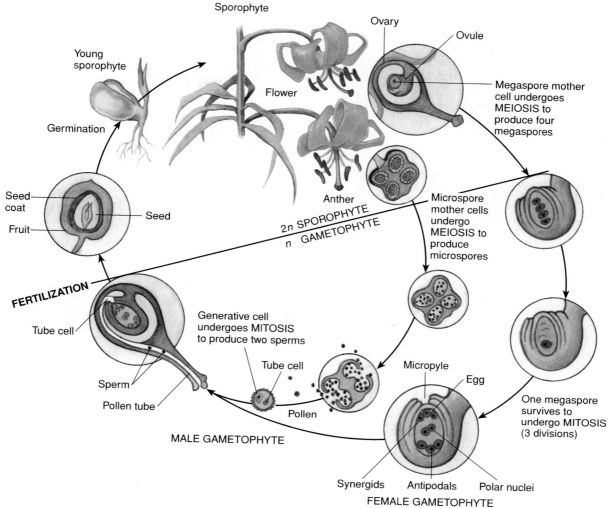

figure 31.1

Generalized life cycle of flowering plants. Although the seeds are enclosed in carpels and the reproductive parts are borne in flowers, the reproductive tissues and organs of angiosperms are similar to those of gymnosperms.

The Koonwarra Angiosperm

The fossil of the world's earliest known flower was discovered in 1986. At first, it was thought to be a fern, but closer examination showed that the "fern" had carpel-bearing inflorescences (fig. 31.3). The fossilized portion of the plant was less than 3 cm long, and overall it resembled a black pepper plant.

The Koonwarra angiosperm had several features that are typical of many modern angiosperms. For example, it had small flowers without petals, a spikelike inflorescence, single-carpel ovaries with short stigmas and no styles, and imperfect flowers with several bracts at their bases. These features occur

in present-day members of the lizard's tail family (Saururaceae), the pepper family (Piperaceae), and the chloranthus family (Chloranthaceae), all of which are dicots (fig. 31.4a). In addition, the leaf venation in the Koonwarra angiosperm resembles that of these families as well as the leaf venation of the birthwort family (Aristolochiaceae), the greenbrier family (Smilacaceae), and the yam family (Dioscoreaceae), the latter two being families of monocots (fig. 31.4b).

The Koonwarra angiosperm shows how the ancestor of flowering plants may have looked: a small, rhizome-bearing herb that had secondary growth, small reproductive organs, and simple, imperfect flowers with complexes of bracts at their

(a)

(b)

(c)

figure 31.2

Angiosperms. (a) *Eucalyptus*, one of the tallest plants. (b) Duckweeds are among the smallest flowering plants; in this photo they are the light green plants floating among the leaves of an aquatic fern (arrow). (c) Bunchberry (*Cornus canadensis*).

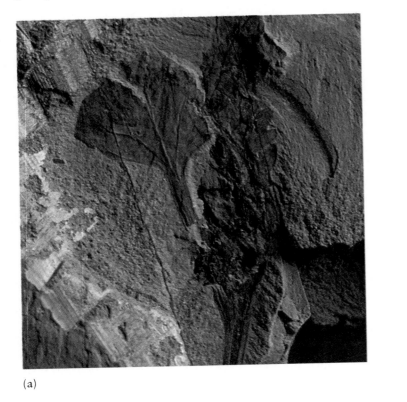

(a)

mm

(b)

figure 31.3

(a) The oldest known fossil flower, which is about 120 million years old, was found in the Koonwarra fossil beds near Melbourne, Australia.
(b) As this drawing shows, the entire fossil resembles a small black pepper plant, less than 3 cm long.

(a)

(b)

figure 31.4

(a) Lizard's tail (*Saururus* sp., Saururaceae), showing small apetalous flowers arranged in a spicate inflorescence. (b) Leaf of *Smilacina racemosa* (American false spikenard), a member of the lily family. Leaf venation of the Koonwarra angiosperm resembles that of the leaves shown here.

bases. Families of living plants that share several features with the Koonwarra angiosperm are believed to be primitive members of the dicots and monocots. Furthermore, the appearance of this plant near the apparent beginning of the evolution of angiosperms and its similarity to dicots and monocots suggest that the Koonwarra angiosperm evolved before the divergence between monocots and dicots. This implies that the monocots and dicots separated into two evolutionary lineages less than 120 million years ago, probably from an ancestor similar to the Koonwarra angiosperm. As explained later in this chapter, however, other evidence points to a much earlier evolutionary split between monocots and dicots.

Characteristics of Other Fossil Flowers

During the time that the Koonwarra angiosperm lived, flowering plants were apparently evolving rapidly. Evidence of rapid diversification among angiosperms is based on fossils from different places around the world that show a variety of floral and vegetative features among different plants. These fossils represent how flowering plants may have evolved from their earliest ancestors into the vast array of species that live today. Some of the more significant of these evolutionary developments are described in the paragraphs that follow.

Flower fragments of angiosperms from the early Cretaceous period already show a diversity of floral features. Some fossils apparently have floral parts arranged in a spiral on their axis, as in the flowers and fruits of the modern *Magnolia* (fig. 31.5a). Other fossils have some or all of their floral parts arranged in circles, or **whorls,** around the floral axis, which is the most common arrangement of floral parts among extant angiosperms (fig. 31.5b). Some early fossils have petals and sepals, while others lack a distinct calyx and corolla.

All of the first flowers were radially symmetrical, like poppies and buttercups; and their petals, when present, were free, (i.e., unattached to each other; fig. 31.5c). Flowers with distinct bilateral symmetry, such as those of modern violets and peas, and flowers with fused petals, such as those of cape primrose, did not appear until the Paleocene period, less than 65 million years ago (fig. 31.5d, e).

The earliest gynoecia had free carpels, and the first fruits derived from them were apparently follicles or nutlets. Fossils of gynoecia with fused carpels are not as old as those with free carpels. The first gynoecia with fused carpels may have developed into capsules. Fused carpels in more recent fossils show the development of nuts, drupes, berries, and pods.

c o n c e p t

The Koonwarra angiosperm is the earliest intact fossil of a flower. It has features that occur in several modern families of flowering plants. The early evolution of certain floral features is known from fragments of angiosperms, beginning at about the time of the Koonwarra angiosperm and continuing through the Cretaceous period and into the Tertiary period.

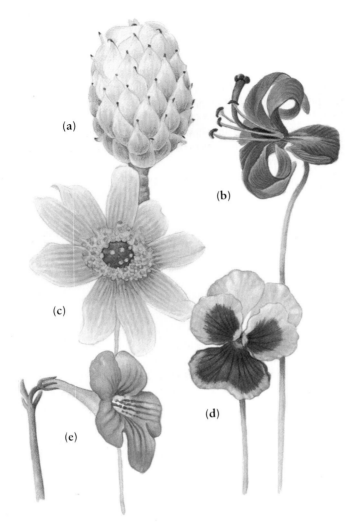

figure 31.5

The earliest flowers had floral parts arranged in spirals (a), as in the carpels of modern magnolia (*Magnolia*), or in whorls (b), as in the petals of the present-day lily (*Lilium*). Many of the oldest fossil flowers also had petal-like sepals, as in the lily, and radial symmetry (c) like the buttercup (*Ranunculus*). More recent flower types include those with bilateral symmetry (d), such as the pansy (*Viola*), and those with petals fused into a tube (e), such as the cape primrose (*Streptocarpus*).

The Origin and Radiation of Angiosperms

Although the Koonwarra angiosperm is the oldest known flower, it is probably not the oldest flowering plant. Nobody knows just how long ago the first angiosperm lived, but fossil pollen from the early Cretaceous period, perhaps 10 million years older than the Koonwarra angiosperm, may have come from angiosperms. Pollen has been found that is even older, but it cannot be distinguished from gymnosperm pollen. Thus, we lack direct evidence for the earliest angiosperms.

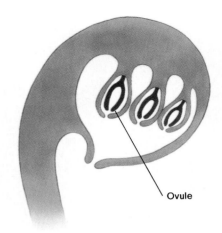

Ovule

figure 31.6

Diagram of a sectioned *Caytonia* cupule containing three ovules.

Nevertheless, several suggestions that intertwine the "when, where, and how" of angiosperm evolution have been made from interpretations of the available fossils.

When Did Angiosperms Evolve?

Assuming gradual evolution, the sudden appearance of a diversity of angiosperms in the Cretaceous period suggests that the evolution of flowering plants began much earlier, perhaps as much as 100 million years before the oldest known angiosperm fossil. If so, then the beginnings of the flowering plants may be found among cycadeoids or other extinct groups that may have shared ancestors with the angiosperms. For this reason, paleobotanists have often focused on comparing the reproductive features of angiosperms with those of different kinds of nonflowering plants.

Of special importance in explaining when angiosperms evolved is determining when the carpel arose. One hypothesis is that the carpel developed from the cupule of a seed fern like *Caytonia* (fig. 31.6). According to this hypothesis, cupule tissue surrounding the seeds fused to form a closed carpel. Seed ferns were prominent in the Carboniferous period, but few persisted into the Mesozoic era. This means that the carpel, or a precarpel, may have originated as early as 200 million years ago. Another hypothesis is that the carpel arose from the longitudinal folding of an ovule-bearing leaf—that is, from a folded megasporophyll (see Chapter 17). This hypothesis lacks support from the fossil record, however, because there are no good examples of such ancient reproductive structures.

Cycadeoids were once considered to be the ancestors of angiosperms, partly because the microsporangia and ovules of cycadeoids occur in the same cone (see Chapter 30). Such an arrangement simulates perfect flowers—that is, flowers with both stamens and carpels on the same receptacle, which are the most common type of flower in angiosperms.[2] (Recall,

---

2. Although the arrangement of microsporangia and ovules on cones simulates a perfect flower, the term *flower* is not used to describe the cones of cycadeoids.

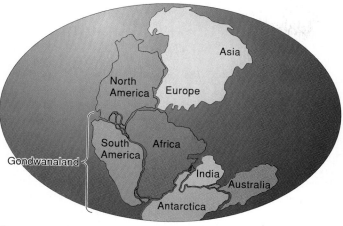

figure 31.7

Positions of the earth's landmasses in the early Triassic period (approximately 220 million years ago).

however, that the Koonwarra angiosperm had imperfect flowers.) Further evidence for a cycadeoid-angiosperm relationship is that the ovules of cycadeoids are believed to have contained linear tetrads (four spores in a row) in the megasporangia, which is a characteristic of angiosperms.

More recently, botanists have used the methods of cladistics to show how angiosperms may have descended rather directly from seed-fern ancestors in a line parallel to cycadeoids. If this were the case, then angiosperms could have originated in the Triassic period at about the same time the cycadeoids first appeared in the fossil record (see reading 31.1, "Angiosperms and Dinosaurs"). This hypothesis is discussed in more detail later in this chapter.

Where Did Angiosperms Evolve?

Botanists believe that pre-Cretaceous angiosperms were well adapted to cool, dry climates. These plants were also probably small, with tough leaves, seed coats, and vessels in their secondary xylem. Most of them were probably deciduous and thus avoided seasonal drying. These hypotheses represent guesswork based on the fossils of more recent Cretaceous angiosperms, and if correct, they suggest that the most likely places for angiosperms to have evolved were in the semiarid central regions of western Gondwanaland (fig. 31.7). Unfortunately, the drier conditions in these upland regions, unlike the wet conditions along shorelines and in lowland basins, did not allow much chance for plants to be preserved in the fossil record.

Plants of lowland basins from the Triassic, Jurassic, and Cretaceous periods were mostly gymnosperms, but angiosperms apparently began to invade these areas by the early Tertiary period, less than 65 million years ago. The more recent invasion of angiosperms into these lowland areas can be explained by climatic and geologic changes at the end of the Mesozoic era. At that time, the average temperature over the earth was dropping, mountains were forming rapidly, and continents were breaking apart and drifting northward. Although these drastic changes drove many kinds of organisms to extinction, they also provided new opportunities for the migration and evolutionary radiation of angiosperms into lowland habitats. Thus, although the first angiosperms probably evolved in cool, dry climates of uplands sometime before the Cretaceous period, the explosive diversification of flowering plants seems to have occurred much later in Tertiary lowlands.

c o n c e p t

Although the first fossils of angiosperms are no older than about 135 million years, the angiosperms probably arose much earlier. Indirect evidence from the possible ancestors of angiosperms indicates that they may have originated as long as 200 million years ago. Angiosperm fossils of that age are unknown, probably because they evolved in dry uplands that were not conducive to fossilization. The rapid evolution of angiosperms into the most diverse group of plants on earth occurred much later in lowlands during the Tertiary period.

How Did Angiosperms Evolve?

Any discussion of how angiosperms evolved must include a topic we have discussed throughout this book: the role of insects. Early in seed-plant evolution, insects became pollen carriers as they searched for food. In turn, plants evolved floral nectar and odors for attracting insects to carry pollen. The earliest, unequivocal angiosperm nectaries are from the late Cretaceous period, but they probably evolved even earlier.

The earliest insects that could have been pollinators were probably beetles. Long before the appearance of angiosperms in the fossil record, cycadeoids were already specialized for pollination by beetles. They had tiny ovules that were protected by scales, which is a feature that is consistent with pollination by insects with chewing mouthparts, such as beetles. By analogy with the cycadeoids, angiosperms that may have lived with the first beetle-pollinated cycadeoids were probably also adapted for pollination by beetles.

Insect pollination was not associated with rapid diversification of angiosperms until the appearance of specialized lepidopterans (e.g., butterflies, moths) and hymenopterans (e.g., bees) during the late Cretaceous and early Tertiary periods. The rise to dominance by angiosperms in the Tertiary, therefore, seems to have been greatly influenced by adaptations for pollination by an increasing diversity of flying insects. Thus, insects probably played an important role in the evolution of angiosperms into the largest and most diverse group of plants.

ANGIOSPERMS AND DINOSAURS

Dinosaurs included some of the most spectacular animals on earth, a few of which reached massive size. They lived throughout the Mesozoic era (Triassic, Jurassic, and Cretaceous periods) in the Age of Dinosaurs, or the Age of Cycads (the term preferred by botanists for the same period), and during the rise of angiosperms. Of these three groups of organisms, only the angiosperms remain as a highly diversified and dominant life-form in modern times. The dinosaurs were all extinct by the early Tertiary period, which coincided with the disappearance of the cycadeoids. By this time, the number of different cycads and other gymnosperms had also begun to diminish. Recent studies by paleobotanists and paleontologists have focused on how the gymnosperms, angiosperms, and dinosaurs may have interacted and even influenced their respective evolutionary histories. The main question asked by botanists is, "How did the angiosperms become dominant at the expense of the gymnosperms?" One possible explanation involves adaptation to insect pollination by angiosperms, as already discussed in the text. Another explanation involves dinosaurs.

The largest herbivores ever to walk the earth were the enormous sauropods, such as *Diplodocus, Brachiosaurus,* and *Apatosaurus* (popularly known as brontosaurus). These animals probably roamed in great herds, browsing on the tops of coniferous trees and other tall gymnosperms. Such browsing would not have killed large trees. Large herbivores did not destroy seedlings by browsing either, because they could not reach down to them. Thus, gymnosperms could flourish without being destroyed by large herbivores.

The coexistence of large, herbivorous dinosaurs and tall gymnosperms began to disintegrate at the beginning of the Cretaceous period, when the body size of dinosaurs mysteriously began to shrink. By the late Cretaceous period, large herbivores had been replaced by smaller, lower browsers of the ornithischian type (e.g., *Ankylosaurus, Parasaurolophus,* and *Montanoceratops*). These dinosaurs had relatively large, muscular heads with flat, grinding teeth for chewing plant tissues. There were more kinds of dinosaurs toward the end of the Cretaceous period than before, although by this time large dinosaurs of the Jurassic period were extinct. The greater diversity of smaller, late Cretaceous dinosaurs may indicate a higher degree of specialization in plant-dinosaur relationships than in earlier periods.

The long-term decrease in body size and increase in diversity of the dinosaurs coincided with the evolutionary radiation of angiosperms and the decline of gymnosperms. Gymnosperms were probably devastated by the low browsers, which would have eaten small seedlings before they were able to reach maturity and produce seeds. Since the first angiosperms were smaller and herbaceous, they probably grew and reproduced more rapidly than woody gymnosperms and therefore stood a better chance of producing seeds before being eaten. Furthermore, the destruction of gymnosperms by small dinosaurs opened up new habitats for the invasion and evolution of angiosperms.

While the dinosaur-gymnosperm-angiosperm connection seems plausible, the suggestion that angiosperms diversified by adapting to insect pollination is equally so. It may be that each of these explanations is correct and that dinosaurs and insects both influenced the rapid evolution of angiosperms in the Cretaceous period. However, the greatest diversification of angiosperms occurred in the Tertiary period, after the dinosaurs were already extinct.

c o n c e p t

One of the main forces behind the evolutionary radiation of angiosperms was adaptation for pollination by insects. The first insect-pollinated flowers probably arose at the same time beetle-pollinated cycadeoids were alive. Pre-Cretaceous, beetle-pollinated angiosperms gave way to a much greater diversity of angiosperms that evolved by adaptation for pollination by butterflies and bees in the Tertiary period.

Angiosperm Diversity

The angiosperms are such a large group that a full appreciation of their diversity would be a monumental undertaking. Systematists normally specialize in some part of the division, such as a group of related species, a genus, a family, or even a group of families. If estimates are correct that even the brightest botanists can know only about 2,000 kinds of plants in significant detail at one time, then what hope is there of keeping track of more than a quarter of a million species? How can we even know all of the species of just one state, such as California, where more than 6,000 species occur, most of which are angiosperms? Part of the answer is to maintain lists, keys, descriptions, and classifications of plants just as botanists have done since before the time of Linnaeus. Books that contain this kind of information are called **floras**. (Floras also include nonflowering plants, but these are almost always a small minority of the species included, unless the floras are devoted to bryophytes or some other special nonflowering group.) The most recent flora of California includes all the relevant information for identifying and classifying plants native to that state in about 1,400 pages. If this proportion held true for all flowering plants, then at least forty books of that size would be needed to cover the entire division.

A flora has not been completed for all species of flowering plants for at least two centuries, nor is one likely ever to be

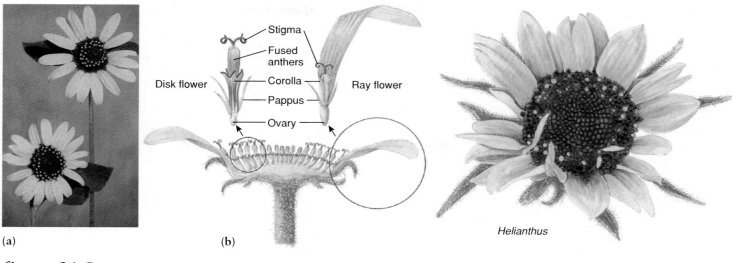

figure 31.8

The "flower" of sunflowers and other members of the family Asteraceae is actually a type of inflorescence called a head. (a) Two inflorescences of the common sunflower, *Helianthus annuus*. (b) Each sunflower inflorescence is comprised of dozens of each of two kinds of flowers: ray flowers around the perimeter of the inflorescence and disk flowers toward the interior.

completed. Most species live in tropical areas, which occur primarily in Third World countries that do not have enough money to finance the work needed to do their floras. Even in the United States, work on an updated flora of North America began only relatively recently. It is expected to be completed by the year 2000, depending on how fast research budgets allow the work to go. It is a 10–20 year project being done by hundreds of taxonomists.

Another way to know about angiosperm diversity is to study it from the top down, that is, by looking at the highest levels of the taxonomic hierarchy first, then examining lower levels one at a time to get a better picture of the whole division. You have already done this to some extent by learning the differences between monocots and dicots, the only two classes in the division. The next step would be to study the characteristic features of each order and then of each family. However, only a few systematists have proposed comprehensive classifications of the orders and families of flowering plants, and their classifications differ not only in the grouping of genera into families and families into orders but also in their various ideas about the phylogenetic relationships of these groups. Table 31.1 gives you an idea of some of these differences by showing how many taxa are included in each category by different systematists.

The features of most importance in classifying plants in floras, comprehensive treatments of divisions, or any group of flowering plants are usually those of the reproductive organs. For example, plants that have small flowers with no sepals, five petals fused into either a tubular or strap-shaped corolla, inferior ovaries, five anthers fused into a cylinder around the style, two stigmas, achenes for fruits, and inflorescences in heads are all classified in the sunflower family (Asteraceae) (fig. 31.8). Sometimes, however, vegetative features are also important because of wide variability in reproductive structures or the similarity

table 31.1

Summary of Different Classifications of Angiosperms

Classification	Number of Orders	Number of Families
Cronquist	83	387
Dahlgren	73	403
Takhtajan	92	410
Thorne	69	440

Note: Systematists identify different systems of classification by the names of the principal authors who wrote them. Those summarized in the table are the four most widely accepted classifications today. More information about them appears in Chapter 24.

of such features among families. For instance, the pineapple family (Bromeliaceae) is unique because of its absorptive leaf scales, nearly spherical crystals of silica, twisted stigmas, and base chromosome number ($n = 25$). These features unite such seemingly disparate plants as the pineapple (*Ananas comosus*) and the Spanish moss (*Tillandsia usneoides*) (fig. 31.9).

c o n c e p t

There are so many kinds of flowering plants that systematists usually study only a few of them at a time, depending on the number of species in a genus or the number of genera in a family. Floras may cover thousands of species in a state or region, but there is no such treatment for all of the world's flowering plants or even for all of the plants in North America. Comprehensive treatments of the flowering plants mostly involve the classification of families and higher taxonomic categories.

figure 31.9

Pineapple plants, showing the stiff, narrow leaves and a pineapple that has developed from an inflorescence of 100 or more flowers whose ovaries and other structures have united into a single, multiple fruit.

figure 31.10

Vellozia is a large genus (approximately 100 species) of tropical Africa and Madagascar.

Testing Taxonomic Hypotheses: The Case of the Velloziaceae

As discussed in Chapter 24, modern classifications attempt to be phylogenetic—that is, to show natural relationships. This approach to explaining diversity automatically presents hypotheses about relationships, which can be tested by cladistic analysis or by the application of new kinds of data (e.g., gene sequencing). Ideally, systematists look at competing taxonomic hypotheses and find evidence to reject all but the best one. Systematists worldwide use this approach regularly, so there are many examples to choose from to explain how it works. Our example involves a family of angiosperms called the Velloziaceae.

The Velloziaceae, tropical monocots, includes about 250 species in six genera (fig. 31.10). The Dahlgren classification includes the Velloziaceae with such families as the pineapple family, the cattail family (Typhaceae), and the ginger family (Zingiberaceae), among others (fig. 31.11a). In contrast, the Cronquist system proposes that the Velloziaceae are more closely related to the families of the Order Liliales (e.g., lilies,

Liliaceae) (fig. 31.11b). This case was resolved by a 1993 study that was based on a cladistic analysis of DNA sequences from the chloroplast gene *rbc*L. The results were used to reject the relationship proposed in the Dahlgren system. This study also corroborated earlier ideas about the occurrence of fluorescent chemicals in the cell walls of some monocots: fluorescent chemicals, such as ferulic acid, bind to cell-wall polymers in the species of some families but not others (fig. 31.12). They include the pineapple, ginger, and cattail families but not the Velloziaceae. The occurrence of such UV-fluorescent cell walls is now considered to be a shared-derived character-state that shows common ancestry of this group of monocots, excluding the Velloziaceae.

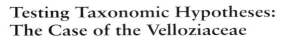

c o n c e p t

Angiosperm diversity is explained by competing taxonomic hypotheses about phylogenetic relationships in different classifications of the flowering plants. Systematists test these hypotheses by comparing DNA sequences or other new kinds of data.

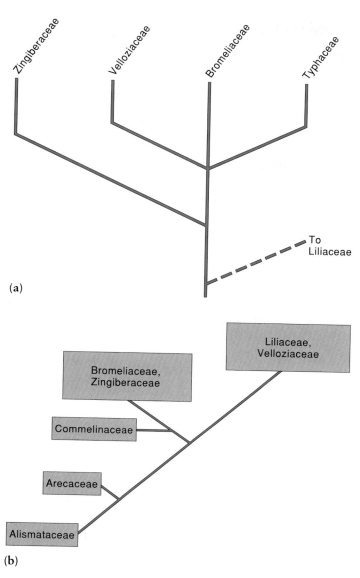

(a)

(b)

figure 31.11

(a) Proposed relationships of certain monocot families based on the Dahlgren system of classification. In Dahlgren's system, the Velloziaceae are closely related to the Bromeliaceae. (b) According to the classification system of Cronquist, the Velloziaceae are more closely related to the Liliaceae than to the Bromeliaceae.

The Phylogenetic Relationships of Angiosperms

A phylogenetic analysis of all of the species or genera of flowering plants cannot be done using available computer technology. Furthermore, the relevant taxonomic information from 250,000 species or 12,000 genera for such an analysis could not be obtained in a reasonable period of time, even if hundreds of

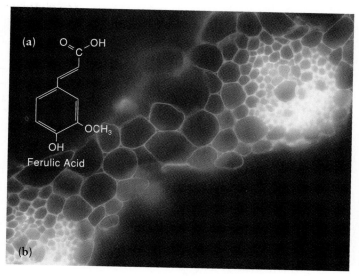

figure 31.12

(a) Chemical structure of ferulic acid. (b) UV-fluorescent cell walls of *Lachnanthes caroliniana*. The linkage of UV-fluorescing compounds to cell-wall polymers is now considered to be a shared-derived character-state for certain monocots.

taxonomists did nothing but gather data. It has recently become possible, however, to construct a complete phylogeny of flowering plants for the 400 or so families and their higher categories. More data are needed to complete the job, but as of 1993 a single gene, *rbc*L, had been sequenced for almost 500 species of seed plants, including representatives of many families of monocots and dicots and all other divisions of seed plants.

Phylogenies of the Flowering Plants

The gene *rbc*L is not quite 1,500 nucleotides long, which means it can potentially provide 1,500 characters, each with four states (A, T, C, G). We can expect, however, that some of these characters are unchanging due to strong natural selection against their mutation. Conversely, other nucleotide positions, which may not be so important for the activity of the protein, may change often enough to show multiple parallel changes (homoplasies) in the phylogeny of a large set of taxa. The analysis of such data, for example, showed that there are at least 3,900 equally parsimonious cladograms, each with 16,305 character-state changes. The high number of character-state changes indicates that there are many homoplasies.

Cladistic methodology does not provide a way to choose the best cladogram from among thousands of apparently equally good ones, so systematists generally look for common branch-points among different phylogenies, from which the strongest suggestions about relationships can be made. The placement of the Velloziaceae, for example, was distant from the Bromeliaceae in all 3,900 cladograms. This provides a strong basis for rejecting the relationships shown by the Dahlgren classification system.

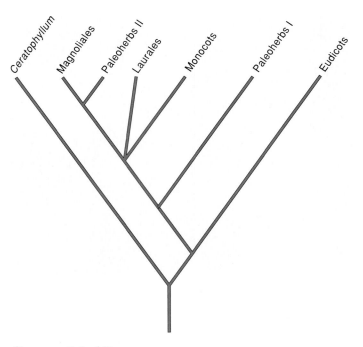

figure 31.13

Simplified molecular phylogeny of flowering plants based on sequences of the chloroplast gene *rbc*L.

figure 31.14

According to a recent analysis based on molecular phylogenies, *Ceratophyllum* may be the most primitive living angiosperm.

Another consistent and interesting result is the relationship of the monocots to the dicots. Although different versions vary somewhat, all results show the monocots to be embedded among the dicots, which means that the dicots are not a unique evolutionary lineage (fig. 31.13). Monocots seem to have diverged with an evolutionary line that gave rise to certain kinds of dicots and to the so-called paleoherbs and moved away from the **eudicots** (most dicots). This result shows that dicots such as magnolias (Magnoliales) and laurels (Laurales) are more closely related to monocots than they are to other dicots. The paleoherbs in this analysis include most of the plants that resemble the Koonwarra angiosperm, which have been suggested as the most primitive angiosperms.

While the molecular phylogeny of angiosperms gives some consistent answers about the relationships of monocots, paleoherbs, and dicots, it also poses a new mystery. The most primitive flowering plant in the latest analysis seems to be *Ceratophyllum*, a weedy, vascular, hornwort (i.e., not a bryophyte) that grows in waterways worldwide (fig. 31.14). You may recognize it as a common aquarium plant. Does its position at the base of the flowering plant cladogram indicate that *Ceratophyllum* is more like the first flowering plant than any other extant angiosperm? If so, does this mean that the angiosperms evolved in an aquatic habitat, not in the cool, dry uplands of the late Cretaceous period as previously thought? Answers to these questions are not yet available, but trying to find them will keep botanists busy for quite a while.

concept

Phylogenies based on a cladistic analysis of DNA sequences provide a basis for supporting or rejecting competing notions about classification. They also present new puzzles that show the relationship between dicots and monocots and the potential origin of angiosperms from aquatic ancestors.

The Origin of Flowers

If a *Ceratophyllum*-like flower is the most primitive type among living angiosperms, then the ancestral angiosperm may also have had simple, monoecious flowers, numerous stamens, and a simple, single-chambered carpel. How did this type of flower arise? Recent cladistic analyses of all seed plants, past and present, have approached this question by trying to find out which nonangiosperm seed plants evolved from the same ancestor as the angiosperms. Results from one of these studies have been used to generate a hypothesis about the origin of flowers from preflowering ancestors.

One possible origin of flowers was suggested from a cladistic analysis of seed plants by James Doyle and Michael Donoghue that was based on the simplified phylogeny showing the relationships among angiosperms, cycadeoids, and gnetophytes (fig. 31.15). These relationships show that the angiosperm-cycadeoid-gnetophyte line shared an ancestor with

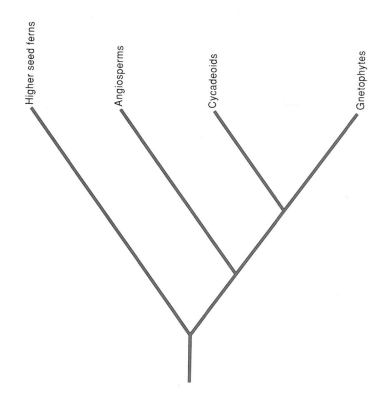

figure 31.15

Simplified phylogeny of these seed plants shows an angiosperm-cycadeoid-gnetophyte evolutionary line arising from the same ancestor that gave rise to the higher seed ferns.

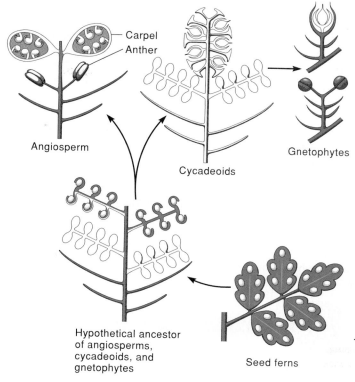

figure 31.16

Proposed origin of flowers based on a cladistic analysis of seed plants.

the higher seed ferns and that the angiosperms evolved earlier than the cycadeoids or the gnetophytes. Based on this phylogeny, Doyle and Donoghue proposed an evolutionary origin of flowers from a bisporangiate compound strobilus that arose from the fertile leaves of seed ferns (fig. 31.16). Furthermore, this ancestral strobilus also gave rise to the bisporangiate strobili of cycadeoids and later to the monosporangiate strobili of gnetophytes. Unfortunately, there are no fossils of such a pre-angiosperm strobilus to provide evidence for this hypothesis. Nevertheless, indirect evidence that angiosperms, cycadeoids, and gnetophytes evolved from a common ancestor with bisporangiate strobili comes from *Welwitschia mirabilis*. The pollen cone of this species has an abortive ovule, which suggests that gnetophytes could originally have been monoecious.

In addition to providing ideas about the origin of flowers, the phylogeny shown in figure 31.16 also gives us a clue as to how old the angiosperms are. If the origin of angiosperms occurred before the cycadeoids appeared in the fossil record, as the cladogram indicates, then flowering plants existed in the late Triassic period, at least 200 million years ago. This suggestion matches that of paleobotanists who have long thought that the angiosperms and cycadeoids arose at about the same time.

c o n c e p t

The most parsimonious evolutionary tree showing the origin of angiosperms suggests that they evolved before cycadeoids and gnetophytes. This phylogeny has been interpreted to mean that flowers arose from an extinct, bisporangiate ancestor that also gave rise to cycadeoids and gnetophytes. Considering the age of cycadeoids in the fossil record, this phylogeny also means that the angiosperms arose at least 200 million years ago.

Dating the Monocot-Dicot Divergence

The **molecular clock hypothesis,** first mentioned in Chapter 11, states that the amount of dissimilarity between two nucleic acid sequences should be proportional to the amount of time through which they have evolved from their common ancestral gene. A nucleic acid sequence can therefore be viewed as a clock: the rate of mutational change in the sequence is like the rate at which the clock ticks. Thus, the "age" of *rbc*L sequences, for example, can be calculated by comparing differences among sequences of various plants to see how long the clock has been ticking. Several biologists have attempted to use this principle to determine when the monocots and dicots evolved from their common ancestor, assuming that this estimate would be close to the time of origin for flowering plants. The best estimate, which was based on comparisons of sequences from *rbc*L and eleven other chloroplast genes, suggests that this divergence occurred approximately 200 million years ago. A different estimate, which was based on comparisons of sequences from two nuclear genes, suggests that the monocot-dicot divergence may have occurred as long as 320 million years ago.

That the monocots and dicots could have diverged as long ago as 320 million years is incomprehensible to paleobotanists, but the history of scientific discovery abounds with surprises that eventually became accepted. However, the estimate of 200 million years is more consistent with the time of origin of angiosperms according to the seed-plant phylogeny proposed by Doyle and Donoghue, and it is currently considered to be the better one.

The Lore of Plants

The informal competition among scientists to find the oldest fossil DNA seemed to be won in 1990, when 17-million-year-old DNA was found in fossils of *Magnolia* leaves. In 1992, however, the record was broken by 30-million-year-old DNA from insects that had been preserved in amber. Even this record was surpassed by a 1993 report of DNA from a 120–135-million-year-old fossil weevil.

Chapter Summary

The angiosperms, which are the most dominant plants on earth, have a flower that includes seeds in a carpel. Fossils of carpels and other parts of flowers are known from Cretaceous deposits that are at least 135 million years old, but the first complete fossil flower is about 120 million years old. This flower resembles several plants in different families of monocots and dicots on the basis of simple features of the flower and its herbaceous type of body.

The main force behind the rapid evolutionary radiation of angiosperms may have been pollination by insects and the availability of habitats left open by the disappearance of many gymnosperms. The first flowers were probably pollinated by beetles; later angiosperms attracted butterflies and bees.

Another hypothesis that explains the rapid evolution of angiosperms involves the influence of dinosaurs. Large, high-browsing dinosaurs mysteriously gave way to smaller, low-browsing dinosaurs by the end of the Mesozoic era. The low-browsing dinosaurs probably devastated gymnosperms by eating young seedlings, which allowed low-growing, herbaceous angiosperms to survive and diversify.

The diversity of angiosperms is so great that most of the knowledge about this group is divided into floras or taxonomic treatments of smaller subsets of taxa. Traditional comprehensive classifications are created primarily by organizing families and higher levels of the taxonomic hierarchy. Modern methods of cladistic analysis and new types of data, such as gene sequences, are now being used to evaluate competing hypotheses from different traditional classifications.

Estimates of rates of mutational change in gene sequences show how old some genes may be and, by inference, how old different plant groups may be. The molecular clocks of dicots and monocots, for example, show that these two groups probably diverged from one another about 200 million years ago, more than 65 million years before the first unequivocal angiosperm fossil.

 ### What Are Botanists Doing?

Molecular phylogenies of plants are being made more often now from comparisons of DNA sequences. How many such phylogenetic studies that were published in the past 12 months can you find in *Biological Abstracts* or other library resources? Which three genes were used most often for inferring phylogenetic relationships?

Writing to Learn Botany

What explanations can you give for the 80-million-year discrepancy between the age of the oldest fossil flower (120 million years) and the age of the monocot-dicot divergence based on the estimated mutational rates of *rbc*L (200 million years)?

Questions for Further Thought and Study

1. What is the fossil evidence that angiosperms originated at least 200 million years ago?

2. What correlation is there between the seed-plant phylogeny estimated by Doyle and Donoghue and the date of the monocot-dicot divergence calculated from the chloroplast gene clock?

3. If angiosperms evolved in the Triassic or Jurassic period, why are there no angiosperm fossils from those periods?

4. Dinosaurs and most other animals went extinct at the end of the Cretaceous period. Some scientists have proposed that such mass extinctions were caused by geological catastrophes. If this is true, how might angiosperms have escaped such catastrophes?

5. In addition to coevolution with insect pollinators, what other kinds of coevolutionary interactions may have occurred between flowering plants and animals?

6. Does the fact that angiosperms do not appear in the fossil record until the Cretaceous period mean that they were not around before then? Explain your answer.

7. The evolution of diversity among angiosperms depended heavily on insects. What other partnerships have angiosperms and other plants established with other organisms to ensure mutual success?

8. Angiosperms usually have a shorter life cycle than do other plants. How might this have affected the evolution of angiosperms?

Web Sites

Review the "Doing Botany Yourself" essay and assignments for Chapter 31 on the *Botany Home Page*. What experiements would you do to test the hypotheses? What data can you gather on the Web to help you refine your experiments?

Here are some other sites that you may find interesting:

http://www.isc.tamu.edu/FLORA/cronang.htm
http://www.isc.tamu.edu/FLORA/gallery.html
Texas A&M University Angiosperm Image Gallery is sponsored by the Department of Biology Herbarium (TAMU) and the Texas A&M University Bioinformatics Working Group.

http://chipmunk.apgea.army.mil/ento/guides.htm
Information from this Guide to Poisonous and Toxic Plants might save your life one day!

http://www.ngdc.noaa.gov/paleo/napd.html
The North American Pollen Database (NAPD) began operation at the Illinois State Museum (Springfield, Illinois, USA) in late 1990; the European database started up shortly thereafter. This database can be a valuable research tool.

Suggested Readings

Articles

Anonymous. 1990. World's earliest flower discovered. *Earth Science* 43(2):9–10.

Baum, D. 1994. *rbc*L and seed-plant phylogeny. *Trends in Ecology and Evolution* 9:39–41.

Clegg, M. T. 1990. Dating the monocot-dicot divergence. *Trends in Ecology and Evolution* 5:1–2.

Crane, P. R., E. M. Friis, and K. R. Pederson. 1995. The origin and early diversification of angiosperms. *Nature* 374:27–33.

Taylor, D. W., and L. J. Hickey. 1990. An Aptian plant with attached leaves and flowers: Implications for angiosperm origin. *Science* 247:702–704.

Books

Fernholm, B., K. Bremer, and H. Jörnvall, eds. 1989. *The Hierarchy of Life: Molecules and Morphology in Phylogenetic Analysis*. New York: Excerpta Medica.

Friis, E. M., W. G. Chaloner, and P. R. Crane, eds. 1987. *The Origins of Angiosperms and Their Biological Consequences*. New York: Cambridge University Press.

Greyson, R. I. 1994. *The Development of Flowers*. Oxford: Oxford University Press.

Heywood, V. H., ed. 1993. *Flowering Plants of the World*. Oxford: Oxford University Press.

Hughes, N. F. 1994. *The Enigma of Angiosperm Origins*. New York: Cambridge University Press.

Ecology

I n the previous unit we discussed the great diversity that has evolved among plants, fungi, algae, and bacteria, and we introduced some of the remarkable adaptations for survival that have developed among these organisms.

In this unit you will examine these and other adaptations in more detail and develop a better appreciation and understanding of the sometimes intricate interrelationships between plants, animals, other organisms, and their environment. The discussions include an examination of ecosystems and the flow of energy within them, the cycling of water and nutrients, and the impact humans have had (and continue to have) on the natural world around them. Also included is an introduction to forest, grassland, desert, and other major natural plant associations.

A tundra community in northern Manitoba, Canada.

Population Dynamics and Community Ecology

32

Chapter Outline

Chapter Overview

This chapter deals with basic ecological concepts and introduces populations, communities, and ecosystems. You'll read about the interactions of producers, consumers, and decomposers, as well as the factors involved in succession. The chapter concludes with discussions of the consequences of human disruption of ecosystems; the topics covered include the greenhouse effect, acid rain, ozone depletion, and water pollution.

Ecology is the study of organisms in relation to their natural environment. Ecologists investigate a host of processes and events, including the growth or decline in numbers of individuals of a particular species, how these species are affected by climate and other environmental changes, how organisms interact within their species and with other species, and how nutrients are recycled within an ecosystem. Virtually all subdisciplines of biology are involved in the study of ecology.

The word *ecology* (from the Greek *oikos*, meaning "home") was first proposed by the German biologist Ernst Haeckel in 1869. Although ecology was recognized as a biological discipline at the beginning of this century, its origins date back to early civilizations when humans, using tools and fire, first learned to modify their environment. Since the 1960s, ecology has become a household word and a vast area of study.

In the past two decades, topics such as the effects of pollution on land, water quality, and people's standard of living have filled thousands of books. During this time, attempts to stop environmental damage have forced major construction projects to file environmental impact reports before proceeding. These reports provide information that helps various agencies to evaluate the wide-ranging effects of development on the flora, fauna, and physical environment and to outline modifications that must be made before a project can be approved. Environmental impact assessment has often produced much controversy and emotional debate between industrial and land developers and those who feel that preservation of the environment should be our highest priority. Effectively resolving these controversies requires an understanding of ecological principles.

Populations, Communities, and Ecosystems

Plants, animals, and other organisms are associated in complex and interdependent ways with each other and with their physical environment. For example, the lichens and mosses on a rock constitute a community, as do the seaweeds in a tidepool. These communities also have animals and other organisms associated with them in larger communities. Therefore, we refer to the unit composed of all the populations of living organisms in a given area as a **biotic community.** Considered together, a community and its physical environment constitute an **ecosystem,** which is interconnected by physical, chemical, and biological processes to other ecosystems. Some populations, communities, and ecosystems may be microscopic, whereas others are much larger—even global.

Populations

A **population** is a group of individuals of a single species living in close proximity and capable of interbreeding. Populations may vary in numbers, density, and the total mass of individuals. Depending on circumstances, a biologist may measure the importance of a population in various ways. If, for example, she is concerned about the preservation of a rare or threatened species, she may simply count the number of individuals, although this may not always be feasible. The biologist may also estimate **population density,** which is the number of individuals in a given area (e.g., five blueberry bushes per square meter). If the individuals in a population have different sizes or distributions, a better estimate of the population's importance to the

ecosystem might be its **biomass** (total mass of the individuals present). Population biologists also study physiology, seed dispersal, germination, survival, and pollination in evaluating the importance of populations to their ecosystem.

Populations within a relatively small geographic area (e.g., whose pollinators have ready access to all individual members) are called **local populations.** If, however, many individuals in several local populations are distributed over a relatively large geographic area, the population may be an **ecological race** within a species. Individuals of a population may be clumped together as a result of uneven distribution of seeds and resources, or plants may be distributed evenly. Some species produce chemicals that inhibit the germination or growth of competitors in their vicinity, resulting in a uniform distribution of plants (fig. 32.1). Uniform distributions are, however, rare in nature. Because dispersal mechanisms often operate more slowly than do changing environmental conditions, most species never completely occupy their potential geographic ranges, which change over time.

Virtually all plants produce seeds or other dispersible reproductive structures. When seeds or spores germinate in a new habitat, we might expect population growth to accelerate. The rate of population growth is, however, slowed by several factors. Annuals complete their life cycles within a single growing season, at the end of which the population stops growing. Some perennials may require several years to complete their life cycles. Also, most seeds and spores either do not germinate, or the seedlings do not survive long enough to reproduce themselves.

The number of offspring under ideal conditions that live long enough to reproduce is referred to as the **biotic potential.** Annuals that produce few seeds per plant often have a greater

figure 32.1

These creosote bushes (*Larrea tridentata*) are evenly spaced because they produce toxins that inhibit competing vegetation. The toxin is effective in the immediate area surrounding each plant.

figure 32.2

A grassland community in spring. Communities are composed of more or less interdependent species living in the same location.

biotic potential than do trees that produce many thousands of seeds because trees have such a long **generation time** (i.e., the time from germination to reproductive maturity). In addition, the number of plants that an ecosystem can support is limited. The maximum number of individuals that can survive and reproduce in an ecosystem is the **carrying capacity** of that ecosystem. As a population increases, competition for nutrients, water, and light increases. When the germination and survival of offspring equal the death rate of mature plants in a population, population growth stops.

Communities

Communities consist of populations of different species living in the same location (fig. 32.2). Similar communities occur under similar environmental conditions, although the actual species composition can vary considerably and can change between the boundaries of an area. A community is difficult to define precisely because members of one community can also occur in other communities. Furthermore, members of one community may be genetically adapted to the biotic and abiotic interactions within that community. If individuals are transplanted to a different community where the same species occurs, they may not necessarily be able to survive alongside their counterparts, which are adapted to their own community. A population adapted to a specific community and environment within the species overall distribution is called an **ecotype.** Ecotypes are often so distinct from one another in their appearance that botanists argue about whether to classify them as separate species.

Ecosystems

Living organisms that interact with one another and with the nonliving environment constitute an ecosystem. The nonliving, or **abiotic, factors** of the environment include light, temperature, oxygen level, air circulation, precipitation, and soil type. The distribution of a plant species is controlled mostly by these abiotic factors and the effects of other organisms (**biotic factors**). Species in arid ecosystems, for instance, are adapted to low precipitation and high temperatures. Such plants are called **xerophytes.** Cacti often have small leaves or nonphotosynthetic leaves in the form of spines, which reduce transpiration and thus adapt them to their particular environments (fig. 32.3). Xerophytes may also have special forms of photosynthesis (such as CAM photosynthesis). Similarly, **hydrophytes** such as water lilies grow in water and have modifications that adapt them to their aquatic environment. The leaves of hydrophytes, for example, have thin cuticles and more stomata on their upper surface than on their lower surface (fig. 32.4).

The mineral content of soils influences the distribution of plant species. The high concentrations of metals such as magnesium, iron, nickel, and chromium and the low amounts of calcium and nitrogen found in serpentine soils characterize a relatively inhospitable environment in which only a few plants can survive. Similarly, the diversity of tree species in rain forests often varies with the levels of phosphorus and magnesium in the soil. Biotic factors such as competition for light, mineral nutrients, water, and space also influence the distribution of plant species, as does the direct removal of plants by grazing animals.

figure 32.3

Barrel cactus (*Ferrocactus covillei*), a xerophyte. The spines are modified leaves that protect the plant and reduce the surface area available for transpiration.

figure 32.4

Water lily (*Nymphaea*), a hydrophyte. The floating leaf blades have thin cuticles; nearly all their stomata are on the upper surface.

c o n c e p t

Ecologists view the interrelationships of organisms as a hierarchy. The lowest level is a population, which is an interacting group of individuals of a single species. Populations of different species make up a community, and communities and their physical environments compose an ecosystem. Ecosystems are the most complex level of organization.

Characteristics of Ecosystems

Trophic Levels Organisms in ecosystems may be grouped into trophic levels as determined by their mode of nutrition. Ideally, ecosystems can sustain themselves entirely by photosynthesis or chemosynthesis and the recycling of nutrients. Autotrophic organisms either capture light energy and convert it, along with carbon dioxide and water, to energy-rich sugars, or they oxidize chemicals as a source of energy. These autotrophic organisms are called **producers.** Heterotrophic organisms such as herbivores (e.g., deer and sheep), which eat only producers, are called **primary consumers. Secondary**

consumers, such as carnivores (e.g., eagles, wolves, and lions), eat primary consumers. Omnivores such as bears may eat plants and animals. **Decomposers** break down organic materials to forms that can be reassimilated as mineral components by the producers. The foremost decomposers in most ecosystems are bacteria and fungi (fig. 32.5).

In any ecosystem, the producers and consumers form **food chains** or interlocking **food webs** that determine the flow of energy through the different levels. Since most organisms have more than one source of food and are themselves often eaten by a variety of consumers, there are considerable differences in the length and complexity of food chains or webs (fig. 32.6).

Flow of Energy Energy enters a food chain at the producer level and flows to subsequent levels of consumers and decomposers in an ecosystem. Only about 1% of the light-energy striking a temperate-zone community is converted to organic material. This organic material and its energy pass to subsequent trophic levels, and as the organisms at each level respire, energy gradually dissipates as heat into the atmosphere. Some energy is stored in organisms that are not consumed and is released when they decompose. For example, the energy in leaves that fall from a plant before a herbivore grazes on them is released when decomposers (bacteria and fungi) degrade the leaves.

Only a small portion of energy stored in one trophic level will flow to the next level. Most energy is lost as heat during growth, maintenance, and decomposition. About 10% of the energy stored in green plants that are eaten by cattle is converted to animal tissue; most of the remaining energy dissipates as heat. When we eat beef, our bodies use about 10% of the beef's stored energy for growth, maintenance, and reproduction. The remaining energy is converted to heat. If 90% of the energy is lost as heat at each level of a food chain, then only about 0.1% of the original energy captured by the producers will be used in a typical food chain with three levels of

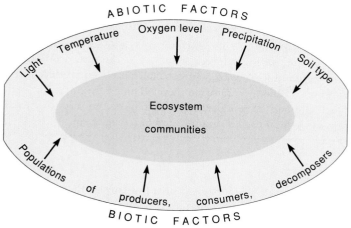

ABIOTIC FACTORS

Light Temperature Oxygen level Precipitation Soil type

Ecosystem

communities

Populations of producers, consumers, decomposers

BIOTIC FACTORS

figure 32.5

The basic components of an ecosystem interact to form an integrated unit.

consumers. Thus, the longer the food chain, the greater the number of producers necessary to provide energy for the final consumer. Conversely, the shorter the food chain, the smaller the number of producers needed to provide energy for the final consumer. Consider the following example:

Assume that for every 100 units of light-energy striking corn plants each day, 10 units are converted to plant tissue. Then suppose that the corn is fed to cattle. Only 10%, or 1 unit, of the original energy is converted to beef. When we eat beef, our bodies, in turn, use only 0.1% of the energy originally converted to plant tissue. If we eat the corn directly instead of the beef, we receive the same amount of energy from 90% fewer plants than it took to produce the beef.

The concept just described has important implications. For example, a vegetarian diet uses energy more efficiently than a diet based on meats. Consequently, where food is scarce or humans are abundant (as in India or Ethiopia), many humans

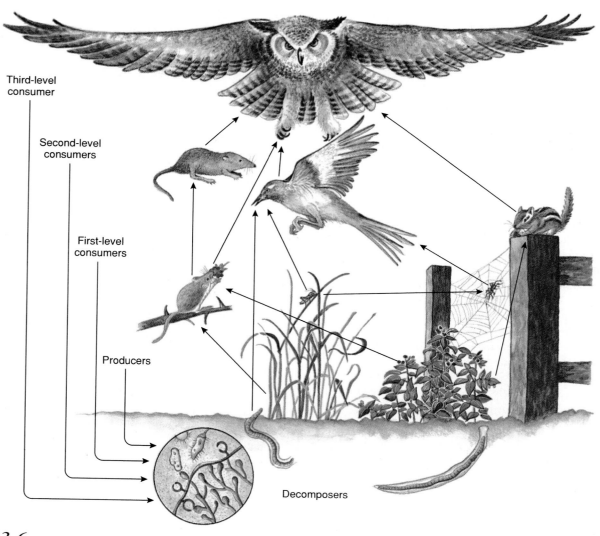

Third-level consumer

Second-level consumers

First-level consumers

Producers

Decomposers

figure 32.6

A hypothetical food web. Although no real food web would be so simple, this diagram shows that plants are producers, animals are consumers, and microorganisms are common decomposers in an interlocking food web. Arrows show the path of energy.

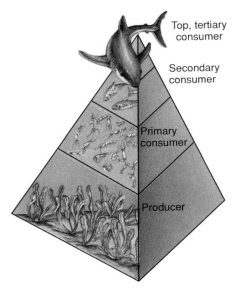

figure 32.7

This pyramid represents how much energy from lower levels of the food web is required to support successively higher levels. In this example, each kilogram of the shark (top carnivore) indirectly requires as much as 1,000 kg of algae (producers).

become vegetarians. In terms of the numbers of individuals and the total mass, there is a sharp reduction of usable energy at each level of the food chain. In a given part of the ocean, for example, billions of microscopic algal producers may support millions of tiny crustacean consumers, which, in turn, support thousands of small fish, which meet the food needs of scores of medium-sized fish, which are finally eaten by one or two large fish. In other words, one large fish may depend on a billion tiny algae to meet its energy needs every day (fig. 32.7).

The interrelationships and interactions among the components of an ecosystem can be complex, but over the long term there is a balance between producers and consumers. An increase in food made available by producers can increase the number of consumers. This increased number of consumers reduces the available food, which then inevitably reduces the number of consumers. The result is sustained self-maintenance of the ecosystem.

Ecosystems exhibit considerable variation in **net productivity,** which is defined as the energy produced by photosynthesis minus that lost in respiration. Productivity (in terms of biomass produced) is usually measured as grams per square meter. Forest ecosystems, for example, produce 3–10 g of dry matter per square meter of land per day, whereas grasslands produce 0.5–3 g, and deserts less than 0.5 g. Crops grown throughout the year (e.g., sugarcane) produce up to 25 g.

Ecosystems are dynamic; they undergo daily changes, seasonal changes, and changes that may take many years. Photosynthesis occurs only during daylight hours, whereas many animals in ecosystems are active during either the day or night but not both. Even in tropical rain forests, some plants lose their leaves for part of the year, and seasonal periods of dormancy and rejuvenation in woody plants are common outside of the tropics.

In any ecosystem, producers, consumers, and decomposers form food chains, with producers at the base. The interrelationships often result in food webs, with a balance between producers and consumers. Energy cannot be recycled and is lost at each level. Huge numbers of producers may be required to meet the needs of a single final consumer at the end of a long chain.

Diversity

Each ecosystem includes a diversity of organisms that are distributed in specific patterns determined by the physical environment and by their relationships to each other. Organisms have diverse growth forms. For example, some herbaceous members of the ecosystem produce rhizomes, tubers, or bulbs, whereas others grow as epiphytes, vines, or parasites. Woody components may be deciduous or evergreen, thin-leaved or thick-leaved, and occur as tall trees, low shrubs, or in many intermediate forms. Ecosystems with similar climates in different parts of the world usually have similar organisms. Ecosystems that include chaparral scrub communities, for example, occur in southern California and in central Chile (see figs. 33.22 and 33.23).

Processes and activities such as expansion of buds, flowering, fruiting, and abscission of leaves are coordinated within an ecosystem. In the Amazon Basin, for example, the fruits of certain trees fall just as the trees' bases become seasonally flooded. Fish that swim under the trees during the flooding often catch the fruits in their mouths as they fall, and the seeds, which pass unharmed through their digestive tracts, are deposited elsewhere.

Species diversity depends on the number of species and the number of individuals per species in an ecosystem. Harsh climates with poor soils, such as those in arctic regions and deserts, limit species diversity, whereas mild or tropical climates with fertile soils permit extensive diversity. Indeed, about half of all living species occur in tropical rain forests, which occur on less than 7% of the earth's landmasses. In rain forests, there is intense competition for light, minerals, and other resources, but each species is well enough adapted to its habitat to allow for great diversity.

Measuring species diversity is no easy task. Thirteen years ago Stephen Hubbell, now at Princeton University, and Robin Foster at Chicago's Field Museum of Natural History Survey counted and marked all the trees in a 0.5 km² area of Panama: 238,000 of them! The plot and data about the growth and death of these trees are now serving as a valuable model for studies of diversity and succession.

How Much Does Diversity Matter? Humans and other organisms depend on ecosystems for a variety of important services (e.g., preservation of soil fertility, control of pest outbreaks). But how can we preserve a working ecosystem? Would the loss of biodiversity—currently a hot topic among environmentalists—change an ecosystem's ability to do those services? The answers to these and many related questions are being sought by researchers in environmental engineering, environmental science, land management, and ecology.

At the heart of the problem is how ecosystem ecologists and population biologists approach their work. Most ecosystem ecologists study how energy flows and materials cycle through an ecosystem and give little thought to the individual organisms involved. Conversely, population biologists study individual species and the food webs they form and generally ignore these organisms' roles in the larger flow of energy and materials. Consequently, we know little about how biodiversity affects the functioning of an ecosystem. Although increased diversity typically increases the production of plant biomass (to a point), there's little evidence about the importance of species diversity in nutrient cycling and decomposition. Ecologists know that we can sustain some loss of species diversity without a loss in the productivity of an ecosystem. This leads to an incredibly tough task: formulating policies for deciding which species really are crucial and which ecosystem functions they're needed for.

Shaping the debate about this issue are two contradictory theoretical predictions about the importance of species diversity. One, presented in 1981 by Stanford University ecologists Paul and Anne Ehrlich, is known as the "rivet popper" hypothesis. According to this hypothesis, the diversity of life is something like the rivets on an airplane; each species plays a small but important role in the working of the whole. The loss of each rivet weakens the plane by a small but noticeable amount; the loss of enough rivets causes the plane to crash.

The opposing idea, called the redundancy hypothesis, was presented in the early 1990s by Brian Walker, an Australian. The hypothesis asserts that most species are superfluous—that is, they are more like passengers than rivets—and that only a few key species are needed to keep the system in motion.

Which is more accurate? Several studies now support the "rivet" hypothesis. One experiment in 1993, in controlled-environment chambers, tested the effect of diversity on productivity by setting up fourteen artificial ecosystems. The result: As species numbers went up, the mean production of plant biomass also increased. A possible explanation is that more species usually generate a more diverse plant architecture, which allows the system to capture more light and produce more plant biomass. This conclusion is supported by studies of multicrop agriculture: the best way to increase the productivity in a cornfield is not to pack in more corn plants but to add other plants such as melons and nitrogen-fixing beans.

Species diversity also increases the ecosystem's resilience: diverse ecosystems recover from stress (e.g., drought) much faster than do ecosystems with less diversity. This is supported by the work of David Tilman and his colleagues at the University of Minnesota, who studied how as many as 250 kinds of plants could thrive in midwestern grasslands. The researchers created 207 4 m² plots distributed among one native prairie and three abandoned fields of different ages. Each season, they clipped a different section in each plot and analyzed its species diversity and biomass. They left some plots alone and added nitrogen fertilizer to others. Nitrogen fertilizer tends to reduce species numbers even as it boosts productivity. During a drought from 1987 to 1988, the productivity in all of the plots fell drastically. However, the drop in species-rich plots was only a fourth of that in species-poor plots. Moreover, the species-rich plots recovered in only one season, rather than in the four seasons required for species-poor plots to recover. Tilman concluded that "Biodiversity is a way to hedge bets against uncertainty, even in managed systems."

Interestingly, the enhancing effects of biodiversity on ecosystem resilience and productivity become saturated after a certain level of diversity. For example, one study showed that the largest gains in stability came with the first ten species; beyond that, adding more species didn't add much stability, perhaps because the essential functional niches had already been filled. Similarly, some evidence suggests that the growth and photosynthetic rates in tropical forests top out with only ten or so tree species. Thus, when it comes to productivity and resilience, some species in ecosystems may be more like passengers than rivets.

Although biodiversity is valuable to a certain point, most ecosystems contain more diversity than is needed to reach peak productivity. This conclusion holds, even when considering large ecosystems. For example, although the temperate forests of the Northern Hemisphere are vastly different in terms of biodiversity (the forests of Europe include 106 tree and shrub species, those of North America 158, and those of East Asia 876), their productivity is virtually identical.

Apparently, the random loss of species does not impair the productivity of an ecosystem, for such extinctions leave behind a few species in each structural category—for example, vines, canopy trees, and understory ferns in a tropical forest. So should we be concerned about the loss of biodiversity and extinctions that human activities are now causing?

Yes, because these extinctions are not random: Logging, burning, and grazing always affect a specific subset of species. Thus, these activities probably have a much greater impact on an ecosystem than do random extinctions.

Habitats

Specific sets of environmental conditions in which organisms live are referred to as **habitats.** Habitats contain organisms adapted to combinations of environmental conditions; the organisms live, reproduce, and die in the specific environments of their habitats. As indicated earlier, living factors such as soil microbes and fungi, plants, and animals are said to be *biotic,* whereas some of a habitat's nonliving physical components such as light, oxygen, elevation, latitude, climate, fire, and avalanches are *abiotic.* Soil components and moisture, shade, associated organisms, and other features of a habitat that affect a plant directly make up its operational habitat; other factors of the habitat may not affect a plant at all. For example, certain species of plants may occur both in an area with a stream running through it and in another area without a stream; the stream apparently has no direct impact on the distribution of specific plants. The stream is, however, a part of the *operational habitat* of aquatic plants within it.

Roles of Biotic Factors in the Habitat As plants respire and photosynthesize, they produce by-products that modify the habitat (e.g., CO_2 that combines with water to form carbonic acid). These plants also absorb minerals, create shade, promote visits by pollinators, produce substances that limit competition by other plants, or promote interaction with other plants in a mutually beneficial way. Plants may also provide shelter and food for other organisms and in other subtle ways create a niche to which they are better adapted than are other plants. An ecological **niche** is the functional role played by a species in its environment, including its activities, location, and relationships to other organisms.

Many organisms occupy ranges with a diversity of features such as topography, available moisture, soil types (referred to as *edaphic factors*), and sunlight. In a given part of the range, a species may compete for water, while in another part, edaphic factors or light may be more critical to competition. If gene flow throughout the range of the species is not widespread and rapid, localized populations may become genetically adapted to local conditions, resulting in ecotypes. Because ecotypes may be morphologically identical to plants in other parts of the range, an experiment would be necessary to determine if natural selection had changed their genetic composition. A simple experiment to answer this question might involve transplanting the suspected ecotypes to a common growing area within the range. Genetic change is evident if the transplanted individuals die or exhibit a different morphology than plants of the same species occurring at the site. If the transplanted individuals show no differences from the local plants, genetic factors are probably not involved.

The extent to which competition is believed to determine the species components of a habitat is controversial. Some biologists believe that the best-adapted species exclude less well-adapted species, whereas others believe that **sympatric** species (i.e., those occupying the same range) overlap in their tolerances of each other and that each uses areas of the range not occupied by the other. Where two species overlap, the better-adapted species dominates the overlap area, while the species that is not as well adapted occurs in only part of its potential niche. If either species is removed, the other species may take over the entire range.

Pollinators and other organisms are important factors in a plant's habitat. Many associations between plants and other organisms are **mutualistic** (mutually beneficial). For example, several species of bleeding hearts (*Dicentra* sp.) have oil-bearing appendages on their seeds. Ants carry the seeds to their nests, strip off the appendages for food, and then deposit the seeds outside the nest, where they germinate. Thus, ants help distribute the plants (fig. 32.8).

Roles of Abiotic Factors in the Habitat Habitat features that are not a function of living organisms are abiotic, and soil mineral content and texture are important abiotic factors. The first plants to become established on new soils are called *pioneers*. Pioneers must often adapt to rapid runoff or leaching of water, lack of protection from wind, high solar radiation, and

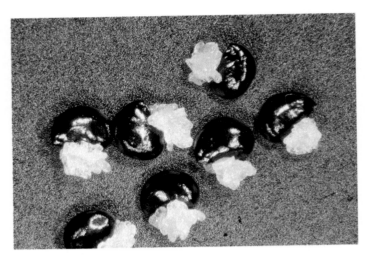

figure 32.8

Seeds of the Pacific bleeding heart (*Dicentra formosa*) have a mutualistic relationship with ants. The glistening white oil-bearing appendages are stripped from the seeds and eaten by ants.

other conditions that many other plants cannot tolerate. The pioneers gradually change the soil's composition by adding organic matter and breaking up bedrock as they penetrate cracks. As already mentioned, carbon dioxide released by respiring roots may combine with water to form carbonic acid, which further breaks down rock particles. Over time, this process, called *succession*, changes a soil's profile. You'll learn more about succession later in this chapter.

Climate plays a vital role in the distribution of nearly all organisms. Tropical plants cannot tolerate freezing or lengthy periods of drought or low humidity. Many temperate-region plants that become dormant in the fall shed their leaves. They may even require periods of freezing to break their bud dormancy. The amount of precipitation and the average temperatures of a region are not, however, as important as their timing and nature. For example, the precipitation in a Mediterranean climate occurs primarily in the winter and may be similar in amount to that of temperate regions that receive precipitation throughout the year. Some plants adapted to Mediterranean climates may not tolerate sustained summer rainfall, whereas plants adapted to precipitation throughout the year can seldom adapt to the long, dry summers of Mediterranean regions.

The total solar radiation available each year for photosynthesis varies considerably with latitude. Cloud cover may reduce the amount of radiation from the sun at the equator, but the sun is directly overhead there for longer each day than anywhere else; sunlight can reach plants at the equator for 12 hours every day of the year. Between the tropics and the Arctic and Antarctic Circles, the differences in daylength between summer and winter can vary by as much as 10 hours, and the plants adapted to these regions usually have photoperiods that determine flowering times. At high latitudes, sunlight is available continuously in midsummer but absent in midwinter. In addition, the sun is overhead only briefly during the year, thereby reducing the amount of radiation reaching the plants. Because of light limitations, cold temperatures, poor soil, and lack of available soil water, the growing season may be less than 3 months long in arctic regions and may also be hampered by high winds and waterlogged soil, which hinders gas exchange and osmoregulation. Similar conditions often prevail at high elevations, regardless of latitude. Landslides, fires, floods, tornadoes, and other disturbances can disrupt an ecosystem by removing large numbers of established organisms.

figure 32.9

Boston ivy (*Parthenocissus tricuspidata*) growing on a wall. Note how each leaf is oriented so that it has maximum exposure to light.

c o n c e p t

An environment in which an organism lives includes biotic factors such as soil microbes and fungi, and other plants and animals. Abiotic (i.e., nonliving) factors include light, oxygen, climate, and physical disturbances.

Interactions between Plants, Herbivores, and Other Organisms

All the plants in a particular habitat share and often compete for nutrients, light, water, and other resources needed for plant growth. Of course, some plants are more efficient and therefore more competitive in their use of a set of resources. The result of this competition may be the elimination of one species by another, a drastic reduction in the number of one species versus its competitors, or the coexistence of competing species if they occupy different *microhabitats* and/or the environment periodically changes. Under such pressures, plants slowly accumulate adaptations to their environment that increase their competitiveness and efficiency of growth.

Competition for light among plants has resulted in the evolution of several mechanisms that allow species to adapt to different light intensities. Depending on the light that is available, different forms of photosynthesis may be involved, leaf orientation may change throughout the day, plants may grow taller, and the thickness of the leaf blades and the number of chloroplasts in the mesophyll may change. The shapes, sizes, arrangements, and orientations of leaves ensure maximum exposure to the light (fig. 32.9). The relationship between light, photosynthesis, and biomass is obvious. Although the total mass of consumers is determined largely by the total mass of food made by the producers, the interactions among producers themselves and among the decomposers and the other members of the ecosystem are usually more subtle. In an ecosystem, the defenses that producers and consumers have against each other have developed through a process of coevolution resulting from natural selection and are maintained in a delicate balance. Flowering plants have evolved a variety of substances that either inhibit or promote the growth of other flowering plants or function as defenses against herbivores. For example, black walnut trees make a substance (juglone) that wilts tomatoes and potatoes and injures apple trees that contact

the black walnut roots. The production of natural antibiotics that kill or inhibit the growth of fungi or bacteria by other plants makes them resistant to various diseases.

Plant-herbivore interactions are widespread. For example, large herbivores such as deer and moose eat a variety of plants, each having a different nutritional value. Each plant species also produces different combinations, types, and amounts of chemicals in addition to proteins, fats, and carbohydrates. Many of these compounds are toxic to consumers, but the animals are not affected because their digestive systems break down and eliminate the toxic compounds (fig. 32.10). The limitations imposed by such compounds cause the consumers to vary their diet, seek familiar foods, and be wary of new ones. If a plant species lacked some natural defense, such as chemical compounds or structural modifications like spines, primary consumers might soon overexploit that species and threaten it with extinction.

Many plants secrete substances that either offend insect consumers or function as insecticides. For example, the volatile oils produced by members of the mint family (Lamiaceae) deter insect larvae and several higher animals from eating the leaves. Some members of the sunflower family (Asteraceae; e.g., chrysanthemums) produce the biodegradable insecticide *pyrethrum*, which is sold commercially. Some plant pests, however, have evolved enzymes that break down these chemicals, enabling the pests to thrive on plants that are toxic to other consumers. Members of the carrot family (Apiaceae), which includes carrots, parsley, and celery, produce aromatic substances that serve as insect repellents. However, larvae of the anise swallowtail butterfly feed exclusively on members of that family; the larvae have specific enzymes that degrade the repellent chemicals (fig. 32.11).

Various inhibitors produced by some bacteria and fungi limit the growth of higher plants, and parasitism by other bacteria, fungi, and flowering plants limits population size. The degree of parasitism varies considerably with the

figure 32.10

A koala eating *Eucalyptus* leaves. The koala's digestive system breaks down substances in the leaves that are toxic to other animals.

organisms involved. About 15% of the 3,000 members of the figwort family, Scrophulariaceae (which includes snapdragons [*Antirrhinum*] and *Penstemon*), exhibit various degrees of parasitism. Species of *Harveya* and *Hyobanche* lack chlorophyll and are entirely dependent on their flowering-plant hosts for their energy and other nutritional needs. *Gerardia aphylla* has pale leaves with about 10% of normal chlorophyll content; additional energy and nutrients come from its host plants. Parasitic species of *Pedicularis* and *Odontites rubra* have green leaves that furnish about half of their energy needs (fig. 32.12). Still other figworts (e.g., *Castilleja* sp.) may parasitize the roots of certain plants but can also live independently. For a more thorough discussion of parasitic plants, see Chapter 20.

figure 32.11

Anise swallowtail butterfly, whose larvae feed exclusively on members of the parsley family (Apiaceae). The larvae have specific enzymes that break down chemicals in the leaves that repel other insects.

figure 32.12

A parasitic Indian warrior (*Pedicularis densiflora*) plant. Although the leaves are green, chlorophyll is insufficient for all the plant's energy requirements. The balance of nutrients needed is obtained through parasitism of the roots of trees and shrubs.

c o n c e p t

Producers and consumers defend themselves in ecosystems by such means as producing substances that inhibit competitive growth or making tissues unpalatable or toxic to consumers. Some consumers produce enzymes that enable them to break down toxic substances, giving them an advantage over competitors. Over time, a delicate balance of defenses has coevolved through natural selection.

The Lore of Plants

Cobra plants (*Darlingtonia californica;* see fig. 20.20) occur in a few swampy places in the mountains of Oregon and northern California. These interesting plants, whose principal leaves form insect-trapping pitchers, have apparently survived because the swamps to which they are adapted have magnesium salts in concentrations high enough to kill most of their competitors.

A pitcher leaf, which may be nearly 1 m long, resembles a cobra head with its hood inflated; it even has "fangs" that insects follow right into the mouth of the trap. Once the insect is in the mouth, it encounters stiff, downward-pointing hairs that facilitate its descent and hinder its escape. Escape is made even more difficult by numerous small patches of transparent tissue on the back of the hood. The false windows mislead the victim in its efforts to find an escape route. Eventually the insect drowns in fluid at the base of the pitcher, where the insect's soft parts are digested by bacteria.

Nutrient Cycles

As you learned in Chapters 5 and 7, nutrients cycle in ecosystems. The cycling of nitrogen, carbon, and water is especially important to plants.

The Nitrogen Cycle

Much of the mass of living cells is protein, and the most abundant element in our atmosphere, nitrogen, constitutes about 18% of this protein. Therefore, the availability of nitrogen is critical to growing plants. Ironically, there are nearly 69,000 metric tons of nitrogen in the air over each hectare of land, but the total amount of nitrogen in the soil seldom exceeds 3.9 metric tons per hectare and is usually much less. This discrepancy exists because the nitrogen of the atmosphere is chemically inert; that is, it does not combine readily with other molecules. Thus, atmospheric nitrogen is largely unavailable to plants and animals.

Most of the nitrogen in plants and, indirectly, in animals comes from the soil in the form of inorganic ions absorbed by the roots. These ions are released by bacteria and fungi that break down the complex molecules in dead plant and animal tissues. Some nitrogen from the air is also *fixed*—that is, converted to ammonia or other nitrogenous substances by *nitrogen-fixing bacteria.* Some of these organisms live symbiotically in plants, particularly in members of the legume family (Fabaceae; e.g., peas, beans, clover, alfalfa), while others live free in the soil (see Chapter 25).

As shown in figure 32.13, nitrogen flows from dead plant and animal tissues into the soil and from the soil back into the plants. Bacteria and fungi can break down huge amounts of dead leaves and other tissues to tiny fractions of their original volumes within a few days to a few months. If their activities were to stop abruptly, the available nitrogen compounds would be exhausted within a few decades, and the supply of carbon dioxide needed for photosynthesis would be seriously depleted as the respiration of decomposers ceased. Forests, jungles, and prairies would die as accumulations of shed leaves, bodies, and debris would bury the living plants and keep the light from reaching their leaves.

Much nitrogen is leached from soil by water. More is removed with each harvest (the average crop removes about 25 kg per hectare per year). This nitrogen can be recycled and used as fertilizer if vegetable and animal wastes are returned to the soil. While bacteria decompose tissues, they incorporate nitrogen, and little is available until they die and release their accumulations into the soil. Accordingly, crops should not be planted in soils to which only partially decomposed materials have been added until bacteria have broken down the organic matter.

Weeds and stubble are frequently controlled by burning which depletes soil nitrogen (fig. 32.14). The annual combined loss of nitrogen from the soil in the United States from fire, harvesting, and other causes is estimated to exceed 21 million metric tons, and only 15.5 million metric tons are replaced by natural means. To offset the net loss, some 32 million metric tons of inorganic fertilizers are applied each year (see reading 20.1, "Putting Things Back"). If organic matter is not added at the same time, the application of inorganic fertilizers combined with the annual burning of stubble may eventually create a *hardpan* soil. Hardpan develops through the gradual accumulation of salt residues, which dissolve humus and disrupt the structure of the soil. Clay particles then clump and produce colloids that are impervious to moisture. In hardpan soils and others low in oxygen (e.g., soils in flooded areas), *denitrifying bacteria* use nitrates instead of oxygen in their respiration, thus rapidly depleting the nitrogen remaining in the soil.

Precipitation returns some nitrogen to the soil from the atmosphere, where it accumulates as a result of the action of light on industrial pollutants, fixation by flashes of lightning, and the diffusion of ammonia released by decay. The activities of nitrogen-fixing bacteria and volcanoes also replenish the nitrogen supply by converting nitrogen to forms that can be used by plants.

c o n c e p t

Microorganisms and fungi are involved in the cycling of nitrogen from dead tissues into the soil and back into plants. Precipitation returns a little nitrogen to the soil from the atmosphere.

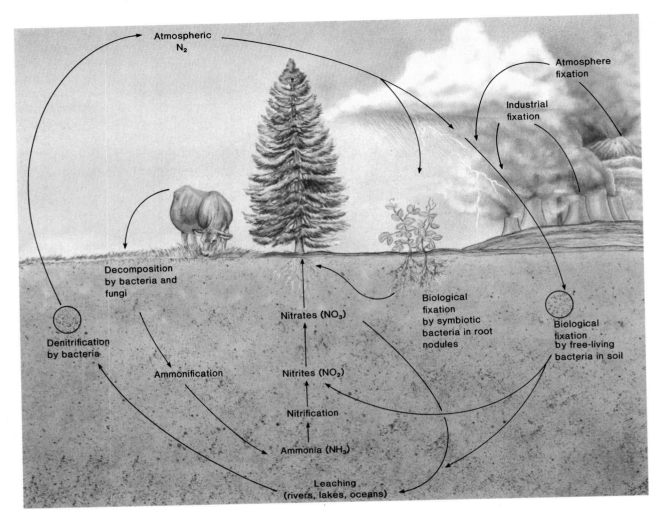

figure 32.13

The nitrogen cycle. Most nitrogen occurs as atmospheric nitrogen (N_2), which is not usable by plants and animals. Nitrogen fixation and decomposition make nitrogen available for plant metabolism as nitrates. Nitrogen from these sources is incorporated into organic compounds in plants and other organisms in the food web.

The Carbon Cycle

Bacteria and fungi also recycle carbon and other substances. As noted in Chapter 7, one of the two raw materials of photosynthesis is carbon dioxide, which constitutes about 0.035% of our atmosphere. It is estimated that all the plants of the oceans and the landmasses use about 14.5 billion metric tons of carbon obtained from carbon dioxide every year. This is replaced by the respiration of all living organisms, with perhaps as much as 90% or more being released by bacteria and fungi as they decompose tissues. Lesser amounts of carbon dioxide are released by the burning of fossil fuels, and a small amount originates from fires and volcanic activity.

All of the carbon dioxide in our atmosphere would be consumed in about 22 years if it were not constantly replenished (fig. 32.15). Oceans act as a buffer by absorbing and storing excess carbon dioxide as carbonates, but their capacity is limited. Some biologists believe that if this storage capacity is

exceeded, the carbon dioxide in the atmosphere will rise dramatically (see reading 32.2, "Curbing Methane Emissions to Curtail the Greenhouse Effect").

c o n c e p t

Decomposers, through respiration, constantly produce most of the carbon dioxide needed by green organisms for photosynthesis. All other living organisms, as well as abiotic processes, are the sources of the remaining carbon dioxide produced. Excess carbon dioxide is stored in oceans.

The Water Cycle

More than two-thirds of the earth's surface is covered by water, which is accumulated in oceans, freshwater lakes, ponds, glaciers, and polar ice caps. Evaporation occurs as

figure 32.14

Burning rice stubble in the San Joaquin Valley of California. Such burns reduce the amount of nitrogen in the soil by altering the bacterial community.

surface waters are warmed by the sun. The water vapor rises into the atmosphere, where it cools, condenses, and returns as precipitation. Air currents move the moisture-laden air around the globe.

When precipitation occurs over land, some of the water returns to the oceans, lakes, and streams in the form of runoff, and some percolates through the soil to underground water tables. When plants absorb water from the soil, all except about 1% of it is transpired through the leaves and stems. Any water retained by the plants is stored in cells, where it is used in metabolism and the maintenance of turgor. The transpired water vapor combines with other evaporated water, and the cycle is repeated (fig. 32.16).

For millions of years, carbon, nitrogen, water, phosphorus, and other molecules have been passing through cycles. Some molecules that were a part of a primeval forest that became compressed and turned to coal may have become part of another plant after the coal was burned. Then the new plant may have been eaten by an animal, which, in turn, contributed molecules to yet another organism. Molecules in our own bodies may have been a part of some prehistoric tree, a dinosaur, a woolly mammoth, or even all three.

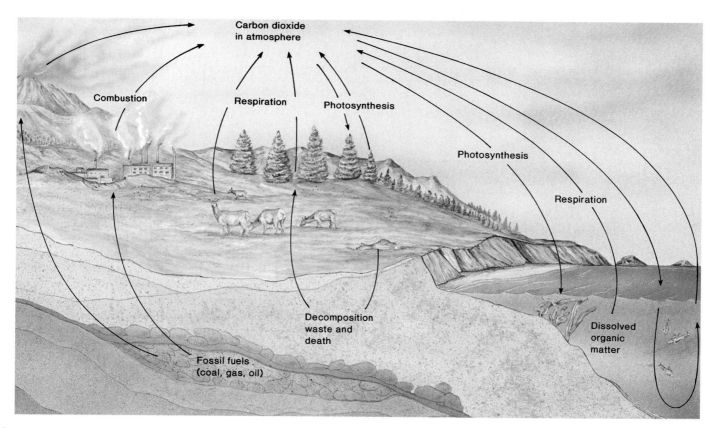

figure 32.15

The carbon cycle. Carbon dioxide is formed from respiration by plants and animals, decomposition of organic matter by bacteria and fungi, burning fossil fuels and wood by humans and natural fires, and volcanic activities. Carbon dioxide is used by plants and other photosynthetic organisms for photosynthesis.

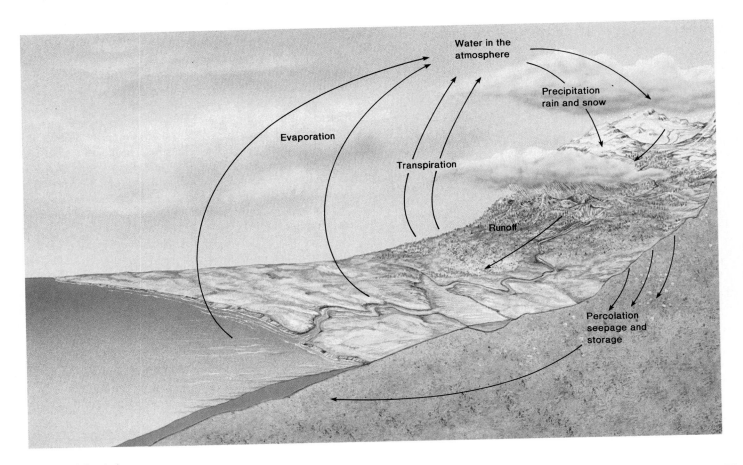

figure 32.16

The water cycle. Atmospheric water falls mostly as rain or snow and either evaporates, runs off, or percolates into the soil. Some of the percolated water reaches underground storage, and some of it is absorbed by plant roots. About 99% of all water taken in by roots is re-released into the atmosphere by transpiration.

Succession

After a volcano spews lava over a landscape or an earthquake or a landslide exposes rocks for the first time, there is initially no life on the lava or rock surfaces. Within a few months, or sometimes within a few years, organisms appear, and a sequence of events known as primary succession occurs. **Succession** is a cumulative, patterned, sometimes directional change in the biotic and abiotic components of a community over time. **Primary succession** refers to those changes occurring on substrate that has never been previously occupied. During primary succession the initial organisms, including pioneer plants, gradually alter their environment as they grow and reproduce. Over time, accumulated wastes, dead organic material, and inorganic debris promote soil formation. Other changes (e.g., changes in shade and water content) favor different species, which may replace the original ones. These new species, in turn, modify the environment further, so that still other species become established (fig. 32.17).

Primary Succession Initiated on Rocks or Lava

Primary succession frequently begins with bare rocks and lava that have been exposed by glacial or volcanic activity or by landslides. Initially, the rocks are sometimes subjected to alternate thawing and freezing, at least in temperate and colder areas. These fluctuations in temperature usually crack or flake the rocks, and lichens often become established on such surfaces. The lichens produce acids that slowly etch the rocks, and as the lichens die and contribute organic matter, they are replaced by other, larger lichens. Certain rock mosses adapted to long periods of desiccation also may become established, and a small amount of soil begins to accumulate, augmented by dust and debris blown in by the wind.

Eventually, enough of a mat of lichen and moss material is present to permit some ferns or even seed plants to become established, and the pace of soil buildup and rock breakdown accelerates (fig. 32.18). If deep cracks appear in the rocks, the larger seeds may widen them further as they germinate and their roots expand. Indeed, seedlings have been known to split rocks that weigh several tons (fig. 32.19).

figure 32.17

Plant succession near Mount St. Helens, Washington. Douglas fir seedlings and other vegetation have reappeared 12 years after complete destruction by volcanic ash.

figure 32.19

Expansion in girth of the roots of this yellow birch has split a large rock.

figure 32.18

Moss and crustose lichens living on rocks. Lichens and mosses are colonizers on rock surfaces in early successional stages.

As soil buildup continues, larger plants take over, and eventually the vegetation reaches an equilibrium in which the associations of plants and other organisms remain the same until another disturbance occurs. Such relatively stable plant associations are referred to as **climax communities.** Climax species are usually more shade-tolerant and larger than species in earlier successional stages; they also tend to grow and become distributed more slowly. For example, the climax vegetation of deciduous forests in eastern North America is dominated by maples and beeches, oaks and hickories, hemlocks and white pines, or other tree associations. In desert regions, various cacti form a conspicuous part of the climax vegetation, while in the Pacific Northwest, large conifers predominate. In parts of the Midwest, prairie grasses and other herbaceous plants form the climax vegetation, and in wet, tropical regions, a complex association of jungle plants constitutes the climax.

Succession, particularly in its more advanced stages, may not proceed in one direction. Sometimes it may appear to become reversed, depending on the influence of neighboring communities, and interpreting what one observes may be difficult. Occasionally when a volcano produces ash instead of lava, the ash buries the landscape and associated vegetation.

As a result, secondary succession may occur; some of the successional stages involving lichens and mosses may be bypassed, and larger plants may become the successional pioneers. This occurred after the series of ash eruptions of Mount St. Helens in the state of Washington during the early 1980s.

Primary Succession Initiated in Water

In the northern parts of midwestern states such as Michigan, Wisconsin, and Minnesota, ponds and lakes of various sizes abound. Many were left behind by retreating glaciers and are often not drained by streams. The water that evaporates from them is replaced annually by precipitation runoff. The ponds and lakes also shrink each year as a result of primary succession. This succession often begins with algae either carried in by the wind or transported on the muddy feet of waterfowl and wading birds. The algae multiply in shallow water near the margin, and with each reproductive cycle the dead parts sink to the bottom. Floating plants such as duckweeds may then appear, often encircling the water. Next, water lilies and other rooted aquatic plants with floating leaves become established. Each group of plants contributes to the organic material on the bottom, which slowly turns to muck. Cattails and other flowering plants that produce their inflorescences above the water often take root in the muck around the edges. As they become established, the accumulation of organic material accelerates.

Meanwhile, the algae, duckweeds, and other plants move farther out, and the surface area of exposed water gradually shrinks. Grasslike sedges become established along the damp margins and sometimes form floating mats as their roots interweave with one another. Dead organic material accumulates and fills in the area under the sedge mats, and herbaceous and shrubby plants then move in. As the margins become less marshy, coniferous trees whose roots can tolerate considerable moisture (e.g., tamaracks or eastern white cedars) gain a foothold, eventually growing across the entire area as the pond or lake disappears. Trees continue to help form soil, and finally the climax vegetation takes over. No visible trace of the pond or lake remains, and the only evidence of its having been there lies beneath the surface, where fossil pollen grains, bits of wood, and other materials reveal the area's history. Such succession may take hundreds or even thousands of years and has never been witnessed from start to finish. The evidence that it does occur, however, is overwhelming (fig. 32.20).

Under natural conditions, stream-fed lakes and ponds eventually fill with silt and debris, although this, too, may take thousands of years. The streams that feed these lakes gradually bring in silt, and the nutrient content of the water slowly rises as dissolved organic and inorganic materials are brought in. This gradual enrichment, called **eutrophication,** stimulates the growth of algae and other organisms, which also add their debris to the bottom of the lake. When sewage and other pollutants enter the lake, eutrophication accelerates. Eutrophication may also accelerate when trees are cleared from land around

figure 32.20

Succession in a pond. The pond basin slowly fills in as algae and rooted plants contribute organic matter.

lakes before the construction of homes and resorts. The cleared land erodes more readily, and precipitation runoff carries soil into the water. Regardless of size, all bodies of water, including rivers, are subjected to these processes.

Secondary Succession

Succession occurs whenever and wherever a natural area has been disturbed. It proceeds at varying rates, depending on the climate, the soils, and the organisms in the vicinity. When an area is disturbed by fire, floods, or landslides, some of the original soil, plant, and animal material may remain. Ecologists refer to the pattern of changes that follow such disturbances as **secondary succession.**

Secondary succession, which has fewer phases and usually arrives at climax vegetation faster than primary succession, may occur if soil is present and some species survive in the vicinity. Indeed, survivors strongly affect subsequent succession. Many secondary successions follow human disturbances, such as the conversion of land to farmland. Other secondary successions follow fires. Grasses and other herbaceous plants become established on burned (or logged) land. These are usually followed by trees and shrubs that have widely dispersed seeds (e.g., aspen and sumac in the Midwest and East, and chaparral plants such as chamise and gooseberries in the West) or that sprout from root crowns. After going through fewer stages than are typical

32–18

of primary succession, the climax vegetation takes over, often within less than 100 years (fig. 32.21). Some biologists contend that in fire-dependent systems such as chaparral, no true climax communities exist.

c o n c e p t

Succession is the progressive change in community composition that occurs as vegetation develops in an area. Soil forms as a result of primary succession, which is initiated on freshly exposed rocks or lava, or in bodies of water. Secondary succession takes place after disturbances occur in areas where soil already exists. The end of a successional series is the climax community.

Fire Ecology

Natural fires started primarily by lightning and the activities of indigenous peoples have occurred for thousands of years in North America and other continents. Trees such as giant redwood and ponderosa pine, although scarred by certain types of fire, often survive, and the dates of fires can be determined by the proximity of the scars to specific annual rings (see fig. 16.8). Growth-ring studies from western North America show that forests of ponderosa pine burned about every 6–7 years. The fires, climate, topography, and soil profoundly affected various biomes. Humans eventually tried to control the fires when they threatened life and property, and this, in turn, altered the vegetation. We have learned that trying to eliminate fires, at least in certain areas, disrupts natural habitats more in the long run than allowing them to occur. Agencies such as the U.S. National Park Service and the U.S. Forest Service now may allow some fires to burn under prescribed conditions.

figure 32.21

Quaking aspen (*Populus tremuloides*) trees often become established in mountainous areas after fires. As succession proceeds, climax vegetation later may replace the aspens.

Fires benefit grasslands, chaparral, and forests by converting accumulated dead organic material to mineral-rich ash, whose nutrients are recycled within the ecosystem (fig. 32.22). If the soil has been burned, some of its nutrients and organic matter may have been lost, and the composition of microorganisms originally present is likely to have changed. Losses are offset by the fact that nitrogen-fixing soil bacteria, including cyanobacteria increase after a fire, and fungi that cause plant diseases decrease.

figure 32.22

Forest fires convert dead organic matter to mineral-rich ash, whose nutrients are recycled within the ecosystem.

At least some of the North American grasslands originated and were maintained by fire. Since grassland fires have largely been controlled, many of these areas have now been invaded by shrubs. In some areas, such as the prairies of the Midwest, grasses are better adapted to fire than are woody plants. These grasses produce seeds within a year or two after germination. Perennial grass buds at the tips of rhizomes usually survive the most intense heat of fires, producing new growth the first season after a fire. Thus, a fire destroys only one season's growth of grass, often after reproduction has been completed.

Shrubs, however, have much of their living tissue aboveground, and a fire may destroy several years' growth. Also, woody plants, unlike grasses, often do not produce seeds until several years after germination. Many shrubs sprout from burned root crowns, particularly in chaparral areas, but repeated burning keeps them small. Most chaparral species, both woody and herbaceous, are so adapted to fire that their seeds will not germinate until fires remove accumulated litter and inhibitory substances produced by the plants during growth.

Fires also play a role in the composition of forests. In the mountains of east-central California, gooseberry and deerbrush are abundant after a fire, but their numbers stabilize within 15–30 years when larger trees return to the area. Ponderosa, jack, and southern longleaf pines, and Douglas firs (which do not tolerate shade) are among the species that repeatedly replace themselves after fires, and the seeds of some species rarely germinate until they have been exposed to fire. In view of the long-range beneficial effects of fires in some ecosystems, wise land-management practices of the future will include guiding succession in many plant communities for the greatest utility and safety.

Humans in the Ecosystem

The total human population of the world was less than 20 million in 6000 B.C. During the next 7,750 years it rose to 500 million; by 1850 it had doubled to 1 billion, and 70 years later it had doubled again to 2 billion. The 4.48 billion mark was reached in 1980, and within 5 years the population had grown to 4.89 billion. The estimate for the year 2000 exceeds 6.3 billion. The earth remains constant in size, but humans have occupied more of it over the past few centuries, and their population density has also greatly increased. In feeding, clothing, and housing themselves, humans have greatly affected their environment. We have cleared natural vegetation from vast areas of land and drained wetlands. We have polluted rivers, oceans, lakes, and the atmosphere. We have killed pests and plant-disease organisms with poisons, which have also killed natural predators and other useful organisms. Since 1955, nearly one-third of the world's cropland (an area larger than India and China combined) has been abandoned because its overuse led to the depletion, degradation, or loss of soil. Finding substitutes for this lost acreage accounts for 60–80% of the world's deforestation. In general, we have disrupted the delicately balanced ecosystems that existed before we began our depredations.

If we are to survive on this planet, we must control the size of our population. Also, many foolish agricultural and industrial practices that have accompanied population growth must be replaced with practices that will restore some ecological balance. Agricultural practices of the future must include returning organic material to the soil after each harvest, instead of adding only inorganic fertilizers. Timber and other crops must be harvested in ways that prevent erosion. The practice of clearing brush with chemicals must be changed or abolished. Industrial pollutants will have to be rendered harmless and recycled whenever possible. Energy conservation will need to become universally practiced, and wasteful packaging and consumption practices must be stopped or sharply curtailed.

Many substances that now are discarded (e.g., organic waste, paper products, glass, metal cans) must be recycled on a larger scale. Biological controls will have to replace the use of toxic chemical controls whenever possible. Water and energy must be conserved. Rare plant species, with their largely unknown genetic potential for medicine and agriculture, will need to be saved from extinction by the preservation of their habitats. The general public must understand the urgent need for wise land management and conservation to resist the influential forces that promote unwise measures in the name of progress—before additional large segments of our natural resources are irreparably damaged or forever lost. The alternative appears to be nothing less than death from starvation, respiratory diseases, poisoning of our food and drink, and other catastrophic events, endangering the survival of humanity.

CURBING METHANE EMISSIONS TO CURTAIL THE GREENHOUSE EFFECT

In an effort to limit global warming, the U.S. government proposed in early 1991 that a comprehensive framework for limiting greenhouse gases would be preferable to concentrating on a single gas. Hogan, Hoffman, and Thompson (*Nature* 354 [1991]:181–82) observed that methane is often ignored in debates about the greenhouse effect. They argued that reducing methane emissions would be relatively easy and could significantly reduce global warming. They point out that cutting methane emissions would be 20–60 times more effective in reducing the potential warming of the earth's atmosphere over the next century than decreasing carbon dioxide emissions. They also noted that methane released by human activities is generally a wasted resource, opening the possibility that reductions might even be profitable.

Much methane is generated in landfills by the anaerobic decomposition of wastes. By collecting this gas, existing recovery systems can reduce emissions by 30%–60%. Coal mining emits similar amounts of methane, although relatively few mines emit most of the gas that is released. If vertical wells are used before and during the mining operations, methane emissions from underground mine areas could be cut by more than 50%, and the cost of mine ventilation would also be lessened.

The production of oil and natural gas also produces much methane. Improved technologies such as leak-proof pipelines can reduce methane emissions cost-effectively. The second largest source of methane related to human activities involves the cattle industry; the cattle themselves produce large amounts of the gas. Some feeds that may reduce methane emissions while increasing produc-

tivity have been identified. One hormone (bovine somatotropine), if approved for use, could reduce methane emissions by 10% while increasing milk production. Recovery systems can profitably recover 50%–90% of the methane generated from animal wastes. Such recovery systems are already in use in India, where they generate enough electricity to power lights for a few hours each evening. However, animal wastes might be more appropriately used as soil fertilizer.

Rice cultivation is considered to be the single largest source of methane emitted by human activities. An integrated approach to irrigation, fertilizer application, and cultivar selection could reduce methane emissions by as much as 30%. Alternative approaches to agricultural practices involving the clearing of land and crop stubble by fire could also cut methane emissions, but trade-offs clearly would be involved.

The Greenhouse Effect

The **greenhouse effect** refers to the maintenance of an equable temperature over the planet and is linked to global warming. Global warming is a global rise in temperature due to the accumulation in the atmosphere of gases that permit radiation from the sun to reach the earth's surface but prevent the heat from escaping back into space. Although the extent and impact of the greenhouse effect is controversial, many biologists believe that it could significantly alter our environment in the future.

Gases such as chlorofluorocarbons are involved in producing the greenhouse effect. These gases are the relatively recent by-products of the manufacture of refrigerants, plastics, and aerosol propellants. Other gases that promote the greenhouse effect have been part of our atmosphere for millions of years—carbon dioxide and methane.

Carbon Dioxide

D. L. Lindstrom of the University of Illinois at Chicago and D. R. MacAyeal of the University of Chicago recently examined records of cores of ancient ice from Siberia, Scandinavia, and the Arctic Ocean. Using computers to simulate the status of ice and atmosphere going back 30,000 years, they found that levels of carbon dioxide had increased enough at the end of the most recent ice age to melt the ice and raise the earth's temperature. Their findings suggest that cycles of

ice ages followed by shorter warm periods may have been caused solely by rising and falling levels of carbon dioxide in the atmosphere.

In 1986 worldwide carbon dioxide emissions from transportation and industrial sources totaled somewhat less than 5 billion tons. By 1987 the total was more than 5.5 billion tons and has continued to rise. The burning of fossil fuels and deforestation (which destroys the major recyclers of carbon dioxide) have caused a 25% increase of carbon dioxide in the atmosphere since 1850. In the last 25 years alone, the increased insulation of carbon dioxide has resulted in the earth's atmosphere becoming 0.4° C warmer, and between 1983 and 1990 the surface temperature of the ocean rose about 0.8° C.

These increases in temperature may seem insignificant, but during the last ice age in North America, when ice covered the northern United States and Canada, the average temperature of the earth at sea level was only 4° C colder than it is now. In 1989, Mostafa Tolba, head of the United Nations Environment Program, estimated that if the current levels of gas release into the atmosphere continue, the earth's temperature will probably rise between 1.4 and 4.3° C in the next 50 years. These higher temperatures will melt the polar ice caps; the released water will raise sea levels and inundate low-lying, often densely populated, coastal areas. The U.S. Environmental Protection Agency estimates that for each 30 cm the ocean rises, it moves 30 m inland. During the past

figure 32.23

Termites in wood. Termites produce a significant amount of methane as they digest wood.

figure 32.24

Damage caused by acid rain to spruce and fir trees.

century, worldwide ocean levels have risen 12.7 cm (5 inches), and it is estimated that more than 1.8 million hectares of land in the United States alone will be flooded if the temperatures rise as predicted. Higher temperatures can also affect winds, currents, and weather patterns, causing droughts and creating deserts in some areas, while causing heavy rainfall in others. The greatest grain-production areas of the world lie in the interiors of continents. If they become warmer and drier, world food supplies may be affected.

Methane

Swamps and wetlands have long been known to be sources of methane produced by anaerobic bacteria. Many animals produce methane during digestion, and large amounts of this gas are released by wood-digesting organisms in the guts of termites (fig. 32.23). The total annual production of methane in the atmosphere has also increased slowly in recent years. A small part of this increase may stem from the increased numbers of termites in cleared tropical rain forest areas, which in 1990 were being destroyed at the rate of more than 50,000 hectares per day.[1] This means that we clear an area of tropical rain forest equivalent to that of more than 1,000 basketball courts *every second*, 24 hours per day.

Acid Rain

Acid rain occurs after the burning of fossil fuels releases sulphur and nitrogenous compounds into the atmosphere. There, sunlight converts these compounds to nitrogen and sulphur oxides, and they combine with water to become acid rain (mostly nitric acid and sulfuric acid). Acid rain changes the pH

of lakes and streams and kills many organisms in them. It also injures plants upon which it falls (fig. 32.24). About half of the Black Forest in Germany has succumbed to its effects. Acid rain has also stunted or killed trees growing downwind from industrial sites. Acid rain also affects nonliving materials. For example, the natural weathering of ancient Mayan ruins in southern Mexico, the Parthenon in Greece, and monuments in Washington, D.C., has been accelerated by acid rain during the past decades.

Acid rain is not responsible for all dead or dying trees in the world's forests. Some trees have perished as a result of insufficient rainfall during successive dry years. Others have succumbed to insect infestations or salt scattered to melt ice and snow on roads, and still others have been weakened by disease.

Ozone

Ozone (O_3) is a form of oxygen in the stratosphere that is more effective than ordinary oxygen (O_2) in shielding living organisms from intense ultraviolet radiation. Ozone is produced from oxygen with the aid of ultraviolet light. Chlorofluorocarbons (CFCs),

---

1. 1 hectare = 11,960 yd^2 = 10,000 m^2 = 2.47 acres

which are inert chemicals used for refrigeration and other industrial purposes, are broken down by sunlight at high altitudes into active compounds that destroy ozone. Destruction of ozone in the stratosphere results in increased exposure to ultraviolet radiation, which increases the incidence of skin cancers, genetic mutations, and damage to vegetation—especially crops.

The accelerating destruction of the ozone shield has been recognized as a serious global problem by both the United States and the European Economic Community. In 1987 the United States proposed a 50% reduction in the production and uses of chlorofluorocarbons by the year 2000, and in 1989 the European Economic Community proposed a total ban on chlorofluorocarbons, also by the year 2000. In 1990, however, developing nations such as India, China, and Brazil had plans to expand the production of chlorofluorocarbons and contended that a ban would place them at an economic disadvantage. Because global cooperation on this matter is urgently needed, the major industrial nations are seeking ways to allay the economic concerns of developing countries by, for example, producing viable alternatives to chlorofluorocarbons. One such alternative was introduced in 1994, when non-CFC refrigerants became standard in the air-conditioning systems of most new automobiles sold in the United States and elsewhere.

Chlorofluorocarbons are not the only danger to the protective ozone layer. Bromine-based compounds called halons, which are commonly found in electronic equipment and portable fire extinguishers, can be up to 10 times more destructive of ozone than are chlorofluorocarbons. Halon concentrations in the atmosphere increased about 20% per year between 1980 and 1986, according to the Environmental Protection Agency. Some scientists believe the concentrations are as much as 50% higher than that and are recommending that powders and other inert gases be substituted for halons in fire extinguishers.

Contamination of Water

Much of the pollution in our lakes and streams comes from the dumping of toxic industrial wastes and runoff from polluted land (fig. 32.25). Other sources include the spraying of pesticides, exhaust and other emissions from aircraft and ships, and airborne pollutants originating from the combustion of fossil fuels. Groundwater supplies become polluted from the infiltration of the soil by pesticides and wastes, septic tank effluents, and garden and farm fertilizers. Even deep wells, which seldom became polluted in the past, are becoming contaminated with substances that damage the health of humans and animals that depend on the water. In response to the problem, many communities are improving their water treatment plants, and many individuals have installed water filtration systems in their homes.

Genetic engineering and bacteria will probably play a major role in reducing water pollution. For example, a bacterium has been bred that can remove more than 99% of 2,4,5-T (a major component of Agent Orange) from a contaminated environment, and several other bacteria are being genetically engineered to break down other toxic wastes.

figure 32.25

Toxic waste pouring into a river.

c o n c e p t

The great increase in numbers of humans has been accompanied by serious disruption or destruction of ecosystems. To ensure human survival, currently widespread individual, industrial, and agricultural practices must be modified or changed. Emissions of gases or other substances that contribute to global warming or destroy stratospheric ozone must be curbed.

Chapter Summary

Ecology examines the relationships of organisms to one another and to their environment. Ecologists view these interrelationships as a hierarchy; at the lowest level are populations, which are groups of individuals of the same species. Populations of different species form a community, and communities and their physical environment form an ecosystem.

Populations vary in numbers, density, and the total mass of individuals. These variations depend on both biotic and abiotic factors.

Ecosystems typically sustain themselves entirely through photosynthesis and interactions among producers, consumers, and decomposers. In any ecosystem, energy flows through the different levels of producers, consumers, and decomposers, and the food chains and webs that they form.

Energy is not recycled in an ecosystem. The producer level of a food chain converts only about 1% of the available light-energy to organic material, and the energy gradually escapes in the form of heat as it passes from one level to another. Variations in food production in an ecosystem are balanced by corresponding variations in consumption, so that the ecosystem is self-maintaining. When two species are grown together in a controlled environment, one of the species often disappears. In nature, if two species occupy the same niche, then at least one is reduced in size and/or number; survival depends on species occupying different microhabitats.

Biotic components influence the composition of an ecosystem through parasitism and by secreting growth-inhibiting substances, growth-promoting substances, and consumer repellents. The chemical compounds or structural modifications coevolved by both producers and consumers are delicately balanced in an ecosystem.

Atmospheric nitrogen is fixed by bacteria found in legumes and other plants. Nitrogen flows from dead plant and animal tissues into the soil and from the soil back to the plants. Water leaches nitrogen from the soil and carries it away when erosion occurs. More nitrogen is lost from harvesting crops, but the loss can be reduced if wastes are decomposed and annually returned to the soil. Replacing lost nitrogen with chemical fertilizers can eventually create hardpan. Carbon, water, and other substances also undergo cycling.

Succession is a pattern of change in the species composition of a given area over a period of time and occurs whenever natural areas on land or in water have been disturbed. Primary succession involves the formation of soil, beginning with either a dry or wet environment. Secondary succession occurs in areas previously covered with vegetation. A stable vegetation (climax vegetation) becomes established at the conclusion of succession and remains until a disturbance disrupts the balance. Fires alter the nutrient and organic composition of the soil and open up areas in which trees such as aspens may become established until slower-growing climax species return.

Human populations have grown rapidly. The disruption of ecosystems by activities directly or indirectly associated with feeding, clothing, and housing billions of people threatens the survival not only of humans but other organisms as well. This survival will ultimately depend on improved agricultural practices, reduction of wasteful activities, conservation, and recycling.

The greenhouse effect refers to the maintenance of an equable temperature over the planet. A global rise in temperature due to the insulating effect of gases has been noted. Acid rain, which damages or kills organisms, occurs when sulphur and nitrogen compounds released by the burning of fossil fuels are converted to nitric and sulphur oxides by sunlight and then mix with water to become acids. In the stratosphere, sunlight converts chlorofluorocarbons into compounds that destroy the ozone shield that protects us from intense ultraviolet radiation. Halons, used in fire extinguishers, also destroy ozone.

Water contamination occurs when toxic wastes, pesticides, effluents from septic tanks, and fertilizers wash or leach into surface and groundwater. Bacteria may soon be used to break down various contaminants.

 ### What Are Botanists Doing?

Go to the library and read an article or research paper that directly or indirectly addresses this hypothesis: A small number of abundant species account for a large fraction of ecosystem function. What do you conclude from this article?

 ### Writing to Learn Botany

If you could introduce congressional legislation to restore disrupted ecosystems, what bills might you propose that could do this without increasing taxes?

Questions for Further Thought and Study

1. Distinguish among populations, communities, and ecosystems.

2. What is meant by the phrase "balance of nature"?

3. Discuss the nitrogen cycle.

4. What is the greenhouse effect? Why should we be concerned about it?

5. What concept or idea discussed in this chapter do you think is most important? Why?

6. Energy and nutrients move differently in ecosystems. What are the consequences of this difference?

7. Today it is fashionable for people to refer to themselves as environmentalists. What does this term mean to you?

8. In industrialized countries, humans have disrupted ecosystems everywhere they have settled. Could humans also improve an ecosystem? If so, how?

9. What vital services do ecosystems provide?

10. If a plant species is unnecessary for the functioning of an ecosystem, should we be concerned if it is eradicated? What might be some economic, moral, or aesthetic reasons for protecting an "unnecessary" plant?

11. How have human activities affected the distribution of plants and animals where you live?

12. Discuss the meaning of the following quotations:

What makes it so hard to organize the environment sensibly is that everything we touch is hooked up to everything else.

—*Isaac Asimov, writer and biochemist*

Eventually, we'll realize that if we destroy the ecosystem, we destroy ourselves.

—*Jonas Salk, medical researcher*

We prefer economic growth to clean air.

—*Charles Barden, environmentalist*

Destroying species is like tearing pages out of an unread book, written in a language humans hardly know how to read, about the place where they live.

—*Rolston Holmes III, philosopher*

13. Recent evidence confirms the biological value of diversity in agriculture, but suggests that diversity does not guarantee the survival of individual species. How would you test this?

14. An ecosystem's productivity and stability—long treated as independent of each other—are, in fact, inextricably linked. How would you test this hypothesis?

15. How might acid rain alter nutrient cycles?

16. How would you best protect biodiversity: by conserving species or by conserving ecosystems? Explain your answer.

Web Sites

Review the "Doing Botany Yourself" essay and assignments for Chapter 32 on the *Botany Home Page*. What experiments would you do to test the hypotheses? What data can you gather on the Web to help you refine your experiments? Here are some other sites that you may find interesting:

http://www.nwf.org/nwf/ed/people/

This site, sponsored by the National Wildlife Federation, discusses how humans affect the environment, as well as things you can do to protect and save the environment. You'll also find a glossary of ecological terms, information about pollution, and links to other sites (e.g., endangered species).

http://www.ncg.ncrs.usda.gov/Welcome.html

This site includes information about issues related to national conservation efforts. You'll be especially interested in the PLANTS national database and the Official Soil Series Descriptions section.

http://www.erl.noaa.gov/

This site, which is sponsored by the National Oceanic and Atmospheric Administration, includes information about technological developments in environmental science and an educational section devoted to improving our understanding of Earth. You can also access links to other ecology-related sites.

http://www.nwf.org/nwf/ed/habitat/

At this site you'll learn about the concept of habitats, as well as the importance of a unique habitat: wetlands.

http://www.gold.net/ecosystem/

At this site you'll find The Ecosystem, which includes updates of news, ecological information, and the latest issue of *The Ecologist*.

http://hawaii-shopping.com/~sammonet/medicalplants1.html

Medicinal Hawaiian plants are more diverse than you might guess. At this site you'll learn all about them.

http://nmnhwww.si.edu/departments/botany.html

The research programs at the Department of Botany, National Museum of Natural History, Smithsonian Institution have something to appeal to all new botanists.

Suggested Readings

Articles

Appenzeller, T. 1993. Filling a hole in the ozone argument. *Science* 262:990–991.

Baskin, Y. 1994. Ecologists dare to ask: How much does diversity matter? *Science* 264:202–203.

Bormann, F. H. 1985. Air pollution and forests: An ecosystem perspective. *BioScience* 35:434–441.

Bushbacher, R. J. 1986. Tropical deforestation and pasture development. *BioScience* 26:22–28.

Castello, J. D., D. J. Leopold, and P. J. Smallidge. 1995. Pathogens, patterns, and processes in forest ecosystems. *Bioscience* 45:16–24.

Finegan, B. 1984. Forest succession. *Nature* 312:109–114.

Holloway, M. 1993. Sustaining the Amazon. *Scientific American* 269 (July):90–99.

la Rivière, J. W. M. 1989. Threats to the world's water. *Scientific American* 261 (September):80–94.

Moffat, A. S. 1996. Biodiversity is a boon to ecosystems, not species. *Science* 271:1497.

Pennisi, E. 1994. Tallying the tropics: Seeing the forest through the trees. *Science News* 145:362–366.

Raloff, J. 1995. When nitrate reigns: Air pollution can damage forests more than trees reveal. *Science News* 147:90–91.

Swetnam, T. W. 1993. Fire history and climate change in giant sequoia groves. *Science* 262:885–889.

Books

Barbour, M. et al. 1987. *Terrestrial Plant Ecology*. 2d ed. Menlo Park, CA: Benjamin/Cummings.

Begon, M. 1990. *Ecology: Individuals, Populations and Communities*. 2d ed. Cambridge, MA: Blackwell Scientific.

Caldwell, M. M., and R. W. Pearcy, eds. 1994. *Exploitation of Environmental Heterogeneity by Plants. Ecophysiological Processes above and below Ground*. San Diego, CA: Academic Press.

Dafni, A. 1993. *Pollination Ecology*. Oxford: Oxford University Press.

Glenn-Lewin, D. C. et al., eds. 1992. *Plant Succession: Theory and Prediction*. New York, NY: Chapman and Hall.

Lovett-Doust, J., and L. Lovett-Doust. 1988. *Plant Reproductive Ecology*. Oxford: Oxford University Press.

Rice, E. L. 1984. *Allelopathy*. 2d ed. San Diego, CA: Academic Press.

Silvertown, J., and J. L. Doust. 1993. *Introduction to Plant Population Ecology*. 3d ed. Cambridge, MA: Blackwell Scientific.

Trona pinnacles and desert plants on a dry lake bed in south central California.

c h a p t e r

Biomes

33

Chapter Outline

Chapter Overview

Ecosystems in different areas of the world are so similar that ecologists classify them into groups called *biomes.* The same biome may therefore span several continents. Although the distribution of biomes is determined mostly by climate, each biome is characterized primarily by its vegetation. Classification of the major vegetation types of the world entails the recognition of anywhere from seven to seventeen biomes, depending on the classifier. In this text, the major ecosystems of the world are grouped into eight terrestrial biomes: tundra, taiga, temperate deciduous forests, grasslands and savannas, deserts and semideserts, mediterranean scrub, mountain forests, and tropical rain forests. This classification includes the major types of vegetation that occur on land, but it excludes aquatic and marine ecosystems.

Why study biomes? Why spend the time and energy to classify biomes, map their boundaries, and list their dominant species? The answer is that biomes are global examples of the association between the physical environment and biological characteristics. This association between moisture, sunlight, and temperature and the adaptations that the extensive vegetation types have acquired over many generations is inseparable. To study biomes is to take a macroview (above the community level of organization) of the results of natural selection within major climates. Each vegetation type has its own **physiognomy** or architecture of characteristics that makes the vegetation as a whole compatible with the surrounding climatic and geological conditions.

Most biomes cover large areas and occur on more than one continent. A biome usually includes several ecosystems, each with its unique combination of plants and animals, that share similar climates and soils. Most of the important biomes in North America also occur on other continents (fig. 33.1).

If you have a keen eye, a little knowledge about plants, and a good imagination, you may already know how the types of vegetation in different parts of the world are similar. For example, desert vegetation in the southwestern United States looks like desert vegetation in northern Africa,[1] and the grasslands of western China look like the grasslands of the central United States. Ecologists recognize such similar types of vegetation as convenient indicators of comparable ecosystems.

A group of similar ecosystems is called a **biome.** Succulent plants such as cacti and low-growing trees and shrubs with spines and small leaves are examples of typical desert vegetation; an area where these plants grow is called a *desert biome.* Grasslands of central North America, western and central Asia, southeastern South America, and northeastern Australia are, collectively, a grassland biome.

Viewing the earth as a collection of biomes is somewhat artificial, because biomes overlap and ecosystems usually include local areas with plants that do not fit the vegetation types of the biome according to our definition. Furthermore, biomes usually exclude aquatic and marine ecosystems, because we do not know enough about them to organize them into groups. Nevertheless, biomes remain a convenient way to organize the terrestrial ecosystems of the world for discussion. The characteristics of individual biomes are influenced mostly by abiotic factors such as precipitation, temperature, and soil type. Furthermore, seasons and climate are influenced by the amount of solar energy that reaches a given area, which changes throughout the year with the angle of the sun's rays.

The wide bands of air currents that occur over the earth (fig. 33.2) and the location of mountains affect vegetation patterns. At the equator, warm air that is more or less saturated with water vapor rises and condenses at higher altitudes, resulting in rain. Moisture-laden air may also be carried by winds toward mountains, where much of the moisture falls on the windward side as the air is forced upward. The leeward side of the mountains usually lies in the "rain shadow," where little precipitation occurs. For example, on the west coast of Hawaii, the largest of the Hawaiian Islands, there are areas that receive less than 22 cm of rain annually, while precipitation only 10 km away on the windward side of the mountains exceeds 250 cm per year.

The organization of abiotic factors and the types of vegetation associated with them have led ecologists to recognize anywhere from seven to seventeen biomes. In this text, we have chosen to be conservative. We emphasize the similarities among eight groups of ecosystems; that is, we group the earth's vegetation types into eight biomes. We exclude aquatic and marine ecosystems because there is no adequate classification of underwater ecosystems on a worldwide scale. Biologists have no idea whether the lakes and oceans of the world can be organized into one biome or a hundred biomes.

---

1. Botanists get extra enjoyment out of movies that are set in certain areas of the world but filmed where the vegetation belies the true location. For example, there is a scene in *The Young Lions* where a German army officer, played by Marlon Brando, peeks over a hill at an enemy encampment in North Africa during World War II. You don't have to look very closely to see the branches of an ocotillo (*Fouquieria splendens*) nearby; this species grows in a southern California desert, where that scene was filmed, but not in the Sahara Desert of Africa, where that scene was supposed to occur.

Tundra—Arctic and Alpine

The word **tundra** is derived from a Russian word meaning "treeless, marshy plain." Tundra is a vast, mostly flat biome whose terrain is marshy in the summer and frozen for much of the remainder of the year (figs. 33.3 and 33.4). It occupies about

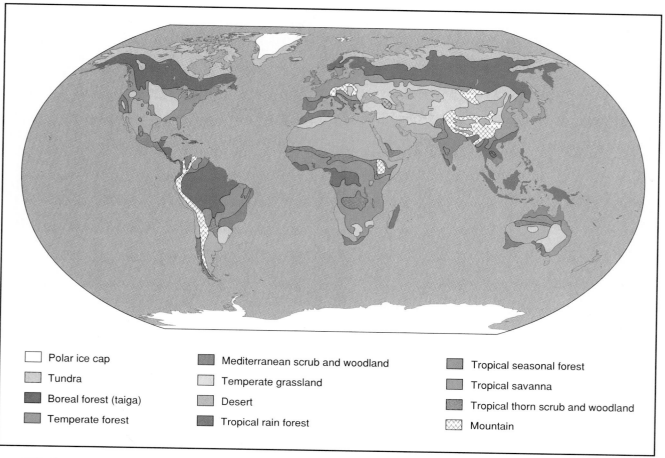

figure 33.1

Major biomes of the world. These biomes are widely distributed, and their distribution is primarily a product of climate and soil type.

Legend:
- Polar ice cap
- Tundra
- Boreal forest (taiga)
- Temperate forest
- Mediterranean scrub and woodland
- Temperate grassland
- Desert
- Tropical rain forest
- Tropical seasonal forest
- Tropical savanna
- Tropical thorn scrub and woodland
- Mountain

25% of the earth's land surfaces, primarily above the Arctic Circle, with some, however, extending farther south. Alpine tundra, which occurs in patches above timberline on mountains below the Arctic Circle, is seldom flat and also differs from arctic tundra in having less annual variation in daylength, less humidity, and more direct solar radiation. The climate and soil of tundra are usually not suitable for agricultural activities.

Fierce, drying winds and freezing temperatures can occur in tundra on any day of the year, but temperatures can also reach 27° C or higher during a midsummer day. In addition to low temperatures, tundra is characterized by shallow (5.0–7.5 cm deep), nutrient-poor clay soils that are waterlogged during the growing season, which lasts only 2–3 months. The cold, anaerobic conditions of the soil prevent any significant recycling of the extensive organic components, which are produced primarily by peat mosses. Although annual precipitation averages less than 25 cm per year, waterlogging occurs primarily because of the flat terrain and because **permafrost** (permanently frozen soil) beneath the surface prevents water from draining into the soil. Permafrost also causes water to accumulate throughout the biome in the form of numerous shallow lakes.

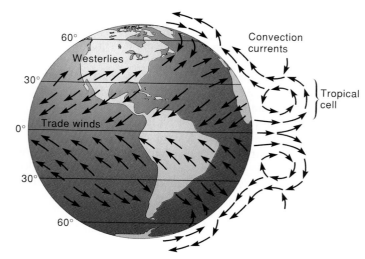

figure 33.2

Global wind patterns are the result of interactions between rising warm air, falling cool air, and the rotation of the earth.

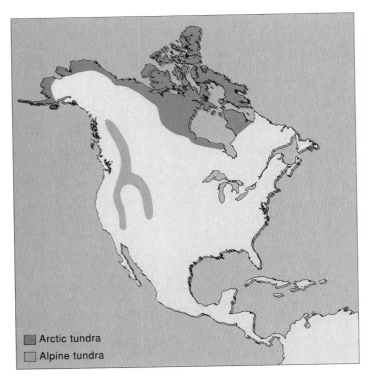

figure 33.3

The distribution of tundra in North America. Arctic tundra occurs only in high, cold latitudes, and alpine tundra occurs in patches on high-elevation, mountainous terrain.

Arctic tundra
Alpine tundra

Permafrost, which usually occurs at a depth of a few cm to about 1 m below the surface throughout arctic tundra, may be up to 500 m deep in northern Alaska and Canada; at Nordvik in northern Siberia it is more than 600 m deep. Permafrost is often absent from alpine tundra.

The vegetation of arctic tundra consists primarily of grasses, sedges, mosses, and lichens, and it is mostly evergreen. Grasses and sedges tend to dominate in alpine tundra, whose environment differs in several ways from those of arctic and antarctic areas. Solar radiation is more intense at high altitudes because humidity is lower, and the sun's rays are more direct than in arctic areas. Daylengths are not as variable, especially in mountains toward the equator.

Tundra is usually treeless, but low shrubs survive and even dominate in many parts. Such plants include willow (*Salix*), birch (*Betula*), and blueberry (*Vaccinium*), which are mostly less than 25 cm tall and rarely more than 1 m tall at maturity. Leaves, which tend to be small, are protected by thick cuticles and dense hairs. The stems are often covered with lichens (which can photosynthesize at −10° C) that protect the plants from drying winds and frigid temperatures. When temperatures drop below freezing, ice crystals form in intercellular spaces. The cells are exceptionally tolerant of the dehydration that occurs as the freezing draws water from the cells. This

figure 33.4

Alpine tundra in Glacier National Park, Montana. Tundra plants have a short but spectacular growing season near and above the timberline, where trees are sparse or absent.

form of protection from cold is so effective that the dormant stems of arctic woody plants can survive immersion in liquid nitrogen, which has a temperature of −196° C.

The flora of tundra also includes low perennials that produce brightly colored flowers during the brief growing season and form brilliant mats over the topsoil (fig. 33.4). Some plants growing in tundra reproduce vegetatively much of the time, producing seeds only during exceptionally mild and long growing seasons (which occur once or twice every century). Their seeds remain viable for long periods—sometimes for centuries. Few of the seeds that germinate encounter the exceptionally long, relatively warm growing seasons necessary for development to maturity; however, just enough do survive to perpetuate the species. Many of these perennials are adapted to the harsh growing conditions by having as little as 10% of their mass aboveground, the remainder being in the form of rhizomes, tubers, bulbs, or fibrous roots below the surface. The level of the permafrost determines the depth to which the roots can grow.

Flower buds usually form by the end of the previous growing season and remain dormant, developing rapidly and opening during the first thaws of the following summer. Some of the bowl-shaped flowers exhibit heliotropic movements, so that the sun's energy is concentrated on the stamens and pistils throughout the day (see Chapter 19). The temperature inside such flowers may be as much as 25° C higher than the air temperatures outside, and the warmth attracts insect pollinators.

During midsummer, photosynthesis can occur throughout all of the 24-hour day, and virtually all of the carbohydrate resulting from photosynthetic activity is produced within a few weeks each year. Most of the carbohydrate is stored beneath the surface, with stem and leaf growth being minimal.

33–5

figure 33.5

A blowout in Alaskan tundra. High winds have torn away the thin layer of vegetation.

Arctic-adapted plants of alpine sorrel (*Oxyria digyna*) flower only after daylengths have exceeded 20 hours, and other arctic species probably have similar photoperiods. The arctic sorrels also reach peak photosynthetic rates at lower temperatures than do plants of the same species adapted to warmer climates. Fruits and seeds of arctic plants mature in as little as 3 weeks. Annuals are rare in tundra because the cold environment usually inhibits the germination of tiny seeds and the subsequent completion of the life cycle in such plants, even during summers that are milder than usual.

Tundra is exceptionally fragile. A truck or car driven across it compresses the soil enough to kill roots, and the tracks remain for many years. Occasionally, sheep grazing on tundra pull up patches of the matted vegetation, leaving exposed edges. High winds catch the exposed edges and rip away larger segments of mat, leaving barren patches called *blowouts* (fig. 33.5).

c o n c e p t

Terrestrial ecosystems are classified into biomes on the basis of the similarities of their vegetation and physical environments. Tundra is characterized by freezing temperatures, permafrost, and the virtual absence of trees and annuals. The tundra biome includes the northernmost ecosystems, which are dominated by low-growing plants. It occupies about one-fourth of the earth's land surfaces.

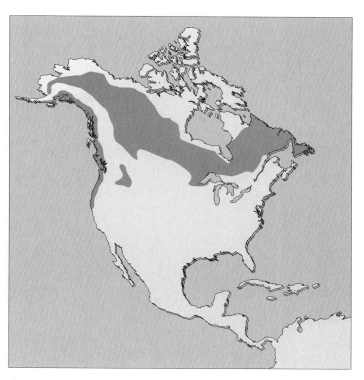

figure 33.6

Distribution of taiga in North America. These northern coniferous forests grow on patchy permafrost and nutrient-poor soils.

Taiga

Taiga, also referred to as *northern conifer* or *boreal forest*, occurs mostly adjacent to and south of the arctic tundra across large areas of North America and Eurasia (figs. 33.6 and 33.7). (Similar vegetation occurs in high mountains of many parts of the world, but these ecosystems are classified as the mountain forest biome.) Permafrost occurs within more than 65% of the taiga, usually less than a meter from the surface. Lightning often starts fires, which melt the permafrost and stimulate plant growth in the burned area for several years.

The soils of taiga are usually acidic and nutrient-poor, making them unsuitable for most agricultural activities other than timber farming. Snow accumulates in the taiga during the winters, which are long and cold. In midwinter there may be only 6 hours of light per day, and temperatures drop to –50° C or lower during the coldest months. In summer, the temperatures may reach 27° C, and daylight lasts for up to 18 hours. Most precipitation occurs in the summer, and ranges from about 25 cm to more than 100 cm in parts of western North America.

The vegetation of taiga is relatively uniform and dominated by a few genera of coniferous trees, including spruce (*Picea*), fir (*Abies*), and pine (*Pinus*). Deciduous trees such as birch, poplar, aspen, willow, alder (*Alnus*), and tamarack (*Larix*)

figure 33.7

A taiga scene. Plants that grow in the taiga are adapted for long, cold winters and heavy snowfall.

often occur in some of the wetter areas, such as the margins of the many lakes, ponds, and marshes. Nitrogen-fixing actinomycetes (*Frankia*) associated with the roots of the alders enable these trees to survive in otherwise comparatively barren soils. A thick-walled epidermis and hypodermis, sunken stomata, and a thick cuticle all adapt the leaves of taiga conifers to the rigors of the harsh winter climate. Many (mostly bulbous) perennials and a few cold-hardy shrubs occur, but annuals, which usually have small seeds and delicate seedlings, are generally prevented from becoming established by the severe climatic conditions.

c o n c e p t

Taiga consists of conifer forest ecosystems that are mostly adjacent to the tundra biome. These hardy plants are adapted to survival in low-nutrient soils and extraordinary ranges of temperature and moisture.

figure 33.8

The distribution of temperate deciduous forests in North America. These broad-leaved trees thrive with warm summers and cold winters.

Temperate Deciduous Forests

Deciduous trees are broad-leaved species that shed their leaves annually during the fall and remain dormant during the winter. These trees tend to be most abundant in areas where the summers are warm and the winters are relatively cold. Like the taiga, nearly all **temperate deciduous forests** occur on large continental masses in the Northern Hemisphere (figs. 33.8 and 33.9). In North America, this type of forest occurs from the Great Lakes region south to the Gulf of Mexico and extends from the general vicinity of the Mississippi River to the eastern seaboard. Temperate deciduous forests also occur in western Europe and Asia. Temperatures within the area vary greatly but normally fall below 4° C in midwinter and rise to above 20° C in the summer. The trees, which usually have thick bark and become dormant before the onset of cold weather, are well adapted to subfreezing temperatures, particularly if the cold is accompanied by snow cover that prevents the ground from freezing down to the root zone. Precipitation averages between 90 and 225 cm per year and occurs mostly during the summer.

The northern limits of the temperate deciduous forests mark a transition from the evergreen conifer forest of the taiga to deciduous trees. Broad-leafed, deciduous trees of the temperate forest require a longer growing season because the slow budding and expansion of newly produced leaves in the spring would be inefficient in the short growing season of a taiga summer. The

figure 33.9

An opening within the eastern deciduous forest.

figure 33.10

Sugar maples (*Acer saccharum*) in the fall. Sugar maples are tapped to make maple syrup (see reading 21.3, "Making Maple Syrup").

short growing season of these harsher climates makes it more cost-effective to have leaves that are established and ready for photosynthesis at the first sign of warmth and extended daylight.

Some of the most beautiful of all the broad-leaved trees occur in temperate deciduous forests. In the upper Midwest of the United States, sugar maple (*Acer saccharum*, fig. 33.10) and American basswood (*Tilia americana*) predominate. Sugar maple also occurs to the Northeast, where it is often associated with the stately American beech (*Fagus grandifolia*). Both maple and basswood are adapted to moist, temperate climates. In the west and west-central part of the forest, oak (*Quercus* sp.) and hickory (*Carya* sp.) are dominant because they tolerate drier conditions. Oaks are also abundant along the eastern slopes of the Appalachian Mountains, where American chestnut (*Castanea americana*) was once a conspicuous part of the forest. The chestnuts have now virtually disappeared, having been killed by chestnut blight disease (*Cryptonectria parasitica*, an ascomycete accidentally introduced from China to North America in 1904). Oak trees extend into the southeastern United States, where they are associated with pine trees and other species such as the bald cypress (*Taxodium distichum*).

Before the arrival of European immigrants, a mixture of large deciduous trees that included maple (*Acer* sp.), ash (*Fraxinus* sp.), basswood, beech, buckeye (*Aesculus* sp.), hickory, oak, tulip tree (*Liriodendron tulipifera*), and magnolia (*Magnolia* sp.) occurred on the eastern slopes and valleys of the Appalachian Mountains. Some of these trees were over 30 m tall and had trunks up to 3 m in diameter. Except in a few protected pockets in the Great Smoky Mountains National Park, the largest trees of this rich forest have been all but eliminated by logging.

American elm (*Ulmus americana*), also once a part of the forest, is rapidly disappearing as Dutch elm disease kills both trees in the wild and those planted along city streets and on college campuses. Several midwestern towns, which had hundreds of elms planted along their streets, were left with few live trees within a year or two after Dutch elm disease was introduced in the vicinity. Dutch elm disease is caused by *Ophiostoma ulmi*, an ascomycete (see Chapter 26) that was introduced to North America from Europe in 1930. The spores are spread by the elm bark beetle, which infects the phloem while boring into the inner bark.

figure 33.11

These trilliums (*Trillium* sp.) have carpeted the floor of part of the eastern deciduous forest before the leaves of the trees have formed a dense canopy that blocks sunlight from reaching the ground.

A mixture of deciduous trees and evergreens occurs on the northern and southeastern borders of the temperate deciduous forest. Hemlock (*Tsuga canadensis*) and eastern white pine (*Pinus strobus*) occur from New England west to Minnesota and south to Alabama along the Appalachians. The once-vast stands of eastern white pine are now almost gone, their valuable lumber having been used for construction and other purposes. Some have been lost to still another tree disease, white pine blister rust (*Cronartium ribicola*, a basidiomycete with two hosts; see Chapter 26), but scattered trees remain. Various pines dominate the Atlantic and Gulf coastal plains from New Jersey to Florida and west to east Texas.

During the summer, the trees of deciduous forests form a relatively closed canopy that prevents most direct sunlight from reaching the floor. Many of the showiest spring flowers of the region, such as bloodroot (*Sanguinaria canadensis*), hepatica (*Hepatica* sp.), Dutchman's breeches (*Dicentra cucullaria*), buttercups (*Ranunculus* sp.), trilliums (*Trillium* sp.), and violets (*Viola* sp.), flower before the trees have leafed out fully, and complete most of their growth within a few weeks (fig. 33.11). Other plants that can tolerate more shade, such as

waterleaf (*Hydrophyllum* sp.), flower after the canopy has formed. Several members of the sunflower family, such as asters (*Aster* sp.) and goldenrods (*Solidago* sp.), flower in succession in forest openings from midsummer through fall.

c o n c e p t

The temperate deciduous forest biome, which occurs where summers are warm and winters are relatively cold, is characterized by broad-leaved trees that lose their leaves annually in the fall.

Grasslands and Savannas

Natural **grasslands** occur toward the interiors of continental masses and along arid coastlines in areas having temperate climates (figs. 33.12 and 33.13). Grasslands tend to intergrade with forests, woodlands, or deserts at their margins, depending on the amounts and patterns of precipitation. On the average

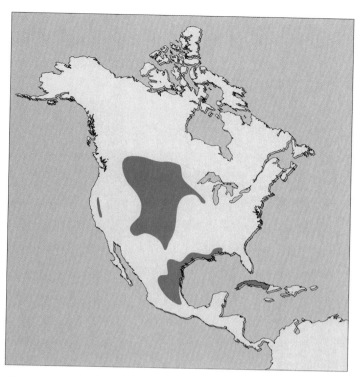

figure 33.12

Distribution of grasslands and savannas in North America. Temperate, arid climates may produce grasslands and savannas with scattered trees.

grasslands receive less precipitation than do deciduous forests. Grasslands may receive as little as 25 cm of rainfall or as much as 100 cm annually in some areas. Air temperatures can range from 50° C in midsummer to –45° C in midwinter.

Savannas, which, like grasslands, are dominated by grasses, occur in areas having subtropical to tropical climates, where long, dry periods are interspersed with seasonal precipitation that may sometimes exceed 150 cm per year (fig. 33.14). In South America, savannas extend from south of the Amazon Forest to near the Uruguayan border; in Africa, they occur both north and south of the tropical forests; in Australia, they occur on both sides of the large interior desert; and in North America, they occur from the drier parts of Texas south to the interior areas of eastern Mexico.

Savannas characteristically include widely scattered, thick-barked deciduous trees that lose their leaves during the dry periods. Baobabs (*Adansonia* sp.; fig. 33.15) are common trees of African and Australian savannas. The larger trees usually give way in the drier areas to smaller scrubby trees (mostly *Acacia* sp.) that typically produce many conspicuous thorns and seldom exceed 15 m in height. Herbaceous perennials with bulbs, rhizomes, and other storage organs are common in grasslands and savannas.

In North America, the natural grasslands, or prairies, were once grazed by huge herds of bison. The bison disappeared as settlers cultivated more and more of the land and hunters slaughtered more and more of the large animals.

figure 33.13

A mixed-grass prairie in South Dakota may also include a variety of plants with showy flowers.

figure 33.14

Savanna in Emas National Park, Brazil.

figure 33.15

A baobab tree (*Adansonia digitata*) in an east-central African savanna. These thick-barked trees lose their leaves during droughts.

By 1889 only 551 bison were left. Large areas of prairie are now used for growing cereal crops (particularly corn and wheat) and for grazing cattle.

Before it was destroyed, the American prairie was a remarkable sight. In Illinois and Iowa, the grasses grew over 2 m tall during an average season and another meter taller during a wet one. A dazzling display of wildflowers began before the young perennial grasses emerged in spring and continued throughout the growing season. Even today, more than fifty species of flowering plants can be observed flowering simultaneously in the middle of spring on as little as 1 hectare of undisturbed natural grassland.

Grasslands in areas with a Mediterranean climate (e.g., the Great Central Valley of California), where most of the precipitation occurs during the winters, may include *vernal pools*. These are temporary pools of rainwater that often accumulate in areas with clay soil or hardpan beneath them. The water evaporates and disappears after the rains stop. Their spring floras include an orderly succession of flowering plants unique to the habitat, some appearing initially at the pool margins, with each species forming a distinct zone or band until the water is gone (fig. 33.16). Some species flower only in the damp soil and drying mud that remains. The seeds of several species germinate underwater.

African savannas are susceptible to how people manage big-game animals such as elephant, giraffe, lion, cape buffalo, zebra, and a large variety of antelope and deer (fig. 33.17).

Elephants, for example, feed almost continuously for much of their lives, with each adult daily consuming between 150 and 175 kg of vegetation. The destruction of vegetation by elephants can become a serious problem when the animals are confined to parks. As the animals multiply, park herds are systematically thinned to prevent devastation of the habitat. If the large animals are not artificially confined, however, they travel large distances in search of seasonal fruits; significant destruction of a restricted habitat is thus avoided.

c o n c e p t

The most common plants in the grassland biome and the savanna biome are grasses. The grassland biome occurs in temperate climates, and the savanna biome occurs in subtropical to tropical climates where there may be long, dry periods each year.

Deserts and Semideserts

Sand, heat, mirages, oasis, and *camels* are features commonly associated with the most well-known of **deserts** but are far from universal (fig. 33.18). These features are typical of the Sahara and other large deserts that occur both north and south of the equator in the interior of Africa and Eurasia, primarily in the vicinity of 20° to 30° latitude. Camels, however, are not

figure 33.16

Meadowfoam (*Limnanthes douglasii*) in flower at the edge of a vernal pool in northern California. Vernal pools are temporary ponds.

typical of the deserts or semideserts that cover roughly 5% of North America, nor are they found in the Australian desert. They are also absent from other smaller deserts such as those of Namibia, Chile, and Peru, which occur in coastal areas where there are cold offshore ocean currents that dry the land.

Deserts can form whenever precipitation is consistently low or where water passes quickly through the soil (fig. 33.19). Many deserts get less than 10 cm of precipitation per year, but some desert areas with porous soil, such as those of the Sonoran and other deserts in the southwestern United States and northwestern Mexico, may receive 25 cm or more annually (see reading 33.1, "Desertification: What Is It and What Causes It?").

The low humidity of deserts causes large fluctuations in daily temperatures. During the summer, for example, daytime temperatures can exceed 35° C and often fall below 15° C the same night. Solar radiation is greater in the dry air than it is in areas where atmospheric water vapor filters out some of the sun's rays. Many desert plants have adapted to these conditions through the evolution of crassulacean acid metabolism (CAM) photosynthesis; C_4 photosynthesis is also more common in desert plants than in those of other areas (see Chapter 7).

Other adaptations of desert plants include thick cuticles, fewer stomata, water-storage tissues in stems and leaves, leaves with a leathery texture and/or reduced size, and even the total absence of leaves. The roots of the bizarre

figure 33.17

Elephants feed continuously on the grasses of a savanna.

Welwitschia plants of southwest African deserts (see fig. 30.14b) get all the water they need from fog drip; these plants thrive in the absence of subterranean water or precipitation. In cacti and other succulents without functional leaves, the stems are photosynthetic. Such plants have widespread, shallow root systems that can absorb water rapidly after the infrequent rains. The water is then stored for long periods inside the stems. Other perennials grow from bulbs that are dormant for much of the year. Some desert trees, such as mesquite (*Prosopis*), have long taproots that can reach several meters down to the water table.

Annuals provide a spectacular display of color and variety, particularly during an occasional season with above-average precipitation (fig. 33.20). The seeds of annuals often germinate after a fall or winter rain and then grow slowly for several months before producing flowers in the spring. Hundreds of different species of desert annuals can occur within a few square kilometers of desert in the southwestern United States.

In the colder parts of North American deserts, sagebrush (*Artemisia* sp.) is common, while in the warmer desert areas creosote bush (*Larrea tridentata*) and many species of cacti predominate. In areas with more precipitation, palo verde (*Cercidium* sp.), mesquite, ocotillo (*Fouquieria splendens*), and junipers (*Juniperus* sp.) are common.

Desert ecosystems, like those of the tundra, are fragile, and recovery from disturbances often takes many years. Large numbers of off-road vehicles and wild donkeys have devastated several desert areas in the southwestern United States; current efforts to curb the destruction thus far have met with very limited success. Unfortunately, our knowledge of desert ecosystems and predictive abilities about their future are poor. However, researchers such as Bill Schlesinger and Jane Raikes from Duke University are exploring the history and mechanics of deserts.

figure 33.18

Sand dunes and an oasis in the Sahara Desert, Algeria.

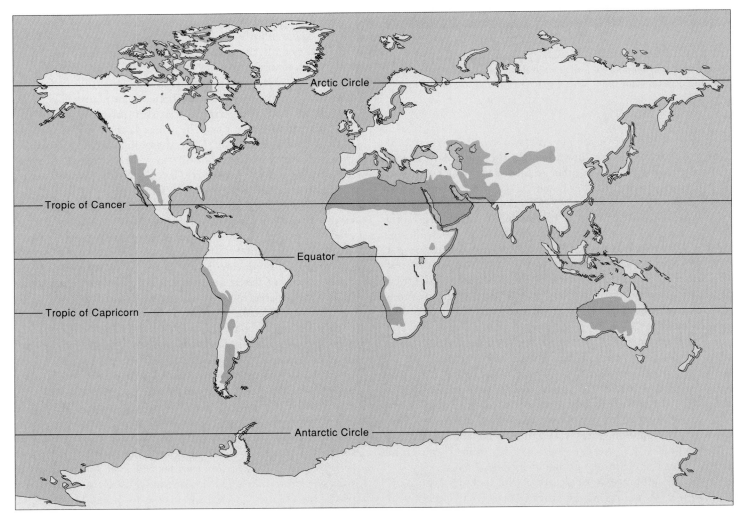

figure 33.19

The distribution of the world's deserts is extensive and growing annually due to global drying and poor land management.

figure 33.20

Spring wildflowers in the Mojave Desert can grow and flower rapidly during brief seasons with rain.

They are examining soil profiles and changes in community structure of a New Mexican desert to develop models that reveal where that desert has been ecologically and where it's going.

c o n c e p t

Deserts occur in hot climates wherever precipitation is consistently low or the soil is too porous to retain water. Plants of the desert biome include succulents and plants with small, tough leaves and deep taproots or extensive, shallow root systems.

Mediterranean Scrub

Areas around the Mediterranean Sea have dry, hot summers and cool, wet winters. Similar areas occur along parts of the west coasts of North and South America, on the southwestern tip of Africa, and in parts of southern and southwestern Australia. Unique scrubby vegetation that is either evergreen or deciduous in summer has evolved in these areas with Mediterranean climates (figs. 33.21 and 33.22). Most growth occurs in the relatively short, wet winters and early spring; plants are dormant for the remainder of the year.

figure 33.21

Distribution of chaparral in North America. Growth seasons in these areas are short with dry, hot summers and cool, wet winters.

figure 33.22

A chaparral scene. California lilac (*Ceanothus integerrimus*) is flowering.

DESERTIFICATION: WHAT IS IT AND WHAT CAUSES IT?

Desertification is the conversion of grasslands or other biomes into deserts. Deserts have been expanding for the past 15,000 years due to worldwide drying since the most recent period of glaciation; thus, desertification is not new. Nevertheless, it has been accelerated in certain parts of the world because of human activities. One of the causes of desertification is the irrigation of arid habitats for farming. Irrigation water always contains salt, in some cases more than a ton (ca. 900 kg) per acre-foot.* As crop plants absorb the water, the salt is left behind. Salt has accumulated in some areas over the years to the point that these areas are about to undergo desertification. For instance, many farmlands in the southwestern United States that are irrigated with water from the Colorado River are becoming polluted with salt and are expected to be useless for farming within a few years.

Desertification in the Sahel region of the southern Sahara Desert is another story (reading fig. 33.1a). Desertification in this semiarid region, which is correlated with the apparent drying out of adjacent savanna ecosystems, has caused tremendous human suffering, most notably the widespread famines of 1972 and 1984. The most widely accepted hypothesis to explain desertification in the Sahel is the drought hypothesis: the Sahel is drying out because of the lack of rain. *Drought* and *famine* are used in the same breath so often that cause and effect are almost fixed in our minds. But let's look a little closer. Widespread famine is a relatively recent phenomenon, yet the low rainfall of the 1970s and 1980s has occurred several times earlier in this century without causing desertification (reading fig. 33.1b). Furthermore, a ranch of more than 1,000 km² in the heart of the Sahel has remained green since it was started in 1968, despite receiving the same low rainfall as adjacent areas undergoing desertification. These observations contradict the drought hypothesis.

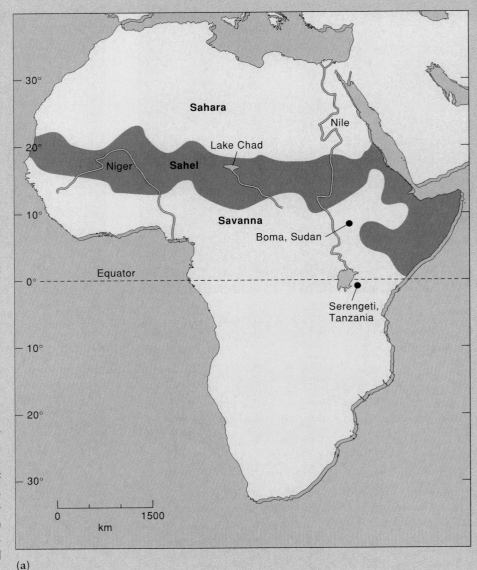

(a)

reading figure 33.1

(a) The Sahel is a semiarid belt that spans Africa between the Sahara Desert and the savanna. (b) Rainfall records from the Sahel show that, since 1900, periods of below-average rainfall have been relatively frequent, although widespread drought and famine in the area have only happened since 1973.

Many of the plants in **Mediterranean scrub** biomes are well adapted to fires, which are frequent. Such plants have thick roots that resprout after the aboveground parts have been burned (fig. 33.23); others have rhizomes that lie close enough to the ground to survive when a fire races through the erect vegetation above them. Still others, such as ear drops (*Dicentra chrysantha; D. ochroleuca*), have seeds that usually do not germinate unless they are seared by fire.

The dense Mediterranean scrub of lower elevations of the Pacific Coast is known as *chaparral*, a word derived from

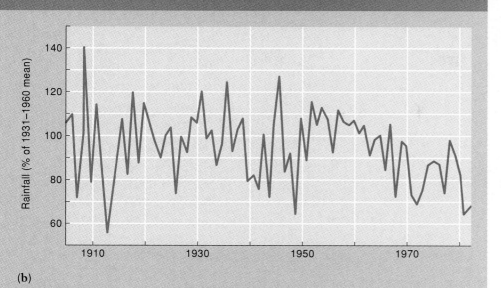

(*Arctostaphylos* sp.), California bay (*Umbellularia californica*), and poison oak (*Toxicodendron diversilobum*).

Ecosystems that have hot, dry summers and cool, wet winters are dominated by scrubby vegetation types, which comprise the Mediterranean scrub biome.

Mountain Forests

Mountain forests occur in widely scattered areas of the world, including Norway, the Himalayan region, the Andes of South America, the Pyrenees and other central European mountains, the Caucasus region, Mongolia and eastern Siberia, the Russian Sikhote Alin and Ural mountain ranges, the Atlas Mountains of Morocco, central African mountain ranges, southwest Saudi Arabia, central New Guinea, and western North America. The **mountain forest** biome is similar to the taiga biome in being characterized by conifer forests. However, permafrost is absent in mountain forests, and unlike taiga, which is restricted to the Northern Hemisphere, this biome occurs in both the Northern and Southern Hemispheres. A brief discussion of North American mountain forests follows.

In the geologic past, deciduous forests extended to western North America. As the climate changed and summer rainfall was reduced, conifers largely replaced the deciduous trees, although some (e.g., maple, birch, aspen, oak) still remain, particularly at the lower elevations. Today, conifer forests occupy vast areas of the Pacific Northwest and extend south along the Rocky Mountains and the Sierra Nevada and California coast ranges (fig. 33.24). Isolated pockets of this biome also occur in other parts of the West, particularly toward the southern limits of the

(b)

What is the difference between the green ranch area and the desertifying areas of the Sahel that might explain why desertification is occurring? Several interrelated factors, all pointing to human activities, are cited as the cause of desertification. This explanation is referred to as the settlement-overgrazing hypothesis. Native people of the Sahel used to migrate seasonally, driving small herds of cattle to the north when rains came and the annual grasses grew, then back to the wetter southern areas as the rainy season passed. The cattle gained weight and reproduced based on a diet of the high-protein annual grasses that grew in the fertile soils of the north. When water became scarce, they returned to the low-protein perennial grasses that grew in the nutrient-poor soil of the south; this diet is not nutritious enough for growth and reproduction.

The pattern of seasonal migration of people and their cattle has been replaced by settlements and agriculture over the years since the Second World War, when Western countries increased their aid for agricultural development in African countries. Unfortunately, crops in the southern areas of the Sahel were unsuccessful because of the poor quality of the soils there. With the exception of the green ranch, cattle were allowed to overgraze the northern annual grasses to compensate for the destruction of these areas by failed farms until the grasses could no longer survive. Both the north and the south are now so severely denuded that they look like they have suffered an extensive drought. However, the effects of the change in lifestyle of the native people from a migratory one to a sedentary one are evidence that overgrazing is the real culprit.

---

*An acre-foot is the amount of water that covers an acre to a depth of 1 ft. It is the preferred unit of measure for water usage in agriculture. There is no metric equivalent to acre-foot, unless you convert acre to hectare and foot to centimeter, but there is no use for such a unit.

(text) Reproduced by permission of the National Research Council of Canada from the *Canadian Journal of Zoology*, Volume 63, pages 987–994, 1985.

chabarro, the Basque name for scrub oak. A similar biome occurs in southern Europe and characterizes the Basque culture. Dominant chaparral species include buckbrush (*Ceanothus cuneatus*), scrub oak (*Quercus dumosa*), silk-tassel bush (*Garrya fremontii*), chamise (*Adenostoma fasciculatum*), manzanita

mountains. The trees tend to be large, particularly in and to the west of the Cascade Mountains of Oregon and Washington and on the western slopes of the Sierra Nevada. Part of the reason for the huge size of trees such as Douglas fir (*Pseudotsuga menziesii*) is the high annual rainfall, which exceeds

figure 33.23

Chaparral shrubs often resprout from the base a year after a fire has swept through the area and destroyed the exposed stems and leaves.

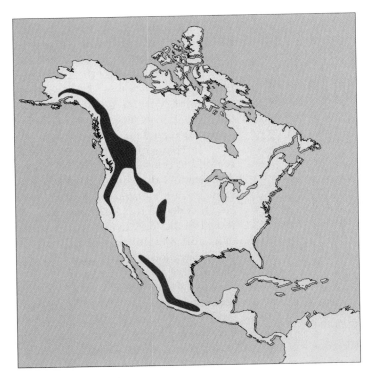

figure 33.24

Distribution of mountain forests in North America. Conifers are most common, but deciduous trees occur at lower elevations.

250 cm in some areas. The world's tallest trees, the coastal redwoods of California (*Sequoia sempervirens;* fig. 33.25), however, apparently depend more on moisture from fog for their size and longevity than on large amounts of rain. The fog, which condenses on the foliage and drips into the soil, also reduces transpiration rates.

As moisture-laden air is forced upward along mountain ranges, annual precipitation changes between sea level and the higher slopes. Temperatures also drop about 1° C for each 100 m of elevation. Thus, the western and southern sides of mountain ranges in the Northern Hemisphere are often warmer and drier than the eastern and northern slopes because the eastern slopes receive direct sunlight only in the mornings when it is cooler and there is less solar radiation; at the same time the northern slopes receive no direct sunlight. It is not surprising, therefore, that even at the same latitudes, different associations of plants are common at different elevations (see reading 33.2 "Biomes: Elevation versus Latitude").

At lower elevations in both the Rocky Mountains and the Sierra Nevada, the predominant conifer is ponderosa pine (*Pinus ponderosa*). At lower elevations in the northern part of the Cascades, Douglas fir, western red cedar (*Thuja plicata*), and western hemlock (*Tsuga heterophylla*) are more common. At intermediate elevations in the Sierra Nevada, the established conifers include sugar pine (*Pinus lambertiana*), white fir (*Abies concolor*), and Jeffrey pine (*Pinus jeffreyi*), while at higher elevations different conifers, such as red fir (*Abies magnifica*), predominate.

figure 33.25

Coastal redwoods (*Sequoia sempervirens*) in northern California rely on consistent moisture from fog and rain.

reading 33.2

BIOMES: ELEVATION VERSUS LATITUDE

High mountain regions have climates and vegetation types that are similar to those of the arctic tundra; mountain areas a little lower in elevation are similar to the high-latitude conifer forests known as taiga. These observations may seem obvious now, but the similarities between ecosystems in northern latitudes and ecosystems in high mountains were not correlated until the early nineteenth century. The correlation between elevation and latitude was first noted by the German explorer and naturalist Alexander von Humboldt,* who observed in 1805 that climbing a high mountain was analogous ecologically to traveling farther north or south from the equator.**

---

*Von Humboldt based his observations on his exploration of Mount Chimborazo in Ecuador, although he could have discovered the same relationships much closer to home—between the high Alps and far northern Europe.

**This observation was first published in Paris in his *Essai sur la geographie des plantes* (*Essay on the Geography of Plants*).

Most of the North American mountain forest biome has comparatively dry summers. Lightning often starts fires and did so long before human carelessness became a leading cause of forest fires. Several tree species are well adapted to survival after being partially burned. Douglas fir, for example, has a thick, protective bark that can be charred without transmitting enough heat to the interior to kill the cambium. Moreover, Douglas fir seedlings thrive in open areas after a fire (fig. 33.26). When the bark of the giant redwoods (*Sequoiadendron giganteum*) of the Sierra Nevada is burned, the trees are rarely killed. This has undoubtedly contributed to the great age and size of many of the trees.

The cones of some pine trees, such as knobcone pine (*Pinus attenuata;* fig. 33.27), remain closed and do not release their seeds until a fire causes them to open. Similarly, the seeds of several other species germinate best after they have been burned. Because of these attributes of members of the mountain forest biome, rangers in some of our national parks occasionally allow fires at higher elevations to run their natural course. The above-average incidence of fires occurring since humans have come in large numbers to the forest, however, has made controlling most fires necessary, even though doing so interferes with natural cycles that would otherwise occur in the biome.

figure 33.26

A large Douglas fir (*Pseudotsuga menziesii*). Note the fire scars on the trunk.

figure 33.27

The closed seed cones of knobcone pine (*Pinus attenuata*) do not open until seared by fire.

The mountain forest biome is similar to the taiga biome in vegetation type; however, the mountain forest biome occurs both north and south of the equator and has no permafrost. North American mountain forests have a fairly conspicuous zonation of species.

Tropical Rain Forests

Tropical areas of the world include many biomes, from seasonally dry areas to grasslands to high mountains. The greatest biological diversity on earth, however, occurs in **tropical rain forests.** The principal tropical rain forests of the world occur in the Amazon Basin of Brazil, the Congo Basin of central Africa, much of Central America, and in or near the equatorial regions of southeast Asia and Indonesia. A narrow band also occurs along the northeast coast of Australia. About 7% of the earth's surface, representing nearly half of the forested

figure 33.28

Distribution of tropical rain forests in North America. Diversity in these rain forests is extraordinary.

areas of the earth and 25% of the earth's species, are included in this biome (fig. 33.28). Tropical rain forests have existed for about 200 million years and, unlike other biomes, were not decimated by glaciation of the ice age.

Rain forests occur throughout areas of the tropics where annual rainfall normally ranges between 200 and 400 cm and temperatures range between 25° C and 32° C, with night temperatures seldom dropping more than 5° C lower than those at noon. Although monthly rainfall varies, there is no dry season, and some precipitation occurs every month of the year, frequently in the form of afternoon cloudbursts. The humidity seldom drops below 80%. Such climatic conditions favor and support a diversity of plants and animals so great that the number of species in tropical rain forests exceeds that of all the other biomes combined.

Rain forests are dominated by broadleaf evergreen trees, whose trunks are often unbranched for as much as 40 m or more, with luxuriant crowns that form a beautiful dark-green and multilayered canopy (fig. 33.29). Small parts of the forest, particularly along rivers, are jungles where the canopy is so dense that little light penetrates to the floor. The few herbaceous plants that survive are generally confined to openings in the forest. There are hundreds of species of trees, each usually represented by widely scattered individuals. Root systems are shallow, and the trees are often buttressed, the broader bases compensating for lack of root depth (fig. 33.30). Organic matter is sparse in tropical soils, which tend to be acidic. The soils are

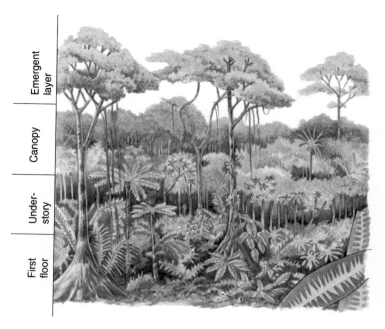

figure 33.29

Levels of plant life in the rain forest. A few tall trees comprise the emergent layer, which overgrows the shorter trees of the dense canopy. Shrubs and low trees occupy the shady understory, and fallen leaves and seedlings are scattered on the forest floor.

figure 33.31

Orchids and other epiphytes on the branches of a tree in a South American rain forest.

figure 33.30

Buttress roots on a tropical tree in Ecuador add stability in moist soils.

often deficient in important nutrients such as potassium, magnesium, calcium, and phosphorus, or the minerals (particularly phosphorus) are in forms that the plants cannot use. Despite the lush growth, there is little accumulation of litter or humus, because decomposers rapidly degrade leaves and other organic material on the forest floor; the nutrients released by decomposition are quickly recycled or leached by the heavy rains.

Most of the plants in rain forests are woody flowering plants; no conifers occur naturally in these ecosystems. Although evergreen plants predominate, some are deciduous species that can shed their leaves from some branches, retain leaves on others, and flower on yet other branches all at the same time, while branches of adjacent trees of the same species are losing their leaves or flowering at different times. Many *lianas* (woody vines that are rooted in the ground) hang from tree branches, and even more numerous *epiphytes* (especially orchids and bromeliads) are attached to limbs and trunks (fig. 33.31). The epiphyte roots, which are not parasitic, do not touch the ground; the plants are sustained entirely by rainwater that accumulates in their leaf bases and by their own photosynthetic activities. Traces of minerals, also necessary to the growth of the epiphytes, accumulate in the rainwater as it trickles over decaying bark, bird droppings, and dust.

figure 33.32

Destruction of tropical rain forest in Belize, Central America. This biome is rapidly disappearing.

Tropical rain forests include broad-leaved, mostly evergreen plants in a hot, humid environment. More species occur in tropical rain forests than in any other kind of ecosystem.

The Lore of Plants

The southernmost recorded flowering plant is the Antarctic hair grass (*Deschampsia antarctica*), which was found at latitude 68° 21′S on Refuge Island, Antarctica. The northernmost plants are the yellow poppy (*Papaver radicatum*) and the Arctic willow (*Salix arctica*), which grow on the northernmost land at latitude 83° N. The highest altitude at which any flowering plants have been found is 6,400 m (21,000 ft) on Kamet in the Himalayas; the plants were *Ermania himalayensis* and *Ranunculus lobatus*.

The Future of the Tropical Rain Forest Biome

In the 1960s major plans were developed to convert the Amazon rain forest into large farms, hydroelectric plants, and mines. At the beginning of the 1990s gold-mining activities were filling the rivers with silt, and the tropical rain forests were being destroyed or damaged for commercial purposes at the rate of more than 35 hectares per minute (fig. 33.32). Stated another way, human activities destroy or damage an area equivalent to that of about 45 football fields *per second*. This damage and destruction has had a devastating effect on wildlife. For example, a 1989 study found that populations of low-flying bird species in a 10-hectare section of the Amazon rain forest fell by 75% just 6 weeks after adjacent land had been cleared. Ten of the forty-eight bird species studied disappeared completely. This situation has been repeated on many occasions. Indeed, dividing large areas of rain forest into smaller pieces separated by as little as 10 m of cleared land can have disastrous effects on the ecology of the entire forest. Most of the bird and other animal species originally present disappear permanently.

Only a small part of the rain forest biome is now protected from commercial development. The rapid loss of biodiversity results in the permanent loss of gene pools with potential for medicine and agriculture. The cleared land, with its poor and eroded soils, often becomes unfit for agriculture after only 2 or 3 years. Even if the land were now to be left fallow, the plant and animal life that has already become extinct will never return. The loss of massive amounts of vegetation directly affects photosynthesis, respiration, and transpiration. If losses continue on a large enough scale, global climate will also be affected. Many more organisms are doomed to extinction before they have even been seen or described for the first time. The rain forest biome will vanish within 20 years if governments and individuals do not stop or slow the large-scale destruction.

Chapter Summary

Biomes are land-based groupings of ecosystems considered on a global or a continental scale. Most biomes are extensive; they include several ecosystems and occur on more than one continent.

Tundra, which occurs primarily above the Arctic Circle, includes shrubs, many lichens and grasses, and tufted flowering perennials. It is also characterized by the presence of permafrost (permanently frozen soil) below the surface. Tundra is fragile and easily destroyed.

Taiga is dominated by coniferous trees such as spruce, fir, and pine, with birch, aspen, tamarack, alder (which has nitrogen-fixing actinomycetes in its roots), and willow in the wetter areas. Many perennials but few annuals occur in taiga. Most precipitation occurs in the summer. Lightning frequently starts fires in this biome.

Temperate deciduous forests are dominated by deciduous trees. In North America such trees include sugar maple, American basswood, beech, oak, and hickory; evergreens such as hemlock and pine occur toward the northern and southeastern borders. In early spring a profusion of wildflowers carpets the forest floor before tree leaves expand.

Grasslands occur primarily in temperate areas toward the interiors of continents. Savannas tend to be in similar locations in areas with subtropical to tropical climates. Savannas also have widely scattered trees. Grasslands located in Mediterranean climatic zones usually include vernal pools that support unique annual floras. Many grasslands have been converted to agricultural use.

Deserts have low annual precipitation and widely fluctuating daily temperatures. Desert plants are adapted both structurally and metabolically to the environment, which has more solar radiation than do other biomes.

Mediterranean scrub has unique, mostly evergreen vegetation adapted to cool, wet winters and hot, dry summers. The vegetation is also adapted to frequent fires.

Mountain forest occupies many mountainous areas on all of the continents. In North America it occurs in the Pacific Northwest and extends south along the Rocky Mountains and the Sierra Nevada and Cascades of California and Oregon. Mountain forests have mostly dry summers, and some of the trees (e.g., Douglas fir, giant redwood) have thick bark that protects them from frequent fires. Other trees (e.g., knobcone pine) depend on fires for the release and germination of seeds.

The tropical rain forests constitute nearly half of all forest land and contain more species of plants and animals than all the other biomes combined. Many woody plants and vines form multilayered canopies, which keep most light from reaching the forest floor. Soils are poor, and nutrients released during decomposition are rapidly recycled. Tropical rain forests are being destroyed so rapidly that they will disappear within 20 years if the destruction is not slowed or stopped.

 What Are Botanists Doing?

Find out what kind of research is being done on biomes by looking through journals such as the *Journal of Environmental Management*, the *Journal of Vegetation Science*, and *Conservation Biology*. Is there any pattern to the kind of research being published about ecosystems, areas of the world, or kingdoms (plant or animal) as the main subjects?

 Writing to Learn Botany

Ecosystems in the grassland biome occur in central and western Asia, eastern Europe, Argentina, New Zealand, central and eastern Mexico, and the central United States. If this biome is so widespread, does it matter that the ecosystems in eastern Europe and in the central United States have been almost totally destroyed by human activities? After all, we still have plenty of the grassland biome in the other ecosystems.

 Writing to Learn Botany

As you have read in this chapter, the tropical rain forest biome is being destroyed at a rate that could render it extinct shortly after we enter the twenty-first century. How could you educate the public about this serious problem, and what could you do to slow or halt the destruction?

Questions for Further Thought and Study

1. Off-road recreational vehicles have devastated parts of the desert biome in recent years; the tracks of any vehicle driven across tundra are visible for many years after. Should we be concerned about such activities? Explain.

2. It has been estimated that at least 25% of all plant species are likely to become extinct in the next 30 years. So far, we seem to be getting along without a number of

plants that have become extinct in the past 10 years. Considering what you learned in this and the previous chapter about biodiversity, why should we try to slow the rate of extinction?

3. The word *deforestation* usually brings to mind chainsaws, bulldozers, and fires in the Amazon rain forest. However, other forests—including those in Central America and Southeast Asia, both of which are biodiversity "hot spots"—are being cleared two to three times faster than are Amazon rain forests. How should this problem be addressed?

4. How do mountain forests differ from temperate deciduous forests?

5. When Mediterranean scrub plants are carefully transplanted to mountain forest habitats with somewhat similar climates, they often die within a year or two. Can you suggest reasons for this?

6. Much of the tropical rain forest biome is in developing countries, which have weak economies. Bearing this in mind, what could be done to halt the destruction of this biome?

Web Sites

Review the "Doing Botany Yourself" essay and assignments for Chapter 33 on the *Botany Home Page*. What experiments would you do to test the hypotheses? What data can you gather on the Web to help you refine your experiments? Here are some other sites that you may find interesting:

http://www.mobot.org/MBGnet/biome/
This site includes information about the world's biomes, as well as the organisms that live in those biomes.

http://www.cnie.org/nle/biodv-6.html
Welcome to the Congressional Research Source report on ecosystems and biomes. At this site, you can learn how Congress defines a biome.

http://iopi.csu.edu.au/iopi/
When you're serious about searching databases concerning biodiversity, check out this site sponsored by the International Organization for Plant Information. Participate in the establishment of a Global Plant Species Information System (GPSIS).

http://www.herbalgram.org/herbalgram.html
The American Botanical Council (ABC) was incorporated in November 1988 as a nonprofit educational organization. ABC's main goal is to educate the public about beneficial herbs and plants.

Suggested Readings

Articles

Monastersky, R. 1990. The fall of the forest. *Science News* 138:40–41.
Perry, D. R. 1984. The canopy of the tropical rain forest. *Scientific American* 251 (November):138–147.
Repetto, R. 1990. Deforestation in the tropics. *Scientific American* 262 (April):36–42.
Zimmer, C. 1995. How to make a desert. *Discover* (February):50–56.

Books

Allan, T., and A. Warren. 1993. *Deserts*. Oxford: Oxford University Press.
Barbour, M. G., and W. D. Billings, eds. 1988. *North American Terrestrial Vegetation*. New York: Cambridge University Press.
Furley, P. et al., ed. 1992. *The Nature and Dynamics of Forest-Savanna Borders*. New York: Chapman and Hall.
Moore, R., and D. S. Vodopich. 1991. *The Living Desert*. Piscataway, NJ: Enslow.
Myers, N. 1992. *The Primary Source: Tropical Forests and Our Future*. Rev. ed. New York: W. W. Norton.
Odom, E. P. 1989. *Ecology and Our Endangered Life-Support Systems*. Sunderland, MA: Sinauer.
Packham, J. R. et al. 1992. *Functional Ecology of Woodlands*. New York: Chapman and Hall.
Richards, P. W. 1994. *The Tropical Rain Forest*. New York: Cambridge University Press.
Riely, J. O., and S. Page. 1990. *Ecology of Plant Communities*. New York: Halsted Press.

Medicinal plants for sale in a market in Brazil. Plants are used for medicinal and superstitious purposes in many parts of the world.

Plants and Society

E

Epilogue Outline

Epilogue Overview

It is difficult to name an area of our existence that is not affected by plants. Plants feed us, house us, clothe us, entertain us, generate oxygen that we breathe, and cure our illnesses. However, plants are not entirely beneficial to humans; plants also damage crops, cause severe allergic reactions, and produce some of the most potent poisons known. Not surprisingly, we have learned to recognize plants that help us and those that harm us. Plants have been integral parts of all civilizations.

(a)

(b)

figure E.1

Plants used for fragrances and spices. (a) Romans used lavender (*Lavandula*) to scent and disinfect their baths. The genus name of lavender, *Lavandula*, is from the Latin word *lavare*, meaning "to wash." (b) A woman in Madagascar cleaning cloves. Cloves are the dried flower buds of *Syzygium aromaticum*.

(a)

(b)

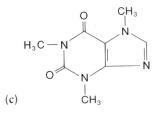

(c)

figure E.2

Tea. (a) Tea comes from *Camellia sinensis*, a plant that originated in subtropical Asia. (b) Tea leaves being processed. (c) Tea and coffee contain caffeine, a compound that stimulates the central nervous system and whose chemical structure is shown here.

E–3

Epilogue

Plants and Society

Throughout this book you've seen many examples of how plants affect our lives. We eat grains, fruits, and vegetables, and are clothed by fibers from stems and leaves. Plants generate the oxygen we breathe, and trees provide us with lumber, paper, and welcome shade on hot days. Many plants make medicines, brilliant dyes, industrial chemicals, and useful oils. The colors and fragrances of flowers and foliage satisfy our aesthetic senses, while plant-derived spices such as black pepper, nutmeg, ginger, cloves, and cinnamon have enhanced our enjoyment of food since before the time of the Roman Empire (fig. E.1). Spices can also preserve foods, which made possible the colonization of the New World. Tea and coffee are the world's most popular beverages (fig. E.2). From the body paints of Amazon Indians to modern cosmetics and from early Egyptian papyrus (fig. E.3) of more than 5,000 years ago to today's pulp mills that produce more than 200 million tons of paper each year, plants affect all aspects of our lives. Here are some of the plant families that we use for other, more unusual, purposes (fig. E.4):

- **Laurel family** (Lauraceae: camphor, sassafras, sweet bay, avocados). We use laurel leaves to crown winners of athletic contests and to bestow academic honors. Sassafras is used to make toothpaste, gum, mouthwash, tea, beer, and gumbo, while sweet bay is used as a spice. Avocados, which contain up to 60% oil, contain more energy per unit weight than does red meat.

figure E.3

Papyrus (*Cyperus papyrus*) growing at the edges of a lake. Papyrus was once used to make paper.

(a) (b) (c)

(d) (e)

figure E.4

Economically important plants. (a) Sassafras (*Sassafras albidum*) is used to flavor gum and mouthwash and in gumbo. (b) Spearmint (*Mentha spicata*) is the source of spearmint oil. (c) Peyote (*Lophophora williamsii*) is the source of various hallucinogens. (d) Monkshood (*Aconitum columbianum*) is the source of aconite, a poisonous drug once used as a sedative. (e) Peppers (*Capsicum* sp.) are favorites of people who like spicy food.

- **Mint family** (Lamiaceae: spearmint, peppermint, *Coleus*). These plants are important sources of oils, such as peppermint oil. Menthol from peppermint is used to make candy, gum, and cigarettes.

- **Carrot family** (Apiaceae: dill, celery, carrot, parsnip, parsley). Poison hemlock contains coniine, the Athenian state poison used to kill Socrates in 399 B.C. Ever the scientist, Socrates insisted on having the stages of his poisoning accurately observed and recorded. Elixir of Celery, advertised in the 1897 Sears and Roebuck catalog, was used to cure nervous ailments, and crushed celery seeds mixed with soda are still marketed as Cel-Ray by Canada Dry. We use some members of the carrot family to make liquor.

- **Cactus family** (Cactaceae: cacti). The fruits of many cacti taste like pears, which explains their common name, prickly pear. Peyote (*Lophophora williamsii*), a small, spineless cactus with carrotlike roots, is the source of mescaline. The Aztecs used dried slices (i.e., *buttons*) of this hallucinogenic plant for religious purposes, as do Indians in Mexico and the southwestern United States.

- **Pumpkin family** (Cucurbitaceae: pumpkin, squash, cantaloupe, watermelon). These plants have been used to make everything from jack-o'-lanterns and dishes to laxatives.

- **Lily family** (Liliaceae: asparagus, sarsaparilla, *Aloe*, chives, garlic, meadow saffron). Sarsaparilla is used to make soft drinks, and *Aloe*, which was once used as a source of phonograph needles, is used to treat burns; *Aloe* is also a common ingredient of many cosmetics. Meadow saffron (fall crocus; *Colchicum autumnale*) is the source of colchicine, an important drug once used to treat gout and now used in biological research. Meadow saffron is different from the true saffron, a member of the iris family and the source of the world's most expensive spice.

- **Buttercup family** (Ranunculaceae: columbine, monkshood, larkspur). Columbine, the state flower of Colorado, has been used as an aphrodisiac. Monkshood and wolfbane contain aconitine, a powerful poison that humans have used to kill a variety of animals, including wolves. Monkshood, an attractive plant grown in many gardens, was used to kill Roman Emperor Claudius and Pope Adrian VI.

- **Spurge family** (Euphorbiaceae: cassava, Pará rubber tree, candelilla, poinsettias). The Pará rubber tree is a source of latex for rubber, and candelilla is the source of several waxes, including candle wax. Cassava, whose roots are used to make tapioca, alcohol, and acetone, is a staple in the tropics.

Despite its widespread use in December, the poinsettia is one of the newest Christmas plants. Joel Poinsett became interested in the plant while serving as American ambassador to Mexico from 1825–29. He brought the plant home, where it was named in his honor.

- **Poppy family** (Papaveraceae: bloodroot, opium poppy). All members of this family produce drugs, the best known of which are morphine and codeine. To learn more about opium poppies, see reading E.1, "Poppies, Opium, and Heroin."

- **Nightshade family** (Solanaceae: tomato, potato, tobacco, petunia, chili pepper). These plants produce drugs such as atropine (once used to dilate pupils and as an antidote to nerve-gas poisoning), hyoscaymine (a sedative), and scopolamine (a tranquilizer). Hot peppers dried and crushed yield cayenne pepper, and sweet peppers are used to make paprika.[1] Fiery chili peppers get their heat from capsaicin, a chemical that's perceived not by our taste buds but rather by pain receptors in our mouth. We can taste it at concentrations as low as 1 molecule per million. That's why only a trace of capsaicin can make an unwary diner grab frantically for water.

Many of the pilgrims considered tomatoes to be evil—on a par with dancing and card playing. Interestingly, tomato juice is said to neutralize butyl mercaptan, the nose-shriveling ingredient of skunk spray. Joseph Campbell brought out his famous tomato soup in 1897, soon after chemist John Dorrance, at a weekly salary of $7.59, figured out how to condense it. In 1962, Andy Warhol's painting of a can of that soup sold for $50,000.

Although modern civilization would collapse without plants, not all plants benefit people. For example, plants such as castor bean, mistletoe, dumb cane, caladium, elephant's ear, philodendron, English ivy, rhubarb, poinsettia, oleander, and yew (the plant used by Robin Hood to make his bows) contain potent toxins. Stinging nettle (*Urtica dioica*), poison ivy (*Toxicodendron radicans*), and the less common poison oak (*T. diversilobum* and *T. toxicarium*) also elicit strong reactions in some people, while plants such as ragweed, grasses, and many fungi cause hay fever and other allergic reactions. Overgrowth of some plants clogs streams and water pipes, and weeds damage crops.

---

1. Vitamin C was purified in 1928 by Hungarian biochemist Albert Szent-Györgyi from peppers in a rejected supper dish of sweet paprika peppers (before then, Szent-Györgyi had tried to purify the vitamin from bovine adrenal glands). The purified vitamin C from those peppers earned Szent-Györhyi a Nobel prize 9 years later.

POPPIES, OPIUM, AND HEROIN

*What shall we do about poor little
Tigger*

*If he never eats nothing he'll never
get bigger;*

*He doesn't like honey and haycorns
and thistles*

*Because of the taste and because of
the bristles*

*And all the good things that an
animal likes*

*Have the wrong sort of swallow or
too many spikes.*

—A. A. Milne, *The House
at Pooh Corner*, 1928

Throughout this book you've seen examples of what Pooh Bear perceptively noted: that mechanical defenses such as thorns are not plants' only defenses. Many species have developed a chemical arsenal that helps deter herbivores. Among them is poppy, *Papaver somniferum*, an annual herb that grows about a meter tall (reading fig. E.1). One of the deterrents in poppy leaves is opium. The euphoric effects of opium were

reading figure E.1

Poppy (*Papaver somniferum*), the source of opium. Opium, the dried latex that oozes from cut fruits, is the source of alkaloids such as morphine and codeine.

known as early as 4000 B.C., and by the 1800s opium was a common ingredient in medications. For example, Godfrey's Cordial, an English concoction of opium, molasses, and sassafras, was used as a cough remedy and a cure for diarrhea.

Today, poppies are often grown innocently in gardens for their pretty purple and white blossoms. Their ripe seeds can be toasted and used as ornaments on crackers or cakes, or crushed and used as a cooking oil. Because the seeds do not contain many alkaloids, their oils produce no euphoric effects. Most of the alkaloids are in latex in the urn-shaped fruits. When the fruits are cut, the latex oozes from the fruits and dries to a brown residue called *opium*. Opium contains more than twenty-five alkaloids, the most abundant of which is morphine (another abundant ingredient is codeine). The illicit sale of heroin, which is made from morphine, is a $6 billion industry.

Worldwide production of opium exceeds 10,000 tons, of which only 400 are used for medicine. India is the largest and only legal producer of opium.

Nature's Botanical Medicine Cabinet

Just as people learned to exploit plants for food, so they learned to use plants as medicine. Plants represent a huge storehouse of drugs: they produce more than 10,000 different compounds to protect themselves from hungry animals. People that search for these compounds are natural-products chemists, such as Isao Kubo:

> Several years ago, I was on a plant collecting trip near Serengeti National Park in East Africa. I was searching for the rare medicinal plant, *Kigeria africana*. After 3 days of combing the Serengeti plain, I came upon an outcrop of *K. africana* trees. Feeling very lucky indeed,

I climbed up onto one of the tree's branches to collect the sausage-shaped fruit.

> But as I busily went about my collecting, I realized I was being watched. Two menacing eyes from an adjacent tree stared at me through the branches. Those eyes belonged to a rather large leopard. I tried with all my might to keep from falling from my perch as my knees knocked and sweat poured from my body. This brief encounter ended abruptly, however, when the leopard decided I was not a suitable meal and disappeared into the bushes. Well, I thought, another day in the life of a natural-products chemist.

—*Isao Kubo, professor of natural products chemistry,
University of California at Berkeley,*
From Medicine Men to Natural Products Chemists *(1985)*

CHEWING ON RESIN

Wounded stems of red or black spruce (*Picea*, Latin for "pitch") release a resin that flows into the wound and hardens upon contact with air, thereby sealing the wound (reading fig. E.2). This resin is called *spruce gum* and was the first commercial chewing gum. The novelty of spruce gum made it popular: rural Europeans have been chewing it for centuries, and in early Sweden spruce gum was considered a present for a potential sweetheart—the Scandinavian equivalent of a box of chocolates. Among the devoted chewers was Mark Twain's Tom Sawyer, who shared some with Becky Thatcher one day after school.

Although the spruce gum business originally thrived—the first factory employed more than 200 people and produced almost 2,000 boxes of gum per day—its popularity soon waned; indeed, only one company in the United States still makes the gum. One

reading figure E.2
Resin oozing from a wounded spruce tree.

reason for spruce gum's decline may have been its unusual taste, which one expert likened to "sinking your teeth into frozen gasoline." Moreover, it turns hard and crumbly when stuck to the bedpost overnight. However, the death of the industry was also due partly to the increased demand for newspapers, which prompted foresters to clear vast forests of spruce to make newsprint (see Chapter 16). Faced with dwindling supplies of spruce, manufacturers started looking for alternatives. Their first choice was paraffin, a by-product of petroleum distillation. However, paraffin never really caught on, despite being the first gum to be sold in packets including picture cards. Today, paraffin is still available in semiedible novelties such as wax fangs, wax lips, and wax buck teeth. A more successful substitute for spruce gum was chicle, the chewable latex of the Mexican sapodilla tree.

DAVID DOUGLAS

Although Isao Kubo's comments earlier in this chapter describe "another day in the life" of a natural-products chemist, they don't come close to describing the life and times of indomitable Scottish botanist David Douglas (reading fig. E.3). Douglas is famous for describing hundreds of plants, including the Monterey pine, digger pine, sugar pine, and ponderosa pine (interestingly, he did not describe Douglas fir, the tree named after him). Under the aegis of the Royal Horticultural Society of London, Douglas led many collecting trips to North America. These trips were fraught with disasters: he was attacked by grizzlies, Indians, and red ants; he braved blizzards and ava-lanches; he suffered all

reading figure E.3
David Douglas.

sorts of diseases; he twice had to eat his horse to avoid starvation; and he was shipwrecked and washed out to sea by a hurricane. Douglas accepted it all, entering only this understated complaint in his diary: "Traveled thirty-five miles, drenched and bleached with rain and sleet, chilled with a piercing north wind; and then to finish the day experienced the cooling, comfortless consolation of lying down wet without supper or fire. On such occasions I am very liable to become fretful." At age thirty-five, Douglas died in Hawaii in 1834, as spectacularly as he lived. While crossing mountains en route to Hilo, he fell into a pit dug as a cattle trap. There, he was gored to death by a wild bull.

(a) (b) (c) (d)

figure E.5

Drugs derived from plants. (a) Many members of the snakeroot family contain alkaloids that help lower blood pressure. (b) Alkaloids from *Ephedra* are used to treat asthma and other allergies. (c) Yams contain a variety of steroidal drugs, some of which are used to make birth control pills. (d) Woman gathering leaves of *Erythroxylon coca*, the source of cocaine.

Natural-products chemists such as Isao Kubo search through the bounty of chemicals made by organisms for substances that people can use. Here are some of the more common drugs derived from plants and a fungal parasite of a plant (fig. E.5):

- Reserpine is extracted from snakeroot (*Rauwolfia serpentina*, a low evergreen shrub) and is used as a sedative and to decrease blood pressure. In India, snakeroot is used to make an antidote for snakebites.[2] Many schizophrenics and others with mental disorders can lead nearly normal lives after being treated with reserpine. In the 1990s annual prescriptions for reserpine in the United States exceeded $100 million.

- Ephedrine, which is extracted from *Ephedra*, is used to ease bronchitis.

- Steroids such as estrogens and testosterone are extracted from yams (*Dioscorea*). The progesterone used to make the first birth control pills was extracted from 10 tons of yams harvested from a Mexican jungle by Russell Marker and his coworkers. Today, more than 60,000 tons of fresh yams are imported into the United States annually for the production of birth control pills.

- Digitoxin, extracted from *Digitalis purpurea* (foxglove, a garden ornamental), is taken as a heart stimulant by more than 3 million Americans each day.

- Cocaine comes from the leaves of *Erythroxylon coca*. This plant, the "divine plant" of the Inca civilization, is native to the eastern slopes of the Andes. Cocaine, an ingredient of Coca-Cola until 1904, is grown mostly in Peru and Bolivia. Cocaine is a stimulant and hunger depressant, which explains why poor laborers can work for 2–3 days without food while chewing coca leaves. The laborers even measure distances in *cocadas*, the distance they can walk on one chew. Today, cocaine is marketed illegally in highly addictive forms, with enormous costs to society.

- Tetrahydrocannabinol (THC) is derived from hemp (marijuana; *Cannabis sativa*) and is used to make hashish. Today, the former Soviet Union is the world's leading producer of hemp. To learn more about hemp, see reading 13.1, "Hemp."

- Lysergic acid (the chemical used to make lysergic acid diethylamide, or LSD) comes from the spore-producing fruiting body of *Claviceps purpurea*, a fungus that infects wheat and rye.

- Nicotine, which comprises 1%–3% of the weight of tobacco, is an ingredient of many insecticides.

2. Most drugs are poisonous when consumed in large amounts. For example, white snakeroot killed thousands of people in the Midwest during the 1800s; in some areas such as in Dubois County, Indiana, the death toll may have been one out of every two people.

table E.1

Some Important Medicinal Plants and Fungi

Plant	Plant Parts Used	Active Compounds	Uses in Medicine
Atropa belladonna (belladonna)	Leaves, roots	Atropine, hyoscyamine	Cardiac stimulant, pupil dilator, antidote for organophosphate poisoning
		Scopolamine	Motion sickness, antiemetic
Cannabis sativa (marijuana)	Leaves, inflorescence	Tetrahydrocannabinol (THC)	Treatment of glaucoma, relief of nausea from chemotherapy
Catharanthus roseus (Madagascar periwinkle)	Leaves	Vinblastine, vincristine	Treatment of leukemia, Hodgkin's disease, and other cancers
Cinchona sp. (fever bark tree)	Bark	Quinine	Treatment of malaria
Colchicum autumnale (autumn crocus)	Corm	Colchicine	Treatment of gout
Digitalis purpurea (foxglove)	Leaves	Digitoxin, digoxin	Cardiac stimulant, diuretic
Dioscorea sp. (yam)	Tubers	Steroids	Production of cortisone, sex hormones, and oral contraceptives
Ephedra sp. (Mormon tea)	Stems	Ephedrine	Decongestant, treatment of low blood pressure and asthma
Erythroxylon coca (coca)	Leaves	Cocaine	Local anesthetic
Hydnocarpus kurzii (chaulmoogra tree)	Seeds, fruits	Ethyl esters of chaulmoogra oil	Treatment of leprosy and related skin diseases
Hydrastis canadensis (goldenseal)	Roots, rhizomes	Hydrastine	Treatment of inflamed mucous membranes
Nicotiana tabacum (tobacco)	Leaves	Nicotine	Stimulant
Papaver somniferum (opium poppy)	Latex from capsule	Morphine, codeine	Narcotic, analgesic, cough suppressant
Penicillium notatum (penicillin)*	Hyphae	Penicillin	Antibiotic
Podophyllum peltatum (May apple)	Roots, rhizomes	Podophyllotoxin	Treatment of venereal warts
Rauwolfia serpentina (rauwolfia)	Roots	Reserpine	Treatment of high blood pressure and psychosis
Salix sp. (willow)	Bark	Salicin	Analgesic, anti-inflammatory, treatment of rheumatoid arthritis and headaches

*Fleming's discovery of penicillin (see Chapter 1) involved *Penicillium notatum*, but most commercially produced penicillin today is derived from X-ray-induced mutants of *Penicillium chrysogenum*, which produce more than 1,000 times the penicillin originally derived from *P. notatum*.

table E.2

Some Important Hallucinogenic Plants and Fungi

Plant	Plant Parts Consumed	Psychoactive Compounds
Acorus calamus (sweet flag)	Rhizomes	α-asarone, β-asarone
Amanita muscaria (fly agaric mushroom)	Basidiocarp	Muscimole, muscazone
Atropa belladonna (belladonna)	Leaves, roots	Scopolamine, hyoscyamine, atropine
Claviceps purpurea (ergot)	Sclerotia	Ergine, isoergine
Datura sp. (jimson weed)	Stems, leaves, seeds	Hyoscyamine, scopolamine
Hyoscyamus niger (henbane)	Leaves, roots	Hyoscyamine, scopolamine
Ipomoea violacea (morning glory)	Seeds	Ergine, isoergine
Lophophora williamsii (peyote or mescal)	Shoots (crowns)	Mescaline
Mandragora officinarum (mandrake)	Leaves, stems	Hyoscyamine, scopolamine
Psilocybe sp. (psilocybe mushroom)	Basidiocarp	Psilocybine
Psychotria viridis (psychotria)	Leaves	Dimethyltryptamine (DMT)
Rivea corymbosa (morning glory)	Seeds	Ergine, isoergine
Trichocereus pachanoi (San Pedro cactus)	Stems	Mescaline

All chemicals in plants are potential drugs. A chemical that oozes from a tree's bark to discourage hungry caterpillars from eating it may also be an effective drug. For example, a chemical in the bark of the Indian neem tree keeps desert locusts off the tree. The people of Serengeti National Park in East Africa chew the twigs of these trees to prevent tooth decay. Other more familiar plants have healing properties as well (tables E.1, E.2).

Today's natural-products chemists are a modern version of the medicine man (or woman) who traditionally explored the healing power of plants (fig. E.6). Herbal medical practices may have begun in prehistoric times, when some individuals became botanical experts by sampling plants themselves. Clay tablets carved 4,000 years ago in Sumeria list several plant-based medicines, as do records from ancient Egypt and China. Roman philosopher Pliny the Elder wrote in the first century A.D., "If remedies were sought in the kitchen garden, none of the arts would become cheaper than the art of medicine." Modern-day medicine men called the *bwana mgana* practice herbal medicine in the region of East Africa explored by Isao Kubo.

figure E.6

Mark Plotkin, a scientist from Conservation International, collecting medicinal plants with the advice of a Wayana medicine man in southeastern Surinam. Despite the importance of such plants (e.g., as sources of medicine), the opportunities for gathering these plants are rapidly disappearing as tribal cultures are lost. Because written records about plants do not exist in many traditional societies, we are losing much valuable knowledge.

A good example of medicines derived from plants used today are the chemicals called *alkaloids,* which come from the periwinkle plant and other species. Alkaloid-derived drugs such as leucocristine and vincaleukoblastine (both from periwinkle; *Vinca roseus*) have helped revolutionize the treatment of some leukemias (blood cancers), and alkaloid narcotics derived from the opium poppy, including morphine, are excellent (but addictive) painkillers. Today, nearly half of all prescription drugs contain chemicals manufactured by plants, fungi, or bacteria, and many other drugs contain compounds that were synthesized in a laboratory but modeled after plant-derived substances. The many medicines that we use from the plant kingdom are a compelling reason why we must stop destroying the world's tropical rain forests, where plant life is so abundant and diverse that all of the species have not even been identified, much less studied.

Case Study: Plants and Malaria

One disease whose spread has been greatly influenced by plants is malaria, which kills more people worldwide each year than any other disease. Malaria starts with chills and violent trembling and progresses to an extremely high fever accompanied by delirium. Finally, the person sweats profusely, is completely

exhausted, and develops a dangerously enlarged spleen. The disease strikes in a relentless cycle, with symptoms returning every 2–4 days. Within weeks, the sufferer either dies or manages to marshall the body's immune defenses against the invading parasite that causes the disease.

Malaria is an interesting example of the relationship between plants and human disease, because plants contribute to both its spread and to its cure. Like many parasitic diseases, the malaria parasite (*Plasmodium falciparum, vivax, malariae,* or *ovale*) must spend part of its life cycle within an intermediate host organism—mosquitoes of the genus *Anopheles.* This insect must bite humans for the disease to spread. The type of vegetation in an area determines whether the mosquito, and the parasite it carries, will thrive. Unfortunately, agriculture often ushers in malaria by replacing dense forests with damp rice fields that are a haven for the mosquitoes. Indeed, in heavily infested areas of Africa, residents are bitten by disease-carrying mosquitoes *every night.*

Some plant products can kill the malaria parasite. In the sixteenth century, natives of Peru gave Jesuit missionaries who were on their way to Europe their secret malaria remedy—bark of the cinchona tree (fig. E.7). However, it was not until 1820 that two French chemists extracted the active

(a)

(b)

figure E.7

Malaria and quinine. (a) Cinchona bark contains quinine, a drug used to combat malaria. (b) Quinine being extracted from cinchona bark.

ingredient from cinchona bark. This substance, which they called quinine, was the standard treatment for malaria until the 1930s. As malaria parasites developed resistance to quinine, malaria spread through the developing nations as more and more land was cleared for farming. New natural antimalarial drugs were sought, and a promising one has come from ancient Chinese medicine.

In 1967 researchers in the People's Republic of China began a systematic study of all plants known to have medicinal properties in search of a new drug to fight malaria. In a document entitled "Recipes for 52 Kinds of Diseases" unearthed from a Mawangdui Han dynasty tomb from 168 B.C., a plant called *qinghao* was described as a treatment for hemorrhoids. A reference from A.D. 340 cited the same plants (also known as *Artemisia annua*, or sweet wormwood, a relative of tarragon and sagebrush; fig. E.8) as a treatment for fever. In 1596 a Chinese herbalist prescribed *qinghao* to combat malaria.

In the 1970s chemists isolated the active ingredient from sweet wormwood and called it *artemisinin*. By 1979 it had been tested on more than 2,000 malaria patients, in whom it cured the fever in 72 hours and eliminated parasites from the blood within 120 hours. The drug is more than 90% effective in treating cerebral malaria, the most severe form of the disease. In animal tests, artemisinin is proving effective against other parasitic diseases as well. Not surprisingly, other animals use plants as drugstores (see reading E.4, "Nature's Medicine Cabinet").

figure E.8

Dried leaves and flowers of *Artemisia annua* contain artemisinin, an antimalarial drug used in China for centuries.

NATURE'S MEDICINE CABINET

To learn more about animal behavior, ecologist Holly Dublin spent most of 1975 doing something rather unusual: tracking a pregnant elephant in Kenya. Dublin noticed that the 60-year-old expectant mother seldom changed her routine of walking about 5 km per day. However, one day the elephant changed her routine: she walked 28 km to a riverbank and began eating leaves from a species of tree that Dublin had never seen an elephant eat before. Before leaving, the elephant ate the entire tree. Four days later, the elephant had her baby.

Dublin was puzzled by the elephant's abrupt change in behavior and unusual meal. Did the tree eaten by the elephant have anything to do with inducing birth? To her surprise, Dublin later learned that pregnant Kenyan women induced labor by drinking tea made from the tree's bark and leaves.

Dublin's observations are among a growing number of studies suggesting that animals use plants as drugstores:

- Chimps often eat leaves of the shrub *Vernonia amygdalin* when they're tired and sick. The plant is used by African tribes to cure the same symptoms. Similarly, chimps eat leaves of *Aspilia*, a member of the sunflower family. These leaves contain thiarurbine-A, a red, sulfur-containing oil that kills pathogenic bacteria and parasitic worms. Humans use extracts of the oil as anticancer drugs.

- Wild rhesus monkeys often eat dirt with their food. That dirt contains much kaolin, a clay that detoxifies many poisons and is the active ingredient in Kaopectate, an antidiarrheal medicine.

Most biologists don't think that these and other examples are coincidence. Rather, they suspect that animals doctor themselves by using plants as preventive medicine.

Agriculture: Plants as Food

Obtaining food today is as simple as a trip to the supermarket or a visit to the garden or farm stand, but for our ancestors living 12,000 years ago, finding a meal was quite a challenge. Tribes of these people moved constantly in search of food (fig. E.9). In what is now the United States, some of our forebearers hunted the large plant-eating mammals that roamed the great plains, while others collected seeds, nuts, roots, fruits, and grasses. These hunter-gatherers were at the mercy of the environment. If there was no food where they were, they had no choice but to move on or starve.

From Hunter-Gatherers to Farmers: The Dawn of Agriculture

People first began to domesticate plants in the Fertile Crescent (today's Lebanon, Jordan, Syria, Israel, Turkey, and Iraq) of the eastern Mediterranean about 12,000 years ago. The first plants to be grown agriculturally in these regions were probably barley and wheat, followed soon thereafter by lentils and peas. Several factors contributed to this first spark of civilization. For example, the last ice age was ending. As the glaciers receded, the land that was revealed became inhabited by a vast assemblage of animals and flowering plants. Humans no longer had to follow herds of animals or gather berries to eat, because food could be found in more areas. People could stay in one place, eating readily available small game, fish, and wild plants.

By about 10,000 years ago, humans learned to take care of certain animals, such as wild sheep and goats, so that they

figure E.9

Hunter-gatherers in central Africa. The woman carries the family dog; the man has caught a turtle. Hunter-gatherers such as these people do not build cities.

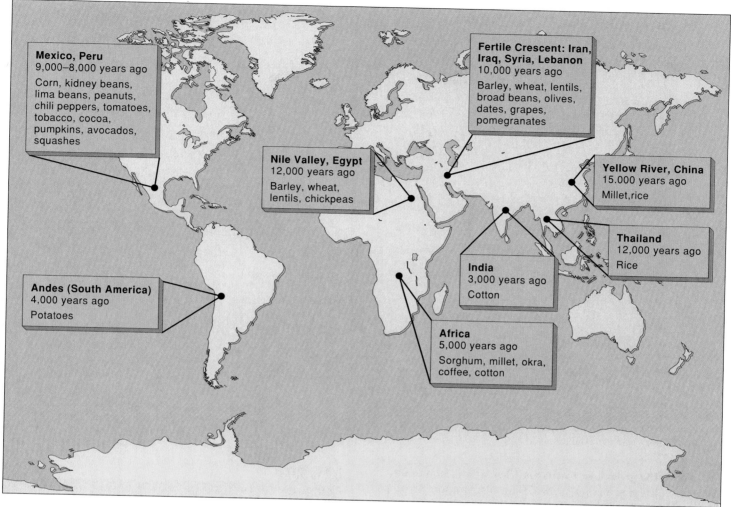

figure E.10

The domestication of agriculturally important plants. Farming arose independently in many places throughout the world.

could have a continual supply of milk and meat. People also learned to manipulate the wild weeds that sustained many tribes. For example, they learned to leave seeds in the ground to ensure a future supply of food. Gradually, these early farmers discovered more about plant reproduction and used their knowledge to improve their food supply. They deduced by trial and error when and how to plant the correct seeds in the right soil and to see that the seeds got enough water to sprout. They learned to recognize when plants were ready to be harvested. People saved the seeds from one season's most useful plants to sow the next year's crop, thereby encouraging certain combinations of traits (a practice called *artificial selection*) that might not have predominated in the wild. Later on, people learned to preserve their harvested food by drying it, thereby reducing even further their dependence on the unpredictable weather and climate. This domestication of animals

and the intentional planting and cultivation of crops marked the birth of agriculture.

Farming probably arose independently in many places around the world (fig. E.10). Archeological evidence indicates that domestication of sheep and goats and cultivation of wheat and barley began in the Fertile Crescent about 12,000 years ago. Gradually, agriculture spread to eastern Europe about 8,000 years ago, and to the western Mediterranean and central Europe about 7,000 years ago. Egyptian and Assyrian tombs, mummy wrappings, paintings, and hieroglyphics from 4,000 years ago depict a rich agricultural society that cultivated pomegranates, olives, grapes, figs, dates, and cereals. A similar spread of agriculture occurred in the Americas, based upon native corn. Yams and cassava were early crops in Africa (fig. E.11). Tomatoes and coffee are the only two crops domesticated in the last 2,000 years.

figure E.11

Cassava (*Manihot esculenta*) is an important food crop in the tropics, particularly in Africa. The large, tuberous roots are peeled, boiled, and mashed. Tapioca is made from cassava.

With the growth of towns and then cities, human influence on cultivated plants spread. The distribution of crops across the planet today indicates that when humans colonized new lands, they often brought their native plants with them and likewise carried new plant varieties from the new land back to the old (for example, corn was introduced to Europe from the Americas by Christopher Columbus in the late fifteenth century).[3] Tracing the origins of crop plants is difficult because archeological evidence is often destroyed over time. The available evidence, however, suggests that different crop plants were domesticated in the Old and New Worlds. Today, thanks to efficient transportation, many different kinds of plants are grown far from where they originated. Moreover, our relatively efficient farming methods free most people to pursue other activities. For example, only about 3% of the U.S. workforce are farmers; these people provide almost all of our food as well as more than 25% of our exports.

The Lore of Plants

Agriculture among Native Americans differs significantly from modern agriculture. Native Americans' agriculture, like the rest of their culture, is based on a harmony with nature. Native American agriculture is effective and sophisticated:

- About 100 of the 120 known food crops were domesticated by Native Americans, including potatoes, avocados, beans, squash, pumpkins, bell peppers, tomatoes, peanuts, pecans, cashews, many berries, and grapes. They made chocolate, flavored their food and drinks with vanilla, grew pineapples (which later were sent to Hawaii), and even coated their popcorn with maple sugar, thus making a treat similar to Cracker Jack.

- Native Americans used selective breeding to develop a wide variety of crops. For example, they developed more than 300 varieties of corn from teosinte, a wild grass.

- Native Americans set up huge farms. Columbus reported cornfields more than 18 mi long that included irrigation canals, crop rotation, and fertilizers.

Most aspects of Native American agriculture were motivated by necessity. Today we continue to learn from Native Americans, the first true environmentalists.

---

3. Columbus brought back the kernels and the native word for the grain: *mahiz*, which survives today as the common name for American corn, *maize*. Only in the United States is the plant called *corn*.

Plants as Food

Our most important use of plants is as food. Today, as much as 90% of the total calories consumed by humans comes from crop plants:

Grains: wheat, rice, corn, sorghum, millet, barley, oats, rye

Tuber and root crops: potato, yam, sweet potato, cassava

Sugar crops: sugarcane, sugar beet

Protein seeds: bean, soybean, pea, lentil

Oil seeds: olive, soybean, peanut, coconut, sunflower, corn

Fruits and berries: citrus, mango, banana, apple

Vegetables: cabbage, lettuce, onion

Most of the calories in our diet come from wheat, rice, corn, potatoes, yams, and cassava (the source of tapioca). Cereals provide much dietary carbohydrate, and the seeds of legumes (beans, lentils, peas, peanuts, and soybeans) are rich in protein. Cereals have different types of amino acids than legumes, so eating these two foods together provides a good balance of proteins. The best plant-derived nutrition combines a cereal (a rich source of carbohydrate) with a legume (a source of protein), a green, leafy vegetable (rich in vitamins and minerals), and perhaps small amounts of sunflower oil, avocado, or olives (which provide fats).

In the tropics, bananas and coconuts are staples of peoples' diets (fig. E.12). Bananas have been cultivated in tropical Africa for the past 2,000 years; there, starchy bananas called *plantains* are a major part of the diet (sweeter varieties are popular in the United States). The 50–100 coconuts that grow on a palm tree each year provide proteins, oils, and carbohydrates, while the leaves and trunks are excellent building materials.

Although we eat many kinds of plants, cereals such as corn and wheat influence our lives more than do other plants. The importance of cereals cannot be understated. All of the world's great civilizations have been based on the cultivation of cereals: maize was the basis of the Incan, Aztecan, and Mayan empires; rice was the staple food in ancient China, Japan, and India; and the civilizations of Egypt, Rome, Greece, and Mesopotamia were based on wheat. Without an abundant and reliable source of these cereals, villages could not have grown into cities, nor cities into empires. Our lives would be very different without cereals.

Cereals: Staples of the Human Diet

Cereals such as wheat, rice, and corn include nine of the ten most economically important groups of plants and provide about half of all the protein in our diets. These plants are all members of the grass family (Poaceae). Rice is the primary source of food in Asia, while corn is a major part of the human diet in South America. In the United States, many of our foods are based on wheat.

figure E.12

(a) Bananas growing in southern Mexico. Bananas are an important crop in the tropics. (b) The 50–100 coconuts that grow on a palm tree each year provide carbohydrates, oils, and proteins.

JOJOBA

Jojoba (*Simmondsia chinensis*) is one of several recently developed crops that has great potential (reading fig. E.5). Its large seeds contain up to 60% of a light yellow, odorless, liquid wax. This wax is used as a lubricant, an ingredient of cosmetics, and even a noncholesterol cooking oil. It is also an excellent substitute for the sperm-whale oil needed in heavy machinery and to make ballistic missles. Jojoba grows in hot deserts that are unsuitable for other crops.

Although jojoba is popular today, before 1980 it was an obscure, low-spreading, grayish green bush growing in the southwestern United States and Mexico. In that year, seeds from wild-growing plants were collected and used to start the first cultivated plantings. The next season, researchers took cuttings from the healthiest bushes and used them to start the next crop. During recent years, the jojoba crop in the United States has grown by at least 40% per year.

reading figure E.5
Jojoba (*Simmondsia chinensis*) and products made from jojoba. Jojoba can grow in hot deserts that are unsuitable for other crops.

Wheat (*Triticum aestivum*) Wheat grows in a range of climates, from the Arctic to the equator, from below sea level to 3,048 m (10,000 ft) above it, and from areas with less than 5 cm (2 in) of annual rainfall to places with more than 178 cm (70 in) of rain per year. Most wheat is used for food. The grain of the "hard" bread wheat contains 11%–15% protein, mostly of a type called *gluten*. When ground and mixed with water, it forms an elastic dough that is excellent for making bread. The grain of the "soft" bread wheat contains 8%–10% protein and is best for making cakes, cookies, crackers, and pastries. Today, the United States produces about 12% of the world's wheat—about 70 million metric tons per year, valued at about $8 billion.

Wheat was probably the earliest domesticated plant; we've cultivated it for at least 12,000 years (fig. E.13). Today there are three kinds of wheat, based upon their number of chromosomes: 14 (diploid), 28 (tetraploid), or 42 (hexaploid) (fig. E.14). Bread wheats are hexaploid, and the durum wheat used to make macaroni is tetraploid. Geneticists study the number of chromosomes in modern wheats to reconstruct the evolution of these important crop plants. It is a fascinating and complex story. Wheat as we know it apparently evolved from an accidental merger between a distant wheat ancestor and a weedy grass. It may have happened as follows: About 10,000 years ago, wandering peoples in what is now Jericho, in Israel, came upon a rich oasis, a spring in the desert ringed by hills

(a)

(b)

(c)

figure E.13

Wheat. (a) Wheat (*Triticum*) fields in Colorado. Grasses such as wheat provide a staple food for humans. (b, c) Harvesting and winnowing of wheat in Tunisia, North Africa. Wheat has been cultivated for at least 12,000 years.

covered with wild grass. While looking for food, the people discovered that the grain held in the grass, when ground, made a fine flour. For many years, they did not know how to encourage the grass's growth intentionally, so each season they would simply forage for whatever nature provided.

Then about 8,000 years ago, something happened that was to have profound effects on agriculture and human civilization. The grass that the people had grown fond of, really an ancient wheat called *einkorn*, crossbred with another type of grass that was not good to eat (fig. E.14). The hybrid grass then underwent a genetic accident (nondisjunction; see Chapter 10) that prevented the separation of chromosomes in some of the developing germ cells. The result was a new type of plant that had twice the number of chromosomes as either parent plant (i.e., twenty-eight chromosomes total). This plant was called *emmer wheat*, and it was a better

source of food than the parent wheat whose grains had become a staple. The doubled number of chromosomes in each cell produced a plant with larger grains. Also, the grains were attached to the plant in such a way that they could be easily loosened and spread by the wind, so that the new wheat was soon plentiful. It was probably about this time that early farmers learned to select seed from the most robust plants to start the crops of the next season.

Then about 6,000 years ago, another "mistake" of nature further improved the quality of wheat. Emmer wheat crossed with another weed, goat grass, and after another fortuitous "accident" of chromosome doubling, led to bread wheat, which

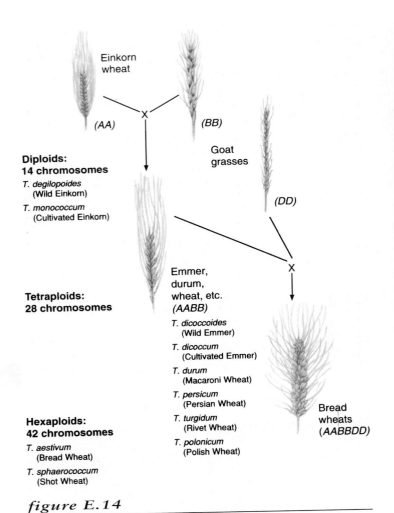

Einkorn
wheat

(AA) X (BB)

Goat
grasses

**Diploids:
14 chromosomes**

T. degilopoides
(Wild Einkorn)

T. monococcum
(Cultivated Einkorn)

(DD)

**Tetraploids:
28 chromosomes**

Emmer,
durum,
wheat, etc.
(AABB)

T. dicoccoides
(Wild Emmer)

T. dicoccum
(Cultivated Emmer)

T. durum
(Macaroni Wheat)

T. persicum
(Persian Wheat)

T. turgidum
(Rivet Wheat)

T. polonicum
(Polish Wheat)

X

**Hexaploids:
42 chromosomes**

T. aestivum
(Bread Wheat)

T. sphaerococcum
(Shot Wheat)

Bread
wheats
(AABBDD)

figure E.14

Evolution of wheat.

figure E.15

Triticale, a popular grain that is a hybrid of wheat (genus *Triticum*) and rye (genus *Secale*). Cells of triticale are polyploid: they each contain a complete set of chromosomes from wheat as well as a set from rye. Triticale can tolerate the harsh environment of rye while producing the high yields of wheat.

has forty-two chromosomes (fig. E.14; also see reading 10.1, "Polyploidy in Plants"). Bread wheat has even larger grains than the emmer wheat that gave rise to it, but at a cost. Its ears are so compact that, on its own, the grain cannot be released. However, farmers collected the rich seed each season for food and kept a certain percentage to be planted the next season. The interdependence of humans and crops that is the basis of modern agriculture was thus born. Today, interesting hybrids, such as triticale, a combination of wheat (*Triticum*) and rye (*Secale*), add variety to our diets (fig. E.15). The cells of triticale are polyploid, containing a complete set of chromosomes from wheat and one from rye. Triticale has the high yield of wheat and can cope with the harsh environments of rye.

Corn (Zea mays) The tasty, sweet, or starchy kernels of the corn plant, *Zea mays*, sustained the Incas, Aztecs, and Mayan Indians of South America and the pilgrims of colonial Massachusetts. Today, corn continues to be a dietary staple from Chile to Canada. The corn crop in the United States presently exceeds 9 billion bushels a year and is used to make food products, drugs, and industrial chemicals, as well as to feed humans

and animals (fig. E.16). Corn is also being used to help solve a troubling environmental problem: plastics. After years of being dumped into our oceans, plastics that were made to last forever are coming back to haunt us. For example, merchant ships dump about 500,000 plastic containers into international waters every day. These and other plastics maim and kill marine life (last year, about 40,000 seals died after becoming entangled in plastics). Other plastics are filling our landfills and polluting our water. In one of the many examples of how plants are used to combat such problems, one company has recently patented a technique for making biodegradable plastics containing corn starch. This technique involves inserting tiny pellets of starch into the plastic, which helps the plastic to decompose to dust in only a few months. If this technology becomes popular, biodegradable containers and cups will soon be common. In the meantime, most products advertised as "biodegradable" fail to live up to their promise.

Like wheat, modern corn probably arose from a naturally occurring wild grass that initially produced a grasslike type of corn with small ears. Fossil evidence exists of such a primitive corn. A clue to the origin of modern corn comes from "teosinte," a group of wild grasses in the genus *Zea* that grow in Mexico today and probably have for thousands of years. The Indians call teosinte *madre de maize*, which means "mother of corn." Unlike other cereals, corn does not look

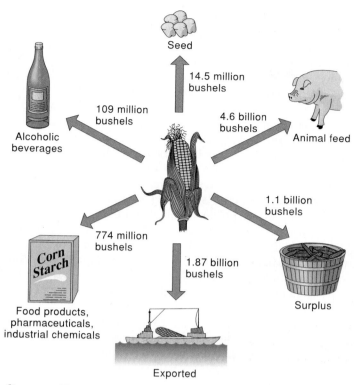

figure E.16

The fate of corn produced in the United States.

like its wild ancestors (fig. E.17a). However, there are a few similarities. Both species have twenty chromosomes and produce fertile hybrids when they are crossed. Wild teosinte sometimes grows on the outskirts of cultivated cornfields.

Evolutionary biologists have suggested hypotheses for corn's origins, based on the interfertility and morphological similarities between modern-day corn and teosinte. The first of these two hypotheses is that the ear of corn evolved directly from the ear of teosinte (fig. E.17). The cause of this change occurred about 7,500 years ago, when a population of wild teosinte faced an environmental stress that was endured by only a few plants. Farmers noticed that the unusually hardy plants had larger and better-tasting kernels, and they chose those plants to cultivate. Part of the teosinte tassel might have enlarged and became the corn tassel. Over the centuries, modern corn, with female ears and male tassels, was artificially selected by farmers seeking plants with plump and tasty kernels. Indirect evidence for this "ear conversion" hypothesis is that a single mutation will change the hard case around a teosinte kernel into the tunicate of corn (fig. E.17b).

The second hypothesis for the origin of corn from teosinte is that the male tassel of teosinte underwent a catastrophic sexual change into a kernel-bearing inflorescence (fig. E.17b). Indirect support for this "sexual transmutation" hypothesis comes from the common appearance of sexual

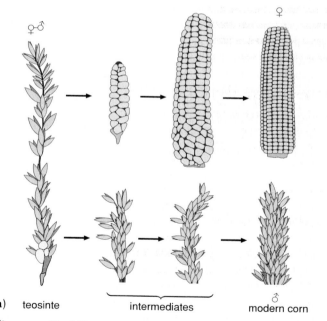

(a) teosinte   intermediates   modern corn

figure E.17

Corn and teosinte. (a) Teosinte (*Zea mexicana*), a wild grass, has structures that may have given rise to modern corn (*Zea mays*). The seeds of teosinte evolved into corn kernels, and the tassels eventually became the familiar tassels of corn. (b) Although no one knows what the earliest corn looked like, this modern ear represents a reasonable guess. (c) Modern corn probably evolved from teosinte. Note the double row of seeds in a single husk.

(b)

(c)

abnormalities in modern corn. These abnormalities include the formation of ears by male inflorescences and the occurrence of tassels in female ears. Moreover, environmental stress is known to induce such sexual reversals in teosinte.

Although modern-day maize may have arisen by the domestication of teosinte, the ears of teosinte and maize differ markedly: teosinte ears produce many fewer kernels that are enclosed by harder fruit cases that make harvest difficult (fig. E.17). The *tgal* gene, which regulates the structure of the fruit case, alone accounts for much of the difference between teosinte and maize. The maize ear was probably derived from the teosinte ear by a series of modifications, each principally controlled by one or two genes.

Charles Darwin, the father of evolution, was also interested in the origin of corn. Darwin used breeding experiments in his greenhouses to show that when corn plants continually self-fertilized, they produced increasingly weaker offspring—that is, inbred plants were more prone to disease and produced ears with fewer rows of kernels. Darwin noted, however, that the offspring of cross-pollinated plants were strong and healthy, with many rows of plump kernels. Although Darwin knew little about the genetic discoveries of his contemporary, Gregor Mendel, what he had demonstrated was the genetic phenomenon of hybrid vigor. Plants with unrelated parents were more vigorous than self-fertilized plants because they had inherited new combinations of genes.

In the first decade of the twentieth century, George Shull, at the Station for Experimental Evolution in Cold Spring Harbor, New York, extended Darwin's greenhouse experiments, thanks to the additional insight provided by the rediscovery of Mendel's laws of inheritance. Shull continually bred plants from the same ears of corn, artificially selecting highly inbred lines. The inbred corn plants were sickly looking, with small ears. But when Shull crossed two different inbred lines to one another, the offspring—hybrids—were exceptionally vigorous. Shull had developed hybrid corn, which revolutionized corn output worldwide. In the United States, corn production has jumped from 21.9 bushels per acre in 1930 to 95.1 bushels per acre in 1979 and well beyond that today, thanks largely to hybrid corn. Indeed, hybrid corn is the largest seed business in the world today. Many of us start our days with flakes made from this tasty grain (see reading 1.2, "Breakfast at the Sanitarium: The Story of Breakfast Cereals").

Today's harvest of corn is about 9 billion bushels per year. We feed about half of that to animals. Two centuries ago, however, much was used to make corn whiskey, which was consumed as a cure for colds, coughs, toothaches, and arthritis. These ailments must have been rampant: between 1790 and 1840, people drank an average of 5 gal of the whiskey per person per year.

Rice (Oryza sativa)   Rice is the most popular food in the world. It's been cultivated in Thailand for more than 12,000 years, and in Asia it comprises 80% of the human diet. In Taiwan, a hungry person is not asked, "Are you hungry?" or "What would you like to eat?" but "Would you like some rice?" In much of the world, as in Taiwan, rice is so much a dietary staple that it is synonymous with food. The 400 million metric tons of rice produced each year are grown in many types of environments, ranging from 53° north latitude to 40° south latitude. Today's rices include twenty wild species plus many cultivars, which are domesticated species produced by agriculture rather than by the evolutionary forces of natural selection (i.e., cultivars do not grow in the wild). The oldest species of rice for which we have fossil evidence, *Oryza sativa*, originated about 130 million years ago in parts of South America, Africa, India, and Australia, which were then joined into one landmass. Unlike wheat and corn, rice is an ancient plant that evolved in a tropical, semiaquatic environment.

Human cultivation of rice began about 15,000 years ago in the area bordering China, Burma, and India. By 7,000 years ago, efforts to control rice growth had spread to China and India; by 2,300 years ago, the crop was growing in the high altitudes of Japan. By 300 B.C., rice was a staple throughout Asia. It has only been over the past 700 years that rice has been eaten in West Africa, Australia, and North America.

As migrating peoples took their native rices with them to new lands, the plants adapted to a wide range of environments, from deep, salty water to the driest of drylands. From the 1930s to the 1950s many nations collected hundreds of native rice varieties, growing small amounts of each type every season, just to keep the collections going.

Other cereals used extensively by humans include barley, millet, oats, and rye. **Barley** was first grown in the eastern Mediterranean and today is grown mostly in Europe, North America, and Australia. Most barley is used as cattle fodder and to make malt used in distilling and brewing. **Millet** was first cultivated in China in about 2700 B.C. Although it is now grown mostly for cattle fodder in the United States, its resistance to drought is also being exploited to make it a food crop in tropical Africa. **Oats** probably originated as a weed growing with other cereals, such as barley or wheat. It was domesticated about 2,500 years ago. Today, it is used primarily as cattle fodder and in breakfast cereals. **Rye** originated as a weed growing among other cereals. It was first cultivated in southwest Asia in about 1000 B.C. and now is used as flour in rye bread and as food for cattle.

Feeding the World

The Struggle to Produce Enough Food

Throughout history, famines have killed millions of people. For example, Indian famines killed 10 million in 1769–1770, 1 million in 1866, 1.5 million in 1869, 5 million in 1876–1878 (as another famine in China killed 10 million), and another million people in 1900. Similarly, the Irish potato famine of 1846–1847 killed 1.5 million people and caused a mass exodus from Ireland. If you're Irish, chances are this famine drove your ancestors to North America.[4]

---

4. Before the famine, a typical Irish family (two adults and four children) ate about 250 lb of potatoes per week. Today, the average American eats about 115 lb of potatoes per year.

OUR FASCINATION WITH PLANTS: A FINAL LOOK

Throughout this book you've read about unusual plants and the equally unusual things we do with those plants. Here's a final look at some of our favorites:

- The world's oldest potted plant is the cycad *Encephalartos altensteinil* brought from South Africa in 1775 and now housed at the Royal Botanic Gardens, Kew, Great Britain.

- The largest hanging basket had a volume of 120 m³ (4, 167 ft³), weighed about 4,000 kg (4.4 tons), and contained 600 plants.

- The longest single unbroken apple peel on record is 52.2 m (172 ft, 4 in) long, peeled by Kathy Wafler in 11.5 hours from an apple weighing 570 g (20 oz).

- The longest distance at which a grape thrown from level ground has been caught in someone's mouth is 99.8 m (327 ft, 6 in).

- The largest jack-o'-lantern in the world was carved from a 375 kg (827 lb) pumpkin.

- The smallest seeds are those of epiphytic orchids; 1.2 million seeds weigh only 1 g.

- The duration record for staying in a tree is more than 26 years by Bungkas, who went up a palm tree in 1970 and has been there ever since. He lives in a nest that he made from branches and leaves.

- The largest rutabaga ever grown weighed 22.1 kg (48 lb, 12 oz), the largest cabbage 56.3 kg (124 lb), and the largest squash 372 kg (821 lb). No word about who ate those whoppers.

- If you've ever wondered how many pickled peppers were in Peter Piper's peck, the answer—in jalapeños—is about 22.5. It's anyone's guess, however, how many seashells he sold down by the seashore.

- The average American eats about 52 kg of potatoes each year, and more than 11 billion kg of potatoes go to make French fries. Luckily for potato lovers, German philosopher Friedrich Nietzsche was wrong when he suggested that "A diet which consists predominantly of potatoes leads to the use of liquor."

There are many factors involved in feeding the world's population, the most important of which is the size of the population. Our population is growing extremely fast. For example, at the beginning of agriculture about 12,000 years ago, only about 5 million people lived on earth. By the time Christ was alive, the population had grown to 250 million. Thereafter, it doubled to 500 million in 1650, doubled again to 1 billion in 1850, doubled again to 2 billion in 1920, and doubled again to 4 billion in 1976. Today's ever-increasing population of more than 5.6 billion demands huge amounts of food.

Botanists have struggled to produce enough food for the population. One of the pioneers in this struggle was Norman Borlaug, a plant geneticist. Borlaug began creating new varieties of plants in 1944 and soon achieved remarkable results. For example, in 1944 Mexico imported wheat to feed its citizens. By 1964, however, Mexico was exporting wheat. Since 1950, production has quadrupled. Similar advances have been made in Pakistan and other countries. This dramatic increase in production was called the Green Revolution. For his work, Borlaug received a Nobel prize in 1970—not so much for the technology that produced the high-yielding crops as for his humanitarianism as he tried to help feed the world. However, even his Green Revolution hasn't been able to keep up with the ever-increasing demand for food. Moreover, critics claim that the Green Revolution actually worsened the problem because it created social and environmental havoc.

The Green Revolution and similar programs have not eliminated world hunger because they have not addressed the problem that drives world hunger: overpopulation. Consider these facts:

- Every year, 100 million people—more than three people per second—are added to the earth's population. That's the equivalent of adding 27 cities the size of Los Angeles.

- Earth's population is now at 5.6 billion people. By the year 2000, it will exceed 6.3 billion. More than 1 billion of those people—that's equivalent to the entire population at the beginning of the Industrial Revolution—will be added in the 1990s.

- More than 95% of the earth's population live in developing countries.

- Each day, the average urban resident of the United States uses about 150 gal of water and 15 lb of fossil fuel while generating nearly 120 gal of sewage, 3.4 lb of garbage, and 1.3 lb of pollutants.

- Earth's population has doubled in less than 35 years and now grows at an annual rate of 1.8%. This means that each day there are 260,000 more people to feed.

The consequences of these facts are devastating. For example, in 1990, more than 1.1 billion people—that's one in every five—lived in abject poverty. More than 500 million people

figure E.18

Deforestation and reforestation of a rain forest. (a) LANDSAT photo showing Rondonia fires. (b) Fire in a Guatemalan rain forest. (c) Destruction of a rain forest by fire. (d) Tree seedlings grown for reforestation of rain forests in Brazil.

were getting less than 80% of the recommended intake of calories; during the hour or so that it takes for us to eat our Thanksgiving Day meal, more than 1,600 people (mostly children) die of hunger. Moreover, our efforts to feed those people are damaging the environment: we're quickly losing topsoil (see reading 20.2, "Losing the Soil") and are polluting the soil and water with herbicides and insecticides. As we bring more land into cultivation, we destroy habitats and threaten many native species of plants and animals (fig. E.18). Earth cannot continue to support such an increasing population.

Prospects for the Future: Beyond the Green Revolution

Although many of the wonders you have learned about in this text have greatly increased crop yields and our ability to produce food, more people are starving than ever before. Our rapidly growing population, combined with ineffective governmental policies and food-distribution methods, has overwhelmed our agricultural system. Solving this problem is a tremendous—perhaps impossible—challenge. However, all hope is not lost.

Throughout this book you've read about one of the most promising tools for helping to feed the world: biotechnology.[5] Just as producing hybrid corn in the 1930s doubled the corn harvest, biotechnology is being hailed as the "second green revolution" that will produce plants that protect and nourish themselves and, in the process, help feed the world. For example, in 1994 a group of researchers used genetic engineering to identify and manipulate a self-incompatibility gene (also called the S gene). A plant having an active S gene recognizes and rejects its own pollen, and therefore cannot fertilize itself. The ability for growers to prevent plants from fertilizing themselves could double the yield (and cut labor costs by half) of many vegetables and flowers.

Other botanists are now using genetic recombination to create high-yielding crops that resist disease, drought, and pests. Still others are improving the caloric and nutritional value of crops. In years ahead, genetically engineered plants will become a leading source for increasing food production. However, biotechnology is not the only tool in the war on hunger.

Another strategy to increase our supply of plant foods is to look for new crops among the many naturally occurring plants. Fewer than 30 of the 240,000 species of flowering plants provide more than 90% of plant-based foods eaten by people. One of the most promising plants that botanists are studying as a new source of food is the majestic-looking amaranth, a member of the pigweed family (fig. E.19). These plants stand about 8 ft (2.4 m) tall and have broad purplish green leaves and massive seed heads. Each plant produces about a half-million mild, nutty-tasting, protein-rich seeds, each the size of a grain of sand. The flowers are a vivid purple, orange, red, or gold. Interestingly, amaranth was cultivated extensively in Mexico and Central America until the arrival of the Spanish conquistadors in the early 1500s, who banned the plant from use as a crop because of its importance in the Aztec religion.

figure E.19

Amaranth is grown for its cereal-like grain and edible leaves. The plants shown here are growing in Pennsylvania.

Amaranth is being grown experimentally at the Rodale Research Center in Kutztown, Pennsylvania, and is available from commercial seed catalogs for home garden use. It grows rapidly, can tolerate a wide range of environmental conditions (high salt, high acid, high alkalinity), yields many seeds, and comes in many varieties. Problems with cultivating amaranth include controlling weeds and pests and harvesting the tiny seeds.

Nutritionally, amaranth is superb. Its seeds contain 18% protein, compared to 14% or less for wheat, corn, and rice. Amaranth is rich in amino acids that are poorly represented in the major cereals. The seeds can be used as a cereal, a popcornlike snack, and a flour to make graham crackers, pasta, cookies, and bread. The germ and bran together are about 50% protein, making them ideal to add to prepared foods and animal feeds. The broad leaves of amaranth are rich in vitamins A and C as well as the B vitamins riboflavin and folic acid. They can be cooked like spinach or eaten raw in salad.

Botanists are also studying a variety of spinachlike plants as new sources of food (fig. E.20). Other plants studied as potential sources of food include saltbush (*Atriplex*) and ironweed (*Bassia scoparia*), both of which contain much

5. Investors are also banking on the promise of biotechnology. For example, Calgene, Inc. of Davis, California, spent $25 million to produce its Flavr Savr™ tomatoes. These tomatoes contain an antisense ("reverse") copy of the gene that codes for the expression of polygalacturonase, an enzyme that softens the cell walls during fruit development. This delays ripening, thereby improving the flavor and doubling the shelf life of the tomatoes to about 2 weeks. The U.S. Food and Drug Administration approved the tomato on 18 May 1994, thereby making it the first food to be genetically altered via molecular biology. (Although foods such as milk and cheese already benefit from biotechnology to enhance production or processing, Calgene's tomato is the first food to have new genes that could not be produced by conventional plant breeding.) Investors expect these tomatoes to capture at least 15% of the $5-billion-per-year tomato market by 1999.

(a)

(b)

figure E.20

Spinachlike plants used as food. (a) A spinach substitute is obtained from *Celosia argentea*, a member of the amaranth family and a close relative of these cockscomb plants (*Celosia cristata*). (b) Cariru (*Talinum*), a locally used spinach in Amazonia.

protein. These plants grow well in harsh conditions (salty soil, hot weather) and are a promising source of food for livestock.

Botanists have also established gene banks and pollen banks to help conserve rare plants and to increase the world food supply (fig. E.21). For example, to offset a potential disaster due to reliance on only a few types of rice, the International Rice Research Institute was founded in 1961 in the Philippines. It soon became a clearinghouse for the world's rice varieties, storing the seeds of 12,000 natural variants by 1970 and of more than 70,000 by 1983. Representatives of other important crops are being banked as well. For example, the National Germplasm System in the United States is the world's largest distributor of germplasm and seeds: in 1994 it distributed more than 300,000 samples to more than 100 nations. Similarly, potato cells are stored at the International Potato Center in Sturgeon Bay, Wisconsin, and wheat cells are banked at the Kansas Agricultural Experimental Station. Pollen and seeds from 250 endangered species of flowering plants have been frozen at a plant gene bank at the University of California at Irvine.

Plant banks offer three priceless services to humanity: a source of variants in case a major crop is felled by disease or an environmental disaster; the return of endangered or extinct varieties to their native lands; and, perhaps most important, a supply of genetic material from which researchers can fashion useful plants in the years to come, even after the species represented in the bank have become extinct.

Establishing a seed or pollen bank is one solution to a growing problem in traditional plant agriculture—the decrease in genetic diversity that makes a crop vulnerable to disease or a natural disaster because it does not have resistant variants. For example, the genetic uniformity of the Irish potato crop made it susceptible to *Phytophthora infestans*, which almost wiped out the crop between 1846 and 1848. (Today in Ireland more than sixty varieties of potatoes are actively farmed to prevent a similar kind of disaster.) However, such lessons are not always heeded. For example, in 1970 more than 70% of the North American corn crop was restricted to just six varieties. A new strain of southern corn leaf blight fungus destroyed 15% of that crop, at a cost of $1 billion. Gene banks and pollen banks provide the diversity to avoid such disasters in the future.

Our ability to feed ourselves will depend largely on our ability to preserve the genetic diversity of our plants, increase the quality of our crops, produce new hybrids, and develop new crops. Although biotechnology will play a major role in this, we will never be able to eliminate world hunger unless governments reform their food-distribution policies and we greatly slow the growth of our population. Plant biotechnology alone is not the answer to world hunger.

figure E.21

Banking plant genes at the cryogenic gene bank at the University of California at Irvine. Pollen is frozen in small, plastic ampules.

Epilogue Summary

Plants help provide us with food, clothing, shelter, and medicine. About 12,000 to 10,000 years ago, groups of people gradually changed from a hunter-gatherer life-style to an agricultural way of life, intentionally saving and planting seeds from the strongest individual plants of crops that could be used as food. Encouraging the propagation of certain individuals artificially selected particular traits; in this way, humans have influenced plant evolution.

Cereals such as corn, wheat, and rice are members of the grass family, whose edible seeds (grains) can be stored for long periods. Modern wheats contain two, four, or six sets of chromosomes, and they probably evolved from a natural cross between ancient einkorn wheat (a diploid) and a wild grass (also a diploid), followed by chromosome doubling in some germ cells about 8,000 years ago. This eventually yielded tetraploid emmer wheat. Emmer wheat also crossed with a wild grass to produce the first hexaploid wheats, which today are used to make bread and other baked goods.

Corn probably evolved from an ancient grass relative, teosinte. Both corn and teosinte have twenty chromosomes, they sometimes grow in the same fields, and they can cross-breed.

About 7,500 years ago, an environmental stress may have selected teosinte plants with large kernels, and early farmers may then have cultivated these individual plants. Breeding experiments led to the development of hybrid corn, which results from crosses between separate inbred lines to yield bountiful crops.

Rice is perhaps the most widely consumed modern cereal crop, and fossils of the plants date back some 130 million years. Today, rice grows in a wide range of environments. In 1961 a rice seed bank was started in the Philippines to preserve different varieties and prevent reliance on a few types. Seed and pollen banks have since been founded for other valuable plant species.

For centuries, humans and other animals have used plants for their medicinal properties. Today, natural-products chemists use a combination of laboratory techniques and information from folklore and herbal medicine to make compounds with effects similar to those of plant-derived compounds in the search for new and more effective drugs.

 ### What Are Botanists Doing?

Some of the biotechnological improvements in dicots have relied on the transfer of a plasmid of *Agrobacterium tumefaciens* into the plant. However, the *Agrobacterium* system does not generally work well with monocots, because they are not usually susceptible to infection by *Agrobacterium*. How, then, are biotechnologists trying to improve monocots such as corn?

Go to the library and read a recent article about biotechnological improvements in monocots. What other approaches do you think should be tried?

 ### Writing to Learn Botany

In an effort to conserve and perhaps increase the genetic diversity of crop plants, many nations donate seeds to gene banks. Sometimes, however, the material donated by a developing nation is used in developed countries to breed or engineer new plant varieties. The new plants—based on the genetic material from the poor nation but the technology of the more wealthy nation—are then sold to the developing nation. Can you think of a more equitable way for nations to cooperate in the development of new plant varieties?

Questions for Further Thought and Study

1. International agricultural officials have debated for a decade about whether Western countries should pay royalties on genes that germplasm banks have given them for free. The genes can be extremely valuable; for example, a barley gene imported for free from Ethiopia protects the $160 million U.S. barley crop from the yellow dwarf virus. Should countries pay royalties for such genetic resources? Why or why not?

2. Describe the ways in which scientists hypothesize that modern varieties of wheat, corn, and rice arose.

3. Why is reliance on only a few varieties of a crop plant dangerous? Cite an example of when such reliance led to devastation.

4. Artificial selection has produced a modern corn plant that has tasty, nutritious kernels. However, modern corn cannot reproduce successfully without the help of people; the kernels are so tightly protected within the ears that a human or animal must disperse the kernels to start the next year's crop. Do you think that this intervention with a natural process is justified? Why or why not? Are there limits to our intervention? If so, what are they?

5. Devise an experiment to test whether the scenario of corn evolving from teosinte is possible.

6. You are planning your vegetable garden and intend to purchase five packets of carrot seeds. The seed catalog describes a new variety of carrot that is resistant to nearly every known garden pest and produces long, highly nutritious carrots in a variety of climates. It sounds too good to be true, but the company that publishes the catalog has had a lot of experience in plant breeding. If you want to obtain as many carrots as possible, would you be better off buying five packets of the new variety or five different types of carrot seeds? Give a reason for your answer.

7. Botanists and others once believed in the so-called *Doctrine of Signatures*, which stated that the shape of a plant part indicated its usefulness in treating a particular ailment. Thus, walnuts, which resemble brains, were used to treat brain disorders, and liverworts, which resemble the liver, were used to treat liver ailments. Some people still believe such claims. How would you test such claims?

8. Early settlers fattened their Thanksgiving turkeys and hogs on beechnuts. Perhaps not surprisingly, writer A. A. Milne gave Piglet a home in a beech tree in *Winnie the Pooh*. In what other fictional accounts do plants play a role?

9. We use different plants for different purposes. For example, wood of poplar does not splinter and therefore is used to make toys, tongue depressors, and Popsicle sticks. How are other plant products suited for their function?

10. Many tropical cultures depend heavily on plants such as sugarcane, bananas, and coconuts. Why haven't these plants had the same worldwide impact as cereals?

Web Sites

Review the "Doing Botany Yourself" essay and assignments for the Epilogue on the *Botany Home Page*. What experiments would you do to test the hypotheses? What data can you gather on the Web to help you refine your experiments?

Here are some other sites that you may find interesting:

http://www.axis-net.com/pfaf

Looking for a plant to remove those grass stains from your new pants? Want to know which plants can be used as disinfectants? Fire retardants? This site is a searchable database for the uses of common plants (e.g., which are best as edibles, medicinals, fuels, etc.). As this site's advertisement says, "This is not your grandmother's botanical database."

http://users.visi.net/~len

"Can I grow palms where I live? How do I do it?" Find the answers on this homepage for Growing Hardy Palms.

http://www.fao.org/waicent/waicent.htm

The World Agricultural Information Centre offers an array of information on agriculture, fisheries, forestry, nutrition, and rural development.

Suggested Readings

Articles

Feldman, M., and E. R. Sears. 1981. The wild gene resources of wheat. *Scientific American* 244(January):102–113.

Kubo, Isao. 1985. The sometimes dangerous search for plant chemicals. *Industrial Chemical News* (April).

Moffat, A. S. 1996. Higher yielding perennials point the way to new crops. *Science* 274:1469–1470.

Pimentel, D. et al. 1989. Benefits and risks of genetic engineering in agriculture. *BioScience* 39:606–614.

Scrimshaw, N. S., and L. Taylow. 1980. Food. *Scientific American* 243 (3):78–99.

Shulman, Seth. 1986. Seeds of controversy. *BioScience* (November).

Tucker, Jonathan B. 1986. Amaranth: The once and future crop. *BioScience* (January).

Vietmeyer, N. D. 1981. New harvests for forgotten crops. *National Geographic* 159(5):702–712.

Woolf, Norma B. 1990. Biotechnologies sow seeds for the future. *BioScience* 40:346–348.

Books

Lewington, A. 1990. *Plants for People*. New York: Oxford University Press.

Simpson, B. B., and M. Conner-Ogorzaly. 1986. *Economic Botany: Plants in Our World*. New York: McGraw-Hill.

Zohary, D., and M. Hopf. 1993. *Domestication of Plants in the Old World*. Oxford: Oxford University Press.

appendix a

Fundamentals of Chemistry

Atoms and Elements

An **element** is a substance that cannot be broken down into a simpler substance by ordinary chemical means. By this definition, there are ninety-two naturally occurring elements in the universe. Each element is represented by a symbol, usually the first one or two letters of its name. (A few of the symbols represent Latin names of elements, such as Au for *aurum*, meaning "gold," and Na for *natrium*, meaning "sodium.") About twenty-two elements are essential to life, but only six—carbon, hydrogen, oxygen, nitrogen, sulfur, and phosphorus—constitute most of what we call living matter. The rest are **trace elements**, those used by plants only in extremely small quantities.

Atomic Weight

The smallest possible amount of an element is an **atom.** Atoms consist of even smaller particles, the most stable of which are **protons, neutrons,** and **electrons.** Atoms with different numbers of these subatomic particles are different elements. For example, an atom with one proton and one electron is elemental hydrogen. A helium atom has two protons, two neutrons, and two electrons.

Each proton or neutron has a mass of about 1.7×10^{-24} gram. For convenience, this mass is defined as 1 **atomic mass unit.** It is also called 1 **dalton,** in honor of John Dalton, who helped develop the atomic theory in the early 1800s. The mass of an electron is about 1/2,000 that of a proton, so it is often disregarded when considering atomic mass. A summary of the features of several common elements is presented in Table A.1.

Structure and Activity of Atoms

Atomic structure depends on how many subatomic particles (protons, neutrons, and electrons) an atom possesses. The smallest atom is hydrogen, which has one proton, one electron, and no neutrons. The electron, which has a negative (−) electric charge, is attracted to the positive (+) electric charge of the proton. (Neutrons, which occur in all elements except hydrogen, have no electric charge.) The opposing electric charges

table A.1

The Symbols, Atomic Numbers, and Atomic Mass of Some Common Elements

Element	Symbol	Atomic Number	Atomic Mass
Hydrogen	H	1	1
Boron	B	5	10.8
Carbon	C	6	12
Nitrogen	N	7	14
Oxygen	O	8	16
Sodium	Na	11	23
Magnesium	Mg	12	24.3
Phosphorus	P	15	31
Sulfur	S	16	32
Chlorine	Cl	17	35.4
Potassium	K	19	39.1
Calcium	Ca	20	40.1
Manganese	Mn	25	54.9
Iron	Fe	26	55.8
Cobalt	Co	27	58.9
Copper	Cu	29	63.5
Zinc	Zn	30	65.4
Molybdenum	Mo	42	95.9

These elements are essential for plant life; each has one or more vital roles. If any one is missing, a plant cannot survive. Atomic number corresponds to the number of protons in the atomic nucleus; atomic mass is the number of protons plus neutrons in each nucleus.

attract each other, but the particles do not touch. This is primarily because the much smaller electron moves at nearly the speed of light, flying around the proton nucleus but never slowing down enough to fall into it. The electron of a hydrogen atom orbits the proton nucleus as though it were a tiny planet orbiting its sun (fig. A.1). Although this planetary model is a useful way to envision the structure of an atom, the actual structure is unknown. Nevertheless, such models help us understand atomic theory and explain the physical properties of atoms.

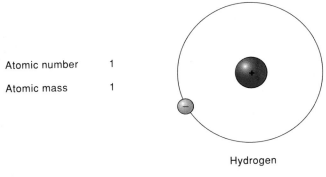

Atomic number   1

Atomic mass   1

Hydrogen

figure A.1
___
An atom of hydrogen has one proton (+), one electron (–), and no neutrons.

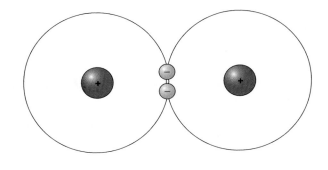

figure A.2
___
The two atoms of hydrogen gas (H_2) are held together by a covalent bond. In covalent bonds, atoms share valence electrons.

Although the exact location of an electron cannot be pinpointed, there is a region around the nucleus where it is most likely to be. This region is called its **orbital.** For the electron of hydrogen, the shape of the orbital is approximately spherical (fig. A.1). Helium has two electrons that also occur in a spherical orbital. Atoms larger than hydrogen and helium have more than two electrons. In these atoms, additional orbitals exist further away from the nucleus to accommodate the added electrons. These outer orbitals are larger, and they occur in different shapes and sizes, in contrast to the relatively small, more-or-less spherical orbitals of hydrogen and helium.

c o n c e p t

An atom is the smallest unit of an element. Each atom is characterized by its size and subatomic composition. These features are determined by the number of protons, neutrons, and electrons that an atom has.

Chemical Bonding: Forces That Hold Molecules Together

One of the most important features of an electron orbital is that it has enough space for only two electrons. An orbital with only one electron can attract another electron to fill the available space. In the case of a hydrogen atom, which has just one electron, the orbital is filled by attracting an electron from another atom. In this way, two hydrogen atoms can share their single electrons to make a combined orbital with two electrons. The combined orbital of two hydrogen atoms makes a **molecule** of hydrogen gas, which is designated by the molecular formula H_2. A molecule of hydrogen gas is held together

by the **chemical bond** between the electrons of two hydrogen atoms. This is one of several types of bonds that are based on the behavior of electrons. These bonds are discussed in the next few paragraphs.

Covalent Bonding

The chemical bond just described for hydrogen gas is called a **covalent bond** (fig. A.2). This is the chemical bond made between **valence electrons,** those electrons occurring in the outermost **electron shell,** or **valence shell,** of the atom. A covalent bond is written as a single line that represents a pair of shared electrons. Using this notation, the structural formula for hydrogen gas is H—H.

Elements with more than one orbital, (that is, all but hydrogen and helium), have more than one electron shell. Unlike the innermost shell, outer shells can contain more than one orbital, up to a maximum of four. For example, a carbon atom has six electrons, two in the innermost orbital and one in each of four outer orbitals in the valence shell. Based on this structure, carbon can share four electrons by covalent bonding. A carbon atom can, for example, bond to four hydrogen atoms. The structural formula of this molecule is:

$$
\begin{array}{ccc}
\text{H} & & \text{H} \\
\backslash & & / \\
& \text{C} & \\
/ & & \backslash \\
\text{H} & & \text{H}
\end{array}
$$

This molecule be written more simply as CH_4, which is the molecular formula for methane. Methane is called a compound molecule, or **compound,** because it consists of more than one kind of element. Except for the simple gases—hydrogen (H_2), oxygen (O_2), and nitrogen (N_2)—virtually all biologically important molecules are compounds.

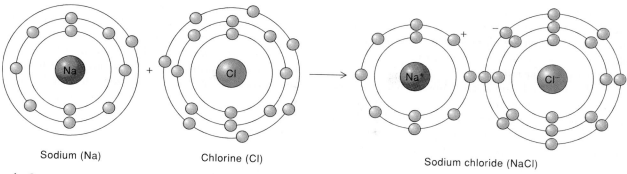

Sodium (Na)   Chlorine (Cl)   Sodium chloride (NaCl)

figure A.3

Sodium and chlorine are bonded ionically to form sodium chloride, or table salt. In ionic bonds, atoms are held together by opposite charges that result from the loss or gain of valence electrons.

Interestingly, a carbon atom can also form a triple covalent bond with a nitrogen atom, or even with another carbon atom. When bonded to nitrogen, a carbon atom still has one free valence electron that must be shared with another atom. If the other atom is hydrogen, the combination of hydrogen, carbon, and nitrogen makes a molecule of a poisonous gas called hydrogen cyanide. Similarly, when a carbon atom forms a triple bond with another carbon atom, each atom still requires one more covalent bond. When both bonds are made with hydrogen, the compound is a molecule of acetylene, the highly flammable gas used in blowtorches.

Ionic Bonding

In extreme cases, valence electrons are not really shared, but instead are completely removed from one atom and transferred to another. This happens between an element that can easily give up an electron, such as sodium, and one that strongly attracts an electron, such as chlorine. The result is that sodium, with 11 protons and 10 electrons, has an extra positive charge; that is, it becomes a positively charged **ion** (*cation*). Chlorine then has 17 protons and 18 electrons, which makes it a negatively charged ion (*anion*). The opposite charges of Na^+ and Cl^- ions attract each other to form the **ionic bond** that makes sodium chloride (NaCl), or table salt (fig. A.3).

Some ions have more than one charge. Magnesium, for example, gives up two valence electrons and is therefore a doubly charged (divalent) cation, Mg^{+2}. It can form ionic bonds to two chlorine atoms to make magnesium chloride ($MgCl_2$).

Besides ionic atoms, entire molecules can become ionic by losing or gaining one or more electrons. Nitrogen and hydrogen, for example, combine to form an ammonium cation (NH_4^+); sulfur and oxygen make a divalent sulfate anion (SO_4^{-2}). Ammonium sulfate, an important ingredient in nitrogen fertilizers, is a compound of these two ions with the molecular formula $(NH_4)_2SO_4$. In this compound, the doubly charged sulfate ion is bonded to two singly charged ammonium ions.

The two strongest types of chemical bonds are covalent bonds and ionic bonds. The forces that make these bonds come from unfilled electron orbitals. When an orbital lacks an electron, it attracts an additional electron to fill it. The orbitals of two separate atoms either share electrons (covalent bond) or transfer electrons (ionic bond).

Hydrogen Bonding and the Special Properties of Water

Nitrogen and oxygen have especially strong attractions for covalent electrons from other atoms. For this reason, they are said to be more **electronegative** than elements such as carbon. When hydrogen is covalently bonded to oxygen, the electronegativity of oxygen forces the hydrogen to become partially positively charged. The partial charge is strong enough to make hydrogen attractive to other electronegative atoms. This attraction is a relatively weak bond, which is neither covalent nor ionic. Rather, this type of bond is called a **hydrogen bond.**

Hydrogen bonds play several significant roles in biological processes. One of these roles is to maintain the shape of large molecules such as enzymes and other proteins that control cellular processes (see Unit Two, "Plants and Energy"). This happens when a bond between the hydrogen of one part of a molecule and an electronegative atom in another part of the molecule holds the two parts together. Molecular shapes are important when different compounds must fit together in a precise way to complete the chemical reactions of cellular processes.

The properties of water, which makes up over 90% of the mass of most plants, are also based on hydrogen bonding. These properties include water's cohesiveness, its high heat of vaporization, and its versatility as a solvent. Each water molecule has the chemical formula H_2O, but the oxygen attracts the hydrogen of neighboring water molecules, and the hydrogen is attracted to other oxygen atoms. Actually, many such bonds are formed and broken and then re-formed in a fraction of a second, but the overall effect is to give water a multimolecular structure. Because of this **cohesion**, plants can pull water upward against gravity, from roots to the tops of the tallest trees.

When molecules of a liquid absorb enough energy, the energy increases their rate of movement. Ultimately, some molecules move fast enough to leave the surface of the liquid in a gaseous state. The amount of energy required to convert 1 g of liquid to its gaseous state is called its **heat of vaporization.** Because water has a cohesive structure, its heat of vaporization is higher than that of most other liquids. Moreover, because heat of vaporization is proportional to the amount of heat that can be removed from the liquid as it vaporizes, water functions as an **evaporative cooler.** Evaporative cooling keeps plants from becoming overheated. This is how a cotton plant, for example, can maintain its leaf temperatures below 30° C even though the air temperature around it exceeds 35° C.

Because water molecules are so much more electronegative at the oxygen end than they are at their hydrogen end, they are said to be **polar.** When salt is added to water, Na^+ ions are attracted to the oxygen pole, and Cl^- ions are attracted to the hydrogen pole. These polar attractions are so strong that the cations and anions are separated and surrounded by water molecules, thereby causing the salt crystals to dissolve and mix into the water. Living cells contain many different kinds of salts that are dissolved in water because of its polarity. The ability of water to dissolve salts also extends to other polar compounds, even though they are not ionic. Ethyl alcohol (CH_3CH_2OH) is one such polar compound that is soluble in water. This property of water contributes to the enormous variety of compounds dissolved in cellular fluids.

c o n c e p t

Hydrogen bonds are very weak but also extremely important in determining the chemical properties of water. Water is a polar solvent because of hydrogen bonding. Its polarity makes it a flexible solvent that can dissolve many substances in cell fluids.

Functional Groups and Biochemical Diversity

The six main elements in biological molecules occur in a variety of combinations as specific subunits of molecules. These subunits are called **functional groups** because they confer chemical properties on the molecules they are part of. Ethyl alcohol is an example of a compound having an electronegative hydroxyl group (OH) and a neutral ethyl group (CH_3CH_2). These functional groups influence the polarity, solubility in water, and reactivity of ethyl alcohol with other compounds.

The great diversity of biochemicals is based on the versatility of carbon, the element that lies at the heart of **organic chemistry** (the chemistry of carbon compounds) and **biochemistry** (the organic chemistry of biological compounds). Most functional groups either contain carbon or are bonded to carbon to make a complete molecule.

Carbon Chains and Rings

The versatility of carbon, as well as its central role in the chemistry of biological processes, comes from the ability of each carbon atom to share four covalent bonds. A carbon atom can be thought of as an intersection of covalent branches in four directions. Covalent bonding occurs in each direction, either with a different element or with another carbon. Because a carbon atom can form a covalent bond with another carbon atom, there is a seemingly infinite variety of possible compounds that have multiple carbons. Known compounds range from the smallest, with one or two carbons (methane, acetylene, ethyl alcohol), to the largest, with up to 30,000 carbons (natural rubber).

Carbon skeletons occur in straight or branched chains or even in closed rings. In each compound, every carbon is fully "satisfied" with four covalent bonds. The simplest compounds are **hydrocarbons** (i.e., those with only carbon and hydrogen). The complexity of compounds increases dramatically when these functional groups are attached to other kinds of elements.

Hydroxyls, Carbonyls, and Carboxylic Acids

Oxygen, which forms two covalent bonds, has multiple roles in different functional groups. The hydroxyl group, which is the functional group of ethyl alcohol and all other alcohols, is one example. Additional roles for oxygen are derived from the double covalent bond between oxygen and carbon, called a **carbonyl** group (C=O). When a carbonyl group occurs at the end of a carbon chain, it is called an **aldehyde;** when it occurs in the interior of a carbon chain, it is called a **ketone.**

Table A.2 shows the structural formulas for several kinds of functional groups, including those containing oxygen. More than one functional group can occur on the same molecule. Glyceraldehyde, for example, is a compound made of a three-carbon chain, two hydroxyl groups, and a carbonyl at the end of the chain. Compounds with both hydroxyl and carbonyl groups are sugars. Glyceraldehyde is the first sugar produced by photosynthesis in plants.

table A.2

Some Functional Groups with Important Roles in Organic Compounds

Group	Name
—OH	Hydroxyl
—NH₂	Amino
—C—CH₃ (with =O below C)	Acetyl
—C—C—C— (with =O above middle C)	Keto
—C—OH (with =O below C)	Carboxyl
—O—P—OH (with OH above and =O below P)	Phosphate

When a carbonyl group and a hydroxyl group occur on the same carbon, the combination is called a **carboxyl group.** In a carboxyl group, the collective electronegativity of the two oxygens is so strong that the hydrogen of the hydroxyl group is stripped of its covalent electron and becomes ionic. Carboxyl groups are therefore **acidic** because they increase the concentration of hydrogen ions (H^+) in solution (see reading A.2, "Moles, Hydrogen Ions, and pH"). Compounds with carboxyl groups are called **carboxylic acids, or organic acids.** The most common organic acid in plants is acetic acid, which is the basic building block of fats and other lipids. It is this acid that gives vinegar its tangy flavor.

Functional Groups Containing Nitrogen

The most biologically significant functional group containing nitrogen is the **amino group** (NH_2). This group is important because it is one of the two functional groups that make up an **amino acid,** the other being a carboxyl group. Amino acids are used for building enzymes and other proteins that are the foundation for almost all biochemical reactions and for a variety of cellular structures.

Functional Groups Containing Sulfur or Phosphorous

Sulfur, like oxygen, has two valence electrons and forms a functional group with a hydrogen atom, which is called a **sulfhydryl group** (SH). Compounds with sulfhydryl groups are called **thiols** and are notable for their foul odors. (Butanethiol is the major ingredient of skunk spray.) Except for this property, sulfhydryls are chemically similar to hydroxyls. They are so similar that plant scientists were able to deduce how plants split water molecules (H_2O) during photosynthesis by analogy with the similar chemical reactions of hydrogen sulfide (H_2S) in bacteria (see Chapter 7).

Sulfur can also form **disulfide** bonds (S—S). These bonds form between two cysteine monomers in proteins and are important for stabilizing a protein's three-dimensional structure.

Of all the biologically important functional groups, the most electronegative is the **phosphate group.** Each phosphate group (PO_4^{-3}) is made of one phosphorous atom and four oxygen atoms. It is derived from a phosphoric acid molecule (H_3PO_4) that, due to the electronegative strength of the oxygens, has released all of its hydrogen atoms as ions. One of the properties of phosphate groups is that they make short-lived, unstable covalent bonds with other phosphate groups. The ease with which these bonds can be broken means that the energy stored in them is easily released. As phosphate bonds break, their energy is used to drive cellular processes. These processes include such functions as building new molecules, moving substances from one place in the cell to another, and controlling what goes through membranes. In Unit Two ("Plants and Energy") you learn about a variety of cellular processes that depend on energy from phosphate bonds. The phosphate group is also important in the sugar-phosphate backbone of the nucleic acids, DNA and RNA, and in the phospholipids of membranes.

concept

The chemical properties of molecules are determined by their functional groups. These are subunits of compounds that have specific combinations of different elements. The most important molecules of biological processes are made of multiple carbons that are attached to an oxygen (hydroxyl, carbonyl, or carboxyl group), a nitrogen (amino group), a sulfur (sulfhydryl group), or a phosphorus (phosphate group). Many compounds have more than one of these functional groups.

Chemical Reactions

Covalent bonds, functional groups, and biological polymers are all products of **chemical reactions**. A chemical reaction is the process of making and breaking chemical bonds, and is written as an equation that has the **reactants** on the left side

ISOTOPES AND RADIOACTIVITY

Each atom is characterized by a specific number of protons, neutrons, and electrons. If an atom changes its number of protons, it becomes another element. However, the same element can have a different number of electrons or neutrons; for example, an element can have different atomic masses because of its variable number of neutrons. Each neutron form of an element is called an **isotope** of the element. Hydrogen has three isotopes, one without neutrons, one with one neutron, and one with two neutrons. To distinguish among them, the isotopes of hydrogen are written ^1H, ^2H, and ^3H, respectively. Similarly, carbon has the isotopes ^{12}C (6 neutrons), ^{13}C (7 neutrons), and ^{14}C (8 neutrons). Like hydrogen, all of the isotopes of carbon have the same number of protons. Notice that the isotope number and the atomic mass are about the same.

The smallest isotope of any element is generally the most abundant in nature. Of all the carbon on earth, 99% is ^{12}C and most of the remaining 1% is ^{13}C; only a trace amount is ^{14}C. The abundant smaller isotopes are also stable, but the larger isotopes of many elements are unstable. They decay by releasing high-energy subatomic particles; that is, they are **radioactive.** (This name has nothing to do with radio waves; radioactive means that an element is like radium, which was the first element discovered to be radioactive.)

Electronic instruments and photographic film can be used to detect the particles of radioactive decay. These detectors and radioactive isotopes are used to study the metabolic processes of a cell. The biosynthesis of many natural products was

	Hydrogen	Deuterium	Tritium
Atomic number	1	1	1
Atomic mass	1	2	3
Number of protons	1	1	1
Number of electrons	1	1	1
Number of neutrons	0	1	2

reading A.1

The three isotopes of hydrogen differ because of the number of neutrons. Because they all have the same number of protons (i.e., 1), they are the same chemical element (i.e., hydrogen) and have identical chemical properties.

first discovered by following the path of radioactive precursors during metabolism. The intermediate products and chemical reactions that lead to sugars, natural rubber, steroids, and many other biological molecules were discovered with the aid of radioactive chemicals. This is still a powerful method for studying the biosynthetic pathways of plant metabolites.

Radioactive isotopes are also useful for dating fossils. This is possible because radioactive decay occurs at a constant rate for each element. For example, as a plant takes in carbon during photosynthesis, the ratio of ^{14}C to ^{12}C in its tissue is the same as the $^{14}C/^{12}C$ ratio of carbon dioxide in the atmosphere. When the organism dies, the ratio

begins to change because ^{14}C decreases in abundance due to its radioactive decay. By knowing the rate of decay, we can calculate how long ago the plant died. Carbon loses half of its radioactivity in 5,600 years, the time period that is called its **half-life.** If a 10-gram sample of a fossil plant has half the amount of ^{14}C as a 10-gram sample of a living plant, that means that the fossil plant died 5,600 years ago. A 11,200-year-old fossil would have one-fourth the amount of radioactivity, and so on. Because of the limitations of electronic detectors, carbon dating is limited to about 50,000 years. Other isotopes, such as potassium-40 (half-life of 1.3 billion years), can be used to date fossils of organisms that lived hundreds of million years ago.

and the **products** on the right side. For example, methane and oxygen (reactants) can react to make carbon dioxide and water (products):

$$CH_4 + 2O_2 \longrightarrow CO_2 + 2H_2O + \text{energy}$$

Notice that all atoms must be accounted for in a chemical reaction; that is, the equation must be balanced. Reactions cannot create or destroy matter, but can only rearrange it. Thus, in this reaction, each methane molecule requires two molecules of oxygen (O_2) to maintain the appropriate proportion of elements in the products.

This is a common reaction that occurs at the burners of gas stoves every day. But why does it occur? That is, what is the driving force that makes the reaction go? The simplest answer is that the stored energy (potential energy) of the covalent bonds in the products is lower than the bond energy of the reactants. Potential energy tends to run "downhill" from higher levels to lower levels. The behavior of potential energy is analogous to that of water at the top of a hill (higher potential energy), which tends to run down to the bottom of the hill (lower potential energy). The difference between the higher potential energy on the left side of the equation and the lower potential energy on the right side of the equation is accounted for by the release of heat energy. This heat energy is equal to the bond energy of the reactants minus the bond energy of the products.

The physical behavior of energy is the driving force underlying all biological processes. Unit 2 ("Plants and Energy") includes 3 chapters devoted to energy and its use by plants. The paragraphs that follow outline some of the most common chemical reactions that are driven by the energy of plant metabolism.

Oxidation and Reduction

The chemical reaction of methane and oxygen presented in the previous section is a combustion reaction; that is, methane and oxygen burn together. In this reaction the covalent electrons of carbon, which are about evenly shared with the electrons of hydrogen in methane, are pulled further away from the carbon by the strongly electronegative oxygen to make carbon dioxide. This shift of electrons away from an element is called **oxidation.** Each hydrogen is also oxidized, since its electron goes from being equally shared with carbon to being pulled away by the oxygen in water.

When one reactant is oxidized, another reactant undergoes **reduction.** In the previous reduction, oxygen is reduced; that is, it partially gains electrons from the carbon in CO_2 and from the hydrogen in H_2O. Although the example reaction involves oxygen, oxidation-reduction (*redox*) reactions do not always require it. Any partial or complete transfer of electrons is a redox reaction. For example, the ionization of sodium and chlorine fits this description (fig. A.3):

$$Na + Cl \longrightarrow Na^+ + Cl^-$$

In this reaction, sodium is oxidized (loses an electron), and chlorine is reduced (gains an electron).

The energy that cells need for growth and metabolism is derived from a redox process called **respiration.** Cellular respiration is a series of many redox reactions that begin with a carbohydrate (glucose) and oxygen and end with carbon dioxide and water. The following equation summarizes the overall process:

$$C_6H_{12}O_6 + 6O_2 \longrightarrow 6CO_2 + 6H_2O + energy$$

The many reactions that occur during respiration, the parts of cells where the reactions take place, and other details about this process are the subject of Chapter 6, "Respiration."

Dehydration and Hydrolysis

Although oxidation and reduction occur during almost all chemical reactions, many reactions are referred to by the products they yield or by the functional groups that are involved. For example, most biological polymers are built by reactions that yield water as a by-product. These reactions are called **dehydration** reactions. The peptide bonds of proteins, the glycosidic bonds of polysaccharides, the sugar-phosphate bonds of nucleic acids, and the glycerol–fatty acid bonds of lipids are all made by dehydration.

Dehydration reactions are reversible. The bond between two glucoses in starch, for example, can be broken to yield glucose monomers. This reverse reaction is called **hydrolysis** because it uses water to break the chemical bond between glucoses. All bonds formed by dehydration can be broken by hydrolysis. In cells, chemical synthesis by dehydration is enhanced by one set of enzymes, whereas chemical degradation by hydrolysis is enhanced by another set of enzymes. The metabolic needs of a cell determine which direction a reaction takes. Both kinds of reactions can occur at the same time but in different parts of the cell.

Decarboxylation and Carboxylation

Covalent bonds in carbon dioxide (O=C=O) are stable and difficult to break. The formation of carbon dioxide, with its low potential energy, is the strong driving force that removes CO_2 from organic acids:

$$R—COOH \longrightarrow R—H + CO_2$$

This **decarboxylation** reaction is common in many cellular processes (e.g., the release of carbon dioxide during respiration). In fact, several decarboxylations occur for each molecule of glucose that is oxidized during respiration. This means that organic acids play a role in the intermediate steps of this process. Organic acids and their decarboxylation are discussed in Chapter 6 ("Respiration").

Although the removal of carbon dioxide has many metabolic roles, the addition of carbon dioxide, or **carboxylation,** is restricted to photosynthesis. This reaction is the foundation for making organic molecules from carbon dioxide in the atmosphere. Energy is required to break the double covalent bonds between carbon and oxygen and to make covalent bonds between carbons. Only green plants, algae, and a few bacteria can transform light energy and use it to carboxylate organic compounds. This process is the topic of Chapter 7 ("Photosynthesis").

Phosphorylation

Reactions that make phosphate bonds are called **phosphorylation** reactions. The most common phosphorylation reaction in any cell is the addition of a phosphate group to an adenine-containing nucleotide, adenosine diphosphate (ADP). The product of this reaction is adenosine triphosphate (ATP). Phosphorylation is accompanied by dehydration. The covalent bond between the second and third phosphate groups of ATP is easily hydrolyzed, thereby releasing energy to another molecule or chemical reaction. Because ATP gives up energy so easily, this phosphate bond is called a high-energy bond. The bond between the first and second phosphate groups is also a high-energy bond. High-energy phosphate bonds are designated by a "~" to distinguish them from ordinary covalent bonds. Thus, phosphorylation can be abbreviated as follows:

$$A—P \sim P + P \longrightarrow A—P \sim P \sim P + H_2O$$

Because of the instability of high-energy phosphate bonds and the demand for this energy to do cell work, each ATP molecule is recycled to ADP in a fraction of a second. ATP is regenerated continuously by the energy derived from respiration and photosynthesis.

c o n c e p t

The most common chemical reactions in cells are the removal of a molecule of water (dehydration) or a molecule of carbon dioxide (decarboxylation), or the addition of a molecule of water (hydrolysis). The function of molecules such as ADP and ATP depends on dehydration and hydrolysis for attaching and removing phosphate groups (phosphorylation).

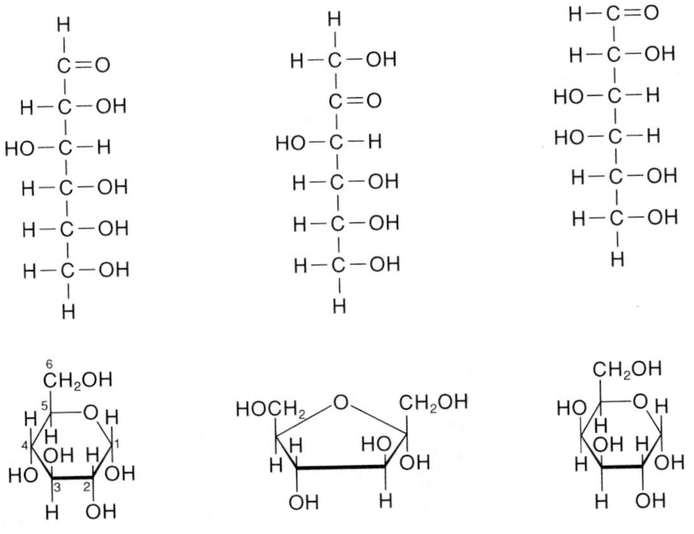

figure A.4

The structures of glucose, fructose, and galactose. Although these sugars are $C_6H_{12}O_6$, they have different chemical properties.

Molecular Shapes and Natural Isomers

Many organic compounds can have different shapes but the same molecular formula. Glucose and fructose, for example, are both $C_6H_{12}O_6$, but they have different chemical properties. Their structural differences are easy to see when their rings are broken open by hydrolysis. The carbonyl group of fructose is a ketone, whereas the carbonyl group of glucose is an aldehyde (fig. A.4). This seemingly small difference is significant. Because of it, fructose is one of the sweetest sugars, whereas glucose has almost no sweetness at all. Glucose and fructose are **structural isomers;** that is, they have the same functional groups bonded to different carbon atoms. Galactose, on the other hand, has the same functional groups on the same carbon atoms as glucose. However, the hydroxyl group on the fourth carbon is in a different orientation on galactose than it is on glucose (fig. A.4). This means that glucose and galactose are **stereoisomers** of each other. They are seemingly identical but, like your left and right hands, are really nonidentical mirror images of each other.

By varying the orientation of hydroxyl groups on carbons 2 through 5 of glucose, there can be 16 possible stereoisomers (2^4). The natural isomers of glucose, however, all have the hydroxyl group to the right of carbon 5. They are called the D forms, from the Latin word *dextro*, which means "right." Thus, natural glucose is D-glucose. The alternative,

L-glucose (from *laevo,* meaning "left") is not made by plants. All other monosaccharides also exist only in the D form in plants. Similarly, the amino group occurs on the same side of its carbon, relative to the carboxyl group, in all amino acids. However, it is on the left side, which means that plants make only L-amino acids.

Optical Isomers

The original designation of D and L forms for sugar and amino acids was an arbitrary one, based solely on the orientation of one functional group. A more useful physical property, one which can be applied to all biological compounds, is the behavior of stereoisomers in light. When a single plane of light, called **plane-polarized light,** is aimed at a substance at one angle, it often emerges from the other side at a different angle. If the angle of the emergent plant of light is to the left of the original plane, the substance is *levorotatory.* If the emergent plane of light is to the right, the substance is *dextrorotatory.* Substances that cause plane-polarized light to rotate are called **optically active.** Optical activity occurs when a carbon atom is asymmetric; that is, when it is attached to four different functional groups. In glucose, for example, carbons 2 through 5 are asymmetric. Carbon 1 is not, because it has "two" oxygens (actually, one oxygen that is double-bonded). Carbon 6 is also inactive. Why?

MOLES, HYDROGEN IONS, AND pH

The molecular weight of a substance is too small to be measured on a normal laboratory scale. Instead, we measure the weight of a chemical in **moles.** A mole is the molecular weight in daltons expressed in grams. For example, the molecular weight of water is 18 daltons, so one mole of water is 18 grams. One mole of table salt (NaCl) is 58.4 grams, based on a molecular weight of 58.4 daltons. Moles are also convenient for expressing the concentrations of substances in water. A solution that contains one mole of a chemical in one liter of water is a **one-molar** (1 M) solution. Therefore, a liter of water with 58.4 grams of NaCl is a 1 M NaCl solution.

One of the most important applications of molarity involves the concentration of hydrogen ions in a solution. Pure water is the standard by which all other solutions are compared, because water is an ionically neutral solution. Its neutrality is not due to the absence of ions, but to the equal concentrations of positive and negative ions. When the oxygen of water pulls hard enough on an electron from one of its hydrogens, two ions form:

$$H_2O \longleftrightarrow H^+ + OH^-$$

This dissociation of water is rare and reversible, but it happens often enough for the concentration of H^+ to be 10^{-7} M. The solution is neutral because the concentration of OH^- is also 10^{-7} M.

Any substance that increases the concentration of H^+ is an **acid;** any substance that decreases the concentration of H^+ is a **base.** For example, hydrochloric acid

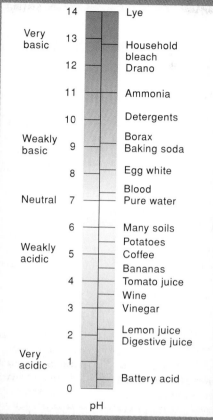

	pH	
Very basic	14	Lye
	13	Household bleach
	12	Drano
	11	Ammonia
	10	Detergents
Weakly basic	9	Borax / Baking soda
	8	Egg white
Neutral	7	Blood / Pure water
	6	Many soils
Weakly acidic	5	Potatoes / Coffee
	4	Bananas / Tomato juice
	3	Wine / Vinegar
	2	Lemon juice / Digestive juice
Very acidic	1	
	0	Battery acid

Source: Data from Essenfeld, et al., *Biology*, copyright © Addison-Wesley Publishing Company.

(HCl) or acetic acid (CH_3COOH) would increase the concentration of H^+. In contrast, sodium hydroxide (NaOH) or ammonium hydroxide (NH_4OH) are bases because they increase the OH^- concentration, thus lowering the proportion of H^+.

When the concentration of one ion increases, the concentration of the other ion decreases proportionally. Thus, if enough acid is added to water to raise the H^+ concentration to 10^{-6} M, the OH^- concentration would decrease to 10^{-8} M. In the extreme, the highest possible concentration of H^+ is 10^0 M and the lowest possible concentration of OH^- is 10^{-14} M. In the other extreme, the highest concentration of OH^- is 10^0 M when the H^+ concentration is at 10^{-14} M. By general agreement, the scale we use is based on the hydrogen ions and is expressed as a negative logarithm of the molarity of H^+. This scale, called the **pH scale,** goes from 0 ($-\log 10^0$) to 14 ($-\log 10^{-14}$). (It would be equally convenient to have a "pOH" scale, but pH was arbitrarily chosen instead.) On this scale, water has a pH of 7. An acid has a pH less than 7 (more H^+), and a base has a pH greater than 7 (less H^+). Note that, because this scale is logarithmic, pH 3 is not double pH 6, but is 1,000-fold higher in H^+ concentration. Small differences in pH mean large differences in the molarity of H^+. Thus, the pH in cells is restricted to a narrow range, usually between 6 and 8.

For more information about acids, bases, and pH, visit http://www.science.ubc.ca/departments/chem/courses/103/tutorials/pH/ and http://www.miamisci.org/pH/default.html on the World Wide Web.

All amino acids, except glycine (H_2N—CH_2—$COOH$), also have at least one asymmetric carbon and are therefore optically active. Nucleotides and other primary metabolites are optically active, as are most secondary metabolites. For most compounds, either the dextrorotatory (+) or the levorotatory (−) isomer occurs naturally, but not both. Chemical reactions, especially enzyme-catalyzed reactions, that depend on the fit of specific isomers to each other will not work if one compound is the "wrong" isomer. For example, when yeasts are fed a mixture of glucose isomers, they metabolize only the (+)-glucose and leave the (−)-glucose behind. In this regard, the respiration of all organisms is the same.

c o n c e p t

Molecules have certain shapes because each carbon forms bonds in different directions. Compounds with the same functional groups are isomers of each other. Structural isomers have the same functional groups on different carbons. Stereoisomers have the same functional groups on the same carbons, but the functional groups are pointed in different directions. When a carbon atom is bonded to four different atoms or functional groups, it is asymmetric, and the molecule is optically active. Optical isomers rotate plane-polarized light in opposite directions.

appendix b

The Metric System

The **metric system** is a standardized system of measurement used by scientists throughout the world. It is also the measurement system used in everyday life in most countries. Although the metric system is the only measurement system ever acknowledged by Congress, the United States remains "out of step" with the rest of the world by clinging to the antiquated English system of measurements involving pounds, inches, and so on.

Metric units commonly used in biology include

meter (m) The basic unit of length

liter (L) The basic unit of volume

gram (g) The basic unit of mass

degree Celsius (°C) The basic unit of temperature

Unlike the English system with which you are already familiar, the metric system is based on units of ten, thus simplifying interconversions (table B.1). This base-ten system is similar to our monetary system, in which 10 cents equals a dime, 10 dimes equals a dollar, and so on. Units of ten in the metric system are indicated by Latin and Greek prefixes placed before the base units:

Thus, we multiply by

0.01	to convert cg to g
0.1	to convert dm to m
1,000	to convert kg to g
0.000001	to convert μL to L
0.1	to convert mm to cm

For example, 620 g = 0.620 kg = 620,000 mg = 6,200 dg = 62,000 cg.

Units of Length

The **meter** (m) is the basic unit of length.

1 m = 39.4 inches (in.) = 1.1 yards (yd)

1 km = 1,000 m = 10^3 m = 0.62 miles

1 cm = 0.01 m = 10^{-2} m = 0.39 in. = 10 mm

1 nm = 10^{-9} m = 10^{-6} mm = 10 angstroms (Å)

470 m = 0.470 km

1 in. = 2.54 cm

1 ft = 30.5 cm

1 yd = 0.91 m

1 mi = 1.61 km

Units of area are squared (i.e., two-dimensional) units of length:

1 m^2 = 1.20 yd^2 = 1,550 $in.^2$ = 1.550×10^3 $in.^2$

1 cm^2 = 100 mm^2 = 0.16 $in.^2$

1 hectare = 10,000 square meters (m^2) = 2.47 acres

1 km^2 = 0.39 mi^2

table B.1

Metric System Divisions and Multiples of the Basic Unit of Length, the Meter

Prefix (Latin)	Division of Metric Unit (meter)	Equivalent	Prefix (Greek)	Multiple of Metric Unit (meter)	Equivalent
deci (d)	0.1	10^{-1} (tenth part)	deka (da)	10	10^1 (tenfold)
centi (c)	0.01	10^{-2} (hundredth part)	hecto (h)	100	10^2 (hundredfold)
milli (m)	0.001	10^{-3} (thousandth part)	kilo (k)	1,000	10^3 (thousandfold)
micro (μ)	0.000001	10^{-6} (millionth part)	mega (M)	1,000,000	10^6 (millionfold)
nano (n)	0.000000001	10^{-9} (billionth part)	giga (G)	1,000,000,000	10^9 (billionfold)
pico (p)	0.000000000001	10^{-12} (trillionth part)	tera (T)	1,000,000,000,000	10^{12} (trillionfold)

The following comparisons will help you appreciate these conversions of length and area:

Height of the Washington Monument	169.1 m	555 ft
Length of a housefly	0.5 cm	0.2 in.
Distance kangaroos can hop in one leap	~7.6 m	~25 ft
Diameter of a pore in a hen's egg	17 μm	0.0007 in.
Surface area of a flu virus	10^{-8} mm^2	1.6×10^{-11} in.^{2}
Skin area of average-sized woman	1.58 m^2	17 ft^2
U.S. paper currency	104 cm^2	16.1 in.^{2}

Measurements of area and volume can use the same units:

1 m^3 = 35.3 ft^3 = 1.31 yd^3

1 cm^3 = 0.000001 m^3 = 0.061 in.^{3}

The following comparisons will help you appreciate these conversions of volume:

Coke can	355 ml	12.0 oz
Legal-sized filing cabinet	0.4 m^3	14 ft^3
Sugar cube	1 cm^3	0.061 in.^{3}

Units of Mass

The **gram** is the basic unit of mass.

1 g = mass of 1 cm^3 of water at 4°C = 0.035 oz

1 kg = 1,000 g = 10^3 g = 2.2 lb

1 mg = 0.001 g = 10^{-3} g

The following comparisons will help you appreciate these conversions of mass:

Penicillin molecule	10^{-18} kg	2×10^{-18} lb
Giant amoeba	10^{-8} kg	2×10^{-8} lb
Human	10^2 kg	220 lb
Nickel	5 g	0.176 oz

Remember that mass is not necessarily synonymous with weight. Mass measures an object's potential to interact with gravity, whereas weight is the force exerted by gravity on an object. Thus, a weightless object in outer space has the same mass as it has on earth.

Units of Volume

The **liter** (L) is the basic unit of volume. A typical thermos bottle holds about 1 liter; a standard flush toilet flushes about 20 liters of water. Units of volume are cubed (i.e., three-dimensional) units of length.

1 L = 1,000 cm^3 = 1,000 mL = 0.001 m^3

1 L = 2.1 pt = 1.06 qt = 0.26 gal = 1 dm^3

1 mL = 0.035 fl oz

1 tsp = 15 mL

1 cup = 0.24 L

Units of Temperature

You are probably most familiar with temperature measured with the Fahrenheit scale, which is based on water freezing at 32°F and boiling at 212°F. Celsius temperatures are synonymous with Centigrade temperatures, and these scales measure temperature in the metric system. Celsius (°C) temperatures are easier to work with than Fahrenheit temperatures since the Celsius scale is based on water freezing at 0°C and boiling at 100°C. You can interconvert °F and °C with the following formula:

$$5(°F) = 9(°C) + 160$$

For example,

45°C (113°F) temperature at which any area of skin of an average person will feel pain

40°C (104°F) typical for a hot summer day

75°C (167°F) hot coffee

37°C (98.6°F) human body temperature

16°C (60.8°F) coldest body temperature a person has been known to have and survive

Celsius versus Centigrade

The Celsius temperature scale was developed in 1742 by Anders Celsius, a Swedish astronomer. Interestingly, Celsius originally set 0°C as the boiling point of water and 100°C as the freezing point of water. Soon thereafter, J. P. Christine revised the scale to its present form—with 0°C as the freezing point and 100°C as the boiling point of water. The Celsius scale is identical to the Centigrade scale.

The following comparisons will help you appreciate the conversions of Fahrenheit and Centigrade temperatures:

Water freezes	0°C	32°F
Butter melts	31°C	87°F
Water boils	100°C	212°F
Fireplace fire	827°C	1,520°F

Hints for Using the Metric System

1. Express measurements in units requiring only a few decimal places. For example, 0.3 m is more easily manipulated and understood than 300,000,000 nm.

2. When measuring water, the metric system offers an easy and common conversion from volume measured in liters to volume measured in cubic meters to mass measured in grams: $1 \text{ mL} = 1 \text{ cm}^3 = 1 \text{ g}$.

3. Familiarize yourself with manipulations within the metric system. Work within one system, and do not convert back and forth between the metric and English systems.

4. The metric system uses symbols rather than abbreviations. Therefore, do not place a period after metric symbols (e.g., 1 g, not 1 g.). Use a period after a symbol only at the end of a sentence.

5. Do not mix units or symbols (e.g., 9.2 m, not 9 m, 200 mm).

6. Metric symbols are always singular (e.g., 10 km, not 10 kms).

7. Except for degree Celsius, always leave a space between a number and a metric symbol (e.g., 20 mm, not 20mm; 10°C, not 10°C).

8. Use a zero before a decimal point when the number is less than one (e.g., 0.42 m, not .42 m).

Web Sites

The following sites contain interesting information and an on-line tutorial about the metric system:

http://edie.cprost.sfu.ca/~rhlogan/metric.html
http://www.essex1.com/people/speer/metric.html

appendix c

Diversity of Living Organisms

The only thing that is constant about the classification of living organisms is that it is always changing. At the turn of the century, biologists were content to view all living organisms as being either plant or animal. By the 1950s, however, biologists worldwide had begun to accept the idea that living things comprised more than two kingdoms. Today, most classifications divide living organisms into at least five kingdoms. **Monera** (bacteria), **Plantae** (plants), **Fungi** (fungi), **Animalia** (animals), and **Protoctista** (algae, protozoans, slime molds, and other organisms that do not fit into the other kingdoms). Other classifications, some of which are beginning to gain acceptance, divide the bacteria into two kingdoms and the protoctists into several kingdoms. At least one classification divides all of life into thirteen kingdoms.

For this text, we have used the five-kingdom system outlined by Lynn Margulis and Karlene Schwartz. We have chosen it mostly for convenience—Margulis and Schwartz have published an easy-to-read, book-length explanation of their classification system (see "Suggested Readings" at the end of this appendix); it is an excellent reference for students and instructors (and authors of textbooks!). However, remember that biological classification systems are dynamic and changing. For example, a recent suggestion that is becoming increasingly popular includes three domains: Eubacteria, Archaebacteria, and Eukaryotes. This classification system is based largely on molecular analysis of organisms' genomes (see discussion on pages 599–601).

Botanical systems of classification differ in certain respects from the Margulis-Schwartz system in categories below the level of kingdom. A botanically oriented hierarchy of classification, for example, includes the main categories of kingdom, division, class, order, family, genus, and species. In contrast, the Margulis-Schwartz system uses the term *phylum* in place of *division,* although phylum is not a formally recognized name in botanical classification. In addition, botanists prefer different names for some divisions of plants other than those used in the Margulis-Schwartz system. In such cases we have adopted the prevailing usages by botanists.

The discussion of classification that follows includes only living plants or organisms that have historically been called plants and that are still presented in most botany texts. This treatment excludes animals and several major groups of protoctists (e.g., protozoans, amoebas). The approximate number of species in each group refers to those that have been named. However, biologists estimate that thousands, or even tens of thousands, of species are yet to be discovered; new species are discovered and named almost every day.

Kingdom Monera

At least 10,000 species of bacteria have been named, although bacteriologists argue that duplicate names for the same organisms might reduce the number of valid species of bacteria to about 2,500. Bacteria are characterized by having a prokaryotic cell type (i.e., no membrane-bound organelles), DNA that is not associated with protein in chromosomes, cell division by fission, and genetic recombination by nonsexual means. Bacteria are the

most metabolically diverse of all groups of organisms. Some species can respire aerobically, others anaerobically. Different bacteria can use sulfide, iron, methane, or carbon dioxide in their energy metabolism. Many bacteria can fix atmospheric nitrogen (N_2) into other molecules (e.g., nitrates, ammonia) that are metabolically useful for eukaryotes such as plants. Certain bacteria undergo photosynthesis like plants, while other bacteria undergo photosynthesis that is unique to prokaryotes.

The Monera are divided into two main groups that are usually referred to as the subkingdoms **Eubacteria** and **Archaebacteria.** Differences between

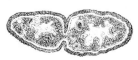

these groups are being discovered continually, to the extent that some biologists see these differences as a basis for classifying each group as a separate kingdom.

The Eubacteria include three divisions that are distinguished mainly by properties of their cell walls. Division **Gracilicutes** contains Gram-negative bacteria, Division **Firmicutes** contains Gram-positive bacteria, and Division **Tenericutes** contains bacteria that have no cell wall.

Division Gracilicutes is divided into three classes, one of which is important in botany because it includes the oxygen-producing photosynthetic bacteria. Bacteriologists call this group Class **Oxyphotobacteria,** but botanists distinguish at least two separate groups in it, either at the class or division level. These are the

Cyanobacteria and the **Chloroxybacteria.** Both groups use chlorophyll *a* for photosynthesis, as do algae and plants. In addition, the Chloroxybacteria use chlorophyll *b*, which is otherwise restricted to plants and some algae.

The Archaebacteria consist of a single division, **Mendosicutes.** Members of this division differ from other bacteria in having eukaryote-like genes with introns, cell walls of glycoproteins and polysaccharides, a eukaryote-like RNA polymerase, ether-linked lipids, and the ability to metabolize methane.

Kingdom Fungi

The fungi are typically filamentous, eukaryotic, spore-producing organisms that lack chlorophyll. They have cell walls made of chitin combined with other complex carbohydrates, including cellulose (chitin is also the main component of the exoskeletons of insects, spiders, and crustaceans). Unlike plants, the main storage carbohydrate of fungi is glycogen, which is also the main storage carbohydrate of animals. All species of fungi are either saprobes or symbionts (i.e., living with other organisms). As symbionts, they may be parasitic, they may provide a benefit to their host, or they may be parasitized by their host.

Nearly 100,000 species of fungi are known, and descriptions of more than 1,000 new species are published each year. Moreover, biologists believe that more than half of all existing fungi have yet to be described. We also suspect that unknown numbers of undiscovered species have already become extinct, due primarily to human activities such as destroying fungal habitats and polluting the air.

Fungi are most often classified into three sexually reproducing groups and one asexually reproducing group, variously treated as divisions, subdivisions, or classes. As divisions, they are the **Zygomycota** (ca. 750 species; e.g., black bread mold, dung fungi, and parasites of amoebas, nematodes, and small animals), the **Ascomycota** (ca. 30,000 species; e.g., yeasts, bread, molds, truffles, morels, and ergot fungi), the **Basidiomycota** (ca. 25,000 species; e.g., mushrooms, stinkhorns, puffballs, jelly fungi, and smut and rust diseases of plants), and the **Deuteromycota** (ca. 17,000 species; e.g., penicillin mold, root-rot fungi, vaginal yeast fungi, and athlete's-foot fungus). The distinguishing features of each of these groups involve differences in reproduction:

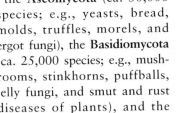

1. Zygospores characterize the Zygomycota.

2. Spore-containing sacs are formed by the Ascomycota.

3. Spores of the Basidiomycota are produced on basidia.

4. Spores of the Deuteromycota are only produced asexually.

Approximately 25,000 species names have been given to lichens, which are organisms that consist of a fungal body that hosts green algae or cyanobacteria or both. The fungus in this symbiotic relationship seems to be parasitic, since it gets nutrients from the algae and cyanobacteria, apparently without providing any benefit in return. The lichens are often classified in existing fungal divisions, the division depending on the type of fungus in the symbiosis. However, Margulis and Schwartz separate all lichens into their own division, the **Mycophycophyta.**

Kingdom Protoctista

The protoctists that have traditionally been of most interest to botanists are the slime molds, the water molds, and the algae. None of the slime molds or water molds can photosynthesize, and these organisms were long considered to be fungi. Conversely, algae usually contain chlorophyll *a* and can photosynthesize like plants. Botanists have historically classified algae as plants, and many botanists still consider most algae to be plants, although this trend is changing.

Slime Molds and Water Molds

These molds are usually classified into four divisions: the **Myxomycota** (plasmodial slime molds), the **Acrasiomycota** (cellular slime molds), the **Oomycota** (biflagellate water molds), and the **Chytridiomycota** (uniflagellate water molds). The diversity of genera in these groups is such that some taxonomists recognize seven divisions instead of the usual four.

The 70 or so species of cellular slime molds are characterized by a phagotrophic mode of nutrition, by carbohydrate storage as glycogen, by cellulosic cell walls, and by amoeboid cell motility. The plasmodial slime molds (ca. 500 species) are also phagotrophic and store glycogen, but differ from the cellular slime molds by the absence of cell walls and by motile cells that swim by using two whiplash flagella. Both kinds of slime molds live in terrestrial habitats.

The water molds absorb nutrients and have flagellated motile cells. The motile cells of the Chytridiomycota (ca. 575 species) have a single whiplash flagellum, whereas those of the Oomycota have two flagella, one of which is whiplash and the other of which is tinsel. The uniflagellate water molds live in freshwater or marine habitats, store glycogen, and form cell

walls of chitin or glucan. Conversely, the biflagellate water molds live only in freshwater habitats, store either glycogen or a glucose polymer called mycolaminarin, and form cell walls of cellulose, chitin, glucan, or some combination of these.

Algae

Algae are an informally defined group of photosynthetic eukaryotes that are classified into seven to eleven divisions. These divisions are at least partially distinguished by their pigments, their energy storage polymers, their cell-wall components, and the number and types of their flagella. The major features of the seven main divisions of algae are summarized next.

Division Chlorophyta: Green Algae

The green algae, which include about 7,500 species, are organisms whose pigments include chlorophylls *a* and *b* and various

carotenoids and xanthophylls. They store starch, contain cellulose in their cell walls, and have either nonmotile cells or motile cells with anywhere from one to about 120 flagella at or near the apex of each swimming cell. Green algae can be unicellular, multicellular, or colonial.

Most green algae live in freshwater, but different species also occur in marine habitats, clouds, snowbanks, or soil, or on the shady moist sides of trees, buildings, and fences. Green algae also live symbiotically with lichen fungi and with several different kinds of animals.

Among the many distinctive groups of green algae, three have received the most attention and are most often classified as classes of the Chlorophyta. Class **Chlorophyceae** includes predominantly freshwater algae that undergo mitosis in a persistent nuclear envelope and cell division by forming a phycoplast. Class **Charophyceae,** which includes mostly freshwater species, and Class **Ulvophyceae,** which consists of marine organisms, both undergo cell division by forming a phragmoplast like that of plants. The Ulvophyceae also have a persistent nuclear envelope during mitosis. However, in the Charophyceae, as in plants, the nuclear envelope disintegrates as mitosis proceeds.

Division Phaeophyta: Brown Algae

The approximately 1,500 species of brown algae derive their color mostly from the xanthophyll fucoxanthin, but they also

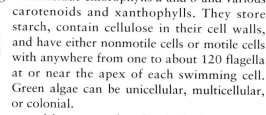

contain chlorophylls *a* and *c* and the orange carotenoid β-carotene. Brown algae store mannitol and a glucose polymer called laminarin. The cell walls of brown algae consist of a cellulosic matrix containing alginic acid, and the motile cells have two lateral flagella, one of which is a forward-projecting tinsel type and the other of which is a

trailing whiplash type. Brown algae are all multicellular and include microscopic forms as well as the most complex of all the algae, the giant kelps. Most brown algae are marine organisms, but a few species live in freshwater. The brown algae are all included in one class, the Phaeophyceae, which contains four orders and five families.

Division Rhodophyta: Red Algae

Like the brown algae, the red algae are mostly marine organisms that are either microscopic filaments or macroscopic forms with complex, leafy branches. The red algae, however, also include some unicellular forms. The 3,900 species of red algae are characterized by proteinaceous pigments called phycobilins. Red algae are red because of their phycoerythrins, which are red phycobilins; they also contain blue-green phycobilins called phycocyanins. Photosynthesis in red algae depends on chlorophyll *a*, but there is no chlorophyll *b*. These algae also contain carotenoids and xanthophylls. The cell walls of red algae usually consist of cellulose and pectins, but the cell walls of some species also form calcium carbonate. Although red algae live almost exclusively in water, mostly marine, they form no cells that swim. All of the red algae are classified into one class, the **Rhodophyceae,** which is divided into five orders and six families.

Division Chrysophyta: Diatoms, Golden-Brown Algae, and Yellow-Green Algae

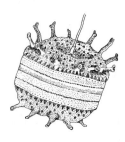

This is the largest division of algae, with more than 11,000 species. The diatoms, golden-brown algae, and yellow-green algae are regarded by some taxonomists to be sufficiently distinct from one another to be classified as separate divisions. At this time, however, most botanists still treat them as three classes of the Chrysophyta: class **Bacillariophyceae** (diatoms), class **Chrysophyceae** (goldenbrown algae), and class **Xanthophyceae** (yellow-green algae). Members of all three classes contain chlorophylls *a* and *c*, carotenoids, and fucoxanthin. The main storage product in these algae is an oil, called chrysolaminarin, that contains a polymer of glucose. Most of the species in this division are unicellular, although some of the golden-brown and yellow-green algae are filamentous, and others are colonial.

Algae that are classified in the three orders and four families of diatoms, which include about 10,000 of the species of the Chrysophyta, are best known for their glass cell walls. Each cell wall consists of two parts that fit together like a pillbox. Although diatoms have no flagellated cells, they apparently swim by a mechanism similar to jet propulsion: each cell secretes mucilage-containing fibrils that propel it forward.

Yellow-green algae (ca. 500 species) are grouped into three orders and four families. These algae either have cell walls of cellulose, or they have silicon-containing scales instead of a well-defined cell wall. Most yellow-green algae have motile cells that have one or two flagella; biflagellate cells have one leading tinsel flagellum and one trailing whiplash flagellum, as do the motile cells of brown algae. Some species lack motile cells, and others have cells that are amoeboid.

Golden-brown algae (ca. 325 species), which are classified into one order with two families, usually have cellulosic cell walls that are rich in pectins, but in certain species the cells are covered with silicon-containing scales. Unlike the other classes of the Chrysophyta, members of the Xanthophyceae lack fucoxanthin. Most golden-brown algae are nonmotile, although some have biflagellated cells like those of the brown algae; still other species have cells that are amoeboid.

Divisions of Unicellular Algae

Three divisions of algae consist exclusively of flagellated unicellular organisms. They are the euglenoids (**Euglenophyta**), the dinoflagellates (**Pyrrhophyta**), and the cryptomonads (**Cryptophyta**). These three divisions are traditionally studied by botanists and zoologists alike because of the plant-like and animal-like features of different species. Each division includes species that are nonphotosynthetic and species that have chlorophyll *a* and carotenoids. In addition, the euglenoids have chlorophyll *b*, and the dinoflagellates and cryptomonads have chlorophyll *c*. Cryptomonads also have phycobilins. A major feature of these divisions is that they are characterized by chloroplasts that are surrounded by an extra membrane, apparently derived from the endoplasmic reticulum.

Euglenoids (ca. 800 species) and cryptomonads (ca. 100 species) lack cell walls. Their cells are bounded by a flexible **periplast,** which is a plasma membrane that has extra inner layers of proteins and a grainy outer surface. Dinoflagellates (ca. 1,100 species) also usually lack cell walls, but most species have armorlike cellulosic plates interior to the plasma membrane. The divisions of unicellular algae are characterized by cells that are motile by one to three flagella. Dinoflagellates have the most distinctive flagellar arrangement among the unicellular algae. Both flagella are lateral, but one coils around the cell and undulates so that the cell spins as it moves forward. The other flagellum trails the cell as a rudder.

The cryptomonads have a nucleus-like organelle between the outer chloroplast membrane and the chloroplast ER. This organelle, called a **nucleomorph,** is absent in other algae. Nuclear division among dinoflagellates is unusual because dinoflagellate chromosomes are permanently condensed and have no histones. Finally, both the euglenoids and the dinoflagellates have a persistent nuclear envelope during mitosis.

Kingdom Plantae

Plants are multicellular eukaryotes having cellulose-rich cell walls, chloroplasts containing chlorophylls *a* and *b* and carotenoids, and starch as their primary food reserve. Plants reproduce sexually by the formation on sporophytes of meiotically derived spores that grow into multicellular gametophytes. In plants, as in some green algae, fertilization produces an embryo that is retained in the parent gametophyte. In all but the bryophytes, the sporophyte is the dominant generation. Plants are informally recognized as three main groups: bryophytes (three divisions), seedless vascular plants (four divisions), and seed plants (five divisions). The main features of each of these twelve divisions of plants are summarized next.

The only swimming cells among plants are the flagellated sperm cells that occur in nine of the twelve divisions. The exceptions are the Pinophyta, Gnetophyta, and Anthophyta (also referred to as Magnoliophyta), none of which forms motile cells.

Division Bryophyta: Mosses

Division Bryophyta is the largest and most familiar division of bryophytes, with more than 10,000 species. Mosses thrive alongside the more conspicuous vascular plants in terrestrial habitats such as on soil, on the trunks of trees, and on shady rock walls. Peat mosses grow more commonly in pools, bogs, and swamps. Some mosses have a central strand of conducting cells that is functionally equivalent to xylem and phloem. This means that, although the bryophytes are referred to as nonvascular plants, mosses do form a simple type of vascular tissue.

Mosses, as well as all other bryophytes, are homosporous. Mosses have leafy gametophytes, multicellular rhizoids, and stomata on the sporophytes. The diversity of these kinds of plants is such that bryologists recognize three classes in the division: the **Sphagnopsida** (peat mosses), which has one order and one family; the **Andreaeopsida** (rock mosses), which also has one order and one family; and the **Bryopsida** ("true" mosses), which includes at least twelve orders and nineteen families.

Division Hepatophyta: Liverworts

Almost 8,500 species of liverworts have been named, ranging in size from tiny, leafy filaments less than 0.5 mm in diameter to a thallus more than 20 cm wide. Liverworts typically have unicellular rhizoids, no cuticle, no specialized conducting tissue, and no stomata. In addition, their spores are shed from sporangia for a relatively short time. Liverworts are the simplest of all living plants. The liverworts are classified in just one class, which is divided into seven orders and twenty-six families.

Division Anthocerotophyta: Hornworts

The hornworts are the smallest of the three groups of bryophytes. There are only about 100 species in six genera, which are divided between two families in the same class and order. The hornworts get their name from the horn-shaped sporophyte. The sporophyte has an intercalary meristem that seems capable of indefinite growth; it also has stomata. The gametophyte is thallose, not leafy, and has no specialized conducting tissue. Female sex organs are embedded in the thallus and contact the surrounding vegetative cells. The thallus also has stomata-like structures, which are absent on gametophytes of all other plants.

Division Psilotophyta: Whisk Ferns

The Psilotophyta are the simplest vascular plants, primarily because they have no apparent roots and most of the species  have no obvious leaves. Instead of roots with root hairs, whisk ferns have rhizomes with rhizoids. Instead of leaves, they have photosynthetic stems and flattened branches that look and function like leaves. Stems are protostelic and are usually dichotomously branched. Like the bryophytes, all members of the Psilotophyta are homosporous. The bisexual gametophytes are small, inconspicuous and nonphotosynthetic.

The only class, order, and family in this division includes two genera: *Psilotum* (3 species), and *Tmesipteris* (7 species). *Psilotum* is widespread in subtropical regions of the southern United States and Asia; it is also a popular and easily cultivated plant that is grown in greenhouses worldwide. *Tmesipteris* is restricted to islands in the South Pacific, where it often occurs as an epiphyte on the trunks of tree ferns. It is rarely cultivated.

Division Lycopodophyta: Club Mosses and Spike Mosses

The Division Lycopodophyta consists of more than 1,100 species worldwide. They are primarily tropical plants that live in terrestrial habitats, either on soil or as epiphytes, but they also form a conspicuous part of the flora in temperate regions. Most of the species are included in two genera: *Lycopodium* (club mosses; ca. 400 species) and *Selaginella* (spike mosses; ca. 700  species). Each genus is classified in a separate class. Their common names derive from their small, mosslike leaves and the club-shaped or spike-shaped cones at the tips of fertile branches.

However, club mosses and spike mosses are differentiated into leaves, stems, and roots, so they are not really mosses. The stems are protostelic, and each leaf has a single, unbranched vein. The club mosses, like bryophytes and whisk ferns, are homosporous. In contrast, the spike mosses are heterosporous. By producing two kinds of spores, the spike mosses always produce unisexual gametophytes—either male or female.

Division Equisetophyta: Horsetails

Equisetum, with about 15 species, is the only living genus in this division. The Equisetophyta, therefore, have only one class, one order, and one family. Some species have branched stems that, with a good imagination, look like horses' tails. Horsetail-type species, as well as unbranched species, are also referred to as scouring rushes because their epidermal tissue contains abrasive particles of glass. They were used by Native Americans to polish bows and arrows and by early colonists and pioneers to scrub pots and pans. *Equisetum* occurs worldwide in most habitats along streams or the edges of forests.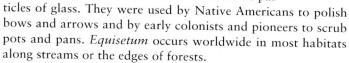

Although horsetails have true leaves, their stems and branches are the dominant photosynthetic parts of the plant body. Horsetails are homosporous, and their gametophytes are photosynthetic.

Division Pteridophyta: Ferns

Ferns include approximately 12,000 living species, making this division by far the largest among the seedless vascular plants.  Ferns are primarily tropical plants, but some species inhabit temperate regions and some even live in deserts.

The classification of ferns is in constant turmoil because of the very different views among fern taxonomists. Neverthe-less, a reasonable system of classification of the ferns groups them into three classes: **Ophioglossopsida, Marattiopsida,** and **Filicopsida.** The Ophioglossopsida, which include but a single order and family, are distinguished by having two kinds of leaves. One kind of leaf is exclusively vegetative and the other is reproductive—that is, it bears sporangia. The Marattiopsida also consist of just one order and one family. This class includes the tree ferns, which have the largest and most complex fronds (leaves) in the division. All members of the Ophioglossopsida and Marattiopsida are homosporous.

Class Filicopsida, known as "true" ferns, is the largest and most diverse group of ferns. The approximately 10,000 species in this class are grouped into three orders and at least ten families. The Filicopsida are distinguished by a type of spore development that is unique to this group of ferns. Such true ferns include homosporous as well as heterosporous species. The heterosporous species are generally aquatic, whereas the homosporous species are terrestrial.

Division Ginkgophyta: Maidenhair Tree

Division Ginkgophyta is one of five divisions of seed plants. Four of these—Ginkgophyta, Cycadophyta, Pinophyta, Gnetophyta—form seeds that are borne on cones or are exposed singly at the tips of fertile branches. Together these four divisions are informally referred to as the gymnosperms (*gymno* = naked, *sperm* = seed). The fifth division, the Anthophyta (i.e., Magnoliophyta), is characterized by seeds that form in fruits. Members of the Anthophyta are informally referred to as angiosperms (*angio* = container). All species of seed plants are heterosporous.

The maidenhair tree, *Ginkgo biloba*, is the only species in this division. Botanists are unsure as to whether this species occurs in natural stands, but it is widely cultivated in temperate regions of the world.

Ginkgo is characterized by distinctive, dichotomously veined, fan-shaped leaves, which are produced on two types of shoots. Relatively fast-growing long shoots and seedlings bear leaves with a distinct apical notch (hence *biloba* in the binomial), while slow-growing spur shoots produce leaves without a notch. The trees are deciduous.

Maidenhair trees are exclusively dioecious. Pollen is born in anthers that occur in loose clusters The ovules of *Ginkgo* are exposed and occur singly at the tips of fertile branches. Ovules develop into seeds with a massive integument that consists of a fleshy outer layer, a hard and stony middle layer, and an inner layer that is dry and papery. The sperm cells of *Ginkgo* are flagellated, although the flagella are not needed for carrying the sperm to the egg.

Division Cycadophyta: Cycads

The approximately 185 species of cycads are divided among three families that are all classified in the same class and order. Cycads live primarily in the tropical and subtropical regions of the world. All species of cycads are dioecious.

Cycads are characterized by their palmlike leaves that bear no resemblance to leaves of other living gymnosperms. Pollen and ovules are formed in sporangia that occur in simple cones on separate plants. Ovules develop into seeds like those of *Ginkgo* in that they have a three-layered integument, but the inner layer of cycad seeds is soft instead of papery. Also like *Ginkgo*, the sperm cells of cycads have flagella that are not needed for carrying the sperm to the egg. The sperm cells of cycads can be up to 400 μm in diameter, which are among the larger sperm cells in the plant kingdom.

Division Pinophyta: Conifers

The informal name of this group, conifers, refers to plants that bear cones, even though some species do not bear cones and other divisions of gymnosperms also include cone-bearing species. This division includes pines, which are the most abundant trees in the forests of the Northern Hemisphere. Most of the approximately 550 species of conifers live in temperate climates, although many also live in alpine habitats or in deserts.

Pines and many other conifers have needle-shaped leaves. Still others have scalelike leaves or leaves with flat blades. Although a few conifers are deciduous, most members of this division are evergreen. Conifers have a wide variety of pollen-bearing and ovule-bearing organs, but none form sperm cells that have flagella.

The Pinophyta consist of two classes of extant species, with a total of two orders and usually six families. The largest class, the **Coniferopsida,** contains pines, firs, cypresses, larches, spruces, hemlocks, and cedars in the Northern Hemisphere, and araucarias and podocarps in the Southern Hemisphere. Members of the single order and family in class **Taxopsida,** such as yews and torreyas, are also native to the Northern Hemisphere.

Division Gnetophyta: Gnetophytes

The gnetophytes include some of the most distinctive, if not bizarre, of all seed plants. The three genera and 71 species of gnetophytes are grouped into one class, but each genus belongs to a separate family in its own order. Gnetophytes are the only gymnosperms that possess vessels, and they are the only gymnosperms that undergo double fertilization. However, unlike double fertilization in angiosperms, double fertilization in gnetophytes is not followed by the formation of endosperm. Instead, the diploid cell from fertilization by the second sperm disintegrates. Like the Pinophyta, the Gnetophyta form nonmotile sperm cells.

Division Anthophyta (Magnoliophyta): Flowering Plants

Botanists estimate that there are at least 260,000 species of angiosperms, which makes the Anthophyta by far the largest division of plants. As a group, the flowering plants are defined by the formation of flowers, by double fertilization that results in a zygote as well as a nutritive endosperm tissue, by the presence of vessels, by the formation of ovules in organs that develop into fruits, and by so-called gametophytes.

Flowering plants are presently divided into two classes, the **Magnoliopsida** (ca. 180,000 species) and the **Liliopsida** (ca. 80,000 species). The Magnoliopsida are informally called dicots, which refers to seeds that have two cotyledon (seed leaves). Dicots are also characterized by flowers whose parts are usually in fours or fives, by netlike venation in leaves, by primary vascular bundles occurring in a ring in the stem, and by the presence of a vascular cambium and true secondary growth in many species. Conversely, the Liliopsida are called the monocots because they form seeds that have a single cotyledon. Also in contrast to dicots, monocots have flowers whose parts usually occur in threes, whose leaves have parallel venation, and whose stems have primary vascular bundles that are scattered. Monocots also lack a vascular cambium and true secondary growth.

There are several comprehensive classifications of flowering plants, which differ according to how genera are lumped into families and families into orders. There is no general agreement among botanists as to which is the best, but four classifications have received the greatest acceptance. These are referred to by the names of the botanists who have proposed them. As summarized in the following table, the different views among these botanists are reflected in the numbers of orders and families that are included in their respective classifications:

Classifier	Number of Orders	Number of Families
Cronquist	83	387
Dahlgren	73	403
Takhtajan	92	410
Thorne	69	440

Suggested Readings

Books

Barinaga, M. 1994. Molecular evolution: Archae and eukaryotes grow closer. *Science* 264:1251.

Beck, C. B. 1988. *Origin and Evolution of Gymnosperms.* New York: Columbia University Press.

Birge, E. A. 1992. *Modern Microbiology: Principles and Applications.* Dubuque, IA: Wm. C. Brown Publishers.

Bold, H. C., and M. J. Wynne. 1985. *Introduction to the Algae.* 2d ed. Englewood Cliffs, NJ: Prentice-Hall.

Carr, N. G., and B. A. Whitton, eds. 1982. *The Biology of Cyanobacteria.* Berkeley, CA: University of California Press.

Chopra, R. N., and P. K. Kumra. 1988. *Biology of Bryophytes.* New York: John Wiley.

Cronquist, A. 1988. *The Evolution and Classification of Flowering Plants.* 2d ed. New York: The New York Botanical Garden.

Hale, M. E. 1983. *The Biology of Lichens.* 3d ed. Baltimore, MD: University Park Press.

Margulis, L., and K. V. Schwartz. 1988. *Five Kingdoms: An Illustrated Guide to the Phyla of Life on Earth.* New York: W. H. Freeman and Company.

Tryon, R. M., and A. F. Tryon. 1982. *Ferns and Allied Plants.* New York: Springer-Verlag.

appendix d

Careers in Botany

Botany is a diverse discipline, ranging from the study of molecules such as DNA to the study of diverse ecosystems such as forests and grasslands. Because of this diversity, botanists can choose from a variety of careers. For example, many cell biologists and geneticists spend most of their time studying cells and genes in laboratories, whereas ecologists, foresters, and taxonomists work outdoors. Many botanists study plant development and pattern formation; others study plants' evolutionary relationships. Still others conserve our natural resources by managing our parks, forests, and wilderness areas. These jobs are enjoyable and rewarding. Moreover, they improve our lives by helping us understand the world around us, while simultaneously increasing and improving our supply of medicines, fibers, and other products derived from plants.

Careers in Botany

Throughout *Botany*, you've seen examples of the many potential careers in botany (e.g., see Reading 16.3 on p. 371). Here are some of the career options you'll have as a botanist:

Agronomy—Agronomists study how soil affects the growth of plants, especially crops. Many agronomists try to increase the yield of crops.

Anatomy—Plant anatomists study the structure and function of plant cells and tissues.

Biochemistry—Plant biochemists study the chemical basis for plant growth and development.

Biophysics—Plant biophysicists study how physical processes affect plant growth and development.

Biotechnology—Plant biotechnologists use plants to produce useful products, including vaccines, antibiotics, and other drugs. Plant biotechnology often involves inserting genes for desirable traits into plants.

Breeding—Plant breeders try to develop new kinds of plants, including those that are disease-resistant.

Bryology—Bryologists study all aspects of bryophytes (mosses, liverworts, and related plants), including their ecology, structure, function, classification, and evolution.

Cell Biology (Cytology)—Plant cell biologists study the structure and function of plant cells. Many of our most important discoveries of plant growth and development are based on discoveries made by cell biologists.

Ecology—Ecologists study how plants interact with the world around them.

Economic Botany—Economic botanists study economically important plants.

Education—Educators (including teachers, museum directors, and science writers) teach others about the biology, cultural impact, and economic importance of plants.

Exploration—Botanical explorers search for rare or undiscovered plants.

Food Science—Food scientists try to develop food from plants and plant products.

Forestry—Foresters manage the production of timber (see Reading 16.3).

Genetics—Plant geneticists study genes, heredity, and variation in plants.

Horticulture—Horticulturists produce ornamental plants and crops. An important subdiscipline of horticulture is landscape design.

Lichenology—Lichenologists study the biology of lichens, which are organisms consisting of a fungus and an alga.

Microbiology—Microbiologists study all aspects of microorganisms, including bacteria and protists such as algae.

Molecular Biology—Plant molecular biologists study the structure and function of macromolecules in plants.

Morphology—Plant morphologists study the evolution and development of flowers, stems, leaves, and roots.

Mycology—Mycologists study all aspects of fungi, including their role in ecosystems and their production of economically important products such as antibiotics.

Natural Resource Management—Natural resource managers are responsible for the use, management, and protection of our natural resources for the benefit of society.

Paleobotany—Paleobotanists study the evolution and biology of fossil plants.

Phycology—Phycologists study all aspects of algae. Phycologists that study marine algae are often referred to as marine botanists.

Physiology—Plant physiologists study the functions of plants, including photosynthesis, respiration, flowering, and the transport of sugars and other solutes.

Plant Pathology—Plant pathologists study the biological aspects of plant diseases.

Pteridology—Pteridologists study all aspects of ferns and related plants (e.g., club mosses).

Systematics—Plant systematists study the evolutionary relationships among plants. Subdisciplines of systematics include taxonomy (identifying, naming, and classifying plants) and chemotaxonomy (the use of chemicals made by plants to classify plants).

Preparing for a Career in Botany

To prepare yourself for a career in botany, be sure that your college career includes a variety of botany and related courses. Equally important to your success will be your ability to solve problems, write and speak effectively, and work well with others. Do all you can to develop these skills; they will benefit you for the rest of your career, regardless of whether you work with plants.

You'll need a college degree—preferably in botany or biology—for most careers in botany. However, don't limit your botanical training to what happens in your classes. Learn about plants in as many ways as you can. For example, learn about the most recent discoveries by reading journals such as *Planta* and *Plant Physiology*. Similarly, go to botany-related meetings in your area. At those meetings, you'll meet other botanists who share your interests.

Seek the advice of botanists. In doing so, remember that most botanists *enjoy* working with students. Ask botanists about what they do and why they study plants. Ask if you can work with them. Indeed, there's no better way of preparing for a career in botany than by working with a botanist. If you're interested in this kind of one-on-one work, ask you professor for information about her/his work, as well as that of other botanists in the area.

Employers

The primary employers of botanists are industry (e.g., Monsanto, lumber and paper companies, breweries, biotechnology companies), educational institutions (high schools, colleges, and universities), and governmental agencies (U.S. Department of Interior, U.S. Department of Agriculture, Public Health Service, NASA, and Environmental Protection Agency). Environmental organizations such as the Sierra Club also employ botanists. If you're interested in working for one of these organizations, contact them directly (see below).

Salaries

Your job and salary will depend on a variety of factors, including your training and experience. Students graduating with a Bachelor's degrees in 1996 received starting salaries of about $27,000 per year; students having Master's degrees started at about $34,000. The average annual salary of botanists in civilian positions was about $39,000; federally employed botanists earned about $44,000. In general, salaries vary with one's expertise, experience, and the cost of living in a particular region or city. These salaries are supplemented by benefits such as rewarding and varied work, individual freedom, and the opportunity to travel.

For More Information

Various aspects of careers in botany can be found at *http://www.ou.edu:80/cas/botany-micro/*. You can also learn about botany and botany-related careers from these organizations:

Botanical Society of America
http://www.falk.ucdavis.edu/bsa.htm[1]

Ecological Society of America
http://sdsc.edu/indx/frameindex/html

America Society of Plant Physiologists
http://www.aspp.org/

American Phytopathology Society
http://www.scisoc.org/aps/overview.html

American Fern Society
http://206.151.68.40:80/fern/

A more comprehensive listing of botany-related organizations (e.g., American Society for Horticultural Science, American Society of Agronomy, American Society of Plant Taxonomists, Society of American Foresters) and their addresses can be found at *http://www.allware.com/kdepew/botany.html#societies*.

---

1. The Botanical Society of America publishes *Careers in Botany: A Guide to Working with Plants* and *Guide to Graduate Studies in Botany in the U.S. and Canada*. You can obtain these useful booklets by writing: Botanical Society of America, Department of Botany, Ohio State University, 1735 Neil Avenue, Columbus, OH 43210, phone (614)292-3519.

glossary

A Horizon (Topsoil) The uppermost layer of soil, usually about 10 to 20 cm. thick

ABA (*see* **Abscisic Acid**)

Abaxial Away from the axis

Abby abbreviation of Abington, which is one of 19 strains of viruses found only gypsy moth caterpillars

Abiotic Something that is non-living and never has been alive

Abrin A seed glycoprotein that is toxic; found in rosary pea

Abscisic Acid (ABA) A plant hormone (growth regulator) associated with water stress and the inhibition of growth; also induces stomatal closing and seed dormancy in many plants

Abscission The detachment of leaves, flowers, or fruits, usually at a weak area termed the abscission zone

Absorption Spectrum The spectrum of light absorbed by a particular pigment

Accessory Fruit A fruit whose flesh is not derived from ovary tissue; for example, an expanded receptacle

Accessory Pigment A pigment that captures light energy and transfers it to chlorophyll *a*; beta-carotene is an example of an accessory pigment

Acetyl Coenzyme A (acetyl-CoA) A two-carbon organic acid whose hydroxyl group has been replaced with coenzyme A

Acid-Growth Hypothesis The hypothesis that acidification of the cell wall leads to the breakage of restraining bonds within the wall, thereby leading to cellular elongation that is driven by turgor pressure

Acropetally Toward the apex

Actin A type of globular protein that makes up actin filaments

Actin Filament The smallest (4–7 nm in diameter) of the three types of filaments that comprise the cytoskeleton

Action Spectrum The spectrum of light that elicits a particular response

Active Transport Movement of solutes across a membrane against their concentration gradient; active transport required energy from cellular metabolism

Acylglyceride Linkage The covalent bond between the organic acid group, such as in a fatty acid, and one of the three hydroxyl groups of glycerol

Adaptive Radiation Evolution of divergent forms of a trait in several species that developed from an unspecialized or primitive common ancestor

Adaxial Toward the axis

Adenosine Triphosphate (ATP) A nucleotide consisting of adenine, ribose, and three phosphate groups; the major source of usable chemical energy in metabolism; when hydrolyzed, ATP loses a phosphate to become adenosine diphosphate (ADP) and releases usable energy

Adenovirus 2 A type of virus that causes human respiratory disease; its role in genetic research involved the discovery of introns

Adventitious Root A root that arises from a leaf or stem (i.e., not from another root)

Aerenchyma A tissue containing large amounts of intercellular spaces

Aerobic Respiration (*see* **Respiration**)

Agar A slimy polysaccharide, consisting mostly of a specific mixture of alpha-galactose sulfates that surround the cell walls of certain red algae; in the United States it is harvested for commerce primarily from *Gelidium robustum*

Akinete A thick-walled dormant cell derived from a vegetative cell

Albuminous Cell Certain ray and axial parenchyma cells in the phloem of gymnosperms; these cells are closely associated with sieve cells, both morphologically and physiologically

Aleurone Layer A group of protein-rich cells located at the outer edge of the endosperm of many grains

Algin An economically important polysaccharide derived from brown algae

Alkaloid A nitrogen-containing base in which at least one nitrogen is part of a ring; examples include nicotine, caffeine, cocaine, and strychnine; alkaloids are often bitter and affect the physiology of vertebrates and other animals

Allele One of the alternative forms of a gene; a gene may have two or more alleles

Allelopathy The production by one organism of chemicals that inhibit the growth of another organism

Allopatric Occurring in different places

Allopatric Speciation Speciation induced by geographical or physical separation of the ancestral population

Allopolyploid A polyploid with multiple sets of chromosomes that originated from more than one species

Allosteric Regulation Regulation that results from a change in the shape of a protein that occurs when the protein binds a nonsubstrate molecule; in its new shape, the protein usually has different properties

Allozymes Enzymes that are coded for by different alleles of the same locus; each form is encoded by different alleles

Alpha-Amylase An enzyme that breaks down starch into smaller units by cleaving the 1,4 linkages between molecules of alpha-glucose

Alpha-Glucose The form of glucose whose structure, when drawn in flat plane, has a hydroxyl group at the first carbon that points down

Alpha-Ketoglutaric Acid A five-carbon organic acid that loses a molecule of carbon dioxide and gains an acetyl-CoA group in the fourth step of the Krebs cycle, thereby being converted to succinyl-CoA; also during this conversion, one molecule of NAD^+ is reduced to NADH

Alpha-Tubulin A type of globular protein that is a main component of microtubules

Amino Acid Acceptor Site A sequence of nucleotides that recognizes and binds to a specific amino acid at the 3′ end of a molecule of transfer RNA

Aminoacyl-tRNA Synthetase A type of enzyme that catalyzes the binding of an amino acid to the amino acid acceptor site on a molecule of transfer RNA

Amylase (see **Alpha-Amylase**)

Amylopectin A highly branched polymer of up to 50,000 molecules of alpha-glucose

Amyloplast A type of plastid that stores starch

Amylose An unbranched chain of up to several thousand molecules of alpha-glucose

Anabolism Biosynthesis; the constructive part of metabolism

Anaerobic Respiration (see **Respiration**)

Anaphase The period of mitosis during which centromeres split and sister chromatids become separate chromosomes that begin to move toward opposite poles of the spindle apparatus

Anaphase I The first anaphase of meiosis; in anaphase I, homologous chromosomes move to opposite poles of the meiotic spindle apparatus, resulting in a halving of the number of chromosomes going to each daughter nucleus

Anaphase II The second anaphase of meiosis; in anaphase II, the centromeres divide, thereby allowing the separation of sister chromatids into independent chromosomes

Androecium (plural, **androecia**) Collectively, all of the stamens of a single flower

Angiosperm A plant whose seeds are born in a fruit; an informal name for flowering plant (Division Anthophyta or Magnoliphyta)

Anisogametes Flagellated gametes of two different sizes; the smaller is designated as male and the larger as female

Anisogamous Organisms that reproduce via anisogametes

Annulus An incomplete ring of cells encircling a fern sporangia; unequal drying of the annulus produces a snapping motion that disperses spores

Anther The pollen-bearing part of a stamen

Antheridogen Substance secreted by young fern gametophytes to stimulate antheridial development

Antheridiophore In some liverworts, the stalk that bears antheridia

Antheridium (plural: **Antheridia**) A unicellular or multicellular structure in which sperms are produced; may be multicellular or unicellular

Anthocyanin Any red or blue pigment that is a flavonoid; anthocyanins are the primary pigments of blue and red plant parts (e.g., flowers, fruits)

Antibody A protein whose formation is induced by an antigen and that binds to the antigen that induced it

Anticodon A sequence of three nucleotides in a molecule of transfer RNA, which is complementary to a codon sequence

Antigen A large, foreign molecule, such as a protein or polysaccharide, that induces its host to form antibodies against it

Antiparallel Refers to double-stranded DNA, in which the direction of each strand is opposite its complementary strand

Antipodal Cell Cells that form at the chalazal of the embryo sac, opposite the micropylar end

Antisense Strand In DNA, the antisense strand of a gene is the one that does not contain a coding sequence for a molecule of RNA; the antisense strand is not transcribed

Apical Dominance The influence exerted by a terminal bud in suppressing the growth of lateral buds

Apical Meristem The meristem at the tip of a root or shoot in a vascular plant

Aplanospores Nonmotile, haploid spores produced from diploid sporangia; develop into the mature haploid phase of the life cycle

Apomictic Asexual production of seeds

Apoplastic Movement The movement of water and solute in the free space of the tissue; the free space includes cell wells and intercellular spaces

Apothecium An open ascocarp; it is usually cup- or saucer-shaped

Archaebacteria Primitive prokaryotes with distinctive chemical and structural features

Archegoniophore In some liverworts, the stalk that bears archegonia

Archegonium A multicellular organ that produces an egg; found in bryophytes and some vascular plants

Aril A fleshy structure that may partially envelop a seed

Artificial Selection Selection by humans of specific traits in organisms being bred to produce desired characteristics

Ascocarp A reproductive structure of ascomycetes, in which asci are formed (sometimes called an **ascoma**)

Ascogenous Hyphae Hyphae with paired male and female nuclei; ascogenous hyphae eventually produce asci

Ascogonium (plural: **Ascogonia**) The female sexual structure of ascomycetes

Ascomycetes A large group of true fungi with septate hyphae; they produce conidiospores asexually and ascospores sexually within asci

Ascospore A spore produced within an ascus

Ascus (plural: **Asci**) A saclike cell of ascomycetes in which, following meiosis, a specific number (usually 8) of ascospores is produced

ATP Synthase A type of membrane-bound enzyme in mitochondria that phosphorylates ADP to ATP by using energy from the diffusion of protons through the enzyme

ATP (see **Adenosine Triphosphate**)

ATP Phosphohydrolase (ATPase) A type of transport protein that uses energy from the hydrolysis of ATP to actively transport ions or other solutes against their concentration gradient

Autopolyploid A polyploid with multiple sets of chromosomes that originated from more than one species

Autotroph An organism that produces its own food, usually by photosynthesis; virtually all plants are autotrophs

Auxin A plant hormone (growth regulator) that influences cellular elongation, among other things; also referred to as indole-3-acetic acid, or IAA

Axial A type of placentation in which the placentae are on the central axis of an ovary having more than one chamber

Axil The upper angle between a twig of leaf and the stem from which it grows

Auxillary Bud Buds that occur in the axil of a leaf

Axillary Placentation Refers to the attachment of ovules along the central axis of an ovary that has more than one ovule-bearing chamber; lily is an example plant that has axillary placentation

b

B Horizon (Subsoil) The layer of soil immediately beneath the topsoil, usually about 25 to 50 cm. thick

Bacteriochlorophyll A modified chlorophyll that is the primary light-trapping pigment in green and purple photosynthetic bacteria

Bacteriophage A type of virus that parasitizes bacteria

Bar A unit of pressure; one bar is the atmospheric pressure of air at sea level and room temperature

Bark The part of the stem or trunk exterior to the vascular cambium

Basidiocarp A reproductive structure of basidiomycetes, in which basidia are formed

Basidioma The fruiting body of basidiomycete fungi; sometimes called a basidiocarp

Basidiomycetes A large and diverse group of true fungi with septate hyphae; they produce basidiospores externally on basidia

Basidium (plural: Basidia) A club-shaped structure upon which, following meiosis, a specific number (usually 4 or 2) of basidiospores is produced

Basipetally Toward the base

Bedrock Solid rock beneath the layers of soil

Beta-Carotene An orange pigment that is made of eight isoprene units; it occurs in most plants as an accessory pigment to photosynthesis

Beta-Glucose The form of glucose whose structure, when drawn in flat plane, has a hydroxyl group at the first carbon that points up

Beta-Oxidation A sequence of biochemical reactions that oxidize fatty acids into a series of two-carbon compounds that are converted to acetyl-CoA

Beta-Tubulin A type of globular protein that is a main component of microtubules

Bilayer In referring to phospholipids, a bilayer is a spontaneously formed double layer of lipid, with an interior of hydrophobic hydrocarbons and an exterior of hydrophilic phosphate groups

Binomial System of Nomenclature A system of applying two-part scientific names to organisms, each name consisting of the genus (generic) name and a species (specific) epithet

Bioassay A quantitative assay of a substance using a part of or an entire organism

Biochemical Organic and inorganic chemicals that occur in living organisms and are involved in the processes of life

Biochemical Cytology Study of the biochemical properties of cell components in conjunction with techniques of microscopy to unravel the details of cell structure and function

Biochemical Reactions of Photosynthesis The temperature-dependent (i.e., "dark") reactions of photosynthesis that reduce carbon dioxide to carbohydrate; occur in the stroma of chloroplasts

Bioenergetics The energy relationships of living organisms

Biogeography The study of geographic distributions of organisms past and present and the mechanisms that caused these distributions

Biological Clock An internal biological timing system that influences cyclic phenomena

Biological Species Concept A species consists of groups of actually or potentially interbreeding natural populations that produce viable offspring

Biomass The collective dry weight of all the organisms in a population, area, or sample

Biotic Pertaining to living organisms

Biotic Community All the populations of interactive living organisms sharing a common environment

Biotic Potential The inherent rate of natural increases, as exhibited by an individual's total number of offspring that survive long enough to reproduce

Bisexual Having antheridia and archegonia on the same gametophyte

Bivalent A pair of synapsed homologous chromosomes in prophase I

Blade The broad, expanded part of a leaf

Bloom For algae, an extensive and conspicuous mass of algae resulting from rapid population growth

Blowout A barren area in arctic tundra caused by wind ripping out part of a vegetation mat whose edges became exposed when pulled up by grazing animals

Body Cell One of two cells produced when the generative cell of a gymnosperm male gametophyte divides; the body cell itself later divides, producing two sperm cells

Bolting Rapid expansion of internodes, increased height, and formation of flowers in otherwise rosette plants in response to cold or hormone application

Bordered Pit A pit in which the secondary wall arches over the pit membrane

Botany The scientific study of plants and plantlike organisms

Bract A structure that is usually leaflike and modified in size, shape, or color

Bracteole Diminutive form of bract

Branch Root A root that arises from another, older root; also called a branch root

Bryophyte Member of a division of nonvascular plants; the mosses, hornworts, and liverworts

Bus Scale Modified leaves that surround and protect a bud

Bulb A short underground stem covered by fleshy leaf-bases that store food

Bulliform Cells Large epidermal cells that occur in groups on the upper surface of leaves of many grasses; loss of turgor pressure in these cells causes leaves to roll up during water stress

Bundle Sheath A layer or layers of cells surrounding the vascular bundle; in C_4 plants, the bundle sheath is photosynthetic and prominent

c

C Horizon (Parent Material) The layer of soil between bedrock and the B horizon. It varies in thickness between about 10 centimeters and several meters, or it may be absent

C3 Plant Plant in which the first fixation of carbon is via the Calvin cycle; the first stable product of C_3 photosynthesis is a three-carbon compound

C4 Plant Plant in which the first fixation of carbon produces a four-carbon acid

Callose A complex carbohydrate in sieve tubes of sieve tube members; callose is especially abundant in injured sieve tubes

Calmodulin A type of protein that is activated when it binds to calcium ions (Ca^{++}); calmodulin activates enzymes in membranes; as much as 2% of the plasma membrane may be calmodulin

Calorie (Cal) 1,000 calories; the amount of heat required to raise the temperature of 1 liter of water 1 °C; a slice of apple pie contains about 365 Cal.

calorie (cal) A unit of heat; one calorie is the amount of heat required to raise the temperature of 1 g of water 1 °C; 1 cal = 4.12J

Calvin Cycle Series of enzymatic reactions in which CO_2 is reduced to 3-phosphoglyceraldehyde (a three-carbon compound) and the CO_2 acceptor (ribulose, 1,5-bisphosphate) is regenerated; also referred to as the reductive pentose cycle

Calyptra The covering that partially or entirely covers the capsule of some species of mosses

Calyx Collectively, all of the sepals of a single flower

CAM (*see* **Crassulacean Acid Metabolism**)

Capillary Movement The movement of water in small tubes such as xylem resulting from its adhesion to surfaces

Capillary Water Water held by surface tension in small pores in the soil

Capsule 1) the sporangium of a bryophyte; 2) a dehiscent, dry fruit that develops from two or more carpels; 3) a slimy layer around the cells of certain bacteria

Carbohydrate An organic compound consisting of a chain of carbons with hydrogen and oxygen attached, usually in a ration of 2:1; glucose, sucrose, and starch are carbohydrates

Carotenoid Any compound in a class of yellow, orange, or red fat-soluble accessory pigments that are derived from eight isoprene units linked together; the most widespread carotenoid in plants is beta-carotene

Carpel The ovule-bearing organ of a flower; the flower of many species has more than one carpel, collectively called the gynoecium

Carpellate Flower A flower whose reproductive parts consist only of carpels; the kernel-bearing flowers on corn cobs are examples of carpellate flower; synonymous with pistillate flower

Carpellate Plant An individual plant whose flowers bear carpels but not stamens; a fruiting mulberry is an example of a plant that is exclusively carpellate (mulberries can form fertile fruits only when pollen is transferred from a staminate plant to a carpellate plant)

Carrageenan A slimy polysaccharide, consisting mostly of a specific mixture of alpha-galactose sulfates that surround the cell walls of certain red algae; the main commercial sources of carrageenan are species of the genus *Chondrus*

Carrion Flower A type of flower that is foul-smelling (carrion odor) and attracts flies or beetles as pollinators

Carrying Capacity The maximum number of individuals in any population of an ecosystem that can survive and reproduce

Casparian Strip The suberized layer covering the radial and transverse walls of endodermal walls

Catabolism The chemical reactions that break down complex materials

Catastrophism The concept that geologic changes result from sudden, violent, large-scale, worldwide catastrophic events

Cation Exchange Process of releasing positively charged ions bound to a surface by replacing them with hydrogen ions

cDNA (*see* **Complementary DNA**)

Cell The structural unit of organisms; plant cells consist of a cell wall and protoplast

Cell Cycle Collectively, the repeating processes of cellular growth and division, including mitosis; the complete cell cycle occurs only in cells that divide, other cells being arrested in development at one of the phases of the cycle

Cell Fractionation The isolation of different organelles or parts of cells by centrifuging a homogenized cell extract in a concentration gradient of, for example, sucrose

Cell Membrane (*see* **Plasma Membrane**)

Cell Plate The disk-shaped structure that forms from the fusion of vesicles at the equator of the spindle apparatus during early telophase in plants and some algae; when mature, the cell plate becomes the middle lamella

Cell Theory A set of postulates describing how cells are the fundamental organizational units of living organisms

Cellulase An enzyme that breaks down cellulose into smaller units by cleaving the 1,4 linkages between molecules of beta-glucose

Central Dogma of Molecular Biology Refers to how genes work to make proteins; each protein-coding gene is transcribed into a molecule of mRNA, which is translated into a sequence of amino acids that comprise a polypeptide (i.e., a protein)

Central Placentation Refers to the attachment of ovules along the central axis of an ovary that has just one ovule-bearing chamber; primrose is an example plant that has central placentation

Centromere A constricted region of a chromosome where sister chromatids are held together

Chalazal Pole The region of the ovule where the stalk of the ovule fuses with the integument; usually the end of the embryo sac that is opposite the micropylar end

Chaparral Dense Mediterranean scrub of lower elevations of the North American Pacific Coast. The unique vegetation, which is either evergreen or deciduous in summer, is well-adapted to fires

Chemiosmosis The coupling of oxidative phosphorylation to electron transport via a proton pump

Chitin A tough, resistant, nitrogen-containing polymer of high molecular weight found in the exoskeletons of arthropods, in the cell walls of many fungi, and in a few other animals and protists

Chlorophyll The pigment responsible for trapping light energy in the primary events of photosynthesis

Chloroplast Organelle specialized for photosynthesis; chloroplasts occur in cells of aboveground parts of plants

Chromatid One of the two threads of a chromosome that has been duplicated in the S phase of the cell cycle; sister chromatids are held together at their centromere

Chromatin A DNA-protein complex that forms chromosomes

Chromatin Fiber A tightly wound coil of chromatin that is believed to consist of six nucleosomes per turn of the coil

Chromosomal Theory of Heredity A set of postulates that accounts for the association of genes with chromosomes

Chymopapain A protease enzyme derived from papaya; used as a drug to dissolve the proteinaceous cartilage of a slipped disc in the spinal column

Circadian Rhythm Endogenous rhythmic changes occurring in an organism on a daily cycle

Circinate Vernation Irregular growth pattern common in fern fronds in which cells on the lower surface of a developing leaf grow faster than those on the upper surface; a "fiddle head" shape results

Cisterna (plural, Cisternae) A set of postulates that accounts for the association of genes with chromosomes

Citrate (citric acid) A six-carbon organic acid that is converted to isocritic acid in the second step of the Krebs cycle

Citric Acid Cycle (*see* **Krebs Cycle**)

Cladistics A method of classifying and reflecting phylogenetic relationships among organisms, based on an analysis of shared features

Cladogram A line diagram portrayal of a branching pattern of evolution, using the concepts and methods of cladistics

Cladophyll A stem or branch that resembles a leaf

Clamp Connection A looplike lateral connection between adjacent cells, occurring in the mycelium of certain basidiomycete fungi

Class A taxonomic category ranking between division and order

Classical Species Concept (*see* **Morphological Species Concept**)

Cleistothecium A closed, more or less spherical ascocarp

Climacteric Rise Point during the ripening of some fruit in which respiratory rates rise to extremely high levels

Climax Community A self-perpetuating community that becomes established at the completion of succession; its composition is strongly influenced by local climate and soils

Climax Vegetation The vegetation of a climax community

Cline Gradual differences in characteristics within a population cross a geographic region

Clone An individual or group of individuals that develop vegetatively from cells or tissues of a single parent individual

cms-T **cytoplasm** (**Texas cytoplasm**) Refers to the phenotype of a variety of corn that is male-sterile due to mitochondrial (cytoplasmic) genes

CO2 Compensation Point Concentration of CO_2 at which the uptake of CO_2 equals the release of CO_2; that is, the point at which photosynthesis equals respiration

Coated Pit A bristle-like structure that occurs in clusters in certain regions of the plasma membrane; these regions form vesicles that pinch off into the cell, thereby removing excess plasma membrane; this process recycles excess plasma membrane in animal cells and it is suspected to do the same in plant cells that have coated pits

Codominance A condition that occurs when both alleles of a heterozygous gene are expressed equally

Codon A sequence of three nucleotides in a gene or molecule of mRNA that corresponds to a specific amino acid or to a stop signal at the end of a gene; of the 64 possible codons, 61 are codes for amino acids and three are stop codons

Coenocytic An organism, or part of an organism, that is multinucleate, the nuclei not being separated by membranes or crosswalls

Coenzyme An organic cofactor of enzyme-catalyzed reactions; NAD^+ and coenzyme A are examples of coenzymes

Coenzyme Q (*see* **Ubiquinone**)

Cofactor A nonprotein substance required by enzymes for proper function

Coleoptile The protective sheath around the embryonic shoot in grass seeds

Coleorhiza The protective sheath around the embryonic root in grass seeds

Colony Hybridization A technique that uses probes to find bacterial colonies that contain a gene of interest

Colonial Algae Algae with a growth form of individual cells loosely attached to each other; cells may or may not be interdependent

Columella Cells Cells in the center of the root cap; characterized by the presence of numerous amyloplasts that sediment in response to gravity

Companion Cell A small cell adjacent to a sieve tube; thought to control the function of sieve tube member

Complementary DNA (cDNA) DNA that is made by reverse-transcribing mRNA into its DNA complement; the collection of vector-cloned c DNA fragments of an organism are its cDNA library

Complete Dominance A condition that occurs when the phenotype of one allele completely masks the phenotype of another allele for a heterozygous gene

Complete Flower A flower that has all four of the main parts: calyx, corolla, androecium, and gynoecium

Compost Partially decayed organic matter used in farming and gardening to enrich the soil and increase its water-holding capacity

Compound Leaf A leaf consisting of two or more independent blades called leaflets

Compression Wood Reaction wood of conifers; compression wood forms along the lower side of leaning stems; compression wood expands and pushes the stem up against gravity

Concentration Gradient The difference in concentration of a substance over a certain distance

Conceptacles Depressions or containers imbedded in the tips of branches of some macroalgae; they contain oogonia and/or antheridia and release eggs and sperm during sexual reproduction

Conidiophore A hypha on which one or more condidia are produced

Conidium (plural: **Conidia**) An externally produced, asexual fungal spore

Conjugation Pilus (*see* **Pilus**)

Continuous Synthesis Refers to the uninterrupted synthesis of DNA in the 5′ to 3′ direction; continuous synthesis occurs in the same direction as a growing replication fork

Convergence The independent evolution of similar structures in organisms that are not closely related

Cork The outermost part of the periderm; the secondary tissue produced by the cork cambium

Corm An elongate, upright, underground stem

Corolla Collectively, all of the petals of a single flower

Cortex Ground tissue located between the epidermis and vascular bundles of stems and roots

Cot Curve A graph of the reassociation of DNA that has been denatured in a salt solution; "cot" comes from initial concentration of DNA (C_o) multiplied by the time required for complete reassociation (t)

Cotyledon Seed leaf; the first leaf formed in a seed; monocots have one cotyledon, and dicots have two

Coupled Reactions Reactions in which energy-requiring chemical reactions are linked to energy-releasing reactions

Coupled Cotransport System A set of active and passive transport proteins that work to actively move ions across a membrane against their gradient, then passively allow the same type of ions to diffuse back down their gradient while coupled to another type of solute that is being transported against its concentration gradient; an example of such a system is the active transport of protons against their concentration gradient by ATPase, followed by the co-transport of protons with sucrose through passive transport proteins back across the membrane

Crassulacean Acid Metabolism (CAM) A type of photosynthesis in which CO_2 is fixed at night into four-carbon acids; during the day, the stomata close and the carbon is fixed via the Calvin cycle; CAM helps plants conserve water and is often characteristic of xerophytic plants.

Cristae (sing., **Crista**) The tubular or vesicle-shaped folds of the inner membrane of mitochondria; cristae contain cytochromes and other components of the electron transport chain that are involved in the synthesis of ATP

Crossing-Over The exchange of genetic material between the chromatids of homologous chromosomes during prophase I of meiosis

Cultivar A variety of plant that is selected for cultivation through hybridization and not found in nature

Cupule Refers to the seed-bearing structure of an extinct group of plants called the seed ferns

Cuticle The waxy coating on the epidermis of all aboveground parts of a plant

Cuticular Wax Wax that is embedded in a cuticle

Cutin The main waxy substance in a cuticle; it consists of hydroxylated fatty acids that are linked together in a complex array

Cyanogenic Glycoside A secondary metabolite; sugar-containing compound that releases cyanide gas when hydrolyzed

Cyanophycin A polypeptide functioning as an energy reserve in cyanobacteria

Cyclic Photophosphorylation The light-induced flow of electrons originating from and returning to photosystem I; cyclic photophosphorylation produces ATP but no reduced NADP

Cyclosis Movement of the cytosol and the cellular components that are suspended in it; cyclosis is usually circular around a central vacuole

Cytochemistry (*see* **Biochemical Cytology**)

Cytochrome Heme-containing proteins that carry electrons in respiration and photosynthesis

Cytochrome b-c₁ Complex A cluster of cytochromes that carry electrons in the electron transport chain; the complex probably also pumps protons across the inner mitochondrial membrane

Cytochrome Oxidase Complex A cluster of cytochrome oxidases that function as the terminal electron carrier in the electron transport chain; this complex donates electrons to oxygen, which is then reduced to form water

Cytokinesis The division of cytoplasm into distinct cells; together with mitosis, cytokinesis comprises the phases of the cell cycle involved in cell division

Cytokinin Group of hormones (growth regulators) that promote growth by stimulating cellular division

Cytology The study of cell structure and function

Cytophotometry A method of studying cells by staining selected parts, such as the nucleus, and measuring how much light they absorb; the absorbance of stained chromatin in a nucleus is proportional to the amount of DNA it contains

Cytoplasmic Inheritance Refers to the inheritance of genes that occur in chloroplasts and mitochondria; it is cytoplasmic because it is not nuclear

Cytoplasmic Male Sterility (*cms*) A male-sterile condition in which sterility is controlled by mitochondrial (cytoplasmic) genes

Cytoplasmic Streaming The active circulation and flow of cytoplasm within a cell and sometimes between cells

Cytoskeleton A network of microscopy filaments that form a mechanical support system in the cell

d

Daughter Colonies Smaller colonies of cells pinched off to the interior of a larger parent colony during asexual reproduction of some algae; disintegration of the parent colony releases the daughter colonies

Day-Neutral Plant Plant whose flowering is not affected by the length of day

Debranching Enzyme A type of enzyme that hydrolyzes the branched linkages of starch

Deciduous Plant Plant that loses all of its leaves during autumn

Decomposer An organism, such as bacterium or a fungus, that facilitates recycling of nutrients in an ecosystem through the breakdown of complex molecules to simpler ones

Deduction The process of devising explanations for observations

Denature To break bonds that maintain the three-dimensional structure of proteins or nucleic acids; also to break the hydrogen bonds that hold DNA together in a double helix

Dendrochronology The study of growth rings of trees to determine past conditions

Dentrifying Bacteria Bacteria that convert nitrates or nitrites to gaseous nitrogen

Deoxyribonucleic Acid (DNA) The nucleic acid containing four different nucleotides whose simple sugar id deoxyribose; genes are made of DNA; DNA exists as a double helix that can be unwound to replicate itself or to make RNA

Desert A biome characterized by low annual precipitation and/or porous soil, low humidity, wide daily fluctuations in temperature, high radiation, and living organisms adapted to these conditions

Desertification The conversion of non-desert biomes into deserts

Desynaapsis The unpairing and separation of homologous chromosomes upon the disintegration of the synaptomeal complex

Deuteromycetes Fungi that have no known sexual reproduction; most reproduce by conidia, and most otherwise have characteristics of ascomycetes. Deuteromycetes are also called Fungi Imperfecti

Dicot A type of angiosperm that belongs to a class whose members are characterized by having two cotyledons (seed leaves) per seed; Class Magnoliopsida

Dictyosome A stack of flattened, membranous vesicles that are often branched; dictyosomes are the sites where precursors of cell wall materials and other cellular components are assembled and prepared for secretion from the cell; dictyosomes are also called Golgi bodies

Dictyostele A complex stele; a type of siphonostele having gaps in the cylinder associated with emergence of leaves (leaf gaps)

Differentiation Physical and chemical changes associated with the development and/or specialization of an organism or cell

Diffuse-Porous Wood Wood in which the vessels are distributed uniformly throughout the growth layers

Diffusion The net movement of particles, either dissolved or suspended, from a region of higher concentration to a region of lower concentration; the energy of diffusion is derived from the random motion of particles that is caused by molecular motion; diffusion tends to cause the distribution of particles to become homogenous throughout a medium

Dihybrid Cross A hybridization experiment that follows the inheritance of phenotypes that are controlled by two different genes

Dikaryotic Fungi whose hyphal cells each have two nuclei, the nuclei usually being derived from two different parents

Dioecious Having the pollen-producing and the ovule-producing organs on different individuals of the same species; mulberry is an example of a dioecious species

Diploid The condition of having two sets of chromosomes in a nucleus

Directional Selection Selection for a phenotype that is either higher or lower in frequency than the most abundant phenotype

Disaccharide A carbohydrate composed of two monosaccharides that are linked by a covalent bond; sucrose and maltose are examples of disaccharides

Discontinuous Synthesis Refers to the synthesis of DNA that occurs in the opposite direction of a growing replication fork; in discontinuous synthesis, DNA polymerase jumps ahead on one strand in the direction of fork movement (in this case, the 3′ to 5′ direction), then builds a new chain "backward" in the 5′ to 3′ direction

Disulfide Bond A type of covalent bond between the sulfur atoms of separate amino acids in the same protein; disulfide bonds strengthen the tertiary structure of proteins

Diterpene A compound that consists of four isoprene units linked together; gibberellins are examples of diterpenes

Diversifying Selection Selection for the low-frequency (extreme) phenotypes above and below the norm of the population; or selection against the high-frequency phenotype (norm)

Division A taxonomic category between kingdom and class

DNA Ligase A type of enzyme that joins adjacent nucleotides together by catalyzing the formation of sugar-phosphate bonds in a strand of DNA

DNA Polymerase A type of enzyme that catalyzes the covalent bonding between nucleotides into a nucleic acid chain

Dolipore A complex central pore occurring in the hyphal septa of many basidiomycete fungi; it is covered by a cap on both sides of the septum

Domain A structural and functional portion of a polypeptide, which may be encoded separately by a specific exon; a portion of a protein that has a globular tertiary structure

Dominant A trait that masks an alternative (recessive) trait when the gene for these traits is heterozygous

Dormancy A condition in which plant parts such as buds and seeds are temporarily arrested in their development; dormancy is typically seasonal and is broken as environmental conditions change during the year

Double Helix The spiral shape of a double strand of DNA

Double Fertilization In angiosperms, the process by which one sperm cell fertilizes the egg to a zygote and another sperm cell fertilizes the polar nuclei to form a primary endosperm nucleus

Dynein A large contractile protein that forms the connecting sidearms and spokes between microtubules in flagella

e

Ecological Race A race composed of many similar variants of the same species in several local populations distributed over a relatively large geographic area

Ecology The study of the interactions of organisms with one another and with their environment

EcoRI An example restriction enzyme that comes from the bacterium *Escherichia coli*; this restriction recognizes the DNA sequence GAATC, then cleaves it between the guanine and the adenine

Ecosystem A major system of organisms that are interacting with one another and with their physical environment

Ecotype An individual taxon of plants adapted to a specific community within its overall distribution

Ectomycorrhizae Mycorrhizae that develop externally and do not penetrate to the interior of the cells they surround

Edaphic Factor A soil factor

Egg Apparatus A group of usually three cells in an embryo sac, one of which is the egg and two of which are synergids

Elater A straplike appendage (usually paired) attached to a horsetail (*Equisetum*) spore; also, a spindle-shaped sterile cell occurring in liverwort sporangia; serves to disperse spores

Electrochemical Gradient The combination of a concentration gradient and an electrical gradient of ions across a membrane

Electrogenic Pump An active transport protein that transports (pumps) ions against their concentration gradient; the main electrogenic pumps in plants are proton pumps

Electron Microscope An instrument that uses an electron beam to magnify images of specimens

Electron Transport Chain A sequence of electron carriers that use the energy from electron flow to transport protons against a concentration gradient across the inner mitochondrial membrane

Embryo In plants, the part of the seed that will form the growing seedling after germination; includes a radicle, apical meristem, and embryonic leaf or leaves

Embryo Sac The common name for the female gametophyte of flowering plants

Enations Flaps of tissue extending from a stem; possibly the origin of microphylls

Endergonic A reaction that requires an input of energy before it will occur; endergonis reactions never occur spontaneously

Endocarp The innermost layer of simple fleshy fruits; the endocarp can be soft, as in tomatoes, or hard and stony, as in peaches

Endocytosis The process by which the plasma membrane invaginates and forms vesicles whose contents from outside of the cell can be brought into the cell

Endodermis Layer of cells inside the cortex and outside the pericycle of roots; radial walls of the endodermis are suberized by the casparian strip

Endomycorrhizae Mycorrhizae that develop within the interior of cells

Endoplasmic Reticulum (ER) An extensive network of sheetlike membranes distributed throughout the cytosol of eukaryotic cells; portions that are densely coated with ribosomes are the rough ER; other regions, with fewer ribosomes, are the smooth ER

Endosperm The nutritive, storage tissue that grows from the fusion of a sperm cell with polar nuclei in the embryo sac

Endospore A thick-walled, stress-resistant structure containing the chromosome and some cytoplasm of a bacterial cell; formed in times of stress and capable of periods of dormancy

Endosymbiotic Hypothesis An explanation for the origin of chloroplasts and mitochondria from the descendants of prokaryotes that lived symbiotically in larger prokaryotic hosts

Entrainment The process by which a periodic repetition of some signal (e.g., light, dark) produces a circadian rhythm that remains synchronized with the same cycle as the entraining (i.e., modifying) factor

Entropy The degree of orderliness in a system

Enzyme A biological catalyst, usually a protein, that can speed up a chemical reaction by lowering its energy of activation; amylase is an example of an enzyme

Epicotyl The region of an embryo above the attachment point of cotyledons

Epicuticular Wax The outermost layer of wax in a cuticle

Epidermis The outermost layer of cells that covers a plant

Epinasty The differential growth of petioles that causes the leaf blade to curve downward

Epiphyte An organism that is attached to another organism without parasitizing it

Epistasis The interaction of two or more genes that act together to make a phenotype; epistasis is best known for serial gene systems that control the multistep biosynthesis of a complex molecule or the sequence of steps in a metabolic pathway

EPSP Synthetase A type of enzyme that catalyzes a step in the synthesis of enolpyruvylshikimic acid-3-phosphate, which is a precursor to aromatic amino acids; the herbicide glyphosate works by inhibiting the activity of this enzyme

Ethylene A gaseous plant hormone (growth regulator) that promotes fruit ripening and other physiological responses

Etiolation The abnormal elongation of stems caused by insufficient light; etiolated stems usually lack chlorophyll

Eubacteria The majority of all bacteria; their cell walls contain muramic acid, certain lipids, and other features that distinguish them from archaebacteria

Euchromatin Lightly staining portion of chromatin, not easily visible by light microscopy

Eudicots An evolutionary line of dicots containing most dicot species; diverging from paleoherbs, more ancient dicots, and monocots

Eusporangiate Having large sporangia arising from multiple fern leaf cells producing many spores collectively surrounded by a multi-layered sporangial wall

Eustele A complex stele common in horsetails and ferns having vascular bundles appearing distinct in cross-section as a ring around the central pith

Eutrophication Gradual enrichment of a lake or stream with nutrients enhancing algal growth

Evolutionary Species An ancestral-descendant sequence of populations evolving separately from others and forming a single unit

Exergonic A reaction that releases energy and occurs spontaneously

Exine The outermost layer of a spore or pollen grain; the exine consists of a resistant polymer that protects the male gametophyte from desiccation

Exocarp The outermost layer (usually the skin) of simple fleshy fruits

Exocytosis The process of expelling the contents of a vesicle from a cell by fusing the vesicle membrane with the plasma membrane and opening the inside of the vesicle to the outside of the cell

Exon A sequence of DNA within a gene that codes for an amino acid sequence

Exon-Shuffling Hypothesis An explanation for how complex new genes arise from the joining of independent exons into new combinations

Extensin A family of related glycoproteins that are structural proteins in cell walls

f

Facilitated Diffusion Passive transport through a transport protein

Fascicle A cluster of pine leaves (needles) or other needlelike leaves of gymnosperms

Fascicular Cambium The part of the vascular cambium that forms between the xylem and phloem within a vascular bundle

Fatty Acid A long, mostly hydrocarbon chain that has an organic acid group at one end; the most common fatty acids in plants are oleic acid, linoleic acid, and linolenic acid

Feedback Inhibition Control mechanism in which the increasing concentration of a molecule inhibits the further synthesis of that molecule

Fermentation A process by which energy is obtained from organic compounds without the use of oxygen as an electron acceptor

Fiber An elongated, thick-walled sclerenchyma cell; helps support or protect the plant

Fibrous Root System Consists of an extensive mass of similar-sized roots

Fiddlehead The curled fern frond prior to unrolling and elongation; also known as a crozier

Field Capacity The water-storage capacity of soil; the amount of water in soil after gravitational percolation stops

Filament The stalk of a stamen; or, the vegetative body of filamentous algae and fungi

Filial Refers to a generation of offspring; the first set of offspring from a hybridization experiment is the first filial generation (F_1), the second set is the second filial generation (F_2), etc.

First Gap (*see* G_1 **phase**)

Fission The asexual division and formation of two similar new cells within mitosis, as seen in prokaryote reproduction

Fitness A measure of an individual's evolutionary success; number of its surviving offspring relative to the number of surviving offspring of other individual's within the population

Flagellum (plural, **Flagella**) A hairlike locomotor organelle that protrudes from the cell into the medium surrounding it; flagella enable cells to swim, but the only swimming cells of plants are the sperm cells of some plant groups; flagella also occur in algae, fungi, bacteria, and animals

Flavin Mononucleotide (FMN) The first electron acceptor in the electron transport chain; FMN takes electrons from NADH in the mitochondrial matrix, plus one proton from NADH and one proton from the matrix to become $FMNH_2$; protons from $FMNH_2$ are released into the mitochondrial intermembrane space

Flavonoid Any phenylpropanoid-derived compound that is linked to three acetate units and condensed into a multiple-ringed structure; the most common flavonoid is rutin; flavonoids also include naringin, which is a bitter substance in grapefruits

Flora The plants or organisms (other than animals) of a particular region; also: a publication devoted to the taxonomy of plants of a particular region

Florigen The hypothetical flowering hormone; florigen has never been identified or isolated

Fluid Mosaic Model A model for the structure of membranes as a fluid phospholipid bilayer through which proteins float in a continually shifting mosaic pattern

Fluorescence The release of energy at a longer wavelength than the energy that was absorbed

Food Web (**Food Chain**) An interlocking flow of energy, involving producers and consumers in an ecosystem

Foot Basal part of a moss sporophyte; the foot is embedded in the gametophyte

Free-Central A type of placentation in which the placentae are on the central axis of an ovary having just one chamber

Free Energy Energy available to do work

Free-Nuclear Embryo An early stage of embryo development in a gymnosperm, in which the zygote nucleus divides repeatedly without walls forming around the nuclei

Frond Photosynthetic leaf-blade of a fern

Fumaric Acid A four-carbon organic that takes on a molecule of water and becomes malic acid in the seventh step of the Krebs cycle

Functional Megaspore The megaspore that, in some types of embryo sac development, is the only one of the four meiotic products to grow into a female gametophyte; the other three spores disintegrate

Fusiform Initials Vertically elongated cells in the vascular cambium that produce cells of the axial system in the secondary xylem and secondary phloem

g

G_1 Phase During interphase, the portion of the cell cycle that occurs between the end of mitosis and the onset of DNA synthesis; G_1 refers to first gap

G_2 Phase During interphase, the portion of the cell cycle that begins at the end of the S phase and lasts until the beginning of mitosis; G_2 refers to the second gap

Gametangium (plural: **Gametangia**) A cell or structure in which gametes are produced

Gamete A haploid reproductive cell that fuses with another gamete to form a zygote; the female gamete is an egg and the male gamete is a sperm; in certain kinds of algae and fungi, however, the gametes are neither male or female

Gametic Meiosis A life-cycle pattern among algae; meiosis produces haploid gametes; resembles life cycles of animals

Gametophyte The phase of the plant life cycle that produces gametes

Gametophytic Self-Incompatibility A type of self-incompatibility that is imposed by gametophyic tissues or organs; an example would be incompatibility that is imposed by the pollen tube, which is a gametophytic structure

Gas Chromatography A technique of separating organic solutes; the samples moved as vapor through a column of liquid or particulate solids, and components of the sample are differentially absorbed into the column

Gas Vacuole A membrane-bound bubble of gas that enables aquatic bacteria to float

Gel Electrophoresis A technique by which nucleic acids or proteins are separated in a gel that is placed in an electric field

Gemmae (sing. **Gemma**) Asexual plantlets in certain liverworts that can form new gametophytes; often form in gemmae cups

Gemmules An erroneous concept of inheritance; described as packets of heritable information produced throughout a mature organism, transported to the reproductive organs, and packaged into gametes before fertilization

Gene A sequence of DNA that codes for a molecule of mRNA, tRNA, or rRNA, or that regulates the transcription of such codes; a gene is the basic unit of heredity

Gene Conversion The change of one allele to another during crossing over

Gene Flow Introduction of genetic material into the gene pool of one population from another population

Gene Gun An instrument that shoots tiny beads coated with DNA directly into cells; some cells treated this way integrate the foreign DNA that is shot into them, thereby becoming transgenic; early models of the instrument used .22 caliber cartridges, hence the name gene gun.

Gene Pool All of the alleles within a population that are available to future generations

Generation Time In plants, the length of time it takes from seed germination to reach sexual maturity

Generative Cell The cell in the pollen grains of angiosperms that divides to form two sperm cells, or the cell in the pollen grains of gymnosperms that divides to form a sterile cell and another cell that divides to form to sperm cells

Genetic Species Concept Two species are considered distinct if their genetic makeup is sufficiently different from one another

Genetic Drift Random changes in gene frequencies within the gene pool of a population

Genetic Distance Measure of the degree of genetic difference between different populations or species

Genetic Code A system of codons (nucleotide triplets) in DNA or RNA that together code for a sequence of amino acids in a polypeptide; 61 of the possible codons are codes for amino acids, the remaining three being stop codons that are not translated

Genetic Engineering The artificial manipulation of genes, or the transfer of genes from one organism to another; synonymous with recombinant DNA technology

Genomic Library The set of fragments of an organism's genome that are cloned in a virus or bacterial plasmid

Genophore A bacterial chromosome; its DNA is not associated with the histone proteins of eukaryotic chromosomes

Genotype An organism's genes, either individually or collectively

Genus (plural: **Genera**) A taxonomic category between family and species; it forms the first part of the binomial of the scientific name of organism

Gibberellin (GA) A type of plant hormone that affects, for example, stem elongation and seed germination

Gill One of the fleshy plates that radiate out from the stipe beneath the cap of a mushroom basidiocarp

Gliadin A storage protein in the grains of wheat

Glucose (*also see* **Alpha-Glucose** and **Beta-Glucose**) A common monosaccharide whose empirical formula is $C_6H_{12}O_6$

Glutelins A complex mixture of storage proteins in the grains of wheat

Glycogen A carbohydrate food reserve similar to starch in many organisms other than plants

Glycolysis The anaerobic metabolic pathway by which glucose is broken down into two molecules of pyruvic acid in the cytosol; the substrate-level phosphorylation of two molecules of ADP to ATP and the reduction of two molecules of NAD^+ to NADH occur for the breakdown of each molecule of glucose

Glycoprotein A type of protein that has sugars attached to it; extension in cell walls is an example of a family of glycoproteins

Glycoside A molecule combining a secondary metabolite with one or more sugars

Glyoxylic Acid Cycle A sequence of biochemical reactions that converts acetyl-CoA into carbohydrate

Glyoxysome A type of microbody that is common is germinating oilseeds and seedlings that arise from them; glyoxysomes contain enzymes that catalyze the breakdown of fatty acids into acetyl-CoA

Glyphosate The generic name of one of the most commonly used herbicides in agriculture

Golgi Body (*see* **Dictyosome**)

Gram Stain A crystal violet stain that is retained by gram-positive bacteria and not retained by gram-negative bacteria, after alcohol or a similar solvent is applied

Gram-Negative (*see* **Gram Stain**)

Gram-Positive (*see* **Gram Stain**)

Grain In wood, the direction of axial cells relative to the longitudinal axis of the tree

Grana (sing., **Granum**) Stacks of thylakoids where the photochemical (i.e., "light") reactions of photosynthesis occur

Grassland A biome characterized by the predominance of grasses

Gravitational Water The water that readily drains from the soil by gravity

Gravitropism The curvature of roots or stems in response to gravity

Greenhouse Effect An increase in temperature due to certain atmospheric gases (e.g., carbon dioxide, chlorofluorocarbons) that allow passing sunlight to heat the earth's surface but prevent the escape of heat back into space

Ground Meristem The fundamental tissue of the apical meristem; produces the cortex

Growth Ring A growth layer in secondary xylem or secondary phloem, as seen in cross section

GTP Cap A molecule of 7-methylguanosine triphosphate (GTP) that is attached to the 5′ end of a molecule of RNA as transcription begins; the GTP cap protects the RNA from degradation as it is being synthesized

Guard Cells Two specialized epidermal cells that form a stomatal apparatus

Gum A hemicellulose that is secreted by plants, which consists of several kinds of monosaccharides; an example is gum arabic, which is a mixture of the monosaccharides arabinose, galactose, glucose, and rhamnose

Gum Arabic A gum produced by the plant species *Acacia senegal*; this gum is a hemicellulose, which in this case is a complex branched chain consisting of arabinose, galactose, glucose, and rhamnose

Guttation The exudation of liquid water from leaves; caused by root pressure

Gynoceium (plural, **Gynoecia**) Collectively, all of the carpels of a single flower

h

Habitat The location, with its own specific set of environmental conditions, where an organism naturally occurs

Habitat, Operational The soil components and moisture, shade, associated organisms, and other habitat features that directly affect an organism

Half-Life Time required in a chemical reaction for half the original reactant material to decay or be consumed

Halon A bromine-based compound that is especially destructive of the ozone layer

Haploid The condition of having only one set chromosomes in a nucleus

Hardpan A hard soil with disrupted structure that may develop through the gradual accumulation of salt residues when inorganic fertilizers are applied annually without the addition of organic matter; it generally restricts the downward movement of water and roots

Hardwood A woody dicot

Hardy-Weinberg Theory Over generations the relative allelic frequencies will remain the same in the absence of mutation, natural selection, and gene flow in small randomly breeding populations

Haustorial Behaving like fungal haustoria; in *Ginkgo* such pollen tubes function as digestive filaments

Haustorium In plants the xylem-to-xylem connection of a parasite to its host

Heartwood Wood in the center of a tree trunk; usually darker due to the presence of resins, oils, and gums; does not transport water and solutes

Helicase A type of enzyme that breaks hydrogen bonds between complementary base pairs of DNA, thereby causing the double strand to split into separate single strands

Helix Anything of a spiral shape; in biology it refers to the shape of DNA molecules, which occur as double helices

Heme A complex organic ring structure, called a protoporphyrin, to which an iron atom is bound; heme occur in the cytochromes of all organisms and in the hemoglobin of animals

Hemicellulose Primarily a cell wall polysaccharide of variable composition and structure; hemicellulose that is secreted by plants is also called a gum (*see* **Gum** and **Gum Arabic**)

Herbarium A systematically arranged collection of dried, pressed, and mounted plant specimens

Heterochromatin A condensed, darkly staining portion of chromatin, easily visible by light microscopy

Heterocyst A relatively large, unpigmented, thick-walled, nitrogen-fixing cell that is produced within the filaments of certain cyanobacteria

Heterogenous Nuclear RNA (hnRNA) The pool of primary RNA transcripts in the nucleus, which are of various, usually large sizes

Heterosis A condition in which crossbred organisms are more fit than inbred organisms because they have more heterozygotic loci

Heterosporous Production of two types of spores; i.e., a large megaspore that becomes a female gametophyte and a smaller microspore that becomes a male gametophyte

Heterotroph An organism that obtains its food from other organisms

Heterozygote Superiority A condition in which individuals heterozygous at one or more loci have higher fitness than an individual with fewer heterozygous loci

Heterozygous A condition in which a gene has two different alleles in a diploid individual

Hill Reaction The photolysis of water and the photoreduction of an artificial electronacceptor by chloroplasts in the absence of CO_2

Histone A type of protein that comprises the protein component of chromatin

hn RNA (*see* **Heterogenous Nuclear RNA**)

Hold Fast A portion of a macroalga that firmly attaches the thallus to the substrate

Homologous Refers to a pair of chromosomes that have alleles for the same genes

Homology A condition in which a common trait possessed by different species was derived from a common ancestor

Homozygous A condition in which both alleles of a gene are the same in a diploid individual

Horizons For soil, the major layers of soil visible in vertical profile; for example, the A horizon is topsoil

Hormone An organic molecule made in one plant that exerts an effect in another part of the plant; effective in small concentrations

Humus The organic portion of soil; derived from partially decayed plant and animal material

Hybrid Vigor (*see* **Heterosis**)

Hybridization Production of offspring from crossing different species or between genetically different populations

Hydrolysis Any chemical reaction that proceeds by the addition of water to break down a molecule; the breakdown of starch by amylase and the breakdown of sucrose by invertase are examples of enzyme-catalyzed hydrolyses

Hydrophilic Refers to chemicals that are freely soluble in water; sugars are examples of hydrophilic compounds

Hydrophobic Refers to chemicals that are not soluble in water by are soluble in nonpolar solvents; lipids and hydrocarbons are generally hydrophobic

Hydrophyte A plant that is adapted to submersion in water or an aquatic environment for at least part of its growing season

Hydrotropism Growth of a root toward water concentration in comparison with one of low solute concentration; a hypertonic solution tends to gain water across a membrane from a solution of lower solute concentration

Hygroscopic Water Water held tightly to the surface of soil particles

Hymenium The layer of asci releasing spores from the ascoma of ascomycete fungi

Hypertonic Refers to a solution of high solute concentration in comparison with one of low solute concentration; a hypertonic solution tends to gain water across a membrane from a solution of lower solute concentration

Hypha (plural: **Hyphae**) A single tubular thread of the mycelium of a fungus or similar organism

Hypocotyl The region of an embryo that is between the radicle and the attachment point of the cotyledons

Hypodermis One or more layers of cells just beneath the epidermis that are distinct from the underlying cortical or mesophyll cells

Hypothesis A proposed solution to a scientific problem that must be tested by experimentation; a working explanation based on evidence and suggesting some principle; if disproved, a hypothesis is discarded

Hypotonic Refers to a solution of low solute concentration in comparison with one of high solute concentration; a hypotonic solution tends to lose water across a membrane to a solution of higher solute concentration

i

Ice-Minus Bacteria Genetically engineered bacteria that contain a foreign gene whose polypeptide inhibits the formation of ice crystals

Imbibition The absorption of water onto the internal surfaces of materials

Immunoassay A technique to detect, identify, and measure quantities of large organic molecules such as hormones; utilizes antibodies to the molecule of interest produced by animals to bind to the molecules and make them evident

Imperfect flower A flower that lacks either an androecium or a gynoecium

Imperfect Fungi (*see* **Deuteromycetes**)

In-Group Analysis The assumption in cladistics that the most prevalent character state is primitive

Inbreeding Mating within the same plant or between the offspring of an inbred parent

Incomplete Flower A flower that has one or more of the parts absent (calyx, corolla, androecium, or gynoecium)

Incomplete Dominance A condition that occurs when the phenotype of one allele only partly masks the phenotype of another allele for a heterozygous gene

Indeterminate Growth Growth that is not inherently limited, as with a vegetative apical meristem that produces an unrestricted number of organs indefinitely

Indole-3-Acetic Acid (IAA) A naturally occurring auxin (*see* **Auxin**)

Indusium An outgrowth covering a fern sorus of a fern leaf

Inferior Ovary An ovary located below the other flower parts on a floral axis

Inflorescence A cluster of flowers that are arranged on their axis in a specific pattern

Inorganic Compound A type of molecule that either lacks carbon or contains carbon but not hydrogen; carbon dioxide and water are examples of inorganic compounds

Integument The layer or layers of tissue that surround the megasporangium (nucellus) in an ovule; the integument becomes the seed coat

Intercalary Meristem Meristem at the base of a blade and/or sheath of many monocots

Intercellular Spaces In leaves, the spaces within a leaf allowing gas circulation between internal photosynthetic cells and the atmosphere via stomata

Interfasicular Cambium The part of the vascular cambium that forms between vascular bundles and connects with the fascicular cambium

Intermediate Filament The middle-sized (8–12 nm in diameter) of the three types of filaments that comprise the cytoskeleton

Internode Part of the stem between two successive nodes

Interphase Collectively, all of the phases of cell growth apart from cell division

Intine The inner layer of a spore or pollen grain; the intine consists of cellulosic and pectic material that is exported from the microspore

Introgression Back-crossing; mating of fertile hybrids with parent populations

Intron A sequence of DNA within a gene that does not code for an amino acid sequence

Inulin A polymer of fructose having beta-2,1 linkages; an alternative to starch as a storage polysaccharide

Invertase A type of enzyme that catalyzes the breakdown of sucrose by hydrolysis into glucose and fructose

Island Biogeography A theory explaining the relationship between defined habitat area (such as an island) available for organisms and the number and diversity of species in that area

Isocitric Acid A six-carbon organic acid that loses a molecule of carbon dioxide in the third step of the Krebs cycle, thereby being converted to alpha-ketoglutaric acid; also during this conversion, one molecule of NAD^+ is reduced to NADH

Isogametes Compatible gametes that appear alike; they are neither male nor female but are derived from compatible strains usually designated as (+) or (−)

Isogamous Organisms that reproduce via isogametes

Isomers Compounds with the same chemical formula but with structural differences

Isoprene The basic five-carbon subunit of terpenoid polymers

Isozymes Enzymes that have the same function but are encoded from different genes

j

Joule (J) The amount of energy needed to move one kilogram through one meter with an acceleration of one meter per second per second; 10^7 ergs; one watt-second; a slice of apple pie contains about 1.5×10^6 J

k

Kilodaltons A unit of mass equal to a molecular weight of one thousand

Kinetic Energy The energy of motion; a solute that moves down its concentration gradient has kinetic energy

Kinetin A purine that acts as a cytokininin

Kinetochore A disc-shaped complex of proteins that is bound on one side to a centromere and on the other side to a spindle fiber

Kingdom The highest taxonomic category

Knots In wood, the bases of branches that have been covered by lateral growth of the main stem

Kranz Anatomy Specialized leaf anatomy characteristic of C_4 plants; characterized by having vascular bundles surrounded by a photosynthetic bundle sheath

Krebs Cycle The metabolic pathway by which acetyl-CoA is oxidized in mitochondria to carbon dioxide; each turn of the Krebs cycle also forms one ATP by substrate-level phosphorylation, reduces one NAD^+ to HADH, and reduces one ubiquinone to ubiquinol; the Krebs cycle is also called the citric acid cycle or the tricarboxylic cycle

l

Lateral Meristem Meristem that produces secondary tissue; the vascular cambium and cork cambium are examples of lateral meristems

Leader Sequence A short non-coding sequence of DNA, immediately upstream from the beginning of a gene, that is transcribed into RNA

Leaf Buttress A lateral protrusion below the apical meristem; the initial stage in the development of a leaf primordium

Leaf Gap Region of parenchyma tissue in the primary vascular cylinder above a leaf trace

Leaf Primordium A lateral outgrowth from the apical meristem that will eventually form a leaf

Leaf Scar A scar left on a twig when a leaf falls from a stem

Leaf Trace The part of a vascular bundle that extends from the base of a leaf to its connection with a vascular bundle of a stem

Leaflet An individual blade of a compound leaf

Lectin A type of protein that binds to carbohydrates on cell surfaces; many lectins are glycoproteins; lectins occur in all parts of the cell but are mostly associated with the endoplasmic reticulum and other membranes, including the plasma membrane

Lenticel Spongy areas in the cock surfaces of stems and roots of vascular plants; allows gas to exchange to occur across the periderm

Leptosporangiate Having small sporangia arising from a single leaf-surface cell of a fern; typically on a delicate stalk and surrounded by a thin wall

Liana A woody vine that is supported by other plants

Light Microscope An optical instrument that uses light to magnify images of specimens

Light-Compensation Point Light level at which photosynthesis equals respiration

Lignin A complex phenylpropanoid polymer that makes cell walls stronger, more waterproof, and more resistant to pests, herbivores, and disease organisms

Lilium-Type Embryo Sac Development A type of embryo sac development that entails all four spores of an ovule; in this type of development, the antipodal cells and one of the polar nuclei are triploid; the other polar nucleus and the egg apparatus are haploid (*also see* **Polygonum-Type Embryo Sac Development**)

Linkage The condition of having genes on the same chromosome (linked); alleles of genes that are linked tend to be inherited together

Loam A soil type consisting of a mixture of sand, silt, and clay

Locus (plural, **Loci**) The position of a gene on a chromosome

Long-Day Plant Plant that flowers when the length of dark is shorter than some critical value; long-day plants flower in spring and summer

Long-Night Plant Plant that flowers only if the uninterrupted dark period exceeds a certain length; sometimes called short-day plant

Looped Domain A fold or loop in a region of packed chromatin fibers, which extends out from the main axis of the chromosome; looped domains may consist of 20,000 to 100,000 nucleotide pairs

m

Macroevolution Evolutionary changes that refer to the development of new species

Macronutrients Inorganic elements required in large amounts for plant growth (e.g., nitrogen, calcium, sulfur)

Magnification Enlargement of an object

Malic Acid A four-carbon acid that is oxidized by the reduction of NAD^+ to NADH in the eighth step of the Krebs cycle; malic acid is also formed by the reduction of oxaloacetic acid that is derived from fixing carbon dioxide to phosphoenolpyruvic acid in C_4 and CAM photosynthesis

Marginal Placentation Refers to the attachment of ovules (placentation) along the edge (margin) of a suture; garden pea pods have marginal placentation

Matric Potential The component of water potential caused by the attraction of water molecules to a hydrophilic matrix

Mediterranean Scrub The often dense, shrubby vegetation that occurs in areas with wet winters and dry summers; it is dominated by evergreen bushes, or those that are deciduous in the summer

Megapascal (MPa) A unit of pressure; one million (10^6) pascals; 1 MPa = 10 atmospheres of pressure; a car tire is typically inflated to about 0.2 MPa, whereas the water pressure in home plumbing is 0.2–0.3 MPa

Megaphyll A leaf type of seedless vascular plants; having multiple, branched veins

Megaspore A spore that will grow into a female gametophyte

Megaspore Mother Cell A cell that will undergo meiosis and cytokinesis to produce megaspores

Megasporophyll Refers to a leaf-like organ that bears megasporangia

Meio-Blastospore A spore that arises by budding from a haploid, meiotically produced spore

Meiosis Nuclear division in which chromosomes are doubled, then divided twice; the daughter nuclei from meiosis have half the number of chromosomes of the parent nucleus; in plants, meiosis forms spores

Meiosis I The first of two nuclear divisions that, in plants, form spores; in meiosis I, homologous chromosomes synapse, cross over, and move to opposite poles of the meiotic spindle apparatus; the separation of homologous chromosomes in meiosis results in a reduction in chromosome number by one-half in daughter nuclei

Meiosis II The second of two nuclear divisions that, in plants, form spores; in meiosis II, centromeres divide and sister chromatids become independent chromosomes that move to opposite poles of the spindle apparatus

Membrane Potential The potential electrical energy of ions across a membrane; membrane potential is measured in volts

Membrane Selectivity The control that a membrane exerts over how much and what kinds of materials pass through it

Membrane System The interconnected membranes of a cell, including the plasma membrane and the various organellar membranes

Mendelian Inheritance Refers to patterns of inheritance that were discovered by Gregor Mendel

Meristem Regions of specialized tissue whose cells undergo cell division

Mesocarp The middle layer (often fleshy) of simple fleshy fruits; the mesocarp occurs between the exocarp and the endocarp

Mesophyll Parenchyma tissue between the epidermal layers of a leaf; is usually photosynthetic

Mesophyte A plant that requires a relatively humid atmosphere and abundant soil water

Messenger RNA (mRNA) A class of RNA that carries the genetic message of genes to ribosomes, where the message is translated into the amino acid sequence of a polypeptide

Metabolism The sum of all chemical reactions occurring in a cell or organism

Metaphase The period of mitosis during which chromosomes become attached to spindle fibers, which align the chromosomes in a circular plane that is perpendicular to the microtubules of the spindle apparatus

Metaphase I The first metaphase of meiosis; in metaphase I, pairs of homologous chromosomes align along an equatorial plane that is perpendicular to the axis of the spindle apparatus

Metaphase II The second metaphase of meiosis; in metaphase II, chromosomes align along an equatorial plane that is perpendicular to the axis of the spindle apparatus

Metaphase Plate The plane of alignment of chromosomes during metaphase; the metaphase plate is perpendicular to the axis of the spindle apparatus

Metaxylem Primary xylem that differentiates after the protoxylem; reaches maturity after the part of the plant in which it is located has stopped elongating

Micelle For soils, the sheet-like structure of a clay particle

Microbody A vesicle-like organelle that is bounded by a single membrane and is generally associated with the endoplasmic reticulum; glyoxysomes and peroxisomes are types of microbodies

Microevolution Evolutionary changes that occur within a population; may eventually lead to the formation of a new species, but not as a one-time event

Microfibril A complex of cellulose molecules that are twisted together into a strong, threadlike component of cell walls

Microhabitat The particular part of a habitat occupied by an individual

Micronutrients Inorganic elements required in small amounts for plant growth (e.g., boron, copper, zinc)

Microphyll A leaf type of seedless vascular plants; having a single unbranched vein

Micropyle The opening in a ovule through which the pollen tube will enter in angiosperms, or through which the pollen grains will enter in gymnosperms

Microsporangium (plural, **Microsporangia**) A microspore-containing sporangium

Microspore A spore that will grow into a male gametophyte

Microspore Mother Cell A cell that will undergo meiosis and cytokinesis to produce microspores

Microsporophyll Refers to a leaf-like organ that bears microsporangia

Microtome An instrument that is used for slicing specimens into microscopically thin sections

Microtubule The largest (18–25 nm in diameter) of three types of filaments that comprise the cytoskeleton; microtubules also move chromosomes during nuclear division and make up the internal structure of flagella

Middle Lamella The pectin-containing layer between cells that probably acts as the glue to hold cells together

Midrib The large central vein of a leaf

Minerals Inorganic chemical compounds usually made with two or more elements

Mitosis The process of nuclear division in which chromosomes are first duplicated, followed by the separation of daughter chromosomes into two genetically identical nuclei; the division of nuclei; together with cytokinesis, mitosis comprises the phases of the cell cycle involved in cell division

Molecular Clock Hypothesis States that the dissimilarity between nucleic acid sequences of two species is proportional to the time since divergence from their common ancestor

Molecular Phylogeny A phylogeny based on molecular data

Monocot A type of angiosperm that belongs to a class whose members are characterized by having one cotyledon (seed leaf) per seed; Class Liliopsida

Monoecious Having the pollen-producing and the ovule-producing organs on the same individuals

Monokaryotic Fungi whose cells each contain a single nucleus

Monophyletic A taxon and all its descendants

Monomer The smallest subunit that is a building block of a polymer

Monosaccharide A simple sugar that cannot be broken down by hydrolysis; glucose is an example monosaccharide

Monoterpene A compound that consists of two isoprene units linked together; menthol is an example monoterpene

Morphological Species Concept Traditional concept of taxonomic species surmising that two species are considered distinct if they are sufficiently different morphologically

Morphological Plasticity Condition in which environmental factors induce different phenotypes from the same genotype

mRNA (*see* **Messenger RNA**)

Mucigel Slimy material secreted by root tips to facilitate growth of the root through soil

Multigene Family A set of duplicated genes; many genes occur in multigene families; an example is the family of genes in which each gene codes for the small subunit of ribulose-1, 5-bisphosphate carboxylase/oxygenase

Mutation A genetic change; mutations include changes in DNA sequences of genes, rearrangements of chromosomes, and the movements of transposable elements

Mycelium (plural: **Mycelia**) Collective term for the hyphae of a fungus

Mycolaminarin A carbohydrate food reserve of water molds (oomycetes)

Mycorrhizae (plural: **Mycorrhizae**) A mutualistic association between a fungus and the roots of a plant

Myxobacteria A group of complex, gram-negative soil bacteria that often form upright, multicellular, reproductive bodies

NADH dehydrogenase complex A complex of enzymes whose function is to transport protons from NADH across the inner mitochondrial membrane

Nastic Movement A movement that occurs in response to a stimulus, but whose direction is independent of the direction of the stimulus

Natural Selection Differential reproduction of phenotypes; genotypes and phenotypes vary among organisms and some of these phenotypes promote reproduction more than other phenotypes

Neck As part of an archegonium, the slender region above the venter surrounding a fluid-filled canal for passage of a sperm cell to the egg within the venter

Nectar A sweet exudate secreted by plants to attract insects (e.g., for pollination)

Nectary A structure in angiosperms that secretes nectar; usually (but not always) associated with flowers

Net Movement The amount of movement that goes in one direction more than another; particles diffuse in all directions, but net movement occurs away from where particles are most concentrated to where they are least concentrated

Net Productivity The energy produced in an ecosystem by photosynthesis minus the energy lost through respiration

Niche The ecological role of a species within a community

Nitrate Reductase System A group of enzymes that catalyze the reduction of nitrate to ammonium

Nitrification The oxidation of ammonium ions or ammonia to nitrate, done by certain free-living bacteria in the soil

Nitrogen Fixation Incorporation of atmospheric nitrogen into nitrogenous compounds; done by certain free-living and symbiotic bacteria

Nitrogen-Fixing Bacteria Bacteria that convert gaseous nitrogen to nitrates or nitrites

Nitrogenase A complex of enzymes that convert atmospheric nitrogen gas into ammonia

Nivea Gene (*niv*) A gene in snapdragons that, when homozygous recessive, blocks the synthesis of flower pigments; plants that are homozygous recessive for this gene have white flowers

Node Point where one or more leaves attach to a stem

Nodule Tumorlike swelling on roots of certain higher plants (e.g., legumes) that houses nitrogen-fixing bacteria

Noncyclic Photophosphorylation The light-driven flow of electrons from water to NADP$^+$ in oxygen-evolving photosynthesis; requires both photosystems I and II

Nonprotein Amino Acids A secondary metabolite; amino acids that are not incorporated into proteins

Nonvascular Plants Plants that lack vascular tissue (e.g., liverworts)

Northern Blotting A procedure by which molecules of RNA are separated by gel Electrophoresis, transferred to a filter, and probed with DNA that is complementary to the RNA sequence of interest; the location of the target sequence is found because it becomes radioactive when the probe anneals to it (*also see* **Southern Blotting**)

Nuclear Envelope The double membrane that surrounds the nucleus

Nucleic Acid An organic acid that is a polymer of mostly four different nucleotides; deoxyribonucleic acid (DNA) and ribonucleic acid (RNA) are the two kinds of nucleic acids

Nucleosome The basic beadlike unit of chromatin in eukaryotes, consisting of DNA that is wound around a core of histone proteins

Nucleotide The subunit of a nucleic acid, consisting of a phosphate group, a simple sugar (either ribose or deoxyribose), and a nitrogen-containing base that is either a purine or a pyrimidine

Nyctinasty The "sleep movements" of leaves in response to changes in turgor of cells at the base of their petioles

Occam's Razor A principle of logic that holds that the best explanation of an event is the simplest, using the fewest assumptions of hypotheses

Oligosaccharins Fragments of plant cell wall known to affect processes, such as growth, reproduction, and differentiation, and therefore functioning as plant hormones

Oogamous Organisms that reproduce via a small, nonmotile or motile, male gamete fertilizing a larger, nonmotile, female gamete

Operculum The lid of the sporangium in mosses

Opposite Phyllotaxis Leaves occurring in pairs at a node

Organella A specialized part of the cell, usually bounded by a membrane; nuclei, chloroplasts, and mitochondria are membrane-bound organelles; ribosomes are membrane-free organelles

Organic Compounds Molecules that contain carbon; usually also include hydrogen and oxygen.

Organic Evolution Changes in the genetic composition of a population of organisms across generations

Organismal Theory A set of postulates describing how whole organisms, not cells, are the fundamental organizational units of living organisms; according to this theory, organisms develop by compartmenting the whole organism into cells, not by building the organism from cells

Organogenesis The formation of organs during plant development

Osmosis The diffusion of water or other solvent through a differentially permeable membrane

Osmotic Potential The potential of solutes to cause osmotic pressure; also called solute potential

Osmotic Pressure The water potential of pure water across a membrane; osmotic pressure is an indicator of how concentrated a solution is on the other side of a membrane from pure water

Osmotically Active Solutes that can cause a change in a cell's osmotic potential; potassium (K$^+$) and other ions are osmotically active

Out-Breeding Mating with unrelated individuals

Out-Group Analysis The assumption in cladistics that the most prevalent character state of plants outside of a given group is primitive

Outcrossing Mating between different individual plants

Ovary The enlarged, ovule-bearing portion of a carpel or of a cluster of fused carpels; after fertilization, an ovary matures into a fruit

Ovule The structure that contains the female gametophyte in seed plants; the female gametophyte is surrounded by a nucellus (megasporangium tissue), which is covered by one or two integuments; when mature, an ovule is called a seed

Oxaloacetic Acid A four-carbon organic acid that is converted to citric acid by the addition of an acetyl group in the first step of the Krebs cycle; oxaloacetic acid is also the product of the carbon dioxide fixation of phosphoenolpyruvic acid in C$_4$ and CAM photosynthesis

Oxidation The loss of electrons from an atom or molecule that is involved in an oxidation-reduction (redox) reaction; oxidation removes energy from one substance, which is coupled with the simultaneous addition of energy to another substance by reduction (*also see* **Beta-Oxidation**)

Oxidative Phosphorylation Phosphorylation of ADP to ATP that uses energy from a proton pump fueled by the electron transport system

Ozone A form of oxygen (O^3) in the stratosphere that, when compared with ordinary oxygen (O^2), more effectively shields living organisms from intense ultraviolet radiation

p

Paleospecies A species defined only by fossil morphology

Palisade Mesophyll The vertical photosynthetic cells below the upper epidermis of a leaf

Panicle A type of inflorescence consisting of a branched main axis with side branches bearing loose clusters of flowers

Papain A protease enzyme; derived from papaya; can digest the muscle tissue of animals and is economically important as a meat tenderizer

Parallelism In cladistics, a pattern of character evolution where the same character state arises from the primitive state more than once

Parapatric Occurring in adjoining places

Parapatric Speciation Speciation that occurs between contiguous populations, often induced by low dispersal range of the individuals

Paraphyletic Term applied to a group of organisms that does not contain all the descendants of a single ancestor

Paraphyses (sing. **Paraphysis**) Sterile filaments that grow among the reproductive cells of certain fungi and brown algae

Parenchyma The tissue type characterized by relatively simple, living cells having only primary walls

Parietal Placentation Refers to the attachment of ovules (placentation) along the wall of an ovary (i.e., parietal); violets are example plants that have parietal placentation

Parsimony In cladistics, the shortest hypothetical pathway that provides the most likely explanation of an evolutionary event

Parthenocarpy Development of fruit without fertilization

Pascal (Pa) The pressure unit (i.e., energy per unit volume) used to measure water potential; one pascal equals the force of one newton per square meter; one atmosphere of pressure equals 1.0×10^5 Pa

Passage Cell Endodermal cells of root that have a thin wall and casparian strip when other endodermal cells develop thick secondary walls

Passive Transport The unrestricted movement of a substance through a biological membrane; the energy for passive transport is the kinetic energy of movement down a concentration gradient; it is passive because it does not require energy from cellular metabolism

Pectin A gluey polysaccharide that holds cellulose fibrils together; pectins are mostly polymers of galacturonic acid monomers with alpha-1,4 linkages

Pedicel The stalk of a flower in an inflorescence

Peduncle The stalk of a flower or of an inflorescence

Pentose Phosphate Pathway A series of chemical reactions that start with glucose-6-phosphate from glycolysis and involve several five-carbon sugars (pentoses); during this pathway, NADP is reduced to NADPH, but no ATP is produced

Peptide Bond A carbon-nitrogen bond that links amino acids together in a chain

Peptidoglycan A large carbohydrate polymer found in the walls of true bacteria; it is composed of long chain molecules interconnected by short chains of peptides

Peptidyl Transferase A type of enzyme in the large ribosomal subunit that catalyzes the formation of a peptide bond between the amino acid at the end of a growing polypeptide and the next amino acid to be added to the chain

Perfect Flower A flower that has an androecium and a gynoecium

Pericarp Refers collectively to the layers of ovary tissue in a fruit; pericarp is the preferred term for fruits whose layers cannot be easily distinguished from one another

Pericycle The layer of cells surrounding the xylem and phloem of roots; produces branch roots

Periderm The protective tissue that replaces epidermis; includes cork (phellem), cork cambium (phellogen), and phelloderm

Peripheral Cells Outermost cells of thee root cap that secrete mucigel; are sloughed from the root cat as the root grows through the soil

Peristome The "teeth" around the opening of the sporangium of mosses

Perithecium A flask-shaped or spherical ascocarp with a terminal opening

Permafrost Permanently frozen soil

Permanent Wilting Point The moisture content of soil at the point when a particular plant's root system cannot absorb water, even when given water and placed in a humid chamber

Peroxisome A type of microbody that occurs primarily in leaves and contains enzymes that metabolize hydrogen peroxide and glycolic acid

Petal One of the parts of the flower that are attached immediately inside the calyx; collectively, the petals of a single flower are called the corolla; the corolla is usually the part of the flower that is conspicuously colored

Petiole The stalklike part of a leaf that connects the blade to the stem

Phagotropic Ingesting solid food particles

Phellem Cork; produced by the phellogen

Phelloderm The inner part of the periderm; forms inside of the phellogen

Phellogen Cork cambium

Phenolic Any compound that contains a fully unsaturated, six-carbon ring that is linked to an oxygen-containing side group

Phenotype An organism's observable features, either individually or collectively; the phenotype results from the interaction of the genotype of an organism with its environment

Phenylpropanoid A complex phenolic that has a three-carbon side chain; phenylpropanoids are generally derived from the amino acids phenylalanine and tyrosine; myristicin, the main flavor ingredient of nutmeg, is a phenylpropanoid

Phloem Vascular tissue that transports water and organic solutes

Phospholipid A lipid that has two fatty acids and a phosphate group bound to a molecule of glycerol; phospholipids are important components of membranes

Photochemical Reactions The "light" reactions of photosynthesis; these reactions occur on the grana of chloroplasts and produce ATP and reduced NADP

Photon The elementary particle of light

Photoperiodism Response to the duration and timing of day and night; the system within plants that measures seasons and coordinates seasonal events such as flowering

Photorespiration The light-dependent formation of glycolic acid in chloroplasts and its subsequent oxidation in peroxisomes

Photosynthesis The production of carbohydrates by combining CO_2 and water in the presence of light energy; occurs in chloroplasts and releases oxygen

Photosystem A complex of chlorophyll and other pigments embedded in the thylakoids of chloroplasts and involved in the photochemical (i.e., "light") reactions of photosynthesis

Phototropism Growth of a stem or root toward or away from light

Phragmoplast A set of microtubules oriented parallel to the axis of the spindle apparatus (perpendicular to the plane of cell division), which will form a cell plate; phragmoplasts occur in plants and in most green algae

Phycobilins Water-soluble accessory pigments occurring in the red algae and cyanobacteria

Phycocyanin A blue photosynthetic pigment in cyanobacteria and red algae

Phycoerythrin A red phycobilin

Phycoplast A set of microtubules oriented perpendicular to the axis of the spindle apparatus (parallel to the plane of cell division), which will form a cell plate; phycoplasts occur only in a few green algae

Phyllotaxis The arrangement of leaves on a stem

Phylogenetic Reflecting evolutionary relationships

Phylum A taxonomic category between kingdom and class in animals; it is equivalent to *division* in plants

Phytochrome A group of proteinaceous pigments involved in phenomena such as photoperiodism, the germination of seeds, and leaf formation; absorbs red and far-red light

Pigment Molecule that reflects and absorbs light at particular wavelengths

Pilus (plural: **Pili**) A minute tube between two bacterial cells, through which transfer of genetic material may occur

Pioneers The first plants to become established on new soil

Pistil A female reproductive structure of a flower; composed of one or more carpels and consisting of ovary, style, and stigma

Pistillate Flower (*see* **Carpellate Flower**)

Pistillate Plant (*see* **Carpellate Plant**)

Pith Parenchyma tissue in the center of a stem; located interior to the vascular bundles

Placenta (plural, **Placentae**) The area inside a carpel where the ovules are attached

Plasma Membrane The semipermeable membrane that surrounds the cytoplasm and is next to the cell wall; also called the cell membrane or the plasmalemma

Plasmid A small, circular fragment of DNA in bacteria; a plasmid can be integrated into and replicated with the rest of the bacterial genome; because of their ability to take up foreign DNA, bacterial plasmids are used as vectors for genetic engineering and research

Plasmodesma (Plural, **Plasmodesmata**) A tiny, membrane-lined channel between adjacent cells

Plasmolysis Shrinkage of cytoplasm away from the cell wall due to the loss of water by osmosis

Plastid A type of organelle that is bounded by a double membrane and is associated with different pigments and storage products; chloroplasts are green, photosynthetic plastids; amyloplasts are storage plastids that contain starch

Pleiotropic Gene A gene that affects more than one phenotypic character; an example of a pleiotropic gene occurs in tobacco, in which a single gene controls the size and shapes of leaves, flowers, anthers, and fruits

Pneumatophore Upward-growing roots of some plants that grow in swamps; contain much aerenchyma and function in gas exchange

Polar Fiber A spindle fiber that does not bind to a kinetochore

Polar Nuclei Nuclei that come from opposite poles of the embryo sac and fuse with a sperm cell to form the primary endosperm nucleus

Polarity Establishment of poles of specialization at opposite ends of a cell, tissue, organ, or organism; for example, polarity leads to the differentiation of roots and shoots

Pollen Grain A male gametophyte that is surrounded by a microspore wall in seed plants

Pollen Tube The germination tube of a pollen grain, which grows from the stigma, through the style, and into the micropyle of the ovule; the pollen tube carries the sperm cells to the embryo sac

Pollination The transfer of pollen from microsporangia to the stigma in angiosperms or to directly to the ovule in gymnosperms

Pollination Droplet A sticky exudate at the mouth of the micropyle of a gymnosperm ovule; pollen grains catching in it are slowly withdrawn to the interior (pollen chamber) as the droplet recedes

Poly-A Tail A chain of adenylic acid molecules that is added to a molecule of RNA immediately after it has been transcribed and cleaved from its DNA template

Polyacetylenes A secondary metabolite; long chain derivatives of fatty acids that contain carbon-carbon triple bonds

Polyembryony, Cleavage The development of multiple embryos in a gymnosperm seed as a result of the differentiation of certain cells of a single embryo

Polyembryony, Simple The development of multiple embryos in a gymnosperm seed as a result of the development of two or more zygotes

Polygene A set of genes that act together, without dominance, to control a continuously variable phenotype; length, width, and oil content are examples of continuously variable phenotypes that are most like to be under polygenic control

Polygonum-Type Embryo Sac Development A type of embryo sac development from a functional megaspore that forms eight free nuclei, three of which become an egg apparatus, two of which are polar nuclei, and two of which become antipodal cells

Polymer A molecule consisting of many identical or similar monomers linked together by covalent bonds

Polymerase Chain Reaction (PCR) A procedure by which free nucleotides are assembled into a nucleic chain in a test tube by enabling the activity of a bacterial DNA polymerase to bind them together; the PCR is cycled 30 or more times to produce a million-fold amplification of the target DNA sequence

Polypeptide A chain of amino acids linked together by peptide bonds

Polyploid A condition in which a nucleus has more than two complete sets of chromosomes

Polysaccharide A carbohydrate polymer composed of many monosaccharides that are linked covalently into a chain; polysaccharides include starch, glycogen, and cellulose

Polysome A cluster of ribosomes on a single molecule of mRNA

Population A group of interbreeding individuals of the same species usually occupying the same territory at the same time

Population Density The number of individuals of a population within a given area

Population Genetics The application of genetic laws and principles to entire populations; assumes that evolution is the result of progressive change in the genetic composition of a population rather than individuals

Population, Local A population within a relatively small geographic area

Postulate A basic or necessary assumption; a set of postulates that address the same phenomenon can be taken together as a theory

Postzygotic Mechanism Barriers to reproductive compatibility that function after the formation of a zygote

Potential Energy The energy stored by matter because of its location or configuration; regarding a solute, the higher its concentration, the steeper is its concentration gradient and the greater is its potential energy; energy available to do work

Preprophase Band A band of microtubules that rings the cell just beneath the plasma membrane in a plane that is perpendicular to the axis of the future mitotic spindle apparatus; the preprophase band also corresponds to the orientation of the future metaphase plate and cell plate

Pressure-Flow Model States that a turgid pressure gradient drives the flow of water and solutes in phloem

Pressure Potential The component of water potential caused by the force created by turgor pressure against a membrane

Prezygotic Mechanism Barriers to reproductive compatibility that function before the formation of the zygote

Primary Cell Wall The usually thin cell wall that forms during cell division; it is part of all but some sperm cells in plants

Primary Consumer A consumer that feeds directly on producers

Primary Growth Growth resulting from the activity of apical meristems

Primary Pit-Field A thin area in a cell wall where clusters of plasmodesmata occur

Primary RNA Transcript A molecule of RNA that includes the GTP cap, the leader sequence, the gene sequence, the trailer sequence, and the poly-A tail

Primary Root (*see* **Radicle**)

Primary Structure The sequence of amino acids in a protein

Primary Succession Succession that is initiated on bare rock or in water after a disturbance has occurred

Primary Thickening Meristem In some monocots, the meristem that increases the thickness of the shoot axis

Primer A short sequence of single-stranded DNA (e.g., 10–30 nucleotides long) that is complementary to one end of a target gene of interest; primers are annealed to their complementary sequences so that DNA Polymerase can begin replicating the target gene in the polymerase chain reaction

Probe In genetic research, a sequence of radioactive DNA or RNA that is used to find the complementary sequence of a gene of interest in a culture of clones or cells

Procambrium A meristem that produces the primary vascular tissues

Producer An autotrophic organism; producers form the base of food chains in an ecosystem

Progymnosperms A group of extinct plants believed to be the ancestors of gymnosperms

Prop Roots Adventitious roots that form on a stem above the ground; help support the plant (as in corn)

Prophase The period of mitosis during which chromosomes condense, first appearing as a mass of elongated threads and later as individual chromosomes

Prophase I The first prophase of meiosis; in prophase I, homologous chromosomes condense, synapse, cross over, and desynapse; chiasmata move to the ends of chromosomes by the end of prophase I

Prophase II The second prophase of meiosis; in prophase II, chromosomes condense, the nuclear envelope disintegrates, and a spindle apparatus is assembled; in many organisms, prophase II is bypassed if telophase I is also bypassed, in which case the meiotic nuclei go directly from anaphase I to metaphase II

Protease Inhibitor Any chemical that inhibits the activity of enzymes that digest proteins (i.e., proteases); protease inhibitors can also be proteins

Protenema (pl. **Protenemata**) The early, filamentous growth of the gametophyte of bryophytes and ferns

Prothallial Cells Two of the four cells produced during the development of a gymnosperm microspore into a pollen grain. The prothallial cells are functionless.

Protoderm The outermost tissue of an apical meristem; produces the epidermis

Protostele The simplest stele; a solid or nearly solid cylinder of xylem and phloem; little or no pith or parenchymatous tissue

Protoxylem The first xylem cells formed in the primary xylem

Pseudoplasmodium A phase of cellular slime molds in which the myxamoebae do not fuse but aggregate into a sluglike body that moves as a unit

Pulvinus Jointlike thickening at the base of a petiole; involved in movements of a leaf (or leaflet)

Punctuated Equilibrium A model stating that long geological time periods with little or no evolutionary change are punctuated by periods of rapid evolution

Punnett Square A gametic grid that is used to show the expected genotypic and phenotypic ratios resulting from a hybridization experiment

Purine A two-ringed nitrogen-containing base that is part of a nucleotide; the most common purines are adenine and guanine

Pyrenoid A spherical protein body imbedded in algal chloroplast; surrounded by starch granules and containing RuBP carboxylase; oxygenase involved in photosynthetic starch production

Pyrimidine A one-ringed nitrogen-containing base that is part of a nucleotide; the most common pyrimidines are thymine, cytosine, and uracil

q

Quaternary Structure The way that different subunits attach to one another in a multisubunit protein

Quiescent Center The relatively inactive region in the apical meristem of a root

r

R Group A general term for the side group of a molecule, such as a methyl group, a hydroxyl group, or a monosaccharide

R-loop A sequence of DNA within a gene that is displaced into a loop-like projection when the gene is annealed to its complementary mRNA; the R-loop does not anneal with the mRNA because it is an intron whose complementary sequence has been spliced out of the mRNA molecule

Radicle The root of an embryo

Ray Initials Cells in the vascular cambium that produce the ray cells of secondary xylem and secondary phloem

Reaction Wood Wood produced in response to a stem that has lost its vertical position; reaction wood straightens the stem

Receptacle The region of the floral shoot where the parts of the flower are attached

Recessive A trait that is masked by an alternative (dominant) trait when the gene for these traits is heterozygous

Recombinant DNA Technology (*see* **Genetic Engineering**)

Recombination Nodule A cluster of enzymes in a synaptonemal complex, which are believed to act in concert to bring matching segments of homologous chromosomes together

Reduction The gain of electrons by an atom or molecule that is involved in an oxidation-reduction (redox) reaction; reduction involves the addition of energy to one substance, which is coupled with the simultaneous removal of energy from another substance by oxidation

Reduction Division A synonym for meiosis, specifically for meiosis I

Reductionism The approach of studying simpler components in order to understand the functions of complex systems

Release Factors A group of cytoplasmic proteins that bind to a stop codon on a molecule of mRNA and interrupt translation by hydrolyzing the bond between the final amino acid in a polypeptide and its transfer RNA

Repetitive DNA Sequences of DNA that occur in many copies in a genome; some sequences of repetitive DNA can occur in a million copies per nucleus

Replication Bubble A region of DNA that has been separated into single strands between opposing replication forks

Replication Fork The region where a DNA double strand is split into separate strands, creating a fork-like appearance in electron micrographs; once replication begins at a replication origin, two replication forks proceed along the double helix in opposite directions from one another

Replication Origin The point of initiation of DNA synthesis along the double helix; two replication forks form at the replication origin and move in opposite directions from one another during DNA synthesis

Replicon A block of DNA between two adjacent replication origins

Reproductive Barriers Various mechanisms that prevent reproduction between individuals, usually from different species

Resin A thick, translucent, combustible, organic fluid usually secreted into resin ducts in pines and many other seed plants

Resin Duct An elongate intercellular space lined with resin-secreting cells and containing resin

Resolving Power The minimum distance necessary to distinguish two points from each other

Respiration The process by which organic compounds are oxidized with the release of energy; respiration is aerobic of oxygen is required as the terminal electron acceptor; it is anaerobic if oxygen is not used as an electron acceptor

Restriction Enzyme A type of enzyme that recognizes a specific sequence of DNA and catalyzes the cleavage of the double helix at that site; most restriction enzymes recognize DNA sequences whose complementary sequence reads the same in the reverse direction; synonym with restriction endonuclease

Reverse Transcriptase A type of enzyme from viruses that catalyzes the synthesis of DNA from an RNA template; in genetics, reverse transcriptase is used for making cDNA of eukaryotic genes

Rhizoids Hairlike extensions of individual epidermal cells anchoring a thallus and increasing surface area for water absorption

Rhizome A fleshy, horizontal, underground stem

Rhizosphere The narrow zone of soil surrounding a root and subject to its influence

Ribonucleic Acid (RNA) The nucleic acid containing four different nucleotides whose simple sugar is ribose; molecules of RNA, which are made as complements of DNA segments called genes, function in protein synthesis

Ribosome An organelle that is responsible for protein synthesis; ribosomes consist of ribosomal RNA (rRNA) and proteins that are arranged into two subunits, one large and one small

Ribosomal RNA (rRNA) The type of RNA that is a component of ribosomes

Ribozyme A sequence of RNA that has enzymatic properties; first named from a self-splicing intron

Ring-Porous Wood Wood having larger vessels in early wood than in late wood, thereby producing a ring when viewed in a cross-section of wood

RisenD A seed glycoprotein that is toxic; found in castor bean

RNA Polymerase A type of enzyme that catalyzes the synthesis of RNA as the complement to a specific sequence of DNA

RNA Processing The trimming of larger primary RNA transcripts in the nucleus into smaller, coding sequences that are exported into the cytosol; synonymous with RNA splicing

RNA splicing (*see* **RNA Processing**)

Root Cap An organ that covers the root meristem; helps the growing root penetrate the soil

Root Hairs Epidermal cells just behind the zone of elongation in roots; increase the absorptive surface area of the root

Root Pressure Positive pressure of roots that forces water up the stem; caused by active movement of minerals in root cells that draws water into the vascular system

Rosin The hard, brittle component of resin remaining after volatile parts have been removed

rRNA (*see* **Ribosomal RNA**)

Rubber A large polymer of up to 6,000 isoprene units

Runner (*see* **Stolon**)

S

S phase During interphase, the portion of the cell cycle in which DNA synthesis occurs; S refers to the synthesis of DNA

Saprobic Obtaining food directly from nonliving organic matter

Sapwood Wood found between the vascular cambium and the heartwood, transports water and solutes

Saturated Refers to fatty acids or other hydrocarbon-containing chemicals whose carbon-carbon bonds are all single bonds; palmitic acid is an example of a saturated fatty acid

Savanna A grassland with scattered trees. Many savannas are located in tropical or subtropical areas.

Scanning Electron Microscopy (SEM) Microscopy that focuses an electron beam that is reflected from a specimen, thereby showing fine details of its surface structure

Scarification The cutting, abrading, or otherwise softening of the seed coat to induce the seed to germinate

Scientific Method A way of analyzing the physical universe; observations are used to construct a hypothesis that predicts the outcome of future observations or experiments; something that cannot be verified cannot be accepted as part of a scientific hypothesis

Sclereids Sclerenchyma cells found in tissues varying from pear fruits to the hard shells of some nuts

Scutellum The cotyledon of a grass seed; the scutellum is specialized for absorbing nutrients from the endosperm as the seed germinates

Second Gap (*see* G₂ **phase**)

Secondary Cell Wall The cell wall that forms interior to the primary cell wall only after cell division is completed; it is restricted to certain cells and often contains lignin

Secondary Consumer A consumer that feeds on primary consumers

Secondary Growth Growth derived from lateral meristems (e.g., the vascular cambium and cork cambium)

Secondary Metabolism The metabolism of chemicals that occur irregularly or rarely among different plants and that usually have no known metabolic role in cells

Secondary Phloem Phloem derived from the outer vascular cambium

Secondary Structure The portion of a protein's shape that is maintained by hydrogen bonds between amino acids

Secondary Succession Succession in habitats where the climax community has been disturbed or removed

Secondary Xylem Xylem formed by the vascular cambium; wood

Seed A mature ovule, consisting of a seed coat that surrounds the embryo and associated tissues

Seed Bank The ungerminated but still viable seeds that occur in natural storage in soil

Seed Coat The outer layer of a seed; the seed coat develops from the integument of the ovule

Seed Ferns An extinct group of plants that were characterized by frond-like leaves and seed-bearing structures; classified together in the Division Pteridospermophyta

Selection Pressures Those environmental factors that promote or retard reproductive success of a phenotype

Selectively Permeable Refers to a membrane that restricts the passage of some solutes through it

Self-Compatible Refers to the potential for successful reproduction between flowers of the same plant or between stamens and carpels of the same flower

Self-Incompatible Incapable of successful reproduction between flowers of the same plant or between stamens and carpels of the same flower

Self-Replication Refers to the ability of DNA to make exact copies of itself

Selfish DNA Refers to DNA that can perpetuate itself by semi-autonomous replication; transposons are considered to be selfish DNA because they can move copies of themselves to several sites in a genome

Semiconservative Replication Refers to the replication of a DNA molecule wherein half of each new double strand consists of one newly synthesized strand and one strand from the parent double helix

Senescence The collective process of aging, decline, and eventual death of plant or plant part

Sense Strand In DNA, the sense strand of a gene is the one that contains the coding sequence for a molecule of RNA and, in the case of mRNA, indirectly for a polypeptide

Sepal One of the outermost parts of a flower; collectively, the sepals of a single flower are called the calyx

Septum (plural: **Septa**) A crosswall in a fungal hypha

Serotype A protein that is a unique antigen; it induces and binds to antibodies that are specific to it alone. Serotypes are used in a classification system applied to viruses

Sessile Leaf Leaf Lacking a petiole; blades of sessile leaves attach directly to the stem

Seta The stalk that supports the capsule of a moss sporophyte

Short-Day Plant Plant that flowers when the length of dark is longer than some critical value; short-day plants usually flower in autumn

Short-Night Plant Plant that flowers only if the uninterrupted dark period is less than a certain length; sometimes called long-day plant

Sieve Area Part of the wall of a sieve element containing many pores through which the protoplasts of adjacent sieve elements are connected

Sieve Cell A long sieve-element having unspecialized sieve areas and tapering endwalls that lack sieve plates; sieve cells occur in the phloem of gymnosperms and lower vascular plants

Sieve Elements Cells in the phloem that transport organic solutes; sieve cells and sieve-tube members are examples of sieve elements

Sieve Plate The part of a wall of a sieve-tube member that has one or more sieve areas

Sieve Tube A series of sieve tube members arranged end-to-end and connected by sieve plates

Simple Leaf A leaf having one blade which may be lobed or dissected

Single-Strand Binding Proteins Proteins that prevent the fusion and rewinding of DNA once the double strands are split apart for replication

Sink Where organic solutes are being transported by the phloem; where metabolites such as sugar are used or stored

Siphonostele A complex stele common in ferns; a central pith surrounded by a tube of vascular tissues

Sisten Chromatids A pair of chromatids in a duplicated chromosome

Sliding-Microtubule Hypothesis An explanation for how chromosomes are moved during anaphase; this hypothesis holds that opposing polar spindle fibers slide past one another, creating a force that pushes the poles of a spindle apparatus apart

Slug (*see* **Pseudoplasmodium**)

Small Nuclear Ribonucleoprotein (snRNP) A complex of small RNA molecules condensed with specific proteins in the nucleus; a snRNP is the basic unit of a spliceosome

snRNP (*see* **Small Nuclear Ribonucleoprotein**)

Softwood Coniferous gymnosperm

Soil Particles The smallest pieces of soil

Solute Potential The component of water potential caused by the presence of solutes in water

Solute A substance dissolved in a solution

Solvent A liquid that dissolves solutes

Sori (singular, **Sorus**); Clusters of sporangia; found especially among the algae and lower vascular plants

Source Where organic compounds such as sugar are being made and loaded into the phloem

Southern Blotting A procedure by which fragments of DNA are separated by gel electrophoresis, transferred to a filter, and probed with DNA that is complementary to the gene of interest; the location of the target gene is found because it becomes radioactive when the probe anneals to it (*also see* **Northern Blotting**)

Speciation Evolutionary formation of a new species

Species (plural: **Species**) A species is a group of similar organisms capable of, or potentially capable of, freely interbreeding. The scientific names of species are binomials consisting of a genus (generic) name and a specific epithet.

Species Diversity The number of species and the number of individuals per species in an ecosystem

Spike A type of inflorescence consisting of an unbranched, elongated main axis whose flowers have very short or no pedicels

Spindle Apparatus Refers to the elliptically shaped collection of spindle fibers in a cell

Spingle Fibers Nearly parallel microtubules that form between the poles of dividing cells; some spindle fibers attach to chromosomes but fibers from opposite poles mostly interact with each other; spindle fibers are believed to move chromosomes both by pulling homologous chromosomes in opposite directions and by pushing poles apart

Spliceosome A cluster of snRNPs; a spliceosome binds to a large primary RNA transcript, cuts out certain parts of the RNA, then splices the rest of the RNA back into a continuous strand

Spongy Mesophyll Leaf tissue consisting of loosely arranged photosynthetic cells

Sporangium A structure within which the protoplasm becomes converted into an indefinite number of spores

Sporopollenin A terpenoid in the exine of a pollen grain that protects the male gametophyte from desiccation

Spore A small reproductive structure, usually consisting of a single cell, which is capable of developing independently into a much larger, mature and often multicellular body

Sporic Meiosis A life-cycle pattern among algae; meiosis by a diploid parent produces haploid spores that develop into mature haploid individuals; resembles life cycle of plants

Sporocarp In heterosporous ferns, a structure bearing several sori consisting of many sporangia

Sporophyll A sporangia-bearing leaf

Sporophyte The phase of the plant life cycle that produces spores

Sporophytic Self-Incompatibility A type of self-incompatibility that is imposed by sporophytic tissues or organs; an example would be incompatibility that is imposed by the stigma, which is a sporophytic structure

Spring (Early) Wood Wood produced in the spring; usually characterized by relatively large cells

Stabilizing Selection Selection for a phenotype within the norm of a population, or selection against extreme phenotypes

Stable Isotope Tracing A technique based on the typical ratio of carbon^{12} to carbon^{13} in tissue samples, which enables ecologists to determine food sources and consumption in food webs

Stalk Cell One of two cells produced when the generative cell of a gymnosperm male gametophyte divides. Immediately before fertilization the body cell divides, becoming two sperms.

Stamen The pollen-producing part of a flower, usually consisting of an anther and a filament; collectively, the stamens of a single flower are called the androecium

Staminate Flower A flower whose reproductive parts consist only of stamens; the tassels at the tops of corn plants are examples of staminate flowers

Staminate Plant An individual plant whose flowers bear stamens but not carpels; a "fruitless" mulberry is an example of a plant that is exclusively staminate (mulberries can reproduce only when pollen is transferred to a carpellate plant)

Starch-Branching Enzyme (SBEI) A type of enzyme that converts straight chains of amylose to the branched polymers of amylopectin; "I" refers to an isoform of the enzyme

Starch Phosphorylase A type of enzyme that cleaves a molecule of glucose from one end of a glucose polymer by phosphorylating the glucose that is removed from the chain

Start Codon The codon at the beginning of a polypeptide-coding gene; the start codon codes for the first amino acid in the polypeptide, which is usually methionine

Stele The central vascular cylinder of roots and stems

Sterol A compound derived from six isoprene units linked together in a multiple-ringed structure; beta-sitosterol is an example of a plant sterol; cholesterol is a widely known example of an animal sterol

Sticky End The single-stranded portion of a DNA sequence after it undergoes a zigzag by a restriction enzyme

Stigma The surface of a carpel that is receptive to pollen grains and upon which the pollen grains germinate; a photosensitive eyespot found in certain kinds of algae

Stipule A leaflike appendage that occurs on either side of the base of a leaf (or encircles the stem) in several kinds of flowering plants.

Stolon A stem that grows horizontally along the ground

Stoma (pl. **Stomata**) The epidermal structure consisting of two guard cells and the pore between them

Stop Codon A codon that occurs at the end of a gene and signals where translation stops

Stratification The exposure of seeds to extended cold periods before they will germinate at warm temperatures

Strobilus, Compound An axis with lateral branches bearing sporophylls

Strobilus, Simple An unbranched axis bearing sporophylls

Stroma The matrix between the grana in chloroplasts; the site of the biochemical (i.e., "dark") reactions of photosynthesis

Structural Polysaccharide A polysaccharide that holds cells and organisms together; cellulose is the most abundant structural polysaccharide in plants

Style A column of carpel tissue arising from the top of an ovary and upon which is the stigma; the style raises the stigma to a receptive position for pollen grains whose pollen tubes must grow through it

Suberin A waxy substance that occurs in cork cells and in the cells of underground plant parts; it consists of hydroxylated fatty acids that are linked together in a complex array

Subsidiary Cella Epidermal cells that are structurally distinct from other epidermal cells and associated with guard cells

Substrate-Level Phosphorylation The transfer of a phosphate group from a substrate, such as phosphoenol pyruvic acid, to ADP, thereby making ATP

Subunit A polypeptide that combines with other polypeptides to comprise a multi-subunit protein

Succession An orderly progression of population replacements in a specific geographic area; it is initiated with pioneer species and completed with the establishment of a stable climax community

Succinic Acid A four-carbon organic acid that is oxidized by the reduction of ubiquinone to ubiquinol in the sixth step of the Krebs cycle; the product of this oxidation is fumaric acid

Succinyl-CoA An acetylated four-carbon acid that is converted to succinic acid by losing its acetyl-CoA group, thereby driving the substrate-level phosphorylation of one molecule of ADP to ATP in the fifth step of the Krebs cycle

Succulent A plant having think, fleshy leaves or stems; succulence is usually an adaptation to water or salt stress

Sucker A sprout on the roots of some plants that forms a new plant

Sucrose Synthase A type of enzyme that catalyzes the reversible breakdown of sucrose from starch by hydrolysis into free fructose and bound glucose; the glucose is bound to a carrier molecule called uridine diphosphate (UDP)

Summer (Late) Wood Wood produced in the summer; characterized by relatively small cells

Superior Ovary An ovary located above the other flower parts on a floral axis

Suspensor Cells A group of cells at the base of the embryo of many seed plants that expands and moves the embryo into the endosperm

Suture The line along which a fruit splits when it is mature

Symbiont One of two (or more) dissimilar organisms that live in close association with each other. The association may be beneficial to both organisms (*mutualism*) or harmful to one organism (*parasitism*)

Sympatric Having the same or overlapping geographic distribution but separated by reproductive or biotic barriers

Sympatric Speciation Formation of a new species entirely within the geographical range of its parental form

Symplast The interconnected living mass of an organism; the symplast is a continuous unit that is comprised of cells that are connected by plasmodesmata throughout the organism

Symplastic Movement of water and solutes through tissues by passing through interconnected protoplasts and their plasmodesmata

Synangium A multi-lobed sporangium at the tip of short lateral branches; possibly derived from fusion of multiple sporangia; common in whisk ferns

Synapsis The pairing of homologous chromosomes by their attachment along a synaptonemal complex; crossing over occurs during synapsis

Synaptonemal Complex A complex of proteins that forms a chromosome-length axis linking homologous chromosomes between the same gene loci

Synegrid A type of cell that occurs next to the egg in an embryo sac; sperm cells entering the embryo sac first pass through one of the synergids

Synonymous Codon Refers to codons that code for the same amino acid

Syringomycin A toxic polypeptide that is secreted by *Pseudomonas syringae*, a species of bacteria that infects corn, beans, and many other kinds of plants

Systematics The classification of organisms into a hierarchy of categories (taxa) based on evolutionary interrelationships

t

Tabomo A group of viruses similar to the tobacco mosaic virus

Taiga A coniferous forest biome adjacent to arctic tundra in large areas of North America and Eurasia.

Tandem Repeat The occurrence of two or more copies of a gene in a row; ribosomal RNA genes typically occur as tandem repeats

Tangential Section A longitudinal section that does not pass through the center of the structure

Tapetum A tissue of sterile cells that surrounds the microspores in a microsporangium; the tapetum acts as a nutritive tissue for the spores and pollen grains while they remain in a sporangium

Taproot A relatively large primary root that produces secondary roots

Tassel The downward-hanging inflorescence of some plants; in corn, tassel refers to an inflorescence of pollen-bearing flowers at the top of the plant

Taxol A drug obtained from the Pacific yew, and also from a fungus that grows on the yew, with potential for treating certain forms of cancer

Taxon (plural: **Taxa**) Any taxonomic category, such as species or family

Taxonomy The classification, description, and naming of organisms

Telomere Sequences of DNA at the tip of a chromosome that counteract its condensation before the onset of nuclear division

Telophase The period of mitosis during which chromosomes seem to mimic prophase in reverse; chromosomes steadily elongate an decondense back into diffuse chromatin, and each new daughter nucleus becomes surrounded by a nuclear envelope

Telophase I The first telophase of meiosis; in telophase I, chromosomes decondense, the spindle apparatus disintegrates, and a new nuclear envelope forms around each daughter nucleus; in many organisms, telophase I is bypassed and the meiotic nuclei go directly from anaphase I to metaphase II

Telophase II The second telophase of meiosis; in telophase II, chromosomes decondense, the spindle apparatus disintegrates, and a new nuclear envelope forms around each of the four new daughter nuclei

Temperate Deciduous Forest A biome dominated by deciduous hardwood trees, and located in areas with temperate climates

Tendril A modified leaf or stem in which only a slender strand of tissue constitutes the entire structure

Tensile Strength The maximum amount of lengthwise pull that a substance can bear without tearing apart

Tension Wood Reaction wood that forms along the upper side of leaning stems; straightens the stem by contracting and "pulling" the stem up

Terpenoid Any compound that is derived from five-carbon precursors called isoprene units; examples include methol (two isoprene units), beta-carotene (eight isoprene units), and rubber (up to 6,000 isoprene units)

Tertiary Structure The portion of a protein's shape that is maintained by disulfide bonds, ionic interactions, or hydrophobic attraction between amino acids

Tetrad Scar A scar on a primitive spore at the point of attachment to three other spores, all four having developed after meiosis; germination takes place in the vicinity of the scar

Thallophytes A term once used to designate fungi and algae collectively

Thallus (plural: **Thalli**) A body not differentiated into roots, stems, or leaves, as seen in lichens and liverworts

Thigmorphogenesis A change in growth and development due to mechanical disturbances; typically a decrease in elongation

Thigmotropism A response to contact with a solid object

Thorn A modified woody stem that terminates in a sharp point

Thylakoid A saclike membranous structure of the grana of chloroplasts; thylakoids house chlorophylls

Tissue Culture A technique for growing and manipulating pieces of tissue in a medium after their removal from the organism

Tonoplast The membrane that surrounds a vacuole; also called a vacuolar membrane

Topoisomerase A type of enzyme that relieves the kinks in DNA that would otherwise block the movement of replication forks; topoisomerases work by breaking one or both strands, thereby allowing the strands to uncoil by swiveling around one another; after uncoiling, the strands are also linked back together by topoisomerases

Totipotent Refers to the notion that every cell has the same genes and therefore the same genetic potential to make all cells other cell types

Trace Elements (*see* **Micronutrients**)

Tracheid Elongated, spindle-shaped cell that transports water in the xylem; also helps support the plant; occur in virtually all vascular plants

Trailer Sequence An extra amount of non-coding DNA that is transcribed into RNA beyond the end of the gene

Transcription The synthesis of a molecule of RNA as a complement to a specific sequence of DNA; transcription occurs in the nucleus

Transduction The transfer of genetic material between bacteria by bacteriophages

Transect, Line A straight line extending between two points, usually established arbitrarily for the purpose of studying vegetation immediately adjacent to it

Transfer RNA (tRNA) A type of small RNA molecule that binds to a specific amino acid and to a codon on messenger RNA; it is called transfer RNA because it is associated with the transfer of amino acids to mRNA in ribosomes; more than 40 different tRNA molecules have been found, at least one for each protein amino acid

Transformation A form of genetic transfer in bacteria by which a fragment of DNA is taken up and incorporated into the DNA of the recipient cell

Transgenic Refers to cells or organisms that contain genes that were inserted into them from other organisms by genetic engineering

Translation The synthesis of an amino acid sequence from a specific sequence of codons along a molecule of mRNA; translation occurs in ribosomes

Transmission Electron Microscopy (TEM) Microscopy that focuses an electron beam through the thin section of a specimen to study its internal structure

Transpiration Evaporation of water from leaves and stems; occurs mostly through stomata

Transport Protein A type of membrane protein that enables the transport of specific solutes across the membrane

Transposable Element A fragment of DNA that is able to multiply and move spontaneously among an organism's chromosomes

Triacylglyceride A combination of a molecule of glycerol with three long-chained organic fatty acids linked to the glycerol by acylglyceride linkages

Tricarboxylic Acid Cycle (*see* **Krebs Cycle**)

Trichogyne A receptive, slender outgrowth for spermatia or similar reproductive cells in red algae and ascomycete fungi

Trichome An epidermal outgrowth (e.g., a hair or scale)

Triplet Refers to a sequence of three nucleotides that together comprise a codon

Triterpene A compound that consists of six isoprene units linked together; sterols, such as beta-sitosterol, are triterpenes

tRNA (*see* **Transfer RNA**)

Tropical Rain Forest An endangered tropical biome with exceptional diversity of species

Tropism A response to an external stimulus in which the direction of the response is determined by the direction from which the stimulus comes; tropisms such as phototropism and gravitropism are produced by differential growth

True-Breeding Refers to purebred strains for a given trait, which means that the gene for that trait is homozygous

Tube Cell The cell in the pollen grains of seed plants that develops into the pollen tube

Tuber A fleshy, underground stem having an enlarged tip (e.g., potato tuber)

Tubulin (*see* **Alpha-Tubulin** and **Beta-Tubulin**)

Tundra A vast biome primarily above the arctic circle, and above the timberlines of mountain ranges further south, whose vegetation includes no typical trees

Tunica-Corpus The organization of the shoot apex of most angiosperms and some gymnosperms; consists of one or more peripheral layers (i.e., tunica layers) and an interior corpus

Turgid Full of water taken in by osmosis

Turgor Pressure The pressure on a cell wall that is created from within the cell by the movement of water into it

Turpentine The volatile, combustible component of resin

Type Specimen A specimen upon which the original description of species is based

u

Ubiquinol The reduced form of ubiquinone; ubiquinol donates electrons to cytochrome *b* in the electron transport chain

Ubiquinone A lipid-soluble quinone whose function is to accept electrons from electron donors like NADH and from the oxidation of fatty acids; also called coenzyme Q

Ultracentrifuge A high-speed centrifuge that is capable of spinning at more than 100,000 revolutions per minute

Unequal Crossing-Over Refers to the exchange of unequal amounts of DNA between homologous chromosomes that are not perfectly aligned

Unisexual Having antheridia and archegonia on separate gametophytes

Unsaturated Refers to fatty acids or other hydrocarbon-containing chemicals whose carbon-carbon bonds include double bonds as well as single bonds; oleic acid (one double bond), linoleic acid (two double bonds), and linolenic acid (three double bonds) are examples of unsaturated fatty acids

Uridine Diphosphate (UDP) A uracil-containing nucleotide that acts as a carrier molecule for glucose and similar monosaccharides; the UDP-sugar complex is also an intermediate compound for the interconversion of one monosaccharide to another (e.g., glucose to galactose)

v

Vacuole A membrane-bound organelle that is filled mostly with water but also may contain water-soluble pigments and other substances; the vacuolar membrane is called the tonoplast

Vascular Bundle Strand of tissue containing primary phloem and primary xylem (and possibly procambrium); often enclosed by a bundle sheath

Vascular Plant A plant having xylem and phloem that conduct fluids throughout the plant

Vascular Tissue Tissue specialized for long-distance transport of water and minerals; xylem and phloem

Vector In genetics, a virus or bacterial plasmid that can take up a foreign gene and integrate it into the genome of a target organism; in plant reproduction, an animal that carries pollen (a pollinator) from a pollen sac to a stigma or to an ovule

Vegetative Cell A cell that is neither sexually reproductive nor divides to form cells that are sexually reproductive; this term particularly refers to the tube cell of angiosperm pollen grains, which is the only vegetative cell of the male gametophyte

Vein Vascular bundle that forms part of the connecting and supporting tissue of a leaf or other expanded organ

Velamen Multiple epidermis covering the aerial roots of some orchids and aroids

Venter The swollen base of an archegonium containing the egg

Vernalization The induction of flowering by cold

Vessel A tubelike structure in the xylem that consists of vessel elements placed end-to-end and connected by perforations; vessel elements conduct water and minerals; found in nearly all angiosperms and a few other vascular plants

Vessel Element One of the cells forming a vessel

Vesticular-Arbuscular (V-A) Mycorrhizae Treelike or bulblike mycorrhizae

Viroid A plant-infecting, viruslike particle with a single circular strand of RNA that is not associated with any protein

Virusoid A particle similar to a viroid but located inside the protein coat of a true virus

w

Water Potential The potential energy of water to move down its concentration gradient; water potential is expressed in units of pressure instead of units of energy, because pressure is simpler to measure

Waxes Complex mixtures of fatty acids linked to a long chain alcohol; also contains free fatty acids with hydroxyl groups and long chain hydrocarbons

Whorl A circular group of at least three leaves or flower parts all attached to an axis at the same level

Wood Secondary xylem

x

Xanthophyll A yellow carotenoid; one xanthophyll, zeaxanthin, in the blue-light photoreceptor in shoot phototropism

Xerophyte A plant adapted for growth in arid conditions

Xylem Vascular system specialized for transporting water and dissolved minerals upward in the plant; characterized by the presence of tracheary elements

y

Yeast Artificial Chromosome (YAC) A yeast chromosome into which large fragments of foreign DNA (millions of base pairs) have been inserted; YACs can be replicated like native chromosomes in yeast cells, thereby cloning large amounts foreign DNA as well

z

Zeatin A natural cytokinin isolated from corn (*Zea mays*)

Zein A storage protein in the kernels of corn

Zoospores Motile, haploid spores produced from diploid sporangia; develop into the mature haploid phase of the life cycle

Zygomycetes A large group of fungi with primarily coenocytic mycelia; they reproduce asexually by spores produced within sporangia; sexual reproduction includes the formation of zygosporangia

Zygosporangium (plural: **Zygosporangia**) A sporangium containing a thick-walled, multinucleate zygospore that develops in zygomycetes after the fusion of isogametes

Zygote The diploid cell that is formed by the fusion of two gametes

Zygotic Meiosis A life-cycle pattern among algae; meiosis of a diploid zygote produces haploid cells that develop into mature haploid individuals; resembles life cycle of fungi

credits

Photographs

Unit Openers

Unit 1: © Bruce Iverson; **Units 2, 3:** © Digital Stock;
Unit 4: © The McGraw-Hill Companies, Inc./Carlyn
Iverson, photographer; **Units 5, 6:** © 1996 Photo Disc,
Inc., Nature, Wildlife and the Environment; **Unit 7:**
© Doug Sherman/Geofile; **Units 8, 9:** © 1996 Photo
Disc, Inc., Nature, Wildlife and the Environment.

Chapter 1

Opener: © Doug Sherman/Geofile; **1.1A** (top left): © Muriel
Orans/Horticultural Photography, Corvallis, Oregon;
1.1B (middle left): © Noboru Komine/Photo
Researchers, Inc.; **1.1C** (bottom left): © Grant Heilman
Photography; **1.1D** (top right): © Verna R. Johnston/
Photo Researchers, Inc.; **1.1E** (middle right): © Walter
H. Hodge/Peter Arnold, Inc.; **1.1F** (bottom right):
© James Shaffer; **Page 6** (top): © J. Villegier/Photo
Researchers, Inc.; **Page 6** (bottom): © Fritz Prenzel/
Peter Arnold, Inc.; **1.2:** © Grant Heilman Photography;
1.3: © Runk/Schoenberger/Grant Heilman Photography;
1.5A: UPI/Bettmann Archive; **1.5B** (all): From *Scientific
American* 250(6):86, 1984. Article by Nina Fedoroff;
Page 12: © John Neubauer; **Page 13:** Aaron Ball;
1.6A: © BioPhot; **1.6B:** © Michael Goodman/
Photo/Nats, Inc.; **1.6C:** © John Kaprielian/Photo
Researchers, Inc.; **1.6D:** © Kjell B. Sandved/Butterfly
Alphabet; **1.6E:** © Walter E. Harvey/National Audubon
Society/Photo Researchers, Inc.; **1.7A, B, C:** Courtesy of
the Monsanto Company.

Chapter 2

Opener: © Leonard Lessin/Peter Arnold, Inc.; **2.3B:** © Biophoto
Assoc./Photo Researchers, Inc.; **2.4B:** © Eric Grave/
Science Source/Photo Researchers, Inc.; **2.9D:** © Scott
Camazine/Photo Researchers, Inc.; **2.12A, B, C:** From D.
Froelich and W. Barthlott, *Mikromorphologie der
Epicuti-cularen Wachse und das System der*

Monokotylen, © 1988; The Academy of Science &
Literature, Maine. Courtesy of Dr. W. Barthlott;
Page 37 (poppy): © Kjell B. Sandved/Butterfly Alphabet;
Page 37 (beet): © Ann Reilly/Photo/Nats, Inc.;
Page 37 (yew): © Ray Pfortner/Peter Arnold, Inc.;
Page 37 (nutmeg): © Walter H. Hodge/Peter Arnold, Inc.;
2.13B: © William E. Ferguson; **2.13C:** © Walter H.
Hodge/Peter Arnold, Inc.; **2.13D:** © Runk/
Schoenberger/Grant Heilman Photography;
2.14B: © Dede Gilman/Visions From Nature;
2.14C: © Betsy Fuchs/Photo/Nats, Inc.; **2.14D:** © Alan
Pitcairn/Grant Heilman Photography; **2.15B:** © Larry
Lefever/Grant Heilman Photography; **2.15C:** © Grant
Heilman Photography; **2.15D:** Courtesy of Hillerich &
Bradsby Co., Inc.

Chapter 3

Opener: © Dr. Jeremy Burgess/SPL/Photo Researchers, Inc.;
3.1B: © Runk/Schoenberger/Grant Heilman
Photography; **3.1C:** W. Dennis Clark; **Page 51** (both):
From Misuzu Baba, Norio Baba, Yoshinori Ohsumi,
Koichi Kanaya and Masaka Osumi, "Three Dimensional
Analysis of Morphogenesis Induced by Mating Pherome
a Factor in Saccharomyces Cerevisiae," *Journal of Cell
Science* 94:207–216, 1989; **3.6A, B:** © Bruce Iverson
Photomicrography; **3.6C, D:** © Richard H. Gross;
3.6 E: © Manfred Kage/Peter Arnold, Inc.; **3.7:** ©
BioPhot; **3.8A:** © Biophoto Assoc./Photo Researchers,
Inc.; **3.8B:** © Runk/Schoenberger/Grant Heilman
Photography; **3.9A:** © Gary T. Cole, U. of Texas,
Austin/Biological Photo Service; **3.10A:** © M. Schliwa/
Visuals Unlimited; **3.10B:** Courtesy of Drs. R. R.
Trelease & Francisco Carrapico; **3.13:** © Biophoto
Assoc./Photo Researchers, Inc.; **3.14:** © E. Vigil/Biology
Media/Photo Researchers, Inc.; **3.15A:** © Biophoto
Assoc./Science Source/Photo Researchers, Inc.;
3.16: Courtesy of W. A. Cote, Center for Ultrastructure
Studies, College of Environmental Science & Forestry,
State University of New York; **3.19A:** © Biophoto
Associates/Photo Researchers, Inc.; **3.20:** Courtesy of

Jean Whatley; **3.22A:** © E. H. Newcomb and W. P. Wergin, University of Wisconsin/Biological Photo Service; **3.23:** From C. M. Kunce, R. R. Trelease and D. C. Doman, "Ontogeny of Slyoxysomes in Maturing and Germinated Cotton Seeds—a Morphometric Analysis," *Planta* 161:156–164, 1984, Springer-Verlag. Courtesy R. R. Trelease, Arizona State University; **3.24 A:** © Dr. Jeremy Burgess/SPL/Photo Researchers, Inc.; **3.25B:** Courtesy of Dr. Keith Porter; **3.26A:** © Knut Norstog; **3.26B:** © Kenneth W. Fink/Nat'l Audubon Society/Photo Researchers, Inc.; **3.27A** (both): Courtesy of Gary L. Floyd.

Chapter 4

Opener: © Bruce Iverson; **4.3:** © Don Fawcett/Visuals Unlimited; **4.5:** R. B. Park & D. Branton, *Brookhaven Symposia In Biology* (19), 1966. Courtesy Albert Pfeifhofer, University of California, Berkeley; **4.7:** © Runk/Schoenberger/Grant Heilman Photography; **4.12:** © Dr. Jeremy Burgess/SPL/Photo Researchers, Inc.; **4.13A:** © Dwight R. Kuhn; **4.13B:** © Alfred Owczarzak/Biological Photo Service; **4.14:** © William E. Ferguson; **Page 84:** © Donald Specker/Animals Animals/ Earth Scenes; **4.19:** Courtesy of L. C. Fowke; **4.25:** © Hugh Spencer/Photo Researchers, Inc.; **4.26:** © Biophoto Assoc./Science Source/Photo Researchers, Inc.

Chapter 5

Opener: © Donald Specker/Animals Animals/Earth Scenes; **5.1:** © Ray Coleman/Photo Researchers, Inc.

Chapter 6

Opener: © Malcolm Wilkins; **6.2:** © William E. Ferguson; **6.12A:** Courtesy of Peter Hinkle, Cornell University; **Page 128:** © Angelina Lax/Photo Researchers, Inc.

Chapter 7

Opener: © Kjell B. Sandved/Butterfly Alphabet; **7.9:** © BioPhot; **7.12B** (left): © Dwight R. Kuhn; **7.12B** (right): © Dr. Jeremy Burgess/SPL/Photo Researchers, Inc.; **Page 146:** Karin Brueschweiler; **7.19B** (all): From J. A. Bassham, 1965 in J. Bonner and J. E. Varner (Eds.) *Plant Biochemistry*, Academic Press, Inc.; **7.22:** © E. H. Newcomb and S. E. Frederick, University of Wisconsin/Biological Photo Service; **7.23A:** © BioPhot; **7.23B:** © Bruce Iverson.

Chapter 8

Opener: © Dwight R. Kuhn; **8.1:** The Bettmann Archive; **8.2:** Photo by Richard E. Ferguson © William E. Ferguson; **8.10:** Photo by Susan S. Whitfield. Courtesy

of C. R. Parks; **8.11:** © John Kaprielian/Photo Researchers, Inc.; **8.13A:** © Muriel Orans; **8.15A:** © Betty Outlaw; **Page 182** (left): © William E. Ferguson; **Page 182** (right): Courtesy U.S.D.A.; **8.16:** © Evelyne Cudel-Epperson, University of California—Riverside, 1992.

Chapter 9

Opener: © Runk/Schoenberger/Grant Heilman Photography; **9.3:** © Ed Reschke; **9.4:** © Bruce Iverson Photomicrography; **9.5:** © Dede Gilman/Visions From Nature; **9.6:** Courtesy of King's College, London; **9.8:** The Bettmann Archive; **9.10G:** © Science VU-NIH/Visuals Unlimited; **9.10H:** © Science VU-NIH/Visuals Unlimited; **9.10I:** From *Cell* 12:817, 1977 from an article by J. R. Paulsen and U. K. Laemmli © Cell Press; **9.10J:** Courtesy of Barbara Hamkalo; **9.10K:** © A. L. Olins, University of Tennessee/Biological Photo Service; **9.10L:** Courtesy of M. M. Schwesinger. From "Plasmids" by Dr. Richard P. Novick, *Scientific American*, Dec. 1980, p. 104.; **9.13A, B:** From *Genetics* 79:137–150, 1975, by T. C. Hus; **9.16B:** From J. Cairns, "The Chromosome of E Coli," *Cold Spring Harbor Symposium on Quantitative Biology*, 28, © 1963 by Cold Spring Harbor Laboratory, Cold Spring, NY. Reprinted by permission of the publisher and author.; **9.19:** © Ed Reschke; **9.20:** Courtesy of Susan M. Wick, University of Minnesota; **Page 205:** © Elliot Meverowitz; **9.21B** (all): © Andrew S. Bajer, Professor of Biology, University of Oregon.

Chapter 10

Opener: © Sheila Terry/SPL/Photo Researchers, Inc.; **10.1B:** © R. F. Evert; **10.1C:** © Ed Reschke; **10.8** (all): © C. A. Hasenkampf, U. of Toronto/ Biological Photo Service; **10.16:** © Science Source/ Photo Researchers, Inc.

Chapter 11

Opener: USDA-ARS photograph by Jack Dykinga; **11.1:** Courtesy of Madan K. Bhattachargya, PhD. From the cover of *Cell* 60(1), January 12, 1990 © Cell Press; **11.4:** © Professor Oscar Miller/SPL/Photo Researchers, Inc.; **11.6A:** Courtesy of O. L. Miller, Jr., Adapted from the *EMBO Journal* 5(13):3591–96, Fig. 1b, 1986; **11.8C:** © Tripos Assoc./Peter Arnold, Inc.; **11.10:** © David M. Phillips/Visuals Unlimited; **11.19A, B, C:** Courtesy of Calgene, Inc. and Bill Santos Photography; **11.27:** © Science VU/Visuals Unlimited; **11.28:** Great Britain M.A.F.F., Crown copyright; **11.31:** Courtesy of Calgene, Inc.; **11.32:** © Runk/ Schoenberger/Grant Heilman Photography; **11.33A:** Research conducted at Oregon State University

Hermiston Ag Research & Extension Center. Photo by Photography Plus, Inc., Umatilla, OR; **11.33B:** © Richard Shade, Purdue University; **11.35A:** Courtesy Madan Bhattacharyya, from *Cell* 60(1):115–122, January 12, 1990 © Cell Press.

Chapter 12

Opener: © Kjell B. Sandved/Butterfly Alphabet;
12.1B: © BioPhot; **12.1C:** © Runk/Schoenberger/ Grant Heilman, Inc.; **12.2A, B, C:** © Bruce Iverson Photomicrography; **Page 265:** © BioPhot; **12.3A:** © BioPhot; **12.3B:** © F. A. L. Clowes; **12.5:** Courtesy J. H. Troughton and L. Donaldson and Industrial Research, Ltd.; **12.6A:** From Susses, "Morphogenesis in Solanum Tuberosum L: Apical Developmental Pattern of the Juvenile Shoot" *Phytomorphology* 5:253–273, 1955; **12.6B:** From Susses, "Morphogenesis in Solanum Tuberosum L: Apical Developmental Pattern of the Juvenile Shoot" *Phytomorphology* 5:253–273, 1955; **Page 269:** From *Nature*, April 11, 1996. Courtesy of Peter Doerner, Ph.D.; **12.7A:** © E. Vigil/Biology Media/Photo Researchers, Inc.; **Page 280:** © BioPhot; **12.13A:** Courtesy J. H. Troughton and L. Donaldson and Industrial Research, Ltd.; **12.13B:** © Kjell B. Sandved/Butterfly Alphabet; **12.13C:** © G. Biittner/Naturbild/OKAPIA/Photo Researchers, Inc.; **12.14:** Courtesy of Paul Green, Stanford University.

Chapter 13

Opener: © Oliver Meckes/Photo Researchers, Inc.;
13.1A: © Ed Reschke; **13.1B:** © Dwight R. Kuhn; **13.2:** © Dwight R. Kuhn; **13.3:** © Randy Moore/ BioPhot; **13.4A:** © R. B. Taylor/SPL/Photo Researchers, Inc.; **13.4B:** Courtesy J. H. Troughton and L. Donaldson and Industrial Research, Ltd.; **13.5:** From Gunning & Pete, "Transfer Cell Plant," p. 420, *McGraw-Hill Yearbook of Science & Technology*, 1971; **13.6:** © BioPhot; **13.7:** © Dwight R. Kuhn; **13.8:** Courtesy of J. D. Mauseth, University of Texas, Austin; **13.9, 13.10:** © BioPhot; **13.11:** © Bruno P. Zehnder/Peter Arnold, Inc.; **13.12:** © BioPhot; **13.13:** © Dr. Jeremy Burgess/SPL/Photo Researchers, Inc.; **13.14, 13.15, 13.16A, B:** © BioPhot; **13.17A:** Courtesy J. H. Troughton and L. Donaldson and Industrial Research, Ltd.; **13.17B:** Courtesy J. H. Troughton and L. Donaldson and Industrial Research, Ltd.; **13.18:** © BioPhot; **13.19A:** Courtesy B. Galatis & K. Mitrakos, provided by James Mauseth; **13.19B:** © Dwight R. Kuhn; **Page 290:** © Runk/ Schoenberger/Grant Heilman Photography; **13.20:** From Yolanda Heslop-Harrison, "SEM of Fresh Leaves of Pinguicula," *Science* 667:173. © 1970 by the AAAS; **13.21:** © Runk/Schoenberger/Grant Heilman Photography;

13.22: © Biophoto Assoc./Photo Researchers, Inc.; **13.23:** © BioPhot; **13.24:** © George Wilder/Visuals Unlimited; **13.25:** © BioPhot; **13.26:** Courtesy Drs. R. A. Meylan & B. G. Butterfield; **13.27:** From Richard Anderson & J. Cronshaw, "Sieve Plate Pores In Tobacco and Bean," *Planta* 91:173–180, 1970 © Springer-Verlag; **13.28:** © BioPhot; **13.29:** © BioPhot; **13.30:** © Bruce Iverson Photomicrography; **13.31:** © Richard F. Trump/ Photo Researchers, Inc.; **13.32, 13.33:** © BioPhot.

Chapter 14

Opener: © Barbara Brundege Photography; **14.2:** © Bruce Berg/ Visuals Unlimited; **14.3A, B:** © Ed Reschke; **14.4:** © Bruce Iverson Photomicrography; **14.5:** © BioPhot; **14.6A:** © Charles Gurche; **14.6B:** © Franz Krenn/OKAPIA 1989/Photo Researchers, Inc.; **14.7A:** © George Gainsburgh/Animals Animals/Earth Scenes; **14.7B:** © Dwight R. Kuhn; **14.8:** © Richard D. Poe/ Visuals Unlimited; **14.9:** © William E. Ferguson; **14.10:** © Alford W. Cooper/Photo Researchers, Inc.; **14.11:** © Ray Coleman/Photo Researchers, Inc.; **14.13:** © Kjell B. Sandved/Butterfly Alphabet; **14.14A:** © Runk/Schoenberger/Grant Heilman Photography; **14.14B:** © Carolina Biological Supply/Phototake; **14.15A:** © Dede Gilman/Visions From Nature; **14.15B:** © Kjell B. Sandved/Butterfly Alphabet; **14.16A:** © Bruce Iverson Photomicrography; **14.16B:** © Runk/Schoenberger/Grant Heilman Photography; **14.17A, B:** © Dwight R. Kuhn; **14.18:** © William E. Ferguson; **14.19:** © Ed Reschke; **14.20A:** © BioPhot; **14.20B:** Courtesy of J. D. Mauseth, University of Texas, Austin; **14.21A, B:** © Kingsley Stern; **14.22A:** © John D. Cunningham/Visuals Unlimited; **14.22B:** © Bruce Iverson Photomicrography; **Page 323** (both): © Richard H. Gross; **14.23:** © E. Webber/ Visuals Unlimited; **14.24A:** © Kingsley Stern; **14.24B:** © BioPhot; **14.24C:** © Kingsley Stern; **14.24D:** © Ghillean Prance/Butterfly Alphabet; **Page 327** (all): © Steven Vogel; **14.25:** © Ed Reschke.

Chapter 15

Opener: © BioPhot; **15.1:** © Dr. Jeremy Burgess/SPL/Photo Researchers, Inc.; **15.2:** © Lynwood M. Chace/Photo Researchers, Inc.; **15.3:** © John D. Cunningham/ Visuals Unlimited; **15.4A, B, 15.5:** © BioPhot; **15.7:** © John D. Cunningham/Visuals Unlimited; **15.8:** Courtesy J. H. Troughton and L. Donaldson and Industrial Research, Ltd.; **15.9A:** © S. Elems/Visuals Unlimited; **15.9B:** © Runk/Schoenberger/Grant Heilman Photography; **15.10:** © BioPhot; **15.11C:** © BioPhot; **15.13A:** © Omikron/Photo Researchers, Inc.; **15.13B:** © Omikron/Photo Researchers, Inc.; **15.13C:** © Dwight R. Kuhn; **15.14:** © Bruce Iverson Photomicrography; **15.17A, B:** © Michael Kienitz;

15.18: © Runk/Schoenberger/Grant Heilman Photography; 15.19A: © Biophoto Associates/Photo Researchers, Inc.; 15.19B: © William E. Ferguson; 15.19C: © Kingsley Stern; **Page 351:** Courtesy Robert E. Eplee, U.S.D.A., APHIS/S & T/WPMC.

Chapter 16

Opener: © Eugene Fisher; **16.1B:** © BioPhot; **16.2:** © The McGraw-Hill Companies, Inc./Kingsley Stern, photographer; **16.4D:** © BioPhot; **16.6A** (left) © Manfred Kage/Peter Arnold, Inc.; **16.6A** (right), **16.7, 16.8:** © BioPhot; **16.9B:** © James Bell/Photo Researchers, Inc.; **16.10:** © Runk/Schoenberger/Grant Heilman Photography; **16.11B, C, D:** © Bruce Iverson Photomicrography; **Page 365** (top left): © F. C. F. Earney/Visuals Unlimited; **Page 365** (bottom left), (top right): © Jeff Gnass Photography; **Page 365** (bottom right): © Henry Ausloos/Animals Animals/Earth Scenes; **16.12B:** © Dwight R. Kuhn; **16.12C** (left): © John J. Smith/Photo/Nats, Inc.; **16.12C** (right): © Dwight R. Kuhn; **Page 370** (left): © E. R. Degginger/Animals Animals/Earth Scenes; **Page 370** (right): © Tim Davis/Photo Researchers, Inc.; **Page 371** (left): Randy Moore; **Page 371** (right): © BioPhot; **16.14A:** © Grant Heilman/Grant Heilman Photography; **16.14B:** © Phillip Hyde; **16.15:** © BioPhot; **16.17A:** © Dwight R. Kuhn; **16.17B:** © Doug Sokell/Visuals Unlimited; **Page 376** (left): © Kirtley-Perkins/Visuals Unlimited; **Page 376** (right): © Porterfield-Chickering/Photo Researchers, Inc.; **16.18:** Courtesy J. D. Mauseth/University of Texas, Austin; **16.19:** © BioPhot; **16.20A** (left): © Biophoto Assoc./Photo Researchers, Inc.; **16.20A** (right), **16.20B, 16.21A:** © Bruce Iverson Photomicrography; **Page 378:** © Christian Grzimek/OKAPIA/Photo Researchers, Inc.

Chapter 17

Opener: © Tom Edwards/Visuals Unlimited; **17.1:** © Scott Camazine/Photo Researchers, Inc.; **17.3A:** © Dwight R. Kuhn; **17.4:** © Professor Malcolm B. Wilkins, Botany Department, Glasgow University; **17.5:** © John J. Smith/Photo/Nats, Inc.; **17.7:** © Dede Gilman/Visions From Nature; **17.8A:** © David M. Stone, 1981/Photo/Nats, Inc.; **17.8B:** © Jane Burton/Bruce Coleman, Inc.; **17.9, 17.12:** © Dede Gilman/Visions From Nature; **17.13:** © Charles Steinmetz/Photo/Nats, Inc.; **17.14:** © David M. Stone/Photo/Nats, Inc.; **17.15A:** © M. P. L. Fogden/Bruce Coleman, Inc.; **17.15B:** © William E. Ferguson; **Page 394** (left): © J. A. L. Cooke/Oxford Scientific Films/Animals Animals/Earth Scenes; **Page 394** (right): © Nuridsany et Perennou/Photo Researchers, Inc.; **17.16A:** © E. F. Anderson/Visuals Unlimited; **17.16B:** © R. E. Litchfield/SPL/Photo Researchers, Inc.; **17.21A:** © Richard Parker/Photo

Researchers, Inc.; **17.21B:** © William E. Ferguson; **17.21C:** © Richard Packwood/Oxford Scientific Films; **17.21D:** © Professor Malcolm B. Wilkins, Botany Department, Glasgow University; **17.23:** © G. I. Bernard/Oxford Scientific Films; **17.24A:** Courtesy Dr. Frank Uasek; **17.24B:** Courtesy of Donald J. Pinkava; **17.25:** © Doug Sokell/Visuals Unlimited; **17.26, 17.27:** © Stephen Dalton/Oxford Scientific Films; **17.28A:** © Greg Crisci/Photo/Nats, Inc.; **17.28B:** © Pat & Tom Leeson/Photo Researchers, Inc.; **17.28C:** © M. J. Coe/Oxford Scientific Films; **17.28D:** © G. I. Bernard/Oxford Scientific Films; **17.28E:** © Sean Morris/Oxford Scientific Films; **17.29A:** © Heather Angel/Biofotos; **17.29B:** © Heather Angel/Biofotos; **17.30A:** © Runk/Schoenberg/Grant Heilman Photography; **17.30B:** © John Gerlach/Animals Animals/Earth Scenes; **17.31A:** © Merlin D. Tuttle/Bat Conservation International/Photo Researchers, Inc.; **17.31B:** © Gerald Thompson/Oxford Scientific Films; **17.32A:** © Sean Morris/Oxford Scientific Films; **17.32B:** © John R. Brownlie/Bruce Coleman, Inc.; **17.33A:** © Fern Stewart/Visions From Nature; **17.33B, C:** © Muriel Orans; **17.33D:** © W. H. Hodge/Peter Arnold, Inc.; **17.34:** © Jane Thomas/Visuals Unlimited; **17.35A:** © Frank T. Awbrey/Visuals Unlimited; **17.35B, C:** © Dwight R. Kuhn; **Page 405** (left): By permission from Webster's Third New International Dictionary © 1986 by Merriam-Webster, Inc., publisher of the Merriam-Webster dictionaries; **Page 405** (right): © Heather Angel/Biofotos; **17.36:** © Dede Gilman/Visions From Nature; **17.37:** © Glenn Oliver/Visuals Unlimited; **17.40:** © William E. Ferguson; **17.41B:** © Harry Engles/Animals Animals/Earth Scenes; **17.42:** © Robert & Linda Mitchell; **17.43A, B:** © Heather Angel/Biofotos.

Chapter 18

Opener: © Dr. Scott Nielsen/Imagery; **18.1A, B:** © Professor Malcolm B. Wilkins, Botany Department, Glasgow University; **18.4C:** Photo and permission supplied by Chun-ming Liu, Oklahoma State University. From *The Plant Cell*, Vol. 5, 621–630, June 1993, from "Auxin Polar Transport Is Essential for the Establishment of Bilateral Symmetry during Early Plant Embryogenesis" by Chun-ming Liu, Zhi-hong Xu, and Nam-Hai Chua, copyright ASPP; **18.5** (all): From J. P. Nitsch, "Growth & Morphogenesis of the Strawberry as Related to Auxin," *American Journal of Botany* 37:212, March 1950.; **18.6:** © Runk/Schoenberger/Grant Heilman Photography; **18.9A:** © R. F. Head/Animals Animals/Earth Scenes; **18.9B:** © Patti Murray/Animals Animals/Earth Scenes; **18.10:** © Sylvan Wittwer/Visuals Unlimited; **Page 430:** © William E. Ferguson; **18.14:** © Arthur M. Greene; **18.15:** Courtesy Cary A. Mitchell; **18.16:** © Runk/Schoenberger/Grant Heilman Photography.

Chapter 19

Opener: © Dr. Scott Nielsen/Imagery; **19.1:** © Runk/Schoenberger/Grant Heilman Photography; **19.3B** (all): © Professor Malcolm B. Wilkins, Botany Department, Glasgow University; **Page 446** (left): © Larry Simpson/Photo Researchers, Inc.; **Page 446** (right): © Kim Taylor/Bruce Coleman, Ltd.; **19.4A, B, C, 19.5A, B:** © BioPhot; **19.6A, B, C:** © Professor Malcolm B. Wilkins, Botany Department, Glasgow University; **19.7:** © Kingsley Stern; **19.8:** © John D. Cunningham/Visuals Unlimited; **19.9A:** © John Kaprielian/Photo Researchers, Inc.; **19.9B:** © Richard H. Gross; **19.9C:** © John Kaprielian/Photo Researchers, Inc.; **19.11A, B:** © Runk/Schoenberger/Grant Heilman Photography; **19.12:** © Fred Bavendam; **19.13A:** © Tom McHugh/Photo Researchers, Inc.; **19.13B:** © Tom McHugh/Photo Researchers, Inc.; **Page 458** (top): Albert Ulrich, Plant Physiologist, Emeritus, Dept. of Soil Science, College of Natural Resources, U. C. Berkeley, 94720; **Page 458** (bottom left): © Larry Lefever/Grant Heilman Photography; **Page 458** (bottom right): © Holt Studios/Animals Animals/Earth Scenes; **Page 459:** © Ed Reschke; **19.17** (all): © Professor Malcolm B. Wilkins, Botany Department, Glasgow University; **19.18:** © Willard Clay; **19.19** (all): © Frank Salisbury; **19.20:** Reprinted with permission from the cover of *Science* Vol. 267, 24 February 1995 and appearing in an article by Steve A. Kay, pp 1065–1236. © 1995, AAAS; **Page 449:** Courtesy of Cora Schmid.

Chapter 20

Opener: © Kjell B. Sandved/Butterfly Alphabet; **20.1:** © William E. Ferguson; **Page 473:** © Walt Anderson/Visuals Unlimited; **Page 474:** © John Colwell/Grant Heilman Photography; **20.5:** © William E. Ferguson; **20.7:** © Jack Wilburn/Animals Animals/Earth Scenes; **20.8A, B, C:** © 1991 Regents University of California Statewide IPM Project; **20.9:** © Harold J. Evans; **20.14:** Courtesy of John G. Mexal, Edwin L. Burke, C. P. P. Reid, © Wadsworth Publishing Co. 1994; **20.15:** © Runk/Schoenberger/Grant Heilman Photography; **20.17A:** © Dwight R. Kuhn; **20.17B:** © E. H. Newcomb & S. R. Tardon, U. of Wisconsin/Biological Photo Service; **20.18A:** © Dwight R. Kuhn; **20.18B:** © William E. Ferguson; **20.18C:** © Kjell B. Sandved/Butterfly Alphabet; **20.19A:** © John Gerlach/Visuals Unlimited; **20.19B:** © Dwight Kuhn; **20.20:** © D. Cavagnaro/Peter Arnold, Inc.; **20.21:** © Dwight R. Kuhn; **20.22** (top): © William E. Ferguson; **20.22** (bottom): © Dede Gilman/Visions From Nature; **20.23A, B:** © Barbara J. Miller/Biological Photo Service; **20.24:** © BioPhot; **20.25:** © Ed Reschke/Peter Arnold, Inc.

Chapter 21

Opener: © Richard Rowan's Collection, Inc./Photo Researchers, Inc.; **21.1, 21.2:** © BioPhot; **21.3:** Courtesy W. A. Cote; **21.4:** Courtesy of B. Meylan and B. G. Butterfield; **Page 502:** Courtesy Frank B. Salisbury; **21.6:** © John D. Cunningham/Visuals Unlimited; **21.10A, B:** © J. N. A. Lott, McMaster U./Biological Photo Service; **21.11:** © Carl W. May/Biological Photo Service; **Page 508** (top): © Steve Kaufmann/Peter Arnold, Inc.; **Page 508** (bottom): © Gregory K. Scott, 1985; **21.14:** © Robert & Linda Mitchell; **21.15:** From I. Raskin and Hans, "Mechanism of Aeration in Rice," *Science* 228:327–329; cover, April 19, 1985. © AAAS, 1985.; **21.16:** Figure from *Botany*, by Peter M. Ray, Taylor A. Steeves and Sara A. Fultz. © 1983 by Saunders College Publishing, reprinted by permission of the publisher; **21.19:** From Gunning and Pate, "Transfer Cells in Angiosperm Line," *Protoplasma* 68:135–156. © Springer-Verlag, 1993; **21.21A, B:** From M. H. Zimmerman, "Movement of Organic Substances in Trees," *Science* 133:73–79 © 1961 American Association for the Advancement of Science.

Chapter 22

Opener: © Runk/Schoenberger/Grant Heilman Photography, Inc.; **22.1:** © Bill Kamin/Visuals Unlimited; **22.2:** © William E. Ferguson; **22.8:** Courtesy of Hunt Institute for Botanical Documentation, Carnegie Mellon University, Pittsburgh, PA; **22.9:** © Science VU-USM/Visuals Unlimited; **22.10:** © Wm. Ormerod/Visuals Unlimited; **22.11A:** © Heather Angel/Biofotos; **22.11B:** © John M. Thager/Visuals Unlimited; **22.12A:** © Bruce Berg/Visuals Unlimited; **22.12B:** © Heather Angel/Biofotos; **22.12C:** © William E. Ferguson; **22.15:** © Robert & Linda Mitchell; **22.19:** © Runk/Schoenberger/Grant Heilman Photography; **22.20:** © Heather Angel/Biofotos; **22.21A, B:** Courtesy of DEKALB Genetics Corp., DeKalb, IL.

Chapter 23

Opener: © Dwight Kuhn; **23.1A:** © Robert E. Lyons/Photo/Nats, Inc.; **23.1B:** © Jon Bertsch/Visuals Unlimited; **23.1C:** © Hans Reinhard/Bruce Coleman, Inc.; **Page 557:** Courtesy of Renee Van Buren; **23.3:** © Dede Gilman/Visions From Nature; **23.4:** © J. Alcock/Visuals Unlimited; **23.5:** © William McPherson/Bruce Coleman, Inc.; **23.10:** Courtesy of Darrell Vodopich; **23.15:** © Jim Strawser/Grant Heilman Photography; **23.16:** © Runk/Schoenberger/Grant Heilman Photography.

Chapter 24

Opener: Courtesy of Hunt Institute for Botanical Documentation, Carnegie Mellon University, Pittsburgh, PA; **24.1:** © Runk/Schoenberger/Grant Heilman Photography; **24.2:** © Tom McHugh/Photo Researchers, Inc.; **Page 577:** © Kingsley Stern; **Page 578:** © Heather Angel/Biofotos; **24.3A, 24.4:** © Robert & Linda Mitchell; **24.8A:** © Thomas Hovland/Grant Heilman Photography; **24.8B:** © Jeff Foott/Bruce Coleman, Inc.; **24.10:** © K. G. Vock/Okapia/Photo Researchers, Inc.

Chapter 25

Opener: © Alfred Pasieka/SPL/Photo Researchers, Inc.; **25.1A:** © David M. Phillips/Visuals Unlimited; **25.1B:** © M. I. Walker/Photo Researchers, Inc.; **25.1C:** W. Dennis Clark; **25.1D:** © P. L. Grilione & J. Pangborn, San Jose State University; **25.2A:** © David M. Phillips/Visuals Unlimited; **25.2B:** © R. Kessel-G. Shih/Visuals Unlimited; **25.2C:** © Runk/Schoenberger/Grant Heilman Photography; **25.3:** © Leon J. LeBeau/Biological Photo Service; **Page 597:** © BPS/T. J. Beveridge/Tom Stack & Associates; **25.4:** © George B. Chapman/Visuals Unlimited; **25.5:** © G. Musil/Visuals Unlimited; **25.6:** © Dr. Dennis Kunkel/Phototake; **25.7:** © Cabisco/Visuals Unlimited; **25.10:** © Peter Gregg/Imagery; **25.11A:** © Philip Sze/Visuals Unlimited; **25.11B:** © Sinclair Stammers/Photo Researchers, Inc.; **25.11C:** © Runk/Schoenberger/Grant Heilman Photography; **25.12:** © Cabisco/Visuals Unlimited; **25.13A:** © Russell Steere/Visuals Unlimited; **25.13B:** © Cabisco/Visuals Unlimited; **25.14A:** Courtesy of Ikuo Kimura, Faculty of Agriculture, Hokkaido University, Japan; **25.14B, C:** Courtesy of Terry Hanzlik, Karl Gordon, Phil Larkin of the Commonwealth Scientific and Industrial Research Organization (CSIRO).

Chapter 26

Opener: © Masana Izawa/Nature Production; **26.1A:** © Ray Coleman/Photo Researchers, Inc.; **26.1B:** © E. R. Degginger/Bruce Coleman, Inc.; **26.1C:** © Richard Walters/Visuals Unlimited; **26.1D:** © Heather Angel/Biofotos; **26.1E:** © Bill Keogh/Visuals Unlimited; **26.1F:** © Hans Reinhard/Bruce Coleman, Inc.; **26.2A:** © Dr. Jeremy Burgess/SPL/Photo Researchers, Inc.; **26.2B:** © M. F. Brown/Visuals Unlimited; **Page 616:** © E. R. Degginger/Bruce Coleman, Inc.; **26.3A:** © Dwight Kuhn; **26.3B:** © R. Hussey/Visuals Unlimited; **Page 620:** © Manfred Kage/Peter Arnold, Inc.; **26.5A:** © John D. Cunningham/Visuals Unlimited; **26.5B:** © Doug Sherman/Geofile; **26.6:** © Gary T. Cole, Ph.D., University of Texas, Austin/Biological Photo Service; **26.8:** © M. Eichelberger/Visuals Unlimited;

26.9: Courtesy of Mylan Laboratories, Inc. Photo by Rienzi & Rienzi Communications, Inc.; **26.10A:** Courtesy of G. S. Ellmore; **26.10B:** © Stephen Sharneff/Visuals Unlimited; **26.11A:** © Robert & Linda Mitchell; **26.11B:** © William E. Ferguson; **26.11C:** © Heather Angel/Biofotos; **26.12:** © John D. Cunningham/Visuals Unlimited; **Page 625A:** © Stephen P. Parker/Photo Researchers, Inc.; **Page 625B:** © Barbara A. Roy, reprinted by permission from Nature 362:57; **26.14B:** © Stanley L. Flegler/Visuals Unlimited; **26.16A:** © M. I. Walker/ Photo Researchers, Inc.; **26.16B:** © Bruce Iverson/ Visuals Unlimited; **26.17:** © Holt Studios, Ltd./Animals Animals/Earth Scenes; **26.19A:** © Bruce Iverson; **26.19B:** © John Shaw/Bruce Coleman, Inc.; **26.19C:** © Cabisco/Visuals Unlimited; **26.19D:** © A. M. Siegelman/Visuals Unlimited.

Chapter 27

Opener: © Randy Morse/Tom Stack & Associates; **27.1A:** © Heather Angel/Biofotos; **27.1B:** © Gary Retherford/Photo Researchers, Inc.; **27.2:** © M. I. Walker/Photo Researchers, Inc.; **27.3A:** © John D. Cunningham/Visuals Unlimited; **27.3B:** © Philip Sze; **27.4:** © Phil Gates, University of Durham/Biological Photo Service; **27.7:** © 1991, W. L. Dentler, University of Kansas/Biological Photo Service; **27.10A:** © Richard H. Gross; **27.10B, 27.11:** © Bruce Iverson; **27.12:** © M. Eichelberger/Visuals Unlimited; **27.13:** © Jeremy Pickett-Heaps; **27.15A:** © Manfred Kage/Peter Arnold, Inc.; **27.15B:** © M. I. Walker/Photo Researchers, Inc.; **27.16:** © Heather Angel/Biofotos; **27.17:** © Carolina Biological Supply/Phototake; **27.18:** © Runk/Schoenberger/Grant Heilman Photography; **27.20:** © Cabisco/Visuals Unlimited; **27.23:** © Tammy Peluso/Tom Stack & Associates; **27.25:** © Sea Studios, Inc./ Peter Arnold, Inc.; **27.26:** Reproduced by permission of the National Research Council of Canada from the *Canadian Journal of Botany*, Vol. 60, pages 85–97, 1982.; **Page 652:** © Phillip Sze; **27.28A:** © Manfred Kage/ Peter Arnold, Inc.; **27.28B:** © Veronika Burmeister/ Visuals Unlimited; **27.29A:** © Peter Siver; **27.29B:** © T. E. Adams/Visuals Unlimited; **27.31:** © R. G Kessel-C. Y. Shih/Visuals Unlimited; **27.32:** © Biophoto Associates/ Photo Researchers, Inc.; **27.33:** © John D. Cunningham/ Visuals Unlimited; **27.35:** © Roland Birke/Peter Arnold, Inc.; **27.36:** © Heather Angel/Biofotos.

Chapter 28

Opener: © Dwight Kuhn; **28.1A:** © Robert & Linda Mitchell; **28.1B:** © E. Webber/Visuals Unlimited; **28.1C:** © Robert & Linda Mitchell; **28.1D:** © David Cavagnaro/Peter Arnold, Inc.; **28.2 (top left):** © Robert & Linda Mitchell; **28.2 (top right):** © George J. Wilder/Visuals Unlimited;

28.4A: © Jeff Gnass; 28.4B: © Heather Angel/Biofotos; 28.4C: © George Holton/Photo Researchers, Inc.; 28.4D: © William E. Ferguson; 28.4E: Kingsley Stern; 28.5: © Bruce Iverson; 28.6: © R. Calentine/Visuals Unlimited; 28.7: © C. Y. Shih-R. G. Kessel/Visuals Unlimited; 28.9A: © Edwin Reschke/Peter Arnold, Inc.; 28.9B: © R. Knauft/Photo Researchers, Inc.; 28.10: © Heather Angel/Biofotos; 28.11B: © J. H. Troughton. Provided by Industrial Research, Ltd.; 28.11C: © Bruce Matheson/Photo/Nats, Inc.; 28.12A: © Kjell Sandved/Visuals Unlimited 28.12B, C, D: Courtesy of Howard Crum, University of Michigan Herbarium; 28.12E: © David M. Dennis/Tom Stack & Associates; 28.12F: Courtesy of Howard Crum, University of Michigan Herbarium; 28.13A: © John D. Cunningham/Visuals Unlimited; 28.13B: © Robert & Linda Mitchell; 28.14: © Triarch/Visuals Unlimited; 28.16A: © Heather Angel/Biofotos; 28.16B: © E. R. Degginger/Bruce Coleman, Inc.; 28.17: © Robert & Linda Mitchell; 28.18A: © William E. Ferguson; 28.18C: Courtesy of Ray F. Evert; 28.18D: Courtesy of Damian S. Neuberger; 28.19: © M. C. F. Proctor; 28.20: Courtesy of Dr. Jonathan Shaw; 28.21A: © M. P. Gadomski/Bruce Coleman, Inc.; 28.21B: © W. Ormerod/Visuals Unlimited; Page 680: © Ira Block; 28.22A: © BIOS/ Peter Arnold, Inc.; 28.22B: © Fred Bavendam; 28.22C: © Dan Budnik/ Woodfin Camp & Associates.

Chapter 29

Opener: © Kjell B. Sandved/Butterfly Alphabet; 29.1: © Grant Heilman/Grant Heilman Photography; 29.4A: © John D. Cunningham/Visuals Unlimited; 29.4B: © Triarch/ Visuals Unlimited; 29.5: © Ed Reschke; 29.6A: © Bruce Iverson; 29.6B: © Bruce Iverson; 29.8: © E. R. Degginger/ Animals Animals/Earth Scenes; 29.10A: © Runk/ Schoenberger/Grant Heilman Photography; 29.10B: © Runk/Schoenberger/Grant Heilman Photography; 29.10C: © James W. Richardson/Visuals Unlimited; 29.12A: © Runk/Schoenberger/Grant Heilman Photography; 29.12B: © Robert & Linda Mitchell; 29.16: © Cabisco/Visuals Unlimited; 29.17: © Barry Runk/Grant Heilman, Inc.; 29.18B: © Patti Murray/ Animals Animals/Earth Scenes; 29.21A: © Dede Gilman/ Visions From Nature; 29.23: © Richard H. Gross; 29.25A: © W. H. Hodge/Peter Arnold, Inc.; 29.25B: © William E. Ferguson; 29.26: © Kjell B. Sandved/ Butterfly Alphabet; 29.27: © R. F. Ashley/Visuals Unlimited; 29.28A: © Runk/Schoenberger/Grant Heilman Photography; 29.28B: © Kjell B. Sandved/Butterfly Alphabet; 29.28C: © Kjell B. Sandved/Visuals Unlimited; 29.29A: © John D. Cunningham/Visuals Unlimited; 29.31A: Kingsley Stern; 29.31B: © Hugh A. Johnson/Visuals Unlimited; 29.33: © Heather Angel/ Biofotos; Page 705: Courtesy of Jeffrey R. Stevens and Robin Gooch; 29.34A: © W. H. Hodge/Peter Arnold,

Inc.; 29.35A, B: Courtesy of Warren H. Wagner, Jr., University of Michigan; 29.36A, B: Courtesy of Spencer Barrett, University of Toronto; Page 709: © Russ Peppers, Ph.D., Geologist, Illinois State Geological Survey; 29.37: © George J. Wilder/Visuals Unlimited; Page 710 (both left): © William E. Ferguson; Page 710 (right): © Stephen J. Krasemann/Peter Arnold, Inc.; 29.38A: The Field Museum, #GEO85637C, Chicago; 29.38B: The Field Museum, #CK-49T, Chicago, painting by Charles Knight.

Chapter 30

Opener: © R. J. Erwin/Photo Researchers, Inc.; 30.3A: Courtesy of Kathleen B. Pigg; 30.3B: © George J. Wilder/Visuals Unlimited; 30.4A: © William E. Ferguson; 30.4B: © Robert & Linda Mitchell; 30.4C: © William E. Ferguson; 30.5A: © Runk/Schoenberger/Grant Heilman Photography; 30.5B: © Grant Heilman/Grant Heilman Photography; 30.5C: © Doug Sokell/Visuals Unlimited; 30.5D: © Alan G. Nelson/Animals Animals/Earth Scenes; 30.6A: © Robert & Linda Mitchell; 30.6B: © Stan W. Elems/Visuals Unlimited; 30.7: © Robert & Linda Mitchell; Page 726: Courtesy of Gary A. Strobel. Photo © William Hess, Brigham Young University; 30.10: © Knut Norstog; 30.11: © Jeff Gnass; 30.12A: © William E. Ferguson; 30.12B, 30.13A, B, 30.14A: © Robert & Linda Mitchell; 30.14B: Science VU/Visuals Unlimited; 30.15A: © Michael S. Thompson/ Comstock; 30.15B: © Rosamond Purcell; 30.17: The Field Museum, #GE085723C, Chicago, photo by Diane Alexander White.

Chapter 31

Opener: © Bob Hasenick; 31.2: © Fritz Prenzel/Animals Animals/ Earth Scenes; 31.2B: © Ed Reschke/Peter Arnold, Inc.; 31.2C: © Hal Horwitz/Photo/Nats, Inc.; 31.3A: Courtesy of the Peabody Museum of Natural History, Yale University; 31.4A: © Robert & Linda Mitchell; 31.4B: © William E. Ferguson; 31.8A: © John Shaw/ Tom Stack & Associates; 31.9: © Wolfgang Kaehler; 31.10: © Kjell B. Sandved/Visuals Unlimited; 31.12B: Courtesy of Michael G. Simpson; 31.14: © E. R. Degginger/Photo Researchers, Inc.

Chapter 32

Opener: © Gary Meszaro/Bruce Coleman, Inc.; 32.1: © Heather Angel/Biofotos; 32.2: © Bruce Iverson; 32.3: © Richard Kolar/Animals Animals/Earth Scenes; 32.4: © William E. Ferguson; 32.8: © Kingsley Stern; 32.9: © E. R. Degginger/ Photo Researchers, Inc.; 32.10: © John Cancalosi/Peter Arnold, Inc.; 32.11: © William E. Ferguson; 32.12: © D. Rintoue/Visuals Unlimited; 32.14: © Jack Wilburn/ Animals Animals/Earth Scenes; 32.17: © Don Johnston/

Photo/Nats, Inc.; **32.18:** © William E. Ferguson; **32.19:** © Runk/Schoenberger/Grant Heilman Photography; **32.20:** © J. Lichter/Photo Researchers, Inc.; **32.21:** © Kim Todd/Photo/Nats, Inc.; **32.22:** © John D. Cunningham/ Visuals Unlimited; **32.23:** © Raymond A. Mendez/Animals Animals/Earth Scenes; **32.24:** © Breck P. Kent/Animals Animals/Earth Scenes; **32.25:** © S. Maslowski/Visuals Unlimited.

Chapter 33

Opener: © Jeff Gnass; **33.4:** © John D. Cunningham/Visuals Unlimited; **33.5:** © B. J. O'Donnell/Biological Photo Service; **33.7:** © Brian Parker/Tom Stack & Associates; **33.9:** © Milton H. Tierney, Jr./Visuals Unlimited; **33.10:** © Jeff Gnass; **33.11:** © Nada Pecnik/Visuals Unlimited; **33.13:** © Ron Spomer/Visuals Unlimited; **33.14:** © Luiz C. Claudio Marigo/Peter Arnold, Inc.; **33.15:** © Don W. Fawcett/Visuals Unlimited; **33.16:** © Jeff Gnass; **33.17:** © David C. Fritts/Animals Animals/Earth Scenes; **33.18:** © Jane Thomas/Visuals Unlimited; **33.20, 33.22:** © Jeff Gnass; **33.23:** © Dede Gilman/Visions From Nature; **33.25:** © Jeff Gnass; **33.26:** © John D. Cunningham/Visuals Unlimited; **33.27:** © D. Newman/Visuals Unlimited; **33.30:** © Wolfgang Kaehler; **33.31:** © Kjell B. Sandved/ Visuals Unlimited; **33.32:** © Kevin Schafer/Tom Stack & Associates.

Epilogue

Opener: © Mark Edwards/Still Pictures; **E.1A:** Photograph supplied by Norfolk Lavender Ltd. © Norfolk Lavender Ltd.; **E.1B:** © Benjamin H. Kaestner, McCormick & Co., Inc.; **E.2A, B:** © Walter H. Hodge/Peter Arnold, Inc.; **E.2C:** © Robert & Linda Mitchell; **E.3:** © L. L. T. Rhodes/Animals Animals/Earth Scenes; **E.4A:** © Grant Heilman/Grant Heilman Photography; **E.4B:** © William H. Allen, Jr.; **E.4C:** © R. Konig/JACANA/Photo Researchers, Inc.; **E.4D:** © Jim Steinberg/Photo Researchers, Inc.; **E.4E:** © D. Cavagnaro/Visuals Unlimited; **Page 812:** © Scott Camazine/Photo Researchers, Inc.; **Page 813** (top): © Alford W. Cooper/Photo Researchers, Inc.; **Page 813** (bottom) Courtesy of Hunt Institute of Botanical Documentation, Carnegie Mellon University, Pittsburg, PA; **E.5A:** © David M. Stone/Photo/Nats, Inc.; **E.5B:** © John D. Cunningham/Visuals Unlimited; **E.5C:** © Heather Angel/Biofotos; **E.5D:** © Brian Moser/ The Hutchison Library, London; **E.6:** © Dr. Mark J. Plotkin *Tales of a Shaman's Apprentice*; **E.7A:** © Michael J. Balick/Peter Arnold, Inc.; **E.7B:** © Walter H. Hodge/ Peter Arnold, Inc.; **E.8:** From Dr. Daniel Klayman, "Quinghausa (Artemysinin): An Anti-malarial Drug From China," *Science* 228:1051, May 3, 1985.

Copyright 1985 by the AAAS; **E.9:** © James P. Blair/ National Geographic Image Collection; **E.11:** © Edward Parker/Oxford Scientific Films/Animals Animals/Earth Scenes; **Page 822** (left): © Runk/Schoenberger/Grant Heilman Photography; **Page 822** (top right): © Alex Kerstitch/Visuals Unlimited; **Page 822** (bottom right): © Walter H. Hodge/Peter Arnold, Inc.; **E.12A:** © Scott Camazine/Photo Researchers, Inc.; **E.12B** (coconut trees): © Walter H. Hodge/Peter Arnold, Inc.; **E.12C** (coconuts): © Patricia Goycolca/The Hutchinson Library, London; **E.13A** (background): © Grant Heilman Photography; **E.13A** (inset): © Cecile Brunswick/Peter Arnold, Inc.; **E.13B:** © Michael Holford; **E.13C:** © E. S. Ross; **E.15:** © Walter H. Hodge/Peter Arnold, Inc.; **E.17B:** © Smithsonian/Carl Hansen; **E.17C:** © Smithsonian/ Antonio Montaner; **E.18A:** NOAA; **E.18B:** © George Holton/Photo Researchers, Inc.; **E.18C:** © 1989 Silvester/ Black Star; **E.18D:** © Luiz C. Marigo/Peter Arnold, Inc.; **E.19:** © Ghillean Prance; **E.20A:** © Ghillean Prance; **E.20B:** © Kjell B. Sandved/Butterfly Alphabet; **E.21:** © Douglas Kirkland/Sygma.

Illustrations

Chapter 1

Figure 1.4: From Gibbs and Lawson, "The Nature of Scientific Thinking" in *The American Biology Teacher*, 54(3):141. Copyright © National Association of Biology Teachers, Reston, VA. Reprinted by permission.

Chapter 2

Reading figure 2.2: From *Diabetes Forecast* Vol. 44(4). Copyright © 1991 American Diabetes Association. Reprinted by permission.

Chapter 3

Figure 3.1a: From Kingsley R. Stern, *Introductory Plant Biology*, 5th ed. Copyright © 1991 The McGraw-Hill Companies, Inc. All Rights Reserved. Reprinted by permission; **Figures 3.21, 3.25a:** From Sylvia S. Mader, *Biology*, 5th ed. Copyright © 1996 The McGraw-Hill Companies, Inc. All Rights Reserved. Reprinted by permission.

Chapter 5

Figure 5.4: From *Biology: The Science of Life*, 3rd Edition by Robert A. Wallace et al. Copyright © 1991 by HarperCollins Publishers, Inc. Reprinted by permission of Addison-Wesley Educational Publishers Inc.; **Figures 5.6, 5.8:** From Purves, Orians, and Heller, *Life: The Science of Biology*, 3rd ed. Copyright © 1992 Sinauer Associates, Inc., Sunderland, MA. Reprinted by permission; **Figures 5.9, 5.12:** From *Biology: The Science of Life*, 3rd Edition by Robert A. Wallace et al. Copyright © 1991 by HarperCollins Publishers, Inc. Reprinted by permission of Addison-Wesley Educational Publishers Inc.

Chapter 6

Reading figure 6.2b: Reprinted by permission from Wivagg, *American Biology Teacher*, 49(2)113. Copyright © 1987 National Association of Biology Teachers, Reston, VA.

Chapter 7

Excerpt, page 146: From Karin Brueschweiler, Arizona State University. Reprinted by permission; **Figure 7.20** From Jensen and Salisbury, *Botany*, 2d ed. Copyright © 1989 Wadsworth Publishing Co. Reprinted by permission.

Chapter 9

Reading figure 9.2a: From *Principles of Genetics* 2/E by Fristrom and Clegg. Copyright © 1988 by W. H. Freeman and Company. Used with permission.

Chapter 11

Figure 11.6b: From Y. N. Osheim, et al., *Cell*, 43:143–151. Copyright © 1985 Cell Press, Cambridge, MA. Reprinted by permission.

Chapter 14

Figure 14.12: From Jensen and Salisbury, *Botany*, 2d ed. Copyright © 1989 Wadsworth Publishing Co. Reprinted by permission.

Chapter 16

Figure 16.11: From Raven, Evert, and Eichhorn: *Biology of Plants*, 5th ed. Worth Publishers, New York, 1992. Reprinted with permission; **16.21:** Figure from *Botany* by Peter M. Ray, Taylor A. Steeves, and Sara A. Fultz, Copyright © 1983 by Saunders College Publishing, reproduced by permission of the publisher.

Chapter 17

Figure 17.10: Courtesy of James E. Canright, *American Journal of Botany*, 39:488, 1952.

Chapter 19

Excerpt and Graph, page 449: From Cora Schmid, Stanford University. Reprinted by permission.

Chapter 21

Figure 21.7 (right): From Ruth Bernstein and Stephen Bernstein, *Biology*. Copyright © 1996 The McGraw-Hill Companies, Inc. All Rights Reserved. Reprinted by permission; **Figure 21.8:** From Jensen and Ross, *Plant Physiology*, 4th ed. Copyright © 1992 Wadsworth Publishing Co. Reprinted by permission.

Chapter 22

Figure 22.4: From Ricki Lewis, *Life*. Copyright © 1992 The McGraw-Hill Companies, Inc. All Rights Reserved. Reprinted by permission; **Figure 22.5:** From Stephen A. Miller and John P. Harley, *Zoology*, 2d ed. Copyright © 1994 The McGraw-Hill Companies, Inc. All Rights Reserved. Reprinted by permission.

Chapter 23

Figure 23.9: From Douglas Futuyama, *Evolutionary Biology*, 2d ed. Copyright © 1986 Sinauer Associates, Inc., Sunderland, MA. Reprinted by permission.

Chapter 24

Figure 24.11: From *Five Kingdoms* 2/E by Margulis and Schwartz. Copyright © 1988 by W. H. Freeman and Company. Used with permission.

Chapter 25

Figure 25.8: Reprinted with permission of Knight-Ridder/Tribune Information Services.

Chapter 31

Figure 31.3b: Courtesy of the Peabody Museum of Natural History, Yale University.

Chapter 33

Figure 33.1: From Eldon D. Enger and Bradley F. Smith, *Environmental Science*, 4th ed. Copyright © 1992 The McGraw-Hill Companies, Inc. All Rights Reserved. Reprinted by permission.

Epilogue

Reading E.1: From *The House at Pooh Corner* by A. A. Milne, illustrations by E. H. Shepard. Copyright © 1928 by E. P. Dutton, renewed © 1956 by A. A. Milne. Used by permission of Dutton Children's Books, a division of Penguin Books USA Inc.; **Figure E.10:** Source: Data from Ricki Lewis, *Life*, The McGraw-Hill Companies, Inc., 1992; **Figure E.16, E.17a:** From Ricki Lewis, *Life*. Copyright © 1992 The McGraw-Hill Companies, Inc. All Rights Reserved. Reprinted by permission.

index

genome, 200, 201–2(box)
root gravitropism in, 447, 448
Arabis holboellil, 625(box)
Arbor Day holiday, 378
Archaea, **599**
Archaebacteria (subkingdom),
599–601, 845–46
characteristics of, vs. Eubacteria, 601
ribosomes of, 601
Archaeopteris sp., 734
Archegonia
of bryophytes, **664,** *666, 671*
of seedless vascular plants, 690
Archegoniophores (liverwort bryophyte),
674, 676
Arctic tundra, 785–87
Arctic willow (*Salix artica*), 802
Arctostaphylos canescens, 576
Argemone albiflora, 542
Argyroxiphium sandwicense, 535
Aridosols, **471**
Aril (gymnosperm), **721**
Aristotle, classification systems of, 521,
524, 574
Artemisia annua, 817
Artemisinin (antimalarial drug), 817
Artificial chromosomes, 247
Artificial selection, **530**
domestication of plants by, 819
for giant vegetables, 537(box)
hybridization and, 168(box)
natural selection compared to, 536–39
Ascetylsalicylic acid, 37 n.
Asci (ascomycetes), **617**
Asclepias tuberosa, 2
Ascogeneous hyphae (fungi), **618**
Ascogonium (ascomycetes), **618**
Ascoma (ascomycetes), **618**
Ascomycetes, **614,** 617–18, *619,* 846
Ascomycota (division), 846
Ascospores (ascomycetes), **618**
Asexual reproduction
in algae, 637–38, 641, 642, 645
in bacteria, 597–98
in bryophytes, 666–67, 669–70,
674, 676–77
in fungi, 612, 614, 615, 622, 623
Ash, white (*Fraxinus americana*), 40,
370(box)
Aspen (*Populus tremuloides*), 350, 775
genetically-based ecotype of, 563, 564
Aspergillosus, 619
Aspergillus, 614, 619
Aspilia sp., 818
Aspirin, 37 n., 438
Asplenium sp., 701
Asteraceae, 157
Astrosclereid, 287
Athlete's foot, 619, 620
Atmosphere
acid rain in, 778

depletion of stratospheric ozone, 778–79
greenhouse effect and global warming
of, 777–78
greenhouse effect and photosynthesis, 157
Atmospheric humidity, transpiration and, 505
Atmospheric pressure, water movement in
plants and, 503
Atmospheric temperature, transpiration
and, 506
Atomic mass unit, **833**
Atoms, **833**
structure and activity of, 833–34
ATP (adenosine triphosphate), **102–5**
chemiosmosis and, 115, 123–25, 143–44
coupled reactions and, 103
as energy currency, 102–3
energy for active transport from
hydrolysis of, 82, 83
phloem loading and requirement for,
516, 517
phosphorylation reaction and production
of, 839–40
production of, in aerobic respiration,
120–21, 123–25
production of, in photosynthesis,
143–44, 148
proton pumps and use or production
of, 85, 86
structure of, 102
synthesis of, 114–15, 120(table), 123–25
ATPase (ATP phosphohydrolases), 82, 83, 86
ATP synthases, **123**
Atriplex occidentalis, 81, 829–30
Austrobaileya, 388
Autopolyploids, **566,** 567
Autoradiography, *149*
Autotrophs, 132
AUX1 gene, 450
Auxiliary cell (red algae), **650**
Auxin, **85, 416,** 417–25
in bryophytes, 680–81
calcium and, 425
discovery of, 417–21
effects of, 423–25
ethylene production stimulated by, 436
functions of, 275(table)
IAA (indole-3-acetic acid) as, 421
influence of, on vascular cambium,
360, 362
as internal cellular signal, 85, 86
name of, 421
plant pathogens and, 430(box)
proton pumping and, 86, 87
stem growth in response to, 308
structure and function of, 275(table), 421
synthesis of, 421
synthetic, 421–22
transport of, 422–23
Auxospores (diatoms), **653**
Avena curvature test, 420
Avery, Oswald, 174

Avery Island, Louisiana, production of
tabasco in, 279
Axillary buds, 190, **263, 311**
apical dominance and growth of, 423–24
Azolla sp., 704, 708
nitrogen-fixing cyanobacteria living
in, 487, 488

b

Bacillariophyceae (class), 847
Bacillus anthracis, 605
Bacillus stearothermophilus, 604
Bacillus subtilis, 596
Bacillus thuringiensis (BT), 15, 16, 249,
604
Bacteria, 593–609
botulism caused by Clostridium, 596,
597(box), 604
chloroxybacteria, 601–3
cyanobacteria, 595, 601
denitrifying, 769
diversity and relationships of, 599–603
economic importance of, 603–4
evolution of photosynthesis and, 152(box)
features of, 586(table), 594–97
fossilized, 533
gene cloning by, 245
generalized life cycle of, 597–98
habitats of, 598–99
human diseases caused by, 597(box),
605(box)
ice-minus, 249
industrial uses of genetically
engineered, 248–50
interactions of, with bryophytes, 681
nitrogen-fixing, 87–88, 350, 352, 486–88,
598, 769
origin and evolution of, 603
photosynthetic, 595, 601–3
proposed higher tax for, 599(table)
shapes and Gram staining of, 595, 596
treatment of Dutch elm disease with
genetically altered, 7 n.2
Bacteriochlorophyll, **595**
Bacteriophages, **174,** *175,* 605
Bailey, Liberty Hyde, 733
Bald cypress (*Taxodium mucronatum*),
363, *365*
Ball, Aaron, 13
Balsa (*Ochroma* sp.), 295
Bamboo, stem growth in, 308, *309*
Banana (*Musa* sp.), 821
Bangia atropurpurea, 652(box)
Bangiophycidae (subclass), 650
Banyan tree (*Ficus benghalensis*), 378(box)
Baobab (*Adansonia digitata*), 791, 792
Bark, 372–74
periderm tissue in, 372–74
secondary phloem tissue in, 372
uses of, 377–78

varieties of, 372
Barley (*Hordeum vulgare*), 84(box), **826**
Barrel cactus (*Ferrocactus covillei*), 762
Bars, water potential expressed in, **78**
Baseball bats, *40*, 370(box)
Bases, 841(box)
Basidia (basidiomycetes), **622**
Basidioma (basidiomycetes), **622**
Basidiomycetes, **614**, 622–23, *624*, 846
Basidiomycota (division), **846**
Bassia scoparia, 829–30
Basswood (*Tilia americana*), 287, 372, 789
Batasins, **438**
Bats, baseball, 370(box)
Bauhin, Gaspard, 575
Bayberry (*Myrica pensylvanica*), 36
Bazzania trilobata, 673
Beagle, Darwin's voyages on H.M.S.,
    527, 528
Bean (*Phaseolus* sp.)
    cotyledon of, *404*
    germination of, *407*
    lectins produced by, *88*
    Mendel's experiments with, *177*
Beer, production of, 84(box), 427
Bees, *172*
Beet (*Beta vulgaris*), *37*, 374, *375*
Belladonna, 330
Bell-fruited mallee (*Eucalyptus pressiana*), 401
Bentham, George, 578
Bergey's Manual of Systematic Bacteriology,
    599(table)
Bessey, Charles, 580
Beta-carotene, **39**, **139**
Beta-glucose, **24**, *25*
Betalain pigment, 533, 582–83, *584*
Beta-oxidation, **126**
Beta tubulin, **54**
Beta vulgaris, *37*, 374, *375*
Betula papyrifera, 373
Beverages, *16*, 329
    beer, 84(box), 427
    chocolate, *5*
    coffee, 6(box), 809
    teas, *5*, 730, 808, *809*
Bhattacharyya, Madan K., 232
B horizon (soil), **470**, *471*
Bidens ctenophylla, *534*, *535*
Bilayer, phospholipid, **74–76**
    selective permeability of, *75*
Bindweed (*Convolvulus* sp.), *450*, *451*
Binomial system of nomenclature, **575–78**
Bioassays, **420**
Biochemical cytology (cytochemistry), **52–53**
Biochemical reactions of photosynthesis,
    144, 148–51
Biochemicals, 22–36
Biochemistry, **836**
Biodiversity. *See* Biological diversity
Bioenergetics, **93**, *94*. *See also* Energy
Biogeography, 533–35

Biological classification, theoretical
    foundation of, 585, 586(table)
Biological clock, **463–64**
Biological diversity
    in ecosystems, 764–65
    locations of great, 560
    of plants, 571, 830
    speciation and, 764, 765
    in tropical rain forests, 764, 800–801
Biological species concept, **555**, *556*
Biomass, **760**
    net productivity of, 764
Biomes, 783–804
    air and precipitation patterns affecting,
        784, 785
    defined, **784**
    deserts and semideserts, 792–95,
        796–97(box)
    effect of elevation and latitude on,
        798, 799(box)
    grasslands and savannas, 790–92
    major world, *785*
    Mediterranean scrub, 795–97
    mountain forests, 797–800
    taiga, 787–88
    temperate deciduous forests, 788–90
    tropical rain forests, 800–803
    tundra, *758*, 784–87
Biophysical controls on plant growth and
    development, 276, 277
Biotechnology, 15–16. *See also* Genetic
    engineering
Biotic community, **760**
Biotic factors, **761**, 763, 765
    roles of, in habitat, 766
Biotic potential, **760–61**
Birch, paper (*Betula papyrifera*), 373
Bird-of-paradise (*Strelitzia reginae*), 547
Birds
    Darwin's finches, 528, *529*, 534
    pollination by, 401
    seed dispersal by, *408*, 409
    seed germination requirements and
        dodo, 405(box)
Bird's foot trefoil (*Lotus corniculatus*), 327
Bird's nest fungus (*Cyathus striatus*), 613
Bisexual colonial green algae, **641**
Bisexual gametophytes, **670**
Bivalent chromosomes, **216**, *217*
Blackberries (*Rubus* sp.), 555
Blackman, F. F., 144
Bladderwort (*Utricularia*), 490
Blade (leaf), 317
Bleeding hearts (*Dicentra* sp.), 766
Blood type, determining, with lectins, 88
Bloom, algal and cyanobacteria, 604, 657–58
Blowouts, tundra, 787
Blue-green algae. *See* Cyanobacteria
Body cells (gymnosperm), **720**
Bogbean (*Menyanthes trifoliata*), *546*, 547
Bolting, **428**, *429*

Bonner, James, 455, 456, 457
Bordered pits, 499
Bordetella pertussis, 605(box)
Boreal forests, 787–88
Borlaug, Norman, 827
Borrelia burgdorferi, 605(box)
Boston fern (*Nephrolepis exaltata*), 186
Boston ivy (*Parthenocissus tricuspidata*),
    324, 767
Botanical Society of America, 554
Botanists
    adventures of David Douglas, 813(box)
    movie-watching and plant recognition by,
        784 n.1
    work of, 8–13
Botany, 3–18
    botanists and use of scientific method
        in, 8–13
    defined, **4**
    goals of, 15–17
    plants, significance of, and, 4–8
    politics and, in Nazi Germany, 118(box)
    unifying themes of, 14–15
Botany students
    A. Ball's research on genetic stability of
        mutants, 13
    K. Brueschweiler's research on
        Photosystem I, 146
    R. Gooch's and J. Stevens's research on
        winter grapefern distribution, 705
    C. Schmid's research on phototropism, 449
Botrychium lunarioides, 705
Botulism (*Clostridium botulinum*), 596,
    597(box), 604
Boundary layer, 505
Boysen-Jensen, Peter, phototropism studies
    by, *417*, 418
Brachysclereids, 288
Bracken fern (*Pteridium aquilinum*), 689, 707
Bracts, 326
Branch (lateral) roots, **336**, 344, *345*
Brassica genus, food plants from, 537(box)
Brassica oleracea, 429
Brassinosteroids, **438**
Bread mold (*Neurospora crassa*), 617
    gene conversion in, 223, *224*
    genetic crossing-over in spore colors
        of, 222, *223*
Bread wheat, 823, 824
Breakfast cereals, 12(box)
Breeding systems, flower diversity and
    function as, 395–97
Brewer's/baker's yeast (*Saccharomyces
    cerevisiae*), 617
Briggs, Winslow, phototropism studies by,
    444, *445*
Bristlecone pine (*Pinus longaeva*), 4 n.1,
    362, 728
British thermal units (Btu), 95 n.2
Broad bean (*Vicia faba*), 69, 190
Bromeliads, 260, 492

Cavities, internal, 299–300
Caytonia sp., 746
Ceanothus integerriumus, 795
Cech, Thomas, 235(box)
Cedar of Lebanon (*Cedrus libani*), 378(box)
Cell(s), 19, 45–71
 competence of, to respond to developmental signals, 277–78
 cytology methods for studying, 48–53
 differentiation of, 270–71
 discovery of, 11–12
 functional organization of, 53–55
 growth and development of. See Development; Growth
 interactions between. See Cellular communication
 laws of thermodynamics and, 98 n.4
 membrane system of, 59–63. See also Membranes; Membrane system
 metabolic pathways in, 99, 100
 microscopic observation of, 46–53
 movements of, 64–67
 nucleus of, 58
 organelles and energy conversion in, 63–64
 origin of chloroplasts and mitochondria in, 66(box)
 preparation of, for microscopy, 51(box)
 ribosomes of, 58–59
 size of, 53, 54
 tissues associated with, 301(table), 302(table)
 wall of. See Cell wall
Cell culture, cloning by, 188, 189
Cell cycle, 187–209
 chromosome movement and mitotic spindle in, 206–7
 control of, 207
 cytokinesis in, 206
 defined, 188
 interphase of, 189–200
 mitosis in, 201–5
 overview of, 188–89
 periods of, 189
Cell division cycle (*cdc*) genes, 207
Cell fractionation, 52
Cell plate, 61, 204, 206
Cell theory, 12, 67, 68, 69
 organismal theory as alternative to, 68, 69
 postulates and problems of, 68, 69
 technology in discovery of, 11–12
Cellular communication, 76, 85–88, 413
 chemical signals, 275–76. See also Hormones
 lectins and glycoproteins in, 87–88
 membrane interactions with other organisms as, 87
 receptors for hormones, 87
Cellular differentiation, plant development and, 270–71, 273–74, 424
 auxin and vascular tissue, 424
 ethylene and, 435
 patterns resulting from, 276, 277

Cellular division
 asymmetric, 273–74
 cytokinins as stimulants of, 431
 plant growth/development and, 268–69, 273–74
 zone of, in roots, 340
Cellular elongation, 269, 270
 effect of auxin on, 423
 proton pumps and, 85
 root zone of, 340
Cellular enlargement, plant growth and, 268–69
Cellular maturation, root zone of, 340, 347
Cellular recognition, epidermal cells and, 294
Cellular respiration, 98, 99, 102, 111
 aerobic, 112–27. See also Aerobic respiration
 anaerobic, 125
 cyanide-resistant, 126–27
 efficiency of, 120–21(box)
 as energy transformation, 107–8
 of lipids, 126, 127, 128
 metabolic relationships of, 113
 of pentose sugars, 126
 photorespiration, 127, 151–53
Cellulases, 27
Cellulose, 24–25
 amylose vs., 27
 chemical structure of, 25
Cell wall, 55–58
 arrangement of cellulose in, 25
 of bacteria, 594
 connections between, 57–58
 connections between primary, 57–58
 primary, 30, 55, 56
 secondary, 55
 structural proteins in, 29–30
Celosia argentea, 830
Celsius, Anders, 843
Celsius temperature scale, 843–44
Centigrade temperature scale, 843–44
Central mother cells, 267
Centric diatom, 651, 652
Centromere, 203, 204
Century plant (*Agave*), 457
Ceratocystis ulmi, 7
Ceratophyllum sp., 752
Cereal plants, 83, 821–26
 amaranth, 829
 as breakfast food, 12(box)
 corn, 824–26. See also Corn (Zea mays)
 rice, 826)
 wheat, 822–24. See also Wheat (Triticum aestivum)
Cereus (*Cereus* sp.), 558
Cerochlamys pachyphylla, 289
Chabarro, 797
Chailakhyan, M. K., 456
Chalazal pole, 392
Chanterelle mushroom, 616(box)
Chaparral biome, 795, 796–97

Character state (cladistics), 582, 583
Chara sp., 643, 644
Charcoal, 377
Charlane Plantation, Georgia, 371(box)
Charophyceae (class), 639, 645, 847
Chase, Martha, experiments of, 174–75
Chelator, 483
Chemical bond, 834–36
Chemical fertilizers, 473(box)
Chemical reactions, 837–41
 carboxylation and decarboxylation, 839
 coupled, 103
 dehydration and hydrolysis, 839
 free energy and, 101
 oxidation and reduction, 101–2, 839
 phosphorylations, 103, 839–40
 photosynthetic, 144–51
 reactants and products of, 837–38
Chemical signals, 87, 275–76. *See also Cellular communication; Hormones*
Chemical structures, 24(box)
Chemiosmosis, 115
 ATP synthesis and, 123–25
 in chloroplasts, 143, 144
Chemistry, fundamentals of
 atoms and elements, 22, 23(table), 833–34
 biochemicals, 22–41
 chemical bonding, 834–36
 chemical reactions, 837–41
 functional groups and biochemical diversity, 836–37
 isotopes and radioactivity, 838
 moles, acids, bases, and pH scale, 841
 reading chemical structures, 24(box)
Cherry (*Prunus serotina*), 513
Chestnut (*Castanea americana*), 789
Chewing gum, 813(box)
Chiasmata, 216, 218, 219, 220
Chicle, 813(box)
China, use of medicinal plants in, 817
Chitin, 612
Chlamydomonas sp., 68, 69, 636, 638
 photosynthesis in, 146, 148
 sexual reproduction in, 640
Chlorella sp., 135, 148, 149
Chlorenchyma cells, 285, 286
Chlorofluorocarbons, 777, 778–79
Chlorophyceae (class), 645, 847
Chlorophylls, 138–40
 in algae, 638, 653, 656
 in bacteria, 595
 chlorophyll a, 138, 141, 142
 chlorophyll b, 138, 139, 142
 chlorophyll c and d, 140
 in choroplasts, 141–42
 as deodorizer, 138
 destruction of, in leaf senescence, 461, 462
 evolutionary relationship of algae containing, 656
 evolution of photosynthesis and, 152(box)
 fate of energy in energized electrons of, 143

Cofactors, **103**
Coffee (*Coffea arabica*), 6(box)
Cohesion, **836**
Colchicine, 55, 56–57
Colchicum autumnale, 55, 811
Cold-hardy plants, 461
Coleoptile, **404**
    phototropism studies using, 417, 418–19,
        444, 445
    polar transport of auxin in, 422
Coleorhiza, **404**
Coleus (*Coleus* sp.), 62, 311, 314, 436, 450
Collenchyma cells, **262**, 285–86, 301(table),
    302(table)
Colonial algae, **634**, *636*
    green, 641
Colony hybridization, **248**
Columbine (*Aquilegia vulgaris*), 811
Columbus, Christopher, 820
Columella (moss bryophyte), **671**
Columella cells, 338
Commercial and industrial uses of plants. *See*
    Economic importance of plants
Common groundsel (*Senecio vulgaris*), *516*
Communication, cellular. *See* Cellular
    communication
Communities, **761**
    biotic, 760
    climax, 773, 774, 775
    in ecosystems, 760, 761
Companion cell, 297, **298**, 302(table), 513
Compartmentalization, **362–63**
Compensation point, **159**
Competition among plants for light, 767
Competitive inhibitors, **107**
Complementary DNA (cDNA), **246**
Complete dominance, **175**
Complete flower, **389**
Composting, 475
Compound(s), **22**, **834**
Compound leaves, **317**
Compound strobili (gymnosperms), **721**
Compound umbel, **395**
Compression fossil, **710**
Compression wood, 369
Concentration gradient, **78**
Conceptacles (rockweed), **648**, *649*
Conchocelis sp., 652(box)
Conduction. *See also* Solutes transport;
    Water transport
    root function of, 338, 347
    structure of cells involved in, 297–98,
        499, 513
Cones (gymnosperm), 720, 723, 729
Conidia (ascomycetes), **618**, *619*
Conidiophores (ascomycetes), **618**, *619*
Coniferopsida (class), 850
Conifers (Pinophyta), 728–30, 850
    extinct, 734–36
    paper produced from, 732(box)
    pine as. See *Pine* (Pinus sp.)
    taiga forest of, 787–88

Coniine, 36, 37(table), 38
Conium maculatum, 38, 327, 789, 811
Conjugation (algae), **642**
Conjugation pilus (bacteria), **597**, *598*
Conocephalum conicum, 673
Conocephalum sp., 665
Conservation, tree farming and, 371(box)
Conservation International, *816*
Consumers in ecosystems, **762**
Continuous synthesis, **198**
Convergence, **533**, *534*
Convolvulus, 450, 451
Cooking, cell biology and, 54
Cooksonia sp., 711
Copa iba (*Copaifera langsdorfu*), 159
Copernicia cerifera, 36
Coprinites dominicana, 627(box)
Corallina sp., 649
Coral reefs, 657
Corchorus capsularis (jute), 47, 288
Cordaitales order, 736
Cork, 55, 372, 376(box)
Cork oak (*Quercus suber*), 55
Corms, **312–13**
Corn (*Zea mays*), 110, 820 n.1, 824–26
    biodegradable plastics containing, 824
    cytoplasmic male sterility in, 182(box)
    endosperm of, 110
    energy transformations by, 91, 92
    ethanol produced from, 604
    fate of, produced in U.S., 825
    gene introns in, 242, 243
    genetic crossing-over in, 222
    germination of, 407
    gravitropism and roots of, 448, 450
    hybrids of, 547, 548
    individual differences in populations of, 538
    interrupted genes in, 243
    knotted-1 gene of, 276
    leaves of, 154, 277, 319, 320, 321
    B. McClintock's research on, 10, 11
    micrographs of, 44, 61
    origins of, in teosinte grass, 548,
        824–26
    roots of, 334, 337, 338, 339, 348, 350,
        448, 450
    root tip, 44, 265, 338
    seedling, 294, 337, 407, 445
    seeds of, 404
    stem structure, 310
    stomata of, 277, 293
    tassels, 182
    transposable elements in DNA of, 10, 11,
        180–81
    zeatin cytokinin in, 430, 431
    zein protein in, 30
Cornus canadensis, 744
Corolla, **384**, *385*
Corpus, **267**
Cortex
    in roots, 341–43
    in stems, 311

Corylus sp., *395*
Corymb, **395**
Corynebacterium diphtheria, 605(box)
Cot curve, **202**
Cotransport across membranes, 82
Cotton (*Gossypium hirsutum*), *16*, *53*, *63*,
    244, 289
Cottonwood (*Populus fremontil*), 306
Cotyledons, 326, **393**
    effects of cytokinins on, 431
    seed germination and, 406, 407
Coupled cotransport system, **82**
Coupled reactions, **103**
Covalent bond, 834–35
Cow pea (*Vigna sinensis*), 293
Cox, Richard, 265(box)
Crassulacean acid metabolism (CAM),
    156(table), 157, **158**. *See also*
    CAM plants
Creighton, H. S., 222
Creosote (*Larrea* sp.), 397, 761
Crick, Francis, 191, *192*
Criminal cases, plants and, 8, 248 n.3
Cristae, **64**, *65*
Cronquist, Arthur, 581, 582
Cronquist system of classification, *581*,
    749(table), 750, 751, 851
Crops
    domestication of, 819–20
    genetically-engineered, 16, 251–54, 829
    plant nutrition and increased yield of, 470
    rotation of, 488
    vernalization response and, 458(box)
Crossing-over of genetic material, **179**,
    216, 218
    *evidence of chromosomal exchange
        in*, 222, 223
    genetic variation promoted by, 546
    repair of DNA and, 227
    unequal, and gene duplication, 224–26
Cross section (wood), 366, **367**
Crown gall disease, *250*
Crown roots, 337
Cryptomonads (Cryptophyta), 635(table),
    653–54, 848
Cryptonectria parasitica, 789
Cryptophyta (division), 635(table), **653**, 848
Cucumber, 292
Cucurbitaceae family, 811
Cucurbita sp., 298
Cultivars, 168(box)
Cuscuta sp., 491, 492
Cushion moss, 668
Cuticle, 69, **289**, **664**, 688
    transpiration and, 507
Cuticular wax, 35–36
Cutin, **36**
Cuvier, Georges, 525
Cyanide, 41
    production of, by clover species, 563
Cyanide-resistant respiration, **126**, *127*,
    128(box)

Freeze-fracture electron microscopy, **52**
Freeze-fracturing of cells, 52
Frithia, 286
Fronds (fern), 313, *314,* **700,** *701*
Fructose, 23, *840*
Fruiting bodies (fungi) 612, *613,* 617
Fruits, *384,* 402–3
  climacteric and nonclimacteric, 433
  development of, 403
  dichotomous key to, 402(table)
  dispersal of, 407–10
  ethylene and ripening of, 433
  evolution in morphology of, 535
  genetically-engineered improvement in
    storage of, 253–54
  gibberellins and formation of, 429
  ovary chambers and, 384, 385
  parthenocarpic, 424
  preventing browning of, 54
  types of, 402–3
Fucales (order), 648
Fucus sp., 272, 648, 649
Fuel, plants as, 330, 377
Fumaric acid, **118**
Functional groups, **836–37**
Functional megaspore, **212, 392, 721**
Fungi (kingdom), **586,** 587, **611–31, 845,**
    846. *See also* Lichens
  characteristics of, 586(table)
  cladistic analysis of, 628
  disappearing species of, 616(box)
  diseases caused by, 7, 613
  diversity of, 614–23
  ecology of, 623–25
  economic importance of, 625–27
  features of, 612
  fossil mushrooms, 627(box)
  generalized life cycle of, 612, 615
  Gibberella fujikuroi, 426
  habitats of, 613
  medicinal, 726(box), 815(table)
  mycorrhizal roots and beneficial, 351,
    485–86
  physiology of, 623
  plant associations with, 485–86
  relationships of, 627
  reproductive strategy of rust, 625(box)
  sensitivity of, to air pollution, 616(box)
  taxol produced from, 726(box)
  turkey tail bracket, 610
Fusiform initials, **358, 360**

g

G_1 phase of interphase, **189,** 190, *191*
G_2 phase of interphase, **189,** 200
GA (gibberellic acid), **426.** *See also*
    Gibberellins
Gaertner, Karl Friedrich, 167
Galactose, *840*
Galápagos Islands, 528
Darwin's study of finches on, 528, *529,* 534
Galium aparine, 287
Gallic acid, 625–26
Gametangia. *See also* Antheridia; Archegonia
  of bryophytes, 666, *671, 681*
  fungi, **617**
  plurilocular, 645
Gametes, **171,** 212, *213. See* Egg (ovum);
    Sperm
  of algae, 636, 637, 649
  of bryophytes, 664–66
  of seedless vascular plants, 688, 689
Gametic isolation, 559
Gametic meiosis, **636,** *637,* 649
Gametophyte, **212,** *213. See also* Pollen; Seed(s)
  of algae, 637, 643, 646, 647
  ancient, 712–13
  of angiosperms, 214, 743
  of bryophytes, 664–66, 667, 669–70, 671,
    674–75, 676–77
  development of female, 214
  development of male, 214
  free-nuclear female (gymnosperm), 721
  gender terminology related to, 670, 704–5
  gymnosperm, 723, 724–25, 726
  of seedless vascular plants, 688, 689–90,
    694, 695–96, 699, 702, 703, 704–5,
    706, 707
Gametophytic self-incompatibility, 397
Gamma plantlets, 268
Gane, R., 432
Gardens, home, 4
Garner, Wightman, 454
Gas chromatography, 420
Gas exchange
  in aerenchyma cells, 285, 286
  in epidermal cells, 291–93
  in hydrophytes, 512
  in periderm, 373, 374
Gasohol, 160
Gas vacuole (bacteria), **595**
Gel electrophoresis, examining proteins
    by, 176(box)
Gemmae (liverwort bryophytes), **674**
Gemmules, **542**
Gene(s), 15, **170,** 175, 232–41
  alleles of. *See* Allele
  AUX1 gene, 450
  cell division cycle (cdc), 207
  crossing-over of. *See* Crossing-over of
    genetic material
  cytoplasmic inheritance of non-nuclear, 180
  DNA as hereditary material in, 174–75.
    See also Deoxyribonucleic acid (DNA)
  engineering of. *See* Genetic engineering
  finding specific, 247–48
  firefly luciferase, 463, 464
  for flower development, 459(box)
  Hardy-Weinberg model for frequencies
    of, 543–44
  interrupted, 237(box), 242–43
jumping (transposable elements), 10, 11,
    180–81, 227
libraries of, 246–47, 248
locus of, 175
mitotic induced (cdc25), 207
multigene families, 226(box)
multiple isozymes of, 177
mutations in, 180, 198(box)
nodD gene, 488
pleiotropic, 177–78, 179
polygenes, 177
protein synthesis and expression
    of, 232–41
psbA gene, 243 n.1
rbcL gene, 737, 751–54
serial systems of, 177
structure of eukaryotic, 242, 243
tandem repeat of, 224, 225
tgal, 826
touch-induced, 454
unequal crossing-over of genetic material
    and duplication of, 224–25, 226
Gene banks, 830, 831
Gene conversion, **223–24,** 225
Gene expression, **232–41**
  forms of RNA, 234–38, 239
  modification of polypeptides, 241
  synthesis of polypeptides (translation),
    239–41
  synthesis of RNA (transcription), 233, 234
Gene flow, **545**
  pollen dispersal and, 562, 563
Gene gun, 254
Gene pool
  defined, 541
  genetic variation in, 544–47
  Hardy-Weinberg model of gene
    frequencies in, 543–44
  population divergence and speciation
    in, 559–69
  reproductive barriers isolating, 556–59
Genera (genus), 575
General Sherman tree, 4
Genera Plantarum, 578
Generation time, **761**
Generative cell, **214, 391**
Genetic code, **241–44**
Genetic controls on plant growth and
    development, 276–77
Genetic distance, **556**
  for evaluating endangered species, 556,
    557(box)
Genetic drift, **545,** 546
Genetic engineering, **244–54**
  cell culture cloning, 188, 189
  development of food crops with, 16,
    251–54, 829
  development of plants resistant to
    drought, disease, and insects using, 15,
    16, 606, 607
  DNA cloning, 245–47

edible vaccines, 277(box)
finding gene of interest in, 247–48
with gene gun, 254
goals of, 253(table)
industrial uses of bacteria made
     by, 248–50
transgenic plants produced by, 250–54
treatment of plant disease with, 7
Genetic recombination, **216**, 220–26
by chromosomal assortment, 220–22, 546
crossing-over in, 222, 223. See also
     Crossing-over of genetic material
gene conversion, 223–24, 225
unequal crossing-over and gene
     duplication, 224–26
Genetics, 163, 231–56
cell cycle and. See Cell cycle
complex inheritance patterns, 175–79
evolution and. See Genetic (synthetic)
     theory of evolution
gene operation and function, 232–41. See
     also Gene(s)
genetic code, 241–44
genetic engineering and, 244–54
genetic recombination, 220–26
meiosis and, 215–18
Mendel and theory of inheritance, 166–73
non-Mendelian inheritance
     patterns, 179–82
of peas, round vs. wrinkled, 232, 254–56
root systems affected by, 349
search for hereditary material, 174–75
sexual reproduction and, 212–15, 226–27
synaptonemal complex, 218–20
Genetic species concept, **556**, 557(box)
Genetic (synthetic) theory of evolution, 541–47
genetic variation, maintenance of, 545–47
genetic variation, sources of, 544–45
Hardy-Weinberg equilibrium and, 543–44
major postulates of, 525(table)
Genetic variation
maintenance of, 545–47
promoters of, 560(table)
sources of, 544–45, 546
Genomes
information stored in, 205
replicons and, 200
size and composition of, 201–2(box)
of yeast, 620(box)
Genomic library, **246**, 247, 248
Genophores, 594
Genotype, **170**–71
Genus (genera), **575**
Geographic distribution of organisms, 533–35
Geographic isolation of organisms, 558
Germination, 404–5, 407
effect of light on lettuce seed, 460
malting barley and, 84(box)
requirements for, 404–6
Giant kelp (Macrocystis sp.), 646, 648
Giant sequoia (Sequoiadendron giganteum),
     4, 6, 7, 365, 496, 576, 728(box), 799

Gibberellins, **39**, **416**, 426–29
discovery of, 426
effects of, 427–29
fiber differentiation due to, 288
functions of, 275(table)
name of, 426
rib meristem affected by, 266
seed germination and, 84(box)
stem growth in response to, 308
structure and function of, 275(table)
synthesis and transport of, 426
vascular tissue differentiation and, 424
Gibbs, Josiah, 100
Gigaspora, 615
Gilbert, Walter, 244
Gills (basidiomycetes), **623**, 624
Ginkgo biloba, 66, 727, 850
fossilized, 522
pollen-bearing cones of, 720
seed-bearing structures of, 721, 722
sperm of, with flagella, 727
Ginkgophyta (division), 727, 850
Glacier National Park, Montana, 786
Gladiolus sp., 250
Gleocapsa, 602
Gliadin, **30**
Gloeotrichia echinulata, 595
Glucose, 22, 23
alpha-, and beta-, 24, 25
breakdown of, to pyruvic acid, 115, 116
harvesting energy from, overview
     of, 114–15
potential energy from, 115, 120
retrieval of, in aerobic respiration,
     112–14
seed germination and, 84(box)
structure of, 24(box), 840
Gluten, 822
Glyceraldehyde, 836
Glycine max, 350
Glycolysis, **112**, **115**, 120(table)
breakdown of glucose to pyruvic acid in,
     115, 116
efficiency of, 120
potential energy of glucose and, 115
ten steps of, 116
Glycoproteins, **29**, **76**, 87–88
determining human blood type with, 88
role of, in sexual reproduction, 88
Glycosides, **36**
Glyoxysomes, **63**
Glyphosate, **252**
Gnetophyta (division), 730–31, 850
Gnetum sp., 730–31
Golden-brown algae (Chrysophyta),
     635(table), 651–53, 847–48
Gold in plant tissue, **482**
Golgi body (dictyosomes), **61**
Gondwanaland, 747
Gonimoblasts (red algae), **650**
Gonorrhea (Neisseria gonorrhoeae), 605(box)
Gooch, Robin, 705

Gossypium hirsutum. See Cotton
     (Gossypium hirsutum)
Gracilicutes (division), **845**
Gradualism, **525**–26
Grafting, 265(box)
Grain (wood), **367**, 368
Gram, **842**, **843**
Gram, H. C., 595
Gram-negative bacteria, **595**, 596
Gram-positive bacteria, **595**, 596
Gram stain, **595**
Grana, **63**, 64, **141**
Grapes (Vitis sp.), 298, 403, 429
Grasses
auxin and cellular elongation in seedlings
     of, 423
C[4] photosynthesis in tropical, 154
fibrous root system of, 337
flowers of, 390
gibberellins and seed germination in, 427
ground tissue of tropical, 320, 321
pollen of, 391
scutellum of, 403, 404, 427
seedling development in, 406
stem growth in, 308, 309
stomata of, 293
Grasslands, 761, 790–92
role of fire in, 775–76
species diversity in, 765
Gravitational potential, 501 n.1
Gravitational water, **475**
Gravitropism, 348–49, **447**–50
Gravity
effect of, on formation of reaction
     wood, 369–70
effect of, on roots, 348–49
growth response to, 447–50
Greece, ancient
classification systems in, 574
philosophy of, and theories of
     evolution, 524
Green algae (Chlorophyta), 635(table),
     638–45
cellular organization of, 68
classification of, 645, 646, 847
colonial, 641
evolutionary relationships of plants and,
     639(box)
filamentous, 642–43, 644
flagella of, 67
lichens and, 620–22
multicellular forms of, 643–45
research on Photosystem I in, 146
unicellular, 638–41
Greenbrier (Smilax), 341
Greenhouse effect, **777**
Green Revolution, 827–28
Griffith, Frederick, 174
Grifola sulphurea, 613
Grimmia sp., 668
Ground meristem, **266**
Groundsel (Senecio vulgaris), 516

plant growth and development influenced
by, 275–76
plant pathology and, 430(box)
receptors for, 87
reproductive, in ferns, 705, 706
vascular cambium formation and, 360
Hornwort (Anthrocerotophyta), 665, 676–77
classification of, 849
gametophyte of, 676–77
sporophyte of, 676, 677
Horsepower, 95 n.2
Horsetails (Equisetophyta), 697–99, 849
gametophytes of, 699, 700
giant, in Carboniferous period, 711, 712
life cycle of, 699
spores of, 700
sporophyte of, 701
stems of, 698
vascular tissue of, 692, 698
House at Pooh Corner, The (Milne), 812
Howard, Alma, 190
Hubbell, Stephen, 764
Human disease. See Diseases of humans
Human impact on ecosystems, 776–80
acid rain as, 778
atmospheric carbon dioxide as, 777–78
atmospheric methane as, 777(box), 778
greenhouse effect as, 777
stratospheric ozone depletion as, 778–79
water supply contamination as, 779
Human population, excessive growth
of, 827–28
Humus (soil), 474–75
Hunter-gatherer cultures, 818
Hutton, James, 525
Hybrid, 168(box)
allozymes of, 177, 178, 565
apomictic, 564
breakdown of, 559
corn, 547, 548
genetic divergence and, 560–61
inviability of, 559
reproductive breakdown of, 559
sterility of, 182, 559
vigor of, 826
Hybridization, 564–65, 566
artificial plant, 168(box)
before Mendel, 166–67
Hydathodes, 299
Hydnora africana, 399
Hydrocarbons, 836
Hydrodictyon sp., 641
Hydrogen (H), 834, 838
ions of, 841(box)
phloem loading and, 517
stomatal opening and, 509
Hydrogen bond, 835–36
Hydroids (bryophyte gametophyte), 670
Hydrolysis, 839
Hydrophilic chemicals, 74
Hydrophobic chemicals, 74

Hydrophytes, 323–24, 349, **512, 761,** 762
Hydroponic farming, 477, 485
Hydroxyl groups, 836, 837(table)
Hygroscopic water, **475**
Hylocomium splendens, 665
Hymenium, **618**
Hyperaccumulators, 481
Hypericum sp., 386, 389
Hypertonic solution, **79**
Hypha (fungi), **612**
Hypotheses, **9**
Hypocotyl, **404**
Hypodermis, root, 341, 342
Hypotonic solution, **79**

IAA (indole-3-acetic acid), 420, **421,** 422–23.
See also Auxin
Ice-minus bacteria, **249**
Imbibition, **427**
Immunoassay, **420**
Impatiens glandulifera, 409
Imperfect flower, **389**
Imperfect Fungi (deuteromycetes), 618–20
Impression fossil, **710**
Inbreeding, **396**
outcrossing vs., 397(table)
Incipient plasmolysis, 501
Incomplete dominance, **175–76**
Incomplete flower, **389**
Independent assortment, law of, **170**
Indian paintbrush (Castilleja sp.), 399
Indian pipe (Monotropa uniflora), 491
Indian warrior (Pedicularis densiflora), 768
Indole-3-acetic acid (IAA), 420, **421,** 422–23.
See also Auxin
Indusium (fern), **701**
Industrial polymers, algae as source of, 658–59
Industrial uses of plants. See Economic
importance of plants
Inferior ovary, **389,** 390
Inflorescence, 394–95, 396
Influenza virus, 607
Infrared radiation (IR), **137**
Ingenhousz, Jan, experiments of, 134
Inheritance, 163, 165–84. See also Genetics
chromosomal theory of, 172–73, 524
complex patterns of, 175–79
cytoplasmic, 180
history of research discoveries in, 174–75
Lamarck's theory of acquired
characteristics, 527
laws of, **170**
Mendel's studies and theory of, 166–73, 524
non-Mendelian, 179–82
polygenic, 177
Inorganic compounds, **22**
Insects
ants. See Ants
aphids, 294, 517–18

butterflies, 294, 399, 768
control of. See Pest control
genetically-engineered resistance of plants
to, 15, 16, 253
methane production by termites, 778
mosquito vector for malaria, 816
plant digestibility adversely affecting, 328
plant hormones affecting life cycle of, 328
plant substances toxic to, 327–29
plant traps lethal to, 489–90, 769
pollination by, 394(box), 399,
400(table), 747
role of, in evolution of angiosperms,
747–48
seed dispersal by, 408
symbiotic relationships between plants
and, 329
Insect-trapping leaves, 325, 326. See also
Carnivorous plants
Insulin, 247, 249
Integument (ovule), **391,** 392
Intercalary meristems, **263,** 308
Intercellular spaces, **498**
Interfascicular cambium, 359
Intermediate-day plants, 455
Intermediate filaments, **55**
International Code of Botanical
Nomenclature, 577(box)
International Potato Center, 830
International Rice Research Institute, 830
Internodes, **308,** 309
Interphase (cell cycle), **188,** 189–200
chromatin formation, 195
chromosome structure during, 193–95
DNA semiconservative replication in,
196, 197
DNA structure and duplication in, theory
of, 191–92
enzymes of replication, 196–200
G_1 phase of, 190, 191
G_2 phase of, 200
replicons, 200
S phase of, 195–200
Interrupted genes, 242
discovery of, 237(box)
evolution of, 243
roles of, 242–43
Intine, **391**
Introgression, **565,** 566
Introns, **236,** 237(box)
gene structure with, and without, 242
RNA splicing by, 235(box)
roles of, 242–43
Inulin, **26**
Inversion loops on synapsed chromosomes,
218, 220
Ionic bond, **835**
Ionizing radiation, **137**
Ion pumps, 76, 84–85. See also Proton pumps
Ions, **835**
hydrogen, and pH, 841(box)

902   I–17   Index

fitness of, 543
interactions between plants, herbivores, and other, 767–68
multicellularity in, 498
plant associations with, 485–91, 766
plants and, 4–8
as soil components, 476, 483
Light, 136–37
absorption of, by pigments, 141, 142, 143
algal reproduction influenced by, 658
chloroplast harvesting system of, 141, 142
competition for, 767
as control factor in photosynthesis, 158–59
effect of, on roots, 348
leaf variation as response to, 321–22
photochemical reactions requiring, 144–48
physiology of bryophytes influenced by, 680
physiology of seedless vascular plants influenced by, 704
plane-polarized, 840
plant response to. See Phototropism
properties of, 136
release of fungal sporangia influenced by, 625
seed germination and, 404, 460
stomatal opening affected by, 509
transpiration and effect of intensity of, 506
wavelengths of. See Wavelengths of light
Light-compensation point, **159**
Light-independent reactions of photosynthesis, **144,** *145,* 148–51
Light microscopy, **46,** 48, 49
different types of, *50*
Light reactions of photosynthesis, **144–48**
Lignin, *40,* **41,** 688
Ligustrum, 319, *320*
Lilac (*Syringa vulgaris*), 319, *499*
Liliaceae family, 811
Liliopsida, **851**
Lily (*Lilium* sp.), *14, 217, 442,* 746, 811
embryo-sac development in, *392, 393*
flower of, *385*
leaf venation, *320*
morphological species concept applied to, *555*
spores and gametes of, *212, 213, 214, 215*
water, 287, 322
Lily family, 811
Limnanthes douglasii, 793
Lindley, John, 576
Lindow man, 680(box)
Lindstrom, D. L., 777
Linkage, genetic, **179,** *180*
Linnaea borealis, 578
Linnaeus, Carolus
binomial system of nomenclature of, 575–78
Species Plantarum of, 572, 575
twinflower named in honor of, 578
Linum sp., 5

Lipids, 33–36
oils, 33–35
phospholipids, 35
respiration of, 126, 127
structure of, 34
waxes and waxlike substances, 35–36
Liriodendron sp., hybridization of, 555, 556
Liter, **842,** 843
Liverworts (Hepatophyta), 665, 672–75, 848
diversity of, 673
gametophyte of, 673, 674–75
Marchantia, 666, 674, 675, 676–77
sporophyte of, 675
Living organisms. See Life and living organisms
Lizard's tail (*Saururus* sp.), 745
Local populations, **760**
Locus of gene, **175**
Lolium sp., 398
Long-day plants, 455
Long-night plants, **457**
Lonicera sp., 387, 399
Looped domain, chromatin, **193,** *194, 195*
Lophophora williamsii, 810, 811
Lore of plants, 6, 23, 31, 54, 59, 88, 98, 114, 172, 200, 205, 254, 278, 279, 294, 349, 378, 410, 424, 428, 438, 454, 511, 534, 560, 578, 603, 613, 622, 627, 658, 733, 754, 769, 802, 820
Lotus corniculatus, 327
Lower leaf zone, **313**
Lumber, 376
Lumpers, classification of, 582
Lupine (*Lupinus* sp.), *395, 406*
Lussenhop, John, *347*
Lycopene, *39, 139*
Lycoperdon sp., *613*
Lycopersicon esculentum. See Tomato (*Lycopersicon esculentum*)
Lycopodium sp., *691, 694, 695,* 705
gametophytes of, *695, 696*
life cycle of, *696*
medicinal qualities of, 709
religious use of, by Druids, 696
sporophyll of, *695*
Lycopodophyta (division), 694–97, 849
Lyell, Charles, 525–26, *528*
Lyginopteris sp., *735*
Lyme disease (*Borrelia burgdorferi*), 605(box)
Lysergic acid (LSD), 814
Lythrum salicaria, 775(box)

MacAyeal, D. R., 777
McClinton, Barbara, 10, *11,* 180–83, 222
Macrocystis sp., 646, *648*
Macroevolution, 547–**49**
Macronutrients, **477,** 480–81(table)
Magnifications, microscopic, **46,** 48
Magnolia (*Magnolia* sp.), *390,* 580, 746, 754

Magnoliophyta (division), 742, 850–51. *See also* Angiosperms; Flowering plants
Magnoliopsida (class), **851**
Maidenhair fern (*Adiantum trapeziforme*), 702
Maidenhair tree. *See Ginkgo biloba*
Malaria, 816–17
Malic acid, **118**
Mallee (*Eucalyptus pressiana*), 401
Malthus, Thomas, influence of, on Charles Darwin, 529–30
Malting, 84(box), 427
Malus. *See* Apple (*Malus* sp.)
Mammoth, Maryland, 454
Mangrove, black (*Avicennia germinans*), 350
Manihot esculenta, 811, 819, 820
Manila hemp (*Musa textilis*), 289
Manna, 518
Maple (*Acer* sp.)
hardwood of, 366
sugar (Acer saccharum), 789
sun leaves and shade leaves of, 321
tapping for maple syrup, 508(box)
Maranta, 453
Marattiopsida (class), 849
Marchantia sp., 674, 675, 676–77
antheridiophores of, *675, 676*
archegoniophores of, *674, 676*
gametangia of, *666*
gametophyte of, *674, 675*
life cycle of, *675*
sporophyte of, *675*
Marginal meristems, **313**
Marginal placentation, **386**
Margulis, Lynn, classification system of, 586, 587, 845–51
Marijuana (*Cannabis sativa*), 295(box), 814
Marker, Russell, 814
Marsilea sp., 704
Mass, units of, 843
Masterson, Jane, 565
Mathematics of leaf arrangement, 316(box)
Mating patterns, genetic variation due to, 546
Matric potential, **501**
Matter, atoms and molecules of, 21–43
Mature region of roots, 341–45
Mauritius, dodo bird and tambalacoque tree on, 405(box)
Maxwell, James, 136
Mayapple (*Podophyllum*), 345
Meadowfoam (*Limnanthes douglasii*), 793
Meadow saffron (*Colchicum autumnale*), 55, 811
Mechanical isolation of species, 558, *559*
Medicinal plants, 5, 37(table), 330, 377, 806, 812–17. *See also* Drugs
animal use of, 818(box)
bacteria, 604
case study of malaria and, 816–17
drugs and medicines derived from, 812–17, 818(box)
fungi, 625–27, 726(box)

Nitrate, 485
Nitrate reductase system, **485**
Nitrification, **485**
Nitrogen
  functional groups containing, 837
  in soil fertilizers, 473(box)
Nitrogenase, **486**, *487*
Nitrogen cycle, *769, 770, 771*
Nitrogen-fixing bacteria, 87, 88, 350, 352,
    486–88, 769
  Anabaena *cyanobacteria, 487, 488, 602,
    708, 709*
  free-living, 486
  on roots, 87, 88, 350, 352, 598
  symbiotic, 487, 488
Nodes, stem, **308**, *309*
Nodules of nitrogen-fixing bacteria, *350,
    352, 487, 598*
  gene for controlling, 488
Nomenclature, binomial system of,
    575–78
Nonclimacteric fruit, **433**
Noncompetitive inhibitors, **107**
Noncyclic electron flow, **146**, *147*
Non-Mendelian inheritance, 179–82
Nonprotein amino acids, **41**
North American biomes
  chaparral (Mediterranean scrub), 795
  grassland and savanna, 791
  mountain forest, 798
  taiga, 787
  temperate deciduous forests, 788
  tropical rain forests, 800
  tundra, 786
Northern blotting, **255**, *256*
Northern coniferous forests, 787–88
Nuclear envelope, **58**
Nucleic acids, 31–32. *See also*
    Deoxyribunucleic acid (DNA);
    Ribonucleic acid (RNA)
Nucleomorph, **654**, 848
Nucleosomes, **193**, *194*
Nucleotide, **31**
  *DNA mutations and gene sequences
    of, 198(box)*
Nucleus (cell), **58**
  daughter, 172, 204, 205
  genetic role of, 174–75
  mitosis and division of, 204–6
  polar, 212, 392
  RNA in, 235–36
Nutrient cycles, 107–8, 769–71
  carbon cycle, 770, 771
  nitrogen cycle, 769, 770, 771
  water cycle, 770–71, 772
Nutrients
  cycling of, 107, 108, 769–71
  essential. See Essential elements
  returning, to soil, 473(box)
  *root absorption of, 338, 339, 346, 347,
    351, 483–84*

soil particles and availability of, 472
*transport of. See Solutes transport;
    Water transport*
trichomes and absorption of, 293–94
Nutrition, 467, 469–95
  essential elements for, 477–82
  obtaining essential elements for, 482–93
  soils and, 470–76
Nyctinasty, 453–54
Nymphaea sp., *287, 322*

O

Oak (*Quercus* sp.), *7, 295, 301, 414, 789*
  cork, 55, 376(box)
  galls on leaves of, 191
  male flowers of, 580
  microhabit isolation of, 558
  vessel and tracheids of, 499
  wood of, 366
Oats, **826**
Obligate photoperiodism, 457
Occam's razor, 583–84
Ocean, rising levels of, 777–78
Ocotillo (*Fouquieria splendens*), *507*
Oedogonium sp., *642*
Oenothera sp., *180, 216, 387, 400*
O horizon (soil), **470**
Oils, 33–35
  tropical, 34(box)
Oleander (*Nerium oleander*), *300, 322, 387*
Oligosaccharins, 437–38
One-molar solution, 841(box)
Onion (*Allium cepa*), *8, 190, 200, 203, 312*
On the Origin of the Species (Darwin), 15,
    417, 524
Oogamous algae, **637**
Oogonium (green algae), *642*
Oomycota (division), 629(table), **846**
Operational habitat, 765
Operculum (moss bryophyte), **669**
Ophioglossopsida (class), 849
Opium, 4, 37(table), 812(box)
Opium Wars, 4
Opposite phyllotaxis, **314**
Optical isomers, 840–41
Optically-active substances, **840**
Opuntia bigelovii, 398
Orange, *403*
  seedless navel, 8, 265(box)
Orbital, electron, **834**
Orchidaceous mycorrhize, 351
Orchids (*Ophrys* sp.), *801*
  epiphytic, 291, 352
  flower of wasp-like, 394
  hybridization of, 566, 567
  *mechanical reproductive isolation of,
    558, 559*
  pollination of, 394(box)
  roots of, 351
  sterility of, 559

Organelles, **46**, 58–64
  energy conversion by, 63–64
Organic acids, **837**
Organic chemistry, **836**
Organic compounds, **22**
Organic fertilizers, 473(box)
Organic soils, 474
Organic solutes, transport of, 297–98, 512–18
Organismal theory, 68–69
Organisms. *See* Life and living organisms
Organized parts, plant, 14
Organogenesis, **431–32**
Orphrys. See Orchids (*Ophrys* sp.)
Orthocarpus pusillus, 399
Orthostichies, **314**
Oryza sativa. See Rice (*Oryza sativa*)
Oscillatoria, 602
Osmosis, **79**, *80*
  inducing, 80–81
  regulation of, in algae, 658
Osmotically active ions, **80**
Osmotic potential, **79**, *80*
Osmotic pressure, **79**, *80*
Outbreeding, **546**, *547*
Outcrossing, **396**
  inbreeding vs., 397(table)
Ovary, **214**, **385**, *386, 387. See also* Fruits
  superior vs. inferior, 389, 390
Overbeek, Johannes van, 429
Overgrazing as cause of desertification, 797(box)
Overpopulation, human, 826–27
Overtopping (leaves), **691**, *692*
Ovules, *212*, **214**, *385, 386*
  development of embryo sac in, 391–93
  gymnosperm, 723, 727
  meiosis and, 173
  seed evolution and, 719
  of seed fern Caytonia, 746
Oxaloacetic acid (OAA), **118**
Oxidases, **63**
Oxidation, **101–2**, *839*
Oxidative phosphorylation, **115**, 119–25, *143*
Oxisols, **471**
Oxygen
  photorespiration caused by increased, 153
  production of, in photosynthesis, 132
  *production of atmospheric, by marine
    algae, 657*
Oxyphotobacteria class, **845**
Oyster mushrooms (*Pleurotus ostreatus*), *613*
Ozone, depletion of stratospheric, 778–79

P

P680 (pigment 680), **142**, *146, 147*
P700 (pigment 700), **142**, *147, 148*
Paabo, Svante, 536
Paál, Arpad, phototropism studies by, *417, 419*
Pacific yew (*Taxus brevifolia*), *37, 372, 377,
    726(box)*
Paleontological species concept, 556

Paleospecies, 556
Palisade mesophyll cells, **319**
Palmately compound leaves, **317**
Palynologists, **709**
Pangenesis, theory of, 524, 542(box)
Panicle, **394**
Panicum, 157
Papain, 30
Papaveraceae family, 811
Papaver radicatum, 802
Papaver somniferum, 4, 37, 812(box)
Papaya (*Carica papaya*), *384*
Paper
   money, 5
   newsprint, 8, 732(box)
   production and consumption of, in
      U.S., 732(box)
   production of, from papyrus, 377 n.2, 809
   production of, pollution and, 17
   wood as source of pulp for, 377
Paper birch (*Betula papyrifera*), 373
Papillae (algae), **642**
Papyrus (*Cyperus papyrus*), 377 n.2, 809
Parallel venation, 318, *320*
Parapatric speciation, **561**
Paraphyses (moss bryophyte), **669**
Parasites, **490**, 767–68
   malaria as, 816
   transgenic plants and bacterial, 250
Parasitic plants, 278, 351(box), 490–92,
      767–68
Parasitism
   populations limited by, 767–68
   roots and, 351
Parastichies, **315**
Parenchyma, **262, 284**–85, 301(table),
      302(table)
   storage cells, in roots, 341
Parietal placentation, **386**
Parsimony, principle of, **584**
Parsley (Apiaceae), *768*
Parthenocarpic fruits, **424**
Parthenocissus tricuspidata, 324, 767
Particle board, 376
Pascals (Pa), **500**
Passage cells, 343
Passiflora sp., 294, 299
Passion flowers (*Passiflora incarnata*), 299
Passive transport, **79**
Passive traps, **489**
Pasteur, Louis, 10
Pauling, Linus, 191
PCR (polymerase chain reaction), **198 n.2,**
      248, 249
Pea (*Pisum sativum*), *164*
   amylose starch of, 26
   artificial hybridization of, 168(box)
   characteristics of, 169(table)
   epidermal cells of, 290
   germination of, 407
   Mendel's studies of, 166–72, 232

round vs. wrinkled, 232, 248, 254–56
   tendrils of, 325
Pear (*Pyrus communis*), *190,* 288
Peat bogs, 672, **679,** 682–83
   ancient humans preserved in, 680(box)
Peat moss (*Sphagnum* sp.), 8, 672(box), 679,
      680(box), 682–83, 848
Pectins, **25**
Pedicels, **394**
Pedicularis densiflora, 768
Peduncle, **394**
Pelc, S. R., 190
Pellicle, **653**–54, 655
Peltate leaves, **317,** *318*
Penicillin, 11, 619, 621
Penicillium sp., *619,* 621
Pennate diatom, **651,** *652*
Pentose phosphate pathway, **126**
Pentose sugars, respiration of, 126
Peperomia, 291
Pepper (*Capsicum* sp.), *810,* 811
   black, 5
   tabasco, 279
Peppermint (*Mentha piperita*), 575, 576, 811
Peptide bonds, **27**
Peptide synthesis, 27
Peptidoglycans, **594**
Peptidyl transferase, **240**
Perfect flower, **389**
Perfoliate leaves, **318**
Pericarp, **402**
Pericycle, 345
Periderm, 295, 301(table), *359,* **372**–74
Peripheral cells (root), 338
Peripheral meristem, 267
Periplast, **848**
Peristome (moss bryophyte), 670, **671,** *672*
Perithecium (ascomycetes), **618**
Permafrost, **785**–86
Permanent wilting point, **475,** *476*
Permian period, mass extinctions of, 627
Peroxisomes, **63**
Persimmon (*Diospyros virginiana*), 284
Pest control
   bacteria useful for, 604
   plant-produced substances for, 767
   transgenic plants and, 253, 254
   viruses useful for, 606, 607
Pesticides, 709, 767
Pests, plants as, 707, 708, 775(box)
Petals, **384,** *385,* 387–88
   origins of, 389
Petiole, 317
Petrifactions, **710**
Petunia, distribution of *rbc*S genes in,
      226(box)
Peyote cactus (*Lophophora williamsii*),
      810, 811
pH, 841(**box**)
   proton pumps and regulation of, 84–85
   of soil, and nutrient availability, 482–83

Phaeophyta (division), 635(table),
      645–48, 847
Phagotrophic organisms, **627**
Phallus impudicus, 613
Phellem (cork), 373–74, 376(box)
Phelloderm, 374
Phellogen (cork cambium), **372**–73
Phenolics, *37(table),* **40**–41
   structure of, 40
Phenotype, **170**–71
Phenylacetic acid, 421
Phenylpropanoids, **40**
Philodendron (*Philodendron*), 94
Phloem, **262,** 297–98, 301(table), 302(table)
   conducting cells of, 297–98, 513–18. See
      also Sieve tube members
   contents exchange between xylem and, 518
   contents of, 517–18
   influence of environment on transport
      in, 517
   in leaves, 318–19
   loading and unloading, 516–17
   models for transport in, 514–17
   primary, 297
   in roots, 345
   secondary, 297, 360, 372, 376–78
   in stems, 310
   transporting organic solutes in, 297–98,
      512–18
Phoenix dactylifera, 166, 167, 284
Phoradendron sp., 278, 491
Phosphate group, **837**
Phospholipids, **35,** 74
   artificial membrane of, 74–76
   structure of, 35
Phosphorous, functional groups
      containing, 837
Phosphorylation, **103,** 839–40
   by chemiosmosis, 115
   oxidative, 115, 119–25
   substrate-level, 114
Photochemical reactions of photosynthesis,
      144–48, *151*
   cyclic electron flow, 145
   noncyclic electron flow, 145–46, 147
   solar tracking (heliotropism), 446(box)
   summary of, 146–48
Photons, **136**–37
   collection of, in photosynthesis, 142
   quantum energy of, 137
Photo-oxidation, carotenoid protection
      against, 139–40
Photoperiod, **454**
Photoperiodism, **454,** 455
   biological clock as timing for, 463
   facultative, 457
   flowering and, 454–55
   leaf variation as response to, 321
   manipulation of, 461
   phytochrome and, 458–61
   types of flowering responses, 455–57

Photophosphorylation, **143**, *144*
cyclic, *145*
Photorespiration, **127**, 151, **153**
Photosensitizers, 328
Photosynthates, fates of, 159–60
Photosynthesis, **96**, 98, 99, 131–62
in bacteria, 595, 601–3
biochemical reactions of, 144, 148–51
C_4 *form of, 154–57*
chloroplasts and, 141–44
control of, 158–59
crassulacean acid metabolism
(CAM), 157–58
efficiency of, 151
as energy transformation, 107–8
evolution of, 152(box)
fate of products produced by, 159–60
genes associated with, 198(box), 226(box)
history of discoveries about, 132–36
nature of light and, 136–37
photochemical reactions of, 144,
145–48, 151
photorespiration and, 151–53
pigments and, 137–41
roots capable of, 352
water-use efficiency, transpiration, and, 506
Photosynthesis-transpiration compromise,
510, *511*
Photosynthetic autotrophs, 132, 152(box)
Photosystem I, 142, *144*, 148
cyclic flow of electrons and, 145
noncyclic flow of electrons and, 147
research on genes for polypeptides of, 146
Photosystem II, 142, *144*, 148
in cyanobacteria, 595, 601
noncyclic electron flow and, 146, 147
Phototropism, 358, 444–45
auxin and, 417–21, 444–45
Cholodny-Went hypothesis on, 444
Darwin's studies of, 417, 418–19, 420
in fungi, 623, 625
heliotropism and, 446(box)
C. Schmid's research on, 449
varying light wavelengths and, 445, 447
Phragmoplast, *204*, **206**
Phycobilins, **601**
in red algae, 648–50
Phycocyanin, *140*, **601**
Phycoerythrin, **601**, 649
Phycoplast, **206**
Phyllotaxis, **314–15**
control of, 315
inhibitor field hypothesis on, 315
mathematical relationships in, 315,
316(box)
patterns of, 314, 315
Phylogenetic classifications, **579–82**. *See also*
Cladograms
of angiosperms, 751–54
of bacteria, 599–600
of gymnosperms, 737

RNA-based molecular, 588
Phylum, **585 n.6**
Physarum polycephalum, 629
Physcia sp., *621*
Physiogomy of vegetation, **784**
Phytoalexins, 328
Phytochelatins, 482
Phytochrome, **458–61**
Phytophthora infestans, 830
Phytoremediation, 482
Picea sp., *731*, 813(box)
Pigments, 62, **137–38**. *See also names of*
specific pigments
accessory, 139–41
in algae, 638, 648–50, 653, 656
anthocyanin, 582–83, 584
betalain, 533, 582–83, 584
chlorophylls, 137, 138. See also
Chlorophylls
cladograms of plants with select,
582–83, 584
complexes of, in chloroplasts, 141–42
in cyanobacteria, 601
evolution of photosynthesis and,
152(box)
flavonoids, 41, 399, 713
leghemoglobin, 350
light absorption by, 141, 142, 143
making and destroying, 140–41
in plants, 137–38
phycobilins, 648
phytochrome, 458–61
rhodopsin, 638
Pigweed (*Amaranthus*), *156*
Pili (bacteria), *594*
Pilobolus, 625
Pine (*Pinus* sp.), *789*
bristlecone, 362
cones of, 716, 720, 723
knobcone, 799, 800
life cycle of, 724–25
mycorrhiza root of, 485
ovule of, 727
pollen grain of, 719
pollen of, 716
resin ducts in, 300, 364, 366
as state trees, 733
tracheid cells of, 296
wood of, 296, 366, 368
Pineapple (*Ananas comosus*), *31*, 158, *750*
Pinguicula sp., *293, 294*, 489–90
Pinnately compound leaves, **317**
Pinophyta (division), 728–30, 850
extinct, 734–36
Pinus longaeva, 4 n.1, 362, 728
Pinus merkusii, 728
Pinus sp. *See* Pine (*Pinus* sp.)
Pioneer species, 766
Piper nigrum, 5
Pistil, **214**, **384**, 385–87
Pisum sativum. See Pea (*Pisum sativum*)

Pitcher plants (*Sarracenia purpurea*), *325,*
489, *490*
Pith
in roots, 345
in seedless vascular plants, 691, 692
in stems, **311**
Pith-rib meristem, 267
Placentae, *385, 386*
Planation (leaves), **691**
Plane-polarized light, **840**
Plankton, role of algae as, 657
Plant(s). *See also* Vascular plants
allelopathic, 707(box)
artificial selection for larger food, 537(box)
basic structure of, 263
botanical themes applying to, 14–15
carnivorous. See Carnivorous plants
cells of. See Cell(s)
characteristics of, 586
characteristics of, and botany's unifying
themes, 14–15
classification of. See Classification
cultivation of, for food. See Agriculture;
Food
defenses of. See Defenses, plant
diseases of. See Diseases of plants
domestication of, 818–20
drought-resistant, 15, 681(box)
economic importance of. See Economic
importance of plants
energy transformations by. See Cellular
respiration; Energy; Photosynthesis
evolutionary relationships of green algae
and, 639(box)
extinct and endangered, 17, 736
form and function of, 259, 262
fossilization of, 710(box)
genetics. See Genetics
growth and development. See
Development; Growth
hallucinogenic, 815(table)
human fascination with, 827(box)
human society and, 807–12
life cycle of, 212, 213, 262–63
medicinal, 812–17. See also Drugs;
Medicinal plants
most common elements in, 23(table)
names of, 574, 575, 577(box)
nutrition of. See Nutrition
reproduction. See Reproduction; Sexual
reproduction
significance of, to all life, 4–8
study of. See Botany
tissues of. See Dermal tissues; Ground
tissue; Meristems; Vascular tissue
Plantae (kingdom), **586**, *587*, **845**, 848–51.
See also Plant(s)
Plantago media, 396
Plantains, 821
Plant hybridization. *See* Hybrid;
Hybridization

Plasma membrane, **46**, 59. *See also*
    Membranes
Plasmids
    cloning in, 245–46
    Ti, 250
Plasmodesmata, 57
    formation of, 206
Plasmodium sp., 816
Plasmolysis, **80**, *81*
Plastics, biodegradable, 824
Plastid, **64**
Plastochron, **315**
Platanus occidentalis (sycamore), *373, 560*
Plato, *524*
Pleiotropic genes, **177–78**, *179*
Pleurisy root, *2*
Pleurococcus sp., *635*
Pleurotus ostreatus, 613
Pliny the Elder, *815*
Plotkin, Mark, *816*
Plurilocular gametangia, **645**
Plywood, *376*
Pneumatophores, *350*
Pneumonia mycoplasma, 595
Poaceae grasses, *156, 157*
Poa sp., *564*
Podophyllum, 345
Poinar, G. O., Jr., *627*
Poinsett, Joel, *811*
Poinsettia, *811*
Poison(s), *810, 811. See also* Toxins
    aconitine (monkshood), 810, 811
    alkaloid, 36, 37(table), 38, 709
    in fungi, 613
    in leaves, 327–28
    in seed proteins, 30
Poison ivy (*Toxicodendron radicans*), *811*
Poison oak (*Toxicodendron diversilobum,*
    Toxicodendron toxicarium), *811*
Polarity in development, **272–73**
Polar molecules, **836**
Polar nuclei, **212**, *392*
Politics and botany in Nazi German, *118(box)*
Pollen, *212*
    development of, 173, 212–13, 390–91
    evolutionary origins of, 718
    gametophytes of, 213
    gene flow and dispersal of, 562, 563
    of grasses, 391
    gymnosperm, 718–20
    interactions of glycoproteins in, 88
    meiosis and, 173
    sterile, 182(box)
Pollen banks, *830*
Pollen grain, **212**, *391. See also* Pollen
    gymnosperm, 718–20, 726
Pollen tube, *88*, **214**, *393, 394*
    haustorial, 727
Pollination, **214–15**
    in angiosperms, 393, 394, 398–401
    in gymnosperms, 726, 727

    mechanical barriers to, 558–59
    mechanisms of, 398–401
    role of insects in angiosperm, 394
    role of nectaries in, 299, 399, 400
Pollination droplet, **726**, *727*
Pollinators, *766*
    deception of, by flowers, 394(box)
Pollution
    sensitivity of bryophytes to, 678
    sensitivity of fungi to, 616(box), 621
    water, 779
Polyacetylenes, **41**
Polyarch roots, *345*
Poly-A tail, RNA, **233**, *234, 236, 237*
Polyembryony, simple and cleavage forms
    of, *721*
Polygenes, **177**, *178*
Polygenic inheritance, *177, 178*
Polygonum-type embryo sac, *391, 392*
Polymerase chain reaction (PCR), **198 n.2,**
    248, *249*
Polymers, biological, **22–36**
    genetically-engineered, 254
Polypeptides, **27**, *241. See also* Proteins
Polyploid plants, **218**, *219(box)*, 565–69, *824*
Polyploids, **565**
Polyploidy, **218**, *219(box)*, 565–69
    aberrant chromosomal separation in,
        566, 567
    large genomes due to, 201
    new types of vegetables formed by, 568(box)
    restoration of fertility by, 568
Polypodium sp., life cycle of, *703*
Polysaccharides, **23**
    storage, 26
    structural, 24–26
Polysiphonia sp., *650, 651*
Polysomes, *58*
Polytrichum commune, 662, 670, 671
Pond, plant succession in, *774*
Poppy (*Papaver* sp.), *4, 802, 811*
    opium, 37, 812(box)
Poppy family, *811*
Population(s), **539, 760**
    biomass of, 760
    biotic potential of, 760–61
    communities of, 761
    density of, 760
    in ecosystems, 760–61
    gene flow in, 545
    genetic drift in, 545, 546
    human, 827–28
    local, 760
    natural selection and, 539–40
    outcrossing and inbreeding in, 396–97
Population density, **760**
Population divergence, *559–69*
    allopatric speciation and, 560–61
    parapatric speciation and, 561
    sympatric speciation and, 561–69
Population genetics, **541**

    genetic variation, 544–47
    Hardy-Weinberg equilibrium and, 543–44
Populus sp., *306, 775*
Populus tremuloides, 350, 563, 564, 775
Porphyridium sp., *649, 650*
Positional controls on plant growth and
    development, *276, 277*
Post, C. W., *12*
Postzygotic barriers, **557**, *559*
Potassium ions (K+)
    loss of turgor and uptake of, 80–81
    stomata, guard cells, and role of, 292,
        293, 509–10
Potato (*Solanun tuberosum*)
    amylopectine in, 26
    Irish famine associated with, 826, 830
    shoot apex of, 267
    tubers of, 312, 313
Potential energy, **78, 96,** *838–39*
    of glucose, 115
P-protein, *298, 513*
Prantl, Karl, *580*
Prayer plant (*Maranta*), *453*
Precipitation
    as abiotic factor, 761, 763
    acid, 778
    biomes and patterns of, 776
    nitrogen cycle and, 769
    water cycle and, 771, 772
Preprophase band, **202**, *203*
Pressure-flow model of phloem transport,
    514–17
Pressure potential, **501**
Prezygotic mechanisms, **557**, *558–59*
Price, Robert J., *594*
Priestley, Joseph, experiments of,
    132–33, 134
Primary body of plant, **284.** *See also* Dermal
    tissues; Ground tissue; Meristems;
    Vascular tissue
Primary cell wall, **55**, *56*
    connections between, 57–58
    interconnections among major
        components of, 30
Primary consumers in ecosystems, **762**, *763*
Primary growth, *263, 264*, **283–304**
    dermal tissue, 289–95
    ground tissue, 284–89
    leaves, 313–30
    roots, 335–54
    secretory structures, 299–300
    stems, 308–13
    terrestrial adaptation and, 300–301
    vascular tissue (xylem, phloem), 295–98
Primary phloem, **297**
Primary pit-fields, *57*
Primary RNA transcript, **233**, *235(box)*
Primary (radicle) root, *336, 337*, **404**
    germination and appearance of, 406, 407
Primary structure of proteins, **27**, *29*
Primary succession, *772–76*

Schlesinger, Bill, 793
Schmid, Cora, research on phototropism by, 449
Schwann, Theodor, 12, 68
Schwartz, Karlene, classification system of, 586, 587, **845–51**
Scientific creationism, 524 n.3
Scientific method, **8–13**
   nonscientists' use of, 13
   use of, 10–12
Sclereids, 287, **288**, 302(table)
Sclerenchyma cells, **262, 286–89**, 301(table)
Scopelophila cataractae, 678
Scouring rushes. *See* Horsetails (Equisetophyta)
Scutellum, **403, 404, 427**
Sea lettuce (*Ulva* sp.), 643, *644, 645*
Sea levels, greenhouse effect and rising, 777–78
Searcher shoots, **312**
Seasons, plant response to changing, 454–62
Secondary body of plant, **358**
Secondary cell wall, **55,** *56*
Secondary consumers in ecosystems, **762,** *763*
Secondary growth, **263,** 357–81
   defined, 358
   from lateral meristems, 263, 264, 358
   secondary phloem and periderm (bark), 372–74
   secondary xylem (wood), 363–70
   unusual, 374–75
   uses of secondary xylem and phloem, 376–78
   vascular cambium, 358–63
Secondary metabolism, **36**
Secondary metabolites, **36–41**
   alkaloids, 36, 38
   examples of well-known, 37(table)
   functions of, 36
   latex as, 300
   minor classes of, 41
   phenolics, 40–41
   terpenoids, 38–40
Secondary phloem, **297,** 372
   differentiation of, 358, 360
Secondary plant body, **358**
Secondary structure of proteins, **27,** 29
Secondary succession, **774–75**
Secondary xylem, **295,** 296, 363–70
   characteristics of wood, 367–69
   differentiation of, from vascular cambium, 358, 360
   kinds of wood, 364–67
   reaction wood, 369–70
   uses of, 376–78
Second law of thermodynamics, **97–98,** 108
Secretory cells, **299**
Secretory structures, 282, **299–300**
   external, 299
   internal, 299–300
Sedum sp., *388*

Seed(s), 403–7
   dispersal of, 407–10
   dormancy of, 404–5, 437, 462
   evolution of, 717, 718
   fruits and, 402–3
   germination of, 84(box), 404–6, 407, 460
   gymnosperm, 720–21
   scarification of, 405
   shape of (round vs. wrinkled), 232, 254–56
   storage proteins in, 30
   structure of, 403–4
Seed bank, **405–6**
Seed-bearing structures, gymnosperm, 721, 722
Seed coat, 403
Seed ferns, 388, 734, *735*
Seedless vascular plants, 687–715
   allelopathic, 707(box)
   club mosses (Lycopodophyta), 694–97
   ecology of, 705–8
   economic importance of, 708–9
   features of, 688–89
   ferns (Pteridophyta), 699–704
   generalized life cycle of, 689–90
   horsetails (Equisetophyta), 697–99
   organization of reproductive structures in, 690, 691
   organization of vascular tissue in, 691, 692
   origins and history of, 709–13
   physiology of, 704–5
   student research on, 705(box)
   whisk ferns (Psilotophyta), 693–94
Seedling
   development of, 406, 407, 423
   etiolated growth in, 459–61
Segregation, law of, **170**
Seismonasty, 451–53
Selaginella sp., 686, 694, *695, 696,* 697, 718
Selection pressures, **539**
Selenium, 478, *479,* 482
Self-compatible flowers, **395–96**
Self-incompatible flowers, **397**
Selfish DNA, **227**
Self-replication of DNA, **195**
Semiconservative replication of DNA, **195,** *196, 197*
   enzymes of, 196–200
   evidence for, 196, 197
Semideserts, 792–95
Seminal roots, 337
Senebier, Jean, experiments of, 134
Senecio vulgaris, 516
Senescence, **424,** 461
   effect of cytokinins on, 432
Sense, RNA, **251**
Sensitive plant (*Mimosa pudica*), 451, *452*
Sepals, **384,** *385,* 388
Septa (fungi), **612**
Sequoiadendron giganteum, 4, 6, 7, 496, 576, 728(box), 799

Sequoia sempervirens, 4, 6, 7, 728(box), 729, 731, 798, 799
Serotype (virus), **606**
Sessile leaves, 317
Seta (moss bryophyte), **671,** 672
Settlement-overgrazing hypothesis of desertification, 797(box)
Sexual reproduction, 211–29
   in algae, 634–37, 640, 641, 642, 643, 645, 647, 648, 649, 650, 651, 653, 654
   in bryophytes, 666–67, 670–71, 675, 677
   effect of hormones on sexual expression in, 435
   in flowering plants, 212–15 743. See also Flowering plants
   in fungi, 612, 615, 618, 619, 622–23, 624
   genetic recombination and, 220–26
   genetic variation promoted by, 546
   in gymnosperms, 718–26
   meiosis and, 172–73, 211, 212, 215–18
   morphology of. See Sexual-reproductive morphology
   in plants vs. animals, 211
   pollination and fertilization in, 214–15
   reasons for, 226–27
   role of proteins in, 88
   in seedless vascular plants, 688, 689, 690, 694, 696, 697, 699, 703
   spores and gametes in, 212
   synaptonemal complex and, 218–20
Sexual-reproductive morphology, 383–412
   of bryophytes, 664, 666
   dispersal of fruits and seeds, 407–10
   flowers, 384–89
   fruits, 402–3
   of gymnosperms, 718–26
   plant diversity and, 389–401
   of seedless vascular plants, 690
   seeds, 403–7
Shade leaves, **321,** 322
Shakespeare, William, 6
Shelf (bracket) fungus (*Grifola sulphurea*), 613
Shoots
   apical meristems of, 264, 267–68
   growth of roots vs., 349–50
   transition between root and, 346
   vascular cambium in, 360, 361
Short and long shoots, **312**
Short-day plants, 455
Short-night plants, **457**
Shull, George, 822
Sieve areas, 297
Sieve cells, 297–98, 302(table)
Sieve elements, **297**
Sieve plates, 513
Sieve pores, 297, 513
Sieve tube members, **298,** 302(table)
   contents of, 517–18
   discovery of, 512

models for transport through, 514–16
phloem loading and unloading
   and, 516–17
structure of, 513
Signal transduction. *See* Cellular
   communication
Silent Spring (Carson), 16
Silicon as essential element, **478**
Silicon metabolism in algae (diatoms), 658
Simmondsia chinensis, 36, 822(box)
Simple leaves, **317**
Simple polyembryony, **721**
Simple strobili (gymnosperms), **721**
Simple umbel, **395**
Sinapis alba, 345
Singer, Rolf, 627
Single-strand binding protein, **196**
Siphonostele (seedless vascular plants),
   691, 698
Siphonostelic roots, **345**
Sisal (*Agave sisalana*), 288
Skoog, Folke, 429
Skunk cabbage (*Symplocarpus foetidus*), 127,
   128(box)
Sleep movements, 463
Sliding microtubules hypothesis of
   chromosome movement, 207
Slime (P-protein), 513
Slime molds, 68, 627–30, 846–47
   characteristics of, 629(table)
Small nuclear ribonucleoproteins
   (snRNPs), **235**
Small RNA (snRNA), **235**
Smilacina racemosa, 745
Smilax, 341
Snakeroot (*Rauwolfia serpentina*), 814
Snapdragon (*Antirrhinum majus*), 177,
   178, 181
snRNA, **235**
Socrates, 811
Sodium as essential element, **478**
Sodium chloride, 835
Soft fibers, 288
Softwoods, **364**, 366
Soil(s), 470–76
   burning rubble and depletion of nitrogen
      in, 769, 771
   components of, 471–76
   erosion and loss of, 474(box)
   formation of, 470–71
   hardpan, 769
   horizons and types of, 470–71
   hydration conditions affecting, 475, 476
   peat moss as conditioner of, 679, 682
   pH of, and nutrient availability, 482–83
   plant distribution and mineral content
      of, 761
   plant nutrition and, 477–93
   primary succession and buildup of,
      772, 773
   properties of, and effect on roots, 349

returning nutrients to, 473(box)
root interface with, 346–47
salty. *See* Saline habitats, plants living in
toxic substances in, 482, 538
transpiration and water content of, 506
Soil particles, **471–72**
   nutrient availability and, 472
   water availability and, 472–73
Solanopteris brunei, 706
Solanun tuberosum. See Potato (*Solanun
   tuberosum*)
Solar energy, 95(box), 766–67. *See also* Light
Solute potential, **501**
Solutes, defined, **76**
Solutes transport, 78, 512–18. *See also* Water
   transport
   contents of, in phloem sieve tubes, 517–18
   exchange between phloem and xylem, 518
   influence of environment on, 517
   membranes and, 78–85
   methods of movement in phloem, 514–17
   parenchyma transfer cells and, 285, 287
   phloem tissue and, 297–98
   roots and, 338, 341–45
   structure of conducting cells, 513
Solution, **78**
Solvent, water as, **76**
Sophora secundiflora, 409
Sori(sorus)
   fern, **701**
   kelp, **648**
Southern, E. M., 255
Southern blotting, **255**
Soybeans (*Glycine max*), 350
Spanish moss (*Tillandsia usneoides*), 493
Spearmint (*Mentha spicata*), 810, 811
Speciation, **549**, 553–70
   allopatric, 553, 560–61
   definitions of species, 554–56
   population divergence and modes
      of, 559–69
   reproductive isolation and, 553, 556–59
   in viruses, 607
Species, **575**
   defining, 554–56
   ecological race within, 760
   endangered, 556, 557(box)
   exotic, 775(box)
   naming new plant, 577(box)
   pioneer, 766
   sympatric, 561–69, 766
Species concepts, 554–56
   biological, 555
   evolutionary, 556
   genetic, 556, 557(box)
   morphological (classical), 554–55
   paleontological, 556
   reproductive isolation and, 556–59
Species diversity, **764**, 765. *See also*
   Biological diversity
Species Plantarum (Linnaeus), 572, 575

Sperm, 173, 212, 214, 393, 394
   algae, 642
   bryophyte, 666, 667, 671
   gymnosperm, 727, 728
   of seedless vascular plants, 688, 689
Spermatia (red algae), **649**
Sphacelaria sp., 658
Sphagnopsida (class), **848**
Sphagnum moss, 8
   peat bogs and, 679, 680(box), 682–83
   shooting spores of, 672(box)
S phase of interphase, **189**
   chromatin synthesis in, 195
   enzymes of replication in, 196–200
Spices, 329, 377, 808
Spike, **394**
Spike mosses (*Selaginella* sp.), 686, 694, 695,
   696, 697, 718, 849
Spinach, 8
Spindle apparatus, **203**
   chromosome movement and, 206–7
Spindle fibers, **203**, 204, 207
Spines, 326
Spiral (alternate) phyllotaxis, **314**
Spirogyra sp., 134, 135, 644
Spirulina bacteria, 603–4
Splachnum luteum, 668
Spliceosome, **235**, 236
Splitters, classification of, 582
Spodosols, **471**
Spongomorpha algae, 68, 69
Spongy mesophyll cells, **319**
Sporangiophores (horsetails), 698, 700
Sporangium
   bryophyte, 696
   of fungi, **612**, 614, 625
   gymnosperm, 721, 724
   sori of, in ferns, 701
   sori of, in kelp algae, 648
   of seedless vascular plants, 688, 690, 691,
      701, 702
   unilocular, in brown algae, 645
Spores, 212, 213
   of algae, 636, 650, 653
   of bryophytes, 662, 672(box), 675
   fossil, 709(box)
   of fungi, 612, 614, **617**, 618, 622, 627
   germination of, from tetrad scar, 718
   germination of, inhibited by plant
      secretions, 707(box)
   of seedless vascular plants, 688, 694,
      695–96, 698–99, 700, 707(box)
Sporic meiosis, **636**, 637, 688
Sporocarps (ferns), **704**
Sporophylls (club mosses), **695**
Sporophyte, **212**, 213
   of algae, 636, 643, 646, 650
   of angiosperms, 743
   of bryophytes, 666, 667, 670–71, 675,
      677, 681
   gymnosperm, 723, 724–25, 727

of seedless vascular plants, 688, 689, 691–92, 704

Sporophytic self-incompatibility, 397

Sporopollenin, **391, 709**

Spring (early) wood, **362,** *363, 366*

Spruce (*Picea* sp.), 8, 731
> chewing gum made from resin of, *813(box)*
> effect of acid rain on, 778

Spurge family, 811

Squash (*Cucurbita*), 298

Stabilizing selection, *540,* **541**

Stahl, Franklin W., 196

Stalk cells (gymnosperm), **718**

Stamens, **384,** *385, 386*
> leaflike, 388
> origins of, 388

Staminate plants, 384

Stanford, Robert, 492

Stanleya elata, 479

Staphylococcus aureus, 596

Starch-branching enzymes (SBEI), **232**

Starches, 26
> retrieval of glucose from, 112–14

Starch phosphorylase, **114**

Star moss (*Tortula ruralis*), 681(box)

Start codon (AUG), **240**

Stele, **691**
> complex, 691, 692
> root, 343–45
> simple, 691

Stelis gemma, 35

Stem(s), 262, 263, 308–13
> axillary buds and branching, 311
> control of growth in, 308–9
> economic importance of, 313
> effect of ethylene on, 435
> functions of, 308
> of horsetails, 698
> modified, 311–13
> secondary growth of dicot, 359
> in seedless vascular plants, 691, 692
> structure of, 309–11

Stemonitis splendens, 629

Stereoisomers, **840**

Steroids, 38, 814

Sterols, **38**

Stevens, Jeff, 705

Steward, F. C., 188

Sticky ends, 245

Stierle, Andrea, 726

Stigma, **214, 385, 638**
> glycoproteins of, 88

Stinging nettle (*Urtica dioica*), 294, 295, 811

Stinkhorn fungus (*Dictyophora* sp., *Phallus* sp.), 613, 622

Stipes (algae), **634,** *636*

Stipules, 326

Stolons, 311–12

Stomata (stoma), *80,* **291,** *292, 293,* **498,** 509–10

differentiation of, in leaves, 274
> diurnal curve of opening, 506
> effect of abscisic acid on closing of, 437
> hypotheses on opening of, 509–10
> leaf orientation and frequency of, 318
> structure and distribution of, 509
> transpiration and, 498, 507, 508
> turgid cells and open, 80

Stonecrop (*Sedum* sp.), 388

Stoneworts (*Chara* sp.), 643, 644

Stop codon, **240,** *241*

Storage function
> in leaves, 326
> parenchyma tissue and, 284, 285
> in roots, 338, 341, 347, 350
> in stems, 308

Storage polysaccharides, 26

Storage proteins, 30
> genetically-engineered, 253

Stradivari, Antonio, wood in violin of, 8

Stratification, **404**

Strawberry (*Fragaria chilensis*), 311, 424, 425, 503

Strelitzia reginae, 547

Streptococcus pneumoniae bacterium, genetic studies of, 174

Streptomyces, 604

Striga asiatica, 351(box)

Strobel, Gary, 7 n.2, 604, 726

Strobili
> in gymnosperms, 721, 723
> in seedless vascular plants, 690, 691

Stroma, **63, 141**

Strontium in plant tissue, **482**

Structural isomers, **840**

Structural polysaccharides, **24,** *25,* 26

Structural proteins, 29–30

Strychnine, 36, 37(table), 38, 293

Strychnos nux-vomica, 38

Style, **214, 385**

Subapical region of roots, 340

Suberin, **36,** 341, 343

Subsidiary cells, 291, 292

Substrate-level phosphorylation, **114,** 115

Subunits, protein, 27

Succession, ecological, 766, **772**–76
> primary, in water, 774
> primary, on rocks or lava, 772–74
> secondary, 774–75

Succinic acid, **118**

Succinyl-CoA, **118**

Succulent stems, **312**

Suckers (roots), 350

Sucrases, **112**

Sucrose, 23
> formation of, 100, 103
> loading of, into phloem, 516–17
> as product of photosynthesis, 159
> retrieval of glucose from, 112
> synthesis and hydrolysis of, 25

Sucrose synthase, **112**

Sugar, maple syrup, 377, 508(box)

Sugarcane, 427

Sulfate, 485

Sulfhydryl group, **837**

Sulfur, functional groups containing, 837

Summer (late) wood, **362,** *363, 366*

Sundew (*Drosera* sp.), 186, 293, 468, 489

Sunflower (*Helianthus* sp.), 176, 177, 288, 767
> flower of, 749
> solar tracking by, 446(box)
> stem structure, 310

Sun leaves, **321,** *322*

Sunlight, 95(box), 766–67. *See also* Light

Superior ovary, **389**

Supernumerary cambia, 374, *375*

Survival of fittest, 539

Suspensor cells (gymnosperm), **721**

Sweet pea (*Lathyrus odoratus*), *179, 180*

Sweet potatoes (*Ipomoea batatas*), 6, 374, *375*

Sweet wormwood (*Artemisia annua*), 817

Swift, Hewson, 190

Sycamore (*Platanus occidentalis*), 373, 560

Symbionts, 612
> fungi and plants as, 705–6
> lichens as, 620–22

Symbiotic nitrogen fixers, 487–88

Sympatric speciation, 561–69, **766**
> clines, ecotypes formation and, 562–64
> hybridization and, 564–65, 566
> polyploidy and, 565–69

Symplast, **68, 343**

Symplocarpus foetidus, 127, 128(box)

Synangium (whisk ferns), 693, **694**

Synapsis, **216**

Synaptonemal complex, **216,** 218–20
> chiasmata, chromosome segregation, and, 219–20
> model of, 219
> prophase I of meiosis and formation of, 216
> recombination nodules on, 219

Synergids, **392**

Synonymous codons, **242**

Syphilis (*Treponema pallidum*), 605(box)

Syringa vulgaris, 319, 499

Systematics, 579(box)

Systemic acquired resistance, 438

Syzygium aromaticum, 808

Szent-Gyorgyi, Albert, 811 n.1

t

Tabasco, 279

Taiga, 787–88

Takhtajan, Armen, 582

Takhtajan system of classification, 582, 749(table), 851

Talinum sp., 830

Tamarisk (*Tamarix ramosissima*), 502

Tambalacoque tree (*Calvaria major*), 405(box)

Tandem repeat of genes, **224**, *225*
Tangential section (wood), **367**, *368*
Tannins, **41**
Tapetum, **391**
Taproot, **336**, *337*
Taproot system, **336**, *337*
Tarahumara tribe of Mexico, 574
Taraxacum sp., 266, *337*, 555
Tassels, corn, **182**
Taxol, 377, **726**
Taxomyces andreanea, 726(box)
Taxon, defined, **585**
Taxonomy, **574**. *See also* Classification
Taxus brevifolia, 37, *372*, 377, 726(box)
Taylor, J. Herbert, 196
Tea (*Camellia sinensis*), *5*, 808
Technology, scientific discovery and use of, 11–12
Tehler, Anders, 627
Telomeres, **247**
Telophase I of meiosis, **215**
Telophase II of meiosis, **215**, *217*
Telophase of mitosis, **189**, *204*, *205*
Temperate deciduous forests, **788–90**
Temperature
    algal reproduction influenced by, 658
    Celsius vs. Centigrade scales of, 843–44
    clover cline along gradient of, 563
    effect of, on roots, 348
    flowering and effect of cold, 458
    of forest interiors, 511
    influence of, on algal reproduction, 658
    plant movement in response to, 454
    seed germination requirements for, 404
    transpiration and atmospheric, 506
    units of, 843–44
Temporal isolation of species, 558
Tendrils
    as modified leaves, 324, 325, 326
    as modified stems, 312
    thigmotropism in, 450–51
Tenericutes (division), **845**
Tension wood, 369
Teosinte (*Zea mexicana*), 548, 824–26
Tepals, 388
Terpenoids, *37(table)*, 38–40
    structure of, 39
Terrestrial plants, 711–13
    cladogram of, 713
    evolution of photosynthesis in, 152(box)
    tissue adaptations for, 283, 300–301
Tertiary structure of proteins, **27**, *29*
Tetracystis, 640
Tetrad scar, **718**, *719*
Tetrahydrocannabinol (THC), 814
Tetraploids, 566
Tetrasporangia (red algae), **650**
Tetraspores (red algae), **650**
Tetrasporophytes (red algae), **650**
Texture (wood), **369**

tgal gene, 826
Thallose liverwort, 673, *674–75*, 676
Thallus (lichen), **620**, *621*
Thelypterin A, **707**
Thelypterin B, **707**
Theobroma cacao, 5
Theophrastus of Eresus, 574
Theory of inheritance, **170**
Thermoactinomyces, 604
Thermodynamics, laws of, 96–99
Thermonasty, 454
Thigmomorphogenesis, **435**, **454**
Thigmotropism, 450–51
Thiols, **837**
Thistle, 59
Thorne, Robert, 582
Thorne system of classification, 582, 749(table), 851
Thorns, **312**
Thottea sp., 317
Thylakoids, *63*, *64*, **141**
    structure of, 77
Thyme (*Thymus vulgaris*), 282
Thymine, *31*, *32*
Tilia americana, 287, *372*
Tillandsia usneoides, 492
Tilman, David, 765
Tinsel flagellum (brown algae), **645**, *646*
Tissue. *See* Dermal tissues; Ground tissue; Meristems; Vascular tissue
Tissue culture, **270**, *271*, **429**
    responses of plant, to kinetin and auxin, 431
Tmesipteris sp., *693*, 849
Tobacco (*Nicotiana* sp.), 177, *178*, 297, 330, 429
Tobacco mosaic virus (TMV), 252, *253*, 605, *606*
Tobamo group of viruses, **606**
Tolba, Mostafa, 777
Toluene, chemical structure, 24(box)
Tomatine, 37(table), 38
Tomato (*Lycopersicon esculentum*), 13, 38, 811
    distribution of rbcS genes in, 226(box)
    effect of ethylene on stem elongation in, 435
    genetically-engineered, 829 n.5
    lycopene pigment of, 39
    stem elongation in, 435
    transgenic, 253
Tonoplast, **62**
Topoisomerases, **198**, *199*
Tortula ruralis, 681(box)
Totipotency, **188**
Touch
    thigmomorphogenesis as response to, 454
    thigmotropism in response to, 450–51
Touch-me-nots (*Impatiens glandulifera*), 409
Tournefort, J. P. de, 575
Toxicodendron radicans, 811

Toxic substances in soil, 482, 538
Toxins, **811**
    botulism, 597(box)
    as chemical defense, 327–28
    growth-inhibiting, produced by plants, 707(box), 767
    plant alkaloids as, 36, 37
    produced by cyanobacteria, 604
    trichome cells and production of, 294, 295
Trace elements, **833**
Tracheids, **295**, *296*, 302(table), 500(box)
    in hardwoods, 364, 366
    transport of water and solutes in, 295, 296, 297, 499, 500, 502–5
Trailer sequence, RNA, **233**, *234*
Trait(s), genetic
    dominant, 167, 175–76
    genotype, phenotype, and, 170–71
    Mendel's experiments using multiple, 171–72
    Mendel's theory of inheritance and, 170
    natural selection theory on, 530
    recessive, 167
Transcription, **232**, 233–34
    steps in, 233, 234
Transduction (bacteria), **597**
Transfer cells, 285, *287*
Transfer RNA (tRNA), **232**, 236–37
    binding of amino acids by, 239
    structure of, 238
Transformation (bacteria), **597**
Transgenic plants, 250–54
    antisense technology and, 250–51
    public issues in production of, 252(box)
    steps in making, 251
    uses of, 251–54
Transitional meristems, **266**
Transition region between root and shoot, **346**
Translation, **232**, 239–41
    amino acid binding prior to, 239
    speed of, 241
    steps in, 240, 241
Transmission electron microscopy (TEM), 49, **50**
Transpiration, **498**
    adaptive value of, 511–12
    factors affecting, 505–10
    leaf architecture and, 318, 498
    in xerophytes, 322–23
Transpiration-cohesion hypothesis, *504*
Transpiration-photosynthesis compromise, 510, *511*
Transpons. *See* Transposable elements (transpons)
Transport. *See* Solutes transport; Water transport
Transport proteins, **81**, *82*
Transposable elements (transpons), 10, *11*, 180–82
    sexual reproduction and, 227

Transposon hypothesis, 227, 548
Travertine, 601, 602
Tree farm, 371(box)
Trees. *See also* Forests; Gymnosperms;
        Wood; *names of specific trees*
    annual harvest of, in U.S., 378
    aspen root system, 350
    Christmas, 424, 732(box)
    commemoration of, 378
    compartmentalization in, 362–63
    generation time of, 761
    giant, 728(box)
    largest, 362
    lore of, 364–65(box)
    sacred, 378(box)
    as state emblems, 733
    tambalacoque, 405(box)
Tremella mesenterica, 613
Treponema pallidum, 605(box)
Triacylglyceride, **33**
Triarch roots, 345
Tricarboxylic acid cycle, **118**
Trichogynes
    ascomycete fungi, **618**
    red algae, **650**
Trichomes, **293–94**
    functions of, 293, 294
    in stem epidermis, 309
    transpiration and, 507
Trifolium sp., 327
Trilliums (*Trillium* sp.), *790*
Trimerophytes, **712**
Trimerophytophyta (division), **712**
Triplet codons, **241–42**
Triterpenes, **38**
Triticale, *824*
Triticum. See Wheat (*Triticum aestivum*)
Tritium, **838**
Trophic levels in ecosystems, 762, *763*
Tropical oils, human health problems and,
        34(box)
Tropical rain forests, **800–803**
    biological diversity in, 764
    deforestation of, 778, 802–3, 828
    plant nutrient gathering in, 492–93
    reforestation of, 828
Tropisms, **444–51**
    gravitropism, 447–50
    phototropism, 444–45, 446(box)
    thigmotropism, 450–51
True-breeding strains, **167**
Truffles (*Tuber melanosporum*), *617*
Tschetverikov, S. S., 543 n.5
Tube cell, **214, 391**
Tuberculosis bacteria, *592*
Tuber melanosporum, 617
Tubers, **313**
Tule tree (*Taxodium mucronatum*), *363, 365*
Tundra, **758,** 784–87
Tunica, **267**
Tunica-corpus model, **267**

Turgor pressure, **62,** 80–81
    nastic movements and, 451–54
    positive, and cellular enlargement, 268
Turpentine, **731,** 732
Twinflower (*Linnaea borealis*), *578*
Twining shoots, **312**

v

Vaccines, production of edible, 277(box)
Vacuoles, **62**
Valence electrons, **834**
Valence shell, **834**
Valerian (*Valeriana sylvatica*), *98*
Vallisneria spiralis, 398
Valves (diatoms), **653,** *654*
Van Helmont, Jan-Baptista, 132, *133,* 477
Van Niel, C. B., 135
Vascular cambium, **264,** 358–63
    controls on activity of, 360–62
    extent of, 362–63
    formation of, 359–60, 361
    hormones and differentiation of, 424
    types of, 358, 359
Vascular plants, **259, 664**
    angiosperm. See Angiosperms; Flowering
        plants
    evolution of first, 711–12
    fossils of, 709, 710(box), 711
    gymnosperm. See Gymnosperms
    seedless. See Seedless vascular plants
Vascular tissue, **262,** 295–98
    auxin and differentiation of, 424
    in leaves, 318–19

    phloem, 297–98
    in roots, 345
    in seedless vascular plants, 691–92
    in stems, 310
    xylem, 295–97
Vaucheria sp., *653*
Vectors, DNA, **244**
Vegetables
    artificial selection for giant, 537(box)
    polyploid, 568(box)
Vegetative cell, **397**
    culturing, 189
Vegetative reproduction, **397–98**
    roots, 336–37, 350
    stolons and runners, 311–12
Vein endings, 318–19, *320*
Veins, leaf, **318,** *319*
    in C₄ plants, 321
Velamen, **352**
Velcro, 8
Velloziaceae, classification of, 750, *751*
Veneers, 376
Venter (bryophyte archegonium), **670**
Venus's flytrap (*Dionaea muscipula*), 451,
        452, 489, 490
Vernalization, **458(box)**
Vernal pools, 792, *793*
Vernonia amygdalin, 818(box)
Vesicular-arbuscular (V-A) mycorrhizae, **625,**
        626
Vessel elements, 295, **296,** 297, 302(table)
    advantages and disadvantages of, 500(box)
    in hardwoods, 364
    *transport of water and solutes in, 296,
        297, 499, 500, 502–5*
Vessels, 499, *500*
Vibrio cholerae, 605(box)
Vicia faba, 69, 190
Vigna sinensis, 293
Virchow, Rudolf, 12, 188
Viroids, **607**
Viruses, **604–7**
    Abby, 607
    adenovirus-2, 237(box)
    bacteriophages, 174–75, 605
    cloning in, 246
    genetic studies of, 232
    HaSV, 606, 607
    origin and evolution of, 607
Virusoids, **607**
Viscum sp., 278 n.4
Visible light, **137**
Vitamins
    biotechnology and creation of, 15
    discovery of vitamin C, 811 n.1
Vitis sp., *298*
Vittaria sp., *690*
Voltziales order, *736*
Volume, units of, 843
Volvox sp., *636, 641*
Von Humboldt, Alexander, 799(box)

Yellow poppy (*Papaver radicatum*), 802
Yellowstone National Park, *602*
Yersinia pestis, 605(box)
Yew (*Taxus* sp.), 37, *372*
    seeds of, *721, 722*
    taxol production from, 377, 726(box)
Young Lions, The (film), 784 n.1

Z

Zamia sp., *65*, 827
    pollen-bearing cones of, 720
    seedbearing structure of, 722

Zea mays. *See* Corn (*Zea mays*)
Zea mexicana (teosinte), 548, 824–26
Zeatin, 430, *431*
Zebra orchid, *559*
Zebrina, *292*
Zein, **30**
Zone of cellular division (root), **340**
Zone of cellular elongation (root), **340**
Zone of cellular maturation (root), **340**, 347
Zoospores (algae), **636**, *637*
Zosterophyllophyta (division), **711**
Zosterophyllophytes, **711**
Zosterophyllum, *711*

Z scheme, 146
Zygomycetes, **614**–17, 846
Zygomycota (division), 846
Zygosporangium (zygomycetes), **614**
Zygote, **188**, *213*, **215**
    of algae, 637
    reproductive isolation mechanisms
        affecting, 557–59
    of seedless vascular plants, 688, 689,
        699, 704
Zygotic meiosis, **637**